Book of Abstracts of the 67th Annual Meeting of the European Federation of Animal Science

EAAP - European Federation of Animal Science

The European Federation of Animal Science wishes to express its appreciation to the
Ministero delle Politiche Agricole Alimentari e Forestali (Italy) and the
Associazione Italiana Allevatori (Italy)
for their valuable support of its activities.

Book of Abstracts of the 67ᵗʰ Annual Meeting of the European Federation of Animal Science

Belfast, United Kingdom, 29 August – 2 September 2016

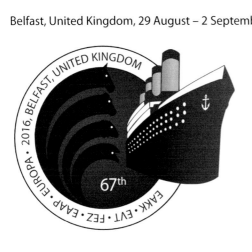

EAAP Scientific Committee:

E. Strandberg
G. Van Duinkerken
H. Spoolder
H. Sauerwein
M. Tichit
M. Klopčič
J. Conington
C. Lauridsen
A. Santos
G. Pollott

OASES
Online Academic Submission and Evaluation System

EAN: 9789086862849
e-EAN: 9789086868308
ISBN: 978-90-8686-284-9
e-ISBN: 978-90-8686-830-8
DOI: 10.3920/978-90-8686-830-8

ISSN 1382-6077

First published, 2016

© Wageningen Academic Publishers
The Netherlands, 2016

Wageningen Academic
P u b l i s h e r s

The individual contributions in this publication and any liabilities arising from them remain the responsibility of the authors.

The designations employed and the presentation of material in this publication do not imply the expression of any opinion whatsoever on the part of the European Federation of Animal Science concerning the legal status of any country, territory, city or area or of its authorities, or concerning the delimitation of its frontiers or boundaries.

The publisher is not responsible for possible damages, which could be a result of content derived from this publication.

Welcome to Belfast and Northern Ireland

On behalf of the Organising Committee and the British Society of Animal Science (BSAS), it is my pleasure to welcome you to the 67th Annual Meeting of the European Federation of Animal Science (EAAP), at the Waterfront Conference and Exhibition Centre, Belfast.

The British Society of Animal Science has hosted four previous annual meetings of EAAP in the UK, but this is the first time the conference has been held in Northern Ireland. This is particularly appropriate in the current year as 2016 is Northern Ireland's 'Year of Food' and August has been designated 'Love Northern Ireland Meat' month. We are extremely proud of our long tradition of producing high quality livestock and livestock products in Northern Ireland, and of our internationally recognised research in livestock production, animal science and animal health. Our social event on the evening of Monday 29 August will provide an excellent opportunity to enjoy local food and culture, and will include a livestock display.

The Annual EAAP meeting is Europe's largest animal scientific conference bringing together around 1,200 delegates from across Europe and worldwide to discuss the issues and challenges associated with efficient and environmentally sustainable livestock production. The overall theme of this year's conference is 'Sustainable Food Production – Livestock's Key Role' and we have worked closely with the local farming, food, and animal health sectors in organising the programme for the meeting.

The conference programme addresses a wide range of animal science topics, with a particular feature of the conference this year including a discussion on the role of grassland and forage as an important global feed resource for farm livestock and sessions on the role of livestock products in human nutrition and health. Delegates are encouraged to attend the Plenary Session entitled 'Sustainable Food Production to 2050: The Key Role of Livestock' on Tuesday morning 30 August with presentations from four internationally renowned speakers.

Networking and interaction between delegates is an essential aspect of any conference and with this in mind we have organised an entertaining and informative social programme. Belfast is renowned as one of the world's friendliest cities and I cordially invite you to enjoy fine food and unique local hospitality during your visit to the conference.

Finally a special thank you to all our sponsors who have very generously supported this conference and are listed on page 10.

Dr Sinclair Mayne
Chair of British Society of Animal Science Organising Committee

National Organisers of the 67th EAAP Annual Meeting

National Organising Committee:

Members of the British Society of Animal Science (BSAS)

Chairperson
- **Dr Sinclair Mayne,** Agri-Food and Biosciences Institute

Members
- **Mr Chris Armour,** Agri-Food and Biosciences Institute
- **Dr Alistair Carson,** Department of Agriculture, Environment and Rural Affairs
- **Dr Jon Day,** British Society of Animal Science
- **Prof Trevor Gilliland,** AgriFood and Biosciences Institute
- **Mr Jim Godfrey,** Farmer, Lincolnshire
- **Ms Bridget Hilton,** British Society of Animal Science
- **Dr Ryan Law,** Dunbia
- **Miss Anna Lavery,** Agri-Food and Biosciences Institute
- **Prof Michael Lee,** University of Bristol and Rothamsted Research
- **Dr Francis Lively,** Agri-Food and Biosciences Institute
- **Dr Elizabeth Magowan,** Agri-Food and Biosciences Institute
- **Dr Steven Morrison,** Agri-Food and Biosciences Institute
- **Mr Mike Steele,** British Society of Animal Science
- **Dr Edward O'Riordan,** Teagasc, Grange, Co Meath, Ireland
- **Dr Geoff Pollott,** Royal Veterinary College, University of London
- **Dr Howard Simmins,** British Society of Animal Science
- **Ms Caroline Stocks,** British Society of Animal Science
- **Prof Eileen Wall,** Scotland's Rural College, Edinburgh
- **Dr Stephen Whelan,** Agriculture and Horticulture Development Board, Stoneleigh
- **Prof Peter Williams,** Independent Consultant

BSAS Presidents

President
- **Prof Liam Sinclair,** Harper Adams University College

Senior Vice President
- **Prof Richard Dewhurst,** Scotland's Rural College, Edinburgh

Vice President
- **Prof Helen Miller,** University of Leeds

Friends of EAAP

By creating the 'Friends of EAAP', EAAP offers the opportunity to industries to receive services from EAAP in change of a fixed sponsoring amount of support every year.
- The group of supporting industries are layered in three categories: 'silver', 'gold' and 'diamond' level.
- It is offered an important discount (one year free of charge) if the sponsoring industry will agree for a four years period.
- EAAP will offer the service to create a scientific network (with Research Institutes and Scientists) around Europe.
- Creation of a permanent Board of Industries within EAAP with the objective to inform, influence the scientific and organizational actions of EAAP, like proposing choices of the scientific sessions and invited speakers and to propose industry representatives for the Study Commissions.
- Organization of targeted workshops, proposed by industries.
- EAAP can represent and facilitate activities of the supporting industries toward international legislative and regulatory organizations.
- EAAP can facilitate the supporting industries to enter in consortia dealing with internationally supported research projects.

Furthermore EAAP offers, depending to the level of support (details on our website: www.eaap.org):
- Free entrances to the EAAP annual meeting and Gala dinner invitation.
- Free registration to journal *animal*.
- Inclusion of industry advertisement in the EAAP Newsletter, in the banner of the EAAP website, in the Book of Abstract and in the Programme Booklet of the EAAP annual meeting.
- Inclusion of industry leaflets in the annual meeting package.
- Presence of industry advertisements on the slides between presentations at selected standard sessions.
- Presence of industry logos and advertisements on the slides between presentations at the Plenary Sessions.
- Public Recognition by the EAAP President at the Plenary Opening Session of the annual meeting.
- Discounted stands at the EAAP annual meeting.
- Invitation to meetings (at every annual meeting) to discuss joint strategy EAAP/Industries with the EAAP President, Vice-President for Scientific affair, Secretary General and other selected members of the Council and of the Scientific Committee.

Contact and further information

If the industry you represent is interested to become 'Friend of EAAP' or want to have further information please contact jean-marc.perez0000@orange.fr or EAAP secretariat (eaap@eaap.org, phone : +39 06 44202639).

The Association

EAAP (The European Federation of Animal Science) organises every year an international meeting which attracts between 900 and 1500 people. The main aims of EAAP are to promote, by means of active co-operation between its members and other relevant international and national organisations, the advancement of scientific research, sustainable development and systems of production; experimentation, application and extension; to improve the technical and economic conditions of the livestock sector; to promote the welfare of farm animals and the conservation of the rural environment; to control and optimise the use of natural resources in general and animal genetic resources in particular; to encourage the involvement of young scientists and technicians. More information on the organisation and its activities can be found at www.eaap.org

Acknowledgements

Thank you

to the 67th EAAP Annual Congress Sponsors and Friends

Principal sponsor

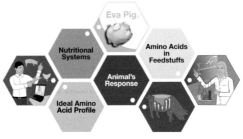

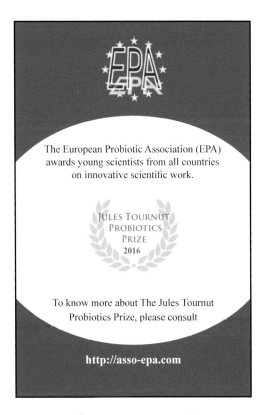

European Federation of Animal Science (EAAP)

President: P. Chemineau
Secretary General: A. Rosati
Address: Via G. Tomassetti 3, A/I
I-00161 Rome, Italy
Phone: +39 06 4420 2639
Fax: +39 06 4426 6798
E-mail: eaap@eaap.org
Web: www.eaap.org

28 August – 1 September 2017

EAAP 2017   eaap2017@agri.ee   +372 731 3420
Ministry of Rural Affairs of the Republic of Estonia
Lai 39 // Lai 41
15056 Tallinn
Estonia

Organised by:

Ministry of Rural Affairs of the Republic of Estonia
- Mr Toomas Kevvai, chairman
- Mr Martin Minjajev, deputy chairman
- Ms Pille Tammemägi
- Ms Sirje Jalakas
- Ms Maria Liisa Luur
- Ms Janika Salev
- Ms Kai Kasenurm
- Ms Inda Vaht
- Ms Eva Lehtla

The Veterinary and Food Board
- Ms Katrin Reili, deputy chairman
- Ms Anneli Härmson

Estonian University of Life Sciences
- Prof Haldja Viinalass, deputy chairman
- Prof Mait Klaassen
- Prof Ülle Jaakma
- Dr Andres Aland
- Ms Krista Rooni

Conference website: www.eaap2017.org

Scientific Programme EAAP 2016

Monday 29 August 8.30 – 12.30	Monday 29 August 14.00 – 18.00	Tuesday 30 August 8.30 – 12.30	Tuesday 30 August 14.00 – 18.00
Session 1 Bioinformatics, quantitative genetics and population genomics with whole genome sequence data Chair: B. Gredler	**Session 12** The use of genetics and genomics to improve disease and welfare traits in cattle Chair: A.W.M. Roozen		**Session 24** Breeding and reproduction in livestock systems in 2030 and beyond: how will science bring us there? Chair: J.G.B. Venneman
Session 2 Creating an enduring dairy sector post quota part 1 Chair: J. Campbell/M. Klopcic	**Session 13** Creating an enduring dairy sector post quota part 2 Chair: R Keatinge/A. Kertz		**Session 25** Young Train: Innovative research and extension in dairy farming, cattle and products Chair: A. Kuipers/ P. Aad
Session 3 Understanding the biology of feed use efficiency in pigs and chickens (linked to EU project ECO-FCE) part 1 Chair: G. Bee/D. Torrallardona	**Session 14** Understanding the biology of feed use efficiency in pigs and chickens (linked to EU project ECO-FCE) part 2 Chair: P. Lawdor/P. Varley		**Session 26** Alternative treatment strategies to address the challenge of antimicrobial resistance Chair: J. Taylor-Pickard
Session 4 Young scientist competition - Genetics Commission part 1 Chair: I. Blunk	**Session 15** Precision breeding: the role of new technologies in enhancing breeding programmes Chair: B. Whitelaw		**Session 27** Young scientist competition - Genetics Commission part 2 Chair: M. Martinez Alvaro
Session 5 Improving the quality and sustainability of beef production part 1 Chair: J.F. Hocquette	**Session 16** Improving the quality and sustainability of beef production part 2 Chair: J.F. Hocquette		**Session 28** Dairy and beef production, breeding and health Chair: M. Vestergaard/ H. Barkema
Session 6 Mixed farming systems - does diversity bring any benefits and at what scale? Chair: J. Ryschawy/ M. Zehetmeier	**Session 17** Livestock research priorities and the global funding challenge Chair: L. Sinclair	**Session 23** **Award Ceremony and Plenary Session** Chair: P. Chemineau	**Session 29** Novel milk-based phenotypes for use in breeding and management applications in dairy production Chair: G. Thaller/M. Crowe
Session 7 New legal landscape guiding access and benefit sharing of animal genetic resources Chair: E. Martyniuk	**Session 18** Free communications in pig production Chair: G. Bee		**Session 30** Adaptation of dairy and dual-purpose cattle to harsh environments Chair: S. König/B. Fürst-Waltl
Session 8 Genomic selection for small ruminants in Europe – how applicable for the rest of the world? Chair: T. Ådnoy	**Session 19** New challenges in environmental assessment: are LCA and ecosystem service assessment frameworks compatible? Chair: T. Turini		**Session 31** Sustainable control of pig and poultry diseases Chair: I. Kyriazakis/S. Millet
Session 9 Healthy livestock production through responsible use of antimicrobials Chair: E. Sossidou	**Session 20** Nutritional physiology Chair: H. Sauerwein		**Session 32** Livestock and climate change: current knowledge and policy challenge Chair: R. Ripoll-Bosch
Session 10 Aquaculture and coastal fisheries Chair: M. Service	**Session 21** Aquafeed and aquabreed Chair: E. Auclair		**Session 33** The future of native horse breeds Chair: J. Katanen
Session 11 Physiological limits of performance due to disproportionate tissue growth Chair: K. Huber	**Session 22** Animal production: the key in a European sustainable circular bio-economy Chair: G. Pollott		**Session 34** Succession planning for European livestock farmers Chair: W. Aston
13.00 – 14.00 **Poster session**		13.00 – 14.00 **Poster session**	

Wednesday 31 August 8.30 – 12.30	Wednesday 31 August 14.00 – 18.00	Thursday 1 September 8.30 – 11.30	Thursday 3 September 14.00 – 18.00
Session 35 Advances in genomic selection Chair: H. Mulder	**Session 45** Free communications in genetics; genetic diversity and disease genetics Chair: J. Fernandez Martin	**Session 55** Free communications in genetics Chair: N. Ibanez-Escriche	**Session 67** Free communications in genetics Chair: J. Lassen
Session 36 Breeding and management aspects of diseases and welfare related traits part 1 Chair: B. Fürst-Waltl/M. Burke	**Session 46** Breeding and management aspects of diseases and welfare related traits part 2 Chair: B. Fürst-Waltl/M. Burke	**Session 56** Amazing grazing part 1 Chair: A. Van den Pol-Van Dasselaar/J.L. Peyraud	**Session 68** Amazing grazing part 2 Chair: J.L. Peyraud/C. Ferris
Session 37 Healthy livestock products for healthy humans Chair: N. Scollan	**Session 47** Meat quality; the role of vertical integration Chair: T. O'Neill	**Session 57** Precision Livestock Farming (PLF): pathways to commercialisation part 1 Chair: I. Halachmi/ R. Dewhurst	**Session 69** Precision Livestock Farming (PLF): pathways to commercialisation part 2 Chair: D. Berckmans/ I. Halachmi
Session 38 Small ruminant mortality: Seeking solutions - genetic and environmental factors that contribute to improved survival Chair: C.M. Dwyer/ J.M. Gautier	**Session 48** Towards 5 ton of pork per sow per year Chair: E.F. Knol/G. Bee	**Session 58** Creating an enduring dairy sector post quota part 3 Chair: A. Kuipers	**Session 70** Free communications in cattle Chair: S. König/ Y.R. Montanholi
Session 39 Longevity: a physiological and management perspective Chair: R.M. Bruckmaier/ M. Vestergaard	**Session 49** Beef and dairy products, quality, human health, substitutes, debate Chair: A. Kuipers/ J.F. Hocquette	**Session 59** Free communications sheep and goat production Chair: J. Conington	**Session 71** Meat and milk quality - highlighting knowledge gaps in the supply chain Chair: O. Tzamaloukas
Session 40 The role of feed evaluation systems in feeding our livestock tomorrow Chair: J. Van Milgen	**Session 50** Genetics of feed efficiency and methane emissions Chair: E. Wall	**Session 60** Free communications in animal health and welfare Chair: G. Das	**Session 72** Optimising animal nutrition at an integral level: improving health, nutrient use efficiency and product quality Chair: G.S. Savoini
Session 41 Multifunctionality of grassland-based livestock farming systems Chair: M. Tichit	**Session 51** Animal nutrition contributing to reduced environmental impact Chair: S. Decampeneere	**Session 61** Genetic and environmental factors to understand dysbiosis in the GI tract of pigs Chair: P. Trevisi	**Session 73** Physiological indicators of animal welfare Chair: C.H. Knight
Session 42 Application of science to improve the management of stable horses Chair: S.J. Wood	**Session 52** Equine tourism and applying science to maximise rider and horse welfare Chair: M. Fradinho	**Session 62** Free communications animal nutrition Chair: G. Van Duinkerken	**Session 74** Equine aging and end of life Chair: A.S. Santos/R. Evans
Session 43 Design livestock systems to meet health and welfare needs of farm animals Chair: V. Ferrante	**Session 53** Animal behaviour and nutritional requirements Chair: L. Boyle	**Session 63** Free communications on equine genetics and breeding Chair: K.F. Stock	
Session 44 Cattle feeding practices and efficiency Chair: V. Cabrera/M. Klopcic	**Session 54** Climate and welfare smart housing systems Chair: T. Amon/I. Halachmi	**Session 64** KT: Using on-farm research and the multi-actor approach to boost effectiveness of knowledge exchange Chair: J. Campbell	
		Session 65 Dog breeding Chair: E. Strandberg	
		Session 66 Managing the food supply chain Chair: S. Durand	
		11.30 – 12.30	
13.00 – 14.00 **Poster session**		**Commission Business meetings**	

Commission on Animal Genetics

Dr Strandberg	President	Swedish University of Agricultural Sciences
	Sweden	erling.strandberg@slu.se
Dr Wall	Vice-President	Reader in Integrative Animal Sciences
	UK	eileen.wall@sruc.ac.uk
Dr Lassen	Vice-President	Aarhus University
	Denmark	jan.lassen@mbg.au.dk
Dr Mulder	Vice-President	Wageningen University
	Netherlands	han.mulder@wur.nl
Dr Ibañez	Secretary	IRTA
	Spain	noelia.ibanez@irta.es
Dr Gredler	Industry rep.	Qualitas AG
	Switzerland	birgit.gredler@qualitasag.ch

Commission on Animal Nutrition

Dr Van Duinkerken	President	Wageningen University
	Netherlands	gert.vanduinkerken@wur.nl
Dr Savoini	Vice-President	University of Milan
	Italy	giovanni.savoini@unimi.it
Dr Apper	Industry rep.	Tereos Syral
	France	emmanuelle.apper-bossard@tereos.com
Dr Auclair	Industry rep.	Phileo
	France	e.auclair@phileo.lesaffre.com
Dr de Campeneere	Secretary	ILVO
	Belgium	sam.decampeneere@ilvo.vlaanderen.be

Commission on Health and Welfare

Dr Spoolder	President	ASG-WUR
	Netherlands	hans.spoolder@wur.nl
Dr Das	Vice-President	University of Goettingen
	Germany	gdas@gwdg.de
Dr Sossidou	Vice-President	Hellenic Agricultural Organization
	Greece	sossidou.arig@nagref.gr
Dr Boyle	Secretary	Teagasc
	Ireland	laura.boyle@teagasc.ie
Dr Ferrante	Secretary	Milan University
	Italy	valentina.ferrante@unimi.it

Commission on Animal Physiology

Dr Sauerwein	President	University of Bonn
	Germany	sauerwein@uni-bonn.de
Dr Quesnel	Vice president	INRA
	France	helene.quesnel@rennes.inra.fr
Dr Silanikove	Vice-President	Agricultural Research Organization (ARO)
	Israel	nsilaniks@volcani.agri.gov.il
Dr Knight	Secretary	University of Copenhagen
	Denmark	chkn@sund.ku.dk
Dr Troescher	Industry rep.	BASF
	Germany	arnulf.troescher@basf.com

Commission on Livestock Farming Systems

Dr Tichit	President	INRA
	France	muriel.tichit@agroparistech.fr
Dr Ingrand	Vice-President	INRA
	France	stephane.ingrand@clermont.inra.fr
Dr Ripoll Bosch	Secretary	Wageningen University
	Netherlands	raimon.ripollbosch@wur.nl
Mrs Zehetmeier	Secretary	Institute Agricultural Economics and Farm Management
	Germany	monika.zehetmeier@tum.de
Dr Turini	Industry rep.	Centre d'Information des Viandes
	France	t.turini@civ-viande.org

Commission on Cattle Production

Dr Klopcic	Secretary/ Industry rep.	University of Ljublijana
	Slovenia	marija.klopcic@bf.uni-lj.si
Dr König	Vice-President	University of Kassel
	Germany	sven.koenig@uni-kassel.de
Dr Halachmi	Vice-President	Agricultural Research Organization (ARO)
	Israel	halachmi@volcani.agri.gov.il
Dr Montanholi	Vice-President	Dalhousie University
	Canada	yuri.montanholi@dal.ca
Dr Fürst-Waltl	Secretary	University of Natural Resources and Life Sciences
	Austria	birgit.fuerst-waltl@boku.ac.at
Dr Vestergaard	President	Aarhus University
	Denmark	mogens.vestergaard@agrsci.dk

Commission on Sheep and Goat Production

Dr Conington	President	SAC
	United Kingdom	joanne.conington@sac.ac.uk
Dr Ådnøy	Vice-President	NMBU
	Norway	tormod.adnoy@nmbu.no
Dr Tzamaloukas	Secretary	Cyprus University of Technology
	Cyprus	ouranios.tzamaloukas@cut.ac.cy
Dr Ligda	Secretary	Hellenic Agricoltural Organisation
	Greece	chligda@otenet.gr

Commission on Pig Production

Dr Lauridsen	President	Aarhus University
	Denmark	charlotte.lauridsen@agrsci.dk
Dr Knol	Vice President/	
	Industry rep.	TOPIGS
	Netherlands	egbert.knol@topigs.com
Dr Bee	Secretary	Agroscope Liebefeld-Posieux ALP
	Switzerland	giuseppe.bee@alp.admin.ch
Dr Velarde	Secretary	IRTA
	Spain	antonio.velarde@irta.es
Dr Millet	Secretary	ILVO
	Belgium	sam.millet@ilvo.vlaanderen.be

Commission on Horse Production

Dr Santos	President	CITAB-UTAD/EUVG
	Portugal	assantos@utad.pt
Dr Evans	Vice president	Norwegian University College of Agriculture and Rural Development
	Norway	rhys@hlb.no
Dr Stock	Vice president	VIT-Vereinigte Informationssysteme Tierhaltung w.V.
	Germany	friederike.katharina.stock@vit.de
Dr Saastamoinen	Vice president	Natural Resources Institute Finland Luke
	Finland	markku.saastamoinen@luke.fi
Dr Holgersson	Secretary	Swedish University of Agriculture
	Sweden	anna-lena.holgersson@slu.se
Dr Cervantes Navarro	Secretary	University of Madrid
	Spain	icervantes@vet.ucm.es

Scientific programme

Session 01. Bioinformatics, quantitative genetics and population genomics with whole genome sequence data

Date: Monday 29 August 2016; 08:30 – 12:30
Chair: B. Gredler

Theatre Session 01

invited Experiences from genomic prediction with imputed sequence data in different species   103
M. Erbe, G. Ni, H. Pausch, R. Emmerling, T.H.E. Meuwissen, D. Cavero, K.-U. Götz and H. Simianer

SNP data management in genomics pipelines   103
E. Groeneveld

Genomic prediction in cattle based on sequence data   104
M. Frischknecht, T.H.E. Meuwissen, B. Bapst, F.R. Seefried, C. Flury, D. Garrick, H. Signer-Hasler,
C. Stricker, Intergenomics Consortium, A. Bieber, R. Fries, I. Russ, J. Sölkner, A. Bagnato and B. Gredler

Accurate sequence imputation enables precise QTL mapping in Brown Swiss cattle   104
M. Frischknecht, H. Pausch, B. Bapst, F.R. Seefried, C. Flury, H. Signer-Hasler, R. Fries, D. Garrick,
C. Stricker and B. Gredler

A family-based imputation algorithm for high and low-coverage sequence data   105
M. Battagin, G. Gorjanc, S. Gonen, R. Antolin and J.M. Hickey

Efficient identification SNPs in high linkage disequilibrium in large genotype and sequence datasets   105
M.P.L. Calus and J. Vandenplas

Selecting animals for whole genome sequencing: methods optimization and comparison   106
A. Butty, M. Sargolzaei, F. Miglior and C. Baes

A hidden Markov model to estimate inbreeding from whole genome sequence data   106
T. Druet and M. Gautier

Accounting for read depth in the analysis of genotyping-by-sequencing data   107
K.G. Dodds, J.C. McEwan, T.P. Bilton, R. Brauning, R.M. Anderson, T.C. Van Stijn, T. Kristjánsson and
S.M. Clarke

SNP-effects underlying GBLUP depend on the definition of the genomic relationship matrix   107
A.C. Bouwman, B.J. Hayes and M.P.L. Calus

Functional annotation of CNV breakpoints in Holstein-Friesian cows   108
J. Szyda, M. Mielczarek, M. Fraszczak, R. Giannico, G. Minozzi, E. Nicolazzi and K. Wojdak-Maksymiec

NGS-based analysis of copy number variations in various cattle breeds   108
M. Mielczarek, M. Frąszczak, E.L. Nicolazzi, G. Minozzi, H. Schwarzenbacher, C. Egger-Danner,
D. Vicario, F. Seefried, A. Rossoni, T. Solberg, L. Varona, C. Diaz, C. Ferrandi, R. Giannico, J.L. Williams,
J. Woolliams and J. Szyda

Poster Session 01

Session 02. Creating an enduring dairy sector post quota (with ADSA, Erasmus+-ISM+ project, EuroDairy project) part 1

Date: Monday 29 August 2016; 08:30 – 12:30
Chair: J. Campbell / M. Klopcic

Theatre Session 02

Session 03. Feed use efficiency in pigs and chickens

Date: Monday 29 August 2016; 08:30 – 12:30
Chair: G. Bee / D. Torrallardona

Theatre Session 03

Session 04. Young scientist competition - Genetics Commission part 1

Date: Monday 29 August 2016; 08:30 – 12:30
Chair: I. Blunk

Theatre Session 04

Session 05. Improving the quality and sustainability of beef production (with Cattle Network WG and Livestock and Meat Commission NI) part 1

Date: Monday 29 August 2016; 08:30 – 12:30
Chair: J.F. Hocquette

Theatre Session 05

Poster Session 05

Session 06. Mixed farming systems - does diversity bring any benefits and at what scale?

Date: Monday 29 August 2016; 08:30 – 12:30
Chair: J. Ryschawy / M. Zehetmeier

Theatre Session 06

Poster Session 06

Session 07. New legal landscape guiding access and benefit sharing of animal genetic resources

Date: Monday 29 August 2016; 08:30 – 12:30
Chair: E. Martyniuk

Theatre Session 07

Session 08. Genomic selection for small ruminants in Europe – how applicable for the rest of the world?

Date: Monday 29 August 2016; 08:30 – 12:30
Chair: T. Ådnoy

Theatre Session 08

Poster Session 08

Session 09. Healthy livestock production through responsible use of antimicrobials

Date: Monday 29 August 2016; 08:30 – 12:30
Chair: E. Sossidou

Theatre Session 09

Effects of antibiotic prophylaxis on the gut microbiome and performance of growing broiler chicks 176
A.A. Tonks, M.J. Woodward and C. Rymer

Session 10. Aquaculture and coastal fisheries

Date: Monday 29 August 2016; 08:30 – 12:30
Chair: M. Service

Theatre Session 10

`invited` Comparing agricultural and urban nutrient loads to coastal systems 176
J.P. Nunes and J. Gomes Ferreira

`invited` Partitioning the faecal load to coastal waters 177
M. Taylor

`invited` Application of a multi-model framework for integrated ecosystem management in lough foyle 177
J.G. Ferreira, H. Moore, P. Boylan, C. Jordan, J.D. Lencart-Silva, C. McGonigle, S. McLean, J.P. Nunes, M. Service and C. Zhu

`invited` The interaction between natural benthic communities and cultivated blue mussels in Belfast Lough 178
H. Moore, A. Boyd, R. Corner, J.G. Ferreira and M. Service

`invited` Achieving a balance – the key to sustainably managing native oyster stocks 178
C. Bromley

`invited` Sustainable Shellfish aquaculture the Ecosystem approach 179
M. Service

Session 11. Physiological limits of performance due to disproportionate tissue growth

Date: Monday 29 August 2016; 08:30 – 12:30
Chair: K. Huber

Theatre Session 11

`invited` Disproportionate growth in farm animals – prerequisite for a disturbed health 179
K. Huber

`invited` Evidence for developmental programming in dairy cattle 180
G. Opsomer, M. Van Eetvelde and M.M. Kamal

`invited` Delayed development in Large White Purebreds than Crossbreds with Meishan born in the Same Litter 180
L. Canario, M.C. Père, H. Quesnel, Y. Billon, W. Hébrard, J. Riquet, P. Mormède and L. Liaubet

Impact of dietary L-arginine supply during early gestation on myofiber development in newborn pigs 181
J.G. Madsen and G. Bee

`invited` Microstructure and function of the thyroid gland may relate to feed efficiency in the bovine 181
J. Ormon, S. Bourgon, J. Munro, A. Macdonald, S. Lam, S. Miller and Y. Montanholi

`invited` Impact of pre-weaning nutritional regimes on mammary gland development in heifer calves 182
S. McCoard, T. Silvestre, A. Molenaar, P. Muir, J. Koolaard, V. Burggraaf, N. Wards and D. Pacheco

Poster Session 11

Session 12. The use of genetics and genomics to improve disease and welfare traits in cattle (with INTERBULL)

Date: Monday 29 August 2016; 14:00 – 18:00
Chair: A.W.M. Roozen

Theatre Session 12

Poster Session 12

Session 13. Creating an enduring dairy sector post quota (with ADSA, Erasmus+-ISM+ project, EuroDairy project) part 2

Date: Monday 29 August 2016; 14:00 – 18:00
Chair: R Keatinge / A. Kertz

Theatre Session 13

Session 14. Feed use efficiency in pigs and chickens

Date: Monday 29 August 2016; 14:00 – 18:00
Chair: P. Lawdor / P. Varley

Theatre Session 14

Session 15. Precision breeding: the role of new technologies in enhancing breeding programmes

Date: Monday 29 August 2016; 14:00 – 18:00
Chair: B. Whitelaw

Theatre Session 15

Session 16. Improving the quality and sustainability of beef production (with Cattle Network WG and Livestock and Meat Commission NI) part 2

Date: Monday 29 August 2016; 14:00 – 18:00
Chair: J.F. Hocquette

Theatre Session 16

Session 17. Livestock research priorities and the global funding challenge (with Agri-Food and Biosciences Institute)

Date: Monday 29 August 2016; 14:00 – 18:00
Chair: L. Sinclair

Theatre Session 17

Session 18. Free communications in pig production

Date: Monday 29 August 2016; 14:00 – 18:00
Chair: G. Bee

Theatre Session 18

Session 19. New challenges in environmental assessment: are LCA and ecosystem service assessment frame-works compatible?

Date: Monday 29 August 2016; 14:00 – 18:00
Chair: T. Turini

Theatre Session 19

Poster Session 19

Session 20. Nutritional physiology

Date: Monday 29 August 2016; 14:00 – 18:00
Chair: H. Sauerwein

Theatre Session 20

Session 21. Aquafeed and aquabreed

Date: Monday 29 August 2016; 14:00 – 18:00
Chair: E. Auclair

Theatre Session 21

Poster Session 21

Session 23. Award Ceremony and Plenary Session

Date: Tuesday 30 August 2016; 08:30 – 12:30
Chair: P. Chemineau

Theatre Session 23

Session 24. Breeding and reproduction in livestock systems in 2030 and beyond: how will science bring us there?

Date: Tuesday 30 August 2016; 14:00 – 18:00
Chair: J.G.B. Venneman

Theatre Session 24

Session 25. Young Train: Innovative research and extension in dairy farming, cattle and products

Date: Tuesday 30 August 2016; 14:00 – 18:00
Chair: A. Kuipers / P. Aad

Theatre Session 25

Poster Session 25

Session 26. Alternative treatment strategies to address the challenge of antimicrobial resistance (with Alltech)

Date: Tuesday 30 August 2016; 14:00 – 18:00
Chair: J. Taylor-Pickard

Theatre Session 26

Session 27. Young scientist competition - Genetics Commission part 2

Date: Tuesday 30 August 2016; 14:00 – 18:00
Chair: M. Martinez Alvaro

Theatre Session 27

Session 28. Dairy and beef production, breeding and health

Date: Tuesday 30 August 2016; 14:00 – 18:00
Chair: M. Vestergaard / H. Barkema

Theatre Session 28

Poster Session 28

Session 29. Novel milk-based phenotypes for use in breeding and management applications in dairy production (with EU GplusE project)

Date: Tuesday 30 August 2016; 14:00 – 18:00
Chair: G. Thaller / M. Crowe

Theatre Session 29

Session 30. Adaptation of dairy and dual-purpose cattle to harsh environments (with Core-Organic projects: 2-ORG-Cows, Organic Dairy Health)

Date: Tuesday 30 August 2016; 14:00 – 18:00
Chair: S. König / B. Fürst-Waltl

Theatre Session 30

Session 31. Sustainable control of pig and poultry diseases (with PROHEALTH)

Date: Tuesday 30 August 2016; 14:00 – 18:00
Chair: I. Kyriazakis / S. Millet

Theatre Session 31

Session 32. Livestock and climate change: current knowledge and policy challenge (with WAAP)

Date: Tuesday 30 August 2016; 14:00 – 18:00
Chair: R. Ripoll-Bosch

Theatre Session 32

Poster Session 32

Session 34. Succession planning for European livestock farmers (with Ulster Farmers Union, Young Farmers Clubs of Ulster and EAAP Cub Club)

Date: Tuesday 30 August 2016; 14:00 – 18:00
Chair: W. Aston

Theatre Session 34

Session 35. Advances in genomic selection

Date: Wednesday 31 August 2016; 08:30 – 12:30
Chair: H. Mulder

Theatre Session 35

Poster Session 35

Session 36. Breeding and management aspects of diseases and welfare related traits (with ICAR) part 1

Date: Wednesday 31 August 2016; 08.30 – 12.30
Chair: B. Fürst-Waltl / M. Burke

Theatre Session 36

Poster Session 36

Session 37. Healthy livestock products for healthy humans (with Devenish)

Date: Wednesday 31 August 2016; 08:30 – 12:30
Chair: N. Scollan

Theatre Session 37

Poster Session 37

Session 38. Small ruminant mortality: Seeking solutions - genetic and environmental factors that contribute to improved survival

Date: Wednesday 31 August 2016; 08:30 – 12:30
Chair: C.M. Dwyer / J.M. Gautier

Theatre Session 38

Poster Session 38

Session 39. Longevity: a physiological and management perspective

Date: Wednesday 31 August 2016; 08:30 – 12:30
Chair: R.M. Bruckmaier / M. Vestergaard

Theatre Session 39

Session 40. The role of feed evaluation systems in feeding our livestock tomorrow

Date: Wednesday 31 August 2016; 08:30 – 12:30
Chair: J. Van Milgen

Theatre Session 40

Session 41. Multifunctionality of grassland-based livestock farming systems

Date: Wednesday 31 August 2016; 08:30 – 12:30
Chair: M. Tichit

Theatre Session 41

Poster Session 41

Session 42. Application of science to improve the management of stable horses

Date: Wednesday 31 August 2016; 08:30 – 12:30
Chair: S.J. Wood

Theatre Session 42

Poster Session 42

Session 43. Design livestock systems to meet health and welfare needs of farm animals

Date: Wednesday 31 August 2016; 08:30 – 12:30
Chair: V. Ferrante

Theatre Session 43

Poster Session 43

Session 44. Cattle feeding practices and efficiency

Date: Wednesday 31 August 2016; 08:30 – 12:30
Chair: V. Cabrera / M. Klopcic

Theatre Session 44

Poster Session 44

Session 45. Free communications in genetics; genetic diversity and disease genetics

Date: Wednesday 31 August 2016; 14:00 – 18:00
Chair: J. Fernandez Martin

Theatre Session 45

Session 46. Breeding and management aspects of diseases and welfare related traits (with ICAR) part 2

Date: Wednesday 31 August 2016; 14:00 – 18:00
Chair: B. Fürst-Waltl / M. Burke

Theatre Session 46

Session 47. Meat quality; the role of vertical integration (with Dunbia)

Date: Wednesday 31 August 2016; 14:00 – 18:00
Chair: T. O'Neill

Theatre Session 47

Session 48. Towards 5 ton of pork per sow per year

Date: Wednesday 31 August 2016; 14:00 – 18:00
Chair: E.F. Knol / G. Bee

Theatre Session 48

Poster Session 48

Session 49. Beef and dairy products, quality, human health, substitutes, debate (with ADSA)

Date: Wednesday 31 August 2016; 14:00 – 18:00
Chair: A. Kuipers / J.F. Hocquette

Theatre Session 49

Session 50. Genetics of feed efficiency and methane emissions

Date: Wednesday 31 August 2016; 14:00 – 18:00
Chair: E. Wall

Theatre Session 50

Session 51. Animal nutrition contributing to reduced environmental impact

Date: Wednesday 31 August 2016; 14:00 – 18:00
Chair: S. Decampeneere

Theatre Session 51

Session 52. Equine tourism and applying science to maximise rider and horse welfare

Date: Wednesday 31 August 2016; 14:00 – 18:00
Chair: M. Fradinho

Theatre Session 52

Poster Session 52

Session 53. Animal behaviour and nutritional requirements

Date: Wednesday 31 August 2016; 14:00 – 18:00
Chair: L. Boyle

Theatre Session 53

Poster Session 53

Session 54. Climate and welfare smart housing systems (with OptiBarn project)

Date: Wednesday 31 August 2016; 14:00 – 18:00
Chair: T. Amon / I. Halachmi

Theatre Session 54

Poster Session 54

Session 55. Free communications in genetics

Date: Thursday 1 September 2016; 08:30 – 11:30
Chair: N. Ibanez-Escriche

Theatre Session 55

Poster Session 55

Session 56. Amazing grazing part 1

Date: Thursday 1 September 2016; 08:30 – 11:30
Chair: A. Van den Pol-Van Dasselaar / J.L. Peyraud

Theatre Session 56

Poster Session 56

Session 57. Precision Livestock Farming (PLF): pathways to commercialisation (with EU-PLF project, BSAS & AGRI-EPI Centre UK) part 1

Date: Thursday 1 September 2016; 08:30 – 11:30
Chair: I. Halachmi / R. Dewhurst

Theatre Session 57

Poster Session 57

Session 58. Creating an enduring dairy sector post quota (with ADSA, Erasmus+-ISM+ project, EuroDairy project) part 3

Date: Thursday 1 September 2016; 08:30 – 11:30
Chair: A. Kuipers

Theatre Session 58

Poster Session 58

Session 59. Free communications sheep and goat production

Date: Thursday 1 September 2016; 08:30 – 11:30
Chair: J. Conington

Theatre Session 59

Session 60. Free communications in animal health and welfare

Date: Thursday 1 September 2016; 08:30 – 11:30
Chair: G. Das

Theatre Session 60

Poster Session 60

Session 61. Genetic and environmental factors to understand dysbiosis in the GI tract of pigs (with PiGutNet)

Date: Thursday 1 September 2016; 08:30 – 11:30
Chair: P. Trevisi

Theatre Session 61

Session 62. Free communications animal nutrition

Date: Thursday 1 September 2016; 08:30 – 11:30
Chair: G. Van Duinkerken

Theatre Session 62

Session 63. Free communications on equine genetics and breeding

Date: Thursday 1 September 2016; 08:30 – 11:30
Chair: K.F. Stock

Theatre Session 63

Poster Session 63

Session 64. KT: Using on-farm research and the multi-actor approach to boost effectiveness of knowledge exchange (with AgriSearch)

Date: Thursday 1 September 2016; 08:30 – 11:30
Chair: J. Campbell

Theatre Session 64

Session 66. Managing the food supply chain (with Institute of Food Science and Technology)

Date: Thursday 1 September 2016; 08:30 – 11:30
Chair: S. Durand

Theatre Session 66

Poster Session 66

Session 67. Free communications in genetics

Date: Thursday 1 September 2016; 14:00 – 18:00
Chair: J. Lassen

Theatre Session 67

Session 68. Amazing grazing part 2

Date: Thursday 1 September 2016; 14:00 – 18:00
Chair: J.L. Peyraud / C. Ferris

Theatre Session 68

Session 69. Precision Livestock Farming (PLF): pathways to commercialisation (with EU-PLF project, BSAS & AGRI-EPI Centre UK) part 2

Date: Thursday 1 September 2016; 14:00 – 18:00
Chair: D. Berckmans / I. Halachmi

Theatre Session 69

Session 70. Free communications in cattle

Date: Thursday 1 September 2016; 14:00 – 18:00
Chair: S. König / Y.R. Montanholi

Theatre Session 70

Poster Session 70

Session 71. Meat and milk quality - highlighting knowledge gaps in the supply chain

Date: Thursday 1 September 2016; 14:00 – 18:00
Chair: O. Tzamaloukas

Theatre Session 71

Poster Session 71

Session 72. Optimising animal nutrition at an integral level: improving health, nutrient use efficiency and product quality

Date: Thursday 1 September 2016; 14:00 – 18:00
Chair: G.S. Savoini

Theatre Session 72

Session 73. Physiological indicators of animal welfare

Date: Thursday 1 September 2016; 14:00 – 18:00
Chair: C.H. Knight

Theatre Session 73

Poster Session 73

Session 74. Equine aging and end of life

Date: Thursday 1 September 2016; 14:00 – 18:00
Chair: A.S. Santos / R. Evans

Theatre Session 74

Poster Session 74

Experiences from genomic prediction with imputed sequence data in different species

M. Erbe[1,2], G. Ni[1], H. Pausch[3], R. Emmerling[2], T.H.E. Meuwissen[4], D. Cavero[5], K.-U. Götz[2] and H. Simianer[1]
[1]Georg-August-University, Department of Animal Sciences, Göttingen, Germany, [2]Bavarian State Research Centre for Agriculture, Institute for Animal Breeding, Grub, Germany, [3]Technische Universität München, Chair of Animal Breeding, Freising, Germany, [4]Norwegian University of Life Sciences, Department of Animal and Aquacultural Sciences, Ås, Norway, [5]Lohmann Tierzucht GmbH, Am Seedeich 9-11, Cuxhaven, Germany; malena.erbe@lfl.bayern.de

Technical progress has made it possible to re-sequence individuals of different species within a reasonable time frame and at acceptable costs. As sequencing all individuals of a livestock breeding population is still too expensive, only a subset of individuals accounting for most of the genetic variation is sequenced and all other individuals are imputed up to all known single nucleotide polymorphisms (SNPs) at sequence level. Results from simulation studies have suggested that sequence data will have a positive effect on the accuracy of genomic prediction and the stability of marker effect estimates especially when using variable selection methods. We thus studied the effect of using sequence data for genomic prediction with data sets from different species, one with over 6,000 Fleckvieh bulls (genotyped with 50k or 777k) and one with almost 900 individuals from a brown layer line (genotyped with 580k), both imputed to sequence level with a reference set of 150 and 25 individuals, respectively. Similar to results from other studies we did not find significantly higher prediction accuracy with imputed sequence data. This was true for usual genomic BLUP models as well as for GBCPP, a fast EM based variable selection method similar to BayesCπ, tested in Fleckvieh and BLUP|GA tested in brown layers. Attempts to reduce noise by using only specific subsets of SNPs (e.g. very accurately imputed SNPs, SNPs from genic regions) generally had a positive effect compared to modelling all SNPs. Possible explanations for the lack of success of genomic prediction with sequence data are inaccuracies in imputed genotypes, especially for variants with small minor allele frequencies, lack of proper models to account for the underlying genetic architecture, and incompleteness of genome maps and structural annotation.

_____

SNP data management in genomics pipelines

E. Groeneveld
Institute of Farm Animal Genetics (FLI), Department of Breeding and Genetic Resources, Höltystr. 10, 31535 Neustadt, Germany; eildert.groeneveld@fli.bund.de

Traditionally, in animal agriculture a relatively small number of phenotypic data points such as milk, protein or fat yield were collected for each animal. With the increase in genotyping the amount of data collected has exploded, adding tens of thousands to millions of SNPs to the phenotypic data for each individual. SNP data are processed in pipelines starting from the initial raw data pool often passing through multiple steps of filtering. The outcome is always a set of SNPs for downstream analysis which is defined by a subset of SNPs from the complete panel and a list of animals. Often these subsets are arrived at in an iterative manner which requires repeated access and processing of the original – possibly huge – full SNP dataset and creating derived and equally sizeable SNP data files. TheSNPpit is a database system that has been developed for the management of high volume SNP data from any genotyping platform allowing mixed storage of SNPs from panels of any size. It implements highly compressed vector storage in a relational database, set-based data manipulation, and a very fast export with PostgreSQL as the database backend being able to handle huge data volumes. Its novel subset system allows the creation of named genotype sets created through the filtering of SNPs (based on major allele frequency, no-calls, and chromosomes) and manually applied sample and SNP lists. Because of the very fast exports of up to 120 mio SNPs/sec export/analyse/delete cycles prevent proliferation of derived data files. As an example, exporting 2,000 samples with 1.5 mio SNPs each took 25 sec, thus, the resultant export file can readily be deleted as it can be quickly recreated for extended analysis. TheSNPpit data repository, with its command-line interface, functions as a feeder to pipelines storing subset definitions in its database at virtually no storage cost. TheSNPpit has been tested for a database of more that 18 mio animals with 3.4 trillion SNPs in 840 GB. Examples of workflows will be given demonstrating the use of subsets and its integration into pipelines. TheSNPpit is freely available as open source software.

Genomic prediction in cattle based on sequence data

M. Frischknecht[1,2], T.H.E. Meuwissen[3], B. Bapst[2], F.R. Seefried[2], C. Flury[1], D. Garrick[4], H. Signer-Hasler[1], C. Stricker[5], Intergenomics Consortium[6], A. Bieber[7], R. Fries[8], I. Russ[9], J. Sölkner[10], A. Bagnato[11] and B. Gredler[2]
[1]*Bern University of Applied Sciences, School of Agricultural, Forest and Food Sciences HAFL, Länggasse 85, 3052 Zollikofen, Switzerland, [2]Qualitas AG, Chamerstrasse 56, 6300 Zug, Switzerland, [3]Norwegian University of Life Science, Arboretveien 6, 1432 Ås, Norway, [4]Iowa State University, Kildee 225, 50011 Ames, USA, [5]agn Genetics, Börtjistrasse 8b, 7260 Davos, Switzerland, [6]Interbull center, SLU, Box 7023, Uppsala, 75007, Sweden, [7]Research Institute of Organic Agriculture (FiBL), Ackerstrasse 113, 5070 Frick, Switzerland, [8]Technische Universität München, Liesel-Beckmann-Straße 1, 85354 Freising-Weihenstephan, Germany, [9]Tierzuchtforschung e.V., Senator-Gerauer-Str. 23, 85586 Poing, Germany, [10]University of Natural Resources and Life Sciences, Gregor-Mendel-Str. 33, 1180 Wien, Austria, [11]University of Milan, Via Celoria 10, 20133 Milano, Italy; mirjam.frischknecht@qualitasag.ch*

One of the factors influencing the accuracy of genomic prediction is the density of SNP data used for prediction. We used sequence genotypes from the 1000 Bull Genomes Project (Run 5) as reference data to impute the whole-genome sequences of around 22,000 Brown Swiss and 15,000 Holstein, Simmental and Swiss Fleckvieh cattle to whole-genome sequences. We used FImpute to obtain HD genotypes and imputed with Minimac from HD to sequence data (16,184,800 variants). We report here the results for non return-rate 56 in heifers which are available already for Brown Swiss. For effect estimation deregressed breeding values of 2,018 Brown Swiss bulls with reliabilities above 0.65 were used. We used the BayesC approach implemented in gbcpp. Accuracy of genomic prediction was calculated as the correlation between the deregressed breeding values and the predicted direct genomic breeding value in a set of 240 of young bulls with accurate breeding values. LD pruned sequence data (5,812,425 SNPs; r=0.412) and sequence data (12,973,772 SNPs; r=0.407) yielded a higher accuracy than 50k data (38,009 SNPs; r=0.400) and missense data (34,184 SNPs; r=0.372). The results will be further evaluated by investigating more traits and breeds.

Accurate sequence imputation enables precise QTL mapping in Brown Swiss cattle

M. Frischknecht[1,2], H. Pausch[3], B. Bapst[2], F.R. Seefried[2], C. Flury[1], H. Signer-Hasler[1], R. Fries[3], D. Garrick[4], C. Stricker[5] and B. Gredler[2]
[1]*Bern University of Applied Sciences, School of Agricultural, Forest and Food Sciences HAFL, Länggasse 85, 3052 Zollikofen, Switzerland, [2]Qualitas AG, Chamerstrasse 56, 6300 Zug, Switzerland, [3]Technische Universität München, Liesel-Beckmann Str 1, 85354 Freising, Germany, [4]Iowa State University, Kildee 225, 50011 Ames, USA, [5]agn Genetics, Börtjistrasse 8b, 7260 Davos, Switzerland; mirjam.frischknecht@qualitasag.ch*

Whole genome sequence data (WGS) are assumed to include causative variants for any trait. With the 1000 Bull Genomes Project, a high number of sequenced bulls and cows became available. Lower density genotypes are available for most individuals with phenotypes i.e. from routine genomic evaluation. Therefore imputation is a widely used approach to increase marker density. In a first step we evaluated accuracy of imputation on BTA 25 from high density (HD, 12,222 variants) to sequence level (642,911 variants) in Brown Swiss cattle. In the 1000 Bulls Genome Project (Run5) sequences of 123 Brown Swiss (BSW) cattle were available. Imputation accuracy was assessed in nine different scenarios varying in composition of reference (e.g. random selection, single vs mixed breed reference) and validation set. For the BSW validation animals we masked sequence genotypes to HD. Imputation was carried out with Beagle4.0, FImpute, Impute2 and Minimac v3. The average correlation between imputed and original dosage was between 0.9409 and 0.982 across all scenarios and imputation programs. In a second step, given the promising imputation results, 1,646 genotyped BSW bulls were imputed genome-wide from HD to WGS (13,938,818 variants) to evaluate the precision of WGS for QTL fine-mapping. Genome wide association studies (GWAS) were carried out for milk fat content (FC) at early (days 8-12) and late (days 298-302) lactation using EMMAX. GWAS revealed one and three QTL ($P<3.84\times10^{-9}$) for early and late FC, respectively. The top variant ($P=2.15\times10^{-16}$) at a QTL for late FC on BTA 20 was a missense mutation in GHR encoding growth hormone receptor. The top association signals at all QTL result from imputed sequence variants demonstrating the enhanced capacities of whole genome sequence data pinpointing candidate causal variants.

A family-based imputation algorithm for high and low-coverage sequence data

M. Battagin, G. Gorjanc, S. Gonen, R. Antolin and J.M. Hickey
The Roslin Institute and Royal (Dick) School of Veterinary Studies, The University of Edinburgh, Easter Bush,
EH25 9RG, Midlothian, Scotland, United Kingdom; mara.battagin@roslin.ed.ac.uk

Sequence data is an increasingly common data type for livestock genetics. Sequence data is expensive and therefore low-cost strategies based on imputation are needed that spread information to whole populations. A widely used low-cost strategy involves: (1) sequencing a small number of key individuals whose haplotypes are shared with many other individuals in a population; (2) resolving the haplotypes carried by these individuals; (3) imputing these haplotypes into their relatives. This study develops and tests a family-based imputation algorithm for sequence data that can maximise the information extracted from a strategy based on sequencing key individuals. The algorithm focuses on resolving the haplotypes carried by key individuals by using high and low-coverage sequence data of the key animals, their parents and grandparents. It can iterate across families connected by pedigree. It builds consensus haplotypes based on the sequence reads of individuals by iterating down the pedigree and across the genome several times. The algorithm accounts for sequencing errors. The performance of the algorithm for imputing sequence data was tested on simulated data. More than 1 million of different scenarios were simulated. Scenarios followed a common structure but differed in one of the following aspects: (1) the number of individuals sequenced within family (from 1 to 7 members); (2) the coverage used to sequence the individuals (from 1× to 30×); and (3) the way in which families were connected. The percentage of corrected phase genotypes in key animals ranged from 0 to 100% depending on the amount of sequence data collected on the focal individual, its parents, and its grandparents. The results show that good accuracy of phasing and imputing sequence data can be achieved even when low cost sequencing strategies are adopted and suggests that the key sires sequencing approach can be optimised from a cost/benefit perspective.

Efficient identification SNPs in high linkage disequilibrium in large genotype and sequence datasets

M.P.L. Calus and J. Vandenplas
Wageningen UR Livestock Research, Animal Breeding and Genomics Centre, P.O. Box 338, 6700 AH Wageningen,
the Netherlands; jeremie.vandenplas@wur.nl

Performance and efficiency of genomic prediction models using whole genome sequence data may be affected by high levels of pairwise linkage disequilibrium (LD). To enable pruning of genotype data for LD, an algorithm is developed that enables rapid detection of any pair of variants in complete or high LD throughout the genome, rather than, for instance, only in a limited window surrounding an individual SNP. The developed algorithm relies on the properties of the squared correlation between phased alleles as a measure of LD. This measure can only reach a value of 1 for two loci when both loci have the same count of the minor allele. Sorting loci based on the minor allele count, followed by comparison of alleles at loci with the same minor allele count provides a quick way to detect loci in complete LD. Likewise, squared correlations close to unity can only be achieved for pairs of loci with a similar minor allele count. Deriving the exact range of differences in minor allele counts that can yield levels of LD above a predefined threshold, enables to considerably reduce the number of pairs of loci for which LD needs to be computed, relative to the total number of (n2-n)/2 pairs. An algorithm was implemented that removes SNPs in complete LD as a first step, and SNPs in high LD as a second step. This algorithm was tested on a simulated dataset containing 10,935,878 SNP and 2,500 animals. The algorithm removed 5,991,192 SNPs that were in complete LD with another SNP, and another 1,341,482 that had a squared correlation greater than 0.99 with an another SNP, leaving 3,603,204 out of the initial 10,935,878 SNPs. To remove high LD SNPs, the algorithm reduced the number of pairs of SNPs for which LD was evaluated from the total of $5.98×10^{13}$ pairwise combinations to $1.34×10^9$. Using a single thread, removing SNP in complete LD required ~20 minutes and removing SNP in high LD required ~11 hours, excluding time to read and write data. Using 10 threads reduced computing time with almost 90% for the second step, which then finished in less than 80 minutes. These results demonstrate that the developed algorithms enable to efficiently remove SNP in high or complete LD throughout the genome.

Selecting animals for whole genome sequencing: methods optimization and comparison

A. Butty[1], M. Sargolzaei[1,2], F. Miglior[1,3] and C. Baes[1]
[1]*Centre for Genetic Improvement of Livestock, Animal Biosciences, University of Guelph, 50 Stone Road East, Guelph, ON, N1G 2W1, Canada,* [2]*Semex Alliance, 5653 Highway 6 North, Guelph, ON N1H 6J2, Canada,* [3]*Canadian Dairy Network, 660 Speedvale Avenue West, Suite 102, Guelph, ON, N1K 1E5, Canada; buttya@uoguelph.ca*

Dairy cattle populations, in particular Holstein animals, have been widely genotyped since the arrival of affordable SNP chip technology. Whole genome sequencing of individuals, however, is still prohibitively expensive. Selection of animals to sequence must be carefully undertaken. Various methods have been proposed to select key animals contributing to the genetic diversity in the population, the first of which relied solely on pedigree information. As the number of genotyped animals increased, methods were adapted to include genomic information. Methods relying on haplotype frequencies have recently been proposed, which emphasize either selection of animals carrying frequent haplotypes, or classify animals based on the rarity of their haplotypes. The algorithms used are commonly iterative in nature. Current methods cover most of the genetic diversity of a population with a reasonable number of animals sequenced. These approaches, however, can be optimized using simulated annealing, a method developed to find the global optimum of a discrete dataset possibly containing several local optima. The simulated annealing method can be used to optimize the haplotype diversity within a group composed of already sequenced and sequencing candidates. Exchanging randomly one candidate at a time, haplotype diversity of each newly formed group will be assessed counting the number of unique haplotypes found for each genomic window. Different window sizes will be tested for each scenario. The proposed method is sensitive to outliers (e.g. crossbred animals) therefore strict removal of outliers is important. This study aims to develop an optimal method to select animals for sequencing. We further develop previously proposed methods using simulated annealing, and a new method based on haplotype diversity is described. Finally, a comparison of the number of animals needed to cover the same amount of genetic diversity and rare haplotypes within the Canadian Holstein population with each method is given.

A hidden Markov model to estimate inbreeding from whole genome sequence data

T. Druet[1] and M. Gautier[2]
[1]*Unit of Animal Genomics, GIGA-R, University of Liège, 1 avenue de l'Hôpital B34 (+1), 4000 Liège, Belgium,* [2]*UMR CBGP, INRA, Campus International de Baillarguet, 34988, Montferrier-sur-Lez, France; tom.druet@ulg.ac.be*

Inbreeding results from the mating of related individuals and has negative consequence because it brings together deleterious variants in one individual. Inbreeding is associated with recessive diseases and reduced production or fitness. Until recently inbreeding was estimated with genealogical data, which have some limitations. Genomic estimates of inbreeding can now be obtained thanks to new technologies. Methods based on genomic relationships assume either that identical-by-state markers are identical-by-descent (IBD) or correct for the allele frequencies estimated in a « founder generation ». Similarly, with runs of homozygosity (ROH), the hypothesis is that stretches of homozygous markers longer than a determined threshold (measured in number of markers or in Mb) are IBD. Although sequencing data are more informative, most of the recent studies used genotyping data. Due to high genotyping errors rates with sequencing technologies, ROH are difficult to use on such data. We herein describe an original method to model inbreeding along chromosomes. It relies on a hidden Markov model that determines locally whether a segment is inbred and gives a measure of inbreeding's age. The method uses a function modeling the probability to observe a genotype in an inbred segments. It is well suited for sequencing data because the function incorporates genotyping errors and uncertainty associated with low or moderate coverage. We first determine the properties of our new method with different simulation scenarii. Then, we use it to estimate inbreeding in the Belgian Blue Beef cattle population with either genotyping arrays (of various densities) or whole genome sequencing data. The estimates will be compared to those obtained with other methods (pedigree, ROH, genomic relationship). Latest results will be presented.

Accounting for read depth in the analysis of genotyping-by-sequencing data

K.G. Dodds[1], J.C. McEwan[1], T.P. Bilton[1], R. Brauning[1], R.M. Anderson[1], T.C. Van Stijn[1], T. Kristjánsson[2] and S.M. Clarke[1]
[1]AgResearch, Invermay Agricultural Centre, Private Bag 50034, Mosgiel 9053, New Zealand, [2]Stofnfiskur, Staðarberg 2-4, Hafnarfjörður 221, Iceland; ken.dodds@agresearch.co.nz

Untargeted genotyping-by-sequencing (GBS) is a method for obtaining SNP genotype information. It does not require the prior construction of SNP sets (e.g. as required for array-based genotyping) and can even be applied to species without a reference genome. The cost of GBS can be reduced by assaying many samples at one time and/or by sampling a smaller fraction of the genome. The former results in less genomic information for a sample (lower average read depth), while the latter reduces the number of SNPs assayed, but increases the read depth per SNP. Analysis of these data should take into account the allelic sampling nature of GBS, rather than treat the GBS results as genotype calls. We show how this can be done for the estimation of linkage disequilibrium and for calculating a genomic relationship matrix (GRM). The GRM can be used for: verification of duplicated samples (positive control samples), breed composition, pedigree, traceability, inbreeding and co-ancestry. It can also be included directly in existing mixed models to estimate genetic parameters and breeding values. The relationship estimation method also accounts for missingness in the data directly, avoiding the need for imputation and resources needed for imputation, such as genome sequence and reference panels. These methods open up opportunities for applications in agricultural species and studies of natural populations, particularly those that currently have poor genomic resources.

SNP-effects underlying GBLUP depend on the definition of the genomic relationship matrix

A.C. Bouwman[1], B.J. Hayes[2,3] and M.P.L. Calus[1]
[1]Wageningen UR Livestock Research, Animal Breeding and Genomics Centre, P.O. Box 338, 6700 AH, the Netherlands, [2]La Trobe University, Applied Systems Biology, 5 Ring Rd, Bundoora 3083, Victoria, Australia, [3]Government of Victoria, Department of Economic Development, Jobs, Transport and Resources, 5 Ring Rd, Bundoora 3083, Victoria, Australia; aniek.bouwman@wur.nl

Genomic prediction is used to predict direct genomic values (DGV) for selection candidates in breeding programs. Besides interest in DGV there is also an increasing interest in SNP-effects estimated using genomic prediction models such as GBLUP. With a GBLUP approach it is straightforward to back-solve the SNP-effects from DGV based on the genotypes of the animals. These SNP-effects can be used for various purposes such as rapid computation of DGV for newly genotyped individuals, for GWAS-type of analyses to get insight on the genetic architecture of a trait, and to estimate DGV based on small genomic regions, so called 'local DGV' for QTL mapping. Genomic prediction models require a genomic relationship matrix (GRM) that can be constructed using different genotype coding methods. Although it has been reported that there is no noticeable effect of genotype coding on DGV, we demonstrate here that the genotype coding used in construction of the GRM does have an effect on the back-solved SNP-effects, in particular for low MAF SNP. This could be particularly important for genomic prediction with use of sequence data, where many variants are at low frequency. In our dairy cattle data set with 643,924 SNP, the correlation between DGV using a GRM built with VanRaden (2008) method 1 (VR1), based on centering genotypes, and method 2 (VR2), based on centering and scaling, was 0.998 (regression coefficient=1), however, the correlation between back-solved SNP-effects using VR1 or VR2 was 0.15. The main difference in SNP-effects between the two methods was found for SNP with a MAF<0.01. In this MAF category, the correlation between SNP-effects was 0.44. Variants with a MAF between 0.01 and 0.05 showed a correlation of 0.90, all higher MAF categories showed a correlation between 0.97 and 0.98. Assuming that causal variants are rare and have a large effect, GRM based on sequence data should be centered and scaled appropriately for a better estimation of SNP-effects.

Functional annotation of CNV breakpoints in Holstein-Friesian cows

J. Szyda[1,2], M. Mielczarek[1,2], M. Fraszczak[2], R. Giannico[3], G. Minozzi[3,4], E. Nicolazzi[3] and K. Wojdak-Maksymiec[5]
[1]*National Research Institute of Animal Production, Krakowska 1, 32-083 Cracow, Poland, [2]Wroclaw University of Environmental and Life Sciences, Biostatistics group, Department of Genetics, Kozuchowska 7, 51-631, Poland, [3]Fondazione Parco Tecnologico Padano, Via Einstein, 26900 Lodi, Italy, [4]University of Milan, Via Festa del Perdono 7, 20122 Milano, Italy, [5]West Pomeranian University of Technology, Aleja Piastów 17, 70-310 Szczecin, Poland; joanna.szyda@up.wroc.pl*

Whole genome DNA sequences of 32 cows were analysed. The total number of raw reads per animal varied between 164,984,147 and 472,265,620. The bioinformatics pipeline included BWA-MEM used for the alignment to the reference genome, Picard and SAMtools used for post alignment editing, and CNVnator used for CNV detection. Using Variant Effect Predictor genomic position of CNV breakpoints were annotated to the reference genome. 487,746 deletions and 61,789 duplications were detected. CNV length ranged between 200 bp and 724,000 bp for deletions as well as 200 bp and 439,300 bp for duplications. The total number of deletions identified per individual was between 13,149 and 22,496. The number of duplications varied between 1,694 and 5,187. The functional annotation showed that the highest numbers of deletion breakpoints were located in intergenic regions (62.84%) and introns (26.84%). 1.35% of the breakpoints were reported within coding regions. The least numbers were reported for non-coding regions (0.32%) and splice regions (0.14%). For duplications, the proportion of breakpoints in each functional group was similar. Most of duplications were placed in intergenic regions (70.37%) and introns (21.63%), intermediate number of breakpoints were found in coding regions, (0.79%) of duplication breakpoints were found in coding regions, the least in splice regions (0.37%) and non-coding regions (0.11%). The highest number of CNV breakpoints (17,315), was observed within the protein kinase, cGMP-dependent, type I gene. CNV detection was performed at the Poznan Supercomputing and Networking Center. Statistical analyses were supported by the National Science Foundation project no. 258836.

NGS-based analysis of copy number variations in various cattle breeds

M. Mielczarek[1], M. Frąszczak[1], E.L. Nicolazzi[2], G. Minozzi[2], H. Schwarzenbacher[3], C. Egger-Danner[3], D. Vicario[4], F. Seefried[5], A. Rossoni[6], T. Solberg[7], L. Varona[8], C. Diaz[8], C. Ferrandi[2], R. Giannico[2], J.L. Williams[9], J. Woolliams[10] and J. Szyda[1]
[1]*Wroclaw University of Environmental and Life Sciences, Kozuchowska 7, Wroclaw, Poland, [2]Fondazione Parco Tecnologico Padano, Cascina Codazza 12, Lodi, Italy, [3]ZuchtData EDV, Dresdner 89/19, Vienna, Austria, [4]Italian Simmental Cattle Breeders Association, Via Nievo 19, Udine, Italy, [5]Swiss Brown Cattle Breeders Federation, Chamerstrasse 56, Zug, Switzerland, [6]Italian Brown Cattle Breeders' Association, Ferlina, Bussoleng, Italy, [7]Norwegian University of Life Sciences, Universitetstunet 3, As, Norway, [8]Universidad de Zaragoza, Pedro Cerbuna 12, Zaragoza, Spain, [9]University of Adelaide, SA 5371, Roseworthy, Australia, [10]Roslin BioCentre, EH25 9PP, Roslin, United Kingdom; magda.a.mielczarek@gmail.com*

Whole genome DNA sequences were determined for 104 bulls representing Brown Swiss (48 individuals) Fleckvieh (30), Guernsey (20), Simmental (16) and Norwegian Red (23) breeds. The total number of raw reads obtained for a single animal varied between 270,678,710 (a Fleckvieh) and 768,980,700 (a Brown Swiss). The average genome coverage ranged from 10 to 28. Alignment to the UMD3.1 reference genome was carried out using BWA-MEM. CNV calling was performed with the CNVnator software. The number of duplications per individual varied between 2,204 and 48,501, while the number of deletions varied between 9,771 and 24,334. Six deletions located on chromosomes 6, 7, 17, 21, 23 and 29 were shared among all animals, while there were no duplications shared among all of them. Length of CNVs varied from 200 bp to 999,300 bp for deletions, and from 200 bp to 925,000 bp for duplications. In conclusion a significant variation in genome structure is observed both within as well as among breeds. The research was carried out within the EU 'Gene2Farm' project (7FP grant No. 289592) and Polish National Science Centre grant No. UMO-2014/15/N/NZ9/03914. Processing of the raw data was performed at the Poznan Supercomputing and Networking Center.

Genome-wide association study in Brown Swiss for udder traits based on sequence data
M. Frischknecht[1,2], C. Flury[1], B. Bapst[2], F.R. Seefried[2], H. Signer-Hasler[1], D. Garrick[3], C. Stricker[4], Intergenomics Consortium[5], A. Bagnato[6], A. Bieber[7], R. Fries[8], I. Russ[9], J. Sölkner[10] and B. Gredler[2]
[1]Bern University of Applied Sciences, School of Agricultural, Forest and Food Sciences HAFL, Länggasse 85, 3052 Zollikofen, Switzerland, [2]Qualitas AG, Chamerstrasse 56, 6300 Zug, Switzerland, [3]Iowa State University, Kildee 225, 50011 Ames, USA, [4]agn Genetics, Börtjistrasse 8b, 7260 Davos, Switzerland, [5]Interbull center, SLU, Box 7023, Uppsala, 75007, Sweden, [6]University of Milan, Via Celoria 10, 20133 Milano, Italy, [7]Research Institute of Organic Agriculture (FiBL), Ackerstrasse 113, 5070 Frick, Switzerland, [8]Technische Universität München, Liesel-Beckmann-Straße 1, 85354 Freising-Weihenstephan, Germany, [9]Tierzuchtforschung e.V., Senator-Gerauer-Str. 23, 85586 Poing, Germany, [10]University of Natural Resources and Life Sciences, Gregor-Mendel-Str 33, 1180 Wien, Austria; heidi.signer-hasler@bfh.ch

Identification of QTL, especially of causative variants within QTL is still challenging. Higher SNP densities aid the identification and fine-mapping of QTL. Based on imputed sequence data we performed GWAS for udder traits in Brown Swiss cattle. The GWAS was performed using a mixed-model approach with deregressed breeding values as phenotypes. The traits investigated included: udder depth (UD), fore udder attachment (FUA), rear udder width (RUW), rear udder height (RUH), fore udder length (FUL), and central ligament (CL). We found significant associations on BTA 3 (UD, FUA), BTA 5 (UD), BTA 17 (FUL, RUW, CL) and BTA 20 (FUA). A single gene was located in the significantly associated regions on BTA 5 (ABCC9) and BTA 20 (HCN1). The region on BTA 17 spans almost 3 Mb and includes 74 genes (maximal region for all the traits combined) and the region on BTA 3 includes 91 genes across 3 Mb. We also looked for associated missense variants in these intervals. Neither for ABCC9 nor for HCN1 we could identify such a variant. On BTA 17 we identified 2 missense variants that were significantly associated with CL. On BTA 3, 11 missense variants were significantly associated with UD and/or FUA. The advantages of using imputed sequence data compared to SNP chip genotypes are mainly through the inclusion of potential causative variants.

Correlation between DNA methylation in CHFR gene promoter and economic traits in J. Black cattle
Y. Suda[1], E. Kobayashi[2], Y. Saito[3], K. Kato[4] and K. Suzuki[5]
[1]Miyagi University, Food, Agricultural and Environmental Sciences, 2-2-1 Hatatate, Taihaku, Sendai, 9820215, Miyagi, Japan, [2]National Institute of Livestock and Grassland Science, Tsukuba, Ibaraki, Japan, [3]Miyagi Prefectural Livestock Experiment Station, Osaki, Miyagi, Japan, [4]Tohoku University, Graduate School of Agricultural Science, Sendai, Miyagi, Japan, [5]Tohoku University, Graduate School of Agricultural Science, Sendai, Miyagi, Japan; suda@myu.ac.jp

Meat quality of Japanese Black cattle (JB) is known to excel in Wagyu, and the breed has become very popular worldwide. However, the main mechanism of developing these characteristic features are not well understood although many researchers have reported SNPs and QTLs which may be related to them. DNA methylation to C in CG rich of the upstream region of start codon controls the expression of many genes on a genome wide level in relation to environmental effects. This study aims to examine a correlation between genome DNA methylation of CG rich in upstream region of CHFR gene and economic traits in JB. Samples were collected from adipose tissues around the kidneys in each 100 JBs produced from same sire. After DNA extraction and purification, EpiXplore Methylated DNA Enrichment Kit, EpiScope Promoter qPCR Array (Human), SYBR Premix Ex Taq GC (Perfect Real Time) of TAKARA and specific software were then used to analyze DNA methylation degree. The DNA methylation degree of CG rich regions in the upstream regions of CHFR gene which encodes E3 ubiquitin-protein ligase correlated significantly with subcutaneous fat thickness and rib thickness (r=0.66). The expression of this gene is required for the maintenance of the antephase checkpoint that regulates cell cycle. So, this gene may play a key role in cell cycle to progress weight gain and lipid accumulation. In this study, although apparent relationships between the known QTL and this gene could not be found, this information might be an effective marker for monitoring meat quality in JB.

Transcriptome profile analysis of ovarian tissues highlights steroids biosynthetic genes in ewes

H. Ghaderi[1], A. Masoudi[1] and B. Arefnejad[2]
[1]Tarbiat Modares University, Animal Breeding, Jalale Ale ahmad Highway, Gisha Bridge, Tehran, 1497713111 Tehran, Iran, [2]OMICS Research Group, Jalale Ale ahmad Highway, Gisha Bridge, Tehran, 1497713111 Tehran, Iran; masoudia@modares.ac.ir

Recently RNA-Seq has been used to investigate the differentially expressed genes in different species. Prolificacy is one of the important traits in sheep production system. Study of the genetic differences between uniparous and multiparous animals may help to improve breeding programs in favor of increasing prolificacy in sheep. Shal sheep is an indigenous Iranian breed, which is capable to have a high prolificacy rate. The aim of this study was to analyze the transcriptomes of ovarian tissues in the groups of the uniparous and multiparous Shal sheep. Using Illumina Hiseq™ 2000 deep sequencing, a total of 46,956,986 and 51,747,856 reads from the samples of uniparous and multiparous Shal sheep were obtained, respectively. Paired-end reads of the sequences were obtained by filtration, and they aligned to the ovis aries reference genome (Ova Version 1.3) using TOPHAT (v2.0.13). After aligning, 26,853 genes were recognized to be expressed in ovarian tissue. To identify significantly differential expression profiles between two groups of the animals, Cufflinks (v 2.1.1), Cuffmerge and Cuffdiff (v 2.1.1) softwares were used. The results shown that 619 genes had significant differences (P<0.05) between uniparous and multiparous animals. Some studies indicated that GDF9 gene has an important effect on prolificacy in sheep. Therefore, this study focused on chromosome 5, which is the host for this gene. In the current study, 1,356 genes were located on chromosome 5. Of these genes, the expression of 28 genes were significantly different between the uniparous and multiparous animals. From these 28 genes, 18 down-expressed and 10 over-expressed were identified in the multiparous animals. Gene ontology terms of KEGG pathway analysis in observed genes showed that the biosynthesis of the steroids pathways is enriched in the ovarian tissue.

The analysis of CNV regions in Holstein-Friesian cows

M. Frąszczak[1], M. Mielczarek[1,2], E.L. Nicolazzi[3], G. Minozzi[3], K. Giannico[3], K. Wojdak-Maksymiec[4] and J. Szyda[1,2]
[1]Wroclaw University of Environmental and Life Sciences, Biostatistics group, Department of Genetics, Kożuchowska 7, Wrocław, Poland, [2]National Research Institute of Animal Production, Krakowska 1, Cracow-Balice, Poland, [3]Fondazione Parco Tecnologico Padano, Haussmann 11/15, Lodi, Italy, [4]West Pomeranian University of Technology, al. Piastów 17, Szczecin, Poland; magdalena.fraszczak@up.wroc.pl

The copy number variation (CNV), one of the important sources of genetic diversity, is defined as the gains (duplications) and losses (deletions) of longer DNA fragments. In this study we concentrated on the CNVs identified based on a read depth (RD) algorithm. Genomic regions with CNVs detected for at least one cow were considered. If two or more CNVs whole or partly overlap, they belong to the same CNV region. The material consist whole genome DNA sequences determined for 29 cows representing the Polish Holstein-Friesian breed. Alignment to the UMD3.1 reference genome was carried out using BWA-MEM and the CNVnator, software was used for CNV detection. The total number of 549,539 CNVs identified in the 29 genomes corresponded to 49,313 CNV regions, after the variant overlap was taken into account. The identified regions of duplication occupied 4.42% of whole genome while the regions of deletions occupied 11.72% of a genome. The shortest duplicated region of 200 bp was located on BTA19, and the longest, identified on BTA27, was equal to 385,300 bp and consisted of 525 single CNVs. For deletions the length of regions varied depending on autosome form 200 bp (BTA25) to 1,374,300 bp (BTA12). The longest region contained 632 single CNVs. Furtheremore overlap between CNV regions and genes was precisely analyzed. In order to assess the accuracy of CNV detection the validation was performed by comparing CNVs detected by using CNVnator, Pindel and GASV (the softwares based on read depth, split read, paired-end mapping algorithm respectively). Finally identified regions were compared to the CNVs from dbVar database. Processing of the raw data was performed at the Poznan Supercomputing and Networking Center. Statistical analyses were supported by the National Science Foundation project no. 258836.

Genome-wide association study in Brown Swiss for fertility traits based on sequence data

M. Frischknecht[1,2], B. Bapst[2], C. Flury[1], F.R. Seefried[2], H. Signer-Hasler[1], D. Garrick[3], C. Stricker[4], Intergenomics Consortium[5], R. Fries[6], I. Russ[7], J. Sölkner[8], A. Bieber[9], A. Bagnato[10] and B. Gredler[2]
[1]Bern University of Applied Sciences, School of Agricultural, Forest and Food Sciences HAFL, Länggasse 85, 3052 Zollikofen, Switzerland, [2]Qualitas AG, Chamerstrasse 56, 6300 Zug, Switzerland, [3]Iowa State University, Kildee 225, 50011 Ames, USA, [4]agn Genetics, Börtjistrasse 8b, 7260 Davos, Switzerland, [5]Interbull center, SLU, Box 7023, Uppsala, 75007, Sweden, [6]Technische Universität München, Liesel-Beckmann-Straße 1, 85354 Freising-Weihenstephan, Germany, [7]Tierzuchtforschung e.V., Senator-Gerauer-Str. 23, 85586 Poing, Germany, [8]University of Natural Resources and Life Sciences, Gregor-Mendel-Str 33, 1180 Wien, Austria, [9]Research Institute of Organic Agriculture (FiBL), Ackerstrasse 113, 5070 Frick, Switzerland, [10]University of Milan, Via Celoria 10, 20133 Milano, Italy; beat.bapst@qualitasag.ch

Fertility in dairy cattle has declined continuously during the last years. In Switzerland low fertility is one of the most important culling reason. Therefore this trait complex is of great economic importance. But fertility traits are functional traits with low heritabilities. We performed GWAS in Brown Swiss and Holstein cattle for the fertility traits: non-return rate after 56 days in heifers (NRH) and cows (NRC); days between first and last insemination in heifers (FLIH) and cows (FLIC); and days to first service (DFS). We used deregressed breeding values as input phenotypes and imputed whole-genome sequence dosages as genotypes. In Holstein (2,336 to 3,257 individuals) we could not identify any significant region for any of the investigated traits. For Brown Swiss cattle (1,392 to 3,278 individuals) however we could identify a QTL on BTA 17 for NRH and FLIH. This locus also showed a suggestive association with NRC and FLIC. The region significantly associated to NRH and FLIH comprises 27 genes. Among the associated variants two non-synonymous variants in the genes GAS2L and ASCC2 could be identified. These two variants could be potentially causative variants.

Genome-wide differential DNA methylation patterns between Iberian & tropical adapted Creole bovines

N. Sevane[1], R. Martinez[2], P. Orozco-Terwengel[1] and M.W. Bruford[1,3]
[1]Cardiff University, School of Biosciences, Museum Avenue, CF10 3AX, United Kingdom, [2]Corporacion Colombiana de Invetigacion Agropecuaria (Corpoica), Centro de Investigaciones Tibaitata, Km 14 via Mosquera, Cundinamarca, Colombia, [3]Cardiff University, Sustainable Places Research Institute, 33 Park Place, Cardiff, CF10 3BA, United Kingdom; orozco-terwengelpa@cardiff.ac.uk

Official assessments of climate change impacts predict a progressive upward trend in average temperatures, leading to a decreased forage production and quality, and an increased disease risk. Therefore, enhancing climate resilience and sustainable production for animals in harsh environments are important goals for the livestock sector. A natural experiment already exists, Columbus' arrival in the Americas, where rapid adaptation to extreme climatic conditions was imposed on a limited number of livestock. This study compares the methylomes of two tropical Creole bovine breeds and their Spanish ancestors to understand epigenetic mechanisms underlying their adaptation. Reduced representation bisulfite sequencing (RRBS) was used to assess differences in methylation between Creole and Spanish samples. Only CpG dinucleotides with 10X coverage (CpG_{10}) where used in downstream analysis. Comparison between groups revealed 3,639 differentially methylated CpG_{10} sites (DMS_{10}, q-value<0.01, differential methylation >25%), annotated to 520 unique features. Gene ontology analysis revealed candidate genes involved in tropical adaptation, such as immune response, tick resistance, skin and hair characters, response to stress, nervous system, vascular physiology, energy management, adaptation to new feeding conditions and reproductive performance. The RRBS results were validated by bisulfite sequencing PCR (BSP). Our results show that the drastic climate change imposed on a reduced number of animals had an impact on their methylome pattern that is currently still measurable, affecting genes implicated in important signalling pathways for adaptation and pointing towards the epigenetic fine-tuning on the regulation of gene activity. This study justifies further work to understand the mechanisms underlying the relationship between epigenetic variation and extensive phenotypic diversity of domestic species.

Functional SNP & splice variants in pigs with divergent lipid phenotypes using genome-wide analysis

T. Cardoso[1], M. Amills[1], R. Gonzalez-Prendes[1], R. Quintanilla[2] and A. Canovas[3]
[1]Center for Research in Agricultural Genomics (CSIC-IRTA-UAB-UB), Universitat Autònoma de Barcelona, Animal Genetics, Bellaterra, 08193, Spain, [2]IRTA, Animal Genetics, Torre Marimon, Caldes de Montbui, 70040-020, Spain, [3]University of Guelph, Animal Bioscience, 50 Stone Road E., Guelph, N1G 2W1, Canada; acanovas@uoguelph.ca

RNA-Sequencing technology was used to examine the structural variation of the gluteus medius (GM) muscle transcriptome of commercial Duroc pigs with divergent phenotypes (HIGH and LOW) for 13 fatness traits. The whole transcriptome of 52 GM samples from HIGH (n=26) and LOW (n=26) pigs were sequenced with a HiSeq2000 platform. An average of 70-million sequence reads were obtained from each sample and mapped to the pig reference genome. In all samples, 80% of the reads were categorized as mapped. In this way, 30 splice variants were differentially expressed between pigs with divergent fatness traits (P<0.01 and Fold-change>1.5). Noteworthy, 8 splicing variants happened to be novel (unannotated in the last pig reference genome available (Sus scrofa 10.2.79)). Pathway analysis evidenced that most of differentially expressed splice variants belong to the glycolysis and gluconeogenesis, oxidative phosphorylation, insulin signaling and fatty acid biosynthesis metabolic routes. We can conclude that in many instances differential expression targets specific mRNA isoforms rather than the global set of transcripts produced by a given gene. Besides, several SNP variants segregated specifically in either the HIGH (2,995 SNPs) or LOW (1,467 SNPs) fatness profiles. Among them, 479 (16.8% of HIGH group-specific SNPs) and 177 (12.0% of LOW group-specific SNPs) were associated with amino acid changes. It would be interesting to correlate SNP frequencies and isoform differential expression data to identify polymorphisms with effects on the mechanism of splicing.

Developments, strategies and challenges for UK dairy sector

G. Jones
Agriculture and Horticulture Development Board, Kenilworth, Warwickshire, United Kingdom;
natalie.reynolds@ahdb.org.uk

Mr. Gwyn Jones is chair of the dairy sector board of the UK Agriculture & Horticulture Development Board (AHDB). He is a dairy farmer and has been farming in West Sussex for over 30 years, running a 350 cow herd with an Anaerobic Digester integrated in the farm business. Mr. Jones is Chairman of RUMA (Responsible Use of Medicines in Agriculture), Chairman of EPRUMA (The European equivalent of RUMA), an active member of the Farm Animal Welfare Council (FAWC), Vice Chairman of the COPA-Cogeca Animal Health and Welfare Working Group, member of the British Veterinary Association (BVA) Welfare and Ethics Group, and a member of the University of Surrey School of Veterinary Medicine's External Liaison and Advisory Committee. He will talk about the challenges and opportunities of the dairy sector in the UK and in a wider perspective. He will also discuss the current situation and issues to deal with.

Analysis of dairy farmers' and stakeholders' expectations and challenges in four European countries

M. Klopčič[1], A. Malak-Rawlikowska[2], A. Stalgiene[3] and A. Kuipers[4]
[1]University of Ljubljana, Biotechnical Faculty, Dept. of Animal Science, Groblje 3, 1230 Domžale, Slovenia, [2]Warsaw University of Life Sciences, Faculty of Economic Sciences, ul Nowoursynowska 166, 02-787 Warsaw, Poland, [3]Lithuanian Institute of Agrarian Economics, V. Kudirkos g. 18-2, 03105 Vilnius, Lithuania, [4]Wageningen UR, Expertise Centre for Farm Management and Knowledge Transfer, P.O. Box 35, 6700 AA Wageningen, the Netherlands; marija.klopcic@bf.uni-lj.si

Data from dairy farmers and stakeholders were collected from 2011 to 2016 concerning their future expectations, as part of a Leonardo da Vinci and an ERASMUS+ project. The goal was to examine if a coherent outlook on future developments and challenges can be found within and between selected European countries. Topics dealt with were choice of development paths, availability of resources, perceived opportunities & threats and farmers' abilities, and economic expectations. In 2011/12, 1,039 questionnaires from farmers were received, spread over Poland, Lithuania and Slovenia. In 2013, 352 of these farmers filled in the questionnaire again as well as 122 farmers from the Netherlands. Those farmers were again questioned in 2016. In 2015, 150 stakeholders in the four countries, composed of 8 groups, i.e. from dairy cooperatives to ministries and NGO's, were asked similar questions. Additionally, they described weak and strong points of the national dairy chain. PCA, cluster analysis and ANOVA methods were used to analyse the dataset. On base of 10 strategies, seven farmer segments were identified. Differences in opinions between these segments were rather limited, while significant differences between countries were found. For instance, farmers in the Netherlands show much more confidence in dealing with the free quota situation and the dairy market than in the other countries. Farmers in Lithuania and Slovenia are more diversified, while the latter farmers are also more consumers oriented. The trend in opinions over years will be discussed. Concerning the stakeholders, preliminary results show only small differences in opinions between the various groups. However, the outlook on the future is quite different between countries. Results of this study provide insights, which can support the development of the dairy sector in the various regions of Europe.

Developments, strategies and challenges for Polish dairy sector

A. Malak-Rawlikowska
Warsaw University of Life Sciences, Faculty of Economic Sciences, Nowoursynowska 166, 02-787, Poland; agata_malak_rawlikowska@sggw.pl

The implementation of the Common Agricultural Policy (CAP) after accession to the EU in 2004 has been a milestone for Polish dairy sector. Easier access to EU markets, the introduction of direct payments, continuing positive price/cost relationship trends and subsidies from the Rural Development Program had a significant impact on the economic situation of dairy farms. Strong concentration of marketable production and an increase in scale led to the decrease in the number of farms. In the period 2004-2015 the number of dairy farms delivering milk to the market decreased by 64% and average milk production per farm almost tripled. The remaining farms continued modernization, investments and breeding improvement. Concurrently, milk yield per dairy cow was steadily increasing. This resulted in an increased production level and increased deliveries to processing plants. Together with the CAP liberalization process, especially abolishment of the milk quota system and reducing export subsidies, dairy market is more exposed to the world market dynamics. In this difficult environment, farmers had to adjust their development strategies in order to survive on the competitive market. Taking into account large investments done in the past years at the farm level, high demand for raw milk expressed by processing industry (resulting from investments in processing powers) and growing demand for dairy products, dairy farmers took the opportunity of abolishment of the milk quota system and continued development towards increased specialisation and production scale. Despite of difficult situation on the global dairy market in 2015/16 (very low prices), Russian embargo, liberalization of CAP (reduced protection of internal market), high penalties for exceeding dairy quota in the last milk quota year 2014/2015, milk deliveries in Poland tend to grew by 6,7% in 2014 and 2,7% in 2015, comparing to the previous year. It is expected that, due to unstable situation on dairy market, low performing farms will withdraw from production, and farm concentration processes will speed up. The upward trend of milk production will continue in the future, but with lower dynamics, depending on the situation on the global market.

Solving the problems facing the dairy sector
D. Dobbin
Northern Ireland Food and Drink Association, Belfast Mills, Belfast, United Kingdom; jason@agrisearch.org

Dr David Dobbin is Group Chief Executive of United Dairy Farmers / Dale Farm, the UK's largest dairy cooperative. He is Chairman of Dairy UK and was appointed as an expert to the EU Agri Markets Task Force in February 2016. He is also a member of the Northern Ireland Food Strategy Board and Chairman of Belfast Harbour Commissioners, Chair of the Management Committee of Ulster Rugby and a past Northern Ireland Chair of the Confederation of British Industry. Dr Dobben will discuss the problems facing the dairy sector from the angle of his industrial and business background, with references to policies and politics.

Developments, strategies and challenges for Netherlands' dairy sector
A. De Veer, I. De Boer and A. Kuipers
Wageningen UR, Livestock Research, De Elst 1, 6708 WD Wageningen, the Netherlands; imke.deboer@wur.nl

In 2014 the Netherlands had 18,000 dairy farms with 1.6 million cows, 12.7 billion kg of milk of which 65% sold abroad, 23 processing companies and 60,000 jobs in the chain. This contributed 1.2% to national economy and 9% to trade surplus. Grassland prices in period 2013-15 varied between €45.000 till 80.000/ha. Recently, milk price is around 30 ct/kg, but for ecological milk 52 ct/kg (2% of total volume). In years preceding the ending of quota system, policy discussions started how to guide the sector in a sustainable way into the quota free era. A land tied dairy sector was envisioned by politics and the sector itself. This was laid down in the so called 'Melkveewet' (Dairy Husbandry Law). Enlargement of no. of cattle was linked to requiring additional land. Moreover, the national produced amount of phosphate from animal manure was already set at 172.9 million kg in agreement with EU commission, of which 84.9 for the dairy cow population. Farmers anticipating the quota free situation enlarged herds in 2013/2014. In 2015, the number of cows was increased by 120.000, + 7, 5% (production + 10%). As result, three month after the abolition of quota, the Netherlands government installed a phosphate quota on farm level, reference July 2015, restricting the sector: 4-8% flat reduction; 10% take away by transfer; price at moment €5,000/cow. Main focus in developing the sector is on increasing cow production efficiency, nutrient use efficiency (by introduction of a nutrient cycle planner; farm specific P-norms are allowed), amazing grazing initiatives, including policy to increase % of cows grazed, cow welfare (by experimenting with bedded pack barns and the innovative cow garden), improving structure of soils, decrease in antibiotic use (partly realized) and cow health initiatives, like increasing resilience of cows. Research in these areas will be presented. Early 2015, a commission was appointed by the government to make a future draft of a sustainable animal sector, which results will be presented as well.

Economic and environmental consequences of milk quota abolition in the Netherlands

C.E. Van Middelaar[1], C.W. Klootwijk[1], P.B.M. Berentsen[2] and I.J.M. De Boer[1]
[1]Wageningen University, Animal Production Systems group, De Elst 1, 6708 WD Wageningen, the Netherlands,
[2]Wageningen University, Business Economics group, Hollandseweg 1, 6706 KN Wageningen, the Netherlands;
corina.vanmiddelaar@wur.nl

This study used a whole-farm optimization model to analyse the impact of the abolition of the milk quota on the economic and environmental performance of an average Dutch dairy farm. The abolition of the milk quota system in the Netherland was accompanied by the introduction of a new manure policy to limit phosphate excretion on expanding dairy farms. The new policy prescribes that any increase in phosphate excretion should be partly processed and partly applied to additional farmland. In addition, phosphate quotas are introduced. Changes in farm structure, management, labour income, nitrogen and phosphate surpluses, and greenhouse gas emissions were assessed by comparing a farm before and after quota abolition and introduction of the new manure policy. Results show that based on current prices, increasing the number of cows after quota abolition is profitable until manure processing or land purchases is required to comply with the new manure policy. Farm intensity increases by about 4%, from 13,578 kg milk per hectare before quota abolition, to 14,130 kg milk per hectare after quota abolition. Labour income increases by €505 per year. When costs of manure processing or land decrease, or when milk prices increase, further farm expansion becomes profitable. Results show that the quota abolition, accompanied by a new manure policy, will slightly increase nutrient losses per ha, due to an increase in farm intensity. Greenhouse gas emissions per unit of milk will hardly change, but total greenhouse gas emissions will increase linearly with an increase in the number of cows.

Developments, strategies and challenges for France dairy sector

V. Brocard[1], C. Perrot[2], G. You[2] and A. Le Gall[1]
[1]Institut de l'Elevage, Techniques d'Elevage et Environnement, BP 85225, 35652 Le Rheu Cdx, France, [2]Institut de l'Elevage, Economie des filières, 149 rue de Bercy, 75795 Paris Cedex 12, France; valerie.brocard@idele.fr

France is the second producing country for bovine dairy sector in Europe. The country probably represents one of the biggest development potentials in the EU together with Germany and Poland. The assets of the dairy sector are the following: the possible combination of high potential yields for forage production (maize silage and grass), the largest agricultural area in Europe, a lower land price than in the neighboring countries, well established know-hows and skills both at production and processing levels, and a powerful and innovative industry. Though the country also presents limiting factors for the expansion of dairy production. The decrease rate of the farm numbers is now quicker than in many other northern European countries. The dairy chain will require farmers and farm managers, and is currently facing issues in relation to the attractiveness of these jobs, to the financial conditions for settling-up, and to the education of the future farmers. The low resort to salaried workers, in particular in large farms, could be a handicap in the search for higher agricultural labour productivity. The grant scheme for farm modernization as well as the readability of the post quota background (with freedom of supply versus contracting and curtailment) may be less favourable for investments than in other countries. Actually the low level of specialization of the French dairy farms, the increase in acreages of cultivated land per agricultural worker, and the competition among agricultural productions within the farms are factors reinforcing the need for increased monitoring of the volatility of the economic efficiency of dairy production, a sector which requires high levels of assets and labour. Finally, in the future, the technical efficiency of the production systems in the farms should lead to an optimal combination of productivity factors (land, labour), production costs and respect of the environment.

Developments, strategies and challenges for Italian dairy sector

A.M. Menghi
Reseacrh Centre on Animal Production (CRPA), Economics, Viale Timavo 43/2, 42122 Reggio Emilia, Italy;
a.menghi@crpa.it

The Italian dairy sector is often mentioned for the highest milk prices paid to the farmer's milk within the European Union. On the other side, the number of Italian dairy farms is rapidly decreasing: more than 1000 farms every year are quitting milk production. The country with the highest number of PDO (Product of Designation Origin) cheeses in Europe, like Parmigiano-Reggiano and Gorgonzola, seems unable to stop this trend. The diagnosis and the critical points of the sector have been detected in several studies. However, the actors of the Italian dairy supply chain are not able to create any strategical national plan to be coordinated with the EU indication, from the 'milk package' to the quota abolishment, to react to the cyclical milk crisis which is affecting the sector. One of the issues since many months on top of the political agenda is the 'price indexing'. This concerns a magic formula to find an impossible automatic price agreement between the strong industrial dairy processing companies and the incredible weak and fragmented farmers associations and co-operatives. Farmers who are not waiting anymore for a political solution of the milk crisis are moving directly to the market with their products, trying to establish a direct contact with the consumers. As a result, an interesting number of new dairy products and processes are emerging, of which some are successful and show an alternative way of gaining added value to their colleagues. Diversification, biodiversity, sustainability and animal welfare are some of the key words for those entrepreneurial farmers.

Developments, strategies and challenges in the German dairy sector

T. Lindena
Thünen Institute for Farm Economics, Bundesallee 50, 38116 Braunschweig, Germany; tomke.lindena@thuenen.de

Among the European member states, Germany is the biggest milk producer in dairy cattle. With 4,284,639 dairy cows in 2015 and 32,395,000 tons of milk yearly production in 2014, Germany holds a share of 18% of the EU dairy cows and 20% of the EU milk produced. The total German milk production has increased about 13% from 2008 until 2014. Thereby, 45% of the milk is produced in two federal states: in Lower-Saxony (northern part of Germany) and in Bavaria (southern part). Taking a look on the quota transfer from July 2007 to November 2014, there is a shift towards dairy production in grassland regions. Moreover, a dichotomy can be observed for the herd size in Germany: In 2014, the average size of farms in the eastern part is 183 cows/farmer. However, in the western part of Germany, dairy farms are smaller (48 cows/farmer). The average herd size throughout Germany is therefore 56 cows/farmer in 2014. Two-thirds of the cows are pure dairy breeds; mainly Holstein Friesian (HF). The average milk yield is 7,352 kg/cow/year. For HF this is 8,865 and for German Fleckvieh 7,243, which is a dual purpose breed, mainly used in the southern part of Germany. The number of dairy farms has decreased about 80% during the past 30 years. In 2015 there were 73,255 dairy farms. Structural change will continue to accelerate as low milk prices pose a particular challenge. The average milk price was below 30 ct/kg in 2015. Lower producer prices, increasing land prices and the availability of qualified labor are particularly challenging up to date. Furthermore, the interest in farming systems by the consumer, society and politics increases. Especially the concept of sustainable dairy farming comes into focus. Retailer set up a list of requirements for the milk production, including aspects such as animal welfare. While in the past sustainability initiatives were focused mainly on dairy factories level, dairy farms are taken into account now. First sustainability concepts are just established on farm level so far or are still in progress. There is growing evidence that dairy farmers, who want to continue their future milk production in a competitive way, have to consider meeting the demands of permanent and comprehensive sustainable requirements.

An analysis of new entrants into the Irish dairy sector during the last years of milk quota

R. McDonald[1], B. Horan[2], K. Pierce[3] and A. Macken-Walsh[4]
[1]Currently: Aurivo Cooperative Society, Farm Profitability Programme, Finisklin Ind Estate, Sligo, Ireland,
[2]Teagasc, Animal and Grassland Research and Innovation Centre, Moorepark, Fermoy, Cork, Ireland, [3]University
College Dublin, School of Agricultural and Food Science, UCD, Belfield, Dublin 4, Ireland, [4]Teagasc, Rural Economy
and Development Research Centre, Aras Ui Mhaoiliosa, Athenry, Galway, Ireland; roberta.mcdonald@aurivo.ie

In the lead up to milk quota abolition in the Europe the dairy industry was preparing for major change on farms and in markets. For the first time in almost 30 years farmers can enter into a quota free environment of milk production. A New Entrant Scheme was established on the back of decisions through the 2008 Health Check which allowed the Irish department of agriculture to create an environment for people to enter the industry prior to the complete removal of milk quotas. As a part of a comprehensive study looking at the financial implications of entering milk production in the lead up to 1st of April 2015, the most viable strategies in establishing a business, as well as more critical questions in this context are how farmers react to the myriad of challenges coming forth from changed policy circumstances and, relatedly what influences will ultimately determine their strategies at farm level? As part of a broader multi-disciplinary research project, this paper presents a narrative analysis of dairy farmers participating in Ireland's New Entrant Scheme, an initiative that has facilitated the establishment of over two hundred new dairy farms in preparation for imminent quota deregulation.

ECO FCE: In ovo manipulation:long term effect of synbiotics on morphology of chicken gut

K. Stadnicka*, J. Bogucka*, G. Elminowska-Wenda, A. Sobolewska, A. Dankowiakowska, A. Szczerba and
M. Bednarczyk
UTP University of Science and Technology in Bydgoszcz, Animal Biochemistry and Biotechnology, Mazowiecka
28, 85-084, Poland; * joint first author; katarzyna.stadnicka@utp.edu.pl

To ensure the best immune protection for the newly hatched chicken, the external supplementation with prebiotics or synbiotics should be given as early as possible. Delivery in ovo on 12th day of egg incubation ensures the efficient transport route of bioactives from the egg air chamber, through highly vascularized allantochorion to the gastrointestinal tract (GI). Upon hatch, the in ovo modulated profile of gut microflora has influence on good health condition of a chicklet, eliminating the need for antibiotics. We presume that this beneficial condition of GI is reflected in morphology of the intestines and might be maintained throughout the chicken life span. Goal of study was to determine microstructure of duodenum and jejunum at 42 day after in ovo manipulation. Two synbiotics (S1, S2) were injected in ovo at 12th day of egg incubation. S1 was composed of Lb. salivarius IBB3154 with galactooligosaccharides and S2 consisted of Lb. plantarum IBB3036 combined with raffinose family oligosaccharides extracted from lupin. After treatment S1, higher weight ($P \leq 0.01$) and increased lenght ($P \leq 0.05$) of duodenum and jejunum were observed than in Control and group S2. This beneficial effect was reflected by the numercially higher villi with larger surface ($P \leq 0.01$) in duodenum in group S1. In jejunum, an increased height of villi ($P \leq 0.01$), smaller crypts ($P \leq 0.01$) and decreased number of neutral goblet cells ($P < 0,05$), which is positive at the end of rearing, was observed in group S2. Overall, a single in ovo treatment with S1 and S2 beneficially affected the structure of intestine at 42 day. The effects differ depending on the formulation of the synbiotic.

ECO FCE: In ovo manipulation – effects on chicken transcriptome and physiology

M. Siwek[1], A. Dunislawska[1], A. Slawinska[1], P. Kolodziejski[2], E. Pruszynska-Oszmalek[2], P. Mackowiak[2] and M. Bednarczyk[1]
[1]*University of Science and Technology, Mazowiecka 28, Bydgoszcz, Poland,* [2]*Univerity of Life Sciences, Wolynska 33, Poznan, Poland; siwek@utp.edu.pl*

Microbiome of chicken gastrointestinal tract can be stimulated in ovo with bioactive substances: prebiotics, probiotics or synbiotics. Goal of this study was to determine molecular and physiological effects of in ovo manipulation in chickens. We analyzed impact of two synbiotics (S1, S2) injected in ovo at 12th day of embryo development on adult chicken transcriptome and selected physiological parameters. S1 was composed of Lb. salivarius IBB3154 with galactooligosaccharide and S2 consisted of Lb. plantarum IBB3036 combined with raffinose family oligosaccharide. Intestinal (jejunum, cecal tonsils), pancreatic, splenic and liver tissues were harvested at 6 time points post hatching (days: 1, 4, 7, 14, 21 and 42). Transcriptome was analyzed in four tissues (cecal tonsils, spleen, jejunum, liver) harvested at 21st day (n=5), using Chicken Gene 1.1 ST Array Strip (Affymetrix, Santa Clara, CA, USA). Metabolic measurements included among others: trypsin and lipase activity in pancreas, alanine transaminase (ALT) activity, total (T3) and free thriiodothyronine (FT3) in blood serum. The highest number of differentially expressed genes (DEG) was induced by S1 in cecal tonsils (159 DEG) and jejunum (95 DEG). S2 had the strongest impact on DEG in liver (159 DEG). Gene ontology analysis indicated activation of immune-related pathways in cecal tonsils (S1) and metabolic pathways in liver (S1 and S2). Beneficial effect of the synbiotics was showed as decrease of ALT on 21st day by both of them and on 42nd day by S2. Trypsin activity in pancreas was diminished on these days after S1 and elevated after S2 treatment. Lipase activity was decreased by S1 and S2 on days 21st and 42nd. Synbiotics hardly changed hormonal profile of metabolism influencing thyroid hormones. Also, no changes were noticed in carbohydrate-lipid indices in blood. In conclusion, S1 proved to have immunomodulatory effect and S2 pro-metabolic effect.

ECO-FCE: Faecal microbiota transplant alters the growth and intestinal bacterial profile of pigs

U.M. McCormack[1,2], B. Metzler-Zebeli[3], S. Buzoianu[2], T. Curiao[2], F. Crispie[4], O. O'Sullivan[4], P. Cotter[4], G.E. Gardiner[1] and P.G. Lawlor[2]
[1]*Waterford Institute of Technology, Department of Science and Computing, Waterford, Waterford, Ireland,* [2]*Teagasc, Pig Development Department, Moorepark, Fermoy, Cork, Ireland,* [3]*University of Veterinary Medicine, Department for Farm Animals and Veterinary Public Health, Vienna, Vienna, Austria,* [4]*Teagasc, Food Research Centre, Moorepark, Fermoy, Cork, Ireland; ursula.mccormack@teagasc.ie*

As the intestinal microbiota plays an important role in energy harvest for the host, there may be potential to improve feed efficiency by manipulating its composition. The objective was to investigate growth and intestinal microbiota composition of pigs following oral inoculation of sows and their offspring with faecal material from pigs screened for low residual feed intake in a previous study. The experiment was a 2×2 factorial design with 2 factors for sow inoculation and 3 factors for the number of piglet inoculations prior to weaning. Pregnant sows (n=22) were blocked by boar used for insemination and body weight and assigned to two treatments: (1) control; and (2) inoculum at day 70 and 100 of gestation. At farrowing, 9 piglets/litter were selected and within litter randomly assigned to 3 treatments: (1) control; (2) inoculum at birth; and (3) inoculum at birth, day 3, 7 and 28 of life. At weaning, 6 pigs per litter (2/trt) were selected and followed to slaughter (~105 kg) as individual pigs. Weight and feed intake were recorded weekly. At slaughter, 36 pigs (n=6/trt) were selected for microbiota profiling (16S rRNA gene sequencing; Illumina MiSeq) on faeces sampled throughout the experiment and on intestinal digesta collected at slaughter. Growth performance was analysed using the Mixed procedure of SAS, and microbiota sequence data were analysed using QIIME. Corrections for multiple comparisons were made using the Benjammani-Hochberg method. Inoculation of sows and/or offspring resulted in reduced offspring body weight throughout the study (P<0.05) and average daily feed intake was lower in pigs from inoculated sows (P<0.05). Intestinal microbiota composition varied between treatments. This study indicates that early-life intestinal microbiota composition can influence lifetime growth in pigs.

ECO FCE: Peri-natal nutrition and lifetime performance of low birth weight pigs from prolific sows

J.G. Madsen[1,2], M. Kreuzer[1] and G. Bee[2]
[1]*ETH Zurich, Institute of Agricultural Sciences, Universitätsstrasse 2, 8092 Zurich, Switzerland,* [2]*Agroscope, 1725 Posieux, Switzerland, Institute for Livestock Sciences, la Tioleyre 4, 1725 Posieux, Switzerland; johannes.madsen@agroscope.admin.ch*

Selection for large litters in commercial dam lines has increased number of piglets with low birth weight (L-BtW), poor survival rate and impaired growth performance. The overall project objective was to investigate rearing and nutritional measures for their efficiency to improve viability and growth of L-BtW piglets. Three experiments were conducted where L-BtW (0.8-1.2 kg BtW) piglets from hyperprolific sows ($\geq$15 total born) were used. In all experiments piglets were weaned in pairs d 7 after birth, allotted to one of three dietary treatments (Control (CON): unsupplemented, L-carnitine (CAR): 0.4 g/piglet/d, L-arginine (ARG): 1.08 g/kg BW/d), and artificially reared (AR) in rescue decks until d 28. In Exp. 1 piglets were fed six times per d according to requirements. In Exp. 2 and 3, they had ad libitum access to the diet. Feed intake and BW was measured daily and weekly, respectively. From the piglets slaughtered at d 28, the Semitendinosus muscle (STM) was removed for analysis of muscle metabolism and protein synthesis pathway activity. Compared to CON, in Exp. 1 and 2, CAR and ARG promoted muscle maturation and activation of protein synthesis pathway. In Exp. 3, one AR piglet per pair was weaned together with a conventionally reared (CR) piglet at d 28 of age. Pre-weaning survival rate and growth performance in the pre and post-weaning period was assessed. All AR piglets survived the nursing period, whereas mortality rate of L-BtW CR pigs reared in loose house sow systems was 56%. Weaned AR and surviving CR piglets displayed weaning weights (5.0 kg) which were below the herd average (6.6 kg). At 88 d of age, BW was comparable between AR and CR piglets (25.8 vs 26.0 kg). In conclusion, Exp. 1 and 2 showed that CAR and ARG could act as bioactive components in L-BtW piglets from hyperprolific sows. In addition, results from Exp. 3 clearly showed that survival rate of L-BtW piglets was markedly increased by applying artificial rearing, and post-weaning growth performance was not compromised when compared to CR piglets.

ECO FCE: Phosphorus nutritional conditioning in broiler chickens

M. Francesch[1], A.M. Pérez-Vendrell[1], E. Fàbrega[2], R. Quintanilla[3] and D. Torrallardona[1]
[1]*IRTA, Animal Nutrition and Welfare, Ctra. Reus-El Morell Km 3,8, 43120 Constantí, Spain,* [2]*IRTA, Animal Nutrition and Welfare, Veïnat de Sies, s/n, 17121 Monells, Spain,* [3]*IRTA, Animal Breeding and Genetics, Ctra. C-59 Km. 12,1, 08140 Caldes de Montbui, Spain; maria.francesch@irta.cat*

A study was conducted to investigate P nutritional conditioning as a feeding strategy to improve lifetime performance and P utilisation in broiler chickens. A total of 600 male broiler chickens (Cobb 500 FF) housed in 24 pens (25 birds per pen) were fed a balanced (6.9 g P/kg; no-CON) or a P deficient diet (4.5 g P/kg; CON) during the first week of live, followed by a common balanced P diet from d 7 to d 21. At d 21, half of the pens from each group continued on a balanced diet (5.8 g P/kg; no-LD) and the other half were subjected to late P deficiency (3.5 g P/kg; LD) until d 43. LD aimed to highlight the possible advantages of CON. This resulted in 4 treatments according to a 2×2 factorial arrangement (CON × LD). A constant Ca:P ratio of 1.3 was used for all diets. Performance (1-42 d), and apparent ileal digestibility (AID), bone mineralisation and gait scores (43 d) were measured. CON reduced daily weight gain and feed intake between d 1-14 (P<0.05) but tended to improve daily weight gain (by 3.3%) and FCR (by 2%) from d 21-42 in no-LD birds. LD impaired performance. CON tended to increase P AID (P<0.10) whereas LD tended to reduced it (P<0.06). At d 43, CON had no effect on bone mineralisation and gait scores, but LD impaired them (P<0.05). In parallel, 24 chickens per treatment were also monitored in individual cages between d 22-30 for the analysis of AID, apparent total tract digestibility (ATTD) and bone mineralisation. In these birds, CON had no significant effects on AID or ATTD, despite a reported increased expression of Type IIb Sodium Phosphate Cotransporter (SLC34A2) in the duodenum of the same animals. There was a CON × LD interaction (P<0.05) for tibia ash concentration, that increased by CON but only for the no-LD birds. LD impaired bone mineralisation (P<0.05). Although not conclusive, results suggest a positive effect of P CON on performance and bone mineralisation; LD did not highlight the possible advantages of CON.

ECO FCE: In vitro evaluation of exogenous enzymes to improve digestibility in pigs and poultry
D. Torrallardona[1], J.I. Badiola[2], A.M. Pérez-Vendrell[1] and M. Francesch[1]
[1]IRTA, Animal Nutrition and Welfare, Mas de Bover, E43120 Constantí, Spain, [2]IRTA-CReSA, Animal Health, Campus UAB, E08193 Bellaterra, Spain; david.torrallardona@irta.cat

The efficacy of exogenous enzymes to improve the nutrient digestibility of different substrates was evaluated with in vitro methods simulating digestion of pigs or poultry at the ileal level. Different combinations between six exogenous enzymes and nine feed ingredients were tested, and these included: endo-1,3(4)-β-glucanase (BGN; with barley or wheat pollard); BGN + endo-1,4-β-xylanase (BGN+XYL; with wheat, wheat pollard, wheat DDGS or maize DDGS); α-galactosidase (GAL; with peas, rapeseed meal, soybean meal or sunflower meal); endo-1,4-β-mannanase (MAN; with rapeseed meal or sunflower meal); protease (PROT; with all nine ingredients); and 6-phytase (PHY; with wheat DDGS or maize DDGS). In general, larger nutrient digestibilities were observed with the pig than with the poultry method. BGN improved NSP digestibility of barley for poultry. BGN+XYL supplementation of wheat improved NSP digestibility, in both models, and DM, CP and starch digestibility in poultry. This enzyme combination also improved NSP digestibility of wheat DDGS in poultry. GAL (but only at large doses of inclusion) improved NSP digestibility of soybean meal in poultry, and DM and starch digestibility of peas in pigs. MAN reduced DM digestibility of sunflower meal in poultry. PROT improved DM and CP digestibility in wheat and soybean meal, and DM in barley and maize DDGS in poultry. PHY improved DM and NSP digestibility of maize DDGS in pigs and poultry. PHY+BGN+XYL improved DM and NSP digestibility of maize DDGS in poultry. The addition of the six enzymes to a mixture of 20% barley, 40% wheat, 10% rapeseed meal, 20% soybean meal and 10% wheat pollard, improved DM, CP and P digestibility in pigs and NSP in poultry. Finally, the undigested residues were incubated with ileal digesta from pigs or chickens. The production of gas suggested a substantial impact of carbohydrases on fermentability, particularly of whole cereals and peas. The in vitro digestibility methods proposed have proven to be a useful tool for the multiple screening of exogenous enzymes on different substrates, as preliminary test for in vivo trials.

ECO FCE: α-galactosidase and xylanase improve performance of pigs fed soybean and rapeseed meals
S. Carné, J. Estévez and J. Mascarell
Industrial Técnica Pecuaria, S.A. (ITPSA), Av. Roma 157, 08011 Barcelona, Spain; scarne@itpsa.com

Legumes and their derived feed materials contain high levels of α-galactosides. These indigestible substances resist thermic treatments and act as anti-nutritional factors that may have an impact on pig feed intake and efficiency. The objective was to investigate the effect of an enzyme preparation (α-galactosidase and endo-1,4-xylanase activities) on the performance of pigs in fattening. In 2 studies, 144 pigs (28 kg BW), individually housed and blocked by BW and sex (12-18 blocks), were fed diets combining soybean (SBM; 10-24%) and rapeseed (RSM; 0-10%) meals until fattening. A sparing of 3% NE of the basal diet and 2 dosages of enzyme preparation (12 and 20 U galactosidase) were assessed, measuring feed intake, growth and F/G. In the first study, final BW was numerically greater for both enzyme dosages vs control (109-113.4 vs 107.8 kg), and differed (P<0.05) when 10% RSM was added. No differences were detected on F/G. In the second study, in which higher levels of SBM (24%) and RSM (10%) were used, final BW was numerically greater for enzyme treatments added on top (102.1 vs 104.3 kg) and in the 3% NE restriction diets (102.1 vs 103.7 kg) with respect to positive control. The enzyme preparation increased the final BW, especially in those pigs fed with 10% RSM. For diets with 3% NE restriction, final BW was unexpectedly greater for the group without enzyme (103.7 vs 108.9 kg; P<0.05) due to an increased feed intake (ADFI: 2,586 vs 2,369 g/d; P<0.05). We hypothesize that the enzyme preparation allowed sparing the 3% NE restriction, whereas in the treatment without enzyme, pigs increased the feed intake to meet their energy requirements. Thus, as protein was not restricted, a higher protein intake and deposition was observed in the control group. F/G was not consistently affected. Further studies of α-galactosidase enzyme preparations in diets with both energy and protein restrictions are warranted to optimise pig performance and feed efficiency.

ECO FCE: The use of enzymes in pig and broiler diets containing low soya and varying levels of energy and ami

M.E.E. Ball[1], V.E. Beattie[2], H. Hayes[2], P.J. Lawlor[3] and E. Magowan[1]
[1]Agri-Food and Biosciences Institute, Hillsborough, BT26 6DR, United Kingdom, [2]Devenish Nutrition Ltd, Belfast, BT3 9AR, United Kingdom, [3]Teagasc, Moorepark, Cork, Ireland; elizabeth.ball@afbini.gov.uk

Ten diets for broilers and ten diets for growing/finishing pigs were used to examine the effect of phytase and protease in broiler diets and the effect of protease and a xylanase/ß Glucanase in growing/finishing pig diets. Diets were RSM and DDGS based except for a soya bean positive control. In the broiler diets, P and Ca levels were reduced to investigate the effect of phytase and amino acid levels were reduced by 1.5 or 3% to test the effects of protease. In the pig diets, phytase was included in all diets (with a P and Ca reduction), available amino acids were reduced by either 3 or 8% and energy (NE) was reduced by either 5 or 10% to test the effects of the protease and xylanase/ß Glucanase respectively. The diets were offered to 600 broilers (0-35 d in pens of 10) and to 800 growing/finishing pigs, (pens of 10 from 17 to 103 kg). In the broiler trial there was no significant effect (P>0.05) on feed intake or weigh gain (average 90.3 g/d (SEM 2.50) and 64.3 g/d (SEM 2.12) respectively between D 0 and 35). When a 3% reduction in amino acids was adopted FCR was poorer in the starter period (1.15, SEM 0.024) and overall between D 0 to 35 (1.42, SEM 0.033) (P<0.05) compared with when a 1.5% reduction was adopted (FCR 1.10 and 1.37 respectively). When phytase was included FCR was improved (P<0.05) in the grower and finisher period. However there was no other significant effects of enzyme addition on bird performance. Likewise there were no strong effects of enzyme addition in the RSM/DDGS based pig diets. Reducing the amino acids and energy in the growing pig diets had no effect on feed intake (average 1,182 g/day) or growth rate (average 656 g/day) (P>0.05) but FCR deteriorated as the level of amino acid and energy reduction increased (P<0.01) (from 1.72 for the RSM/DDGS control to 1.95 when amino acids were reduced by 8% and energy by 10%).

ECO-FCE: Effect of carbohydrase use and soaking of cereals on the growth of liquid fed finisher pigs

A. Torres-Pitarch[1,2], E.G. Manzanilla[2], G.E. Gardiner[3], R. Lizardo[4], D. Torrallardona[4], J.V. O'Doherty[1] and P.G. Lawlor[2]
[1]University College of Dublin, School of Agriculture and Food Science, Belfield, Dublin, Ireland, [2]Teagasc, Pig Development Department, Moorepark, Fermoy, Co Cork, Ireland, [3]Waterford Institute of Technology, Dep. of Science, Waterford, Co. Waterford, Ireland, [4]IRTA, Monogastric Nutrition, Constantí, Tarragona, Spain; peadar.lawlor@teagasc.ie

Mixing the cereal fraction of pig diets with water prior to feeding (soaking) may increase the activity of exogenous NSP-degrading enzymes. A 2×2 factorial design was used to assess the effect of soaking the cereal fraction of the diet with or without supplementation of a carbohydrase enzyme (xylanase and β-glucanase; XB). A total of 252 pigs (31.0 kg; ±0.65 SEM) housed in same sex pens of 7 pigs/pen were allocated to 1 of 4 dietary treatments: (1) Fresh liquid diet (FLD) where the diet was mixed with water immediately prior to feeding, (2) FLD + XB, (3) Soaked liquid diet (SLD) where the cereal fraction (35% wheat, 38% barley and 11% pollard) of the diet was soaked for 3 h before feeding the diet and (4) SLD + XB. The experiment lasted 71 days during which growth and feed intake were recorded. At slaughter, cold carcass weight, muscle depth and fat depth were recorded. The data was analysed by the MIXED procedure of SAS. All pigs finished the experiment at a similar live weight (98.4±0.65 kg SEM; P>0.05). Pigs fed the soaked cereal diets had higher ADG than pigs fed the fresh liquid diets (731 vs 669±24.9 g/day SEM; P<0.10) and a 4% numerical improvement in FCR during the first week of the experiment. For the entire experiment, ADG, ADFI and FCR were not significantly affected by enzyme supplementation. At slaughter, fat depth was increased by enzyme supplementation (13.0 vs 12.3±0.20 mm SEM; P<0.05) and lean meat yield was reduced (57.3 vs 56.6±0.17% SEM; P<0.01). In conclusion, carbohydrase supplementation reduced lean meat yield in finisher pigs and whilst differences were not significant, tendencies indicate that soaking cereals before feeding may increase ADG. Both of these effects warrant further investigation.

ECO FCE: Combined use of cereal extrusion and enzymes in chickens

M. Francesch[1], I. Riahi[1], A.M. Pérez-Vendrell[1], E. Fàbrega[1], M. Gispert[2] and D. Torrallardona[1]
[1]IRTA, Animal Nutrition and Welfare, 43120 Constantí and 17121 Monells, Spain, [2]IRTA, Product Quality, 17121 Monells, Spain; maria.francesch@irta.cat

The aim of this study was to evaluate the effect of extrusion on wheat (W) or barley (B), in combination with carbohydrases on performance, digesta viscosity, nutrient apparent ileal digestibility (AID), welfare and meat quality of Cobb 500 FF chickens. A 2×2 factorial arrangement of treatments was used for each cereal, with extrusion (EXT; 105-135 °C and 13-15% water) and enzymes (ENZ; Xylanase for W and β-glucanase for B) as main factors. Cereals were included in the diets at 63-69% (W) and 55-62% (B), and 6 replicates of 40 birds were used for each treatment from d 1-38. EXT had no significant effect on performance for W, but it increased feed intake (FI) and impaired FCR for B (P<0.05). ENZ reduced FI and average daily gain for W and B, and improved FCR for B (P<0.05). EXT increased digesta viscosity (P<0.05) for B and W (2.7 and 1.6-fold, respectively), and there was an EXT×ENZ interaction (P<0.05): larger viscosity reductions with ENZ were observed for extruded B than raw cereal, whereas ENZ reduced viscosity only for extruded W. EXT increased energy AID for W (P<0.05), but had no effect on protein AID and reduced that of some amino acids (P<0.05). Barley extrusion reduced AID for energy, protein and all amino acids (P<0.05). ENZ increased amino acid AID in W diets (except Lys), and Gly, Thr, Ala, Ile, Leu, Val, Met and Phe AID in the B extruded diet (P<0.05). EXT×ENZ interactions (P<0.05) were observed for fat AID in both cereals. For B, EXT reduced fat AID but only in absence of ENZ. For W, the increase in fat AID was larger if W was extruded. Barley EXT impaired litter quality and feather cleanness and increased the severity of pododermatitis (P<0.05), but these were improved by ENZ (P<0.05). Extrusion increased carcass yield for W and intramuscular fat, skin a* and muscle L* for B; and reduced pH 24 h post mortem for B and muscle L* for W (P<0.05). ENZ increased carcass yield for B and surface area for W, and reduced muscle b* for W (P<0.05). It is concluded that EXT did not improve productivity of broiler chickens, and that in combination with ENZ did not bring any additional improvement, despite of some positive responses on AID and welfare.

Heritability of udder morphology and colostrum quality traits in swine

A. Balzani[1], H.J. Cordell[2], E. Sutcliffe[3] and S.A. Edwards[1]
[1]Newcastle University, AFRD, Agriculture building, Newcastle University, NE1 7RU, Newcastle upon Tyne, United Kingdom, [2]Newcastle University, Institute of Genetic Medicine, Centre for life, NE1 3BZ Newcastle upon Tyne, United Kingdom, [3]ACMC Ltd., Upton House, Beeford, YO258AF Driffield, United Kingdom; a.balzani@ncl.ac.uk

The heritability of udder quality traits, defined as morphology and colostrum Immunoglobulin G (IgG) concentration at farrowing, was estimated together with the genetic and phenotypic correlations of these traits with other production and reproduction criteria. Udder morphology traits were record in 988 Meidam sows and colostrum samples were collected from 528 sows. Teat length (len), diameter (dia), inter-teat distance within the same row (samer), and teat distance from the abdominal mid line (aml) were recorded to the nearest millimetre. For each sow, a record was also made of udder development score (dev), the proportion of teats oriented perpendicular to the udder (OR) and the proportion of non-functional teats (NoFun). Colostrum IgG concentration (COL) was estimated with the Brix refractometer. Heritability of udder morphology traits varied from high (len: h2=0.46; dia h2=0.56) to moderate (samer h2=0.37; aml h2=0.22; dev h2=0.25; NoFun h2=0.3; OR h2=0.1 and COL h2=0.35). The trait samer was negatively genetically correlated with the number of stillborn (rg=-0.48), and positively with the number of live-born piglets (NBA rg=0.69). The highest genetic correlation with productive traits was estimated between aml and average daily gain during rearing (ADG rg=0.42), although this had a negative phenotypic correlation (rp=-0.13). Teat length was also moderately correlated with ADG (rg=0.27). Back fat thickness at 100 kg was positively correlated with dia and TT (rg=0.28; rg=0.36), and negatively only with dev (rg=-0.22). The same results were found for the phenotypic correlation between BFAT and dev and TT (rp=-0.09; rp=0.10). Udder quality traits can be included in the breeding goal and weighted appropriately with other important traits in the breeding objectives to enhance maternal performance.

Feeding of Enterococcus faecium NCIMB 10415 to piglets affects B-cells on cell, gene and miRNA level

S. Kreuzer-Redmer, J. Bekurtz, D. Arends and G.A. Brockmann
Humboldt-Universität zu Berlin, Züchtungsbiologie und molekulare Genetik, Invalidenstr. 42, 10115 Berlin,
Germany; susanne.kreuzer.1@agrar.hu-berlin.de

Feeding of the probiotic Enterococcus faecium NCIMB 10415 (E. faecium) has been described to promote growth performance and health in pigs. Here we analyzed the response of B cells via flow cytometry and differential expression analyzes of miRNAs and potential target genes in gut associated lymphatic tissues from E. faecium fed piglets versus untreated controls. Piglets from each feeding group (probiotic/control) were sampled at day 12±1 (n=6), 34±1 (n=6) and 54±2 (n=8) of age balancing for litter and gender. Relative proportions of B cells were obtained via IgM antibodies by flow cytometry. RNA-Seq and miRNA-Seq of a pool of total RNA of lymphocytes from gut associated lymphatic tissues of three individuals of every feeding group was performed at the age of 34 days. Target prediction of miRNAs was done with miRanda and PITA. A lower relative frequency of IgM+ B cells was detected at days 34 and 54 in the probiotic fed group as compared to the control group. Analysis of RNA-Seq data revealed an up-regulation of miRNA-423-5p and down-regulation of 31 putative target transcripts with an immune relevant function. Validation by RT-qPCR confirmed a significant up-regulation of miR-423-5p (2.11 fold; P=0.03) and, additionally a down-regulation of selected, immune relevant target gene IGLC (0.61 fold; P=0.03) and IGKC (0.69 fold; P=0.04). In a further step, we investigated the expression pattern of miR-423-5 and its putative target transcripts over the course of time during the feeding experiment. We identified an anti-correlation between miR-423-5p and IGLC transcript amounts. The transcript data suggest IGLC as potential target of miR-423-5p. Luciferase reporter assays in a HeLa cell line verified IGLC as a target of miR-423-5p. Hence, the probiotic E. faecium led to lower relative B cells counts and a lower relative transcript amount of B cell associated genes after weaning. These effects are likely, partially mediated through miRNA-423-5p.

Alternative consideration of social genetic effects models in Duroc pigs

M. Ragab[1,2], M. Piles[2] and J.P. Sánchez[2]
[1]Fac. of Agriculture, Kafr El Sheikh University, Poultry Production, El Giesh street, 33516, Kafr El Sheikh, Egypt,
[2]Institut de Recerca i Tecnologia Agroalimentàries (IRTA), Animal Breeding, Torre Marimón, 08140, Caldes de
Montubi, Barcelona, Spain; moha.ragab@hotmail.com

There are important statistical difficulties associated with social interaction models, basically deriving from the collinearity between direct and social interaction effects. We present an alternative implementation of these models to alleviate these problems. Bi-weekly average daily gain records (ADG) were recorded on 663 animals between 105 and 182 d of age. The animals were reared in pens with between 10 and 14 mates. Three types of models were fitted: (1) repeatability animal model (RAM); (2) repeatability animal model with traditional social interaction effect (RSM); and (3) repeatability animal model with a social interaction effect function of the degree of competition between each pair of mates (RS1M). All proposed models included the fixed effects of batch, pen size and age, and the interaction between batch and age, and also the random effect of pen. In RS1M pair-specific competitions were defined as standardized Euclidean distance between each pair of animals based on two alternative feeding behavior traits: feeding rate (FR) or time between two consecutive feeding (FT). Estimated total heritability was 0.34 (0.09) and 0.42 (0.17) using RAM and RSM, respectively. IUnder RS1M at the average distance (0), the estimated total heritability was 0.38 (0.09) regardless the behavior trait considered for defining distance. Estimated social interaction heritability for RSM was 0.004 (0.003), this parameter under RS1M reached 0.06 (0.09) when FT was used to define the distance between competitors and the parameter was evaluated at the 3rd quartile of the distance distribution. Using feeding behavior traits to account for variability in the intensity of competition between animals generated parameter estimates with lower errors than the traditional social interaction model, which would be expected to yield higher accuracies in the predictions derived from such models.

Patterns of sexual size dimorphism in various chicken breeds
J. Geibel[1], S. Weigend[2], A. Weigend[2], A.R. Sharifi[1] and H. Simianer[1]
[1]Georg-August-Universität Göttingen, Department of Animal Sciences, Animal Breeding Group, Albrecht-Thaer-Weg 3, 37075 Göttingen, Germany, [2]Friedrich-Loeffler-Institut, Institute of Farm Animal Genetics, Höltystrasse 10, 31535 Neustadt-Mariensee, Germany; johannesgeibel@online.de

Rensch's Rule (RR) states that sexual size dimorphism (SSD) will increase with increasing body size. This is mostly explained by sexual selection in wild populations, hence there is the question whether this mechanism is also present in domestic populations. To answer this question, we had data available on a set of 1,387 adult male and female individuals out of 89 chicken breeds, which were further classified into 12 different groups depending on breed characteristics. For every animal there were measurements of body weight, length of breast bone, wing length and length and thickness of legs. Calculating the Sexual Dimorphism Index (SDI, giving the ratio males are larger than females) of every breed provided first indications of the presence of SSD. The SDI of body weight (0.35) is higher and varies much more than SDI of skeleton measurements (0.16-0.25). We also calculated the slope S of the standardised major axis regression of the log of the male measurement against the log of the corresponding female measurement to estimate allometric effects. A slope of S>1 shows positive allometry, which would be in agreement with RR. We did not find the overall slope of body weight to be different from 1. Hence, our data does not support the statement of RR in domesticated chickens. However, the slope seems to differ across breed groups, but differences hardly reached significances due to limited sample sizes. For skeleton measurements the slopes were found to be higher than 1 (1.027-1.068), but also differing over groups of breeds. This is the first study trying to evaluate RR in a large data set in a domesticated species based on individual phenotypes. Overall, we conclude that RR cannot be confirmed in domestic chicken for body weight, but in tendency for some skeleton measurements.

Improving gait in ducks and chickens: effects of selection on leg morphology and function
B.M. Duggan, P.M. Hocking and D.N. Clements
The Roslin Institute and Royal (Dick) School of Veterinary Studies, The University of Edinburgh, Easter Bush, Midlothian, EH25 9RG, Scotland, United Kingdom; brendan.duggan@roslin.ed.ac.uk

Gait problems have been well-documented in broiler chickens yet similar problems have not been reported in Pekin duck, which has undergone a similarly intense selection process. In order to understand how selection for greater meat yields has affected walking ability, bone morphology and gait was assessed objectively in divergent lines. The Pekin duck was compared to its slower growing ancestor, the mallard, and the broiler chicken was compared to the slower growing layer chicken. For each line, at 3, 5 and 7 weeks, 12 birds were allowed to walk over a pressure-sensing walkway. The leg bones of these birds were scanned by computed tomography to assess morphology. Results were analysed by ANOVA, accounting for age and sex. Duck lines walked faster than chickens and, within each species, lighter lines moved faster than their heavier conspecifics (P<0.005). Lighter lines also walked with a narrower step width than their heavier conspecifics (P<0.001). The angle of the foot when walking differed between species (P<0.001); the feet of both duck lines were internally rotated and those of both chicken lines were sagittally aligned with the direction of travel; this is partly explained by bone torsion. Morphologically, leg differences between the species were clear, notably a cranial curvature in the tibiotarsi of chickens compared to a caudal curvature in those of ducks (P<0.001). Duck tibiotarsi were also more laterally curved than those of chickens (P<0.001), which may be an adaptation for swimming. Duck legs reached adult size earlier, allowing them greater opportunity to remodel their bones to handle loads imposed on them. Selection for high meat yield and rapid growth has led to changes in morphology and gait in both species. The growth rates in both birds are similar; however, Pekin ducks ambulate with relative ease compared to broiler chickens. Objectively measuring components of gait rather than traditional visual gait assessment may lead to greater breeding success.

Modulating birth weight heritability in mice

N. Formoso-Rafferty[1], I. Cervantes[1], N. Ibáñez-Escriche[2,3] and J.P. Gutiérrez[1]
[1]Universidad Complutense, Producción Animal, Avda. Puerta de Hierro s/n, 28040 Madrid, Spain, [2]University of Edinburgh, The Roslin Institute, Edinburgh, Edinburgh, United Kingdom, [3]IRTA Lleida, Genètica i Millora Animal, Centre IRTA Lleida, Lleida, Spain; n.formosorafferty@ucm.es

Expected genetic response is proportional to the heritability of the trait, and this parameter is considered inherent of a specific trait in a particular population. Modifying the environmental variance of the birth weight by artificial selection has been shown to be feasible in a divergent selection experiment in mice. In addition, models assuming heterogeneity in residual variance lead to different estimates of heritabilities across combinations of systematic (environmental) effects. The objectives of this work were to (1) estimate the evolution of the heritability of birth weight in mice in the mentioned experiment, as well as (2) estimating different heritabilities regarding systematic effects. Data belong to eleven generations of a divergent selection experiment conducted to modify the environmental variability of birth weight in mice. A total of 15,431 birth weight records from 959 females and 1,641 litters in combination with 14,786 pedigree records were used. The model used in the analysis included generation, litter size, sex and parity number as systematic effects. Each record of birth weight was assigned to the mother of the pup in a model which assumes that the environmental variance is heterogeneous and partially under genetic control. Differences in heritabilities between lines reached values of 0.06 in the last generations. Choosing the most extreme values of systematic effects, the birth weight heritability ranged from 0.04 to 0.22. According to these results, the possibility of modulating the heritability for this trait could be explored in two ways: first selecting to decrease the environmental variability, or second choosing the appropriate levels of the systematic effects.

Local vs global ancestry: regions deviating from genome wide admixture in a composite cattle breed

N. Khayatzadeh[1], G. Mészáros[1], Y.T. Utsunomiya[2], J.F. Garcia[2], U. Schnyder[3], B. Gredler[3], I. Curik[4] and J. Sölkner[1]
[1]University of Natural Resources and Life Sciences (BOKU), Department of Sustainable Agricultural Systems, Augasse 2-6, 1090 Vienna, Austria, [2]UNESP-Univ, Estadual Paulista, Departamento de Medicina Veterinária, Jaboticabal and Aracatuba, São Paulo, Brazil, [3]Qualitas AG, Chamerstrasse, Zug, Switzerland, [4]University of Zagreb, Department of Animal Science, Svetošimunska cesta, Zagreb, Croatia; negar.khayatzadeh@students.boku.ac.at

Swiss Fleckvieh is a composite breed of Simmental (SI) and Red Holstein Friesian (RHF), established around 1970 in Switzerland. For a set of 300 admixed animals investigated in this study, the global ancestry proportions were estimated 0.68 RHF and 0.32 SI (SD 0.19). We searched through the 29 autosomes of these individuals to find regions showing extreme deviations (excess or deficiency) in local versus global ancestry (Δ ancestry). Sources for Δ ancestry may be demographic processes such as gene flow, genetic drift and selection. Two different approaches were employed to establish significance thresholds for Δ ancestry. With the first method, extreme outliers from normal distribution, with Bonferroni correction assuming 5,000 or 1000 independent segments, were considered significant. The second approach was a permutation test. Local ancestry estimates were concatenated across all autosomes for each animal separately and the genome was permuted by cutting at random location. SD for 20,000 permutations was calculated and the 5% and 1% quantiles were defined as thresholds. Both methods yielded similar significance thresholds. Considering 1% genome-wide threshold of the permutation test, we found a significant region on chromosome 18 (18.7-25.9 Mb).Based on the 5% genome-wide as well as Bonferroni thresholds with 1000 hypotheses, a region on chromosome 13 (46.3-47.3 Mb) was also significant. Both observed signals are in direction of increased SI ancestry by 0.16-0.18. The peaks may be interpreted as admixture signals of selection. Since the composite population has undergone crossbreeding for only few generations (~10) so far, the signals are not narrow enough to point at selection for any single candidate gene causing the ancestry deviation.

Effects of heat stress on the transcriptomic profile of blood cells in lactating dairy goats

A. Contreras-Jodar[1], A.A.K. Salama[1,2], G. Caja[1], S. Hamzaoui[1], M. Vailati[3] and J.J. Loor[3]
[1]*Universitat Autònoma de Barcelona, Group of Ruminants Research (G2R), Edifici V, Travessera dels Turons s/n, Cerdanyola, Barcelona, 08193, Spain,* [2]*Animal Production Research Institute, Nady El-Said Street, 12816; Dokki, Giza, Egypt,* [3]*University of Illinois, Animal Sciences, 1207 W. Gregory Drive, 61801, Urbana, IL, USA; alexandra.contreras@uab.cat*

High temperature is a major stress that negatively affects welfare, health, and production of dairy animals. Heat-stressed animals are also more prone to diseases, suggesting that their immunity is hindered. Although productive and physiological responses of dairy animals to heat stress are well known, there is still limited information about the response at the transcriptomic level. Our objective was to evaluate the changes in blood transcriptomics of dairy goats under heat stress. Eight adult Murciano-Granadina dairy goats in mid-lactation were submitted to 2 climatic treatments for 35 d. Treatments and temperature-humidity index (THI) were: (1) thermal neutral (TN: 15-20 °C, 40-45%, THI=59-65); and (2) heat stress (HS: 12 h at 37 °C -40%, THI=86; 12 h at 30 °C -40%, THI=77). Blood samples were collected at d 35 and RNA was extracted for microarray analyses (Affymetrix GeneChip Bovine Genome Array). The signal intensity of globin genes was low and did not affect the detection of gene expression. The analysis revealed that 55 genes were up-regulated, whereas 88 were down-regulated by HS. The Dynamic Impact Approach revealed that 31 biological pathways were impacted by HS. Pathways of leukocyte transendothelial migration, cell adhesion molecules, hematopoietic cell lineage, calcium signaling, and PPAR signaling were negatively impacted by HS, whereas nucleotide metabolism increased. Overall, these changes indicate alteration in the functionality and efficiency of immune cells, which might be linked to a greater immune susceptibility to diseases of dairy goats under HS conditions.

Comparing methods to analyse raw, large-scale methane data

M. Szałański[1,2], G.F. Difford[2,3], P. Løvendahl[2] and J. Lassen[2]
[1]*Poznan University Of Life Sciences, Department of Genetics and Animal Breeding, Wołyńska 33, 60-637 Poznan, Poland,* [2]*Aarhus University, Department of Molecular Biology and Genetics, Center for Quantitative Genetics and Genomics, Blichers Allé 20, 8830 Tjele, Denmark,* [3]*Wageningen University, Animal Breeding and Genomics Centre, P.O. Box 338, 6700 AH Wageningen, the Netherlands; mszalans@mbg.au.dk*

The enteric fermentation of feed by ruminant animals contributes significantly to methane emissions and the global greenhouse effect as well as lowering feed efficiency. This can result in significant, negative changes to global and local environment and financial losses for agriculture. Consequently, large-scale emission mitigation strategies are now under evaluation. Two predominant methods of mitigation are: nutrition modifications and genetic selection. While nutrition often gives only a temporary effect, the genetic selection provides permanent and cumulative results thereby being a long term strategy. Among all ruminants, dairy cattle are most suitable to collect large amount of data, necessary to start genetic evaluation. This is due to technological progress, i.e. introduction of Automatic Milking Systems (AMS) or Automatic Feeding Systems (AFS), as well as to intensive and concentrated production environments. Thus there is a need to compare different types of methodologies and especially data analysis, resulting in different phenotype estimates. Enteric methane concentrations in the breath of ~100 lactating Holstein cattle was collected by means of 'sniffer' technologies in AMS during milking at the Danish Cattle research Center and commercial dairy herd. Through the use of various mixed models on various methane trait shapes and an alternative approach, the eructation peaks analysis, methods were contrasted. Each method has its pros and cons. Results in terms of repeatability and rank correlations obtained were not completely equal, but somewhat similar. The phenotype which would be used to encourage farmers to breed for methane emission mitigation should be repeatable, heritable, and easy to estimate and analyze.

Fertility and health traits correlations with methane emissions

L. Zetouni[1], M. Kargo[1,2] and J. Lassen[1]
[1]Aarhus University, Molecular Biology and Genetics, QGG, Blichers Allé 20, Postboks 50, 8830, Denmark, [2]SEGES Cattle, Agro Food Park 15, 8200, Aarhus N, Denmark; lzetouni@mbg.au.dk

There has been an increasing interest, over the last decade, on understanding methane's relationship with traits of interest in dairy cattle, as a way to improve mitigation strategies. A lot has been uncovered about methane and the physiology of the rumen's microbes and nutrition features of cows, but so far little is known about how methane relates to health and fertility traits. This information is extremely important to incorporate into a total merit index. Therefore, our goal was to investigate the genetic correlations between methane and health and fertility traits. Data was collected on 1,500 Holstein cows from 11 commercial herds in Denmark. Methane emission was measured during milking in milking robots, and then quantified using information on milk production, weight and days carried calf to predict carbon dioxide production and multiplied by the ratio between methane and carbon dioxide. The health and reproduction traits analyzed were: days between calving and first insemination, first to last insemination, number of inseminations, udder diseases and 'other diseases', which summed up reproduction, digestion and fertility treatments during the corresponding lactation. Bivariate linear models were used in the analysis to estimate the genetic correlations between methane and each one of the traits analyzed, Preliminary results have shown methane and the traits analyzed seem, in general, to be favorably correlated. Heritabilities for methane were around 0.3, while for the other traits studied the estimates were low, ranging around 0.05 (for other diseases), mostly due to the limited population size. Further investigations with a bigger dataset are being performed in order to establish more accurate results on methane relationship with health and fertility traits in dairy cattle.

Bayesian analysis of selection response on linear or ratio feed efficiency traits in pigs

M. Shirali[1], P.F. Varley[2] and J. Jensen[1]
[1]Center for Quantitative Genetics & Genomics, Aarhus University, Blichers Allé 20, 8830, Tjele, Denmark, [2]Hermitage Genetics, Sion Road, Kilkenny, Ireland; mahmoud.shirali@mbg.au.dk

To propose a selection criterion for feed efficiency either as linear (residual feed intake (RFI)) or ratio (feed conversion ratio (FCR)), a Bayesian approach of calculating response to selection was used. Response to selection was defined as the difference in additive genetic mean of the selected top 10% population and the total population after adjusting for genetic trends. The Bayesian method integrates over unknown population parameters and takes proper account of traits defined as a ratio. A trivariate animal model were used to analyse 3,724 MaxGro terminal line pigs with records for average daily feed intake (ADFI) and body weight gain (ADG) during 60 to 110 kg test period and lean meat percentage (LMP) at the end of test period. Breeding values and genetic parameters of genetic and phenotypic RFI (RFI_g and RFI_p, respectively) were estimated from conditional distribution of ADFI given ADG and LMP using genetic or phenotypic partial regression coefficients obtained from genetic or phenotypic (co)variance matrices of the trivariate model, respectively. Breeding value for FCR was obtained using breeding values of ADFI and ADG after adjusting for the mean of each trait. The posterior means of heritability (standard deviation) were obtained for RFI_g (0.15 (0.03)), RFI_p (0.20 (0.03)), ADFI (0.32 (0.04)), ADG (0.26 (0.04)) and LMP (0.56 (0.06)). Selection for RFI_g showed direct response of -0.16 (0.01) kg/d in RFI_g and correlated responses of -0.16 (0.01) kg/kg on FCR, -0.15 (0.02) kg/d on ADFI and no effect on production traits. However, selection for FCR resulted in direct response of -0.17 (0.01) kg/kg on FCR and correlated responses of -0.14 (0.02) kg/d on RFI_g, -0.18 (0.03) kg/d on ADFI and 0.98 (0.18)% on LMP. Direct selection for FCR or RFI_p resulted in disproportional selection on production traits. Application of RFI_g to breeding programs would involve joint selection on residual feed intake and its component traits of production.

Genotype-Environment interaction for bulls used in expanding dairy herds
S. McParland
Teagasc, Animal and Grassland Research and Innovation Centre, Moorepark, Fermoy, Co. Cork, Ireland;
sinead.mcparland@teagasc.ie

A genotype-environment interaction (G×E) implies different animal genotypes do not perform equally in different environments. The aim of this study was to test for and quantify G×E of bulls across (1) different herd sizes and (2) herds differing in rate of expansion. Data were obtained from the national database for 3,040 milk-recorded herds. Linear robust regression was used to quantify the rate of herd expansion; herds were classified as (1) not expanding, (2) expanding slowly (3 cows/year), or (3) expanding rapidly (9 cows/year). Predicted herd size in 2012 was obtained from the regression and used to stratify herds into small, medium and large herds (average=46.5, 72.6 and 125.3 cows, respectively). Sire predicted transmitting ability (PTA) from the December 2011 national genetic evaluation were obtained from the national database. All PTAs were deregressed and retained where the reliability exceeded 50%. Holstein-Friesian cows which calved for the first time between 2012 and 2014 were retained for analysis. Milk, fat and protein yield were analysed using a linear mixed model adjusted for the fixed effect of parity, herd class (expansion or size), the interaction between sire PTA and herd class, and the random effect of herd-year-season of calving. Regression coefficients of daughter performance on sire PTA were poorer than expected and deviated significantly from 1. However 56% of the data used in this study represented first parity animals. As cows matured, their performance was closer to that expected from their sire PTA. Sire milk, fat and protein PTA interacted ($P<0.001$) with both herd size and rate of expansion. Bull performance was poorer in large herds ($b_{fat}=0.55$) relative to small herds ($b_{fat}=0.66$). Furthermore bull performance was greatest in herds expanding slowly ($b_{fat}=0.69$) relative to either static herds ($b_{fat}=0.60$) or herds expanding rapidly ($b_{fat}=0.56$). Bull performance is not equal in herds of different size or rate of expansion and this should be considered when purchasing bull semen.

B-spline Basis Functions for Modelling Marker Effects in Backcross Experiments
M. Reichelt, M. Mayer, F. Teuscher and N. Reinsch
Leibniz Institute for Farm Animal Biology (FBN), Institute of Genetics and Biometry, Wilhelm-Stahl-Allee 2, 18196 Dummerstorf, Germany; reichelt@fbn-dummerstorf.de

The identification of chromosome sections affecting quantitative traits (QTLs) is essential in the field of quantitative genetics. There are various Bayesian approaches to identify these QTLs. These approaches are based on linear mixed models that include random effects of all markers simultaneously. In experimental populations – such as for example backcross type – higher marker densities result in more parameters than observations ($p>>n$). One of the consequences are inflated estimates of the genetic variance. B-splines offer the opportunity to model the genetic effects of any number of markers by a limited number of basis function effects. Models with basis function effects were compared to marker models in a simulation study. For that reason we simulated 18 different scenarios for a backcross population. 12 QTLs with different effects were either independent or linked in repulsion and coupling in each case. For each scenario 200 experiments with 500 individuals were examined and results averaged over repeated experiments. Markers were equally spaced with 1 cM ($p>>n$) and 5 cM ($p≈n$) distances. Two different degrees for the basis functions and two different number of equidistant knots, which has to be defined in advance, were chosen. B-spline basis functions improved the precision of estimated marker effects, genetic variances and genetic predictions compared to a standard Bayesian model. Computation time was immensely decreased. In conclusion B-splines offer a suitable way for adapting the number of model parameters to the size of the genome, irrespective of marker density.

Random regression analyses of carcass traits yield useful herd-year profiles for management purposes
T.M. Englishby[1,2], D.P. Berry[2], M.P. Coffey[1], K.L. Moore[1] and G. Banos[1]
[1]Scotland Rural College (SRUC), Animal and Veterinary Sciences, Easter Bush, Midlothia, EH15 9RG, United Kingdom, [2]Teagasc, Animal and Bioscience, Animal & Grassland Research and Innovation Centre, Teagasc, Moorepark, Fermoy, Cork, Ireland; tanya.englishby@sruc.ac.uk

Routinely collected abattoir data provides an important source of information for the genetic evaluations of carcass traits. Such information may also be useful to inform management decisions. The objective this study was to quantify the genetic and non-genetic contributions to carcass characteristics in steers with particular emphasis on generating herd-specific profiles for carcass characteristics across age. Slaughter records from 93,338 steers, aged between 360 and 900 days, from 6,117 sires, raised in 1,582 finishing herds were available from the UK national database. Genetic and herd-year of slaughter (co)variance components for each trait were generated using sire random regression models that included cubic polynomials for fixed and quadratic polynomials for random effects; heterogeneous residual variances were assumed across slaughter ages. Average heritability estimates for carcass weight, conformation and fat score across all ages at slaughter were, 0.37, 0.41 and 0.37, respectively. The proportion of the phenotypic variance attributable to differences in herd-year of slaughter ranged from 34.50% (637 days) to 46.80% (540 days) for carcass weight, 35.60% (840 days) to 46.80% (540 days) for conformation and 22.73% to 25.60% (540 days) for fat score. The correlation between herd-year effects wihin trait weakened as the distance between ages increased, indicating different management practices may be more applicable for differently aged cattle. Eigenvalues and eigenfunctions of herd-year covariance matrices revealed variation among herds in the shape of their growth profiles for carcass traits. Positive correlations were evident between the herd-year of slaughter effects for carcass weight and conformation, for carcass weight and fat score, and for conformation and fat score. Herd-year of slaughter parameters presented indicates that herd environment is an important factor in the variability in carcass traits across ages and the output from routine test-day model genetic evaluations may be useful in management and decision support tools.

Composition and sensory qualities of bull beef from different breeds and production systems
G.B. Mezgebo[1,2], F.J. Monahan[2], M. McGee[1], E.G. O'Riordan[1], D. Marren[1], R.I. Richardson[3] and A.P. Moloney[1]
[1]Teagasc, Grange, Co. Meath, Ireland, [2]School of Agriculture and Food Science, University College Dublin, Dublin 4, Ireland, [3]University of Bristol, Bristol BS40 5DU, Bristol, United Kingdom; gebrehawerya.mezgebo@ucdconnect.ie

The aim of this study was to compare composition and sensory qualities of beef from early-maturing (EM) and late-maturing (LM) bulls from contrasting production systems (PS). Fifty six weaned spring-born suckler bulls were assigned to a two breed types (B) × two PS factorial arrangement of treatments. The two B were EM (Aberdeen Angus and Hereford-sired) and LM (Charolais and Limousin-sired). The two PS were: either ad libitum concentrates to slaughter (C) or grass silage ad libitum plus 2 kg concentrate daily during the winter (123 days) followed by 99 days at pasture and then C (GSPC). The duration of the concentrate finishing period was 230 and 85 days for C and GSPC, respectively. Age at slaughter was 16 and 18 months for C and GSPC, respectively. The bulls were slaughtered on reaching the mean live weight to achieve a target carcass weight of 380 kg. At 48 h post-slaughter, Longissimus thoracis muscle was excised (from the 10th rib position), aged for 14 days at 2 °C and stored at -18 °C prior to compositional and sensory analyses. Data were statistically analysed using the General Linear Model procedure of SPSS. Intramuscular fat content was higher (P<0.001) for EM than for LM (39.9 vs 18.2 g/kg), and for C than for GSPC (39.1 vs 18.9 g/kg). Muscle moisture content was lower (P<0.01) for EM than for LM (729 vs 748 g/kg), and for C than for GSPC (734 vs 743 g/kg). Sensory characteristics were similar (P>0.05) between EM and LM. Tenderness, flavour liking and overall liking were higher (P<0.05) for C than for GSPC (tenderness 4.7 vs 4.4, flavour liking 5.5 vs 5.1 and overall liking 5.1 vs 4.7; scale 1-8 (8 extremely tender/liked)). Abnormal flavour was lower (P<0.05) for C than for GSPC (2.3 vs 2.5; scale 1-8 (8 highest abnormal flavour). Results show that the different breed types had a different muscle composition but this did not result in different sensory characteristics. The results also illustrate that dietary inclusion of grass silage followed by pasture resulted in a lower rating for sensory quality by trained panellists.

Prediction of carcass and live weights of some beef cattle breeds

Y. Bozkurt, C.G. Tuzun and C. Dogan
Suleyman Demirel University, Faculty of Agriculture, Department of Animal Science, Cunur Kampus, Isparta, 32260, Turkey; yalcinbozkurt@sdu.edu.tr

This research aimed to develop prediction models for accurate estimation of performance and carcass features of beef cattle grown in feedlot beef system by using Digital Image Analysis (DIA) and Artificial Neural Networks (ANN). For this purpose, 40 animals were used in total and composed of 20 animals of the Brown Swiss breed and 20 animals of the Holstein breed with the age of about 4-5 months at the beginning of the experiment. Animals were fed the same dietary rations throughout the experimental period of 12 months. Liveweights and some body measurements such as body length, wither height, chest circumference, body depth, hip height and hip width were determined fortnightly and also at the end of the finishing period. When the animals reached 500-550 kg liveweights, they were slaughtered and the hot carcass weight, the carcass length, the carcass depth were determined as metric measurements. The digital images of each live animal and their carcasses were taken and the same parameters were also determined from the images. Then, prediction models were developed by DIA and ANN. There were no significant differences (P >0.05) in the predicted values between breeds. Therefore, their data were combined to develop models. The results showed that liveweights can be predicted from body length (R^2=95.8% and 90% by DIA and ANN respectively) and chest girth (R^2=95.7% and 95% by DIA and ANN respectively). The best predictor variables were the same to estimate hot carcass weight. The best predictive variable was found to be carcass length by ANN (R^2=92%). ANN can also be used as an accurate prediction method.

Predicted phenotypic carcass meat yield and cut yields in cattle differing in genetic merit

S.M. Connolly[1,2], A.R. Cromie[3] and D.P. Berry[2]
[1]*Anglo Beef Processers, Castle Street Ardee Co Louth, Ireland,* [2] *Teagasc, Moorepark, Animal & Grassland Research and Innovation Centre, fermoy co cork, Ireland,* [3]*Irish Cattle Breeding federation, Highfield House Bandon Co Cork, Ireland; stephen.connolly@teagasc.ie*

Previous studies documented that animals of superior genetic merit for carcass weight, conformation and fat score, subsequently, express superior phenotypic performance. What is not clear is if genetic merits for these 'macro' terminal traits translate into superior phenotypic performance for more detailed carcass attributes. The objective of the present study was to determine if genetically elite animals, based on the Irish terminal index, are, in fact, superior phenotypically for carcass meat yield and other detailed carcass characteristic. Estimated breeding values (EBV) for carcass weight, fat, conformation, feed intake, calving difficulty, gestation length and mortality were calculated as the average of the sire and dam EBV for each trait based on the April 2010 national genetic evaluation. Following edits, 43,189 animals from 3,494 finishing herds remained. Animals were categorised into four terminal index groups based on individual genetic merit as: very high, high, low, and very low groups. The association between terminal index EBV and phenotypic performance was quantified using mixed models. The dependent variables were vhvc (very high value cuts), hvc (high value cuts), mvc (medium value cuts), lvc (low value cuts), total bone and total bone percentage of carcass weight as determined from video image analysis (VIA). Animals in the highest genetic merit group had, on average, a greater (P<0.05) yield of vhvc, hvc, mvc, lvc, and total meat yield compared to the very low genetic merit group even after adjusting to a common carcass weight. Animals in the very high genetic merit group yielded, on average, 75.37 kg bone (19.50% of carcass weight) while the very low genetic merit animals yielded 73.44 kg bone (21.11% of carcass weight). Animals of greater genetic merit had greater meat yield compared to their lower genetic merit contemporaries. This suggests that higher genetic merit animals have superior performance at both producer level and at a meat industry level, thereby increasing revenue across the entire production chain.

Effect of hang, cut, cook and doneness on eating quality
J. Tollerton, N. Gault, E. Tolland, D. Devlin, L. Majury and H. Lewis
AFBI, Food Research Branch, 18A Newforge Lane, Belfast, BT9 5PX, United Kingdom; joan.tollerton@afbini.gov.uk

Beef is a luxury product whose eating quality still challenges the beef industry worldwide. Several factors including breed, age, sex and muscle type are known to affect the consumers' preference. Indeed studies have indicated that 20% of cooked beef does not meet the consumers' expectation and can affect when next they purchase beef. A consumer study comprising 1,400 people from Northern Ireland and Australia assessed the eating quality of Northern Irish and Australian beef. Consumers were asked to assess the eating quality of four muscles (striploin, rump, knuckle and topside) from carcases both Achilles and Hip hung. Consumers were also asked to evaluate the effect of different cooking methods and degrees of 'doneness'. Consumers scored samples for tenderness, juiciness, flavour liking, satisfaction and overall liking. Results showed that Northern Ireland and Australian consumers judged beef in a very similar way. However, Northern Irish consumers did give slightly higher scores for juiciness and satisfaction. Data also indicated that the Northern Irish consumer places greater importance on flavour when compared to their Australian counterpart. The most important factor influencing eating quality was the cut (muscle) where striploin scored best when grilled and rump scored best when roasted. As expected, an interaction between hanging method and muscle showed that tenderness of striploin but not fillet can be improved by hip hanging the carcase. Cooking method significantly affected eating quality for some attributes. On comparison with grilling, roasting improves tenderness and overall liking of rump meat though has a negative effect on the juiciness of sirloin. The degree of doneness had small effects on the overall liking of grilled and roast steak, it did however, have differing effects on different muscles. Knuckle scored higher when cooked 'medium' and rump when cooked 'well done'. 'Doneness' did not affect the sensory scores for sirloin or topside.

Beef eating quality: a European journey
L.J. Farmer
Agri-Food and Biosciences Institute, Food Research Branch, Newforge Lane, Belfast BT9 5PX, United Kingdom; linda.farmer@afbini.gov.uk

The European beef industry is exceedingly diverse, both in terms of breeds of cattle and rearing regimes. There are numerous regional breeds, from Limousin and Piedmontese to Aberdeen Angus and the 'Black and White' cattle of Poland. In some countries, extensive production on pasture is common while in others most cattle are housed indoors and reared intensively on concentrate diets. The beef produced is a highly priced and valued food item. However, there is evidence that it does not always deliver the expected eating quality. Data from more than 15,000 consumer assessments of 700 cattle from Ireland (north and south), Poland and France shows that, across Europe, 19% of sirloins and more than 50% of topsides are regarded as unsatisfactory by consumers. This is not a new problem and nor is it unique to Europe. Evidence shows that EUROP grade, used across Europe as a standard for carcase quality and by which farmers are paid, bears no relationship to the final quality of the meat. Various initiatives have attempted to solve this problem in countries such as Japan, S Korea, Australia, New Zealand, USA and UK and research has been conducted across Europe and further afield. This paper will summarise the approaches used to quality assure beef eating quality. The European beef industry faces criticism for its apparent high carbon footprint and unhealthy image. Therefore, it is important that solutions to the eating quality issue do not adversely impact on these issues. In Europe there is an initiative to capitalise on recent research to devise tools which will help the beef industry to deliver beef that meets consumer requirements for palatability. The aim is for the industry and research communities to work together to identify a practical and effective way forward that also delivers an environmentally friendly and nutritional product.

Growth performance and meat characteristics of Awassi sheep that holds the Callipyge gene
K.I.Z. Jawasreh, A.H. Al-Amareen and A.Y. Abdullah
Jordan University of Science and Technology, Animal Production, 00962, Irbid, Jordan; kijawasreh@just.edu.jo

Frozen semen of four Rambouillet ram's (R) (homozygous for the mutation of the Callipyge gene – CLPG) was imported from USA (Utah University). The introgression of the CLPG into Awassi has been initiated by producing F1 following crossing of R with Awassi (AW) ewes, while the first backcross (FBC; 75% (AW) and 25% (R)) was formed by using the F1 Callipyge carriers with AW ewes. In order to examine the performance of the FBC (heterozygous for CLPG) compared to the AW, fattening trial was designed using 16 weaned male lambs (eight from AW and FBC lambs) that were kept in individual pens for 98 days. A well balanced ration was offered, the feed intake was recorded in daily basis and weights were monitored weekly. At the end of the fattening period, the lambs were slaughtered in order to investigate the carcass cuts and meat characteristics. Callipyge carrier lambs (CAW) exceeded AW lambs in slaughter weight (P=0.0002). The average daily gain for AW and CAW were 0.189 and 0.332 kg/ day (P<0.0001), respectively. Awassi lambs consumed 1.217 kg feed/kg live weight more than CAW (P=0.0254). The slaughter weight of the CAW was higher than that of the AW (P=0.0001): CAW weight was 50.9 kg at slaughter age (176.5±1.34 days), while AW was 37.07 kg only. The hot and cold carcass weights were higher for CAW (P<0.0001). Furthermore, dressing percentage as well as weights of shoulders, legs, rack and loin were significantly higher for CAW, Fat tail weight was non-significantly affected by the genotype (P<0.298). The heart, liver, kidney and kidney fat weights were significantly higher in CAW. Weights of Longissimus, of total leg muscle, of Intermuscular and subcutaneous fat and of bones were higher in CAW compared to AW. The ratios of muscle and bone weights to total leg weight were higher in CAW, while the intermuscular fat/ leg weight and subcutaneous fat/ leg weights were non-significantly affected by the genotype. Eye muscle area was 14.8099 in AW while 25.3521±1.4086 in CAW (P<0.0001). Eye muscle weight was higher in CAW than in AW (+0.168 kg, P=0.0003). Shear force values were 7.28 in CAW compared to 3.22 in AW (P<0.0002). In conclusion, CLPG rams can be used to improve meat quantity and quality of AW.

Effect of muscle type and ageing on eating quality of beef
J. Tollerton, N. Gault, E. Tolland, L. Farmer, D. Devlin, L. Majury and H. Lewis
AFBI, Food Research Branch, 18A Newforge Lane, Belfast, BT9 5PX, United Kingdom; joan.tollerton@afbini.gov.uk

There are several variables that can affect the consistency and eating quality of beef. Research has highlighted that different muscles have different eating quality and that a lower than expected eating experience can seriously affect the customer's choice regarding their next beef pruchase. In order to investigate the relationship between the eating quality of different beef muscles, sensory evaluations were carried out using topside, silverside, fillet, sirloin, rump and knuckle muscles. Investigations included studying the effect of electrical stimulation, carcase hanging, ageing, cooking method and degree of 'doneness'. Results showed highly significant differences (P<0001) in eating quality across the evaluated muscles. Differences in overall liking score (0-100) had a range of 30 points when fillet and topside muscles were included. While the order of sensory scores for grilled beef was fillet > striploin > rump > knuckle > topside, the difference between parts of the same muscle were surprisingly large. The eating qualtiy of striploin taken from the anterior end of the loin scored between 6-8 units higher than the posterior, depending on hanging method. Similar variations in sensory score were observed for roasted rump. Analysis highlights that tenderstretch hanging improved the acceptability of all muscles apart from fillet, which still remaned most liked. Ageing improved sensory scores for anterior and mid sirloin and topside as well. An interaction between muscle and cooking method showed that while some muscles gave similar sensory scores when grilled and roasted, rump received much higher scores when roasted. Investigations of the correlations between sensory scores for different muscles from 32 animas showed that here was no consistent relationship between the eating qualty of one muscle with that of another muscle taken from the same carcase. For example, the sensory evaluation of striploin cannot reliably be used to predict the eating quaity of rump or topside. This lack of correlaton can be explained by the fact that muscles respond differently to factors such as hanging method, electrical stimulation and ageing.

Factors affecting Eating Quality of Northern Irish Beef

D.J. Devlin, N.F.S. Gault, B.W. Moss, E. Tolland, J. Tollerton and L.J. Farmer
Agri-Food and Biosciences Institute, Food Research Branch, Newforge Lane, BT95PX Belfast, United Kingdom;
declan.devlin@afbini.gov.uk

The beef industry in Northern Ireland (N.I.) comprises a vital part of the agri-food sector making a significant contribution to the overall economy. The production of N.I. beef with consistently high eating quality has the potential to increase domestic sales and improve marketability abroad. Beef eating quality as defined by the consumer is predominantly made up of texture (tender or tough), juiciness, flavour and overall palatability with tenderness deemed the most important. To investigate the factors affecting beef eating quality that could have an enhanced effect for N.I. beef production, AFBI has conducted a number of experiments using tightly controlled consumer sensor y panels in accordance with the Meat Standards Australia protocols. More than 6,000 consumers tasted roast beef or grilled steak prepared from over 900 beef primals from 239 animals. On a quality scale of 0-100, findings indicate that the variation in eating quality associated with muscle type is 30 points. Hanging method (Hip or Achilles suspended) contributes between 8 and 18 points. Cooking method accounts for 5-12 points of variation, with animal breed, aging process and marbling fat accounting for a further 5-10 consumer points. Optimised processing techniques to ensure consistent quality was found to be of considerable importance. The rate of carcase chilling together with level of electrical stimulation applied must be within defined limits in order avoid an adverse effect on eating quality. Many of these factors were found to interact with each other. The successful management of these factors to assure the eating quality of beef presents an opportunity for the N.I. beef industry.

The variability of European beef can be reduced by predicting consumer satisfaction

S.P.F. Bonny[1,2], J.-F. Hocquette[1,3], D.W. Pethick[2], I. Legrand[4], J. Wierzbicki[5], P. Allen[6], L.J. Farmer[7], R.J. Polkinghorne[8] and G.E. Gardner[2]
[1]INRA, Recherches sur les Herbivores, UMR1213, 63122 Saint Genès Champanelle, France, [2]Murdoch University, School of Veterinary and Life Sciences, 60 South st, Murdoch, 6150, Australia, [3]Clermont Université, VetAgro Sup, UMR1213, 63122 Saint Genès Champanelle, France, [4]Institut de l'Elevage, Service Qualité des Viandes, MRAL, 87060 Limoges Cedex 2, France, [5]Polish Beef Association, Ul. Kruczkowskiego 3, 00-380 Warszawa, Poland, [6]Teagasc Food Research Centre, Ashtown, Dublin 15, Ireland, [7]Agri-Food and Biosciences Institute, Newforge Lane, Belfast BT9 5PX, United Kingdom, [8]Polkinghornes, 431 Timor Road, Murrurundi, NSW 2338, Australia; spfbonny@gmail.com

Delivering quality beef to the consumer relies upon both quantifying beef palatability and then accurately predicting that from information available at slaughter. Recent work has outlined the ability of carcass traits such as sex, age and breed to predict beef eating quality across a diverse range of countries and production systems in Europe. However, the variation of consumer responses to beef within Europe has yet to be quantified. Following Meat Standards Australia testing protocols, over 19,000 consumers from Northern Ireland, Poland, Ireland, France and Australia tasted cooked beef samples, scored the sensory characteristics and then allocated them to one of four quality grades; unsatisfactory, good-every-day, better-than-every-day and premium. A total of 22 different muscles, cooked by four different cooking methods and to three different degrees of doneness were tested. Linear discriminant functions were calculated for each experimental group, and for the dataset as a whole, using the sensory scores which were used to predict the consumer scored quality grade. Overall 26% of the beef was unsatisfactory. The discriminant analysis allocated 68% of samples to the correct quality grade, similar to previously reported values. Importantly, only 7% of the beef unsatisfactory to consumers was misclassified as acceptable. These results demonstrate that a MSA-like grading scheme could be used to predict beef eating quality and underpin a commercial eating quality guarantee these European countries.

Effect of carcass weight on composition and sensory qualities of beef from bulls

G.B. Mezgebo[1,2], F.J. Monahan[2], M. McGee[1], E.G. O'Riordan[1], D. Marren[1], R.I. Richardson[3] and A.P. Moloney[1]
[1]Teagasc, Grange, Co. Meath, Ireland, [2]School of Agriculture and Food Science, University College Dublin, Dublin 4, Ireland, [3]University of Bristol, Bristol BS40 5DU, Bristol, United Kingdom; gebrehawerya.mezgebo@ucdconnect.ie

The aim of this study was to determine the composition and sensory qualities of beef from suckler bulls from different carcass weights (CW). Sixty spring-born (mean birth date 8 March) late-maturing (Charolais and Limousin-sired) weaned suckler bulls were housed in slatted floor pens (groups of five) and provided with a finishing diet of ad libitum concentrates (870 g/kg rolled barley, 60 g/kg soya bean meal, 50 g/kg molasses and 20 g/kg minerals/vitamins) and grass silage (dry matter digestibility: 700 g/kg) until they reached treatment mean live weights to achieve target CW of 340, 380 and 420 kg. The duration of the concentrate finishing period was 145, 180 and 230 days for 340, 380 and 420 kg CW, respectively. Age at slaughter was 14, 15 and 17 months for 340, 380 and 420 kg CW, respectively. At 48 h post-slaughter, Longissimus thoracis muscle was excised from each carcass for the compositional and sensory analyses. Data were statistically analysed using the General Linear Model procedure of SPSS. Intramuscular fat content tended ($P<0.06$) to be higher for 420 kg CW than for 380 and 340 kg CW, which did not differ (38.4, 26.2 and 25.3 g/kg for 420, 380 and 340 kg CW, respectively). Muscle moisture content was lower ($P<0.01$) for 420 kg CW than for 380 and 340 kg CW, which did not differ (725, 747 and 744 g/kg for 420, 380 and 340 kg CW, respectively). Flavour liking was higher ($P<0.05$) for 420 and 380 kg CW (which did not differ) than for 340 kg CW (5.4, 5.3 and 5.1 for 420, 380 and 340 kg CW, respectively; scale 1-8 (8 extremely liked)). Abnormal flavour was lower ($P<0.05$) for 420 and 380 kg CW (which did not differ) than for 340 kg CW (2.3, 2.3 and 2.5 for 420, 380 and 340 kg CW, respectively; scale 1-8 (8 highest abnormal flavour)). Tenderness, juiciness, beefy flavour and overall liking ratings were similar ($P>0.05$) for all CW. Results suggest that an increase in CW tended to increase intramuscular fat content but this increase did not result in a change in tenderness and juiciness of the beef, as evaluated by trained panellists.

Growth and carcass traits of purebred Simmental and Simmental×beef breed crossbred bulls

M. Pesonen and A. Huuskonen
Natural Resources Institute Finland (Luke), Tutkimusasemantie 15, FI-92400 Ruukki, Finland; maiju.pesonen@luke.fi

The objective of the present research was to study the potential for improvement of growth and carcass traits through Simmental (Si) × beef breed crossbreeding compared to purebred Si bulls. The data from Finnish slaughterhouses included observations of 2,152 pure Si, 805 Si×Aberdeen Angus (Si×Ab), 255 Si×Blonde d'Aquitaine (Si×Ba), 980 Si×Charolais (Si×Ch), 1,033 Si×Hereford (Si×Hf) and 999 Si×Limousin (Si×Li) bulls. The carcasses were classified for conformation and fatness using the EUROP classification. Statistical analyses were performed using the SAS Mixed procedure. Differences between the breed groups were compared using a Dunnett's test so that pure Si was used as a control breed. The average slaughter age was 565 days. Mean carcass weights for the Si, Si×Ab, Si×Ba, Si×Ch, Si×Hf and Si×Li bulls were 402, 396, 414, 414, 393 and 394 kg, and daily carcass gains were 686, 675, 709, 725, 664 and 669 g/d, respectively. All crossbreds differed significantly from Si bulls in both carcass weight and carcass gain. The conformation score of the pure Si bulls was 8.3, and improved 10, 7 and 2% by using Ba, Li and Ch crosses, respectively. Si×Ab crossbreds produced 7% and Si×Hf crossbreds 8% poorer conformed carcasses compared to pure Si bulls. The carcass fat score of the Si×Ab, Si×Hf and Si×Li bulls was 22, 22 and 4% higher than that of the Si bulls. There were no differences in fat score between Si and Si×Ba or between Si and Si×Ch bulls. The yields of loin, tenderloin, inside round and outside round were lower with Si×Ab bulls compared to pure Si bulls. With Si×Hf crosses the yields of inside round and outside round were lower compared to Si bulls. There were only few differences in the yield of valuable cuts between Si bulls and Si×late maturing crossbreds. However, the yield of corner round was higher with Si×Ba and Si×Li bulls compared to Si bulls. It can be concluded that crossbreeding with Ch and Ba breeds improve both gain and carcass conformation compared to pure Si bulls.

Performance of bulls offered diets based on whole crop silages with or without protein inclusion

M. Pesonen, E. Joki-Tokola and A. Huuskonen
Natural Resources Institute Finland (Luke), Tutkimusasemantie 15, FI-92400 Ruukki, Finland; maiju.pesonen@luke.fi

A 3×2 factorial design with growing bulls was used to study the effects on performance of (1) forage type and (2) inclusion of rapeseed meal (RSM) in the barley-based concentrate. The three forage types were grass silage (GS), at the milk stage harvested whole crop barley silage (BSM), and at the dough stage harvested whole crop barley silage (BSD). The experiment comprised in total 30 Hereford bulls. At the beginning of the experiment the bulls were divided into five blocks of six animals by live weight (LW). Within the block, the bulls were randomly allotted to one of the six treatments. The bulls were placed in adjacent tie-stalls and offered silages ad libitum. The concentrate used was rolled barley alone or barley plus RSM. The amount of the concentrate supplementation was 37 g/metabolic LW/animal/day. The crude protein (CP) content of barley was 119 g/kg dry matter (DM). RSM was given so that the CP content of the concentrate was raised to 140 g/kg DM. The data were subjected to ANOVA using the SAS GLM procedure. The statistical model included the fixed effects of forage type, RSM supplementation and their interaction and the random effect of the block. Mean total DM intakes for the GS, BSM and BSD bulls were 9.04, 8.74 and 9.46 kg/d, and carcass gains were 883,770 and 867 g/d, respectively. The BSD bulls differed significantly from the BSM bulls in both DM intake and carcass gain. Mean carcass weights of the GS, BSM and BSD bulls were 395, 369 and 392 kg, respectively, and tended to be lower in the BSM bulls compared to the other forage treatments. The forage type had no effects on the carcass conformation or fat score. RSM supplementation increased total DM intake (8.64 vs 9.52 kg/d, P<0.01), carcass gain (784 vs 896 g/d, P<0.001) and carcass weight (371 vs 400 kg, P<0.01) of the bulls. There were no effects of RSM on carcass conformation score or fat score. The interactions between the forage type and RSM supplementation were not significant.

Effect of using sustained-release non-protein nitrogen on feed conversion efficiency in beef cattle

M. Agovino[1], A. Linscott[2] and F. Aubry[1]
[1]Alltech Biotechnology Centre, Summerhill Road, Dunboyne, Co. Meath, Ireland, [2]Alltech UK, Ryhall Road, Stamford, Lincs, PE9 1TZ, United Kingdom; magovino@alltech.com

This study investigated the effect on feed conversion efficiency and average daily gain (ADG) of two different sustained-release (SR) ruminal non-protein nitrogen (NPN) source in fattening and finishing beef cattle. 143 beef finishers, mainly Limousin cross with a small number of British Blue X's (heifers and steers), were split into 3 feeding groups. Control C (n=47; basal diet based on maize silage, grass silage, fresh potatoes, rolled wheat, pot ale syrup, minerals and urea), OP group (n=47; basal diet reformulated to include SR NPN – Optigen®, Alltech Inc. and replace urea), OS group(n=47; basal diet reformulated to include a newgeneration of SR NPN – Optisync®, Alltech Inc and replace urea). Animals were on treatment for 130 days in average. Dry matter intake (DMI), feed conversion efficiency (FCE) and daily gain (ADG) were measured. Cattle were weighed at the start, middle and end of the trial. Dung sieve analysis was carried out. Dung samples were sieved twice during the trial period using the Nasco's Digestion Analyzer (3 sieves). The digestion analyzer evaluates rapid by-product passage and performance by evaluating livestock manure. Cattle on Control diet showed mucin casts. Mucin casts were not obvious in cattle of OP and OS group and there was less undigested feed in the top sieves. ADG was 1.185 kg/day on the control group, 1.415 in OP group and 1.441 in OS group. FCR (Feed Conversion Rate) was 10.43, 8.71 and 8.55 respectively in control, OP and OS group. Both OP and OS groups showed that sustained-release (SR) ruminal non-protein nitrogen (NPN) source could be used to replace feed grade urea in beef finishing diets in order to improve FCR.

The effect of suckler cow condition score in pregnancy on calving parameters and progeny performance
D.E. Lowe, F.O. Lively and A.W. Gordon
AFBI Hilsborough, Large Park, Hillsborough, BT40 2NQ, United Kingdom; denise.lowe@afbini.gov.uk

In the UK, cow body condition score (BCS) is often increased at grass and then those reserves mobilised over the winter period in order to lower winter feed costs. The objective of this study was to evaluate the impact of changes in BCS of spring-calving suckler cows from weaning to calving on calving parameters and growth of their progeny to weaning. Data from 3 winter periods of the project were combined (mean of 88 cows/ year). Cows were penned according to breed and body condition score at weaning and subsequently fed to either gain, lose or maintain BC so as to achieve a calving BCS in the region of 2.5 at the point of calving. At calving, cows were scored for calving difficulty and a sample of colostrum was collected and subsequently analysed for casein, fat, lactose, protein and urea N. Calves were weighed and scored for vitality at birth and subsequently weighed every 3 weeks to weaning. For statistical analysis, each cow was categorised according to the change in their BCS as follows: BCS at weaning <2.5 and stayed the same or ±0.25 by calving; BCS at weaning ≥2.5 and stayed the same or ±0.25 by calving; BCS at weaning <2.5 and increased by calving; BCS at weaning ≥2.5 and increased by calving; BCS at weaning <2.5 and decreased by calving or BCS at weaning ≥2.5 and decreased by calving. The continuous type variables were analysed under the linear mixed model methodology with cow ID used as a random effect and BCS category as a fixed effect in all models. The score-type variables were each assessed using a random permutation test for a 2D contingency table with score variable and BCS category as classification factors. BCS change from weaning to calving had no significant effect on colostrum composition, calving difficulty score ($P=0.288$) or calf vitality ($P=0.226$). There was a trend that progeny from cows that had BCS≥2.5 at weaning and lost condition by calving, which is currently considered best practice, weighed lightest at weaning. This negative impact on progeny performance would offset any potential saving in winter feed costs of this system. Additional research is required to valid this finding and to investigate the relationship between cow body condition score on lifetime progeny performance.

Growth of bull calves at the dairy farm does not affect their performance as rosé veal calves
A. Jensen, M. Johansen and M. Vestergaard
Aarhus University, Foulum, Department of Animal Science, Blichers Allé 20, 8830 Tjele, Denmark;
mogens.vestergaard@anis.au.dk

The rosé veal production in Denmark is based on dairy bull calves, which are shipped from the dairy farms to the veal calf producers at 2-8 weeks of age and slaughtered before 10 months of age. Recent Danish studies on farm data suggest that a low calf mortality level in the dairy herd is positive for the later performance of veal calves. However, rosé veal farmers usually pay for LW of the calves at time of purchase as the expectation is that a high LW favours the later performance. But no solid data are available to verify if a high daily gain in a dairy herd is positive for the later performance. The objective was to test if daily gain from birth to entry (ADG1 (Mean (SD)): 621 (123) g/d), LW and(or) age at entry (LW1: 66 (7) kg and Age1: 35 (9) d) affected the daily gain during the 62 (6) days in the starter unit (ADG2: 992 (71) g/d), daily gain during the 192 (7) days in the finisher unit (ADG3: 1,430 (33) g/d) and daily gain during both starter and finisher units (ADGtot: 1,320 (33) g/d). The effect of ADG2 on ADG3 and ADGtot was also evaluated. Data included 2,436 calves coming from 41 different dairy farms. The regression analysis was based on mean data for the 41 dairy farms and included Stepwise regression. Results showed that ADG2 was best described by a model including LW1 and Age1 ($R^2=0.45$) and that ADG2 was negatively affected by LW1 ($\beta=-7.6$ g/kg; $P<0.001$) and positively by Age1 ($\beta=9.1$ g/d; $P<0.001$). ADG3 was not affected by Age1, LW1, ADG1 and(or) ADG2 (all P values >0.3). Finally, ADGtot was best described (model $R^2=0.96$) by LW1, ADG2 ($\beta=0.23$ g/g; $P<0.001$) and ADG3 ($\beta=0.75$ g/g; $P<0.001$). These results suggest that veal farmers do not benefit from buying calves from dairy farms with high growth rate and high LW relatively to the age at shipping and that overall growth performance instead is highly dependent on growth during the first 2 months in the starter unit.

The impact of animal movements on lifetime performance of beef cattle

V.S. Murphy[1,2], D.E. Lowe[2], F.O. Lively[2], F.M. Titterington[2] and A.W. Gordon[3]
[1]Queen's University Belfast, University Road, Belfast, BT7 1NN, United Kingdom, [2]Agri-Food and Biosciences Institute, Large Park, Hillsborough, BT26 6DR, United Kingdom, [3]Agri-Food and Biosciences Institute, Newforge Lane, Belfast, BT9 5PX, United Kingdom; victoria.thompson@afbini.gov.uk

The aim of this study was to evaluate the impact of animal movements between producers on lifetime performance of beef cattle in Northern Ireland (NI). Each movement, which results in cattle mixing, may increase social stress and aggression, and expose animals to a greater risk of disease which could have implications on lifetime performance. Data were sourced from the Bovine Information System (BovIS), which assimilates slaughter data from local abattoirs and phenotypic information from the government database, Animal and Public Health Information System (APHIS). The lifetime performance of beef cattle which had been finished on the farm of origin was compared with those which have had two or more residencies in their lifetime. Only prime animals were selected i.e. steers and heifers under 30 months of age and young bulls under 24 months of age. Breed was divided into three categories; continental, native and dairy. Animals were removed where breed was recorded as 'unknown'. A total of 1,604,563 prime animals were available and analysed using linear regression. Although there was a significant effect of the number of farm residencies on daily carcass gain (DCG), cold weight and age at slaughter in all animal and breed types from 2009 to 2015 ($P<0.05$), the variance accounted for all variables ranged from 0.1 to 0.9 for heifers, 0.3 to 1.9 for steers and 0.4 to 7.6 for young bulls. Therefore, only a small percentage of the variance in DCG, cold weight and age at slaughter can be attributed to the number of farm residencies. Predicted DCG of continental steers from 2009 to 2015 was 0.51 kg for those which have been finished on the farm of origin and 0.48 kg for those which have had 4 farm residencies in their lifetime. The findings from this study indicate that although there is a significant effect of the number of farm residencies on the lifetime performance of beef cattle, the variance accounted for is low and therefore may not have a biological significance.

Effect of feed ration composition on sustainability

L. Mogensen[1], M.T. Knudsen[1], I.S. Kristensen[1], N.I. Nielsen[2] and T. Kristensen[1]
[1]Aarhus University, Department of Agroecology, Blichers Alle 20, 8830 Tjele, Denmark, [2]SEGES, Agro Food Park 15, 8200 Aarhus N, Denmark; lisbeth.mogensen@agro.au.dk

Livestock production has a major environmental impact and feed production contribute considerably to this. To lower the environmental impact of the feed, farmers should choose feed items that have a lower environmental impact. The goal is that the farmer in the future can formulate a feed ration that at the same time meet the nutritional recommendations, and minimize the effects on climate, environment and biodiversity. We have developed a method based on life cycle assessment (LCA) to document the sustainability of the whole chain of producing each feed item taking into account cultivation, processing, and transport as well as contribution from soil carbon changes. The new sustainability parameters are given per kg dry matter (DM) feed and include: carbon footprint (CO_2-eq.), land use (m^2), impact on biodiversity (potential disappeared fraction), fossil energy consumption (MJ) and eutrophication (NO_3-eq.). These sustainability parameters were quantified for 17 different home-grown feed items and 26 types of purchased feed. In general, the values for carbon footprint, eutrophication and energy consumption were highly correlated. Whereas the correlation between these sustainability parameters and land use or biodiversity is low. The sustainability of producing 1 kg DM of roughage are generally higher than for producing 1 kg DM of cereals, and grass has a more positive effect on biodiversity and soil carbon sequestration than other types of roughage. The sustainability parameters can be used for investigating effect of changes in feed rations, for example clover grass vs maize silage. Including clover grass silage in the ration had a positive effect on biodiversity and did not affect carbon footprint. Using local protein in the form of rapeseed cake vs imported soybean meal had a positive effect on carbon footprint, land use as well as biodiversity.

Isotope ratios of bioelements for inferring beef origin and zebu feeding regime in Cameroon

M. Perini[1], B.M. Nfor[2], F. Camin[1] and E. Piasentier[3]
[1]Fondazione E. Mach, via Mach 1, 38010 San Michele all'Adige (TN), Italy, [2]Livestock Development Corporation SODEPA, P.O. 1410, Yaoundé, Cameroon, [3]University of Udine, via Sondrio 2, 33100 Udine, Italy; edi.piasentier@uniud.it

The aim of the study was to deal with a lack of knowledge regarding the stable isotope composition of beef from zebu cattle reared in tropical Africa. Sixty beef carcasses belonging to the most popular zebu breeds (Goudali, White Fulani and Red Mbororo) were selected and sampled at the slaughterhouse of Yaounde. The carcass fat colour was visually evaluated on three-level qualitative scale (white, cream or yellow). The stable isotope ratios of five bio-elements – H, O, C, N and S – were analysed in protein and fat muscle fractions, together with the fatty acid composition of meat samples from Longissimus dorsi muscle. Zebu beef from Cameroon has a particular isotope profile, characterised by higher $\delta^{13}C$, δ^2H and $\delta^{18}O$ values than those reported in other areas of the world, as a consequence of the almost exclusive use of tropical C4 pasture grasses for cattle feeding and of the geographic and climatic gradient in the isotope composition of precipitatation water. Moreover, the isotopic composition of the fat fraction of muscle was significantly affected by the subcutaneous fat colour. Zebu with white subcutaneous fat ('white type') showed a clear tendency to be more enriched in ^2H isotopes and more depleted in ^{13}C isotopes than the 'yellow type', while the 'cream type' represented an intermediate condition. These trends correlated with fat composition: ^2H enrichment and ^{13}C depletion were significantly correlated with a high PUFA content, while ^2H depletion and ^{13}C enrichment were correlated with a high SFA content. It was argued that, as a consequence of better nutritional status, the 'yellow types' had a more diluted phospholipid (PL) content and a higher proportion of neutral lipids (NL) in their fat in comparison with the 'white types', and thus a less negative $\delta^{13}C_{FAT}$ value, in agreement with the findings that PL fraction displays a more negative diet-tissue fractionation than NL fraction. Within Cameroon, multi-element analysis give promising results for tracing the regional origin of beef and some aspects of the cattle breeding system, such as the animal's nutritional status.

Variation in M. longissimus texture and cooking loss from cattle within a commercial supply chain

S. Mackintosh[1], J. Draper[2], A. Scott[3], S. Morgan[1], H. Powell[1], I. Richardson[4] and N. Scollan[1]
[1]Aberystwyth University, IBERS, Gogerddan, Aberystwyth, SY23 3EE, United Kingdom, [2]ABP Food Group, 6290 Bishops Court, Solihull Parkway, Birmingham, B37 7YB, United Kingdom, [3]Sainsbury's Supermarkets Ltd, 33 Holborn, London, EC1N 2HT, United Kingdom, [4]University of Bristol, School of Clinical Veterinary Science, Langford, Bristol, BS40 5DU, United Kingdom; sim17@aber.ac.uk

Texture is a key factor in consumer perception of meat quality at the point of consumption. Variation in beef eating quality, particularly texture, is a primary cause of both failure to re-purchase and consumer complaints regarding beef products. Factors affecting beef tenderness are multifactorial, and despite standardisation of slaughter and carcass processing, wide variability still exists. This in part may be due to the diversity of cattle production systems that underpin the supply chain. This study assessed strip loin texture from Limousin (LIM, n=261) and Aberdeen Angus (AA, n=238) steers from a total of 25 farms, within a single supply chain. Cattle of a defined market specification were selected from a range of finishing diets to represent a range of production systems within the supply chain. Production data was confirmed by telephone interviews. Longissimus dorsi steaks were cooked to 72 °C and then chilled overnight (4 °C). Cooking loss was assessed on raw vs cooked-chilled steaks. Meat texture was assessed using Warner-Bratzler shear force. Cooking loss ranged from 16.5-34.8% and 17.3-31.3% for LIM and AA cattle respectively. Shear force measurements ranged from 1.7-4.0 kg and 1.5-4.4 kg for LIM and AA cattle respectively. Understanding the extent of variation in beef texture in a supply chain is an important step to facilitating improvements in overall meat eating quality. Furthermore, identifying the impact of production systems on texture and cooking loss may offer opportunities for improvement.

Genetic variability of the MHC class II (DRB.3) in South African and Namibian beef cattle breeds

L. Haikukutu[1], T.O. Itenge[1], L. Bosman[2], C. Visser[2], J. Lyaku[3], F. Mausse[3] and E. Van Marle-Köster[2]
[1]University of Namibia, Department of Animal Science, Faculty of Agriculture and Natural Resources, 9000, Windhoek, Namibia, [2]University of Pretoria, Animal & Wildlife Science, Lynnwood Road, 0002, Pretoria, South Africa, [3]University of Namibia, School of Veterinary Medicine, Faculty of Agriculture and Natural Resources, 9000, Windhoek, Namibia; evm.koster@up.ac.za

Tick infestation is a limiting factor in beef production in the sub-tropical areas of both South Africa (SA) and Namibia. The consequences of tick infestation include blood loss, supressed growth and tick-borne diseases. Acaricides for livestock dipping has been the primary method of control, but the resistance of ticks to these chemicals is of major concern. Genetic variation in the MHC class II genes has been shown to be associated with resistance to tick-borne diseases. In this study the genetic variability in the DRB3 region was studied in three SA and two Namibian beef cattle breeds. Blood was collected from 40 SA Bonsmara, 41 Nguni and 39 Hereford cattle from three different geographical locations in SA, and 42 Afrikaner and 30 Sanga types from Namibia. Ticks under the tail area were counted and converted into six classes ranging from class one (0-10) to class six (>50). Samples were analysed for four DRB3 loci (DRB3, DRBP1, RM185, BM1815) following standard DNA extraction and PCR procedures. Statistical analysis was done using Msat Toolkit and Structure. PIC values ranged from 0.508 to 0.847. Unbiased Heterozygosity varied between 0.76 (SA Bonsmara) to 0.66 (Namibian Afrikaner). The highest allele number was found in the SA Bonsmara and Namibian Sanga (7.7), with the lowest number in the Namibian Afrikaner. Tick scores were low (counts between 11 and 20) for all the breeds except for the Bonsmara with counts of 31-40 ticks per animal. Structure analyses grouped the five populations into three indistinct clusters. For the four loci, limited variation was observed between the Sanga types, composite Bonsmara and Taurine breeds. Despite the high polymorphic nature of loci, genetic variability between the populations was lower than expected. It will be important to investigate the infection status of individuals within these populations for a better understanding of the mechanisms for tick resistance.

Grass-based production systems for late-maturing sired suckler bulls

K. McMenamin[1,2], M. McGee[2], A.P. Moloney[2], A.K. Kelly[1] and E.G. O'Riordan[2]
[1]University College Dublin, School of Agriculture and Food Science, Belfield, Dublin 4, Ireland, [2]Teagasc, AGRIC, Grange, Dunsany, Co. Meath, Ireland; kevin.mcmenamin@teagasc.ie

In temperate climates including grazed grass in the diet of cattle is economically desirable. The aim of this experiment was to determine the effects of contrasting grass-based production sytems (PS) on liveweight and carcass weight and carcass traits of suckler bulls. Sixty weaned spring-born Charolais and Limousin sired bulls (live-weight 425 kg, s.d. 35.2; age 390 days (d), s.d. 39.2), previously offered grass silage ad-libitum (GS) + 2 kg of a barley-based concentrate (C) daily over 145 d, were blocked on age, weight and sire breed and assigned to one of four PS. The four PS were: (1) Grazed-grass (GG) only for 200 d (G0); (2) GG (100 d) then GG + 0.5 expected dry matter intake (DMI) as C (100 d) (G0GC); (3) GG + 0.5 DMI as C (200 d) (GC); and (4) GG (100 d) then housed indoors on ad-libitum C + GS (100 d) (G0AL). Bulls rotationally grazed Lolium perenne-dominant swards to a target post-grazing sward height of 4.5 cm. Paddock areas differed for G0, G0GC and GC such that residency time and, pre and post-grazing sward heighs were similar for all PS. Slaughter age was 19.3 months. Data were analysed using mixed models with treatment included as a fixed effect and block as a random effect. Average daily live-weight gains (kg/d) over 200 d were: G0 1.25, G0GC 1.30, GC 1.25 and G0AL 1.49, respectively. Slaughter (699 vs 650 kg) and carcass (407 vs 374 kg) weights (P<0.05), and carcass fat (7.7 vs 5.2, P<0.001) and conformation (10.6 vs 9.5, P<0.05) scores (1-15) were greater for G0AL than the other PS, which did not differ (P>0.05). Kill-out proportion (g/kg) was highest for GC (587) and lowest for G0 (564) with G0GC (570) and G0AL (584) being intermediate. In conclusion, only G0AL produced carcasses acheiving current market specifications (greater than or equal to 6 on 1-15 scale) for fat score.

Comparison of methods to evaluate the chemical composition of carcass from beef cattle
M. Al-Jammas, J. Agabriel, J. Vernet and I. Ortigues-Marty
INRA, UMR 1213 Herbivores, Theix, 63122 St Genès Champanelle, France; ortigues@clermont.inra.fr

In cattle, carcass quality is an important criterion in the grid of payment to producers. The most accurate determination of beef carcass quality involves dissection of the entire carcass. This, however, is very costly and cumbersome. Alternative methods were developed from cut carcass or indirect measurements to calculate carcass composition using prediction equations: (1) dissection of the 9th, 10th and 11th rib followed by analysis of its chemical composition; (2) specific gravity of the whole carcass; and (3) combination of easily obtained measures (EOM) such as kidney-pelvic-heart fat, USDA yield grade (YG), quality grade, subcutaneous fat thickness (SFT), ribeye area, marbling and hot carcass weight. These alternative methods have become reference methods in research but have not been directly compared. Current objectives were to evaluate if the carcass composition calculated from the different reference methods matched. For this purpose, equations were developed between single proxy traits (YG, SFT) and reference carcass composition value, for each of the reference method. A meta-analysis was applied on 30 studies from 25 publications that reported carcass quality data from both proxy traits and reference methods (53, 31 and 16% of data based on rib dissection, specific gravity and EOM respectively). Within-study variance-covariance models were developed (Minitab 16) according to Sauvant et al. Results showed that the amounts of carcass lipids calculated from EOM are significantly lower than those calculated from rib dissection or specific gravity (-25 kg and -19 kg respectively, representing up to 17% of total carcass lipids). Similarly, amounts of carcass proteins calculated from rib dissection were significantly higher than those obtained from specific gravity or EOM (+3 kg and +2 kg respectively, up to 6% of total carcass protein). In conclusion, the three reference methods considered differ in their degree of accuracy to determine the chemical composition of the carcass.

Feed efficiency, carcass traits and selection for yearling weight in Nellore cattle
J.N.S.G. Cyrillo[1], H.A. Fidelis[1], L.O. Tedeschi[2], S.F.M. Bonilha[1], R.H. Branco[1] and M.E.Z. Mercadante[1]
[1]Instituto de Zootecnia, Centro APTA Bovinos de Corte, CP 63, 14174000 Sertãozinho, SP, Brazil, [2]Texas A&M University, Department of Animal Science, 101 Kleberg Center, 77845, College Station, TX, USA; cyrillo@iz.sp.gov.br

Data from five studies were compiled to assess the relationships between residual feed intake (RFI), carcass traits and selection for yearling weight (YW) in Nellore cattle. Three selection lines were established by selection experiment: Selection Nellore (NeS) and Traditional Nellore (NeT), based on maximum selection differential on YW and Control Nellore (NeC), selected for the contemporary group mean of YW. Five feed efficiency tests (minimum duration of 70 days) were carried out to classify NeS, NeT and NeC bulls in low and high RFI levels. Next, a sample (n=127; initial age=534 days) with 0.9620 kg DM/d^{-1} of average difference between RFI levels was finished and slaughtered to evaluate carcass traits (slaughter weight, kg; hot carcass weight, kg; rib eye area, cm^2; and backfat thickness, mm) and internal organs expressed per 100 kg of hot carcass weight (liver; kidneys; and kidney-pelvic-inguinal fat). A meta-analysis in a mixed model included RFI level, selection line, their interactions and age as covariate as fixed effects; diet within year and year as random effects. No significant differences in carcass traits or internal organs were found between low and high RFI animals. Significant differences (P<0.05) between selection lines were detected for all carcass traits (e.g. slaughter weight: 405; 481; 484±12.1 kg) and for kidneys (0.23; 0.34; 0.28±0.13 kg) for NeC, NeS and NeT, respectively. Significant interactions between RFI levels and selection lines were detected for liver (P<0.05), having low RFI animals in NeC line heavier liver than high RFI animals (1.99±0.11 vs 1.85±0.11), whereas in NeS line low RFI animals had lighter liver than high RFI animals (1.80±0.11 vs 1.94±0.11), respectively. These results suggest that efficient and not efficient animals do not differ in carcass traits. There is evidence that RFI, growth performance and liver weight are related.

Indicator-based sustainability assessment of smallholder beef cattle production in South Africa

T. Marandure[1], C. Mapiye[1], G. Makombe[2] and K. Dzama[1]
[1]Stellenbosch University, Animal Science, P. Bag X1, 7602 Matieland, South Africa, [2]University of Limpopo, Turfloop Graduate School of Leadership, P.O. Box 756, 0787, Fauna Park, South Africa; tawamarandure@yahoo.co.uk

Ninety-five farmers were involved in deriving a set of social, environmental and economic sustainability indicators which were used to assess sustainability of the smallholder cattle production system in Ncorha and Gxwalibomvu communities in the Eastern Cape, South Africa. The derived indicators were scored on a five-point Likert-type scale and aggregated to provide a score for each of the three dimensions of sustainability and the net sustainability score. Aggregated sustainability scores were grouped into three categories; non-sustainable (<33%), conditionally sustainable (33-65%) and sustainable (>65%). Most respondents indicated good to excellent operational levels for social indicators including access to information (67%) and gender balance (66%). For environmental sustainability, respondents indicated very good to excellent operational levels for air quality (100%) and chemical use (85%). Social grants (54%) were the dominant economic indicator with income levels of less than R1000/month. Cattle income was the least common economic indicator but had the highest income levels (>R3,000/month). Aggregate sustainability scores showed that cattle production systems in Ncorha and Gxwalibomvu, respectively, were socially (48.2% and 56.6%) and environmentally (54.2% and 57%) conditionally sustainable but economically (15.7% and 10.8%) non-sustainable. Overall, cattle production systems in Ncorha (39.4%) and Gxwalibomvu (41.5%) were conditionally sustainable. Interventions to improve the economic dimension of sustainability while, simultaneously maintaining or improving the social and environmental dimensions of sustainability of the system were recommended.

Pasture-based finishing of early-maturing sired suckler beef bulls at 15 or 19 months of age

C. Lenehan[1,2], A.P. Moloney[2], E.G. O'Riordan[2], A. Kelly[1] and M. McGee[2]
[1]University College Dublin, School of Agriculture and Food Science, Belfield, Dublin 4, Ireland, [2]Teagasc, Animal and Grassland Research and Innovation Centre, Dunsany, Co. Meath, Ireland; mark.mcgee@teagasc.ie

Compared to high-concentrate systems, grass-based finishing of suckler beef bulls is economically attractive. Achieving a commercially acceptable carcass fat score (6, scale 1-15) with young bulls finished at pasture is difficult. The performance of early-maturing sired suckler bulls finished at pasture, with or without concentrate supplementation, at 15 or 19 months of age was evaluated. Sixty Aberdeen Angus-sired bulls (initial live weight 399 kg s.d. 38.1; age 389 days s.d. 24.6) were blocked by weight and assigned to a 2 (slaughter age, SA: 15 (S15) or 19 (S19) months) × 2 (finishing strategies, FS: grass only (G0) or grass + 3.2 kg dry matter (DM) barley-based concentrate daily (GC)) factorial arrangement. They were turned out to pasture on 7 April. For GC, concentrates were introduced immediately for S15 and 104 d later for S19; slaughter occurred 63 and 192 d post-turnout, respectively. Bulls rotationally grazed Lolium perenne-dominant swards. To avoid confounding with variances in herbage nutritive value, G0 and GC paddocks differed in area such that residency time and, pre and post-grazing herbage mass were similar for both. Data were analysed using mixed models with SA and FS as fixed effects and, block as a random effect. There were no (P>0.05) SA×FS interactions. Increasing SA significantly increased carcass weight (265 vs 355 kg), kill-out proportion (542 vs 561 g/kg) and conformation (6.7 vs 8.3, 1-15) (P<0.001) and carcass fat (5.8 vs 6.8) scores (P<0.01). Supplementation reduced estimated herbage intake by 0.71 and 0.32 kg DM/kg DM of concentrates for S15 and S19, respectively. Supplementation increased carcass weight (P<0.001) – S15 (257 vs 273 kg; 79 g carcass/kg DM) and S19 (342 vs 368 kg; 92 g carcass/kg DM) – and kill-out proportion (547 vs 557 g/kg, P=0.06) but had no effect (P>0.05) on carcass fat or conformation scores. In conclusion, carcasses were adequately finished, with or without concentrates at S19, but not at S15.

Comparison of rolled barley with citrus pulp as a supplement for growing cattle offered grass silage
C. Lenehan[1,2], A.P. Moloney[2], E.G. O'Riordan[2], A. Kelly[1] and M. McGee[2]
[1]University College Dublin, School of Agriculture and Food Science, Belfield, Dublin 4, Ireland, [2]Teagasc, Animal and Grassland Research and Innovation Centre, Dunsany, Co. Meath, Ireland; mark.mcgee@teagasc.ie

A variety of feedstuffs of varying carbohydrate composition are available to rectify deficiencies in nutrient supply for growing cattle fed grass silage-based diets. The objective was to examine the effects of replacing rolled barley (starch) with citrus pulp (digestible fibre) in a supplement on intake and performance of young growing cattle offered grass silage (dry matter digestibility, 681 g/kg) ad libitum for 101 days. Weaned, early and late-maturing breed, male suckled beef calves (n=120, initial live weight 357 kg, s.d. 39.3) were blocked by sire breed, gender and weight and from within block randomly assigned to one of two concentrate supplements based on, starch (862 g rolled barley, 60 g soya bean meal, 50 g molasses, 28 g vitamin and minerals/kg) (RB) or digestible fibre (855 g citrus pulp, 80 g soya bean meal, 53 g molasses, 12 g vitamins and minerals/kg) (CIT). Concentrates, formulated to have similar concentrations of PDIE (107 g, true protein digestible in the small intestine when energy limits microbial protein synthesis) /kg DM) were offered (1.6 kg DM) once daily. On day 87, blood samples were taken before and 2 h after feeding, and rumen fluid samples were collected (trans-oesophageal) 2 h post-feeding. Statistical analysis was carried out using mixed models with diet included as fixed effect, block as a random effect and, with appropriate covariates. Supplement type did not affect (P>0.05) grass silage intake (4.45 kg DM), live weight gain (0.670 kg/day), final live weight, ultrasonically assessed body composition or measurements of skeletal size. Rumen pH (6.64 vs 6.79), ammonia (51 vs 81 mg/l) and acetate-to-propionate ratio (2.7 vs 3.2) (P<0.001) were lower for CIT than RB. In conclusion, citrus pulp can replace barley in concentrate supplements for growing cattle without negatively affecting performance.

Restricted or ad libitum creep feeding of nursing beef calves grazing native pastures
A. Simeone, V. Beretta, J. Olaizola, J. Piegas and X. Silveira
University of the Republic, Ruta 3, km 363, 60000 Paysandu, Uruguay; asimeone@adinet.com.uy

A study was conducted in Uruguay (32°22'S, 58°03'W) to evaluate the effect of creep feeding (CF) and of creep-feed offer, on growth, supplement and milk intake of nursing calves grazing on native pastures. Fifty-one spring calving cow-calf pairs were randomly allotted to six paddocks and three treatments (2 replicates per treatment): control (C) with no CF; limited CF (LCF) with supplement (18% CP, EM: 2.9 Mcal/kg) offered daily at 1 kg/100 calf liveweight (LW); or ad libitum CF (ACF) with same supplement. At the start of CF, calves were 69±8.6 days old and weighed 78±5.7 kg. CF lasted until weaning (182±8.6 days old). Pastures were continuous grazed at 1.1 cow-calf pair/ha. LW and milk intake (MI) were recorded at the beginning and every 28 days. Supplement DM intake (SI) was measured weekly, and conversion ratio (SCR) was calculated as SI/ LW gain response with respect to C treatment. Data were analyzed according to a randomized plot design with repeated measures. Statistical model included treatment (T), week (W) and T×W effects, and initial LW or MI as covariates. Independent of T, sward biomass and height decreased with W (W1: 2,967 kg/ha, 19.8 cm; W17: 1,200 kg/ha, 5.7 cm; P<0.01). CF improved LW gain, with higher response for ACF compared to LCF (C: 0.61[c], LCF: 0.85[b], ACF: 1.11[a] kg/d; SE 0.03 P<0.01). SI was higher for ACF (1.6 vs 1.0% LW, P<0.01). Creep feeding did not affect mean MI (C: 3.5, LCF: 2.9, ACF: 2.7 kg/d; P>0.10) but a significant T×W interaction was observed for relative contribution of MI expressed as percentage of total (SI+MI) intake. Interaction indicated that differences between treatments appeared after W3 and increased towards W17 (C: 100[a], LCF: 42[b], ACF: 9.4[c]%, P<0.01). Although a better SCR value was observed for ACF, difference was not significant (5.6 vs 7.5 kg/ kg, P>0.10). Results suggest that CF is a useful technique for increasing LW gain of nursing calves. Decision upon restricted or ad libitum feeding would depend on target LW at weaning.

The effects of breed, month of parturition and progeny gender on beef cow fertility

F.M. Titterington[1], S.J. Morrison[1], F.O. Lively[1] and S. Dawson[2]
[1]Agri-Food and Biosciences Institute, Large Park, Hillsborough, BT26 6DR, United Kingdom, [2]Agri-Food and Biosciences Institute, Newforge Lane, Belfast, BT9 5PX, United Kingdom; frances.titterington@afbini.gov.uk

Reproductive performance is a key factor in the efficiency and profitability of a beef cow enterprise. The objectives of the current study were to evaluate current levels of beef cow fertility and investigate the effects of breed, season, year, and progeny gender. Data were sourced from the Bovine Information System (BovIS), which assimilates slaughter data from local abattoirs and information from the government database (APHIS). The calving interval data (CI; measured in days, d) included 273,764 records collected between 1997 and 2012 and included the seven most common breeds (and their crosses) in Northern Ireland (Charolais, CH; Limousin, LIM; Belgian Blue, BB; Simmental, SIM; Blonde D'Aquitaine, DAQ; Aberdeen Angus, AA; and Hereford, HER), accounting for 94.1% of beef dams recorded. CI data were restricted to between 250 and 600 days. Predicted CI was analysed as a mixed linear model using the REML commands in GenStat. Year of birth was included as a random effect in the analysis and the variance component due to year of birth was constrained to be 7.24%. Mean CI for all cows was 395 d, 30 days longer than the optimum 365 d. CI for each dam breed was CH 399 d, LIM 394 d, BB 399 d, SIM 395 d, DAQ 394 d, AA 392 d, and HER 396 d; with CH and BB dams having the longest CI ($P<0.05$). Cows older than 144 months had a longer CI ($P<0.05$) compared to cows younger than 144 months. Using a CH sire led to a shorter subsequent CI of 392 d ($P<0.05$) compared to LIM 395 d, BB 396 d, SIM 396 d, DAQ 396 d, AA 394 d, and HER 395. Cows calving in June had the shortest subsequent CI (376 d; $P<0.05$) whilst cows calving in November had the longest subsequent CI (410 d). Progeny gender did not significantly affect CI. This study established the level of beef cow fertility in Northern Ireland is sub optimal and there is room for improvement; and the factors identified as influencing fertility included dam breed, sire breed, and month of parturition. This knowledge can be used to direct breeding programs and inform knowledge transfer protocol to improve sustainability of beef production.

Animal welfare index (AWI): an on-farm welfare evaluation of beef farms in Ireland and Belgium

P. Lawrence, M. McGee and B. Earley
Teagasc, AGRIC, Grange, Dunsany, Co. Meath, Ireland; bernadette.earley@teagasc.ie

Issues relating to the welfare of farm animals are becoming increasingly important internationally. The study objectives were to: (1) examine the welfare status of Irish beef production farms using an animal welfare index (AWI); and (2) benchmark the AWI with a European beef production system: 122 farms (105 in Ireland and 17 in Belgium) were visited once during the winter housing period. The Irish beef production systems visited were suckler calf-to-beef (n=30), suckler calf-to-weaning/store (n=63) and dairy calf-to-beef (n=12). The Belgian farms comprised of calf-to-beef (n=14) and suckler calf-to-weaning/store-to-finish production systems (n=3). The AWI grouped 51 indicators into six categories: locomotion (6), social interactions (8), flooring (4), environment (9), stockpersonship (13) and husbandry management (11). The 6 category scores were integrated to give an overall AWI. There was no difference in AWI among the three Irish beef production systems, whereas the AWI of the Belgian system was lower ($P<0.001$) compared with the Irish systems. The mean AWI for the Irish farms were greater ($P<0.001$) than for the Belgian farms (70.3% for the suckler calf-to-beef system; 71.9% for suckler calf-to-weaning/store-to-finishing; 66.6% for dairy calf-to-beef, and 56.7% for Belgian farms, respectively). The environment (0.66), stockperson (0.50), husbandry management (0.48) and social (0.46) categories were correlated ($P<0.001$) with the AWI for the suckler calf-to-beef systems. The social (0.63), environment (0.55), stockperson (0.51) and husbandry management (0.32) categories were correlated ($P<0.001$) with the AWI for suckler calf-to-weaning/store-to-finishing systems. The stockperson (0.67) and environment categories (0.65) were correlated ($P<0.001$) with the AWI for dairy calf-to-beef systems. The environment (0.88), stockperson (0.69), social (0.60) and locomotion (0.55) categories were correlated ($P<0.001$) with the AWI for the Belgian beef production system. This study demonstrated the very high welfare standard of Irish beef production systems when compared to the intensive beef production systems in Belgium.

Effect of cytoplasmic lineage in genetic evaluation for economic traits in Nellore cattle

L. Grigoletto[1], E.C. Mattos[1], M.H.A. Santana[1], F. Baldi[2], J.P. Eler[1], L.G.G. Figueiredo[3] and J.B.S. Ferraz[1]
[1]College of Animal Science and Food Engineering, USP, 225 Duque de Caxias Norte Av., 13635-900 Pirassununga, Sao Paulo, Brazil, [2]School of Agricultural Science and Veterinarias, UNESP, Acess way Prof. Paulo Donato Castellane s/n, 14884900 Jaboticabal, Sao Paulo, Brazil, [3]National Association of Breeders and Researchers, ANCP, 463 João Godoy St., 14020-230 Ribeirão Preto, Sao Paulo, Brazil; jbferraz@usp.br

Maternal effects and it's inheritance have been suggested to have a permanent influence on selection and expression potential traits in animals. This study aimed to evaluate the impact of the incorporation of cytoplasmic lineage effect on the estimation of (co)variance components and genetic parameters of growth and reproductive traits in Nellore cattle (Bos indicus). Records of birth weight (BW, n=243,391), weaning weight (WW, n=431,681), post weaning weight gain adjusted to 345 days (PWG, n=172,131) and scrotal circumference at 18 months of age (SC, n=78,438) were analyzed. We used pedigree data from 496,190 Nellore animals, progeny of 2,688 bulls and 134,728 cows. The number of cytoplasmatic lines found, using LinMat software, was 28,459. Two single-trait models were used to obtain estimates of (co)variance components using REML method by BLUPF90 program. Model 1 (M1) included Lc as a random effect. Model 2 (M2) did not includ the Lc effect. Estimates of cytoplasmic effects accounted for 0.07, 0.15, 0.05 and 0.3% of total variation of BW, WW, PWG and SC, respectively. Variance of Lc effect did not affect the estimates of direct and maternal heritability coefficients. For PWG and SC, the weaning management group (WMG) was considered as an uncorrelated random effect. Such approach contributed to maintain animals in the analysis and increase the number of observations per CG. The likelihood ratio tests (LRT) suggested that M1 significantly (P>0.01) better fits to the data than M2 for BW and could therefore be recommend for genetic evaluation of this Nellore population. These results suggest that the cytoplasmic effect exists and permanently influence the generations of each ancestral line even if the effect has small magnitude.

Effect of SNP polymorphisms on the intramuscular fat content in Hungarian Simmental cattle

I. Anton[1], B. Húth[2,3], I. Füller[2], A. Zsolnai[1,3] and G. Holló[3]
[1]NARIC-Research Institute for Animal Breeding, Nutrition and Meat Science, Gesztenyés u.1, 2053 Herceghalom, Hungary, [2]Association of Hungarian Simmental Cattle Breeders, Zrinyi u. 3, 7150 Bonyhád, Hungary, [3]University of Kaposvár, Guba S. u. 40, 7400 Kaposvár, Hungary; anton.istvan@atk.naik.hu

In the last few years, advances in molecular genetics have enabled the application of MAS (Marker Assisted Selection) in achieving different breeding objectives. Some SNPs (single nucleotide polymorphisms) have been demonstrated to affect intramuscular fat content and meat quality traits in farm animals. Genome-wide association study (GWAS) based on typing of 777,000 SNPs by DNA chip technique is suitable for the improvement of beef quality in different cattle breeds. Correlations among DNA-chip data and intramuscular fat content (IMF) can be highlighted by statistical analysis. Sixty Hungarian Simmental bulls were selected for this study. During slaughtering, rib samples have been taken from m. longissimus dorsi (LD) cut between the 11-13th rib. Chemical percentage intramuscular fat was determined from LD. After DNA extraction from LD, SNP typing was performed on high-resolution SNP chips developed for cattle (Illumina Bovine HD Chip, 777,000 SNPs). Statistical analyses were performed by using the SVS software, which is suitable to fit various models, accounting for population stratification or environmental interaction. According to the analysis outcome, two loci have been identified to be associated with IMF. These loci (-log10P>5) seem to be useful in selection program and are located on chromosome 13 and 17, respectively. Frequency of favorable alleles gives the possibility to assist MAS by molecular tools. Selection for favorable alleles at reported loci on chromosomes mentioned above might be performed, if increased IMF is desirable. The project was supported by the Hungarian Scientific Research Fund (Project 111643).

Metabolic and endocrine status in heifers from two beef breeds submitted to different rearing diets

J.A. Rodríguez-Sánchez, I. Casasús, J. Ferrer and A. Sanz
Centro de Investigación y Tecnología Agroalimentaria de Aragón, Avda. Montañana, 930, 50059, Spain;
asanz@aragon.es

Thirteen Parda de Montaña (PA) and 12 Pirenaica (PI) autumn-born-heifers were used to evaluate the consequences on the onset of puberty and metabolic and endocrine status of 2 dietary treatments calculated to promote different body weight (BW) gains. Two feeding management (0.8 [HIGH] vs 0.6 [LOW] kg/d) were applied on beef heifers from 2 breeds (PA vs PI) during the rearing phase, from weaning at 6 months of age to breeding at 15 months of age. The heifers were bled weekly to determine their progesterone level, to estimate the onset of puberty, and every 3 months to assess the glucose, NEFA, cholesterol, urea, IGF-I and leptin plasma levels. All the heifers reached puberty at similar BW (322 ± 38 kg; 55% mature BW), but at different ages depending on the breed, since PA heifers were more precocious than PI ones (9.1 vs 10.7 months, respectively, $P<0.05$). Heifers from the HIGH treatment tended to be pubertal earlier than those from the LOW one (9.4 vs 10.5 months, $P<0.09$). Glucose level was affected by the feeding management (5.04 vs 4.85 mmol/l, in HIGH and LOW respectively, $P<0.05$) and NEFA by the breed (0.11 vs 0.15 in PA and PI heifers respectively, $P<0.05$), whilst cholesterol (3.40 ± 0.86 mmol/l), urea (6.13 ± 1.13 mmol/l), IGF-I (206.4 ± 85.2 mmol/l) and leptin (2.68 ± 1.42 mmol/l) were not affected by any treatment. The nadir of cholesterol level was found at 12 months of age (2.89 ± 0.44 mmol/l), when the heifers had started cycling. The heifers with greater level of urea ($r=0.58$, $P<0.01$) at 6 months, and IGF-I ($r=0.43$, $P<0.001$) in the whole trial, were more precocious. The circulating leptin increased through the rearing phase (2.39 vs 3.38 ng/ml, at 6 and 15 months of age, $P<0.01$), along with fat deposition, but it was not related with the age at puberty. In conclusion, the heifers from both breeds reach the puberty early enough to be bred at 15 months, even with gains of 0.6 kg/day during the rearing phase. The urea and IGF-I levels could be used to identify the more precocious heifers.

Abundance of beige and brown adipocyte markers in different adipose depots of 26-month-old cattle

K. Komolka, E. Albrecht and S. Maak
Leibniz Institute for Farm Animal Biology, Institute for Muscle Biology and Growth, Wilhelm-Stahl-Allee 2, 18196
Dummerstorf, Germany; maak@fbn-dummerstorf.de

Brown and beige adipocytes were recently detected in white adipose tissue depots of adult cattle. Activity of these cell types during the fattening period could contribute to breed differences in fat deposition. Therefore, we compared the abundance of markers for adipocyte types in steers from three cattle breeds largely differing in fat deposition (Japanese Black n=6, Holstein n=5, and Charolais n=6) under high energy feeding and thermoneutral conditions. Markers for white (leptin, LEP), beige (transmembrane protein 26, TMEM26, uncoupling protein 1, UCP1), and brown adipocytes (Zic family member 1, ZIC1, UCP1) were analyzed by RT-qPCR in subcutaneous (SCF), intramuscular (IMF), intermuscular (IRMF), perirenal (PF), and visceral fat (VF). As expected LEP mRNA was abundant in all samples with significantly lower amounts in VF compared to all other depots ($P<0.05$). Brite marker TMEM26 was weakly but evenly expressed in all depots in all animals whereas UCP1 mRNA showed higher individual variation in some depots. ZIC1 was not detectable in VF and PF but abundant in SCF, IRMF and IMF. Expression was significantly lower in SAT and IMF compared to IRMF ($P<0.05$). For none of the markers, significant breed differences could be established. In addition to mRNA abundance, investigations on protein abundance and tissue localization of beige and brown adipocytes were conducted in bovine adipose tissue depots using commercial antibodies against UCP1 and ZIC1. We demonstrated that fat depots of cattle at an age of 26 months still comprise different adipocyte types. However, our results did not indicate that higher energy expenditure because of higher abundance or activity of beige or brown adipocytes contributed to differences in fat deposition among the investigated breeds.

Response of early-weaned beef calves in feedlots to non protein nitrogen level in the diet
V. Beretta, A. Simeone, J. Clerc, M. Fonseca and J. Rocco
University of the Republic, Ruta 3, km 363, 60000 Paysandu, Uruguay; beretta@fagro.edu.uy

An experiment was conducted to evaluate the effect of non protein nitrogen (NPN) level in the diet on feedlot performances of spring born early-weaned beef calves. Twenty-five castrated Hereford calves (98.1±12.5 kg) were randomly allocated to five total mixed rations (20% alfalfa hay, 80% concentrate) differing in substitution rate of CP from soybean meal (SBM) for NPN: 0%, 15%, 30%, 45% or 60%. A mixture of urea: slow release urea was used (varying from 100:0 in 0% to 26:74 in the 60% treatment). Concentrates were formulated for similar energy intake (ME 3.05 Mcal/kg) and CP supply (CP 17%) across treatments. Animals were fed ad libitum in individual pens outdoors during 11 weeks, and weighed every 14 days. Final subcutaneous back fat (SBF) and Longissimus dorsi area (LDA) were determined by ultrasonography. Dry matter intake (DMI) was determined daily and the feed to gain ratio (FGR) was calculated. Apparent organic matter digestibility (OMD) and blood urea concentration (BU) were determined on week 6. Data was analyzed according to a randomized plot design, using the Mixed Procedure of SAS for live weight (LW) and DMI repeated measures, and the GLM procedure for FGR, SBF, LDA, OMD and BU. Linear and quadratic effects associated to NPN level were tested. Animal LW increased linearly with time ($P<0.01$) at different rates depending on treatment (1.42[a], 1.40[a], 1.32[a], 1.22[a] and 0.87[b] kg/d for 0, 15, 30, 45 or 60% replacement, respectively; $P<0.01$). A quadratic response to increasing level of NPN was observed for live weight gain, final LW, LDA ($P<0.01$) and DMI ($P=0.05$), while no differences were detected ($P>0.1$) for SBF (2.8 mm, SE: 0.17), FGR (3.76 kg/kg, SE: 0.06), or OMD (79.4%, SE: 5.1). A quadratic effect was also observed for BU (BU, mg/dl $-0.0063x^2-0.064x+15.3$; $P<0.01$). Results suggest that when early-weaned calves are fed with high grain rations, performance and N utilization would be simultaneously optimized when NPN substitutes supplemental CP from SBM in the concentrate within a range between 5 and 10%.

Identification of genomic regions related to tenderness in Nellore beef cattle
M.E. Carvalho[1], F.R. Baldi[2], M.H.A. Santana[1], R.V. Ventura[1,3], G.A. Oliveira Junior[1], R.S. Bueno[1], M.N. Bonin[4], F.M. Rezende[5], J.P. Eler[1] and J.B.S. Ferraz[1]
[1]University of Sao Paulo, Medicine Veterinary, Duque de Caxias, 225, Pirassununga, SP, 13635-900, Brazil, [2]Sao Paulo State University, Animal Science, Via Prof. Paulo Donato Castellane s/n, Jaboticabal, SP, 14884-900, Brazil, [3]Beef Improvement Opportunities, 294 Mill St East Suite 209, Elora, Ontario, ON N0B 1S0, Canada, [4]Embrapa Beef Cattle, Av. Radio Maia, 830, Campo Grande, MS, 79106-550, Brazil, [5]Federal University of Uberlandia, Av. Pará, 1720, Uberlandia, MG, 38405-320, Brazil; jbferrraz@usp.br

The aim of this study was to identify, by ssGWAS, genomic regions that potentially have association with tenderness in Nellore cattle. Phenotypes were obtained according to standard USDA Quality Grade (1999), from longissimus thoracis muscle between the 12th and 13th ribs on the right half-carcass and aged for 7 days. Data of 909 Nellore bulls were analyzed. Those animals were genotyped with Illumina Bovine beadchip HD® GGPi (74K). Based on another Nellore population genotyped for Illumina beadchip BovineHD® (777K), genotypes were imputed by FImput software. Analyses were performed using a pedigree composed by 6,276 animals and, assuming contemporary group (farm and slaughter batch) as fixed effect and age at slaughter as a covariate. Single step analyses were realized by Blupf90 program considering windows of 10 markers (SNP) to estimate their effects, this procedure enables the identification of regions associated with tenderness along the chromosomes. After quality control (MAF<0.05%, call rate<90%), 463,995 SNPs in autosomal chromosomes were used in the association analyses. Based on that, 18 regions in 14 different chromosomes (1, 4, 6, 7, 8, 10, 18, 19, 20, 21, 22, 25, 26 and 29), that explained more than 1% of the additive variance, were explored and some genes were identified in these regions, as AVEN, SHISA7, UBE2S, CDC42EP5, C16orf96, SNORA1 and FAM119A. In conclusion, with ssGWAS method using high density panel, it was possible to identify regions related to tenderness in Nellore cattle. Posteriorly, those genes and their pathways will be investigated to evaluate their importance for meat quality traits.

Intramuscular connective tissue properties in longissimus thoracis muscle of Hungarian Simmental

G. Holló[1], B. Húth[1], E. Egri[1], I. Holló[1] and I. Anton[2]
[1]Kaposvár University, Guba S. street 40., 7400 Kaposvár, Hungary, [2]National Agricultural Research and Innovation Centre, Gesztenyés str 1, 2053 Herceghalom, Hungary; hollo.gabriella@sic.ke.hu

The intramuscular connective tissue plays an important role in determining meat tenderness. Our objective was to compare the collagen/hydroxyprolin content and computed tomographic (CT) connective tissue proportion in longissimus thoracis (LT) muscle of Hungarian Simmental (HS) bulls and cows. Animals were slaughtered at the same live weight (n=24, bulls: 530.6±44.7 kg, cows: 527.33±53.54 kg) at commercial abattoirs in Hungary. The average age of bulls and cows were 696 days (d) and 3,042 d respectively. After 24 h chilling, LT sample at 12th rib were taken from the right half carcass. The CT examinations were carried out with the usage of a 16-slice CT system (Siemens Somatom Sensation Cardiac, slice thickness: 5 mm). The samples were scanned at different user-selectable tube voltages e.g. low:80 kV and high: 140 kV. The CT value at LT muscle area of each mixed scan (80±140 kV) was obtained. Volumetric connective tissue content was determinated (above 200 CT value). After CT, the hydroxyproline/ collagen content of LT was determined. The intramuscular fat content was analysed by Soxhlet method. For statistical analysis, IBM SPSS 20.0 program was used. Cows had lower carcass weight (247 kg v. 295 kg), EU muscle (3.5 v. 5.5) and fat score (4.2 v. 5.9) than bulls (P<0.01). Bulls had higher LT area, but the intramuscular fat content is same for bulls (2.8±1.9) and for cows (2.7±1.9). On the other hand bulls had lower CT intramuscular connective tissue proportion in LT compared to cows (0.41±0.21% v. 0.66±0.28% P<0.01). The same tendency can be seen for collagen content (0.5±0.21% v. 0.67±0.11% P<0.01). The correlation between CT connective tissue proportion of LT and collagen content was r=0.8. There was a weak positive correlation of slaughter age with CT connective tissue or collagen content (r=0.3-0.4). Conclusion: Intramuscular connective tissue properties increased with slaughter age in LT of HS breed. Older cows had higher collagen and connective tissue proportion than bulls. Mixed CT scans can be used for the analysis of intramuscular connective tissue content. This work was supported by grant of OTKA (13729).

On-line prediction of beef muscle chemical composition and texture using near infrared spectroscopy

J. Cafferky[1,2], V. Campos[1,2], A. Sahar[2], R. Hamill[2], T. Sweeney[1], G. Downey[2], A. Cromie[3] and P. Allen[2]
[1]University College Dublin, School of Veterinary Medicine, Belfield, Dublin 4, Ireland, [2]Teagasc Food Research Centre, Ashtown, Dublin 15, Ireland, [3]The Irish Cattle Breeding Federation, Shinagh House, Bandon, Co. Cork, Ireland; jamie.cafferky@teagasc.ie

Chemical composition and shear force values of beef are critical for sensory quality, nutritional value and product formation. However, analysis for these traits is both time consuming and destructive to the meat. Visible-Near Infrared Spectroscopy (Vis-NIRS) has been proposed as a rapid, non-destructive analytical technique with the potential to predict chemical composition values in beef carcasses on-line with a reasonable level of accuracy. Hence, the aim of this study was to calibrate a chemometric model to rapidly predict beef LTL chemical composition and shear force values on-line using Vis-NIRS. Spectra were collected on-line from the exposed surface of the neck and rump muscles at 1 h and 2 h post-mortem (day0), from the cut surface of M. longissimus thoracis et lumborum (LTL) at 0 h and 1 h after quartering (day after slaughter, day1) and offline in the laboratory on the cut surface of the LTL muscle immediately and after 1 h blooming (2 days after slaughter, day2). LTL steaks aged for two days were then analysed in the laboratory for moisture and fat content (n=222 for fat, n=206 for moisture) and fourteen-day aged steaks were analysed for Warner-Bratzler Shear Force (WBSF) (n=145). Chemometric analysis gave R-squared validated prediction values of 0.41 for fat (quartering day1 0 h), 0.32 for moisture (quartering day1 0 h), and 0.50 for WBSF (loin day2 1 h). While useful models were not identified for the early post mortem (neck and rump) spectra, potentially useful predictive models for fat, moisture and WBSF were identified from spectra collected on days 1 and 2 post-mortem.

Growth paths for suckler bulls slaughtered at 19 months of age: A meat quality perspective

L. Moran[1], M.G. O'Sullivan[2], M. McGee[3], K. McMenamin[3], E.G. O'Riordan[3] and A.P. Moloney[3]
[1]Teagasc, Ashtown Food Research Centre, Dublin 15, Ireland, [2]UCC, School of Food and Nutritional Sciences, Cork, Ireland, [3]Teagasc, Animal Grassland Research and Innovation Centre, Co. Meath, Ireland; lara.moran@teagasc.ie

The aim was to assess the influence of different winter growth rates (WGR) followed by supplementation at pasture on selected meat quality characteristics of spring born bulls. Bulls (n=90) were blocked (breed, age and weight), and assigned to a 2 (WGR) × 3 (pasture supplementation) design. During the first winter (127 days), bulls were offered grass silage (GS) ad libitum (AD) and either 3 (WGR3) or 6 (WGR6) kg of a barley-based commercial concentrate. After turn-out to pasture, bulls were offered: grass without supplementation (PO), grass plus 0.2 predicted dry matter intake (DMI) as concentrates or grass plus 0.4 predicted DMI as concentrates (PL). After 98 days, the bulls were housed and offered concentrates and GS AD for 76 days prior to slaughter at ~19 month of age and collection of the Longissimus thoracis (LT) muscle. The colour, chemical composition (unaged), instrumental texture and sensory characteristics (14 days of ageing) of LT were measured and the data analysed using the Glimmix procedure of SAS. On average, WGR6 bulls had greater (P<0.05) carcass weight than WGR3 bulls, but no differences were found due to pasture supplementation (carcass weight PO=412 kg, PL=418 kg, PH=432 kg). An interaction (P<0.05) between WGR and pasture supplementation was found for LT shear force (WBSF). Within WGR3, WBSF was lower (P<0.05) for PO compared to PL and PH which did not differ, whereas within WGF6, WBSF was lower (P<0.05) for PL compared to PH and PO which did not differ. However, these differences were not detected by the trained sensory panel (P>0.005). An interaction (P<0.05) between WGR and pasture supplementation was also found for redness (a). Within WGR3, redness was lower (P<0.05) for PL compared to PO and PH which did not differ, whereas within WGF6, redness was lower (P<0.05) for PO compared to PH and PL which did not differ. Chemical composition was similar for all treatments. In conclusion, variations in the growth path exerted minor effects on appearance and eating quality, and these differences did not affect the sensory perception of tenderness.

Effect of Stress, Fasting, Lairage and interaction with bulls and steers on Beef Eating Quality

D.J. Devlin, N.F.S. Gault, B.W. Moss, E. Tolland, J. Tollerton and L.J. Farmer
Agri-Food and Biosciences Institute, Food Research Branch, Newforge Lane, BT95PX Belfast, United Kingdom; declan.devlin@afbini.gov.uk

A wide range of factors influence consumer acceptability of red meat. The pH in meat 24 hours post slaughter (ultimate pHu) is an indicator for meat quality attributes e.g. colour and tenderness. High pH meat is a result of an inadequate supply of muscle glycogen in the animal. Depletion of muscle glycogen occurs when an animal is exposed to stress both physical (hunger, fatigue) and psychological (social regrouping). The sex of an animal and its stage of development can also affect the incidence of dark cutting meat. A number of experiments have been conducted by AFBI to investigate the effects of both gender and stress on eating quality. In one experiment the eating quality of bulls versus steers (32 animals of each gender) under different conditions of holding (time in lairage) and pre-slaughter stress (clipping) was determined. Another experiment looked at the effects of physical stress (mixing and fasting) on the eating quality of Holstein steers (n=32). No statistically significant overall effects of sex type (bull vs steer) on eating quality attributes were found. When held in lairage overnight, beef from bulls had higher pHu values than that from steers. There were few interactions of sex type with duration of lairage and clipping and where these did occur may reflect the differences in pHu between bulls and steers in overnight lairage. Fasting steers up to 72 hours prior to slaughter had no effect on meat quality. Mixing of steers for one hour prior to transport when coupled with short journey and lairage, except for juiciness in the strip loin, had no effect on eating quality. The unexpected absence of any effect in this experiment may be related to the temperament of the cattle or an initial high nutritional status of individual cattle with only a small proportion adversely affected in terms of eating quality.

Evaluation of marbling traits by X-ray computed tomography in Hungarian Simmental breed

G. Holló[1], B. Húth[1], I. Holló[1] and I. Anton[2]
[1]Kaposvár University, Guba S. street 40., 7400 Kaposvár, Hungary, [2]National Agricultural Research and Innovation Centre, Gesztenyés út 1, 2053 Herceghalom, Hungary; hollo.gabriella@sic.ke.hu

Marbling is an economically important factor in many beef carcass classification systems. The aim of this study was to evaluate marbling traits of longissimus thoracis muscle with different methods: USDA marbling score, intramuscular fat content (IMF) by image analysis of X-ray computed tomography (CT) scans and by conventional analytical method. Bulls (n=46) were slaughtered at an average weight of 535.94+126 kg and an average age of 645.66+437 days. After 24 hrs, chilled LT sample at the 12th rib were taken from the right half carcass. USDA marbling score was assessed visually at the cut surface of the ribeye. The CT examinations were carried out using a 16-slice CT system (Siemens Somatom Sensation Cardiac, slice thickness: 5 mm; tube voltage: 120 kV). With simple thresholding, pixels between -19 and +19 CT value were included in the evaluation of IMF. After CT, the intramuscular fat content of LT was determined by chemical extraction. For statistical analysis, the IBM SPSS 20.0 program was used. The average growth rate of bulls was 954 g/day. The chemical fat content of LT varied between 0.5 and 7.9%, whereas CT fat content varied between 0.1 and 8%. IMF in the 12th rib and in longissimus muscle on CT-scans closely correlated with chemical fat content (r=0.69 and 0.80 respectively). The highest frequency of USDA marbling score was 'small' (53.2%), followed by 'slight' (25.5%), 'modest' (17%) and 'moderate' (2.1%). The same frequency tendency has been observed for IMF measured by chemical and CT method. It was concluded that IMF was affected by slaughter weight, and that it decreased with higher growth rate (r=-0.46 – -0.49). This work was supported by grant of OTKA (13729).

Muscle development as characteristic for beef production in 4 local breeds of cattle

P. Polák[1], J. Tomka[1], R. Šveistienė[2] and A. Račkauskaitė[2]
[1]National Agricultural and Food Centre, RIAP, Hlohovecká 2, 951 41 Lužianky, Slovak Republic, [2]Institute of Animal Science of LUHS, R. Zebenkos 12, 82317 Baisogala, Radviliškis district, Lithuania; polak@vuzv.sk

Because local breeds carry special traits to tolerate harsh condition and use seasonal pasture for milk production, the hypothesis was that local breeds can be also used for beef production on pasture. The aim of this study was to assess this possibility with 4 breeds of cattle from central Europe currently used in dairy system. Muscle thickness measured by ultrasound muscle and body development evaluated by trained classificatory as information of body conformation as well as milk production were taken into consideration. Totally 58 heifers (23 Pinzgau – P, 19 Lithuanian whitebacked – LWB, 13 Lithuanian ashgray – LAG and 3 Lithuanian red breed of old genotypes analysed – LR) were used in the study. Age of heifers was between 14 and 28 months. Muscle and fat thickness were measured by Aloka PS 2, with probe UST-5044-3.5 MHz, 172 mm on back on last thoraces vertebra and on rump on os ischii. Average daily gain, body and muscle development were evaluated by means of linear evaluation of exterior used for beef cattle in Slovakia done by a certified person. Average daily gain in all breeds was 695 g for P, 730 g for LR, 740 g for LWB and 751 g for LAG. Total layer of muscle was higher for P (199.02 mm) than for LR (166.50 mm), LWB (155.47) and LAG (162.21). The highest score for linear evaluation of muscularity was 16.63 points for P but 14.07 for LAG, 14.00 for LR and 13.05 for LWB. Also score for muscle development was the highest in P. The intra breed variance in each breed was higher than variance among breeds. The results could be affected by the low number of animals and level of breeding in herds of origin. Because all 4 breeds are considered as rare endangered breeds within the scheme of genetic resources conservation, the number of animals in breeds was low and the range of age was high. However, all breeds shown potential to be used in cow–calf system due to good milk production in pasture and moderate muscle development. Focus should be put to body weight and overall muscle development. More data are needed in order to increase number of animals as well as to harmonize age of heifers in each group.

Effect of diets containing sunflower cake on the meat quality of young bulls
V.S. Oliveira, R.L. Oliveira, T.C. Rocha, J.S. Trajano, R.D.X. Ribeiro, N.G. Nascimento Júnior and P.A. Oliveira
Federal University of Bahia, Av. Adhemar de Barros 500, 40170110 Salvador Bahia, Brazil; ronaldooliveira@ufba.br

This study was conducted to determine the optimal amount of sunflower cake in the diet of young bulls, based on chemical and physical characteristics of the meat. We hypothesize that the inclusion of SC in the diet will not affect the meat quality parameters. Thirty-two young Zebu bulls (374.4 ± 42.5 kg), were distributed in a completely randomized design, and fed with a total mixed ration that contained 40% of Cynodon sp. chopped hay and 60% of concentrate composed of corn bran, soybean meal, mineral premix, and the amounts of sunflower cake (SC): 0; 9; 18 and 27% of total DM. Those amounts were the treatments. The feed trial lasted 90 days. The data were subjected to analysis of variance and regression testing. Significance was declared as $P<0.05$. There are no effects of SC inclusion on moisture (70.1%), crude protein (25.5%), and ash (1.2%) contents of meat ($P=0.99$, 0.26 and 0.51, respectively), but there was a quadratic effect on the total lipid content ($P=0.07$), and the inclusion of 2.8% of SC had the lowest value for this parameter (4.2%). A quadratic effect ($P=0.04$) was observed in cholesterol content, and the inclusion of 17.1% of CS promoted the highest value (69.0%). The inclusion of SC caused a linear increase ($P=0.03$) in the water retention capacity, with means of 92.2, 96.1, 94.5 and 95.3%, at the inclusion levels of 0, 9, 18 and 27%, respectively. A quadratic effect ($P=0.02$) was observed in shear force measurement, with highest value for 12.3% (2.8 kgf) of SC inclusion. The inclusion of SC on diet did not affect meat luminosity and color indexes a, b and c ($P=0.45$, 0.95, 0.91, 0.99, respectively), with means of 40.6, 21.7, 9.0 and 23.5, respectively. Sunflower cake can be added up to 27% (DM basis) without compromise meat quality parameters.

Integrated crop-livestock systems may increase micronutrient content of foodstuffs
S.L. Kronberg[1], G.F. Combs[2], M.A. Liebig[1] and J. Ryschawy[3]
[1]USDA-Agricultural Research Service, Northern Great Plains Research Laboratory, P.O. Box 459, Mandan, ND, USA, [2]USDA-Agricultural Research Service, Grand Forks Human Nutrition Research Center, (retired), Grand Forks, ND, USA, [3]INRA, UMR 1248 AGIR, Chemin de Borderouge, 31324 Castanet-Tolosan, France; scott.kronberg@ars.usda.gov

Plants provide essential mineral nutrients to people and animals, and these minerals are derived from soil. Some minerals are under-consumed by many people and this contributes to costly health problems. Several factors influence mineral uptake by plants including crop rotation, soil management before seeding and fertilization. This study evaluated the mineral content of a legume (pea, Pisum sativum) and a cereal (wheat, Triticum aestivum) in response to soil management, nitrogen fertilization and grass management before annual crop production. Peas were produced: (1) using no tillage, minimum tillage or conventional tillage; (2) with or without nitrogen fertilization; and (3) with the perennial grass, which grew on the land for several years before annual crop production, either left on soil to decay or removed as hay. For wheat production, only no-till was used, but the other treatments were evaluated. Dry peas contained greater concentration ($P<0.05$) of calcium and zinc when grown with no-till techniques versus minimum or conventional tillage and contained more ($P<0.05$) magnesium and manganese when grown using no-till or minimum tillage. Zinc and manganese concentrations in dry pea were higher with nitrogen fertilization ($P<0.05$) while magnesium concentrations were lower ($P<0.05$). Copper and iron concentrations were greater ($P<0.05$) in dry pea and wheat when the perennial grass was not removed as hay. Magnesium concentration in wheat was elevated ($P<0.05$) when perennial grass was removed. These results suggest that concentrations of some micronutrients may be increased in annual crops by disturbing soil less, rotating perennial grassland with annual crops and grazing or feeding grass on the farmland rather than offsite.

Mixed crop-livestock systems across scales: toward new agroecological models?

J. Ryschawy[1,2]
[1]Université de Toulouse, INPT-ENSAT, CASTANET-TOLOSAN, 31324, France, [2]INPT-ENSAT INRA UMR 1248
AGIR, CASTANET-TOLOSAN, 31324, France; julie.ryschawy@toulouse.inra.fr

Integrated crop-livestock systems (ICLS) are recognized worldwide as models for the agroecological transition. Interactions between crops, livestock and grasslands potentially provide multiple ecosystem services. Soil quality improvement and recoupling of nitrogen and carbon cycles would in particular result from organic fertilization from livestock waste and crop-grassland rotation. Increased landscape heterogeneity through the integration of grasslands within diversified crop rotations would enhance biological regulation. According to these assumptions, ICLS would be a key agroecological model for rethinking agriculture. However, ICLS are decreasing at the farm level due to workforce and skills constraints and the strong specialization trend linked to the economic and politic contexts. A new opportunity would be to develop ICLS at local scale based on exchanges between specialized farmers. Conceiving ICLS at the local scale goes beyond farm-scale workforce constraints while providing comparable economic and environmental benefits. Spatial and temporal coordination would provide landscape heterogeneity and soil quality improvement. Collective organization allows farmers to combine and share their technical skills while limiting qualitative and quantitative workforce constraints. Crop farmers could diversify their rotations while integrating grasslands and obtaining manure. Livestock producers could benefit from local traceable and lower-cost feed production. Still, ICLS at the local scale would require new research effort to design relevant technical and social practices and evaluation indicators. Design and assessment tools should integrate local knowledge and be easily out-scaled. We will illustrate in this presentation some French case-studies considering these different scales of interest: from the ICL farm to the local level integration.

Mixed farming and the sustainability of beef production systems in Wales, UK

P.K. Nicholas-Davies[1], A. Olaizola[2,3], O. Barrantes[2,3] and G. Evans[4]
[1]Aberystwyth University, Institute of Biological, Environmental and Rural Sciences, Gogerddan Campus, SY233EE, Aberystwyth, United Kingdom, [2]Agrifood Research and Technology Centre of Aragon (CITA), Avda. Montañana 930, 50009 Zaragoza, Spain, [3]Universidad De Zaragoza, C/ Pedro Cerbuna, 12, 50009 Zaragoza, Spain, [4]Celtic Pride, Cross Hands Food Park, SA146SX, Llanelli, Camarthenshire, United Kingdom; pkn@aber.ac.uk

The aim of this study was to assess the sustainability of a group of Welsh case study farms by means of farm typologies and to explore the relationship between farms sustainability and types of mixed farming systems. The MESMIS sustainability framework was applied to 38 farms (with varying levels degrees of mixing of cropping and livestock) supplying the Celtic Pride Premium Beef brand in mid and south Wales. To classify the farms based on their sustainability, a cluster analysis was performed on five attributes (productivity; stability, reliance and resilience; adaptability; equity; self-sufficiency) for sampled farms of the case study group (n=36) and four groups of farms were identified, Group 1 (n=10) 'Highest equity', Group 2 (n=10) 'Best productivity', Group 3 (n=8) 'Least self-sufficient' and Group 4 (n=8) 'Least stability and adaptability'. The farms comprised in Groups 1 and 2 were on average more sustainable than the farms of Groups 3 and 4. To study the relationships between beef farms sustainability and types of beef farming systems, the typology of the Celtic Pride farms sustainability was crossed with a classification of beef farming systems. Mixed beef farming systems were distributed in Group 1 of sustainability 'Highest equity' and Group 2 'Best productivity'. These groups showed the highest adaptability and stability and self-sufficiency and high productivity, respectively. Specialised grassland beef farming systems were distributed in Group 2 and Group 4 'Least stability and adaptability'. Conversely, Fodder cropping beef farming systems were distributed in all the sustainability groups, but 53% belonged to Group 3 'Least self-sufficiency'.

Does the diversity of integration practices enhance the resilience of mixed crop-livestock systems

F. Stark[1], H. Archimède[2], R. Poccard-Chappuis[3], E. Gonzalez-Garcia[4] and C.-H. Moulin[5]
[1]IAVFF Agreenium, 42 rue Sheffer, 75016 Paris, France, [2]INRA, URZ, Prise d'Eau, 97170 Petit Bourg, Guadeloupe, [3]CIRAD, UMR SELMET, Baillarguet, 34000 Montpellier, France, [4]INRA, UMR SELMET, Place Pierre Viala, 34060 Montpellier, France, [5]Montpellier SupAgro, UMR SELMET, Place Pierre Viala, 34060 Montpellier, France; charles-henri.moulin@supagro.fr

The mixed crop-livestock systems (MCLS) are characterized by the diversity of productions and the crop-livestock integration (CLI) practises. From an agro-ecological perspective, four main emergent properties are expected from MCLS: resilience, productivity, efficiency and self-sufficiency. The objective was to assess, at farm level, the resilience of MLCS and explore if CLI could confer more resilience. We performed the study on 17 family farms in three territories of the humid tropics, Guadeloupe (8), Brazilian Amazonia (4) and Cuba (5), contrasted from a socioeconomic point of view, thus looking to cover a range of CLI configurations. The farms range from 2 to 100 ha, 1 to 5 workers and combine cattle or pigs, pasture or forage crops, and crops for food and cash. We used the ecological network analysis (ENA) method to study the flow networks of nitrogen (N) for a sample of seventeen farms. We characterized CLI in terms of intensity and organization of flows, and examined the resilience of the N flow networks, from indicators of development capacity, ascendancy and overhead proposed by Ulanowicz and Norden (1990). The results show that both the intensity and organization of CLI flows are consistent in characterizing CLI in a range of situations i.e. from low CLI intensity based on few flows to high CLI intensity based on complex flows. The resilience indicator is independent of the intensity of the CLI practices at farm level. On the other hand, the resilience exhibits a strong correlation (r=0.91) with the organization of flows. The CLI seems effective in promoting resilience, viewed as the stability provided by the homogeneous distribution of flows among the farming system compartments. This study shows the interest of ENA quantifying CLI and resilience at farm level. The diversity of CLI flows provides a certain type of resilience, besides other interest like the nutrient recycling, leading to a better efficiency of N use.

ArchiMod: a metamodel of farming systems functioning to address future livestock challenges

L. Puillet[1], O. Martin[1], B. Méda[2] and F. Garcia-Launay[3]
[1]INRA, UMR 791 MoSAR, 75005 Paris, France, [2]INRA, UR 83 URA, 33780 Nouzilly, France, [3]INRA, UMR 1348 PEGASE, 35590 Saint Gilles, France; laurence.puillet@agroparistech.fr

Livestock farming systems (LFS) are facing different challenging issues, questioning our viewpoint: rather than technical objects to be optimized, LFS are components of the agro-ecosystem (AES). This change in our viewpoint involves: (1) integrating dynamical, multidimensional and multifunctional aspects; (2) facilitating collaboration among disciplines; and (3) developing and re-using computer models. In this work, we developed ArchiMod, a metamodel describing structure and functioning of any LFS based on a set of concepts, used in a protocol of description and a graphical language. The graphical language represents the biophysical functioning of an AES, seen as made of entities (non-living components) and biomasses of living organisms, in a given life stage of a given species, interacting by way of processes within 4 contexts of interactions. The concept of context for processes enables progressively organizing the diversity of processes, each of them corresponding to a defined level of organization, where a given type of processes occurs. The Organism Stage context describes input and output flows at the interface of a living organism with its external environment (e.g. feed intake). The Species pathway context describes transitions between organism stages within each species life cycle (e.g. non-lactating to lactating for dairy female). The Environmental Interactions context describes biotic (e.g. pollen dispersal by insects) and abiotic (e.g. N leaching) processes within a physical environment. Finally, the Management operation context describes processes driven by the farmer (e.g. drying-off). Key principles of ArchiMod are: (1) each process is a building block used to describe the whole system dynamics; (2) the system's functioning is described by defining the contextualization of all elementary processes; (3) it gives a graphical display of the whole set of components to be defined in a mathematical and/or computer model; (4) it provides the rules to articulate elementary models, independent of the mathematical formalism chose to describe processes. ArchiMod was tested on contrasted LFS with various species (fish, rabbit, goat, sheep, cattle, chicken and pig).

The revival of Mixed Farming Systems – will dreams finally become true?

P.L. De Wolf, W. Sukkel, P.A. Galama, J. Verloop and M. Pleijte
Wageningen University and Research Centre, Edelhertweg 1, 8219 PH Lelystad, the Netherlands;
pieter.dewolf@wur.nl

Mixed farming systems seem to be a very attractive concept: Every now and then, projects and initiatives are started to revitalise the mixed farm. The linked concept of closed cycles for nutrients and carbon is also a returning issue on the agendas. However, the trend of specialisation just goes on in European agriculture, despite all efforts to re-introduce mixed farming systems. We will first review several initiatives in Europe could reveal the drivers behind this returning interest in mixed farming and closed cycles. What are the motives of such initatives, what is the expected outcome? Who are the initiators and the main actors? Is there anything in common over these initiatives, or do they define the concept differently? The institutional context could possibly explain why specialisation is more dominant than mixed farming systems. The paper will then explore some macro-trends in the context of the agricultural system, such as the availability of global feed and fertiliser inputs, the agricultural policy of the EU and European member states, the structure of the agricultural knowledge and innovation system and the role of suppliers. Are there any signs of changes that could favour mixed farming, on farm, region or even European level? This part of the paper is based on the work of the EIP Focus Group Mixed Farming Systems. The last part of the paper concentrates on the current situation at farm level. Several examples from various European countries are analysed: could specialised arable or livestock farms develop towards mixed farming systems? What model of mixed farming system is most suitable: on farm level or cooperative models? What are be the main benefits and drawbacks and the enabling or hindering conditions? For instance, at farm level, specialisation has many advantages, because it makes efficient use of various resources and can bring the highest economy of scale. Can these advantages be maintained, for instance in a cooperative mixed farming system, or compensated by other advantages of mixed farming systems at farm level?

How much animal-source food can we produce while avoiding feed-food competition?

H.H.E. Van Zanten, O. Van Hal and I.J.M. De Boer
Wageningen University, Animal Production Systems group, De Elst 1, 6708WD, the Netherlands;
hannah.vanzanten@wur.nl

Livestock directly contribute to food supply by providing essential nutrients to humans, and indirectly support cultivation of food crops by providing manure and draft power. Livestock, however, also consume human-edible food or graze on land suitable for cultivation of food crops. As we face feeding 9.7 billion people by 2050, preferably without expanding the amount of agricultural land, there is an increasing need to avoid competition for land between animals and humans. We performed a review on studies that provide insight into the amount of animal-source food (ASF) produced without feed-food competition. So called default livestock are only fed with co-products, food-waste, crop-residues, or biomass from grazing land. Results showed, that between 7 g and 27 g of animal source protein per person per day can be produced from default livestock. Considering feed-food crops and feeding food-waste had an important contribution in this. Considering feed-food crops implies that choices have to be made between different crops, based on their contribution to feed and food production. Oil production from soy cultivation, for example, resulted in the co-product SBM. The practice of feeding food-waste to livestock is currently prohibited but shows potential in extensively reducing the environmental impact of livestock production. Considering feed-food crops and feeding food-waste are examples of mitigation strategies that currently can be implemented to reduce the environmental impact of the livestock sector. In mixed crop-livestock systems, particularly in developing countries, considering feed-food crops and using food-waste are embedded in production system, and can, therefore, be an example. In general a paradigm shift is needed: research should no longer focus on increasing efficiency of the animal or the animal production chain, but on increasing efficiency of the entire food system. Although ASF produced from default livestock, does not fulfil the current demand for ASF, about one third of the protein each person needs can be produced without competition for land between feed and food production. Livestock, therefore, does have an important contribution to future nutrition supply.

Livestock farming system diversity and resource use efficiency – What the history tells for France?
J.P. Domingues[1], A.H. Gameiro[1], T. Bonaudo[1], B. Gabrielle[2] and M. Tichit[1]
[1]INRA, UMR SADAPT, 16 rue Claude Bernard, 75231 Paris Cedex 05, France, [2]AgroParisTech, UMR ECOSYS, Thiverval Grignon, 78850, France; dominguessantos@agroparistech.fr

Recent research suggests that crop-livestock integration can occur at higher levels such as landscape and regions through the combination of different specialized farms. Nonetheless there is lack of knowledge at these more aggregated levels on how more diverse systems could provide better efficiency in the use of resources. Objective of this study was to assess the production and environmental performance in terms of nitrogen use in regions with contrasted livestock farming systems and intensification trajectories. Nitrogen flows related to animal and crop components were calculated for 14 departments located in Great Western (GW) and 11 departments in Massif Central (MC). Using multi-metrics, we calculated indicators of nitrogen conversion efficiency (NCE), the balance of human edible protein (HEP) and the protein self-sufficiency (PSS). All metrics were computed over 1970 and 2010. The analysis shows that overall NCE was higher in GW than in MC, although MC had a better balance of HEP and higher PSS. Greater NCE was partly driven by livestock population composition, with 44% of monogastrics in GW against 8% in MC. Livestock density in GW was roughly twice as much as in MC. Land use diversity, measured with Shannon index, was positively linked to NCE across the 25 departments. Even if at present a better NCE in GW backs higher provisioning of livestock products, it introduces a high environmental cost on the long run. It further exacerbates the competition with human nutrition (by allocating large amounts of crops to animal feeding); it also increases dependence on globally sourced protein (with massive imports of soy beans). Overall the system with greater land use diversity provided higher production output. More research is needed to understand the benefits of diverse mixing strategies at a given level of intensification, including on biodiversity and ecosystem service provision.

Mixed farming systems in the beef value chain of the city of Malang, Indonesia
S.J. Oosting[1], J.J.M. Boekhorst[1], M.G. Kok[1] and I. Subagiyo[2]
[1]Wageningen University, Animal Production Systems group, P.O. Box 338, 6700 AH Wageningen, the Netherlands, [2]Brawijaya University, Jalan Veteran, Malang 65145, Indonesia; simon.oosting@wur.nl

Beef demand is high on the densely populated island of Java, Indonesia and local production does not meet the demand. We studied the beef value chain (BVC) of the city of Malang on East Java, where 1 million inhabitants predominantly consume beef in Bakso, a meatball soup. Three farming system are important in the BVC: breeder farms (mixed crop-livestock farms with reproductive cows, starter bull calves and replacement female calves), feeder farms (with 1 to 10 bulls as a side activity) and feedlots. We conducted on farm interviews (n=65) and expert interviews to study the role of these farming systems in Malang city's BVC and input use and production performance at the farming systems. During farm visits live weight and age of 240 bulls were recorded, as well as age at first calving, and length of calving interval if reproductive animals were present. Average daily gain (ADG) was estimated for bulls for which live weight was recorded more than once. We estimated that for a daily supply of 100 finished bulls of 500 kg live weight to the city of Malang, around 100,000 smallholder breeder farms supplied the starter calves, around 30,000 feeder farms were involved in the intermediate to late fattening stage, and around 40 feedlots were involved in finishing the bulls. Feedlots purchased bulls preferably at a weight of 400 kg from feeder farms or directly from breeder farms and such bulls could vary in age between 1.5 and 5 yrs. ADG of the bulls at a feedlot was approximately 1 kg, independent of age (P>0.05). An essential principle of the BVC was that production of starter calves at the breeder farms was at a low input basis which makes the BVC resilient. The BVC could reach higher output by improving the reproductive and growth performance at breeder farms. This would, however, imply a transition of a low input to a relatively high input farming system which would not be feasible under present price levels of inputs and outputs and which would reduce the number of smallholders benefitting.

Self-sufficiency for animal feed: a multi-level framework to promote agroecological farming systems
V. Thenard, A. Charmeau, P. Triboulet and G. Martin
INRA, UMR1248 AGIR, Chemin de borde rouge, 31326 Castanet Tolosan, France; vincent.thenard@toulouse.inra.fr

Faced to global change and negative externalities of agricultural systems, agroecology is a conceptual, methodological and operational framework using ecologic principles to redesign farming systems towards improved sustainability. Self-sufficiency of farming systems is a key principle of agroecology, and especially the self-sufficiency for animal feed in livestock systems. Although some farms combine livestock and crop production, the production of fodder and concentrates is usually insufficient to feed animals. Developing agroecological farming systems requires a better understanding of the stakes of feed self-sufficiency at different levels (farms, territory, supply chain). The purpose of this communication is to present a framework characterizing self-sufficiency for animal feed at farm level in order to discuss the issue of this self-sufficiency at different levels. The study was based on the analysis of data from 41 livestock farms in southern France. Farm data was collected for three types of livestock systems: dairy cattle, beef cattle and dairy sheep. First we developed indicators to calculate livestock farm self-sufficiency for animal feeding on a dry matter, energy and protein basis respectively. Three other indicators were calculated and provided information on livestock farm structures, intensification and performances (environmental, economic). Analysis of the results allowed distinguishing the different livestock systems according to their level of self-sufficiency. We showed relations between the different types of self-sufficiency and the farm performances: intensifying production and animal productivity increase energy and protein requirements for animals. Farmers need to supply animal feed with purchased concentrates, limiting the self-sufficiency at farm level. Results were aggregated at the regional level to assess the scope for self-sufficiency for animal feeding among farms. Secondly we interviewed different types of local actors involved in feed (farmers, agricultural network, and feed factories) to better understand the issues linked to feed self-sufficiency at farm and territory scales. We discussed constraints and levers which could be used to promote a located self-sufficiency for animal feeding.

Crop livestock integration at landscape level in Mediterranean: a comparative analysis
J. Lasseur[1], M. Napoleone[1], M. Mohamed[2], C.H. Moulin[1] and V. Alary[2]
[1]INRA-UMR SELMET, 2 place Viala, 34060 Montpellier, France, 2 place Viala, 34060 Montpellier, 34060 Montpellier, France, [2]CIRAD-UMR SELMET, TA C/112 Baillarguet, 34398 Montpellier, France; charles-henri.moulin@supagro.inra.fr

Previously sheep farming in French Mediterranean relied on forage resources combination from contrasted component of landscape (Sylva, Saltus and Ager), mixed farm was the pivot. Specialization at farm and local levels reshape crop livestock integration (CLI): Cash crop specialization for plains, urbanization of the coast, withdrawal in hinterland for specialized sheep farms. Crop and livestock integration is now at stake at territory level to increase forage self-sufficiency, efficiency in resource consumption and adaptation to global change. We distinguish three ways for CLI: vicinity grazing on cash crop fields (V.G.), distant winter grazing (D.G.) in vineyards of coastal zone, distant production of winter forage (D.F.) on removed peri-urban vineyard of plain. VG rely on inter-individual relationships inherited from traditional social organization, decreasing and of limited perspectives due to disintegration of rural communities and local specializations. DG is based on mobility and a wide social network. It is an adaptive way of integration with a high ability to catch opportunities, even ephemeral ones. Large size flocks are concerned allowing in return their enlargement. Farmers get access to a wider range of social relations and forage resources. It doesn't contribute in explicit territorial development schemes. D.S. limit purchases in forage input for small size mountain livestock farms. This cooperative action connects a group of farmers with stakeholders of coastal area to initiate actions based on win-win interest. For farmers it redefine activity in line with the stakes of agro-ecological transition. Sustainability relies on the evidence that farmer's activities on coastal zone are more than a service delivery, for which others could be more efficient. It needs to be aware of equilibrium between the two poles of farming (mountain and plain). The sustainability for recoupling at landscape level relies on territorial and non-sectorial policies that have to be more strongly stated.

Opportunities and challenges of diversified crop-livestock system in new reclaimed lands, Egypt
A. Aboul Naga[1], V. Alary[2], M. Osman[1] and X. Juanes[2]
[1]Animal Production Research Institute, Nadi Elseed str., Dokki, Cairo, Egypt, [2]CIRAD, Centre International de baillarguet, TA C-112 A, 34398 Montpellier cedex 5, France; adelmaboulnaga@gmail.com

In 2013-14, a farm survey was conducted over 175 small-scale farms to assess the level of diversification and integration of crop-livestock activities in the new reclaimed land (NRL) of Egypt, over a gradient of land reclamation from 1960-2010. In the first years of settlement, crop-livestock systems were mainly developed according to soil type, water availability, accumulated experience and lesson learned from neighbors. Mostly the crop system based on traditional crops; wheat and berseem in the Winter and maize, and vegetable(beanuts and potato) in the Summer. Berseem and maize were for animal feeding. Livestock included large ruminants (buffaloes and cattle) for milk and meat and small ruminants as a cash flow complement and for social and religious events. In the areas where common beneficiaries farmers are dominant, they continue to allocate 30-50% of the land to fodder crops. Besides, 46% of university graduates, have replaced their traditional crop system with fruit tree plantations (mainly oranges); they reduced significantly their livestock activity, as soon as they started to collect fruits; livestock having been the main financial basis for the development of tree systems. The new opportunities of diversification of the crop system through trees in the NRL have supplanted the livestock activity, considered it as a labor intensive activity for graduates, with economic constraint related to feed costs and health issues. In parallel, the diversification through livestock activities remains a major pillar of farm sustainability for common beneficiaries, in order to face the problems of water availability, and to maintain soil fertility. Behind these different strategies, different challenges can be observed: increase farm income or/and reduce family labor load and/or reduce water and soil risks at the local level. The farm survey shows that the different scale challenges evolve during the settlement according to farm land tenure, past experience and level of education.

Which animal do farmers need for tropical mixed farming systems?
N. Mandonnet[1] and T. Ceresita[1,2]
[1]INRA, UR143, Unité de Recherches Zootechniques, Petit-Bourg, 97170, Guadeloupe, [2]INRA, UE 1294, Plateforme Tropicale d'Expérimentation sur l'Animal, Petit-Bourg, 97170, Guadeloupe; nathalie.mandonnet@antilles.inra.fr

In the Global South, improvement of agricultural outputs is necessary because while world population will double by 2050 areas devoted to agriculture will decrease. Efficiency in animal productions is essential to allow coverage of protein nutritional needs of people, both in quality and quantity. Moreover, the demand for fresh locally-produced animal products is not satisfied yet in the tropics. In the Caribbean territories mixed farming systems are the most common farming systems (about 80%) and can constitute a solution to reach food sovereignty in such limited and isolated spaces. Lessons to be learned from these systems involve improving animal performance while respecting the natural balance with the environment and maintaining the multi-functionality of plants and animals. Enhancement of adaptation to stresses (biotic, abiotic and socio-economic) in species or animal genotypes implies their equilibrium with the environment without notion of maximizing the production functions, ensuring their survival and well-being. This approach underpins for the farmer an integrated management of health, nutrition, genetics, reproduction, in close relationship with other compartments of the farming system. In this context, balanced enhancement of mechanisms and genetics of adaptive traits is a key element. The animal must be efficient that is to say must reconcile different physiological functions of production, reproduction and adaptation. This optimization leads to decreased inputs and finally overall efficiency of mixed farming systems. The aim of the breeder is to choose the best balance between output-reproduction-adaptation, at the individual, the breed or the flock scale combining genetic and physiological diversity. The main thrust of this breeding policy is to give to humans and animals their right place in the food chain taking into account the farmers' skills and wills. This idea is included in the agroecological approach and may give guidelines for food sovereignty worldwide.

Test various strategies to reduce Nitrogen (N) Phosphorus (P) and Carbon (C) losses in mixed farms

H. Chambaut[1], J.L. Fiorelli[2], S. Espagnol[3], P. Mantovi[4], J. Verloop[5], S. Maignan[6], E. Ruane[7] and P. Nicholas[8]
[1]Institut de l'Elevage, Angers, 49000, France, [2]INRA, SAD ASTER, Mirecourt, 88500, France, [3]Institut du porc, Le Rheu, 35650, France, [4]FCSR, Reggio Emilia, I-42100, Italy, [5]Wageningen University, Wageningen, 6700 AA, the Netherlands, [6]Coopedom, Domagné, 35113, France, [7]Teagasc Research Centre, Moorepark, Co. Cork, Ireland, [8]Aberystwyth University, Gogerddan Campus Aberyswyth, SY233EE, Wales, United Kingdom; helene.chambaut@idele.fr

The Framework 7 European Project Cantogether (2012-2015) hypothesised that complementarity between animals and crops can reinforce sustainability of agricultural systems at farm or territory level. A participatory approach to enhance the degree of mixing in farms resulted in new practices and techniques to be tested at farm level (i.e. more crop and animal types, diversification in farm functions, but also more interactions beween farms enterprises). The benefits on NPC farm gate balance (FGB), direct nitrogen emissions (NO_3, NH_3, N_2O, NO), P run off risk, and carbon (CO_2 and CH_4) losses were estimated using IPCC Tier2 and EMEP methodology on 9 farms located in Europe (649-1,385 mm rainfall/year). Baseline systems covered a wide range of production intensity, from organic to high production farming (3.2-11.7 TDM/Ha, 0.98-4.57 LSU/Ha); NPC fluxes displayed significant ranges per hectare: FGB N surplus 49-359 kg, P surplus 0-36 kg, nitrate-N leaching 26-135 kg, ammonia-N losses 12-122 kg and methane emission 2.9-19.4 t eq. CO_2. Emission related to energy consumption ranged from 0,2 to 1,6 t eq CO_2/ha. After new MFS practices were implemented, primary production of agriculture area was rather stable (-0.1 TDM/ha), and required less mineral fertilizer (-42% N mineral fertilizer and -84% P used per TDM produced). 56% of farms decreased nitrate leaching and 78% lowered direct contribution to climate warming per hectare. Nevertheless, only 44% of farms improved emissions per unit of protein product outputs, and 55% of farms reduced permanent grassland area (less sink of C). These results show that improved nutrient cycling on the farm associated with mixed farming was sometimes impaired with a loss of farm outputs.

Are farming systems a key to understand dairy sheep breed management in Thessaly (Greece)?

L. Perucho[1], C. Ligda[2] and C.H. Moulin[3]
[1]INRA, UR LRDE, Quartier Grossetti, 20250 Corte, France, [2]H.A.O., Veterinary Research Institute, P.O. Box 60 272, 57001 Thessaloniki, Greece, [3]Montpellier SupAgro, UMR SELMET, 2 place Pierre Viala, 34060 Montpellier cedex 1, France; chligda@otenet.gr

Thessaly is a region of central Greece with a significant sheep production dispersed across interior lowlands, hills and surrounding mountainous ranges. In this contrasted geography, the diversity of land management and access to production means leads to a diversity of farms characteristics and farming practices. These practices include the breed management, characterized by the use of pure breeds (registered under selection scheme or part of a conservation program), local populations (not registered in breed management program) and crossbred animals at two different scales: the farming system and the region. We applied Multiple Correspondence Analysis (MCA) and Hierarchical Classification on Principal Components (HCPC) on structural farm data collected from 41 semi-structured interviews to assess and qualify the link between identified farming systems and the choice of genetic material. Three clusters of farms were distinguished on 8 main variables: land use, genetics, flock size, length of milking period, milking equipment location, cultural community and existence of other sources of income. Results show that local breeds, raised as pure breeds, are facing the expansion of crossbreeding and remain in specific farming systems, not connected with specific add-value processes. The Karagouniko sheep breed, historically bred in Thessaly, is raised as pure breed under milk recording in western lowlands, in flocks of 100 to 200 ewes grazing collective grasslands, by farmers usually combining other source of income. Although the use of local pure breeds remains minority in Thessaly, farmers stress their interest for the adaptive traits of local breeds in terms of valorization of spontaneous resources, robustness to rough weather conditions and diseases, longevity and walking on mountain pastures. These results could contribute to build regional genetic management programs adapted to the diversity of breeding objectives and farms contexts.

ABC of the Nagoya Protocol

E. Martyniuk
Warsaw University of Life Sciences, Faculty of Animal Sciences, Department of Genetics and Animal Breeding, ul Ciszewskiego 8, 02-786 Warszawa, Poland; elzbieta_martyniuk@sggw.pl

The Nagoya Protocol, a treaty negotiated under the Convention on Biological Diversity was adopted in 2010 and came into the force on 12th October 2014. The objective of the Nagoya Protocol is the fair and equitable sharing of the benefits arising from the utilization of genetic resources and associated traditional knowledge. By 1st April 2016, 73 countries ratified this treaty, 31 from the African Region and 10 + EU from the European region. The three key elements of the Nagoya Protocol include: Access, Benefit Sharing and Compliance. Parties to the Protocol are obliged to provide for legal certainty, clarity and transparency of their domestic access and benefit-sharing legislation or regulatory requirements regarding Access to their genetic resources. Users have to obtain PIC (Prior Informed Consent) from a provider country before accessing their genetic resources. Also, traditional knowledge associated with genetic resources held by indigenous and local communities has to be accessed with PIC or approval and involvement of these communities. However, in exercising sovereign rights over natural resources, Parties to the Nagoya Protocol may decide not to regulate access and to wave PIC requirements, as is the case of some European countries. As exchange of genetic resources is based on bilateral basis, the Benefit Sharing arrangements should be negotiated between user of genetic resources and the provider country established in a contract (MAT: mutually agreed terms). Such arrangements may include both monetary and non-monetary benefits. The Protocol provides also for the possibility of establishing a Global Multilateral Benefit Sharing Mechanism, but the need for and modalities of such mechanism have not yet been determined. The third key element: Compliance is considered at two levels. Parties to the Nagoya Protocol have to comply with provisions of the Protocol (e.g. develop domestic access benefit sharing legislation; establish competent national authority and checkpoint, issue PICs, provide information to ABS Clearing House and monitor utilisation of genetic resources), while users within their jurisdiction have to comply with domestic laws as well as the laws of a provider country. The scope of the Protocol includes genetic resources of all livestock species.

The implementation of the Nagoya Protocol in the EU

A. Kozłowska
European Commission, DG Environment, Office BU9 3/124, Brussels, Belgium; alicja.kozlowska@ec.europa.eu

Nagoya Protocol is implemented in the EU by Regulation No 511/2014 on compliance measures for users from the Nagoya Protocol on Access to Genetic Resources and the Fair and Equitable Sharing of Benefits Arising from their Utilization in the Union (the EU ABS Regulation). The Regulation entered into application on 12 October 2014, date on which the Nagoya Protocol itself entered into force for the Union. Some important obligations set up by the Regulation entered into application only a year later thus on 12 October 2015. The cornerstone of the EU ABS Regulation is due diligence. In line with Article 4 of the Regulation the user (a natural or legal person that utilises genetic resources or traditional knowledge associated with genetic resources) is under the obligation to ascertain that genetic resources and the associated traditional knowledge which they utilise have been accessed in accordance with applicable access and benefit-sharing legislation or regulatory requirements, and that benefits are fairly and equitably shared on mutually agreed terms. The EU ABS Regulation establishes two checkpoints, i.e. one at the stage of receiving research funding and the other one at the stage of final development of the product. In both situations the user is expected to declare either that they exercise due diligence in accordance with Article 4, or that they have fulfilled the obligations under Article 4. The modalities of the application concerning due diligence declarations are defined in the Commission implementing regulation – Regulation (EU) 2015/1866 laying down detailed rules for the implementation of Regulation (EU) No 511/2014 as regards the register of collections, monitoring user compliance and best practices. Furthermore, the EU ABS Regulation provides for checks on user compliance to be carried out by the Member States. The Member States need to prepare risk-based plans for implementation of the checks and establish dissuasive and proportionate penalty system for infringement of the ABS rules. The EU ABS Regulation provides also for two voluntary mechanisms facilitating compliance with obligations of the Regulation, namely best practices and registered collections. The procedural aspects of these voluntary mechanisms are also defined in the Commission implementing regulation referred above.

Exchange of animal genetic resources in research

M. Tixier-Boichard

GABI, INRA, AgroParisTech, Université Paris-Saclay, 78350 Jouy-en-Josas, France; michele.boichard@jouy.inra.fr

Originally, the concept of animal genetic resources was related to the management of populations of domestic animals. Exchanges of reproductive material are handled by breeders' associations or breeding companies, whereas researchers are rather involved in studies of reproductive biotechnologies. Genomics has strongly increased the use of biological samples in animal sciences. DNA banking started in the 90s, with European projects on genetic diversity of a large range of breeds. At that time, getting the samples was straightforward and providers were acknowledged in the first paper. The principles introduced by the Nagoya Protocol are virtuous since traceability of resources used in research is a recommended practice to ensure the relevance of results. Furthermore, offering breeders a feedback on the results obtained with their samples is a way to facilitate their cooperation to a project. However, setting a complex administrative procedure for ABS raises a number of issues which are likely to have counterproductive effects: (1) cost and, particularly, time of collecting samples will increase, staff training and additional administrative work will increase expenses of research institutions which may decrease their capacity to conserve genetic resources;(2) time from the research idea to the start of the work will increase because of the need to obtain permits from foreign administrations, so that reactivity of research will decrease, projects aiming at a large geographic coverage will have to develop dedicated partnerships; (3) an unfair competition will arise for research and exploitation of research results in EU as compared to the USA which are not part of the CDB; (4) heterogeneity of national laws within EU will delay the realization of research. At least, a simplified access for researchers to the resources of their own country could be recommended within the EU. The starting IMAGE project on animal gene banks could study options for a specific access framework to animal genetic resources, weighting all costs and benefits for research and stakeholders.

Genetic resources in commercial animal breeding

J.G.B. Venneman

European Forum of Farm Animal Breeders, Rue de Trèves 61, 1040 Brussels, Belgium; jan.venneman@effab.info

Breeding is selecting the best parents for creating the next generation. The latest implemented development is genomic selection; this tool makes it possible to predict the breeding value of the animal and its offspring at a younger age and more precisely than in the decades before. All breeding programs are based on the aforementioned principle but the systems differ from sector to sector. Breeding in the poultry sector is based on improvements of purebred lines (grand-grand-parents) and crossing these lines (from grand-parents via parents to the end product for the commercial sector (layers, broilers, turkeys, ducks, etc.) to fully utilize heterosis. In the commercial pig breeding sector the system is based on crossing of two purebred lines to create an F1 sow line that is crossed with a purebred boar line to create an F2 pig for the commercial sector. Both the poultry and pig breeding sectors are operating in closed production systems. Companies are using the genetic variation within their own populations, but sometimes they buy genetic material from other sources. Breeding in the aqua sector is more or less comparable to breeding in the pig and poultry sector. Breeding in the ruminant sector is done in open systems. Everybody has access to the genepool of other companies by buying genetic material (semen, embryos, oocytes). However also in the ruminant sectors only the variability within the sector is used. Input of genetic material from developing countries does not take place, also due to veterinary restrictions. Crossing between breeds is getting more common. Different sectors are using different ways to conserve their genepool. In some species this is done in situ (poultry), in some species this is done ex situ. In the recent past conserved gene pools were needed if a genetic variant were to be retrieved and introgressed. With current state of the art the precise description of the genetic variant might be enough to bring it quickly, clean and efficient (back) in a population of choice. The EU commercial breeding sector is not using genetic resources from non OECD countries. For this reason the Nagoya Protocol and the EU Regulation is not applicable. However, when the use of New Breeding Techniques like trans-genesis and cis-genesis would become possible in Europe, the need and wish to import Genetic Resources from non-OECD countries cannot be excluded.

FAO contribution to support implementation of the Nagoya Protocol in agriculture

I. Hoffmann[1] and R. Baumung[2]
[1]FAO, Commission on Genetic Resources for Food and Agriculture, Viale delle Terme di Caracalla, 00153 Rome, Italy, [2]FAO, Animal Genetic Resources Branch, Viale delle Terme di Caracalla, 00153 Rome, Italy; roswitha.baumung@fao.org

The Commission on Genetic Resources for Food and Agriculture, at its Eleventh session in 2007, approved its first Multi Year Programme of Work, which included access and benefit-sharing (ABS) as a crosscutting issue. The Commission realized the need to contribute to ongoing intergovernmental discussions on ABS in order to ensure that it move in a direction supportive of the special needs of the agricultural sector, in regard to all components of biological diversity of interest to food and agriculture. Since the adoption of the Nagoya Protocol on Access to Genetic Resources and the Fair and Equitable Sharing of Benefits Arising from their Utilization to the Convention on Biological Diversity, ABS work intensified within the Commission with a milestone meeting of the Ad Hoc Technical Working Group on ABS in Svalbard in September 2012. In April 2013, the Commission decided on a process to advance work on ABS, and established a Team of Technical and Legal Experts on ABS (TTLE-ABS). The TTLE-ABS was mandated to prepare Draft Elements to Facilitate Domestic Implementation of Access and Benefit-Sharing for Different Subsectors of Genetic Resources for Food and Agriculture (ABS Elements), drawing on studies, reports and other inputs. ABS Elements were meant as voluntary tools to assist national governments to develop their domestic ABS legislation. In January 2015, the Commission welcomed the ABS Elements, which were subsequently welcomed by the Thirty-Ninth Session of the FAO Conference. The Conference also noted the complementarity between the work of the Commission and the Nagoya Protocol in regard to access and benefit-sharing for genetic resources. The Commission further requested its sectoral Intergovernmental Technical Working Groups (on plants, animal, forest, and aquatic genetic resources), to continue elaborating subsector-specific elements for ABS, including consideration of the role of traditional knowledge associated with GRFA and their customary use. This work is currently in progress, and any outcome of this work will be considered by the TTLE-ABS. This important work is meant to help the Commission to move its ABS agenda forward in light of the special needs of agriculture and the rapidly evolving field of ABS.

Case studies performed in preparation of the EU Sectorial Guideline on Animal Breeding

B. Visser
Wageningen University, Centre for Genetic Resources, P.O. Box 16, 6700 AA Wageningen, the Netherlands; bert.visser@wur.nl

The Nagoya Protocol is implemented in the European Union through EU Regulation 511/2014. In order to support the various sectors in the EU utilizing genetic resources and to facilitate compliance by users, the EU Commission is developing a series of eight sectorial guidelines, in addition to a horizontal guideline addressing generic issues, in relation to the EU Regulation. One of these sectorial guidelines is focussing on the animal breeding sector, including aquatic species and companion animals. In particular, these sectorial guidelines will provide guidance in determining which activities qualify as research and development in the context of the Nagoya Protocol, and therefore should be considered as within scope of the EU Regulation, assuming specific other conditions have been fulfilled as well. In addition to providing an overview of the sector in the EU and a description of its use of genetic resources, each sectorial guideline will present a number of case studies, describing research and development activities which are considered to be in scope or out of scope of the EU Regulation. Proposals for the sectorial guidelines have been developed by small drafting groups and will be presented to a wide array of stakeholders in dedicated workshops. The EU Commission will evaluate the outcome of this process and finalize each of the sectorial guidelines. The workshop on the draft sectorial guideline for animal breeding already took place on July 1. A selection of case studies incorporated in the draft sectorial guideline on animal breeding will be presented.

Genomic selection in sheep: where to now?

J. McEwan, K. Dodds, S. Rowe, R. Brauning and S. Clarke
Animal Genomics Team, AgResearch, Invermay Agricultural Centre, PB50034, Mosgiel, New Zealand;
john.mcewan@agresearch.co.nz

Use of molecular information in New Zealand sheep breeding became widespread 15 years ago with adoption of DNA 'mix and match' parentage testing coupled with single gene tests including those affecting fertility and carcass muscling. An international effort to sequence the genome and create high density SNP chips suitable for genomic selection then commenced a decade ago and resulted in the introduction of genomic selection in New Zealand five years ago. This has expanded into a monthly across flock and breed national genomic evaluation for more than 20 traits in selected maternal breeds and crosses. A similar development programme is underway in terminal breeds and includes meat quality traits. Research is also underway into using genomics to select animals that emit less methane and this has renewed focus on key input traits such as feed efficiency. While a significant fraction of the industry uses this technology, the cost currently limits its use to elite ram lambs via two stage selection and also depends on imputation from lower density chips. Current research is now focussed on reducing the cost of genotyping via the use of genotyping by sequencing. The target outcome is parentage, breed prediction, co-ancestry and genomic selection for the equivalent of the current cost of DNA parentage so that it will becomes a routine measurement for all animals in ram breeding flocks.

The potential for genomic selection for small ruminants (SR) in developing countries

R. Mrode[1], G. Mekuriaw[2], J. Mwacharo[3] and A. Djikeng[4]
[1]International Livestock Institute, Animal Biosciences, P.O. Box 30709, Nairobi 00100, Kenya, [2]Addis Ababa University, Microbial Cellular and Molecular Biology, P.O. Box 1179, Addis Ababa, Ethiopia, [3]International Centre for Agricultural Research in the Dry Areas, Small Ruminant Genetics and Genomics Group, P.O. Box 5689, Addis Ababa, Ethiopia, [4]Biosciences eastern and central Africa (BecA-ILRI Hub), Director, P.O. Box 30709, Nairobi 00100, Kenya; r.mrode@cgiar.org

The versatility of indigenous SR's to adapt to diverse agro-ecologies and production systems imply that they play a key role in sustaining livelihoods of (agro)-pastoralists and small holder farmers in many developing countries (DC). In most DC's, phenotypic and pedigree information remains problematic to collect, cohorts of contemporaries are often of inadequate size, breeding infrastructures are non-existent and advanced reproductive technologies are difficult to apply. These drawbacks can however be circumvented through community–based breeding programmes (CBBP) which can provide a framework to implement basic recording and design mating schemes. CBBP together with case-control protocols for use in genome-wide association analyses offer great opportunity for identifying genomic regions with major gene effects that could be used for genomic selection (GS) and/or introgression. The availability of genomic data can facilitate the implementation of GBLUP using genomic relationship matrices; determine breed composition and admixture in the absence of pedigree records, assessment of genetic biodiversity and the identification of genomic fingerprints of positive selection. Such data can provide a platform to design mating schemes that optimize productivity and adaptability in a diverse genepool of SR's and the development of suitable synthetic breeds. In the context of the CBBP, the propagation of alleles, underlying production and adaptation traits, via gene/genome/haplotype block editing could be combined with GS in developing appropriate synthetic breeds that optimize productivity and adaptability.

Implementation of single step genomic selection in UK dairy goats

S. Mucha[1], R. Mrode[1,2], M. Coffey[1], S. Desire[1], I. Maclaren-Lee[3] and J. Conington[1]
[1]Scotland's Rural College, Animal & Veterinary Sciences, Easter Bush, EH259RG Midlothian, United Kingdom, [2]Animal Biosciences, International Livestock Institute, 00100 Nairobi, Kenya, [3]Illumina Cambridge Ltd., Great Abington, CB216GP Cambridge, United Kingdom; sebastian.mucha@sruc.ac.uk

Despite increasing goat milk consumption in the UK, until very recently there have been no breeding value estimations. Therefore, a research project was undertaken aiming to implement routine breeding value estimation incorporating genomic technology. This was done in collaboration with two goat farms in the UK owned by a single farming business. The population was created in 1985 by crossing 3 breeds: Alpine, Saanen, and Toggenburg which resulted in a new synthetic breed. Data provided by the participating farms comprised 590,409 milk yield records on 14,453 dairy goats kidding between 1987 and 2013 with a pedigree of 30,139 individuals. Animals were genotyped commercially with the 50K Caprine Illumina SNP chip. The reference population was composed of a mixture of 1,474 males and females and the validation set contained 305 animals. A total of 47,306 SNP were used after QA. Various methods for estimation of genomic breeding values were evaluated and single step GBLUP resulted in the highest accuracy of GEBVs for milk yield (0.61). Apart from breeding value estimation, SNP information was also applied for parentage verification and parent discovery. Initially the project was focused only on milk production but in subsequent years this was extended to body conformation, mastitis, longevity, feed efficiency, phantom pregnancies and milk composition. The number of genotyped animals has increased accordingly to over 7,100 individuals with a target of close to 12,000 individuals in the coming year. This study has demonstrated that genomic selection can be successfully implemented in small populations. Moreover it proves that the initial investment can be outweighed by economic benefits from improved accuracy of breeding value estimation and lower costs associated with only rearing sires that will be used for breeding.

Practical challenges of implementing Genomic Selection and breeding British Texel Sheep

J. Yates
British Texel Sheep Society, National Agricultural Centre, CV83LG, United Kingdom; johnyates@texel.co.uk

Dairy industries in many countries are realising increases in the rate of genetic progress through the application of genomics. The greatest opportunity is likely to be in hard to measure traits (HTM) maternal traits, including disease resistance/tolerance, and ease of care (cost-saving) traits. 13 M ewes are mated annually by 365,000 rams in the UK. 27% of were Texel rams being used on 17,700 UK sheep flocks, which make the Texel breed the most common ram breed, so any genetic improvement will have the greatest impact on sheep performance at a National level. Financial benefits of genetic improvement between 1996 and 2015 amount to £197.0 million, or £4.8 million annually. The benefits can be attributed to a combination of an increasing genetic trend, increased use of rams from pedigree registered flocks (as opposed to rams from commercial flocks), and increased use of the Texel rams to breed replacements. By 2016 the Society will have 4,000 HD genotypes within its reference population. Genome Wide Association Study will be carried out using the Multi-Locus Mixed Model (MLMM) algorithm. Genomic estimated breeding values (gEBVs) will be predicted using the single-step method. Integrating pedigree, genotype and phenotype data for hard to measure health traits, mastitis and footrot. Imputation from 50K to 700K will also enhance the outputs. The inclusion of genomic information in sheep breeding programmes is expected to result in a large (~10-32%) increase in economic response to selection in terminal sire sheep, relative to a conventional breeding programme. Genotyping costs are expected to fall over time however cost benefit and the speed of which genomic selection can be adopted in sheep breeding programmes remains a challenge.

Implementation strategies for genomic selection in Norwegian sheep breeding

M. Lillehammer[1], A.K. Sonesson[1] and T.H.E. Meuwissen[2]
[1]Nofima, P.O. Box 210, 1431 Ås, Norway, [2]Norwegian University of Life Sciences, Department of Animal and Aquacultural Sciences, P.O. Box 5003, 1432 Ås, Norway; marie.lillehammer@nofima.no

The breeding goal for Norwegian sheep includes large weight on maternal traits and slaughter traits, and genomic selection should hence be a useful strategy to increase total genetic gain as well as genetic gain for traits currently evaluated by a time-consuming progeny test scheme. In this study, different implementation strategies of genomic selection with different male generation intervals were tested by stochastic simulations. The breeding goal was assumed to have 30% weight on traits measured on candidates, 20% weight on traits measured on slaughter lambs and 50% weight on maternal traits. Currently, the artificial insemination (AI) rams serves 15% of the ewes. They are 3 years old and selected among rams progeny tested for 2 years within local breeding units. To genotype the local rams (300 2-year olds or 1,500 1-year olds) each year and use genomic selection on 3 year old AI rams, increased genetic gain for maternal traits, but reduced genetic gain for other traits, causing no change in total genetic gain and a 56% increase in rate of inbreeding, compared to the conventional breeding scheme. More optimal was to genotype one year old local rams and use genomic selection to select 2 year old rams for AI and rams to use for a second year locally. This scheme increased total genetic gain by 18% and genetic gain for maternal traits by 65%, compared to the conventional scheme. Rate of inbreeding increased by 75% compared to the conventional scheme and by 12% compared to genomic selection with 3 year old AI rams. A reduction of the generation interval for AI rams to 1 year, achieved by genotyping newborn lambs and use genomic selection to select AI rams directly, increased genetic gain to 31%, compared to the conventional scheme, with a similar distribution of genetic gain between traits as the other genomic selection schemes. Rate of inbreeding, however, increased by 357% per year, and this scheme would be logistically challenging.

Genomic selection for a multi-breed sheep population in Ireland

N. McHugh[1], A. O'Brien[1], E. Wall[2], K. McDermott[2], T. Pabiou[2], S. Randles[2] and D.P. Berry[1]
[1]Teagasc, Animal & Bioscience Department, Animal & Grassland Research & Innovation Centre, Moorepark, Fermoy, Co. Cork, Ireland, [2]Sheep Ireland, Bandon, Co. Cork, Ireland; noirin.mchugh@teagasc.ie

Genomic selection has revolutionised animal breeding internationally and is now at a low enough cost to be exploited by the Irish sheep industry. The pipelines and logistics for the implementation of genomic evaluations are available from previous research conducted for the Irish dairy and beef industries. Research funding from the Irish Department of Agriculture, Food and the Marine provided the necessary funding for the initial research on the feasibility of genomic evaluations in Irish sheep. Genotype information is available on approximately 15,500 animals with recorded phenotypes from the six main breeds: Texel, Suffolk, Charollais, Vendeen, Belclare and Beltex; these breeds were chosen based on the volume of phenotypic information available in the national genetic evaluations. Genotyping was conducted on various platforms including high density (606,006 SNPs; n=192), medium density (51,135; n=2,895) and a custom low density (16,301; n=12,000) panels. Ten sheep have also been sequenced at 10X. All low density genotypes are imputed with an accuracy (i.e. correlation) of 0.983 to medium density for genomic analysis. Individual animal call rate of <90% existed for ~7% of genotypes. Initial results reveal 5% parentage errors, with 32% of the genotype errors relating to dam misidentification and 68% to sire misidentification. One-step multi-breed genomic predictions have been developed. To date all genotyping costs have been incurred through competitive research funding. In Autumn 2016, however, a genotype subsidy will be put in place using the remaining research funds to incentivise breeders to genotype all replacements entering the flock and from spring 2017 onwards a streamlined genotyping service will be available to breeders for genotyping all lambs at their own cost. Future research will focus on development of low cost DNA parentage options for commercial and hill breeds, and the possible inclusion of more breeds into the genomic evaluations. It is envisaged the routine genomic evaluations will be available for the Irish sheep population from spring 2017 onwards.

Imputation of medium density genotypes from custom low density genotype panel in sheep

D.P. Berry[1], A. O'Brien[1], S. Randles[2], K. McDermott[2], E. Wall[2] and N. McHugh[1]
[1]Teagasc, Moorepark, Fermoy, Co. Cork, Ireland, [2]Sheep Ireland, Bandon, Co. Cork, Ireland; donagh.berry@teagasc.ie

A custom low density genotype panel has been developed with 16,351 single nucleotide polymorphisms (SNPs), 12,118 of which are on the medium density Illumina Ovine50 Beadchip which has 51,135 SNPs. The objective of the present study was to quantify the accuracy of imputation from the low density to medium density panel in five different sheep breeds. Medium density genotypes were available on 2,375 sheep from the breeds Suffolk (n=566), Texel (n=318), Vendeen (n=461), Charollais (n=559) and Belclare (n=471). The youngest 75 animals per breed were used as the validation population; the 37,278 autosomal SNPs on the medium density panel that are not on the low density panel were masked for the validation population. Imputation was undertaken across the entire genome simultaneously using both family and population wide linkage (disequilibrium) information. Concordance rates and the correlation between the true and imputed genotypes were estimated for the validation animals which included the low density SNPs in the calculation. Across all genotypes, the correlation between the actual and imputed genotype was 0.983; the mean genotype (allele) concordance rate was 0.979 (0.989). The mean genotype and allele concordance rate per individual varied from 0.864 to 0.997 and from 0.929 to 0.999, respectively. The individual with the poor concordance rate was an outlier and the minimum genotype (allele) concordance rate excluding this individual was 0.920 (0.958). Mean genotype concordance rate per breed was 0.984, 0.972, 0.982, 0.969 and 0.989 for Belclare, Charollais, Suffolk, Texel and Vendeen, respectively. Imputation accuracy not accounting for pedigree was marginally better than when pedigree was accounted for in the imputation process. Imputation accuracy with a reference population of only the breed of animal to be imputed was also marginally better than when multiple breeds were included in the reference population; imputation accuracy of breeds not represented in the reference population were considerably worse. The low density panel is therefore a useful, lower cost, strategy to achieve genomics evaluations in these sheep breeds.

Sheep breeding, environmental and non genetic factors affecting lamb growth under low imput systems

I. Ben Abdallah and M. Djemali
INAT, Animal Genetic and Feed Resources Lab, 43 Avenue Charles Nicole, Mahrajene, 1082, Tunisia; ichrak25121988@gmail.com

Sheep owners in Tunisia are approximately 270,000; 65% of them raise the fat tail Barbarine sheep breed known for its adaptation to different environments. Even though a national recording program to produce improved young rams has been established since the sixties, sheep productivity is still low. The objectives of this study were: (1) to identify main sources of variation of the different growth traits; (2) compute adjustment factors for non genetic effects; and (3) analyse the selection scheme to produce improved rams. A total of 117,522 Barbarine lambs; enrolled in the National Sheep Recording Program during the period 2004-2014 was used in this study. A linear model including Flock ownership type (State or Private), Farm-Flock-year-Season of lambing, Age of dam and Sex-Type of birth was developed to identify major sources of variation. Least Square Means were used to derive adjustment factors for sex-type of birth and age of dam. Pedigree information was used to identify lambs with known pedigree. Main results showed that average lamb weights at birth, 10 d, 30 d, 70 d, and 90 d of age were respectively 4±0.5; 6±1.4; 9±2.2; 15±3.8 and 18±4.1 kg, respectively. Average lamb weight gains between 0-30 d; 10-30 d; 30-70 d and 30-90 days, were 145±62; 171±60; 157±52 and 148±45 g/d, respectively. New adjustments factors were computed which will help flock owners improve their ability to cull poor producing animals and get better female replacements. Only 0.4% of lambs had their pedigree known and only 254 male lambs are with known pedigree showing that the national sheep improvement program is in need for adjustments to benefit better from the breed genetic diversity available. An alternative breeding scheme is developed and could inspire many developing countries under similar conditions.

Candidate mutation responsible for inherited skeletal developmental abnormalities in Cheviot sheep

O. Matika[1], M. Davey[1], J. Del-Pozo[2], H.A. Finlayson[1], C. Farquharson[1], D.J. Headon[1], J.W. Kijas[3], Z.H. Lu[1], L. McTeir[1], V. Riggio[1], J. Schoenebeck[1], T. Schwarz[2], K.A. Staines[1], J.A. Woolliams[1], A.L. Archibald[1] and S.C. Bishop[1]
[1]*The Roslin Institute and Royal (Dick) School of Veterinary Studies, University of Edinburgh, Easter Bush, EH25 9RG Midlothian, United Kingdom,* [2]*Royal (Dick) School of Veterinary Studies, University of Edinburgh, Easter Bush, EH25 9RG Midlothian, United Kingdom,* [3]*CSIRO Agriculture, St Lucia, QLD 4067, Australia; oswald.matika@roslin.ed.ac.uk*

Achondroplasia in Cheviot sheep, which was first described 45 years ago, is characterised by abnormalities of the head, forelimbs and hooves and surviving affected lambs are generally culled shortly after birth. The condition has been observed in certain flocks and progeny from specific rams and ewes which was indicative of a recessive genetic condition. In addition, cases can be co-twinned to normal lambs which is strong evidence for a genetic mechanism. As the incidence suggested an autosomal recessive mode of inheritance, 9 cases and 27 control samples were scanned for regions of homozygosity with the OvineSNP50 BeadChip and then a subset with the Sheep HD BeadChip. Finally, DNA from a pool of 5 cases and single dams were sequenced. The homozygosity mapping identified a region on chromosome (OAR) 3. Sequencing identified two genes as possible candidates within the region of interest: ADAMTS20 and PRICKLE1 genes. We found a 10 bp deletion in the open reading frame of the PRICKLE1. Knock-outs of the mouse PRICKLE1 gene yield similar phenotypes. Thus, we have identified a strong candidate for the mutation responsible for achondroplasia in Cheviot sheep and have characterised the associated phenotypes in greater detail. DNA tests for the mutation should allow the elimination of this genetic disease.

Genetic parameters for breech strike indicator traits in South African merino sheep

A.J. Scholtz[1], S.W.P. Cloete[1,2], J.B. Van Wyk[3], J.J.E. Cloete[2,4] and A.C.M. Kruger[1]
[1]*Directorate Animal Sciences, P/Bag X1, Elsenburg 7607, South Africa,* [2]*University of Stellenbosch, Department of Animal Sciences, Matieland 7602, South Africa,* [3]*University of the Free State, P.O. Box 339, Bloemfontein 9300, South Africa,* [4]*Elsenburg Agricultural Training Institute, Private Bag X1, Elsenburg 7607, South Africa; schalkc@elsenburg.com*

Blowflies are important external parasites of woollen sheep in South Africa. Animal welfare concerns resulted in the abolishment of the Mules operation as a control measure for breech strike. Challenge-based selection of animals for resistance to flystrike is furthermore not ethically acceptable. Data of a divergent selection experiment of Merino sheep for reproduction were recorded to study indicator traits for breech strike as potential indirect criteria in yearling animals (2004-2012). Traits considered were autumn dag score (ADS), spring dag score (SDS), breech cover score (BCS), crutch cover score (CCS), belly and points score (BPS), breech wrinkle score (BWS), and the diameters of the vertical (VBA) and horizontal (HBA) bare areas around the perineum. The number of records used ranged from 910 (SDS) to 1,854 (BWS). Yearling progeny selected for reproduction had lower ADS and SDS, poorer BPS and longer VBA and HBA dimensions that than their contemporaries selected against reproduction. Heritability estimates amounted to 0.18 for ADS, 0.49 for SDS, 0.21 for BCS, 0.41 for CCS, 0.31 for BPS, 0.47 for BWS, 0.09 for VBA and 0.17 for HBA. ADS was positively related to SDS (0.87), BCS (0.44), BWS (0.51) and HBA (0.48) on the genetic level. BCS was positively related to CCS (0.37) and negatively related to VBA (-0.99) and HBA (0.85), while CCS was positively related to BPS (0.67), BWS (0.50) and negatively to VBA (-0.72). The genetic correlation of BPS with BWS amounted to 0.38 and that of VBA with HBA 0.68. Genetic correlations indicated that selection for reduced BCS and CCS will result in larger dimensions for VBA and HBA. The former subjective scores may thus be used as a proxy for the more time-consuming physical measurements of the bare area. Selection for lower scores for CCS and BWS may compromise BPS, which represents unfavourable correlations. Indirect selection for breech strike indicator traits seems feasible.

Breeding value estimates of fertility traits in Swiss Red Engadine sheep

A. Burren[1], C. Gazzarin[2], K. Steiner[2], A. Spengler[2] and H. Joerg[1]
[1]Bern University of Applied Sciences, School of Agricultural, Forest and Food Sciences HAFL, Länggasse 85, 3052 Zollikofen, Switzerland, [2]Schweizer Engadinerschaf Zuchtverein SEZ, Rietstrasse 17, 9016 St. Gallen, Switzerland; hannes.joerg@bfh.ch

In 2015 breeding value estimates of fertility traits were introduced for the Red Engadine sheep, following the same principle as for the other four Swiss sheep breeds namely White Alpine, Black-Brown Mountain, Valais Blacknose, and Brown Headed Meat sheep. We selected the following four traits to assess the breeding value (BV) using REML and BLUP methods: age at first lambing (BV1, n=5,913), lambing interval (BV2, n=4,597), litter size one (BV3, n=6,427) and litter size two (BV4, n=4,794). The last two traits are the number of lambs at the first and second lambings. The pedigree data comprised 39,621 sheep. The statistical model accounted for fixed effects of herd×year of the first lambing (BV1, 2, 3 and 4), season of the first lambing (BV1 and 3) and random effects of animals and residuals (BV1, 2, 3 and 4). Estimated heritabilities were 0.32, 0.08, 0.19 and 0.15 for BV1, BV2, BV3 and BV4, respectively. The phenotypic and genetic correlations varied among traits. Estimated values for these parameters were 0.17/0.84 (BV3-BV4), 0.17/0.18 (BV1-BV3), 0.02/0.10 (BV2-BV3), 0.08/0.29 (BV1-BV4), 0.07/0.22 (BV2-BV4) and 0.00/0.20 (BV2-BV1). These estimates correspond to the values known for the other four Swiss sheep breeds and results presented in the literature for similar traits. Litter sizes of lambings three and above are highly correlated with litter size two and were therefore not used directly for breeding value estimates. The average breeding values per year of birth reflect the varying genetic trends followed by the four traits. These trends provide information on the selection practised within a breed. In the case of the Red Engadine sheep, the estimated breeding values for age at first lambing, lambing interval, litter size one and litter size two are well suited to improve the fertility of the breed.

Association of the ACAA2 gene with dairy traits in Chios sheep

S. Symeou[1], A. Hager-Theodorides[2], A. Psifidi[3], O. Tzamaloukas[1], G. Banos[4] and D. Miltiadou[1]
[1]Cyprus University of Technology, Department of Agricultural Sciences, Biotechnology and Food Science, P.O. Box 50329, 3603 Limassol, Cyprus, [2]Agricultural University of Athens, Faculty of Animal Science and Aquaculture, Iera Odos 75, 11855 Athens, Greece, [3]The Roslin Institute, Easter Bush, EH25 9RG Midlothian, Scotland, United Kingdom, [4]SRUC, Animal & Veterinary Sciences, Roslin Institute Building, Easter Bush, EH25 9RG Midlothian, Scotland, United Kingdom; sa.symeou@edu.cut.ac.cy

The Acetyl-CoA acyltransferase 2 catalyses the last step of mitochondrial fatty acid β oxidation, playing a key role in meeting animal energy needs. A SNP (HM537015:g.2982T>C) located in the 3' UTR of the ACAA2 gene has been linked with milk yield in dairy sheep. In this study, the association of the g.2982T>C SNP with milk yield, fat, protein content and prolificacy was studied on 1,374 Chios sheep from 25 flocks in Cyprus and Greece. A cost effective sequencing and a Taqman RT-PCR assay were developed for genotyping all animals. Allele frequencies were similar in Cyprus and Greece with a respective T frequency of 0.54 and 0.60. The mixed model for the association analysis included the fixed effects of flock, lactation by lambing age interaction, year by season of lambing interaction, lactation duration for milk yield, and genotype; animal was fitted as a random effect. The SNP was significantly associated with milk yield in the Cypriot population with an additive effect of 10.61 kg (P<0.01) on first and 7.62 kg (P<0.05) on all lactations, and a positive dominance effect of 13.02 kg (P<0.01) and 9.49 kg (P<0.01), respectively. In contrast, a significant negative dominance effect was revealed in the Greek population amounting to -12.63 kg (P<0.01) on all lactations. The SNP also had a significant additive effect of -0.05 (P<0.01) on protein content, and significant additive (0.06, P<0.05) and dominance (0.09, P<0.05) effects on prolificacy in the Greek and Cypriot populations, respectively. Founder effect putatively accounting for the observed differences are discussed.

Genotype concordance rate between Illumina and Affymetrix ovine genotype panels

D.P. Berry[1], E. Wall[2], A. O'Brien[1], K. McDermott[2], S. Randles[2], T. Pabiou[2], P. Flynn[3], S. Park[4], R. Weld[3], D. Kelly[4] and N. McHugh[1]
[1]Teagasc, Moorepark, Fermoy, Co. Cork, Ireland, [2]Sheep Ireland, Bandon, Co. Cork, Ireland, [3]Weatherby's Ireland, Naas, Co. Kildare, Ireland, [4]Identigen Ltd., Dublin, Dublin, Ireland; donagh.berry@teagasc.ie

The objective was to quantify the concordance between genotypes of 89 sheep, genotyped on both the Illumina and Affymetrix platforms. DNA extracted by a single company was used to generate genotypes on 51,135 biallelic SNPs using the Illumina platform. A custom Affymetrix was developed to mimic the Illumina panel and genotyping was undertaken by a separate service provider using the same DNA; 67 of the SNPs on the Illumina panel were not present on the Affymetrix panel. Furthermore alternative probe sets for 7,827 of the Illumina SNPs were included on the Affymetrix panel; only the probe that had the greatest concordance with the Illumina genotype was retained. The mean call rate per individual for the Affymetrix genotype (0.970) was lower ($P<0.001$) than the mean call rate per individual on the Illumina platform (0.989). No strong obvious relationship was evident between the individual animal call rates on both platforms. A total of 771 SNPs (i.e. 1.51%) had a call rate of <0.90 on the Illumina platform with 4,649 SNPs (i.e. 9.10%) having a call rate <0.90 on the Affymetrix platform; 147 of the 771 SNPs with a poor call rate on the Illumina platform had a call rate of <0.90 on the Affymetrix platform. There was almost a 1 percentage unit lower mean SNP call rate from the Illumina platform for SNPs with a call rate <0.90 on the Affymetrix platform compared to SNPs with a call rate ≥0.90 on the Affymetrix platform. Across all individuals, the genotype (allele) concordance rate between both platforms was 0.964 (0.981). Across all SNPs, the genotype (allele) concordance between both platforms was 0.961 (0.980). Of the 745 SNPs that had a concordance rate <0.50, 91 (i.e. 12%) had a GC score of <0.55 which is the threshold commonly used in association studies. Excluding the SNPs with a concordance rate of <0.80; the mean genotype (allele) concordance rate per SNP for the remaining 48,326 SNPs was 0.982 (0.991). It is not possible, however, to determine which platform generated the true genotype. Good concordance exists between genotypes generated using either platform.

Runs of homozygosity profiles in five sheep breeds

D.C. Purfield[1], E. Wall[2] and D.P. Berry[1]
[1]Animal & Grassland Research and Innovation Centre, Teagasc, Moorepark, Fermoy, Co. Cork, Ireland, [2]Sheep Ireland, Bandon, Co. Cork, Ireland; deirdre.purfield@teagasc.ie

Runs of homozygosity (ROH) defined as contiguous lengths of homozygous genotypes are a common feature of mammalian genomes. Because inbreeding increases autozygosity of loci, ROH are now widely used as a predictor of whole genome inbreeding levels and also in the identification of genomic regions subjected to recent and/or ancient selective pressure. The objective of this study was to profile ROH in five sheep breeds: Belclare, Charollais, Suffolk, Texel and Vendeen. A total of 2,364 Illumina OvineSNP50 genotypes were available. Single nucleotide polymorphism (SNP) edits applied included the removal of animals with genotype call rates <95%, genotypes on the sex chromosomes, with a call rate <95%, a minor allele frequency of <0.01, or that deviated from Hardy Weinberg equilibrium ($p<1\times10^{-16}$). After edits, 39,487 SNPs from 2,320 animals remained. ROH were defined using a sliding window approach, but allowing for up to one heterozygous SNP and a maximum of two missing SNPs per window. Short ROH (i.e. ROH length <10 Mb) were common across all breeds but to varying frequencies. The Suffolk breed had a greater mean proportion of their genome (185.02 Mb), covered in short ROH; mean coverage per breed for the other breeds ranged from 76.98 Mb to 118.77 Mb. Long ROH (i.e. ROH length >20 Mb), which are indicative of recent consanguineous matings, were not a common feature of the ovine genome; 1.68% of animals exhibited long stretches of homozygosity. The proportion of the autosome per individual covered in ROH (i.e. a measure of homozygosity) ranged from 0.00 to 0.24. The most homozygous chromosome across all breeds was OAR16 with, on average, 7.05% of the chromosome covered in ROH. A genomic region on OAR2 at 132-134 Mb in the Belclare and Texel breeds was frequently found in ROH in 31.88% and 50.80% of the populations, respectively. Genes identified in this region include MTX2, KIAA1715, ATF2 and CHN1.

Genomic selection in Romanian indigenous sheep breeds: preliminary results

S.Z. Kusza[1,2], D.E. Ilie[1], M. Sauer[1], I. Patras[1] and D. Gavojdian[1]
[1]*Research and Development Station for Sheep and Goats Caransebes, Drumul Resitei km 2, 325400, Caransebes, Romania,* [2]*University of Debrecen, Böszörményi str 138., 4032 Debrecen, Hungary; kusza@agr.unideb.hu*

Aim of the current preliminary study was to identify the base for the genetic polymorphism of β-casein gene in two indigenous dairy sheep breeds belonging to the South-Eastern European Zackel group (Racka and Turcana), through the use of TaqMan genotyping assay, in order to provide information for future selection schemes of the Romanian indigenous sheep breeds reared for milk, with the ultimate purpose to produce hypoallergenic sheep milk and derived dairy products in order to meet new market demands. The polymorphism of the beta-casein gene in the two local sheep breeds (Racka n=98 and Turcana n=111, respectively) was identified. Genomic DNA was extracted from hair follicles and beta-casein genotypes were determined by the rapid TaqMan (Applied Biosystems, USA) genotyping assay. Homozygote genotypes GG were not detected in any of the two studied populations. In both Racka and Turcana breeds, the A variant had the highest frequency, 0.98% and 0.97%, respectively. The most frequent genotype was AA for 96.94% in Racka and 94.59% in Turcana, followed by AG, with an incidence of 3.06% and 5.41%, respectively. Genotype GG was not found in the studied breeds. In the current study, the fast DNA tests for genotyping ovine CSN2 was successfully optimized, however, further samples and correlations of genomic results with milk quality characteristics and production data are needed for the development of future breeding schemes of the Romanian indigenous dairy and dual-purpose sheep breeds.

Genomic study of horn morphology in Italian goat populations

A. Talenti, G. Rota, S. Frattini, B. Coizet, G. Minozzi, G. Pagnacco and P. Crepaldi
University of Milan, Via Celoria 10, 20133, Milan, Italy; giulietta.minozzi@unimi.it

Horn shape is a morphological trait showing high phenotypic variability in goat. The genetic basis of this morphological variability is still poorly described. The purpose of this study was to deepen the genetic basis of horn shape in Italian goat populations. Studied samples included 503 animals of 16 populations, classified in 5 major clusters of breeds base on their horn shape. Data were quality checked for minor allele frequency (MAF>0.1%), genotype and individual call rate (5% and 10%, respectively). To avoid dimorphism-related bias, only females were retained in the working dataset. Marker selection was performed using a combination of LOSITAN and case-control GWAs on pairwise comparison among groups. Only markers monomorphic in at least one out of five clusters were retained. Finally, panel were thinned retaining only markers showing a significant difference in genotype frequencies comparison among groups. After final panel thinning procedure, markers showing a relevant difference in allele frequencies in at least two groups were analyzed, and genes close to the markers (+ 100 Kb) were identified. After data quality check and male exclusion, the working dataset included information of 50,989 markers and 326 female goats belonging to 12 breeds. After combined LOSITAN/GWAs selection, followed by monomorphic analysis, only 102 SNPs were retained. This panel was further thinned to 66 markers showing a significant difference in genotype frequencies in at least one comparison among groups. Of these, 3 showed a significant and biologically relevant difference (>0.8) in minor allele frequencies in at least two groups. Three genes (RXFP2, KIT and RAP2B) were identified close to the 3 significant markers, RXFP2 is particularly interesting due to its role in sheep polledness, KIT is involved in the migration of neural crest cell and RAP2B, belonging to RAS family involved in the activation of epidermal and fibroblast growth factor receptor. Future work will be focused on the validation of these results in wider scenario.

Influence of non-genetic factors in growth and reproductive traits of Santa Ines sheep

R.L. Aguirre[1,2], J.P. Eler[2], E.C. Mattos[2] and J.B.S. Ferraz[2]
[1]Universidad Nacional de Loja, Area Agropecuaria y de Recursos Naturales Renovables, Av. Pio Jaramillo Alvarado y Reinando Espinosa, La Argelia, Loja, Ecuador, [2]University of Sao Paulo/FZEA, ZMV/NAP-GMABT, Rua Duque de Caxias Norte, 225, Campus da USP, 13635-970 Pirassununga, SP, Brazil; jbferraz@usp.br

The objective of this study was to investigate the influence of some non-genetic factors on growth and reproductive performance of Santa Ines sheep, a meat hair sheep breed, very adapted to tropical and subtropical environments, raised in different regions of Brazil. Records from 13.186 animals belonging to 33 farmers from 11 states of Brazil were analyzed. The reproductive characteristics analyzed were age at first lambing, lambing interval (LI), prolificacy, female productivity, besides the growth traits weights at birth, 60 days, 180 days and 270 days of age. The non-genetic effects considered were sex, year (2003-2014), season (dry and rainy), herd, region, type of parturition (single or twins) and maternal age were included in the evaluation as fixed effects. The average age at first lambing was 22 ± 1.5 months, suffering influence ($P<0.05$) of herd and region. The lambing interval was 363 ± 160 days and was significantly affected by year, season, type of lambing, herd, region and ewe's age. The average of prolificacy was of 1.36 lambs/birth, being lower in primiparous ewes (1.2 ± 0.04) than in ewes with two or more parities (1.4 ± 0.03) and it was influenced by the herd, region and season while ewe productivity at an average age at 87.8 ± 32.8 months, in the entire productive life was 3.2 ± 0.5 lambs/ewe. Birth weight and weights at 60, 180 and 270 days in this population had averages of 3.64 ± 0.8, 14.6 ± 5, 31.7 ± 10.7 and 38.5 ± 13.7 kg respectively. Birth weight was significantly influenced by sex, ewe's age, type of lambing, herd, region and year of birth. The other evaluated weights were affected by sex, type of lambing, herd and region. All information analyzed lead to the conclusion that non-genetic factors affect growth and reproductive performance of Santa Ines ewes and must be included in genetic models analysis.

Evaluation of the inbreeding effects in a small sheep population and reducing it by rotating rams

H. Khalfaoui and M. Djemali
INAT, Animal Genetic and Feed Resources Lab, 43 Avenue Charles Nicole, Mahrajène, 1082, Tunisia; khalfaoui.hajer@yahoo.fr

The majority of sheep owners in developing countries have a small size flocks with almost no access to outside rams. The objectives of this study were to: (1) evaluate the inbreeding effect in a small sheep flock; and (2) propose a ram rotating scheme to keep inbreeding in the flock at a low level. Growth and reproduction traits were recorded in a small size flock of the Balck Thibar breed in Northern Tunisa during 2011-2014. The number of active rams and breeding females was used to evaluate the effective population size (Ne) and the inbreeding rate (ΔF).A linear model including year-month of lambing, lamb sex-type of birth and ram effects were used to monitor lambs growth trends. Main results showed: the Ne and ΔF varied from (Ne=7.7 with ΔF=0.064) in 2013 and (Ne=15.31 and ΔF=0.032) in 2012 showing a real inbreeding case. Active rams and breeding females were 4 and 90; 2 and 70 and 3 and 160, in 2012, 2013 and 2014, respectively. Six groups were formed on an index computed based on the average lambs weaning weight for each ewe. The best producing group will serve the following one and the next one will serve the following one until the 6th Group.

Democratization of genomic tools to improve sheep productivity in developing countries
F. Gazbar, I. Ben Abdallah and M. Djemali
INAT, Animal Genetic and Feed Resources Lab., 43 Charles Nicole 1082 Mahrjène, Tunis, Tunisia, 1082, Tunisia;
fatmagazbar06@hotmail.fr

While the gap is widening between developed and developing countries in the field of classical animal genetics and breeding, genomic tools are becoming the method of choice in developed countries. The objectives of this study were to: (1) demonstrate by a real case how the lack of suitable breeding strategies can be a major threat to autochthonous sheep breeds improvement; and (2) show how genomic tools could be an efficient alternative tool to reverse the decrease of native breed productivity. A total of 54,476 fat tail Barabrine lambs from 77 flocks recorded during the period 2010-2014 were used in this study. Lamb growth traits were evaluated based on an additive linear model including flock ownership type (state or private), flock-season within flock ownership type, age of dam at lambing and lamb's sex-type of birth. An approximate estimation of lamb weights at weaning by 2030 was made. Main results showed that average weights of Barbarine lambs at birth, 30 d., 70 d. and 90 days of age were 4±0.37, 8±1.88, 14±3.04 and 16±3.5 kg, respectively. Average daily gains between 10-30, 30-70 and 30-90 d were 138±51; 135±44 and 132±41 g/d, respectively. These performances were lower than results reported in previous studies for the same breed since the sixties until 2000 indicating the negative effect due to a lack of reliable strategies to genetically manage the breed. If the situation continues without change, estimated lamb weaning weight will decrease by 200 g/year to be 13 kg by 2030 (R^2=0.85). The paper presented different uses and conditions to allow genomic tools to reverse the decreasing trend if they become democratized, their cost is lowered and the technology is transferred to low input production systems.

Novel phenotype management using molecular characterization in Prolific sheep 'W' line in Tunisia
S. Bedhiaf-Romdhani[1], B. Jemmali[2] and Y. Ben Sassi[3]
[1]INRA-Tunisia, PAF, Rue Hédi Karray, 1004, El Menzeh, Tunisia, [2]ESAM, Mateur, Mateur, Tunisia, [3]INAT, Av Charles Nicolles, 1082, Tunisia; bedhiaf.sonia@gmail.com

The nucleus of the prolific Barbarine sheep 'W' line was raised in INRA-Tunisia (Ouesslatia research station), created in 1979 and managed through 7 mating families to avoid inbreeding. Fertility and prolificacy in 'W' line varied from 82 to 91% and from 141 to 190%, respectively. On the average one ewe produces 1.6 lambs/litter. Average lamb weights were 4.5, 12 and 25 kg at birth, 1 month and at weaning, respectively. Heritability and repeatability estimates of prolificacy were evaluated to be 0.13 and 0.62, respectively. Molecular characterization's results showed that using ISSRs technique, the 'W' line was found genetically distant from all Barbarine ecotypes. Using the PCR-RFLP to detect the FecGH[+] mutation in the Barbarine 'W' line, we obtained a total of 93% of 'W' individuals already indexed prolific were heterozygous to the GDF9 gene and were carrying the mutation (FecGH[+]). Being able at an early age to identify future replacements will help sheep owners to improve the prolificacy of their flocks at lower management and feeding costs. At the national level, this prolific 'W' line could strengthen the national selection nucleus providing improved prolific animals for multipliers flocks helping to cover the base flocks' demand.

Inference of population structure of 16 sheep breeds using SNP genotype data

A.C. O'Brien[1,2], D. Purfield[2], N. McHugh[2], E. Wall[3], T. Pabiou[3], K. McDermott[3], S. Randles[3], S. Fair[1] and D.P. Berry[2]
[1]Department of Life Sciences, UL, Limerick, Ireland, [2]Moorepark Research Center, Fermoy, Co. Cork, Ireland, [3]Sheep Ireland, Bandon, Co. Cork, Ireland; aine.obrien@teagasc.ie

Information on the genetic diversity and population structure of sheep breeds is useful when deciding the most optimal, for example, crossbreeding strategies to improve phenotypic performance by exploiting heterosis. Knowledge of individual animal breed composition can also be useful for breeding programs in composite breeds. Furthermore, such information can also be useful in determining the possible transferability of genomic predictions across breeds. To objective of the present study was to quantify the genetic diversity and population structure of the most predominant sheep breeds used in Ireland. Single nucleotide polymorphism (SNP; n=49,392) data were available on 394 animals representing unrelated animals from the 16 'breeds' Belclare (n=35), Beltex (n=28), Bluefaced Leicester (n=21), Charollais (n=39), Donegal Cheviot (n=19), Wicklow Cheviot (n=20), Easy Care (n=14), Galway (n=30), Kerry Blackface (n=10), Waterford Blackface (n=10), Lleyn (n=39), Rouge (n=25), Suffolk (n=29), Texel (n=34), Vendeen (n=60) and Zwarbtle (n=31). The Bluefaced Leister had the least mean observed heterozygosity (HO=0.251) between individuals within breed while the Kerry Blackface had the greatest mean heterozygosity (HO=0.371). The pair-wise breed fixation index (Fst) ranged from 0.066 (Donegal Cheviot and Wicklow Cheviot) to 0.283 (Bluefaced Leicester and Suffolk).The Fst between the Beltex and Texel was 0.092. The Bluefaced Leicester was most distant from the other breeds with a mean breed pairwise Fst of 0.236. Some animals in the Belclare breed, which is a composite, developed in the 1970, still retained up to 18% Lleyn and up to 25% Texel although no Galway bloodline (based on the 30 animals in the study) was evident in the 35 animals included in the study; the Galway breed was one of the breeds in the original development of the Belclare.

Advancement of Dairying in Austria (ADDA): Farmer surveys on mastitis, DCT, and antibiotic use

C.L. Firth[1], W. Obritzhauser[1], C. Egger-Danner[2] and K. Fuchs[3]
[1]University of Veterinary Medicine, Institute of Veterinary Public Health, Veterinaerplatz 1, 1210 Vienna, Austria, [2]ZuchtData EDV-Dienstleistungen GmbH, Dresdnerstrasse 89/19, 1200 Vienna, Austria, [3]AGES – Austrian Agency for Health and Food Safety, Zinzendorfgasse 27/1, 8010 Graz, Austria; clair.firth@vetmeduni.ac.at

ADDA is a three-year research project covering topics relevant to dairy farming and milk production. The project covers seven research areas, namely: cattle nutrition, reproduction, udder health, milk quality, integrated tools for farm management, reducing the use of antibiotics, and increasing economic efficiency. ADDA is unique in that it combines expertise from the fields of academia, agriculture, commercial dairies, veterinary agencies and government authorities. This study is researching how dairy farm management and antibiotic use affect mastitis incidence. Data are currently being collated from 18 veterinary practices throughout Austria with respect to antibiotic use on 283 dairy farms. In addition, each veterinarian providing data on antibiotic use will be requested to fill in a questionnaire for each farm enrolled in the project. Commercial dairy employees and milk recorders will also be asked to complete a survey, primarily regarding milking technique and hygiene on each of these farms. Furthermore, the farmers themselves are being questioned on general farm management, dry cow therapy, housing systems, mastitis management, etc. As of February 2016, links to surveys had been sent via email to 220 farmers who provided their email address and consented to the study. To date (7th March), 26.0% have completed the questionnaire. Initial results show that 67.2% of these farms have loose housing systems, while 32.8% tether their dairy cows. Blanket dry cow therapy was implemented on 49.1% of farms, with 50.9% preferring to be selective in their use of dry cow udder tubes. Only 10.5% of farmers regularly cultured milk samples from all animals prior to drying off, whereas 89.5% carried out this precaution in cows with current, or a history of, udder problems. Final conclusions regarding the relationship between mastitis, dry cow therapy, antibiotic use and management factors will be made once all the data are available and have been analysed.

Prevalence of health and welfare conditions in growing pigs on a farm with high antibiotic usage

A. Diana[1,2], E.G. Manzanilla[2], J.A. Calderon Diaz[2], N. Leonard[1] and L. Boyle[2]
[1]UCD, School of Veterinary Medicine, Dublin, Ireland, [2]Teagasc, Moorepark, Fermoy, Co. Cork, Ireland; alessia.diana@teagasc.ie

Antibiotics (AB) are used to treat infectious disease but are unlikely to be of benefit to other production diseases and welfare issues where improved management may be more beneficial. The aim of this study was to establish the prevalence of such health and welfare conditions at each production stage on a farm with high AB usage. All pigs born (n=1,050) during 1 wk were weaned at 28 ± 2 d into 20 pens of 55 pigs. Health and welfare checks were conducted on a weekly basis during the 1st and 2nd weaner stages and the first 5 wks of the finisher stage for a total of 15 wks. Pens were assessed for 10 min and the proportion of pigs affected (mean±SE) by these conditions was recorded: tail (TL), ear (EL) and flank lesions (FL), lameness, hernias, poor body condition (PBC), sickness, bursitis, limb injuries and skin disorders. Conditions were ranked according to the 3 most common health and welfare conditions per stage. The most common health condition during the 1st stage was PBC ($3.7\pm0.4\%$) with bursitis 2nd ($1.48\pm0.2\%$) and hernias 3rd ($0.71\pm0.1\%$). In the 2nd stage bursitis was the 1st condition ($8.6\pm1.2\%$) with PBC 2nd ($5.0\pm0.9\%$) and lameness 3rd ($3.2\pm0.7\%$). Bursitis was also the most common condition ($12.2\pm0.8\%$) observed in the finisher stage, followed by lameness ($9.9\pm0.8\%$) and PBC ($8.2\pm0.7\%$). EL ($40.8\pm3.4\%$; $53.8\pm4.2\%$; $12.2\pm1.1\%$, 1st, 2nd stage and finisher respectively) was the most common welfare condition at all stages, followed by TL ($10.8\pm1.5\%$; $33.3\pm3.1\%$; $8.9\pm0.8\%$) and FL ($0.15\pm0.06\%$; $0.56\pm0.1\%$; $3.3\pm0.4\%$). In spite of the high level of in-feed AB, indicators of poor health and welfare were detected at all production stages. The link between these indicators, the treated disease and their efficacy as a tool to monitor the effect of corrective practices should be further explored.

Phenotypic and genotypic identification of antibiotic resistance genes of staphylococci isolated fro

R. Saidi
Laghouat University, Agronomy, Department of Agronomy, Telidji Amar University, BP 37G, Ghardaïa Road, 03000 Laghouat, Algeria, 03000, Algeria; saidi.radhwane@yahoo.fr

The study was carried out to investigate the phenotypic and genotypic identification of in vitro antimicrobial susceptibility of 21 Staphylococci (10 Staphylococcus aureus and 11 Coagulase Negative Staphylococci) isolated from bovine mastitis to 12 antimicrobial drugs frequently using in veterinary medicine in Algeria. Isolates of staphylococci from bovine mastitis were tested for antibiotics with disc-diffusion method according to the National Committee for Clinical Laboratory Standards guidelines in the Mueller-Hinton agar, and resistant genes mecA, blaZ, aac-aph, ermA, ermC, tetK and tetM were detected by PCR. Staphylococci isolates showed high resistance to penicillin (95.23%), oxacillin (80.95%), and clindamycine (80.95%). Among 21 isolates of Staphylococci, 20 were found to be methicillin and multidrug resistant. The distribution of antibiotic-resistant genes was mecA (100%), tetM (100) followed by blaZ (42.85%). The finding of methicillin-resistant staphylococci (MRS) from bovine mastitis is the first report in Algeria and revealed the status of resistant isolates in herd that might be helpful in treatment, controlling of resistant strains and for deciding culling of cows.

Antibiotic and synthetic growth promoters in animal diets

M. Gonzalez Ronquillo[1], J.C. Angeles Hernadez[2] and O.A. Castelan Ortega[1]
[1]Universidad Autonoma del Estado de México, Facultad de Medicina Veterinaria y Zootecnia, Departamento de Nutricion Animal, Instituto Literario 100, 50000, Toluca, Mexico, [2]Universidad Nacional Autonoma de México, Facultad de Medicina Veterinaria y Zootecnia, Programa de Maestría y Doctorado en Ciencias de la Producción y de la Salud Animal, Circuito Exterior, Ciudad Universitaria, 04510, Delegacion Coyoacan, Mexico; mrg@uaemex.mx

Food quality and safety have been a significant and pressing issue in recent years. In light of the FAO's definition of food security – the physical, social and economic access to sufficient and nutritious food – food safety plays a fundamental role. Animal feed and feeding is pivotal to the livestock industry, but the use of veterinary antibiotics (VAs) and synthetic growth promoters (SGP) diminishes the sustainability of the diets and can cause an accumulation of residues in animals (meat, milk and eggs) and the environment (water and soil pollution). Wastewater systems are another major pathway through which antibiotics and hormones can enter the environment, with negative consequences. Recent modeling suggests that the global consumption of antimicrobials will increase by 67% in 2030, mainly for China, USA, Brazil, India, and Mexico. Antimicrobial consumption has a heterogenic geographic distribution, depending on the legislation in each continent and country, level of industrialization of animal production and characteristics of the market for food animal products. Analytical methods play a crucial role in food analysis, to determine the presence of antibiotics and other additives. In general, a monitoring program put in place to educate the population on the hazards of residues in animal products is necessary, in conjunction with a continuous decrease in the use of antibiotics and synthetic growth promoters in animal diets.

Establishing a monitoring system for the use of antibiotics at animal level in Austria

W. Obritzhauser[1], M. Mayerhofer[2], C. Firth[1], K. Fuchs[3] and C. Egger-Danner[2]
[1]University of Veterinary Medicine, Veterinärplatz 1, 1210 Vienna, Austria, [2]ZuchtData, Dresdner Straße 89/19, 1200 Vienna, Austria, [3]Austrian Agency for Health and Food Safety, Zinzendorfgasse 27/1, 8010 Graz, Austria; w.obritzhauser@dairyvet.at

Animal welfare and food safety, particularly with respect to the use of antibiotics on farm, are increasingly important to consumers. The occurance of antibiotics resistant bacteria in humans is much discussed and the reasons for increasing resistance are unclear. An excessive use of antibiotics in animal production (e.g. growth promoters, prophylactic use) is suspected to be one of the main causes. In Austria, surveillance systems on antimicrobial consumption in veterinary medicine have been put in place. Within the Austrian central cattle database, a voluntary system of recording diagnoses has been set up earlier, presently the system is extended to monitor medicinal products used at animal level. As a precursor for monitoring drug use, a standardised list of licensed pharmaceuticals has been prepared by the Austrian medical authority. An observational study is being performed to investigate mastitis and antibiotic use on 280 dairy farms in Austria. Data are provided by the attending veterinarians. Bacteriological analyses are carried out on milk samples from all cows treated for udder disease. Isolates of mastitis pathogens, as well as commensal bacteria, are tested for resistance. Detailed information is also being collected via questionnaires, which are sent to farmers, veterinarians and commercial dairy consultants. The experiences gained by the farmers and veterinarians in this study will be used to implement a routine recording system for drug use on Austrian dairy farms. Herd level data will allow an evaluation of management practices regarding the use of antibiotics on farm. Such detailed information on the current situation of antimicrobial use in livestock production is vital for the development of the One Health guidelines to reduce antimicrobial resistance, affecting both humans and animals.

Associations between mycobacterial purified protein derivative hypersensitivity and MAP ELISA

A.E. Kennedy[1,2], N. Byrne[2], J. O'Mahony[1] and R.G. Sayers[2]
[1] Cork Institute of Technology, Department of Biological Sciences, Cork, Ireland, [2]Teagasc, Animal & Bioscience Research Department, Moorepark, Fermoy, Co Cork, Ireland; aideen.kennedy@teagasc.ie

Intradermal testing, involving administration of purified protein derivative (PPD) to elicit a delayed hypersensitivity (DTH) response, is used as a diagnostic tool for bovine tuberculosis (bTB). The single intradermal cervical comparative test (SICCT) involving the use of avian and bovine PPD is administered at least annually to every bovine over six weeks of age in the Republic of Ireland. It has been suggested that administration of PPD may allow identification of bovines infected with Mycobacterium avium subspecies paratuberculosis (MAP), the causative agent of Johne's disease (JD). The aim of this study was to investigate if animals showing DTH reactions to PPD following SICCT had an associated increase in MAP ELISA response, thereby identifying potential cases of sub-clinical JD. During mandatory annual SICCT of a 139-cow dairy herd, skin thickness measurements (mm) were recorded at the site of avian and bovine PPD administration. Cows were assigned to DTH categories based on increases in skin thickness measurement at 72 hours post-PPD administration. Blood samples were also collected pre and post-SICCT (day 10, 16, 42, 57 & 91) for ELISA testing. Linear regression was performed to identify associations between DTH measurements and MAP ELISA results with logistic regression used to analyse categorical variables (ELISA positive vs negative and DTH categorisations). Significant associations were identified between increased skin thickness due to PPD and MAP ELISA readings, most notably on day 10 post-PPD, when a 1 mm skin thickness increase at the site of bovine PPD administration was associated with a MAP ELISA increase of 10.5 S/P (P=0.012). Animals with DTH responses at both avian and bovine PPD sites were most likely to test ELISA positive in the post-PPD period relative to other categories. Whether this indicates improved sensitivity at identifying subclinical MAP positive cows or represents decreased test specificity due to cross reacting mycobacterial antigens is unknown. Further research is required to identify whether skin thickness increases post-PPD can identify animals that have been previously exposed to MAP.

Managing antibiotic use in the Netherlands

A. Kuipers[1] and H. Wemmenhove[2]
[1]Expertise Centre for Farm Management and Knowledge Transfer WUR, P.O. box 35, 6700 AA Wageningen, the Netherlands, [2]Livestock Research WUR, P.O. box 338, 6700 AH Wageningen, the Netherlands; abele.kuipers@wur.nl

Antibiotics use in animals has become part of the societal discussion. National goal was to reduce use by 50% in 2013 compared to 2009. Herd health and treatment plans have been introduced. From 2012 on, 3/4 generation drugs (3/4GE) are only allowed in exceptional cases. Selective dry cow treatment was introduced in 2014 as a recommended practice based on the cell count of cows. Since 2014, also veterinarians are classified according to total quantity of drugs applied. To gain insight, antibiotic use and farm factors were examined on 94 dairy farms during 2005-2012. The level of use (in no. of daily dosages) was rising in period 2005-2007, followed by a period of growing societal interest in animal antibiotic use resulting in a reduction in use in 2011-2012. On average, 68% was applied to the udder: mastitis 25% and dry-cow therapy 43%. Drugs other than applied to udder health tended to decrease the most, while farmers were reluctant to lower use of dry-cow therapy tubes. Use of 3/4GE minimized from 18% of NDD in period 2005-2010 to 1% in 2012. The drop in use varied between 3 groups of farmers studied of which one was guided for some years. A logistic function fitted nicely to the adaptation process. Farm and herd factors affecting antibiotics use were studied. Especially use for dry cow therapy could be well explained by health status, cell count and calving interval. The 'more successful' farmers seemed to use somewhat more antibiotics than the other colleagues. They were also able to adapt more easily to the new conditions. Since 2012, all farms are monitored by the Veterinary Medicines Authority (SDa). Nationally, a 58% reduction was achieved through 2014. However, the decrease levelled off in recent years. The dairy sector reduced less than the pig and poultry sectors. Also insight in variation in use by veterinarians will be given. SDa experts pinpointed at the relation between less use and a reduction in prevalence of antibiotic resistant E. Coli bacteria.

Chicken embryo lethality assay for determining the virulence of Enterococcus faecalis

A.E. Blanco[1], M. Barz[1], W. Icken[1], D. Cavero[1], A.R. Sharifi[2] and M. Voss[1]
[1]Lohmann Tierzucht GmbH, Am Seedeich 9-11, 27472 Cuxhaven, Germany, [2]Animal Breeding and Genetics Group,
Georg-August-University Goetting, Animal Sciences, Wilhelmsplatz 1, 37073 Goettingen, Germany; blanco@ltz.de

Enterococcus faecalis (E. faecalis) is the major pathogen found in field cases of amyloid arthropathy (AA) in chickens, resulting in a great negative impact on pullet growth. Consequently, the Embryo Lethality Assay (ELA) is proposed in the present study as a model to evaluate the virulence of E. faecalis strains, specifically the pathogenic avian strain K923/96 which has been used in former studies related with AA and which will be used as the reference strain in further analysis. The ELA was repeated four times on a sample of 3,443 eggs of White Layers in total. 0.2 ml of five different doses of the same strain, varying from 2.5 to 2,500 colony-forming units (cfu) per ml, were inoculated into the allantoic cavity of 10-day old embryos. Additionally, one control group was included in each trial, which was inoculated with 0.2 ml of sterile phosphate buffered saline. The mortality rate of the embryos was determined by candling the eggs over a period of 7 days. Statistical analysis of mortality data were carried out using a survival model as well as by applying a linear logistic model for repeated measurements with a binary response variable, which was modeled as a binomial random variable (y_i). The infectious dose showed a significant effect on the embryonic mortality rate. The estimated average mortality rate of the control group was low (3%) and differed significantly from all other doses. The embryonic mortality rate with the doses of 2.5, 5, 25, 250 and 2,500 cfu/ml was 41, 46, 66, 89 and 92% respectively. As expected, the higher the infectious dose, the greater the embryo mortality rate and the lower the embryonic survival time. The highest embryonic mortality rate was recorded 2 and 3 days post inoculation in all doses and trials. The results suggest that the ELA can be used as a rapid and reliable tool to establish the pathogenicity of E. faecalis strains dependent on the doses. In order to corroborate its capacity to predict the virulence of this bacterium, more ELA with different E. faecalis strains are required.

On farm risks associated with the prevalence of resistant strains of Escherichia Coli: A pilot study

A. Xexaki[1], E. Sossidou[2], G. Filioussis[1] and E. Petridou[1]
[1]Aristotle University of Thessaloniki, Faculty of Veterinary Medicine, Laboratory of Microbiology and Infectious Diseases, University Campus, 54124 Thessaloniki, Greece, [2]Hellenic Agricultural Organization-DEMETER, Veterinary Research Institute, NAGREF Campus, 57001 Thessaloniki, Greece; sossidou.arig@nagref.gr

The aim of this pilot study was to investigate the prevalence of resistant strains of Escherichia coli in broiler flocks as related to on farm risk factors. The study was carried out on 132 broiler houses, randomly selected from farms sited in regions which represent more than 70% of the Greek broiler production. A structured questionnaire was used to collect data by personal interviews with the farm manager, including biosecurity measures, harvesting system, feeding and watering method, staff, microclimate in broiler house, antimicrobial therapy and information about the establishment and equipment of the farm as well as the genotype, age and population of the flock. Two samples per house of fecal swabs were collected from the floor, then one Escherichia coli strain per sample, isolated from TBX medium, was analysed using the Kirby-Bauer Disk Diffusion Susceptibility Test for 16 antimicrobials. Preliminary results derived from 71 samples were statistically related with 6 risk factors i.e. bedding material, season, type of drinkers, flock genotype, hatchery and site of the farm. The risk for the prevalence of resistant strains of Escherichia coli strains was significantly (P≤0.05) increased in winter/spring, when rice hulls is used as bedding material and when nipple drinkers with tray is used as drinking equipment. More specifically, the high prevalence of resistance in chloramphenicol (84.6%) and in nalidixic acid (92.3%) was significantly (P≤0.05) associated with the use of rice hulls. Moreover, the high prevalence in nalidixic acid (100%) and in chloramphenicol (91.7%) was significantly related to winter and spring season, respectively. Finally, when nipple drinkers with tray are used, the prevalence in nalidixic acid (100%) was significantly (P≤0.05) higher compared with drinkers without trays.

Effects of antibiotic prophylaxis on the gut microbiome and performance of growing broiler chicks

A.A. Tonks[1], M.J. Woodward[2] and C. Rymer[1]
[1]Animal, Dairy and Food Chain Science, School of Agriculture, Policy and Development, University of Reading, Reading, RG6 6AR, United Kingdom, [2]Food Nutritional Science, School of Chemistry, Food and Pharmacy, University of Reading, Reading, RG6 6AP, United Kingdom; a.tonks@pgr.reading.ac.uk

The routine administration of antibiotic prophylaxis to broiler chicks has been known to alter the development and composition of a healthy gut microbiome with consequences on health and performance. This study aimed to characterise the effects of Linco-spectin on the caecal microbiota of Ross 301 broilers and investigate the subsequent influence on live weight (LW) and feed conversion ratio (FCR). Linco-spectin was administered in drinking water from 1-3 days of age and the caecal microbiome was characterised through Roche 454 DNA sequencing at 4, 9, 15 and 29 days (D) of age. Birds were weighed at each sampling point with feed weighed at D15, 22 and 20 for the calculation of FCR. The relative abundance of each bacterial phylum, class and order were calculated and statistical analysis was carried out using ANOVA in Minitab 17. Antibiotic-treated birds (AB) weighed less than control (CON) birds at every time point. Statistical significance was noted at: D4 ($P=0.002$) where Bacillales was absent from CON; D15 ($P=0.10$) where Lactobacillales were more abundant ($P=0.025$) and Enterobacteriales less abundant ($P=0.026$) in AB; D22 ($P=0.021$) though DNA was not analysed at this point. FCR was similar at D15 and D29 but 0.1 higher in AB at D22 – though not at any level of statistical significance ($P>0.1$), this would have a significant economic impact. At D29, Clostridiales and Burkholderiales were significantly more abundant in AB ($P=0.015$ and $P=0.005$ respectively), though this had no effect on performance. These differences show that Linco-spectin significantly affects the gut microbiome of growing broiler chicks and antimicrobial prophylaxis can influence bird performance, perhaps through the modulation of biochemical pathways, immune responses and gut development. Further investigations into the effects on Linco-spectin on the gut metabolome will be carried out through 1H NMR.

Comparing agricultural and urban nutrient loads to coastal systems

J.P. Nunes and J. Gomes Ferreira
CESAM & Dept. Environment and Planning, University of Aveiro, Campus Universitário de Santiago, 3810-193 Aveiro, Portugal; jpcn@ua.pt

Terrestrial nutrient loads interact with aquaculture, both supporting primary production and disrupting water quality. Nutrient management is essential to achieve the objectives of the EU's Water Framework Directive, and normally focuses on urban wastewater loads; however, diminishing returns from wastewater treatment has shifted the focus towards agricultural exports, spatially dispersed and irregular in time, complicating quantification and management. This work analysed urban and agricultural loads to two coastal systems with important aquaculture activities, Lough Foyle in northern Ireland and the Ria Formosa in southern Portugal. Watershed modelling was combined with measurements to quantify long-term nutrient loads from agriculture and wastewater, identify sources, and analyse temporal patterns. Results indicate that agricultural loads are more important than wastewater loads. The most common landuses are the major sources even with lower per area exports. Agricultural loads show seasonal and inter-annual patterns driven by rainfall, becoming concentrated in shorter periods than wastewater discharges, particularly in the drier Portuguese system. Aquifer contamination could continue to release agricultural after fertilization is controlled. This highlights that managing agricultural nutrients is more complex than wastewater management, requiring solutions for both amounts and timing of loads which can be implemented throughout the agricultural landscape.

Session 10

Theatre 2

Partitioning the faecal load to coastal waters

M. Taylor
AFBI, New Forge, Belfast, United Kingdom; malcolm.taylor@afbini.gov.uk

Unexpected changes in the classification of commercial shellfish beds can have severe economic consequences. Similarly there is a requirement by regulatory authorities to target scarce resources at the appropriate management measures which will be effective. As point sources of pollution have become under stricter controls the relative contribution of diffuse or non point sources to drops in water quality have become more significant. This paper explores the use in conjunction with traditional sampling programmes of source apportionment by means of Microbial Source Tracking (MST) to enable identification of the most dominant trend of pollution that is impacting on overall water quality. Northern Ireland has a high proportion of land dedicated to farmland and it is recognised that agricultural pollution may severely impact coastal water quality and shellfish status. The use of MST will enable a risk based model that will advise and dictate targeted remediation towards improvement of only those wastewater assets which are directly contributing to poor water quality is undertaken.

Session 10

Theatre 3

Application of a multi-model framework for integrated ecosystem management in lough foyle

J.G. Ferreira, H. Moore, P. Boylan, C. Jordan, J.D. Lencart-Silva, C. McGonigle, S. McLean, J.P. Nunes, M. Service and C. Zhu
Universidade Nova de Lisboa, Dept. Environmental Sciences and Engineering, Faculty of Science and Technology, Monte de Caparica, 2829-516, Portugal; joao@hoomi.com

The Enhanced Application of the SMILE Ecosystem model to Lough Foyle (EASE) project, provides an example of the use of an integrated multi-model framework. Integrated catchment management, and the implementation of legislative instruments such as the Water Framework Directive (2000/60/EC) and Marine Strategy Framework Directive (2008/56/EC) have created a need for management frameworks to be extended beyond coastal systems to integrate catchment processes, especially the drivers of particulate organic, and dissolved nutrient discharge to coastal systems. The EASE framework provides an improved understanding not only of the effects of nutrient control on shellfish production, but on the analysis of the role of multiple uses (and sources), which may include finfish cage culture. This approach may be used to analyse changes to the use of the catchment, and how those changes affect primary and secondary production within Lough Foyle. The linkage of estuarine dynamics with mesoscale shelf processes also helps managers understand the extent to which particular activities may be managed. EASE has an additional focus on ecosystem services, with particularly on top-down eutrophication control by bivalves, and their potential role in watershed-scale nutrient credit trading. By providing standalone solutions e.g. for catchment or for shelf exchanges, managers can use components of the EASE framework to address specific issues, while relying on the combination of these to provide integration as required.

The interaction between natural benthic communities and cultivated blue mussels in Belfast Lough
H. Moore, A. Boyd, R. Corner, J.G. Ferreira and M. Service
Agri-Food and Biosciences Institute, Fisheries & Ecosystems Branch, New Forge, Belfast, United Kingdom;
heather.moore@afbini.gov.uk

Belfast Lough is one of the main areas for bottom cultivation of mussels in Northern Ireland. Shellfish act as a strong environmental biofilter, and can play an important role in moderating phytoplankton concentration, therefore reducing the potential for eutrophic conditions to develop. In 2013 the European Commission published its Strategic Guidelines for the sustainable development of EU aquaculture which states that aquaculture development must be balanced against improving the EU water environment, controlled legislatively through the Water Framework Directive (WFD) and the Marine Strategy Framework Directive (MSFD). This paper presents a modelling approach to support decision-making for Belfast Lough in the context of the WFD and MSFD, with specific focus on (1) Interaction between natural benthic communities and cultivated shellfish (blue mussel, Mytilus edulis) and (2) Effects of the above components on primary production and phytoplankton biomass, particularly with respect to top-down control of eutrophication. Shellfish aquaculture within Belfast Lough was shown to provide an ecosystem service while simultaneously contributing to local employment and food security.

Achieving a balance – the key to sustainably managing native oyster stocks
C. Bromley
Queen's University Belfast, IBIS Project, University Road, Belfast BT7 1NN, United Kingdom; cassbrom@gmail.com

The story of oyster production is one of 'boom and bust' cycles. Native oysters (Ostrea edulis) once occurred in large populations throughout European waters. Following a long history of wild harvesting and aquaculture, the industrial revolution gave rise to an industry which harvested millions of oysters each year for consumption and re-stocking. Over the past 150 years, factors such as overexploitation, disease, and invasive species have led to massive population declines. In recent years there has been increased interest in regeneration, owing to renewed market demand and growing recognition of the environmental importance of oysters. This presentation will discuss the results of recent work in Lough Foyle, one of the few remaining wild native oyster fisheries in European waters. The project has synthesised knowledge from 150 years of literature, information from today's producers, and data collected from field surveys and experiments to provide advice to managers, producers and restorers. The results show that key to sustainably managing native oyster stocks is to achieve a balance – a balance between commercial and environmental considerations; a balance between stakeholders; and a balance between natural and aquaculture production. Adopting aquaculture techniques has assisted with stock regeneration in the past and could play a role in native oyster restoration today.

Session 10

Theatre 6

Sustainable Shellfish aquaculture the Ecosystem approach

M. Service
AFBI, New Forge, Belfast, United Kingdom; matt.service@afbini.gov.uk

Although shellfish culture is generally regarded as relatively benign concerns have been raised by regulatory authorities across Europe about the potential of these activities to damage or impact the features and species of Natura 2000 sites. Either through: (1) physical presence of aquaculture structures; (2) human presence within the licensed sites; (3) smothering of benthic communities caused by faeces and pseudo faeces; and (4) the removal of potential food sources or the introduction of non native species. This paper will present examples of management practise and tools adopted in Northern Ireland to ensure the sustainable management of Bays and Inlets used for shellfish culture. The Geographic Information Systems (GIS) programme ArcGIS are used to analyse spatially the licensed aquaculture sites in relation to the designated features and species of Special Areas of Conservation (SACs) and Special Protection Areas (SPAs). Carrying capacity modelling and benthic deposition modelling were used to determine the level of impact of the current licensed aquaculture sites within Strangford Lough on the designated features of the SAC, which includes the biogenic reef feature. Finally proposals for Bay Scale and Individual Farm Scale monitoring will be discussed.

Session 11

Theatre 1

Disproportionate growth in farm animals – prerequisite for a disturbed health

K. Huber
Institute of Animal Science, Faculty of Agricultural Sciences, University of Hohenheim, Anatomy and Physiology of Farm Animals, Fruwirthstrasse 35, 70599 Stuttgart-Hohenheim, Germany; korinna.huber@uni-hohenheim.de

Growth of organisms is a concerted symmetric development of all organs and tissues to fulfill physiological functions during ontogeny and in adult life. If a tissue or organ is not able to grow proportionally, this can be defined as asymmetric or disproportionate growth (DG) with potentially detrimental health consequences for the organism. Prenatal challenges regarding nutritional and oxygen supply and high metabolic performance early in postnatal life might be main reasons for developing DG. In this paper, indications for DG in high performance farm animals will be reviewed. Most of our farm animals are used early in their ontogeny for production of foodstuff such as milk, eggs and meat. In dairy cows, an early first calving age is pursued. Calving at 24 month of age is accompanied by immature mammary gland tissue. Furthermore, since heifers need to allot their metabolic energy not only to milk production but also to their own growth, this may lead to less developed organs and tissues. Immature tissues and cells are overburdened when requested for high metabolic performance, thereby developing mitochondrial and endoplasmatic reticulum stress. These cellular mechanisms are able to generate a generalized pro-inflammatory situation and to shorten productive lifes due to occurrence of production diseases. Similarly to the dairy cow, high-performing laying hens were genetically forced to lay eggs at early age. Important cellular structures were not fully developed at this age leading to severe health issues. Fatty liver hemorrhagic syndrome is a well-known production disease in laying hens which are lacking collagen type III fibres in the liver and thereby, liver tissue integrity. In growing broilers, breast muscle growth is disproportionately higher compared to other body tissues. Tibia dyschondroplasia is a well-known production disease in broilers which are lacking capillary spreading in the metaphysis of their bones. Thus, the remodelling of hyaline cartilage to mineralized bone tissue is not possible, leading to immature bones and joints.

Evidence for developmental programming in dairy cattle
G. Opsomer, M. Van Eetvelde and M.M. Kamal
Faculty of Veterinary Medicine (Ghent University), Department of Reproduction, Obstetrics and Herd Health,
Salisburylaan 133, 9820 Merlbeke, Belgium; mieke.vaneetvelde@ugent.be

We hypothesized that (heavy) lactation during gestation in modern dairy cows affects growth and development of the calf, potentially affecting its health in later life. We have performed several epidemiological studies to test this hypothesis. In a first study in which 1,594 Holstein Fresian calves born out of primi (n=540) and multiparous (n=1,054) dams were included, we demonstrated that multiparous dams produced on average 6,193.1 (±1,352.8) kg milk during gestation. Weight of the newborn calf was used as a proxy for its intra-uterine development, and was in first parity mothers shown to be significantly associated with the dam's age while with gestational milk yield in multiparous cows. Blood samples of newborn calves were taken to measure insulin levels and to assess insulin resistance of their glucose metabolism. Calves born following a higher gestational yield tended to be more insulin resistant. Besides gestational milk yield level, also the length of the dry period had a significant effect, with calves born after a longer dry period being heavier and more insulin resistant. Gender of the calf was associated with most of the above mentioned parameters, female calves being lighter and more insulin resistant. In a second study, placentas were collected to count and measure cotyledons as a proxy for surface availability in terms of exchange between mother and calf. Growth and lactation during gestation increased the number of cotyledons, while total cotyledonary surface seemed to be positively influenced by the final nutrient demand of the calf. Remarkably, season of calving was significantly associated with most of the traits examined. Calves born during the warmer months were lighter and less insulin resistant, while concomitant placentas had a significantly larger cotyledonary surface. The latter suggests season of birth potentially being an important characteristic in terms of health and production for dairy cattle. As a conclusion, our data illustrate evidence for developmental programming in dairy cattle. The latter warrants for more fundamental studies to decipher the underlying mechanisms which at their turn may indicate preventive and curative strategies in order to increase life expectancy in dairy cattle.

Delayed development in Large White Purebreds than Crossbreds with Meishan born in the Same Litter
L. Canario[1], M.C. Père[2], H. Quesnel[2], Y. Billon[3], W. Hébrard[3], J. Riquet[1], P. Mormède[1] and L. Liaubet[1]
[1]INRA UMR1388 / INPT ENSAT / INPT ENVT GenPhyse, Animal Genetics, 24 chemin de Borde Rouge, Auzeville, 31326 Castanet-Tolosan, France, [2]INRA, Animal Physiology, UMR1348 Pegase, 35590 Saint-Gilles, France, [3]INRA, Animal Genetics, UE1372 Genesi, 1770 Surgères, France; laurianne.canario@toulouse.inra.fr

A crossbreeding design based on the use of mixed semen from Large White (LW) and Meishan (MS) boars was set up to study differences in late development of purebred and crossbred fetuses developed in the same uterine environment, i.e. either LW or MS. Caesareans were performed at the end of the second gestation to study characteristics of fetuses at 90 d and 110 d of development. A total of 312 fetuses from 9 LW and 10 MS sows were dissected. The four genetic types of piglets were compared according to body, bones and organs traits measured on the experimental farm or from image analyses. The model for statistical comparison included the fixed effect of sex, the interaction between genetic type and age of development, the percent of purebreds in the litter and litter size as covariates, and the sow random effect. Purebred (PB) and crossbred (CB) fetuses differed for body weight and body length at 90 d and 110 d of development in MS sows (P<0.05) but not in LW sows. LW PB were heavier and longer than MS PB at 110 d but not 90 d (P<0.05). The body mass and ponderal indexes of LW PB were lower than those of LW CB at 90 d and 110 d, and of MS PB and MS CB at 90 d only. These indexes were lower in MS CB than MS PB at 90 d but the difference disappeared at 110 d. Proportionally to body length, the femur was the shortest in LW PB, intermediate in the 2 reciprocal CB types, and the longest in MS PB (at 90 d). Proportionally to body weight, liver was the lightest in LW PB, of intermediate weight in the 2 CB types, and the heaviest in MS PB (P<0,05). Ratios of the brain weight above weight of several organs at 90 d were the highest in LW PB, intermediate in the 2 CB types, and the lowest in MS PB, indicative of sparing effects in LW PB. Our findings outlined a delay of late fetal development in LW PB fetuses. The complementary comparison of born alive LW PB that died or survived the first 24 h after birth confirmed that disproportionate development is correlated with lower survival.

Impact of dietary L-arginine supply during early gestation on myofiber development in newborn pigs
J.G. Madsen[1,2] and G. Bee[2]
[1]ETH Zurich, Universitätstrasse 2, 8092 Zürich, Switzerland, [2]Agroscope, Route de la Tioleyre 4, 1725 Posieux,
Switzerland; johannes.madsen@agroscope.admin.ch

Prenatal survival and development of offspring is positively affected when sows with average litter size are fed a gestation diet supplemented with arginine above recommendation. The objective of this study was to determine whether a similar arginine effect can be expected in high prolific sows (litter size >15 pigs born) where intrauterine crowding impairs myofiber hyperplasia. In all, 16 sows were used in the present study: 10 unilateral oviduct ligated (UO; non crowded) and 6 intact hyperprolific sows (IN; crowded). Except from d 14 to 28 of gestation, sows were offered daily 2.8 kg of a standard gestation diet. From d 14 to 28 of gestation, sows were randomly allotted to receive daily, in a cross-over design (2 periods: 5th and 6th parity), either 25 g L-arginine·HCl (ARG) or 43 g L-alanine (Ctrl) both as top dressing for 14 d. At farrowing, litter birth weight (BtW), number of total-, live- and stillborn were assessed. Four piglets per sow were selected according to sex and BtW (lightest and average BtW), euthanized, and subsequently complete Semitendinosus muscle (STM) were excised. Organs were weighed and myofiber traits of the light (STM$_l$) and dark (STM$_d$) portion of the STM were assessed by ATPase staining. Litter size and average litter BtW was neither affected (P>0.25) by intrauterine crowding nor by arginine supplementation. Selected piglets were 29% lighter (P<0.01) and tended (P=0.07) to be 6% heavier when born from IN and ARG sows, respectively. While intrauterine crowding had no effect on myofiber characteristics, newborn pigs born from the ARG group had a 17% greater (P<0.01) STM area and 12% greater (P<0.05) myofiber number in the STM$_l$. Consequently, these piglets had 12% more (P<0.01) myofibers in the STM than those of the Ctrl group. Results strongly indicate a positive effect of a surplus of L-arginine during early gestation on myogenesis of the STM. The lack of significant interactions between the extent of intrauterine crowding and ARG supplementation on the traits of interest suggest that arginine is effective in an non crowded as well as in a crowded intra uterine environment.

Microstructure and function of the thyroid gland may relate to feed efficiency in the bovine
J. Ormon[1], S. Bourgon[1], J. Munro[1], A. Macdonald[2], S. Lam[2], S. Miller[2,3] and Y. Montanholi[1]
[1]Dalhousie University, 58 River Road, Truro, NS, B2N5E3, Canada, [2]University of Guelph, 50 Stone Road E,
Guelph, ON, N1G2W1, Canada, [3]AgResearch, Invermay Agricultural Centre, Private Bag 50034, Mosgiel, 9053,
New Zealand; yuri.r.montanholi@gmail.com

Thyroid hormones are major determinants of metabolic rate. Evidence supports a direct association between levels of thyroxine (T4) and triiodothyronine (T3) with oxygen uptake and feed efficiency in cattle. The objective was to further understand bovine thyroid physiology by including the evaluation of thyroid-stimulating hormone (TSH) levels and features of glandular histomorphometry. Cattle were tested for productive performance and residual feed intake (RFI; kg/d) was determined. Blood plasma and thyroid glands were harvested from 79 crossbred feedlot cattle shortly after stunning. Blood plasma was submitted to determine hormonal concentrations (T3, T4 and TSH). Fragments of thyroid gland were processed for histomorphometry, which involved photomicroscopy and image analysis to determine follicle size, epithelial cell size and ratio of epithelial layer to follicle size. Cattle were categorized as feed efficient (Low-RFI=39) and feed inefficient (High-RFI=40). Least square means of hormonal and microstructural measures were compared between the two categories. Preliminary results from a subset of animals indicate no difference (P>0.05) in levels of T3 (3.69 vs 3.80 nmol/l) when comparing Low-RFI vs High-RFI cattle. Differences in thyroid microsctructure between feed efficiency groups were suggested. Dimensions of the follicles, epithelial cells and ratio of epithelial layer to follicle size were 3.6, 19.7 and 4.2% greater in Low-RFI than High-RFI cattle. These results may be indicative of an increased workload in the thyroid gland in relation to improved efficiency of feed utilization in cattle. Upon completion of the analysis, this research may have implications in the assessment of phenotypic diversity to benefit genetic selection and in the assessment of diversity to improve herd resilience.

Impact of pre-weaning nutritional regimes on mammary gland development in heifer calves

S. McCoard[1], T. Silvestre[1], A. Molenaar[1], P. Muir[2], J. Koolaard[1], V. Burggraaf[3], N. Wards[1] and D. Pacheco[1]
[1]AgResearch Limited, Private Bag 11008, Palmerston North 4442, New Zealand, [2]On-Farm Research Ltd, State Highway 2, Hastings 4178, New Zealand, [3]AgResearch Limited, Private Bag 3123, Hamilton 3240, New Zealand; sue.mccoard@agresearch.co.nz

The aim of this study was to evaluate the effect of contrasting nutritional regimes pre-weaning on the growth and development of the mammary gland (MG), which may have important implications for future milk production. Hereford × Friesian heifer calves (4-6 days of age) were reared using one of four contrasting diets created by modifying the amount of fat and the volume of the milk replacer (MR) offered, and the type of grain in the starter: group 1 (control) fed standard MR (21% fat, 23% protein, fed at 12.5% of body weight (BW)) plus a standard pelleted meal (SM); group 2 fed high fat MR (31% fat and 23% protein) at 12.5% BW plus SM; group 3 fed standard MR plus high fermentable starch meal, and group 4 fed standard MR at 20% BW. Both starter meals had similar crude protein contents (19% DM) and metabolisable energy (12.9 MJ/kg DM). Post-mortem samples were collected at either 6 weeks or 6 months of age to examine MG development using a histological approach. At 6 weeks, calves in group 4 had 1.5 times heavier MGs compared to all other groups (P=0.04), larger average fat pad (FP) cell size (P<0.001) and 2.5 to 3.5 times more large fat cells (P<0.001) compared to all other groups. At 6 months, while MG parenchyma or FP weight did not differ between groups (P<0.10), average FP cell size differed (group 4>1>3>2; P<0.001) and group 4 calves had 1.4 to 3 times more large fat cells and fewer small fat cells (P<0.001) with the group 2 calves having the least small (fat cells (P<0.001). In summary, increased nutrient intake through milk increased the weight of the MG at weaning, the average size of fat cells and the proportion of large fat cells in the MG at both weaning and 6 months of age, which may have future implications for milk production. This study suggest that both the amount of nutrients from milk, and the fat content of the milk replacer influences mammary development in the growing heifer.

Effect of intra uterine position on corporal composition of rabbit fetuses

R. Belabbas[1], M.L. García[2], H. Ainbaziz[3], A. Berbar[1], G.H. Zitouni[1], N. Benali[3], M. Lafri[1], Z. Boumahdi[1] and M.J. Argente[2]
[1]Biotechnology Laboratory of Animal Reproduction (LBRA), University of Saad Dahleb Blida I, 09000, Algeria, [2]Departamento de Tecnología Agroalimentaria, Universidad Miguel Hernández de Elche, 03312 Alicante, Spain, [3]Health and Animal Production, National Veterinary School, Algeria, 16200 Alger, Algeria; r_belabbas@yahoo.fr

The objective of this work was to examine the effect of intra uterine position of rabbit fetuses at 25 d of pregnancy on their corporal composition. Fifty seven unilaterally ovariectomized multiparous rabbit does were used in this study. The females were mated and sacrificed at 25 d of pregnancy. Immediately after, the uterine horn was examined in order to count the number of blood vessels reaching each implantation site, the position of each fetus within the uterine horn (oviduct, middle or cervix) and their sex by direct examination of gonads. Only the females with litter size near to the mean of the population were used (n=14). For each female, three fetuses were chosen (one from the oviduct position, one from the cervix position and the last one from the middle position). The fetuses (n=42) were frozen and stored at -20 °C until analysis. Representative samples of ground matter were freeze dried and analysis for dry matter, humidity, protein, ash, lipids and energy contents. The intra uterine position did not affect the ash content, dry matter and rate of humidity. Conversely, the protein, energy and lipid contents were significantly higher for the fetuses in oviduct position compared to the other positions (+15%; P<0,05). The fetuses with implantation sites receiving equal or more than 6 blood vessels showed higher percentage of protein (+9%, P<0,05), lipids (+13%; P<0,05) and energy contents (+16%; P<0,05) that those receiving less. Finally, there were no differences for all the parameters measured between fetuses of sex male and female. In conclusion, at birth, the fetuses from the oviduct position or receiving higher number of blood vessels should have high probability of live related to higher fat and protein reserves.

Growth, carcass and feed efficiency traits of lambs born to ewes restricted during mid gestation

L. Piaggio[1], G. Ferreira[1], M.J. Marichal[2], R. San Julián[3], F. Baldi[4] and G. Banchero[3]
[1]Secretariado Uruguayo de la Lana, S. Gomez 2408, Mdeo, 12100, Uruguay, [2]Facultad de Agronomía, UdelaR, Av. Garzón 780, Mdeo, 12900, Uruguay, [3]Instituto Nacional de Investigación Agropecuaria, Ruta 8 km 28, Treinta y Tres, 33000, Uruguay, [4]FCAV Unesp, Jaboticabal, São Paulo, 14884-900, Brazil; mariadejesus.marichal@gmail.com

The effect of energy restriction during dam's gestation on lamb's birth (BW), weaning (WW) and slaughter weight, carcass and meat traits was evaluated. Eighty three adult Polwarth ewes sired with Texel, bearing single (n=54) or twin (n=29) lambs were allotted to two treatments by day 45 of gestation: T60 ewes offered a TMR (137.2 g CP/kg DM; 2.7 Mcal ME/kg DM) to provide 60% of the energy requirements for gestation from day 45 to day 115 of gestation and T100 ewes with 100% energy requirements. On day 116 all ewes were shorn and grazed oat grass (1,396 kg DM/ha; 180 g CP/kg DM; 2.4 Mcal ME/kg DM) until weaning (104 days of age). After weaning, 40 lambs were fed individually ad libitum to record feed intake (FI), weight gain (WG) and feed efficiency (FE=FI/WG) and 72 lambs were fed in 3 collective pens to determine ADG. The finishing TMR had 207 g CP/kg DM and 2.6 Mcal ME/kg DM. At 150 days of age, lambs were shorn and slaughtered. All traits were analyzed by ANOVA using SAS 9.2. The model included the type of birth, restrictions, and the interaction between these factors. For WBW, FBW and carcass characteristics age were included as covariate. Means were compared by Tukey's. Feeding restriction affected (P<0.01) preweaning ADG (T60=164 vs T100=183 g/d/lamb), WBW (T60=23.2 vs T100=25.3 kg) and FBW (T60=34.7 vs T100=37.2 kg) but not BBW or post weaning ADG. Litter size and sex affected all growth traits. T60 lambs FI tended (P=0.10) to be higher than T100 lambs (59.7 vs 54.9 kg/finishing period/lamb) and FI was higher (P<0.01) for T60 vs T100 (3.82 vs 3.21%). FE tended (P=0.16) to be better for T100 vs T60 lambs (5.5 vs 6.3 kg/LW).Treatment, sex and litter size affected (P<0.05) slaughter weight. Differences between treatments were due to different preweaning ADG and consequently different final and slaughter weight.

Effect of nutritional restriction in late gestation on beef female calves live weight and placentas

C. Batista[1], J.I. Velazco[2], G. Banchero[2], F. Baldi[3] and G. Quintans[2]
[1]Facultad de Agronomía-UdelaR, Garzón 780, 12900, Uruguay, [2]National Institute for Agricultural Research, Ruta 8, km 281, 33000, Uruguay, [3]FCAV Unesp, Via de Acceso Prof. Paulo Castellane, 14884-900, Brazil; gquintans@inia.org.uy

The effect of maternal undernutrition during late gestation was investigated in beef multiparous cows. All cows got pregnant by fixed time artificial insemination and all gestated female calves. The placenta structure and live weight (LW) of female calves were analyzed. Twenty two pregnant British cross-bred cows were assigned at day 195.5±1.60 of gestation (mean ± sem) to one of two treatments: (1) cows fed rations calculated to provide 125% of the energy requirements (high, H, n=11); (2) cows fed rations calculated to provide 75% of the energy requirements (low, L, n=11) where crude protein was not limiting. The duration of the experimental period was 79.5±0.73 days (last trimester of gestation). Cows weighed 479.5±11.5 kg and presented a body condition score (BCS, scale 1 to 8) of 4.0±0.05 at the onset of treatments. At birth all the placentas were collected, weighed, the total number of cotyledons was counted and LW of calves registered. Statistical analyses was performed using the restricted likelihood method, applying a mixed model. The model included the random effects of sire (4 sires) and dam breed composition (percentage of Angus and Hereford), and the fixed effects of dam age (from 4 to 9 years old) and treatment. At birth, calves in H tended to be heavier (P=0.09) than those from L treatment (34.9±2.59 vs 32.9±2.49 kg). Placenta weight was similar (P=0.18) for calves from H and L treatment (4.6±0.78 vs 4.2±0.76 kg). Also, no differences were found in the total number of cotyledons (97.3±9.75 and 88.3±8.23 for H and L, respectively; P=0.31). Under the conditions of this experiment the different nutritional levels tended to induce differences in calves LW at birth. However, the greater calves LW in H treatment cannot be explained neither by the weight of the placenta nor by the total number of cotyledons. Further research is needed to better understand the mechanisms underlying this process.

Genetic parameters for growth and carcass traits of Pekin ducks

M. Waehner[1], H. Pingel[2], R. Fischer[3] and R. Wehlitz[3]
[1]Anhalt University of Applied Sciences, Agriculture, Strenzfelder Allee 28, 06406 Bernburg, Germany, [2]University Halle-Wittenberg, Animal Breeding, Th.-Lieser Str. 11, 06120 Halle/Saale, Germany, [3]Saxon State Office for Environment, Agriculture and Geology, Animal Breeding, Am Park 3, 04886 Köllitsch, Germany; m.waehner@loel.hs-anhalt.de

In the breeding of Peking ducks a remarkable progress was observed. Therefore the current performance potentials has to be analyzed. From a heavy strain of Peking ducks in the Chinese company SAIFEIYA 1,376 pedigreed birds (full and half-sibs) could be used for calculation of heritability as well as phenotypic and genetic correlations. The ducklings were kept from the age of 15 to 42 days in single boxes (2.0×0.70 m) with individual feeding of standard duck grower pellets. Growth, feed efficiency and breast muscle thickness were measured. At day 42 all ducks were slaughtered and the carcass composition was measured. The genetic parameters were estimated by software packages PEST and VCE-6. The average body weight at day 42 of 3.72 kg with h^2=0.47 characterizes a heavy strain of Peking ducks. There is a high feed efficiency (gain divided by feed consumption) from 15 to 42 days of age with 0.45. The heritability for feed efficiency is h^2=0.36. The breast muscle thickness amounts to 1.98 cm with h^2=0.56. The same heritability was estimated for breast muscle percentage. The genetic correlation between breast muscle thickness and the percentage of breast muscle to carcass with 0.78 is high enough for practical selection. The genetic correlations of feed efficiency to breast muscle thickness and to breast muscle percentage are low with 0.04 and 0.17, respectively, but is high to leg muscle percentage (rg−0.66). There is also a high negative genetic correlation between feed efficiency and amount of skin with rg=-0.53. It means the higher the feed efficiency the lower the amount of skin with subcutaneous fat. High feed efficiency is a presupposition for optimum biological efficiency and sustainability in duck production. The late start of breast muscle growth is a limit for the reduction of slaughter age. The main objective of successful duck breeding for meat production is (beside of the percentage of breast muscle) the improvement of feed efficiency.

Influence of diet on carcass characteristics of broiler chicken

R. Dominguez[1], J. Méndez[1], D. Franco[1], M.V. Sarriés[2], P. Cachaldora[1] and J.M. Lorenzo[1]
[1]Centro Tecnolóxico da Carne (CTC), Rua Galicia, 4, Parque Tecnológico de Galicia, San Cibrao das Viñas, Ourense, 32900, Spain, [2]Universidad Pública de Navarra, Producción Agraria, ETSIA-Edif. Los olivos-Campus de Arrosadía-Pamplona-Navarra, 31006, Spain; jmlorenzo@ceteca.net

This research was conducted to study the effect of diet on carcass characteristics of Broiler chicken. For this study a total of 309 chicken were used. Animals were randomly divided into three groups: CO (commercial feeding), WG (wheat grains feeding) and WF (wheat flour feeding). The left side of the carcass was quartered (wing, breast, drumstick, and thigh). In general, diet had great impact in carcass measurements. The results obtained showed that the inclusion of wheat in the diet (both, as flour and as grain) significantly (P<0.01) decreased the live (3.06 and 3.01 kg for WG and WF groups, respectively) and carcass (2.24 and 2.20 kg for WG and WF groups, respectively) weight compared to CO batch (3.24 and 2.39 kg for live and carcass weight, respectively). These differences could be due to the animals from CO treatment presented significantly (P<0.001) higher drumstick weights (420 g vs 404 and 382 g for CO, WG and WF batches, respectively) and breast weights (818 g vs 726 and 724 g for CO, WG and WF batches, respectively). However, the wing and thigh weights were no significantly (P>0.05) affected by diet. Regarding commercial cuts, the inclusion of wheat in the diet had a positive effect on thigh, drumstick and wing percentages. However, breast percentage showed the highest values in CO batch. Finally, as a general conclusion, the inclusion of wheat in the diet decrease the live and carcass weights and the weight of the main cuts.

Response of bovine jejunal transcriptome to dietary restriction and subsequent compensatory growth

K. Keogh[1], S.M. Waters[1], P. Cormican[1], A.K. Kelly[2] and D.A. Kenny[1]
[1]Teagasc, Animal and Bioscience Research Department, Grange, Dunsany, Co. Meath, Ireland, [2]University College Dublin, School of Agriculture and Food Science, Belfield, Dublin 4, Ireland; david.kenny@teagasc.ie

The objective of this study was to evaluate the effect of dietary restriction and subsequent re-alimentation induced compensatory growth (CG) on the global gene expression profile of jejunal epithelial tissue. Holstein Friesian bulls (n=60) were assigned to one of two groups: (1) restricted feed allowance (RES; n=30) for 125 days (Period 1) followed by ad libitum access to feed for 55 days (Period 2) or (2) ad libitum access to feed throughout (ADLIB; n=30). At the end of each period, 15 animals from each treatment group were slaughtered and jejunal epithelial cells collected. Messenger RNA was then isolated from all jejunal epithelium samples and cDNA libraries prepared and subsequently sequenced. Resultant sequence reads were first checked for quality and then aligned to the bovine reference genome. The number of mapped reads was counted and differentially expressed genes (DEG) determined. DEGs were defined as having a P<0.05 and false discovery rate of 10%. Average daily gain (ADG) for Period 1 was 0.6 kg/d for RES and 1.9 kg/d for ADLIB. During re-alimentation an ADG of 2.5 and 1.4 kg/d was observed for RES and ADLIB, respectively. 24 genes were differentially expressed in RES animals compared to ADLIB at the end of Period 1, with only one gene differentially expressed at the end of Period 2. When analysed within treatment (RES, Period 2 vs Period 1), 26 genes were differentially expressed in animals undergoing CG. Genes identified as differentially expressed represented those involved in processes including digestion, metabolism, immune function, cellular growth and differentiation. This study provides an insight into the molecular mechanisms underlying the expression of CG in jejnunal epithelium in cattle; however the results suggest that most of the CG has occurred by day 55 of re-alimentation.

Effect of plane of nutrition on age at puberty and blood metabolites in Holstein Friesian bulls

C. Byrne[1], A.M. English[1,2], S. Fair[2], P. Lonergan[3] and D.A. Kenny[1]
[1]Teagasc, Animal and Bioscience, Dunsany, Meath, Ireland, [2]University of Limerick, Department of Life Sciences, Limerick, LK, Ireland, [3]University College Dublin, School of Agriculture and Food Science, Belfield, Dublin 4, Ireland; colin.byrne@teagasc.ie

The aim of this study was to examine the effects of plane of nutrition during the first and second 6 months of life on age at puberty and metabolites in dairy bulls. Autumn-born Holstein-Friesian bulls (n=83) with a mean (±S.D.) age and body weight of 17 (4.4) days and 52 (6.2) kg, respectively, were assigned to a high [H] or low [L] plane of nutrition for the first 6 months of life. At 24 weeks of age, bulls either remained on the same diet or were switched to the opposite diet, until puberty, resulting in four groups: HH; HL; LL and LH. Bulls were weighed weekly pre-weaning and fortnightly post-weaning. Puberty was defined as an ejaculate containing 50×10^6 sperm with >10% progressive motility. Blood samples were collected monthly until puberty. Data were analysed using mixed models ANOVA and orthogonal contrasts (SAS, version 9.3). All results are presented as mean±s.e.m. H bulls grew at 0.96±0.04 kg/day and L bulls at 0.60±0.05 kg/day during the first 6 months of life (P<0.001). There was a pre × post-6 month diet interaction for average daily gain; growth slowed in HL to 0.70±0.05 kg/day and accelerated in LH to 1.6±0.03 kg/day (P<0.001). Bulls on a H and L plane of nutrition pre-6 months reached puberty at 291±4.8 and 323±4.6 days, respectively (P<0.001), irrespective of diet offered post-6 months; leading to effect of plane of nutrition (P=0.35) post- 6 months of life on age at puberty. Urea and BHB levels were higher in bulls on a L diet pre-6 months (P<0.05). NEFA were higher in bulls on a H diet pre-6 months of age (P<0.01). Bulls on a L diet post-6 months had higher BHB and NEFA levels than H (P<0.05). There were higher urea and triglyceride levels in bulls on a H diet post-6 months (P<0.01). In conclusion, feeding an increased plane of nutrition pre-6 months of age hastens puberty onset in dairy bulls. Bulls on a high plane of nutrition post-6 months did not reach puberty earlier than bulls restricted throughout life, despite reaching a growth rate, in line with unrestricted bulls. Overall metabolite levels reflect level of nutrition.

Dietary's effect of protein levels on digestive organs and small intestine histologic of rabbits
N. Benali, H. Ainbaziz, Y. Dahmani, H. Tighiouart, R. Kaddour, A. Zouambi, R. Bellabes, M. Cherrane and S. Temim
High National School veterinay, Alger, Rue Issad Abbes, Oued Smar, 16000, Algeria; na.benali@yahoo.fr

An experiment was conducted to evaluate the effect of different dietary crude protein levels on the development of digestive organs and the morphology of the intestinal mucosal. Seventy two rabbits of the local population (42 days old) were allotted to three experimental groups of 24 rabbis each and fed ad libitum three diets that were iso-energetic (2,500 Kcal/kg/DE) and had the same fiber content (13.5%) but contained 3 levels of crude protein: low protein diet (LP 16% CP), medium protein diet (MP, 18% CP) and high protein diet (HP, 20% CP). The weight and length of the digestive organs and morphometric indices (villus height and crypt depth) of duodenum, jejunum and ileum were measured at the end of fattening period (92 days old). Data collected for digestives organs and histological examination were submitted to analysis of variance Anova 1 and statistically analyzed by StatView of SAS (Abacus Concepts, 1996, Inc., Berkeley, CA, USA). The medium protein diet (MP 18% CP) gave a significantly higher effect on the weight of stomach (106 g vs 79.8 and 67.3, $P<0.05$), small intestine (52.5 g vs 42.3 and 45.1, $P<0.05$), proximal colon (15.5 g vs 10.00 and 11.03, $P<0.05$) and the length of the caecum (50.0 cm vs 43.2 and 44.7, $P<0.05$) when compared with the digestive organs of rabbits fed the LP and HP diets, respectively. The MP diet resulted in greater villus heights (1,941 μm vs 1,267 and 1,677, $P<0.05$) and crypt depths (378 μm vs 292 and 277, $P<0.05$) in the jejunum, and villus heights and crypt depths of the ileum (976 μm vs 706 and 725-299 μm vs 269 and 270, $P<0.05$) compared with the LP and HP diets, respectively. However, the hHP diet increased the villus heights (1,927 μm vs 808 and 1,379, $P<0.05$) and crypt depths (344 μm vs 236 and 319, $P<0.05$) in the duodenum. In conclusion, a medium protein diet, could positively influence the development of digestive organs and the morphometric features of the intestine in rabbits.

Testicular development and the attainment of puberty in Icelandic male horses
O.R. Dyrmundsson, G.G. Gunnarsson and I. Sveinsson
Agricultural Consultant, Jorusel 12, 109 Reykjavik, Iceland; oldyrm@gmail.com

The native Iceland breed of horses is widely known for the smooth gait, tölt, as well as for its strength, endurance and good fertility. These characteristics are well documented but there has been a lack of scientific information, both in Iceland and abroad, on reproductive traits such as sexual development in males of this breed. In former times it was normally stated in Iceland that colts attained puberty by two years of age. Until 2013 the Icelandic Livestock Management Act stipulated that male horses could remain intact until 16 months of age, without confinement, unless individual, local Horse Breeding Societies applied for a derogation for their district demanding castration not later than at 12 months of age. Late in the 20th century, when the level of nutrition had generally improved, it was evident that some fast growing males were attaining puberty by one year of age and this prompted our small study of testicular development, carried out in W-Iceland. Firstly, the results indicated that both testes and epididymides of 45 males, castrated at 10-14 months of age, were substantially smaller than those of a 48-month old stallion studied for comparison, as would be expected. Secondly, out of the 45 males, 16 (35.6%) had mature spermatozoa in either one or both of their epididymides and thus had attained physiological puberty. Their age ranged from 12-14 months (mean=12.5) while those not having epididymal spermatozoa were in the range of 10-13 months (mean=12.0). Thirdly, looking at individual testicles and epididymides, those with mature spermatozoa (n=22) weighed 32.6 g and 9.8 g, respectively, while the immature testicles and epididymides (n=68) weighed only 13.9 and 7.6 g (n=68), respectively. All testicular weight measurements were highly correlated with both size and volume measurements. These findings led to the conclusion that at least some Icelandic male horses could, as young as 12 months of age, impregnate mares, provided they had the physical and behavioural ability to mate with them. This is in agreement with casual observations and herd records. Now most colts in Iceland, not selected for breeding, are castrated by 12 months of age, mainly in the spring.

GénoSanté: Improve the productive health of dairy cows by genomic selection and management

H. Leclerc[1,2], A. Barbat[1,3], M. Philippe[4], S. Fritz[1,5], J.-B. Davière[6], L. Manciaux[7], F. Guillaume[4], T. De Bretagne[4] and D. Boichard[1,3]
[1]UMT 3G, Bat 211, 78350 Jouy-en-Josas, France, [2]Institut de l'Elevage, Bat 211, 78350 Jouy-en-Josas, France, [3]GABI, INRA, AgroParisTech, Université Paris-Saclay, 78350 Jouy-en-Josas, France, [4]Evolution, rue de la Motte Brûlon, 35706 Rennes, France, [5]ALLICE, Bat 211, 78350 Jouy-en-Josas, France, [6]Seenergi, Bvd des loges, 53942 Saint Berthevin, France, [7]BCEL-Ouest, rue Curie, 22195 Plérin, France; helene.leclerc@idele.fr

GénoSanté is a collaborative project bringing together French companies (milk recording organisations and herd support with BCEL-Ouest and Seenergi; AI cooperatives with Evolution, Elitest and Midatest; and a dairy industry with Agrial) and research institutions (INRA, ALLICE and IDELE). The aim of this project is to strengthen the selection tools used to improve productive health through the development of new indices, to reduce antibiotic use, and improve the profitability of dairy herds. The first results deal with ketosis, one of the most common disorders in dairy cows due to energy deficit during the early stages of lactation. Its prevalence reaches 5% and 15-25% for clinical and subclinical forms, respectively. Data included milk beta-hydroxybutyrate (BHB), acetone and a ketosis indicator (Cetodetect®) from 887,000 Holstein (3 millions of samples) and 148,000 Normande cows (487,000 samples) collected between 2012 and 2015 on parity 1 to 5, from 7 to 120 days in milk. BHB and acetone measurements were log-transformed to improve. Genetic breeding values indicators and permanent environment effects were estimated using a single trait repeatability BLUP animal model applied to multiple lactations. Heritability estimates of BHB were 0.12 and 0.15 for Holstein and Normande breeds, respectively, and 0.10 and 0.16 for acetone. Genetic correlation estimates between the 3 traits ranged from 0.72 to 0.85 in Holstein and from 0.85 to 0.98 in Normande. Genetic indicators for the 3 traits were favorably correlated with protein percent index (0.18 to 0.28) and fertility index (0.08 to 0.17). A genomic evaluation was carried out for 31,213 Holstein and 6,870 Normande males and females with both genotype and phenotypes. In this same project, genetic and genomic evaluations for claw disorders are in preparation.

Quantitative genetic analysis of the bTB diagnostic Single Intradermal Comparative Cervical Test

S. Tsairidou[1], S. Brotherstone[2], M.P. Coffey[3], S.C. Bishop[1] and J.A. Woolliams[1]
[1]The Roslin Institute and R(D)SVS, University of Edinburgh, Easter Bush, Midlothian, EH25 9RG, Edinburgh, United Kingdom, [2]Institute of Evolutionary Biology, University of Edinburgh, King's Buildings, West Mains Road, EH9 3JT, Edinburgh, United Kingdom, [3]SRUC, Easter Bush Campus, EH25 9RG, Edinburgh, United Kingdom; smaragda.tsairidou@roslin.ed.ac.uk

Bovine Tuberculosis (bTB) is a disease of significant economic importance, being one of the most persistent animal health problems in many countries including the UK. Genetic selection for cattle more resistant to bTB can offer a complementary control strategy. Such selection will be partially informed by the diagnostic Single Intradermal Comparative Cervical Test (SICCT). The aim of this study was to conduct a quantitative genetic analysis of SICCT measurements collected during bTB testing, investigating the genetic control of the response to SICCT and exploring an important question, namely whether in addition to increasing bTB resistance, genetic selection might also genetically alter the performance of SICCT. The genetic basis of variation in SICCT response was explored by means of fitting linear mixed models in ASReml, to obtain heritability estimates, genetic and environmental correlations and genetic regressions, for SICCT and its components. Further, the correlation between SICCT in healthy animals and test positivity, and the impact of age on the heritability were assessed. SICCT was found to be a robust test at the genetic level. Its hierarchical, comparative design provides substantial protection from correlated responses in its components arising from selection, and controls for initial skin thickness and for age-dependent differences. SICCT was found to be lowly heritable, with a weak correlation among healthy animals with test positivity. These results demonstrate that although all theoretical risks cannot be excluded, the benefit from selection for bTB resistance is unlikely to be compromised by any adverse correlated response in SICCT, which is expected to be weak and slow, allowing it to be monitored and managed.

Host genetics of resistance to bovine tuberculosis infection in dairy cattle
S. Wilkinson[1], S.C. Bishop[1], A.R. Allen[2], S.H. McBride[2], R.A. Skuce[2,3], M. Bermingham[1], J.A. Woolliams[1] and E.J. Glass[1]
[1]The Roslin Institute and R(D)SVS, University of Edinburgh, Easter Bush EH25 9RG, United Kingdom, [2]Agri-Food and Biosciences Institute, AFBI Stormont, Stoney Road, Belfast BT4 3SD, United Kingdom, [3]Queen's University Belfast, Medical Biology Centre, Belfast BT9 7BL, United Kingdom; adrian.allen@afbini.gov.uk

Genetic variation for bovine tuberculosis (bTB) resistance exists in cattle, thus breeding for resistance is a viable control option. bTB infection may be described using a spectrum of phenotypes as not all infected cattle have detected signs of disease. This study investigated the host genetics under-pinning different bTB infection outcomes. 1,966 dairy cows, defined as confirmed cases (visible lesions detected), unconfirmed cases (no visible lesions detected) and controls, were genotyped at 538,231 SNPs. The contribution of each chromosome (BTA) to heritability (h^2c) was estimated using a linear mixed model. Regional heritability mapping using 100-SNP overlapping windows was carried out to identify genomic regions of variation (h^2r). Analyses considered controls against each case phenotype. 15 chromosomes contributed to heritability ($1.7 \leq h^2c \leq 7.7\%$). By case phenotype, 4 of 5 chromosomes explaining variation in confirmed cases were also found for unconfirmed cases, but an extra 10 chromosomes were unique to unconfirmed cases. Four genomic regions with variation were found: 71.5-71.8 Mb on BTA13 (h^2r=2.1%; confirmed cases), 19.3-19.8 Mb on BTA17 (h^2r=5.3%; unconfirmed cases), 57.2-57.5 Mb on BTA22 (h^2r=3.9%; unconfirmed cases) and 6.6-7.1 Mb on BTA23 (h^2r=3.5%; unconfirmed cases). Genes associated with these regions were identified. Of interest was the bovine leukocyte antigen (BoLA) class IIb region located at 6.97-7.53 Mb near the QTL found on BTA23. BoLA genes have a major role in immune response to infection, where they encode cell surface proteins that recognise foreign molecules. The results suggest there are at least two bTB infection outcomes controlled by distinct and overlapping genetic variants. Breeding values can incorporate both bTB infection states as they share underlying chromosomal variance.

Genome-wide association identify regions underlying bovine tuberculosis resistance in dairy cattle
K. Raphaka[1], O. Matika[1], E. Sanchez-Molano[1], R. Mrode[2], M. Coffey[2], E.J. Glass[1], J.A. Woolliams[1], V. Riggio[1], S.C. Bishop[1] and G. Banos[1,2]
[1]The Roslin Institute and Royal (Dick) School of Veterinary Studies, Genetics and Genomics, Easter Bush, Midlothian, Edinburgh, EH25 9RG, United Kingdom, [2]Scotland's Rural College, Roslin Institute Building, Edinburgh, EH25 9RG, United Kingdom; kethusegile.raphaka@roslin.ed.ac.uk

The significant social disruption and economic losses from bovine tuberculosis (bTB) due to infection from Mycobacterium bovis present an ongoing challenge to the cattle industries in the UK and worldwide. Genetic variation in cattle provides an opportunity to select for more resistant animals. The objective of this study was to identify genomic regions associated with resistance to bTB in dairy cattle using Genome-Wide Association Study (GWAS) and Regional Heritability Mapping (RHM). In the latter, regions consisting of 100 SNPs and overlapping by 50 SNPs were defined and analysed. De-regressed estimated breeding values (dEBV) for susceptibility to bTB were calculated for Holstein Friesian bulls based on bTB skin test records of their daughters and then used as phenotypes. A total of 529 bulls with dEBV and 50K SNP genotypes were considered. Of the 56,134 SNPs, 36,625 SNPs were retained for analyses following quality control. GWAS identified two SNPs on chromosomes 4 and 17 which were suggestive (i.e. one false positive per genome scan) of association with bTB resistance at $-\log_{10}$(P-value) of 4.56. The estimated SNP-based polygenic heritability for resistance to bTB was 0.45±0.11. RHM identified genomic regions on chromosomes 10 and 17, all of which were significant at a suggestive level of $-\log_{10}$(P-value) of 2.57. The proportion of total phenotypic variance accounted for by these regions was 0.05 and 0.10-0.11 for chromosomes 10 and 17, respectively. The genomic regions identified in this study provide an opportunity for exploration of plausible candidate genes and contribute to the understanding of the pathways involved in cattle resistance to bTB.

GWAS of resistance to paratuberculosis in French Holstein and Normande cattle

M.P. Sanchez[1], R. Guatteo[2], A. Davergne[3], C. Grohs[1], A. Capitan[1,4], P. Blanquefort[5], A. Delafosse[6], A. Joly[7], C. Fourichon[2] and D. Boichard[1]
[1]GABI INRA, AgroParisTech, Université Paris-Saclay, Domaine de Vilvert, 78350 Jouy en Josas, France, [2]UMR1300 BioEpAR, Oniris, INRA, Atlanpole-Chantrerie CS 40706, 44307 Nantes, France, [3]GDS, Haute-Normandie, 76000 Rouen, France, [4]Allice, 149 rue de bercy, 75012 Paris, France, [5]GDS, Pays de la Loire, 49800 Trélazé, France, [6]GDS, Orne, 61000 Alençon, France, [7]GDS, Bretagne, 56000 Vannes, France; marie-pierre.sanchez@jouy.inra.fr

A case-control genome-wide association study (GWAS) was applied to identify QTL of resistance to Mycobacterium avium ssp. paratuberculosis (MAP) infection in French Holstein (HOL) and Normande (NOR) cattle. Cases were infected shedder cows (confirmed with both positive blood ELISA and fecal PCR) or clinical cases. Controls included animals with 3 repeated negative blood ELISA, negative fecal PCR test, at least 72 months old and born in the same herd and at the same time period as cases. A total of 405 NOR (210 cases / 195 controls) and 989 HOL (437 cases / 552 controls) cows were genotyped with the Illumina BovineSNP50 BeadChip. GWAS was conducted within breed with GCTA, accounting for the population structure through a 50K-based genomic relationship matrix. The most significant region associated with infectious status was found on chromosome (BTA) 12 at 69.8 Mb ($P<2E^{-8}$) in HOL cows. Another association was identified on BTA23 in the region of the major histocompatibility complex (MHC) in both breeds. Other chromosomal regions ($P<7E^{-5}$) were found on BTA1 (55 Mb), BTA10 (91 Mb) and BTA13 (56.6 Mb) in HOL and on BTA1 (92.6 Mb), BTA9 (39.3 Mb), BTA15 (58.4 Mb) and BTA17 (8 Mb) in NOR cows. Seven additional regions were found with more moderate significance level ($P<1E^{-4}$) in NOR (BTA13 and 25) or HOL (BTA3, 9, 11, 25, 26 and 28) cows. This on-going study presents encouraging results. After collecting and genotyping additional cases and controls, whole genome sequences of the cows will be imputed using the 1000 bull genome reference population and GWAS will be carried out directly on whole genome sequence data to pinpoint candidate mutations. This study was funded by INRA (GISA), Apisgene, and GDS France. The authors thank the farmers and GDS who contributed to this project.

Unravelling the contribution of host genetics to infectious disease outbreaks

A. Doeschl-Wilson[1], O. Anacleto[1], S. Tsairidou[1], G. Lough[1], R.D. Houston[1], J.A. Woolliams[1], H.A. Mulder[2], S. Cabaleiro[3], M. Saura[4] and B. Villanueva[4]
[1]The Roslin Institute and R(D)SVS of the University of Edinburgh, Genetics and Genomics, Roslin Institute Building, Easter Bush, Midlothian, EH26 8NT, United Kingdom, [2]Wageningen University, P.O. Box 338, 6700 AH Wageningen, the Netherlands, [3]CETGA, A Coruna, 15965 Ribeira, Spain, [4]INIA, Ctra. De la Coruña, Km 7., 28040 Madrid, Spain; andrea.wilson@roslin.ed.ac.uk

Genetic analyses of infectious disease data usually focus on disease resistance. Increasing evidence however shows that risk and severity of disease outbreaks also depend on host infectivity (ability to transmit infections) and tolerance to infection. Estimating genetic parameters for these traits has proven difficult, because current quantitative genetics methods fail to account for the infection dynamics and dependence between traits. Our studies aimed to develop statistical methods to estimate additive genetic risk and SNP effects associated with resistance, tolerance and infectivity. These tools were applied to simulated and real data to gain novel insights into the host genetic contribution to infectious disease spread and impact in livestock populations. When applied to epidemiological data, our dynamic social effects models produced reliable estimates of additive genetic risks and SNP effects for both host susceptibility and infectivity. Prediction accuracies in simulations were above 0.5 and 0.4 for susceptibility and infectivity, respectively, even when information on time-to-infection is imprecise. Informed by these simulations, a large-scale infection experiment in turbot was designed, and revealed that infectivity is heritable and likely genetically correlated with resistance and tolerance. Furthermore, in order to disentangle the relative importance of resistance and tolerance to host responses to infection, we derived a theoretical expression to estimate relative contributions of both traits to the overall genetic variance in host responses. When applied to growth data in pigs infected with PRRS virus, resistance and tolerance were estimated to have similar contributions to the genetic growth variance under infection. Our results offer new opportunities and challenges for implementing novel disease traits into genetic disease control strategies.

Estimating gene effects on host susceptibility and infectivity from individual disease status (0/1)

F. Biemans[1,2], P. Bijma[2] and M.C.M. De Jong[1]
[1]Wageningen University, Quantitative Veterinary Epidemiology Group, Radix Building, 6708 PB Wageningen, the Netherlands, [2]Wageningen University, Animal Breeding and Genomics Centre, Radix Building, 6708 PB Wageningen, the Netherlands; piter.bijma@wur.nl

Breeding against infectious diseases in livestock is desirable, but challenging. Current approaches attempt to explain an individual's disease status from its own pedigree or genotype, and thus capture genetic effects on host susceptibility only. Epidemiological theory, however, demonstrates that disease prevalence depends also on host infectivity, i.e. the propensity to infect others. Moreover, phenotypic studies suggest large differences in infectivity among hosts, such as the existence of supershedders that excrete disproportionally more pathogens than average hosts. If such differences are genetic, heritable variation in disease prevalence may be considerably greater than currently believed. Thus our aim was to develop methods to estimate gene effects on both susceptibility and infectivity of hosts. We considered an endemic disease, with time-interval data on individual disease status (0/1) generated by a Susceptible-Infected-Susceptible (SIS) model, in herds of 100 members. The number of susceptible and infectious individuals at the start of each interval, the number of cases within each interval, and the genotypes of individuals were assumed known. Simulations were used to investigate bias and precision of the estimates, and the impact of the length of the time interval. Results show that a Generalized Linear Model (GLM) with a complementary log-log link function can be used to estimate susceptibility and infectivity effects of alleles at single loci. Susceptibility estimates were more accurate than infectivity estimates. Infectivity estimates were biased downwards, i.e. conservative. For infectivity, the optimum recording interval was ~1/3 of the average infectious period of individuals. The GLM developed here can be used to quantify genetic effects on host infectivity, which are currently overlooked. Application to digital dermatitis in dairy cattle is ongoing.

Interest and limits of a bovine MD chip to study Bos taurus × Bos indicus crossbred animals in India

D. Cruz Hidalgo[1], M. Swaminathan[2], P.D. Deshpande[2], M. Boussaha[1], R. Saintilan[1], D. Laloe[1] and V. Ducrocq[1]
[1]GABI, INRA, AgroParisTech, Université Paris-Saclay, 78350 Jouy-en-Josas, France, [2]Central Research Station, BAIF, Uruli Kanchan, Maharashtra, India; vincent.ducrocq@jouy.inra.fr

BAIF is an Indian NGO serving 4 million poor rural families in 16 states. Besides improvement of indigenous breeds, BAIF promotes the use of Holstein/Jersey purebred and crossbred bulls to improve Bos indicus non-descript cows with limited production. BAIF runs a big bull stud producing 8 million semen doses a year. A total of 288 animals from BAIF stud and nucleus farm were genotyped with the Illumina Bovine SNP50 Beadchip® (MD chip). These included Holstein (59 males (M), 23 females (F)) Jersey (39M), Holstein-Bos indicus crossbreds (78M, 7F), Jersey-Bos indicus crossbreds (28M) and Bos indicus animals (54M) mainly of Gir and Sahiwal breeds. After quality control, 43,167 SNP were kept but 18,191 of these were monomorphic in Gir and Sahiwal. These genotypes were analyzed using Principal Component Analysis (PCA) which clearly separated Holstein, Jersey and Bos indicus breeds with their crosses in intermediate positions. PCA revealed a large heterogeneity among Holstein animals, due to varied procurement sources over time. However, SNP information appeared limited to distinguish between indigenous breeds. Breed composition was studied using the Admixture software. It showed good agreement with known pedigree for Jersey crossbreds but the proportion of Holstein genes among Holstein crossbreds was frequently overestimated and Bos indicus breed composition was often incorrect. The results illustrated the necessity to genotype a larger number of local Bos indicus animals. More importantly, there is a need for the development of a more adequate MD chip with more polymorphic SNP. Being more cost efficient than the current bovine HD chip, such a MD chip could be used for determination of breed composition of crossbreds and Bos indicus animals, for genetic diversity studies but also for genomic selection of animals adapted to the local environment.

Detection of QTL for mastitis resistance and related functional traits in Bos taurus and Bos indicus

J. Jardim[1], G. Sahana[2], C. Quirino[1], M. Peixoto[3], G. Santos[3] and M. Lund[2]
[1]University of North Fluminense Darcy Ribeiro (UENF), Laboratory Of Animal Breeding, Avenida Alberto Lamego,2000, sala 54, 28013-602, Brazil, [2]Aarhus University, Quantitative Genetics and Genomics, Blichers Allé 20 building K21, 8830 Tjele, Denmark, [3]Brazilian Agricultural Research Corporation (EMBRAPA), Centro Nacional de Pesquisa em Gado de Leite, Rua Eugênio do Nascimento, 610, 36038-330 Juiz de For a, MG, Brazil; juliagazzoni@mbg.au.dk

The quantitative traits loci (QTL) which are segregating both in Bos taurus and Bos indicus can be mapped precisely if information from both population are joined because they diverged at least one hundred thousand years ago. This study aimed to make genetic associations of milkability traits, udder conformation and udder health in Danish Holstein cattle and to examine if QTL identified in this Bos taurus cattle are segregating in a Bos indicus population, Guzera breed by multi-trait meta-analysis. First we carried out association analysis for milking speed and udder conformation traits (~5,000 deregressed-proof breeding values) in Danish Holstein cattle using imputed whole genome sequence data (~15 million SNPs). Significant QTL regions on 16 chromosomes were identified for udder index trait and 18 genomic regions showed significant association with milking speed in Danish Holstein cattle. Association results for clinical mastitis, milk, fat and protein yields in Danish Holstein cattle were available from previous studies. High-density (770k) genotype data of 25 Guzera sires were used to impute 900 cows genotype with 50k SNP chip. A genome scan for somatic cell count, milk fat and protein yields were carried out in Guzera cattle using imputed HD genotype data. The genomic regions identified in Holstein cattle are being studied in Guzera cattle. Meta-analysis across taurus and indicus populations for individual traits will be carried out using weighted Z-score model. We also plan to do multi-trait meta-analysis within population for closely related trait. The work is in progress. The authors acknowledge FAPEMIG for financing the study, CAPES for granting scholarship and Aarhus University for the support and shared knowledge.

Using whole genome sequences to identify QTL for udder health and morphology in French dairy cattle

T. Tribout[1], M. Barbat[1,2], A. Govignon-Gion[1,3], A. Launay[3], R. Lefebvre[1], A. Barbat[1], M. Boussaha[1], P. Croiseau[1], M.P. Sanchez[1] and S. Fritz[1,2]
[1]GABI, INRA, AgroParisTech, Université Paris-Saclay, Domaine de Vilvert, 78350 Jouy-en-Josas, France, [2]Allice, 149 rue de Bercy, 75012 Paris, France, [3]Idele, 149 rue de Bercy, 75012 Paris, France; thierry.tribout@jouy.inra.fr

Genome-wide association studies (GWAS) at the sequence level were performed in Montbéliarde (MO), Normande (NO) and Holstein (HO) breeds for somatic cells counts, clinical mastitis scores, and 9 to 11 udder morphology traits by breed. The number of bulls considered by trait varied from 1,857 to 2,515 in MO breed, from 624 to 2,203 in NO breed, and from 4,959 to 6,321 in HO breed. The considered response variables were the bulls' daughter yield deviations (DYD), derived from the national genetic evaluations. The DYD reliability of all the bulls considered in the analyses exceeded 0.2 (clinical mastitis) or 0.5 (other traits). Genotypes of the bulls for 27,754,235 sequence variants were imputed in 2 steps, using FImpute software: first from 50K level to HD level using 522 MO, 546 NO, and 776 HO HD genotyped bulls as a reference, and then to the sequence level using 1,147 sequenced bulls from the 1,000 bull genomes project. GWAS were done independently within each breed and for each trait, using GCTA software, accounting for the population structure through a HD-based genomic relationship matrix. A total of 49, 17 and 45 significant QTL (-logP>6) were detected in MO, NO and HO breeds, respectively. Among them, 9, 2 and 11 QTL in MO, NO and HO breeds, respectively, were highly significant (-logP>9). Most of the QTL affected only 1 trait in 1 breed, but locations on chromosomes 4, 6, 17, 19 and 29 showed significant results for 2 to 5 traits within a breed and/ or similar traits in 2 breeds. Multi-markers analyses (BayesC method) were realized on targeted regions around the QTL (from 2 to 9 Mb length) using GS3 software, to reduce the effect of long distance linkage disequilibrium and to narrow the location of the potential causative mutations. Combining these results with functional annotations led us to several good candidate genes, such as RBM19, GC, NPFFR2, RASSF6 and LIFR. Authors acknowledge the financial support from APIS-GENE and the contribution of the 1,000 bull genomes consortium.

Including cow genotypes improved the genetic predictions of claw disorders in Norwegian Red
C. Ødegård[1], M. Svendsen[1] and B. Heringstad[1,2]
[1]Geno Breeding and AI Association, Storhamargata 44, 2317 Hamar, Norway, [2]Norwegian University of Life Sciences, Department of Animal and Aquacultural Sciences, P.O. Box 5003, 1430 Ås, Norway; cecilie.odegard@geno.no

Genotyped cows could be an important source of information when predicting breeding values of low heritable traits, such as claw disorders. The aim of the study was to validate the effect of including genotypes of cows in prediction of breeding values using single-step GBLUP (ssGBLUP). The three claw health traits included in the total merit index of Norwegian Red were analyzed: corkscrew claw, infectious claw disorders (group of heel horn erosion, dermatitis and interdigital phlegmon) and laminitis-related claw disorders (group of sole ulcer, white line disorder and sole and white line hemorrhage). The dataset consisted of 410,292 claw health records from 263,518 cows after 2,412 sires and from 6,992 herds, and was analyzed using a multivariate animal model. The frequency of corkscrew claw, infectious claw disorders and laminitis-related claw disorders were 11%, 8% and 9%, respectively. Number of genotyped cows with at least one claw health record was 2,416. Number of genotyped sires was 2,235, where 1,396 sires had at least 30 daughters with claw health records. These latter sires were randomly assign to 10 groups and included in the 10-fold cross-validation. Three relationship matrices based on pedigree (A) and genomic (G) information combined to a H-matrix in ssGBLUP were compared: (1) a matrix only; (2) G-matrix for genotyped sires and A-matrix for the rest; and (3) G-matrix for genotyped sires and cows and A matrix for the rest. The predicted breeding values from the 10 validation sets, with reduced data, were correlated to breeding values predicted using ssGBLUP including all claw health records and a relationship matrix consisting of all available genotypes and pedigree information. The results showed that including genotypes of sires were beneficial when predicting breeding values of claw disorders and including cow genotypes achieved a slightly further improvement.

A genome wide association study for Bovine Digital Dermatitis using M-stages
G. Kopke[1], R. Jungnickel[1], H. Dressel[1], B. Waurich[1], M. Wensch-Dorendorf[1], F. Rosner[1], H.H. Swalve[1] and D. Döpfer[2]
[1]Martin-Luther-University Halle-Wittenberg, Institute of Agricultural and Nutritional Sciences, Th.-Lieser-Str. 11, 06120 Halle, Germany, [2]University of Wisconsin, School of Veterinary Medicine, 2015 Linden Drive, Madison 53706-1102, USA; grit.kopke@landw.uni-halle.de

Bovine Digital Dermatitis (BDD) is an infectious disease with global importance. Besides the influence of Treponema spp. and environmental factors, host genetic factors could play an important role in the etiology of this claw disease. The objective of this study was to identify chromosomal regions contributing to a cow's resistance against BDD, or affecting the course of disease. A total of 6,735 Holstein cows were scored for BDD on 6 large dairy farms. Each farm was visited three times for repeated scoring. Every cow received a score for M-stage (0, 2 or 4 as described in Döpfer et al. in 1997) to classify BDD lesions and one score for signs of chronicity for the hind legs. Chronicity was defined as none (healthy skin), hyperkeratotic (thickened skin) or proliferative (diffuse, filamentous or mass-like overly grown epidermal tissues). Distinct traits were defined and a genome wide association study (GWAS) was performed for 1,366 cows with genotypes (Illumina bovine beadchip 50K and 10K imputed to 50K) using Plink v.1.07 and SAS 9.4. 15 SNPs were detected as significant (Bonferroni adjusted p-value<0.05) and 29 SNPs as suggestive (Bonferroni adjusted p-value<0.1) for the different BDD traits. Several SNPs were significantly related to acute BDD events. A further set of SNPs partly were significantly linked to chronic BDD incidences and partly were associated with skin proliferation. For binary traits reflecting resistance to BDD significant SNPs were located on chromosomal regions different from those associated with categorical traits describing severity status. Data analysis is ongoing as more genotypes for the cows scored will be available from an associated project.

A benefits model for a maternally focused beef breeding program in Ireland incorporating genomics
F. Hely[1], P. Amer[1], C. Quinton[1], T. Byrne[1] and A. Cromie[2]
[1]AbacusBio Ltd, P.O. Box 5585, Dunedin 9058, New Zealand, [2]Irish Cattle Breeding Federation, Shinagh, Bandon, Co. Cork, Ireland; fhely@abacusbio.co.nz

Genetic improvement for maternal traits has recently been identified as a major focus for improvement of the suckler beef industry in Ireland. While selection within Ireland and importation of foreign bulls and semen has led to genetic progress, this progress has been in terminal traits to the detriment of suckler cow efficiency. A benefits model was developed using the Limousin sire breed as a case study to quantify the industry benefits from a breeding programme focusing on improving the replacement index over a 20-year period. The model considered the current flows of genetic merit from bulls that are progeny tested as part of the national maternal beef breeding programme, foreign sources, and pedigree herds through to commercial suckler beef and dairy herds, and the impact of modifying these flows. The benefits model also considered the impact genomics may have on the flow of genetic merit, through earlier use of bulls awaiting full progeny test results, and improved accuracy of the index with genomic selection. Increasing the usage of the top progeny tested maternal AI (GI AI) bulls in both pedigree and commercial suckler herds, by displacing foreign AI usage, proved to be a key strategy for improving the average replacement index over the 20-year period considered. The net present value (NPV) of the benefits from increasing GI AI to 30% usage in pedigree herds and 20% usage in commercial herds was €145M in replacement index units after 20 years. Compounding this with selecting the GI AI bulls from all pedigree herds using genomic selection, and identifying these bulls for widespread use earlier lead to a further increase by €71M in the NPV of total benefits after 20 years. In addition to the total benefits from improving replacement index, the benefits model used the expected change in replacement index component traits to estimate the reductions expected in greenhouse gas emissions from improving the average replacement index of the suckler herd. This gave an expected reduction of 220,000 CO_2 equivalent tonnes annually after 20 years of genetic gain when gains in the Limousin breed (which makes up 45% of the national population) were extrapolated to industry level.

Genomic breeding values of carcass, female fertility and calf survival traits for UK Limousin cattle
K.L. Moore, A. Moran and M. Coffey
Animal & Veterinary Sciences, SRUC, Roslin Institute Building, Easter Bush, Midlothian EH25 9RG, United Kingdom; kirsty.moore@sruc.ac.uk

Carcass traits, female fertility and calf survival are difficult to select for. They are expressed at end of life and in the case of fertility and survival have low heritability making them ideal for genomic selection. The first Limousin UK GEBVs became available in 2016 resulting from a four year project using Visual Image Analysis (VIA) carcass traits records on commercial multi breed animals. A SNP key was developed based on 64,453 VIA carcass records and 4,152 Limousin genotypes imputed to 50k. GEBV accuracy was 0.68 for carcass weight and slaughter age and ranged 0.72 to 0.75 for primal yields; topside, silverside, knuckle, striploin, rump and fillet. This represents a ~10% increase in accuracy compared to conventional EBVs of Limousin sires for VIA carcass traits – and a much reduced generation interval. When compared to the wider Limousin population the increase was much larger at around 40%. A 12 month research project starting in 2016 will expand the suite of GEBVs to include female fertility and calf survival traits. Traits being considered are; age at first calf, calving interval and cow reproductive lifespan. These traits currently receive EBVs but for GEBVs the national British Cattle Movement Service (BCMS) will be used to enrich the existing data to include commercial herds. In BCMS it is compulsory to record all calves that survive past 2 weeks of age and this will enable a GEBV for calf survival post 2 weeks of age. The new female fertility and calf survival GEBVs will be included in the 2017 official Limousin genetic evaluations allowing Limousin breeders to use genomics to make early selection decisions for both carcass and maternal traits and this will accelerate the genetic improvement for these currently difficult to breed for traits.

Underlying mutation(s) for a deformed cleft lip in a Holstein-Friesian sire family
M.M. Judge[1], B.W. Kirkpatrick[2], J.F. Kearney[3] and D.P. Berry[1]
[1]Teagasc, Moorepark, Co. Cork, Ireland, [2]University of Wisconsin, Madison, WI, USA, [3]ICBF, Bandon, Co. Cork, Ireland; michelle.judge@teagasc.ie

A deformed cleft lip phenotype was detected at apparent low frequency among the offspring of a phenotypically normal Holstein-Friesian bull. The objective of the present study was to identify the region(s) of the genome possibly underlying this phenotype. Pedigree analysis of affected calves indicated no obvious inbreeding to a common ancestor thus essentially ruling out the likelihood of a recessive mutation. Bovine50 Beadchip genotypes were available on the phenotypically normal Holstein-Friesian sire and high density genotypes were available on a selection of both his affected (n=23) non-affected (n=44) paternal half-sib progeny. Phased genotypes were used to predict paternally inherited haplotypes in the offspring. In one approach used, the association of the paternally inherited haplotype with the affected phenotype utilised both affected and non-affected offspring in a typical linkage analysis. In the second approach, only offspring with the affected phenotype were used; this approach quantified the significance of the deviation of the inheritance of a given paternal haplotype from the expectation of equal representation of both haplotypes under the null hypothesis of no association. Linkage analysis results, based on all offspring, identified associations for regions on both distal BTA13 ($P<2\times10^{-5}$) and BTA17 ($P<2\times10^{-4}$). When only including affected offspring in linkage analysis, the significance of the association observed on BTA13 increased considerably ($P<1\times10^{-24}$). Analysis of just the non-affected animals revealed no such patterns. No obvious candidate gene existed in this region but several yet uncharacterised genes did exist. Two modes of action could be likely contributors to this phenotype: incomplete dominance of an allele or a mosaic mutation. Whole genome sequence of two affected half-sibs is being undertaken to locate candidate causal mutations in the heterozygote state.

The estimation of additive and dominance effects underlying lameness in Fleckvieh and Braunvieh cows
J. Szyda[1,2], T. Suchocki[1,2], M. Fraszczak[2], H. Schwarzenbacher[3] and C. Egger-Danner[3]
[1]National Research Institute of Animal Production, Krakowska 1, 32-083 Cracow, Poland, [2]Wroclaw University of Environmental and Life Sciences, Biostatistics group, Department of Genetics, Kozuchowska 7, 51-631, Poland, [3]ZuchtData GmbH, Dresdner Str. 89, 1200 Wien, Austria; joanna.szyda@up.wroc.pl

Data available for the analysis includes lameness scores, production information, pedigree and 76,934 SNP genotypes for 2,981 cows representing Brownvieh and Fleckvieh breeds. The major goal of the project was to estimate additive and dominance effects of the SNPs. A series of mixed linear models with a fixed additive and dominance effect of a single SNP and a random additive polygenic effect of a cow were used for this purpose. Correction for multiple testing was performed by a method accounting for an inter-correlation between SNPs expressed by a pairwise linkage disequilibrium. SNPs which showed significant additive or dominance effects on lameness were functionally annotated to the UMD3.1 bovine reference genome. Furthermore Gene Ontology (GO) terms were analysed in order to identified functional clustered significantly enriched within the set of significant SNPs. Finally KEGG pathways were tested for their under/over representation in the significant set as compared to the metabolic pathway constellation represented by the whole bovine exome. Linkage disequilibrium calculation was performed at the Poznan Supercomputing and Networking Center.

Genomic selection: the added value for the industry in the Netherlands

I.J.P. Vermeer, W.M. Stoop and G. De Jong
CRV, Animal Evaluation Unit, P.O. Box 454, 6800 AL Arnhem, the Netherlands; irma.vermeer@crv4all.com

The introduction of genomically enhanced breeding values (GEBV) in The Netherlands and Flanders has led to huge changes in breeding program management, due to improved reliabilities of young breeding stock and the decreasing need for progeny-testing schemes for young bulls. Moreover, the availability of GEBV has also led to a multitude of industry and market opportunities. It has created new developments in herdbook registration, genetic disorders, and on-farm genetic management, both in selecting breeding stock and bull choice. The process from request to breeding value is intricate. The goal is to finish the process from genotype request until GEBV within 21 days. This includes processing the request, sending sample material (ID-labelled containers for hair, nose-swabs, ear-tag samples or blood), sampling by herd owner, CRV technician or veterinarian, sending the sample tissue to a general collection point, and from there to the appointed laboratory for DNA extraction and chip analysis, and finally the processing of the raw genotypes and the genomic evaluation and blending procedure within CRV. Within the process challenges lie in the fields of: (1) consistently maintaining the level of accuracy obtained for GEBV and genotypic information in general; and (2) transparency to adopt this new technology in the market. Within the market, the dominance of young bulls in the breeding value lists has caused a shift from the traditional view on breeding values as being daughter-based predictions, to the new GEBV era with fast moving markets where the genetic progress is higher and bulls are much faster replaced. Due to the speed of developments, it is important to keep your market involved. Clear images can help visualise the mechanics behind GEBV.

Challenges and new developments in dairy cattle nutrition

G. Van Duinkerken and R.M.A. Goselink
Wageningen UR Livestock Research, De Elst 1, 6708 WD Wageningen, the Netherlands; gert.vanduinkerken@wur.nl

Global dairy production will grow about 2% per year with the majority of this extra production coming from developing countries. Farm profitability is highly dependent of input prices (e.g. feed) and output prices (milk), and these prices will be volatile. The increasing dairy production enhances feed demands, and urges the need to increase feed efficiency. Resource use efficiency will be a hot topic in dairy cattle nutrition, with at least three major components: (1) precision livestock feeding concepts; (2) feeding strategies to increase animal health and longevity; and (3) novel feed materials. Precision livestock feeding concerns an integration of emerging data collection technologies, mathematical modelling and our ability to add value to such concepts in common practice. There is an increasing amount of (real time) phenotypical data available which enables us to describe individual variation as well as temporal variation. Examples are automated recording of body weight, (individual) feed intake, body condition scoring, and monitoring of rumen conditions and rumination activity. The specific challenge is to integrate new data and models to improve feeding programmes and develop them into precision feeding concepts. Furthermore, it is a challenge to apply these data for the development of 'early warning' systems, which can be the basis for dietary interventions. Improving animal health and longevity will further increase resource efficiency; it will also have a direct impact on animal welfare, the reduction of antibiotics use as well as societal acceptance of the dairy sector. Focus areas to increase lifespan are the transition period and young stock raising. There is an increasing interest in novel sources of feed energy and protein. Although most business cases for large scale production of such novel feeds are not (yet) valid, it is likely that research efforts will further expand because of a worldwide increasing competition for resources for feed, food and fuel. Research and development on novel feed materials for dairy cattle will most likely focus on feed materials like aquatic proteins and products from biorefinery of biomass. Moreover, there will be a renewed research interest in protein crops, forages, grass-based production and the use of crop residues and by-products.

Is dairy crossbreeding a profitable way for Holstein farms?

C. Dezetter[1,2,3], N. Bareille[3], D. Billon[3], C. Cortes[2], C. Lechartier[2] and H. Seegers[3]
[1]PASS'SAS, 4 rue des Epicéas, 25640 Roulans, France, [2]ESA Angers, 55 rue Rabelais, 49007 Angers, France, [3]INRA-Oniris-UMR BIOEPAR, La Chantrerie, 44307 Nantes, France; c.dezetter@groupe-esa.com

The objective of this study was to assess ex-ante the effect of introducing dairy crossbreeding in Holstein dairy operations on farm profitability. A three-breed crossbreeding scheme Holstein × Montbéliarde × Scandinavian Red was introduced in Holstein dairy operations and compared to keeping on Holstein pure breeding scheme. Two initial operations were simulated according to the prevalence of reproduction and health disorders (average or high). Two management strategies were set to keep either a constant number of cows or a constant volume of milk sold during the 15 years of simulation. Milk prices and concentrates costs were set to the French prices for 2014. Economic effects from the changes in performance for milk yield and solids, reproduction, health and longevity induced by different genotypes were assessed with an individual-based mechanistic, stochastic and dynamic simulation model. After 5 years of simulation, crossbreeding scheme reduced average milk yield per cow-year compared to Holstein scheme. However, changes in milk contents, reproduction, udder health and longevity were always in favour of the crossbreeding scheme. When keeping a constant number of cows, averaged margins over 15 years were quite similar between crossbreeding and pure breeding schemes in operations with average prevalence of disorders. However, in operations with high prevalence of disorders, the crossbreeding scheme increased margins up to 90€/cow-year. When keeping a constant volume of milk sold, the crossbreeding scheme improved margins up to 10€/1000 l for both prevalence levels of disorders. As the global economic context for dairy operations is quite volatile with an unfavourable trend, crossbreeding appears to be a possible way to improve Holstein farms profitability.

Challenges, new developments, techniques, and knowledge gaps in cow herd health

H.W. Barkema
University of Calgary, Faculty of Veterinary Medicine, HS 3330 Hospital Drive NW, Calgary, AB, T2N 4N1, Canada; barkema@ucalgary.ca

In response to economic realities, innovations, consumer expectations and regulations, the dairy chain is changing rapidly worldwide. Changes include a reduction in number of farms, herd size, and increased numbers of organic farms. As a result of increasing herd size, but in some countries also in response to animal welfare regulations, the proportion of dairy herds housed in tie-stalls has decreased considerably. Although in some countries access to pasture is regulated, in countries that traditionally practiced seasonal grazing, fewer herds let their dairy cows graze. Concurrently, to remain competitive farmers have been required to adopt increased standards for food safety and biosecurity, become less reliant on the use of antimicrobials and hormones and provide assurances regarding animal welfare. Additionally, there has been wide adoption of technologies that resulted in dramatic changes in management (e.g. automated milking systems, cow activity monitors, and automated calf feeders) and allow for automatic collection of data, as well cow-side tests (e.g. milk-based tests for pregnancy determination). However, utilization of data and action lists that these systems generate for health and welfare of livestock is still largely unrealized, and more training of dairy farmers, their employees and their advisors is necessary. Finally, possibilities of using milk for disease diagnostics and monitoring are considerable and dairy herd improvement associations will continue to expand the number of tests offered to diagnose diseases and pregnancy. Undoubtedly, these changes will have a profound impact on health and welfare of dairy cattle. In this presentation, an overview is presented of some of the most important recent changes in the dairy sector that affect health and welfare of dairy cows, and the science associated with these changes. Additionally, knowledge gaps are identified where research is needed to guide the dairy sector through changes that are occurring, or that we expect will occur in the future. There is every expectation that changes in the dairy sector will be further accentuated.

Challenges and developments in nutrient use efficiency with a focus on land and manure management
D.A. McConnell .
Agriculture and Horticulture Development Board, Animal Science Department, Stoneleigh Park, Kenilworth,
Warwickshire, CV8 2TL, United Kingdom; debbie.mcconnell@ahdb.org.uk

With global food demand expected to double by 2050 and changing dietary consumption patterns, the long term expansion opportunity for the dairy sector is significant. However, since 1960 the agriculture land area per capita has halved to 0.7 ha, requiring more efficient farming systems and increased productivity per hectare. Concurrent to increased demand for dairy products, are strengthening societal and legislative pressures to reduce environmental footprint and improve the sustainability of our food chain. This has translated into additional demands on land function, requiring not only primary food production but water and carbon regulation, nutrient cycling and provision of habitats for functional biodiversity. Since 2000, both nitrogen and phosphorus use efficiency have increased in European agriculture. However, within the European dairy sector there is continued opportunity to improve nutrient use efficiency at farm level. This can be achieved though the uptake of existing best practice guidance (for example, 58% of all UK grassland soil samples remain below target for pH, phosphorus and potassium content) and through advances in research and innovation. New developments in plant breeding such as identifying genetic markers for improved nutrient use, advances in technology for precision application of nutrients and novel methods of facilitating nutrient recovery from animal manures also provide exciting opportunities to further increase nutrient use efficiency on farm. In addition, precision technologies have a key role to play in data collection to allow for more accurate estimation and improved understanding of nutrient distribution and nutrient cycling on farms. This presentation will explore the key barriers to improving nutrient use efficiency on European dairy farms and will consider the knowledge exchange mechanisms and technologies required to overcome these barriers.

Role of Dilution of Maintenance in reducing the environmental impact of animal protein production
R.A. Cady
Elanco, Global Sustainability, ScientificAffairs, 101 Thresher Dr., Unit 51, Eureka, Missouri, 63025, USA;
cadyra@elanco.com

Origin of the term 'dilution of maintenance' appears to be lost to history. Earliest mentions of the concept go back to the turn of the 20th century and 1st appearance of the term appears to be in 1970s animal nutrition textbooks. Maintenance is defined as the obligatory nutritional requirement necessary for healthy survival at basal metabolism. Primary factors influencing maintenance requirements are metabolic size, gender, physical activity level, immunological status, and ambient temperature. Maintenance is an overhead nutritional requirement that must be met before any non-obligatory productive function can be sustained (i.e. growth, reproduction, lactation) unless body reserves are mobilized (a temporary state). Once maintenance requirements are met, nutritional intake can be partitioned to productive activities until the requirement for such activities are met or the activity ceases. When all requirements are met, excess nutrients are either excreted or stored as body reserves. Maintenance requirement is not, for the most part, influenced by the level of productive functions. Thus, maintenance requirement of an open 700 kg dairy cow producing 20 kg milk daily is essentially equivalent to her 700 kg cohort producing 25 kg daily. The higher producing cow requires more nutrients than the cohort to meet the nutrient requirements for lactation but not for maintenance. Thus, measured as a percentage, the maintenance requirement per kg of milk produced is smaller for the higher producing cow, allowing a higher percentage of consumed feed to support milk production. This is an example of 'dilution of maintenance'. There are two choices for increasing animal source protein production; increase animal numbers or increase productivity. Increasing animal numbers increases the environmental footprint per unit of protein produced because of increasing the population maintenance requirement. Increasing productivity reduces the environmental footprint per unit of protein produced by spreading the fixed cost of maintenance over more units of production. Other methods also exist for reducing maintenance of the production sector by improving health and cow comfort which reduce waste and the need for additional animals in the population.

Helping dairy farmers to improve economic performance utilizing data-driven DSS tools
V.E. Cabrera
Department of Dairy Science, University of Wisconsin-Madison, 1675 Observatory Dr., Madison, WI 53706,
USA; vcabrera@wisc.edu

Dairy farming is a highly dynamic and integrated production system that requires continuous and intense, data-driven, decision-making. Several dairy farm components that include cattle, crops, soils, weather, management, economics, and environment are extremely and dynamically interrelated. Therefore, successful dairy farming management and decision-making for improving economic performance can only be achieved by effective and efficient use of meaningful data together with decision support system (DSS) tools within an integrated systems approach. This presentation will describe the development and application of a suite of more than 40 computerized DSS tools aimed to assist dairy farm managers and dairy farm advisors to improve their continuous economic decision-making and problem solving abilities. No single or special methodology was used to develop each or all of these DSS tools, but instead a combination and adaptation of methods and empirical techniques with the overarching goal that these were: heavily farm specific data-driven, highly user-friendly, grounded on the best scientific information available, remaining relevant throughout time, and providing fast, concrete, and simple answers to complex farmers' questions. These DSS are becoming innovative tools by applying high quality farm data and expert information to useful and farm-specific management decisions taking advantage of latest computer technologies. All the DSS tools object of this presentation are hosted at http://DairyMGT. info, Tools section and are categorized within dairy farming management areas such as: nutrition and feeding, reproductive efficiency, heifer management and cow replacement, production and productivity, price risk management and financial analysis, and environmental stewardship. The presentation will reveal practical and real-life applications of a number of these DSS tools to demonstrate satisfactory system assessment, acceptable future predictability, adequate scenario evaluation, and, consequently, satisfactory decision-making. The presentation will include discussion of the trade-offs among data amount and quality, user-friendly design, computational detail, accuracy of calculations, and bottom line performance and effective decision-making.

Outlook fo dairy markets, price volatility and farm economic performance in Ireland
T.H. Hennessy and T. Donnellan
Teagasc, Athenry, G, Ireland; thia.hennessy@teagasc.ie

This paper will examine the development of the Irish dairy sector in the period leading up to milk quota removal and the developments over the last two years. Particular attention will be paid to the impact of price volatility on farm income and farmers' expansion plans. The economic performance of expanding dairy farms will be reviewed and in particular the ability of farmers to cope with milk price volatility will be discussed. Future investment requirements will be estimated and stress tested for alternative milk prices.

ECO-FCE: Growth and intestinal microbiota differ in pigs ranked on RFI on three European sites

U.M. McCormack[1,2], T. Curiao[2], S. Buzoianu[2], E. Magowan[3], B. Metzler-Zebeli[4], D. Berry[2], F. Crispie[2], O. O'Sullivan[2], P. Cotter[2], H. Reyer[5], G.E. Gardiner[1] and P.G. Lawlor[2]
[1]Waterford Institute of Technology, Department of Science and Computing, Waterford, Waterford, Ireland, [2]Teagasc, Research Centre, Moorepark, Fermoy, Cork, Ireland, [3]Agri-Food and Biosciences Institute (AFBI), Pig Department, Hillsborough, NI, United Kingdom, [4]University of Veterinary Medicine, Department for Farm Animals and Veterinary Public Health, Vienna, Vienna, Austria, [5]Leibniz Institute for Farm Animal Biology (FBN), Institute for Genome Biology, Dummerstorf, 18196 Dummerstorf, Germany; ursula.mccormack@teagasc.ie

Residual feed intake (RFI) is the difference between a pig's actual feed intake and its predicted feed requirements for maintenance and growth. The objective here was to investigate the growth performance and intestinal microbiota composition of pigs ranked on RFI on three sites (Ireland, Northern Ireland, Austria). Sows on each site were inseminated using MAXGRO (Hermitage Genetics, Ireland) semen and the resultant 409 progeny were reared to slaughter using a standardised management and dietary regime. Feed intake, weight, back-fat depth and muscle depth were recorded weekly for individual pigs between day 42 and day 91 post-weaning (pw) for the calculation of RFI. RFI was calculated as the residuals from a least squares regression model of average daily feed intake on average daily gain (ADG), metabolic live-weight, gender, and all two way interactions as well as the effects of back fat, muscle depth and lean. Pigs within litter and across site were stratified into high (HRFI; n=60) and low (LRFI; n=60) RFI, with a minimum spread of 2 std. deviations from the mean between the high and low RFI. Microbial profiling was performed on faeces collected at day 42 and 105 pw and on ileal and caecal digesta collected at slaughter, using high-throughput 16S rRNA gene sequencing (Illumina MiSeq). Growth performance was analysed using the Mixed procedure of SAS, and microbiota sequence data were analysed using QIIME. Weight and ADG did not differ between RFI ranks or between sites (P>0.05). ADFI was lower in LRFI pigs (P<0.001) and FCE was improved in the LRFI pigs (P<0.001). Intestinal microbial composition differences were found between RFI ranks and they could be exploited as biomarkers for feed efficiency.

ECO-FCE: Assessing differences in intestinal function in pigs of low and high residual feed intake

B.U. Metzler-Zebeli[1], P.G. Lawlor[2], E. Magowan[3], M. Hollmann[1] and Q. Zebeli[1]
[1]University of Veterinary Medicine Vienna, Veterinaerplatz 1, 1210 Vienna, Austria, [2]Teagasc, Moorepark, Fermoy, Co. Cork, P61 C996, Ireland, [3]Agri-Food and and Biosciences Institute, Large Park, Hillsborough, N. Ireland BT26 6DR, United Kingdom; barbara.metzler@vetmeduni.ac.at

A plethora of physiological processes take place in the small intestine that may influence nutrient demand and digestive capacity and thereby contribute to diverging feed efficiency in pigs. Here, we aimed to investigate the small intestinal morphology, duodenal disaccharidase activity, jejunal permeability and gene expression in relation to sugar and short-chain fatty acid transport, tight-junction proteins and innate immune response in finishing pigs selected for diverging residual feed intake (RFI) at d 98 post-weaning. Mucosal permeability of the jejunum was determined by the Ussing chamber technique and gene expression by RT-quantitative PCR. Low RFI pigs had shorter crypts in the duodenum and a 2-fold greater duodenal lactase activity than high RFI pigs (P<0.05), whereas duodenal maltase and sucrose activity did not differ between RFI groups. Low RFI was further associated with an increased luminal mucosa-to-blood flux of horseradish peroxidase in the distal jejunum compared to high RFI (P<0.05), whereas gut electrophysiological parameters and relative expression of ZO1 and OCLN were similar for both groups. The absorptive response to mucosal addition of D-glucose tended to be greater in high RFI than in low RFI pigs, whereas relative expression of SGLT1 in jejunal mucosa did not differ between RFI groups. Low RFI pigs had a lower jejunal expression of TLR4 (P<0.1) and TNFA (P<0.05) but higher expression of ALPI than high RFI pigs. Differences found in the activation of the jejunal immune response may be due to different microbial profiles in low and high RFI pigs. Results further indicated that jejunal permeability and sugar digestion and uptake by the small intestine may explain some of the variation in finishing pigs of diverging RFI.

ECO FCE: Holistic transcriptome-profiling of feed conversion efficiency (FCE)-divergent pigs

H. Reyer[1], M. Oster[1], E. Magowan[2], R. Quintanilla Aguado[3], J. Horodyska[4], R.M. Hamill[4], P. Lawlor[5], B. Metzler-Zebeli[6], E. Murani[1], S. Ponsuksili[1] and K. Wimmers[1]
[1]Leibniz Institute for Farm Animal Biology (FBN), Wilhelm-Stahl-Allee 2, 18196 Dummerstorf, Germany, [2]Agri-Food and Biosciences Institute, Large Park, Hillsborough, Co. Down BT26 6DR, United Kingdom, [3]IRTA, Animal Breeding and Genetics, Torre Marimon, E08140 Caldes de Montbui, Spain, [4]Teagasc, Food Research Centre, Ashtown, Dublin 15, Ireland, [5]Teagasc, Pig Development Department, Moorepark, Fermoy, Co. Cork, Ireland, [6]University of Veterinary Medicine Vienna, Veterinärplatz 1, 1210 Vienna, Austria; reyer@fbn-dummerstorf.de

For economic and ecological reasons, FCE is of particular interest in livestock. However, there are still knowledge gaps about underlying host-based molecular mechanisms. Thus, FCE relies on biological processes affecting the whole organism. Specifically, regulatory circuits including the sensation and perception of intrinsic requirements, appropriate adaptive systemic responses, and processes of resource allocation are involved. Transcriptomic analyses within ECO-FCE focused on tissues of the gut-brain axis (duodenum, jejunum, ileum, caecum, hypothalamus) and metabolic tissues (muscle, liver) of closely related pigs divergent in FCE. Moreover, individual molecular differences linked to gut microbiota and various environmental stimuli (feeding regime, ingredients and food additives, nutritional condition) were examined. The holistic analyses showed the systemic make-up of FCE with slight transcriptional changes in the hypothalamus which were cascaded and amplified in the body resulting in considerable alterations in posterior responsive tissues like liver and muscle. Interestingly, the analyses highlighted the thyroid signalling pathway as affected due to divergent FCE. Moreover, alterations in muscle tissues were classified into three categories: accumulation of fatty acids, adhesion of connective tissue and apoptosis. The systemic transcriptome-profiling provides potential candidate genes and regulated themes for FCE and, ultimately, identifies strategies to improve efficiency while reducing the ecological footprint of livestock.

ECO-FCE: Feed efficiency related gut microbiota profiles vary in chickens raised at two locations

S.-C. Siegerstetter[1], R.M. Petri[1], E. Magowan[2], Q. Zebeli[1], P.G. Lawlor[3] and B.U. Metzler-Zebeli[1]
[1]University of Veterinary Medicine Vienna, Veterinärplatz 1, 1210 Vienna, Austria, [2]Agri-Food and and Biosciences Institute, Large Park, Hillsborough, N. Ireland BT26 6DR, United Kingdom, [3]Teagasc, Moorepark, Fermoy, Co. Cork, P61 C996, Ireland; sina-catherine.siegerstetter@vetmeduni.ac.at

The gut microbiota is co-responsible for harvesting energy from feed and thus may influence the chicken's residual feed intake (RFI). Information is scarce on whether RFI-related bacterial profiles are similar or specific for different rearing environments. This study aimed at characterizing the gut bacterial microbiome and predicted metagenome of broiler chickens of diverging RFI raised at 2 locations. Two identical broiler chicken trials were conducted in Austria (L1) and UK (L2). Feces, ileal and cecal digesta of low (n=34) and high (n=35) RFI chickens were used for Illumina MiSeq sequencing and data were analyzed using PROC MIXED in SAS. Sex and RFI did not affect total bacterial 16S rRNA gene copies per gram digesta and feces. For many highly abundant OTUs, results for RFI-effects at L1 differed from those at L2. Low RFI males and females at L1 comprised more Turicibacter (one of the highly abundant OTUs) in the ileum than high RFI birds, whereas high RFI males at L2 had more Turicibacter than low RFI birds. Additionally, high RFI males at L1 had more Lactobacillus in ileal digesta compared to low RFI chickens, whereas this genus was associated with low RFI in males and females at L2. In cecal digesta, mainly unclassified species of the order Clostridiales were differentially abundant between low and high RFI chickens, showing again diverging results for RFI at L1 and L2. Predicted metagenomics using KEGG pathway analysis supported location-specific gut microbiota profiles as indicated by RFI × location interactions for important metabolic pathways (e.g. amino acid-, carbohydrate-, vitamin-, and lipid metabolism, bacterial chemotaxis and cell motility). Thus, clearly distinguishable RFI-related gut bacterial profiles and imputed metabolic functions could be established at the 2 locations. However, location-specific effects made it difficult to clearly associate chicken's RFI with specific bacterial taxa and metabolic functions.

ECO-FCE: Improved feed efficiency in chickens is associated with intestinal size and function

B.U. Metzler-Zebeli[1], M. Hollmann[1], E. Magowan[2], P.G. Lawlor[3] and Q. Zebeli[1]
[1]University of Veterinary Medicine Vienna, Farm Animals and Veterinary Public Health, Veterinärplatz 1, 1210 Vienna, Austria, [2]Agri-Food and and Biosciences Institute, Large Park, Hillsborough, N. Ireland BT26 6DR, United Kingdom, [3]Teagasc, Pig Development Department, Moorepark, Fermoy, Co. Cork, Ireland; barbara.metzler@vetmeduni.ac.at

In chickens, conflicting results exist regarding the impact of digestion on feed efficiency (FE). Additionally, information regarding the impact of rearing environment on the FE-related variation in gut function is scarce. Here, we investigated differences in nutrient digestion, intestinal size, jejunal permeability and related targeted gene expression in chickens of diverging residual feed intake (RFI) raised at 2 locations (L1, Austria; L2, UK; n=9/sex, RFI and location). Irrespective of RFI, chickens from L1 gained on average 350 g more while having the same FI as birds from L2. Different hepatic energy needs were indicated by linearly increasing liver weight from low to high RFI (P<0.05). Fecal protein digestibility was decreased from low to high RFI (P<0.05). RFI did not affect disaccharidase activity at the mid-jejunal brush border, but L1 chickens had greater maltase and sucrase activity than L2 birds (P<0.05). Gut electrophysiological data and mucosal flux of the permeability marker fluorescein isothiocyanate showed greater jejunal permeability in low vs high RFI females (P<0.05). Conversely, expression of tight-junction proteins was similar for RFI groups but greater in chickens from L2 than L1 (P<0.05). Also, jejunal expression of MCT1 was lower and that of IL1B, ALPI and TLR2 higher in L2 birds than L1 birds (P<0.05). Higher jejunal MCT1 expression in low vs high RFI males may indicate greater SCFA absorption (P<0.05). Differences in microbial activity may have led to trends for RFI × location interactions for jejunal expression of ALPI in females, TLR4 in males, and IL1B in both sexes (P<0.1). Thus, diverging energy needs for first pass nutrition together with increased gut permeability and differential activation of the mucosal immune response in the jejunum may in part explain FE differences in chickens. Location-related variation in digestive enzyme and jejunal gene expression may have contributed to the different growth of chickens from L1 and L2.

ECO FCE: Molecular alterations of broilers differing in feed conversion efficiency (FCE)

H. Reyer[1], N. Trakooljul[1], E. Magowan[2], R. Quintanilla Aguado[3], B. Metzler-Zebeli[4], E. Murani[1], S. Ponsuksili[1] and K. Wimmers[1]
[1]Leibniz Institute for Farm Animal Biology (FBN), Wilhelm-Stahl-Allee 2, 18196 Dummerstorf, Germany, [2]Agri-Food and Biosciences Institute, Large Park, Hillsborough, Co. Down BT26 6DR, United Kingdom, [3]IRTA, Animal Breeding and Genetics, Torre Marimon, 08140 Caldes de Montbui, Spain, [4]University of Veterinary Medicine Vienna, Veterinärplatz 1, 1210 Vienna, Austria; reyer@fbn-dummerstorf.de

An efficient conversion of feed depends on both extrinsic (i.e. diet, housing) and intrinsic factors (i.e. gut morphology, resource allocation) which will determine growth rates and health status. To improve FCE, it is necessary to understand molecular mechanisms of endogenous factors at a tissue specific level. Most FCE-related tissues/organs are specialised but also multi-functional with an array of distinct biological functions. Thus, the intestines facilitate digestion and absorption of nutrients while mediating mucosal immunity. The liver as a major metabolic organ is involved in partitioning and metabolising of macronutrients. These ultimately influence target tissues like muscle and fat, which largely contribute to growth related aspects and gain in live weight. Accordingly, broiler trials were conducted to identify key molecular differences in these efficiency-related tissues. Specifically, the trials addressed molecular adaptations to nutrient challenges like early phosphorus restriction and environmental influences on lifetime performance of broilers. Holistic approaches including RNA-seq were used to generate transcriptome profiles of liver, small intestine, and muscle tissues of FCE-divergent broilers (selected based on residual feed intake). Analyses revealed 172, 120 and 81 differentially expressed genes (DEG) in duodenum, jejunum and ileum between high and low FCE animals. For liver, leg and breast muscle, 159, 186, and 54 DEG were identified. In the gut, enriched molecular pathways were related to growth, immune function, glucose uptake, transport, and lipid metabolism. Altogether, unique expression patterns in distinct tissues emphasise complex mechanisms contributing to the variation of FCE.

ECO FCE: Residual feed intake and feeding behaviour in group housed pigs

M. Palmer[1], E. Magowan[2], B. Metzler-Zebeli[3], P. Lawlor[4] and N.E. O'Connell[1]
[1]Queen's University Belfast, Institute for Global Food Security, 18-30 Malone Road, Belfast, Northern Ireland, BT9 5BN, United Kingdom, [2]Agri-Food and Biosciences Institute, Sustainable Agri-Food Sciences, Large Park, Hillsborough, Northern Ireland, BT26 6DR, United Kingdom, [3]University of Veterinary Medicine, Vienna, University Clinic for Swine, Veterinärplatz 1, 1210 Wien, Austria, [4]Teagasc, Pig Development Department, AGRIC, Moorepark, Fermoy, Co. Cork, Ireland; m.a.palmer@qub.ac.uk

Residual feed intake (RFI) is a measure of feed efficiency (calculated as the observed feed intake minus intake predicted from observed growth). This study aimed to examine the relationship between feeding behaviour and RFI in finishing pigs, focusing on within-litter variation. The study used 336 finishing pigs from 33 litters, housed in pens of 7-13 littermates and fed through single space automatic feeders. Data were collected from four batches of pigs on three sites. Start and end time of feeding bouts and weight of feed eaten was recorded continuously. Data were selected from 2×24 h periods per week during weeks 11-20 of life. Weekly values of each parameter were averaged to give one overall value for each pig. RFI was calculated for the period 10-20 weeks of age. Data were analysed using a REML linear mixed model in Genstat. The fixed model for each feeding behaviour parameter included RFI, gender and gender by RFI interaction, the random model included litter within site and boar. There was no relationship between RFI and the mean size or duration of individual feeding bouts or the number of feeding bouts per kg of feed eaten. As might be expected, there were positive relationships between RFI and feed intake ($P<0.01$), duration of feeding in 24 h ($P<0.01$) and number of feeding bouts in 24 h ($P<0.01$), indicating that more efficient pigs had a lower feed intake, fed for a shorter time and had fewer feeding bouts in 24 h. A positive relationship was also found between RFI and rate of feeding ($P<0.01$); pigs with better feed efficiency had a slower rate of feeding. The finding of differences in feeding rate between littermates with varying RFI suggests that there may be scope to improve RFI by altering feeding practices, however further research is required.

ECO FCE: An industry view on the future focus of feed use efficiency

R.C. Borg
Cobb Europe BV, Koorstraat 2, 5831 GH Boxmeer, the Netherlands; randy.borg@cobb-vantress.com

Feed efficiency is a key breeding objective goal for most livestock genetic improvement programs. Although individual performance data is important for the breeding program, within a commercial environment feed utilization is benchmarked based on group performance. Defining feed efficiency as the ratio of production:intake is logical but does not account for individual components of feed utilization throughout the production period. Therefore it is important to not only consider an individual's feed intake and production but also their contributions relative to health and robustness within the production system. Components of feed efficiency are influenced by both biological (i.e. physiologic stage) and production environment (ration, feed form, management practices). Therefore, to ensure further genetic progress in feed use efficiency it is critical to invest in phenotyping technologies that can capture the component traits of an individual that influence how the individual's performance contributes to overall feed use and total production within the group. In this presentation we will address some specific components of feed efficiency and highlight opportunities where individual phenotypic data recording could be considered within a breeding objective.

ECO FCE: Understanding the genetic architecture of feed efficiency traits in monogastrics

H. Reyer[1], M. Graczyk[2], A. Borowska[2], T. Szwaczkowski[2], P.F. Varley[3], E. Murani[1], S. Ponsuksili[1], R. Hawken[4] and K. Wimmers[1]
[1]Leibniz Institute for Farm Animal Biology (FBN), Wilhelm-Stahl-Allee 2, 18196 Dummerstorf, Germany, [2]Poznan University of Life Sciences, Departament of Genetics and Animal Breeding, Wolynska 33, 60-637 Poznan, Poland, [3]Hermitage Genetics, Sion Road, Kilkenny, Ireland, [4]Cobb-Vantress Inc., 412 East Siloam Springs, AR 72761, USA; reyer@fbn-dummerstorf.de

The sustainable and efficient usage of resources is a major concern in agri-food production. The biodiversity of livestock species regarding their feed conversion efficiency (FCE) provides a huge potential for selective breeding. Genetic analyses are able to show links between genomic variation and variation in efficiency and growth related traits. Therefore, the genetic work undertaken within ECO FCE aimed to elucidate the genetics of FCE in pig and chicken to provide results significant for both science and animal breeding. Specifically, different methodical approaches including GWAS and entropy analyses were employed to identify genomic regions and plausible candidate genes building the molecular foundation of FCE. Analyses were conducted based on populations of each 1000 pigs and broilers, respectively, which were genotyped using moderate density SNP arrays (60K). The analyses of a commercial Cobb broiler line, which was already under selection for FCE traits and preselected based on body weight, revealed only low heritability in feed conversion ratio, feed intake and weight gain. Nevertheless, this population provides the opportunity to focus on genes with low to moderate effects on the traits, as depicted by the identification and evaluation of positional and functional candidate genes AGK and GTF2I. Regarding pigs, the analyses of a terminal sire line population (Maxgro; Hermitage) revealed a high potential for genetic adaptation regarding FCE and feeding behavioural traits. The estimated heritability for these traits ranged from 0.3 to 0.4. Potential candidate genes are involved e.g. in water homeostasis, glycolysis, mitochondria assembly, and beta-oxidation. Consequently, as exemplified by the divergent molecular themes affected in both species, FCE-relevant candidate genes will be further investigated.

ECO FCE: Value of RFI vs FCR in genetic selection programs

M. Shirali[1], P.F. Varley[2] and J. Jensen[1]
[1]Aarhus University, Center for Quantitative Genetics & Genomics, Blichers Allé 20, 8830, Tjele, Denmark, [2]Hermitage Genetics, Kilkenny, Sion Road, Ireland; mahmoud.shirali@mbg.au.dk

Feed utilization efficiency due to its effect on economic and environmental sustainability is one of the main breeding objectives in livestock breeding programs. Here we present new approach using Bayesian inference on direct and correlated response to selection to examine different selection criteria for feed efficiency either as linear (residual feed intake (RFI)) or ratio (feed conversion ratio (FCR)). Response to selection was defined as the difference in additive genetic mean of the selected top 10% population and the total population after adjusting for genetic trends. Bayesian method integrates over unknown population parameters and 'fixed' environmental effects and handles properly ratio traits. Population parameters of component traits (average daily feed consumption (ADFI), body weight gain (ADG) and lean meat percentage (LMP)) of RFI were obtained from a tri-variate model for 3,724 MaxGro terminal line pigs. RFI was estimated genetically (RFI_g) or phenotypically (RFI_p) from conditional distribution of ADFI given ADG and LMP using either genetic or phenotypic partial regression coefficients, respectively. The posterior distribution of breeding value for FCR was derived from posterior distribution of environmental and additive genetic effects on ADFI and ADG. Selection for RFI_g showed direct response of -0.16 (0.01) kg/d in RFI_g and correlated responses of -0.16 (0.01) kg/kg on FCR, -0.15 (0.02) kg/d on ADFI and no effect on production traits. However, selection for FCR resulted in direct response of -0.17 (0.01) kg/kg on FCR and correlated responses of -0.14 (0.02) kg/d on RFI_g, -0.18 (0.03) kg/d on ADFI and 0.98 (0.18)% on LMP. Direct selection for FCR or RFI_p resulted in disproportional selection on production traits. Application of RFI_g to breeding programs would involve joint selection on residual feed intake and its component traits of production. The methodology is applicable to livestock breeding programs having feed intake and production records.

ECO-FCE: Integrative genomic models for FCE in monogastrics

P. Sarup[1], J. Jensen[1], M. Shirali[1], T. Ostersen[2], H. Reyer[3], P. Varley[4] and P. Sørensen[1]
[1]Center for Quantitative Genetics and Genomics, Aarhus University, Blichers Allé 20, 8830 Tjele, Denmark,
[2]SEGES Danish Pig Research Centre, Axeltorv 3, 1609 København V, Denmark, [3]Leibniz Institute for Farm Animal
Biology, Wilhelm-Stahl-Allee 2, 18196 Dummerstorf Mecklenburg-Vorpommern, Germany, [4]Hermitage Genetics,
Sion Road, Kilkenny, Ireland; pernille.sarup@mbg.au.dk

Genetic variance for complex traits is often estimated using linear mixed models that incorporate information from dense single nucleotide polymorphism (SNP) markers using a realized genomic relationship matrix. In such models, individual genetic markers are usually weighted equally and genomic variation is treated as a 'black box.' This approach is useful for selecting animals with high genetic potential, but it does not generate or utilise knowledge of the biological mechanisms underlying trait variation. Here we propose a linear mixed-model approach that can evaluate the collective effects of sets of SNPs and thereby open the 'black box.' The described Genomic Feature Best Linear Unbiased Prediction (GFBLUP) model has two components that are defined by genomic features. We analysed data on average daily gain, feed efficiency, and lean meat percentage of boars from different pig breeds, utilizing genotypes from a 60 K SNP chip. In addition, information on known quantitative trait loci (QTL), gene ontology, or KEGG pathways from public databases and differentially expressed genes were integrated in the GFBLUP as genomic features. Our results showed that the most significant QTL categories were indeed biologically meaningful. Additionally, for high heritability traits, prediction accuracy was improved by the incorporation of biological knowledge in prediction models. A simulation study using the real genotypes and simulated phenotypes demonstrated challenges regarding detection of causal variants in low to medium heritability traits. The GFBLUP model showed increased predictive ability when enough causal variants were included in the genomic feature to explain over 10% of the genomic variance, and when dilution by non-causal markers was minimal. Predictive ability was increased by the inclusion of prior QTL information obtained outside the training data for high heritability traits.

ECO FCE: Improving protein efficiency of livestock

E.D. Ellen, T. Veldkamp and Y. De Haas
Wageningen UR, P.O. Box 338, 6700 AH Wageningen, the Netherlands; esther.ellen@wur.nl

Livestock convert crops and crop products into high quality human edible proteins. However, some protein sources used in feed for livestock could also be used directly for human consumption. Due to the increase in world population and increase in wealth, it is expected that the demand for human edible protein sources (from animal as well as plant origin) will increase. This will result in competition between humans and animals for high-quality protein sources. Therefore, increasing protein efficiency of livestock is an important challenge in livestock production. To improve protein efficiency of livestock, the animal production sector may: (1) use animals that are more efficient with protein sources; and/or (2) efficiently use alternative protein sources or protein sources with a higher percentage human inedible (such as sunflower seed meal, rapeseed meal, peas, vicia faba, lupines and their concentrates, chick peas, lucerne, leaf proteins from grass or sugar beet leaves, algae, duckweed, or insects). The aim of this study was to investigate different future directions to improve protein efficiency of livestock. The focus in this study was on pig and laying hen production, where we focussed on improvement within species. In this study, protein efficiency was defined as 'the amount of protein input to produce one kg of protein output'. To take into account competition between humans and animals for protein sources, protein input is corrected for whether the protein is suitable for human consumption, also known as human edible proteins. In this definition, land use has not been taken into account. To improve protein efficiency, it is important to consider both the animal output (for instance producing more eggs) and the protein input (use of human inedible protein sources). The most promising solutions are: (1) alternative protein sources; (2) precision feeding, taking into account genotype × nutrition interaction and nutrigenomics; and (3) selective breeding. A challenge will be the limited availability of datasets with phenotypic observations for protein efficiency. Close collaboration between geneticists and nutritionists is recommended in order to define a phenotype that can be used for protein efficiency and to optimise selection procedures for protein efficiency.

ECO FCE: Key advances from ECO FCE and how they will shape the future

E. Magowan[1] and N.E. O'Connell[2]
[1]Agri-Food and Biosciences Institute, Hillsborough, BT26 6DR, United Kingdom, [2]Queens' Univeristy Belfast, Institute of Global Food Security, Belfast, BT9 5BN, United Kingdom; elizabeth.magowan@afbini.gov.uk

ECO FCE has developed new ideas, tools and strategies to advance further improvement in the FCE of pigs and chickens and their environmental footprint and overall has advanced the understanding of the factors which drive feed use efficiency (FCE) in monogastrics. A major task in ECO FCE was to collate and utilise the wealth of information that already existed in peer review journals. A key output from this has been the development of a protocol which could be adopted for the future automated collation of information from animal production experimentation reported through peer reviewed literature. The development of nutritional tools, both in vitro and in vivo, to improve FCE was also of key focus in ECO FCE. In vitro 'ileal digestion' methods (adapted to be used on an Ankom system) have proven useful for the multiple screening of exogenous enzymes on different substrates and as preliminary tests for in vivo trials. Also, the nutritional conditioning of chickens has been found to increase the efficiency of absorption or metabolic utilisation for this nutrient later in life for example, results suggest that a positive effect of P nutritional conditioning on performance and bone mineralisation. Specific symbiotics injected in ovo have also been developed which positively affect the ability of the chicken to convert feed to lean gain. A novel aspect of ECO FCE was to identify key factors driving good and poor feed use efficiency in pigs and broiler chickens, when reared across different environments. Strong environmental effects have been observed and the work has suggested that the physiological mechanisms underlying divergence differ for males and females. Interesting correlations with feeding behaviour, gut microbiota as well as macronutrient transport, endocrine regulation, osmohomeostasis and mitochondrial function have been observed. At a genetic level ECO FCE has identified biomarkers of divergent feed efficiency and has development various modelling methodologies to advance the ability to breed better for improved feed use efficiency.

Editing allelic variation into the porcine genome

B.A. Whitelaw
The Roslin Institute and R(D)SVS, University of Edinburgh, Easter Bush Campus, Midlothian EH25 9RG, United Kingdom; bruce.whitelaw@roslin.ed.ac.uk

Animal breeding relies on the availability of genetic variation in the target breeding population. The new molecular tools termed genome editors now allow single generation introgression of alleles that are not present in a given population. This presentation will illustrate this for the porcine RELA gene. We have used two type of genome editor; zinc finger nuclease and transcription activator-like effector nuclease. Using these molecular scissors to produce sequence specific double strand DNA breaks we have produced indels and, when used in conjunction with a DNA template, specific DNA base changes in the porcine RELA gene. These genetic events result in either a truncated RELA gene product or a different allelic version of RELA to be produced. These animals will now be breed to produce cohorts for virus challenge to test the hypothesis that genetic variation in the porcine RELA gene underlies the dramatic differences in African Swine Fever Virus induced pathology observed between different pig species. We apply the genome editors to the zygote but they can be used in conjunction with somatic cell nuclear transfer. Currently we achieve editing in 30% of founder animals born in both pigs and sheep. Elsewhere editors have been used in other livestock species. Challenges for the future is to increase the frequency of editing and to multiplex enabling several loci to be edited simultaneously in the same animal. These new molecular tools enable new strategies to be added to livestock breeding regimes.

Potential of promotion of alleles by genome editing in breeding programs

J.M. Hickey[1], G. Gorjanc[1], M.A. Cleveland[2], R.K. Varshney[3], C.B.A. Whitelaw[1], J.A. Woolliams[1] and J. Jenko[1]
[1]The Roslin Institute and Royal (Dick) School of Veterinary Studies, The University of Edinburgh, Easter Bush, EH25 9RG Midlothian, Scotland, United Kingdom, [2]Genus plc., 100 Bluegrass Commons Blvd., Suite 2200, Hendersonville, TN 37075, USA, [3]International Crop Research Institute for the Semi-Arid Tropics, ICRISAT, Patancheru, India; john.hickey@roslin.ed.ac.uk

Genome editing (GE) is a method that enables specific nucleotides in the genome of an individual to be changed. The aim of this study was to evaluate the potential of GE for improving quantitative traits that are controlled by many quantitative trait nucleotides (QTN), termed herein as promotion of alleles by genome editing (PAGE). Scenarios differed in the (1) number of QTN defining the trait, (2) number of edits per sire, (3) number of edits per generation, and (4) the strategy for the use of PAGE. A base line scenario involved selecting individuals on true breeding values (i.e. GSonly – genomic selection with perfect accuracy) for ten generations. Alternative scenarios complemented this base line with PAGE (GS+PAGE). The effect of different PAGE strategies was quantified by comparing the response to selection, the change in the allele frequency, the number of distinct QTN edited, the sum of absolute effect of the edited QTN, and the inbreeding. Response to selection after ten generations was between 1.04 and 4.24 times higher with GS+PAGE than with GSonly. When the total resources for PAGE were limited editing a few sires for many QTN resulted in more genetic gain and inbreeding compared to editing many sires for a few QTN. Between the scenarios GSonly and GS+PAGE there was little difference in the average change in QTN allele frequency but there was a major difference for the 20 QTN with largest effect. This study showed that PAGE has a great potential for application in breeding programs.

Modelling energy partitioning and milk production performance of suckler beef cows

D. Sapkota[1,2], A.K. Kelly[1], M. McGee[2] and P. Crosson[2]
[1]School of Agriculture and Food Science, University College Dublin, Belfield, Dublin 4, 1, Ireland, [2]Animal & Grassland Research and Innovation, Teagasc, Dunsany Co. Meath, C15PW93, Ireland; deep.sapkota@teagasc.ie

In pasture-based suckler beef systems, beef cows are often faced with changes in feed resources' availability over the annual production cycle. The energy intake from these ressources are partitioned to maintain homeostasis and homeorhesis functions. Thus, the knowledge and understanding of energy partitioning is of interest in respect of its impact on productive outputs in suckler beef cows. The objective of this work was to develop a dynamic model of energy partitioning and performance of suckler beef cows based on Irish Net Energy feeding system. A dynamic model, operating with a daily time step, was developed in Stella (version 9.0) considering a time frame from the establishment of pregnancy in a beef heifer to the end of second lactation. Feed energy consumption is prioritised towards maintenance and pregnancy functions with the remaining energy available for lactation and growth. Where energy intake is in excess or in deficit of these functions, animals have the capacity to deposit or mobilise respectively, this energy in the form of body reserves. A scenario representing a typical feeding system for spring-calving cows in Ireland was simulated. Results indicated that maintenance requirements were approximately 60% of the total energy consumed by the cow. Feed energy consumption was insufficient to meet demands from three months pre to two months post-calving. At this time, the energy deficit was met by the mobilisation of body reserves. Results showed that the model was capable of representing energy partitioning, milk production and individual cow performance. Therefore, the proposed model provides a basis to simulate alternative feeding practices for suckler beef cows in pasture-based systems.

Session 16

Theatre 2

Does suckler cow genotype matter at calving?

D.E. Lowe and F.O. Lively
AFBI Hillsborough, Large Park, Hillsborough, Co. Down, Northern Ireland, BT26 6DR, United Kingdom;
denise.lowe@afbini.gov.uk

The aim of this three year study was to compare two suckler cow genotypes, namely Limousin × Holstein (LH) (sourced from the dairy herd) and Stabiliser (a composite breed), in terms performance at calving. Both dam genotypes were bred to a Stabiliser sire. At calving, cows were scored for calving difficulty, cow temperament, mothering ability and milk supply. In addition, a sample of colostrum was collected as soon after birth as possible in Years 1 and 2 and analysed for milk composition. The continuous variables were analysed as a one-way ANOVA, looking at the effect of breed. Cow date of birth was used as covariate. The scoring variables were fitted in turn with breed as a 2-dimensional contingency table using a random permutation test. There was no significant effect of dam genotype on concentrations of casein, lactose, protein or urea nitrogen in the colostrum. Colostrum from LH dams had a numerically higher fat concentration in both years but it only approached significance in Year 2 (mean concentrations of 13.1 (S) and 48.8 (LH) g/kg; P=0.050). Dam genotype had no significant effect on incidence of calving difficulty. There was no significant effect of dam genotype on cow temperament scores in Year 1 and 2, but S cows tended to have a better temperament score in Year 3 than LH cows (P=0.058). When the three years of data was combined, dam genotype had no significant effect of cow temperament or mothering ability scores. There was also no significant effect of dam genotype on the vitality score of the calves. There was a trend for a higher proportion of LH cows to have a plentiful supply of milk compared to S cows when the 3 years of data were combined (P=0.084), as assessed by the milk supply score. However it was not consistent over the 3 years (P=0.088 in Year 1; P<0.001 in Year 2; ns in Year 3). This study demonstrated that both dam breeds exhibit high maternal attributes at calving. However further work is required to investigate if LH cows have a more plentiful milk supply since this has potential to influence growth rate of progeny.

Session 16

Theatre 3

Net Feed Efficiency (NFE) measured in finishing Stabiliser steers

J.J. Hyslop[1], R. Fuller[2], U. Taylor[2], D. Thirlwell[3], D. Dreux[4] and D. Pullar[2]
[1]SAC Consulting Ltd, BSS, Edinburgh, EH26 0PJ, United Kingdom, [2]BIG, Southburn Offices, Driffield, Yorkshire, United Kingdom, [3]JSR Farming Ltd, Southburn Offices, Driffield, Yorkshire, United Kingdom, [4]Richard Keenan (UK) Ltd, Stoneleigh Park, Warks, England, United Kingdom; jimmy.hyslop@sac.co.uk

Selecting for improved efficiency (lower NFE expressed in DMI/d) is recognised as a viable method of improving producer sustainability and reducing Greenhouse Gas (GHG) emissions from beef production systems. Whilst NFE is usually measured in young breeding bulls, the objective of this study was to measure feed intake, animal performance and NFE in finishing Stabiliser steers prior to slaughter. Individual feed intake, growth and backfat depth were determined in 80 Stabiliser steers during a 56 day measurement period immediately prior to slaughter where a mixed forage:concentrate complete diet (50:50 DM basis) was offered ad libitum. Residuals of the regression of dry matter intake on mid test metabolic liveweight, daily liveweight gain (DLWG), backfat depth and killing out proportion were used to compute individual NFE values. Steers were ranked by NFE into three groups where low NFE (efficient), mid NFE and high NFE (inefficient) were delineated on the basis of 0.5 of the s.d. of the mean NFE value respectively. Differences between these three groups were then tested using the residual maximum likelihood (REML) facility in Genstat 15. Average DLWG did not differ significantly (P>0.05) across the three NFE groups as expected (1.82, 1.87, 1.80 kg/d for the low, mid and high NFE groups respectively, s.e.d.=0.042). Alternatively, NFE values were highly significantly different (P<0.001) between the low, mid and high NFE steer groups (-0.79, -0.04 and +0.83 kg DMI/d respectively, s.e.d.=0.122). Results indicate that NFE calculations can be determined in finishing steers prior to slaughter and may have utility in NFE selection programmes aimed at improving the feed efficiency of Stabiliser cattle.

Associations of feed efficiency and animal category with rumen functional parameters in cattle

S. Lam[1], J. Munro[2], L. Guan[3], M. Steele[3], F. Schenkel[1], S. Miller[1,4] and Y. Montanholi[2]
[1]University of Guelph, Guelph, ON, Canada, [2]Dalhousie University, Truro, NS, Canada, [3]University of Edmonton, Edmonton, AB, Canada, [4]Invermay Agricultural Centre, Agresearch, Mosgiel, New Zealand; slam02@uoguelph.ca

Approximately 75% of total metabolic energy in ruminants is derived from ruminal metabolism involving microbial fermentation and absorption of energetic compounds across the rumen wall. The aim was to identify associations of rumen volatile fatty acids (VFA), microbiology, and histology profiles with feed efficiency (residual feed intake, RFI; kg/d) and across animal categories (calf vs yearling). Grass-fed heifer calves (HC=107) and yearling heifers (YH=3 6) and grain-fed cattle (n=48) were assessed for RFI. Rumen fluid was sampled for VFA and microbial population analysis. Grain-fed cattle were harvested and rumen tissue was collected for histomorphometric analysis. Least square means were compared between HC and YH groups and RFI groups (H=50% efficient; L=50% inefficient per population) using the SAS GLM procedure. Results indicated higher total VFA (44.8, 38.2 µmol/ml), and lower acetate (71.9, 74.7% of total VFA) concentration and bacterial abundance (4.4×10^{10}, 7.1×10^{10} 16s rRNA copy/ml) in HC compared to YH. The comparison of RFI groups showed higher total VFA concentration (45.0, 41.2 µmol/ml), methanogen (1.8×10^{17}, 1.1×10^{17} 16s rRNA copy/ml) and fungi abundance (1.4×10^{12}, 5.9×10^{11} 16s rRNA copy/ml) in H compared to L grass-fed animals. Comparisons in the grain-fed cattle revealed lower acetate (63.9, 73.9% of total VFA) and butyrate concentrations (12.6, 14.7% of total VFA), and greater rumen papillae width (143.7, 124.4 µm) in H vs L groups, with no differences in microbial abundance. Therefore, rumen function is influenced by animal category and results suggest higher methane emissions from feed efficient grass-fed cattle. Additionally, greater papillae width in efficient grain-fed cattle suggests higher ruminal absorptive capacity. This study provides further understanding of biological processes underlying efficiency of feed utilization in beef cattle.

The effect of increased production efficiency in beef production I: Cow population size

L. Aass, B.A. Åby and O.-M. Harstad
Dept. of animal and aquacultural sciences, Norwegian University of Life Sciences, P.O. Box 5003, 1432 Aas, Norway; laila.aass@nmbu.no

In a system with milk quotas, increased milk yield/cow from genetic gain and/or improved management will inevitably lead to a reduced number of dairy cows, consequently leading to reduced dairy beef production. To maintain domestic beef production to match market demands under import restrictions, an increase in suckler beef production is needed. However, beef production contribute significantly to greenhouse gas (GHG) emissions, which, in general, is higher from suckler beef compared to dairy beef. To optimise sustainability, a highest possible volume of dairy beef is thus preferable. A key influencing factor to this is the milk yield per dairy cow. A simulation study involved four scenarios for strategies in Norwegian dairy and beef production to meet the demand for food towards 2030. All scenarios included three prognoses for increase in milk yield/ dairy cow (2%, 1%, 0%). To stipulate market demands of milk and beef in 2030, domestic prognoses for human population growth (1% per year) and future changes in import restrictions were included. This gave varying production goals of milk (1,230 to 1,770 mill. litres) and beef (80' to 110' tonnes), respectively. Production statistics from official and livestock databases for year 2012 and population sizes of 233' dairy cows and 70' suckler cows were used as basis for the simulations. The results revealed that a yearly increase in milk yield/cow had substantial effect both on number of dairy cows, dairy beef production and the need for suckler cows. Dependent of the increase in milk yield per dairy cow, the number of dairy cows decreased up to 42% within the 2% alternative, while the need for suckler cows increased up to 247%, dependent of the 2030 beef production goal. An increase in kg carcass/cow of 1%/year towards 2030 from improvements in population means of functional cow traits and carcass weights reduced the need for suckler cows with 30' to 45' heads, given constant beef production goals.

The effect of increased production efficiency in beef production II: Greenhouse gas emissions

B.A. Åby[1], P. Crosson[2], L. Aass[1] and O.M. Harstad[1]
[1]Norwegian University of Life Sciences, Box 5003, 1430 Ås, Norway, [2]Animal & Grassland Research and Innovation Centre, Teagasc, Dunsany, Co. Meath, Ireland; bente.aby@nmbu.no

In a system with milk quotas, increased milk yields per cow lead to a reduction in the dairy cow population. Thus, fewer dairy-origin calves are available for fattening, resulting in lowered beef production. Where demand for beef increases, beef production from the suckler cow population can compensate for the lowered beef output from the dairy herd. This may increase the GHG emissions, as emissions are generally higher for suckler vs dairy beef. However, increased efficiencies in beef production can increase beef output per cow thus reducing the number of cows required to produce a given beef demand. Aass et al. (this conference) estimated the number of dairy and suckler cows needed given several scenarios for domestic production targets in 2030 and prognoses for increases in milk yield per dairy cow. In addition, the potential for increased efficiencies in beef production on Norwegian dairy and beef cattle farms were identified and the reduced need for suckler cows estimated. The aim of this study was therefore to quantify the effects of increased production efficiencies on GHG emissions (per kg beef and total emissions). GHG emissions were estimated using two whole farm system models, HolosNor and BEEFGEM. Using data for current production performance (250 kg and 277 kg per cow for dairy and suckler cows, respectively) GHG emissions were 18.5 and 25.5 CO_2-equivalents per kg carcass for dairy and suckler beef, respectively. Preliminary analysis showed that improved efficiencies (i.e. increased carcass weights, reduced no. of stillborns, calving intervals and age at first calving) in dairy and suckler cow herds decreases the GHG per kg of beef and thus, total GHG emissions. We therefore conclude that increased efficiencies in beef production both in dairy and suckler cow herds may be an important mitigation option. Further analyses will be performed and results will be presented at the conference.

Environmental sustainability of integrated France-Italy specialized beef chain using the LCA method

M. Berton[1], M. Lherm[2], J. Agabriel[2], L. Gallo[1] and E. Sturaro[1]
[1]University of Padova, DAFNAE, viale del l'Unviersità 16, 35020 Legnaro (Padova), Italy, [2]INRA, UMR1213 Herbivores, Theix, 63122 Saint-Genès-Champanelle, France; marco.berton.4@studenti.unipd.it

Cradle-to-farm-gate Life Cycle Assessment (LCA) method was applied to assess the environmental impact and land competition of the France-Italy beef sector, based on geographically separated suckler cow-calf (French Massif Central) and fattening (North East Italy) phases. The herd activities, as well as production and transport of feedstuffs and materials were included. The batch (i.e. animals homogenous for breed, sex, origin, fattening farm and finishing period) was used as reference, whereas the kg of body weight (BW) sold as functional unit. The study involved 73 Charolais male batches (4,882 heads) reared during 2014. A cluster analysis was performed to connect suckler cow-calf herds with fattening batches. Birth date, BW and age at sale were the variables. For each of the three clusters obtained, a mean French farm was twinned within farm observations of groups of farms in the INRA Charolais Network database. For each batch, number of animals, BW (405±13 kg) and age (316±36 d) at the sale to Italy and at the sale to the slaughterhouse (729±23 kg and 542±36 d, respectively) were collected. During fattening, diet samples were monthly collected and chemically analysed to calculate the dry matter (DM) intake (10.6±0.5 kg DM/head/d) and composition. Impact categories were (expresses as mean ± SD per 1 kg BW): global warming (13.0±0.7 kg CO_2-eq), acidification (199±14 g SO_2-eq) and eutrophication (60±4 g PO_4-eq) potentials, cumulative energy demand (32±4 MJ) and land occupation (LO, 19.2±0.7 m^2/year). Land use competition with potential human edible crops was calculated by considering the permanent grassland (located mainly in mountainous areas) as not competitive. The land occupation without grassland was 3.9±0.3 m^2/year per 1 kg BW sold, implying a competitive LO comparable to monogastric meat production. The inclusion of indicators of land competition should be included in the LCA analyses aimed at evaluate the land occupation and the global efficiency of ruminant farms.

Evaluation of production efficiencies among primiparous suckler cows of diverse genotype at pasture

S. McCabe[1,2], N. McHugh[3], N. O'Connell[1] and R. Prendiville[2]
[1]*School of Biological Sciences, NI Technology Centre, Queens University Belfast, Belfast, United Kingdom,* [2]*Teagasc Grange, Dunsany, Co. Meath, Ireland,* [3]*Teagasc Moorepark, Fermory, Co. Cork, Ireland; simone.mccabe@teagasc.ie*

The replacement index was derived in order to try and identify animals suitable for breeding or selecting replacement on the basis of their genetic merit for maternal traits. The aim of this study was to determine estimates of grass dry matter intake (GDMI) and production efficiency among primiparous suckler cows of diverse genotype. Data were available from 52 high and 32 low index primiparous cows. Cows were managed under a rotational grazing system with a pre and post-grazing height of 11.4 (s.d. 1.56) and 4.4 (s.d. 0.74) cm. Cow live weight (LW) and body condition score (BCS) were recorded every three weeks. Estimates of milk yield were measured at 120 and 156 days in milk using the weigh-suckle-weigh technique and GDMI was established using the n-alkane technique at 126 and 162 days in milk. Cow LW was 555 kg and 566 kg ($P>0.05$) with a BCS of 2.65 and 2.82 ($P<0.001$) for high and low index cows, respectively. High merit cows had significantly greater ($P<0.001$) milk yield than low merit cows; 7.8 and 6.7 kg/d respectively. No difference in GDMI was observed between the genotypes; 11.8 kg and 12.0 kg ($P>0.05$) for high and low index animals, respectively. Estimated milk yield per unit GDMI was similar for high and low animals; 0.58 and 0.54 kg ($P>0.05$), respectively. No difference in GDMI per 100 kg LW was observed between high and low merit cows; 2.20 and 2.11 kg ($P>0.05$) respectively. Milk yield per 100 kg LW for high merit animals was comparable to low merit animals at 1.26 and 1.13 kg ($P>0.05$). No difference in residual feed intake was observed ($P>0.05$) between high (-0.03) and low (0.22) merit animals. Results suggest that with the exception of BCS and a 1.1 kg/d increase in milk yield for high merit cows, genotype had no effect on LW, GDMI or production efficiency for primiparous cows at pasture during mid-lactation.

A survey of commercial beef farms: investigating the impacts of farm management on herd fertility

F.M. Titterington[1], S.J. Morrison[1], F.O. Lively[1], A. Ashfield[2], A.W. Gordon[2] and A. Johnston[3]
[1]*Agri-Food and Biosciences Institute, Large Park, Hillsborough, BT26 6DR, United Kingdom,* [2]*Agri-Food and Biosciences Institute, Newforge Lane, Belfast, BT9 5PX, United Kingdom,* [3]*College Agriculture Food and Rural Enterprise, 45 Tirgracy Road, Co. Antrim, BT41 4PS, United Kingdom; frances.titterington@afbini.gov.uk*

A high level of reproductive performance is essential to maximise profitability in a beef cow enterprise. To assess the impact of farm management on herd fertility, a survey of 105 Northern Ireland beef cow farms was conducted comprising of 441 variables; the relationship between management variables and fertility was analysed. Each herd's average calving interval (CI) and the proportion of cows with a CI greater than 450 days (extended calving interval; ECI) was calculated to establish herd fertility. The relationship between each response variable (CI and ECI) and each explanatory variable (farmers' answers to questionnaire) was examined using univariate linear regression analyses. All response variables found to be associated with the explanatory variables were modelled against each group in turn using a fully automated multivariate stepwise regression algorithm employing the method of forward selection with backward elimination; all analyses were carried out in GenStat. It was found that farmers who vaccinated cows had a 5% reduction in the proportion of cows with ECI ($P<0.05$), and as the number of vaccines administered to a cow increased the CI decreased ($P<0.05$). Regular vaccination of breeding bulls reduced CI by 9 days ($P<0.001$). Bull selection had several interactions with herd fertility, most notable was farmers who used visual selection rather than estimated breeding values (EBV) to select bulls were found to have a 15 day longer CI ($P<0.001$) and 6.7% higher proportion of cows with ECI ($P<0.001$). When asked to rate their herd fertility, farmers who considered herd fertility as good or very good had higher levels of fertility ($P<0.05$) compared with those who did not. In conclusion, the survey found a number of important variables linked to improved fertility including selecting sires based on EBVs and using a robust vaccination programme.

Analysis of suckler cow reproductive performance on 37 Irish beef farms
R.F. Taylor[1,2], M. McGee[2], A.K. Kelly[1] and P. Crosson[2]
[1]*School of Agriculture & Food Science, University College Dublin, Belfield, Dublin 4, Ireland, [2]Animal & Grassland Research and Innovation Centre, Teagasc, Grange, Dunsany, Co. Meath, Ireland, C15 PW93, Ireland; rachel.taylor@teagasc.ie*

As the calf is the primary output generated in beef suckler cow systems, profitability is largely determined by the ability of the cow to produce a calf annually and to remain in the herd for a number of years. A further consideration for pasture-based systems is the requirement to match calving dates with the onset of the grazing season and thus, a calving to calving interval (CI) of close to 365 days is important. Therefore, reduced reproductive efficiency in suckler beef cows results in decreased profitability of suckler beef production systems. In the context of suckler farms being predominantly loss-making, before non-market support payments are included, improving the reproductive performance of the suckler cow herd is a critical element of improved profitability. In addition to CI, key measures of reproductive efficiency include calving rate (number of calves produced per cow per year), age at first calving, mortality rate and dystocia. In Ireland, national figures from the Irish Cattle Breeding Federation (ICBF) reported that in 2015 CI, calving rate and percentage of heifers calving at 22-26 months of age averaged 407 days, 0.82 calves per cow per year and 18% of total first calving heifers, respectively. In this paper we will present the findings of an in-depth analysis of the reproductive performance of 37 Irish suckler beef farms participating in a knowledge transfer programme in which reproductive performance is one of the key areas of focus. We will identify the relationship between farm factors and reproductive efficiency. Furthermore, a key objective is to relate herd level reproductive performance to farm financial records to determine their effect on farm profitability.

Improving efficiency and reducing waste in the beef supply chain
N. Scollan[1], S. Mackintosh[1], I. Richardson[2], A. Singh[3], N. Mishra[3] and J. Draper[4]
[1]*Aberystwyth University, Institute of Biological Environmental and Rural Sciences, Gogerddan Campus, Abersywtyth SY233EB, United Kingdom, [2]University of Bristol, School of VeterinarySciences, Langford, Bristol BS40 5DU, United Kingdom, [3]University of East Anglia, Norwich Business School, Norwich, NR47TJ, United Kingdom, [4]AngloBeef Group, Solihul Parkway, Birmingham B377YB, United Kingdom; ngs@aber.ac.uk*

New innovations in the beef supply chain that improve eating quality, increase colour shelf life and increase efficiency in the production and distribution chain will reduce wastage. The Waste and Resources Action Programme (WRAP, 2008) estimated that 161,000 tonnes of meat and fish were wasted in UK homes, which changed little by 2012, with 56,000 tonnes of beef being discarded, 13% of all beef purchased, 10,000 tonnes of which was 'not used in time'(WRAP, 2012). For beef, it is estimated that retailer 'back of store' waste was ~1,600 animal equivalents, each tonne of waste being equivalent to 4.5 tonnes of carbon. Premature discolouration and 'unusual' odours that consumers find off-putting on opening vacuum packs are two of the most important areas addressed. This paper will review opportunities to reduce waste by taking a participatory approach from retailer to primary production. Identifying critical control points which underlie waste in the beef supply chain is essential and will deliver innovative approaches to reduce. In addition to optimising antioxidants in the meat to protect colour, and modifying boning and packaging practices, novel supply chain management processes are important. The latter employs a variety of operational research and artificial intelligence techniques to minimise waste and enhance the efficiency of beef supply chain, using Decision Support Systems to assist in making improved decisions in real time. This research will be capable of application throughout the beef industry supply chain and also produces knowledge and approaches that may be applied to other meat supply chains.

Alternative finishing strategies for dairy steers

B. Murphy[1,2], A.K. Kelly[1] and R. Prendiville[2]
[1]University College Dublin, Belfield, Dublin 4., Ireland, [2]Teagasc, Animal and Grassland Reseach & Innovation Centre, Grange, Dunsany, Co. Meath, Ireland; brian.murphy@teagasc.ie

The aim of this study was to investigate alternative finishing strategies for dairy steers. Forty five steers were assigned to one of three strategies: slaughtered at 21 months of age and finished at pasture plus 5 kg of concentrate dry matter (DM) per head daily for 60 days (21S) or 110 days (21L) or slaughtered at 24 months of age and finished indoors on grass silage plus 5 kg DM of concentrate per head daily (24MO). During the second season at pasture 21L, 21S and 24MO were rotationally grazed at pasture on a pasture only diet for 118, 173 and 235 days, respectively. Steers were adapted to their finishing diet over a 10 day period. Both 21L and 21S were slaughtered on 3 November 2015 and 24MO was slaughtered on 9 February 2016. Data were analysed using PROC MIXED of SAS. A Tukey test was included to compare the pairwise comparisons between the treatment means. By design, live weight at the start of finishing was greater for 24MO than 21L, 21S was intermediate ($P<0.001$); 538, 432 and 483 kg, respectively. Live weight at slaughter ($P<0.001$) and carcass weight ($P<0.001$) were greater for 24MO than 21S and 21L; 612, 537, 535 kg and 308, 276 and 275 kg, respectively. Kill out proportion tended ($P=0.06$) to be greater for 21L than 24MO, 21S was intermediate; 514, 502 and 513 g/kg, respectively. Conformation score was greater ($P<0.05$) for 24MO than 21L and 21S was intermediate; 4.07, 3.29 and 3.40, respectively. Fat score was greater ($P<0.001$) for 24MO than 21L and 21S; 7.93, 5.93 and 5.33, respectively. Average daily gain (ADG) during the finishing period and lifetime ADG were similar for all treatments. Concentrate DM intake during the finishing period was lower ($P<0.001$) for 21S than 21L, 24MO was intermediate; 283, 542 and 436 kg per head, respectively. These results indicated that steers finished at pasture for 60 days at the end of the second season represent a viable alternative to traditional indoor winter finishing.

Willingness to pay for beef is similar between different consumer groups

S.P.F. Bonny[1,2], J.-F. Hocquette[1,3], D.W. Pethick[2], I. Legrand[4], J. Wierzbicki[5], P. Allen[6], L.J. Farmer[7], R.J. Polkinghorne[8] and G.E. Gardner[2]
[1]INRA, Recherches sur les Herbivores, UMR1213, 63122 Saint Genès Champanelle, France, [2]Murdoch University, School of Veterinary and Life Sciences, 90 South St, Murdoch, 6150, Australia, [3]Clermont Université, VetAgro Sup, UMR1213, 63122 Saint Genès Champanelle, France, [4]Institut de l'Elevage, Service Qualité des Viandes, MRAL, 87060 Limoges Cedex 2, France, [5]Polish Beef Association, Ul. Kruczkowskiego 3, 00-380 Warszawa, Poland, [6]Teagasac Food Research Centre, Ashtown, Dublin 15, Ireland, [7]Agri-Food and Biosciences Institute, Newforge Lane, Belfast BT9 5PX, United Kingdom, [8]Polkinghornes, 431 Timor Road, Murrurundi, NSW 2338, Australia; spfbonny@gmail.com

Quantifying consumer willingness to pay for beef will enhance the development of beef quality systems globally. Over 19,000 consumers from Northern Ireland, Poland, France and Australia participated in a sensory panel where they allocated beef samples to one of four categories: unsatisfactory, good-every-day, better-than-every-day or premium quality. Consumers were then asked to detail their willingness to pay for these four categories, and complete a short demographic questionnaire which recorded their age, income bracket, occupation, gender, composition of the household and their attitude towards beef. Consumer willingness to pay for beef of different quality levels was found to be remarkably consistent between different demographic groups. Consumers were willing to pay between 150-200% more for premium beef and there was a 50% penalty in value for unsatisfactory beef. By far, the greatest influence of willingness to pay was country of origin. This difference was unable to be explained by the other demographic factors examined in this study, such as occupation, gender, frequency of consumption and the importance of beef in the diet. Consumer age had a small negative relationship with willingness to pay and the importance of meat in the diet had a small positive effect. Therefore, we can conclude that the willingness to pay for beef is highly transferrable between different consumer groups.

Suckler cow efficiency: G×E interaction in British and continental beef cattle breeds

M. Wetlesen[1], B.A. Åby[2] and L. Aass[2]
[1]Nord University, Kongens gate 42, 7713 Steinkjer, Norway, [2]Norwegian University of Life Sciences, The Department of Animal and Aquacultural Sciences, Universitetstunet 3, 1432 Ås, Norway; marit.wetlesen@nord.no

Currently, Norwegian consumers have a higher demand for beef meat than the domestic production covers. To increase the production, we need to extend our knowledge about suckler cow efficiency in different breeds related to different management regimes and resource availability. Production intensity varies due to huge differences in geographical and climatic conditions. Therefore, this study aims to investigate the significance of the genetic × environment interaction (G×E) on suckler cow efficiency in British and continental beef cattle breeds in various management systems. Among important parameters affecting suckler cow efficiency is weaning weight, calving interval and age at first calving. Data from the Norwegian Beef Cattle Herd Recording System (SFK) were used in the analyses. In addition, data were collected by interviewing farmers from 31 beef cattle herds about the available resources and the applied management system on each farm. Both British (Aberdeen Angus and Hereford; 18 herds) and continental breeds (Charolais, Limousin and Beef Simmetal; 13 herds) were represented. These herds were located all over Norway, having unique production environments. The production environments varied relatively in areas of cultivated/uncultivated pastures, feeding and management strategies. The herds have been ranked and analyzed according to a production intensity scale, combined with suckler cow efficiency parameters to estimate G×E interactions. The analyses will show if there are any differences in suckler cow efficiency between British and Continental breeds when exposed to varying production intensity. Preliminary results showed that for British breeds the amount of feed used for the cow from one calving to the next varied from 16,754 to 25,859 MJ NE, and for the heifer breeding from 21,692 to 31,470 MJ NE. The calving interval varied from 359 to 426 days and age at first calving varied from 731 to 913 days for all breeds. Weaning weight for heifer calves for British and Continental breeds was 232 and 272 kg, respectively.

Comparative carcass characteristics of progeny from high and low replacement index cows

S. McCabe[1,2], N. McHugh[3], N. O'Connell[1] and R. Prendiville[2]
[1]School of Biological Sciences, NI Technology Centre, Queens University Belfast, Belfast, United Kingdom, [2]Teagasc Grange, Dunsany, Co. Meath, Ireland, [3]Teagasc Moorepark, Fermory, Co. Cork, Ireland; simone.mccabe@teagasc.ie

The aim of this study was to compare the performance of progeny from primiparous cows of diverse genotypes. Data were available from 69 progeny from high and low replacement index dams; 26 heifers and 43 steers, respectively. The dams were sourced from both the dairy and the suckler herd and bred with terminal Aberdeen Angus or Limousin bulls. All animals were similarly managed under a rotational grazing system while at pasture. A 2 (gender) × 2 (slaughter ages) factorial arrangement of treatments was assigned. Heifers at 19 months of age were finished on a pasture only diet and slaughtered off pasture at the end of the second grazing season while those at 21 months were finished indoors during the second winter on a grass silage ad-libitum diet supplemented with 5 kg concentrates DM/day for 72 days pre-slaughter. Steers were slaughtered at 21 months and 24 months and followed the same finishing system as 21 month old heifers. Live weight at slaughter was 570 kg and 568 kg for progeny from high and low merit cows (P>0.05), with a carcass weight of 310.2 kg and 309.8 kg (P>0.05), respectively. Progeny from high merit cows had a similar kill out proportion to low merit cows; 545 and 547 g/kg (P>0.05). Carcass conformation of 6.7 and 6.8 (P>0.05) along with carcass fat scores of 7.7 and 7.6 (P>0.05) was obtained for progeny from high and low merit cows, respectively. Results show that of all traits investigated carcass performance was similar for progeny from high and low genetic merit cows for maternal traits.

Influence of population structure on the compilation of the Bonsmara genomic reference population

L. Bosman[1], R.R. Van Der Westhuizen[2], C. Visser[1] and E. Van Marle-Köster[1]
[1]University of Pretoria, Department of Animal and Wildlife Sciences, Private Bag X20, Hatfield, Pretoria 0028, South Africa, [2]SA Studbook, P.O. Box 270, Bloemfontein 9300, South Africa; evm.koster@up.ac.za

The largest beef breed in South Africa is the Bonsmara, a locally developed composite breed, consisting of approximately 81,000 registered cows. The Bonsmara was developed specifically for adaptability to sub-tropical conditions using objective measurements between 1937 and 1963, and as a breed is subjected to mandatory performance recording. Biological samples have also been stored for the last 10 years. The Bonsmara is therefore well-placed to engage in genomic selection, and is in the process of compiling a genomic reference population. To date, 583 Bonsmara cattle (388 bulls and 195 cows) were genotyped with the GeneSeek® Genomic Profiler Bovine HD™ Chip (GGP-HD) 80K chip (Neogen, Lincoln, NE, USA), and a preliminary study into the population structure of the reference population was conducted. The average minor allele frequency observed for the Bonsmara was 0.280 across 56,248 autosomal SNPs after quality control, while the observed and expected heterozygosities were 0.361 and 0.365 respectively. After pruning the dataset for SNPs in linkage disequilibrium (LD), 19,119 SNPs were retained, averaging 659 SNPs per autosomal chromosome. An average SNP density of 1 SNP per 90 kb was observed. Structure analysis revealed a non-homogenous population with a high amount of introgression, which may negatively influence genomic breeding value prediction accuracy. Genotyping of approximately another 990 Bonsmara cattle are pending, using the GeneSeek® GGP-HD 150K chip. As more animals will be added to the reference population in the future, the profile of the reference population may change in such a way to ensure better GEBV accuracies.

An evaluation of synchronisation methods for suckler herds

F.O. Lively[1], D. Anderson[2] and D.E. Lowe[1]
[1]Agri Food and Bioscience Institute, Agriculture Branch, Large Park, Hillsborough, BT26 6 DR, United Kingdom, [2]AgriSearch, Large Park, Hillsborough, BT26 6DR, United Kingdom; francis.lively@afbini.gov.uk

Whilst there are a wide range of synchronisation programmes used throughout the world, farmer feedback from a pilot study undertaken by AFBI indicated a strong preference for programmes which required minimal handling and fixed time AI. Consequently, an on-farm study was undertaken on 12 herds, to evaluate the role of different synchronisation programmes for beef cows and heifers. The study involved 402 cows and 219 heifers, with two synchronisation programmes for each evaluated during summer 2015. The heifer synchronisation programmes included a 3 handling programme (Day 0 insertion of progesterone device plus gonadotrophin (GnRH) injection, Day 5 progesterone device removal plus prostaglandin injection and 72 hours later (Day 8) fixed time AI (FTAI) and GnRH injection) (H1) and a 4 handling programme (Day 0 insertion of progesterone device, Day 7 prostaglandin injection, Day 8 progesterone device removal and 56 hours later (Day 10) FTAI) (H2). The cow synchronisation programmes included a 3 handling programme (Day 0 insertion of progesterone device plus GnRH injection, Day 7 progesterone device removal plus prostaglandin injection and 72 hours later (Day 11) FTAI plus GnRH injection (C1) and a 5 handling programme (Day 0 insertion of progesterone device plus GnRH injection, Day 7 prostaglandin injection, Day 8 progesterone device removal, Day 10 (36 hours after progesterone device removal) GnRH injection and Day 11 (56 hours after progesterone device removal) FTAI) (C2). The conception to first service was significantly (P<0.05) lower for H1 compared to H2 (0.47 and 0.68, respectively) but there was no significant different between C1 and C2 (0.56 and 0.64, respectively). Combined conception to service 1 and service 2 (which was either AI or stock bull) and overall in-calf rate did not significantly differ across treatments. The conclusion from this study would suggest that there is potential to reduce the number of handlings within cow synchronisation programmes without any significant impact on herd fertility but not with heifers.

Using consumer based assessments to derive economic values for quality of pork meat
A.D.M. Biermann, E. Rommelfanger, J. Anthe, H. Frevert and S. König
University of Kassel, Nordbahnhofstraße 1a, 37213 Witzenhausen, Germany; annedmbiermann@gmx.de

The aim of this study was to evaluate economic values (EV) for meat quality traits in pork by including consumer based assessments. This approach was realized within a consumer survey including a total of 101 respondents. In the first part, respondents tasted meat samples from different animals and stated their willingness to pay (WTP) by comparison to the price they usually pay for this kind of meat. All samples were also analyzed in consideration of different traits: shear force (SF), intramuscular fat content (IMF), meat color (COL), and drip loss (DL). In the second part, different characteristics concerning the appearance of meat were assessed from photographs, i.e. optical color (O_COL), and optical drip loss (O_DL). According to part one, respondents stated their WTP for each characteristic/class. To calculate EV for quality, classes were defined and the averaged WTP was used to derive prices for each class. EV per phenotypic standard deviation (SD) based on WTP for tasted meat samples were -1.91€ for SF, and 28.18€ for IMF at the population mean of 10.10 kg/cm^3 (SD=2.45), and 1.57% (SD=0.49). Implementation of EV in breeding programs will decrease SF and increase IMF through selection and finally improves meat quality in the population. For traits concerning the appearance of meat, EV based on WTP for tasted meat samples were -2.99€ for COL, and 41.90€ for DL at the population mean of 74.18 (0 = bright; 90 = dark) (SD=6.96) and 4.80% (SD=2.86). In contrast, EV for same appearance traits, but based on the WTP using photographs (O_COL and O_DL) were -18.93€ and -14.81€. Those divergent results for same traits by using different methods suggest utilization of photographs for the derivation of economic weights for traits which are not ascertainable by tasting (e.g. color and drip loss). Including consumer based assessments into economic evaluations of traits can be advantageous, especially for breeding programs targeting on niche production with main focus on specific product and meat quality. Future studies might include assessments of a trained consumer panel for more standardization of this approach.

Response of early-weaned beef calves in feedlots to protein level and grain source in the diet
V. Beretta, A. Simeone, F. Pérez and S. Risso
University of the Republic, Ruta 3, km 363, 60000 Paysandu, Uruguay; beretta@fagro.edu.uy

An experiment was conducted to evaluate the effect of dietary protein level and starch source on feedlot performances of spring born and early weaned (81.6±32.4 days old) beef calves. Thirty Hereford calves (111±40.3 kg) were randomly allocated to five total mixed rations (20% alfalfa hay, 80% concentrate). Four rations were sorghum grain based, formulated for similar ME supply (2.6 Mcal/kg) but differing in crude protein level (CP) 13, 16, 19 (S19) and 22% DM; a fifth ration was formulated for 19% CP but substituting sorghum grain for maize grain (M19). Calves were fed ad libitum in individual pens outdoor during 10 weeks, and liveweight (LW) was recorded every 14 days. Dry matter intake (DMI) was determined daily and the mean feed to gain ratio (FGR) was calculated. Apparent OM digestibility (OMD) and blood urea-N concentration (BU) were determined on week 4. The experiment was analyzed according to a randomized plot design with repeated measures. Linear and quadratic effects associated to CP level were tested. LW gain increased linearly with CP (b=0.033 kg/d, P<0.01) with no effect due to starch source (M19:1.31 kg/d vs S19: 1.26 kg/d; P>0.10). Treatments did not affect DMI (kg/d) (P>0.10), consequently CP intake increased linearly for sorghum diets (b=0.074 kg/d, P<0.01). When expressed as percentage of LW, DMI in M19 tended to be lower compared to S19 (3.2 vs 3.6%, P=0.06). As a result, FGR improved linearly with CP level (b=-0.07 kg/kg, P<0.01) and with maize compared to sorghum as starch source (M19: 3.74 vs S19: 4.42 kg/kg; P<0.01). OMD tended to increase linearly with CP (b=0.53, P=0.08) while a quadratic response was observed for BU (maximum 28 g/l for CP=21%). Substituting sorghum for maize did not affect OMD (P>0.10) but it reduced BU (0.17 vs 0.29 g/l; P<0.01). Results suggest that when feeding high grain sorghum diets to early-weaned calves, growth, FGR and N utilization would benefit from increasing CP up to 22%. Given the same CP level, substituting sorghum for maize would improve FGR and N utilization as well.

A parsimonious imprinting model applied to Simmental slaughterhouse data

I. Blunk[1], M. Mayer[1], H. Hamann[2] and N. Reinsch[1]
[1]Leibniz Institute for Farm Animal Biology (FBN), Genetics and Biometry, Wilhelm-Stahl-Allee 2, 18196 Dummerstorf, Germany, [2]Landesamt für Geoinformation und Landentwicklung Baden-Württemberg (LGL), Stuttgarter Straße 161, 70806 Kornwestheim, Germany; blunk@fbn-dummerstorf.de

The expression of genomically imprinted genes depends on the parental origin of their alleles due to epigenetic DNA-modifications. So far, imprinting receives little consideration in animal breeding programs although its pertinence for important traits is widely known in agricultural species. To consider imprinting, a model (imprinting model) with a genetic effect as sire and a genetic effect as dam has been proposed that provides an estimate of the imprinting variance. When large datasets are available for genetic evaluation purposes, their utilization is hardly feasible as the dimension of mixed model equations becomes too large, especially when advanced models such as the imprinting model are used. To reduce the number of equations, we modified the imprinting model by replacing the genetic effect as dam by the maternal grandsire's effect who now represents the maternal side of inheritance. To demonstrate its practical use, we applied the modified model to 1,366,160 fattening bulls accompanied by pedigrees of up to 2,637,761 ancestors to investigate the influence of imprinting on beef performance in dual-purpose Simmental. Carcass conformation, carcass fatness, killing out percentage and net BW gain were analyzed using the modified model in a linear and generalized linear form. Applying the linear model led to relative imprinting variances ranging between 8.56% and 16.64% whereas the generalized linear model revealed slightly higher proportions. For three of the four traits the maternal gamete turned out to be the greater contributor. A model equivalent to the modified model allowed the direct prediction of highly reliable parent-of-origin effects which can be interpreted as imprinting effects in the absence of X-chromosomal variation.

Towards more efficient use of performance test data for Norwegian beef cattle

K. Haugaard
TYR, Storhamargata 44, 2317 Hamar, Norway; katrine@tyr.no

Since 1998, the Norwegian Beef Breeders Association, TYR, has tested beef cattle bulls at a performance test station located in Stange, Norway. The five largest breeds are included in the national breeding scheme: Aberdeen Angus, Charolais, Hereford, Limousin and Simmental. The testing capacity is 80 bulls each year (60 bulls in the years 1998-2003), and the number of bulls from each breed varies from year to year. Currently the performance test starts when the bulls are approximately 230-day-old and lasts for 147 days. During the test, the bulls' weight and intake of forage and grain are registered daily. The information is then used to calculate the bulls' roughage utilization potential with the aim of finding the bulls with the highest daily weight gain on the least amount of roughage. Based on conformation scoring, breeding values from the routine breeding value estimation and the results from the performance test, the best bulls are selected for AI (Artificial Insemination) and are potentially the sires of the next generation of bulls to enter the performance test. To the authors knowledge, this is the only performance test station for beef cattle that register the daily feed intake of roughage and grain as well as daily weight gain of the animals. So far, the information from the test station have been used only for phenotypic selection of new AI bulls after the end of each test. However, with 15 years of data on roughly 1,000 bulls, there is a potential to include the information into the breeding evaluation and breeding value estimation. Selection of AI-bulls based on estimated breeding values rather than phenotypic performance would possibly increase the genetic gain. The aim of the study is to: (1) find the optimal utilization of the performance test data; (2) predict variance components and heritabilities for breeding value estimation; and (3) estimate genetic correlations between performance test data and production traits measured on field data.

Ultra-low density genotype panels for breed assignment of Hereford cattle
M.M. Judge[1], M.M. Kelleher[2], J.F. Kearney[2,3], R.D. Sleator[3] and D.P. Berry[1]
[1]Teagasc, Moorepark, Co. Cork, Ireland, [2]ICBF, Bandon, Co. Cork, Ireland, [3]CIT, Cork, Co. Cork, Ireland;
michelle.judge@teagasc.ie

Breed assignment of a biological sample is advantageous as a means of delivery on consumer expectation of traceable food products. The objective of this study was to accurately quantify the Hereford proportion in biological samples using ultra-low density genotype panels. Medium density genotypes (13,306 single nucleotide polymorphisms [SNPs]) were available on 54,703 commercial and 4,042 purebred animals. The breed proportion of the commercial animals was generated based on the medium density genotypes and this estimate was regarded as the gold standard breed composition. Ten genotype panels (100 to 1,000 SNPs in increments of 100 SNPs) were developed from the medium density genotypes; methods used to identify the most informative SNPs for breed assignment included the Delta statistic, the Fixation (Fst) statistic, and an Index statistic made up of a combination of the Delta and the Fst statistic. Breed assignment analyses were undertaken for each breed, panel density, and SNP selection method separately. Breed assignment was undertaken separately for all commercial animals (n=54,703) and animals deemed to contain some proportion of Hereford based on available pedigree (n=5,187). The predicted breed proportion of animals was then compared to the gold-standard breed prediction. Panel density and SNP selection method affected the correlation between predicted and actual breed proportion. The Index method of SNP selection numerically (though not significantly) outperformed all other selection methods in accuracy when panel density was ≥300 SNPs. The correlation between actual and predicted breed proportion increased as panel density increased. Using 300 SNPs (selected using the Global Index method), the correlation between predicted and actual breed proportion was 0.995. It is necessary to include at least 300 SNPs on genotype panels for the accurate prediction of the Hereford breed proportion of biological samples when chosen from commercially available panels.

Cortisol response to ACTH challenge in beef heifers divergently ranked on phenotypic RFI
A.K. Kelly[1], P. Lawrence[2], B. Earley[2], D.A. Kenny[2] and M. McGee[2]
[1]School of Agriculture and Food Science UCD, Belfield Dublin 4 Ireland, NA, Ireland, [2]Teagasc Grange, Dunsany Co. Meath, NA, Ireland; alan.kelly@ucd.ie

This study aimed to determine the physiological stress response to an exogenous adrenocorticotropic hormone (ACTH) challenge in beef heifers divergently ranked on phenotypic residual feed intake (RFI). Individual intake and growth was measured on 85 heifers (initial live weight 311 kg, s.d. 48.8) offered grass silage ad libitum and 2 kg of concentrate daily for 84 d. The 12 highest [0.48 kg DM/d] and 12 lowest [-0.50 kg DM/d] ranking animals were used for this study. Heifers received dexamethasone (20 µg/kg BW i.m.) 12.40 h prior to the ACTH (1.98 IU/kg metabolic BW0.75) challenge. Blood samples were collected at -40, -20, 0, 20, 40, 60, 80, 100, 120, 150, 180, 210, 270, 300, 330, and 390 min relative to ACTH administration (0 min) for plasma ACTH and cortisol concentrations. Haematology variables were determined intermittently. Data were analysed using a repeated measures mixed models ANOVA (PROC MIXED) in SAS incorporating terms for RFI group, sample time and their interaction, as appropriate. Heifers differing in RFI did not differ (P>0.10) in ACTH concentration. Concentration of ACTH peaked (P<0.001) in both RFI groups at 20 min post-ACTH administration, following which concentration declined to baseline levels by 150 min. Similarly, cortisol systemic profile peaked at 60 min and concentrations remained continuously elevated for 150 min. A RFI × time interaction was detected for cortisol concentrations (P=0.06). High RFI heifers had a greater cortisol response than Low RFI from 40 to 150 mins relative to ACTH administration. Cortisol response was positively associated with RFI status (r=0.32; P<0.01). No effect of RFI or RFI × time (P>0.10) was evident for neutrophil, lymphocytes, monocyte, eosinophils and basophil count. Plasma RBC number (6.07 v. 6.23, P<0.05) and HCT % (23.2 v. 24.5, P<0.05) were greater for low than high RFI animals. Evidence is provided that feed efficiency is associated with HPA axis function and stress response in beef heifers.

GWAS studies reveals genomic regions related to early pregnancy in Nellore (Bos indicus) heifers

G.A. Oliveira Jr.[1,2], A.S.M. Cesar[1,2], J.B.S. Ferraz[2] and D. Garrick[1]
[1]Iowa State University, Animal Science, Ames, IA, USA, Ames, USA, [2]University of Sao Paulo/FZEA, ZMV/NAP-GMABT, Rua Duque de Caxias Norte, 225, Campus da USP, 13635-970 Pirassununga, SP, Brazil; jbferraz@usp.br

Reproductive traits influences the number of animals produced and the generation intervals in livestock production. The understanding of the genomic regions involved could help in identifying more efficient animals, reducing the generation interval and increasing profitability of producers. Two research projects of the University of Sao Paulo, Brazil, in cooperation with the Univ. of Copenhagen and Iowa State Univ. studied early pregnancy in Bos indicus heifers. The database gathered early pregnancy information from Nellore heifers that were exposed to breeding at about 15 months of age (the normal time in Brazil is over 24 months). The gestation diagnosis were made by ultrasound examination 40 days after the breeding. The trait was analyzed as categorical 1 as the success, for heifers that were diagnosed pregnant and 0 as a failure, for those not pregnant. Genotyped were obtained by GeneSeek GGP-HDi technology that holds around 78,000 SNPs. A total of 1,254 animals and 64,800 SNPs passed from quality control and were analyzed using Bayes-B methodology and GenSel software with non-overlapping one megabase (Mb) windows to estimate the genetic variance. The genomic heritability was estimated at 0.09. The results showed three regions spread at chr 5, 14 and 18, that explained more than 1% of the genetic variance, especially chromosome 5 that harbor a region that explained 4.89%. The genes TOM1 and HO-1, which were known as important genes in the 73 Mb window related with pregnancy in humans, are in similar regions with strong signals in this research. Similarly, genomic regions between 20-30 Mb on chr14 were reported in literature to influence reproduction traits in cattle. The regions identified also embrace the gene SAMP14, which has been suggested to have a role in sperm-egg interaction in humans, through acrosomal membrane. In conclusion, these results suggest that the use of genomic can enhance the ability to identify better animals for selection in early pregnancy in Nellore cattle.

An international perspective on research funding

A.R. Peters
Royal (Dick) School of Veterinary Science, University of Edinburghand Scotland's Rural College, Edinburgh, EH26 ORZ, United Kingdom; andy.peters@sruc.ac.uk

Given the well-publicised challenges of population growth, environment, climate change and food security there has been renewed interest in the efficiency of global agriculture in general and livestock productivity in particular. Thus many institutions in the agri-veterinary field are well placed to contribute to solutions for these issues on a global scale. International donors e.g. USAID, the Bill and Melinda Gates Foundation (BMGF) and DFID UK, have seen significant increases in their budgets for food security. There have been several initiatives in Edinburgh to leverage these opportunities. The Global Alliance for Livestock Veterinary Medicines (GALVmed) was recently incorporated to develop and deliver livestock vaccines for developing countries, funded by DFID, BMGF and the EC. More recently BMGF has funded two partnerships between UoE and SRUC: (1) to establish a completely novel Centre for Tropical Livestock Genetics and Health (with the International Livestock Research Institute; ILRI) to focus on integration of tropical adaptations with productivity traits in cattle and poultry in sub-Saharan Africa (SSA); and (2) to create a new team to address cattle mortality, technology evaluation and harmonisation of livestock data collection in SSA. It is increasingly apparent that such complementary partnerships are the key to future success in tackling these global issues.

H2020: influening it's development and implementation
L. Mahy
European Commission, DG Agriculture and Rural Development, Brussels, Belgium; louis.mahy@ec.europa.eu

Horizon 2020, the EU's flagship programme for research and innovation, is by definition an international collaborative programme. It aims for a first class knowledge base built on excellent science, and an environment conducive to innovation in Europe. It also includes the agricultural sector and aims at jobs, growth and investment in rural areas. To filter the most important needs from the sector down to a limited number of research topics, the process via which the latter are formed involves an intensive dialogue with the stakeholders. This presentation will give an outline of some key aspects of this dialogue. An overview will be given of the different channels via which the dialogue takes place and on how livestock research community and – industry can voice their needs. Also the broader context in which these research needs are to be expressed, will be discussed. What is to be expected from the agricultural research community? How does the political context, and more specifically the priorities related to the H2020 programme, relate to them? How does this translate into research topics? What is the relation to the multi-actor approach? In short, this presentation aims at informing the livestock research community about how to further integrate into the H2020 programme.

A better return for dairy farmers on funds for research and knowledge transfer by more international cooperation
W. Koops
ZuivelNL, P.O. Box 93453, 2509 AL The Hague, the Netherlands; koops@zuivelnl.org

Dairy farmer's organizations and industry-led levy bodies in different EU countries work together in ECIP (European Cattle Innovation Partnership) to get a better return on industry and public funds. Besides a short description of members and activities of ECIP more information will be provided how dairy farming research in the Netherlands is financed by public and by industry, with special attention how Dutch dairy farmers collectively contribute to research by ZuivelNL (in English: DairyNL) and the decision making process. An important objective of international collaboration is to reduce duplication and improve the efficiency of resource utilization. Another objective is to support the European Commission in their ambition to reduce the gap between research and farming practice by a more end user driven knowledge development. A reason why ECIP members initiated the 3 year project EuroDairy, a pan-European network for dairy farming funded by the EU Horizon 2020 programme.

The UK agricultural strategy: a platform for sustainable innovation and commercial success
J. Bainbridge
UKTI, UKTI AgriTech Organisation, EH26 0RZ, United Kingdom; janet.bainbridge@ukti.gsi.gov.uk

The UK Government has developed a long-term Agri-tech Strategy in partnership with industry to ensure that the outputs from the UK's world class science base are translated into benefits for society and the economy worldwide. The strategy brings together Government departments, farming and industry representatives, the agricultural science community and research funders. The agri-food sector contributed 7.6% (£103 billion) of national GVA in 2013 and 13% of national employment in 2014 (3.8 million employees) – of which agriculture and fishing contributed £9.8 billion and food and drink manufacturing contributed £26.5 billion GVA in 2013. Government has committed £160million, which will be co-invested with industry to address agri-tech challenges: (1) £70 million the Agri-Tech Catalyst that supports collaborative industry/research projects to take agricultural innovations from the laboratory to the marketplace. (98 projects funded); (2) £90 million to establish world class Centres for Agricultural Innovation to support the wide scale adoption of innovation and technology, developing skills and capability in the food and farming supply chain. The role of the UKTI Agri Technology organisation is to provide a focus for businesses in terms of Inward Investment and Trade. We consist of civil servants and specialists and operate around the world across agri tech innovation. Specifically we have specialist experts in Animal health, Plant science, Precision agriculture and also Trade.

An introduction to the UK's new Centres of Excellence – CIEL, Agri-EPI, Agrimetrics
S. Hoste[1], D. Ross[2] and D. Flanders[3]
[1]CIELivestock Ltd., Sand Hutton Innovation Campus, Sand Hutton, York, YO41 1LZ, United Kingdom, [2]Future Farming Systems, SRUC, Roslin Institute Building, Easter Bush, Midlothian EH25 9RG, United Kingdom, [3]Agrimetrics, Rothamsted Research, Harpenden, Hertfordshire, AL5 2JQ, United Kingdom; enquiries@cielivestock.co.uk

Four new 'centres of excellence' have recently been established in the UK using an £80M investment from the UK government (BIS) through Innovate UK. The centres, namely CIEL, CHAPS, Agri-Epi and Agimetrics, will uplift the capability and capacity of the UK Agri food research arena and will focus on livestock production, crop production, precision technology and 'big data' respectively. They all represent large scale consortia between industry and academia with the key goal of delivering solutions for industry to ultimately improve the profitability of UK Agri-Food. CIEL will focus on research from cell to whole farm systems across all the main farmed livestock species e.g. genomics and physiology through to digestion and reproduction and the resulting health and welfare as well as product quality and safety of all farm animals. CIEL will conduct industry driven projects through its academic base as well as industry partners' facilities and its wide network of commercial farms. Agri-Epi will conduct a wide range of industry-led activities in applied research and development, demonstration, training and education. They aim to become world-leading in engineering and precision agriculture for the livestock, arable, aquaculture and horticulture sectors. Initially Agri-Epi will explore cutting edge technologies e.g. automated vehicles, unmanned aerial vehicles and new instrumentation to monitor both livestock production efficiency and in-field performance of cropping systems. Agrimetrics is the world's first Big Data Centre of Excellence for the whole agri-food industry. It is supporting a revolution in the use of data science and modelling in the agri-food industry and contributing to a more intelligent and therefore more productive, efficient, resilient and sustainable food system.

Pig carcass tail lesions: associations with record keeping and farm performance parameters

N. Van Staaveren[1,2], A. Hanlon[1] and L. Boyle[2]
[1]University College Dublin, School of Veterinary Medicine, Belfield, Dublin 4, Ireland, [2]Teagasc Animal and Grassland Research and Innovation Centre, Pig Development Department, Moorepark, Fermoy, Co. Cork, Ireland; nienke.vanstaaveren@teagasc.ie

Detailed record keeping of financial and farm performance figures is fundamental to improve pig farm productivity. However, it is not known how this relates to aspects of pig welfare. The aim of this study was to examine the influence of record keeping on prevalence of tail lesions in pigs. Associations between prevalence of tail lesions and production parameters at farm level were also investigated. Pig carcasses were observed in 2 abattoirs in Ireland after scalding/dehairing. For each carcass, tail lesion score (0-4), sex and farm number were recorded. Tail lesion scores were collapsed into none/mild (score ≤1), moderate (score 2) and severe lesions (score ≥3). The prevalence of tail lesion outcomes was calculated and data were analysed at batch level. Batches of farms which kept records (ePM batch) were identified by farm number. The difference between ePM batches (n=27) and non-ePM batches (n=46) was analysed using generalized linear mixed models (PROC GLIMMIX). Farm performance data were obtained from the ePM system (n=14) and related to the average tail lesion score (TS) of the batch. On average, 26.8% of pigs in a batch had moderate tail lesions while 3.4% of pigs in a batch showed severe tail lesions. A lower percentage of pigs with moderate tail lesion was observed in ePM batches (23.2±0.92%) compared to non-ePM batches (38.2±0.96%, P<0.001). No difference between ePM and non-ePM batches was found for the prevalence of severe tail lesions (P>0.05). A negative correlation was found between TS and weight at transfer of weaners (r=-0.5, P<0.05). TS was positively correlated with the number of finishing days (r=0.5, P<0.05) and negatively associated with average daily gain in the finisher period (r=-0.4, P=0.07). This study provides the first indication that record keeping is associated with lower risk of carcass tail lesions and that such lesions are associated with farm based productivity data. Further research is needed to examine possible underlying factors which differentiate farms which keep records (ePM) from those which do not.

What do maternal tests actually test?

C.G.E. Grimberg-Henrici[1], I. Czycholl[1], O. Burfeind[2] and J. Krieter[1]
[1]Institute of Animal Breeding and Husbandry, Christian-Albrechts-University, Olshausenstr. 40, 24098 Kiel, Germany, [2]Futterkamp, research farm of the Chamber of Agriculture of Schleswig Holstein, Gutshof 1, 24327 Blekendorf, Germany; cgrimberg@tierzucht.uni-kiel.de

Several studies use behavioural tests to characterise sows with regard to their maternal performance. These studies choose usually a selection of tests and the combinations vary between studies. However, it is unknown which combinations of tests are advisable. Besides, there is rarely valid information about the significance of these tests. In the present study 47 sows were tested in week 2 and 4 of lactation in six successive maternal tests which were conducted in their home pens and in a test arena. The test arena (14.4 m^2) had a piglet nest in a corner and the sows were only able to hear and smell their piglets. The sows' reaction to piglet distress calls, separation from and reunion with their piglets was tested. Sows' behaviour as willingness to protect their piglets, responsiveness to their piglets and searching for their piglets were documented. The observed parameters in these tests were analysed using the statistical procedures principal component analysis (PCA). PCA identifies redundancies in parameters and investigates underlying domains of behaviour. Eight principal components were extracted by PCA which explain 75.9% of the total variance. Component loadings greater than or equal to 0.40 were interpreted as highly positive and component loadings less than or equal to -0.40 as highly negative. Preliminary results show different underlying domains of maternal behaviour of sows (communication, care, contact and local attachment with piglets). Maternal tests investigate different maternal domains and the importance of these tests varies. There is incidence that the experimental environment in which the tests were placed triggered different maternal domains, although the experimental setup was obviously the same. These results show that several maternal tests are needed to cover all domains of maternal behaviour to get valid information about maternal performance of sows.

Chemical Causes of Pig-House Odour

L.J. Farmer[1], J.T. Kennedy[1], T.D.J. Hagan[1] and E. Magowan[2]
[1]Agri-Food and Biosciences Institute, Food Research Branch, Newforge Lane, Belfast BT9 5PX, United Kingdom,
[2]Agri-Food and Biosciences Institute, Agriculture Branch, Large Park, Hillsborough BT26 6DR, United Kingdom;
linda.farmer@afbini.gov.uk

The odour from pig-houses is often a problem both to nearby residents and to pig farmers. Often this forms a barrier to business expansion. A number of researchers have investigated the composition and dynamics of pig-house odour, with a view to devising mitigation strategies. Nevertheless, the chemical components of the odour responsible are not unambiguously understood. Researchers have reported up to five hundred different volatile organic compounds emitted from pig production facilities. Of these, perhaps 20-30 are 'key odour impact compounds'. We have determined which odour compounds contribute most to the characteristic pig-house odour through the calculation of 'odour activity values' (OAV). The OAV is the concentration of a single compound divided by its odour threshold concentration. Compounds with an OAV greater than 1.0 are likely to contribute to the overall odour. OAVs were calculated using odour detection thresholds which had been selected for consistency of their methods of determination. Analysis of published papers providing concentrations of volatile compounds from pig-houses show that the most important odour compounds contributing to pig-house odour are short chain acids, sulphur compounds, phenols, indoles (e.g. skatole) and amines. These compounds have the highest odour activity values and often have pungent and offensive odours. Ammonia, while a contributor and an important greenhouse gas, is not the major cause of pig-house odour. Studies on pig-houses in Northern Ireland confirm these findings and suggest there are factors affecting their concentrations which are not fully understood. The parameters affecting the concentrations and OAVs of these compounds will be reviewed, together with potential mitigation strategies. Evaluation of OAVs suggests that mitigation factors related to slurry management or pig diets can have differential effects on these classes of odorous compounds.

The impact of the sanitary environment on health, pubertal development and sex odour of male pigs

S.P. Parois, A. Faouen, N. Le Floc'h and A. Prunier
INRA, Agrocampus Ouest, UMR 1348 PEGASE, Domaine de la Prise, 35590 Saint-Gilles, France, Metropolitan;
severine.parois@rennes.inra.fr

Previous studies have shown that the sanitary environment can affect the health and growth of pigs. The objective of the present study was to evaluate the effect of a chronic inflammation induced by a degraded sanitary environment on the pubertal development of boars and on boar taint. Boars from a commercial cross line were allocated at 139 ($\pm$0.1) days of age (D0) and 81.3 ($\pm$0.6) kg of live weight either to good (n=61) or poor sanitary conditions (n=54) for 4 weeks. The poor environment was previously soiled by non-contemporary pigs, poorly ventilated, not cleaned during treatment; the good environment benefited from reinforced hygiene measures. Inflammatory status and pubertal development were evaluated on D0 and D27 by measures on blood samples (blood formula, inflammatory proteins; testosterone, estradiol). Every week, boars were weighed, sampled for saliva (estrone, inflammatory protein CRP) and their rectal temperature was measured. Fat was collected on D27 by biopsy and at slaughter (116.7$\pm$0.4 kg live weight) to measure boar taint compounds (androstenone, skatole, indole). Pigs in the poor environment were dirtier (P<0.001). The treatment had a slight effect on rectal temperature (D7: 38.4$\pm$0.04 vs 38.6$\pm$0.06 °C, P<0.001; and D14: 38.3$\pm$0.04 vs 38.6$\pm$0.04 °C, P<0.001) and a marked effect on fat skatole (0.037 vs 0.082 µg/g of pure fat) and indole (0.039 vs 0.14 µg/g) with higher values for boars in the poor environment (P<0.0001). Overall, treatment had no clear impact on the inflammatory status, the pubertal development and androstenone of the boars. However, the soiled increased markedly the dirtiness and fat skatole.

Breed and feeding effects on boar taint under organic conditions

H. Brandt[1], K. Höinghaus[2], R. Bussemas[2] and F. Weißmann[2]
[1]Institut für Tierzucht und Haustiergenetik, Ludwigstr. 21 B, 35390 Giessen, Germany, [2]Thünen-Institut für Ökologischen Landbau, Trenthorst 32, 23847 Westerau, Germany; horst.r.brandt@agrar.uni-giessen.de

A total of 280 boars descending from Duroc and Piétrain sires were fed under organic conditions in the research station in Trenthorst. All boars in the control group were fed with a 100% organic diet to a final weight of 115 kg live weight while half of the boars got the same diet supplemented with 10% native potato starch for the last 30 days of fattening. All carcasses are evaluated for boar taint during the slaughter process and are all analyzed for skatole and androstenone content in the lab. On average 1.045 µg/g androstenone and 0.0483 µg/g skatol with standard deviations of 0.0918 and 0.107, respectively, were measured. Significant differences ($P \leq 0.01$) between sire breed were found for androstenone and skatole. A significant interaction ($P \leq 0.01$) between sire breed and feeding regime was found for skatole. The progeny of Duroc sires showed higher androstenone contents than the progeny of Piétrain sires. Within the boars descending from Duroc sires, there was no effect of the native potato starch feeding on skatole while the progeny of Piétrain sires under the feeding regime with native potato starch showed the same level of skatole like the Duroc progeny but a three times higher skatole level in the control group without native potato starch.

Are testis volume and boar taint related: a trial in entire and immunocastrated boars

M. Bernau[1], S. Schwanitz[1], L.S. Kreuzer[1], P.V. Kremer-Rücker[2] and A.M. Scholz[1]
[1]Ludwig Maximilians University Munich, Livestock Center Oberschleissheim, St. Hubertusstrasse 12, 85764 Oberschleissheim, Germany, [2]University of Applied Sciences Weihenstephan-Triesdorf, Steingruberstrasse 2, 91746 Weidenbach, Germany; maren.bernau@lmu.de

The aim of this study was to evaluate, whether it is possible to predict boar taint in boar carcasses by evaluating different variables at the boar in vivo. A total of 68 boars (Piétrain × German Landrace; 34 entire (EB) and 34 immunocastrated boars (IB)) were scanned during their growth (ø 60 & 90 kg live body weight). Magnetic resonance imaging (MRI) was used to scan the whole body. Different variables were created to determine fat and muscle composition of the shoulder, loin, belly and ham region. Additionally the testis volume was determined. After slaughtering cheek and salivary gland probes were sampled for the sensory analysis. A general linear model procedure of SAS 9.3 was used, with boar group (EB or IB), result of sensory analysis and their interaction as fixed effect. The significance level was set with $P < 0.05$. EB with a positive result had larger testis volumes (scan 90: 523.10 cm^3) than negative ones (437.82 cm^3) or IB. Significant testis volume differences ($P \leq 0.0017$) between EB and IB were detected at the first and second test stage at scan 90. No significant differences could be obtained for the testis volume of sensory test negative and positive IB (315.22 cm^3 and 286.36 cm^3). Regarding the MRI variables, significant differences between the boar groups could be achieved for the variable shoulder fat volume at scan 90 (EB: 59.29±2.73 cm^3; IB: 67.32±2.69 cm^3), the shoulder outline at scan 60 (EB: 1,095.42±12.87 cm^3; IB: 1,048.84±12.87 cm^3) and the testis measurements (volume and area) at scan 90. EB tend to have smaller back fat volumes than IB ($P=0.0535$). MRI allows the detection of differences between EB and IB without tedious dissection. A wide range of possible variables can be created and compared with sensory test results. For EB, testis volume showed a significant relationship to sensory testing.

Consumer evaluation of different cooking methods to deal with boar taint in fresh meat
F. Borrisser-Pairó[1], N. Panella-Riera[1], M. Egea[2], M.B. Linares[2], M. Gil[1], M.A. Oliver[1] and M.D. Garrido[2]
[1]IRTA-Monells, Product Quality, Finca Camps i Armet, 17121, Girona, Spain, [2]University of Murcia, Food Science and Technology, Espinardo, 30071, Murcia, Spain; francesc.borrisser@irta.cat

The presence of boar taint in fresh pig meat may affect its sensory quality and the acceptability of consumers due to the presence of Androstenone (AND) and skatole (SKA). These two compounds are present in meat from some boars. Masking strategies or different cooking methods might be a good way to deal with boar taint in fresh meat. The aim of this study was to compare the acceptability of fresh meat from boars and castrated male pigs cooked with different cooking methods: (1) vacuum cooking (in a bath until a core temperature of 65-70 °C); (2) fried/breaded (with garlic and parsley in olive oil). Two types of meat were used in this study: (1) Type BM: loins from boar with high levels of AND (2.3 μg/g in fat) and medium levels of SKA (0.14 μg/g in fat); (2) Type CM: loins from castrated male pigs with no detectable levels of AND and SKA. One hundred and fifty consumers evaluated the samples by making a comparison between BM and CM for each cooking method. For each comparison, consumers evaluated which sample they preferred in terms of 'odour' and 'taste'. For each sample, they also assessed 'odour and taste liking' rated on a Likert scale (from 1: 'dislike extremely 'to 9: 'like extremely'). Results were analysed according to the consumers' preferences using GLIMIX procedure of SAS®. Regarding the vacuum method, 51% of the consumers preferred the odour of CM vs BM, and 52% preferred the flavour of CM vs BM. Considering the fried/breaded method, 49% preferred the odour of CM, and 49% preferred flavour from meat of CM vs BM. Results of liking are expressed as percentage of consumer scoring ≥5 in the Likert scale. Considering vacuum method, 73% of the consumers scored CM over 5, and 76% BM. Regarding the fried/breaded method, the percentages were 92% for CM and 95% BM. No significant differences (P>0.10) were found between BM and CM in both cooking methods in both analyses. These results confirm the hypothesis that the cooking methods used may be useful to mask and counteract the negative effect that the presence of boar taint may have in the sensory quality in fresh pig meat.

The relationship between feed efficiency and the expression of genes associated with appetite
S. Vigors, T. Sweeney, A.K. Kelly and J.V. O'Doherty
University College Dublin, Veterinary Medicine, School of Veterinary Medicine, University College Dublin, Belfield, Dublin 4, D4, Iraq; stafford.vigors@ucd.ie

The objective of this study was to examine the association of residual feed intake (RFI) and feeding behavior with both hypothalamic and gut peptides involved in appetite control. 75 Male pigs (initial BW 22.4 kg (SD=2.03)) were fed a standard finishing diet (13.8 MJ DE/kg and 9.5 g standard ileal digestible (SID) lysine /kg) during a 42 day period for the purpose of calculating RFI and evaluating feeding behaviour. Following the calculation of RFI, 8 high RFI (HRFI) and 8 low RFI (LRFI) pigs (average weight 85 kg, sem 2.8 kg), were slaughtered (115 d.o). Tissue was then collected from the hypothalamus and small intestine to analyse the gene expression of gut peptides and neuropeptides associated with appetite control. Behavioural analysis confirmed that LRFI pigs ate less (P<0.0001), spent less time eating per day (P<0.05), had smaller meals (P<0.05), and spent less time eating each meal than the HRFI pigs (P<0.10). In the jejunum and ileum, the HRFI pigs had increased expression of glucagon-like peptide 1 receptor (GLP-1R) (P<0.05), with no changes in the other measured gut or hypothalamic peptides (P>0.10). While RFI was unrelated to neuropeptide gene expression, the amount eaten per feeder visit was correlated with cocaine and amphetamine related transcript gene expression (CART) in the periventricular nucleus (r=0.63; P<0.01). The expression of proopiomelanocortin (POMC) was negatively correlated with eating rate (r=-0.62; P<0.011). In the duodenum the amount eaten per visit was positively correlated with the expression of cholecystokinin (CCK) (r=0.56) and GLP-1R (r=0.54). In conclusion HRFI pigs had increases in activity related to feeding behaviour and increased gene expression of GLP-1R. The results from this study also potentially identify a number of neuropeptides involved in the control of feeding behaviour in pigs.

Genotype by diet interactions in selection lines for residual feed intake in pigs

E.D. Mauch[1], J.M. Young[2], J.F. Patience[1], N.K. Gabler[1] and J.C.M. Dekkers[1]
[1]Iowa State University, Animal Science, Kildee Hall, Ames, IA, USA, [2]North Dakota State University, Animal Science, Hultz Hall, Fargo, ND, USA; edmauch@iastate.edu

In 2001, selection of Yorkshire pigs for feed efficiency based on residual feed intake (RFI) began at Iowa State University and continued for ten generations, resulting in a high and a low RFI line. During the selection process, pigs (n=2,309) were fed a standard corn and soybean-meal diet that was high in energy and low in fiber content (HELF). In four parties of generations 8, 9 and 10, half of the pigs (n=311) were fed a diet containing 18% less net energy and 175% more neutral detergent fiber (LEHF) by including by-products such as corn bran, wheat middlings, and soybean hulls. The objective of this study was to evaluate the impact of feeding the LEHF diet on genetic parameters and responses to selection. Individual feed intake was recorded using FIRE© feeders, along with biweekly body weights from ~40 to 118 kg and end of test ultrasonic backfat depth (BF) and loin muscle area (LMA). From these data, average daily feed intake (ADFI), average daily gain (ADG), feed conversion ratio (FCR), and RFI were calculated. Bivariate analyses were conducted to estimate genetic parameters across diets. Heritability estimates ranged from 0.24 to 0.59 across traits, with traits having similar estimates across diets. Growth related traits (ADG, BF, and LMA) had high positive genetic correlations of 0.99 ± 0.23 across diets. Feed intake related traits, RFI, ADFI, and FCR, had lower genetic correlation estimates across diets of 0.82 ± 0.28, 0.86 ± 0.17, and 0.92 ± 0.45, respectively. However, the observed line difference for RFI on the LEHF diet was only 40% of correlated response predicted based on genetic parameter estimates. In conclusion, genetic parameter estimates were consistent across diets, and genetic correlations across diets were close to one for growth related traits but slightly lower for feed related traits. Observed correlated response in RFI under the LEHF diet to selection for RFI under the HELF diet was less than expected, suggestive of genotype by diet interactions. Funded by AFRI-NIFA grant #2011-68004-30336.

Transglycosylated starch beneficially shifts digestion to the hindgut in growing pigs

M.A. Newman, Q. Zebeli, A. Ladinig and B.U. Metzler-Zebeli
University of Veterinary Medicine Vienna, Veterinaerplatz 1, 1210, Vienna, Austria; monica.newman@vetmeduni.ac.at

Dietary inclusion of resistant starch (RS) can reduce overall starch digestibility, which can increase hindgut fermentation and beneficially alter short-chain fatty acid (SCFA) profiles. These effects can be beneficial to intestinal health, however, each RS source often yields differing results. Therefore, the objective of this study was to investigate the effects of transglycosylated cornstarch (TGS) on apparent ileal (AID) and total tract digestibility (ATTD) and SCFAs under different feeding regimens in two experiments with growing barrows. Pigs were fed 2 diets containing 72% purified starch [waxy cornstarch (CON) or TGS]. The first experiment (E1) utilized 8 restrictively fed ileal-cannulated pigs in a crossover design, which allowed for continuous collection of both feces and ileal digesta. The second experiment (E2) utilized 16 pigs in a serial slaughter model fed ad libitum in a completely randomized design, which allowed for collection of feces as well as digesta throughout the gastrointestinal tract. In E1, TGS resulted in reduced ($P<0.05$) AID of energy (GE) and dry matter (DM), as well as reduced ($P<0.05$) ATTD of GE, DM, and protein (CP) compared to CON. This reduction in digestibility at the ileum led to an increase ($P<0.05$) in available substrate to the large intestine for microbial fermentation, and increased ($P<0.05$) hindgut disappearance of GE and DM with TGS compared to CON. Likewise, in E2, TGS showed an overall reduction ($P<0.05$) in ATTD of GE, DM, starch, and CP compared to CON. Differences in SCFA concentrations between diets progressively increased through the large intestine from no difference in the cecum, to a trend ($P<0.10$) in the proximal colon, to significantly higher ($P<0.05$) in the mid colon for TGS compared to CON. In conclusion, TGS demonstrated characteristic effects by reducing nutrient digestibility regardless of feeding regimen. This led to increased SCFA production via a more distal hindgut fermentation, which is beneficial to overall gut health and makes TGS particularly advantageous compared with other RS sources.

Home grown rapeseed meal as soya bean meal alternative for grower and finisher pigs

J.G.M. Houdijk[1], S.P.J. Kightley[2], M.M. Kasprzak[3], J. Wiseman[3], P. Carre[4] and O.A. Olukosi[1]
[1]SRUC, West Mains Road, Edinburgh EH9 3JG, United Kingdom, [2]National Institute of Agricultural Botany,
Huntingdon Road, CB3 0LE Cambridge, United Kingdom, [3]University of Nottingham, Sutton Bonington, LE12 5RD
Loughborough, United Kingdom, [4]CREOL, Rue Gaspard Monge, 33600 Pessac, France; jos.houdijk@sruc.ac.uk

Canadian studies suggest that rapeseed meal (RSM) upper inclusion of 15% in UK pig diets may be out of date. We hypothesized that RSM from modern UK oilseed rape 'double zero' varieties may be used at greater inclusion rates to replace soya bean meal (SBM) in grower/finisher pig diets. RSM from oilseed rape varieties DK Cabernet and PR46W21, with glucosinolates of 2.7 and 4.6 μmole/g, respectively, were incorporated at 0, 5, 15 and 25%, gradually and completely replacing SBM in nutritionally complete grower/finisher pig diets, formulated for NE of 9.5 and 9.3 MJ/kg, and standardized ileal digestible lysine levels of 0.98 and 0.89%, respectively. After one adaptation week, and for separate batches of grower pigs (initial body weight of 39±0.5 kg) and finisher pigs (initial body weight of 62±0.9 kg), RSM diets were fed ad libitum to 2 groups of 3 male and of 3 female pigs for 3 weeks; control diets were fed to 4 groups of 3 males and of 3 females. Diet and sex effects were analysed by 7×2 factorial ANOVA using growth performance during adaptation as covariate, and orthogonal contrasts to locate diet effects. RSM did not affect grower pig feed conversion ratio but reduced feed intake (P=0.004) and body weight gain (P=0.048) at RSM inclusion levels >15%, with smaller feed intake for PR46W21 than DK Cabernet (P=0.008). In finisher pigs, RSM did not affect feed intake. However, male finishers grew faster than female finishers (P=0.022), although only in the absence of RSM (P=0.066). Males had better feed conversion than females, both for growers (P=0.023) and finishers (P<0.001), although finisher feed conversion improved at 25% RSM inclusion (P=0.04). These data support the view that RSM from UK oilseed rape may be used up to 25% inclusion in nutritionally complete finisher pig diets to replace SBM. However, for grower pigs, even when heavier than the typical 30 kg initial body weight, RSM inclusion should not exceed 15%. AHDB Cereals and Oilseeds (RD-2012-3812) funded this work.

Faba beans to pigs: properties and possibilities of different cultivars

E. Ivarsson and M. Neil
Swedish University of Agricultural Sciences, Department of Animal Nutrition and Management, P.O. Box 7024,
750 07 Uppsala, Sweden; emma.ivarsson@slu.se

In diets for pigs white-flowered faba beans (WF) are preferred to colour-flowered (CF), due to higher digestibility and fear for negative effects of tannins which are more abundant in CF. However, CF cultivars have higher and safer yields than WF. Since there is variation in tannin contents among CF cultivars it is of interest to explore the variation and see if some are more useful than others in pig feed. In an ongoing study 84 samples from 11 CF and 5 WF cultivars grown in 3 different regions in Sweden were analysed for dry matter, ash, crude protein (CP), amino acids, condensed tannins, vicine, convicine and enzymatic digestibility in vitro. The digestible (d) organic matter (OM) and dCP were estimated by multiplying the OM and CP content with the corresponding in vitro digestibility. The dOM varied between 56.3-61.7% and the dCP between 27.4-30.5% in the CF. The dOM was between 63.2-67.1% and the dCP between 28.2-33.2% in WF. The condensed tannins were 5.67-7.38 (g/kg) in the CF and were in most WF below the detection limit. Based on these results, 2 CF cultivars (1 promising and 1 less promising) and 1 WF were chosen for a feeding trial. Julia had the highest dCP, second highest dOM and the lowest content of condensed tannins in the CF and was selected as promising. Fuego was selected as the less promising CF. In WF, Gloria had the highest dOM and dCP and was selected. The CF cultivars were fed both with 10 and 20% inclusion, whereas the WF was fed with 20% inclusion. The faba bean diets were compared to a commercial piglet feed, given a total of 6 diets. Each diet was fed to 6 pens with 10 piglets per pen during a 4 weeks period following weaning at around 33 days of age (average weight 11.4 kg). Preliminary results show that the pigs increased to 15.6 kg in average at 7 weeks of age, the weight gain during the 2 first weeks did not differ between cultivars which indicate that weaning pigs can perform well on faba beans diets independently of cultivar.

Effect of hemp products on gastric ulcers in growing pigs

C. Lauridsen[1], M. Curtasu[1], M.L.M. Pedersen[2], J.Ø. Lindegaard[2], M.B.F. Nielsen[2] and N. Canibe[1]
[1]Aarhus University, P.O. Box 50, 8830 Tjele, Denmark, [2]SEGES, Axeltorv 3, 1609 København V, Denmark;
charlotte.lauridsen@anis.au.dk

Impaired gastric health, including gastric ulcers, is a well-known problem in pig production. Gastric ulcers are difficult to detect because the pig does not show signs of impaired liveliness before severe pathogenesis. Dietary inclusion of industrial hemp can reduce inflammation and increase the consistency of the gastric content, and may therefore potentially reduce gastric ulcer development. The objective was to study the effect of including hemp products in the feed on the prevalence of gastric ulcer in finishers, and to search potential biomarkers of gastric ulcer. A total of 428 pigs enrolled the experiment from 30 to 110 kg live weight, and divided into four dietary treatments: (1) meal; (2) pellets; (3) pellets including 4% hemp cake (before pelleting); (4) pellets including 4% hemp husk (after pelleting). Two days before slaughter, a fecal and a saliva sample were obtained from each pig. After slaughter, each stomach was characterized and clinically scored for presence of ulcer. Gastric content and epithelium were sampled from 20 stomachs without ulcer and 20 stomachs with severe ulcer. Results showed that the Meal group had significantly lower prevalence of gastric ulcers (scored 6-10) than the three pelleted groups (10% versus 92-95%). Pigs on diet 4 had a lower prevalence of severe gastric ulcers (score 8-10) than pigs fed diet 2 and 3, but had still much higher prevalence than pigs fed diet 1. In parallel, the proportion of stomachs with increased consistency of the content was 48-52% from pigs on diet 2 and 3, but 67% from those on diet 4, and 99% from pigs on diet 1. Pigs with severe gastric ulcers and pigs with no gastric ulcers could be discriminated by subjecting saliva to a metabolomics analysis using LC-MS/MS and intestinal epithelial tissues to transcriptome analysis of inflammatory genes. However, inflammatory fecal markers were not different. In conclusion, hemp husk, but not hemp cake, could reduce the proportion of severe gastric ulcers. However, feeding meal remains to be the best strategy to prevent the development of gastric ulcers in pigs.

Nurse sow strategies: an effective way to rear super-numerous piglets?

O. Schmitt[1,2,3], E. Baxter[1], L. Boyle[3] and K. O'Driscoll[3]
[1]SRUC, Animal Behaviour and Welfare, Animal and Veterinary Science Research Group, West Mains Road, Edinburgh EH9 3JG, United Kingdom, [2]The University of Edinburgh, Department of Animal Production, Royal (Dick) School of Veterinary Studies, Roslin, Midlothian, United Kingdom, [3]Teagasc, Pig Development Department, Teagasc Moorepark Research Centre, Fermoy, Co. Cork, Ireland; oceane.schmitt@teagasc.ie

Using nurse sows to rear super-numerous piglets from large litters (>14 born alive) is a fostering strategy intended to optimise pre-weaning survival. However, there is little information about the growth of fostered compared to non-fostered piglets. This study measured survival and growth rate of piglets subject to different fostering strategies. At one day-old, the heaviest 3-4 piglets born into large litters (removal [R], n=9) were fostered either onto a sow 21 days into lactation (N1: one-step strategy, n=10); or onto a sow 7 days into lactation (N2: two-step strategy, n=9). Piglets from N2 were fostered onto a sow 21 days into lactation (N3: two step strategy, n=9). Cross-fostering was allowed when up to 25% of remaining R piglets were <1 kg (RC, n=10). All treatments were balanced for sex. Piglets were weighed at D0 (foster day), D1, D3, D10, D17, D24, D31, at weaning (W), one week post-weaning (W7) and at transfer to the 2nd stage weaner accomodation (S2). Data were analysed using mixed model approach, accounting for repeated measures. Treatment had no effect on piglets' weight or overall growth rate, until weaning. However, R and RC piglets had greater average daily gain (ADG) than N1 piglets until D11 (P<0.05), and N2 piglets until D4 (P<0.05). At transfer to S2, R piglets were lighter than N1 piglets (N1=13.22±0.46 kg, R=11.06±0.48 kg; P<0.005), due to N1 piglets having greater ADG than R piglets post-weaning (W-W7: N1=0.21±0.10 kg, R=0.10±0.09 kg, P<0.05; W7-S2: N1=0.43±0.02 kg, R=0.36±0.02 kg, P=0.05). Treatment did not affect pre-weaning survival. Fostering of heaviest piglets did not compromise growth or pre-weaning survival, compared to lighter littermates which remained with their dam.

Oral fluid samples in group-housed sows: a welfare-friendly method alternative to blood samples?
C. Fablet, V. Dorenlor, F. Eono, S. Eudier, E. Eveno, D. Liegard-Vanhecke, N. Rose and F. Pol
Anses, Epidemiology and Welfare Unit, BP 53, 22440 Ploufragan, France; christelle.fablet@anses.fr

Blood is usually used as diagnosis fluid for pigs. Sampling is done by an invasive method which can stress and hurt the animal. Furthermore, it requires at least two trained operators and may be laborious especially in group-housed sows. This study aimed at assessing the feasibility of oral fluid (OF) sampling on a chewing device already used in growing pigs, as an alternative to blood sampling in group-housed gestating sows. The study was carried out in 30 French herds (1,359 sows) selected on the gestation housing type (straw bedding vs slatted floor). In each herd, individual OF and blood samples were taken from at least 30 sows and three pens selected at random. From these pens, pen-based OF samples were collected on a chewing device provided for 45 minutes. Sampling and chewing times per sow and OF quantity were recorded for each sample; individual information was collected for each sow. Every 15 minutes, the lying sows were counted (to measure the pen activity level). Factors associated with individual sampling time and the probability that a sow chew the pen-based sampling device were identified by logistic regression models. Individual OF sample took 2:50 min (one operator, 4 ml) while blood sample took 1:15 min (at least two operators). Sampling time was significantly higher when straw bedding and varied according to the chewing device design and the operator. 45.8% of the sows from 78.8% pens chew the pen-based device (49 ml). Sows were less attracted by some kinds of collective device, when straw bedding, when multiparous and when the activity of the pen was low. OF sampling is a safe and easy technique performed by a single operator. This is a promising welfare friendly sampling technique for group-housed sows based on the animal cooperation. The type of sampling (individual or pen-based) should be adapted according to animal housing.

Alterations in metabolites and hormones of pigs following anaesthesia
G. Das, A. Vernunft, S. Görs, E. Kanitz, J. Weitzel, K.-P. Brüssow and C.C. Metges
Leibniz Institute for Farm Animal Biology, Wilhelm-Stahl-Allee 2, 18196, Dummerstorf, Germany;
gdas@fbn-dummerstorf.de

Experimental setups, in which acute surgical interventions are performed, require sedation and/or general anaesthesia. However, these can interfere or confound with the effects of the experimental factors of interest on measured physiological variables. In search for a physiologically-friendly anaesthesia regime for such setups, we investigated effects of general anaesthesia induced by propofol (Prop) or pentobarbital (Pent) or propofol plus isoflurane (Prop + Isof) on plasma concentrations of commonly measured metabolites and hormones. In two experimental repetitions, 6 female pigs fitted with jugular vein catheters were used. On the first and last experimental days (ED: 1-12 d) basal measurements were made on plasma samples collected at regular intervals for a period of 6 h (Control). On the following EDs, each pig was rotationally anaesthetized either with Prop or Pent or Prop + Isof on different days, separated with washout periods (2-3 days). Plasma concentrations of cholesterol, urea, glucagon and cortisol remained unaffected by any of the treatments ($P>0.05$) with no significant interaction by time ($P>0.05$). Glucose and lactate concentrations were increased ($P<0.05$) by Prop and Pent within the first hour of anaesthesia ($P<0.05$). Prop and Pent caused reduced non-esterified fatty acids concentrations, which were more pronounced during the last 2 h of the studied period. Triglycerides concentrations were increased by all three anaesthetic agents within the first 45 min with Pent exerting a less stronger effect than Prop and Prop + Isof. Prop and Pent caused the highest elevation in insulin levels 30 min after the application. Our data demonstrate altered levels of metabolites and hormones induced by all three treatments with propofol plus isoflurane exerting less pronounced effects, suggesting varying degree of alterations in kinetics of carbohydrate and lipid metabolism induced by anaesthesia.

Association of polymorphism in MC4R gene with muscle and backfat thickness in different ages of pigs
J. Tomka, K. Vašíčková, I. Bahelka, D. Vašíček, M. Oravcová and M. Bauer
NAFC, Research Institute for Animal Production Nitra, Hlohovecká 2, 95141 Lužianky, Slovak Republic;
tomka@vuzv.sk

The missense mutation in melanocortin-4 receptor (leading to the Asp298Asn substitution) has been associated with several production traits in pigs. The objective of this work was to study the association of this polymorphism with muscle and backfat depths and average daily gain in different ages of Landrace pigs. During the two years period 75 animals were weighed at the age of 40, 60, 80 and 100 days. The backfat and muscle thickness were measured at the time of weighing. The frequencies of G and A-allele were 0.71 and 0.29. The effect of polymorphism was analyzed using linear models with year, sex of animal, dam, age at weighing and genotype fitted as fixed effects. The results showed significant association of this polymorphism with backfat thickness in the age of 60, 80 and 100 days. There was no association showed with muscle thickness in any stage of life. There was significant association of this polymorphism with average daily gain in the period from 40 to 60 days. To conclude the A-allele has been associated with thicker layer of backfat during the whole experiment and higher daily gain in the period from 40 to 60 days. The study was performed during the realization of the project 'MARKERY No. 26220220190'.

Gene co-expression networks and profiles reveal potential biomarkers of boar taint in pigs
M. Drag[1], R. Skinkyte-Juskiene[1], D.N. Do[2], L.J.A. Kogelman[3] and H.N. Kadarmideen[1]
[1]University of Copenhagen, Department of Large Animal Sciences, Grønnegårdsvej 7, 1870 Frederiksberg C, Denmark, [2]McGill University, Department of Animal Science, Ste-Anne-de-Bellevue, QC, Canada, [3]Danish Headache Center & Department of Neurology, Rigshospitalet Glostrup, Nordre Ringvej 67, 2600 Glostrup, Denmark; markus.drag@sund.ku.dk

Boar taint (BT) is an offensive odour or taste of porcine meat which may occur in entire male pigs due to skatole and androstenone accumulation. To avoid BT, castration of young piglets is performed but this strategy is under debate due to animal welfare concerns. The study aimed to reveal potential BT biomarkers for optimized breeding. Male pigs (n=48) with low, medium and high genetic merit of BT were selected and tissues from liver and testis were subjected to transcriptomic profiling by RNA-Seq. The reads were mapped to the Sus scrofa reference genome (Ensembl, ver. 79) which resulted in ~87% uniquely mapped reads. Quality control by Qualimap revealed ~51% of reads mapped in the exonic. Differential expression (DE) comparison of low, medium and high BT using Limma revealed a 10-fold difference in numbers of DE genes (FDR<0.05) between liver and testis, with testis being the highly active tissue. GOseq was used to find enriched gene ontology (GO) terms and REVIGO was used to filter semantic similarities. In both liver and testis, a GO termed 'oxidoreductase activity' was enriched (P<0.05) due to high amounts of 5α-reductases involved in steroid metabolism, including androstenone synthesis. In testis, >80 DE genes were functionally classified by the PANTHER tool to 'Gonadotropin releasing hormone receptor' and 'Wnt signaling' pathways which play a role in reproductive maturation and proliferation of spermatogonia, respectively. WGCNA was used to build co-expression modules and enrichment analysis and semantic filtering revealed the GO terms 'catalytic activity' and 'transferase activity' to be overrepresented (P<0.05) in liver and testis, respectively. Transferases include prenyltransferases which are involved in catalysis of the precursor of steroid hormones. Extraction of hub genes from important modules and integration with DE results revealed potential biomarkers for BT.

Transglycosylated starch alters lipid metabolome and improves insulin response in growing pigs

M.A. Newman, Q. Zebeli, E. Eberspächer and B.U. Metzler-Zebeli
University of Veterinary Medicine Vienna, Veterinaerplatz 1, 1210, Vienna, Austria; monica.newman@vetmeduni.ac.at

Dietary inclusion of resistant starch (RS) has been shown to have beneficial effects on blood glucose, insulin, and lipids. However, results between varying RS types has shown to be inconsistent. Therefore, this study evaluated the impact of transglycosylated cornstarch (TGS) on blood glucose, insulin, and lipids, and serum metabolome in the pre and postprandial phase in growing pigs. Eight jugular vein catheterized barrows (initial body weight, 26.1 kg) were fed 2 diets containing 72% purified starch [waxy cornstarch (CON) or TGS] in a crossover design. A meal tolerance test (MTT) was conducted with serial blood samplings, and samples were analyzed for glucose, insulin, and blood lipids. Serum was additionally analyzed for metabolites using a targeted mass spectrometry-based approach. Pigs fed TGS had reduced immediate postprandial serum insulin (P<0.05) and glucose (P<0.1) response compared to pigs fed CON diet. Serum metabolome profiling also showed reduced (P<0.05) sum of hexoses 30 min postprandially compared with CON. The MTT indicated increased (P<0.05) serum cholesterol postprandially in TGS-fed pigs, but no differences in the fasting state compared to CON-fed pigs. Furthermore, serum urea was increased (P<0.05) at all time points with TGS compared to CON, indicative of increased protein catabolism. Additionally, serum metabolome profiling showed reduced (P<0.05) amino acids such as alanine and glutamine with TGS, suggesting increased gluconeogenesis compared to CON, likely due to a reduction in available glucose. Of the metabolites affected by dietary treatment, sphingomyelins and akyl-acyl-phosphatidylcholines were generally increased (P<0.05) pre-prandially, whereas diacyl-phosphatidylcholines and lysophosphatidylcholines were reduced (P<0.05) postprandially compared with CON. In conclusion, TGS did not elicit a reduction in blood cholesterol levels, but led to characteristic changes in postprandial insulin and glucose metabolism, which may have caused the alterations in serum amino acid and phospholipid metabolome profiles.

Use of phytase and xylanase to improve the nutritional value of rice co-products fed to pigs

G. Casas[1] and H. Stein[2]
[1]Universidad Nacional de Colombia, Produccion Animal, Carrera 30 45 03, Colombia, [2]University of Illinois, Animal Science, 1207 W Gregory Drive, 61801, USA; gacasasbe@gmail.com

Three experiments with pigs were conducted with the objectives of: (1) determining effects of microbial phytase and microbial xylanase on digestibility of phosphorus (P) and on DE and ME in rice co-products; and (2) determining effects of inclusion of full fat rice bran (FFRB) or defatted rice bran (DFRB) without or with supplementation of xylanase on growth performance of weanling pigs. In Exp. 1, 80 barrows (initial BW: 13.6±0.8 kg) were allotted to 10 diets to determine the DE and ME in FFRB, DFRB, brown rice, and broken rice without or with microbial xylanase (Econase XT-25, AB Vista, Marlborough, UK). In Exp. 2, 96 barrows (initial BW: 19.4±1.4 kg) were allotted to 12 diets to determine the standardized total tract digestibility (STTD) of P in rice co-products without or with microbial phytase (Quantum Blue, AB Vista, Marlborough, UK). In Exp. 3, 532 pigs (initial BW: 9.3±0.5 kg) were allotted to a basal diet or diets containing 10, 20, or 30% FFRB or 10, 20, or 30% DFRB. All diets were formulated without and with 16,000 units per kg of microbial xylanase. Addition of xylanase increased (P<0.05) concentrations of DE and ME (DM basis) in diets containing FFRB and DFRB, but not in diets containing brown rice or broken rice. Addition of microbial phytase increased (P<0.05) the STTD of P and decreased (P<0.05) the excretion of P from pigs fed diets containing all rice co-products. Average daily gain increased as 10% FFRB or 10% DFRB was included in diets for weanling pigs, but then started to decline as 20 or 30% were used (quadratic, P<0.05). The gain to feed ratio was not affected by the inclusion of DFRB in diets fed to weanling pigs, but increased (quadratic, P<0.05) as inclusion of FFRB increased. However, microbial xylanase did not affect growth performance of pigs. Overall, it is concluded that microbial enzymes may be used to increase digestibility of nutrients and energy in rice co-products fed to pigs.

Sex and extruded linseed diet +/- concentrate in fiber effect on the fatty acid composition of pork

A. De Tonnac[1], J. Mourot[1] and M. Guillevic[2]
[1]National Institute for Agricultural Research, Domaine de la Prise, 35590 Saint-Gilles, France, [2]Valorex, La Messayais, 35210 Combourtillé, France; auriane.de-tonnac@rennes.inra.fr

The aim of this study is to compare the effect of diets containing extruded linseed +/- concentrate in fiber on the deposit of fatty acids in castrated and entire male pig's tissues. Thirty six crossbreed castrated and entire growing-finishing male pigs (51 kg initial live weight, LW) received one of three experimental diets including a control diet (C), extruded linseed diet (EL) and EL poor fiber diet (EL-). Diets respectively contained 0.57, 3.90 and 4.35 g/kg of feed of ALA. Pigs were stunned at 115 kg LW and their backfat, adipose tissue of internal ham (ATH), liver and longissimus dorsi muscle were sampled to determine their fatty acids composition. Data was analysed by GLM procedure of SAS with the diet and the sex as main factor followed by a Bonferroni test. Results were expressed in g/100 g of total fatty acids (FA). There was less C18:2n-6 in the liver of castrated than in entire male pigs fed EL and EL- diets, in the backfat of pigs fed C and EL- diets and in ATH of pigs fed only with EL-. There was also a lower percentage of C18:3n-3 in the backfat and ATH of castrated compared to entire male pigs fed EL- and/or EL. The total of polyunsaturated fatty acids (PUFA), n-6 and n-3 FA were lower in the backfat of castrated than entire male pigs fed EL- diet. It was also the case for n-6 FA and n-3 FA in the backfat of castrated pigs fed the C and EL diet, respectively. The percentage of C18:3n-3 and total n-3 FA were lower in the muscle of castrated than entire male pigs fed the EL and EL- diets or EL- only, respectively. Regardless of the sex ALA deposition was statistically increased in the backfat and the muscle of pigs fed the EL- diet compared to the EL diet and castrated pigs had lower PUFA levels than entire male pigs, whatever the diet and the tissue.

Indices of economic efficiency the weaner production at Croatian piggeries to be sustainable

M. Sviben
Freelance consultant, Siget 22B, HR10020 Zagreb, Croatia; marijan.sviben@zg.t-com.hr

The indices of economic efficiency of the sustainable piglet production were published at the 63rd EAAP Annual Meeting in 2012 as follows: 30.21 for the number of weaners per sow a year ($E_S = n_W/n_S$) and 151.5 for the number of weaners per farrowing pen a year ($E_P = n_W/n_P$). Those two coefficients are connected as follows: $E_P = E_S \times n_S/n_P$. The ratio of the number of sows exploited on an average during a year (n_S) and of the number of farrowing pens at the piggery (n_P) requires to be 5.015. Since the data, collected at the piggeries over Croatia during 2014, were published by the Croatian Agricultural Agency in 2015, it was possible to see achieved indices of economic efficiency of the weaner production and to compare them to required magnitudes. At ten piggeries, having 294-2,602 sows, 84-780 farrowing pens and 8,428-68,818 produced weaners, the coefficients E_S varied from 22.77 to 30.74 (75.37-101.75% of 30.21 piglets), on an average 27.25 or 90.21%. The coefficients E_P varied from 81.75 to 140.71 (53.96-92.88% of 151.5 piglets), on an average 100.86 of 66.58%. The number of weaners per farrowing pen a year was 3.50% determinated by the number of weaners per sow a year. The coefficient of economic efficiency per farrowing pen was 70.22% determinated by the ratio of the number of sows exploited on an average during the year and of the number of farrowing pens at the piggery. The ratios varied from 3.011 to 5.000 (60.04-99.70% of 5.015), on an average 3.725 or 74.28%. Suchlike ratio n_S/n_P was the product of the number of sows used per farrowing a year (n_S/n_F – the reciprocal farrowing index, required to be 0.3847) and the number of farrowings (litters, shifts) per pen a year (n_F/n_P – required to be 13.035). Between the number of shifts and the number of weaners per farrowing pen a year (E_P) the correlation coefficient ($r_{X/Y}$) was 0.931 and the determination coefficient ($r_{X/Y}^2$) was 0.8668 (86.68%). Economic efficiency of the weaner production at Croatian piggeries could be improved increasing the number of shifts per farrowing pen a year.

The efficiency of the weaner production determinated by the sows' ability, the use of sows and pens
M. Sviben
Freelance consultant, Siget 22B, HR10020 Zagreb, Croatia; marijan.sviben@zg.t-com.hr

Since the data, collected in 2014, were published by The Croatian Agricultural Agency in 2015, it was possible to study the interrelations of four factors (n_W/n_F – the indicator of the sows ability, n_F/n_S – the farrowing index, n_S/n_F – the reciprocal farrowing index, n_F/n_P – the number of farrowings (litters, shifts)/pen/year) which gave the product known as the index of economic efficiency of the weaner production per farrowing pen – E_P. The interrelations of four factors and the measure of economic efficiency of the weaner production per farrowing pen, which includes the index of economic efficiency of the sows' exploitation ($E_S = n_N/n_F \times n_F/n_S$), were studied elaborating the data from 10 Croatian piggeries in 2014 and the data collected at the farm in Rokovci during 10 years (1994-2003). The means were found out as follows: weaners/litter 11.597 at 10 piggeries and 8.881 in Rokovci, farrowings/litter 2.346 at 10 piggeries and 2.165 in Rokovci, sows/farrowing 0.427 at 10 piggeries and 0.465 in Rokovci, shifts/pen/year 8.725 at 10 piggeries and 8.803 in Rokovci. The numbers of weaned piglets per sow a year (E_S) were on average 27.25 at 10 piggeries and 19.23 in Rokovci. The indices of economic efficiency per farrowing pen a year (E_P) were on an average 100.86 weaners/pen/year at 10 piggeries and 78.03 weaners/pen/year in Rokovci. The determination coefficients ($r_{X/Y}^2$) between the number of shifts per farrowing pen a year and the index od economic efficiency E_P were found out as follows: 0.8668 at 10 piggeries and 0.8010 in Rokovci. The prediction equations of the index of economic efficiency per pen were $Y_C = 0.986X_1 + 1.073X_2 - 105.868$ at 10 piggeries and $Y_C = 0.922X_1 + 1.084X_2 - 99.986$ in Rokovci. The magnitudes of X_1 were the percentages of minimum number of shifts per pen a year (7.185 at 10 piggeries, 7.665 in Rokovci). The magnitudes of X_2 were the percentages of the number of weaners per litter at the piggery with the least number of shifts per pen a year (2014 – 11.506) or in Rokovci in 1997 (9.596). In two equations the regression coefficients were almost equal. It did not matter how the sows' ability had been expressed (normal in past times or hyperprolific nowadays).

Attempt of SPME technique for analysis quality of boar's semen
A. Zwyrzykowska[1], K. Śpitalniak[1], W. Bielas[2], R. Kupczyński[1], A. Szumny[3] and B. Jarosz[3]
[1]*Wroclaw University of Evironmental and Life Sciences, Department of Hygiene, Environment and Animal Welfare, Chelmonskiego 38c, 51-630 Wroclaw, Poland,* [2]*Wroclaw University of Evironmental and Life Sciences, Department of Animals Reproduction, Pl. Grunwaldzki 49, 50-366 Wrocław, Poland,* [3]*Wroclaw University of Evironmental and Life Sciences, Departament of Chemistry, C. K. Norwida 25, 50-375 Wrocław, Poland; anna.zwyrzykowska@up.wroc.pl*

Evaluation of boar semen quality for measures of fertility is an important component to success with reproductions in pigs. The macroscopic examination of sperm routinely evaluates also the smell of the ejaculate. The composition of volatile compounds is used not only as a food quality, but also as a diagnostic method for assessment of animal health nutrition status as well as animal welfare. Semen samples from seven boars were assessed by computer assisted semen analysis and flow cytometry. The semen samples were divided into three groups with different sperm motility, with the values from 30 to 85%. Next all semen were analysed on gas chromatograph coupled with mass detector using microextraction for solid phase. The grey fiber coated with divinylbenzene carboxen and polydimethylsiloxane polymer in head-space vials was used for volatile compounds analysis. We found about twenty volatile compounds. Predominated signals, on GC-MS chromatograms were: n-aldehydes (hexanal, octanal, nonanal) together with aromatic benzaldehyde were visible. We found also short and medium alcohols, such as hexanol, 1-octen-3-ol, octanol, nonoanol. From the other hand, on chromatogram triethylamine, N-formyl dibuthylamine were presented. Tentatively tetramethylguanidine was identified. Interestingly, dehydrojasmone – the product of degradation of prostaglandin occurred in all samples. In investigated samples, we found also natural terpenes, such as linalool, limonene, geranyl acetone and ionones, originated from the animal diet. To the best of our knowledge, obtained results of volatile compounds analysis are first in scientific literature. In our opinion, the SPME technique could be a novel approach for sperm quality assessment.

Weaner pig performance using a wet or dry feed system

P. McMullen[1], R. Wregor[2], W. Henry[3], A. O'Connell[2] and E. Magowan[1]
[1]Agri-Food and Biosciences Institute, Hillsborough, BT26 6DR, United Kingdom, [2]JMW Farms, Armagh, BT60 1DL, United Kingdom, [3]Rektify Ltd, Gilford, BT63 6HJ, United Kingdom; elizabeth.magowan@afbini.gov.uk

Scientific data comparing the use of modern liquid feed systems with dry feed systems for weaned pigs is lacking. A total of 3,120 pigs were used across two trials to investigate pig performance when: (1) starter 1 and 2 diets were offered in wet or dry form; and (2) when only starter 1 diet was offered in wet or dry form (starter 2 diet was offered in wet form). At weaning (26 d) pigs were allocated to treatment based on weight and gender and placed in groups of 52. Pigs were fed the starter 1 diet for 10 d, and then the starter 2 diet to 26 d post weaning. The 'ad libitum' probe wet feed system (2.9 cm/pig of feeder space) delivered a 3:1 liquid:dry mix. Dry feed was pelleted and offered through dry multi space feeders (5.2 cm/pig of feeder space). Pig weight and feed intake was recorded at wean, 10 d and 26 d post weaning. Average Daily Gain (ADG), Average Daily Feed Intake (ADFI) and Feed Conversion Ratio (FCR) were calculated. When both the starter 1 and 2 diets were offered in wet form, pig weight at 26 d was similar ($P>0.05$) to when the diets were offered in dry form (average 17.7 kg). Between weaning and 26 d, the ADFI was higher ($P<0.001$, 568 g/day) and the FCR of pigs was poorer ($P<0.01$; 1.54) when the starter 1 and 2 diets were offered in wet form compared with when they were offered in dry form (423 g/day and 1.19 respectively). When only the starter 1 diet was offered in dry form pig weight at 10 d and 26 d post weaning was similar ($P>0.05$) compared with when it was offered in wet form (average 10.5 and 18.1 kg respectively). The intake of pigs in the first 10 days after weaning was higher ($P<0.05$; 315 g/day) when starter 1 diet was offered in wet form compared with dry form (242 g/day). Wet feeding increased pig feed intake post weaning but this did not translate into increase growth and as such the feed use efficiency was negatively affected by offering either or both the starter 1 and 2 diets through a wet feed system compared with a dry feed system.

Investigating the association of porcine PCSK9 genotype and muscle fatty acid composition

W. Chański[1], R. González-Prendes[1], A. Castelló[1,2], J. Jordana[2], A. Manunza[1], R. Quintanilla[3] and M. Amills[1]
[1]Center for Research in Agricultural Genomics (CSIC-IRTA-UAB-UB), Animal Genetics, Universitat Autònoma de Barcelona, E08193 Bellaterra, Spain, [2]UAB, Ciència Animal i dels Aliments, Universitat Autònoma de Barcelona, E08193 Bellaterra, Spain, [3]IRTA, Animal Breeding and Genetics, Torre Marimon, E08140 Caldes de Montbui, Spain; raquel.quintanilla@irta.cat

A GWAS in a Duroc pig population allowed us detecting a genomic region displaying significant associations with serum cholesterol, triglyceride and low-density lipoprotein concentrations on pig chromosome 6 (141-147 Mb). This region contains the proprotein convertase subtilisin-like kexin type 9 (PCSK9) gene (SSC6, 145 Mb), which is a key regulator of lipoprotein metabolism. In the current work, we have genotyped by pyrosequencing a missense PCSK9 c.1222G>A mutation (E408K) in 273 Duroc pigs. Herewith, we have genotyped by pyrosequencing, in 273 Duroc pigs distributed in five half-sib families, a non-synonymous PCSK9 substitution (c.1222G>A) that involves a non-conservative replacement of a glutamate by a lysine at position 408 of the mature protein. Subsequently, we have carried out an association analysis with several phenotypes related to lipid metabolism and muscle fatty acid composition using the mixed-model based GEMMA software. Results evidenced that this polymorphism is not the causal mutation of the previously mentioned SSC6 QTL, but additional associations with muscle fatty acid composition were detected. The most relevant result has been the association between PCSK9 genotype and palmitelaidic content at the gluteus medius skeletal muscle, that was highly significant at the nominal level ($P=0.008$). These findings make sense from a biological perspective because the increased expression of PCSK9 enhances the catabolism of CD36, one of the main long-chain fatty acid transporters, and there is also evidence that it may lower the absorption of palmitate.

Covariance components for growth of boars during on-farm test

A. Ule, M. Klopčič, M. Kovač and Š. Malovrh
University of Ljubljana, Biotechnical Faculty, Department of Animal Science, Groblje 3, 1230 Domžale, Slovenia;
anita.ule@gmail.com

The aim of this paper was to assess the dispersion parameters for growth of boars. Longitudinal measurements of body weight were analyzed by random regression models (RRM) and compared with results from single trait (SMT) and multiple-trait analysis (MTM). Data included 3586 records of Pietrain boars from on-farm performance testing. Dam parity and year-month interaction as season on test-day were treated as fixed. The covariates were litter size and age. Random part of the model contained direct additive genetic, common litter and permanent environment effect. Legendre polynomials were used to model random effects where polynomials of the second order were found to be sufficient. Covariance components were estimated by REML as applied in VCE6. The estimated heritability in RRM ranged from 0.26 at weaning to 0.67 at the end of the test and was lower than in STM (0.34-0.84) and MTM (0.36-0.84). The proportion of common litter effect was the greatest at weaning (0.36) and decreased with age. The proportion of variation due to the permanent environment was the lowest at weaning (0.05) and increased with age. Additive genetic correlations estimated by RRM were the highest among adjacent age classes and gradually decreased with time. The smallest correlations were between weights at birth and weights at other ages (0.07-0.17). The estimates did not differ much among methods applied, some variation was only at the end of test period because of non-random samples of slow growing boars in the last two age groups. RRM estimates were less sensitive to non-random sampling and less numerical instable than MTM.

The role of life cycle assessment in environmental decision making in livestock production

I.J.M. De Boer
Wageningen University, Animal Production Systems group, Department of Animal Sciences, P.O. Box 338, 6700
AH Wageningen, the Netherlands; imke.deboer@wur.nl

Life cycle assessment (LCA) is nowadays a generally accepted method to gain insight into the environmental impact along the life cycle of an animal-source food product, and to identify options to reduce this impact. Besides food provision, however, animal production systems also contribute to other ecosystem services, such as production of bio-energy or clean drinking water, carbon sequestration, maintenance or enhancement of biodiversity, or preservation of the landscape. These other ecosystem services are generally ignored in LCAs of animal-source food, which might favour system with a high animal intensity per ha of land. LCA studies that do address ecosystem services besides food production can be divided in two categories. The first category considers other ecosystem services, such as the enhancement of biodiversity or preservation of the landscape, as another output of the system. In such a multiple-output situation, the environmental impact of the animal production system is allocated to the multiple outputs (e.g. milk, and other ecosystem services), using one of the main allocation methods (system expansion, physical or economic allocation). The second category considers an ecosystem service as another environmental impact category, besides climate change, eutrophication, etc., and, therefore, aims to incorporate ecosystem services into the LCA framework. These studies face the complexity of determining characterization factors (modelling the cause-effect chains) of various ecosystem services. This presentation addresses pros and cons of both approaches, and, more generally, the strengths and weaknesses of LCA to assess the environmental impact of animal production systems.

Emerging challenges and opportunities in ecosystem service assessment in the livestock sector
M. Tichit
INRA, UMR SADAPT, Paris, 75005, France; muriel.tichit@agroparistech.fr

The term ecosystem service (ES) refers to the benefits humans derive from ecosystems and thus formalize the dependence of humankind on the biosphere. This dependence is particularly apparent in agroecosystems. We review emerging applications in ES approaches aimed at supporting environmentally informed decisions. We review strength and weakness of ES approach when applied to livestock systems. The ecosystem service framework offers an integrated viable screening tool that can pinpoint environmental hotspots in rural areas where livestock systems are key players. Future advances of ecosystem service approach in broadening the assessment to economic and social performance of livestock systems will make it more relevant for both farmers and policy makers. Finally, we discuss the conditions at which the ecosystem service framework could also support product development, and consumer choices through the use of labels.

Interactive debate with audience on the two keynote talks
T. Turini
CIV, Rue de Bercy, Paris 75012, France; t.turini@civ-viande.org

A 45 mn discussion will follow the two keynote presentations. The discussion will be introduced by a reviewer. The reviewer will propose a few statements about strengths, weaknesses, complementarities of ecosystem service and LCA frameworks for assessing livestock sustainability. Each statement will be discussed with the audience who will be invited to express support or disagreement and propose new arguments. Our final objective is to achieve a significant list of statements envisioning new strategies for assessing sustainability of livestock farming systems.

Evaluating ecosystem services in the life cycle assessment framework
T. Nemecek
Agroscope, Institute for Sustainability Sciences, Reckenholzstrasse 191, 8046 Zurich, Switzerland;
thomas.nemecek@agroscope.admin.ch

Several approaches have been proposed to include ecosystem services (ESS) in LCA: (1) using multiple functional units, (2) using allocation to ESS, (3) using system expansion and (4) introducing additional impact categories. Using an area-related functional unit (1) can reflect some of the ESS; however, the area used by agricultural production is only a poor indicator, not allowing a differentiation of various levels of ESS. Allocating a part of the environmental impacts to ESS (2) requires an economic valuation of the latter, generally quantified by the subsidies. However, subsidies do not necessarily represent the true value of ESS and this method does not allow for a detailed assessment of different ESS. Furthermore, this approach means that the provision of ESS is associated with environmental impacts such as climate change or eutrophication and it is not possible to distinguish between different ESS, since they all have to be summed up. The system expansion approach (3) requires the definition of an alternative system for ESS provision, which is often difficult to find and also debatable. It is concluded that including ESS as separate impact categories (4) is the best and most flexible approach allowing for a detailed and differentiated assessment, without the need to quantify the economic value of ESS.

Fitting ecosystem services assessment into the LCA framework
C. Cederberg
Chalmers University of Technology, Energy & Environment, Maskingränd 2, 412 96 Göteborg, Sweden;
christel.cederberg@chalmers.se

The last decade has seen an increase in environmental systems analysis of livestock production often based on life cycle assessment (LCA) methodology. The impacts of land occupation and land transformation on ecosystem services, an issue highly relevant to LCAs of milk and meat products as the global livestock sector is a large land user, have been neglected due to lack of consistent methodology. Here, the latest method development on how to include ecosystem service assessment in the LCA framework is presented and challenges concerning data availability, scales and setting the reference situation are discussed.

Utility of ecosystem services to inform regional livestock management: beyond a choice experiment

L. Goti[1], A. Bernués[2] and R. Ruiz[3]
[1]Thünen Institute of Sea Fisheries, Economic Analysis, Palmaille 9, 22767 Hamburg, Germany, [2]Centro de Investigación y Tecnología Agroalimentaria de Aragón, Avda. Montañana 930, 50059 Zaragoza, Spain, [3]NEIKER Tecnalia, Granja Modelo de Arkaute, Carretera N-1, Km 355, 01192 Arkaute, Alava, Spain; leyre.goti@thuenen.de

The OVIPRINT research project had as objectives the measurement of GHG as well as the ecosystem services (ES) related to traditional livestock systems. Beside GHG measurements the project provided the opportunity to perform a choice experiment on different ecosystem services delivered by livestock production systems in the Gorbea natural park, in the Basque Country region of Spain. There has been some modelling efforts in the literature on the interaction between livestock and ecosystem services, but in a different methodological approach to multifunctionality the set up of a choice experiment brings the opportunity to take advantage of a complementary view to that of modelling. This consists in in the identification of the relevant ecosystem services through a literature review, the qualitative analysis of a set of focus groups with diverse stakeholders (shepards, scientists and technicians, consumers and mountaneers) as well as interviews with managers to collect data on their perception of ecosystem services and finally the gathering of information on subsidies and governance structure. The qualitative and quantitative data were used to undertake a governance analysis that included identification of uncertainties and conflicts. The development of a choice experiment thus allowed for a parallel collection of data on e.g. subsidies and governance structure of livestock systems and protected areas in the Basque Country region, that through an alternative analysis provided useful insights on what was blocking the delivery of certain ecosystem services as fire prevention.

Livestock cultural and territorial vitality services are key to unlock the agroecological transition

J. Beudou[1,2], G. Martin[1,2] and J. Ryschawy[1,2]
[1]Université de Toulouse, INPT-ENSAT, Castanet-Tolosan, 31324, France, [2]INRA, UMR 1248 AGIR, Castanet-Tolosan, 31324, France; julie.ryschawy@toulouse.inra.fr

Little research has been dedicated to the cultural and territorial vitality services provided by livestock farming to society. These services also are largely ignored in discussions regarding the transition of livestock farming towards agroecology. This article aims to show why these services must be taken into account to favour the agroecological transition of livestock farming. We examined the nature and interrelationships of cultural services (relating to gastronomy, landscape, etc.) and territorial vitality services (relating to employment generated, tourism, social bonds, etc.). To do so, we drew from the perceptions of local actors on two contrasted French territories: Aubrac and Pays de Rennes. Livestock farming is prominent in both territories, but the orientations of their livestock farming differ. Twenty semi-structured interviews were conducted on each territory. Our work revealed the diversity of the cultural and territorial vitality services provided by livestock farming, and their importance for local actors. These services could act as obstacles to an agroecological transition. For example, jobs linked to livestock farming in Pays de Rennes and the vitality of tourism in Aubrac tend to reinforce the already existing livestock farming. However, the desire of actors to provide or to benefit from cultural and territorial vitality services also could act as a lever to reconfigure agri-food systems, and consequently livestock farming systems. Our results indicated that to better take these services into account, the framework within which agroecological transition is considered, both scientifically and politically, needs to be adjusted. Our study demonstrates the need to develop further interdisciplinary approaches to include social services and thus consider the food system and territorial scale.

Analysis of the net food production of different livestock categories in Austria

P. Ertl[1], A. Steinwidder[2], W. Knaus[1] and W. Zollitsch[1]
[1]BOKU–University of Natural Resources and Life Sciences, Department of Sustainable Agricultural Systems, Gregor Mendel Strasse 33, 1180 Vienna, Austria, [2]Agricultural Research and Education Centre Raumberg-Gumpenstein, Institute of Organic Farming and Farm Animal Diversity, Trautenfels 15, 8951 Pürgg-Trautenfels, Austria; werner.zollitsch@boku.ac.at

Depending on the proportion of potentially human-edible feeds in livestock diets, animal production either increases or decreases the quantity of food available for human consumption. The aim of the present study was to investigate the amount of potentially human-edible feeds in diets of different livestock categories and to assess their net contribution to human protein and energy supply. National data (annual feed balance and official animal production data) for the main livestock categories (cattle, dairy cows, fattening bulls, swine, broilers, laying hens, turkey, sheep, and goats) were used for this case study. With the exception of intensive growing-finishing bulls, diets of ruminants included 9-20% potentially human-edible protein and energy, whereas for monogastric animals, values ranged from 46 to 84%. Under current feeding conditions in Austria, cattle are the only species that are net contributors to both human protein and energy supply, with dairy cows currently producing about 2 and 1.5 times more human-edible protein and energy, respectively, than they consume via feeds. When considering the differences in the protein quality between human-edible plant inputs and animal products, not only cattle, but also laying hens, sheep and goats increased the value of protein available for human consumption under current conditions. Results of this study suggest that the general critique on the low efficiency of animal production in terms of net food production needs a clear differentiation. Especially ruminants have a great potential to increase the total quantity of food available for humans.

Land occupation for livestock feed production in Austria

P. Ertl[1], A. Steinwidder[2], W. Knaus[1] and W. Zollitsch[1]
[1]BOKU–University of Natural Resources and Life Sciences, Department of Sustainable Agricultural Systems, Gregor Mendel Strasse 33, 1180 Vienna, Austria, [2]Agricultural Research and Education Centre Raumberg-Gumpenstein, Institute of Organic Farming and Farm Animal Diversity, Trautenfels 15, 8951 Pürgg-Trautenfels, Austria; werner.zollitsch@boku.ac.at

Livestock compete with humans for different resources. One of the biggest issues is the competition for arable land, as the production of food of animal origin from feedstuffs cultivated on cropland is much less efficient than the utilization of arable land for the direct production of human food of plant origin. The aim of the present case study was therefore, to determine the percentage of feed energy and protein (% of total) originating from different land categories (i.e. arable land where feedstuff is the main crop; arable land where feedstuff is a co-product; grasslands; other origin) which are utilized in the main Austrian livestock systems. Data on feed inputs were taken from the national feed balance for the years 2011-2013. For cattle, sheep, and goats, over 50% of the feed energy and protein was derived from grasslands, which clearly highlights the role of ruminants in utilizing land unsuitable for growing food crops for supplying human foods. For monogastric livestock (swine, broilers, laying hens and turkeys), feed energy mainly originated from arable land, where the feedstuffs were the main crop (67-76%), whereas feed protein mainly originated from arable land where feedstuffs were co-products (48-59%). This indicates that the energy supply, especially for monogastric animals, might be more challenging in an attempt to improve sustainability and land use efficiency than protein supply, which is mainly provided from co-products from crop production. In conclusion, substantial amounts of livestock feed is derived from land where feedstuffs are not grown as the main crop and therefore, the general criticism on livestock systems regarding its competition for land needs some differentiation.

Life cycle assessment models for prediction of dairy farm GHG emissions using Tiers 2 and 3 methods
T. Yan
Agri-Food and Biosciences Institute, Agriculture Branch, Hillsborough, County Down BT26 6DR, United Kingdom;
tianhai.yan@afbini.gov.uk

Life cycle assessment models for prediction of dairy farm GHG emissions using Tiers 2 and 3 methods T. Yan, S. Morrison, C. P. Ferris and A. Aubry Agri-Food and Biosciences Institute, Agriculture Branch, Hillsborough, County Down BT26 6DR, UK Greenhouse gas (GHG) emissions from enteric fermentation and manure management account for over 60% of total emissions within the farm gate for milk production systems. Thus it is important to have accurate estimates of emissions from these two sources to develop appropriate mitigation strategies. The objective of the present study was to use our respiration calorimeter data, and other information to develop life cycle assessment models for prediction of GHG emissions from grassland based dairy farms. Our models estimate feed intake for each type of dairy herds using the UK energy feeding system (Feed into Milk) developed from energy metabolism data obtained in this Institute and Reading University. These feed intake data, together with our CH_4 emission and digestibility data, are then used to calculate enteric CH_4 emissions and manure CH_4 and N_2O emissions for grazing and confined cattle. Our models also use Northern Ireland information to estimate carbon sequestration for land use change. The GHG emissions from 'other sources' are calculated using emission factors from the literature for farm conditions similar to Northern Ireland, and default factors from IPCC 2006. These include emission sources from concentrate production and transportation, fertiliser manufacture and application, plant residue, electricity and fossil fuel use, veterinary medication, and other minor sources within farm gate. These models were validated using data from 9 local dairy farms operating a range of feeding systems (e.g. low vs high concentrate input; short vs long grazing period). The results revealed that GHG emissions associated with the production of one kg energy corrected milk ranged from 0.89 to 1.07 kg CO_2 equivalent, when carbon sequestration was included. These emission data are comparable with estimates from a range of models used in European countries.

Assessing sustainability in chicken supply chains: OVALI, a methodology shared by stakeholders
B. Méda[1], J. Protino[1], L. Dusart[2], P. Lescoat[3], P. Magdelaine[2], C. Berri[1] and I. Bouvarel[2]
[1]INRA, UR83 Recherches Avicoles, 37380 NOUZILLY, France, [2]ITAVI, 7 rue du Faubourg Poissonnière, 75009 PARIS, France, [3]AgroParisTech, UMR 1048 SADAPT, 75231 PARIS, France; bertrand.meda@tours.inra.fr

Chicken supply chains (CSC) are facing many challenges regarding the three pillars of sustainability (environmental, economic and social). The objective of this study was to develop a methodology: (1) to assess the sustainability of CSC in their territory; (2) identify their weak points and strengths; and (3) propose innovations to improve their sustainability. The method was built with a participatory approach, to take into account the diversity of standpoints of stakeholders involved in the CSC (e.g. feed producers, farmers, slaughterhouses, civil society, research and education…). The final assessment grid is composed of 9 objectives, 28 criteria and 45 indicators. Each criterion is assessed using one or more indicators, that are converted into score (180 points per pillar). The indicators were evaluated using data from surveys, literature, expertise or complementary tools (production costs and LCA simulators). The analysis of criteria allow the identification of critical points and for each pillar, a grade from A+ to E was attributed. A French CSC producing 'standard' chicken was evaluated, showing in particular a need to increase competitiveness and protein autonomy. Thus, a scenario combining practices (heavier birds, energy-saving houses, changes in feeding…) was simulated and assessed. Such a scenario allows improvements on the three pillars with for example, lower production costs (live weight, breast meat), lower environmental impacts (greenhouse gases emission and energy use) and better product image thanks to the local feedstuffs used in animal feeding. These results also show that improving sustainability in CSC is possible with no antagonism between pillars. The OVALI Sustainability for poultry® method is expected to help stakeholders in their initiatives to improve sustainability in CSC.

Elaboration of an experimental model of the oxidative stress in weaned piglets

E. Royer[1], F. Barbé[2], D. Guillou[2], Y. Rousselière[1] and E. Chevaux[2]
[1]*Ifip-institut du porc, 34 bd de la gare, 31500 Toulouse, France,* [2]*Lallemand SAS, 19 rue des briquetiers, 31702 Blagnac, France; eric.royer@ifip.asso.fr*

An experimental model was established as tool for the study of the oxidative stress in weaning pigs. In two experiments, 360 weaned piglets were randomly allocated to eight groups in 2×2×2 factorial designs. In Exp.1, the factors were the sex, a vaccination at weaning against porcine circovirus type 2 (PCV2) or not, and phase 1 diets with NRC (2012) levels for vitamin E and selenium (SA) or extra supplementation in vitamin E, selenium yeast and superoxide dismutase-rich melon supplement (HA). In Exp.2, a double vaccination against PCV2 and porcine influenza, heat stress at d 9-10, 23-24 and 37-38 (36.5 °C over 6 h period) or controls, and SA or HA diets were applied. Blood samples were taken from 6 piglets per treatment at d 13, 28 and 40 in Exp.1 and d 13 and 40 in Exp.2. Serum haptoglobin, glutathione peroxidase activity (GPx), blood lipid peroxides and protein carbonyls were determined. Half-hemolysis time (HT50) of whole blood (WB) and red blood cells (RBC) exposed to a controlled free radical attack were determined. The HA supplementation increased HT50 of WB and RBC (P<0.02) in Exp.2, as well as HT50 of WB at d13 (interaction, P=0.04) and RBC of non-vaccinated pigs (interaction, P=0.05) in Exp.1. In Exp.2, GPx increased (P=0.01), whereas lipid peroxides (P=0.01) and protein carbonyls (P=0.05) decreased as a result of HA diet. Vaccinations increased haptoglobin in Exp.1 and 2, as well as lipid peroxides in Exp.2 (P=0.05). Moreover, the double vaccination decreased GPx activity for HA fed pigs at d 40 (interaction, P=0.05) in Exp.2, and protein carbonyls in Exp.1 and 2 (P<0.06). Vaccination × heat stress × time interactions (P<0.05) were observed on HT50 values in Exp.2. At d 40, heat stress decreased WB (P=0.01) and RBC HT50 (P<0.01) for the vaccinated piglets, whereas such effects were not observed for the non-vaccinated piglets. In conclusion, a model based on heat stress and vaccination may be efficient to assess strategies limiting oxidative stress.

Mitochondrial abundance and respiratory chain activities in tissue of cattle divergent in RFI

C.E. McKenna[1,2], D. Kenny[2], R. Porter[1] and S. Waters[2]
[1]*Trinity College Dublin, Biochemistry, Dublin 2, Dublin 2, Ireland,* [2]*Teagasc Grange, Animal Bioscience, Dunsany, Co. Meath, Co. Meath, Ireland; mckenncl@tcd.ie*

It has been hypothesised that mitochondrial function contributes to variation in feed efficiency between animals. The objective of this study was to determine the effect of phenotypic ranking for residual feed intake (RFI) on mitochondrial number and functionality of respiratory chain complexes (I-IV) in liver and muscle tissue of purebred Simmental heifers and bulls. Individual dry matter intake (DMI) and growth were measured in heifers (n=24) and bulls (n=28) with an initial live weight (SD) of 367 kg (42.4) and 381 kg (51.5), respectively. Animals were offered concentrate ad libitum plus 3 kg grass silage for 70 days. Liver and muscle biopsies were harvested at the end of the feed intake recording period and spectrophotometric assays performed. Residuals of the regression of DMI on ADG and mid-test $BW^{0.75}$, were used to compute individual RFI coefficients, within gender. Animals were ranked by RFI into high (inefficient) and low (efficient) groups. This resulted in 10,9,10,9 high and low RFI heifers and bulls, respectively. Statistical analysis was performed using the MIXED procedure of SAS. Overall ADG and daily DMI for heifers were 1.26 kg (SD=0.4) and 9.14 kg (SD=0.5), respectively and 1.86 kg (SD=0.3) and 9.56 kg (SD=1.02) respectively, for bulls. High RFI heifers and bulls consumed 10% more (P<0.05) than low RFI heifers and bulls. At the mitochondria level, we observed that there was no significant difference in mitochondrial abundance in a comparison of high and low RFI animals in muscle or liver. However, we observed that mitochondrial complex I activity was 1.5-fold lower in muscle from high RFI animals (P<0.0027) when compared to low RFI animals. By contrast there was no significant difference in complex 2, 3 or 4 activities in the equivalent comparison. We also observed that complex IV activity was higher in liver from less efficient bulls and heifers. These data may indicate a relationship between RFI and mitochondrial electron transport chain in muscle and liver.

Muscle and serum acylcarnitine profiles in dairy cows during the periparturient period

Y. Yang[1], J. Rehage[2], S. Dänicke[3], C. Prehn[4], J. Adamski[4], H. Sauerwein[1] and H. Sadri[1]
[1]Institute of Animal Science, Physiology and Hygiene Group, University of Bonn, Bonn, Germany, [2]Clinic for Cattle, University of Veterinary Medicine Hannover Foundation, Hannover, Germany, [3]Institute of Animal Nutrition, Friedrich-Loeffler-Institute, Braunschweig, Germany, [4]Genome Analysis Centrer, Molecular Endocrinology Group, Helmholtz Zentrum München, Germany; hsadri@uni-bonn.de

Free fatty acids (FA) are released into the circulation in response to the negative energy balance (NEB) occurring during early lactation when skeletal muscle turns into the major site for use of fat-derived fuels, such as non-esterified FA and ketone bodies. Our objective was to characterize the concentrations of carnitine and its acyl esters (acylcarnitines; ACC) both in blood serum and in skeletal muscle of dairy cows to address potential changes in the capacity for mitochondrial β-oxidation of FA in skeletal muscle in context with the NEB typical for early lactation. ACC are formed when FA-CoA enter mitochondria for β-oxidation and TCA cycle by the carnitine shuttle. Levels of ACC may vary depending on the metabolic conditions, but can accumulate when rates of β-oxidation exceed rates of TCA cycle which may in turn impair mitochondrial function and provoke insulin resistance and are thus linked with the risk for production diseases in dairy cows. Biopsies from M. semitendinosus and blood were collected at day (d) -21, 1, 21, and 70 relative to calving from 11 Holstein dairy cows. The ACC concentrations were assessed by FIA-MS/MS profiling through targeted metabolomics. Time-dependent changes ($P \leq 0.02$) in the serum concentrations of C2, C3, C4, C4:1, C7-DC, C8, C16, C18, and C18:1-OH were observed. The concentrations of C3, C5-OH, C4-OH, C5:1, C6:1, C16, C16:1, C18, C18:1, and C18:2 in muscle changed over time ($P \leq 0.02$). Muscle carnitine remained unchanged despite a decline ($P \leq 0.01$) in the serum concentrations, suggesting that the decrease in serum concentrations likely results from increased carnitine uptake to maintain the intracellular concentrations. Serum long chain-ACC and muscle long chain-, monounsaturated-, and polyunsaturated-ACC were greater ($P \leq 0.05$) on d 1 than other time-points, likely reflecting increased lipid flux and consequently incomplete or overloaded FA oxidation in the skeletal muscle around calving.

Health and metabolism as reflected by calves' plasma redox state

A. Shabtay, A. Asher, V. Sibony, R. Agmon, S. Weyl-Feinstein and M. Cohen-Zinder
Agricultural Research Organization, Beef Cattle Unit, Newe Ya'ar Research Center, Ramat Yishay 30095, P.O. Box 1021, Israel; shabtay@volcani.agri.gov.il

Neonatal calves are often exposed to significant immunological and environmental stressors which may impose deleterious effects on their health and production, leading to a decrease in animal welfare and profitability of the beef and dairy enterprises. Bovine respiratory disease complex (BRD) is a major health concern with a significant economic impact on the beef cattle industry. Accumulating circumstantial evidence indicate the involvement of oxidative stress in the development of BRD. With respect to livestock production, the cardinal scientific concern would be to improve feed conversion efficiency either by genetic selection or environmental manipulations. In two independent experimental systems, we exposed neonatal Holstein calves to either transportation or artificial light (at night) stress. While in the transportation experiment we looked for biomarkers in the blood to early predict BRD susceptibility, the aim of light manipulation was to reveal a metabolic profile that would shed light on the biochemical basis of production efficiency. In both experimental systems it turned out that the redox state of plasma may: (1) predict susceptibility to BRD three months ahead; (2) indicate metabolic disorders and explain, at a certain level, the physiological and biochemical mechanisms that underlie the environmental influence on production efficiency. It is suggested that disease susceptibility and feed conversion efficiency are both reflected by plasma redox state and lactate:pyruvate ratio, which might be indicative of cytosolic free NAD+/NADH ratio and affect maintenance of energy production.

Effect of nutrient supply on mammary gland development and gene expression in pre-weaned calves

L.N. Leal[1,2], G.J. Hooiveld[1], M.A. Steele[3], M.V. Boekschoten[1], F. Soberon[4,5], M.E. Van Amburgh[4] and J. Martín-Tereso[2]

[1]*Wageningen University, Droevendaalsesteeg 4, 6708 PB Wageningen, the Netherlands,* [2]*Trouw Nutrition Research and Development, Veerstraat 38, 5831 JN Boxmeer, the Netherlands,* [3]*University of Alberta, Department of Agricultural, Food, and Nutritional Science, Edmonton, T6G 2P5, Canada,* [4]*Cornell University, Ithaca, NY 14853, USA,* [5]*Shur-Gain USA, Nutreco Canada Inc., Guelph, Ontario, N1G 4T2, Canada; leonel.leal@trouwnutrition.com*

Mammary gland growth in calves was assumed to be isometric during the first 2 months of life. However, recent data suggests that pre-weaning mammary development can follow an allometric path. The present study aimed to identify genes and biological pathways meaningful to mammary gland development that are affected by pre-weaning nutrient supply. Twelve new-born female Holstein calves were randomly assigned to two different planes of nutrition: HIGH; (1.26 MJ intake energy per kg $BW^{0.75}$ per day; n=6) or LOW (11.7 MJ intake energy per day; n=6) until harvest at 54 days of age. Mammary gland mass of the HIGH fed calves was 3.4 fold larger than LOW calves (0.41 vs 0.12% of body weight, respectively). Analysis of the mammary transcriptome profiles revealed that upon HIGH feeding, fat metabolism, triglyceride biosynthesis, extra cellular matrix interactions, branched-chain amino acid metabolism and respiratory electron transport pathways were expressed, whereas genes associated with RNA translation, muscle contraction and apoptotic processes were found to be suppressed. The array data also show that signal transduction pathways were affected by the dietary treatments. HIGH diet activated ErbB1 signalling and TGF-β; and supressed NF-kB and IL-6 mediated events. Potential activation of transcription factors TP53, TP63, NFKB1, coupled with the suppression of ESR2 and BLC6, were identified as the main upstream regulators of the expression profiles found in the mammary gland. Elevated planes of nutrition substantially modified gene expression profiles and subsequent mammary organogenesis in pre-weaned calves.

Parathyroid hormone response to long-term phosphorus supplementation in steers

D.M. McNeill[1], S.T. Anderson[2], L.J. Kidd[1], S.P. Quigley[3], A.T. Lisle[3] and D.P. Poppi[3]

[1]*University of Queensland, School of Veterinary Science, Gatton Campus, 4343 Gatton, Australia,* [2]*University of Queensland, School of Biomedical Science, St Lucia campus, 4072, St Lucia, Australia,* [3]*University of Queensland, School of Agriculture and Food Sciences, Gatton Campus, 4343, Gatton, Australia; d.mcneill@uq.edu.au*

Plasma inorganic phosphorus (PiP) concentration is a common indicator of phosphorus (P) status in cattle. However, it can be misleading. Bone biopsies are considered a better indicator but the technique is invasive. To find novel and less invasive indicators of P status, our aim was to explore the metabolic markers linked to responses in bone tissue and PiP. Thirty Bos indicus cross weaner steers were individually fed, ad libitum, at one of five levels of P, n=6 per level, from inadequate to in-excess of P requirements (P1-P5: 0.9, 1.3, 1.8, 2.1 and 2.4 g P/kg DM). After 6 months, biopsies were collected from the 12th rib to determine cortical bone thickness (ribCT) and plasma for the determination of bone and calcium metabolism markers, including Parathyroid hormone (PTH), plasma inorganic phosphorus (PiP) and plasma total calcium (Ca). Responses at each P level were compared by ANOVA and curves were fitted to the means to define pattern of response. Plasma concentrations of PiP (1.06, 1.21, 1.53, 2.17, 2.54; sem=0.124 mmol/l; P<0.001) and ribCT (2.53, 2.63, 3.24, 3.29, 3.65; sem=0.300 mm; P<0.01) increased linearly with dietary P level. LogPTH showed a sigmoidal response from very low, undetectable concentrations to plateau between the third and fourth level of dietary P (0.420, 0.678, 1.306, 2.002, 1.945; sem=0.2975 log pg/ml; P<0.001). Using PiP as a better indicator of P intake than dietary P concentration and Ca/log PTH as a hypothesised expression of PTH sensitivity, Ca/logPTH (8.18, 5.91, 3.10, 1.36, 1.42; sem=1.064; P<0.001) showed an exponential decline relative to the increases in PiP with dietary P previously mentioned. The lower asymptote of this association was achieved somewhere between 1.8 to 2.1 g P/kg DM, which is approximately the dietary P requirement for steers. PTH and Ca warrant further investigation as metabolic predictors of P requirement in cattle.

Novel physiological responses in phosphorus deficient beef cows

S.T. Anderson[1], D.M. McNeill[2], L. Castells[3], J.G. Spiers[1], L.J. Kidd[2], K. Goodwin[4], M.T. Fletcher[3] and R.M. Dixon[3]
[1]University of Queensland, Biomedical Sciences, St Lucia, Australia, [2]UQ, Veterinary Science, Gatton, Australia, [3]UQ, Queensland Alliance for Agriculture and Food Innovation, St Lucia, Australia, [4]Department of Agriculture and Fisheries, Brian Pastures Research Station, Gayndah, Australia; stephen.anderson@uq.edu.au

Phosphorus (P) deficiencies are common in beef cattle grazing seasonally dry tropical rangelands such as in northern Australia, but little is known about the underlying physiological mechanisms that regulate bone metabolism in such animals. Forty heifers calving at 3 years of age were fed either P deficient (0.08 g P/kg DM, 'Low') or P adequate (0.22 g or 0.28 g P/kg DM, 'High') diets during late pregnancy and lactation respectively, in a factorial design with four dietary treatments: HH, HL, LH and LL. Animal production responses have been reported (1). Here we report on plasma concentrations of Parathyroid Hormone (PTH) and 1,25-dihydroxycholecalciferol (1,25-$(OH)_2$-D_3) and markers of bone deposition (Osteocalcin; OCN) and resorption (C-terminal telopeptides of Type I collagen; CTX-1). PTH increased over time (P<0.05) in cows ingesting high P diets (HH and LH), whilst PTH was markedly decreased in cows ingesting low P diets (LL and HL) particularly in lactation. These diet effects on PTH were negatively correlated with plasma total calcium concentrations. In contrast, 1,25-$(OH)_2$-D_3 in pregnancy was markedly increased (P<0.05) in animals ingesting the low P diet. OCN levels, indicative of bone deposition, decreased (P<0.05) as pregnancy progressed and reached a nadir at calving, before increasing (P<0.05) in lactation. The high P diet increased (P<0.05) OCN during pregnancy and lactation, although this effect was limited. Low P diets were always associated with higher (P<0.05) CTX-1 concentrations, indicative of marked bone resorption. Overall beef cows ingesting severely P deficient diets attempt to achieve P homeostasis through pronounced bone resorption together with activation of Vitamin D3. Both processes appear to be independent of PTH. Such physiological responses in cattle with P deficiency warrant further investigation.

Hydroxy-selenomethionine: an efficient source of selenium in fattening pigs

S. Green[1], K. Coppens[2] and P.A. Geraert[1]
[1]ADISSEO, Innovation & Marketing, 10 Place du Général de Gaulle, 92160 Antony, France, [2]Denkavit Ingredients BV, Swine Nutrition, Tolnegenweg 65, 3780 Voorthuizen, the Netherlands; pierre-andre.geraert@adisseo.com

Selenium (Se) is an essential trace element for animals, mainly represented through two seleno amino acids: selenomethionine (SeMet) and selenocysteine (SeCys). The latter being specifically incorporated into selenoproteins, a group of protein involved in various functions such as thyroid hormone activation, antioxidant defense or immunity. This trial aimed to compare the selenium deposition and antioxidant biomarkers in practical conditions using different dietary organic selenium sources. The basal diets contained 0.25 ppm of selenium as sodium selenite and were enriched either with 0.2 ppm of selenium as selenomethionine (SeMet) or with hydroxy-selenomethionine (OH-SeMet). Pigs (PIC × Belgian Piétrain) received the dietary treatments from 20 weeks of age (approx. 65 kg) until 28 weeks of age before slaughtering (115 kg). Commercial breeding conditions were used for this experiment, with farms supplemented with one of the two additional selenium sources. At the end of the experiment, 10 pigs per farm were selected (five castrated males and five females) for blood sampling and tissue collection. Glutathione peroxidase activity and total glutathione (tGSH) quantification were determined on blood as well as total selenium deposition in liver and diaphragm muscle. Statistical analyses did not indicate Sex or Treatment×Sex significant interaction (P>0.05) for the different parameters measured. Muscle and liver Se concentrations were significantly increased in the OH-SeMet treatment compared to SeMet treatment: 1.07 vs 0.52 mg Se/kg dry matter product for muscle and 2.07 vs 1.83 mg Se/kg dry matter product for liver, respectively (P<0.05). Blood glutathione peroxidase activity was similar between the two treatments: 216 vs 221 U/l, (P>0.05); whereas increased tGSH was observed in the OH-SeMet treatment: 916 vs 720 µmol/l, (P<0.05). Taking into account the limitations of such a field trial, results indicate increased Se tissue deposition and higher tGSH levels with OH-SeMet compared to SeMet. Such higher tGSH levels and total Se deposition would support an enhanced antioxidant capacity as well as better meat quality in those OH-SeMet fed pigs.

Added dietary cobalt and vitamin B12 in the periparturient period in dairy cows

D.V.W. Achchillage[1], A.H. Brassington[2], K.D. Sinclair[2] and L.A. Sinclair[1]
[1]Harper Adams University, Dept. Animal Production, Welfare and Veterinary Sciences, Edgmond, Newport, Shropshire, TF10 8NB, United Kingdom, [2]Nottingham University, School of Biological Sciences, Sutton Bonington, LE12 5RD, United Kingdom; lsinclair@harper-adams.ac.uk

Vitamin B12 (B12) is required by the dairy cow for the regulation of energy metabolism and liver health particularly during the transition period, and is normally produced in sufficient quantities by the rumen microbes from dietary cobalt (Co). Recent European Union (EU) legislation has however, limited the maximum inclusion rate of Co. This study determined the effect of added dietary Co, B12 or injectable B12 on energy metabolism and performance during the transition period in high-yielding dairy cows. Fifty six Holstein-Friesian cows received one of four treatments from drying off until 8 weeks postpartum: C, no added Co (background of 0.2 mg Co/kg DM); DC, additional 0.2 mg Co/kg DM; DB, additional 0.68 mg B12/kg DM or IB, injected B12. Cows were weighed, condition scored (1-5 scale) and blood sampled at drying off, calving and at 2, 4 and 8 weeks post calving, with additional blood samples collected at 2 weeks prior to calving and live weight/condition score at 4 weeks prior to calving. Liver biopsy samples were collected at 4 weeks post calving. Cows were group fed prior to calving, with individual intake and milk yield recorded post-calving. There was no effect (P>0.05) of treatment on live weight pre or post calving (mean values of 664 kg and 3.0 units respectively). Similarly, there was no effect of treatment on DM intake or milk yield post calving (mean values of 21.6 and 39.6 kg/d respectively). Plasma β-hydroxybutyrate decreased from drying off until calving, then increased at calving and in early lactation (P<0.001), but was not affected by treatment. Post calving, mean liver triacyglyceride concentration was 31.3 mg/g fresh weight, but there was no effect (P>0.05) of treatment. In conclusion, the recently introduced EU limit on the inclusion of Co will not impair energy metabolism or performance during the transition period, and there is no benefit to added dietary B12.

Relationship between mitochondria and compensatory growth in Holstein Friesian bulls

C.E. McKenna[1,2], K. Keogh[2,3], S. Waters[2], D. Kenny[2], R. Porter[1] and A. Kelly[3]
[1]Trinity College Dublin, Biochemistry, Dublin 2, Dublin 2, Ireland, [2]Teagasc Grange, Animal Bioscience, Dunsany, Co. Meath, Co. Meath, Ireland, [3]University College Dublin, Department of Agriculture and Veterinary Sciences, Belfield, Dublin 4, Dublin 4, Ireland; mckenncl@tcd.ie

Compensatory growth is an accelerated growth phenomenon which occurs following a period of dietary restriction. This trait is exploited worldwide in animal production systems as a method for lower feed costs. The physiological and molecular control of compensatory growth remains to be elucidated. Previous studies have hypothesised a link between mitochondrial function and compensatory growth; thus the aim of this study was to investigate a potential link between mitochondrial function and compensatory growth in hepatic tissue. Hepatic tissue was collected from Holstein Friesian bulls following 125 days of dietary restriction and again following 55 days of subsequent re-alimentation during which the animals exhibited significant compensatory growth. The data were compared with those of control animals offered the same on an ad libitum basis throughout the 180 day study. Spectrophotometric assays to assess mitochondrial abundance and function of the respiratory chain enzyme complex; complex 1 were performed on hepatic tissue. Mitochondrial abundance tended to be higher in restricted animals when compared with ad libitum animals (P=0.07) while no significant difference were observed between groups once the restricted animals underwent re-alimentation (P=0.17). No difference was observed in Complex 1 activity between restricted and ad libitum animals (P=0.6) however there was a tendency (P=0.06) for complex 1 activity to be higher in restricted animals compared to ad-libitum animals once they undergo re-alimentation. These results suggest a potential relationship between compensatory growth and mitochondrial abundance and function.

Tissue mRNA expression of branched-chain amino acid catabolism enzyme in early lactating dairy cows

H. Sadri[1], A.R. Alizadeh[1,2,3], J. Rehage[4], S. Dänicke[5] and H. Sauerwein[1]
[1]Institute of Animal Science, Physiology & Hygiene Unit, University of Bonn, Germany, [2]Department of Embryology, Reproductive Biomedicine Research Center, Royan Institute for ReproductivBiomedicine, Tehran, Iran, [3]Department of Animal Science, Saveh Branch, Islamic Azad University, Saveh, Iran, [4]Clinic for Cattle, University of Veterinary Medicine, Hannover, Germany, [5]Institute of Animal Nutrition, Friedrich-Loeffler-Institute, Braunschweig, Germany; hsadri@uni-bonn.de

The branched-chain α-ketoacid dehydrogenase complex (BCKDH) catalyzes an irreversible oxidative decarboxylation of branched-chain α-keto acids, which is considered as the rate-limiting step in the overall branched-chain amino acids (BCAA) catabolic pathway. Skeletal muscle is a major site for BCAA catabolism; in addition, adipose tissue (AT) has been appreciated in humans as being active in BCAA metabolism. Thus, to test for the potential involvement of AT in BCAA metabolism in dairy cows, 25 primiparous Holstein cows were slaughtered on either DIM (days in milk) 1 (n=5), 42 or 105 (each n=10), and samples from M. semitendinosus and 3 visceral AT (mesenterial, omental, and retroperitoneal) and 3 subcutaneous AT (sternum, tail head, and withers) were obtained and assayed for mRNA abundance of BCKDHA and BCKDHB. The mRNA abundance of BCKDHA in mesenterial, sternum, and tail head AT was greater (P≤0.03) on DIM 105 compared with DIM 1 and 42, but remained unchanged in omental and retroperitoneal AT and in muscle. The BCKDHB mRNA abundance in mesenterial, sternum, tail head, and withers AT and muscle was greater (P≤0.04) on d 105 compared to the other time-points. Across all tissues, the mRNA abundance of BCKDHA was lower (P<0.05) in retroperitoneal AT than in omental and withers AT, whereas that of BCKDHB was greater (P<0.05) in retroperitoneal AT than tail head AT. The BCKDHA and BCKDHB mRNA expression was detectable in different fat depots in a comparable abundance with that of muscle, suggesting that AT may be an important contributor to the adaptive processes associated with whole body BCAA metabolism during early lactation.

OH-Selenomethionine: an efficient source of functional selenium in dairy cows

M. Briens, L. Bahloul, L. Saibi, P.A. Geraert and Y. Mercier
ADISSEO, R&D, 10 Place du Général de Gaulle, 92160 Antony, France; pierre-andre.geraert@adisseo.com

Selenium (Se) is an essential trace element in all animal species. In ruminants, its deficiency causes various pathologies such as myopathies, decrease of fertility and depressed immune status. Selenomethionine (SeMet) represents a non reactive storage form of selenium. Conversely, selenocysteine (SeCys) is specifically included in selenoproteins involved in various biological mechanisms particularly oxidoreduction pathways. Dietary selenium supplementation is currently done through mineral (sodium selenite) or organic (selenized yeasts or SeMet based products) forms. The present studies compare the bioefficacy of different selenium sources in dairy cows. In the first study, 24 prim'Holstein cows (3 groups of 8 animals) were distributed according to their lactation, stage, production level, milk fat and protein contents. Cows were fed during 61 days a control (C) non Se supplemented diet; a diet supplemented in sodium selenite (SS) or in OH-SeMet (SO) (2.4 mg Se/cow/day). Those dietary treatments did not affect the production performance (P>0.05): 35.4, 34.6 and 35.6 kg milk per day respectively; neither the milk composition: 35.8, 36.4 and 37.0 g fat/kg milk or 31.1, 31.7 and 32.0 g protein/kg milk respectively. Significant differences in milk selenium contents were observed (P<0.05) 9.1, 14.1 and 16.9 mg Se/kg milk for C, SS and SO respectively. Such values confirm the higher bioavailability of OH-SeMet. In a second study, nine dry prim'Holstein cows (3 groups of 3 animals) were used to compare selenized yeast (SY) (containing 65% of its Se as SeMet) and OH-SeMet (SO) to a control group (C). The supplemented groups SY and SO received 10 mg Se/cow/day during 21 days. During that period, blood samplings were performed at 8 times. Plasma selenium content showed a level double compared to the C group in SY and SO fed cows. Moreover, the SO supplemented cows exhibited a higher SeCys plasma content compared to SY group. Those results confirm the better efficacy of OH-SeMet as a source of functional Se.

Endotoxin concentrations in the gut increase from the duodenum to the ileum in broiler chickens

C. Emsenhuber, N. Reisinger, E. Mayer and G. Schatzmayr
BIOMIN Holding GmbH, Technopark 1, 3423 Tulln, Austria; caroline.emsenhuber@biomin.net

Endotoxins play an important role during inflammation processes and are continuously released in the gastrointestinal tract by Gram-negative bacteria. Several factors (e.g. wrong feed formulation, mycotoxins, heat stress…) can impair the intestinal barrier function and thus endotoxins from the intestine can enter the blood stream. Once there, small amounts can lead to strong immune responses and weaken the birds' immune system. Therefore, endotoxin concentration in different regions of the intestinal tract of broiler chicken were assessed. Intestinal content from six healthy broilers was sampled and stored at -20 °C until analysis. Samples were thawed at room temperature, weight-in and diluted 10-fold with endotoxin free water followed by incubation for 1 hour at 37 °C while shaking at 300 rpm. Afterwards, samples were centrifuged for 10 min at 4,500×g and heat inactivated for 15 min at 100 °C. Supernatants were diluted either 10-, 50-, or 100-fold with endotoxin free water. Endotoxin concentration was measured with the limulus amoebocyte lysate (LAL) assay. Glucan buffer was used for reconstitution of the LAL reagent to avoid interferences of the sample matrix. Spike recovery was measured for every sample to avoid false positive or false negative results. Endotoxin concentration in the duodenum (370±388 EU/g) and jejunum (480±435 EU/g) was significantly lower than in the ileum (4,270±3,802 EU/g). Endotoxin spike recovery was valid for all samples (100-200%). Our results show that a certain concentration of endotoxins is also present in the intestinal tract of healthy broilers. In addition, we could see that the endotoxin concentration increases from duodenum to the ileum. As long as the intestinal barrier is intact, endotoxins might not harm the animal. However, during stress events, endotoxins might enter the blood stream and negatively impact animal health.

Immunogenic properties of Lactococcus and Lactobacillus probiotics in chicken HD11 cell line

A. Slawinska[1], A. Plowiec[1], A. Dunislawska[1], M. Siwek[1] and H. Bluyssen[2]
[1]UTP University of Science and Technology, Department of Animal Biochemistry and Biotechnology, Mazowiecka 28, 85-084 Bydgoszcz, Poland, [2]Adam Mickiewicz University, Department of Human Molecular Genetics, Umultowska 89, 61-614 Poznan, Poland; slawinska@utp.edu.pl

Probiotics are commensal bacteria, which – once ingested – colonize guts of the host and exert beneficial properties on its metabolism and health. Some probiotic strains interact with immune system. This study aimed to determine immunogenic properties of probiotic bacteria in chicken macrophage-like cell line (HD11). Four probiotic strains, i.e. Lactococcus lactis subsp. lactis IBB2955 (P1), Lactococcus lactis subsp. cremoris IBB477 (P2), Lactobacillus salivarius IBB3154 (P3) and Lactobacillus plantarum IBB3036 (P4) were used to stimulate immune-related signaling pathways in HD11 cell line. This cell line comprises of chicken monocytes and macrophages, which belong to innate immune system. In vitro stimulation of HD11 cells with probiotics lasted for 6 hours and was followed by transcriptome analysis of host with Affymetrix GeneAtlas System. All bacteria strains triggered gene expression regulation in HD11 cells. In P1-P3 groups 2,861 genes on average were differentially expressed; 38% up-regulated vs 62% down-regulated. In P4 group only 514 genes were regulated; 55% up-regulated vs 45% down-regulated. Metabolic pathways triggered in HD11 by all probiotic strains (P1-P4) included Toll-like receptor/NOD-like/Jak-STAT/MAPK and mTOR signaling pathways. Common pathways regulated by Lactococcus spp. (P1, P2) included propionate metabolism, endocytosis and natural killer cell mediated cytotoxicity. P2 specifically triggered pentose phosphate pathway and Notch signaling. In all groups (P1-P4), genes with the highest up-regulation included INHBA, CCL4, CCL20, IL12B, IL6, IL8 and TNFAIP6; whereas genes with the highest down-regulation were GPR34, GPR65, FUT7, CCR4 and PTPN6. In conclusion, all bacteria strains (P1-P4) triggered immune-related signaling pathways in HD11. Distinct molecular responses to different strains have been determined. Acknowledgements: National Science Centre in Poland (grant no: UMO-2013/11/B/NZ9/00783)

Effect of essential oils on NK cell activity in poultry

H. Al-Khalaifa
Kuwait Institute for Scientific Research, Shuwaikh, 13109-Safat-Kuwait, Kuwait; hkhalifa@kisr.edu.kw

There has been interest in enrichment of poultry meat with unsaturated fatty acids to increase its consumption by humans. However, there is claim that these fatty acids have immunomodulatory effects in chickens. The purpose of this study was to investigate the effect of various dietary essential oils on natural cytotoxicity in poultry. One day old male Cobb 500 broilers (n=20) were fed on one of four essential oils containing n-3 PUFA: flaxseed oil-, sunflower oil-, fish oil (FO) or canola-enriched diets until slaughter. At slaughter, samples of blood were collected from each bird and natural cytotoxicity test was conducted. The overall differences between dietary treatments were analysed using one-way analysis of variance (ANOVA) using the general linear model in Minitab. Differences between the treatment groups were considered statistically different at $P \leq 0.05$. There was a clear fatty acid modulation effect based on the different essential oil used across all tissues. Natural cytotoxicity was investigated. It was highest in splenocytes and PBMCs from broilers fed flaxseed oil, followed by those fed canola or sunflower oil, and lowest for those from broilers fed FO. These results suggest that canola oil may enrich chicken meat with n-3 PUFA without significant detrimental effects on chicken natural cytotoxicity.

Duodenal morphometry and digestibility of broilers fed with dry whey alone or combined with butyrate

C. Pineda-Quiroga, R. Atxaerandio, R. Ruiz and A. García-Rodríguez
Neiker-Tecnalia, Animal Production, Campus Agroalimenario de Arkaute, 01080, Vitoria-Gasteiz, Spain;
cpineda@neiker.eus

Dry whey is a co-product of the dairy cheese industry characterized by high lactose content (Approx 70% of dry matter). Poultry have no β-galactosidase required to digest lactose, therefore lactose was proposed as prebiotic. Prebiotics could have a favourable effect on intestinal development and digestion of nutrients. Moreover, butyric acid and their salts are recognized as development promoter of enterocytes and enhancer of mucosa integrity. Thus, the aim of this study was evaluate the effect of supplementing broiler diets with dry powder whey (W), fat-coated calcium butyrate (B) and a mixture of both (WB) on duodenal histomorphometry (DH) and coefficient of ileal apparent digestibility of nutrients (CIAD). Four treatments with 10 replicates of 30 one day-old chickens Ross 308 were randomly allocated in pens of 2.75 m^2. Animals were fed with one of the following corn-soybean based diets: control diet (0W0B), 60 g W/kg (60W0B), 1 g B/kg (0W1B) and 60 g W/1 g B/kg mixture (60W1B). Chromium oxide (Cr_2O_3, 50 g/kg) was added to diets as inert marker. At day 21 of feeding, 3 birds by treatment were randomly slaughtered to obtain the duodenum for histological examination with haematoxiline and eosine staining. Ileal digesta content of sampled animals was collected, pooled and freeze dried to determine Cr_2O_3. As a result, animals fed with 60W1B showed an increase on duodenal villus surface area compared to 0W0B (83.4 vs 53.6 μm^2), 60W0B (53.5 vs 53.6 μm^2) and 0W1B (49.3 vs 53.6 μm^2) (P<0.05). As consequence, feeding with 60W1B increase CIAD of crude protein compared to 0W0B, 60W0B (0.815 vs 0.707) and 0W1B (0.815 vs 0.757). Also, 60W1B increment CIAD of calcium and phosphorus compared to 0W0B (0.764 vs 0.628; 0.702 vs 0.505), 60W0B (0.764 vs 0.628; 0.702 vs 0.525) and 0W1B (0.764 vs 0.723; 0.702 vs 0.613), respectively (P<0.05). Feeding with 60W0B did not affect DH or CIAD of nutrients, while 0W1B reduced villus height and villus:crypt ratio compared to the remaining treatments (P<0.05). The presented results show that feeding broilers with a mixture of dry whey and fat-coated calcium butyrate may act synergistically, by increasing villus surface area and CIAD of nutrients.

The effects of various dietary vegetable oil sources on villi morphology & liver aldehydes in layers

A. Hajibabaei[1], S.H. Golzar Adabi[1], N.H. Casey[1] and A.G. Bayraktaroglu[2]
[1]University of Pretoria, Department of Animal and Wildlife Sciences, Pretoria, 0002, South Africa, [2]Ankara University, Department of Histology and Embryology, Ankara, 3642, Turkey; ali.hajibabaei@up.ac.za

Twenty-six-week-old laying hens (90 Lohmann Classic Brown) received dietary treatments containing canola and linseed oil at 2% inclusion levels and a control diet for 21 days. The 2% dietary linseed oil, which had higher total polyunsaturated fatty acids (PUFA) and linolenic fatty acid contents, resulted in higher liver malondialdehyde and wider villi, while the 2% canola and linseed oil treatments recorded smaller reproductive organs relative to live weight. It was concluded that young layers might have an anatomical response to high levels of dietary PUFA, which might improve nutrient absorption and cause a biochemical response, resulting in potentially negative cytotoxic aldehydic lipid peroxidation products.

Fate of a brominated flame retardant (HBCD) in chickens and pigs: a generic modelling approach

B. Méda[1], E. Royer[2], S. Guérin[1], A. García De Salazar[2], E. Dominguez-Romero[3,4], J. Henri[5], A. Travel[4], E. Baéza[1] and C. Jondreville[3]
[1]INRA, UR83 Recherches Avicoles, 37380 Nouzilly, France, [2]IFIP, Institut du Porc, 31500 Toulouse, France, [3]INRA, USC 340 AFPA, 54500 Vandoeuvre-les-Nancy, France, [4]ITAVI, UMT BIRD, 37380 Nouzilly, France, [5]ANSES, Laboratoire de Fougères, 35306 Fougeres, France; bertrand.meda@tours.inra.fr

Hexabromocyclododecane (HBCD) is an endocrine disruptor both for humans and animals. It is widely used as a flame retardant in building materials of livestock housing and has been detected in animal products. Thus, there is a great need for methods to investigate the contamination of animal products by HBCD. This study proposes a generic Physiologically Based PharmacoKinetic (PBPK) model for growing chickens (C) and pigs (P) simulating HBCD accumulation in tissues. It consists in the connection of two sub-models. The physiological sub-model represents different compartments (total weight and lipid weights): plasma, liver, two muscles with different lipid contents (breast and leg for C; loin and ham for P), one adipose tissue (abdominal fat for C; backfat for P) and one compartment representing the rest of the animal. In the ADME sub-model (absorption, distribution, metabolism, excretion), HBCD is absorbed in digestive tract, before being distributed in the organism by the plasma considering differences in blood irrigation between tissues. Since HBCD is a lipophilic and apolar compound, a partition coefficient (Kp), defined as the ratio between neutral lipid concentrations in a tissue and in plasma, was used to predict the distribution of HBCD in each tissue. In the liver, a fraction of HBCD is eliminated by metabolization (hepatic clearance). Model was calibrated to take into account differences in tissue growth (allometric growth) and lipid contents according to species, genetics (e.g. fast or slow growing strain for C) and sex (male, female or castrated male for P). For each species, the model was validated using experimental data where animals were fed: (1) with a control diet (i.e. no HBCD); (2) with a contaminated diet; or (3) with a contaminated diet (exposition) and then a non-contaminated diet (decontamination). This generic model will be used to assess the risk of meat contamination by HBCD in different pig and chicken production systems.

Effect of feeding dietary AlgamuneTM AM on the performance of piglets

N.S. Muley[1], J.E. Zerrahn[2], J. Busch[1] and A. Lemme[1]
[1]Evonik Industries AG, Nutrition & Care, Rodenbacher Chaussee 4, 63457 Hanau-Wolfgang, Germany, [2]Evonik Industries AG, Nutrition & Care, Damhaven 3 C, 7100 Vejle, Denmark; nitin.muley@evonik.com

A series of 2 experiments were carried out to test the efficacy of Algamune™ AM (microalgae-based biomass containing linear beta-1,3-glucan, dose 100 g/ton feed) in weaned piglets. Piglets were not offered antibiotic in either experiment. Feed and water were fed ad libitum. In the first experiment, piglets were fed diets (isocaloric and isonitrogenous) with and without Algamune™ AM. Piglets (n=640, Danbred sow × Pietrain boar, ≈5 kg initial body weight) were separated on the basis of body weight and sex, and placed in 40 pens (16 piglets per pen) with a waterer (1 per pen) and feeder (1 per 2 pens; 32 piglets per feeder). Performance data were analyzed using Proc mixed model (SAS Inc, NC, USA) with pen and sex as covariates. Weight gain (d0-25:4.14 kg and d0-51:13.69 kg) and feed conversion (d0-25:1.59 and d0-51:1.53) in piglets fed Algamune™ AM was similar (P>0.05) as compared to weight gain (d0-25:4.11 kg and d0-51:13.30 kg) and feed conversion (d25:1.59 and d51:1.56 kg) in piglets fed control diets. Piglets fed dietary Algamune™ AM had lower (3.13%) mortality than piglets fed control feed (5.11%). In the second experiment, piglets (≈7.7 kg initial body weight, Landrace/Yorkshire × Duroc) were separated on the basis sex and placed in 42 pens (20-22 piglets per pen of either sex) with a waterer (1 per pen) and feeder (1 per 2 pens; 40-44 piglets per feeder). Slight variation in CP content (18.1% CP for control and 18.8% CP for treatment) of feeds was observed. Zinc oxide (2,500 ppm) was added to both diets at d14. Experimental feeds were offered till day 21. Data were analyzed using GLMM (mixed effect linear model) procedure of R. Results (d21) indicated that average daily gain between pigs fed control (0.27 kg) and treatment (0.27 kg) diets was similar (P>0.05). However, piglets fed Algamune™ AM had significantly better (P=0.02) feed conversion (1.45) than piglets fed control diets (1.53).

Effect of spay-dried porcine plasma in weaned piglets against an E. coli challenge

X. Guan[1], F. Molist[1], C. Van Vuure[2] and P.J. Van Der Aar[1]
[1]Schothorst Feed Research, Swine Nutrition, Meerkoetenweg 26, 8200 AM Lelystad, the Netherlands, [2]Darling Ingredients International, Sonac, Kanaaldijk Noord 20-21, 5690 AA Son, the Netherlands; xguan@schothorst.nl

Weaning is the most challenging period for piglets and post-weaning diarrhoea caused by Escherichia coli (E. coli) is still one of the main problems. Spray-dried porcine plasma (SDPP) is frequently used as an alternative protein source of dry skim milk powder (DSMP) especially during the first 2 weeks after weaning. The objective of the trial was to compare the likely effect of SDPP compared to DSMP after an E. coli challenge. Thirty-two piglets weaned at 26 days of age with average body weight of 6.9 kg entered the experiment and were allocated to 2 different treatments DSMP and SDPP. The experimental period lasted for 22 days and piglets received 2 different diets. During the first 15 days piglets received either a diet with 5% DSMP or a diet with 5% SDPP. From day 15 till 22, all the piglets received the same diet. On day 10, piglets were orally challenged with 5 ml of 8.7 Log of E. coli. During days 11 to 15, 18, 20 and 22, faecal samples were collected to analyse the excretion of E. coli. On days 8, 14 and 21, blood samples were taken to analyse the level of haptoglublin and natural antibodies. Piglets were weighed at the beginning of the trial and at days 15 and 22. Feed intake and feed conversion ratio was measured for the same periods. Results were analysed as repeated measurements with dietary treatment, day and its interaction as main factors. During days 11 to 15, piglets fed the diet with SDPP showed 0.6 Log E. coli (P=0.01) lower in the faeces. Piglets fed with SDPP also showed a significant transitory increase (P=0.025) of haptoglobulin level in the blood after the challenge (day 14), together with a significant (P=0.003) improvement of the average daily gain. In conclusion, the addition of 5% SDDP during the first two weeks in weaned piglets challenged with E. coli resulted in lower E. coli counts after the challenge with improved growth performance.

Effect of zinc oxide sources and dosages on intestinal bacterial counts and gut integrity of piglets
J. Michiels[1], N. Van Noten[1], J. Degroote[1], A. Romeo[2] and W. Wang[1]
[1]Faculty of Bioscience Engineering, Department of Applied Biosciences, Ghent University, 9000 Gent, Belgium, [2]Animine, 335 Chemin du Noyer, 74330 Sillingy, France; aromeo@animine.eu

Zinc oxide (ZnO) can be supplied at pharmacological dosage (2,400 ppm of Zn) in piglet diets in order to improve gut health and performance. In this study, the effects of a potentiated ZnO source (HiZox®) at low dose were compared to the regular ZnO at nutritional and pharmacological levels. The conventional ZnO was evaluated at 110 and 2,400 ppm of Zn, vs 110 and 220 ppm of Zn for the potentiated ZnO. High iron (Fe) level was used to induce gastro-intestinal disturbances: diets including regular ZnO contained 100 or 500 ppm of iron from FeSO4, vs 500 ppm of Fe for diets with the potentiated ZnO. The study consists of two identical time replicate experiments, with 48 weaned piglets, 20-21 d of age. In each experiment 12 pens were used. Each trial lasted 15 days. Animal performance, bacterial counts in intestinal contents and Ussing chamber measurements in distal jejunum were recorded. Regarding the growth performance, groups fed with 2,400 ppm of Zn were better than the other groups whatever the iron content. They had also the lowest bacterial counts in distal small intestine and were significantly better than the groups fed with 110 ppm of Zn from regular ZnO: for example with E. coli, 4.9 $\log_{10}$ cfu/g vs 7.2 $\log_{10}$ cfu/g with 100 ppm of Fe, 5.6 $\log_{10}$ cfu/g vs 7.0 $\log_{10}$ cfu/g with 500 ppm of Fe (P<0.05). Groups fed with potentiated ZnO were also significantly better (P<0.05) than the groups fed with 110 ppm of standard ZnO for coliforms, and numerically better for E. coli. Concerning the measurements in Ussing chambers, transepithelial electrical resistance (TEER) of jejunal mucosa was significantly (P<0.05) higher for groups fed with potentiated ZnO, compared with groups fed with 110 ppm of standard ZnO, showing a better integrity of the tissue. In conclusion, the potentiated ZnO at low dosage showed positive effects on intestinal health, similar to the effects of pharmacological dosage of standard ZnO.

Vitamin E and selenium injections improved the reproductive performance of Awassi ewes
M.S. Awawdeh, A.H. Eljarah and M.M. Ababneh
Jordan University of Science and Technology, Faculty of Veterinary Medicine, P.O Box 3030, Irbid 22110, Jordan; mawawdeh@just.edu.jo

The objective of this study was to investigate the effects of vitamin E and selenium (Se) injections on the reproductive performance of estrus-synchronized Awassi ewes. Estrus was synchronized in 74 ewes and ewes were randomly assigned to one of two groups where ewes received: 0 (control, n=36) or 13.6 mg/kg BW of vitamin E plus 0.045 mg/kg BW of Se (vitamin E/Se, n=38). Vitamin E/Se injections were given at the time of insertion, withdrawal, and 19 days after withdrawal of intra-vaginal sponges. Serum samples were collected from ewes (n=20 per group) at injection times to measure Se contents. Pregnancy rates were determined by progesterone assay (at day 19) and by ultrasound (at day 40) after sponge removal. Ewes were regularly observed until lambing where lambing rate and litter size, sex, and weight were recorded. Vitamin E/Se injections had no (P>0.25) effects on live weight of ewes at lambing or BW change from breeding to lambing. Vitamin E/Se injections positively improved (P<0.05) pregnancy rates determined by progesterone assay (from 80.6 to 97.4%) and ultrasound examination (from 63.9 to 86.8%). Vitamin E/Se injections tended (P=0.08) to reduce pregnancy losses from 44.8 to 24.3%. In addition, vitamin E/Se injections markedly increased the percentage of ewes that lambed after only one service (from 64.0 to 93.3%) but did not affect overall lambing percentage. Vitamin E/Se injections did not (P>0.20) affect litter size, sex, or weight. Under conditions of the current study, injecting Awassi ewes with vitamin E/Se concurrent with estrus-synchronization program improved their reproductive performance.

Reducing rumen fermentation of wheat with NaOH can ameliorate heat stress in grain-fed sheep
P.A. Gonzalez-Rivas, P.A. Giraldo, K. Digiacomo, B.J. Leury, J.J. Cottrell and F.R. Dunshea
The University of Melbourne, Faculty of Veterinary and Agricultural Sciences, Royal Parade, Parkville, Victoria
3010, Australia; fdunshea@unimelb.edu.au

The metabolic heat increment in ruminants and the risk of heat stress (HS) are closely related to the rumen fermentation characteristics of feed. Therefore, reducing the rumen fermentation of wheat with 3% sodium hydroxide (NaOH) treatment may reduce heat load and the impact of HS in grain-fed sheep. An in vitro gas production technique was used to determine the gas production kinetics of 20 replicates of wheat, corn and 3% NaOH treated wheat grain incubated in buffered rumen fluid at 39 °C during 24 h. Then, 31 Merino × Poll Dorset wethers were housed in 2 climate-controlled rooms and were fed either crushed corn grain plus forage (CD), crushed wheat grain plus forage (WD) or 3% NaOH treated whole wheat plus forage (TWD) during 3 experimental periods: P1) 7 d of thermoneutral conditions (18-21 °C/40-50% relative humidity [RH]) and 1.8 times maintenance feed intake; P2) 7 d of HS (28-38 °C/30-50% RH) and 1.8 times maintenance feed intake; and P3) 7 d of HS as P2 and 2 times maintenance feed intake in a in a randomized control experiment. Water was offered ad libitum. Rectal temperature (RT), respiration rate (RR), and flank skin temperature (FT) were measured four times a day. Gas production curves were fitted to a modified Hill model and gas kinetic and physiological parameters were analysed by the Restricted Maximum Likelihood (REML) in GenStat v15. Wheat grain treated with 3% NaOH had a slower rate of gas production than wheat (-23%, $P<0.001$) and corn (-7%, $P<0.05$) reaching maximum fermentation activity later than wheat (7.9 vs 9.7 h $P<0.001$). All physiological parameters were elevated during HS, especially during P3 ($P<0.001$). Sheep fed CD had lower RR and RT than WD and TWD especially during HS ($P<0.001$). Sheep fed TWD had lower RR than WD ($P<0.001$). Flank temperature was higher for WD, while no differences were observed in FT between CD and WTD ($P<0.001$). These data confirm the slower in vitro fermentability of 3% NaOH treated wheat and corn compared to wheat and that reducing the rate of fermentation with TWD or CD improves tolerance to heat stress in grain-fed sheep.

Comparing the efficiency of intravaginal progesterone dispenser to CIDR for Awassi sheep
R. El Balaa[1], E. Mousaoumai[2], P. Aad[3], K. Houchaimi[4] and S. Abi Saab[5]
[1]University of Balamand, Balamand, Koura, 100 Tripoli, Lebanon, [2]Lebanese University, Faculty of Agriculture and Veterinary Medicine, Dekwaneh, Beirut, Lebanon, [3]Notre Dame University, Zouk Michael, Keserwan, Lebanon, [4]Lebanese Agricultural Research Institute, Tal el Amara, Bekaa, Lebanon, [5]USEK, Université Saint Esprit, Kaslik, Lebanon; paad@ndu.edu.lb

Due to its fat tail, Awassi sheep create a hotter microclimate around the progesterone sponge, which could lead to speeding its deterioration and thus decreasing its efficiency. The objective of this study is to test the efficiency of CIDR (Controlled Internal Drug Release) and the amelioration it could bring to the heat synchronization process 51 Awassi females were distributed into 3 groups according to age and weight: GI included 19 low weight females 35.68 kg (SD=2.88) aged between 6 and 8 months. GII included 16 medium weight females 44.85 kg (SD=5.50) aged between 8 and 12 months. GIII included 16 high weigh females 57.96 kg (SD=6.89) aged between 12 and 16 months. Each group was divided into two subgroups, one receiving the intravaginal apparatus and the second the CIDR apparatus. Progesterone dispensers were placed for 14 days during which two ultrasound tests were performed. After the removal of the progesterone devices, an intramuscular injection of 500 IU (International Units) of PMSG (Pregnant Mare Serum Gonadotropin) was administered. 48 hours later, adult males were introduced for heat detection, followed by ultrasound check for confirmation. Data was analysed through χ^2 and ANOVA tests using SPSS statistical program. The success rate of synchronization regardless of the used device was significantly high with 75.5% compared to a 24.5% failure rate ($P<0.05$). Success rate at GIII 93.7% was significantly higher than that of GI 58.8% ($P<0.05$) which shows a possible effect of age and weight. Over all, No significant difference was noted between the success rates of the two treatments with 84% for the sponge, against 66.7% for CIDR ($P<0.05$). Results showed that the performance of intravaginal sponges and CIDR are similar for Awassi sheep heat synchronization; age remained a key parameter for the start of the oestrous cycle.

Can metabolomics help to predict amino acid fluxes across animal tissues?

G. Cantalapiedra-Hijar[1], A. De La Foye[1,2], D. Durand[1], D. Gruffat[1], E. Pujos-Guillot[2], C. Chantelauze[1], S. Makhlouk[1] and I. Ortigues-Marty[1]
[1]Institut National de la Recherche Agronomique, UMR 1213 INRA-VetAgro Sup, Unité Mixte de Recherches sur les Herbivores, 63122 St Genès Champanelle, France, [2]Institut National de la Recherche Agronomique, UMR 1019, Plateforme d'Exploration du Métabolisme, Nutrition Humaine, 63122 St Genès Champanelle, France; ortigues@clermont.inra.fr

The study of nutrient fluxes across organs and tissues is important for animal physiologists to better define animal nutritional requirements and strategies to improve animal production. Standard methods for determining nutrient fluxes across organs and tissues involve the use of multi-catheterized animals. However, their use raises ethical issues and implies laborious and expensive experiments. Consequently, alternative biomarkers are being explored. Our hypothesis was that concentrations of metabolites present at the peripheral level (easy to sample vessels) could reflect amino acid (AA) fluxes across organs and tissues in dairy cows. To test this hypothesis, a non-targeted metabolomic approach (LC-MS) was conducted on 73 plasma samples from 4 vessels irrigating and draining blood of gut tissues, liver and mammary gland in 5 dairy cows fed each 4 different diets. A total of 999 different ions, representing metabolites with their isotopes and some in-source fragments, were extracted from all plasma metabolic profiles. Around one third (n=307) of these ions showed an intensity significantly (P<0.05) changed between the entry and exit of at least one of the organs, and thus could be considered as representing molecules with an endogenous rather than a dietary origin. Only endogenous ions present at the artery, an easy to sample vessel, were further statistically analyzed. A Sparse partial least square model (S-PLS) was used to predict measured AA fluxes across the three organs from the intensity of arterial endogenous ions. AA fluxes across the liver was not correctly predicted by variables kept by the S-PLS model (cross validation criterion Q2<0.0975). In contrast, a total of 52 ions allowed predicting the AA fluxes across gut tissues (Q2=0.50; R2 cross valid=0.45) and mammary gland (Q2=0.37; R2 cross valid=0.30). In conclusion, metabolomics may help to predict amino acid fluxes across gut and mammary gland but not across liver.

Effect of Marjoram and Chamomile extracts on ram semen characteristics after freeze-thawing process

H. Karimi[1], H. Daghigh Kia[2], F. Sadeghi Sadegh Abad[2] and H. Vaseghi[2]
[1]University of Tabriz, 29 Bahman Blvd, 5166614766, Iran, [2]University of Tabriz, Animal Science, 29 Bahman Blvd, 5166614766, Iran; karimi@tabrizu.ac.ir

During exposing sperm to cold shock and osmotic pressure, membrane oxidation increases due to the production of free radicals. The aim of this research was to investigate the effect of extracts of Chamomile and Marjoram herb as natural antioxidants on frozen semen of ram. Marjoram herb contains high levels of phenolic compounds such as gallic acid, polyphenols such as Rosmarinic acid and Quercetin. Chamomile contains chamazulene and flavonoids. Semen was collected by artificial vagina from 5 mature Ghezel breed rams during the breeding season. Semen samples, which were diluted with a Tris-based extender containing the antioxidants Marjoram 66 µl/ml and Chamomile 100 µl/ml extracts, and an extender containing no antioxidants (control), were cooled to 5 °C and frozen in 0.25-ml French straws in liquid nitrogen. Frozen straws were thawed individually at 37 °C for 30 s in a water bath for evaluation. After freezing-thawing, the sperm motility parameters were evaluated using the CASA system, the viability of sperms using eosine-nigrosin stain, membrane integrity using the hypo osmotic swelling test, the sperm abnormalities using Hancock solution and lipid peroxidation by measuring the malondialdehyde concentration. The results showed that adding 66 µl/ml Chamomile and 100 µl/ml Marjoram significantly improved sperm motility parameters, viability and plasma membrane integrity after freeze-thawing compared with the control group.

Effect of eCG on amino acids concentrations in follicular fluid of repeat breeder cows
M.M. Ababneh and S. Manasrah
Jordan University of Science of Technology, Veterinary Clinical Sciences, Irdid 22110, Jordan; ababneh@just.edu.jo

A Repeat breeder (RP) cow is defined as any cow that failed to get pregnant after three services or artificial insemination during estrus. Inclusion of equine chorionic gonadotropin (eCG) in ovulation synchronization and fixed-time AI Protocols is reported to improve fertility in RB cows. The aim of this study was to determine the effect of eCG on individual amino acids (AA) concentrations in follicular fluid (FF) collected from RB cows. Cows were synchronized to estrus with presynch-ovsynch protocol. At the time of the last prostaglandin (PG) $F_2\alpha$ injection, cows were either injected with 500 IU eCG (eCG group; n=9) or normal saline (control; n=9). Trans-vaginal aspirations of FF and blood collection were performed 18 hours after the last gonadotropin-releasing hormone injection. AA concentrations were determined by reverse-phase HPLC. Concentrations of most AA including Aspargine, Threonine, Alanine, Valine, Tryptophan, Isoleucine, Leucine, Orinthine and Tyrosine in FF were correlated with their concentrations in plasma in both groups. AA concentrations in the FF were not affected by eCG except for Cystine (Cys) and Serine (Ser). Cys decreased (P=0.019) but Ser tended to increase (P=0.07) in eCG treated cows. This difference in Ser and Cys concentrations in FF may be due to eCG effect on metabolism or secretion of those AA in the preovulatory follicle. The findings of altered concentrations of Cys and Ser in FF after eCG administrations is interesting. This is because eCG might have an indirect effect on antioxidant status and reactive oxygen species (ROS). Cystine is a precursor for glutathione synthesis. Glutathione concentration is a critical part of oocyte cytoplasmic maturation and protects cells against the effects of the ROS. It is plausible the eCG had shifted the cellular uptake toward increasing cystine transport leading to increased intracellular Cys and therefore decreased Cys concentration in the FF. Ser, on the other hand, was not transported and therefore accumulated in the FF. Further studies are needed to determine if these changes in FF Cys concentrations reflect changes in the intracellular oocyte concentrations and oocyte quality.

Effect of forage type fed with or without molybdenum and sulfur inclusion on copper status of lambs
A. Hussein
Harper Adams University, Edgmond, Newport, Shropshire, TF108NB, TF10 8NB, United Kingdom;
ahussein@harper-adams.ac.uk

Liver Copper (Cu) has been shown to be influenced by forage type. This may be due to their fermentation characteristics influencing thiomolybdate production in the rumen. The effect of forage type with and without Cu antagonist inclusion on the rumen pH and Cu status of lambs was investigated. Sixty Scottish Blackface store lambs with a mean live weight of 23 kg were used in a 10 week study. Lambs were offered ad libitum grass silage (GS), fermented whole crop wheat (FWCW) or urea-treated whole crop wheat (UWCW), with 300 g of concentrate, formulated either no antagonists or 5 mg/kgDM molybdenum or 4 g/kgDM Sulfur, to reduce the efficiency of Cu absorption by half. The dry matter intake (DMI) and live weight gain (LWG) were monitored weekly, with blood samples being collected by venepuncture. At the end of the experiment lambs were slaughtered and rumen fluid and liver samples collected. Blood plasma and liver samples were analysed for mineral content. The experiment was analysed by ANOVA as a 3×2 factorial design. There was no significant interaction (P>0.05) between forage type and Cu antagonists on any parameters. The DMI of lambs offered diet UWCW and FWCW was higher (P=0.007) than those offered diet GS (0.75, 0.74, and 0.67 kg/d respectively). The LWG of lambs fed UWCW was higher (P<0.001) compared to FWCW or GS diet (0.15, 0.12, and 0.11 kg/d respectively). Lambs offered GS tended (P=0.059) to have a higher rumen pH than those fed UWCW or FWCW diet (6.38, 6.22, and 6.21 respectively). Liver Cu concentration was lower (P<0.05) in lambs offered FWCW diet than those offered GS or UWCW (119, 172, and 189 mg/kg DM respectively). Adding antagonists reduced (P<0.05) liver Cu concentration compared to no added antagonists (127 and 192 mg/kg DM respectively). Dietary treatments had not effect on plasma Cu concentration. The results indicate that there was no relationship between rumen pH and the liver Cu status therefore some other characteristics is likely to influence Cu availability.

Effect of artificial rearing on fatty acid metabolism and adipogenesis in lambs

G. Theodorou[1], E. Tsiplakou[1], G. Papadomichelakis[1], D. Sparaggis[2], I. Politis[1], V. Christodoulou[3] and G. Zervas[1]
[1]Agricultural University of Athens, Animal Sciences and Aquaculture, 75 Iera Odos, 11855, Athens, Greece, [2]Agricultural Research Institute, P.O. Box 22016, 1516 Nicosia, Cyprus, [3]Hellenic Agricultural Organization 'Demeter', Animal Research Institute, Paralimni, 58100 Giannitsa, Greece; gzervas@aua.gr

Although artificial rearing is a widespread practice in dairy farms there is not enough scientific data to evaluate possible negative effects in lamb development. The aim of the study was to examine the effect of artificial rearing in the transcriptome of the adipose tissue of lambs. Forty lambs of the Chios breed were divided into two groups of artificial and natural rearing. Lambs were slaughtered at 35 and 90 days of age. Gene expression data showed a marginally diverse pattern between the two groups especially in the visceral fat depot. However, the most interesting results proved to be the differences observed between genders where male lambs displayed higher levels of expression for fatty acid synthase (FASN), stearoyl-CoA desaturase (SCD), acetyl-CoA carboxylase alpha (ACACA), hormone-sensitive lipase and peroxisome proliferator-activated receptor gamma and the differences in gene expression between the three different fat depots examined (omental, visceral and tail) where the tail fat depot exhibited significantly higher expression of FASN, SCD, ACACA and leptin when compared to the respective levels of the visceral and omental adipose tissue depots. Finally as expected there were major differences in gene expression levels between the age of 35 and 90 days with a notable difference in LPL expression where there was a differential shift in the tail fat depot when compared to the other two indicating once more that the tail fat depot plays a significant role in fatty acid metabolism during the first months postnatal.

Historical developments and current issues in fish nutrition

J.W. Schrama[1] and S.J. Kaushik[2]
[1]Wageningen University, Aquaculture and Fisheries Group, P.O. Box 338, 6700 AH Wageningen, the Netherlands, [2]INRA, UR 1067, NuMeA, Nutrition, Metabolism and Aquaculture, BP 3, 64310 St Pée sur Nivelle, France; johan.schrama@wur.nl

For many decades, the yearly growth rate of cultured fish is above 7%. Currently, more than 50% of human fish consumption comes from aquaculture. This presentation addresses the historical developments in terms of nutrition. It includes: changes in nutrient levels and energy content of fish feeds; changes in feed processing; progress in knowledge on requirements; alteration in ingredients used in fish feeds (fishmeal & –oil replacement). Comparisons are made to show similarities and differences to land animals. Aquaculture is characterized by a large variability in many aspects. This variability is a challenge for fish nutritionist to formulate/develop optimal fish feeds. More than 100 species are commercially farmed and their number is still increasing. Differences in feeding ecology, trophic levels, anatomical differences (e.g. gastric vs agastric species), differences in aquatic environment (fresh water, brackish or marine; cold, temperate or tropical), all can have implications on meeting nutrient requirements. Next to species diversity, culture/husbandry conditions exhibit large diversity (pond, race way, cages; water quality management/water refreshment). Examples are given on impacts of husbandry conditions on meeting nutrient/energy requirements. In culture systems with minimal water exchange, the diet can indirectly influence fish by its impact on water quality (feed waste/spillage, faeces and metabolite excretion). In other words, feed ration size and composition are important factors, which determine the quality of the environment in which the fish live. This presentation addresses the implications of the biotic and abiotic factors as well as husbandry conditions in relation to nutrition such as: nutrient requirements; impact of physical condition feed/food type consumed (e.g. size & water content); interaction between nutrition and water quality; impacts of fishmeal/oil replacement.

No fish meal and fish oil in aquafeed: a challenge for the sustainable development of aquaculture

S. Skiba-Cassy, V. Lazzarotto, G. Corraze and F. Médale
INRA, UMR INRA UPPA, Nutrition, Metabolism, Aquaculture (NUMEA), Aquapôle INRA, Quartier Ibarron, 64310 Saint Pée sur Nivelle, France, France; skiba@st-pee.inra.fr

Within the past 30 years, global consumption of fish and shellfish has increased three-fold. Since capture fisheries have remained stable, the increased demand for seafood has been met by aquaculture whose global production has dramatically grown, from less than 4 Mt in 1980 to 67 Mt in 2012 and is expected to further increase by 33% until 2021. For a long time, intensive fish farming has exclusively relied on the use of complete feeds formulated with fish meal (FM) and fish oil (FO) that adequately cover nutritional requirements of fish. The increased demand for FM and FO for aquaculture combined with regulatory measures on fisheries captures have led to shortage in FM and FO. In this context, the use of alternative raw materials in aquafeeds has become essential for the sustainable development of aquaculture. The proportion of FM and FO has been reduced in aquafeed and replaced by a combination of plant ingredients providing sufficient quantity of proteins to cover high protein requirement and specific needs in essential amino acids and fatty acids. However, whereas plant ingredients offer several advantages (diversity, availability and lower prices), they also have limitations related to lower protein content and amino acid profile for protein sources, absence of long chain polyunsaturated n-3 fatty acids in vegetal oil and relatively high levels of starch. Our recent studies indicated that rainbow trout is able to survive, grow and produce viable offsprings when fed a totally plant based diet throughout the whole life cycle. However total replacement of FM and FO leads to decreased feed intake, feed efficiency and growth rate as well as metabolic changes, even if diets correctly cover fish nutritional requirement. Attention is now paid on microalgae as provider of long chain polyunsaturated n-3 fatty acids and new alternative sources of protein such as insect meal and yeast to improve growth and flesh quality of fish fed plant-based diets with no FM and FO.

Evaluation of dietary wheat gluten and scFOS on gut health and growth performance of rainbow trout

S. Voller[1], A. Rodiles[1], M. Rawling[1], S. Davies[2], E. Apper[3] and D. Merrifield[1]
[1]Plymouth University, School of Biological Sciences, Drakes Circus, PL4 8NG Plymnouth, United Kingdom, [2]Harper Adams University, Flatt Rd, TF10 8NB Shropshire, United Kingdom, [3]Tereos Starch & Sweeteners s.a.s, Z. I. et Portuaire, 67390 Marckolsheim, France; samuel.voller@plymouth.ac.uk

The aquaculture industry is now valued at >$150 billion, exceeding beef production volumes, and providing approx. 50% of seafood consumed by humans. Increasing the sustainability of aquafeeds is paramount to improving the long term sustainability and productivity of the aquaculture industry. The replacement of fishmeal with alternative protein sources, such as those derived from plants, has received much attention and has enabled the expansion of aquaculture production despite the stagnated levels of availability and increased cost of fishmeal. Identification of plant protein sources with greater health benefits and fewer antinutritional factors than soybean derived products is a high priority. Gluten meals are high-protein co-products from the removal of starch from cereal grains. They are highly digestible proteins rich in glutamine. The gut microbiota is increasingly recognised as an important contributor to gut, and systemic level, health. As such, gut modifying feed additives, such as probiotics and prebiotics, are also topics of intensive research efforts in aquaculture. Therefore, the effects of the prebiotic scFOS and wheat gluten meals were assessed through in vivo rainbow trout feeding trials. Over the course of a number of feeding trials wheat gluten products at inclusion levels of 20% to 30%, at the expense of soy protein concentrate, have resulted in significantly improved SGR, weight gain, and FCR (P<0.05) of rainbow trout. High throughput sequence analysis revealed that wheat gluten, with or without scFOS, can modulate gut microbiota promoting potentially beneficial bacteria. Gene expression analysis of intestinal tissue has revealed an increase in glutathione-s-transferase in response to specific wheat gluten products. Overall, the results obtained thus far indicate that the utilisation of wheat gluten and scFOS in aquafeeds can have beneficial impacts on some health and production parameters of rainbow trout, and in turn positive impacts on the sustainability of salmonid production.

Feeding ragworm (N. virens) or mussel (M. edulis) to sole (S. solea) alleviates nutritional anaemia

J. Kals
Wageningen UR Livestock Research, Animal nutrition, P.O. Box 338, 6700 AH Wageningen, the Netherlands;
jeroen.kals@wur.nl

Sole fed commercial pellets (P) suffer from an anaemia with haematocrit (Hct) levels of ±11%. Sole in ponds stocked with ragworm (RW) show Hct levels of 24±5%. Based on the feeding ecology of sole we hypothesise that feeding RW or Mussel (M) increases Hct and haemoglobin (Hb) levels. (1) test if a diet change from P to RW or M increases Hct and Hb in sole; (2) estimate the rate of increase of Hct and Hb; (3) determine if the anaemia is due to an inflammatory response or a nutritional deficiency using expression of marker genes related to iron homeostasis and infection. The 1st trial had 2 treatments: in the 1st, sole was fed RW and sampled every 2-3 days for 26 d. In the 2nd treatment, sole was fed P and only sampled at d 26. Sole used were anaemic. In a 2nd trial we tested if M had an equal effect as RW. Sole were fed 3 diets; P, M, and RW. Feeding was restricted and equal for all diets. In a 3rd trial sole were fed P, P treated with RW extract, or RW. Fish were fed to satiation. All fish were sampled for blood. Data were analysed using ANOVA. A p-value<0.05 is considered significant. When significant, means were compared using the Fishers LSD test. Hct and Hb levels at start were 9.8±3.6% and 16.3±6.6 g/l. After 26 d the Hct level of sole fed RW rose to 19.0% and the Hb level doubled (P<0.05). Hct and Hb levels of sole fed P stayed low with 8.8±3.3% and 11.7±5.5 g/l (P>0.05). The rate of increase of Hct and Hb fitted with the Morgan-Mercer-Flodin growth model. Hct and Hb levels of sole fed M or RW rose strongly and are not different (P>0.05). Again, sole fed P stayed anaemic (P<0.05). On DM, sole fed RW had an equal feed intake (FI) as sole fed P with RW extract. Despite equal FI, sole fed RW grew 37% faster. Sole fed P grew 71% slower compared to sole fed RW. Gene expression indicated the anaemia in sole fed P is a nutritional anaemia. The studies show a clear effect of RW or M on Hct and Hb in sole. Feeding RW or M alleviates the anaemia in sole.

Effect of iron source on the alleviation of nutritional anaemia in common sole (Solea solea)

J. Kals[1], R.J.W. Blonk[1], H.W. Van Der Mheen[1], J.W. Schrama[2] and J.A.J. Verreth[2]
[1]Wageningen UR Livestock Research, Animal nutrition, P.O. Box 338, 6700 AH Wageningen, the Netherlands,
[2]Wageningen University, Aquaculture and Fisheries Group, P.O. Box 338, 6700 AH Wageningen, the Netherlands;
jeroen.kals@wur.nl

Sole fed commercial pellets (P) suffer from a nutritional anaemia. Hypotheses: (1) the nutritional anaemia in sole fed P is due to an iron deficiency; (2) the assumed iron deficiency is due to inadequate absorption of iron; (3) an increase in absorption due to a higher bioavailability of heme or iron chelates will alleviate anaemia in sole; and (4) haematocrit (Hct) and haemoglobin (Hb) are expected to follow iron absorption patterns. Further, we estimated the absorption of copper (Cu), cobalt (Co), chromium (Cr), manganese (Mn), molybdenum (Mo) and zinc (Zn) to evaluate interaction between iron source and absorption of other minerals. Sole (220±45 g) were fed 4 diets, each with a different iron source: iron sulphate, iron methionate, iron proteinate or heme, in triplicate for 23 days. Feeding level was restricted and equal for all diets. Data were analysed using ANOVA. Homogeneity was tested using the Levene's test. Tank was used as experimental unit. For all tests a p-value<0.05 was considered significant. When significant, means were compared using the Fishers LSD test. At the start of the trial, sole reared on P had values of Hct, Hb and hepatosomatic index of 12.5%, 19.6 g/l and 1.13%, respectively. Values at the end of the trial did not differ from the values at the start and were not affected by the source of iron. The apparent absorption coefficient (AAC) of iron, Mn, Zn, Co, Cr and Mo, except for Cu (P<0.05), were unaffected by iron source. Yet, the iron absorption was high for all sources. The AAC of Cu was 15-22% higher in sole fed the diet with heme. The use of different iron sources, including heme, did not affect Hct and Hb in anaemic sole. The high absorption of iron and Cu in sole fed heme did not affect Hct and Hb, suggesting the nutritional anaemia in sole is not an iron nor a Cu deficiency anaemia.

The effects of dietary inulin on growth performance and health status of juvenile Nile tilapia
N. Tiengtam, S. Khempaka, P. Paengkoum and S. Boonanuntanasarn
Suranaree University of Technology, School of Animal Production Technology, 111 University Avenue, Muang,
Nakhon Ratchasima, 30000, Thailand; nattanan_tga@hotmail.com

The present study investigated the effects of prebiotic inulin on growth performance and health status of juvenile Nile tilapia (Oreochromis niloticus). Three dietary treatments (each diet in four replicates) were formulated to incorporate inulin at 0 (control), 2.5, and 5 g kg^{-1}. Fish were reared in concrete ponds for 8 weeks. Fish fed the inulin diets exhibited better growth performance than fish fed the control diet. There were no significant differences in the moisture, crude protein, crude lipid, and ash contents among experimental treatments. Dietary inulin increased red blood cell number. Among the fourteen blood chemicals examined, dietary inulin led to increased glucose, albumin, and magnesium content. Inulin supplementation at 5 g kg^{-1} improved lysozyme activity and alternative complement haemolytic 50 (ACH50) activity. However, no significant difference was observed in total immunoglobulin content among treatment groups. Dietary inulin at 5 g kg^{-1} increased the height of intestinal villi and goblet cell number. Taken together, these results indicated that inulin at 5 g kg^{-1} had beneficial prebiotic effects on growth performance and health status of juvenile Nile tilapia.

Random regression analysis of feed efficiency in families of Turbot (Scophthalmus maximus)
W.M. Rauw[1], A. Kause[2], S. Cabaleiro[3], R. Caamaño[3], L.A. Garcia-Cortes[1] and L. Gomez-Raya[1]
[1]INIA, Departamento de Mejora Genética Animal, Ctra. de La Coruña km 7.5, 28040 Madrid, Spain, [2]LUKE,
Biometrical Genetics, Myllytie 1, FI-31600 Jokioinen, Finland, [3]CETGA, Punta de Couso s/n, 15965 Aguiño (A
Coruña), Spain; rauw.wendy@inia.es

We investigated the use of random regression models to analyze variation in growth, feed efficiency, and in-group dominance dynamics in families of Turbot (Scophthalmus maximus). A total of 84 full-sibs in each of eight families where allocated across 24 tanks, each family replicated in three tanks. Body weight was measured individually four times between approximately 250 and 370 days of age. Feed intake was recorded and weight gain calculated for the three corresponding time periods. Feed conversion ratio (FCR), residual feed intake (RFI), and the coefficient of variation in body weight (CV, as a measurement of inter-individual group competition) were calculated for each tank in each period. Data were analyzed with a random regression model that estimated components of variation attributable to families. Results showed that variation attributable to families explained the intercept of the regression for BWG (19%) and FCR (18%) (P<0.001), and the slope of the regression (the trend over time) for CV (5%; P<0.001). These differences between families suggest that a genetic component may exist and that it may be possible to select for these traits. FCR was highly phenotypically correlated with RFI (r=0.91; P<0.0001; n=24). Families with higher feed efficiency did not have more uniform body weights, therefore, superior feed efficiency in families may not have resulted from decreased inter-individual group competition. The use of hierarchical random regression models are proposed for the analysis of variation of feed efficiency in groups of fish where multiple measures of feed intake and growth are recorded.

Session 21

Theatre 8

Genetic evaluation with uncertain parentage in aquaculture

R. De Paz[1], B. Villanueva[1], M. Herlin[2], A. Millán[3], P. Martínez[3], M.A. Toro[4] and J. Fernández[1]
[1]INIA, Mejora Genética Animal, Ctra. Coruña Km 7,5, 28040 Madrid, Spain, [2]ABSA-Culmarex, Crta. Can Pastilla, 07007 Es Coll D'en Rabassa, Mallorca, Spain, [3]Facultade de Veterinaria, USC, Departamento de Xenética, Avda. Carballo Calero, 27002 Lugo, Spain, [4]ETS Ingenieros Agrónomos, Universidad Politécnica de Madrid, Departamento de Producción Agraria, Av. Puerta de Hierro, 2, 28040 Madrid, Spain; jmj@inia.es

The use of molecular markers for assigning paternity and reconstructing genealogies is a standard approach in aquaculture breeding programs when mass spawning is applied or when families are reared together. The reconstructed genealogy is needed for obtaining BLUP (best linear unbiased prediction) estimates of breeding values, and for controlling the increase in inbreeding. When few microsatellite markers are used, there will be a proportion of individuals not assigned to any parent or assigned to multiple parents. In this situation there are three possibilities: (1) using exclusively the offspring assigned to a single mother and a single father (discarding thus more selection candidates); (2) genotyping the dubious individuals (and the potential parents) for more markers, with an extra cost; and (3) calculating an additive relationship matrix that accounts for such uncertainty. However, the advantages of including more individuals in the evaluation through uncertain paternities or the balance between the extra cost of genotyping more markers and the gain in accuracy and selection intensity is unknown. In this study we simulated a sea bream breeding program for comparing three different strategies in terms of the accuracy of the estimated breeding values and selection intensity: (1) use only those individuals assigned unambiguously to both parents; (2) use only those individuals assigned to at least one single parent; (3) include also individuals with both multiple paternity and maternity. In the two last strategies, the probabilities of each putative parent were taken into account. Two different numbers of microsatellites were considered in the paternity analysis. Results show that the inclusion of uncertain paternities gives no consistent advantages. Selection intensity when using only perfectly assigned offspring was already high enough and thus there was little margin for improvement.

Session 21

Theatre 9

Quantitative genetics of susceptibility of Atlantic salmon to the salmon louse

B. Gjerde
Nofima, Genetics and breeding, Osloveien 1, 1430 Ås, Norway; bjarne.gjerde@nofima.no

Salmon louse (Lepeophtheirus salmonis) is the largest problem for the Atlantic salmon industry. In Norway the lice cost for 2015 was estimated to NOK 5 billion (Euro 540) or NOK 4/kg salmon produced. During the last years the louse have developed resistance towards all available chemoterapeutants and current control methods are all short term since the fish often become reinfected and thus have to be deloused at regular intervals. Recently some farmers have been forced to slaughter the fish before desired marketing size due to lack of lice control, and further growth of the industry is not to be expected before the lice problem is solved. The economic weight of susceptibility to lice is therefore high. Lice count data from controlled challenge tests of 827 families from four year-classes show that there is substantial genetic variation in susceptibility to the salmon louse. High genetic correlations has been found between susceptibility in replicated tests, in repeated test with delousing in between, of sessile and adult louse, in challenge test and field test, and close to zero genetic correlations of susceptibility of lice with other traits (growth rate, furunculosis, ISA and IPN). Predictions show that with a total number of 300 tested families selection for only increased resistance to the louse may reduce lice count per fish with 24% per generation, or 75% over five generations. The prospects are therefore good that selection for increase resistance to the salmon louse will reduce the need for delousing. Additional genetic gain may be possible through MAS or GWS selection. However, competition between breeding companies makes it difficult for the companies to put a lot of weight on this trait relative to other traits in the breeding objective with higher short-term benefit. A solution to this is to use government regulations to force all the breeding companies to prioritize and document increased resistance to the salmon louse.

258

EAAP – 67th Annual Meeting, Belfast 2016

Effects of feeding polyphenols extracted from grapes on flesh quality in farmed sea bass

F. Giannico[1], M.A. Colonna[1], P. Rotondi[1], A.M. Facciolongo[2], F. Toteda[1] and A. Caputi Jambrenghi[1]
[1]*University of Bari Aldo Moro, Department of Agricultural and Environmental Science, Campus, Via Orabona, 4, 70126 BARI, Italy,* [2]*Italian National Council of Research, Institute of Biosciences and Bioresources, Via Giovanni Amendola, 165/A, 70126 Bari, Italy; francesco.giannico@uniba.it*

Polyphenols contained in grapes, among which proanthocyanidins, have a wide range of biological effects including antioxidant, bactericid and modulatory effects against several enzyme systems. They can be used for minimizing or preventing lipid oxidation in pharmaceutical products, retarding the formation of toxic oxidation substances, mantaining nutritional quality and prolonging the shelf life of food. Furthermore, polyphenols may help to protect animla health and welfare, especially in intensive farming systems common in aqualculture. The research was conducted in order to evaluate the influence of dietary supplementation with a polyphenol complex extracted from Nero di Troia small berry grapes (14058 EXTRAUVA, Farmalabor, Italy) in farmed sea bass (Dicentrachus labrax L.). Three groups received a control feed (C) or an experimental one supplemented with two levels of polyphenols: 100 or 200 mg/kg feed, respectively. Ten basses of about 393 days of age for each group were analyzed for: biometric measurements (lengths of: total body, head, abdominal, trunk, caudal peduncle, body muscle and fork; height; thickness; weight of: total body, head, tail, skin, edible bones and viscera) and related indexes; pH; colour; rheological properties (TPA-Texture Profile Analysis) and muscle lipid oxidation (TBARS). There were no significant differences between groups with regard to somatic measurements, pH and colour. Fish fillets of the control group showed a higher gumminess in comparison to the two groups supplemented with grape polyphenols (14.82 vs 7.38-7.91; P<0.01). The TBARS test showed that the same MDA concentraion was recorded in polyphenol groups, markedly lower than the control one (0.032 vs 0.044 mg MDA/kg fillet; P<0.05), thus confirming the positive effects of dietary polyphenols on preservation of lipid peroxidation.

The replacement of fish meal with rice wine residual on growth performance in diets of Thai Panga

K. Taveewannaboon[1] and S. Ponchunchoovong[2]
[1]*Suranaree University of Technology, School of Animal Production, 111 university road, 30000, Thailand,* [2]*Suranaree University of Technology, School of Animal Production Technology, 111 university road, 30000, Thailand; samorn@sut.ac.th*

This study was conducted to assess the efficacy of the replacement of fish meal with rice wine residual (RWR) in the diets of Thai Panga (Pangasianodon hypopthalmus×Pangasius bocourti). Fish with an average initial weight of 220.38±1.60 g were randomly stocked in cages (1×1×1.5 m3) at a rate of 40 fish per cage. Six experimental diets were formulated to provide isonitrogenous (320 g/kg CP) and isocaloric (15.00 kJ/g) diets, by increasing RWR levels to replace 10, 20, 30 and 40% of dietary fish meal. The control groups were fed a commercial diet (CA) and the basal diet without RWR (CB). Fish were reared for 8 months. The final weight, weight gains, daily growth rate, specific growth rate and FCR showed that RWR can be used effectively as a substitute for up to 30% of the fish meal. Increasing the replacements to 40% resulted in a significant lowest growth performance compared to the other treatments (P<0.05). The replacement of fishmeal by RWR did not affect survival rate (P>0.05). Replacing 30% of RWR can improve the growth performance of Thai Panga.

Genetic variation in microenvironmental plasticity of body weight in Tilapia (Oreochromis niloticus)

S. Agha[1,2], A. Doeschl-Wilson[2], W. Mekkawy[1,3], C.E. Lind[3], J. Benzie[3] and N. Ibanez-Escriche[2]
[1]Faculty of Agriculture, Ain Shams University, Animal Production Department, Haddayek Shoubra, Cairo, Egypt, 11241, Egypt, [2]The Roslin Institute, University of Edinburgh, Easter Bush, EH25 9RG, Edinburgh, United Kingdom, [3]WorldFish, Penang, 11960, Malaysia; safsoof23@hotmail.com

In animal breeding, genetic differences in environmental variation can be detected as genetic heterogeneity of residual variance which is also a measure of the micro-environmental plasticity (sensitivity). The objectives of this study were to (1) estimate the genetic variation in micro-environmental plasticity for body weight at harvest (BW) in Genetically Improved Farmed Tilapia strain (GIFT), and (2) investigate the effect of selective breeding for BW across multiple generations on its environmental variance. The data consist of 74,253 individuals representing 1,131 full-sibs families selected for BW over 13 generations in Malaysia. Three different animal models were implemented using Bayesian inference. Model HOM assumes the homogeneity of environmental variance, while in model HET1 is assumed heterogeneous environmental variance and in model HET2 is assumed that part of the environmental variance is under genetic control. Deviance Information Criterion values favored HET2 model. The analysis revealed the presence of the additive genetic variance in both the mean (0.45) and the variance of BW (0.34). The posterior mean of the genetic additive correlation between the additive genes affecting the mean BW and its variance (95% highest posterior interval) was -0.53 (-0.47, -0.59). Negative correlation implies that genotypes with high BW tend to be less sensitive to local environmental perturbations (more homogeneous) i.e. robust. The genetic trends showed that the BW was improved by selective breeding, although not a clear correlated genetic change occurred in the environmental variance during the selection period. Including the micro-environmental sensitivity information in aquaculture breeding programs may improve both the performance and robustness of the farmed fish.

Recombinant Saccharomyces cerevisiae expressing delta 6 desaturase of Nile tilapia

S. Boonanuntanasarn
Suranaree University of Technology, Institute of Agricultural Technology, 111 University Avenue, Muang, Nakhon Ratchasima, 30000, Thailand; surinton@sut.ac.th

The enzyme delta 6 desaturase ($\Delta6$) converts linolenic acid (C18:2n6) and α-linolenic acid (C18:3n3) into γ-linolenic acid (C18:3n6) and stearidonic acid (C18:4n3). This study was aimed to produce the recombinant Saccharomyces cerevisiae expressing delta 6 desaturase from Nile tilapia (Oni-$\Delta6$). PGK promoter was cloned from genomic DNA of S. cerevisiae by PCR. The expression vector (pPGK) was constructed by joining PGK promoter to CYC 1 poly (A) signal. Recombinant yeast carrying pPGK (RY-pPGK) was generated by transformation of pPGK into S. cerevisiae. The Oni-$\Delta6$ cDNA was cloned from the liver of Nile tilapia by RT-PCR. The plasmid vector which expressed Oni-$\Delta6$ driven by PGK was constructed and designated as pPGK-Oni-$\Delta6$. Later, recombinant yeast carrying pPGK-Oni-$\Delta6$ (RY-pPGK-Oni-$\Delta6$) was generated. The expression of RY-pPGK-Oni-$\Delta6$ was analyzed by RT-PCR. Non-transformed S. cerevisiae (NT), RY-pPGK and RY-pPGK-Oni-$\Delta6$ were grown in yeast extract peptone-dextrose medium (YPD) at 30 °C for 24 h. Each culture was added C18:2n6 substrate. Subsequently, the NT, RY-pPGK and RY-pPGK-Oni-$\Delta6$ were harvested, and fatty acid contents were determined. RY-pPGK-Oni-$\Delta6$ contained a high level of C18:3n6, compared with that of NT and RY-PGK, demonstrating that recombinant S. cerevisiae expressing Oni-$\Delta6$ could have the $\Delta6$ activity by converting C18:2n6 into C18:3n6.

Livestock, land use, global change and food security

T.D. Searchinger

Princeton University, Woodrow Wilson School of Public and International Affairs, 405 Robertson Hall, Princeton, NJ 08544, USA; tsearchi@princeton.edu

Global agriculture is on a course to contribute around 12 gigatons or more of greenhouse gas emissions per year by 2050, including land use change. This level of emissions would by itself fill up 70% of the total, annual emissions budget that may be consistent with acceptable warming. Meat and milk production already contribute a majority of agriculture's global emissions, and use around three quarters of global agricultural land. The growing global demand for meat and milk is the main driver of rising emissions along with rising population. Although holding down the growth in global consumption of milk and meat is critical to meeting climate targets, improving the efficiency of livestock production globally is also critical to reduce emissions. Dairy production in some countries, for example, contributes around five-times the emissions per litre of milk as in other countries. Better feed and better grazing practices are particularly critical to meeting global targets. Analyses that characterize national livestock systems and their improvement potential at a high level of detail will be a necessary tool for driving these improvements. Some major technological innovations will also be necessary to hold down emissions as well as other environmental threats. This talk will summarize the challenges, and the best immediate and long-term opportunities for progress.

Grasslands, future food demand and environmental impact

J.F. Sousanna[1] and P. Dillon[2]

[1]INRA, 147 rue de l'Université, Paris Cedex 07, France, [2]Teagasc, Moorepark, Fermoy, Co. Cork, Ireland; jean-francois.soussana@paris.inra.fr

The grassland biome covers about one fourth of the ice-free land area and provides about half of the global proteins used by ruminants with a relatively low conversion efficiency, estimated at 12 and 6% in 2010 for dairy production and for ruminant meat production, respectively, on a protein basis. It is likely that these global conversion efficiencies have improved by less than 1% per decade since the 1960's. Since half of global grassland above-ground net primary productivity is not ingested by domestic herbivores, global grasslands are often perceived as the last frontier for a sustainable intensification of animal agriculture. Indeed, improving pasture productivity and quality could avoid tropical pasture expansion and enhance the conservation and development of carbon rich landscapes. Nevertheless, the livestock sector is increasingly competing with human consumption for edible grains and crop by-products, while the average herbage use efficiency is unlikely to have increased by more than one third since the 1960's. By the 2050's, grasslands could be relatively more resilient to climate change than arable crops, which could increase their role both as a feed source and as a global carbon stock. Reaching two major goals of our times, global food security and climate stabilization, will require increased investments in grassland improvement and restoration, in livestock health and production efficiency and a global convergence towards nutritionally balanced diets that have potential to reduce both chronic undernourishment and the incidence of metabolic diseases related to obesity.

Improving the lives of farm animals: science in the development of valid welfare indicators
C.J. Nicol
University of Bristol, School of Veterinary Science, Langford House, BS40 5DU, United Kingdom;
c.j.nicol@bris.ac.uk

It is generally accepted that the moral legitimacy of keeping sentient animals for food production depends upon such animals leading lives that are worth living. Indeed, the long-term sustainability of the animal production sector depends on consumer trust that this precept is both endorsed, and actively pursued, by the farming community. However, assessing the quality of an animal's life is not an easy task. It requires a blend of science, extrapolation and philosophical argument. My talk will focus on the scientific part of this complex exercise. I will review the advances that have been made in recent years in developing valid indicators of animal welfare and explain the role of animal-centred information in providing the ground-truth for validation. Animal-centred information can be obtained from carefully-designed preference tests and from assessments of animal emotion and mood. Valid welfare indicators are those that are most strongly linked to the animal-centred data. By additionally taking steps to avoid redundancy and duplication sets of measures can be devised for use in practical assessments and audits on commercial farms. Then, by combining information obtained at different points in the life of a farm animal, an overall assessment of quality of life can be constructed. Practical illustrations of all stages of this process – from theory to commercial application – will be given using my work on laying hens as an example.

The role of mixed crop-livestock farming systems in ensuring food security
P. Thornton[1] and M. Herrero[2]
[1]CGIAR Research Program on Climate Change, Agriculture and Food Security, ILRI, P.O. Box 30709, Nairobi 00100, Kenya, [2]Commonwealth Scientific and Industrial Research Organization (CSIRO), 306 Carmody Road, St Lucia, QLD 4067, Australia; p.thornton@cgiar.org

Mixed crop-livestock systems, in which crops and livestock are raised on the same farm, occur very widely across the globe. These systems provide half of the world's cereal production and more than 80% of the ruminant meat and milk; they also provide most of the staples consumed by many millions of poor people in developing countries. The mixed systems will be critical for future food security as well: demand for food is projected to increase by 70% to the middle of the century as a result of population growth, rising incomes, changing diets and urbanisation. These systems face severe challenges, particularly in the developing world in relation to shrinking farm sizes, climate change and increasing climate variability, and weak institutional and market support. Mixed farms may need to consolidate and expand, diversify or specialise in the future. Nevertheless, considerable potential exists to sustainably intensify these systems through improving crop and livestock productivity and, often, strengthening the interactions between them. At the same time, the mixed systems have a pivotal role to play in mitigating greenhouse gases from the agriculture, forestry and land use sectors. Despite a wealth of information on the technical possibilities, there is much less information on the costs, benefits and trade-offs involved, particularly at the local level. In all cases, technical solutions need to be implemented within an enabling environment that provides the necessary incentives, policies and markets for systems to intensify sustainably at scale. Understanding how the mixed systems may evolve in the coming decades in response to global change is crucial for identifying appropriate development pathways that can balance national policy objectives with enhanced livelihoods and food security for vulnerable and resource-poor communities. Donor support, public-private partnerships, market and infrastructure development, appropriate monitoring mechanisms and large-scale behavioural change will all be required if the developing-country productivity challenge is to overcome.

Animal-derived foods for sustainable diets across the life stages: do they have a future?
D.I. Givens
University of Reading, Reading, RG6 6AR, United Kingdom; d.i.givens@reading.ac.uk

Public health nutrition is facing many major challenges and three of these will shape food-related policy for decades to come. These are the rapidly increasing burden of obesity, the increasing age of populations and the challenge of increasing world food production by some 50% by 2030 to meet the increasing demands whilst minimising the effect on the environment. Animal-derived foods in particular have been highlighted as having a high environmental cost but much less attention has been paid to their role in healthy diets that have impact from childhood to old age. Whilst most people know that milk/dairy products are very important sources of dietary nutrients such as calcium and iodine and that red meat is an excellent source of iron and zinc, there is more uncertainty about whether or not these foods contribute to increased risk of cardiometabolic and other chronic diseases. The evidence from long term cohort studies that high milk consumption does not increase CVD risk and indeed may provide some benefit is now pretty unequivocal, although the effects of butter as well as benefits of fat reduced milk and saturated fat reduced milk are less certain. Recent data point to a beneficial association between fermented dairy products and type 2 diabetes. The situation with red meat is complex although the evidence linking red meat with colorectal cancer is mainly a function of processed meat and this needs to be considered alongside risks from other lifestyle issues such as alcohol consumption. There is now good evidence that specific health issues related to children (e.g. calcium, vitamin D intake), pregnant women (e.g. iodine and iron status) and the elderly (e.g. loss of muscle mass, vitamin B_{12}) can be alleviated by animal-derived foods or components of them and these effects are not all explained by traditional nutrition. They must however be considered when debating future sustainable food production; for example, a simplistic replacement of milk proteins by plant proteins may not provide the same long term health benefits.

FABRE TP: Promoting research and innovation activities in the animal breeding and reproduction sector
J. Venneman and M. Van Straten
FABRE TP secretariat, Dreijenlaan 2, 6703 HA, Wageningen, the Netherlands; jan.venneman@effab.info

The Farm Animal Breeding and Reproduction Technology Platform (FABRE TP) is the meeting point for professionals from the industry, knowledge institutes and funding organisations in the field of animal genetics, genomics, breeding and reproduction. FABRE TP: (1) develops research and innovation strategies and provides a coherent business-focused analysis of bottlenecks and opportunities related to societal challenges and industrial leadership actions; (2) mobilises industry and other stakeholders within the EU to work in partnership and deliver on agreed priorities (3) shares information and enables knowledge transfer to a wide range of stakeholders across the EU. FABRE TP was launched in 2006 and is an EU recognised Technology Platform. In 2008 FABRE TP published its first Strategic Research Agenda followed by an Implementation Plan. A second Strategic Research Agenda was published in 2011. During the year 2015 the structure of FABRE TP was adjusted. The Board formulated a new Mission, Internal Rules and an action Plan. The official re-launch of FABRE TP was at the EAAP 2015 Conference in Warsaw. Currently (April 2016) FABRE TP has almost 60 members, partly from the breeding and reproduction industry (via EFFAB) and partly knowledge institutes.

Genome editing in animal breeding

B. Whitelaw

The Roslin Institute and R(D)SVS University of Edinburgh, Developmental Biology, Easter Bush Campus, Midlothian EH25 9RG, United Kingdom; bruce.whitelaw@roslin.ed.ac.uk

Man had been breeding farm animals for millennium. At first selection was based on simple characteristics and individual preferences, this strategy to continually produce more productive animals now benefits from genetics. Initially we had individual genetic markers which through association with trait phenotype enabled selection of desired genotypes, breeding regimes are now entering the genomic selection era for many livestock species. To say the least, this is an exciting time for animal breeding – both academically and commercially – and is being fuelled by the relentless expansion of genomic sequence data. Regardless of the depth of genomic knowledge available for a given species or breed and how clever the animal breeder is, animal breeding relies on the availability of genetic variation. New molecular scissors called genome editors now enable the animal breeder to access genetic variation otherwise unavailable. This could be variation not present at all in a given breeding population, breed or species – or present but not at a higher enough frequency within the target population to allow rapid fixation. The genome editors are site-specific endonucleases what can produce indels at the target locus or when delivered with a DNA template can produce sequence changes, for example allele swaps. Thus genome editors enable use of desired genetic variation in the animal of choice.

Biotechnologies and precision farming: a new era for the ruminants breeding sector

X. David

Allice, 149 rue de Bercy, 75595 Paris cedex 12, France; xavier.david@allice.fr

In a few years, the ruminants breeding sector has been facing 3 main technological breakthroughs: genomic selection, semen sexing and internet of monitoring tools. They generated quantities of data, encouraged genetic progress, changed practices and relationships between stakeholders. Now, let's imagine what will be the next future. How knowledge on genome interacting with environment, and especially knowledge on genome expression can be increased? From lab to milking parlour, how will it change the breeding and management practices? Finally, what kind of benefits can we expect for the farmers, the animals, the environment and the society?

Poultry breeding towards 2030
V. Olori and A.M. Neeteson-Van Nieuwenhoven
Aviagen, 11 Lochend Road, Newbridge Midlothian, United Kingdom; victor.olori@aviagen.com

Poultry breeding is a highly responsive, adaptive and sophisticated industry. Its evolution will be influenced by socio-economic, natural and technological factors affecting demand and hence poultry production in 2030. These include global population growth and development; availability of agricultural production resources; consumer attitude and demand for safe 'green' and ethical production systems and advances in technology for poultry production and health including the application of genomics in poultry breeding. Future demand and constraints of poultry production creates challenges as well as opportunities, driving current breeding programs decisions. The nature of the birds of the future can thus be gleaned from today's breeding objectives and contemplated selection criteria. Technological advancement, especially its impact on data recording and genomic selection, provides additional tools for breeding program efficiency and accurate selection on a broad range of traits. This will facilitate precision breeding to meet the needs of the future. With the world population projected to reach 8.5 billion in 2030, more poultry products will be needed from birds reared in a broad range of environments and production systems. The key challenge is how to increase output sustainably while reducing cost, improving safety and maintaining high welfare standards. Ongoing evolution of poultry breeding objectives is a proactive response to provide farmers with a wide variety of efficient poultry birds which will be resilient in a range of production systems to meet the diverse demands of different consumers in the future. This is attainable with improvement in health, robustness and reproductive efficiency with emphasis on fertility and hatchability to improve chick and minimise waste. Improved feed efficiency will result in massive reduction in agricultural land and resources needed for feed production reducing the environmental footprint of poultry farming. This approach is imperative given the future outlook on costs and availability of inputs such as feed, water and energy arising from population growth, climate change, and human efforts to mitigate it.

The perspectives of genetically modified livestock in agriculture and biomedicine
H. Niemann
Institute of Farm Animal Genetics, Mariensee, 31535 Neustadt, Germany; heiner.niemann@fli.bund.de

The first transgenic livestock were reported in 1985 via microinjection of foreign DNA into zygotic pronuclei. This technology has been the method of choice for more than 20 years, but has recently been replaced by more efficient protocols based on somatic cell nuclear transfer (SCNT) which are compatible with targeted genetic modifications. Although the efficiency of transgenic animal production by microinjection technology is low, farm animals with agriculturally important transgenic traits were produced. Typical agricultural applications included improved carcass composition, improved lactational performance, and wool production; enhanced disease resistance and reduced environmental impact. Transgenic animal production for biomedical applications has already found broad acceptance. In 2006, the European Medicines Agency (EMEA) approved the commercialization of the first recombinant protein drug produced by transgenic animals. Recombinant antithrombin III, produced in the mammary gland of transgenic goats, was launched as ATryn® for prophylactic treatment of patients with congenital antithrombin deficiency. Pigs with transgenic expression of human immunomodulatory genes have contributed to significant progress in xenotransplantation research with survival periods of non-human primates receiving transgenic porcine hearts or kidneys approaching six months. As the genome sequencing projects for various farm animal species (cattle, horse, pig, poultry, sheep) near completion, it has become increasingly practical to target the removal or modification of individual genes or even single base pairs with the aid of gene editing tools (ZFNs, TALENs, CRISPR/Cas). It is anticipated that this approach to animal breeding will be instrumental in meeting global challenges in agricultural production in the future and might ultimately contribute to satisfying the increasing demand for valuable animal derived proteins for an ever-increasing global human population and will also open new horizons in biomedicine.

Pig breeding into the 21st century a view towards 2030 and beyond
C. Lewis
Genus PLC, Matrix House Basing View, Basingstoke Hampshire RG21 4DZ, United Kingdom;
craig.lewis@genusplc.com

During the history of intensive animal breeding there have been three major technological breakthroughs, selective breeding, BLUP evaluations, and use of genomic technologies. The pig sector in general has always been at the forefront of driving technological advances in the breeding sector due to the efficient biological process of pig production and industry structure. As we move toward 2030 it will arguably be the most rapidly moving period of innovation in animal agriculture with new breeding technologies changing the way in which potentially the pig industry can provide animal protein to a growing population. We are already seeing the impact of genomics allowing us to create more kgs of meat produced per sow than ever before and new advances in the ways the industry can utilize directly the genome to increase welfare, food safety, reduce environmental impact of production while also increasing efficiency for producers is of significant societal good. What is clear is that the technology is being rapidly developed to deliver positive changes in the industry, this coupled with our increase understanding of the underlying biology of the pig and development of novel trait recording will enable the industry to utilize new technologies within the major production systems for rapid uptake in the industry. What is less clear is the political and social acceptance of the new technologies, and key to implementation is clear, open, and honest dialogue with all global stakeholders during the coming years.

Salmon aquaculture 2030: what difference will new breeding technologies make?
A. Norris
Group Genetics, Marine Harvest ASA, Sandviksboder 77AB, 5035 Bergen, Norway; ashie.norris@marineharvest.com

Salmon aquaculture, as a relative newcomer, has seen tremendous change in the past 15-20 years in all areas of production from farming technology to breeding. Huge advances in particular have been made in the freshwater production of juveniles and broodstock so much that the modern production systems are barely recognisable today to a salmon-farmer-time traveller from 20 years ago. Selection programs, started in 1970s-1990s, have only recently started using tools and technologies of modern genomics. On the other hand the sector is dealing with far greater challenges in the marine environment than it dealt with in the early days of salmon aquaculture most likely as a result of climate change and increasing production pressures in some parts of the world. There are two approaches to dealing with these challenges; partially or fully remove the fish from the marine environment or produce a more robust fish better equipped to withstand it. New breeding technologies will be crucial to meet either or both of these challenges. Genomics, in particular, will allow for greater selection efficiency for complex traits, such as resistance to parasites and other pathogens, by allowing for exploitation of the large family and population sizes available in fish species.

Agreement of two Laser Methane Detectors with respiration chambers and in a dairy barn

D. Sorg[1], S. Mühlbach[1], F. Rosner[1], B. Kuhla[2], M. Derno[2], S. Meese[3], A. Schwarm[3] and H.H. Swalve[1]
[1]Martin-Luther-University Halle-Wittenberg, Institute of Agricultural and Nutritional Sciences, Th.-Lieser-Str. 11, 06120 Halle, Germany, [2]Leibniz Institute for Farm Animal Biology (FBN), Institute of Nutritional Physiology, Wilhelm-Stahl-Allee 2, 18196 Dummerstorf, Germany, [3]ETH Zurich, Institute of Agricultural Sciences, Universitätstr. 2, 8092 Zurich, Switzerland; diana.sorg@landw.uni-halle.de

Laser Methane Detectors (LMD) can be used to quantify CH_4 emissions in the air exhaled by dairy cows. Aim of the present study was to validate the new version of the LMD (LMM mini-g, Crowcon Detection Instruments) with respect to measuring concentrations of CH_4 comparing two devices among each other and with measurements in respiration chambers at two locations, taken as the gold standard. Ambient air in the chamber was measured by LMD and the chamber's gas analyser while recording the cow's activity. LMD devices were compared measuring background CH_4 concentration in a dairy barn. Data was analysed using Pearson's coefficient of correlation, Bland-Altman methodology and inverse regression. The two LMDs agreed very well with each other in the chamber (r>0.99) and in the dairy barn (r=0.97). Agreement of LMD and chamber was higher (r≥0.86) than reported with the older model (r=0.80). Differences in CH_4 emissions due to cow's activity were both detected by chamber measurements and LMD. Differences in measurements among the methods increased with increasing concentrations of CH_4 but only were of a relevant magnitude for concentrations above normal values taken at the mouth of a cow. In conclusion, the new generation of LMD is suitable to measure fluctuating CH_4 emission profiles from dairy cows under farm conditions. Further research is needed to evaluate the use of such profiles as a phenotype for genetic selection.

Factors associated with IgG concentration and specific antibody inhibition in bovine colostrum in NI

A. Dunn[1], S. Morrison[1], A. Ashfield[2], M. Welsh[3] and A. Gordon[2]
[1]Agri-food and Biosciences Institute, Sustainable Livestock, Large Park, Hillsborough, BT26 6DR, United Kingdom, [2]Agri-food and Biosciences Institute, Economics, Newforge Lane Belfast, BT9 5PX, United Kingdom, [3]Sisaf, Animal health, Queens Rd, Queen's Island, Belfast, BT3 9DT, United Kingdom; amanda.dunn@afbini.gov.uk

A survey was conducted on 21 commercial dairy herds in Northern Ireland, during a 1 year period. Colostrum samples were collected (n=1,239) and analysed for IgG, and a sub-set of samples were analysed for ecoli, rotavirus, and coronavirus inhibition. IgG and specific antibodies were quantified using bovine specific ELISA kits from Bio-X Diagnostics (Jemelle, Belgium). Mean IgG concentration was 55 mg/ml; there was significant variation between samples, ranging from 1.4 to 165 mg/ml with 44% of colostrum samples less than the satisfactory 50mg/ml recommendation for good quality. Mean ecoli, rotavirus and coronavirus inhibition were 85.9%, 32.5% and 53.9%, respectively. Cows immunized against ecoli, rotavirus and coronavirus had a greater (P<0.05) E. coli and corona inhibition score than non-immunized cows. Timing of vaccination also had a significant effect (P=0.003) on colostral ecoli inhibition levels; the greatest probability (97%) of achieving ecoli inhibition >60% was if the dam was immunized at 9 to 12 weeks before parturition, whereas there was only a 50% prospect of achieving this if immunized at 0 to 3 weeks prior to parturition. Colostral IgG concentration within a uni-variate analysis of vaccinated cows was shown to be associated with the timing of BVD, leptospirosis, IBR, ecoli, rotavirus and coronavirus immunization (P<0.05). Season of calving was associated with ecoli and rotavirus inhibition, cows calving in the winter months had the greatest prospect (55%) of achieving an ecoli score >60% inhibition. Length of dry period also had a significant effect (P<0.001) on ecoli inhibition, the greatest probability of achieving ecoli inhibition >60% was from cows with a dry period of 8 to 16 weeks. REML was used to perform uni and multi-variate analysis within Genstat 16th edition. The findings from the study will help inform best practice guidelines on vaccine and colostrum management on commercial dairy farms helping reduce dairy calf morbidity and mortality.

Metabolic adaptation of periparturient dairy cows characterised by changes of the blood metabolome

Á. Kenéz[1], S. Dänicke[2], M. Von Bergen[3] and K. Huber[1]
[1]University of Hohenheim, Institute of Animal Science, Fruwirthstr. 35, 70599 Stuttgart, Germany, [2]Federal Research Institute for Animal Health, Institute of Animal Nutrition, Bundesallee 50, 38116 Braunschweig, Germany, [3]Helmholtz Centre for Environmental Research, Department of Metabolomics, Permoserstr. 15, 04318 Leipzig, Germany; akos.kenez@uni-hohenheim.de

High-yielding dairy cows experience great metabolic stress during the transition from late pregnancy to early lactation. This is due to the complex adaptation processes affecting energy homeostasis in support of milk production. According to their adaptation efficiency some cows develop severe metabolic diseases while others are able to maintain metabolic health. This study aimed to characterise metabolic pathways affected the most during the transition period, and to identify individual metabotypes within the cow population. Twenty-six German Holstein cows were used to collect blood samples repeatedly during the transition period: 42 and 10 days before calving and 3, 21 and 100 days after calving. Blood serum samples were subjected to a targeted liquid chromatography-mass spectrometry (LC-MS) based metabolomics analysis using the AbsoluteIDQ p180 Kit of Biocrates Life Science AG (Innsbruck, Austria). Processed metabolomics data were evaluated by principal component analysis (PCA) and by heatmap visualisation. The PCA revealed a clear separation of the blood samples according to the sampling day, indicating a notable shift of the metabolic composition during the studied period. According to the heatmap the acylcarnitines provided a clear and consistent separation of cows within the herd. The concentration of glycerophospholipids and sphingolipids was remarkably decreased 10 day before and 3 days after calving than earlier and later in the transition period. Amino acids and biogenic amines showed a more homogenous pattern with less variation in concentrations over time. Analysing longitudinal changes of the blood metabolome and identifying new biomarkers by this approach can help understanding the multifaceted metabolic adaptation of transition cows. The biological interpretation of the differences in blood acylcarnitine concentration may serve as a source for characterising and predicting healthy and diseased metabolic phenotypes in dairy cows.

An automated warning system for lameness in dairy cows

D. Piette, T. Norton and D. Berckmans
KU Leuven, Biosystems, kasteelpark arenberg, 3001 Heverlee, Belgium; daniel.berckmans@kuleuven.be

Lameness in dairy cows is a leading cause for production loss and reduced animal welfare. Nowadays farmers have to invest a lot of time and money in manually detecting the lame cows in their herd. Consequently, a tool that detects lameness in real time and in an automated way can create added value for the farmer. In previous work, an algorithm has been developed that uses 3D top view images from the back of the cow to calculate a back posture score that reflects the lameness of the cow. The scope of this work is to generate an automated warning system for lameness based on the back posture scores of the cows. For this purpose, the back posture scores of 209 Holstein Friesen cows were collected over a period of one year in order to establish a historical back posture dataset for each cow. The gold standard for lameness, i.e. manual scoring, was collected for all cows by the same observer on a weekly basis over the final six weeks of data collection. This work presents the prototype of the warning system developed for use in a real-time lameness monitoring on a dairy farm in Sweden with 1,500 cows. This algorithm developed performs with a sensitivity of 79% and a specificity of 82.3%. Four specific aspects of the warning system will be addressed in detail in the paper. The first one being how individualising the warning list generation on a cow level can improve algorithm performance. The second aspect deals with how to define the baseline of each cow for warning generation based on the cow's historical dataset. Thirdly the effect of the length of the historical data used for baseline calculation on the performance of the algorithm will be discussed. Finally, the feedback of the farmer who is using the prototype warning system will be reviewed.

The use of mid-infrared spectrometry to estimate the ration composition of lactating dairy cows

M. Klaffenböck[1], A. Steinwidder[2], C. Fasching[2], G. Terler[2], L. Gruber[2], G. Mészáros[1] and J. Sölkner[1]
[1]University of Natural Resources and Life Sciences (BOKU), Division of Livestock Sciences, Gregor-Mendel-Strasse 33, 1180 Vienna, Austria, [2]Federal Agricultural Research and Education Centre Raumberg-Gumpenstein, Raumberg 38, 8952 Irdning-Donnersbachtal, Austria; klaffenboeckmichael@gmx.at

Milk composition of dairy animals is influenced by the composition of the ration fed. Since milk composition is routinely determined using mid-infrared (MIR) spectrometry, MIR spectra may potentially also be used to predict ration composition. The objective of this study was to determine if, and how well the percentages and absolute amounts of hay, grass silage, pasture feed, maize silage and concentrate in the feed ration can be estimated using MIR spectrometry. A total of 10,200 milk samples from 90 dairy cows from two experimental farms of the federal research centre in Raumberg-Gumpenstein, Austria, were collected, and the intakes of all ration components were measured. Pasture feed intake was estimated using energy requirement and in-barn energy intake. Using partial least square regression (PLS), equations were developed to estimate ration compositions corresponding to each milk sample. In separate analyses, other factors like milk yield or concentrate intake were included as additional predictors. To evaluate accuracy, the square of the correlation between observed and estimated values (R^2; coefficient of determination) and ratio to performance deviation (RPD) were used. The highest R^2 values were observed for the ration proportion of pasture feed (%), maize silage (%) and concentrate intake (kg), with 0.76, 0.56 and 0.56 respectively. Estimation of groups of feedstuffs (all forages, energy-dense feedstuffs) also resulted in R^2 values of >0.6. Including parameters milk yield and/or concentrate intake improved R^2 values by 0.01-0.15. The results of this study indicate a potential use of MIR spectra as a promising predictor for ration composition of dairy cows.

Myoepithelial cell contraction participates in mammary epithelial cell exfoliation in cow milk

L. Herve[1,2], H. Quesnel[1,2], S. Wiart[1,2], P. Lamberton[1,2], C. Mustière[1,2], V. Lollivier[1,2] and M. Boutinaud[1,2]
[1]Agrocampus Ouest, UMR PEGASE, 35590 Saint-Gilles, France, [2]INRA, UMR PEGASE, 35590 Saint-Gilles, France; lucile.herve@rennes.inra.fr

A varying proportion of mammary epithelial cells (MEC) is exfoliated through shedding from the mammary epithelium into milk, which might influence MEC number in the udder and thus milk yield. We hypothesized that myoepithelial cell contraction induced by oxytocin (OT) release at milking might stimulate MEC exfoliation through mechanical forces. Therefore, to investigate the role of myoepithelial cell contraction in MEC exfoliation, we inhibited or induced myoepithelial cell contraction by injecting the OT receptor antagonist Atosiban (Ato) or exogenous OT, respectively. Eight multiparous cows were assigned to 2 treatments during 2 milkings according to a crossover experimental design of 2 periods: (1) Control + OT; and (2) Ato + OT. For the Control + OT treatment, cows were first milked to recover standard milk and then received an i.v. injection of OT (5 IU) to collect residual milk through a second milking. For the Ato + OT treatment, cows were injected with Ato (50 µg/kg of BW), milked to recover the cisternal milk, and then received 5 IU of OT before a milking to collect the alveolar milk. For each type of collected milk, MEC were purified using an immuno-magnetic method and the percentage of apoptotic MEC was assessed by flow cytometry. Tight junction opening was assessed by plasma lactose concentration determination. Data were analyzed using an ANOVA procedure. Inhibiting myoepithelial cell contraction by Ato injection induced a decrease in MEC number in milk (P<0.01). In contrast, OT injection after standard milking increased the concentration of MEC in milk (P<0.05) and decreased the percentage of apoptotic MEC in milk (P<0.01) while OT injection during Ato + OT treatment increased the number of MEC in milk (P<0.01). During the milking after the Ato injection, plasma lactose concentration was on average lower than during standard milking whereas in both treatments, OT injections increased the lactose concentration in plasma (P<0.001). These results together showed that myoepithelial cell contraction associated with a loss of mammary epithelium integrity stimulates MEC exfoliation into milk.

Evaluation of genetic beta casein variations in a dairy cattle herd in Germany

A.-S. Kraus[1], M. Bernau[1], E. Märtlbauer[2] and A.M. Scholz[1]
[1]Livestock Center Oberschleissheim, St. Hubertusstraße 12, 85764 Oberschleissheim, Germany, [2]Institute for Hygiene and Technology of Food of Animal Origin, Veterinary Faculty, Munich, Schönleutnerstraße 8, 85764 Oberschleissheim, Germany; ann-sophie.kraus@lvg.vetmed.uni-muenchen.de

Caseins belong to major proteins in cow milk, which can be differentiated into alphaS1-, alphaS2-, beta and kappa-caseins. There are several subtypes of beta-caseins including the most important types A1 and A2 that differ from each other because of a mutation localized in bovine chromosome 6 in the amino acid at position 67: histidine in A1, proline in A2. Further subtypes are A3, B, C, D, E, F, G, H1, H2, and I. The casein genotype underlies large genetic variability. Old breeds with lower milk yield are supposed to have a higher A2 frequency, like e.g. Guernsey with 0.88-0.9; while in contrast modern high performance dairy cattle have a higher A1 frequency, like e.g. Holstein-Friesians with 0.31-0,66. A very few studies suggest differences in gastrointestinal responses in adult humans consuming milk containing either beta-casein A1 or A2 type. Some individuals react to A1 beta-casein with higher faecal calprotectin values and increased intolerance in comparison with A2 beta-casein. During human digestion, the histidine in A1 milk generates beta-casomorphin-7, a bioactive peptide with opioid activity that is also linked to heart disease, Type 1 Diabetes and neurological disorders. Due to the A1 intolerance in some human individuals, a few countries like Australia or New Zealand started to launch pure A2 milk. The aim of this study is to identify the genetic beta-casein variations in the research dairy cattle herd of the Livestock Center Oberschleissheim. So far, a number of 142 cows of varying breeds like Brown Swiss, Gelbvieh, German Fleckvieh, German Holstein and there hybrids has been examined. Milk samples were taken and analyzed by PCR. Totally, 52% of the dairy cattle show homozygous A2/A2, 15% homozygous A1/A1 and 33% heterozygous A1/A2 genotypes. Unexpectedly, A2/A2 had in tendency a higher milk yield with 30.68 kg/day in comparison with 30.55 kg/day for A1/A2 and 29.25 kg/day for A1/A1.

Replacing grass silage with forage pea silages in the diet of dairy cows on milk fatty acid profile

C.E.A. Campbell, J.A. Huntington and L.A. Sinclair
Harper Adams University, Animals, Newport, TF10 8NB Shropshire, United Kingdom; ccampbell@harper-adams.ac.uk

Increased global demand for soya bean, and associated fluctuations in its availability and price has resulted in greater interest in the utilisation of home grown, protein-rich forage sources for dairy cows, although their effect on milk fatty acid (FA) profile is unclear. The effects of the inclusion of forage peas differing in flower colour (and therefore tannin content) as a replacement for grass silage on the intake, performance, and milk FA profile were therefore studied. Eighteen multiparous dairy cows that were 100 ± 23.1 days in milk received one of 3 diets in each of 3 periods of 28 day duration, in a Latin square design with measurements taken in the final 7 days of each period. The proportion of grass (G), white (W) or red (R) flower pea silage to maize silage was 40:60 (DM basis), and all treatments contained 55:45 forage to concentrates (DM basis). Diets were formulated to be isonitrogenous. Cows were fed once daily at 08:00 h and milked twice daily. Following extraction, milk FA were analysed by gas chromatography. Data were analysed as a Latin square design using Genstat (v. 16). Dry matter (DM) intake and milk yield were highest ($P<0.01$) in cows when fed C (mean values of 21.4, 19.9 and 20.7 kg DM/d, and 40.6, 38.4 and 38.3 kg/d for C, W and R respectively), but there was no effect ($P>0.05$) on milk fat content, with a mean value of 35.2 g/kg. Compared with C, cows receiving W or R had a lower milk protein content ($P<0.05$). Dietary treatment had little effect on the majority of milk FA, but C18:3n-3 was higher ($P<0.001$) in cows when fed R than C, while C18:2n-6 was lower ($P<0.001$) in cows when fed W or R than C (mean values of 0.16, 0.16 and 0.18 g/100 g for C18:3 and 0.59, 0.45 and 0.47 g/100 g for 18:2n-6 for C, W and R respectively). Compared to grass silage the inclusion of forage peas in the diet of high yielding dairy cows reduced intake and milk yield, and had a minor effect on milk FA profile.

Comparative performance of Holstein-Friesian dairy cows of contrasting Economic Breeding Index
M. O'Sullivan[1,2], S. McParland[1], K.M. Pierce[2] and F. Buckley[1]
[1]*Teagasc, Animal and Grassland Research and Innovation Centre, Moorepark, Fermoy, Co. Cork, Ireland,* [2]*University College Dublin, School of Agriculture and Food Science, Belfield, Dublin 4, Ireland; morgan.osullivan@teagasc.ie*

The objective of this study was to evaluate the phenotypic performance of two genetic groups of Holstein-Friesian cows based on the Irish national dairy index, the EBI (economic breeding index). Data were available from 2013 to 2015. Ninety ELITE cows (mean EBI €249) and 45 cows representing national average genetics (NA; mean EBI €133) were evaluated annually across 3 contrasting seasonal pasture-based feeding treatments; Control, High Concentrate and Low Grass Allowance. All cows were parity 1 in 2013, 33.3% parity 1 and 66.6% parity two in 2014, and 27% parity 1, 27%parity 2 and 46% parity 3 in 2015. Milk volume (5,499 vs 5,704 kg) was significantly lower (P<0.001) but milk fat (44.3 vs 41.6 g/kg) and milk protein content (37.5 vs 35.8 g/kg) was significantly higher with ELITE compared to NA. Milk solids (fat plus protein) yield was numerically higher and milk value (P<0.05) was significantly higher with ELITE cows (448 kg and €1,891) compared with NA cows (442 kg and €1,862). Elite cows were lighter (490 vs 498 kg; P<0.001) but maintained higher body condition score (2.94 vs 2.76; P<0.001) throughout lactation. Pregnancy rate to first service (58 vs 47%; P=0.07), six-week in-calf rate (70% vs 58%; P<0.02) and 13-week in-calf rate (90 vs 81%; P<0.008) were higher, and calving to conception interval (83 vs 86 days; P=0.18) was lower for ELITE compared with NA. Based on the three years data there was no genotype × feeding treatment interaction implying a similar milk response to additional concentrate and similar reduction in milk yield due to a lower grass allowance expressed by both genotypes. The results obtained in the current study provide confidence to the Irish dairy chain that genetic selection based on EBI delivers more productive and more fertile dairy cows. In addition the study provides leadership, that the decline in fertility evidenced in the Holstein-Friesian population caused by aggressive selection almost solely for milk production, may be reversed when appropriate selection pressure on suitable fertility traits is applied while, simultaneously improving production.

Profiles of mammary adaptability and associated traits in dairy cows using a 24-h milking interval
C. Charton[1,2], H. Larroque[1] and J. Guinard-Flament[2]
[1]*INRA / Université de Toulouse INP ENSAT / Université de Toulouse INP ENVT, UMR 1388 GenPhySE, Chemin de Borde Rouge, 31326 Castanet-Tolosan, France,* [2]*INRA / Agrocampus Ouest, UMR 1348 PEGASE, 65 rue de Saint Brieuc, 35042 Rennes cedex, France; clementine.charton@rennes.inra.fr*

Understanding the dairy cows' adaptability is of interest to secure dairy systems relative to changes of animals' environment and breeding. Mammary gland is a key component of cows' adaptability. To characterize its adaptive potential, its functioning was perturbed by carrying out a single milking interval of 24-h (24h-MI). This study aimed to determine: (1) whether cows' responses to the 24h-MI could be gathered into typical adaptive profiles; and (2) whether those profiles were influenced by traits describing milk yield and composition, mammary distension, milk flow rate, inflammation, and epithelium integrity. The 24h-MI challenges (n=93) were performed on 56 Holstein-Friesian cows. They consisted of 3 successive periods: 1 wk of twice daily milking (TDM) as a control, 1 single 24h-MI and then 13 d of TDM. PCA followed by clustering revealed 3 profiles of mammary adaptability. Cows in cluster 1 lost on average 4.5 and 6.8 kg/d more milk yield than cows in cluster 2 and 3. Cows in cluster 1 were characterized by showing no increase in milk flow rate and by having the lowest increase in udder teats distance between TDM and 24h-MI day. Thus cows in cluster 1 either lost more milk due to a less compliant udder or had a lesser udder distension because of their substantial milk yield reduction. Cows in cluster 3 recovered on average 63.9% of the milk they had lost, whereas cows in cluster 1 recovered almost completely (90.5%) and cows in cluster 2 recovered completely (100.9%). The lesser recovery observed in cluster 3 might be associated with a lower estimated milk yield potential or an initially inflamed udder (as indicated by a higher occurrence of initial milk interleukin-8 (IL-8), a small cytokine playing a key role in mediating inflammation as it triggers neutrophil recruitment) or both. To conclude, profiles of mammary gland adaptability were discriminated on both milk yield loss and recovery responses and were correlated to some new specific traits: udder compliance, initial udder inflammation status and cow milk yield potential.

Discovery of new αs2-casein isoforms suggests 2 phosphorylation pathways

Z.H. Fang[1,2], M.H.P.W. Visker[2], G. Miranda[1], A. Delacroix-Buchet[1], H. Bovenhuis[2] and P. Martin[1]
[1]INRA, AgroParisTech, Université Paris-Saclay, GABI, Domaine de Vilvert, 78350 Jouy-en-Josas, France,
[2]Wageningen university, Animal breeding and genomics center, P.O. Box 338, 6700 AH Wageningen, the
Netherlands; zih-hua.fang@wur.nl

Phosphorylation is an important post-translational modification of casein (CN), and is one of the key factors responsible for constructing casein micelles. Variation in the degree of αs2-CN phosphorylation is of great interest because it might be associated with variation in the size of casein micelles among milk of individual cows, thus might affect milk technological properties. This study aimed to investigate the variation in the phosphorylation degree of αs2-CN among milk of individual cows and to explore relationships among different phosphorylation isoforms of αs2-CN. For this purpose, morning milk samples from 529 French Montbéliarde cows were analyzed using Liquid Chromatography coupled with Electrospray Ionization Mass Spectrometry (LC/ESI-MS). We detected the presence of αs2-CN-9P, -14P, and -15P in bovine milk, in addition to the known isoforms αs2-CN-10P, -11P, -12P, and -13P. To our knowledge, this is the first study that shows the presence of these three new αs2-CN isoforms. The relative concentrations of each αs2-CN phosphorylation isoform varied considerably among milk of individual cows. Furthermore, the phenotypic correlations and hierarchical clustering suggest two regulatory systems for the phosphorylation of αs2-CN, one responsible for isoforms with lower levels of phosphorylation (αs2-CN-10P and -11P), and the other responsible for isoforms with higher levels of phosphorylation (αs2-CN-12P, -13P and -14P). Identifying all phosphorylation sites of αs2-CN and investigating the genetic background of the different αs2-CN phosphorylation isoforms may provide further insight into the phosphorylation mechanism of caseins.

Dietary inclusion of marine algae and its impact on milk composition and performance in dairy cows

B.E. Till[1], J.A. Huntington[1], J. Taylor-Pickard[2] and L.A. Sinclair[1]
[1]Harper Adams University, Animals, Newport, TF10 8NB, Shropshire, United Kingdom, [2]Alltech, Summerhill
Road, Dunboyne, Ireland; btill@harper-adams.ac.uk

It has been suggested that marine products that are enriched with very long chain polyunsaturated fatty acids can be fed to dairy cows to modify milk composition and improve human intake of these nutritionally beneficial fatty acids. The objectives of the current study were to determine the effect of different inclusion rates of marine algae on milk yield, composition, and performance of dairy cows. Twenty Holstein dairy cows yielding 40 kg/d at the beginning of the study were randomly allocated to one of four diets, in a 4×4 latin square design, with four periods each of 28 days duration. The basal ration contained maize and grass silages and straight feeds, and was supplemented with one of four levels of algae oil to provide four treatments; 0 (Control; C); 29 (Low; L); 58 (Medium; M) and 87 (High; H) g/cow/day. Individual milk yield and intake were recorded daily during the final 7 days of each period, with milk samples taken on four occasions for subsequent analysis. The cows were weighed and body condition scored (1-5 scale) at the beginning of the study and end of each sampling period. Data were analysed using Genstat (v17). Supplementation with algae had no effect on dry matter intake (P=0.849), milk yield (P=0.759), body condition score (P=0.406), live weight (P=0.467) or live weight change (P=0.177), with mean values of 23.1 kg/d, 38.5 kg/d, 2.94 units, 667 kg and 0.34 kg/d respectively. There was an effect of algal oil supplementation on milk fat content; (P=0.011) which was lowest at the highest level of algae oil inclusion (mean values of 39.6, 38.4, 37.1 and 35.9 g/kg for C, L, M and H respectively). There was no effect of dietary treatment on milk protein content (P=0.282), with a mean value of 32.4 g/kg. These results demonstrate that marine algae can be added to the diet of high yielding dairy cows diet without any adverse effect on performance, except for milk fat content that will be reduced at the highest level of inclusion.

Comparative efficiency of lactation curve models using Irish experimental dairy farms data
F. Zhang and M. Murphy
Cork Institute of Technology, Process, Energy and Transport Engineering, Bishopstown, Cork, T12 P928, Ireland;
zhangensi@gmail.com

The objective of this study was to find a well-fitting model to represent the lactation curves of Irish dairy cows over various prediction horizons. Ten representative models from multiple categories including; curve fitting, static regression models and auto-regressive models were used to model forecast total daily milk production. The experiment was carried out on a total of six years of daily milk yield records from farms situated in the south of Ireland. The model simulation was set at herd level and a sample herd of 100 cows was randomly selected from the regional herd population. The parity composition of the sample herd was kept identical to that of the regional herd population. The SSE (Sum of Squared Errors) value of the ten lactation models varied substantially, except for the Surface Fitting model. Based on these criteria, the Surface Fitting model had the lowest SSE value (1,714,385 litres and 1,610,491 litres) over the 365-day horizon and the 30-day horizon. R^2 value of Surface Fitting model (0.97) was the highest and the RMSE (Root Mean Squared Error) value of the Surface Fitting model was the lowest (68.5 litres and 66.4 litres), in comparison to the other lactation models. However, in the short-term prediction (10-day horizon), the NARX model proved to be the most accurate forecasting model with the lowest SSE value (1,198,267 litres) and the lowest RMSE value (57.3 litres). The results showed the Surface Fitting model and the NARX model provided the most accurate milk production forecast for the sample herd, for long-term (365-day and 30-day) and short-term (10-day) forecast horizons, respectively.

American and German citizen attitudes towards cow-calf separation on dairy farms
G. Busch[1], D.M. Weary[2], A. Spiller[3] and M.A.G. Von Keyserlingk[2]
[1]Free University of Bolzano, Faculty of Science and Technology, Universitätsplatz 5, 39100 Bozen-Bolzano, Italy,
[2]University of British Columbia, Animal Welfare Program, 2357 Main Mall, V6T 1Z4 Vancouver, Canada, [3]Georg-
August-University Göttingen, Department für Agrarökonomie und Rurale Entwicklung, Platz der Göttinger Sieben
5, 37073 Göttingen, Germany; gesa.busch@unibz.it

The incorporation of public values into the design of farming practices is a necessary aspect of achieving a sustainable agriculture sector. In dairy farming, the separation of cow and calf immediately following birth is arguably one of the most emotional topics. Scientific evidence can be found which reveals both advantages and disadvantages of this practice. However, results from qualitative opinion research indicate that this practice is not supported by the public once it is made aware of it. This study analyzes the views of American and German citizens regarding the separation of cow and calf in a quantitative segmentation approach (K-Means Clustering) to provide insight into public opinion. 476 US and 491 German citizens took part in the study. All participants were confronted with two handling practices, namely early and late separation of cow and calf on dairy farms. Although the majority of participants favors late separation (USA=55.1%, Germany=68.7%), a smaller proportion supports early separation (USA=27.3%, Germany=20.2%). Three different clusters could be identified according to participants' attitudes towards separation practice. Cluster A (38.7% of participants) favors late separation, Cluster B (43.6%) is unsure whether to favor late separation and Cluster C (17.9%) favors early separation (P≤0.000). Cluster A (pro late) and C (pro early) participants' opinion towards separation practice is reinforced after reading balanced arguments for both, early and late separation, while Cluster B participants' opinion is not influenced by reading these arguments. For dairy industries, the results reveal challenges for developing accepted farming practices due to heterogeneity in public opinion and confirmation bias in citizen information search.

Tasting acid coagulated, ripened and blue goat cheese combined with tocte, honey and blackberry jam

V. Inca Guerrero[1], E. Ureña Ureña[1], I. Barba Cuji[1], J. Palmay Paredes[1], C. Hernández Maya[1], R. Remache[1], J. Espinoza Castro[1], J. Erazo Solines[2] and D. Sánchez Macías[1]
[1]Universidad Nacional de Chimborazo, Dpto. of Animal Production and Processing, Ingeniería Agroindustrial, Avda. Antonio José de Sucre s/n., 060150, Riobamba, Ecuador, [2]Mondel Company, KM 2,5 Vía Sangolqui, Amaguaña. Los Chillos, 170501, Sangolquí, Ecuador; dsanchez@unach.edu.ec

Local or regional products have authentic, original, seasonal, durable and other features important for the local or regional community; as well promote social cohesion. Tocte (an Andean nut), honey and blackberries are three typical and much consumed products in the Andean region. The objective of this study was to analyze how Andean people accept acid coagulated, ripened or blue goat cheese, combined with tocte, honey or blackberry jam. Three expert panelists previously tasted acid coagulated, ripened and blue goat cheese combined with tocte, honey and blackberry jam, with the aim to choose the six best combinations to give inexpert people: acid coagulated goat cheese with honey and blackberry jam, ripened goat cheese with blackberry jam and tocte, and blue goat cheese with honey and tocte were selected. Sensory testing was conducted with 280 Andean region people from Ecuador, who only 10% have tasted goat cheese previously. They ranked the products using a 9-point hedonic scale for overall acceptance, where 1=dislike extremely and 9=like extremely. Also, into the same taste cheese type, they had to choose the best combination and were instructed to write down the reasons why they liked or disliked each product. The 90% of consumers liked or liked extremely acid coagulated goat cheese with honey or blackberry jam, mainly because the sour-sweet taste and palatable sensation, but the best punctuation was for the combination with honey (7.22 vs 7.09). Respect to ripened goat cheese, 76.5% and 68% liked or liked extremely the cheese combined with blackberry jam or tocte, respectively. Consumers (73%) preferred the blackberry jam combination versus tocte (5.83 vs 5.34). And finally, when consumers tasted the blue cheese, 49% and 41% liked or liked extremely the blue cheese with honey or tocte, respectively. However, 65% preferred the honey combination versus tocte with blue goat cheese (4.46 vs 3.92).

Characterizing the mammary epithelial cell exfoliation process during milking in dairy cows

L. Herve[1,2], V. Lollivier[1,2], P. Lamberton[1,2], S. Wiart[1,2], S. Philau[1,2], H. Quesnel[1,2] and M. Boutinaud[1,2]
[1]Agrocampus Ouest, UMR PEGASE, 35590 Saint-Gilles, France, [2]INRA, UMR PEGASE, 35590 Saint-Gilles, France; lucile.herve@rennes.inra.fr

The presence of mammary epithelial cells (MEC) in milk indicates that some MEC are shed from the mammary epithelium. The MEC exfoliation process in dairy cows likely participates in the regulation of MEC number in the udder and thus in milk yield variations. The mechanisms that regulate the exfoliation process are not known. However, we can hypothesize that milking, through its mechanical pressure on the mammary epithelium, is a special time for exfoliation of the MEC. In order to characterize MEC exfoliation during milking, 9 multiparous cows were milked and milk samples were collected each minute for composition analysis. The samples collected at the beginning (after 1 and 2 min), in the middle and at the end of milking were used for MEC purification using an immuno-magnetic method. Tight junction opening was assessed by plasma lactose concentration and Na:K ratio in milk. Data were analyzed using the mixed procedure of SAS. Milk flow increased until the 3rd minute and then decreased as the milking progressed (P<0.001). Milk composition changed during the course of milking: the milk fat content increased (P<0.001) while the milk protein content decreased (P<0.001). At the end of milking, the somatic cell count increased (P<0.001) whereas the percentage of MEC in total somatic cells decreased (P<0.05), suggesting an influx of immune cells. The concentration of MEC in milk decreased between the 1st and the 2nd minute of milking suggesting that the milk of the 1st minute of milking contained MEC which were exfoliated before milking started. The milk MEC concentration then increased from the 2nd minute to the end of milking. A decrease in the Na:K ratio in milk (P<0.01) and an increase in plasma lactose concentration (P<0.01) during milking indicated a disruption of mammary epithelium integrity. The results of this study show that MEC are exfoliated before milking but also during milking. The exfoliation process during milking is concomitant with the disruption of mammary epithelium integrity.

Actual purchase dairy products and expectative after introducing the goat cheese benefits to Andeans
V. Inca Guerrero, E. Ureña Ureña, I. Barba Cuji, R. Remache, J. Palmay Paredes, C. Hernández Maya, J. Espinoza Castro and D. Sánchez Macías
Universidad Nacional de Chimborazo, Dpto. of Animal Production and Processing, Ingeniería Agroindustrial, Avda. Antonio José de Sucre s/n., 060150, Riobamba, Ecuador; dsanchez@unach.edu.ec

Globally, the consumers are seeking for food products which inspire them, surprise them and offer them an emotional connection with the goods they consume. However, Riobamba is an Andean city with people with food tradition deeply rooted, having little experience with goat cheese. Almost 80% of habitants from Riobamba are considered indigenes, as Quichuan. The objective of this study was to analyze their purchasing profile of dairy products, especially goat cheese, explain the goat milk and cheese benefits, and meet purchase expectations after knowing such benefits. Through questionnaires, 280 people from Riobamba (Chimborazo, Ecuador) were asked about how often they purchase normal or processed and enriched dairy products (with calcium, omega 3, probiotics, lower in fat, etc.), ripened, blue or goat cheese. As well, they were asked about how often they consume those products. Then, they were explained about the benefits of milk, cheese and more specifically acid coagulated, ripened and blue goat cheese. Also, a table with different attractive dishes was shown and a recipes book was given. After that, they were asked again if they knew such benefits and how are now their expectative to buy goat dairy products. Descriptive statistical analysis was used to analyze the data with SAS program (v.11). The results show that few people buy enriched milk (32%), dairy products with immunological properties (18%), ricotta (8%) or ripened and blue cheeses (13%). However, more than 90% of the respondents consume 2 or more dairy portions per day, and the two main motivations to buy cheese are the inherent quality and tasty. Near 90% of the participants have never consumed goat cheese, and 82% did not know its benefits. However, the 67% of the respondents are now disposed to buy goat cheese. To conclude, if they have to choose, they will purchase acid coagulated goat cheese (41%), ripened goat cheese (36%) or blue goat cheese (23%).

Milking interval and cisternal udder evaluation in Egyptian Maghrebi camels (Camelus dromedarius L.)
E.B. Abdalla[1], G. Caja[2], A.F. Seioudy[3], O.A. Salama[3] and M.H. Farouk[4]
[1]Fac. Agric., Ain Shams Univ., Anim. Prod. Dept., 68, Hadayek Shubra St., 11241, Cairo, Egypt, [2]Univ. Autònoma de Barcelona, Grup de recerca en remugants, department de ciència animal i dels aliments, Campus de la UAB, Plaza Cívica, s/n, Bellaterra, 08193, Barcelona, Spain, [3]Anim. Prod. Res. Institute, Agric. Res. Center, Camel Res. Dept., 5, Nadi-El Said St., Dokki, 12611, Giza, Egypt, [4]Fac. Agric., Al-Azhar Univ., Anim. Prod. Dept., Nasr City, 11884, Cairo, Egypt; ahmedfathy_86@hotmail.com

Effects of six different milking intervals (4, 8, 12, 16, 20 and 24-h) on milk yield and milk composition were studied in Egyptian Maghrebi dairy camels (n=10) at mid-late lactation [281±41 days in milk, 4.2±0.3 l/d]. Intravenous injection of oxytocin (OT) was administered before milking at each experimental milking to induce complete milk ejection. Cisternal and alveolar milk were measured before and after inducing milk letdown with OT, respectively for all milking intervals and udder cisterns were explored by ultrasonography. Milk secretion rate decreased exponentially (R^2=0.90) from 4- to 24-h intervals (168±28 to 105±10 ml/h). Comparing with 12-h milking interval (3.51 l/d), estimated daily milk yield was 115, 106, 75, 75 and 72% for 4, 8, 16, 20 and 24-h intervals, respectively. Total milk solids and milk fat content decreased with increasing milking interval, showing the greatest value at 4-h intervals (12.91±0.29 and 4.36±0.10%, respectively) and the lowest at 24-h intervals (10.19±0.29 and 2.44±0.10%, respectively). Milk protein (3.54±0.19%), lactose (3.83±0.19%) and ash (0.65±0.02%) did not change significantly (P>0.05) for all milking intervals. Camel udders showed small cisterns (132±30 ml; 7.7% of total milk in the udder) and did not vary with increase in milking interval. Alveolar milk contained more fat (3.26 vs 2.22%; SEM, 0.10%) and protein (3.39 vs 2.96%; SEM, 0.05%) than those observed in cisternal milk. In conclusion, this short-term study proved the low storage capacity of the Egyptian Maghrebi camel udder but also showed their moderate ability to adapt to extended milking intervals at mid-late lactation.

Relationship between in-line recorded milk flow rate and longevity in Estonian Holstein cows

T. Kaart, A. Tänavots, M. Liiva and H. Viinalass
Institute of Veterinary Medicine and Animal Sciences, Estonian University of Life Sciences, Department of Animal Genetics and Breeding, Kreutzwaldi 62, 51014 Tartu, Estonia; alo.tanavots@emu.ee

In nowadays industrial world, the homogeneity of cows-related farm procedures (incl. milking process) has become an essential economic factor. Another aspect, which has a major impact on the profitability of the dairy industry, is cows' functional longevity. The relationships between in-line recorded average milk flow rate (AFR, kg/min) expressing the efficiency of the milking process, and survival rate and culling reasons (CR) characterising the longevity, in Estonian Holstein (EHF) cows were investigated. AFR was recorded in-line from 2010 to 2015 in total of 20,484 first parity EHF cows belonging to 58 owners. The number of culled cows at the end of 2015 was 5,190 (25.3%). The Cox proportional-hazards regression model was built and regression tree with package 'partykit' in R was constructed to study the effect of AFR on censored longevity. The correspondence analysis (CA) was used to discover the common patterns among final nodes of regression tree and culling reasons. The overall AFR was 2.16 kg/min (SD=0.79) and the AFR of culled cows was 2.08 kg/min (SD=0.77). The survival rate was the highest on cows with AFR>3.16 kg/min and the lowest on cows with AFR<1.24 kg/min, indicating that the cows with faster AFR survive longer than cows with moderate or slower AFR. This tendency follows also the Cox model. According to the CA the slower AFR (≤1.66 kg/min) was mainly related with culling due to the low milk yield and various udder and teat defects, and especially with poor milkability. However, the culling due to mastitis was associated mostly with moderate and culling due to the metabolic and reproductive disorders with faster AFR. The results reveal that faster milking EHF cows have longer productive life and the culling rate due to mastitis does not increase.

Feed intake of dairy cows ante partum and the effect of milk production and metabolism post partum

P. Kühne[1], H. Scholz[1] and T. Engelhard[2]
[1]Anhalt University of Applied Sciences, Faculty LOEL, Strenzfelder Allee 28, 06406 Bernburg, Germany, [2]State Institute for Agriculture and Horticulture Saxony-Anhalt, Lindenstraße 18, 39606 Iden, Germany; p.kuehne@loel.hs-anhalt.de

The performance of dairy cows raised up in the last years quickly. They have to reach a high feed intake to get quickly over the point of the negative energy balance. From this point of view it seems to be of great importance to set the focus on the level of feed intake ante partum (a.p.) to consequently get a higher stability in the lactation. The study is based on the dry matter intake (DMI) – five days before calving – of 151 dairy cows of the State Institute of Agriculture and Horticulture Saxony-Anhalt in Iden. Multiparous cows (n=116) were allocated in three groups: animals with an average of <11.5 kg DMI per day [group 1], the second group with an average of 11.5-14.3 kg DMI per day [group 2] and the group 3 with an average of ≥14.3 kg/d DMI. The present study showed that dairy cows with the highest feed intake a.p. [group 3] had a higher milk yield in the first 105 days of lactation. Cows of group 3 produced a significant higher amount of milk protein at a comparable level of milk fat compared to group 1 and group 2. Cows of all groups changed body weight and back fat thickness at a comparable level. The dairy cows with daily feed intake ante partum of <11.5 kg DMI had the significant highest level of the metabolic parameters free fatty acids (NEFA) and beta hydroxybutyrate (BHB) during lactation. Especially in the first week after calving the NEFA were with an average of 1.51 mmol/l and the BHB in the 2nd/3rd week p.p. (1.78 mmol/l) and at the 7th/8th week p.p. (1.79 mmol/l) above the reference value. The results illustrated that scoring the rumen filling for identification of cows with problems of dry matter intake ante partum is useful for a perfect start to the following lactation.

The use of soy lecithin in fatty dairy cows during perinatal period

E. Wojtas and A. Zachwieja
Wroclaw University of Environmental and Life Sciences, Department of Cattle Breeding and Milk Production, Chelmonskiego 38c, 51-630 Wroclaw, Poland; edyta.wojtas@up.wroc.pl

Nutrition is a key factor in determining the state of health of animals. The aim of the study was to investigate the effect of the addition of soy lecithin in the ration of fatty cows in the perinatal period on the value of selected biochemical indices of blood. Twenty cows of Polish Holstein-Friesian were included in the analysis. At the beginning of the dry period, cows were evaluated on their condition using a five-point scale (BCS). For the experiment were selected animals, which condition exceeded a value of 4 points (fatty cows). The cows were kept in the freestall system and fed TMR (total mixed ration). In the experimental group (10 cows), soy lecithin was administrated in an amount of 40 g/pcs. /day from 3 weeks before calving until the end of the 2nd week of lactation. Blood samples were collected from the tail vein in four periods: (1) 3 weeks before calving; (2) immediately after calving; (3) in the 2nd week of lactation; and (4) in the 4th week of lactation. Levels of alanine aminotransferase (ALT), aspartate aminotransferase (AST), glucose, free fatty acids (FFA) and β-hydroxybutyric acid were determined in the serum of the cows. The results were analyzed by using a one-way analysis of variance test. The largest differences were observed in the case of glucose. In the serum cows from the experimental group, the levels remained within the normal range (2.2-4.5 mmol/l) and was significantly greater at all dates of collection of blood as compared to the control group of cows. The level of alanine aminotransferase, aspartate aminotransferase, free fatty acids and β-hydroxybutyric acid stayed within the normal limits in cows from the experimental group, but in comparison to the cows from the control group, there were no statistically significant differences. The results found stabilization of biochemical markers in the blood of cows that received soy lecithin in the ration in the perinatal period.

Genetic relationships between fertility and milk coagulation in two Estonian dairy cattle breeds

M. Liiva[1,2], T. Kaart[1,2], M. Vallas[1,2] and H. Viinalass[1,2]
[1]Bio-Competence Centre of Healthy Dairy Products LLC, Kreutzwaldi 1, 51014 Tartu, Estonia, [2]Institute of Veterinary Medicine and Animal Sciences, Estonian University of Life Sciences, Department of Animal Genetics and Breeding, Kreutzwaldi 62, 51014 Tartu, Estonia; tanel.kaart@emu.ee

In dairy cattle farming, the genetic evaluation of fertility traits has had increasing importance in last decade. Meanwhile in dairy industry the improvement of milk technological properties has become one of the basis of more efficient producing. In present study the genetic relationships between heifers' fertility traits (age of first insemination (AFI), number of inseminations (NI) and non-return rate at 56 days after first insemination (N56)) and milk coagulation properties (rennet clotting time (RCT) and curd firmness 30 min after addition of the clotting enzyme (A30)) were estimated. Fertility data from 5,822 heifers from 105 farms and two breeds, Estonian Holstein (5,011) and Estonian Red (811), having first calving between years 2004 and 2009, and 24,031 records of RCT and A30 from first lactation were analysed (incl. pedigree information over 24,000 animals). Bivariate animal models were fitted with ASREML. For all variables the random herd effect and fixed breed and calving year-season effects were considered, for log-transformed RCT and A30 the 3rd order Legendre polynomial for additive genetic animal and permanent environmental effects plus fixed effects of calving age and sampling year-season were included, and for fertility traits the random technician effect was included. For N56 models the logit link function was applied. The coagulation traits both had quite high heritability (around 0.40) and quite weak negative genetic correlation between them (-0.14). Conversely, the heritability of fertility traits was between 0.02 and 0.04 and the genetic correlations between them were high (absolute values between 0.62 and 0.83). The genetic correlations between fertility traits and RCT and A30 were between -0.1 and 0.1 over the lactation (except genetic correlation between N56 and A30 was up to 0.3).

Measurement of rumen filling and relation to feed intake of German Holstein cows

P. Kühne[1], H. Scholz[1] and T. Engelhard[2]
[1]Anhalt University of Applied Sciences, Faculty LOEL, Strenzfelder Allee 28, 06406 Bernburg, Germany,
[2]State Institute for Agriculture and Horticulture Saxony-Anhalt, Lindenstraße 18, 39606 Iden, Germany;
p.kuehne@loel.hs-anhalt.de

Changes in feed intake are useful in early detection of disease in dairy cows. The rumen filling score can give information about actually feed intake of dairy cows. This score can evaluated visually from score 1 with a low feed intake until score 5 for high feed intake in different stages of lactation. First measurements between rumen filling score and the deepness of the paralumbar fossa was evaluated by Burfeind and showed some good accordance. Investigation 1 contains 62 dairy cows (German Holstein) until the first 100 days of lactation and was measured on six dairy farms in Germany. Therefore was used a special measuring instrument with one specific measuring point which was taken of each paralumbar fossa. At the same time the rumen filling score was estimated by visual evaluation according to the system of Zaaijar et al. The research represents the score 2, 3 and 4. There was a negative correlation between the measurement and the rumen filling score of r=-0.785 (P≤0.001). For the rumen filling score (RFS) was found the following deepness: RFS 2 with 6.3 cm; RFS 3 with 4.6 cm and RFS 4 with 3.0 cm. The values obtained are significantly different from each other. Further individual traits (e.g. body deepness, body condition score, body girth, pin wide) don't correlate with deepness of paralumbar fossa or show significant differences between the rumen filling score. Investigation 2 contains the visual examination of rumen filling score and the dry matter (DM) feed intake of dairy cows in the 5 days ante partum (n=64). The highest dry matter intake (average 14 kg DM per cow and day) achieved cows with RFS 4. Cows with Score 2 or 3 of rumen filling had a mean daily dry matter intake of 10 kg. The present results show that the rumen filling evaluation of dairy cows in the time before calving is very useful for an estimation of the feed intake.

The effect of n-3 fatty acid supplementation on oxidative status in calves

K. Śpitalniak, R. Kupczynski, K. Pogoda-Sewerniak and A. Zwyrzykowska
Wrocław University of Environmental and Life Sciences, Department of Environment, Hygiene and animal Welfare,
Chelmonskiego 38 c, 51-630 Wroclaw, Poland; kinga.spitalniak@up.wroc.pl

As one of the major components of the cell membrane polyunsaturated fatty acids are responsible for the protection of cells against oxidative stress. The balance between omega 3 and omega 6 lets the organism to keep control of the intensity of oxidative stress and inflammation. Oxidative stress in the newborn animals might have a negative impact on growth and metabolic processes, which are especially important in dairy farming. The aim of the research was to evaluate the influence of polyunsaturated fatty acids on total antioxidant status (TAS), superoxide dismutase (SOD) and glutathione peroxidase (GPx) and fatty acid profile. The experiment was performed on 27 Holstein-Friesian calves divided into 3 groups: control group, group receiving formulation with esters of α-linolenic acid (10 g/d) and group receiving formulation with flax oil (10 g/d) from 14 to 56 days of age. The formulations were added to the milk replacer in the amount of 1% DM/calf/day. Blood samples for laboratory tests were taken from the external jugular vein at 7,14, 28,35 on the 45th day of their life. The following parameters have been determined: total antioxidant status (TAS), superoxide dismutase (SOD) and glutathione peroxidase (GPx). The analyzes were performed using the multi-detection microplate reader Synergy HT BIO-TEK, Randox reagents. The study confirmed that the analyzed formulations have significant impact on the value of TAS (total antioxidant status). Intensive increase of TAS was found in the group receiving add esters of α-linolenic acid (0.76 to 1.66 U/l). GPx activity in both experimental groups presented an upward trend. At the beginning of the experiment concentration of glutathione peroxidase in the control group was high, further was systematically decrease with age.

Combatting AMR – new antibiotics and new approaches
B.F. Gilmore
Queen's University Belfast, School of Pharmacy, Belfast, United Kingdom; b.gilmore@qub.ac.uk

The majority of clinically useful antibiotics were discovered during the 'Golden Age' of antibiotic discovery, from the 1940s to the early 1970s. Since then, the discovery and approval of new antibiotic entities has not kept pace with the emergence of resistance and loss of efficacy of conventional antibiotics. This presentation will focus on the potential approaches to combat AMR through the reinvigoration of the antibiotic discovery pipeline, the quest to bring antibiotic alternatives from the laboratory to clinical reality and will discuss the key findings of the O'Neill Review.

Antimicrobial resistance and its impact on cattle production now and in the future
D.C. Barrett
University of Bristol, Production and Reproduction, School of Veterinary Sciences, Langford, Bristol, BS40 5DU, United Kingdom; david.barrett@bristol.ac.uk

Antimicrobial resistance (AMR) is a growing concern world-wide both in medicine and veterinary medicine. Individuals, scientific bodies and lobby groups may argue over the importance of antimicrobial use in livestock production, but the fact remains that human health is at risk due to AMR globally. While as scientists we must argue for evidence based policy making, there is already enough evidence to support the need to reduce antimicrobial usage in all sectors. The scientific arguments have now in part been overtaken by a political drive for change at the very highest levels. In 2016 AMR will be discussed by leaders of the G7, G20 and United Nations and it is likely that all three bodies will call for substantial changes on a global scale as outlined by Lord O'Neill. The responsibility for antimicrobial stewardship on farms rests ultimately with the veterinary profession as it is only veterinary surgeons that have the right to prescribe such medications; however livestock farmers, their other advisors and those controlling every step in the food supply chain also have a responsibility, as do those responsible for the production, promotion and supply of veterinary medicines. Milk buyers, processors and retailers of all animal products are increasingly monitoring, benchmarking and seeking to influence medicine usage on farm. After giving an insight into medicine use on farms, this presentation will outline some of the research undertaken by members of the AMR Force research group at the University of Bristol (@AMRForce). This group is seeking to answer fundamental questions about AMR in livestock, the environment and in human patients, as well as show that it is possible to keep healthy and productive cattle without the need for critically important antimicrobials such as fluoroquinolones and 3rd and 4th generation cephalosporins. The author will also briefly discuss more general future implications of AMR on farming practises.

Shifting the balance: breaking the antibiotic resistance cycle
R. Murphy
Alltech, Sarney, Summerhill road, Dunboyne, Co. Meath, Ireland; rmurphy@alltech.com

Antibiotic resistance has the potential to become one of the greatest problems of our generation, given the ever-increasing rise in bacterial strains that are less and less sensitive to existing treatments. While abuse of antibiotics in humans is probably the major contributor, policy makers have turned the spotlight on agricultural use as a way to control the problem. Treatment of animals with antimicrobials can cause drug resistance to zoonotic pathogens (e.g. Salmonella, Campylobacter) that can be transmitted to humans. Globally it is recognized that there is no so-called 'silver bullet' to replace antibiotic use in animal production and producers will have to continually improve hygiene and husbandry to address the issue. Products that will assist the move to antibiotic-free production status include many that are designed to regulate and support the gut environment and its microflora: coccidial vaccines, probiotics, competitive exclusion products, feed enzymes, functional nutrients, organic acids, organic minerals, plant-based products, and yeast cell wall derivatives. Of the functional ingredients currently in use for microbial control, mannan-oligosaccharides (MOS) and Mannose Rich Fractions (MRF) are widely used in animal nutrition. Given their ability to bind and limit the colonization of gut pathogens, MOS and MRF have proven to be an effective solution for antibiotic-free diets, as well as providing support for immune function and digestion.

Tackling antibiotic resistance
P. Wall
School of Public Health, Physiotherapy and Population Science, Woodview House, Belfield Dublin 4, Ireland; patrick.wall@ucd.ie

Increasing adverse publicity globally about antibiotic resistance is leading to policy responses from regulators and industry that maybe more in proportion to the media interest rather than in proportion to the risk to public health. The conventional media is feeding off the social media and vice versa compounding consumer anxiety. Many of the major global food service and retail giants have responded by pledging to deliver animal protein from antibiotic free production systems. Often the claims are made by their marketers rather than their technical experts, who are now faced with honoring the promises made to their customers. The public perception of modern farming is that the animals are intensively reared and are filled with antibiotics. Large-scale production is considered factory farming and bad, and small scale is considered good. However it is not the scale of the farm that is the issue rather it is how it is managed. Antibiotics are not a substitute for good husbandry practices. The simplest way to reduce the amount of antibiotics being used is to reduce the amount of disease. Antibiotic resistance is not a pathogen problem, a poultry problem, or a pig problem, it is a people problem. No food animal demands antibiotics and a change of behavior is required from the people tending the livestock. Alternative strategies to control the spread of disease must be adopted. These include robust biosecurity, genetic selection of disease resistant stock, immunization strategies, good husbandry practices and optimal nutrition.

Sparse single-step genomic Best Linear Unbiased Prediction in crossbreeding schemes
J. Vandenplas, M.P.L. Calus and J. Ten Napel
Wageningen UR Livestock Research, Animal Breeding and Genomics Centre, P.O. Box 338, 6700 AH Wageningen,
the Netherlands; jeremie.vandenplas@wur.nl

The single-step genomic Best Linear Unbiased Prediction (ssGBLUP) is becoming the method of choice to predict genomic breeding values (GBV) for livestock. The main reason is that ssGBLUP enables simultaneous use of data from genotyped and non-genotyped animals by combining genomic and pedigree relationship matrices. However, ssGBLUP could become infeasible with more than 150,000 genotyped animals, mainly due to the expensive creation and inversion of the genomic relationship matrix. Recently, the so-called algorithm for proven and young animals (APY) was proposed for computing a sparse inverse of the genomic relationship matrix. The APY is based on genomic recursions, where a GBV of a new genotyped animal, called non-core animal, is conditioned on GBV of all the previous genotyped animals, called core animals. Accurate computations of GBV using ssGBLUP with the APY for simulated and real single-breed populations were obtained by several authors. However, many livestock production systems are based on crossbreeding schemes, involving animals from multiple breeds and their crosses, and the APY still requires validation for such complex situations. The aim of this study is the validation of the APY for situations involving crossbreeding schemes, and the implementation of an efficient ssGBLUP using the APY for a large number of purebred and crossbred (genotyped) animals. Data were simulated to mimic a three-way crossbreeding scheme. Different distances between the breeds, as well as different strategies to select core animals, were considered.

Empirical comparison of different methods for genomic evaluation in pigs
A. Fangmann[1], A.R. Sharifi[1], M. Erbe[2] and H. Simianer[1]
[1]Georg-August-University Goettingen, Animal Breeding and Genetics Group, 37075 Goettingen, Germany,
[2]Institute for Animal Breeding, Bavarian State Research Centre for Agriculture, 85586 Poing-Grub, Germany;
anna.fangmann@agr.uni-goettingen.de

Initially, methods for using genomic information to predict breeding values (BV) were multi-step methods, which involve many assumptions which, if violated, may result in loss of information, inaccuracies and bias. To overcome this, single-step genomic best linear unbiased prediction (ssGBLUP) was proposed combining pedigree, phenotype and genotype of all individuals for genetic evaluation. Our objective was to implement ssGBLUP for genomic predictions in pigs and to compare the accuracy of ssGBLUP with multi-step methods in practical data. For this purpose, we used data from a German Landrace population. In total, the trait 'number of piglets born alive' was available for 182'054 litters of 41'090 sows, the pedigree comprised 147'461 individuals and 526 animals were genotyped. After quality control, 495 animals with 44'368 SNPs on 18 autosomes remained for analysis. We compared different methods: conventional parent average (PA), the direct genomic value (DGV) calculated with genomic BLUP (GBLUP), a genomic enhanced breeding value (GEBV) obtained by blending the DGV with the conventional PA, and ssGBLUP. Predictive abilities, i.e. correlations between deregressed proofs and genomic BVs, were calculated with forward prediction for young genotyped validation animals, which were born in 2011 to 2014. Predictive abilities were similar for GEBVs (0.19) and ssGBLUP (0.18) but higher than for GBLUP (0.08). Generally, predictive abilities were rather small which might be due to a poor relationship between reference and validation animals or the low number of genotypes. Assessment of reliabilities from MME for young genotyped animals indicated that genomic prediction outperforms PA, while ssGBLUP provided the highest reliability and blending yielded lower reliabilities. To asses a possible bias, regression of DRP on different genomic BV based on five-fold cross-validation was performed. As expected, ssGEBV was less biased than DGV or GEBV. Although, ssGBLUP did not outperform GBLUP or GEBVs, there are indications for a better performance when a reasonable number of genotypes becomes available.

Power and precision of mapping genes in simulated F2 crosses using whole genome sequence data

M. Schmid, R. Wellmann and J. Bennewitz
University of Hohenheim, Institute of Animal Science, Farm Animal Genetics and Breeding, Garbenstrasse 17,
70599 Hohenheim, Germany; markus_schmid@uni-hohenheim.de

In the past, mapping experiments in pigs often relied on F2 crosses, which were established with distantly related founder breeds (e.g. one European and one Asian type breed) or closely related breeds (e.g. two European type breeds). In this study we investigated the power and precision to map QTN in simulated F2 crosses using maximum marker density. The focus was on founder segregating genes, because these are of interest to improve genomic selection within the founder breeds. We analyzed simulated whole-genome sequence data of porcine F2 crosses with closely (EU1xEU2) and distantly (ASxEU2) related founder breeds as well as their pooled data (JOINT design). As a reference, a sample of the breed of interest (EU2), which connected both crosses, was evaluated. The founder breed genomes (2M) were simulated for 10,000 generations using an idealized population drift model. To create F2 designs, crossing schemes followed a mating plan. QTN were inserted after simulations had been finished. GWAS was conducted for 50 replicates applying a single marker regression model and power and precision were calculated. Our results showed, that mapping power is high in the cross ASxEU2, but lower in EU1xEU2. Power was highest in the JOINT design. In contrast, precision was highest in EU1xEU2 and lowest in ASxEU2. Notably, precision was similar in the cross EU1xEU2 and in the founder breed EU2. Pooling data resulted in an intermediate precision. Compared to purebred EU2, mapping power was strongly increased. Moreover the number of SNPs can be substantially increased in F2 designs compared to purebred populations, which strongly depended on the number of founder individuals within a design and their phylogenetical relatedness. To conclude, the existing porcine F2 crosses established in the past are so far underused experimental populations to map founder breed segregating genes precisely, if the founder breeds are closely related or several F2 crosses are pooled or both.

Using phenotypes of three-breed cross to improve breeding value estimation of purebred animals

C.A. Sevillano[1,2], J.W.M. Bastiaansen[2], J. Vandenplas[3], R. Bergsma[1] and M.P.L. Calus[3]
[1]Topigs Norsvin Research Center B.V., P.O. Box 43, 6640 AA Beuningen, the Netherlands, [2]Wageningen University, Animal Breeding and Genomics Centre, P.O. Box 338, 6700 AH Wageningen, the Netherlands, [3]Wageningen UR Livestock Research, Animal Breeding and Genomics Centre, P.O. Box 338, 6700 AH Wageningen, the Netherlands; claudia.sevillanodelaguila@wur.nl

Genomic breeding programs for pigs and poultry mainly use training populations of purebred animals (PB), while targeting to improve performance of crossbred animals (CB), which are usually a three-breed cross. Traits as feed efficiency with a low genetic correlation between PB and CB performance stress the need of using CB in the training population. When using CB, effects of SNPs may be breed-specific because of difference in Linkage Disequilibrium (LD) patterns between a SNP and a QTL, and differences in allele frequencies and allele substitution effects of QTL in the parental breeds. Estimation of breed-specific effects of alleles in a CB population, requires knowing their breed-of-origin. We investigated the performance of an approach for assigning breed-of-origin of alleles (BOA) in real data of three-breed cross pigs (finishers). BOA consists of three steps: (1) phase the haplotypes of PB and CB; (2) determine unique haplotypes among the PB; and (3) assign the breed-of-origin for each allele carried on haplotypes of CB. Genotypic and phenotypic data was available for 14,187 PB from the three parental breeds, and 1,723 finishers. On average 93.0% of the alleles of a finisher were assigned a breed-of-origin without using pedigree, and 94.6% using pedigree information. Using these derived breed origin of alleles, we built three (i.e. one for each breed) partial genomic relationship matrices. These matrices are used in a model to disentangle the genetic effect of a finisher in three genetic values, one from each breed origin, and assume zero correlations between the performance of PB from different breeds. This model will be tested for daily gain, feed intake, back fat, and loin depth.

Genomic evaluation of carcass traits of young bulls: comparison of two and single-step approaches

I. Croué[1,2] and V. Ducrocq[1]
[1]INRA, UMR1313 GABI, 78352 Jouy-en-Josas Cedex, France, [2]Institut de l'Élevage, UMR1313 GABI, 78352 Jouy-en-Josas Cedex, France; iola.croue@jouy.inra.fr

Meat production has a large economic impact for dual purpose breeds farms, partly through the production of young bulls, slaughtered between 12 and 24 months. In order to implement an appropriate genomic evaluation for carcass traits of Normande and Montbeliarde young bulls, breeding values were estimated for young candidates to selection using different approaches. Records of 159,115 Normande and 154,119 Montbeliarde were used. Evaluation approaches were compared measuring their accuracy (correlation between preadjusted performances and estimated breeding values) and bias (regression coefficient of preadjusted performances on estimated breeding values) on two validation populations: the 20% youngest genotyped young bulls and the 20% youngest AI sires. Genomic approaches were more accurate than BLUP (+0 to +7 points of correlation, except for age at slaughter in the sire validation population evaluated using a GBLUP, where the accuracy was lower), but had worst regression coefficients (regression coefficient of 0.676 to 1.078 with genomic approaches, vs 0.806 to 1.141 with BLUP). Multiple weighting of genomic and pedigree information in the single-step relationship matrix were tested in order to obtain the best single-step evaluation possible. With appropriate correction of the relationship matrix, a single-step evaluation led to similar accuracies than a two-step procedure, but slightly better regression coefficients, except for the low heritability trait (age at slaughter), where the single-step procedure was highly preferable. These results will be used for the development of a routine genomic evaluation of carcass weight, conformation and age at slaughter of young bulls in the Normande and Montbeliarde breeds in France.

Joint Estimation of Additive, Dominance and Polygenic Variances Using Female Reference Population

H. Alkhoder, Z. Liu, F. Reinhardt and R. Reents
vit Germany, Heinrich-Schröder-Weg 1, 27283 Verden, Germany; hatem.alkhoder@vit.de

Routine genomic evaluations in dairy cattle are based on corrected phenotypes of bulls, e.g. deregressed estimated breeding values or daughter yield deviations (DYD). As more and more cows have being genotyped, genomic reference population may be changed from the bulls with daughters to cows with own phenotypes. Analyzing cows' own phenotypes, e.g. original test-day yields or yield deviations, allows an estimation of dominance effects in addition to all model effects in the genomic evaluation based on the bull reference population. About 12,000 genotyped cows with phenotypes were used to jointly estimate additive and dominance variances as well as the residual polygenic variance. A GBLUP model was applied to the genotype and phenotype data for the simultaneous estimation of the three genetic variances. The new genomic model was compared to the current genomic model, based on DYD of bulls, without dominance effects via genomic validation based on youngest validation bulls with daughter information. Results of this study are presented.

TagSNP selection and genotype imputation in Atlantic Salmon

S.A. Boison, M. Lillehammer, A.K. Sonesson and M. Baranski
NOFIMA, Aquaculture and Genetics, Osloveien 1, Ås, Norway; solomon.antwi.boison@nofima.no

Reducing the cost of genotyping is crucial for the successful implementation of genomic selection in several Atlantic salmon selective breeding populations. The aim of the study was to develop a low-density marker panel (LDP) from a 37K Affymetrix SNP array and to quantify the accuracy of imputation for the developed LDP. Selection of markers from the high-density 37K Affymetrix SNP array to constitute the LDP was based on minor allele frequency (MAF), Linkage disequilibrium (LD) and the product of LD and MAF (LDMAF). Genotype data of 4,263 animals from three Atlantic Salmon populations were used to develop the LDP with ~2,000 (2K) SNP markers. Imputation was undertaken with FImpute and accuracy was computed as the correlation between true and imputed genotypes averaged across samples. The highest accuracy of imputation was obtained when markers were selected using both MAF and LDMAF (0.958; SD=0.012). Using LD only to select markers for the LDP gave the lowest accuracy (0.943; SD=0.014). Our results show that, developing a 2K low-density marker panel for imputation to 37K is feasible. Furthermore, imputation accuracy was reasonably high and thus we expect very little effect of the use of imputed genotypes for genomic prediction in these three aquaculture selective breeding populations.

SNP discovery in the sheep milk casein genes using transcriptome sequencing

A. Suárez-Vega[1], B. Gutiérrez-Gil[1], C. Klopp[2], G. Tosser-Klopp[3] and J.J. Arranz[1]
[1]Universidad de León, Producción Animal, Facultad de Veterinaria, Campus de Vegazana s/n, 24071 León, Spain, [2]INRA, Plateforme bioinformatique Toulouse Midi-Pyrénées, UR875 Biométrie et Intelligence Artificielle, 24 Chemin de Borde Rouge, Auzeville, CS 52627, 31326 Castanet Tolosan, France, [3]INRA, UMR1388 GenPhySE (Génétique, Physiologie et Systèmes d'Elevage), 24 Chemin de Borde Rouge, Auzeville, CS 52627, 31326 Castanet-Tolosan, France; asuav@unileon.es

Caseins, encoded by the CSN1S1, CSN1S2, CSN2 and CSN3 genes, are the major proteins in sheep milk and constitute about 76-83% of total milk protein. Genetic variants of all major milk proteins have been demonstrated to influence, from greater to lesser extent, the composition and/or technological properties of milk. In this study we have used the transcriptome sequencing technology (RNA-Seq) on sheep milk somatic cells (MSC) to identify the genetic variability of the genomic region encoding sheep caseins, a 250 kb cluster located on the ovine chromosome 6, in two common dairy sheep breeds farmed in Spain, Assaf and Churra. We have found 735 variants, most of which were already described on the Ensembl database and positioned in introns. Among the variability identified in coding regions, five missense variants were found on genes CSN1S1, CSN2 and CSN1S2. Among the missense mutations detected on CSN1S2, two of them (rs430397133 and rs424657035) are present in the protein variant classically known as CSN1S2*B, which is related to higher milk, fat, and protein yield and protein content when compared with the most common form CSN1S2*A variant. Further research is required to estimate the allele frequencies and determine the phenotypic effects of the variants identified through this RNA-Seq approach in the Churra and Assaf sheep populations.

Whole liver transcriptome analysis for the metabolic adaptation of dairy cows

N.T. Ha[1,2], C. Drögemüller[3], F. Schmitz-Hsu[4], R.M. Bruckmaier[1], H. Simianer[2] and J.J. Gross[1]
[1] Veterinary Physiology, Vetsuisse Faculty, University of Bern, Bremgartenstr. 109a, 3001 Bern, Switzerland, [2]Animal Breeding and Genetics Group, Department of Animal Sciences, Georg-August-University Goettingen, Albrecht-Thaer-Weg 3, 37075 Göttingen, Germany, [3] Institute of Genetics, Vetsuisse Faculty, University of Bern, Bremgartenstr. 109a, 3001 Bern, Switzerland, [4]Swissgenetics, Meielenfeldweg 12, 3052 Zollikofen, Switzerland; nha@gwdg.de

During early lactation, dairy cows experience a severe metabolic load often resulting in the development of various diseases. The inevitable deficiency in nutrients and energy at the onset of lactation requires the adjustment of the hepatic metabolism to overcome metabolic stress. In this study, we conducted a whole liver transcriptome analysis for the transition cow to identify novel factors crucial for the metabolic adaptation. Liver samples were taken from 6 cows at 3 points of time: week 3 ante-partum; week 2 and 3 post-partum. Using RNA-seq, we compared the transcriptomic profile of the 6 cows before and after calving. We performed a differential gene expression (DGE) and a combination of the gene-set enrichment analysis and perturbation analysis, incorporating gene network information, to identify significant genes (Ensembl) and pathways (KEGG). Among the ~10,000 expressed genes, we discovered ~1,000 genes to be significantly differentially expressed (FDR<5%), of which ~43% are linked to lipids, ~16% to oxidative stress and ~6% to the glucose metabolism (GO). The combined pathway analysis further revealed 7 significant pathways, including the pathways 'adipocytokine signaling pathway' and 'steroid hormone biosynthesis'. The DGE and pathway analysis have demonstrated that major hepatic changes from late pregnancy to early lactation are related to gluconeogenesis and fat mobilization. We further find indications for immunological changes that may contribute to the impaired immune system of dairy cows during the transition period. The discovered factors are a valuable contribution to a better understanding of the metabolic adaptation and hence, should be more closely investigated in future studies.

Genes and microbes, the next step in dairy cattle breeding

G.F. Difford[1,2], J. Lassen[1] and P. Løvendahl[1]
[1]Aarhus University, Center For Quantitative Genetics and Genomics, Blichers Alle, 8830, Denmark, [2]Wageningen University, Animal Breeding and Genomics Centre, Wageningen, 6700, the Netherlands; gareth.difford@mbg.au.dk

The performances of dairy cattle are viewed as the combination of their genes and the environment they are raised in. It has become apparent that the numbers of microbial cells on and in the dairy cow, far outnumber the actual cow's cells by a factor of ~120. The sheer magnitude of microbial numbers indicates microbes may influence a cow's performance, particularly where ruminal microbes play a crucial role in digestion and metabolism. Interestingly, studies have shown that for some ruminal microbes the cow has an influence on their presence and abundance, whilst other microbes are only found in animals raised in certain parts of the world or fed particular diets. This begs the question 'How does the cow's genes and environment affect the microbial composition and abundance and how do these microbes in turn affect the host's performance?' The objective of this study was to identify production traits under varying levels of genetic and microbial influence. Thereafter, evaluate the changes in predictive accuracy, with the inclusion of host genetic information and bacterial and archaeal diversity and abundance. To this end, rumen samples were collected from ~1000 lactating Holstein cows on commercial farms in Denmark and the 16s rRNA gene sequenced and assembled to obtain ruminal diversity and abundance estimates. Additive genetic and microbial variances were estimated simultaneously using univariate animal models and the most contrasting traits selected. Bayesian variable selection models were fitted for genetic and microbial information independently and jointly and the predictive accuracy evaluated through cross validation. Results show interactions between variance captured by additive cow genetics and microbial diversity and abundance. The increase in predictive accuracy of traits varied from 0-40%, with the largest increases associated with lowly heritable traits. Results show some cow performance traits are influenced by ruminal microbial diversity and abundance outside of the host; however, more research is required to fully understand the mechanisms driving the observed cow and microbial abundance and diversity dynamics.

Copy number variations associated with insect bite hypersensitivity in Friesian horses

A. Schurink[1], V.H. Silva[2,3], B.D. Velie[3], B.W. Dibbits[1], R.P.M.A. Crooijmans[1], L. François[4], A. Stinckens[4], S. Blott[5], N. Buys[4], G. Lindgren[3] and B.J. Ducro[1]
[1]Animal Breeding and Genomics Centre, WU, P.O. Box 338, 6700 AH Wageningen, the Netherlands, [2]Netherlands Institute of Ecology, NIOO-KNAW, 6708 PB Wageningen, the Netherlands, [3]Department of Animal Breeding and Genetics, SLU, P.O. Box 7023, 75007 Uppsala, Sweden, [4]KU Leuven, Department of Biosystems, Livestock Genetics, P.O. Box 2456, 3001 Heverlee, Belgium, [5]Reproductive Biology, Faculty of Medicine and Health Sciences, The University of Nottingham, LE12 5RD Leicestershire, United Kingdom; anouk3.schurink@wur.nl

Copy number variations (CNVs) are the largest source of genetic variation that has been identified in the equine genome so far. CNV-based GWAS could therefore facilitate detection of genetic variation for traits of interest. The aim of our study was to identify, characterize and associate CNVs in Friesian horses with insect bite hypersensitivity (IBH), a common equine allergy to insects that causes intense itch and seriously reduces welfare. After pre-processing of raw intensity signals from 670,796 SNPs for 138 cases and 131 controls, CNVs were identified using PennCNV and merged into CNV regions (CNVRs) using CNVRuler. A SNP-based GWAS performed beforehand (chi squared test with 1df) showed a clear association between the ELA region on ECA20 and IBH in Friesian horses. Regions identified on ECA9, 11 and 20 have rather consistently been associated with IBH across equine breeds investigated so far. Currently, the CNV analysis is at an early stage and different strategies to improve the SNP array overall signal quality are being tested. The preliminary CNV-based GWAS, which analysed gains and losses separately by logistic regression, indicated a loss in the ELA region significantly more frequent in affected horses. CNVRs significantly associated with IBH could point to genes involved in this complex polygenic disease. Our study will increase the knowledge on CNVs in horses and will highlight which genes and pathways are affected by CNVs. Moreover, it will facilitate in our attempt to improve selection against IBH.

Copy number variations reveal metabolic pathways involved with feed conversion ratio in beef cattle

M.H.A. Santana[1], G.A. Oliveira Junior[2], M.E. Carvalho[1], L. Grigoletto[1] and J.B.S. Ferraz[1]
[1]Faculdade de Zootecnia e Engenharia de Alimentos, Universidade de São Paulo, Duque de Caxias Norte 225, 13635900, Brazil, [2]Iowa State University, Ames, IA 50011, USA; jbferraz@usp.br

Feed conversion ratio (FCR) is a feed efficiency trait that measures the animal's capacity to convert feed consumed into the desired output (e.g. meat deposition). There are alternative genomic frameworks such as copy number variants (CNV) that can be useful to explain the variability and unveil the molecular architecture of complex traits like FCR. The study aimed to identify candidate genes and putative pathways involved with FCR from CNV associations. FCR data of 1,475 Nellore cattle (574 ± 95-d old and 381 ± 45 kg) were used from feed efficiency tests conducted in Brazil. Genotypic information (777,962 SNPs) from 2,253 animals were used to detect CNV (PennCNV). The CNV regions (CNVRs) were determined merging overlapping CNVs (CNVRuler). Linear regression was performed to associate FCR and CNVRs and false discovery rate (FDR) was considered significant. The CNVRs associated with FCR were considered as promising QTLs and candidate genes within these regions were used in enrichment analysis (Biomart, Ensembl Genes 81, DAVID 6.7 and KEGG). We identified 139,089 CNVs and 2,667 CNVRs (1,112 loss, 938 gain and 617 mixed regions) in autosomal chromosomes (Chr). From those, 16 CNVRs were non-fixed and 12 CNVRs (Chr 1, 5, 7, 10 and 12) had significant association (FDR<0.05) with FCR and were constituted by gain and loss (mixed) of CNV. Fifty-one genes were found over these regions and the metabolic pathways found in KEGG were endocytosis, glycerophospholipid metabolism, insulin signaling pathway, histidine metabolism, olfactory transduction, oxidative phosphorylation and retinol metabolism. Our results showed that the CNV approach found distinct genomic regions, but the identified genes are part of the same biological processes, mainly lipid, protein and energetic metabolism. These findings contribute to the knowledge of the genetic basis of feed efficiency.

Dietary supplementation of Saccharomyces cerevisiae on production and health status in dairy cattle

S. Ghazanfar[1], M. Qubtia[2], F. Hassan[2], A. Muhammad[3], I. Ahmed[1] and M. Imran[2]
[1]National Agriculture Research Centre, National Agriculture Research CentrE, Park Road, Islamabad, Pakistan, 44000, Pakistan, [2]Department of Microbiology, Department of Microbiology, Faculty of Biological Sciences, Quaid-i-Azam University Islamabad-45320, 44000, Pakistan, [3]Livestock Research center, Livestock Research center, National Agricultural Research Centre, Park Road, Islamabad-45500., 44000, Pakistan; shakira_akmal@yahoo.com

Imbalanced diet is considered a major constraint for livestock promotion in developing countries. The uneven dietary pattern leads to poor growth and production efficiency. Probiotics have a beneficial effect on GIT health by stimulating the development of a healthy microbiota, and preventing enteric pathogens from colonizing the intestine, increasing digestive performance resulting in enhanced growth rate and milk yield. Many commercial yeast products available are imported and are not suitable for local breed. Therefore, this study is planned to compare laboratory and commercial yeast supplemented to dairy cattle. Nine lactating dairy cattle were randomly divided into three equal groups. In Group I, cows fed on 3 kg concentrate, 8 kg maize silage and 30 kg oats fodder per animal. In Group II, cows fed on control diet plus commercial (COM) yeast (Yac-Sac1026; 10 g/day/animal) corresponding to 2.5×10^7 cfu/g, while group III was fed control diet plus laboratory (LAB) yeast (8 g/day/animal; corresponding to 3.13×10^7 cfu/g) for 60 days. Results revealed that the LAB group produced more (P<0.05) milk with 18% high fat content than other groups. The nutrient detergent fibre and acid detergent fibre digestibility were significantly (P<0.05) better in LAB fed group. The outcomes of the ruminal gut microflora showed that the average, Lactococcus (cfu/g) counts were increased (P<0.05) while Enterococcus and coliform (cfu/g) counts were (P<0.05) decreased in LAB groups than other groups which leads to improved GIT microbial balance. It is concluded that LAB yeast improved the health and production performance of lactating dairy cattle. Locally isolated yeast strain may be adopted well in the cattle gut than exotic probiotics.

Accuracy of genomic breeding values from endocrine and traditional fertility traits in dairy cows

A.M.M. Tenghe[1,2], B. Berglund[3], R.F. Veerkamp[1,2] and D.J. De Koning[3]
[1]Wageningen UR Livestock Research, Animal Breeding and Genomics Centre, Droevendaalsesteeg 1, P.O. Box 338, 6700 AH Wageningen, the Netherlands, [2]Wageningen University, Animal Breeding and Genomics Centre, Droevendaalsesteeg 1, P.O. Box 338, 6700 AH Wageningen, the Netherlands, [3]Swedish University of Agricultural Sciences, Department of Animal Breeding and Genetics, Ulls väg 26, P.O. Box 7023, 750 07 Uppsala, Sweden; amabel.tenghe@wur.nl

Endocrine fertility traits defined from progesterone concentration levels in milk have been suggested as alternative indicators for fertility in dairy cows because they are less biased by farm management decisions and more directly reflect a cow's reproductive physiology than traditional fertility traits. With the aim of enabling the use of endocrine fertility traits in genomic selection, we evaluated the added value (accuracy) of using endocrine fertility traits in genomic prediction of fertility. Endocrine and traditional fertility records were available for 2,447 Holstein cows with 5,339 lactations from Ireland, the Netherlands, Sweden, and the United Kingdom. The endocrine traits were commencement of luteal activity (CLA) and proportion of samples in luteal activity (PLA), and the traditional trait was calving to first service (CFS). Genomic estimated breeding values (GEBV) were derived using genomic BLUP in univariate and bivariate analysis, with 85,485 single nucleotide polymorphisms. The accuracies of GEBV were evaluated by 5-fold cross-validation. Accuracies of GEBV ranged from 0.04 to 0.15 across all traits for univariate analysis, and 0.02 to 0.49 for bivariate analysis, indicating low to modest predictive ability. Improved accuracies of GEBV for CFS were achieved in bivariate analysis where endocrine and traditional fertility traits were used, and there was a better predictive ability of CFS in bivariate analysis with CLA than with PLA. This first study on genomic predictions for fertility using endocrine traits suggests some improvement over using only the traditional traits. Further studies with larger training populations may show bigger improvements.

Bone Alkaline Phosphatase as an indicator of phosphorus status in breeder cows

S.T. Anderson[1], R.M. Dixon[2], D.M. McNeill[3], J.G. Spiers[1], L. Castells[2], L.J. Kidd[3], K. Goodwin[4] and M.T. Fletcher[2]
[1]*University of Queensland, Biomedical Sciences, St Lucia, QLD 4072, Australia,* [2]*UQ, Queensland Alliance for Agriculture and Food Innovation, St Lucia, QLD 4072, Australia,* [3]*UQ, Veterinary Science, Gatton, QLD 4343, Australia,* [4]*Department of Agriculture and Fisheries, Brian Pastures Research Station, Gayndah QLD 4625, Australia; stephen.anderson@uq.edu.au*

Plasma inorganic phosphorus (PiP) concentration is useful in diagnosing P deficiency in cattle grazing rangelands with P deficient soils. Additional markers of bone metabolism might improve diagnosis of the P status of cattle. Bone alkaline phosphatase (BAP) is an enzyme secreted by osteoblasts. Whilst BAP increases during bone deposition in growing animals, in mature animals BAP concentrations are usually stable. We examined the effects of dietary P on BAP concentrations in mature breeder cows. Mature mid-pregnant Bos indicus cross cows (n=40) were housed in individual pens and for 13 weeks fed ad libitum P-deficient diets with either low or high metabolisable energy content (8.6 MJ or 9.7 MJ/kg DM; LE and HE) with/without dicalcium phosphate (0.08 or 0.28 g P/kg DM; HP and LP) in a 2×2 factorial design (HE-LP, HE-HP, LE-LP and LE-HP). Plasma BAP concentrations were measured by MicroVue EIA (Quidel). From a starting point of moderate P status, P deficient diets (LE-LP and HE-LP) reduced PiP to <1.0 mmol/l, whilst HP diets (LE-HP and HE-HP) increased PiP >2.0 mmol/l. Total calcium concentrations were higher (P<0.05) in LP than HP diets Over time plasma BAP concentrations increased (P<0.05) on LP diets and decreased on HP diets. On both LP and HP diets BAP also increased (P<0.05) with HE intake. The results show BAP exhibits an inverse relationship to dietary P intake and may be a sensitive marker of P status in mature cows. Increased BAP in a P deficient state is unexpected given it is a marker of bone deposition. However BAP acts to dephosphorylate organic phosphate esters and upregulation of BAP in P deficiency is a likely homeostatic response. Research supported by Meat and Livestock Australia and Queensland Department of Agriculture and Fisheries.

The impact of the rumen microbiome on the detailed milk fatty acid profile in Danish Holstein cattle

A.J. Buitenhuis[1], N.A. Poulsen[2], S.J. Noel[3], D.F. Plichta[4] and J. Lassen[1]
[1]*Aarhus University, Dept. of Molecular Biology and Genetics, Center for Quantitative Genetics and Genomics, Blichers Allee 20, 8830 Tjele, Denmark,* [2]*Aarhus University, Dept. of Food Science, Blichers Allee 20, 8830 Tjele, Denmark,* [3]*Aarhus University, Dept. of Animal Science, Blichers Allee 20, 8830 Tjele, Denmark,* [4]*Denmark Technical Univeristy, Center for Biological Sequence Analysis, Kemitorvet, Building 208, 2800 Lyngby, Denmark; bart.buitenhuis@mbg.au.dk*

The microbial community makes the rumen to function as a forestomach where the vegetable feed fiber is partially digested by the microbes before it is passed to the stomach. The volatile fatty acids which are formed by rumen fermentation serve as an important nutrient resource for the host and thereby could potentially influence the milk fatty acid composition. To investigate this relation a total of 310 milk samples and rumen samples were taken from Danish Holstein cows from three different herds. The detailed fatty acid profile of the milk samples was determined by gas chromatography covering C6:0, C8:0, C10:0, C12:0, C13:0, C14:0, C14:1, C15:0, C16:0, C16:1, C18:0, C18:1c9, C18:1t11, C18:2n6, C18:3n3, and C18:2c9t11 (CLA). The rumen samples were sequenced using the 16S approach. In total 8,515 bacterial and 168 archaeal OTUs (97% similarity) were detected. A genomic relationship matrix was generated based on the bovine 50K SNP array for all cows. Hierarchical clustering of the bacteria and archaea separately revealed a cluster of 402 bacterial OTUs and a cluster of 11 archaeal OTUs. The impact of these clusters on the individual fatty acids was evaluated by random regression models accounting for fixed effects of herd, days in milk and parity together with random effects of bacteria, animal and error. The archaea cluster only explained variance for C13:0 (0.41±0.35) and C15:0 (0.16±0.11), whereas the bacteria cluster explained a large part of the variance of C14:0, C18:1n11t, C18:1n9c, and C18:3n3 (≥0.90). These results demonstrate a strong link between specific fatty acids and archaea or bacteria in the rumen.

Impacts of energy level of the diet and nature of forage on milk spontaneous lipolysis in dairy cows

E. Vanbergue[1,2], J.L. Peyraud[1] and C. Hurtaud[1]
[1]INRA, Agrocampus Ouest, UMR PEGASE, Domaine de la prise, 35590 Saint Gilles, France, [2]Institut de l'élevage,
Service Productions Laitières, Monvoisin, Le Rheu, France; elise.vanbergue@rennes.inra.fr

Spontaneous lipolysis (SL) is one of the criteria of milk technological and organoleptic quality. Few studies have been done to investigate the impact of feeds on SL although milk quality is hardly related to feeding systems. This trial was aimed to evaluate the effects of the energy level of the diet and the nature of forage on SL. Thirty two cows were divided into 4 groups according their diet. Diets were based on maize silage or haylage and dried grass. Two levels of feed restriction were applied: 'non-restricted' which cows were fed at 100% of ad libitum dry matter intake (DMI) and 'restricted' which cows were fed at 75% of ad libitum DMI. The trial was conducted in a reverse design on the energy level of the diet. Cows were followed during 11 weeks: 3 control weeks and 6 experimental weeks divided in 2 periods. Milk samples were collected from morning and evening milkings at the end of each period for fat, protein, fat globule size and SL determination. Data were analyzed with SAS GLM procedure. Milk fat was lower when cows were fed with maize silage (3.53 vs 3.91%) and was not affected by feed restriction. Milk fat globule size was lower when cows were fed with maize silage (average diameter: 3.25 vs 3.55 µm) and was not affected by feed restriction. SL was sharply higher in milk from evening milkings compared to morning milkings (0.72 vs 0.29 mEq/ 100 g of fat). In morning milks, SL was higher when cows were fed with maize silage (0.40 vs 0.20 mEq/100 g of fat) and restricted (0.35 vs 0.26 mEq/100 g of fat). In evening milks, SL was higher when cows were restricted (0.88 vs 0.60 mEq/100 g of fat) and there was no significant effect of the nature of forage. We supposed that the mechanisms of SL are different in morning and evening milks. Higher impact of feed restriction on evening milks would be explained by higher energy requirements during the day. Further investigation will be done to clarify the mechanisms behind.

Effects of a by-pass Omega-3 source on milk yield, milk composition and fertility

A.J. Escribano[1], J.J. Mallo[1] and M. Quintela[2]
[1]NOREL S.A., Technical Department (Ruminants), C/Jesus Aprendiz, 19, 1 – A y B, 28007. Madrid, Spain,
[2]Progando, S.L., Technical and Quality Department, Polígono Industrial de Sabón, Parcela 7ª, 15142. Arteixo,
Spain; ajescribano@norel.net

Nowadays, consumers' demands towards healthier food are increasing. In this sense, milk products with a healthier fatty acid profile (PUFAs) are on the spotlight. Omega-3 fatty acids' popularity is remarkable due to their positive effects on health. From the sector side, there is a need to increase the low milk prices paid in origin as a consequence of the interactions among several facts: (1) the timid pruchses of extra milk by the industry after the end of EU milk quota; (2) the expected milk consumption by developing countries does not take off; (3) increase in production costs. Technically, the increase in production levels came along with reductions in fertility rates, which is basic for milk production. To help farmers to deal with this context, NOREL has developed a highly-energetic source of by-pass Omega-3, whose effects on milk production and quality were studied in comparison with extruded linseed (the main source of Omega-3 in Spanish dairy farms). 102 milking cows (one farm) of 33.94 (SD=9.27) liters/head/day average was selected. The study was divided in three periods: Control period 1 (the main source of Omega-3 was extruded linseed); Experimental period (the main source of Omega-3 was product under study); Control period 2 (the same as Control period 1). Both diets were isoenergetic and isocost. Control periods lasted 4 weeks, while experimental period length was 6 weeks. Results showed that the use of the product under study lead to an increase of 1.06 liters/cow/day (P<0.001), 0.12% milk fat (3.73 vs 3.61; P=0.001), and no protein dilution effect (+0.02%; P>0.05) during the experimental period. Moreover, the transfer rate of C18:3n3 from diet to milk was increased from 8.18% to 13.48% (which means a 164.42% increase). These results will allow achieving desired Omega-3 levels in milk products while increasing milk yield and milk fat, and in turn milk price. Thus, this product constitutes a solution for the sector and avoid the negative interactions between feeding PUFAs, healthy rumen environment and achieving the levels of milk fat required by the industry.

Modeling heterogeneous co-variances for genomic regions in prediction for milk protein compositions
G. Gebreyesus[1,2], M.S. Lund[2], L. Janss[2], H. Bovenhuis[1] and A.J. Buitenhuis[2]
[1]Wageningen University, Animal Breeding and Genomics Centre, P.O. Box 338, 6700 AH Wageningen, the Netherlands, [2]Aarhus University, Center for Quantitative Genetics and Genomics, Department of Molecular Biology and Genetics, Blichers Allé, 8830 Tjele, Denmark; grum.gebreyesus@mbg.au.dk

Milk protein compositions are scarcely recorded and have a low correlation with routinely collected traits such as milk protein yield, limiting the potential gain in reliability from multi-trait predictions. However, the co-variance between traits may not be consistent across the genome with some genomic regions explaining larger proportion of the covariance with the indicator trait than others. This study aimed to develop multivariate genomic prediction models taking into account heterogeneous co-variances across genomic regions. Studied traits included, protein percentage, αS1-CN, αS2-CN, β-CN, κ-CN, α-LA, β-LG, glycosylated κ-CN and αS1-CN-8P measured on 650 Holstein cows. De-regressed proofs were also available for milk protein yield from 5,200 bulls. Bayesian Factor models were developed assuming different genomic sections: each chromosome, 100 or 200 consecutive SNPs on a chromosome fitted as genomic sections. Co-variances between each milk protein and milk protein yield were estimated for each section instead of the total genome. Alternatively, different GBLUP models assuming uniform co-variance across the genome were run to compare the performance of the novel model with five-folds cross validation. The Bayesian factor models resulted in the highest prediction reliability for most proteins studied with reliabilities as high as 0.57 for β-LG, 0.67 for glycosylated κ-CN and 0.70 for κ-CN using 100 SNPs as genome sections. There was a gain in reliability of up to 50 percentage points for κ-CN and β-LG compared to the highest reliable of the whole-genome based GBLUP models. The results indicate that prediction for scarcely recorded traits might benefit enormously by taking into account heterogeneous co-variances per genomic regions in multi-trait predictions instead of the whole genome approach assuming uniform co-variance across the genome.

Partitioning of rumen-protected n-3 and n-6 fatty acids in bovine muscles with different metabolism
C. Wolf, K. Giller, S.E. Ulbrich, M. Kreuzer and J. Berard
ETH Zurich, Department of Environmental Systems Science, Universitätstrasse 2, 8092 Zurich, Switzerland; christina.wolf@usys.ethz.ch

Novel coating techniques to protect unsaturated fatty acids (FA) from microbial biohydrogenation now allow investigating the partitioning of n-3 and n-6 FA to different bovine muscles. In the present experiment, 18 Angus heifers received a standard diet composed of straw, hay, wheat, molasses and soybean meal, (ratio 40:6:10:7:7) for 8 weeks until slaughter which was supplemented daily with 0.45 kg coated lipids in 7.45 kg total feed dry matter. Half of the heifers received coated fish oil rich in n-3 FA (C3), the other half coated sunflower oil rich in n-6 FA (C6). The Extensor carpi radialis (ECR), Biceps femoris (BF) and Longissimus thoracis (LT), muscles characterised by very high, intermediate and low involvement in locomotion, were sampled. Physicochemical meat quality and profile of the intramuscular FA (IFA) were investigated. Data were analysed using the mixed model of SAS considering supplements, muscles and their interaction as fixed effects, and slaughter date as random effect. Supplement effects were mostly non-significant for muscle pH, colour, water-holding capacity, shear force and gross chemical composition. Although all muscles of the C3 heifers had a higher proportion (g/100 g total IFA) of n-3 IFA and a lower proportion of n-6 IFA compared to those of the C6 heifers (overall average n-3: 7.7 vs 4.4, SEM: 0.85, P<0.001; n-6: 7.5 vs 8.4, SEM: 0. 93, P<0.10), there were differences between the muscles in the level of FA and their relative incorporation rate. In general, ECR had a higher proportion of n-3 and n-6 IFA compared to BF and LT (n-3: 9.6, 4.8, 3.9, SEM: 0.85, P<0.001; n-6: 12.0, 6.9, 4.9, SEM: 0.96, P<0.001). In addition, interactions (P<0.01) of supplement and muscle were found for the fish oil specific very long-chain n-3 FA. In conclusion, the results show that individual n-3 FA are differently allocated to muscles. This suggests a specific need for these FA which may be connected to locomotion.

Potential use of MIR spectroscopy to assess the origin of milk produced in a PDO area

F.G. Colinet[1], J.A. Fernández Pierna[2], F. Dehareng[2], C. Bertozzi[3], D. Veselko[4], M. Sindic[1], M. Tielemenans[1] and N. Gengler[1]
[1]University of Liege, Gembloux Agro-Bio Tech, 5030, Gembloux, Belgium, [2]Walloon Agricultural Research Centre, 5030, Gembloux, Belgium, [3]Walloon Breeding Association, 5590, Ciney, Belgium, [4]Milk Committee, 4651, Battice, Belgium; frederic.colinet@ulg.ac.be

Several European local dairy products are currently labeled with the Protected Designation of Origin (PDO) trademark based on their production area. One of the conditions to obtain this designation is that the raw material (I.e. the milk) needs to be produced in the region delimited within the PDO's specifications. The aim of this study was to discriminate the milk produced in the PDO area 'Ardenne' from the milk produced in other parts (non-PDO) of the Walloon Region using mid infrared (MIR) spectroscopy as analytical tool. More than 1,800,000 standardized MIR spectra from the Walloon bulk milk spectral database generated during the process of milk quality analysis for payment between January 2012 and December 2015 were used. In order to simplify the discrimination analysis model on the whole dataset which required taking into account of several effect (e.g. seasons, months and years), 1,450 sub-datasets were generated using a 5 days' time window approach throughout the 4 years. In each sub-dataset, only one MIR spectrum per farm was selected randomly, the other spectra were discarded of the sub-dataset. Afterwards, a first derivative pretreatment of the spectrum was applied using Savitzky-Golay derivative with a window size of 5 wavelengths. Then, partial least square discrimination analysis was performed on PDO and non-PDO milk samples of each calibration sets. In average, a correct classification rate of 79% (±2%) was obtained. The average sensitivity and specificity were 77% (±2%) and 80% (±2%), respectively. Given the existing variability between farms (due in particular to the breeds and the feed), the obtained results were promising and highlighted the usefulness of MIR spectroscopy to assess the origin of milk produced in a specific PDO area.

Effect of regime on vitamin E, A and D in beef

S. Stewart, A. Fearon, C. McRoberts, L. Farmer, N. Gault, C. Ferris and F. Lively
Agrifood and Biosciences Institute, Food Research, Newforge Lane, Belfast BT9 5PX, United Kingdom; sharon.stewart@afbini.gov.uk

There is considerable interest in the contribution of diet to our health and reducing the risk of disease. Supplementation of food for human consumption has been suggested as a method of addressing dietary deficiencies. This is expensive and many surveys show that consumers prefer their food to be natural with few, if any, additives. Grass is a source of vitamin E and beta-carotene which is converted in the body into vitamin A (retinol). It is also a source of vitamin D2, but D3 is considered a more useful vitamin in the body and its uptake and use can be impaired by too much D2. In cattle UVB radiation in sunlight naturally converts 7-dehydrocholesterol in skin into pre-vitamin D3 and thence to cholecalciferol (vitamin D3). These vitamins can be transferred into milk and meat. Because humans consume a large amount of both meat and milk the small amount of D3 present becomes a useful source of this vitamin. However little information is available on vitamin content, particularly D3, in NI meat and milk although it is known that any cattle that are housed on no concentrates use up D3 reserves by late winter. Meat samples were collected from producers that had been farmed in a range of conditions (housed, outdoor, grass fed, concentrate fed and a mix). Levels of vitamins E, A, and D3 were determined using solvent extraction followed by HPLC. Significant differences were observed in the vitamin levels in meat from animals which were reared indoors and outdoors, or on grass versus concentrates. Vitamin A is significantly higher for animals that are indoors, and again for indoor fed concentrate versus no concentrate. Significant differences were also observed in grass versus silage. This research also suggests that pasture was comparable to concentrates which may be assumed to be supplemented. To derive marketing benefits from optimal red meat vitamin content processors will need to identify diet and production practices for producers to employ in order to consistently achieve the requisite vitamin profile.

A meta-analysis of the effect of the diet fed to Holstein cows on their milk calcium content

A. Boudon[1,2], T. Havard[1,2], M. Gelé[3], B. Rouillé[3] and C. Hurtaud[1,2]
[1]Agrocampus Ouest, UMR 1348 PEGASE, Rue de St-Brieuc, 35000 Rennes, France, [2]INRA, UM 1348 PEGASE (Physiologie Environnement et Génétique pour l'Animal et le Système d'Elevage), Domaine de la Prise, 35590 Saint-Gilles, France, [3]Idèle, Service Productions Laitières, 149 rue de Bercy, 75595 Paris cedex 12, France; anne.boudon@rennes.inra.fr

Milk Ca is an important determinant of milk coagulation and cheese-making abilities. Milk and dairy products in westernized diets also represent an optimum source of Ca. This meta-analysis aimed to determine the effect of the diet fed to the cows on the calcium content of their milk. The 611 measurements of milk calcium contents and diet composition, obtained during 31 experiments realized at the INRA experimental farm of Méjussaume, were gathered in a data basis. Each data represents an average of measurements obtained on a same individual during a period of 3 to 5 days. The experiments aimed to determine whether milk quality was affected by the nature of the forage (148 data), the nature and the amount of dietary N (108 data) or the nature and the amount of dietary energy (223 data). Milk calcium content was on average 1.19 g/kg (±0.13). It was not affected by the dietary calcium balance, i.e. the difference between calcium intake and calcium requirement, as long as the balance remained lower than 20 g/d. Above this limit, milk calcium contents declined with dietary calcium balance. In experiments comparing natures of forage, total, soluble and colloidal calcium contents of milk were lower when diets were based on pasture rather than maize silage while milk calcium contents obtained with diets based on grass silage or hay were intermediary. The dietary calcium and phosphorus balances were also higher with the diet based on pasture. Both the dietary PDI and net energy balance had slight positive effect on colloidal calcium content of milk (P<0.05) but they did not affect soluble or total calcium content. It could not be concluded from this study if the lower milk calcium content of grazing cows was explained by high dietary calcium and phosphorus balances or by other intrinsic differences between maize silage and grazed pasture.

The challenge of detecting cheese-damaging clostridia in milk

J. Brändle, K.J. Domig and W. Kneifel
BOKU-University of Natural Resources and Life Sciences, Institute of Food Science, Department of Food Science and Technology, Muthgasse 18, 1190, Austria; johanna.braendle@boku.ac.at

Butyric acid producing clostridia are known to cause late-blowing, a severe quality defect in hard and semi-hard cheese. In the course of the cheese ripening process, clostridia produce tremendous amounts of gas and butyric acid, which causes rancid off-flavours and cheese blowing defects such as slits, cracks and irregular eyes. Bacterial spores enter the raw milk during milking and due to their high resistance to environmental stress, they are able to survive the cheese production process. As options to eliminate or reduce clostridial spores from milk are limited, the selection of suitable cheese milk prior to production is crucial for cheese producers. However, it remains difficult to determine the clostridial spore count in milk because of the lack of an international standard method. Routinely applied methods differ among countries, regions and even laboratories and detailed performance data on real milk samples are scarce. In order to evaluate the most suitable method, we selected the four most frequently used applications for the quantitative determination of clostridia in milk (Bryant-and Burkey-, RCM-, modified RCM- and NIZO method in MPN procedures) and analysed various milk samples from Austrian dairies in parallel. A subset of positive tubes was spread onto a non-selective medium and the isolates were identified by 16s rDNA sequencing to obtain information about method selectivity. We concluded that the detected spore numbers are extremely method-specific and that none of the hitherto applied procedures provides the required selectivity to detect only relevant cheese-damaging clostridia. These results underline the need for an innovative method to detect clostridial spores in milk.

Assessment of passive immunity in Irish dairy and suckler beef calves

C.G. Todd[1], K. Tiernan[1,2], M. McGee[3], P. Crosson[3], I. Lorenz[2] and B. Earley[1]
[1]Teagasc, Animal and Bioscience Research Department, AGRIC, Grange, Dunsany, Co. Meath, Ireland, [2]University College Dublin, School of Veterinary Medicine, Belfield, Dublin 4, Co. Dublin, Ireland, [3]Teagasc, Livestock Systems Research Department, AGRIC, Grange, Dunsany, Co. Meath, Ireland; cynthia.todd@teagasc.ie

Transfer of passive immunity occurs when a newborn calf absorbs colostral immunoglobulins into circulation. Calves that have failure of passive transfer are at greater risk of morbidity and mortality. The study objective was to assess passive immunity in Irish dairy and suckler beef calves. Dairy (n=84) and suckler beef (n=111) farms throughout Ireland were visited. A total of 1,040 dairy and 923 suckler calves between 1 and 21 days of age were blood sampled by jugular venipuncture. The following indirect test methods were used to assess passive immunity: Zinc Sulphate Turbidity (ZST) test, total protein (TP) concentration, globulin (GLOB) level, and gamma-glutamyl transferase (GGT) activity. Cut-point values from the literature were applied to categorize the results; ZST: Low (L) $\leq$ 10, Medium (M) = 10-20 and High (H) $\geq$20 units, TP: Failure (F) <52 and Adequate (A) $\geq$52 g/l, GLOB: F<16 and A$\geq$16 g/l, and GGT: F<50 and A$\geq$50 U/l. Mean ZST units, TP concentration, GLOB level and GGT activity for dairy calves were 18.1 (SD=7.9), 63.0 (SD=8.4), 35.4 (SD=9.3) and 352.8 (385.3), respectively. Mean ZST units, TP concentration, GLOB level and GGT activity for suckler calves were 16.1 (SD=7.4), 60.6 (SD=8.0), 33.1 (SD=8.9) and 226.7 (295.0), respectively. Dairy calves were less likely to have ZST results in the lower categories than suckler calves (L: 13.9 vs 21.1%, M: 50.3 vs 51.0%, H: 35.9 vs 27.8%, P<0.01). Similarly, dairy calves were less likely than suckler calves to be classified in the failure category for TP concentration (10.3 vs 14.4%, P<0.05) and GLOB level (0.2 vs 1.0%, P<0.05). Dairy and suckler calves did not differ for GGT activity (P=0.31). These results suggest that Irish suckler calves are at greater risk for failure of passive transfer than dairy calves. Research is on-going to elucidate which indirect test method is most robust for the assessment of passive immune status in calves.
Acknowledgement: We acknowledge the DAFM Stimulus Fund (11/S/131) for financial support.

Effect of production area, season and farm dimension on milk production and quality

C. Lazzaroni and D. Biagini
University of Torino, Department of Agricultural, Forest and Food Sciences, Largo P. Braccini 2, 10095 Grugliasco, Italy; carla.lazzaroni@unito.it

Milk amount and quality are the parameters taken into account for the payment, but both are affected by several factors that can impact heavily on the profitability of the production. Knowledge on the effects of these factors may allow farmers to keep them under control and thereby reduce the negative effects or exploit the positive ones, always taking into account the cost-benefit relation. This is particularly important in areas where there is the higher milk production, as in the Po Valley where more than 75% of Italian milk is produced. To verify the situation in two areas of North-West of Italy, a study was carried on dairy farms located in zones characterised by different land morphology (10 located in the irrigated plan, and 14 located on the not irrigated hill), to point out differences in milk production. Effects of production area, production season and month (over the whole year), farm dimension (small: <50 cows, as average; medium: 50-100 cows; large: >100 cows), and farm management were analysed by ANOVA on milk production (kg/d/head), content of fat (%), crude protein (%), lactose (%), and somatic cells count (n×1000). Differences in the productive parameters were found for all studied effects: for production area (higher milk production and quality in farms on the plain: 30.4 vs 24.9 kg/d/head, P<0.000; 3.79 vs 3.68% fat, P<0.01; 3.35 vs 3.31% crude protein, P<0.05; 4.86 vs 4.76% lactose, P<0.000; 230.75 vs 318.42 n×1000 somatic cells count, P<0.000), for production season and month (higher quality in winter than in summer: 3.82 vs 3.67% fat, P<0.05; 3.39 vs 3.26% crude protein, P<0.000; 246.02 vs 342.06 n×1000 somatic cells count, P<0.05), and for farm dimension (higher milk production and quality, except for crude protein, in larger farms: 30.82 kg/d/head, P<0.000; 4.85% lactose, P<0.000; 253.47 n×1000 somatic cells count, P<0.001).

Effect of breed, housing, milking system, farm dimension and season on milk production and quality
C. Lazzaroni and D. Biagini
University of Torino, Department of Agricultural, Forest and Food Sciences, Largo P. Braccini 2, 10095 Grugliasco,
Italy; carla.lazzaroni@unito.it

Milk amount and quality are the parameters taken into account for the payment, but both are affected by several factors that can impact heavily on the profitability of the production. Knowledge on the effects of these factors may allow farmers to keep them under control and thereby reduce the negative effects or exploit the positive ones, always taking into account the cost-benefit relation. To verify the situation in the North-West of Italy, a study was carried on 200 dairy farms located in the Saluzzo's plain, to point out differences in milk production. Effects of breed (Holstein, Piemontese, crossbred Piemontese × Friesan, mixed), housing systems (tied stall or loose system), milking systems (hand or machine), farm dimension (small: <50 cows, as average; medium: 50-100 cows; large: >100 cows), production season and month (over the whole year) were analysed by ANOVA on milk production (t/head/year), content of fat (%), crude protein (%), somatic cells count (n×1000), and bacterial count (cfu). Differences in the productive parameters were found for all studied effects: for breed (higher milk production in Holstein, followed by mixed breed, crossbred Piemontese × Holstein, and Piemontese cows: 6.0 vs 3.5 vs 2.7 vs 1.4 t/head/year, P<0.000), for housing systems (higher milk production in loose system: 7.5 vs 3.1 t/head/year, P<0.000; but lower quality: 3.7 vs 3.8% fat, P<0.001, and 3.2 vs 3.3% crude protein, P<0.05), for milking systems (higher milk production with milking machine: 3.9 vs 1.8 t/head/year, P<0.000), for farm dimension (higher milk production in larger farms: 4.5 t/head/year, P<0.000; with higher fat 3.6%, P<0.000; but also higher somatic cells count 577 0 n×1000, P<0.05), for production season and month (higher quality in autumn than in summer: 3.9 vs 3.7% fat, P<0.000; 3.4 vs 3.2% crude protein, P<0.000; but not for somatic cells and bacterial count, P<0.000).

Individual dairy curd production and association analysis with 14 candidate genes in Italian buffalo
G. Cosenza[1], L. Zicarelli[2], N.P.P. Macciotta[3], L. Ramunno[1] and A. Pauciullo[4]
[1]University of Naples, Via Università, 100, 80055 Portici (NA), Italy, [2]University of Naples, Via F. Delpino, 1,
80137 Napoli, Italy, [3]University of Sassari, Viale Italia, 39, 07100 Sassari, Italy, [4]University of Turin, Largo P.
Braccini, 2, 10095 Grugliasco (TO), Italy; alfredo.pauciullo@unito.it

The genetic merit of Italian buffaloes is mainly related to the ability of producing mozzarella cheese. This trait is measured by the estimated mozzarella production (PKM index) according to Altiero et al. Several studies have shown that PKM based only on milk yield, fat (FP) and protein percentage (PP) may over- or under-estimate the real mozzarella yield. In this study, the individual curd production (ICP) is proposed as an index of the mozzarella making ability. Values were compared with estimated PKM. Moreover, associations between ICP and 14 loci related to milk traits (ACACA, BLG, CSN1S1, CSN1S2, CSN2, CSN3, LEP, LPL, LTF, OXT, OXTR, PRL, PRLR and SCD) were tested. Individual milk samples of 233 Italian buffaloes were taken monthly along the whole lactation. A total of 36 variables were measured. Among them: FP, PP and lactose percentage, milk coagulation properties (r, clotting time; k_{20}, curd firming time; a_{30}, curd firmness at 30 min), pH, ICP at 24 h (ICP_{24}), etc. Cows were genotyped for published and new SNPs found in the explored genes. Associations between phenotypes and candidate gene polymorphisms were tested using a mixed linear model that included fixed effects of the genotype, farm, calving season, days in milk, parity, date of the test, and the random effect of the animal. The association study showed that 2 out of 14 SNPs (OXTR g.129C>T and SCD g.-461A>C) were associated with ICP_{24} and fat percentage. The CC vs TT genotype at OXTR showed an increase of 18.08 g/kg milk (P=0.044) of ICP_{24} and +1.01 FP (P=0.001). AC genotype at SCD showed an over-dominance effect, with a higher increase vs AA of 13.15 g/kg milk (P=0.015) for ICP_{24} and +0.54 FP (P=0.010). The estimated/real PKM ratio showed that the actual PKM underestimates the prediction of mozzarella yield (on average 0.929). These data confirm the findings of Parlato & Zicarelli on the necessity of a new prediction model, and offer useful indication for the application of MAS programs in buffalo.

Factors affecting the freezing point of milk of Polish Holstein-Friesian cows
A. Otwinowska-Mindur and E. Ptak
University of Agriculture, Department of Genetics and Animal Breeding, al. Mickiewicza 24/28, 30-059 Kraków,
Poland; rzmindur@cyf-kr.edu.pl

The objective of this study was to estimate the influence of lactation number, month of milk sample, lactation stage and herd size on the freezing point of milk of Polish Holstein-Friesian cows. Data comprised 4,719,787 milk samples from the first seven lactations of 752,770 Polish Holstein-Friesian cows. The data were made available by the Polish Federation of Cattle Breeders and Dairy Farmers. Milk freezing point (FP), and milk, fat and protein yields were analyzed. Cows calved in 20,057 herds, in 2013 and 2014. Depending on the number of cows, herds were divided into 6 herd-size classes: up to 9, 10–15, 16–25, 26–50, 51–150, and more than 150 cows. Ten lactation stages were created as 30-day intervals throughout the lactation. The fourth to seventh lactations were treated as one 4+ class of lactations. Four-factor analysis of variance using the GLM procedure in SAS was applied to each trait (milk freezing point, and milk, fat and protein yields). The mean FP of milk samples (0.5326 °C) as well as more than 92% of all milk samples did not exceed the maximum limit for the freezing point of cow's raw milk, which, following Polish standards, was taken to be 0.52 °C. Low and negative correlations between milk FP and milk, fat and protein yields within lactations were found. The lowest correlations within lactations were between FP and milk yield (-0.067 to -0.021) and the highest between FP and protein yield (-0.271 to -0.191). The freezing point of milk samples was highly significantly (P<0.001) affected by all examined factors: month of sample, successive lactations, stage of lactations and herd-size. FP was lowest for milk samples taken from January to March, and highest for samples from the end of the year. Milk FP increased with lactation number. Mean FP decreased with time within lactation, except in the first 30 days (first stage). Generally, FP was highest in small herds (up to 9 cows) and lowest in large herds (more than 150 cows).

Data sharing to support breeding decisions in a segmented beef production chain
F. Maroto-Molina[1], A. Gómez-Cabrera[1], J.E. Guerrero-Ginel[1], A. Garrido-Varo[1], J.A. Adame-Siles[1], R. Santos-Alcudia[2] and D.C. Pérez-Marín[1]
[1]*University of Cordoba, Department of Animal Production, Campus de Rabanales, Ctra. Madrid-Cadiz km 396, 14071, Cordoba, Spain,* [2]*COVAP, Ctra. Industrial Dehesa Boyal s/n, 14400, Pozoblanco, Spain; g02mamof@uco.es*

Beef production is conducted through distinct production segments (farms, feedlots, etc.) that may be characterized by differences in management practices, location and ownership. This is a challenge for the efficiency of the production chain. Most farms are cow-calf farms that produce calves to be sold to feedlots. Economic return of each animal depends on feedlot performance and carcass quality and farmers can affect them through breeding decisions, but they do not have the information. This study used a database containing full records of 384 calves to explore the opportunities and challenges of sharing data about feedlot and slaughterhouse performance (daily gain, dressing percentage, conformation and fat cover scores) with farmers in order to support breeding decisions. Real-world data present some drawbacks, especially that variability factors are not as controlled as they are in experiments. A correlation between average daily gain (ADG) and slaughter weight and significant differences between sexes and between animals entering the feedlot in the first and in the second half of the year have been found. This information was used to standardize individual records and relative indexes were calculated by dividing the performance value of each animal by the average data of similar animals. Small differences were observed among breeds. Limousin purebred calves stand in carcass conformation (6% above average) and Berrenda crossbred in fat cover (8%). Differences among cows were greater than among breeds (20% for ADG, 18% for conformation and 13% for fat cover), which is coherent with the strategy of using individual data in order to support breeding decisions. The data management practices designed, allow the support of breeding decisions at farm level, although there is a lack of data of high impact in the performance, e.g. feed intake. Besides, there are other important data, e.g. cow fertility, which must be obtained at farm level.

Evolution of milk composition in cow's milk during the time course of milking
C. Hurtaud[1,2], E. Vanbergue[1,2,3], C. Cirot[1,2,3], L. Hervé[1,2] and M. Boutinaud[1,2]
[1]Agrocampus Ouest, UMR 1348 PEGASE, 65 rue de Saint-Brieuc, 35000 Rennes, France, [2]INRA, UMR 1348 PEGASE, Domaine de la Prise, 35590 Saint-Gilles, France, [3]Institut de l'Elevage, Monvoisin, 35650 Le Rheu, France; catherine.hurtaud@rennes.inra.fr

Our objective was to study the evolution of milk composition during the time course of milking. We used 9 dairy cows averaging 56±6 days in milk. Yield, composition (fat, protein, calcium) and free fatty acids (determined by copper soap method) were measured on milk samples collected every min during morning milking (on one day per dairy cow). Statistical analysis was carried out using the Proc Mixed procedure in SAS software with repeated statement. Milk fat content increased during milking (from 6.9 to 62.6 g/kg). Lactose and protein contents slightly but significantly decreased during milking (from 51.8 to 47.7 g/kg and from 28.1 to 26.4 g/kg respectively). Milk free fatty acids (Meq/100 g fat) measured after milking decreased during the first 3 minutes and after reached a plateau. Milk free fatty acids measured after 24 hours of storage at 4 °C followed the same trend. Milk lipolysis remained unchanged during milking in our trial. Milk fat globule diameter measured as $d_{4,3}$ increased at the beginning of milking, but reached a plateau very quickly. Measured as $d_{3,2}$, it also increased at the beginning of milking and decreased at the end of milking that could suggest a reduction in milk fat globule membrane availability towards the end of milking. Milk calcium decreased during milking (from 1,205 to 1,147 mg/kg). Intra animal, there was a strong curvilinear relationship between milk fat content and milk fat globule diameter, $d_{4,3}$ (R^2 from 0.67 to 0.98), a positive linear relationship between milk protein content and milk calcium and lactose contents (R^2 from 0.32 to 0.95 and from 0.50 to 0.99, respectively) and a negative linear relationship between milk fat globule diameter and milk free fatty acids measured after 24 hours of storage at 4 °C (R^2 from 0.13 to 0.97). In conclusion, there was a great change in milk composition during milking and a complete milking is necessary to obtain a representative picture of the secreted milk.

Effect of Spirulina supplementation in heat stressed dairy cows' ration on milk fatty acid profile
M.A. Karatzia, M. Ioannidou, G. Samouris and E.N. Sossidou
Hellenic Agricultural Organization-DEMETER, Veterinary Research Institute, Ktima Thermis, 57001, Thessaloniki, Greece; mkaratz@vet.auth.gr

Heat stress conditions are directly associated with a decline of overall production performance and specifically milk yield in dairy cows. Additionally, exposure to heat stress strongly modifies milk quality, altering its fatty acid profile. Different strategies regarding nutritional manipulations are proposed in order to ameliorate the negative effects of heat stress on production. This study attempts to estimate the effect of Spirulina sp. – a microalga, rich in proteins, vitamins and polyunsaturated fatty acids- supplementation in dairy cows' ration, during heat stress, on milk fatty acid profile. The trial was undertaken in a commercial farm in Northern Greece. Twelve heat stressed (average T.H.I.=73,86±0,042) Holstein cows in early lactation were assigned to 2 groups (Controls-C, n=6 and Spirulina-S, n=6). Controls were fed a standard ration and S group consumed an extra 100 g of Spirulina powder, daily for 61 days during June and July 2015. Milk was sampled on days 0, 31 and 61 and fatty acid (FA) composition was determined by gas chromatography. Saturated FA (SFA) (C6:0-C15:0) in the controls increased significantly (P≤0.05) between samplings. On the contrary, in S group, unsaturated FA C16:1 and C18:1 increased significantly (P≤0.05) between samplings. Moreover, polyunsaturated FA:SFA ratio increased significantly (P≤0.05) between 1st and 3rd sampling (1st:0.038±0.004 to 3rd:0.072±0.016) in S group. Spirulina powder supplementation enhances heat stressed cow milk with health-associated unsaturated FA, an observation that requires further investigation. This research project is funded under the Action 'Research & Technology Development Innovation Projects'-AgroETAK, MIS 453350, in the framework of the Operational Program 'Human Resources Development'. It is co-funded by the European Social Fund through the National Strategic Reference Framework (Research Funding Program 2007-2013) coordinated by the Hellenic Agricultural Organization-DEMETER.

Survey of mycotoxins in grain, feed and diary products in Croatia
S. Zjalic[1], V. Brlek[2], I. Sabljak[2], M. Grubelic[2], L. Coso[1,3] and M. Matek Saric[1]
[1]University of Zadar, Mihovila Pavlinovica 1, 23 000 Zadar, Croatia, [2]Croatiakontrola, Karlovacka cesta 4L, 10000 Zagreb, Croatia, [3]Croatia Control Ltd., Rudolfa Fizira 2, 10150 Zagreb Airport, Croatia; szjalic@unizd.hr

Mycotoxins, secondary metabolites of some fungi, are unavoidable contaminants of food and feed stuff. Due to their toxic effects the concentration of different mycotixns in food and feed commodities are limited by low regulations in almost all the countries in the world. The influence of climatic conditions on presence of mycotoxins in food and feed stuff is well known and documented. In this work we monitored the presence of fusarium toxins (fumonisins, DON, zearalenon) and aflatoxins on corn, corn based feed and diary products in Croatia in last 5 years and compared them with climate data (humidity, rain and temperature). More than 100 samples of grains produced in central Croatian region of Slavonia, cattle feed and 1000 samples of milk were analysed per year. Our data shows that the major control measures and better agriculture practice, adopted as a consequence of the major consciousness of the problem, after the outbreak of aflatoxin contamination in Croatian corn and milk in 2012 gave some results. Even if in certain years the climatic conditions were favourable for mycotoxin contamination, their concentration in feed and diary products were always in the limits imposed by EU regulation.

Effect of dietary olive pomace supplementation on cow milk quality
F. Castellani, S. Marchetti, L. Grotta, F. Palazzo, E. Marone and G. Martino
University of Teramo, Faculty of Bioscience and Technology for Food, Agriculture and Environment, C. Lerici 1, 64023 Mosciano Sant'angelo (TE), Italy; fpalazzo@unite.it

Olive oil industry by-products are hardly biodegradable for the presence of antimicrobial and phytotoxic organic substances and their disposal represents an environmental problem with non-negligible costs, so they can be used in livestock sector as supplements or substitutes of conventional feed. The study aimed to evaluate the effect of dietary olive pomace supplementation on cow milk, in order to identify an alternative solution that might alleviate the environmental pollution caused by this by-product and to improve the dietetic-nutritional characteristics of the milk. The study was conducted using 24 omogeneous Holstein Friesian dairy cows, according to age, number of calves and lactation period, divided into two groups (control and experimental). During the 50-days-trial, both groups were fed with the same diet while experimental group received 2 kg DM of olive pomace/cow/die. Milk yield, milk chemical composition (protein, casein, lactose, lipids, fatty acids profile, urea and somatic cells), milk coagulation properties (rennet coagulation time: RCT; curd firming time: k_{20}; curd firmness: a_{30}) and olive pomace and milk total phenols content were determined. Statistical analysis was carried out using GLM procedure of SAS. Milk yield was monitored during the trial and it did not differ between two groups. Oleic acid ($P<0.01$) and MUFA ($P<0.01$) were higher in experimental milk while no differences were found for the other chemical parameters. RCT ($P<0.01$) and k_{20} ($P<0.01$) were lower in experimental milk compared with the control one while there were no differences for curd firmness between two groups. Phenols were found in olive pomace but they did not switch in milk: total phenols content was similar in control and experimental groups. Results indicate that dietary olive pomace supplementation influenced the content of oleic acid and MUFA and also the cheese-making ability.

Variability of MIR predicted cheese yield in Walloon Region of Belgium

F.G. Colinet[1], F. Dehareng[2], C. Bertozzi[3], D. Veselko[4], T. Troch[1], M. Sindic[1] and N. Gengler[1]
[1]University of Liege, Gembloux Agro-Bio Tech, 5030, Gembloux, Belgium, [2]Walloon Agricultural Research Centre, 5030, Gembloux, Belgium, [3]Walloon Breeding Association, 5590, Ciney, Belgium, [4]Milk Committee, 4651, Battice, Belgium; frederic.colinet@ulg.ac.be

Economically, cheese yield is very important. At the level of cheese manufacturers, an improvement of cheese yield by 1% increases the benefit without increasing milk consumption. Recently, methods to predict cheese yield based on the mid infrared (MIR) spectrum were described. Based on individual bovine milk samples collected in Walloon Region during 4 years, an equation to predict cheese yield, expressed as g of coagulum per 100 g of processed milk, was developed. This equation was applied on more than 850,000 standardized MIR spectra from the Walloon bulk milk spectral database generated during the process of milk quality analysis for payment between January 2014 and December 2015. The data were analyzed using mixed model procedures in order to study the variability of predicted cheese yield at the level of the Walloon Region. As expected, first results highlighted the effect of the seasons, with a difference of about 3% between least square means of summer and winter. Independently of the season effect, the provincial origin of the milk samples had a highly significant effect. Differences between provincial origins could be related to the main feed managements practised in each Walloon Province, e.g. in the Provinces of Liege and Luxembourg, cattle have more often and longer access to the pasture than in the other Walloon provinces. Furthermore, maize silage is incorporated more often and with a higher level in the ration in the Province of Hainaut. This first investigation highlighted the usefulness of MIR spectroscopy to predict cheese yield from milk bulk sample, which is of great interest to all cheese manufacturers. However, further researches on the variability at the Walloon level will be conducted applying innovative modeling approaches.

Effects of intensification of grazing systems on meat quality of Nelore steers in Brazil

R.R.S. Corte[1], S.L. Silva[1], A.F. Pedroso[2], R.T. Nassu[2], R.R. Tullio[2], A. Berndt[2], L.S. Sakamoto[1], P.H.M. Rodrigues[1] and P.P.A. Oliveira[2]
[1]University of Sao Paulo, Duque de Caxias Norte, 225, Pirassununga, Brazil, [2]Embrapa Southeast Livestock, São Carlos, SP, Brazil; rscorte@usp.br

The intensification of livestock production in tropical grazing areas should be based on the best use of the potential of pasture growth with an environmental sustainability focus. To assess the effects of intensification of different grazing systems on the meat quality of Nelore steers (n=24;271±2.2 kg of live weight – LW;15 months old) were allotted, for two consecutive years, to four grazing systems with 2 area replications (block): (1) DP: degraded pasture (1.1 AU/ha; *Brachiaria decumbens*); (2) IHS: irrigated with high stocking rate (5.9 AU/ha; *Panicum maximum*); (3) DHS: dryland with high stocking rate (4.9 AU/ha; Panicum maximum); (4) DMS: dryland with moderate stocking rate (3.4 AU/ha; *Brachiaria brizantha*). With exception of the degrade pasture, all pastures were manage in a rotational grazing system. Each paddock was grazed by three steers as testers and regulating animals were used to adjust sward heights. Data were analyzed as completely randomized block design using PROC MIXED and the year was used as a random effect. Animals were slaughtered with approximately 450 kg LW The left half-carcass was cut between the 12th and 13th rib and 2.5 cm steaks were removed for meat quality (shear force, water holding capacity, pH, cooking losses, and objective color) analyses. The grazing systems tested produced meat with similar quality: shear force (DP=7.34, IHS=7.51, DHS=7.11, DMS=7.16 kgfcm^{-2}), water holding capacity (DP=77.97, IHS=78.62, DHS=76.88, DMS=77.88%), pH (DP=5.46, IHS=5.41, DHS=5.42, DMS=5.43) cooking losses (DP=25.67, IHS=26.30, DHS=26.86, DMS=23.72%) and color, DP (L*=35.49, a*=14.55, b*=11.37) IHS (L*=35.03, a*=15.4, b*=11.93) DHS (L*=36.2, a*=15.81, b*=12.4) and DMS (L*=34.38, a*=14.76, b*=11.26). Results indicated no adverse effects of intensification of grazing systems on meat quality. Although, all grazing systems resulted in acceptable meat quality composition, future methane emissions and performance results of this study must also be used to determine the most efficient grazing system for beef cattle in Brazil.

Potential of milk MIR spectra to develop new health phenotypes for dairy cows in the GplusE project

A. Vanlierde[1], C. Grelet[1], N. Gengler[2], C. Ferris[3], M.T. Sorensen[4], J. Höglund[4], F. Carter[5], A. Santoro[5], K. Hermans[6], M. Hostens[6], P. Dardenne[1] and F. Dehareng[1]
[1]CRA-W, 5030, Gembloux, Belgium, [2]ULg-GxABT, 5030, Gembloux, Belgium, [3]AFBI, BT9 5PX, Belfast, United Kingdom, [4]AU, 8000, Aarhus, Denmark, [5]UCD, Belfield, Dublin, Ireland, [6]UGent, 9000, Gent, Belgium; a.vanlierde@cra.wallonie.be

Animal production systems, including dairying, must become more efficient. This is addressed within the 'Genotype plus Environment' project (GplusE), an objective of which is to record novel cow phenotypes and to develop predictors for these novel phenotypes using milk mid infrared (MIR) spectra. MIR allows the 'status' of cows to be predicted using a rapid, cost effective, routine process. Data (liveweight, body condition score, uterine health, residual feed intake, lameness) and samples (milk, blood, liver, feed) were collected from 135 dairy cows on 3 European farms (AFBI-UK, UCD-IRL, AU-DK) from calving until day 49 post calving. Classification models have been developed using PLSDA technique to link phenotypes of interest and potential biomarkers to the MIR spectra of milk. For example, a model has been developed using 60 observations which allows us to distinguish animals with or without lameness with a predicted classification of 68 and 71% respectively. Regression models have also been developed which allow us to predict biomarkers using milk MIR spectra. Some of these can be on the basis of threshold levels (eg. milk NAGase, which is associated with udder inflammation), while others can be predicted quantitatively (eg. IGF1, which is linked to uterine health). This database will allow us to develop additionnal tools to predict new health indicators from milk MIR spectra, that can be easily implemented at a large scale. Predictions will be validated through new data collected with the same protocol from 3 other European university farms.

Potential for novel metabolite measurements in milk as biomarker phenotypes for dairy traits

J. Höglund[1], M.T. Sorensen[1], T. Larsen[1], L. Foldager[1], C. Ferris[2], M. Bell[2], F. Carter[3], A. Santoro[3], M. Crowe[3] and K.L. Ingvartsen[1]
[1]Aarhus University, Animal Science, AU-Foulum, 8830 Tjele, Denmark, [2]Agri-Food and Biosciences Institute, Hillsborough, BT26 6DR, United Kingdom, [3]University College Dublin, School of Veterinary Medicine, Belfield, Dublin 4, Ireland; kli@anis.au.dk

In the EU-project 'Genotype plus Environment' (GplusE), a major objective is to develop predictors for health phenotypes which can be used to help make farm management decisions, and in breeding programmes. The predictors are based on information derived from milk, namely the content of metabolites and enzymes, mid infrared (MIR) spectra and content of glycans. This abstract describes work with milk metabolites and enzymes. The health phenotypes are based on data describing milk production, feed intake, live weight, body condition score, uterine health, lameness, and disease diagnoses and treatments. In addition, residual feed intake (RFI) was calculated from milk production and feed intake data. The data were collected from 135 dairy cows on 3 European university/research farms (AFBI-UK, UCD-IRL, AU-DK), from calving until 50 days in milk (DIM). Seven different diets were offered across the three farms. An analysis was undertaken to examine the impact of including milk metabolites (beta-hydroxy butyrate, isocitrate, free glucose, glucose-6-phosphate, uric acid and urea) and enzymes (N-acetyl-beta-D-glucosaminidase (NAGase) and lactate dehydrogenase) into a general model describing the variance of phenotypes such as RFI and milk somatic cell count (SCC). The general mixed effects model included diet, parity (first, second, ≥third), diet × parity, DIM, DIM^2, parity × DIM, parity × DIM^2 as fixed effects. An initial analysis has shown that from the 8 metabolites and enzymes examined, beta-hydroxy butyrate, isocitrate and free glucose possess the highest descriptive potential with respect to RFI, while NAGase, lactate dehydrogenase and free glucose possess the highest descriptive potential with regard to SCC.

Potential for novel glycan measurements in milk as biomarker phenotypes for dairy traits

A. Santoro[1], J. Vandepitte[2], M. Hostens[2], F. Carter[1], E. Matthews[1], W. Waegeman[2], A.G. Fahey[1], K. Hermans[2], C. Ferris[3], M. Bell[3], M.T. Sorensen[4], J. Höglund[4] and M.A. Crowe[1]
[1]University College Dublin, School Of Veterinary Medicine Veterinary Science Centre, Belfield, D4, Ireland, [2]Ghent University, Salisburylaan 133, D4, 9820 Merelbeke, Belgium, [3]Agri-Food and Biosciences Institute, 12 Stoney Rd, Belfast, United Kingdom, [4]Aarhus University, Blichers Allé 20, Building P20, 8830 Tjele, Denmark; andreia.santoro@ucdconnect.ie

In the EU-project 'Genotype plus Environment' (GplusE), a major objective was to develop predictors for health phenotypes. The predictors were based on milk metabolites and enzymes, mid infrared (MIR) spectra and glycomics. Glycans are small carbohydrate molecules that in conjugation with proteins play important roles in cell signalling, molecular recognition, immunity, and inflammation. N-glycans are attached to the side chain of some asparagine residues within the protein scaffold. This project studied glycans on the immunoglobulin (IgG) fraction in milk. The project involved135 Holstein Friesian cows from three European universities/ research farms (AFBI – UK, AU-DK and UCD-IE), in the first 50 days of lactation. Milk samples collected at 7, 14, 21 and 35 days were analysed in a robotised UPLC-based platform in 96-well format. Traits recorded related to production (lactation number, daily milk yield, live weight, daily intake), health (disease diagnostics, uterine health, lameness score, body condition score) and fertility (heat detection, service, sire used, and hormones). Classification and regression trees were trained using data from 2 herds, and cross tested for the remaining herd. Preliminary results showed both high sensitivity and specificity using N-linked glycans released from IgG to predict the occurrence of retained placenta and clinical mastitis (amongst other traits) in dairy cows. The models will be validated using data collected from three other partner herds. IgG glycomic profiles can be used as a milk-based phenotype to substitute for other difficult to measure traits in dairy production. Supported by FP7-KBBE-613689.

OptiMIR: Use of MIR spectra to predict multiple cow status as advisory tools for dairy farms

C. Grelet[1], N. Gengler[2], C. Bastin[3], A. Vanlierde[1], S. Smith[4], M. Gelé[5], H. Soyeurt[2], X. Massart[3], P. Dardenne[1] and F. Dehareng[1]
[1]CRA-W, 5030, Gembloux, Belgium, [2]ULg, GxABT, 5030, Gembloux, Belgium, [3]AWE, 5590, Ciney, Belgium, [4]SRUC, EH9 3FH, Edimburgh, United Kingdom, [5]IDELE, 49105, Angers, France; c.grelet@cra.wallonie.be

Considering the current increasing of herd size, there is a need for precise and rapid information on individual cow state. Mid infrared (MIR) technology is already used worldwide for milk analysis; it allows rapid and cost effective determination of milk composition. The objective of OptiMIR project was to optimize the use of MIR spectra in order to produce indications on cow status thereby providing advisory tools to dairy farmers. Hence phenotypes of interest were collected in several countries and linked to MIR spectra. Since the OptiMIR network comprised 65 MIR instruments in 6 countries, standardisation of MIR data was necessary, allowing the collation of spectral databases and the use by all milk recording organizations (MRO) of the models developed. Using chemometric tools (like PLS regression), predictive models were developed to provide indicators on fine milk composition, on milk biomarkers of physiological imbalance, and directly on status of the cows. Equations predicting fine milk composition such as fatty acids and minerals were consolidated through the OptiMIR network, providing indirectly information on technological properties of milk and cow status. As biomarker of early physiological imbalance, an equation predicting citrate in milk was developed with good accuracy ($R^2cv=0.86$); and as milk biomarkers of ketosis, BHB and acetone were calibrated with fair results ($R^2cv=0.63$ and 0.67 respectively). Direct classification of spectra regarding low vs high risk of ketosis was also performed (84.5% sensitivity and 84.2% specificity). Direct regressions were realized for various negative energy balance criteria (r from 0.43 to 0.57) and enteric methane ($R^2cv=0.7$). All equations are available to be used by MRO on field and converted into advisory tools for the dairy sector.

Developing a bovine ketosis risk indicator using spectral analysis and animal phenotype data
S. Smith[1], V. Hicks[2], R. Cooper[3], J. Forrest[4], A. McRae[4] and E. Wall[1]
[1]SRUC, Midlothian, EH93JG, United Kingdom, [2]NMR, Chippenham, SN151BN, United Kingdom, [3]EBVC Ltd, Cumbria, CA110DT, United Kingdom, [4]Royal (Dick) School of Veterinary Studies, Midlothian, EH259RG, United Kingdom; stephanie.smith@sruc.ac.uk

Lactating dairy cattle have high energy requirements, with lactational demands commonly 3.0-4.5 maintenance. Whey they occur, energy deficits are compensated by fat mobilisation (measured by non-esterified fatty acids; NEFAs) and their subsequent metabolism. If NEB is severe or prolonged, beta-oxidation of fat may result with ketone bodies produced, including betahydroxybutyrate (BHB). Very elevated concentrations of NEFA and BHB represent a maladaptive response to NEB, associated with poor metabolic health, production and fertility. The objective of this study was to use reference serum BHB and NEFA records from nine herds to calibrate concurrent mid-infra red (MIR) spectral data from milk and generate a prediction tool for these metabolic indicators. Prediction tools, developed using MIR, have been successful for numerous difficult-to-measure traits, providing cost and time benefits over 'gold-standard' methods. A final dataset of 1,469 blood metabolite and concurrent spectra were obtained from two research herds (SRUC's Dairy Research Herd at Crichton, 956; University of Edinburgh's herd at Langhill Farm, 224) and seven commercial farms (289). Prior to alignment spectral data were standardised according to Grelet et al. Partial least squares regression was used to calibrate the spectral data with the reference blood measurements and split-sample cross validation performed in SAS. Subject to additional edits where necessary, the optimal square root of the coefficient of determination (R) was 0.74 for NEFA and 0.72 for BHB (log transformed). The optimum sensitivity and specificity were 0.73, 0.95 for NEFA and 0.80, 0.67 for BHB. Whilst NEFA tool is deployable, the BHB will likely benefit from additional sampling we have secured. Ultimately the tool will be applied to national herd data to help indicate prevalence as part of routine milk recording and used in combination with other health indicators e.g. MIR-predicted energy balance.

Genetic and environmental effects on individual wavenumbers of bovine milk infrared spectra
Q. Wang, A. Hulzebosch and H. Bovenhuis
Wageningen University, Animal Breeding and Genetics group, Droevendaalsesteeg 1, Radix (building nr. 107), 6708 PB Wageningen, the Netherlands; qiuyu.wang@wur.nl

Fourier transform infrared (FTIR) spectroscopy is widely used to determine milk composition. In this study 1,060 milk FTIR wavenumbers ranging from 925 to 5,008 cm^{-1} of 1,759 Holstein Friesian cows on 354 herds in the Netherlands were available. The objective is to investigate the genetic and environmental factors on milk FTIR wavenumbers. Single trait analysis was performed for each wavenumber in order to assess the significance of fixed effects and REML estimates of variance components. The fixed effect lactation stage had significant effects on 712 wavenumbers. Most of the wavenumbers had moderate to high heritability ranging from 0.20 to 0.60. Differences between herds explained 10 to 30% of the total variance for most wavenumbers. This suggests that milk FTIR wavenumbers are indicative for differences in feeding and management between herds. The diacylglycerol O-acyltransferase 1 (DGAT1), stearoyl-CoA desaturase (SCD1), kappa-casein and beta-lactoglobulin polymorphisms are known to have large effect on milk composition and therefore we studied the effects of these polymorphisms on milk FTIR. The DGAT1 polymorphism had highly significant effects on many wavenumbers. In contrast, the SCD1 polymorphism did not significantly affect any wavenumber. SCD1 is known to have a strong effect on the content of C10:1, C12:1, C14:1, and C16:1 fatty acids. Therefore, these results suggest that FTIR spectra contains little direct information on these mono unsaturated fatty acids. The kappa-casein and beta-lactoglobulin polymorphisms had significant effects on a few wavenumbers. Some wavenumbers are located near absorption peaks of water, therefore they are usually excluded in the prediction equations for milk composition. However, we found that some of these wavenumbers are affected by DGAT1 polymorphism and lactation stage. This suggests that these wavenumbers contain useful information regarding milk composition.

Novel milk phenotypes based on a biological model of lactation

G.E. Pollott
RVC, PPH, Royal College Street, London NW1 0TU, United Kingdom; gpollott@rvc.ac.uk

Milk production throughout lactation follows a characteristic curve. It has been postulated that three basic processes determine the level of milk production on any given day of lactation: (1) the number of mammary cells produced up to that day; (2) the number of cells dying off through apoptosis; and (3) the secretion rate of the cell. Lactation curves have been analysed with a biological model of lactation to investigate the relations ship between these three factors. One conclusion from this work was that if it were possible to measure two these factors then the third could be estimated and the results used to improve the selection and management of cows for milk production. Previous experiments to measure these factors have been reported and in this paper a synthesis of results is attempted in order to explain milk production in terms of the relationships between the three components. The two measured factors were the rate of apoptosis, measured by microvesicle production, and the secretion rate of the cell, as indicated by changes in gene expression throughout lactation. This left the number of mammary cells as the unestimated factor of the three. Data derived from 20 cows was integrated using a biological model of lactation. Information on lactose gene expression levels was used in conjunction with microvesicle measurements derived from monthly sampling throughout lactation. Combining these factors to model milk production indicated that the number of active mammary cells probably had the greatest influence on daily milk yield (0.63 of variation). This factor varied between cows to a much greater extent than the other two measured variables. Further research is required to verify this and a method to measure the number of active cells would probably be the best way to improve the methodology. The biological model of lactation has also been used to model fat, protein and lactose production during lactation. The current methodology was extended to include these three milk components. Not surprisingly, lactose production mirrored that of milk yield per day but fat and protein both had different but with characteristic patterns of synthesis throughout lactation. This approach looks promising to deconstruct milk production into some of its underlying processes.

Prediction of sheep milk chemical composition using its pH, electrical conductivity and refraction

A.I. Gelasakis[1], R. Giannakou[1], A. Kominakis[2], G. Antonakos[3] and G. Arsenos[1]
[1]Veterinary Faculty, A.U.Th., Panepistimioupoli, P.O. Box: 393, GR 54124, Thessaloniki, Greece, [2]Agricultural University of Athens, Iera Odos 75, GR 11855, Athens, Greece, [3]Frizarta Cooperative, 13th km Agrinio-Ioannina, GR 30100, Agrinio, Greece; gelasakis@vet.auth.gr

The objective of the study was to assess the predictive value of pH, electrical conductivity and refractive index of milk regarding its fat, protein and lactose concentration. A total of 308 purebred Frizarta ewes, were used. Milk samples were obtained from individual ewes and assayed for fat, protein and lactose concentration, as well as for milk electrical conductivity (MEC), pH and refractive index (RI). Data were analysed by means of descriptives and analytical statistics using the statistical software package IBM SPSS Statistics 21. Descriptives included the calculation of mean and standard deviation for each one the continuous variables in the dataset. Stepwise multiple linear regression models were developed to estimate the predictive value of pH, MEC and RI of milk regarding its content of fat, protein and lactose. The internal validity of the prediction model was evaluated by split-half cross validation and bootstrapping techniques. Average values for pH, MEC, RI, fat, protein and lactose concentration were 6.7, 4.0 ms/cm, 13.4, 5.5%, 5.4% and 4.7%, respectively. Milk fat concentration increased ca 0.8% and decreased ca 1.3% for each unit increase of RI and pH, respectively. Milk protein concentration increased ca 0.6% and 0.2% and milk lactose concentration decreased ca 0.1% and 0.5% for each unit increase of RI and MEC, respectively. The developed predicting equations were significant (P<0.001) and equal to 3.96+0.77×(RI)−1.3×(pH), -3.27+0.58×(RI)+0.24×(MEC) and 8.41−0.13×(RI)−0.50×(MEC), for fat, protein and lactose concentration, respectively. Among the models the highest validity was observed for the model predicting milk protein concentration followed by the model predicting milk lactose concentration. Although, automatic infrared milk-analysers remain the golden standard for the measurement of milk fat, protein and lactose concentration, it can be concluded that the prediction of these traits using RI, MEC and pH as predictors is a feasible and economical option for dairy ewes' milk.

Representative sample selection using FT-IR milk spectra to predict dry matter intake in dairy cows
N. Shetty, P. Løvendahl, M.S. Lund and A.J. Buitenhuis
Aarhus University, Molecular Biology and Genetics, Center for Quantitative Genetics and Genomics, 8830 Tjele,
Denmark; nisha.shetty@mbg.au.dk

Dry matter intake (DMI) is proxy trait for energy intake and thereby production efficiency in cattle fed forages. However, recording of DMI is very costly and not applicable to commercial dairy farms. Alternatively, milk Fourier transform mid-infrared (FT-IR) profile may be useful as indicator trait for DMI. The effect of using FT-IR profiles to select representative samples to predict DMI in dairy cows was explored. The data set consisted of 1,044 records from 97 Danish Holstein and 43 Jersey cows. Weekly averaged DMI (in kg), milk yield (MY in kg), live weight (LW in kg) and FT-IR spectra were used. Partial least squares regression method was used to develop the prediction models. The sample selection method was based on infrared spectra. The procedure begins with sorting the samples according to the Mahalanobis distances to the center of the data matrix and then selecting the most extreme sample. A limiting distance is defined and all the samples that are closer to the selected sample than this distance are made redundant and the procedure is repeated. By using only MY as predictor, the validation showed a coefficient of determination (R^2Val) of 0.59 and the root mean square error of prediction (RMSEP) of 2.24. Adding LW improved the model to R^2Val of 0.72 and RMSEP of 1.83. Including FT-IR profile further improved the prediction results to R^2Val of 0.79 and RMSEP of 1.59. Sample selection method reduced 1,044 records into a representative 245 records (104 cows, 71 Holstein and 33 Jersey). The prediction results based on this reduced set (23% of original records), were similar to using full data with R^2Val=0.76 and RMSEP=1.66, when MY, LW and FT-IR were used as predictors. Such reduced representative calibration sets are very useful in practice, as it substantially reduces the effort and cost of performing analysis of traits in large number of samples with no significant loss in prediction accuracy.

Milk mid-infrared spectral data as a useful tool to predict feed intake in dairy cows
S.E. Wallén[1], T.H.E. Meuwissen[1], S. McParland[2] and D.P. Berry[2]
[1]Norwegian University of Life Sciences, Dept. of Animal and Aquacultural Sciences, Arboretveien 6, 1430 Ås,
Norway, [2]Teagasc, Moorepark, Co. Cork, Ireland; sini.wallen@nmbu.no

Having access to (predicted) feed intake data can be advantageous not only for breeding for improved feed efficiency but can also be useful in herd management decisions. The objective of the present study was to quantify the ability of mid-infrared spectroscopy (MIRS) of milk to predict feed intake in lactating Norwegian Red dairy cows. Data for this study originated from six different feeding experiments from a single dairy research herd between the years 2007 and 2015. Records were available on energy intake (EI), milk yield, and milk concentration of fat, protein and lactose. Splines were used to interpolate daily values where the MIR spectra were also available but only where phenotypes were available within 10 days of the MIR observation. The number of records with both a phenotypic value and MIR was approximately 1,000 on 92 to 184 cows depending on the trait. Partial Least Squares (PLS) regression analysis was used to relate the milk MIR to the milk concentration of fat, protein and lactose as well as for energy intake using one-at-a-time cross validation; for the prediction of energy intake milk yield itself was also considered as a predictor trait. The number of fitted factors in the PLS varied between 14 and 19. The proportion of the variability explained by the MIR in milk fat, protein and lactose concentration as well as energy intake was 0.74, 0.88, 0.81 and 0.60, respectively. When milk yield was also included in the model, the proportion of the variation in energy intake increased to 0.64. The proportion of variability in energy intake predicted by milk yield and milk concentration combined was just 0.22 thereby indicating that additional information in the MIR is being exploited in predicting energy intake.

Going beyond current limits in defining and using milk infrared spectra based phenotypes
N. Gengler, H. Hammami and A. Lainé
ULg, GxABT, Passage des Deportes 2, 5030, Belgium; nicolas.gengler@ulg.ac.be

Mid-infrared (MIR) spectral data has been recognized as a major source for novel milk-based phenotypes for breeding and management. Current limits exist however in defining and using these phenotypes that are related to: (1) type of infrared technology and the associated sampling methods; (2) scope of reference traits; (3) used prediction models; (4) portability of prediction equations; and (5) use of results for breeding and management. The first limit is currently falling as on-farm in-line near-infrared (NIR) tools become available. This will also add real-time capacity to many novel applications pioneered in MIR. But due to sampling frequency, and ioff-line nature, MIR is lacking this capacity. Recently the scope of reference traits was considerably enlarged extending beyond milk components, the prediction of methane emissions, being an example. Also milk MIR spectral data is in general an excellent tool to check for the status of the cows (e.g. body energy). Two other recent examples reported here are the ongoing research on identification of cow's heat stress status and the detection of early pregnancy in milk composition. Work on heat stress is starting to detect milk spectra based signatures of reduced intake and increased body reserve mobilization. Results based on pregnancy validated test-day records showed that in the period up to 90 days pregnant the relative reaction of certain MIR wavelengths were at least 2× larger than that of milk yield. The recently finished OptiMIR project has been instrumental in pushing back boundaries allowing improved portability of equations and novel applications. A final issue is the correct use of indirect (M)IR predictors, distinguishing (i.e. genetic correlations <1) MIR based indirect index traits from breeding goal (e.g. direct) traits. If the required (genetic) parameters become available, direct, and therefore optimal, use of MIR spectral data to complement direct traits will become possible.

Phenotypic characterization of milk processing traits predicted by mid-infrared spectroscopy
G. Visentin[1,2], M. De Marchi[2], A. McDermott[1,2], D.P. Berry[1], M. Penasa[2] and S. McParland[1]
*[1]Teagasc, Animal and Grassland research and Innovation Center, Moorepark, Fermoy, Co. Cork, Ireland,
[2]University of Padova, Department of Agronomy, Food, Natural Resources, Animals and Environment (DAFNAE), Viale dell'Università 16, 35020 Legnaro (PD), Italy; donagh.berry@teagasc.ie*

Milk processing characteristics are important quality traits, yet little is known about factors governing their variability due to the resources required to measure these characteristics in a sufficiently large population. Milk mid-infrared spectroscopy prediction models were recently developed for rennet coagulation time (RCT), curd-firming time (k20), curd firmness (a30 and a60), heat coagulation time (HCT), casein micelle size (CMS), and pH in dairy cows. These prediction models were applied to 182,002 spectra collected from 10,122 Irish dairy cows from both research and commercial herds between the years 2013 and 2015. Sources of variation were investigated using linear mixed animal models which included the fixed effects of contemporary group (herd-year-season of calving), breed, stage of lactation, parity, milking time, heterosis, and recombination. Within and across parity cow effects were included as random terms. Mean RCT, k20, and a30 were 20.03 min, 5.26 min and 29.87 mm, respectively. These traits were most favourable for cheese manufacturing in early lactation, concurrent with the lowest values of both pH and CMS. All traits deteriorated in mid-lactation but improved towards the end of lactation, since strong correlations existed between them (from |0.51| between RCT and a30 to |0.66| between RCT and pH). In direct contrast, HCT was the greatest in mid-lactation, but the least in both early and late lactation. Relative to multiparous cows, primiparous cows, on average, produced milk more suitable for cheese and milk powder production. Results from the present study may aid in decision-making for milk manufacturing, especially in countries characterised by seasonal supply of fresh milk.

Traits associated with protein composition and colour of bovine milk

A. McDermott[1,2], G. Visentin[1,2], M. De Marchi[1], D.P. Berry[2], M.A. Fenelon[3] and S. McParland[2]
[1]University of Padova, Viale dell'Università 16, Animals and Environment, 35020 Legnaro, (PD), Italy, [2]Teagasc, Animal and Grassland Research Centre, Moorepark, Fermoy, Co. Cork, Ireland, [3]Teagasc, Food Research Centre, Moorepark, Fermoy, Co. Cork, Ireland; audrey.mcdermott@teagasc.ie

The aim of this study was to determine the associations of parity, stage of lactation and milking time with protein composition and colour of bovine milk. Mid-infrared spectral (MIRS) data from 2013 to 2015, inclusive were obtained from the milk samples of 3 research herds. Spectral outliers were detected using mahalanobis distance and removed. Data were further edited to remove records from cows which were greater than 305 days in milk or were greater than tenth parity. Equations developed recently to predict individual protein fractions and milk colour directly from the MIR spectrum were applied to the data yielding 64,606 records from 1,102 Holstein-Friesian cow lactations of alpha s1, beta, and kappa casein as well as the yellowness of milk. Factors associated with each of the milk quality traits were quantified using linear mixed models. All models were adjusted for milking time, parity (1, 2, 3, 4 and 5+) and stage of lactation (5 equal groups of 60); and the random effects of within and across lactation permanent environmental effects and herd_year_season. Older animals had yellower milk with a higher level of kappa casein and lower level of beta casein (P<0.001). As lactation progressed, milk got yellower in colour (P<0.001) and had higher levels of kappa, beta and alpha s1 casein (P<0.001). Futhermore, although evening milk had a yellower colour than morning milk and higher levels of alpha s1 casein, there was no difference in beta casein across milking time, and the difference in kappa casein was small.

Towards preventive health management in native dual-purpose cattle via novel breeding strategies

S. König
University of Kassel, Animal Breeding, Nordbahhofstr. 1a, 37213 Witzenhausen, Germany;
sven.koenig@uni-kassel.de

This contribution reflects the major scientific objectives of the collaborative EU-project 2-ORG-COWS. The objective of the project is to provide an in-depth evaluation of novel functional traits and associated environmental descriptors as a prerequisite: (a) for the implementation of breeding strategies for local dual-purpose cattle kept in grassland systems; and (b) for preventive livestock health management. Based on an interdisciplinary research design involving the disciplines animal breeding, animal husbandry, animal nutrition, and precision livestock farming, the most relevant European dual-purpose cattle breeds will be evaluated with regard to internal and cross breed comparisons focusing on: Novel traits, economic evaluation criteria, selection signatures and biodiversity, and different breeding strategies. The central theme for identical and harmonized health trait recording across country borders is based on sensor technology, which will be implemented in dual-purpose cattle experimental and contract herds. Sensor technology allows longitudinal trait recording of cow behavior, health and welfare traits (e.g. grazing activity, rumination) in close intervals. Management oriented sensor software delivers a large amount of validated data (health, behavior, etc.) from all cows in the experimental herds, defined as level I for trait recoding. An evaluation of indicator traits within a limited set of practical (contract) herds with dual-purpose breeds addresses level II. Most reliable indicator traits will be used for a population wide recording scheme (level III), and also including conventional control breeds for studies on genotype × environment interactions (G×E). Studies aiming on the identification of G×E imply a detailed characterization of environmental descriptors (rating of the grassland system, ingredients and quantities of the feed ration, meteorological data) on an identical scale all across Europe.

Animal-human-technology interactions: novel means of phenotyping cattle health and welfare

U. König V. Borstel

University of Göttingen, A.-Thaer-Weg 3, 37075 Göttingen, Germany; koenigvb@gwdg.de

With increasing availability and affordability of technology for use at the animal level, possibilities for automated recording of data related to animal production, health and behaviour have skyrocketed. In particular with dairy cattle, sensor-based technology at the individual animal level is well established and provides an abundance of data mainly intended for use in production, reproduction and health management. For example, automated detection or measurement of body temperature, heat, mastitis, metabolic disorders, and birth prediction is already commonplace in modern dairy cattle farming. Applications of automated monitoring of e.g. rumination, lameness, heat stress or methane emission detection are under development. Besides its use in management decisions, these data provide with its longitudinal structure on an individual animal level, invaluable information also for breeding purposes. However, in particular with sensors recording behaviour-based measurements, such as locomotion sensors for heat detection, interpretation of data remains challenging, and results from automated processing may require verification by human personnel. With regard to welfare-related traits, it is important to note that so far, only indicator traits are available that can provide at best a rough estimate on an animal's internal state. Development of automated behaviour tests that can provide a closer insight into the animals' needs and desires would be desirable to obtain more detailed information on how animals cope with given husbandry and management conditions. In other areas such as early detection of lameness, mastitis or upcoming parturition, technology already proves to be clearly superior to trained personnel. These facts raise the question, if future dairy farmers will increasingly be technology rather than animal experts, and if so, what this means to animal welfare. The present paper reviews prospects and limitations of existing and evolving technology for simultaneous use in dairy cattle breeding and selection as well as husbandry and management.

Characteristics of organic dairy farm types in seven European countries

A. Wallenbeck[1], A. Bieber[2], A. Spengler Neff[2], B. Fuerst-Waltl[3], C. Winckler[3], S. Ivemeyer[4], C. Simantke[4], S. March[5], J. Brinkmann[5], T. Rousing[6], J. T. Sorensen[6], J. Walczak[7], P. Wójcik[7] and V. Ribikauskas[8]

[1]SLU, Box 7023, 750 07 Uppsala, Sweden, [2]FiBL, Ackerstrasse 113, 5070 Frick, Switzerland, [3]BOKU, Gregor Mendel-Str.33, 1180 Vienna, Austria, [4]Uni-Kassel, Nordbahnhofstraße 1a, 37213 Witzenhausen, Germany, [5]Thuenen-Institute, Trenthorst 32, 23847 Westerau, Germany, [6]AU, Postboks 50, 8830 Tjele, Denmark, [7]NRIAP, 32-083 Balice, Cracow, Poland, [8]LSM-uni, Tilzes 18, 47181 Kaunas, Lithuania; anna.wallenbeck@slu.se

Development of suitable management and breeding strategies for organic animal production depends on the characteristics of the production systems, which vary between countries, regions and over time. The importance of this variation is frequently discussed among researchers, but often based on inadequate information. An important step in the Core Organic project ORGANICDAIRYHEALTH is to develop a data base on characteristics of major organic dairy farm types in Europe. The process of creating this database include three steps: (1) identify major organic dairy farm types in the partner countries (AT, CH, DE, DK, LT, PL, SE); (2) describe characteristics of the production system in each farm type; (3) Establish and deliver a database with key information on these farm types. Major organic dairy farm types in each country were identified according to a standardised protocol including four criteria; Herd size, Production level, Geographical location and Housing type. The farm type identification protocol was completed using data from national milk recording and organic certification bodies. The number of major farm types identified per country was: AT-5, CH-4, DE-4, DK-1, LT-4, PL-5, SE-4. The production system in each farm type was investigated by a profound farmer questionnaire including detailed questions on: Farm location, structure and size, Housing and milking system, Production level, Animal health and management, Feeding strategy, Breeding and reproduction. We aimed for complete protocol information from at least 10 farms per farm type and ended up with in total 419 completed questionnaires (AT-50, CH-20, DE-41, DK-10, LT-40, PL-198, SE-60). The data base will be used in other steps of the current project and as a basic resource for future research and decision making.

Comparison of native and commercial dairy cattle breeds on organic farms in five European countries

A. Bieber[1], A. Spengler[1], B. Fuerst-Waltl[2], S. Ivemeyer[3], C. Simantke[3], C. Stricker[1], J. Walczak[4], A. Wallenbeck[5], C. Winckler[2] and P. Wójcik[4]
[1]Research Institute for Organic Agriculture (FiBL), Ackerstr. 113, 5070 Frick, Switzerland, [2]University of Natural Resources and Life Sciences Vienna (BOKU), Gregor-Mendel-Str. 33, 1180 Vienna, Austria, [3]University of Kassel, Nordbahnhofstr. 1a, 37213 Witzenhausen, Germany, [4]National Research Institute of Animal Production (NRIAP), Department of Technology, Ecology and Economics, Krakowska 1, 32-083 Balice near Cracow, Poland, [5]Swedish University of Agricultural Sciences, Department of Animal Breeding and Genetics, Box 7023, 75007 Uppsala, Sweden; anna.bieber@fibl.org

Local/ native breeds are indirectly or explicitly recommended in organic standards. They may be better adapted to local, and especially organic, conditions characterized by a stronger dependency on local feed resources, higher amounts of pasture-based feeding systems and/or lower amounts of concentrate feeding, as well as stronger restrictions regarding medicine use. Nevertheless, little knowledge is available regarding health status, medicine use and phenotypic functional characteristics of local/native breeds in organic dairy production in Europe. Therefore, local and commercial dairy breeds from organic farms were compared regarding direct health traits, health associated functional and conformation traits as well as production traits with the aim to map the suitability of local breeds for organic farms in the respective countries. The study included data from organic herds from Austria, Germany, Poland, Sweden and Switzerland for the period 01.07.2011 to 30.06.2014. Preliminary results reveal lower milk yield for local breeds, but in many cases better fertility (shorter calving intervals, lower number of inseminations). Limited data on useful lifetime indicates a superiority of local breeds for this trait, whereas commercial breeds showed a higher lifetime production (kg ECM).

Genetic relationships and trait comparison between and within lines of local dual purpose cattle

M. Jaeger, K. Bruegemann and S. König
University of Kassel, Animal Breeding, Nordbahnhofstr. 1a, 37213 Witzenhausen, Germany; maria.jaeger@uni-kassel.de

The aims of the present study were: (1) to analyse the genetic structure of the local black and white cattle breed (DSN); (2) to compare different lines of DSN and Holstein (HF) cows for production and functional traits; and (3) to analyse genotype by environment interactions based on defined clusters. Clusters were created according to geographic, environmental and phenotypic descriptors (e.g. longitudes, altitudes, breeding organisation, average herd production level). The study considered 4,251 DSN and 4,187 HF cows from 38 herds keeping both breeds. Farms mostly reflected pasture based low input production systems. The average generation interval considering the years 1950 to 2010 was 5.8 years for the DSN population. Based on the average increase of inbreeding per generation, the effective population size comprised 125 animals, and the current average inbreeding coefficient in 2013 was 2.13%. The most influential DSN sires with more than 100 daughters were identified. Genetic relationships between those sires with the active cow population ranged between 6 and 9%. The genetic relationship between the current DSN and HF populations was quite small (0.06%), but reflecting identical genealogies. Due to artificial insemination, genetic relationships among different cluster combinations were on a similar level. Linear and generalized linear mixed models were applied for breed and line comparisons. The energy efficiency indicator fat-to-protein ratio (FPR) was higher for HF early in lactation in the first parity ($P<0.0001$), but opposite results were found with increasing day in milk. Interestingly, least square means for somatic cell scores were slightly higher for DSN compared to HF. A multiple trait approach, i.e. defining same traits in different clusters as two different traits, was applied to prove genotype by environment interactions based on genetic correlation estimates. Genetic correlations between low heritability functional traits were lower than for production traits, and partly below the threshold of $rg=0.80$.

Possibilities and requirements for organic dairy breeding lines

M. Kargo[1,2], L. Hjortø[2] and J.R. Thomasen[3]
[1]SEGES, Agro Food Park 15, 8200 Aarhus, Denmark, [2]Aarhus University, Center for Quantitative Genetics and Genomics, Blickers alle 20, 8830 Tjele, Denmark, [3]VikingGenetics, Ebeltoftvej 16, 8960 Randers, Denmark; morten.kargo@mbg.au.dk

Until now, improvements in the efficiency of organic dairy production have mainly occurred through better management, feeding and production strategies, using the same dairy cattle breeds as in conventional production systems. The potential of breeding lines that are genetically adapted to organic milk production is essentially unexploited even though specific lines offer the potential to breed for cows with characteristics specific for organic production systems. The main reason why organic lines of dairy cattle have not been established so far is most likely that the correlation between the breeding goals for organic and conventional production systems is above 0.8 in Northern Europe. A correlation higher than that has been shown to cause loss of genetic gain when dividing populations with less than a million cows in lines. However, that was under the assumption that progeny testing of bulls was the overall driver of genetic gain. With the recent implementation of genomic selection, the number of genotyped cows with phenotypes is the overall driver of genetic gain, so even in a small population (<50,000 cows), genetic gain can be maintained at the same level. Therefore, the critical limit regarding the correlation between breeding goals is likely higher now. Mainly two factors affect this correlation. Firstly, the economic values in the breeding goal and secondly the presence of biological genotype-by-environment interactions. On the condition that the economic values differ between the two production systems and genotype-by-environment interactions exist, it may be beneficial to establish different lines of conventional and organic dairy cattle and thereby further improve the efficiency of organic dairy production through selective breeding. Possibilities and requirements for setting efficient organic dairy breeding lines will be discussed in this presentation.

Preferences for Breeding Goal Traits for Danish Red and Jersey Cattle

M. Slagboom[1], M. Kargo[1,2], D. Edwards[1], A.C. Sørensen[1], J.R. Thomasen[1,3] and L. Hjortø[1]
[1]Aarhus University, Molecular Biology and Genetics, Blichers Allé 20, 8830 Tjele, Denmark, [2]SEGES Cattle, Agro Food Park 15, 8200 Aarhus N, Denmark, [3]VikingGenetics, Ebeltoftvej 16, 8960 Randers SØ, Denmark; margotslagboom@mbg.au.dk

The aim of this study was to characterize preferences of farmers with Red Danish Cattle (RDC) or Jersey herds. A breed-specific survey was established to characterize preferences of farmers for improvements in ten traits, by means of pairwise rankings using the online software 1000Minds. These pairwise rankings were based on equal economic worth of trait improvements. The RDC-survey was filled in by 87 farmers and the Jersey-survey by 76 farmers. Both RDC and Jersey herds had the highest preference for improvements in mastitis, followed by milk production. The lowest preference was given to calving difficulty. By means of a cluster analysis distinct clusters of farmers were identified and named according to the trait improvements that were the most preferred per cluster. RDC herds were divided into three different clusters: Robustness, Production & Health, and Production & Fertility. The Jersey herds were divided into the clusters Production & Fertility, Production & Robustness, and Survival. For RDC herds we found that the cluster Production & Fertility had the highest percentage of organic farmers (50%), the lowest percentage of crossbreeding and the lowest yields and herd size. The cluster Robustness had the highest yields and the highest prevalence of udder disorders, and these farmers ranked production traits very low and mastitis very high. For Jersey herds, the percentage of dead cows was highest for the cluster Survival. This cluster also ranked mastitis the highest compared to the other clusters, and contained herds with the highest prevalence of udder disorders. These results suggest that farmers choose to improve traits that are more problematic in their herds. This study shows that heterogeneity exists in farmers' preferences for trait improvements and that some herd characteristics can be linked to the different farmer clusters. The results from this study give a strong basis for setting up customized indexes or breeding goals for the different farmer clusters, which might increase the uptake of genetic merit.

Exploring the genetic background of parasite resistance in selected lines of black and white cattle

K. Brügemann[1], K. May[1,2], C. Scheper[1], C. Strube[2] and S. König[1]
[1]University of Kassel, Animal Breeding, Nordbahnhofstr. 1a, 37213 Witzenhausen, Germany, [2]University of Veterinary Medicine Hannover, Institute for Parasitology, Bünteweg 2, 30545 Hannover, Germany; kerstin.bruegemann@uni-kassel.de

Regaining importance of keeping dairy cows in grassland systems implies a detailed evaluation of breeding strategies on genetic resistances against endoparasite infections. The present study aimed on: (1) a comparison of different black and white cattle selection lines for three endoparasite traits; and (2) the estimation of genetic parameters for parasite resistances. A research design was implemented to create three different genetic lines within herds on the basis of a German Holstein cow (GHC) population: line 1 = GHC × New Zealand (NZ) sires, line 2 = GHC × GH sires, line 3 = GHC × GH pasture sires. GH pasture sires represent bulls of German origin with high breeding values for the 'grassland traits' being important in NZ. A forth genetic line (line 4) in the cross-classified research experiment included local black and white dual-purpose cows (DSN). 1,995 faecal samples were taken during two farm visits from 1,139 cows kept in 17 grassland farms located in North-west Germany. Endoparasite traits were the faecal egg counts for gastrointestinal nematodes (FEC_GI) and for trematodes (FEC_FP), and faecal larvae counts for Dictyocaulus viviparus (FLC_DV). Lowest values for nematode infections were identified for the pasture adopted genetic lines 1 and 3. Heritabilities for endoparasites were 0.05±0.04 (FEC_GI), 0.33±0.06 (FEC_FP), and 0.05±0.04 (FLC_DV). The genetic correlation between FEC_GI and FLC_DV was almost one (rg=0.99), but close to zero between FEC_GI and FEC_FP (rg=-0.06), and between FLC_DV and FEC_FP (rg=0.05). Antagonistic genetic relationships were identified between endoparasites and milk yield. The genetic correlation between FEC_GI and SCS was 0.36, and also positive between endoparasites and the energy deficiency indicator fat:protein-ratio. Infections with endoparasites contribute to physiological imbalances, and other health problems as well.

Genotype × environment interactions in dual purpose cattle in harsh environments

B. Bapst
Qualitas AG, Chamerstrasse 56, 6300 Zug, Switzerland; beat.bapst@qualitasag.ch

In Switzerland cattle are kept in a range of different production systems and environments: from lowland to mountain area, from intensive to extensive feeding systems and from conventional to organic production systems. Some of these environments can be classified as harsh environments. It is assumed that some genotypes affect traits differently in diverging environments. If that would be the case, genotype × environment (G×E) interactions would cause re-rankings of the cattle's corresponding breeding values in the breeding value list. First steps of our investigations addressed the question whether dual purpose cattle are affected by G×E interactions. Specifically, we treated the yield traits (milk kg (mkg), fat kg (fkg) and protein kg (ekg)) of Swiss Original Braunvieh cattle in lowland and mountain area as different traits. We analyzed 104,984 lactation records of 35,614 Original Braunvieh cows by applying an animal model with repeated measurements. These records were divided into different environments. The extent of the genetic correlation (r_a) between the traits in opposite environments indicates whether G×E interaction exists. In a preliminary analysis r_a of the 3 split traits was found to be between 0.98 and 1. This indicates almost no G×E interaction exists for yield traits in Original Braunvieh cattle in these two environments. Further investigations applying different methodology and environment definitions will be carried out to verify these results.

G×E in organic and conventional production systems in Austrian Fleckvieh cattle
C. Pfeiffer[1], C. Fuerst[2], H. Schwarzenbacher[2] and B. Fuerst-Waltl[1]
[1]*University of Natural Resources and Life Sciences (BOKU), Department of Agricultural Systems, Gregor-Mendel-Strasse 33, 1180 Vienna, Austria,* [2]*ZuchtData EDV-Dienstleistungen GmbH, Dresdner Strasse 33, 1200 Vienna, Austria; chri_pfeiffer@hotmail.com*

Improvement of breeding and management resulted in a considerable increase of production traits in Austrian dairy cattle. Apart from production systems with a high intensity farm management, sustainable and organic dairy production systems are an integral part in Austria. Possible G×E depending on the method of production and intensity of farm management might exist and lead to a re-ranking of animals. Thus, an adaptation of the breeding programs would be required. Therefore, G×E were estimated between milk yield, persistency, functional longevity, somatic cell score, clinical mastitis, early fertility disorders, cystic ovaries, milk fever, non-return rate at 56 days of cows and interval from first to last insemination in Austrian Fleckvieh cows milked in three different production systems (organic, conventional low and high level of farm intensity). Genetic correlations (ra) between production systems were estimated using an approximate multivariate two-step approach applied to yield deviations and de-regressed estimated breeding values. In general, no severe G×E were found. Genetic correlations between the production systems organic and conventional low level of farm intensity, organic and conventional high level of farm intensity, conventional low level of farm intensity and conventional high level of farm intensity ranged from 0.952 to unity, 0.890 to unity and 0.886 to unity, respectively. The lowest ra were observed for non-return rate at 56 days of cows (0.890) between organic farms and farms with high intensity management and for functional longevity (0.886) between farms with low and high intensity management. For all other traits, ra were close to unity. From the breeding perspective, autonomous breeding programs for different production systems within Austria are currently not needed.

Defining consensus desired gains for a Kenyan Holstein-Friesian breeding goal
C.M. Kariuki[1,2], J.A.M. Van Arendonk[3], A.K. Kahi[4] and H. Komen[1]
[1]*Wageningen University, Animal Breeding and Genomics, P.O. Box 338, 6700 AH Wageningen, the Netherlands,* [2]*Chuka University, Animal Sciences, P.O. Box 109, 60100 Chuka, Kenya,* [3]*Hendrix Genetics, Research Technology and Services, P.O. Box 114, 5830 AC Boxmeer, the Netherlands,* [4]*Egerton University, Animal Sciences, P.O. Box 536, 20115 Egerton, Kenya; charles.kariuki@wur.nl*

The Kenyan dairy cattle industry contributes to food and nutrition security and is a source of income for numerous households. Selective breeding can enhance efficiency in this industry. The current approach that rely on semen importation is sub-optimal and a local breeding program is needed. The Kenyan industry is characterized by diverse production and marketing systems. Desired genetic gains approach can be used to define consensus breeding goal to optimize production in all systems. We used Analytic Hierarchy Process to determine individual preference values for milk yield (MY), calving interval (CI), production life time (PLT), mature body weight (MBW), and fat yield (FY). Results show that classical classification of production systems into largescale and smallholder systems failed to capture differences in trait preferences. These differences became apparent when classification was based on productivity at the individual animal level, with high intensity (HIP), low intensity (LIP) producers and processors (Pr) as the main important groups. Highest preferences were PLT and MY for HIP, CI and PLT for LIP, and MY and FY for Pr. Highest disagreements between the groups were observed for FY, PLT, and MY. Individual and group preferences were aggregated into consensus preferences using Weighted Goal Programming. Desired gains were obtained as product of consensus preferences and percentage genetic gains (G%). These were 2.42, 0.22, 2.51, 0.15 and 0.87% foMY, CI, PLT, MBW and FY, respectively. Consensus breeding goal can play a major role in the establishmet and acceptalityof a local breeding schemefor the highly diverse production and marketingcircumstancesin Kenya.

Genotype by environment interaction for activity based estrus traits in Danish Holstein

A. Ismael[1,2], E. Strandberg[1], B. Berglund[1], M. Kargo[2,3], A. Fogh[3] and P. Løvendahl[2]
[1]Swedish University of Agricultural Sciences, Department of Animal Breeding and Genetics, Uppsala, P.O. Box 7023, 750 07, Sweden, [2] Aarhus University, Dept. of Molecular Biology and Genetics, Tjele, 8830, Denmark, [3]Knowledge Center for Agriculture (SEGES), Aarhus, 8000, Denmark; ahmedismaelsayed@gmail.com

The objective of this study was to investigate whether genotype by environment interaction (G×E) exists for female fertility traits with regards to production level. The fertility traits considered were: (1) interval from calving to first high activity (CFHA); (2) duration of first estrus (DFE); and (3) strength of first estrus (SFE). Estrus traits were derived from electronic activity tags for 11,522 first parity cows housed in 125 herds. Data were analyzed using random regression animal model (RRM) by regressing the phenotypic performance on the average herd energy corrected milk yield as an environmental gradient. For validation of the random regression model results, data were further analyzed using a bivariate model (BM) considering the same trait in high and low production environments as different trait. The heritability estimate for CFHA decreased from 0.25 to 0.10 with increasing production level. For DFE, heritability estimate obtained by BM for the low producing herds were more than twice the estimates for high producing herds, this trend was however not detected by the RRM results. For SFE, heritability estimates decreased from 0.12 in the lowest environment to the minimum estimate of 0.03 before increasing gradually to 0.13 in the highest environmental scale. The genetic correlation of the same trait in low and high production environments was around 0.74 for CFHA, 0.22 for SFE and 1.00 for DFE using RRM with similar estimates using BM. These results indicated no evidence of G×E interaction effect on CFHA, DFE, while there was strong G×E for SFE, which indicate that G×E interaction should be included in the evaluation of this trait.

Statistical analysis of yield milk by AMMI analysis of genotype-environment interaction

A. Hamrouni and D. M'naouer
Institut National Agronomique de Tunis, Production Animale, rue de kornich Chott El salem Gabès 6061, 6061, Tunisia; abirturki@yahoo.fr

Genotype-by-environment interaction (GEI) is an important phenomenon in animal breeding. Our objective was to examine the existence of a genotype × region interaction on milk yield production in Holstein dairy cows in Tunisia and look for genotypes that have general adaptability, or look for genotypes that have specific adaptability for specific environments. Data were classified as belonging to three regions of the country: Northern, Central and Southern. Only data from sires with daughters in three regions were used in each analysis. We evaluated 26 common sires sires in this three environments. The adaptability of genotypes were evaluated based on the additive main effects and multiplicative interaction (AMMI) method. The AMMI method allowed for easy visual identification of superior genotypes for each set of environments. A preliminary ANOVA shows large differences due to environment and genotype, and a large GE interaction (P<0.01). AMMI analysis showed that milk yield was significantly (P<0.01) affected by environments (E), genotypes (G) and genotype × environment interaction (GEI). The first two multiplicative axis terms (PCA1 and PCA2) explained 78% of GEI sum of squares. According to AMMI1 biplot G24 and G9 exhibited specific adaptablity for Northern environments. Genotypes G4 and G16 revealed specific adaptation for Central environments. Genotypes G25 and G11 exhibited specific adaptablity for Southern environmentes. Genotypes G10, G10, G7 and G15 revealed general adaptability.

Adaptation of Holstin Dairy Cattle to Southern Mediterranean Environmental Conditions
M. Khalifa and M. Djemali
INAT, Animal Genetic and Feed Resources Lab., INAT, 43 Avenue Charles Nicolle 1082 Mahrajène, 1082, Tunisia;
moniakhlifi3h@gmail.com

The effect of month or season of calving on production traits has been reported in dairy cattle as a main source of variation of milk traits. The dairy cow produces differently in relation to the period of calving. The shape of its lactation curve based on seasonal effects represents a reliable tool to evaluate its adaptation to cliamte variation within and across years. The objectives of this study were to: (1) identify major environmental and non genetic factors of Holsteins production traits under Southern Mediterranean production systems; and (2) describe their milk, fat and protein lactation curves based on season of calving. A total of 1,107,314 test days recorded in 2,926 Hosltein herds during 2003-2014 was used in this study. A linear model was developed including Herd-year, month of calving, lactation number and test day control number for first lactations and lactations ≥2. Least square solutions of month of calving allowed to identify 3 different seasons of calving with season1 including five months (September to January), season 2 incuding four months (February to May) and season 3 including Summer months. The same linear model, without month of calving, was used within season. Main results showed that herd-year, month of calving, lactation number, milk test control number were all significant sources of variation for milk, fat and protein. Average test days for milk, fat and protein varied from 17±7 to 11±5, 0.6±0.28 to 0.4±0.21 and 0.52±0.23 to 0.36±0.18 kg/d, respectively. A total of 18 curves were obtained. One curve was established for each season, each lactation and each trait. The trends of the lactation curves showed that cows calving in season one (automn-winter) gave more milk than cows calving in other seasons until the 5th month after calving. Cows calving in summer season gave less milk until the 5th control, after wich their production became greater than cows calving in other seasons. The highest differences among seasons varied from 3 to 4 kg/d and 4 to 6 kg/d, respectively for primparious and multiparious. Lactation curves trend translated how Holsteins are adapting to Southern Mediterranean Environmental conditions.

Organic and Conventional Dairy Farmers Prefer Different Improvements in Breeding Goal Traits
M. Slagboom[1], M. Kargo[1,2], D. Edwards[1], A.C. Sørensen[1], J.R. Thomasen[1,3] and L. Hjortø[1]
[1]Aarhus University, Molecular Biology and Genetics, Blichers Allé 20, 8830 Tjele, Denmark, [2]SEGES Cattle, Agro Food Park 15, 8200 Aarhus N, Denmark, [3]VikingGenetics, Ebeltoftvej 16, 8960 Randers SØ, Denmark; margotslagboom@mbg.au.dk

In dairy cattle breeding, breeding goals (BG) are developed and subsequently a selection index that farmers want to use. Therefore it is important to take their preferences for BG traits into account. Two production systems that are expected to influence farmer preferences for BG traits are organic and conventional systems. The aim of this study was to characterize preferences of organic and conventional Danish dairy farmers for improvements in BG traits for Holstein cattle. A survey was established to characterize preferences for improvements in ten traits, by means of pairwise rankings using the online software 1000Minds. These pairwise rankings were based on equal economic worth of trait improvements. The survey was filled in by 106 organic and 290 conventional farmers. The most preferred trait improvement for both production systems was in cow fertility, and the least preferred improvement was in calving difficulty. Organic farmers ranked milk production and feed efficiency higher than conventional farmers. The reason for this may be that the average milk yield is lower and the milk price is higher in organic herds compared to conventional herds. Organic farmers also ranked calf mortality higher. Conventional farmers ranked mastitis, leg and claw diseases, other diseases and cow mortality higher compared to organic farmers. By means of a cluster analysis distinct clusters of farmers were identified and named according to the trait improvements that were the most preferred per cluster. The cluster analysis for organic farmers identified two production-based clusters ('Production & Fertility' and 'Production & Mastitis') and one robustness-based cluster, whereas a cluster analysis for conventional farmers found one production-based cluster, one health-based cluster and one survival-based cluster. This study shows that organic and conventional farmers differ in their preferences for improvements in BG traits, and the results can be used to set up specific breeding goals for organic and conventional farming systems.

Dairy cows for an efficient milk production based on roughage

H. Joerg[1], A. Wenger[1], C. Aebersold[1], A. Burren[1], M. Schwarzenberger[2] and M. Huber[2]
[1]Bern University of Applied Sciences, School of Agricultural, Forest and Food Sciences, Länggasse 85, 3052 Zollikofen, Switzerland, [2]Bildungs- und Beratungszentrum Arenenberg, Arenenberg, 8268 Salenstein, Switzerland; hannes.joerg@bfh.ch

In Switzerland, the dairy cows are fed with a lot of roughage. Most dairy cows go to pasture and eat grass during the growing season. In winter the diet is based on hay and silage. Therefore a total selection index, called Swiss index, was created for breeding dairy cows that are suitable for milk production based on roughage. The main goals of the breeding index are progress in the fitness traits, less milk yield and cows with a mean hip height. The weights of the individual traits in the Swiss index are therefore on fat and protein content, cell count, persistence, fertility, calving is and against hip size. We analysed the Swiss index the four dairy breeds Brown Swiss, Swiss Fleckvieh, Red Holstein and Holstein for the correlations between the index and the breeding values of the single traits. The number of bulls was in Brown Swiss, Swiss Fleckvieh, Red Holstein and Holstein of 14,393; 14,623; 18,160 and 11,370; respectively. Since the bulls are born in the last 60 years, the course of the breeding values and the correlations to an index such as the total index can be tracked. The tracking in the four breeds showed that progress has been made in the milk yield, in the body size and in the udder traits. Comparison of correlations between single traits and the total selection index as well as the Swiss index revealed interesting differences in milk yield, fertility and hip height. The correlations between the breeding value for milk yield and the total index are approximately 0.8, but the correlations between the breeding value for milk yield and the Swiss index are between -0.16 and 0.2. The correlations between the breeding value for fertility and the Swiss index are in the range of 0.6 and much higher than with total index. The correlations between the breeding value for hip height and the Swiss index are negative. Breeding according to Swiss index will lead to cows with no increased milk yield, better fertility and medium body size.

Preliminary study on the meat quality of Albera bulls, an endangered cattle breed from the Pyrenees

N. Panella-Riera, M. Vitale, M. Gil and M.A. Oliver
IRTA, Product Quality, Finca Camps i Armet s/n, 17121 Monells, Spain; mauro.vitale@irta.cat

The Albera is an endangered local bovine breed from the Albera massif, in the eastern Pyrenees, characterized by a high degree of rusticity. It represents an important part of the genetic biodiversity of Mediterranean breeds. The Albera breed is suitable for extensive production systems due to its adaptation in the Pyrenees and Pre-Pyrenees area. The aim of this study was the characterization of the quality of carcass and meat from bulls of the Albera breed for the first time. Ten Albera bulls (age at slaughter: 2.7±0.4 yrs) were reared in 2 types of farms: (1) a mountain farm (MF; n=5) in the Pre-Pyrenees area, with access to wooded areas and mountain meadows; and (2) a saline meadows farm (SF; n=5), near to the protected Natural Park 'Aiguamolls' in the Empordà region. After slaughter, carcass and meat quality were measured. Two aging methods were applied during 12 days post mortem (pm): wet aging of a portion of the longissimus dorsi (LD) muscle or dry aging of the whole carcass (and subsequent sampling of the same proportion of the LD). Instrumental color, expressible juices, cooking loss and shear force were recorded, and a trained panel assessed the aged meat. The bulls differed in carcass traits: carcasses of MF bulls were heavier than SF (MF=340.4 kg vs SF=273.6 kg; P=0.02) and had lower fatness (MF=1+ vs SF=2; P=0.04) and the same average conformation (R). The intramuscular fat content was lower than 2%. The aging method as well as the interaction between farm production system and aging method did not significantly affect the technological and sensory attributes of meat. The farm production system did not affect instrumental color, expressible juices or cooking losses of aged meat (P>0.05). MF aged meat was significantly harder than SF one (WBSF−6.5 vs 3.6; and sensory hardness=6.4 vs 3.2; P<0.01), more chewy (7.3 vs 4.6; P<0.01) and with a less strong beef flavor (3.4 vs 4.5; P=0.01)). Results underlined the importance of feed strategy to improve technological and sensory traits of Albera's bull meat and offered an interesting alternative to carcass aging for its producers.

Review on environmental benefits of organic milk production
G. Pirlo
Council of Agricultural Research and Economics, via A. Lombrado 11, 26900 Lodi, Italy; giacomo.pirlo@entecra.it

About 2.7% of the milk produced in Italy in 2012 was organic and it met only a small portion of the consumers' demand. Increase of organic milk (OM) production could satisfy it, but it could also contribute to reduce some environmental pressures on this agricultural sector. In the framework of a national project on organic milk production (VaLatteBio), the results of eight research papers have been reviewed to verify what the extent of the effect of OM production could be on global warming (GW), acidification (AC), and eutrophication (EU) compared with conventional milk (CM) production system. Average milk production was 7,193 (CV 12.9%) and 6,291 (CV 11.8%) kg of Energy Corrected Milk (ECM) for CM and OM respectively. When expressed per kilogram of ECM, average GW was 1.24 (CV 16.9%) and 1.35 (CV 17.0%) kg of CO_2eq for CM and OM respectively; AC was 0.11 (CV 63.1%) and 0.09 (CV 82.1%) kg PO3-4eq for CM and OM respectively; EU was 0.025 (CV 92%) and 0.0093 (CV 80%) kg SO2eq for CM and OM respectively. If expressed per ha of land occupied, average GW was 9,709 (CV 51%) and 6,556 (CV 38%) kg of CO_2eq for CM and OM respectively. Too few data are available for AC and EU per ha of land occupied. Studies showed large variability in AC and EU due to differences in production systems and in methodologies adopted. Organic milk production does not appear to contribute to reduce GW per unit of milk produced, but the studies did not consider soil carbon sink. On the contrary, it can reduce pressure on local environmental impacts, such as those caused by acidification of soil and water. Further data on environmental impact of OM production could stimulate an increase of this economic sector.

PROHEALTH, production diseases and piglet mortality
S.A. Edwards[1], S. Matheson[1], F. Pandolfi[1], R. Thompson[1,2], T. Ploetz[2] and I. Kyriazakis[1]
[1]Newcastle University, School of Agriculture, Food & Rural Development, Newcastle upon Tyne NE1 7RU, United Kingdom, [2]Newcastle University, Open Lab, Newcastle upon Tyne NE1 7RU, United Kingdom; sandra.edwards@ncl.ac.uk

PROHEALTH is a project funded under EU FP7 (no. 613574) to address the prevalence, causes and alleviation of production diseases of pigs and poultry raised in the wide variety of EU intensive systems. Production diseases are those which persist in intensive systems, and whose prevalence or severity tends to increase with production level. Piglet mortality is a good example of a production disease which, as a result of the combined effect of losses from stillbirths and pre-weaning mortality ranging in our survey data from 21-24% of total born, costs the industry €4-9 per pig sold according to PROHEALTH economic analysis. Like all production diseases, piglet mortality is a consequence of complex interactions between individual predispositions of the animals, diverse aspects of their production environment, including housing, nutrition and management, and the level of pathogen challenge to which they are subject. Work within PROHEALTH explores the different causes of mortality and their individual risk factors. It highlights the influence of increasing prolificacy on piglet maturity and viability, the important role that the gestation period plays in subsequent viability, and the sources of variability in maternal health and behaviour which are key for survival. These investigations harness novel digital approaches to quantify key aspects of maternal behaviour, and explore the influence of genetics and management on these behaviours. Based on this understanding, interventions will be developed and tested for their efficacy to reduce mortality, which will benefit animal health and welfare, farm economy and the societal perception of the industry.

Effects of gestation housing system on maternal stress, piglet maturity at birth and early survival
H. Quesnel[1], H. Pastorelli[1], E. Merlot[1], I. Louveau[1], L. Lefaucheur[1], F. Robert[2], M.C. Père[1] and F. Gondret[1]
[1]INRA, UMR1348 PEGASE, Domaine de la Prise, 35590 Saint-Gilles, France, [2]Deltavit, CCPA group, 35150 Janzé, France; helene.quesnel@rennes.inra.fr

Sow housing and management system can generate maternal stress which could influence piglet development during gestation and postnatal survival. We investigated the effects of two housing systems for gestating sows on piglet maturity at birth and early survival. Gestating sows were group-housed in a conventional system on slatted floor (C, n=49) or in larger pens enriched with straw bedding (E, n=57). On gestation day (DG) 105, all sows were transferred into farrowing pens on slatted floor. Piglet mortality was recorded in all litters and saliva cortisol in all sows, while neonate and lacteal traits were investigated in a subset of litters and sows (C n=18, E n=19). Concentrations of cortisol were greater (P<0.001) in C than in E sows on DG 35 and 105 but no longer after the transfer to farrowing crates on DG 107 and day 4 of lactation. Mortality rates were greater amongst C piglets, especially within 3 days post-partum (13.6 vs 6.3%, P<0.001). System did not alter piglet weight at birth but gut weight was lower (P<0.05) in C piglets. The C piglets also had greater plasma concentrations of fructose (+13%), lower concentrations of glucose (-9%) and lower glycogen content in muscle (-8%) (P<0.05), suggesting a lower maturity of C piglets. However, other tissue indicators of physiological maturity did not differ between C and E neonates. The housing system neither influenced piglet weight gain over 24 h from birth nor altered nutritional composition of colostrum and milk collected at day 4 of lactation, the exception being protein content that tended to be greater in colostrum from C sows (P=0.07). To conclude, the conventional system during gestation induced maternal stress and greater neonatal mortality, without major effects on piglet physiology at birth. Research was funded by the EU FP7 Prohealth project (no. 613574).

Relationships between sow conformation and crushing events in commercial piglet production
S.M. Matheson[1], G. Walling[2] and S.A. Edwards[1]
[1]Newcastle Univeristy, AFRD, NE1 7RU, United Kingdom, [2]JSR Genetics Limited, Southburn, YO25 9ED, United Kingdom; stephanie.matheson@newcastle.ac.uk

Pig selection maximises length of back and lean tissue growth rate for meat production. The resulting change of sow body shape may increase difficulty with control of posture change and consequently crushing of piglets. The aim of this study was to investigate the relationship between sow conformation, farrowing environment and piglet crushing by the sow. Piglet births (n=11,752) were recorded for 21 weeks in a population of Landrace sows crossed with either White Duroc or Large White sires. All sows (n=874) were scored for front and hind leg conformation and the parity and floor surface of their farrowing crate was recorded (FAR-FLOOR). Piglet data gathered at first processing (18-24 hours after birth) were weight, gender, IUGR-status determined by head morphology and reason for death. All piglets remained in their birth litters until processing, but were fostered thereafter. Gilts were individually tagged but boars were unidentifiable after processing. In total, 495 piglets were crushed (LTOT) – 349 boars and gilts crushed before processing (LPROC), with a further 146 gilts crushed between processing and weaning. Non-significant effects (Proc mixed model with sow-week as a random factor) on both LTOT and LPROC included sow parity, piglet sire breed, gender and sow leg conformation scores. Significant effects on LPROC were an interaction between piglet weight and IUGR-status (P=0.014), the interaction between FAR-FLOOR and IUGR-status (P=0.01). When looking at LTOT, the piglet weight/IUGR status interaction loses significance, although the interaction between FAR-FLOOR and IUGR-status remains significant (P=0.005). In conclusion, sow limb conformation does not directly influence the risk of piglet crushing, though the interactions of IUGR-status and farrowing floor require more investigation. This research was funded by the EU FP7 Prohealth project (no. 613574).

A framework for automatic quantification of posture state and transitions in farrowing sows

R.J. Thompson[1,2], S.M. Matheson[2], T. Plötz[1], S.A. Edwards[2] and I. Kyriazakis[2]
[1]Newcastle University, Open Lab, Floor 3, 89 Sandyford Road, Newcastle upon Tyne, NE1 8HW, United Kingdom,
[2]Newcastle University, School of Agriculture, Food, and Rural Development, Agriculture Building, Newcastle
University, Newcastle upon Tyne, NE1 7RU, United Kingdom; r.j.thompson3@ncl.ac.uk

Understanding the behaviour of farrowing sows can highlight the need for intervention or lead to the prediction of movement patterns that are potentially dangerous for their piglets, such as crushing when lying down. We present a novel approach to automated classification and quantification of sow postures and posture transitions that enables large scale and accurate continuous behaviour assessment on farm. We use data recorded using tri-axial accelerometers attached to the hind-end of the sow, from a deployment that involved six sows over the period around parturition. We automatically classify the posture state of the sows for the full dataset with a mean F_1 score (a measure of predictive performance between 0 and 1) of 0.78. Posture transitions were detected with an F_1 score of 0.79. We automatically extract and visualize descriptive features that characterise the manner in which the sows change posture in order to provide clear and intuitive comparison of sow activity and lying behaviours. The methodology presented in this paper can be readily applied in large scale deployments with substantial potential for enhancing animal welfare and productivity on farm.

Responses to weaning in two pig lines divergently selected on residual feed intake depending on diet

H. Gilbert[1], J. Ruesche[1], N. Muller[2], Y. Billon[3], F. Robert[4], L. Roger[4] and L. Montagne[2]
[1]INRA-INPT, UMR1388 GenPhySE, 31326 Castanet-Tolosan, France, [2]INRA-AgroCampus Ouest, UMR1348
PEGASE, 35000 Rennes, France, [3]INRA, UE1372 GenESI, 17700 Surgères, France, [4]CCPA group, Quartier du
Haut-Bois, 35150 Janzé, France; helene.gilbert@toulouse.inra.fr

Two divergent pig lines for residual feed intake (RFI) (LRFI, pigs eating less than predicted for their production level and HRFI, pigs eating more than predicted) have been selected for 10 generations. Selecting for improved feed efficiency might impair animal ability to cope withstress. Weaning is a challenge that generates digestive troubles in pig production. This study tested the effects of a protective dietary regime after weaning on growth and health from weaning to slaughter in the RFI lines. Two batches of 66 females and castrated males from each line (264 pigs total) were tested in conventional post-weaning units (22 pigs per pen). Half were fed a conventional two-phase dietary sequence after weaning (transition 12 days after weaning), the other half were fed a three-phase protective feed sequence (transitions 10 and 19 days after weaning). The two first diets of the protective sequence contained a higher diversity of ingredients, including cooked cereals. Piglets were weighed at weaning (D0; 28 days of age), D1, D2, D6, D12, D19, D26 and when exiting the post-weaning unit (D42, 10 weeks of age). Diarrhoea and pen feed intake were recorded. No antibiotics were used. The LRFI piglets were heavier at weaning, with lower levels of diarrhoea at D1 and D2 than HRFI piglets (P<0.01). At D42, there was no BW difference between lines. The LRFI pigs had a lower feed intake and growth rate from D0 to D18 (P<0.001), with higher levels of diarrhoea at D6 (22% vs 14%). The protective sequence improved feed intake and growth and reduced diarrhoea, mainly in the LRFI line (P<0.001), suggesting a positive effect on the piglets that are more negatively affected by weaning. Very limited effects of the feeding sequences were subsequently observed on growth rate and body composition until slaughter. Thus, the protective diet after weaning had positive effects during the post weaning period but no consequence thereafter. This work has received funding from the EU-FP7 PROHEALTH project (grant agreement no. 613574).

Effects of weaning conditions on immune parameters of piglets

A. Buchet[1,2,3], C. Belloc[1] and E. Merlot[3]
[1]INRA UMR 1300 BioEpAR, BP 40706, 44307 Nantes, France, [2]COOPERL ARC Atlantique, BP 60238, 22403 Lamballe, France, [3]INRA, UMR 1348 Pegase, 35590 Saint Gilles, France; arnaud.buchet@rennes.inra.fr

Weaning leads to social, nutritional and spatial changes for piglets inducing very high adaptive requirements. The aim of this study was to investigat the effects of weaning and management practices at weaning on the immune stimulation of piglets. To disassociate weaning from the age effect, a 2×2 factorial design was used, where 64 piglets born from 12 litters were weaned in optimal (OC) or deteriorated (DC) farming conditions (n=16 per group) at either 21 (W21) or 28 (W28) days of age. In OC post-weaning pens, animals originating from 2 litters were housed in groups of 4 (0.39 m^2/pig) in clean and disinfected pens. In DC post-weaning pens, animals originating from 4 litters were housed in groups of 8 (0.20 m^2/pig) in dirty pens after a waiting time of 4 h at 20 °C. One week after weaning, DC pigs were mixed again. From 12 to 61 days of age, pigs were blood sampled weekly for leucocyte count and immunoglobulin G (IgG) and M (IgM) levels. Concentration of IgG decreased from 12 to 40 days of age and stabilized thereafter ($P<0.001$). Concentration of IgM and lymphocyte count increased from 12 to 47 days of age ($P<0.001$). IgG and IgM concentrations were positively ($P<0.05$) while lymphocyte count was negatively ($P<0.01$) influenced by body weight. Neutrophil counts were stable uo to 5 days after weaning, but increased 12 days after weaning ($P<0.001$). This rise was greater in DC than OC pigs ($P<0.001$) and in W28 compared to W21 pigs ($P<0.01$). DC pigs had also greater IgM concentrations 12 days after weaning compared to OC pigs ($P<0.001$). In conclusion, the greater neutrophil count and IgM production in DC pigs 12 days after weaning, regardless the age at weaning, indicates that challenging conditions lead to higher stimulation of the immune system compared to optimal weaning conditions. This might explain the metabolic and growth differences observed between OC and DC pigs described elsewhere in this meeting.

This little piggy went to market: applications of genomics for sustainable pig health

M.F. Rothschild[1], F. Bertolini[1], J. Harding[2], D. Ciobanu[3], B. Mote[3] and G. Plastow[4]
[1]Iowa State University, Department of Animal Science, 2255 Kildee Hall, Ames, IA 50011, USA, [2]University of Saskatchewan, Department of Large Animal Clinical Sciences, Western College of Veterinary Medicine, Saskatoon, SK, Canada, [3]University of Nebraska-Lincoln, Department of Animal Science, Lincoln, NE, USA, [4]University of Alberta, Livestock Gentec, Department of Agricultural, Food and Nutritional Science, Edmonton, Alberta, Canada; mfrothsc@iastate.edu

Losses in pig production due to disease, reduced productivity and increased health costs exceed $2 billion annually in the US and $20 billion worldwide. Usual approaches to disease management include vaccination, drug therapy and biosecurity. One often overlooked approach is the application of genetics and genomics to select for pigs that are generally healthier or less susceptible and, in some cases, resistant to particular diseases. Recent advances in genomic technologies and sequencing of the pig genome have provided excellent clues as to genetic differences for susceptibility to diseases such as PRRS, PCV2 and possibly PEDV and to environmental factors, such as heat stress, that affect pig health and productivity. Future approaches to combating disease may include genotype-dependent therapies and vaccinations as well as genome editing. Opportunities to apply these approaches and their sustainability will be examined.

Can we select for both resistance and tolerance of pigs to PRRS?

G. Lough[1], H. Rashidi[2], I. Kyriazakis[3], N. Deeb[4], J.C.M. Dekkers[5], J. Lunney[6], R. Rowland[7], H.A. Mulder[2] and A. Doeschl-Wilson[1]
[1]Roslin Institute, Genetics & Genomics, Easter Bush, EH25 9RG, United Kingdom, [2]Wageningen University and Research Centre, Wageningen, 6708 PB, the Netherlands, [3]Newcastle University, Newcastle upon Tyne, Tyne and Wear, NE1 7RU, United Kingdom, [4]PIC North America, 100 Bluegrass Commons Blvd., Ste. 2200, Hendersonville, TN 37075, USA, [5]Iowa State University, 239 Kildee Hall, Ames, IA 50011, USA, [6]USDA, Animal Parasitic Diseases Laboratory, 10300 Baltimore Avenue, Beltsville, MD 20705, USA, [7]Kansas State University, K-231 Mosier Hall, Manhattan, KS 66506, USA; graham.lough@roslin.ed.ac.uk

A host can adopt two strategies when faced with infection: resistance, the ability to limit pathogen replication, and tolerance, the ability to limit the impact of infection on performance, with or without reducing pathogen load. Both strategies may be under host genetic control, but evidence for genetic variation in tolerance and the relative contributions of resistance and tolerance to infection-induced performance losses are poorly understood. Using infection and growth data from 1,320 weaner pigs challenged with the same dose of Porcine Reproductive and Respiratory Syndrome (PRRS) virus, previous studies have established significant genetic variation in resistance (serum viral load, VL) and average daily gain (ADG) of infected piglets. Our aims were to determine whether there was significant genetic variation in tolerance of pigs to PRRS and contributions of resistance and tolerance to overall genetic variation in growth under infection. Using multi-variate linear mixed models, we found evidence for genetic variation in tolerance of pigs to PRRS. However, random regression sire models for ADG with a tolerance slope fitted for each sire, did not result in a better fit to the data than models without genetic variance in tolerance. Furthermore, we found a strong positive genetic correlation between resistance and tolerance (0.76), indicating that selection on one trait would favourably affect the other. Resistance and tolerance contributed 4 and 5% to the overall genetic variance in ADG, respectively. In contrast, the level of ADG (ADG at average VL) absorbed most of the genetic variation in ADG (84%). Implications of these findings for genetic control of PRRS and other infectious diseases will be discussed.

Are the consequences of coccidiosis affected by selection improved performance?

P. Sakkas[1], I. Oikeh[1], D.P. Blake[2] and I. Kyriazakis[1]
[1]Newcastle University, School of Agriculture Food and Rural Development, King's road, Newcastle on Tyne, NE1 7RU, United Kingdom, [2]Royal Veterinary College, University of London, Department of Pathology and Pathogen Biology, Royal Veterinary College, University of London, AL9 7TA, Hatfield, United Kingdom; panagiotis.sakkas@ncl.ac.uk

Coccidiosis reduces mineral absorption and bone mineralization in broilers. We hypothesized that the effects of a coccidian infection will be more pronounced on broilers selected for higher ADG and lower FCR, including reduced bone mineralization. 144 chicks of either a fast (F) or a slow growing (S) genotype were orally inoculated with 0 (Control), 2.5×10^3 (Low), or 7×10^3 (High) E. maxima oocysts at d 13 post hatch. Each treatment included 8 replicate pens with 6 birds per pen. The right femur and tibia of 1 bird/pen were dissected at d6 and d13 post infection (pi) to assess dry defatted bone (DDB), ash weight and bone breaking strength (BBS), as a proportion of BW on the day of dissection and percentage bone ash (PBA). Host genotype, parasite dose, and their interaction were treated as factors and data were analysed with GLM. On D6pi birds of the S genotype had higher ($P<0.05$) tibia DDB and ash, and higher femur DDB. On D13pi birds of the S genotype still had a higher degree of bone mineralization having higher tibia DDB and ash ($P<0.05$), as well as higher femur DDB ($P<0.01$), ash ($P<0.001$) and BBS ($P<0.05$). Infection caused a reduction in femur PBA and BBS ($P<0.05$) on d6pi. Infection effects were more pronounced on d13pi; as infected birds tended to have reduced tibia ash ($P<0.1$), and reduced tibia PBA and BBS ($P<0.01$), and femur PBA ($P<0.05$). At the same time infection reduced both tibia and femur DDB ($P<0.01$) and tended to reduce tibia ash. In conclusion, S growing genotypes showed a higher degree of mineralization over the growing period both for tibia and femur. Infection with E. maxima affected mineralization at peak of parasite replication (d6pi), but effects were more pronounced at the recovery period. However, our hypothesis that the effects of infection would be more pronounced in F genotypes was rejected, as no genotype and dose interactions were detected.

Effect of breeding and sex on performance and incidence of osteochondrosis in fattening pigs

M. Karhapää[1], H. Siljander-Rasi[1], T. Kortelainen[1] and M.-L. Sevón-Aimonen[2]
[1]Natural Resources Institute Finland, PL 18, FI-01301 Vantaa, Finland, [2]Natural Resources Institute Finland, Myllytie 1, FI-31600 Jokioinen, Finland; maija.karhapaa@luke.fi

Leg problems impair welfare and cause economic losses in pig production. The objective of this study was to determine the effect of breeding line and sex on susceptibility to leg disorders, lameness and occurrence of osteochondrosis (OC) in pigs. Finnish Landrace gilts and boars (n=137) of initial body weight (BW) 40 kg and final BW 120 kg were used. Litters from the same population were differentiated into a high maternal fertility line (HF) and high performance line (HP) on the basis of the average breeding index values of their parents. The pigs from different lines were group housed and had 5-phase individual feeding (ad libitum) planned to fulfill the nutrient requirements of growing breeding boars. Lameness was tested a week before slaughter. Joint surfaces of a total of 270 femurs or humeri were evaluated for OC lesions using a rating scale from 0 to 5 (0=no changes). Two 12 mm thick slices were cross-sectioned from the distal end of each femur and humerus and the severity OC was evaluated from slice surfaces. The statistical model included fixed effects of breeding line, sex, line×sex and the random effects of batch, sire and litter within sire. The ADG was 901 and 870 g/d, FCR 26.1 and 25.7 MJ NE/kg and carcass lean was 62.8 and 63.5% for the HF and HP lines, respectively (P>0.10). The leg disorders or lameness tests showed no differences between the breeding lines or sexes. The evaluation of distal femoral joint slices (sum of the OC scores) revealed no differences in OC severity between breeding lines but boars had more severe OC lesions than gilts (P<0.01). Only 1.2% of the surfaces of distal humerus joints were considered as normal. The majority (47.1%) had irregularity in the joint surface (score 2). OC lesions were more severe in humerus joints compared to femur joints. Results indicate that OC is common in Finnish Landrace pigs and boars have more severe OC lesions than gilts. This research was funded by the EU FP7 Prohealth project (no. 613574).

Dynamic response to weaning of two lines of pigs divergently selected on residual feed intake

N. Muller[1], H. Gilbert[2], F. Robert[3], L. Roger[3] and L. Montagne[1]
[1]INRA, Agrocampus Ouest, UMR1348 PEGASE, 65 rue de Saint-Brieuc, 35000 Rennes, France, [2]INRA, INPT, UMR1388 GenPHySE, 31326 Castanet-Tolosan, UMR1388 GenPHySE, /, 31326 Castanet-Tolosan, France, [3]CCPA group, ZA du Bois de Teillay, 35150 Janzé, France; lucile.montagne@agrocampus-ouest.fr

Breeding efficient pigs is a way to reduce dietary costs and environmental waste. However, optimisation of feed efficiency must not be linked to a decrease in the ability of animals to cope with challenges or stress, such as during the weaning period. This study aimed to characterize the response of two lines of pigs divergently selected for residual feed intake (RFI), an indicator of feed efficiency, calculated as the difference between observed and theoretical feed intake estimated from maintenance and production requirements. Animals of the RFI- line (- for under-consumption) are more efficient than animals from the RFI+ line (+ for over-consumption). Thirty six piglets from each line were weaned at 28 days of age and housed in individual pens. Their zootechnical performance and physiological characteristics were followed during three weeks. Piglets from the more efficient line (RFI-) appeared more affected by weaning during the first week than piglets from the RFI+ line. Daily feed intake and growth rate of pigs from the RFI- line were 65% and 25% lower (P<0.01) than RFI+. RFI- piglts suffered more from undernutrition, had more diarrhoea (10 vs 4/36 pigs), and had a higher inflammatory status in the first week after the weaning when compared with pigs from the RFI+ line. These piglets then seemed to adapt to the weaning conditions and to recover during the second and third week. Both lines had similar zootechnical performance and physiological characteristics at the end of the post-weaning period. To conclude, the dynamics of the post-weaning responses differed between lines, despite similar performance by the end of the weaning period. Piglets from the RFI- line were more negatively affected in the first week after weaning than those from the RFI+ line. This work has received funding from the EU-FP7 PROHEALTH project (grant agreement no. 613574).

The effect of prebiotics to control artificial infection of Salmonella
R. Breitsma[1], D. Parfitt[1] and B. Rouleau[2]
[1]Micron Bio-Systems, BFF Business Park, Bath Road, TA6 4NZ, Bridgwater, Somerset, United Kingdom,
[2]University of Guelph, Ontario Agricultural College, 50 Stone Road East, N1G Guelph, Ontario, Canada;
renata.breitsma@micronbio-systems.co.uk

Bacterial disease outbreaks impose significant constraints on poultry production. Poor digestion and nutrient absorption caused for example by Salmonella adversely affect animal welfare and productivity. This, in turn, can lead to significant losses for the farmer and can increase the potential for the contamination of poultry products marketed for human consumption. It is known that dietary supplementation with natural growth promoters such as prebiotics can assist in Salmonella prevention. To evaluate the effect of two different prebiotics on Salmonella shedding 80 day-old broiler chicks (Ross) from the same origin were randomly divided into 4 treatment groups. The control group received no feed additives, whereas three trial groups received dietary supplementation with two different prebiotics. Treatment groups 2 and 3 were supplemented with mannan-oligosaccharide (MOS) derived from Saccharomyces cerevisiae cell wall at inclusion levels 0.5 and 2 kg per ton of feed (kg/t), whereas treatment group 4 with 5 kg/t of fruchto-oligosaccharide (FOS). At 4 days of age all chicks were orally inoculated with challenge dose 8×10^5 cfu/bird of Salmonella ser. Enteritidis. At 8 days post infection (dpi) till the end of the trial (56 dpi) twice a week the cloacal swabs were taken and analyzed qualitatively for Salmonella. The results of the present study showed that in the groups supplemented with 0.5 kg/t MOS and 5.0 kg/t FOS Salmonella was detected qualitatively in cloacal swabs till the end of the trial. In the group supplemented with 2.0 kg/t of MOS Salmonella was not detected starting from 45 dpi ($P<0.05$) till the end of the trial. It might be assumed that due to adhesion effect of MOS to pathogenic bacteria the colonization of the gastro-intestinal tract with Salmonella was reduced consequentially minimizing the risk of shedding Salmonella into environment.

Effect of feed level and source of vitamin D on bone mineralization and performance of broilers
P. Sakkas[1], R.M. Gous[2], T.R. Hill[1] and I. Kyriazakis[1]
[1]Newcastle University, School of Agriculture Food and Rural Development, King's rd, NE1 7RU, Newcastle upon Tyne, United Kingdom, [2]University of KwaZulu-Natal, Animal and Poultry Science, Carbis Rd, Pietermaritzburg, 3201, South Africa; panagiotis.sakkas@ncl.ac.uk

We hypothesize that: (1) broiler performance and bone mineralization will benefit from a higher vitamin D (vit D) supplementation level and increased 25-OH-D_3 ($25D_3$): D_3 (D_3) ratio in the feed than current recommended levels and; (2) The effects of increasing $25D_3$: D_3 will be more pronounced at lower levels of vit D supply. 2,880 d-old Ross 308 chicks, were sexed and reared separately in 48 floor pens, each pen containing 60 birds. Birds were allocated to 9 diets offered from d1-d21, and were allocated one of 3 different levels of vit D supplementation, (Low (L): 2,000, Medium (M): 4,000 and High (H): 8,000 IU/kg of diet respectively) and one of 3 combinations of sources of vit D: 100% D_3, 50% D_3: 50% $25D_3$ or 100% $25D_3$. At d21 tibia breaking strength, tibia and femur dimensions, ash weight as a proportion of BW, and tibia ash percentage and ash Ca and P percentage were measured in 2 chickens per/pen. Results were analysed with GLM with pen as the experimental unit and level, source of vit D and sex as factors. Sex affected BW ($P<0.0001$) and FCR ($P<0.02$) being higher and lower for males than for females, respectively. FCR tended to be affected by the interaction between level and vit D source ($P<0.1$); birds on the L level tended to have lower FCR when fed $25D_3$ than D_3. Sex affected tibia and femur length, both being higher for female than male birds ($P<0001$). Ash weight was affected by vit D level ($P<0.019$), being higher for birds on the H than the M or L levels of vitamin D. There was an interaction between level and sex tibia Ca content ($P=0.032$) being higher for male than female birds on M. Level and source of vit D tended ($P<0.1$) to interact for tibia width and ash. Our results suggest that at low supplementation levels, $25D_3$ tended to improve FCR while mineralization increased beyond 4,000 IU/kg.

An ELISA test for the diagnosis of the two most common nematodes in chickens
G. Das[1], M. Hennies[2] and M. Gauly[3]
[1]Leibniz Institute for Farm Animal Biology, Institute of Nutritional Physiology 'Oskar Kellner', Wilhelm-Stahl-Allee 2, 18196, Dummerstorf, Germany, [2]TECOdevelopment GmbH, Marie-Curie-Straße 1, 53359, Rheinbach, Germany, [3]Faculty of Science and Technology, Universitätsplatz 5, 39100, Bolzano, Italy; gdas@fbn-dummerstorf.de

Infections of chickens with Ascaridia galli (Ag) and Heterakis gallinarum (Hg) are currently re-emerging. The aim of this study was to compare diagnostic ability of a newly developed ELISA test with faecal egg counts (FEC) in chickens experimentally or naturally infected with either of the nematodes. An ELISA based on Ag body antigens was developed to measure antibodies (IgY) against both Ag and Hg, in plasma and egg yolk samples, which were collected from experimentally (n=154) or naturally-infected chickens (n=28) or from un-infected controls (n=34). Worm burdens were determined 28 or 30 weeks post infection in two separate experiments (Exp) performed with either nematode. Area under ROC curve (AUC) was significantly (P<0.05) higher with the egg yolk samples (0.98) than with plasma (0.90) or FEC (0.90) in Ag-Exp, while all three tests performed similarly (≥0.94) in the Hg-Exp. Both the ELISA and FEC tests had the highest specificity (100%) in the Ag-Exp, whereas 7-10% of uninfected control birds were false classified in the Hg-Exp with no difference between ELISA and FEC. The ELISA test was able to detect Ag-infected animals more accurately with egg yolk samples (sensitivity 97%; CI: 90-100%) than with plasma (82%; CI: 71-90%) or with FEC (79%; 68-88%). ELISA with plasma or egg yolk and the FECs resulted in similar test accuracy with sensitivities ranging from 82 to 93% in the Hg-Exp. The ELISA had a sensitivity of 87.5 and 95% using plasma samples to identify naturally-infected animals harbouring either Ag or Hg, respectively. When egg-yolk samples were used, the test sensitivity improved for Hg (100%). As the egg yolk antibodies provide highest test accuracy with both nematodes and are able to be collected from the animals in a non-invasive, host-friendly way, we suggest the ELISA system with the egg yolk samples as the most accurate diagnostic test.

Effects of oregano, attapulgite, benzoic acid on gut health of broiler chickens
I. Skoufos[1], I. Giannenas[2], N. Papaioannou[2], A. Tsinas[1], S. Skoufos[2], J. Anastasiou[3] and A. Tzora[1]
[1]TEI of Epirus, Agriculture Technology, Division Animal Production, Kostakioi Artas, 47100, Arta, Greece, [2]Aristotle University of Thessaloniki, School of Veterinary Medicine, Thessaloniki, 54124, Thessaloniki, Greece, [3]Trinity Nutrition Ltd. & Invesco Feeds Technology Ltd., Aeta Place, Gortnakesh, Cavan, Co. Cavan, Ireland; jskoufos@teiep.gr

The aim of the present study was to investigate the effects of oregano, attapulgite, benzoic acid and their combination on microflora composition of jejunum and caecum, the intestinal architecture and the breast and thigh meat quality. A total of 400 one-day-old broiler chicks were used in a 42-day trial to evaluate the efficacy of oregano essential oil (ORE), attapulgite (ATT), benzoic acid (BEN), and their mixture (MIX) on intestinal microbiota, composition and intestinal architecture. Chickens were randomly distributed into five treatments with four replicates of twenty chickens per pen: CON, control diet; ORE, addition of 0.5 g/kg Ecodiar® containing oregano essential oil; ATT, addition of 3 g/kg Ultrafed® containing attapulgite; BEN, addition of 1 g/kg Vevovital® containing benzoic acid; MIX, addition of 0.5 g/kg Ecodiar, 3 g/kg Ultrafed® and 1 g/kg Vevovital. At the end of the trial, total counts of bacteria, Enterobacteriaceae, Lactobacilli, and Clostridium perfringens were enumerated by real time PCR at both jejunum and cecum at the day of slaughter. Intestinal morphology was carried out in duodenum, jejunum and ileum, for villus height and crypt depth. Cell proliferation was also evaluated in the small intestine and the cecum. No significant (P>0.05) differences were found in meat quality. Enterobacteria populations in the jejunum, were significantly higher (P=0.009) in the MIX group. Crypt depth was significantly lower (P<0.019) and villus height to crypt depth ration was significantly higher (P<0.020, respectively) for the BEN group. Cell proliferation for ORE and MIX groups was significantly higher compared to the other three groups in the duodenum (P=0.001) and the jejunum (P=0.012). The results showed that dietary inclusion of oregano, benzoic acid and their mixture increased lactobacilli in cecum. In conclusion, dietary supplementation with oregano EO and benzoic acid could be a potential additive to improve intestinal microflora and functionality.

Chestnut tannins reduce coliform post-weaning diarrhoea in piglets
M. Girard, S. Thanner, A. Gutzwiller and G. Bee
Agroscope, ILS, Tioleyre 4, 1725 Posieux, Switzerland; marion.girard@agroscope.admin.ch

Post-weaning diarrhoea (PWD) in piglet, sometimes lethal, is one of the most widespread disease in pork production. Weaning is a stressful period during which digestive tract physiology rapidly changes. Hence, piglets are more susceptible to gastrointestinal pathogens, like enterotoxigenic E. coli F4 (ETEC) responsible for diarrhoea. The large use and misuse of antibiotics has led to increased occurrence of resistances, so alternative solutions are needed urgently. Hydrolysable tannins (HT) are known to have antimicrobial properties. The present experiment aimed to study whether a standard diet (C; CP: 17%; DE: 14 MJ/kg) supplemented with HT from chestnut (T; Silvafeed Nutri P/ENC for Swine, Silvateam, Italy) could reduce the incidence of PWD in piglet artificially infected with ETEC. At 23 to 31 d of age, 72 piglets were weaned and allocated in a 2×2 factorial design balanced for weaned body weight and litter. From the day of weaning, piglets had ad libitum access to water, electrolytes and either C or T diet. In each of the 2 groups, 4 d after weaning, 18 piglets were infected with 5 ml of ETEC (K88ac/heat labile and stable toxins) suspension at 10^8 cfu/ml orally, whereas the remaining 18 received 5 ml of saline solution. Feed intake and body weight were recorded weekly. The faecal score was determined daily for 14 d using a 5-level scale (1=dry to 5=watery diarrhoea). In the first week after infection, the faecal score and the number of days in diarrhoea were reduced (P<0.01) in the T group (2.7 ± 1.24 and 2 ± 1.9 d respectively) compared with the C group (3.1 ± 1.32 and 3 ± 2.4 d respectively). However, average daily gain and feed intake were similar (P>0.05) between the groups during the experimental period. No differences in the frequency of antibiotic treatment between the T and C group (5.6% of piglets received antibiotic treatment after 4 d of watery diarrhoea) was recorded and no piglet died. Consequently, in the first week after infection, T diet did not improve growth performance but reduced the severity of PWD.

Effect of tannin structures as antimicrobial agents against pathogenic bacteria
M.M. Dakheel[1], F.A.H. Alkandari[2], M.J. Woodward[2], I. Mueller-Harvey[1], C. Drake[1] and C. Rymer[1]
[1]University of Reading, Animal Science, School of Agriculture, Policy and Development, University of Reading, RG6 6AT, Reading, U.K, United Kingdom, [2]University of Reading, Food and Nutritional Science, School of chemistry, Food and Pharmacy, University of Reading, RG6 6AR, Reading, Berkshire, UK, United Kingdom; m.m.dakheel@pgr.reading.ac.uk

Plants contain an abundance of natural substances that hold promise as feed additives. Tannins are an important group of these plant compounds. Tannins can link with proteins and other components to make complexes that, depending upon their chemical nature, alter their activities. Recently, condensed tannins have been illustrated to be toxic against several pathogenic organisms, such as gastrointestinal parasites. The aim of this study was to investigate the effect of tannin compositions on antimicrobial activity against Gram-negative and Gram-positive bacteria. Tannin structures ranged from procyanidins to prodelphinidins and also included mixed tannins (procyanidin/prodelphinidin mixtures) and the mean degree of polymerisation (i.e. average tannin size) of these tannins ranged from dimers to decamers. The minimum inhibitory concentrations (MIC) of these tannin extracts were determined against two pathogenic models, Gram-negative (avian pathogenic Escherichia coli, APEC) and Gram-positive (Staphylococcus epidermidis), by using the agar diffusion and broth microdilution methods. Stepwise linear regression showed that the MIC for E Coli (which ranged from 5.0 to 10.0 mg/ml) was negatively related to mDP and positively related to PC% (MIC=7.83-0.591 mDP+0.0468 PC, R^2=0.781, P=0.005) while for S. epiderdimis (range 1.25 to 5.00 mg/ml) the relationship was: MIC=2.720-0.3100 mDP+0.03822 PC (R^2=0.848, P=0.001). These results show that PD has greater antimicrobial activity than PC, and that Gram positive bacteria may be more susceptible to tannins than Gram negative bacteria, perhaps because of the structure of their cell membranes. Higher mDP values were also important in reducing the MIC.

Efficacy of a novel efficient phytate degrading bacterial 6-phytase in weaned piglets

D. Torrallardona[1] and P. Ader[2]
[1]Institut de Recerca i Tecnologia Agroalimentàries (IRTA), Monogastric Nutrition, Mas de Bover, E43120 Constantí, Spain, [2]BASF SE, G-ENL/MT – F 31, 68623 Lampertheim, Germany; david.torrallardona@irta.cat

The objective of this study was to investigate the efficacy of a newly developed bacterial 6-phytase (6-Phy) on performance and Ca and P apparent total tract digestibility (ATTD) in weaned piglets. 312 newly weaned piglets (6.1±0.86 kg) in 104 pens were offered 13 treatments consisting of a basal negative control (NC), the NC supplemented with 4 levels of monocalcium phosphate (MCP), and the NC supplemented with 8 levels of 6-Phy. NC was formulated to contain 4.2/4.0 g P between 0-21/21-42 d post-weaning, respectively. MCP was added to provide 0.6/0.5, 1.2/1.0, 1.8/1.5 or 2.4/ 2.0 g additional P/kg feed between 0-21/21-42 d. 6-Phy was added to provide 125, 250, 375, 500, 750, 1000, 1,500 or 2,000 FTU/kg feed. A Ca:P ratio of 1.3 was used for all diets. Body weight and feed intake were controlled at 21 and 42 d. Faeces were sampled at 39-42 d to measure Ca and P ATTD using TiO_2 as indigestible marker. At day 43 piglets were killed by pentobarbital injection and dry weight and ash of left Os metacarpale III were measured. Over the 42 d trial, linear responses to 6-Phy and MCP were observed for weight gain (ADG) and feed efficiency (G:F; P<0.01). ADG (g/d) was increased from 282 in NC to 356 in the 1.8/1.5 g MCP-P diet and to 360 in the 1000 FTU 6-Phy diet. G:F was improved from 0.61 in NC to 0.69 in the 2.4/2.0 g MCP-P diet and to 0.69 in the 1000 FTU 6-Phy diet. ATTD of Ca and P (%) increased linearly with 6-Phy (P<0.001) from 58.7 and 9.2 in NC to 72.0 (1000 FTU) and 57.4/59.8 (1000/2,000 FTU), respectively. Finally, linear responses to 6-Phy and MCP were also observed for Os metacarpale III dry weight and ash content (P<0.01). Bone weight (g) increased from 2.98 in NC to 3.91 in the 1.8/1.5 MCP-P diet and to 3.84 in the 1000 FTU 6-Phy diet. Ash content (%) increased from 30.0 in NC to 38.4 in the 2.4/2.0 MCP-P diet and to 41.6 in the 750 FTU 6-Phy diet. It is concluded that the novel bacterial 6-phytase improves P ATTD in weaned pigs and can replace MCP without affecting performance and bone mineralization.

Effect of dietary mannan-rich fractions on bacterial communities, a new approach with old data

A. Corrigan and R. Murphy
Alltech, European bioscience centre, Meath, A86X006, Ireland; acorrigan@alltech.com

The understanding and description of nutritional interventions is important to help develop and understand their use in poultry allowing for the appropriate manipulation of diets to improve poultry health and performance. In this study we analysed the results of four previous broiler experiments to characterise consistent changes in the bacterial community structure of the broiler caecum in response to dietary supplementation with a mannan oligosaccharide. High throughput sequencing of the V4-V6 regions of the 16 S rRNA gene had been carried out previously to assess the caecal microbial alterations in response to MOS supplementation in a standard commercial production setting across four separate broiler trials. The datasets from each individual trial were then pooled and binned as one dataset and analysed for differences between control and MRF supplemented diets at day 35 posthatch. Results showed a trend that Phylum Bacteroidetes was increased and Phylum Firmicutes was decreased across all four trials compared to the control. We then used the random forest bioinformatics approach to identify a highly relevant set of microbial OTU's which are indicative of MOS supplementation in the broiler caecum. Results showed that using this approach to analyse 16 S rRNA gene sequences has allowed us to identify a novel set of candidate biomarkers for the identification of an MRF specific caecal microbiome. This work has given us the capability to begin understanding the link between how the GI microbes may influence the phenotypic effects typically associated with MRF supplementation in broilers such as improved FCR, immune modulation and improved body weight. This knowledge will be instrumental in developing future nutritional strategies and will be beneficial to the poultry industry.

Livestock and climate change: can we steer a path between the devil and the deep blue sea?
P. Thornton
CGIAR, Research Program on Climate Change, Agriculture and Food Security, International Livestock Research Institute, P.O. Box 30709, Nairobi 00100, Kenya; p.thornton@cgiar.org

Although climate change is just one of the drivers affecting livestock production systems globally, its impacts may be severe, especially in developing countries where hundreds of millions of people depend wholly or in part on livestock for their livelihoods. An overview of the current state of knowledge regarding climate change impacts on livestock production is presented. From an adaptation perspective, many options are available, ranging from changes in animal species and breeds to improved grazing and feeding strategies, for example. The impacts of many of these options on household incomes and food security are not that well understood at present. The technical mitigation potential of the livestock sector is considerable, via options that include sustainably intensifying livestock production, promoting carbon sequestration in rangelands, reducing emissions from manures, and changes in consumption patterns. The economic potential is much less than what is technically possible because of adoption constraints, costs and numerous trade-offs. Both adaptation and mitigation in the livestock sector at the appropriate scale will require widespread behavioural change on the part of all participants. Such behavioural shifts in turn will require highly effective enabling policy environments. The livestock-climate change nexus is complex, and the contrasts between developed and developing countries are profound. The ways in which the technical and policy-related challenges might be addressed in different places need to be appropriately contextualised: what is appropriate for vulnerable livestock-keeping communities in developing countries will be very different for what would apply to intensive livestock producers serving highly-regulated markets in Europe, for instance. A nuanced approach to livestock and climate change issues is greatly needed. Considerably more research and investment are needed to increase the affordability and adoption of adaptation and mitigation practices while avoiding negative impacts on livelihoods, economic development and the environment, in the pursuit of reducing poverty and enhancing food security for all.

Global comparisons of migration and adaptation features in pastoralist communities
A. Jenet, N. Buono, K. Van Troos, S. Mason, S. Di Lello, R. Saavedra and M. Gomarasca
Vétérinaires Sans Frontières International, Av Paul Deschanellaan 36-38, 1030 Brussels, Belgium; coordinator@vsf-international.org

Concerns about increased marginalization have let institutions to put pastoralism on the agenda such as the International Fund for Agricultural Development. A consultation process to evaluate the global political integration and the enabling environment of pastoralists was carried out in 26 countries distributed in 5 subcontinents and in depth in 8 selected pastoralist hotspots. On aspects of enabling environment, a minimum of 3 interlocutors per country were interviewed, whereas 315 pastoralists have been surveyed in the hotspots in respect to practices. Multiple comparisons of means were carried out using Scheffe function of SPSS statistical package. Mobility is a critical livelihood feature that enables pastoralist to adapt to harsh conditions. Results revealed that Afar (East Africa, annual migration distance 85±14.1 km), Arkhangai (Asia, 67±13.4 km), Chaco (South American, 55±12.7 km) and Altiplano (67±8.1 km) were characterized by limited mobility, while pastoralists in Tiris Zemmour (North Africa, 100±13.3 km), Gourma (West Africa, 168±14.1 km) and Wagadou (105±14.1 km) reported migrations from significant higher distances and were only exceeded by pastoralists from the East African Chalbi territory reporting 345±14.1 km annual herd migration. Impeded mobility was probably the reason of why in this survey migration and herd splitting were only mentioned by 50 and 29% of the pastoralist when asked about drought adaptation mechanisms while selling of livestock even at low prizes was for 62% of the pastoralists the main coping mechanism during periods of stress. It is remarkable that pastoralists chose distressful coping mechanisms which require longer periods to recover over adaptive mechanisms that do no harm. Perhaps there are constraints that limit full mobility? If mobility played a decreasing role as coping mechanism, market access becomes important.

Towards the sustainable transformation of cattle value chains in Nicaragua

R. Van Der Hoek, B.K. Paul, M.A. Mena, C.A. Birnholz, M.A. Mora and A.M. Notenbaert
CIAT, International Center for Tropical Agriculture, Tropical Forages Program, Km 17 Recta Cali-Palmira, 763537 Cali, Colombia; r.vanderhoek@cgiar.org

In Nicaragua the cattle sector is a major pillar of the economy and presents an important opportunity for the improvement of farmer livelihoods, through dairy and beef production, processing and marketing. The CGIAR Research Program on Livestock and Fish is implementing activities to capture this opportunity and transform the dual-purpose cattle value chain with a specific focus on inclusiveness and gender equality. These efforts need to be balanced with long-term sustainability and environmental stewardship. Due to its extensive character, current cattle production leads to soil degradation, deforestation and a shift of the agricultural frontier towards the vulnerable Caribbean region. The sector also contributes 63.3% of total greenhouse gas (GHG) emissions. Interventions on forages, breeding and husbandry have a great potential to increase cattle productivity and carbon accumulation, mitigate GHG emissions and recuperate degraded soils. Improved pasture systems, for example, contribute to carbon accumulation, and reduce GHG emissions per unit of product. Based on standard calculations, such as the Revised Universal Soil Loss Equation (RUSLE) and the guidelines of the Intergovernmental Panel on Climate Change (IPCC), this study assesses in two contrasting sites (rainfall) in Nicaragua the environmental sustainability of three intervention scenarios: (1) traditional extensive system; (2) silvopastoral system with improved pastures and fodder banks; (3) semi-intensive system based on scenario (2) with improved cattle breeds. Compared to scenario (1), scenario (3) shows highest increase in milk productivity (from 350 to 1,300 kg/ha/year), highest carbon accumulation (2 t/ha/year of CO_2 equivalent), as well as a decrease in GHG emissions from 10.7 to 5.1 kg CO_2 equivalent per kg of Fat-Protein Corrected Milk (FPCM). The methodology allows for rapid feedback to farmers and other value chain actors on interventions and encourages the various value chain actors to apply alternative and inclusive development pathways.

LCA results on a low emission farming concept in highly integrated pig and poultry production

T. Kaufmann[1], M. Binder[1] and C. Haasken[2]
[1]Evonik Nutrition & Care GmbH, Rodenbacher Chaussee 4, 63457 Hanau, Germany, [2]Evonik Technology & Infrastructure GmbH, Rodenbacher Chaussee 4, 63457 Hanau, Germany; thomas.kaufmann@evonik.com

Livestock value chains have emissions of green house gases like CO_2, CH_4 and N_2O (Global Warming Potential (GWP) in kg CO_2e) and nutrient emissions like N, P, and S, responsible for eutrophication (kg PO_4e) and acidification (kg SO_2e). Whereas life cycle analyses (LCA) according to ISO standard have more environmental impact categories, there is general agreement that these are the 3 most critical ones in evaluating the environmental impact of animal farming and food products thereof. This study aimed at showing the mitigation potential under typical European conditions of poultry and pig operations for the named impact categories by combining best practises for low emission feeding, manure management and renewable energy production in form of elecricity, heat, gas or biofuel. System boundaries where cradle to farm gate includig the effects during manure storage and application, the functional unit (FU) selected was 1000 kg live weight (LW) for pigs or broilers. 5 LCA scenarios using GaBi datasets were calculated. 1: reference (no supplementation of amino acids (AA), no biogas plant (BG) using manure as feedstock (-/-); 2: AA, no BG: (AA/-); 3: AA, BG for heat and electricity: (AA/CHP); 4: AA, BG upgraded to pure bio methane fed into gas grid: (AA/CH_4); 5: AA, BG upgraded to pure compressed biomethane used as biofuel to replace diesel fuel: (AA/CMG). In the feedmixes, soybean meal has been assumed to come from South America with a 53% share of land use change origin. For broilers (pigs) GWP could be reduced from 4,970 (4,293) kg CO_2e/1000 kg LW in sc. 1 to 2,854 (2,581), 2,645 (2,076), 2,689 (2,109) and 2,632 (2,068) in sc 2-5, respectively. For AP (kg SO_2/1000 kg LW) the respective values were 65,3 (50,1), 31,7 (34,3), 15,3 (16,3), 16,3 (17,0) and 16,2 (16,9). EP (kg PO_4/1000 kg LW) was reduced from 24,6 (23,9) to 14,3 (17,6) and 11,0 (13,4) in sc. 2 and 3 with no further improvement. It is concluded, that AA supplementation alone already has a significant mitigaton potential, which can be further increased by the different biogas options. Lower feed protein content using more AA and digestate treatment offer further reduction potential.

GHG emissions, production and economics of typical French beef and dairy farms in the horizon 2035

C. Mosnier[1], A. Duclos[1], M. Lherm[1], B. Lelyon[2], J. Agabriel[1] and A. Gac[2]
[1]Inra, UMR 1213 Herbivore, Theix, 63122 Saint-Genès Champanelle, France, [2]Institut de l'élevage, Monvoisin, 35652 Le Rheu cedex, France; cmosnier@clermont.inra.fr

In order to mitigate climate change, France committed during the COP21 to reduce its emission by 12% between 2013 and 2028. Would the cattle production sector meet this objective? Four contrasted scenarios have been developed at national level in order to explore possible futures: S1 'trend', S2 'milk production increases to answer the global demand', S3 'local oriented, environmentally friendly production', S4 'tough policy to reduce GHG emissions'. The bioeconomic model Orfee has been created and used to assess the impacts of the main drivers of these scenarios on the evolution of typical beef and dairy farms. For all scenarios we assume that some technological progresses would be available (increase in milk yield up to 25%, younger age at first calving, silage of cereal-protein crops, alfalfa, higher efficiency of fertilizers, organic to intensive production), but not necessary chosen by the model optimization. Two levels of prices are tested for milk and meat (favorable or not). In addition, labour productivity is doubled in S2 to account for the assumed high level of technology available on farms; in S3, organic farming with low level of concentrate feeds is imposed; in S4 a tax on carbon emission is introduced. The typical farms simulated encompass four beef cattle farms ranging from a farm specialized in suckler cow production based on permanent grasslands in mountainous areas to a cash crop farms with a male fattening unit. The four dairy farms have contrasting soil potential, forage systems and milk yield. We show that dairy cows in plain areas with cash crops and forage crops are the most sensitive to changes between scenarios since they could intensify their production more easily or quit dairy production to produce cash crops only. Technological progresses enable to increase production profit and to reduce GHG emission. Scenario S2 is the most favorable to production and income but emits more GHG. Scenario S3 induces a reduction of profit and has contrasted effects on GHG emissions. Cattle production is highly impacted by carbon taxes.

Pasture-based livestock production in the context of climate change

F.P. O'Mara
Teagasc, Research Directorate, Head Office, Oak Park, Carlow, Ireland; frank.omara@teagasc.ie

Livestock are significant contributors to climate change, and are responsible for 10-12% of global anthropogenic greenhouse gas emissions. Most of these emissions come from methane produced from rumen fermentation, and also from emissions associated with manure, either deposited directly at pasture, or during storage and subsequent spreading. Another important source of emissions is from fertilisers used in pasture and feed production for livestock. The EU funded AnimalChange project concluded that the potential to reduce emissions from European livestock production was less than for regions outside the EU as EU production systems are already relatively efficient. Nevertheless, overall emission intensity could be reduced by 15-20% between 2005 and 2025 by marginal adjustments such as feed efficiency, biogas and energy use efficiency. The potential reduction is higher in the longer term (30 to 50% by 2050) as new technologies and practices such as legume introduction into grassland, geographical relocation and bio-refineries become more widespread. Adaptation will be increasingly important and will likely focus on extreme weather events and emerging diseases in the short term, while geographical distribution of livestock and feeding strategies will also evolve in the longer term. A technology Foresight recently conducted by Teagasc has identified technologies that will be important for agriculture over the next 20 years. Those relevant to livestock production include plant and animal genomics and related technologies, animal and soil microbiota, and digital technologies. New advances in these technology areas will be important if pasture and livestock production is to intensify sustainably in the coming years, and contribute to the twin challenges of food security and climate change.

Key factors influencing the carbon footprints of Northern Ireland dairy farms

A. Aubry[1], S.J. Morrison[1], T. Yan[1], P. Caskie[2], P. Keatley[2] and C. Ferris[1]
[1]Agri-Food and Biosciences Institute, Agriculture branch, Hillsborough, BT26 6QE, United Kingdom, [2]Department of Agriculture and Rural Development, Upper Newtownards Road, BT4 3SB, Belfast, United Kingdom; aurelie.aubry@afbini.gov.uk

Accurate estimates of greenhouse gas (GHG) emissions from livestock production systems are crucial to the development and evaluation of effective GHG mitigation strategies. The objectives of this study were: (1) to estimate the carbon footprints of commercial dairy farms using a GHG calculator developed using a partial lifecycle assessment approach and Tier II and III emissions factors, and (2) to identify farm characteristics that account for variability among these farms. The carbon footprints of 100 dairy farms were estimated, using technical and financial data collected for DARD's Farm Business Survey (FBS)/Farm Accountancy data network (FADN) system in Northern Ireland. Greenhouse gas emissions per kg of energy corrected milk (ECM) sold were on average 1.22 ± 0.15 kg CO_2e/kg ECM (range, 0.89 to 1.69 kg CO_2e/kg ECM). The large farm-to-farm variability illustrates the scope that exists to reduce emissions. Stepwise linear regressions indicated the best equation to predict emissions, as follows: carbon footprint (kg CO_2e/kg ECM) = $1.28 - 0.0815$ M $+ 0.32$ PH $+ 0.66$ C $+ 0.486$ N (R^2=0.64), where M is ECM sold (kg ECM/cow/yr) $\times 10^{-3}$, PH is the proportion of heifers, C is concentrate used per kg of ECM (fresh kg/kg ECM), N is the inorganic fertiliser N applied (kg N/ha/yr) $\times 10^{-3}$. Relationships between carbon footprints and other physical and financial indicators were also explored. Overall, results indicate that the carbon footprint tended to decrease with increased milk production per cow, reduced concentrate feed levels, reduced electricity consumption and fertiliser use, and reduced replacement rates, the latter often being overlooked in carbon footprint studies. Our approach also demonstrates the usefulness of FBS/FADN as a source of representative data to calculate average emission levels and ranges, thus providing indicators to monitor progress towards GHG reduction strategies.

Weather influences milk yield, feed intake and feed efficiency in dairy cows

D.L. Hill and E. Wall
Scotland's Rural College, Animal and Veterinary Sciences, King's Buildings, West Mains Road, Edinburgh, EH9 3JG, United Kingdom; davina.hill@sruc.ac.uk

Climate models predict an increase in mean ambient temperatures and the frequency of extreme events, such as heat waves, over the 21st century. When dairy cows experience conditions above their thermoneutral zone they are expected to decrease dry matter intake (DMI) to reduce metabolic heat production, but reduced feed intake can have negative effects on milk yield. Here, we tested how milk yield, DMI and feed efficiency (FE) vary in response to natural weather conditions in an experimental herd of Holstein Friesians in the UK. Cows belonged to 2 lines: those selected for high genetic merit for milk solids (Select) and those at the UK average (Control). We predicted that: (1) select cows would have higher milk yield, DMI and FE than Controls; and (2) milk yield, DMI and FE would vary with an index of temperature and humidity (THI). Animals received a mixed ration, distributed ad libitum, into automatic feed measurement gates, and were milked 3x a day. Milk yield was expressed as fat and protein corrected milk yield (FPCMY), and FE as FPCMY per unit dry matter. We analysed >73,000 daily records from 328 cows and 8 years using Restricted Maximum Likelihood. Select cows had greater FPCMY, DMI and FE than Controls. As THI increased, cows of both lines decreased FPCMY and DMI. This suggests that dairy cows under current temperate conditions are more sensitive to ambient heat than previously recognised. Importantly, there was a positive relationship between FE and THI. Improvements in FE with increasing temperature could partially offset the future cost of reduced milk yield under a warming climate. Understanding how efficiency and productivity are likely to change under future weather conditions can enable us to develop policies and management practices that minimise economic loss. This will be an important aspect of adaptation to climate change.

Animal health and greenhouse gas intensity: the paradox of periparturient parasitism

J.G.M. Houdijk, B.J. Tolkamp, J.A. Rooke and M.R. Hutchings
SRUC, West Mains Road, EH9 3JG Edinburgh, United Kingdom; jos.houdijk@sruc.ac.uk

We present the first direct measurements of pathogen challenge impacts on greenhouse gas (GHG) emissions for small ruminants. Twin-rearing ewes were fed pelleted lucerne ad libitum and infected with Teladorsagia circumcincta (PAR, n=16) or sham-infected (CON; n=16). Methane (CH_4) and carbon dioxide (CO_2) emissions were measured in respiration chambers on paired ewes (n=8). Assessed dry matter (DM), digestible organic matter (dOM) and nitrogen intake informed manure CH_4 and nitrous oxide (N_2O) emissions. Compared to CON ewes, PAR ewes had reduced DM intake (4.43 vs 4.16 kg/day; sed 0.10), dOM intake (2.17 vs 1.72 kg/day; sed 0.13), litter weight gain (718 vs 669 g/day; sed 24) and bodyweight gain (-69 vs -162 g/day; sed 37; all P<0.05). Compared to CON ewes, PAR ewes produced per day less enteric CH_4 (55 vs 49 g; sed 2.8; P<0.05) and CO_2 (3,463 vs 3,241 g; sed 26; P<0.10). Whilst yields per kg DM intake were similar, averaging 10.5±0.56 g/kg for enteric CH_4 and 675±111 g/kg for CO_2, yields per kg dOM intake for PAR ewes were greater than for CON ewes, i.e. 22.0 vs 25.2 g/kg (sed 1.54; P<0.05) for enteric CH_4 and 1,413 vs 1,664 g/kg (sed 111; P<0.05) for CO_2. In contrast, CON and PAR ewes produced per day similar amounts of manure CH_4 (2.98±0.14 g) and N_2O (2.03±0.06 g). However, yields per kg DM intake for PAR ewes were greater than for CON ewes for manure CH_4 (0.56 vs 0.65 g; sed 0.03; P<0.05) and for NO_2 (0.40 vs 0.42 g; sed 0.01; P<0.10), whilst these effects were more pronounced for yields per kg dOM intake for manure CH_4 (1.42 vs 2.07 g; sed 0.29; P<0.01) and for NO_2 (0.85 vs 1.12 g; sed 0.12; P<0.05). Extra feed needed for delayed weaning to similar lamb weight and greater maternal body weight loss compensation increased GHG intensity of PAR over CON ewes by 11% for enteric CH_4, 15% for CO_2, 32% for manure CH_4 and 30% for N_2O, resulting in combined GHG intensity of 9.28 vs 10.74 CO_2-eq/kg lamb weight gain (sed 0.39; P<0.01) for CON and PAR ewes, respectively. Our data support the view that ewe parasitism increases GHG intensity, implying that ewe worm control can improve productivity and reduce environmental footprint of sheep production systems. We thank Scottish Government (RESAS) for funding this work.

Equations to predict methane emissions in pasture-based systems

S. Stergiadis[1,2], C. Zou[2,3], X. Chen[2,4], M. Allen[5], D. Wills[2] and T. Yan[2]
[1]University of Reading, School of Agriculture, Policy and Development, P.O. Box 237, Reading, RG6 6AR, United Kingdom, [2]Agri-Food and Biosciences Institute, Large Park, Hillsborough, Co Down, BT26 6DR, United Kingdom, [3]Chinese Academy of Agricultural Sciences, Buffalo Research Institute, Nanning, Guangxi, 530001, China, P.R., [4]Lanzhou University, College of Pastoral Agriculture Science and Technology, Lanzhou 730020, China, P.R., [5]Agri-Food and Biosciences Institute, 18a Newforge Lane, Belfast, Co Antrim, BT9 5PX, United Kingdom; s.stergiadis@reading.ac.uk

The aim of this study was to develop predictions of methane emissions (CH_4), suitable for cows fed at low feeding levels in pasture-based systems, using dry matter (DM) intake (DMI) and/or liveweight (LW), grass nutrient and metabolisable energy (ME) concentration and digestible organic matter in DM (DOMD). Solely fresh-cut grass of various quality was offered at maintenance energy level to 25 nonpregnant nonlactating dairy cows. Grass nutrient composition, methane emissions in calorimetric chambers and nutrient intakes/outputs were measured daily and 3-day average of 300 data were used (n=100). Residual maximum likelihood analysis was used to develop prediction equations. Using DMI or organic matter intake (OMI), as sole predictors, showed lower prediction error than using digestible DMI or OMI. Adding concentrations of grass nutrients, ME and DOMD, in equations already using DMI or OMI as primary predictors, reduced prediction error and the best accuracy was shown in the equation: CH_4 (g/d) = $-74.28 + 25.72$ DMI (kg/d) $- 19.43$ ME (MJ/kg DM) $+ 426.9$ DOMD (kg/kg DM), R^2=0.73. Equations using only LW and grass nutrient composition as predictors, to be used in pasture-based farms where DMI measurements are not accurate, performed satisfactorily. A cross validation of the present equations and those currently used in greenhouse gas inventories, indicated that using the latter may under-predict methane emissions from pasture-based systems. The new equations showed improved accuracy and may be recommended for grazing animals at low feeding levels.

Development of mathematical models to predict enteric methane emission from cattle in Latin America

M. Benaouda, O.A. Castelán-Ortega and M. González-Ronquillo
Universidad Autónoma del Estado de México, Faculty of Veterinary Medicine, Carretera a Tlachaloya SN, Cerrillo Piedras Blancas, 50090 Toluca de Lerdo, Mexico; moh_ben44@hotmail.com

Ruminants contribute to global warming by the production of huge amounts of methane (CH_4) from enteric fermentation. A large part of world's cattle population is found in Latin America (LA) where little work has been done to quantify CH_4 emission. Expensive facilities and large number of animals required to measure in vivo CH_4 emission largely explain the lack of data. Thus, interest has increased to develop mathematical models to predict CH_4 emission. The objective of the present study is to collect a data set from in vivo studies conducted in LA and to develop statistical models, using mixed regression analysis, to predict CH_4 production by cattle from dietary and animal variables. A database from 20 publications was constructed, which included 66 mean observations of CH_4 outputs measured on 490 cattle. The best-fit (resulting in the lowest Root Mean Square Predicted Error; RMSPE and lowest Akaike Information Criterion; AIC) simple linear equations were CH_4(l/day) = $8.76_{(\pm2.73)}$ + $27.19_{(\pm3.53)}$×Dry Matter Intake (DMI; kg/day)[RMSPE=22%, AIC=651 and R^2=78%] and CH_4 (l/day) = $-13.74_{(\pm8.63)}$ + $4.78_{(\pm1.79)}$×Dry Matter Digestibility (DMD;%) [RMSPE=45%, AIC=597 and R^2=55%]. Multiple regression equation that predicted CH_4 slightly better than simple prediction equations was CH_4(l/day) = $-59.22_{(\pm5.47)}$ + $26.52_{(\pm4.01)}$×DMI + $1.11_{(\pm0.11)}$×DMD [RMSPE=21%, AIC=489 and R^2=92%]. However, further inclusion of other variables such as neutral detergent fiber (NDF;kg/day) and acid detergent fiber (ADF;kg/day) intakes improved the model predictability: CH_4(l/day) = $130.08_{(\pm8.15)}$ + $38.83_{(\pm6.81)}$×DMI − $0.8_{(\pm0.13)}$×DMD − $76.62_{(\pm23.47)}$×NDF intake + $70.68_{(\pm33.87)}$×ADF [RMSPE=23%, AIC=464 and R^2=93%]. Cross-validation of the last model [Observed CH_4(l/day) = $1.13_{(\pm0.1)}$ × predicted CH_4 (l/day) + $4.89_{(\pm30.62)}$; AIC=509 and R^2=74%] indicated that differences in CH_4 production between the diets could be predicted accurately. We concluded that feed intake, NDF, ADF intake and digestibility are the main predictors of total CH_4 production by cattle in LA. The models developed in the present study would be useful for regional inventory preparation.

Nongenetic factors affecting methane emissions from dairy cattle

M. Pszczola, K. Rzewuska and T. Strabel
Poznan University of Life Sciences, Department of Genetics and Animal Breeding, Wolynska 33, 60-637 Poznan, Poland; strabel@man.poznan.pl

Methane (CH_4) and carbon dioxide (CO_2) concentration were measured during milking in automated milking system (AMS) using Fourier transform infrared method. Data were collected at one commercial farm of Polish Holstein-Friesian cows in Western Poland with average milk production of 11,000 kg. Measurements were collected between 02/12/2014 and 09/02/2016 from 290 cows. Air samples were taken every 5 seconds from three AMS units in sequences of 24 hours per AMS. Measured CH_4 and CO_2 concentrations were averaged per milking. CO_2, as related to breathing rather than to eructation was used as a reference for data editing. Air samples taken when no cows were present in AMS, were used to quantify background CO_2 concentration. Observations were discarded whenever average CO_2 concentration was lower than the mean CO_2 background concentration (~1,810 ppm). Due to herd management system resulting in a lack of observations with gas measurements from the beginning of lactation, only observations starting at 25 day of lactation onwards were analyzed. Lactation length was limited to 305 days. After data edits, 34,387 observations from 290 cows each with average number of 119 average records remained for further analyzes. Methane emission was quantified in grams per day (CH4_GRAMS) based on CH_4 to CO_2 ratio and heat production units. The mean CH4_GRAMS was 415 g (SD 95.2 g) ranging from 52 to 970 g. The obtained phenotype was used for further analysis to determine factors influencing methane emission. The significance of the sampling time of a day, month of the year, week of observation, lactation number, lactation phase, AMS unit and age of calving was tested using mixed model approach with random cow effect. All the analyzed factors were highly significant (P-value<0.001). Based on these analyses, a model to estimate genetic variance of methane emissions will be built.

Ambient temperature and its effect on performance and udder health

V. Auer[1], C. Egger-Danner[2], H. Formayer[3], D. Leidinger[3], E. Ofner-Schroeck[4], F. Steininger[2], E. Zentner[4], K. Zottl[5] and B. Fuerst-Waltl[1]
[1]Univ. Nat. Res. Life Sci. Vienna (BOKU), Gregor Mendel-Str. 33, 1180 Vienna, Austria, [2]ZuchtData EDV-Dienstleistungen GmbH, Dresdner Str. 89, 1200 Vienna, Austria, [3]Univ. Nat. Res. Life Sci. Vienna (BOKU), Peter-Jordan-Str. 82, 1190 Vienna, Austria, [4]HBFL Raumberg-Gumpenstein, Raumberg 38, 8952 Irdning-Donnersbachtal, Austria, [5]LKV Austria, Dresdner Str. 89, 1200, Austria; steininger@zuchtdata.at

Prolonged periods of high ambient temperatures, particularly in connection with high humidity, not only affect the farms' production resources but may also have a direct impact on dairy cows. Rising ambient temperatures result in problems with regard to the regulation of the internal body temperature as the cows' ability to dissipate body heat is reduced. In high-yielding cows this effect is additionally enhanced due to higher metabolic activity. Certain construction and husbandry measures in dairy cattle farms may however alleviate the negative effect of high ambient temperature and humidity. Thus, data of the husbandry system (barn type tie stall/loose housing; forced ventilation yes/no; pasture yes/no) were collected in 150 Austrian dairy cattle farms covering the breeds Fleckvieh, Brown Swiss and Holstein (Projects StartClim2014.C and EfficientCow). Those data were merged with performance data from the central cattle data base and meteorological data from the Central Institute for Meteorology and Geodynamics. The effect of temperature (average of the maximum temperature and temperature-humidity-index, THI, up to and including three days before performance testing) on dairy performance traits (milk yield, fat and protein content) and udder health (Somatic cell score, SCS) was analyzed excluding and including husbandry systems. Taking husbandry system not into account, performance traits were significantly negatively affected by higher temperatures and higher THI while no effect on SCS was observed. However, including the husbandry system, significant but varying effects of temperature and THI within husbandry systems were found for SCS as well. Particularly with regard to the expected increase of days with high ambient temperature, the results illustrate the necessity to reduce dairy cows' heat stress in specific husbandry systems.

Evaluating climate change through month of calving effect on milk yield of dairy cattle

M. Djemali
INAT, Animal Genetics and Feed Resources Lab., 43 Avenue Charles Nicolle, 1080, Tunisia; mdjemali@webmails.com

Climate change is becoming a global concern. The specific objective of this study was to identify a biological method to quantify climate change effects. The effect of month of calving on dairy cattle milk yield in subsequent years could be a reliable indicator of climate change quantification. Milk data of Friesian dairy cattle, raised under South Mediterranean production conditions and recorded during two periods (1983-1987) with 74,201 lactations from 327 herds and (1998-2009) with 25,000 lactations from 156 herds were used in this study. In the first period, cows were North American Holsteins (2%), European Friesian (28%) and their progeny born and raised in Tunisia (70%). In the second period North American sired cows were (19%) and European sired cows were (14%). Statistical analyses of collected data identified significant sources of variation by region, herd-year within region countries of origin, month of calving and days in milk for dairy cattle milk yield. The effect of month of calving on milk yield was quantified by the least squares solutions obtained with the last month restricted to zero. In both periods, cows calving in the fall and early winter (September-January) yielded more milk than cows calving in spring (February-May) or summer (June-August). In the first period, cows calving in January yielded 384 kg more milk than cows calving in August. In the second period, the difference between January and August was reduced to be only 220 kg. Least squares means differences of milk yield from January to December (as a basis of comparison) were, respectively, in the first period (113, -34, -76, -100, -189, -238, -246, -271, -160, 7, 35 and 0 kg). These differences were, respectively, in the second period (-47, 6, -30, -86, -203, -226, -267, -268, -177, -76, -47 and 0 kg). Differences among month of calving effects on milk yield within periods showed a similar trend. However, differences in the second period showed a clean tendency of a decrease in differences between months especially between what is known in the first period (1983-1987) as traditional cool months and hot months which might be a climate change translation.

Use of archaeol as a proxy for methane emission in dairy cattle

L. Sandberg[1], N. Krattenmacher[1], S. Görs[2], B. Kuhla[2], J. Tetens[1], C.C. Metges[2] and G. Thaller[1]
[1]Institute of Animal Breeding and Husbandry, CAU, Olshausenstr. 40, 24118 Kiel, Germany, [2]Leibniz Institute for Farm Animal Biology (FBN), Wilhelm-Stahl-Allee 2, 18196 Dummerstorf, Germany; lsandberg@tierzucht.uni-kiel.de

Methane (CH_4) is a potent greenhouse gas. Genetic variation in CH_4 output from dairy cattle has been reported by several studies. Implementation of novel breeding approaches to mitigate enteric CH_4 requires the evaluation of methods for phenotyping a sufficient number of individuals. Archaeol, a membrane lipid from methanogens and detectable in faeces, might provide a useful proxy for quantification of CH_4 emissions from cattle on a large scale. For evaluation of archaeol, gas emission of four dairy cows in 1st and six in 2nd lactation was measured in respiration chambers in two consecutive trials with high and low starch diets at various days in milk (high starch HS; 178 g starch/kg DM at 100±3 DIM vs low starch LS; 99 g starch/kg DM at 135±3 DIM). The aim was to investigate the relationship of archaeol concentration and CH_4 output considering various aspects such as DM or starch intake with regression analysis. It was observed that cows fed HS diet had lower CH_4 output (520.4 l CH_4/d, P=0.4) and also excreted less concentrations of fecal archaeol (37.1 µg/g DM, P=0.001), compared to the LS trial (559.3 l CH_4/d and 47.6 µg/g DM). Cows fed HS have lower CH_4 outputs as starch in the diet shifts short-chain fatty acid production from acetate towards propionate accompanied with reduced availability of hydrogen for methanogenesis. Positive linear relationship between faecal archaeol concentration (µg/g DM) and CH_4 production relative to DM intake (l CH_4/kg DMI; R^2=0.53, P=0.001, n=20) as well as CH_4 production relative to starch intake (l CH_4/kg starch intake; R^2=0.94, P<0.0001) was found. Analysis of variance revealed a highly significant effect for diets. Results showed that faecal archaeol excretion is diet dependent. The use of archaeol as a linked trait for CH_4 emission in cattle needs further investigations.

The establishment of female line of alternative meat chicken for smallholder farmer's occupation

A. Molee[1], W. Molee[1], S. Khempaka[1], C. Hormta[1], P. Mernklatoak[1], T. Chormai[2], R. Bunnom[1], N. Tanpol[1] and B. Likitdejcharoj[1]
[1]Suranaree University of Technology, School of Animal Production Technology, 111 University Avenue, Muang District, Nakhon Ratchasima Province, 30000, Thailand, [2]Department of Livestock Development, Kabinburi Animal Husbandry and Research Center, Kabin Buri District, Prachin Buri, 25110, Thailand; amonrat@sut.ac.th

Alternative meat chicken is an answer for food security, food sustainability, and for smallholder's occupation. The establishment of female line of the chicken which has the ability to produce the low cost of one day old chick, their progeny has a suitable growth performance, and has the potential to improve their genetic with regard to egg production, are very need to do. Therefore the aim of the study is to evaluate the potential of crossbred chicken that can be female line of meat chicken. The criteria which were used to evaluate are: 1 egg production of the crossbreds, 2. growth performance of their progenies, and 3. the potential to improve the genetic of the crossbred. Sixty males exogenous crossbred chickens, and 60 males of Shiang Hai breed were randomly crossed with 900 female high performance egg production chicken to produce 1,500 for SUTT1 (Suranaree University of Technology tester1) crossbred, and 1,500 for SUTT2. Individual data of age at first egg (AFE), body weight at first egg (BWFE), egg weight at first egg (EWFE), monthly egg number (EN) (from the first 5% of egg production to 56 weeks of age), were collected. The cost of one day old chick was calculated. Genetic parameters, additive variance, residue variance, heritability, and genetic correlations were estimated by REML method with REMLF90. The results show that both SUTT1 and SUTT2 crossbred chickens have the potential to be in the female line for alternative broiler chickens. However, SUTT1 has more opportunity of selection, since the heritability, and additive variance of SUTT1 were higher than SUTT2 in most period.

Nurturing the livestock farmers of the future

K. Poppe
Wageningen UR, LEI, P.O. Box 29703, 2502 LS The Hague, the Netherlands; krijn.poppe@wur.nl

Key to the future of any industry in an aging Europe is the need to attract high calibre human capital. However, falling numbers of livestock farmers, often linked to increasing efficiency and greater professionalization in farming, can signal to young people that leaving the sector is better (and less risky) than joining or staying. Developments in ICT and precision farming techniques tend to accelerate this trend as capital investment and increasing scale become ever more important to farm business viability. At the same time current agricultural policies increase land prices and expose farmers to increasing price volatility in agricultural commodity markets. Not surprisingly, research shows that young farmers and new entrants see access to land and finance as the most significant barriers to entering the industry and growing farm businesses. This raises the question if, and how, agricultural policies could better support young farmers, and the industry as a whole, to attract its next generation of farm managers. The answer is not simple, as the succession process can be long and is influenced by a range of factors, not least taxation, of which –at a European level – little is known. The paper critically reviews the problem of the declining number of farms in the livestock industry and discusses how the CAP and national policies influence farm succession decisions.

Farm family succession planning

M. Lobley
Environment, Economics and Policy Institute (LEEP), University of Exeter, Exeter, EX4 4PJ, United Kingdom;
m.lobley@exeter.ac.uk

With most farms still being considered to be family farms of one form or another, the process of succession can be seen as the key process by which the business as well as an array of intangible assets is passed between generations of farming families. Indeed, it is often an objective of the family business to pass it on to one or more familial successors. The process of succession can be a powerful influence on the development of the farm business and can be a difficult and taxing time for farm family members. Against this background this paper draws on international research to look at rates of succession, some of the difficulties associated with successful succession, the benefits of succession planning and the dangers of getting it wrong.

Farming at the periphery of Europe: challenges and policy opportunities for livestock farms
J. Dwyer
University of Gloucestershire, Countryside and Community Research Institute, Oxstalls Campus, Oxstalls Lane,
Gloucester, Gloucestershire, GL2 9HW, United Kingdom; jdwyer@glos.ac.uk

Much policy attention has focused upon the challenges of maintaining viable farming in marginal areas of Europe, in past decades. Today, the support paid for this purpose under the Common Agricultural Policy's second pillar is the second largest spending measure in the Rural Development budget (17% of the total), signifying its continuing relevance. And yet farming in these areas remains beset with problems which threaten the long-term prospects for sustainable agriculture. In this context, succession planning is as much about successful business innovation as it is about careful management of fiscal and other liabilities. Recent work in upland areas of the UK, as well as some parts of central, southern and eastern Europe, offers ideas for how to make marginal farming an attractive, fulfilling and realistic career choice for the next generation. Marginal area farmers and farming families in future need to be multiple asset managers, engaged in diverse markets and enterprise types and building strong connections with wider public interest. New forms of collaboration and partnership, between farmers, but also with communities and other stakeholders, could help to foster better conditions for this kind of development.

Barriers to new entrants to farming: a Northern Ireland perspective
C. Jack[1], C. Miller[2] and D. Anderson[1]
[1]Agri-Food and Bioseciences Institute, Agricultural and Food Economics, 18a Newforge lane, Bt9 5PX, United Kingdom, [2]Queens University, School of Medicine, Dentistry and Biomedical Sciences, Belfast, BT7 1NN, United Kingdom; claire.jack@afbini.gov.uk

Throughout Europe small scale family farm businesses have long been considered the backbone of rural society. The nature of these family farm businesses are such that they tend to pass down through the generations and it is this custom of family succession which can create structural difficulties within farming. The trend across all EU-27 countries has been one of farm operators staying in the business longer alongside a decline in young farmers entering the sector. The aim of this study was to explore, at a Northern Ireland level, key barriers to new entrants to farming. An online survey was undertaken of new or potentially new entrants, focusing on socioeconomic characteristics' of new entrants, attitudinal responses to farming as a career and the real or perceived barriers and issues encountered in establishing a sustainable farm business. Most of the 420 respondents anticipated that their entry into farming would be through succession (taking over management of the existing family farm) and then ultimately inheritance of some/or all of the farm business and assets. Respondents indicated that broader economic and social factors (beyond the cost and accessibility of land, availability of capital and working capital), were also important. There was an acknowledgement of the role and impact of farm family household dynamics and the wider rural economy. From a policy perspective, and within the context of overall household well-being, there may be a need for support measures which aid succession transition through professional mediation and facilitation, supporting shared/partnership farming and different land-tenure arrangements. The results reflected the strong interaction between 'family' and 'business' highlighting the difficulties encountered in family farming, particularly in relation to the sustainability and viability of the farm as it moves from one generation to the next.

Access to land and land mobility
A. Jagoe
European Council of Young Farmers (CEJA), Carrigaline, Co, Cork, Ireland; pressofficer@ceja.eu

Alan Jagoe is a 33 year-old dairy and arable young farmer from Carrigaline, Co. Cork in Ireland He is President of the European Council of Young Farmers (CEJA) and before that as President of Macra na Feirme – Ireland's young farmer and rural youth organisation. Alan has a keen interest in Land Mobility, conducting studies and writing reports as part of CAP reform in Ireland. CEJA acts as a forum for communication and dialogue between young farmers and European decision makers, promoting a younger and innovative agricultural sector across the EU 28 and creating good working and living conditions for young people setting up in farming and those who are already 'Young Farmers'. The main concerns for young farmers that CEJA seeks to address are the access to land, credit and production rights whilst strengthening education and training facilities for young people in rural areas. CEJA represents 2 million Young Farmers in Europe with regular contact with young farmers' organisations, agricultural institutions and associations throughout Europe and worldwide. CEJA is independent of any political ideology.

Improving the connectivity between young farmers and young scientists
S. Finan
Ballinlough, Co, Roscommon, Ireland

Seán Finan is a beef farmer from Ballinlough Co Roscommon in the Republic of Ireland. Seán is the 35th National President of the Irish young farmers organisation Macra na Feirme from 2015-2017. He has held various roles within Macra na Feirme at club, county and national level including Vice President and Treasurer before being elected National President. He is farming on the home farm with his father and brother and believes there is a future in the beef and livestock sector for technically efficient, educated, business minded operators. Seán has a Degree in Civil Engineering (Hons) from the National University of Ireland, Galway and a Green Certificate in Agriculture. In his role as National President Seán represents Macra na Feirme and young Irish farmers at a National and also at a European level through CEJA. Sean, during his presidency is working to further grow and develop the Macra na Feirme Land Mobility Services, roll out a further training programme for young farmers and continues to lobby and represent young farmers on the issues which they face at the moment. These issues include income and viability, access to land, access to affordable credit, access to education, income volatility and implementation of CAP.

Succession at a Northern Ireland level

B. Bell
Ulster Farmers' Union, 475 Antrim Road, Belfast, Antrim BT15 3DA, United Kingdom; info@ufuhq.com

To help get more young people into farming in Northern Ireland, the Young Farmers Clubs of Ulster (YFCU) and the Ulster Farmers' Union (UFU) are planning to launch a new land mobility scheme for Northern Ireland. Last year YFCU and the UFU developed plans for a land mobility initiative, similar to the pilot scheme already in place at that time in the Republic of Ireland. The plan is to put older farmers facing this challenge in touch with younger farmers wanting to get into the industry. Both can then jointly develop long-term operational and financial plans for the farm – a classic mix of experience and youth coming together to the benefit of both. As a starting point a UFU funded survey of several hundred UFU members has recently been concluded. The information about the scale of the problem is vital if Northern Ireland is to make progress with a land mobility scheme. The intention is to create a broader range of succession options that would not otherwise be available to retiring or retired farmers. This will create a win/win situation for both the older farmer and the developing young farmer. Now that the survey is completed, the YFCU and UFU hope to launch the full land mobility initiative at the earliest opportunity.

Session 35 Theatre 1

Multi-breed genomic evaluations for 1 million beef cattle in Ireland

A.R. Cromie[1], R. Evans[1], J.F. Kearney[1], M. McClure[1], J. McCarthy[1] and D.P. Berry[2]
[1]Irish Cattle Breeding Federation, Highfield House, Cork, IE7, Ireland, [2]Teagasc, Moorepark, Fermoy, IE7, Ireland; acromie@icbf.com

Key stakeholders in Ireland (Irish Cattle Breeding Federation, Teagasc and the Department of Agriculture, Food and Marine) are currently developing multi-breed genomic evaluations for some 1 million beef cattle. The project is co-funded through the EU's Rural Development Program, with the overall objective of increasing rates of genetic gain for key traits related to profitability and environmental sustainability within the Irish suckler beef herd. A total of €300m has been allocated to the project over the 6-year period (2015-2020), of which some 15% will be allocated towards the cost of genotyping and related genomic evaluations. Phase 1 of the project is underway, with 300k beef animals genotyped in 2015. This is in addition to a further 100k animal's which were genotyped in 2013 and 2014, as part of an initial Irish government and industry funded initiative to help establish the required infra-structure (phenotypes and genotypes) for large scale multi-breed genomic evaluations. All animals have been genotyped on the International Dairy and Beef chip (IDB), with the latest version (the IDBv3) being a customised 54k chip developed in conjunction with Teagasc and Illumina Inc. It is anticipated that a further 300k animals/year will be genotyped in 2016 and 2017, resulting in a total of 1 million beef animals required for routine genomic evaluations by end 2017. Phase 2 of the project, will result in a further 1 million animals being genotyped in 2018-2020, bringing the total requirement for routine genomic evaluations to in excess of 2 million animals. Analysis to-date has been based on a sub-sample of some 100k sires and cows with reliable evaluations for key profit traits. Single-step genomic evaluations using Mix99 software have been applied to the dataset, with an almost doubling of reliability from the current 20% to almost 40% for individual traits and the relevant economic indexes. Initial feedback from industry has been very positive, with an expectation that the evaluations will become official from April 2016, after which we anticipate running routine evaluations, based on the increasing genotype and phenotype data every 2-3 months.

APY inverse of genomic relationship matrix: theory, analyses and questions
I. Misztal, I. Pocrnic, D. Lourenco and Y. Masuda
University of Georgia, Animal and Dairy Science, GA 30602, USA; ignacy@uga.edu

Genomic relationship matrix (GRM) can be inverted by Algorithm for Proven and Young (APY) based on recursion on a random subset of animals. While a regular inverse has a cubic cost, the cost of the APY inverse can be close to linear, allowing inexpensive computations with millions of genotyped animals. Theory proposed for APY assumes that optimal size of the subset (maximizing accuracy of genomic predictions) is due to a limited rank of GRM, which is a function of independent chromosome segments (Me) and subsequently of effective populations size (Ne). Simulation studies have shown that: (1) the dimensionality is almost a linear function of Ne but for large Ne can be depressed by limited number of genotyped animals and SNP markers; (2) accuracy of predictions with APY inverse is higher than with a regular inverse; (3) the distribution of independent chromosome segments is skewed. Tests using commercial data sets confirmed results by simulation. Comparisons of eigenvalue plots between simulated and commercial populations indicated an effective population size of 157 for Holsteins, 115 for Angus, 107 for Jerseys, 41 for broiler chicken and 30 for pigs. Experiences with the APY inverse raise a few questions. Can the rank provide information on the minimum SNP chip size that eliminates the polygenic component (or missing heritability)? Is the rank of GRM for a large two-breed population twice that of a single population? In simulation studies where QTL are on SNP markers, the best correlation of a simulated QTL effect is not with the actual SNP effect but with an average of adjacent SNPs. Is the optimum number (window size) of adjacent SNP a function of Ne and dictates the maximum resolution in GWAS? With all causative SNP are identified and their variances known, and appropriate weighted GRM (with APY inverse applicable) has the rank of the number of causative SNPs. Is the rank of weighted GRM with incomplete identification of QTLs (e.g. via GWAS or BayesB) smaller than that of a regular GRM? The APY inverse solves the problem of large-scale genomic computations and provides new insight into the genomic information.

Single-step genomic evaluation with many more genotyped animals
E.A. Mäntysaari and I. Strandén
Natural Resources Institute Finland (Luke), Biometrical Genetics / Green Technology, Humppilantie 7, 31600 Jokioinen, Finland; ismo.stranden@luke.fi

Recent innovations in algorithms have made single-step genomic evaluations feasible for complicated models, large populations and enormous number of genotyped animals. In state of the art implementations, the genomic relationship inverse is replaced by much sparser G_{APY} –inverse, and the inverse of pedigree based genetic relationship matrix A^{-1}_{22} (the inverse of relationship matrix for genotyped animals) is solved with help of Cholesky decomposition of A^{11} (the part of the A^{-1} for non-genotyped animals). For good performance, APY (algorithm for proven and young) requires right number and well chosen set of core animals. If the markers are not modeling 100% of the genetic variance, desired fraction w of A_{22} has to be added to G matrix before approximation of the inverse. We introduce two alternative approaches to implement single-step models, for evaluations where the number of genotyped animals (N_g) is much more than SNPs. In ssSNPBLUP, genotyped animals are modeled using $u_2=a_2+Zg$, where Z is the standardized genotype matrix of marker covariables, g is a vector of SNP-effects, and a_2 is a vector of residual polygenic effects independent from markers. While in the GBLUP the genomic (co)variance matrix $var(u2)= G+wA_{22}$ is inverted directly, in ssTBLUP we replace the G^{-1}_w by $(wA_{22})^{-1}+TT^T$, where T is a rectangular matrix, with N_g rows and r columns. As in $ssG_{APY}BLUP$ with w=0, the matrix A_{22} is not formed explicitly. The $r \leq$ column rank of Z, and is never more than number of SNPs. The new approaches give equivalent results with full implementation of ssGBLUP when w>0. Computationally both of the models are more demanding than ssGBLUP or $ssG_{APY}BLUP$ when the number of genotyped animals is less than number of markers, but for implementations with $N_g >> r$, both models are faster than ssGBLUP.

Test-Day single-step genomic evaluation using APY algorithm
M. Koivula[1], I. Strandén[1], G.P. Aamand[2] and E.A. Mäntysaari[1]
[1]Natural Resources Institute Finland (Luke), Green Technology, Myllytie 1, 31600 Jokioinen, Finland, [2]Nordic Cattle Genetic Evaluation, Agro Food Park 15, 8200 Aarhus N, Denmark; minna.koivula@luke.fi

Number of genotyped animals is increasing rapidly in dairy cattle creating computational challenges for genomic evaluation. The APY algorithm for core and young animals is one approach to overcome the computational challenges. We compared the single-step BLUP (ssGBLUP) using APY algorithm and regular ssGBLUP for the joint Nordic Red dairy cattle (RDC) evaluations for milk, protein, and fat using test-day model. We used test-day (TD) data containing ca. 4 million cows with 90 million records. There were 56,017 genotyped RDC animals of which 27,095 had either records or offspring in the full TD data. We tested different sets of core animals in APY to study how it affects the solutions. For the validation, four latest years of observations were removed from the TD data. The reduced data was used to solve GEBVs and EBVs for all animals in the pedigree and the full data was used to solve current EBVs and GEBVs for testing purpose. Deregressed proofs for 305 d milk, protein and fat yields were used for a posteriori validation. According to the results, the TD ssGBLUP model gives considerably higher validation reliabilities (r^2) than TD model without genomic information, e.g. for milk ssGBLUP r^2 was 0.46 and EBV r^2 was 0.36. No significant differences were found in validation reliability between the regular and APY TD ssGBLUP models. Also the GEBVs for the genotyped reference animals from the full TD model were 0.999 the same with the different ssGBLUP. With sufficiently large core group (more than 10,000 animals) also correlation between GEBVs for the young genotyped animals was over 0.99. Thus, the use of APY algorithm is feasible and can overcome some of the computational problems.

Efficient computations for single-step genomic evaluation
I. Strandén[1], K. Matilainen[1], G.P. Aamand[2] and E.A. Mäntysaari[1]
[1]Natural Resources Institute Finland Luke, Myllytie 1, FI-31600 Jokioinen, Finland, [2]Nordic Cattle Genetic Evaluation, Udkaersvej 15, 8200 Århus, Denmark; ismo.stranden@luke.fi

Genomic evaluations by single-step BLUP (ssGBLUP) allow combining genomic and pedigree information. Inclusion of genomic information requires dense matrices of size number of genotyped animals in the coefficient matrix of mixed model equations (MME). Because rest of the coefficient matrix in MME is very sparse, computations due to the dense matrices start to dominate computing time as number of genotyped animals increase. ssGBLUP has two dense inverse matrices for genotyped animals: genomic and pedigree based relationship matrix. We present an approach for solving ssGBLUP where these dense matrices will be replaced by computationally less demanding steps. For the inverse of pedigree relationship matrix of genotyped animals, the matrix is not explicitly build but required computations are made by a matrix equivalent formula. We present three approaches: iteration on pedigree (IOP), iteration in memory (IM), and cholmod (CM). For the inverse of the genomic relationship matrix, the approximate method of proven and young (APY) approach through matrix decompositions is formulated. These approaches were tested on a joint Nordic fertility evaluation for Holstein. Genomic data totaled 81,031 genotyped animals including bulls from the Eurogenomics collaboration. The pedigree had 9.73 million animals of which 7.5 million had fertility record. The fertility model had 11 traits. ssGBLUP were solved by preconjugate gradient (PCG) iteration. According to the results, computing time per iteration was 1.04 min. by regular ssGBLUP, 1.44 min. by IOP, 1.22 min by IM, and 1.06 min. by CM. When APY using 10,000 (20,000) animals in the core was used, the computing times were 0.86 (1.00) min. for IOP, 0.57 (0.74) min. for IM, and 0.46 (0.60) min for CM. Thus, ssGBLUP computations can be reduced by using a combination of approximate and direct computing approaches.

Dimensionality of genomic information and APY inverse of the genomic relationship matrix

I. Pocrnic[1], I. Misztal[1], D.A.L. Lourenco[1], Y. Masuda[1] and A. Legarra[2]
[1]University of Georgia, Animal and Dairy Sciences, Athens GA 30605, USA, [2]INRA, UMR1388, Castanet Tolosan, 31326, France; ipocrnic@uga.edu

The objective of this study was to evaluate by simulation the dimensionality of genomic information in closed populations and its effect on genomic predictions using regular or APY inverses of the genomic relationship matrices (GRM). Six populations were simulated with approximate effective population size (Ne) from 20 to 200. Each population consisted of 10 non-overlapping generations, with 25,000 animals per generation and phenotypes available for generations 1 to 9. The last three generations were fully genotyped assuming genome length L=30. GRM were constructed for each population and analyzed for distribution of eigenvalues. The number of the largest eigenvalues explaining x% of variation (EIGx, x=90, 95, 98, 99) in GRM ranged from 814; 1,611; 3,701; 6,253 (Ne≈20) to 5,512; 9,245; 15,483; 20,786 (Ne≈200), respectively. Genomic estimated breeding values (GEBV) were computed by single-step GBLUP using either a direct inverse of GRM or an inverse with the Algorithm for Proven and Young (APY) that is based on recursion on a random subset of animals; APY inverse has an approximately linear cost as opposed to cubic for the regular inverse. The sizes of the subset in APY were set to EIGx, with × as before. Accuracies of genomic estimated breeding values (GEBV) for the last generation with APY inverse peaked at EIG98 and were slightly lower with EIG95, EIG99 or the direct inverse. Most information in GRM is contained in about NeL largest eigenvalues, with no information beyond 4NeL. Genomic predictions with APY inverse of GRM are more accurate than by a regular inverse.

'Gene dropping A': Avoiding classical computations of the numerator relationship matrix in genomics

C. Edel, E.C.G. Pimentel, R. Emmerling and K.U. Götz
Institute of Animal Breeding, Bavarian State Research Center for Agriculture, Prof.-Dürrwaechter-Platz 1, 85586 Poing-Grub, Germany; christian.edel@lfl.bayern.de

In many applications of genomic selection the additive-genetic variance cannot be completely described by (high density) SNP-markers because of incomplete marker-QTL linkage disequilibrium. In applications of genomic BLUP this unexplained part of the genetic variance is commonly accounted for by adding an arbitrary proportion of the numerator relationship matrix A to the genomic relationship matrix G. Additionally, in standard formulations of single-step GBLUP the inverse of the numerator relationship matrix of genotyped animals is required to compute the so called inverse H matrix. However, computing the numerator relationship matrix for genotyped animals can be time consuming even in genomic selection applications of moderate size. With the advent of large amounts of genotyped females it might even become infeasible at some time in the near future. In this investigation we propose to calculate a numerical approximation for the numerator relationship matrix by dropping virtual SNPs, randomly assigned to the pedigree base, through the pedigree. We illustrate how a virtual genomic relationship matrix, calculated from an increasing number of SNP genotypes generated independently by gene dropping, converges towards the true numerator relationship matrix. By a small simulation we illustrate the performance of this approximation in explaining variance not covered by the observed marker genotypes for varying amounts of simulated SNPs and different basic conditions. Finally we demonstrate the feasibility of the approach by using it in a large two-step GBLUP routine application for dual purpose Fleckvieh using a forward prediction scenario. We discuss the findings in terms of validation reliability, bias and computing time in comparison to standard implementations using the numerator relationship matrix.

Single-step SNP-BLUP with on-the-fly imputed genotypes and residual polygenic effects

M. Taskinen, E.A. Mäntysaari and I. Strandén
Natural Resources Institute Finland (Luke), Myllytie 1, FI-31600 Jokioinen, Finland; matti.taskinen@luke.fi

Single-step genomic BLUP (ssGBLUP) evaluation combines relationship information from pedigree and genomic marker data. The inclusion of the genomic information into mixed model equations (MME) requires inverse of combined relationship matrix H that has a dense matrix block for genotyped animals. To avoid inversion of dense matrices, ssGBLUP can be transformed to Single-Step SNP-BLUP (ssSNP-BLUP) having observed and imputed marker coefficients. Simple block LDL' type decompositions of ssGBLUP matrix H were derived to obtain different types of linearly equivalent ssSNP-BLUP MME. For non-genotyped animals the imputed marker coefficient terms in the ssSNP-BLUP were calculated on-the-fly using sparse matrix decompositions without storing the imputed genotypes. Usual assumption is that the genomic information can not account all the genetic variation in animals. Therefore the ssSNP-BLUP was extended to account residual polygenic effects. These were added to genotyped animals and transmitted to non-genotyped animals using relationship coefficients similar to imputed genotypes. The relationships can be further orthogonalized to improve convergence of iterative solutions. The accuracy and iteration convergence of the derived MME were tested with a small data having 73,579 animals of which 2,885 were genotyped with 37,526 markers. The fully orthogonalized, non-preconditioned ssSNP-BLUP formulation was found to behave comparable to diagonally preconditioned original ssGBLUP.

Single-step GBLUP using APY inverse for protein yield in US Holstein

Y. Masuda[1], I. Misztal[1] and T.J. Lawlor[2]
[1]University of Georgia, 425 River Road, Athens GA 30602, USA, [2]Holstein Association USA, Inc., 1 Holstein Place, Brattleboro, VT 05302, USA; yutaka@uga.edu

Single-step genomic BLUP (ssGBLUP) is a method for genomic evaluations supporting millions of genotyped animals with the inverse of a genomic relationship matrix (G^{-1}_{APY}) calculated using an algorithm 'APY'. The objective of this study was to provide initial results in an application of ssGBLUP with G^{-1}_{APY} to 305-day protein yield for US Holsteins. Two G^{-1}_{APY} were tested; one was from 139,057 genotyped bulls with 12,895 core animals (APY140K) and the other one was from 764,029 genotyped animals with 12,913 core animals (APY760K). The predictor data set consisted of phenotypes recorded after 1989 and pedigrees limited to 3 generations back from recorded or genotyped animals. Genomic predictions (GPTA2011) were calculated for predicted bulls that had no recorded-daughters in 2011 but had at least 50 such daughters in 2015. We used the official DYD in 2015 (DYD2015) for the predicted bulls (n=3,797). We also used the official GPTA in 2011 with a multi-step method as a comparison. Coefficient of determination (R^2) and slope (b_1) were calculated from a linear regression of DYD2015 on GPTA2011. Using APY140K, the R^2 was 0.50 compared to 0.51 from the official GPTA. The b_1 was much better (0.98) compared with 0.81 from the official GPTA. With APY760K, the R^2 slightly decreased and the b_1 was still close to 1. The small decrease (up to 0.04) in R^2 could be due to several reasons: use of different data sets from the official ones, inclusion of highly imputed genotypes, and use of unweighted SNP markers in G. Also the validation could be biased due to selection when using all genotypes with incomplete phenotypes. All the computations finished within 11 hours including 4.2 hours to set up G^{-1}_{APY} with APY760K. We will finish the computation within 14 hours using 1 million genotyped animals with the same model. Single-step GBLUP can provide genomic predictions for all genotyped bulls and cows while accounting for pre-selection. Further research will determine the impact of various factors affecting the reliability such as validation methodology, weighting SNP markers, and quality of genotyped data.

Single step Marker Assisted Selection in breeding value estimation: do markers add value?

B. Zumbach, M.S. Lopes, S. Wijga, R. Bergsma and E.F. Knol
Topigs Norsvin Research Center B.V., Schoenaker 6, 6641SZ Beuningen, the Netherlands;
birgit.zumbach@topigsnorsvin.com

In pig breeding the estimation of breeding values including both genotyped and non-genotyped animals in a single step approach (ssGBLUP) has become standard. A way to further increase accuracies of breeding values could be the addition of markers with large effects discovered by Genome Wide Association studies (GWAS). The aim of this study was to verify if GWAS findings (QTL) add value to ssGBLUP under a marker assisted approach (MA-ssGBLUP) at the example of number of teats (NTE). Records of ~730,000 Large White pigs were available, of which ~15,000 animals were genotyped. The single step relationship matrix (H) was built with the software calc_grm. Breeding values for NTE were estimated using the MiXBLUP software with H as relationship matrix. In MA-ssGBLUP the GWAS SNPs were included as a fixed regression in the model. To provide all animals with a genotype for the selected GWAS SNPs, genotype probabilities were estimated. The analysis was multi-trait including 14 SNPs, of which 4 were specific for NTE. For comparison of SNP effects a single trait MA-GBLUP run with the 4 NTE SNPs and all the 14 SNPs was performed. Predictive ability of the different predicted breeding values (ssGBLUP: û; MA-ssGBLUP: $\hat{u}_g$, ûSNP, $\hat{u} = \hat{u}_g + \hat{u}SNP$) was estimated on a validation data set including ~1/3 of the youngest genotyped animals which had a phenotype and their contemporaries. The SNP effects of the single trait and 13-trait MA-GBLUP runs were in agreement. The by far largest SNP effect was for MARC0038565 on chromosome 7, pos ~103 kB which had been identified as the site of a major QTL. Predictive ability of MA-ssGBLUP $\hat{u}_g$, ûSNP and û were 0.428, 0.208 and 0.469, respectively, while û-ssGBLUP was 0.462. Although the confirmed QTL for NTE has a considerable effect the added value of MA-ssGBLUP over ssGBLUP is only marginal and not significant. Also for other traits and lines the MA-ssGBLUP confirmed the SNP effects detected by GWAS, however predictive ability was in general the same as for ssGBLUP. From a certain level of genotyped animals in the reference population onwards there is no evidence of added value in MA-ssGBLUP as H seems to account for the SNP effects.

Single-step genomic evaluation for fertility in Nordic Red dairy cattle

K. Matilainen[1], I. Strandén[1], G.P. Aamand[2] and E.A. Mäntysaari[1]
[1]Natural Resources Institute Finland (Luke), Myllytie 1, FI-31600 Jokioinen, Finland, [2]Nordic Cattle Genetic Evaluation, Agro Food Park 15, Skejby, 8200 Århus N, Denmark; kaarina.matilainen@luke.fi

Joint Nordic (Denmark, Finland, Sweden) genetic evaluations of female fertility traits were updated in 2015. Because the number of genotyped cows has increased considerably in recent years, the fertility model was found to be re-evaluated with genomic information included. The single-step genomic evaluation (ssGBLUP) was based on the same multi-trait multi-lactation animal model and variance components as the routine breeding value evaluation (BLUP). Heifer traits were non-return rate and length of service period. Cow traits in first, second and third parities were non-return rate, length of service period, and days from calving to first insemination. All 11 traits had low heritabilities (0.015-0.04), but moderately high genetic correlations between the parities (0.60-0.88). Phenotypic data contained 4 million animals with records, and pedigree consisted of 5,3 million animals. There were also 315 phantom parent groups (PPG), which were included into the pedigree or both into the pedigree and genomic information using QP-transformation to calculate breeding values directly. Genotypes were available for 23,520 animals. Validity of BLUP and ssGBLUP were studied using regression of deregressed proofs for full data on estimated breeding values (EBV) and genomic enhanced breeding values (GEBV) for reduced data. Both BLUP and ssGBLUP were solved with MiX99-program. With the PPG in the pedigree only, the BLUP and ssGBLUP converged after 3,052 and 29,063 iterations, respectively, while the ssGBLUP with the PPG in both the pedigree and genomic information converged after 2,453 iterations. Genetic trends based on males' EBV and GEBV followed well each other. For EBV, validation reliabilities were 0.16-0.32 and regression coefficients were 0.88-1.14 depending on the trait. The corresponding figures for GEBV were 0.22-0.45 and 0.72-0.99. In conclusion, the joint Nordic fertility evaluation using ssGBLUP was feasible after the genetic groups were correctly accounted. Especially for the cow fertility traits, the ssGBLUP model gave higher validation reliabilities than BLUP.

Identification of key contributors to increase phasing accuracy in complex population structures

M. Neuditschko[1,2], H.W. Raadsma[1], M.S. Khatkar[1], E. Jonas[1,3], E.J. Steinig[4], C. Flury[5], H. Signer-Hasler[5], M. Frischknecht[5,6], R. Von Niedrhäusern[2], T. Leeb[6] and S. Rieder[2]
[1]University of Sydney, Reprogen Animal Bioscience Group, 425 Werombi Road, 2570 Camden, Australia, [2]Agroscope, Swiss National Stud Farm, Les Longs-Prés, 1580 Avenches, Switzerland, [3]SLU, Department of Animal Breeding and Genetics, Box 7023, 75007 Uppsala, Sweden, [4]James Cook University, College of Marine and Environmental Sciences, 1 James Cook Drive, 4811 Townsville, Australia, [5]Bern University of Applied Sciences, Länggasse 85, 3052 Zollikofen, Switzerland, [6]University of Bern, Institute of Genetics, Bremgartenstrasse 109a, 3001 Bern, Switzerland; markus.neuditschko@agroscope.admin.ch

Knowledge of the genetic contribution of individuals is essential to select informative individuals for genotype imputation. Based on the Eigen Value Decomposition (EVD) of a relationship matrix we describe a novel approach to determine the genetic contribution of individuals within populations. The approach was applied and validated in four disparate datasets including simulated population, a highly structured experimental sheep population and two large complex pedigreed populations namely horse and cattle. In the simulated and sheep datasets, we identified all known key contributors within the populations, whilst in the horse and cattle dataset we applied the method to select small reference populations that increased phasing accuracy. Compared to commonly applied strategies to select informative individuals for genotype imputation including the identification of marginal gene contributions (PEDIG) and the optimization of genetic relatedness (REL) the selection of key contributors provided the highest phasing accuracies within the selected reference populations. Therefore, this method provides a valuable complement to common applied tools to select individuals for re-sequencing.

Multi-varietal genomic selection in French pig populations

C. Carillier-Jacquin[1], L. Tussel[1], J. Riquet[1], M.J. Mercat[2] and C. Larzul[1]
[1]INRA-GenPhySE, Animal Genetics, INRA, 24 Chemin de Borde Rouge, Auzeville CS 52627, 31326 Castanet Tolosan cedex, France, [2]IFIP/BIOPORC, Pôle Génétique, La Motte au Vicomte, 35651 Le Rheu cedex, France; celine.carillier@toulouse.inra.fr

Reference population size is a limiting factor to succeed in genomic selection. Multi-breed genomic evaluations were proposed as a solution to increase reference population size but did not always lead to genomic breeding value (GEBV) accuracy improvement. The aim of this study is to investigate multi-varietal genomic evaluation using varieties of Piétrain pigs. The data consisted in a main reference population (V1) of 96 genotyped boars, their 752 genotyped and phenotyped offspring and two additional populations of 13 (V2) and 17 (V3) genotyped boars with their 118 and 177 genotyped and phenotyped offspring, respectively. We study the interest of adding the largest reference population to each of the two small populations on GEBV accuracies. Sixty traits were studied including production traits, skin lesions, hormonal and blood parameters. Single-step genomic evaluations were carried out either separately for each variety or by grouping the varieties (V1 and V2 or V1 and V3). Theoretical accuracies obtained in the single variety genomic evaluations were compared to the ones obtained in multi-varietal evaluations for the sires and the offspring. The breeding value predictability in single variety and multi-varietal genomic evaluations were also compared using a 4-fold cross-validation. Average genomic kinship coefficient between the main reference population and the other two populations was of 8% (V1 and V2) and 12% (V1 and V3). GEBV theoretical accuracies and predictive ability of animals from the smallest reference populations (V2 and V3) improved when considering multi-varietal evaluation. These improvements were higher for V2 which was most related to the main reference population. Slightly higher theoretical accuracies were obtained for the offspring than for the boars. This preliminary study suggests that increasing the reference population size with animals from another variety enhances the GEBV accuracy and the predictive ability of animals from small size populations. This enhancement is proportional to the degree of relatedness between varieties.

Justification for setting individual animal genotype call rate threshold at 85%

D.C. Purfield[1], M. McClure[2] and D.P. Berry[1]
[1]Animal & Grassland Research and Innovation Centre, Teagasc, Moorepark, Fermoy, Co. Cork, Ireland, [2]Irish Cattle Breeding Federation, Bandon, Co. Cork, Ireland; deirdre.purfield@teagasc.ie

Data quality controls are essential to ensuring accurate genomic analyses including genomic predictions. One such quality control measure commonly employed is individual animal call rate, defined as the proportion of SNPs per animal where a genotype was called. No consensus exists on the minimum animal call rate that should be imposed; animals failing the reach these thresholds are discarded from analyses and often, at the expense of the breeder, the animal is resampled and re-genotyped. The objective of the present study was to propose the minimum call rate that could be applied without jeopardising data quality. A total of 144,672 samples genotyped on a custom Illumina genotype panel on 143,827 dairy and beef cattle were available. The genotyping panel comprises of 11,086 single nucleotide polymorphisms (SNPs) on either the Illumina Bovine SNP50 or high density genotyping panels. All genotypes were called using the Illumina GenCall method. Lab-dates (n=4) where >15% of the samples genotyped had a call rate <90% were not considered further. Of the remaining 142,433 samples, 493 animals had both a poor call rate (<90%) and a subsequent high call rate (>99%) after re-sampling and re-genotyping. The mean call rate for all samples was 98.77% (range: 15.81%-99.97%). The genotype and allele concordance rate was estimated for the 493 animals with both a low call rate and subsequent high call rate. Genotype concordance between low and high-call genotypes increased as call rate increased; 39.06% for a low call rate <40%, 61.10% for 50-60%,75.16% for 60-70%, 83.54% for 70-75%, 87.77% for 75-80%, 95.99% for 80-85% and 97.17% for 85-90%. Allelic concordance also increased linearly, suggesting that SNPs were unlikely to be called as opposing genotypes with an allele concordance of 98.45% for call rates between 85 to 90%.

Detection of pig genome regions determining performance traits

A. Borowska[1], H. Reyer[2], K. Wimmers[2], P.F. Varley[3] and T. Szwaczkowski[1]
[1]Poznan University of Life Sciences, Department of Genetics and Animal Breeding, Wolynska st. 33, 60-637 Poznan, Poland, [2]Leibniz Institute of Farm Animal Biology, Institute of Genome Biology, Wilhelm-Stahl-Allee 2, 18196 Dummerstorf, Germany, [3]Hermitage Genetics, Sion Road, Kilkenny, Ireland; tszwaczkowski@gmail.com

Recently, several statistical approaches to detect the effects of SNPs on livestock performance traits have been described. One of them is entropy analysis (the information theory) that measures the uncertainty connected with values of the traits. Conditional entropy indicates the most informative SNPs, whereas mutual information identifies the relationship between SNPs. The aims of this study were to verify the validity of the SNPs in relation to swine production traits and indicate interactions between the most informative SNPs using the analysis of entropy. The records of 1,296 pigs were included. Molecular data contained 61,565 SNPs. After filtering for low call rate and minor allele frequency, 50,951 SNPs were included in the entropy analysis. The following traits were analysed: backfat, days to 110 kg, feed conversion ratio (from 60 kg to 110 kg liveweight), per cent lean meat (achieved using an ultra-sonic scanner 'piglog 105') and terminal line index for Maxgro. The values of traits were divided into quartiles. The entropy H(SNP), conditional entropy H(Trait/SNP) and mutual information MI(SNP1\SNP2) were estimated for SNPs. The computations were performed using the R statistical package program. The performed study indicates important genome regions determining pig performance traits, distributed across a number of chromosomes. The most informative SNPs were mainly located on chromosomes: 1, 4, 5, 7, 9, 12, 14 and 17. The largest number of SNPs with high H (T/S) values were on chromosome 1. The most informative pairs of SNPs connected with all traits were located on chromosomes: 1, 14, 15 and 16. High mutual information was registered for SNPs located nearby.

Characteristics of haplotypes in a multi-breed dairy and beef cattle population
T.R. Carthy[1], D.C. Purfield[1], M. McClure[2] and D.P. Berry[1]
[1]Teagasc, Animal and Grassland and Research Innovation Centre, Moorepark, Fermoy, Co Cork, Ireland, [2]Irish Cattle Breeding Federation, Bandon, Cork, Ireland; tara.carthy@teagasc.ie

Knowledge of the extent of haplotype sharing among breeds could provide useful insights in the potential of effective across-breed genomic predictions. The objective of this study was to characterise the haplotypes present in a large multi-breed dairy and beef cattle population and to quantify the commonalities that exist between breeds. Genotypes were available from two different Beadchip panels, the Illumina BovineHD beadchip and the International Dairy and Beef (IDBv2) SNP chip on 113,562 animals. The animals included 69,777 cross-bred animals and 22,172 Holstein-Friesian, 373 Jersey, 5,412 Charolais, 6,732 Limousin, 1,612 Simmental, 3,861 Angus, 2,961 Hereford, and 661 Belgian Blue pure-breed animals as determined from pedigree analysis. Missing genotypes were imputed and linkage phase of haplotypes were inferred. Haplotypes consisting of 50 adjacent SNPs were constructed resulting in blocks of approximately 157 kb in length. A large number of haplotypes existed at each genomic position. Limiting to common haplotype (>1% frequency in the population), considerably reduced the number of haplotypes at each genomic position. The mean number of common haplotypes at each genomic position in the multi-breed population was 22.9%. The mean number of common haplotypes per breed across the genome was 16.9% varying from 11.4% in the Jersey to 17.9% in the Charolais. Across all pure-breed animals, 2.3% of the common haplotypes appeared in all breeds whereas 64.3% of the common haplotypes were breed-specific. The two dairy breeds, Holstein-Friesian and Jersey only shared 1.3% of common haplotypes. In the beef breeds, all the beef breeds shared 3.6% of the common haplotypes, whereas the three continental beef breeds (Charolais, Limousin, and Simmental) and the two British Isle breeds (Angus and Hereford) shared 11.6% and 16.4% of the common haplotypes, respectively.

SNP clustering associated with performance traits in chicken: entropy analysis
M. Graczyk[1], H. Reyer[2], K. Wimmers[2], R. Hawken[3] and T. Szwaczkowski[1]
[1]Poznan University of Life Sciences, Department of Genetics and Animal Breeding, Wolynska st. 33, 60-637 Poznan, Poland, [2]Leibniz Institute of Farm Animal Biology, Institute of Genome Biology, Wilhelm-Stahl-Allee 2, 18196 Dummerstorf, Germany, [3]Cobb-Vantress Inc., 4703 U.S. Highway 412 East Siloam Springs, AR 72761, USA; graczykmag@gmail.com

Body weight and its changes over time as well as feed efficiency can be perceived as main traits in genetic improvement programs in meat-type chicken. Recently, entropy analysis (based on the so-called 'information theory') is becoming more and more popular to search for SNPs affecting performance traits. It enables the exploration of large data sets, e.g. genomic SNP data, while indicating which part of the information (of SNPs) is more useful than the other. The objective of this study was to select the most informative SNPs and indicate chromosomal regions involved in the characteristics of performance traits. In total, 855 individuals were genotyped using the 60K Illumina iSelect chicken array. Analysed phenotypes comprised body weight at 36 days, body weight at 39 days, body changes between 36-39 days, feed intake between 39-46 days, and feed conversion ratio. Entropy of trait, conditional entropy and mutual information were calculated to indicate the association between SNPs. The results show that for all analysed traits the most informative SNPs are located on chromosome 1, 2, 4, 8, 12 and the sex chromosome Z. Large mutual information was registered for SNPs located nearby. Clusters of the most informative pair of SNPs connected with all recorded traits were located on chromosome: 1, 2, 3, 4, 6, 12, 20 and Z. Additionally, SNPs interacting with all analysed subsets of SNPs were observed. It can be concluded that entropy analysis is a useful method to estimate an association between SNPs and denote the specific chromosomal regions connected with the development of a trait.

A genome-wide association study using CNVs for production traits in Brown Swiss dairy cattle

R.T.M.M. Prinsen[1], M.G. Strillacci[1], F. Schiavini[1], A. Rossoni[2], B. Gredler[3], M.A. Dolezal[4], A. Bagnato[1] and A. Bieber[5]
[1]*Università degli Studi di Milano, Veterinary Medicine, Via Celoria 10, 20133 Milano, Italy, [2]Associazione Nazionale Allevatori Razza Bruna, Loc. Ferlina, 37012 Bussolengo (VR), Italy, [3]Qualitas AG, Chamerstrasse 56, 6300 Zug, Switzerland, [4]University of Veterinary Medicine, Veterinärplatz 1, 1210 Wien, Austria, [5]Research Institute of Organic Agriculture, Ackerstrasse 113, 5070 Frick, Switzerland; raphaelle.prinsen@unimi.it*

Detecting genetic variation such as Copy Number Variation (CNV) in cattle provides the opportunity to study their association with productive traits. The aim of this study was to investigate CNVs in 1,410 samples of the Brown Swiss cattle breed using Illumina BovineHD Genotyping BeadChip data and to perform a genome-wide association analysis for three production traits: milk yield (MY), fat yield (FY) and protein yield (PY). After severe quality control, CNVs were called with the Golden Helix SVS 8.3.1 and PennCNV software and were summarized to CNV regions (CNVRs) at a population level, i.e. overlapping CNVs, using BEDTools. Additionally common CNVRs between the two software were set as consensus. CNV-association studies were executed with the CNVRuler software using a linear regression model. Genes within significant associated CNVRs for each trait were annotated with a GO analysis using the DAVID Bioinformatics Resources 6.7. The quality control filtered out 294 samples. The GoldenHelix software identified 25,030 CNVs summarized to 398 CNVRs while PennCNV identified 62,341 CNVs summarized to 5,578 CNVRs. A total of 132 CNVRs were identified to be significantly associated with one or more of the three evaluated traits. The greatest association signal was given by a CNVR on chromosome 4 for MY. The associated regions overlap with 256 genes annotated on the Bos taurus genome assembly (UMD3.1). The result of this study is a comprehensive genomic analysis of the Brown Swiss breed, which enriches the bovine CNV map in its genome. Finally, the results of the association studies deliver new information for productive traits considered in selection programs of the Brown Swiss breed.

Analysis of FSHB gene polymorphism in Polish Landrace and Polish Large White × Polish Landrace sows

D. Polasik[1], M. Kumalska[1], Y. Sawaragi[1], G. Żak[2], M. Tyra[2], P. Urbański[3] and A. Terman[1]
[1]*West Pomeranian University of Technology in Szczecin, Al. Piastów 45, 70-311 Szczecin, Poland, [2]National Research Institute of Animal Production, ul. Sarego 2, 31-047 Krakow, Poland, [3]Institute of Genetics and Animal Breeding, Postępu 36A, 05-552 Magdalenka, Poland; grzegorz.zak@izoo.krakow.pl*

FSHB gene is considered as a marker for reproductive traits in human as well as in domestic animals. The aim of this study was to estimate associations between polymorphism in FSHB gene and reproductive traits in Polish Landrace breed (L, n=260) and Polish Large White × Polish Landrace crossbreed (LWxL, n=381) sows. The following traits were analyzed: total number of piglets born (TNB), number of piglets born alive (NBA) and number of piglets weaned (NW). The polymorphism in FSHB gene was determined by PCR-RFLP method using HaeIII restriction enzymes. A mixed linear model was applied for association analysis which considered effects of genotype, litter order and year season. Two alleles of FSHB gene were identified – A and B with following frequency: 0.058 and 0.942 (L), 0.150 and 0.850 (LWxL) respectively. The distribution of genotypes was in Hardy-Weinberg equilibrium in both of analyzed breeds. Analysis of relationship between FSHB/HaeIII polymorphism and reproductive traits in sows showed that genotype BB was favorable for all analyzed traits in first and second parity but differences were small and not confirmed statistically. The significant effect (P≤0.01) of litter order on analyzed traits was however observed.

Genotyping-by-sequencing application to different livestock species
A. Gurgul, A. Miksza-Cybulska, I. Jasielczuk, T. Szmatoła and M. Bugno-Poniewierska
National Research Institute of Animal Production, Krakowska 1, 32-083 Balice, Poland; artur.gurgul@izoo.krakow.pl

Together with reduction of costs and availability of next generation sequencing platforms, its routine application in livestock species genomics becomes more viable. In comparison with standard genotyping microarrays, it gives a more comprehensive view into the genome of the species, allowing not only for reproducible genotyping but also for detection of new or rare variants and direct analysis of causative mutations. Apart from the whole genome sequencing, a few other approaches relying on genome portion enrichment have been proposed. One of them is genotyping-by-sequencing (GBS), based on reduction of genome complexity with restriction enzymes. This approach seems to be inexpensive, simple to carry, repeatable and may be applied to variety of species (even with unknown genome sequences), and it does not require customized genotyping microarrays production. In this study, we tested the use of GBS approach for whole-genome genotyping in several livestock species, including cattle, sheep, horses and others. Libraries were generated using PstI-HF endonuclease and 48 custom-designed barcoded adapters. The obtained results demonstrated that GBS libraries can be easily generated for variety of species with only a little optimization steps and produce high quality sequences which can be then used for polymorphism discovery and genotyping.

Extent of linkage disequilibrium and haplotype block structure in Limousine and Simmental cattle
I. Jasielczuk, A. Gurgul, T. Szmatoła, T. Ząbek and M. Bugno-Poniewierska
National Research Institute of Animal Production, Department of Genomics and Animal Molecular Biology, Krakowska 1, 32-083 Balice, Poland; artur.gurgul@izoo.krakow.pl

Knowledge on the extent of linkage disequilibrium (non-random connection of alleles at different loci) between markers is essential to determine the number of markers required for applications based on genotyping technologies. In this study, we characterized the extent of LD based on the pairwise r2 statistic of SNPs at a distance up to 5 Mb and defined haplotype block structure including SNPs located at a maximum distance of 500 kb in Limousine (LM) and Simmental (SM) cattle breeds maintained in Poland. A panel of 40,158 (LM) and 40,117 (SM) high quality SNPs from Illumina BovineSNP50 v2 BeadChip was utilized for the analysis. We evaluated levels of LD in nine inter-marker distance thresholds. As expected, the maximum average r2 of 0.30 was obtained at a short distance up to 25 kb in both studied breeds. The minimum average values of r2 (0.01) was observed at the distances of 1,500-3,000 kb and 3,000-5,000 kb in LM and 3,000-5,000 kb in SM. The extent of LD estimated for individual autosomes was diversified. Average r2 per autosome reached the highest value on BTA7 – 0.036 (±0.072) in LM and 0.046 (±0.083) on BTA6 in SM cattle. The lowest mean LD was observed on BTA12 (0.026±0.055) and BTA25 (0.026±0.052) in LM and BTA10 (0.026±0.037) in SM cattle. In total, 828 (LM) and 667 (SM) haplotype block structures spanning 89,781 kb (LM) and 72,582 kb (SM) of the genome were detected. In LM cattle 3,181 (7.92% of all SNPs used) and in SM cattle 2,580 (6.43%) SNP markers formed blocks with a range of 2-15 SNPs per block. Mean block lengths were estimated at 108.4±99 kb for LM and 108.8±97 kb for SM. For both studied breeds BTA1 showed the longest haplotype block structures, having 66 blocks spanning 8,120 kb in LM cattle and 57 blocks spanning 6,754 kb in SM. The shortest haplotype blocks were observed on BTA28 in both breeds with 6 blocks in LM and 4 blocks in SM covering in total 243 kb and 29 kb, respectively.

Effect of QTL number and minor allele frequency, and marker density on genomic evaluation

E. Karaman and M.Z. Firat
Akdeniz University, Faculty of Agriculture, Department of Animal Science, Antalya, 07058, Turkey;
emrkaraman@gmail.com

Genomic prediction relies on the prediction of genomic estimated breeding values of animals utilizing thousands of markers distributed to whole genome. The extent of linkage disequilibrium (LD) between quantitative trait loci (QTL) and single nucleotide polymorphisms (SNPs), which depends on the differences in minor allele frequencies (MAFs) of QTL and SNPs, has an impact on the reliability of genomic prediction. Marker panels tend to include SNPs with high MAF, and empirical studies showed little improvement in reliability of genomic predictions when using low or high density SNP panels. Here, we examine, utilizing 52,842 SNP genotypes of 3,534 pigs and simulated phenotypes with a heritability of 0.5, how QTL number and MAF, and marker density affect genomic evaluation. Beside, some genomic prediction methods, BayesB, BayesCπ and GBLUP were compared in terms of reliability of predictions. Phenotypes were simulated for 50 or 500 QTL, and the number of markers was varied (5K, 20K and 50K). In order to evaluate the impact of QTL MAF, two scenarios were created: (S1) QTL and markers were randomly selected, and (S2) QTL were selected among the SNPs with MAF<0.01, while markers were selected at random. Our results showed that a great extent of heritability was missing (~40%) in S2 when the trait was controlled by 50 QTL and 5K markers were used. Differences in reliability of predictions for different genomic prediction methods were more pronounced in S2 than in S1, and as marker density increased, the gap between the reliabilities of variable selection methods and GBLUP increased.

Hoof health and housing conditions: a never ending story?!

A. Fiedler, C. Kroeger and K. Schindhelm
HoofHealth Practice, Heerstr.3, 81247 Munich, Germany; dr.andrea.fiedler@t-online.de

Modern dairy cows demand high standards concerning housing conditions. Animal welfare is the basis for high fertility rates, high milk yield, good hoof health and longevity. It is often the simplest essentials, the basic knowledge and the willingness that are lacking in order to break out of the vicious circle. Concerning temperature, airflow, humidity and incidence of light free stalls can bring together a variety of requirements under one roof. Knowledge about light regimens and light intensity for milking cows versus young stock and dry cows can improve reproduction and milk yield. Cows need animal-friendly walking alleys as a prerequisite for high productivity and longevity: slip-resistant with adequate abrasion, durable and easy to clean for a high hygiene standard. Hoof health is negatively influenced by poor hygiene and slippery alleys. Standing and walking comfort can be improved by rubber mats on the floor. Slatted floors as well as paved/concrete floors should be cleaned regularly. Avoiding blind alleys and offering broad passageways for good cow flow minimizes stress and can improve performance. Stock density is another factor involved in hoof health. A cow-cubicle ratio of 1:1 (and animal-feeding place ratio as well) is acceptable. Cubicle dividers and brisket board must not impede the cow's normal movements, but should guide her to lie in the right place. Excellent lying comfort is offered by well managed deep litter cubicles. Mattresses have to be combined with a littering depth of 3-5 cm. Consistent feeding improves the health condition of the cows and a higher feed intake for all cows results in better performance of the low ranking cows. Empty troughs are strictly prohibited. And appropriate feeding management is mandatory in the young stock. Young stock needs a consistent and well controlled feeding from the first day of life and during the complete rearing period. Leftover feed or inedible feed must not be fed to young stock. Digital dermatitis in particular can be a great problem in herds with badly managed young stock. High standards of hygiene and 'cow comfort' set the stage for healthy dairy cows. The quality and functionality of every single component is like a piece of a puzzle that contributes to the success of the whole system.

ICAR Claw Health Atla: one year of implementation

N. Charfeddine[1], G. Thomas[2], A. Fiedler[3], K.E. Müller[4], K.F. Stock[5], A. Steiner[6], A.M. Christen[7], V. Daniels[8], P. Nielsen[9], D. Döpfer[10], B. Heringstad[1], C. Egger-Danner[1], J. Pryce[1] and J. Kofler[1]
[1]Conafe, ctra de andalucia km 23,600, 28340 Madrid, Spain, [2]Institut de l'Élevage, Maison Nationale des Eleveurs, 149 rue de Bercy, 75595 Paris Cedex 12, France, [3]Association of Certified Hoof Trimmers, Heerstr. 3, 81247 Munich, Germany, [4]Freie Universität Berlin, Königsweg 65, 14163 Berlin, Germany, [5]VIT, Heinrich-Schröder-Weg 1, 27283 Verden, Germany, [6]University of Bern, Bremgartenstr. 109a, 3012 Bern, Switzerland, [7]Valacta, 555 boulevard des Anciens-Combattants Sainte-Anne-de-Bellevue, H9X 3R4 Quebec, Canada, [8]Vic's Custom Clips est. 1984, P.O. Box 43 2016 Arva St., Arva, ON, N0M 1C0, Canada, [9]SEGES P/S, Agro Food Park 15, 8200 Aarhus, Denmark, [10]University of Wisconsin in Madison, 2015 Linden Dr, Madison, WI 53706, USA; noureedine.charfeddine@conafe.com

Claw and foot disorders causing lameness are gaining in interest worldwide. Their importance for the whole cattle sector and their role in animal welfare have motivated worldwide engagement for effective breeding for improved claw health which requires trait definitions that are comparable between countries. Therefore, in 2014 the ICAR Working Group on Functional Traits (ICAR WGFT) started interdisciplinary collaborative work involving international claw health experts, which resulted in the publication of the ICAR Claw Health Atlas in June 2015. The descriptions in the Atlas provide a universal tool for claw trimmers and practitioners for accurate classifications of claw disorders. The current focus is the implementation of harmonized claw data recording in national programs. This included translating the Atlas to multiple languages and raising awareness of stakeholders, as well as providing support if needed. The presentation will give an overview about the status of implementation with its challenges and highlight further planned developments and activities.

The impact of claw health and lameness on fertility in Austrian dairy herds

J. Burgstaller[1], S. Guggenbichler[1], B. Fuerst-Waltl[2], F. Steininger[3], J. Kofler[1] and C. Egger-Danner[3]
[1]University Clinic for Ruminants, Food Animal and Veterinary Public Health, Veterinaerplatz 1, 1210 Vienna, Austria, [2]University for Natural Resources and life Science, Gregor-Mendel Straße 33, 1190 Vienna, Austria, [3]ZuchtData, Dresdner Straße 89, 1200 Vienna, Austria; johann.burgstaller@vetmeduni.ac.at

The Austrian cattle breeding associations were searching for efficient dairy cows, which produce milk, calve without complications every year and can be fed with farm grown foodstuff. To find those cows data was collected from 170 breeding herds consisting of 5,500 dairy cows. The first step included the evaluation of farm specific factors, including animal husbandry, feeding, claw trimming practice and management. At every milk yield recording event, which was performed 8 to 11 times during the year of 2014, BCS, lameness scoring, body weight and body measures were evaluated. The ration was analysed and documented, animal health and claw trimming data was collected during the observation period. The project gathered data from different sources; milk and fertility data from animal recording, additional documented data as mentioned at every milk sampling event, claw health data from claw trimming events and animal health data from the veterinarians. Data on claw health and lameness was used to evaluate the impact on fertility. Lameness groups were created, according to the data on lameness scoring, to be able to compare cows that have never been lame, to cows that were slightly, moderately or severely lame. Fertility parameters used in this study were calving interval, days open, non-return rate, insemination rate, first service conception rate. Non-lame Fleckvieh cows were 97 days open on average compared to chronically lame cows which were 113 days open. Calving interval in non-lame Holstein cows was 392 days compared to 425 days in severely lame cows. First service conception rate was 50% for all non-lame cows, whereas first service was successful in 35.4% in moderately lame cows. A significant increase in days open, prolonged calving intervals and decrease of the first service conception rate was observed for cows detected as lame when compared with those considered non-lame.

Linear type traits show pronounced phenotypic relationships to foot and claw health

B. Fuerst-Waltl[1], C. Fuerst[2] and C. Egger-Danner[2]
[1]Univ. Nat. Res. Life Sci. Vienna (BOKU), Gregor Mendel-Str. 33, 1180 Vienna, Austria, [2]ZuchtData EDV-Dienstleistungen GmbH, Dresdner Str. 89, 1200 Vienna, Austria; birgit.fuerst-waltl@boku.ac.at

Worldwide, the awareness of the relevance of a properly functioning locomotor system in cattle has been raised lately. In Austria, problems relating to feet and legs rank third among the most important disposal reasons in cows. Apart from economics, welfare related aspects have to be considered as these diseases may cause severe pain. Besides veterinarian diagnoses and hoof trimmers' records, data on linear scorings of feet and leg traits may also contribute towards improved cow health. As a first step, phenotypic relationships between linearly scored feet and legs traits, claw diagnoses and observations, and lameness were analysed in about 4,000 Fleckvieh (FV) and 1,500 Brown Swiss (BS) cows kept in herds across Austria. Within the project EfficientCow, all veterinarian diagnoses and observations of professional hoof trimmers were recorded during the year 2014. Additionally, cows' lameness was scored on a scale of 1 (not lame) to 5 (severly lame) in the course of each performance testing (approx. 8-11 times/year and cow). Once a year, linear scoring took place independent from lactation number. After pre-correction of several conformation traits, their partly non-linear relationship to claw health and lameness was analysed by additionally fitting the random effect of herd and the fixed effects lactation number, calving year and month, and alternatively type of recording (diagnoses), hoof trimmer or evaluator (lameness). In both breeds, animals with higher overall feet and legs scores were less prone to claw diseases and lameness. Within single linear type traits, animals of both breeds having a somewhat straighter hock angle (scores around 4; 1 = straight, 9 = extremely sickled) had significantly less feet and legs problems while no effect of hock quality could be found. Slightly steeper foot angles also resulted in reduced lameness. Next steps will include genetic analyses of all traits involved. However, non-linear relationships may complicate the interpretation of genetic correlations.

Genetic parameters for hoof disorders in dairy cattle

S.C. Ring[1,2], A.J. Twomey[1,2], N. Byrne[1], E. O'Brien[1], R.G. Sayers[1], A. Geoghegan[1], M.L. Doherty[2] and D.P. Berry[1]
[1]Teagasc, Moorepark, Fermoy, Co. Cork, Ireland, [2]UCD, School of Veterinary Medicine, Belfield, Dublin 4, Ireland; siobhan.ring@teagasc.ie

Lameness is a major factor influencing animal welfare as well as the productivity, profitability and competitiveness of the Irish dairy industry. The objective of this study was to quantify the contribution of additive genetics to the incidence of different hoof disorders and their genetic correlation with mobility score (MS). Hoof trimming was undertaken by professional trimmers on 7,533 lactating Irish dairy cows from 51 dairy herds during the year 2015; the presence and intensity (i.e. score of 0 to 3 or 0 to 5) of overgrown (OG), sole hemorrhage (SH) and white line (WL) was recorded for both back feet by two operators. Prior to hoof trimming, MS (i.e. score of 0 to 3) was undertaken on 4,700 cows. Animals purchased during 2015 and animals with a most recent calving event >365 days prior to hoof trimming were discarded. Furthermore, only animals with a known sire in a herd-year-season of calving contemporary group (CG) of ≥5 animals were retained. After all data edits, 7,087 animals in 222 contemporary groups from 51 herds remained. (Co)variance components were estimated using animal linear mixed models. Fixed effects included in the models were CG, parity, age relative to parity median, stage of lactation at trimming, hoof disorder recorder × trimming date interaction and heterosis and recombination coefficients; animal was included as a random effect. The direct heritability (standard error in parenthesis) for MS and susceptibility to OG, SH and WL as an ordinal scale trait was 0.08 (0.03), 0.12 (0.02), 0.22 (0.03) and 0.19 (0.03), respectively; the genetic standard deviation for MS, OG, SH and WL was 0.16, 0.65, 0.27 and 0.60 standard deviation units, respectively. Considering OG, SH and WL as a binary trait, the respective heritability estimates were 0.12 (0.02), 0.18 (0.03) and 0.11 (0.02); the genetic standard deviation for OG, SH and WL was 0.12, 0.19 and 0.15 standard deviation units, respectively. The genetic correlation between MS and hoof disorders, ranged from 0.10 to 0.22 but the standard errors were large (0.18 to 0.20). This study indicates that genetic variation exists in susceptibility to hoof disorders.

Breeding for healthier livestock

C. Maltecca, J. Howard and F. Tiezzi
North Carolina State University, Department of Animal Science, Box 7621, Raleigh, NC 27695, USA;
cmaltec@ncsu.edu

Managing inbreeding and genomic diversity is an often forgotten path to ensure healthy livestock. Management of diversity rests on three pillars. Understanding the basis and consequences of genetic diversity. Controlling the population effective size. Optimize genetic variability deployed through mating plans. Inbreeding management often rely on the implicit assumption that individuals with the same inbreeding share the same genomic load. Marker information allows instead for regions-specific homozygosity causing inbreeding depression to be identified. Yet, these regions are expected to be at a low frequency so that traditional association methods based on estimating dominance effects lack statistical power. Methods that exploit the fact that longs runs of homozygosity (ROH) are enriched with deleterious variants can have greater power in identifying haplotypes linked to inbreeding depression. We have employed a combined coalescence and forward-in-time simulator to investigate the use of alternative similarity metrics in identifying and curtail the propagation of deleterious haplotypes in selected cattle populations. As the genomic load is spread over variants of increasingly small effect, measures that aggregate the overall effect of a region become more effective than estimating dominance effects or the use of overall genomic similarity in curtailing the accumulation of inbreeding. Using the same principle we employed a heuristic method to identify haplotypes with negative effect for variants of small effect. When employed in simulated data the heuristic is better able to detect regions of the genome associated with inbreeding depression regardless of the effect size of the single mutations. This might have implications both in reducing and managing the genomic load of breeding populations, as well as providing a basis for the dissection of complex traits related to fitness and/or disease resistance.

Use of somatic cell count patterns for selection against mastitis

J.J. Vosman, G. De Jong and H. Eding
CRV, Animal Evaluation Unit, Wassenaarweg 20, 6843 NW Arnhem, the Netherlands; jorien.vosman@crv4all.com

Since the incorporation of the breeding value for somatic cell count (SCC) in the Dutch total merit index, an increase in udder health has been observed. This positive trend for udder health was strengthened by replacing the breeding value SCC with the udder health index (UHI) in 2010. The UHI consists of two breeding goal traits: subclinical mastitis (SCM) and clinical mastitis (CM). SCM is based on observations on test day SCC pattern. CM is based on the correlation with SCM and with the five indicator traits derived from SCC: early SCS150, late SCS400, Infection, Severity and Peaks in SCC. Since 2016 direct observations of CM are incorporated in the genetic evaluation CM. As a practical validation of the method based on correlations with indicator traits, direct observations of CM are included. Observations are based on animals with CM, but also on the herd mates present at the time of infection. Daughters of bulls with high breeding value for CM show lower CM incidence. In general only a slight increase in reliability can be observed, without any change in breeding value. For young bulls with more than 100 observations the added reliability was 1-3%. This confirms that the use of SCC patterns is the basis for an accurate method to predict CM breeding values which can be used to select against mastitis.

Selection for disease resistance in Canadian dairy cattle

F. Malchiodi[1], A. Koeck[1], J. Jamrozik[1,2], A.M. Christen[3], D. Kelton[4], F.S. Schenkel[1] and F. Miglior[1,2]
[1]University of Guelph, CGIL, Dept of Animal Biosciences, N1G2W1 Guelph, Ontario, Canada, [2]Canadian Dairy Network, 660 Speedvale Avenue West, N1K1E5 Guelph, Ontario, Canada, [3]Valacta, 555 boul. des Anciens-Combattants, H9X 3R4 Sainte-Anne-de-Bellevue, Quebec, Canada, [4]University of Guelph, Dept of Population Medicine, N1G2W1 Guelph, Ontario, Canada; miglior@cdn.ca

Since 2007, eight different diseases are routinely recorded in Canada: mastitis, displaced abomasum (DA), ketosis (KET), milk fever, retained placenta, metritis, cystic ovaries, and lameness. The overall goal is to develop a series of genetic evaluation systems for resistance to diseases in Canadian dairy cattle. In August 2014, a new index for resistance to mastitis was officially implemented and included in the national selection indices. In addition to the presence of clinical mastitis, the index includes patterns of SCS and conformation traits (body condition score (BCS), udder depth and fore attachment). A genomic evaluation was also implemented, with an increase in reliability for young and proven bulls. The second evaluation system (resistance to metabolic diseases) will be implemented later this year. In addition to the direct information of the presence of KET or DA in the first 100 DIM, three predictors of metabolic diseases are included: fat:protein ratio and milk β-hydroxybutyrate (BHBA) at the first test-day, and BCS. BHBA is a very a strong indicator of KET. The third genetic evaluation system will be for hoof health. Besides the presence of lameness, data on hoof lesions are collected by hoof trimmers, through a centralized data collection system. Hoof lesions had low heritability, but the distribution of the estimated breeding values showed exploitable variation among sires. In addition to the presence of a lesion, conformation traits routinely recorded by professional classifiers were considered. Some feet and leg traits may be included in the index for hoof health, together with the presence of lameness as recorded by the producers. Selection for enhanced health in Canadian dairy cattle is feasible by using an efficient combination of producer recorded traits, together with routinely recorded traits from milk recording and classification system.

Postpartum diseases, health and fertility traits in Norwegian Red crosses and Holstein in Israel

E. Rinell and B. Heringstad
Norwegian University of Life Sciences, Animal and Aquacultural Sciences, Arboretveien 6, 1430 Ås, Norway; ellen.rinell@nmbu.no

In Israel, veterinarians routinely check dairy cows weekly after calving until they are healthy. Therefore, extensive records on postpartum diseases are available. Norwegian Red has been used for crossbreeding in Israel since 2005. The aim of this study was to compare health and fertility of Norwegian Red × Holstein crosses (NRX) with their Holstein (HO) counterparts in Israel. A large number of unique health traits and a high milk yield despite the warm climate makes it interesting to investigate NRX performance in Israel. The data consisted of records from 13,092 cows from 29 herds in Israel with an average of 146 NRX and 780 HO records each and calvings from 2007-2014. Seven postpartum disorders (mean incidence in NRX vs HO) were analyzed: mastitis after calving (both 0.5%), milk fever (0.6% vs 0.2%), retained placenta (5% vs 8%), metritis (34% vs 40%), ketosis (1.5% vs 1.4%), udder edema (0.5% vs 0.7%) and anestrus (both 0.3%). Use of prostaglandin for ovulation induction was also analyzed (1% in NRX vs 2% in HO). These were binary traits registered as '1' indicating that the disorder was present or a treatment took place, and otherwise '0.' Days open and body condition scores were also analyzed. The analysis was done in SAS 9.4 using proc logistic for binary traits and proc glm for linear traits. The model for all traits included breed, herd, calving year, calving month, and parity as fixed effects. There were fewer cases of milk fever, metritis, retained placenta ($P<0.01$) and udder edema ($P<0.05$) in NRX compared with HO. Israeli Holstein had fewer cases of ketosis than NRX ($P<0.01$). There were no significant breed differences found for mastitis after calving and anestrus. Usage of prostaglandin was lower in NRX ($P<0.01$) which could signify fewer fertility issues in crosses. Likewise, NRX had fewer days open than HO ($P<0.01$). Body condition score before and after calving and at peak lactation was significantly higher for NRX compared with HO ($P<0.01$).

Genetic variation of seasonal tick counts in South African Nguni cattle

N.O. Mapholi[1,2], A. Maiwashe[2,3], C.B. Banga[2], O. Matika[4], V. Riggio[4], M.D. Macneil[2,3,5] and K. Dzama[1]
[1]University of Stellenbosch, Department of Animal Sciences, P Bag X1, Matieland, 7602, South Africa, [2]Agricultural Research Council, Animal Breeding and Genetics, Private Bag X2, Irene, 0062, South Africa, [3]University of the Free State, P.O. Box 339, Bloemfontein, 9300, South Africa, [4]The Roslin Institute and R(D)SVS, University of Edinburgh, Easter Bush Midlothian, EH25 9RG, United Kingdom, [5]Delta G, 145 Ice Cave Road, Miles City, MT 59301, USA; ntanga@arc.agric.za

Tick and tick-borne disease are a major challenge in livestock production in the sub-tropics, with the advent of climate changes. The objective of this study was to assess tick species distribution and their colonisation on animals' body sites, and to estimate genetic parameters for these tick traits in Nguni cattle. Data comprised seasonal tick counts on various anatomical regions of 586 Nguni cattle under natural infestation, in four herds located in four provinces of South Africa. Heritabilities and correlations for log-transformed counts were estimated using bivariate analysis, fitting a sire model, accounting for fixed effects. Six different tick species (Amblyomma hebraeum, Boofilids, Rhipicephalus evertsi evertsi, Rhipicephalus appendiculatus, Hyalomma marginatum, Rhipicephalus simus) were identified attached to eight anatomical regions (i.e. head, ear, neck, back, legs, belly, perineum and tail) on the animals. Heritability estimates ranged from 0.01 ± 0.01 to 0.26 ± 0.01 across seasons and traits (anatomical region or tick species). Genetic correlations ranged from -0.79 ± 0.33 to 1.00 ± 0.00 among counts for different anatomical regions and 0.00 ± 0.01 to 0.99 ± 0.01 among tick species. Phenotypic correlations were low to moderate, ranging from 0.06 ± 0.01 to 0.72 ± 0.01 among anatomical regions and 0.01 ± 0.02 to 0.44 ± 0.01 for tick species. The largest genetic variation in tick count was observed in the summer (September to January). High genetic correlations were observed between whole body count and most of the anatomical region counts, with belly and perineum counts being the most suitable indicator traits for whole body count. Our results would allow identifying the most appropriate season and anatomical region for tick sampling in a cost-effective selection programme for tick resistance in Nguni cattle.

Preliminary results from a largescale lameness study in Irish dairy herds

N. Byrne, S. Ring, A. Twomey, E. O'Brien, D. Berry and R. Sayers
Teagasc, Animal and Biosciences, Moorepark, Fermoy, Co. Cork, Ireland; noel.byrne@teagasc.ie

Abolition of EU milk quotas has led to increased herd sizes, enlargement of grazing platforms and an increase in the distances cows walk for milking. Existing research indicates that longer walking distances predisposes dairy cows to hoof disorders. The objectives of this study were to: (1) document the prevalence of mobility abnormalities and foot lesions in a large population of dairy cows; and (2) investigate associations between mobility scoring and the presence of a foot lesion. Dairy cows on 46 Irish dairy farms were mobility scored (MS) using the DairyCo four point scale on two occasions (spring and autumn 2015). Mobility score was dichotomised for statistical analysis; cows recording a MS of 0 or 1 were categorised as 'none or minor mobility impairment' and those recording a MS of 2 or 3 were categorised as 'moderate to severe mobility impairment'. Hoof-trimming was also completed on over 6,000 cows and each individual categorised on the basis of how many lesions were simultaneously present (e.g. 0, 1, 2, etc.). Univariable logistic regression was used to examine associations between MS in spring and MS in autumn, MS and the presence of at least one hoof disorder, and MS and the number of foot lesions present. Foot lesions were characterised as mechanical (e.g. white line disease) or infectious (e.g. digital dermatitis). In spring, 4% of all cows recorded MS of two or greater. The prevalence of moderately to severely lame cows increased significantly to 6.7% in autumn ($P<0.001$), with cows significantly more likely to record a MS of two or greater compared to spring (OR=9.83; $P<0.001$). Mechanical lesions (n=9,413) were more highly prevalent than infectious lesions (n=156). In both spring and autumn, cows with multiple lesions were more likely to record a MS of two or greater. Cows with a MS of two or greater were over three times more likely to have at least one foot lesion than cows with a lessor MS ($P<0.001$). The prevalence of moderate to severe lameness in the Irish dairy system compares favourably with more intensive, indoor systems internationally. The increased prevalence of cow lameness in autumn, however, is a cause for concern, and most likely reflects the cumulative effect of daily road walking as the lactation progresses.

Inter and intra-observer reliability of different methods for recording temperament in cattle

A. Vogt, I. Schlechter, E.L. Aditia, S. Schütze, K. Hille, M. Gauly and U. König Von Borstel
University of Göttingen, Department of Animal Science, Albrecht-Thaer-Weg 3, 37075 Göttingen, Germany;
anina.vogt@agr.uni-goettingen.de

Breeding of cattle with docile temperament can facilitate routine practices and enhance welfare of cattle in common production systems. The aim of the study was to compare and validate assessment methods for temperament traits of cattle for their potential use in breeding programmes. Calves of five different beef and dairy cattle breeds were video-recorded during a 2 min tethering (n=159) and a 2 min crush test (n=162) at the age of 38±25 days. Subsequently, two observers blinded to the identity of records each analysed the video clips three times with the software Interact in intervals of 1-2 days between viewings. Frequency and duration of behaviour patterns such as head movement, tail-movement or defecation were recorded. In addition, each behaviour pattern was assessed on a 10 cm visual analogue scale (VAS) and at the end of each observation an overall behaviour score on a numerical scale from 1-5 and on a VAS was assigned to each animal. Using variance components from mixed model analysis, results revealed that both inter and intra-observer reliability were superior for the majority of behaviour features when assessed with the VAS compared to the more time-consuming assessment of exact frequencies and duration with Interact software (e.g. Intra-OR 'tail movement': Observer I: r_{VAS}=0,83 and r_{Freq}=0,26; Observer II: r_{VAS}=0,91, r_{Freq}=0,37 and r_{Dur}=0,87; Inter-OR 'head movement': r_{VAS}=0,68 and r_{Freq}^2=0,36; Inter-OR 'tail movement': r_{VAS}^2=0,65 and r_{Freq}^2=0,22 for the tethering test). Especially the overall behaviour score showed a high intra-observer reliability when assessed with the VAS (Observer I: r_{VAS}–0,97 and r_{Num}–0,77, Observer II: r_{VAS}=0,99 and r_{Num}=0,84 for the crush test). However, not all traits proved to be appropriate for recording with the VAS, particularly rarely occurring ones. In conclusion, a VAS that is based on carefully selected features can be considered a reliable and practical method for assessment of temperament in cattle.

Genetic relationship between clinical mastitis, somatic cell score and milk in Czech Holstein

L. Zavadilová, J. Bauer, L. Krpálková, Z. Krupová, M. Michaličková and M. Štípková
Institute of Animal Science, Přátelství 815, 104 00 Prague, Czech Republic; zavadilova.ludmila@vuzv.cz

Cases of mastitis (CM) from 35 845 lactations of 15 822 cows were recorded on 7 farms in the Czech Republic from 1996 to 2014. The data were used to estimate the genetic parameters of clinical mastitis, considered in the entire lactation or in the first part of lactation (150 days), and to determine the genetic relation of clinical mastitis with the somatic cell scores (SCS305) and milk production (MY305). The CM considered as an all-or-none trait (CM1) with values of 0 (no CM case) and 1 (at least 1 CM case), the number of CM cases per lactation (CM2) and the number of days of CM per lactation (CM3) were analyzed with bivariate linear animal models, either for the whole lactation (CMx_305) or only for the first 150 days of lactation (CMx_150). Factors included in the linear model were the parity, effect of herd, year of calving, calving season, fixed linear and quadratic regression of age at first calving, permanent environmental effect of the cow, and additive genetic effect of the cow (pedigree 27 902 animals). Heritability of CM traits was in the range 0.05–0.08. Genetic correlations were 0.96±0.018 between CM1_305 and CM1_150, 0.98±0.009 between CM2_305 and CM2_150 and 0.99±0.002 between CM3_305 and CM3_150. Genetic correlations of SCS305 with the CM305 traits were 0.85±0.037, 0.81±0.037, and 0.78±0.040 for CM1_305, CM2_305, and CM3_305, respectively. Genetic correlations of SCS305 with the CM150 traits were 0.84±0.038, 0.80±0.037, and 0.77±0.048 for CM1_150, CM2_150, and CM3_150, respectively. Genetic correlations of MY305 with the CM traits were in range 0.19-0.29 (±0.060). Our findings show that using the first part of lactation to the 150th day in milk is sufficient for genetic evaluation instead of evaluating the incidence of CM for the whole lactation. The work was supported by the project QJ1510144 and project MZERO0714 of the Ministry for Agriculture of the Czech Republic.

Comparative study on fitness related traits in Fleckvieh and Brown Swiss cattle breeds

R. Neamt[1], L.T. Cziszter[1,2], F. Neciu[1], D. Gavojdian[1,3] and D.E. Ilie[1]
[1]Research and Development Station for Bovine Arad, Arad, Calea Bodrogului 32, 310059, Romania, [2]Banat University of Agricultural Sciences and Veterinary Medicine 'King Michael I of Romania', Timisoara, Calea Aradului 119, 300645, Romania, [3]Research and Development Station for Sheep and Goats Caransebes, Caransebes, Drumul Resitei km2, 325400, Romania; gavojdian_dinu@animalsci-tm.ro

The main objective of the current comparative study was to evaluate mastitis and lameness incidences and attrition rates in the dairy cattle breeds Fleckvieh and Brown Swiss. Work was carried out at the Research and Development Station for Bovine Arad (46°10'N 21°19'E). The study herd consisted of 309 Fleckvieh and 35 Brown Swiss cows, with age and parity balanced between genotypes and representing a diverse sampling of genetic lines for each breed. Clinical mastitis incidence was 3.28±1.02% in Fleckvieh and 8.58±4.80% in Brown Swiss cows, respectively (P≤0.05). Genotype did not influence the lameness incidence (P>0.05), with averages of 2.96±0.97% and 5.71±3.98% in Fleckvieh and Brown Swiss, respectively. Attrition rates were 19.67±2.28% in Fleckvieh and significantly higher (P≤0.01), 28.57±7.75%, in Brown Swiss. For all three studied traits, the Brown Swiss cows proved to have lower organic resistance, compared to that of the Fleckvieh breed. Based on the current findings, it would be advisable to include new traits into the selection schemes of the Brown Swiss breed reared in Romania, with special focus on mastitis resistance and cow stay-ability, in order to improve animal welfare and the overall herd productivity and farm profitability.

Relationship between dairy herd performance and exposure to bovine herpesvirus-1

R.G. Sayers
Teagasc, Animal and Biosciences, Moorepark, Fermoy, Co. Cork, Ireland; riona.sayers@teagasc.ie

Assessing the potential costs associated with a particular infection disease allows the risk and potential benefits of introducing control programs at farm and national level to be quantified. Additionally, cost-benefit analyses are useful in prioritising and promoting on-farm disease control programs. The objectives of the current study were to use bovine herpesvirus 1 (BoHV-1) prevalence data in Irish dairy herds to document associations between milk production, fertility performance, mortality and viral status in Irish dairy herds. Bulk milk samples were collected from 305 herds on four occasions (March, June, August, November) over the 2009 lactation. These data were used to classify farms in terms of their endemic BoHV-1 status. Cow level (milk parameters only) and herd level performance data were sourced from the Irish Cattle Breeding Federation. Vaccinated herds (n=36) were excluded from statistical analysis, as were herds for which complete datasets were not available. Ordinary linear regression was used to investigate associations between milk, fertility and mortality performance, and herd level BoHV-1 results in 274 herds. Additionally, a hierarchical mixed-effect regression was used to investigate associations between predicted individual cow milk yield, milk fat yield, and milk protein yield and BoHV-1 bulk milk herd status. No fertility or mortality related associations were identified, with the exception of calving rates in multiparous cows in univariable linear regression models. Multiparous cows in herds positive for exposure to BoHV-1 recorded reduced milk yields per cow per year of almost 250 L (P=0.05) in the multivariable linear model. The mixed effect model documented a loss of 0.91 L per cow per day (P=0.60). Milk fat and protein yields were also impacted by herd BoHV-1 status again highlighting sub-optimal milk production in BoHV-1 bulk milk positive herds. Limited data is available relating to the impact of endemic BoHV-1 in dairy herds. The current study supports the growing evidence that sub-clinical BoHV-1 infection results in sub-optimal performance in dairy herds. Eradication of BoHV-1 should be considered, therefore, in an era of expanding milk demand globally.

A survey of miscarriage causes in dairy herds in Poland

E. Bauer, J. Żychlińska-Buczek and K. Merda
University of Agriculture in Krakow, Animal Science, Al. A. Mickiewicza 24/28, 31-120 Kraków, Poland;
e.bauer@ur.krakow.pl

The diagnosis of miscarriages often presents a challenge to the herd owner and the herd veterinarian. Although a gradual increase in the abortion rate in a herd may be noted over a period of many years, a sudden and dramatic increase is more commonly seen. For this reason, prompt and thorough action is required when abortions occur. Well-kept records will often be of benefit during the investigation of abortion problems. At present it is believed that reproductive problems are major health problems for dairy cattle. Among them miscarriage is a significant cause of economic losses incurred by cattle farmers. The aim of this study was to determine the frequency of miscarriage causes in Polish dairy herds and to compare them with the most common cause of abortion in cattle worldwide. Farm data were collected with a direct detailed survey on: farm characteristics, composition and diets of each animal category, crop data and management, facilities and milking parlour. The study was based on 1,500 polish Holstein-Friesian cows and heifers with known pedigree, with data collected between 2005 and 2010, on two commercial dairy farms. Diseases and welfare traits were measured daily and expressed in %. Miscarriage rates for cattle were different. The results indeed clearly suggest significant differences in the causes of miscarriage in the selected herds. Abortion as a form of impaired fertility in cows is caused by infectious agents and nutrition in most cases. The process of miscarriage recognition in cattle herds represents one of the biggest problems and is challenging for veterinarians. The medical-veterinary program of care creates an opportunity to control the breeding cows. It allows evaluation of the abnormality and thus an effective prevention and treatment. Measures should be taken to minimize abortions' occurrence. Good sanitation, proper feeding, use of vaccination programs will help to reduce the incidence of abortion and increase profit.

An All-Ireland study of sero-prevalence for reproductively important diseases on beef cow herds

M.H. Parr[1], F. Randi[1], D. Barrett[2], M.G. Diskin[3] and D. Kenny[1]
[1]Teagasc, Animal and Grassland Research and Innovation Centre, Grange, Dunsany, Co. Meath, Ireland,
[2]Department of Agriculture, Food and Marine, Regional Veterinary Laboratory, Fawcett's Bridge, Doonally, Co.
Sligo, Ireland, [3]Teagasc, Animal and Grassland Research and Innovation Centre, Mellows Campus, Athenry, Co.
Galway, Ireland; mervyn.parr@teagasc.ie

Numerous bacterial, viral and protozoan diseases have been associated with poor reproductive performance in cattle. Leptospirosis (hardjo-bovis and -prajitno genotypes; Lepto), bovine viral-diarrhoea (caused by bovine viral-diarrhoea virus; BVDV), infectious bovine rhinotracheitis (IBR) (caused by bovine herpesvirus-1) and neosporosis (caused by Neospora caninum) are all transmissible diseases that are considered to be of economic importance to both national and international cattle production systems. The aim of the study was to quantify the sero prevalence of the above listed pathogens according in spring calving beef cow herds, across the island of Ireland. The study included both vaccinated and unvaccinated herds only. For the purpose of this summary, finding will only include unvaccinated herds. During the summer months of 2014 and 2015, a total of 5,908 cows from 169 herds were blood sampled. Herds were ranked on size and subdivided into quartiles based on the number of cows per herd (4-20; 21-29; 30-46 and 47-94) resulting in similar numbers of herds per quartile grouping (n=42, 41, 42 and 41, respectively, for quartiles 1 through 4). Sero-prevalence of Lepto was 59, 58, 72 and 80% for the four quartiles, respectively. Sero-prevalence of BVD was 75, 79, 78 and 81% in each quartile respectively. Sero-prevalence of IBR was 30, 39, 45 and 53% in each quartile respectively. Sero prevalence of neosporosis was 6, 7, 6 and 5% in each quartile respectively. Overall herd pregnancy rate at the end of the breeding season was 89 and 88%, for 2014 and 2015, respectively. This study provides, for the first time, sero-prevalence data for Lepto, BVDV, IBR, and neosporosis on beef cow herds across the island of Ireland. Further investigation will be required to determine the relationship between exposure to these pathogens and herd reproductive and performance efficiency.

Genetic relationship between type traits and number of lactations in Czech Fleckvieh cows

L. Novotný[1], L. Zavadilová[2] and J. Bauer[2]
[1]*Czech Moravian Breeding Corporation, Hradištko 123, Hradištko 123, 252 09, Czech Republic,* [2]*Institute of Animal Science, Genetics and breeding of farm animals, Přátelství 815, 104 00 Prague, Czech Republic; zavadilova.ludmila@vuzv.cz*

The genetic relationships between conformation traits and longevity expressed as number of lactations (NL) were analysed in 91 486 Czech Fleckvieh cows first calved from 2003 to 2009. Data collection was completed in 2015. Cows were scored for conformation during the first lactation. Genetic correlations were estimated by bivariate runs using the DMU program for variance component estimation. For type traits, the linear model included fixed effects of herd and date of scoring, fixed effect of classifier, fixed linear and quadratic regression on the age at first calving. For the number of lactations, the linear model included fixed effects of herd, year and season of calving and fixed linear and quadratic regressions on the age at first calving and those on milk production in first lactation (for NL expressed as functional longevity, NL_f). Animal models included random effect of animal with pedigree (334 322 animals). The values of heritability for conformation traits were in the range from 0.06 to 0.58, heritability of NL was 0.06 and this of NL_f was 0.05. Low or intermediate genetic relationships between recorded type traits and NL traits were found. The highest genetic correlations were found between NL and udder (0.41), udder depth (0.21), rear and fore udder length (0.23; 0.26), rear udder attachment (0.28) and feet and legs (0.24), for the NL_f, udder depth (0.33), feet and legs (0.26), udder (0.25) and muscularity (0.16). Negative genetic correlations between NL and type traits were observed for stature (-0,20), rump width and rump length (-0.25, -0.24), for NL_f, stature (-0.18), rump width and rump length (-0.19, -0.17). We can conclude that selection of cows with higher breeding values for udder, feet and legs and udder depth results in selection of cows that achieve the high number of lactation. The work was supported by the project QJ1510144 and project MZERO0714 of the Ministry for Agriculture of the Czech Republic.

The influence of pricing systems on the economic value of traits in dual-purpose cattle

M. Michaličková[1], Z. Krupová[1], E. Krupa[1], L. Zavadilová[1] and J. Bujko[2]
[1]*Institute of Animal Science, Přátelství 815, 104 00 Prague 10, Czech Republic,* [2]*Slovak University of Agriculture, Trieda A. Hlinku 2, 949 76 Nitra, Slovak Republic; michalickova.monika@vuzv.cz*

Economic values for milk traits and somatic cells score (SCS) were calculated for the dairy production system of the Slovak Simmental breed. For the first time, an economic value for clinical mastitis incidences (CM) was defined. The bio-economic model of the program package ECOWEIGHT ver. 6.0.4 was used for calculation. On average, the 305-d milk yield was found to be the most important trait (21%) followed by SCS (12%). Lower economic values were reported for fat and protein content (9 and 8%, respectively). The relative importance of fat and protein content in milk increased when more emphasis was put on milk components and on qualitative parameters in the milk pricing system. Rather low relative economic values for CM incidence (0.9%) were observed. Generally, an increase in the mean value for SCS and CM traits is economically unfavourable (negative marginal economic values). Results found in this study provide first important information about potential traits to be considered in a breeding goal for dairy cattle in Slovakia. In Slovak conditions, the milk pricing system varied highly among the farms. Generally, fat and milk protein content and SCS are the key parameters in the milk pricing system. This study was funded by projects QJ1510217 and MZERO0714 of the Czech Republic.

Livestock nutritionists or human health professionals?
P.G. Wall
University College Dublin, School of Public Health, Physiotherapy and Population Science, Woodview House, Belfield Dublin 4, United Kingdom; patrick.wall@ucd.ie

Many professionals in the agri-food sector work within their confined areas without realising the real objective of their activities. Phenomenal advances have been, and continue to be made, in the genetics of food-producing animals and fish. Relentless selection for production traits has delivered us very different livestock from those our forefathers tended. Animal nutritionists are far ahead of their human counterparts when it comes to diet formulation and performance. Genetics creates the potential and nutrition delivers on it, but suboptimal animal health or welfare can undermine any gains these may offer. Producing safe food is a key objective but is not the final end game. Food is the fundamental fuel for human health and 'you are what you eat' is a true dictum. Diet-related disease in people is a major public health issue. Increasingly, primary agricultural output is coming under the spotlight in both the scientific and general media for contributing to human health problems with headlines such as 'Red meat causes cancer' and 'Dairy products clog up your arteries'. However, it is now possible to modify the composition of the final output from farms by altering the rations fed, whether the desired outcome is less saturated fat, more omega 3 or any number of other constituents. Human nutrition is key to health, so the final objective for most activities in the agri-food sector should be improved human health, and all engaged in activities along the food chain should consider themselves in the 'human health business'. Doctors and nurses are not in the health business; rather they are in the sickness business.

Omega chicken: a case study
H. Hayes[1], J. Coleman[2] and U. Lavery[2]
[1]Devenish Nutrition Limited, Belfast, N. Ireland, United Kingdom, [2]Moy Park, Craigavon, N. Ireland, United Kingdom; heather.hayes@devenishnutrition.com

Many chronic diseases such as cardiovascular disease are the result of poor diet and lack of exercise. They present a huge drain on government funds and cause a high level of stress to individuals & families affected. The diet of an animal has a significant impact on food quality. With the correct animal nutrition, primary agricultural products such as chicken & milk have the potential to deliver specific nutrients into our diet which can have a positive impact on human health. One such product is Omega enriched chicken developed by Devenish Nutrition in partnership with poultry supplier Moy Park & currently retailed through Waitrose. Many populations within Europe consume well below the EFSA daily recommendation of 250 mg long chain omega 3 fatty acids (eicosapentanoic acid EPA C20:5n3 & docosahexanoic acid DHA C22:6n3). This is often due to a dis-taste for oily fish which traditionally has been the only significant dietary source of these essential omega 3 fatty acids. This intake shortfall is especially apparent in young children & teenagers. Chicken is a popular meat especially with the younger population & the partners within this project identified the opportunity to naturally enhance EPA & DHA in chicken, through feed, to a level which could make a positive contribution to heart, brain & vision function. As a case study, aspects of the scientific development of Omega chicken will be considered along with practical application and potential health benefits to the consumer.

Consumption of omega 3 enriched chicken and cardiovascular benefits in humans

A.V. Stanton
Royal College of Surgeons in Ireland, 123 St Stephens Green, Dublin 2, Ireland; astanton@rcsi.ie

Greater consumption of oily fish, and elevated plasma levels of long-chain omega-3 polyunsaturated fatty acids (omega-3-PUFAs), has been strongly and consistently associated with reduced incidences of heart attacks and strokes. Despite international guidelines recommending at least 1 serving/week of oily fish, many people do not eat fish at all, and worldwide, deficiencies are common. Unfortunately, Omega-3-PUFA supplements do not appear to reliably provide the same beneficial protection. We recently studied an alternative to oily fish or supplementation, namely chicken-meat naturally enriched with algae-sourced omega-3-PUFAs. In an open sequential study, 30 healthy participants ate 3 servings/week of omega-3-PUFA-enriched chicken-meat for 5 weeks. Plasma omega-3-PUFAs levels increased from 82 mg/g at baseline to 93 mg/g (12% increment, $P=0.006$). Furthermore two cardiovascular health biomarkers improved – clinic systolic BP reduced from 116 to 113 mmHg ($P=0.014$), and urinary thromboxane (measure of platelet turnover) reduced from 1,435 to 1,024 pg/mg creatinine ($P=0.037$). Whether further increments in plasma omega-3-PUFA levels, and greater cardiovascular benefits, can be achieved with more prolonged eating of omega-3-PUFA-enriched chicken-meat is currently being tested in a larger, longer, double-blind, randomized, controlled, nutritional trial. If so, omega-3-PUFA-enriched chicken-meat may offer consumers an attractive alternative to eating oily fish or to taking supplements.

Long-chain omega-3 polyunsaturated fatty acids and cognitive performance

J.K. Virtanen
University of Eastern Finland, Institute of Public Health and Clinical Nutrition, Kuopio, Finland; jyrki.virtanen@uef.fi

Cognitive impairment is a considerable burden as people age. Brain tissue membranes contain long-chain omega-3 polyunsaturated fatty acids (PUFA) and particularly docosahexaenoic acid (DHA), which is an important structural component. Therefore, intake of long-chain omega-3 PUFA may prevent or delay age-related cognitive decline. However, fish, a major source for the long-chain omega-3 PUFA, is also a major source for methylmercury, a neurotoxicant. Therefore, mercury exposure could attenuate the potential beneficial effects of fish consumption on cognitive decline. We investigated the associations of the serum long-chain omega-3 PUFA and pubic hair mercury, both established biomarkers for intake, with five cognitive performance tests among 768 older men and women from the population-based Kuopio Ischaemic Heart Disease Risk Factor Study (KIHD) in Finland. We found statistically significant associations of the serum long-chain omega-3 PUFA, and especially DHA, with better performance in the Trail-Making Test and the Verbal Fluency Test, both assessing frontal lobe functioning. Mercury was associated only with worse performance in the Trail-Making Test and it did not markedly modify the associations between the long-chain omega-3 PUFAs and cognitive performance. In conclusion, higher intake of fish and the long-chain omega-3 PUFA could be beneficial for the aging brain.

The effect of selenium enriched silage on the selenium status of finishing lambs and their products

M.R.F. Lee[1,2], H.R. Fleming[2], C. Hodgson[2], R.I. Richardson[1] and D.R. Davies[3]
[1]*University of Bristol, School of Veterinary Science, Langford, Somerset, BS40 5DU, United Kingdom,* [2]*Rothamsted Research, North Wyke, Okehampton, Devon, EX20 2SB, United Kingdom,* [3]*Silage Solutions Ltd, Bwlchyblaen, Ponthrydygroes, Ceredigion, SY25 6DP, United Kingdom; michael.lee@rothamsted.ac.uk*

Selenium (Se) is a trace element essential for cellular function, which has been linked with reduced risk of cancer, cardiovascular disease, cognitive decline and thyroid disease in humans. In Europe Se is often deficient in livestock diets due to the low Se status of soil. Supplementation of diets with organic selenomethionine is common due to its greater bioavailability over inorganic Se. Silage L. plantarum inoculants have been shown to convert inorganic Se into organic selenocysteine and elemental nano-Se. Here we report the effect of feeding inoculated silage enriched with Se versus control inoculated silage on the Se status of finishing lambs and their products. Charollais × Suffolk lambs (20 male and 20 female, mean weight 42 kg±1.7 kg, mean age 217±4.3 days) were paired according to weight, sex and previous diet, and one of each pair randomly allocated to treatment. Se status of biological samples was determined by HPLC-UV-HG-AFS during a 10 week trial. DM intake was comparable 0.8±0.03 kg/d with Se intake significantly different 0.14 vs 1.60 mg/d on the control and Se enriched silage, respectively. This was reflected in higher Se concentrations in faeces (0.84 vs 4.17 ug/g DM); wool (1.01 vs 1.87 ug/g DM); blood (0.19 vs 0.26 ug/ml) and muscle (0.31 vs 0.41 ug/g). TBARS as a measure of rancidity during shelf life trails were higher on the control (4.6 vs 3.8 mg/g) although not significantly different, shelf life of the meat was significantly higher on the Se treatment (9.05 vs 10.2 days). This study shows that silage inoculum L. plantarum enriched with Se can be used to deliver Se to livestock, further work to determine the relative bioavailability to livestock of Se enriched silage vs inorganic (selenite) and selenomethionine (organic) is required.

Vitamin D3 and 25-hydroxyvitamin D supplementation in laying hen diets, on egg vitamin D content

S.K. Duffy[1], G. Rajauria[1], P.O. McAlpine[1], K.D. Cashman[2] and J.V. O'Doherty[1]
[1]*University College Dublin, School of Agriculture and Food Science, Belfield, Dublin 4, Ireland,* [2]*University College Cork, School of Food and Nutritional Sciences, University College Cork, Cork, Ireland; sarah.duffy@ucdconnect.ie*

Biofortification of eggs with vitamin D may be one way to counteract the vitamin D insufficiency in humans in northern Europe. The objective of this study was to assess the effects of vitamin D3 (D3) versus 25-hydroxyvitamin D (25-OH D3) enriched diets fed to laying hens, on egg vitamin D concentration and egg quality parameters. Sixty Hy-line Brown hens, 90% in lay with an initial body weight of 1.68 kg (SD=0.020) were randomly assigned to four houses, consisting of fifteen birds per house. Experimental design was a 4×4 Latin square consisting of four experimental treatments (T1) 1,500 IU D3, (T2) 3,000 IU D3, (T3) 3,000 IU 25-OH-D3, (T4) 1,500 IU D3 and 1,500 IU 25-OH-D3 per kg of diet and four experimental periods (8 weeks). Each treatment was offered in each period. Estimation of Vitamin D3 and 25-OH-D3 content of egg yolks for each treatment was carried out by analytical High Performance Liquid Chromatography. Albumen height, Haugh unit, yolk height and yolk colour was analysed for fresh, 5 day and 10 day old eggs. Statistical analysis was carried out using MIXED procedure of SAS. Experimental treatment had a significant effect (P<0.01) on total vitamin D3 activity of biofortified eggs (T1; 2.92, T2; 3.88; T3; 5.06 and T4; 4.85 µg/egg (SD=0.022)) respectively. Treatment had no effect (P>0.05) on albumen height, Haugh unit and yolk height, throughout the four experimental periods. Hens offered T3 had a significantly (P<0.01) darker egg yolk colour compared to other treatments. In conclusion, results indicate that biofortification of laying hen diets with 25-OH-D3 may be a useful approach for tackling low vitamin D intakes in northern Europe without having any negative impacts on egg quality.

Transcriptional regulation of lipid metabolism in transition dairy goats by fish oil and stearate

G. Farina[1], G. Invernizzi[2], J.M. Caputo[1], A. Agazzi[1], V. Dell'Orto[1], J.J. Loor[3] and G. Savoini[1]
[1]University of Milan, Department of Health, Animal Science and Food Safety, via Celoria 10, 20133 Milan, Italy, [2]Università Telematica San Raffaele Roma, via di Val Cannuta 247, 00166 Rome, Italy, [3]University of Illinois, Department of Animal Sciences, 1207 W Gregory Dr, 61801 Urbana, IL, USA; greta.farina@unimi.it

To better understand the interaction between saturated or unsaturated fatty acids and genes involved in lipid metabolism in liver and subcutaneous adipose tissue, a nutritional study involving twenty three second parity alpine dairy goats fed either a non fat-supplemented basal diet (C; n=8), or a basal diet supplemented with stearic acid (ST; n=7) or a basal diet supplemented with fish oil (FO; n=8) was set up. 30 g/head/d extra fatty acids during the dry period and 50 g/head/d during lactation were supplemented starting one week before parturition up to 21 days in milk. Liver and subcutaneous adipose tissue samples were harvested at day 7, 7 and 21 relative to kidding and immediately snap frozen in liquid nitrogen. mRNA levels of CPT1A, MSMO1 and SLC27A2 in liver and ACACA, FASN, PLIN2, LEP and IL6 in adipose were measured by quantitative real-time RT-PCR. Data obtained were analyzed using the MIXED procedure of SAS. Milk production, milk composition, body weight and body condition score did not differ between treatments. CPT1A had a similar pattern between treatments and was up regulated at day 7 and down regulated at day 21. No differences were observed for MSMO1 expression. Long-Chain-Fatty-Acid-CoA Ligase (SLC27A2), was down regulated before kidding and, at day 7, reached the highest up regulation in all treatments. At day 21, was then down regulated in C and ST but still up regulated in FO (P<0.05). The enzyme encoded by this gene plays a key role in lipid biosynthesis and fatty acid degradation and could be linked to the delayed fat mobilization in FO subcutaneous adipose tissue observed at day 21. These results support the idea that dietary unsaturated and saturated fatty acids are able to differentially regulate lipid metabolism in liver and adipose tissue.

Vitamin D3 supplementation in heifers, on blood serum 25-OH-D3 and beef vitamin D3 concentration

S.K. Duffy[1], J.V. O'Doherty[1], K.D. Cashman[2] and A.K. Kelly[1]
[1]University College Dublin, School of Agriculture and Food Science, Belfield, Dublin 4, Ireland, [2]University College Cork, School of Food and Nutritional Sciences, UCC, Cork, Ireland; sarah.duffy@ucdconnect.ie

Beef biofortification is a potential food-based strategy for increasing vitamin D dietary intake. The objective of this study was to examine the effect of vitamin D_3 (D_3) inclusion rates in beef finishing heifers on beef D3 content, animal performance and carcass characteristics. Thirty continental (Charolais + Limousin cross) heifers were randomly allocated to 3 dietary treatments (T1) basal + 0 IU D_3, (T2) basal + 2,000 IU D_3 and (T3) basal + 4,000 IU D_3. Dietary treatments were offered for the final 28 days of an 80 day intensive finishing period. The basal diet consisted of a standard ad-libitum finishing regime of concentrates and forage (straw) offered at a ratio of 90:10. Animal growth was recorded weekly and individual dry matter intakes were recorded using Calan Broadbent Controlled feeding system. Blood samples were taken prior to slaughter for blood serum 25-OH-D_3 concentration. Longissimus dorsi (LD) muscles were excised 10 days post slaughter for D3 analysis carried out by analytical High Performance Liquid Chromatography. Data was analysed using mixed model ANOVA in SAS. Increasing dietary D_3 resulted in a linear response (P<0.05) in systemic 25-OH-D3 blood concentrations ((T1) 88.63 vs (T2) 132.52 vs (T3) 165.45 n/mol/l sem 10.692)). Similarly, D3 (P<0.05) content of LD beef also showed a linear response to dietary D3 inclusion. Indeed, D3 content in the highest inclusion treatment (T3), showed 33% enhanced D3 content compared to (T2) and 72% from (T1), ((T1) 0.019 vs (T2) 0.045 vs (T3) 0.067 µg/100 g sem 0.0240)). During the supplementation period heifer intake, performance and efficiency was not effected by treatment (P<0.10). Caracas parameters such as weight, kill out%, fat and conformation were also not altered by treatment offered (P>0.10). In conclusion results indicate that D3 content of beef can be successfully enhanced through short term dietary supplementation.

The effect of insect meal on egg fatty acid profile

E. Chatzidimitriou[1], V. Maurer[2], F. Leiber[2], M. Baranski[1], S. Stergiadis[1,3] and G. Butler[1]
[1]Newcastle University, Nafferton Ecological Farming Group, School of Agriculture, Food and Rural Development, Newcastle Upon Tyne, NE1 7RU, United Kingdom, [2]Research Institute of Organic Agriculture (FiBL), Ackerstrasse 113, 5070 Frick, Switzerland, [3]University of Reading, School of Agriculture, Policy and Development, P.O. Box 237, Earley Gate, Reading, RG6 6AR, United Kingdom; eleni.chatz@ncl.ac.uk

Replacing soy in poultry diets will lower the dependency on high-protein feed imports, thus improving the overall sustainability of egg production systems. Insects, which are a staple feed for many wild birds and poultry, have high-quality protein and are rich in other nutrients, such as fat, minerals and vitamins, thus being a promising alternative protein source. The objective of this study was to investigate the effects of replacing soybean cake with Hermetia illucens meal in layers diet of different flocks on egg fatty acid (FA) profile. A 3-week feeding trial was carried out with 30 organic Lohman Selected Leghorn layers at 64-74 weeks old, randomly distributed to three feeding groups: (1) control; (2) Hermetia meal at 12% of diet; and (3) Hermetia meal at 24% of diet. The experiment was subsequently repeated three times, using different flocks of the same genotype. Eggs were sampled on the last day of the feeding period and FA analysis of yolks was carried by gas chromatography. Analysis of variance was performed using diet and flock as fixed factors. Increasing dietary insect meal (to 12% and 24% of diet) significantly increased saturated FA content (+6% and +9%) and decreased polyunsaturated FA (-13% and -25%), including both omega-6 (-14% and -27%) and omega-3 FA (-17%, only at the high-insect diet). Similarly the resulting omega-6/omega-3 ratio was significantly lower (-10 and -12%). The flock also showed significant effects on FA profile in all FA groups except saturated FA. Interactions between flocks and dietary treatments were also significant for polyunsaturated FA and omega-3 FA.

In vitro antioxidant, ACE-inhibitory and citomodulatory properties of whey and casein proteins

C. Giromini[1], A.A. Fekete[2,3], J.A. Lovegrove[3], D.I. Givens[2], R. Rebucci[1] and A. Baldi[1]
[1]University of Milan, Health, Animal Science and Food Safety, via celoria 10, 20133 Milan, Italy, [2]School of Agriculture, Policy and Development, University of Reading, Food Production and Quality Research Division, Whiteknights Campus, P.O. Box 226, RG6 6AR, Reading, United Kingdom, [3]Hugh Sinclair Unit of Human Nutrition, University of Reading, Department of Food and Nutritional Sciences, Whiteknights Campus, P.O. Box 226, RG6 6AR, Reading, United Kingdom; antonella.baldi@unimi.it

Bovine milk proteins represent an important source of bioactive peptides. Whey (W) and casein (C) proteins exhibit an extensive range of bioactivities. In order to gain insight into their bioactive potential, isolated W and C proteins were digested in vitro using pepsin and pancreatic enzymes. Soya protein (S) was included in the study as non-animal protein control. The total digesta was filtered using 3 kDa Sartorius centrifuge tubes, in order to mimic the intestinal absorption. A permeate (mimicking the absorbed fraction) and a retentate (mimicking the intestinal fraction) were obtained. The antioxidant capacity (TEAC) and ACE-inhibitory (ACE-I) activity were determined before and after digestion. Protein retentate fractions were also tested for their citomodulatory activity on the human intestinal cell epithelium. All undigested proteins showed TEAC values below 0.5μmol trolox equivalents/mg proteins. Following in vitro digestion, a significant increase (P<0.05) was seen in TEAC values for W, C and S permeate to 2.32, 2.49 and 2.26 μmol, respectively. Whereas, the TEAC values obtained for W, C and S retentate were 0.81, 0.69 and 0.71 μmol, respectively. W protein permeate exhibited a significant ACE-I activity, compared with C and S permeate and compared with undigested W protein. All protein retentate, at specific concentrations, were able to stimulate the viability of HT29-MTX-E12 cells. The comparison of the in vitro effect of the two major intact milk proteins performed in this study can provide valuable knowledge regarding their bioactivities, promoting their use in the formulation of functional food ingredients.

Health beneficial omega-3 fatty acids levels in lambs of dairy breeds raised in Slovakia

M. Margetín[1,2], M. Oravcová[2], O. Debrecéni[1], L. Luptáková[1] and J. Tomka[2]
[1]*Slovak University of Agriculture Nitra, Tr. Andreja Hlinku 2, 949 76 Nitra, Slovak Republic,* [2]*National Agricultural and Food Centre, Research Institute for Animal Production Nitra, Hlohovecká 2, 951 41 Lužianky, Slovak Republic; tomka@vuzv.sk*

The meat quality of 80 light lambs of both genders from traditional rearing of four dairy breeds raised in Slovakia (Improved Valachian – IV, Tsigai – Ts, Merino – M and synthetic population of Slovak dairy sheep – SD) were assessed on the basis of essential and health beneficial fatty acids of intramuscular fat (IMF). The average weight of lambs before slaughter was 17.79±2.63 kg and age of lambs was 56.3±5.86 days. Lambs suckled dam milk; after 7 days from birth lambs were offered commercial concentrates and hay. Fatty acid (FA) profiles in M. longissimus dorsi were determined by gas chromatography. Differences between IV, TS, M and SD lambs were calculated using GLM model with effects of breed (B), sex (S) and interaction R×S. Highly significant differences in IMF were observed between IV, Ts, M and SD lambs (3.25±0.193; 3.57±0.193; 3.52±0.193 and 3.70±0.193 mg/100 g meat; P<0.001). The highest content of essential FA (linoleic, α-linolenic) were found in M lambs (8.68±0.550 and 54.09±2.131 mg/100 g meat). The differences between breeds were highly significant (P<0.001). The level of CLA was highest in IV lambs (28.11±1.93 mg/100 g meat) but only in lambs of this breed was observed unfavorable ratio of omega-6/omega-3 FA (4.15±0.142). The differences in health beneficial long chain omega-3 FA (EPA, DPA, DHA) between breeds were also highly significant (P<0.001). The level of EPA+DHA was highest in M and Ts lambs (56.21±3.40 and 38.57 mg/100 g meat) and smallest in SD lambs (20.44 ± in 3.400 mg/100 g meat). The content of EPA+DPA+DHA in individual lambs ranged from 19.48 to 162.18 mg/100 g meat. Thus, meat of traditionally reared lambs in Slovak dairy breeds appears to be a rich source of omega-3 FA. The study was supported by the projects APVV 0458-10 and VEGA 1/0364/15.

Producing lower saturated fat dairy products by changing the dairy cow diet: the RESET study

K.E. Kliem[1], D.J. Humphries[1], O. Markey[1,2], D. Vasilopoulou[1,2], C.C. Fagan[2], A.S. Grandison[2], R. Morgan[1], J. Guo[1], D.I. Givens[1] and J.A. Lovegrove[2]
[1]*University of Reading, Animal, Dairy and Food chain Sciences, School of Agriculture, Policy and Development, RG6 6AR Reading, United Kingdom,* [2]*University of Reading, Food and Nutritional Sciences, Whiteknights, RG6 6AP Reading, United Kingdom; k.e.kliem@reading.ac.uk*

It has been shown that replacing saturated fatty acids (SFA) with cis-unsaturated fatty acids in the human diet can reduce the risk of cardiovascular disease, and yet there are few studies that have focused on dairy products in particular. This study was part of a larger project involving a human intervention study. The main objective was to identify whether milk and dairy products could be produced with a consistently lower SFA/higher cis-monounsaturated fatty acid (MUFA) concentrations when compared with the control products (obtained from conventional production methods) for use in the human study. Production was divided into two periods to correspond with stages of the human study, with period 1 (Dec 13 to Sept 14) involving a total of 58 Holstein-Friesian cows (mean±SEM; parity 4.0±0.12; DIM 181±7.7; yield at start 35.0±0.77 l/d) and period 2 (Oct 14 to May 15) involving 41 cows (parity 4.0±0.12; DIM 205±7.2; yield at start 33.4±0.91 l/d). Cows were fed total mixed rations (50:50 forage:concentrate DM basis), and supplemented with 1 kg high oleic acid sunflower oil (AAK Ltd) per cow per day. Milk was used to make butter, cheddar cheese and UHT milk for the human study (treatment) and these modified products were compared with equivalent products obtained via a commercial partner (control). Full fatty acid (FA) profile of both treatment and control products during both periods were analysed. There was a difference (P<0.05) in total SFA (70 vs 54 and 70 vs 52 g/100 g FA control vs treatment, for periods 1 and 2, respectively), cis-MUFA (23 vs 32 and 22 vs 32 g/100 g FA control vs treatment for periods 1 and 2) and total trans FA (4.0 vs 10.7 and 3.8 vs 11.8 g/100 g FA control vs treatment for periods 1 and 2). Results demonstrate that it is possible to produce milk with a consistent FA profile, using different cows, over periods of time.

Genetic approaches to improving lamb survival under extensive field conditions
F.D. Brien[1] and M. Young[2]
*[1]University of Adelaide, School of Animal and Veterinary Sciences, Roseworthy Campus, Roseworthy SA
5371, Australia, [2]Beef and Lamb New Zealand, Genetics, P.O. Box 39-085, Christchurch 8545, New Zealand;
forbes.brien@adelaide.edu.au*

Providing good ewe nutrition and effective shelter at lambing to improving lamb survival has long featured in advice given to sheep producers. This paper discusses genetic approaches to improving lamb survival under extensive field conditions. Lamb survival has been included as a trait in genetic evaluation in New Zealand since 2001 and more recently in Ireland, with similar approaches currently being implemented in the United Kingdom and under consideration in Australia. However, in general, lamb survival has or is being indirectly evaluated as a component of net reproduction rate or evaluated as a trait of the ewe (rearing ability). Effective genetic evaluation for lamb survival requires accurate and comprehensive reproduction and pedigree records, the availability of which is variable between and within countries and breeds, with the Merino breed notably lacking pedigree information in most recorded flocks. The ability to determine pedigree from DNA testing and association testing between lambs and their dams using electronic ear tags is improving this situation, but is also reducing the opportunity to collect data near the time of lambing on traits useful for indirect selection. In New Zealand and Ireland and starting in the United Kingdom this year, both direct and maternal breeding values for lamb survival are provided to breeders. In Australia, the intention is to initially provide a breeding value for rearing ability and then work towards issuing direct breeding values for lamb survival. The actual and prospective genetic gains in lamb survival from using various selection strategies will be discussed at the EAAP conference. Finally, recent studies in sheep and goats suggest that genetic evaluation of lamb survival and potential genetic gains may be enhanced by considering survival in singles, twins and multiple born neonates as separate traits.

Small ruminant neonatal mortality: What are the important phenotypes?
C.M. Dwyer
*SRUC, Animal Behaviour and Welfare, Roslin Institute Building, Easter Bush Campus, Edinburgh EH25 9RG,
United Kingdom; cathy.dwyer@sruc.ac.uk*

Neonatal mortality of small ruminants has remained an intractable problem across all sheep and goat producing countries. Survival as a genetic trait has very low heritability, partly because of issues with accurate on-farm recording, and partly because it is a complex trait where many genes and conflicting factors can be in play. However, it is also likely that the trait may be largely determined by management and environmental factors, and opportunities for genetic gain are small. Recently there has been some evidence from looking in more detail at specific animal characteristics that improve survivability (such as lamb behaviour), or by looking at post-mortem evidence for vulnerability, that the heritability of these specific phenotypes can be greater than that of survival as a whole. This raises two issues to identify the important phenotypes for survival: (1) what are the important traits that increase survivability; and (2) can these phenotypes be readily measured in a commercial setting, in both intensive and extensive conditions? Identification of these phenotypes may lead to improved possibilities to increase our ability to achieve genetic progress in survival, but may also be relevant to achieving improvements in management by targeting the most appropriate changes to increase survival. In identifying these phenotypes, the perinatal period seems to be most important, since the majority of lamb and kid pre-weaning mortality occurs during this short time window. In this paper I will review what we currently know about phenotypic traits that are associated with improved survivability (e.g. dystocia, birth weight, lamb vigour, maternal behaviour) and consider the feasibility of measuring these on farm. The aim of this paper is to stimulate discussion about recording traits on farm that can be used to improve the survival of small ruminant neonates.

Genetic parameters for lamb survival

A. McLaren, S. Mucha, K. Kaseja, K. Moore and J. Conington
Scotland's Rural College, Animal & Veterinary Sciences, Roslin Institute Building, EH25 9RG, Midlothian, United
Kingdom; ann.mclaren@sruc.ac.uk

A key factor affecting the profitability of sheep farms, as well as animal welfare, is lamb survival. The topic is complex due to the large number of different factors that can be involved. In addition to farm environment influences, there are a wide range of biological factors that can have an impact on the survival of lambs. One aspect worthy of consideration is the genetic component of the trait. In order to investigate the potential for including a lamb survival trait in future breeding programmes, the heritability was investigated using data collected from commercial Dorset, Lleyn and Texel flocks. Lamb survival was considered as a binary trait, measured as a trait of the lamb, not as a trait of the ewe. Lambs were coded 0 if they were born dead or else did not have an 8 or 21-week weight recorded, or coded 1 otherwise. The data were from performance recording flocks throughout the UK and contained survival records for 15,433 (from 20 flocks), 51,174 (from 47 flocks) and 48,995 (from 108 flocks) Dorset, Lleyn and Texel lambs, born between 2003-2013, respectively. Univariate analyses were carried out to determine the genetic, maternal and permanent environmental effects. The models used were threshold animal models with binary distributions and a LOGIT link function. They also included the fixed effects of litter size at birth, sex, farm, birth year, dam age and the interaction between birth year and month. Lamb birth weight was also included and treated as a covariate. The overall proportion of lambs unaccounted for, and therefore assigned to code 0, were 13.0%, 10.6% and 13.4% for the Dorsets, Lleyns and Texels respectively. Direct heritability estimates were low for all three breeds ranging from 0.04-0.08. Maternal heritability and permanent environment estimates ranged from 0.004-0.02 and 0.01-0.02 respectively. Overall, the results indicate that lamb survival has a low genetic basis for all three breeds. However, it would still be of benefit to include lamb survival as a future breeding goal.

Genetic determinism of maternal behaviour in sheep and relationship with lamb mortality

D. Hazard[1], A. Kempeneers[1], D. Foulquié[2], F. Carriere[2], E. Delval[3] and A. Boissy[3]
[1]INRA UMR1388 Genphyse, 24 chemin de Borde Rouge, 31326 Castanet Tolosan, France, [2]INRA UE321 La
Fage, Saint Jean Saint Paul, 12250 Roquefort sur SQoulzon, France, [3]INRA UMR1213 Herbivores, Theix, 63122
Saint-Genès-Champanelle, France; dominique.hazard@toulouse.inra.fr

Increasing behavioural autonomy of animals ensures their adaptation to various environments and contributes to the sustainability of livestock production. More particularly, maternal behaviour is vital for the offspring survival. Breed differences in maternal behaviour are well known but genetic variation within breed has been poorly investigated. The aim of the study was to estimate in sheep genetic parameters for maternal reactivity and search QTL associated with the related behavioural traits. Ewes (n=851) reared outside were individually phenotyped just after lambing in two behavioural tests based on the reactivity to separation from their lambs and/or to a human. Ewes were tested for one or two lambing. In the first test performed outdoor 2 hours after lambing, the ewe's reactivity was assessed by measuring flight distance from lambs, time to restore contact with lambs and facility to follow transport of lambs. In the second test performed indoor 24 h after lambing, the ewe's reactivity was assessed by measuring vocalizations, locomotion and proximity to lambs with or without human presence. Lamb mortality was recorded until weaning. Heritability estimates were moderate (h^2=0.11 to 0.23±0.05) for flight distance from lambs, time to restore contact with lambs and facility to follow the transport of lambs. Genetic correlations between these traits ranged from 0.78 (±0.21) to 0.95 (±0.06). For the second test, heritability estimates ranged from 0.11 (±0.05) for proximity to 0.50 (±0.05) for vocalizations. Genetic correlations ranged from 0.13 (±0.06) to 0.44 (±0.07). Lamb mortality was significantly lower in ewes showing a higher maternal reactivity. A genome wide association study of maternal reactivity is currently carried on in ewes genotyped using the Illumina ovineSNP50 beadchip. The moderate to high heritabilities and potential SNPs polymorphisms for the behavioural traits of ewes may offer opportunities in genetic selection for improving maternal behaviour and reducing lamb mortality.

Survival rates and growing performance of lambs born from Assisted Reproductive Technologies

E. Emsen[1], M. Kutluca Korkmaz[2], B. Odevci[3] and H. Demirezer[4]
[1]Ataturk University, Animal Sicence, Ataturk University, Department of Animal Science, 25240, Turkey, [2]Ataturk University, Ispir Hamza Polat Vocational School, Ataturk University, Ispir Hamza Polat Vocational School, 25900, Turkey, [3]Imona Technologies, Imona Technologies, Istanbul Technical University, Technopark, ARI-3, 34450, Turkey, [4]Mustafa Kemal University, Reproduciton and Artificial Insemination, Mustafa Kemal University, Veterinary Faculty, 31030, Turkey; ebruemsen@gmail.com

Assisted Reproductive Technologies (ART) provides opportunities to introduce new genetics through embryo transfer and artificial insemination. Recently, a great deal of research has been performed concerning behavior characteristics due to both economic causes and animal welfare. This study was carried out using native fat tailed native Tuj breed (2-4 ages) of ewes received two frozen-thawed embryo genotypes while crossbreed and native purebred lambs obtained via laparoscopic insemination. Data were collected from 51 lambs (Romanov-R: 10, Charollais-C: 10, CharollaisXTuj-F1: 13 and Tuj-T: 18). The roles of dam and lamb in determining the growth performance and survivability of offspring based on specific behavioral differences were investigated using inter-breed embryo transfer and LAI. Charollais lambs were recorded with higher birth (4.9±0.2 vs 3.5±0.2 kg; P<0.001) weights compared to Romanov lambs. Weaning weights (75 d) were found higher in C lambs (29.4±1.9 kg; P<0.001) than those observed in R (22.4±1.3 kg), F1 (19.6±1.2 kg) and Tuj (19.4±1.1 kg) lambs. Tuj ewes had a relatively but not significantly longer duration of the licking/grooming event (17.7 min) for its own breed of lambs than purebred (R: 15.3 and C: 10.2 min) and crossbreed (F1: 12.7 min) exotic breed of lambs. Romanov and Tuj lambs were significantly more (P<0.01) active than C and F1 lambs in the first 2 h after birth and required less birth assistance. At weaning C lambs had significantly lower (50%; P<0.05) survival rates than R (100%), F1 (92%) and T (83%) lambs. It was concluded that survival and weaning performance of lambs born from embryo transfer and artificial insemination more likely depends on lamb rather than maternal behavioral characteristics.

An on-farm experiment to assess new phenotypes: how to involve farmers?

A. Cheype[1], F. Tortereau[2] and J.M. Gautier[3]
[1]Institut de l'élevage, Boulevard des Arcades, 87060 Limoges, France, [2]INRA, GenPhySE, 24 chemin de Borde Rouge, Auzeville Tolosane, CS52627, 31326 Castanet-Tolosan Cedex, France, [3]Institut de l'élevage, BP 42118, 31321 Castanet-Tolosan Cedex, Finland; agathe.cheype@idele.fr

Lamb vigor has been studied in different countries and genetic analyses showed that selection on these traits was possible. In these studies scoring systems have been developed to assess lamb vigor on farm. In France, the introduction of these new traits in selection will be completely dependent on the feasibility and the farmers' willingness to collect such data in commercial farms. In order to tackle this issue, a two-year project started in 2015 to develop suitable scoring systems for French farmers. This project has been set up with a bottom-up approach based on strong involvement of farmers and technicians. At first, six farmers and four technicians highly motivated by the topic have experimented the feasibility to collect those new traits during one lambing season over more than 950 lambs. The aim of first the study was to get feedback from farmers in order to define which traits was recordable and relevant, and to gather practical way to do it. The six farmers had fully tried to record all the different traits for each lamb where the information was available. During a focus group, all farmers agreed to keep five over nine traits proposed at the beginning: birth weight, birth assistance, lamb activity at birth, suckling ability and when appropriate, death date. The first positive step has allowed us to follow up with twenty more farmers in order to verify, with a broader group, if those new farmers were able to collect the selected traits and to get new ways to do it.

Ease of data collection on farm; reality or myth

H. Davies[1] and T. Canning[2]
[1]Llandre farm, Pumsaint, Llanwrda, SA198UT, United Kingdom, [2]FARMWIZARD, 54 Elmwood Avenue, Belfast, BT9 6AZ, United Kingdom; huwllandre@hotmail.co.uk

The 'TAG' project was funded by the Rural Development Plan for Wales (RDPW) 2007-13 Supply Chain Efficiencies Scheme (SCES). It ran from June 2009 to September 2015 and initially involved 80 farmers. The aim of the project was to improve the efficiency and quality of Welsh sheep production systems through better flock data collection and analysis. http://www.menterabusnes.co.uk/en/tag A key finding of the TAG project was the importance of user friendly data collection devices which allow the farmer to scan an EID tag and record multiple associated events e.g. weighing or lambing events. Historically, the most popular EID data collection device among sheep farmers was the MS Windows based Psion Workabout pro™ fitted with an Agrident EID reader. This solution was generally recognised as being user friendly, although the device is too large to fit in a pocket and is expensive (£1,500), creating a barrier to the adoption of new EID data performance management technology. The introduction and proliferation of Smartphone devices have become increasingly available to consumers and most of the farmers we deal with already regularly use a smart phone device. These devices have seen great advancements in User Interface Design and Internet Browser capabilities and now support HTML 5. This new standard offers exciting new functionality such as an offline database, presenting a great opportunity for developers to produce data access and recording applications. An enhanced work stream was developed in the final year of the TAG project along with the cloud based software provider Farm Wizard (http://www.farmwizard.co.uk) in order to investigate the on farm practicalities of a Smartphone Web Application for sheep farmers. The application stores an offline flock database, including historic events and performance data on their animals. The app has Bluetooth capability, enabling interaction with stick readers and Bluetooth enabled weigh scales. Voice recognition technology and real time device-to-cloud data transfer technology was also developed. Working with Farm Wizard enabled valuable access to commercial partners who are prepared to invest the resources needed to capture future opportunities as they arise.

Panel discussion on how improving data quality related to Small ruminant mortality

C.M. Dwyer[1] and J.M. Gautier[2]
[1]SRUC, Animal Behaviour and Welfare, Roslin Institute Building, Easter Bush Campus, Edinburgh EH25 9R, United Kingdom, [2]Institut de l'Elevage, Ruminant Health and Welfare and Dairy Products Quality Department, BP 42118, 31321, France; cathy.dwyer@sac.ac.uk

This challenge session will explore ways to reduce lamb and kid mortality in extensively and intensively reared sheep systems. Specifically, it will address genetic and phenotypic approaches to improving lamb and kid mortality, and address ways to standardise recording of perinatal survival for indoor and outdoor lambing. Issues around improving data quality, tools to record lamb and kid survival and ways to do this will be considered, as will small ruminant mortality before, during and after birth and including abortion and stillbirth. To do so, a panel of discussion will be organised and moderated by Cathy Dwyer. Don't hesitate to join in order to share your experiences and knowledge.

Definition of a total merit index in Austrian dairy sheep and goats

B. Fuerst-Waltl[1] and C. Fuerst[2]
[1]Univ. Nat. Res. Life Sci. Vienna (BOKU), Gregor Mendel-Str. 33, 1180 Vienna, 1180 Vienna, Austria, [2]ZuchtData EDV-Dienstleistungen GmbH, Dresdner Str. 89, 1200 Vienna, Austria; birgit.fuerst-waltl@boku.ac.at

Routine genetic evaluations for dairy and fitness-related traits are going to be implemented in Austrian dairy goats and sheep in September 2016. A milk, fitness and total merit index (TMI) will be defined. The milk index consists of the traits milk, fat and protein yield while for fitness related traits somatic cell score, number of lambs born and number of lambs born alive, the latter to account for stillbirth, are considered. While dairy breeding values are calculated in multivariate runs, fitness traits are univariately estimated. Economic values had already previously been derived for sheep but needed to be estimated for goats. The relative economic values (in %) of dairy: fitness traits are 91 : 9 and 89 : 11 in dairy goats and sheep, respectively. Within fitness traits, the relative economic values (in %) for number of lambs born : number of lambs born alive : somatic cell score = 31 : 14 : 55 in goats and 35 : 21 : 44 in sheep. However, the weighting in the TMI does not correspond to the expected selection response as the latter is also influenced by heritabilities, reliabilities and genetic correlations. When selecting for a TMI purely based on derived economic values, 99% of the expected monetary selection response would be achieved by means of dairy traits in both species. For all fitness traits, only a marginal, albeit positive, genetic gain may be expected. To illustrate the effect of a stronger weight on fitness traits, the economic value of milk-kg was set to zero in alternative scenarios. As a result, the relative weights (in %) dairy : fitness were 60 : 40 for goats and 50 : 50 in sheep corresponding to expected selection responses (in %) of 87 : 13 and 79 : 21, respectively. The derivation of economic values revealed that rearing losses and longevity would also be traits of interest. However, due to lack of data, currently neither of them can be included in routine evaluations. The final decision on the weights in the TMI will be made by representatives of the breeding organizations.

Lamb serum IgG concentration is reduced when lambs are fed using a stomach tube compared to suckling

T.M. Boland[1], F.P. Campion[1] and F.M. McGovern[2]
[1]University College Dublin, School of Agriculture and Food Science, Belfield, Dublin 4, Ireland, [2]Teagasc, Animal and Grassland Research and Innovation Centre, Mellows Campus, Athenry, Co. Galway, Ireland; tommy.boland@ucd.ie

Feeding colostrum to neonatal ruminants using a stomach/oesophageal tube is frequently practiced and recommended to ensure sufficient and timely intake of colostrum. This study aimed to characterize lamb serum immunoglobulin G concentration at 24 hours post-partum in lambs fed colostrum via stomach tube, or allowed to freely suckle during the first 18 hours of life. sixty four twin bearing ewes were individually fed to 100% of ME requirements during late gestation and at parturition ewes were allocated to one of two treatments as follows stomach tube (ST) or suckling (S; n=32 per treatment). Ewes in the ST group had an udder cover placed over the udder from parturition to 24 hours post-partum. These ewes were hand milked at 1, 10 and 18 hours post-partum following oxytocin administration and lambs received colostrum to a maximum rate of 50 ml per kg birth weight at each time point. Progeny of the S ewes were allowed to suckle freely during the first 24 hours post-partum. All lambs were blood sampled at 24 hours post-partum and udder covers removed at this stage. Lamb birth weight did not differ with treatment (P>0.05), but lambs from the S group were heavier from day 14 to 70 post-partum (P<0.01) and tended (P=0.09) to be heavier at weaning (day 98). Lamb serum IgG concentration was lower in lambs fed via stomach tube from 24 hours post-partum to day 70 post-partum (P<0.05). Lamb serum IgG concentration declined from 24 hours post-partum to two weeks post-partum and subsequently increased to a peak at 70 days post-partum. While stomach tubing is a valuable technique for colostrum management in the ruminant neonate, these findings indicate that lamb serum IgG concentration is reduced when lambs are fed colostrum via stomach tube compared to natural suckling.

The influence of ram on number of born and weaned lambs in Suffolk

J. Schmidova[1,2], M. Milerski[2], A. Svitakova[2], H. Vostra Vydrova[2] and L. Vostry[2]
[1]Czech University of Life Sciences in Prague, Faculty of Agrobiology, Food and Natural Resources, Faculty of Agrobiology, Food and Natural Resources, Kamýcká 129, 165 21, Prague, 6 Suchdol, Czech Republic, [2]Institute of Animal Science, Genetics and breeding of farm animals, Přátelství 815, 104 01, Prague-Uhříněves, Czech Republic; schmidova.jitka@vuzv.cz

The objective of the present study was to quantify the service sire effect in terms of (co) variance components of number lambs born and weaned. The database with 11,311 lambings in purebred Suffolk sheep from 1995-2013 was used. The basic model equation for the analysis of variance of litter size contained effects of ewe's age at lambing, contemporary group, permanent environmental effect of ewe and direct additive genetic effect of ewe. The other model was extended by contemporary group of ewes during mating (harem), and additive genetic and permanent environmental effect of service ram. Variance components and genetic correlations were estimated by the Gibbs sampling method. The proportions of variance for the service sire effect for number of lambs born and weaned were 4.1% and 2.6%. Including service sire, harem, and ram's permanent environmental effect in the model for number lambs born favourably decreased deviance information criterion. The annual genetic trends estimated as regression coefficient of additive genetic values on animal's birth year were 0.4% of lambs born and 0.2% of lambs weaned for female fertility. In male contribution on litter size it was 0.2% of lambs born and 0.1% of lambs weaned. Results from present study demonstrate that service rams in Suffolk sheep have low but detectable influence on litter size of their mates. Genetic parameter estimates indicate that direct selection on the service ram effect could increase litter size and achieve genetic gain through ram selection.

Impact of early weaning on growth performance and small intestine development in lambs

S. McCoard[1], N. Hennes[1], F. Knol[1], O. Cristobal-Carballo[1], A. Khan[1] and D. Stevens[2]
[1]AgResearch Limited, Animal Nutrition & Physiology Team, Private Bag 11008, Palmerston North 4442, New Zealand, [2]AgResearch Limited, Farm Systems South Team, Private Bag 50034, Mosgiel, New Zealand; sue.mccoard@agresearch.co.nz

Artificial lamb rearing is used in dairy sheep farming systems to maximise the amount of milk produced. Cost effective rearing systems designed for pastoral-based dairy sheep systems that do not limit the growth and health of the lambs are required. We investigated the impact of early weaning of artificially-reared lambs on growth and small intestine development. Lambs (n=32) were reared using milk replacer with ad libitum concentrate pellet and chopped meadow hay indoors and weaned off milk at either 4 or 6 weeks using a gradual step down system. Following weaning, lambs were grazed outdoors until 16 weeks. Live weight was recorded weekly and duodenum samples collected post-mortem at 4 and 16 weeks (n=8/group at each time point), weighed and processed for histological evaluation of morphology. Live weight and duodenal morphology data were analysed using REML variance components analysis. At 4 weeks, lambs weaned at 4 weeks were 10% lighter than lambs weaned at 6 weeks (P=0.009), however, by 16 weeks, no differences in live weight were evident between groups (P=0.284). Small intestine weight did not differ between the weaning groups either at 4 or 16 weeks of the trial (P>0.05). At 4 weeks, duodenal villi width was 17% greater in the 4 week versus the 6 week weaning group (P=0.03) but there were no differences in villus height, crypt depth, muscularis propria thickness, goblet cell number or villus height:crypt depth ratio (P>0.05). Similarly, no differences in duodenal morphology were evident between the groups at 16 weeks of age. To our knowledge this is the first study evaluating the effect of early weaning on lamb performance and small intestine development. Weaning at 4 weeks of age using a gradual step down milk feeding system does not adversely affect longer-term growth performance or small intestine development in artificially-reared lambs.

Impact of early weaning on rumen development and fermentation profiles in artificially-reared lambs
O. Cristobal-Carballo[1,2], F.W. Knol[2], S. Muetzel[2], M.A. Khan[2], D. Stevens[2] and S.A. McCoard[2]
[1]*Massey University, IVABS, Tennent Dr, 4474 Palmerston North, New Zealand,* [2]*Agresearch Limited, Animal Nutrition & Physiology Team, Grasslands Research Centre, Tennent Drive, 4442 Palmerston North, New Zealand; omar.cristobal@agresearch.co.nz*

Early weaning of artificially-reared lambs offers the potential to decrease rearing costs of commercial dairy sheep systems. However, early weaning may impact functional development of the rumen and therefore transition off milk and onto solid feed. The objective was to investigate the impact of early weaning of artificially-reared lambs on rumen development and fermentation. Lambs (n=32) were fed milk replacer at 20% of body weight with ad libitum grain-based meal and chopped hay indoors for 6 weeks followed by transition to pasture. Lambs were weaned off milk at either 4 (early) or 6 weeks (control) using a gradual step down system. Blood samples were collected fortnightly for β-hydroxybutyrate (BOH). Rumen wall samples (dorsal (DS), ventral (VS), dorsal blind (DBS) and ventral blind sacs (VBS)) and rumen contents were collected at slaughter (n=8/group) for evaluation of morphology and volatile fatty acids respectively. Data were analysed using a linear mixed model. At week 4 and 16, rumen weight, fermentation profiles, papillae number and width were not different between weaning groups (P>0.05). Early weaned lambs had 55% (P<0.035) and 60% (P<0.0002) higher BOH at 4 and 6 weeks respectively, which may indicate improved activity of the rumen wall to metabolise butyrate. At week 4, compared to controls, early weaned lambs had 11% shorter DS rumen papillae (P<0.046), but 16% thicker DBS mucosal epithelial cell layer (P<0.001), which may favour the increased absorptive capacity of the rumen epithelium. Early versus control weaned lambs also had 17% and 13% thicker muscular layers in the VBS (P<0.0001) and VS (P<0.005), respectively at 16 weeks indicating greater physical development. Weaning at 4 weeks of age using a step down milk feeding system improves some aspects of the morphological and functional development of the rumen to support the weaning transition and growth in artificially-reared lambs.

Pre-lambing metabolic profile of twin-bearing ewes in a hill environment
P. Zhou[1,2], N.R. Lambe[2], C.R. Morgan-Davies[2], N.D. Sargison[1] and T.G. McEvoy[2]
[1]*University of Edinburgh, Easter Bush Campus, EH25 9RG, United Kingdom,* [2]*Scotland's Rural College, West Mains Road, EH9 3JG, United Kingdom; ping.zhou@sruc.ac.uk*

The aim of this study was to investigate pre-lambing metabolic status of twin-bearing Scottish Blackface ewes (UBF: unimproved; or IBF: genetically improved Scottish Blackfaces) and Lleyn ewes farmed as one flock in a hill environment (56°N, 4°W). Two supplementary feeding levels (low/high) were applied from early January 2015 to lambing in April, with ewes assigned to whichever was appropriate on the basis of actual liveweight (LWT) or subjective body condition score (BCS) changes. Blood samples from 20 healthy 3 to 5-year-old twin-bearing (according to scanning results) ewes per genotype were collected on 18th March. The metabolites measured were β-hydroxybutyrate (BOHB), albumin, urea N, Cu and Mg. Only ewes that gave birth to twins and lambed 25-45 days after sampling were used in Generalized Linear Model analysis. Terms tested in the models included genotype, ewe LWT and BCS at scanning, days prior to parturition, litter weight, feeding level and ewe age. Terms used in the final model for each metabolite were determined by stepwise regression. The estimated mean plasma concentrations of BOHB, albumin, urea N, Cu and Mg for Lleyn (n=16), UBF (n=17) & IBF (n=16) ewes (0.41, 0.43 & 0.43 mmol/l, 31.6, 31.3 & 31.7 g/l, 3.9, 3.4 & 3.3 mmol/l, 13.5, 13.3 & 13.6 μmol/l and 0.99, 0.89 & 0.90 mmol/l, respectively) were all within standard reference ranges. Feeding level was negatively associated with BOHB (P<0.05) and urea N (P<0.001) status. Mean Mg plasma concentration in Lleyn ewes exceeded that in UBF and IBF ewes (P<0.05), which indicates that the Lleyn ewes had higher dry matter intakes than the Scottish Blackface ewes. The other metabolites did not differ significantly among genotypes. To conclude, in 2015, twin-bearing Lleyn ewes seemed as competent as Scottish Blackface ewes in this harsh hill environment, albeit perhaps while consuming greater amounts of feed.

Potential new indicators predicting longer productive lives in dairy cows

K. Huber
Institute of Animal Science, Faculty of Agricultural Sciences, Universtiy of Hohenheim, Anatomy and Physiology
of Farm Animals, Fruwirthstrasse 35, 70599 Stuttgart-Hohenheim, Germany; korinna.huber@uni-hohenheim.de

The failure to adapt metabolism to the homeorhetic demands of lactation is considered as a main factor in reducing the productive life span of dairy cows. Classical plasma variables such as non-esterified fatty acids, beta-hydroxybutyrate, glucose and insulin are not sufficiently indicating mal-adaptation to onset of lactation. In addition, a plenty of inflammatory and oxidative stress markers are known to indicate the pro-inflammatory situation in transition cows. However, the origin of these cytokines is not fully understood. It is hypothesized that dysfunctions of basic cellular pathways are responsible for the deviation of metabolism in the mal-adapting dairy cow. Secondly, it is hypothesized that these cellular dysfunctions occur before classical indicators and inflammatory markers show metabolic disorders. Nineteen multiparous dairy cows were used in this study, which were not different in performance, and classical biochemical profiles were obtained throughout the study period. To identify novel pathways and biomarkers potentially related to productive life in dairy cows serum samples were examined by means of (targeted) metabolomics. From 42 days before up to 100 days after parturition 188 metabolites were quantified at each time point using the AbsoluteIDQ® p180 Kit (Biocrates Life Science AG, Innsbruck, Austria). This panel is related to human metabolic disturbances such as diabetes type II and systemic inflammation and includes metabolites from 5 compound classes, namely acylcarnitines, proteinogenic and modified amino acids, glycerophospho and sphingolipids and biogenic amines. In the dairy cows, metabolites were identified such as long-chain acylcarnitines and biogenic amines being associated with extended productive life span. These metabolites might be mainly secreted by the liver and depend on the functionality of hepatic mitochondria. The concentrations of biogenic amines and some acylcarnitines differed already before the onset of lactation thus indicating their predictive potential for continuation or early ending of productive life.

Extended lactation in dairy cows

J. Sehested, C. Gaillard, T. Kristensen, J.O. Lehmann, L. Mogensen, G.M. Maciel and M. Vestergaard
Aarhus University, Foulum, Depts. of Animal Science, Agroecology and Food Science, Blichers Allé 20, DK 8830
Tjele, Denmark; jakob.sehested@anis.au.dk

Lactation can be extended to last at least 16 to 18 month by delayed rebreeding in dairy cows. An extended lactation strategy holds potential to improve dairy cow longevity and efficiency due to fewer calvings with associated health risks and fewer dry periods over an expected lifetime. In a typical Danish dairy production the lactation period will be 10 to 11 month based on early rebreeding after calving. However, rebreeding in this period is often challenging and with poor outcome in terms of pregnancy rate, and it is also not clear if early rebreeding is the optimal strategy in relation to milk production, reproduction, cow health, milk quality, environmental impact or economic outcome of the production. Results on some of these aspects from recent Danish experiments and from investigations on private Danish farms on extended lactation strategy will be presented and discussed. The feeding level and the feeding strategy imposed will affect lactation yield and persistency for the dairy cows on extended lactation strategy. An extended lactation strategy might increase the variation between cows and challenge the commonly applied TMR feeding strategy with one feed ration used for all lactating cows. Likewise, an individualised or group-based feeding strategy might be more favorable in relation to increased lactation length and the implied demand for increased lactation persistency for the extended lactation strategy to become successful. Results on individualized and live-weight based feeding strategies in relation to extended lactation performance and metabolic load of the dairy cows will be presented and discussed. Finally, extended lactation strategies will be discussed in relation to dairy cow lifetime performance and herd scenarios.

Production, health and welfare characteristics of cows selected for extended lactation

J.O. Lehmann, L. Mogensen and T. Kristensen
Aarhus University, Department of Agroecology, Blichers Allé 20, Dk-8830 Tjele, Denmark;
jespero.lehmann@agro.au.dk

Decision support software developed to assist dairy farmers with their daily decision-making typically do so by processing a large amount of cow specific information. Managing selected cows for extended lactation may be a way of utilizing their milk production potential for longer. The objective of our study was to describe the available information in early lactation for a decision to either delay or inseminate a cow. Data came from 422 completed extended lactations from 4 herds that differed in breed, herd size and average milk yield. Lactations were grouped in 3 equal sized milk performance groups (MPG) according to their relative milk performance within herd and parity (primi and multiparous). The peak yield in high MPG were 24 and 17% greater for primi and multiparous, respectively, and the persistency in the period between 60 and 305 days in milk (DIM) was 7 percentage points greater for both parity groups when compared with the respective cows in the low MPG. Primiparous cows in the high MPG were 1.8 months older at first calving, required 0.3 inseminations more per conception as heifers, produced 5 kg ECM more per day at 2nd and 3rd milk recording, and fewer tended (P<0.1) to have had an assisted calving when compared with cows in the low MPG. Multiparous cows in the high MPG had a higher persistency between 60 and 305 DIM, 6 days longer dry period and 0.4 inseminations more per conception in the previous lactation, and they produced around 8 kg ECM more per day at 2nd and 3rd milk recording as well as tended (P<0.1) to have fewer assisted calvings when compared with the low MPG. Numerically, fewer cows in the high MPG were treated for a disease. A principal component (PC) analysis showed that the total variation of 29 variables with early lactation cow specific information could be clustered into 3 PC; fertility, disease and milk production for primiparous cows and 4 PC; milk yield, fertility, peak yield and persistency for multiparous cows.

Customising dry period length: consequences for energy balance and yield over multiple lactations

A.T.M. Van Knegsel[1], A. Kok[1,2], R. Van Hoeij[1], J. Chen[1], N. Mayasari[1] and B. Kemp[1]
[1]Wageningen University, Adaptation Physiology group, De Elst 1, 6708 WD Wageningen, the Netherlands,
[2]Wageningen University, Animal Production Systems group, De Elst 1, 6708 WD Wageningen, the Netherlands;
ariette.vanknegsel@wur.nl

It is well known that a shorter or no dry period (DP) improves the energy balance (EB) in early lactation, mostly due to a reduced milk yield. Knowledge on effects of DP length on health and milk yield over multiple lactations is limited. It can be hypothesized that, when a shorter or no DP improves fertility or affects lactation persistency, long-term consequences for milk yield are less severe. Recently, we focused in several studies on the long-term value of shortening the DP to improve cow health. The results indicate that shortening or omitting the DP shift milk yield from the critical period shortly after calving to the period before calving, when energy needs can be met more easily. Shortening the DP from 60 to 30 days resulted in a limited reduction in milk yield, but improved the EB in the subsequent lactation and had no consequences for somatic cell count in milk, colostrum quality, or development of the calves. Omitting the DP resulted in a strong reduction in milk yield, but also had a large positive impact on the EB, metabolic status and resumption of ovarian cyclicity in the subsequent lactation. Effects of DP length on lactation persistency were absent. Milk yield losses, but also improvement of the EB, were reduced when this management strategy was applied over successive lactations. Effects of DP length on milk yield and EB were depended on parity, genotype, and udder health status. Studies are ongoing to determine cow characteristics related to the individual differences in response to a short or no DP, aiming at the development of a 'Customised Dry Period' strategy for individual cows. This 'Customised Dry Period' will not only be evaluated for consequences on cow health and productivity, but also for the long-term effects on net herd returns and environmental impacts.

Dry period plane of energy: effects on glucose tolerance in peripartum dairy cows
S. Mann, F. Leal Yepes, T. Overton, J. Wakshlag, B. Cummings and D. Nydam
Cornell University, Department of Population Medicine and Diagnostic Sciences, Tower Road, 14853 Ithaca, NY,
USA; sm682@cornell.edu

This abstract was presented at the 2015 Graduate Student Competition: ADSA Production Division Graduate Student Oral Competition, PhD. Overfeeding energy during the dry period may affect peripartal insulin sensitivity of dairy cows. The objectives were to describe the effect of different dry period dietary energy strategies on glucose tolerance and on concentrations of insulin, glucagon, glucose, β-hydroxybutyrate (BHBA) and nonesterified fatty acids (NEFA) in the peripartum period. To test our hypothesis that dry period plane of energy leads to changes in glucose tolerance, cows (n=84) were blocked by expected calving date and randomized into 3 treatment groups 57 d before expected parturition: a TMR formulated to supply 100% of energy requirements (C), or 150% of requirements (H), or 100% for the first 28 d of the dry period followed by a TMR supplying 125% of energy requirements until calving (I). At 28 and 10 d pre-, as well as 4 and 21 d postpartum, intravenous glucose tolerance tests (IVGTT; 0.25 g glucose/kg) were performed and the area under the curve (AUC) of glucose and insulin were estimated. Baseline samples were analyzed for concentration of insulin, glucose, NEFA, glucagon and BHBA. Mixed-effects ANOVA were carried out (SAS, v. 9.3). The AUC for glucose and insulin was not different among groups at any time point (P>0.10). Differences in baseline values were most notable on d 4 postpartum and indicated increased postpartum lipolysis and ketogenesis in group H whereas we detected a tendency for lower glucose concentration despite higher glucagon concentration in this group. In conclusion, these findings suggest that prepartum plane of energy had an effect on the hormonal regulation of gluconeogenesis and ketogenesis (as reflected by the concentrations of insulin and glucagon), but did not lead to a detectable effect on peripartal glucose tolerance as assessed by IVGTT.

The effects of offering concentrates during the dry period on dairy cow performance
R.A. Law, M. Romero Oiartzun, A. Brown, A. Gordon, A.F. Carson and C.P. Ferris
Agri-Food and Biosciences Institute, Agriculture Branch, Large Park, Hillsborough, BT26 6DR, United Kingdom;
r.law@dunbia.com

Feeding concentrates during the dry period is often advocated to improve the body condition score of cows prior to calving, and to 'prepare the rumen' for high concentrate diets post calving. However, the actual benefits of feeding concentrates during the dry period are less clear. This study was conducted on 10 Northern Ireland dairy farms over a two year period, and involved 1,217 dairy cows. Cows were categorised as having either a 'low' (1.0-2.5) or 'moderate/high' (2.75-5.0) body condition score (BCS) at drying off (1-5 scale, where 1 = emaciated and 5 = overly fat). Cows in the 'low' BCS group were managed on one of three treatments during an 8 week dry period: concentrates offered for the full eight weeks, concentrates offered for the final three weeks, and no concentrates offered during the dry period. Cows within the 'moderate/high' BCS group were managed on the latter two treatments only. Concentrate feed levels during the dry period averaged 2.7 kg per cow per day. 88% of cows had a BCS of 2.5 or less at drying off (Low). While cows on all treatments lost BCS during the dry period, those within the moderate/high BCS group had a greater BCS loss than those within the low BCS group. Although cows offered no concentrates during the dry period had a greater BCS loss than those offered concentrates (P<0.001), this effect was small and of no practical importance. BCS at drying off had no effect on milk yield, milk fat + protein yield, or conception to first or second AI during the subsequent lactation (P>0.05). Similarly, offering concentrates during the dry period had no effect on milk production or fat + protein yield during the subsequent lactation, or on overall fertility. Thin cows (BCS<2.25) offered no concentrates during the dry period had an increased risk of being culled during the subsequent lactation (P=0.086). With the exception of 'thin cows', few benefits were observed when cows were offered concentrates during the dry period.

A novel method to prevent hypocalcemia: Can we improve cow health and longevity in the herd?

L.L. Hernandez

University of Wisconsin-Madison, Dairy Sciene, 1675 Observatory Dr, Madison, WI 53706, USA;
llhernandez@wisc.edu

Dairy cattle losses due to premature death and culling cause major financial strain to individual dairy producers and the industry, have welfare implications, and are increasing over time. Periparturient diseases are responsible for substantial economic losses for dairy farmers. Approximately 25% of culling and 75% of injuries, diseases, and other health disorders occur within the first 60 d postpartum. The first insult to the dairy cow at parturition is hypocalcemia. Cows that are affected by either clinical (CH) or subclinical hypocalcemia (SCH) are at increased risk for development of other metabolic and reproductive complications. Approximately 25% of primiparous cows and 50% of multiparous cows in the US dairy cow population suffer from subclinical hypocalcemia SCH, and between 5 and 10% of cows develop CH. Currently accepted practices for treatment and prevention of SCH and CH include oral Ca supplementation post-calving and anionic salt supplementation pre-calving. However, post-calving Ca supplementations, while critical for treatment, are not sufficient to prevent maternal hypocalcemia and its associated peripartum disorders. Use of anionic salts in the pre-partum period has reduced the incidence of SCH and CH. However, approximately 25% of cows will still be afflicted with SCH. The lack of adequate therapies leaves a large percentage of the US dairy cow population unprotected and new therapies are lacking because the physiological mechanisms of SCH are not fully understood. Therefore, our research group has been working on novel methods for the prevention of CH and SCH. We are currently using 5-hydroxy-L-tryptophan (5-HTP), a serotonin precursor, to improve calcium homeostasis during the early peripartum period. Preliminary data indicate that administration of 5-HTP prepartum improves calcium metabolism postpartum. We have performed experiments both in the US model for feeding and housing dairy cows as well as repeated the experiment in a European system. Our data present a potential novel route to prevent SCH and CH in dairy systems. Overall, particularly in the US system, this could reduce the removal of cows from the herd during the first 60 days in lactation.

How longevity drives greenhouse gas emissions of milk produced by dairy cows in two feeding regimes

F. Grandl[1,2], S. Peter[3], M. Kreuzer[1] and M. Zehetmeier[4]

[1]ETH Zürich, Institute of Agricultural Sciences, Universitätstr. 2, 8092 Zürich, Switzerland, [2]Qualitas AG, Chamerstr. 56, 6300 Zug, Switzerland, [3]Flury&Giuliani GmbH, Sonneggstr. 30, 8006 Zürich, Switzerland, [4]Bavarian State Research Center, Agricultural Economics, Menzinger Straße 54, 80638 München, Germany; florian.grandl@qualitasag.ch

Feed production, enteric fermentation and rearing of replacement heifers are the main drivers of greenhouse gas (GHG) emissions from dairy production systems. Promoting longevity has a great potential to reduce GHG emissions, as a longer productive lifetime dilutes the emissions from replacement heifers. However, changing enteric methane (CH_4) emissions of cows during lifetime could counteract this advantage. We therefore investigated emissions of milk (kg CO_2-equivalents per kg energy corrected milk, kg CO_2-eq./ kg ECM) produced from 30 dairy cows with an age ranging from 2.4 to 10.1 years fed a diet either with or without concentrate. The cows were part of an experiment, where CH_4 emissions and feed intake were measured at around 110 days in milk. A GHG balance was calculated for current milk production of these cows by aggregating emissions from the feeds consumed, estimated heifer rearing emissions, and the measured CH_4 emissions. The emissions were related to the actual milk yield. Emissions from feed production did not change with the age of the cows but were higher for the cows fed the concentrate diet (0.43 vs 0.35 kg CO_2-eq./kg ECM). CH_4 emissions showed a maximum in cows of approx. 4 and 5.5 years of age and were lower for the cows fed concentrate (0.39 vs 0.49 kg CO_2-eq./kg ECM). Proportional emissions from the rearing phase decreased steeply in early productive life and began to level off when the 4th to 5th lactation was completed. Combining these main drivers for GHG emissions of milk production systems showed that average emissions of milk produced decreased markedly in cows from the 1st to the 4th lactation or up to approx. 15 kg milk yield per day of life, and decreased only slightly for older cows or for higher milk yield per day of life, respectively. Therefore, increasing the longevity of cows in dairy herds could reduce the overall environmental impact of milk production.

Once recorded metabolic adaptation does not allow to predict longevity in dairy cows

J.J. Gross[1], L. Grossen-Rösti[1], F. Schmitz-Hsu[2] and R.M. Bruckmaier[1]
[1]*Veterinary Physiology, Vetsuisse Faculty University of Bern, Bremgartenstrasse 109a, 3012 Bern, Switzerland,*
[2]*Swissgenetics, Meielenfeldweg 12, 3052 Zollikofen, Switzerland; rupert.bruckmaier@vetsuisse.unibe.ch*

Health disorders occurring during lactation account distinctly for culling in dairy cows. This study investigated if metabolic status obtained once in life during a negative energy balance (NEB) in early lactation allows to predict age and lifetime performance animals achieved at culling. In total, 200 multiparous and metabolically stressed cows (milk fat-protein-ratio > 1.5 in previous early lactation) entering at least their 3rd lactation (parity: 5.0±2.1, mean ± SD) were selected. Blood was sampled in week 4 post-partum (p.p.) and plasma analyzed for concentrations of free fatty acids (FFA), β-hydroxybutyrate (BHBA), glucose, insulin, cholesterol, triglycerides, and insulin-like growth factor-1 (IGF-1). Age at culling, reasons for culling, days lactating, and total lifetime performance in milk yield were recorded. Relationships between number of achieved lactations, lifetime performance, milk yield per day of life and per DIM, metabolic, and endocrine parameters were evaluated. Age of cows at culling ranged from 4.7 to 20.2 yr with parities from 3 to 17. A retrospective classification of the one-time recorded metabolic adaptation in week 4 p.p. did not differ between animals of different parities at culling. Furthermore, there was no relationship neither between the metabolic adaptation recorded in a preceding lactation and the number of lactations achieved, nor to the lifetime milk production. However, milk yield per d of life and per d in milk, respectively, were correlated with plasma concentrations of FFA and IGF-1. Contrary to the wide spread assumptions, an inadequate adaptation due to a high metabolic load during a NEB as shown by high BHBA and FFA plasma concentrations did not necessarily result in an earlier culling of dairy cows, although they are obviously more prone to metabolic disorders.

Poster presentations: longevity: a physiological and management perspective

R.M. Bruckmaier[1] and M. Vestergaard[2]
[1]*Vetsuisse Faculty, University of Bern, Veterinary Physiology, Bremgartenstr. 109a, 001 Bern, Switzerland, [2]Aarhus University, Department of Animal Science, Blichers Allé 20, 8830 Tjele, Denmark; rupert.bruckmaier@vetsuisse.unibe.ch*

Please join this 'Short presentation of the posters' event as the final theatre presentation of session 24. Poster presenters are offered the possibility to shortly present their poster during the final 15 min of the theatre session. As a total of 8-9 poster authors are offered this possibility, each poster presenter will be given 1-2 min to present the poster in 1 or 2 slides. In case the poster author wishes to use this possibility, the slide(s) has to be sent to the chair (Rupert M Bruckmaier) no later than Sunday 28 August 2016. If you do not want to utilize the option of presenting your poster this way, it is also OK. Based on the incoming slides from poster authors, a common slide show will be compiled and made available for this presentation. At start of the presentation, all poster authors will be asked to line up in a queue so everything can be kept within the 15 min time slot available. Rupert Bruckmaier will ring the bell when times run out, so be sharp with your slides and your words. There will be no time for questions from the audience. Questions have to be put directly to the poster authors when you visit them during the poster session. Titles and contents of the posters can be viewed in the following abstracts.

Impact of litter size in first parity on sow lifetime performance and health

E. Andersson[1], J. Frössling[1,2], B. Algers[1], L. Engblom[3] and S. Gunnarsson[1]
[1]Swedish University of Agricultural Sciences, Department of Animal Environment and Health, P.O. Box 234, 532 23 Skara, Sweden, [2]National Veterinary Institute, Department of Disease Control and Epidemiology, 751 89 Uppsala, Sweden, [3]Swedish University of Agricultural Sciences, Department of Animal Breeding and Genetics, P.O. Box 7023, 750 07 Uppsala, Sweden; emma.andersson@slu.se

In the last decades, the average litter size of sows has been steadily increasing. The negative consequences on the welfare of piglets born in large litters are well known, but studies on the effects of large litter sizes on sow lifetime performance and health are limited. This was therefore the focus of two observational studies, which were both based on retrospective analysis of available production data. In our first study, lifetime performance (number of litters produced in a lifetime and stayability) of 48 707 crossbred sows from 24 herds was investigated. In our second study, disease frequency during gestation and lactation in 939 purebred Yorkshire sows from one herd was investigated. The unit of interest was sow and litter size in the first parity was the main exposure of interest in both studies. The potential association between litter size and the different outcomes were analysed using multivariable regression models. Results from the first study indicated that sows having a medium sized litter in their first parity produces more litters in a lifetime and that the probability to produce at least 4 litters in a lifetime is also higher for such sows, compared to sows having a small or large litter size. The second study found a significant higher disease frequency during gestation and lactation in first parity sows with a large litter size. Our results indicated that aiming for keeping sows giving birth to a medium-sized litter, with approximately 11-14 piglets born in total, may improve sows lifetime production, stayability and health; and this should be considered when planning breeding strategy and annual removal in piglet producing herds.

Polish Simmentals: early predictors of longevity

M. Morek-Kopec[1], W. Jagusiak[1] and A. Zarnecki[2]
[1]University of Agriculture in Krakow, Department of Genetics and Animal Breeding, Al. Mickiewicza 24/28, 30-059 Krakow, Poland, [2]National Research Institute of Animal Production, ul. Krakowska 1, 32-083 Balice near Krakow, Poland; rzzarnec@cyf-kr.edu.pl

Genetic correlations between length of productive life and body conformation were estimated to identify type traits that can be used as early predictors of longevity in a small Polish Simmental dual-purpose cattle population with about 10,000 milk-recorded cows. Breeding values of 166 Simmental sires from December 2015 routine genetic evaluations were used. EBVs for functional longevity, defined as length of productive life corrected for production, were estimated using a Weibull proportional hazard model. EBVs for 18 linear type traits (scored from 1 to 9) and stature and chest girth (measured in cm) were estimated by a single-trait BLUP animal model. Genetic correlations were approximated based on the correlation between EBVs of body conformation traits and longevity, adjusted by reliabilities. Generally, the estimates of genetic correlations were low to moderate. Positive genetic correlations with functional longevity were found for udder width (0.25) and udder depth (0.11), as well as for rear leg set side view (0.27) and rear leg set rear view (0.16). Unfavourable correlations were obtained for traits characterizing capacity, especially body depth (-0.36) and chest girth (-0.24). There were smaller negative correlations with stature (-0.151), rump width (-0.16), muscularity of front end (-0.14), rear end muscularity (-0.13) and chest width (-0.12). Weak negative correlations were found for rear udder height (-0.18) and udder support (-0.18), and near-zero correlations for rear teat placement (0.04), front teat placement (0.01), fore udder (-0.06) and foot angle (-0.05). The traits with the largest correlations, which could be used as correlated predictors of longevity, were body depth, chest girth, udder width and rear leg set side view. These traits seem to indicate that cows with smaller body capacity have a lower relative risk of culling.

Genetic relationships between age at first calving, its underlying traits and survival of Holsteins

J. Heise[1,2], K.F. Stock[1], F. Reinhardt[1] and H. Simianer[2]
[1]IT Solutions for Animal Production (vit), Heinrich-Schröder-Weg 1, 27283 Verden, Germany, [2]Georg-August-University Göttingen, Department of Animal Sciences, Albrecht-Thaer-Weg 3, 37075 Göttingen, Germany; friederike.katharina.stock@vit.de

Models for genetic evaluation of longevity in dairy cattle are often corrected for age at first calving (AFC), implying absence of genetic relationships between AFC and survival. There are reasons to reconsider this hypothesis, because AFC is mainly determined by age at first insemination (AFI) and the interval from first to last insemination (FLI). In our study, we analyzed phenotypic and genetic relationships between AFC, AFI, FLI and survival (1/0) of different periods of the first lactation (S1: 0 to 49 d, S2: 50 to 249 d, S3: 250 d from calving to 2nd calving). Data consisted of 10 samples of in total 1,731,571 German Holsteins born between 2000 and 2009. Phenotypic relationships were analyzed using univariate logit threshold models with either S1, S2 or S3 as dependent variables and AFC, AFI or FLI as independent variables. The model for genetic analyses was a linear multiple trait sire model. Considered fixed effects were herd×year of calving for S1 to S3 and AFC, herd×year of insemination for AFI and FLI and additionally type of semen (fresh, frozen, sexed frozen or information not provided) for FLI. Phenotypic effects of AFC, AFI and FLI on S1 to S3 were mainly negative and strongest on S3. Mean heritability estimates from the 10 samples were 0.020 (S1, SD: 0.007), 0.023 (S2; SD: 0.006), 0.026 (S3; SD: 0.005), 0.561 (AFC; SD: 0.208), 0.424 (AFI; SD: 0.129) and 0.010 (FLI; SD: 0.002). Estimated genetic correlations between age traits (AFC, AFI) and S1 to S3 ranged from -0.40 (AFC, S3; SD: 0.09) to -0.24 (AFI, S3; SD: 0.14), between FLI and S1 to S2 they were close to zero, and between FLI and S3 it was -0.43 (SD: 0.11). Our results suggest that correcting for age at first calving in genetic evaluations for longevity removes important functional genetic variance and should therefore be avoided.

Simulations to define the optimum lifetime management for Holstein cows

C. Gaillard[1], H.N. Phuong[2], P. Blavy[3] and J. Sehested[1]
[1]Aarhus University, 20 Blichers Allée, 8830 Tjele, Denmark, [2]Department of Economic Development, Jobs, Transport and Resources, Bundoora, 3083 Victoria, Australia, [3]AgroParisTech & INRA, 16 rue Claude Bernard, 75231 Paris, France; charlotte.gaillard@anis.au.dk

Interest in extended lactation has increased this last decade. However, little is known about the long term effects of extended lactation on the lifetime productive and reproductive performance. The two main objectives of this study were: (1) to determine the optimum duration of extended lactation to enhance efficiency and pregnancy rate compared with a 10 mo lactation; and (2) to define if the lactation duration should vary with the age of the cow to reach an optimum lifetime performance. A lifetime performance model taking into account the changing physiological priorities of an animal during its life and through repeated reproduction cycles, was used to simulate different lifetime scenarios. The model was tested to fit 16 mo lactations. For each scenario, a total of 300 cows were randomly generated. Simulation of cows managed for repeated 16 mo lactation all their life (EL) had the highest lifetime efficiency (ratio of energy in milk to energy intake), and the longest lifetime compared with shorter (10, 12, or 14 mo) or longer (18, 20, or 22 mo) lactations. Pregnancy rates were not significantly improved with EL compared with a 10 mo lactation (N). Simulations of cows managed for a 16 mo lactation during their first lactation, followed by 10 mo lactations for the rest of their life (EL-N), had a lifetime efficiency similar to that of the EL scenario. Simulations of cows managed for a 10 mo lactation during their first lactation, followed by 16 mo lactations for the rest of their life (N-EL), had a lifetime efficiency similar to the N scenario. To conclude, these simulations indicated that: (1) the 16 mo lactation is the optimum extended lactation length in terms of productive and reproductive performance; and (2) that managing the primiparous cows with a 16 mo extended lactation, followed by 10 mo lactations, allows the lifetime efficiency to increase being similar to cows managed for 16 mo lactations all their life.

Genetic correlations between energy balance and lactation persistency of Holsteins in Japan

A. Nishiura[1], O. Sasaki[1], M. Aihara[2] and H. Takeda[1]
[1]NARO Institute of Livestock and Grassland Science, Tsukuba, Ibaraki, 305-0901, Japan, [2]Livestock Improvement Association of Japan, Koto-ku, Tokyo, 135-0041, Japan; akinishi@affrc.go.jp

A postpartum energy deficit is caused by the inability of feeding to compensate for the high milk production during the early stage of lactation. The negative energy balance (EB) is assumed to be the reason for reproductive problems, as well as metabolic diseases, including ketosis and mastitis. The objective of this study was to estimate genetic correlations between EB and lactation persistency (LP) in the first three parities of Holstein cows in Japan, with the goal of improving EB. The data consisted of test-day milk records of Holstein cows in Japan that were collected from 2008 to 2013. All lactations were from 6 to 305 days in milk (DIM) and were required to have a minimum of 9 records. The data set for analysis consisted of 3,457,401 records of 228,741 cows. EB were calculated using the multiple regression equation of Friggens et al. or Løvendahl et al. Genetic parameters of EB and LP were estimated using a multiple-trait random regression model, in which the records in different lactations were treated as separate traits. LP was defined as the difference in estimated milk yield between 240 and 60 DIM. Genetic correlation estimates between EB and LP in the same lactations were negative soon after parturition and positive after 80 DIM. Genetic correlation estimates in 1st lactation were higher than in later lactations. Genetic correlation estimates between EB in one lactation and LP in the subsequent lactation were smaller than in the same lactations. Genetic correlation estimates between LP in one lactation and EB in the subsequent lactation were -0.2 to 0.5. The shape of the curve of genetic correlation estimates between EB in 3rd lactation and LP in 2nd lactation was different from each other. In conclusion, energy deficit in the early stage of lactation is genetically related to low LP, that is, low milk yield in the middle to late stage of lactation.

Voluntarily delayed rebreeding and double insemination effects on pregnancy rates of Holstein cows

C. Gaillard, M. Vestergaard and J. Sehested
Aarhus University, 20 Blichers Allée, 8830 Tjele, Denmark; charlotte.gaillard@anis.au.dk

During the last decades, the milk production per cow has increased while fertility and mounting behavior have decreased in Holstein cows. Voluntarily delaying rebreeding to a period where the cow is in positive energy balance might help improving pregnancy rates. To overcome the problems associated with estrus detection, the use of a double artificial insemination (AI) technique, where cows are inseminated 1 or 2 days apart, have previously been applied but with inconsistent results. The two main objectives of the present study were: (1) to determine if the pregnancy rates were improved when delaying rebreeding to 8 months after calving compared with cows inseminated at around 2 months after calving; and (2) to evaluate the effect of a double AI technique around estrus on the pregnancy rates compared with the use of a single AI per estrus. The dataset comprised 235 observations, each representing one AI, from 62 cows followed for consecutive lactations of 10 months (early rebreeding at 2 months) and one lactation of 16 months (late rebreeding at 8 months). The double AI technique was applied in 41% of the AI recorded. The results showed that even though more cows tended to express mounting behavior at the late rebreeding compared with the early rebreeding (63.3 vs 45.9% respectively, P=0.06), the pregnancy rate at first AI was not improved (40.0 vs 38.7% respectively, P=0.27). The cows receiving the double AI technique had a lower pregnancy rate compared with the cows receiving a single AI per estrus (45.8 vs 63.8 respectively, P<0.01). Moreover, use of the double AI technique induced confusion as how to report the reproductive performance in the herd, taking into account the total number of insemination/success or considering only one value per estrus. To conclude, delaying rebreeding to 8 months increased mounting behavior at AI time, but did not increase pregnancy rate at first AI. The double insemination technique had a negative effect on the pregnancy rate at first AI.

Herd dynamics and economics of different strategies for managing a herd for extended lactation

J.O. Lehmann[1], L. Mogensen[1], S. Østergaard[2], J.F. Ettema[3] and T. Kristensen[1]
[1]Aarhus University, Department of Agroecology, Blichers Allé 20, 8830 Tjele, Denmark, [2]Aarhus University, Department of Animal Science, Blichers Allé 20, 8830, Denmark, [3]Aarhus University, Department of Molecular Biology and Genetics, Blichers Allé 20, 8830, Denmark; troels.kristensen@agro.au.dk

The objective of this study was to investigate the effect of different extended lactation management strategies on herd dynamics and economics. Herd dynamics, milk production and feed consumption were estimated with simulations using the SimHerd-model (www.simherd.com) where lactation curve and intake capacity were adjusted for calving interval. Managing all or first parity cows for calving intervals of 15 (CI15) or 17 (CI17) months increased the number of milking days in the herd and reduced days dry when compared with a calving interval of 13 (CI13) months. However, milking days were unchanged and reduced when only older cows were managed for CI 15 and CI17, respectively; partly, due to a greater proportion of cows being dried off earlier. Managing all or older cows for extended lactation reduced the proportion of first parity cows and reduced the number of young stock per annual cow. This was particular pronounced when all or older cows were managed for CI17 instead of CI13 (-24 and -16%, respectively). Reduction in ECM per annual cow ranged from 2 to 486 kg with the largest reductions seen when all and older cows were managed for CI17, while the reduction was only 2 and 13 kg ECM when only first parity cows were managed for CI15 and CI17, respectively. Revenues from milk and meat were reduced in strategies with extended lactation, but feed costs and other costs were also reduced as extended lactation leads to fewer calvings, which leads to fewer costs associated with diseases and reproduction. Also, the reduction in total number of animals reduced total work hours, and the sum of all changes led to an increase in herd contribution margin of up to 6%, which was achieved when first parity cows were managed for CI17. Besides that, extended lactation reduced the number of livestock units by 1.2 to 7.6% and area required to grow the feed ration by 0.9 and 7.7%.

Influence of the first lactation on the productive life of Holstein cows

M. Grayaa[1,2,3,4], H. Hammami[4], C. Hanzen[2], B. Rekik[1], A. Gillon[5] and N. Gengler[4]
[1]UC, ESAM, Mateur, 7030, Tunisia, [2]ULg, FMV, Liège, 4000, Belgium, [3]UC, INAT, Tunis, 1082, Tunisia, [4]ULg, GxABT, Gembloux, 5030, Belgium, [5]AWé, Ciney, 5590, Belgium; mriwa8@hotmail.fr

Length of productive life (LPL) defined as the time from first calving to culling is an ambiguous trait resulting from many factors. This preliminary study aims to link LPL to several of these factors including lactation curve traits. Lactation curve traits for milk were provided by the Walloon Breeding Association for 20,766 primiparous Holstein cows calving from 2003 to 2014 in 395 larger herds (>50 cows). A linear model that included fixed effects of herd (H), calving year (CY), 4 classes of calving season (CS), 4 classes of milk production adjusted to 305 days (M305), and 9 classes of age at first calving (AFC) together with linear regressions on LPL expressed in days in milk (DIM), peak production (PK), persistency (PS) defined as a post-peak slope, was used to study variations of LPL. Effect of censored LPL was reduced by the CY effect. Observed means were 7,427 kg (M305), 393.7 days (DIM),45.9 months (LPL) and 27.7 months for AFC. As expected LPL was affected by herd (P<0.01) linked to management strategies among herds, which culling decisions, and by calving year (P<0.01) as animals born earlier are expected to be able to achieve a higher LPL. Season had a significant effect on LPL (P<0.01), as autumn seemed the least favorable season. Cows with intermediate yields tended to stay longer in the herd than low (M305<5,000) or high (M305>10,000) producing ones, a result that can be explained by voluntary culling of low producing cows and of involuntary culling for reasons such as diseases for very high producing cows. Results for AFC were less clear, later calvings (>28 month) being linked to lower LPL. For linear regressions of DIM on LPL, significance was close (P<0.06) for a positive relationship, but this could be an artifact due to the definition of LPL. Finally, the lack of significance for the regressions of PK and PS on LPL in this study should be considered a preliminary result for two reasons: (1) the simultaneous presence of other correlated effects as milk yield in the model; and (2) the use of a linear regression when non-linear relationships are more likely (intermediate optimum).

Lost in transition: a reaction norm model to breed cows that can better cope with metabolic stress

H. Simianer[1], N.T. Ha[1,2], J.J. Gross[2], U. Schnyder[3], F. Schmitz-Hsu[4] and R.M. Bruckmaier[2]
[1]Georg-August-University Goettingen, Albrecht-Thaer-Weg 3, 37075 Goettingen, Germany, [2]University of Bern, Bremgartenstr. 109a, 3001 Bern, Switzerland, [3]Qualitas AG, Chamerstr. 56, 3052 Zollikofen, Switzerland, [4]Swissgenetics, Meielenfeldweg 12, 3052 Zollikofen, Switzerland; hsimian@gwdg.de

Dairy cows often suffer a severe metabolic stress in the transition phase, caused by the discrepancy of high energy demand for rapidly increasing milk production and limited feed intake. We hypothesize, that some cows are genetically less well suited to cope with this metabolic stress than others, leading to adverse follow-up effects on longevity. To test this, we designed a reaction norm model in which the functional longevity is linked to the metabolic challenge in the early part of the first lactation. As challenge variables we used either the sum of the milk yield or the accumulated fat/protein ratio of the first three test days, pre-adjusted for herd-test day variance. We defined a random regression sire model, in which for each sire a random regression coefficient was estimated to reflect the genetic component of the reaction to the challenge. The model was fitted to data of ~1.4 million daughters of ~5,000 Brown Swiss bulls with suitable observations available (>9 daughters per bull). We found with both challenge variables the sire variance for the slope of the random regression to be significantly different from zero, suggesting a genetic component for the ability to cope with metabolic stress. In other words, some bulls have daughters in which metabolic stress causes a reduction of longevity, while in other bulls the longevity of the daughters remains largely unaffected by an energy deficit in the transition phase. A bootstrap procedure was used to obtain confidence bands for the individual reaction norms of bulls, which differed quite substantially in slope. Genome-wide association studies for the slope coefficient revealed similar results for both the milk yield and the accumulated fat/protein ratio, with 7 markers showing genome wide significant (α=0.05, Bonferroni correction) effects. The results of the study show that the ability to cope with metabolic stress in the transition phase clearly has a genetic component and could be used to breed metabolically robust dairy cows.

Good legs means good stayability of the sows

T.H. Le[1,2], E. Norberg[1], B. Nielsen[3], P. Madsen[1], K. Nilsson[2] and N. Lundeheim[2]
[1]Aarhus University, Dept. of Molecular Biology and Genetics, Blichers Alle 20, 8830 Tjele, Denmark, [2]Swedish University of Agricultural Sciences, Dept. Animal Breeding and Genetics, Ulls väg 26, 750 07 UPPSALA, Sweden, [3]Danish Agriculture and Food Council, Pig Research Centre, Axeltorv 3, 1609 København V, Denmark; le.hongthu@mbg.au.dk

Leg weakness is one of the major reasons for culling of sows. Improvement of leg conformation is expected to improve both animal welfare and producers' profit. We summarize here the possibilities to breed for better legs/gaits and the potential influence on reproduction and stayability, based on data on Landrace and Yorkshire sows in Sweden and Denmark. Purebred young pigs were scored for leg conformation traits at performance testing (5 months of age) in our studies. Leg conformation traits were analyzed together with litter size in parities 1 and 2 and stayability traits (culled or not culled after parity 1 or parity 2), recorded on purebred sows in nucleus and multiplier herds. Genetic (co)variances between traits were estimated. We estimated heritabilities for leg conformation traits ranging from 0.02 to 0.20 depending on breed and population. Estimated genetic correlations between leg conformation and litter size were favorable for Swedish Yorkshire (+0.15 to +0.42), but unfavorable in the Danish populations (-0.17 to -0.46). Estimated correlations between leg conformation and stayability were favorable in both Swedish Yorkshire and the Danish populations (0.12 to 0.42). Our results indicate that leg conformation can be improved by selection. Indirect selection for improved stayability based on leg conformation traits would be possible but influenced by the size of the genetic correlations between traits. Our ongoing project on genome wide association study using high density SNP chip is expected to identify chromosome regions associated with leg traits in three Danish pig populations (Landrace, Yorkshire and Duroc). Results from this study could be used as prior information in genomic selection, helping to more effectively improve leg conformation and accordingly improve sow longevity.

Novel NIR predictions of the nutritive value of cereal and protein meals across animal species
S. Parker-Norman and H. Graham
AB Vista, 3 Woodstock Court, Marlborough, Wilts SN8 4AN, United Kingdom; hadden.graham@abagri.com

Nutritionists traditionally use proximate analysis, where ingredients and feeds are analysed for their basic nutritional content, to ensure accuracy and consistency in diet formulation. These analyses are time-consuming and expensive, restricting the number of samples that can be analysed and creating delays between sampling and receiving analytical results. An alternative is to use a Near Infra-red Reflectance (NIR) spectrophotometry, a non-destructive, cost effective and fast technology, for predicting chemical and physical composition of feedstuffs. Several thousand feedstuff samples (maize, sorghum, wheat, barley, soyabean meal and canola meal) were collected worldwide and subjected to novel NIR calibrations to predict proximate composition, available energy for swine, poultry and ruminants, and digestible amino acids. Energy value varied within and between cereal types and animal species, with the variation in available energy within a cereal type ranged from 1-4 MJ/kg DM. It is estimated that a 1 MJ/kg difference in energy would be worth between €10-20/t in animal feeds. Further, for canola meal the lysine content ranged from 16-23 mg/kg and the reactive lysine (ie. the lysine present in a form that can be utilized by the animal following digestion) from 70-100%. The use of novel on-line and hand-held NIR units with unique NIR calibrations based on in vivo data offer the feed industry the potential to purchase feedstuffs and formulate feeds based on accurate nutritional data, offering considerable cost savings, more consistent animal performance and reduced environmental impact of animal production.

INRA Feed Unit System for Ruminants: a new 'Red Book'
P. Nozière[1], J.L. Peyraud[2] and D. Sauvant[3]
[1]INRA, UMR Herbivores, Theix, 63122, France, [2]INRA, UMR Pégase, St-Gilles, 35590, France, [3]INRA, UMR MoSAR, Paris, 75005, France; pierre.noziere@clermont.inra.fr

The INRA feed unit system for ruminants, widely adopted in France and several countries, has been regularly updated. But some limits appeared regarding emerging challenges for animal nutrition: prediction of feed efficiency, product quality, effect of feeding on animal health and emissions to the environment. It also did not properly cover the use of low-quality or high-concentrate diets. To address these issues, a group of INRA researchers involved in ruminant nutrition was estabished in 2010 to work on the 'Systali' project. The system has been renewed using meta-analyses (large data bases of digestion, metabolic balance and animal performance, representative of the major feeding practices), and modelling with evaluations and simulations. The dietary supply model has been renewed to account for digestive interactions and flows of individual nutrients (VFA, glucose, AA, LCFA). Feed values are no longer additive and depend on the final ration. Requirements have been updated to account for variability in metabolic efficiency and new ration values, and various productive and non-productive animal responses to diets have been modeled. The main advances are presented in a new 'Red book', the 'INRA Feed Unit System for Ruminants' (Wageningen Academic Publishers), organised in 4 parts. The first two parts present biological concepts and equations (with their field of application and statistical precision) used to predict intake (including at pasture) and nutrient supply (Part 1), animal requirements and responses to diets (Part 2). They apply to net energy, metabolisable protein and amino acids, but also water, minerals and vitamins. Part 3 presents the use of concepts and equations in rationing with two purposes: (1) diet calculation of a given performance objective; and (2) prediction of the multiple responses of animal to diet changes in the case of dairy and meat, large and small ruminant productions. Part 4 provides tables of feed values, and their prediction. For each part, specificities for tropical and Mediterranean areas are presented. All the equations and concepts will be embedded in the INR@tion software available on the internet for practical use.

Comparative prediction of digestive interactions dairy cows

D. Sauvant[1], J.B. Daniel[1] and P. Nozière[2]
[1]*AgroParisTech-INRA UMR MoSAR, Life Sciences, 16 rue C. Bernard, 75231 Paris Cedex 05, France, [2]INRA UR Herbivores, Phase, Center of Theix, 63122 St Genès Champanelle, France; sauvant@agroparistech.fr*

Modeling digestive interactions (DI) is a major difficulty in developing of feed unit systems, espcially for high yielding ruminants. DI generally lead to underestimation of the nutritive value of the ration. In the INRA Systali project, Sauvant and Nozière quantified by meta-analysis the DI induced by dry matter intake as a % of live weight (0.5<DMI<4.0), proportion of concentrate (0<PCO<0.8) and the rumen protein balance (-70<RPB<100, = CP − 6.25×(non NH_3-N at duodenum), g/kgDM). The target was to compare, on a same database, the impacts of DI proposed by INRA, NRC (2001), and Finnish research workers. The comparison was performed on the MoSARCO database for dairy cows (874 treatments, 288 experiments focused on energy or protein supply). Expressed as discount factors<1 multiplicative of theoretical digestibility, the correction of NRC (2001) was a bit less than that of Systali (0.93±0.02 vs 0.89±0.03). In contrast, as the method to calculate NE from digestibility was not the same between the two systems, the correction in Mcal of Net Energy for Lactation (NEL/d) (CorrNEL) was larger for NRC than for INRA (-2.65±1.22 vs -2.03±1.31 Mcal/d). The correction for Finland (CorrNEL$_{FIN}$) was calculated from the dietary ME at maintenance (tabulated values), dietary CP, and DMI. Mean CorrNEL$_{FIN}$ is close to CorrNEL$_{INRA}$ (-1.99±1.13 vs -2.03±1.31 Mcal/d). The intra-experiment regressions between systems are: CorrNEL$_{NRC}$ = 0.29 CorrNEL$_{INRA}$ − 2.04, RMSE=0.56 and CorrNEL$_{FIN}$ = 0.49 CorrNEL$_{INRA}$ − 1.03, RMSE=0.46. The influence of DMI on CorrNEL was similar for the 3 systems; in contrast for NRC, the impact of PCO was lower, and NRC also provided an inverse sign for the correction for RPB, which led to less accurate results. In conclusion, calculated DI are different from one feed system to another.

Evaluation of structural effectiveness of diets for dairy cows: an on farm comparison of systems

A. Behrendt[1], D. Albers[2], A.R. Sharifi[3] and J. Hummel[1]
[1]*University Göttingen, Animal Science, Kellnerweg 6, 37077 Göttingen, Germany, [2]Agricultural chamber Lower Saxony, Albrecht-Thaer-Str. 1, 26939 Ovelgönne, Germany, [3]University Göttingen, Animal Science, Albrecht-Thaer-Weg 3, 37075 Göttingen, Germany; jhummel@gwdg.de*

A sound prediction of structural fibre supply is important to prevent subacute rumen acidosis in dairy cows. Different evaluation systems like 'Structure Value' (SV), 'structure effective Crude Fibre' (seCF) and 'physically effective Neutral-Detergent Fibre' (peNDF$_{>8mm}$ and peNDF$_{>1.18mm}$) exist. In this study, systems were compared based on estimated fulfilment of recommendations by a diet (% of recommendation), and based on their relation with potential indicators of adequateness of diet. Mixed rations of 40 German farms were evaluated. Intake of the group was quantified and diets analysed for particle size distribution (Penn State Separator) and chemical composition. Milk yield plus nutrients and energy balance were evaluated for the herds. For 10 individuals per herd, further indicators for appropriateness of diet were evaluated (faecal score; urinary net-acid-base-excretion; rumination rate as chews/min). Data were statistically evaluated with a mixed model and comparison of means. The relation to indicators was evaluated via correlation. Test diets had an average milk yield of 34.0 l/d, milk fat was 4.04%, concentrate proportions 32-49% and maize silage proportions of roughage 0-82%. Fulfilment of requirements was estimated highest by SV and lower by seCF and both peNDF. Some correlation with indicators like milk fat/protein ratio (proportion of animals <1.0) was found for peNDF (r ~ -0.39) and SV (r=-0.55), but not for seCF. The study confirms results that SV tends to estimate a higher fulfilment of recommendations for structural fibre than other systems. Assuming approximately appropriate diets, this points to an over-estimation of structural fibre supply. The peNDF system resulted in reasonable estimates and showed correlation with indicators like milk fat/protein ratio. In this combination, peNDF shows some advantage over seCF and SV in our study.

Conceptual model of digestion for pigs and poultry and its factors of variation

S. Roger[1], A. Narcy[1], J. Van Milgen[2], M.J. Duclos[1] and E. Recoules[1]
[1]INRA, UR0083, 37380 Nouzilly, France, [2]INRA, UMR1348, 35590 Saint-Gilles, France; sonia.roger@tours.inra.fr

Improving the efficiency of pig and poultry husbandry systems requires a better understanding of nutrient digestion. Mechanistic modelling is a useful tool to identify and quantify the impact of different factors affecting the digestion. Up to now, despite the genericity of physiological processes, models of digestion developed for monogastrics have been species-specific or focused on individual nutrients. In addition, practical nutrition has been mostly based on table values but ignores interactions among nutrients and among nutrients and the animal. The aim of this study was to develop a conceptual model of digestion for pigs and poultry while accounting for different factors of variation (e.g. species, sex, and physiological stage). The digestive tract of the animal is considered as a system where inputs correspond to the ingested nutrients, the compartments are the stomach, the small and large intestine, and outputs are the quantity of absorbed and excreted nutrients. For each compartment, the input and output kinetics were determined using literature knowledge. Preliminary results suggest that the output of the stomach was represented by a decreasing exponential function. For the small and large intestines, inputs of nutrients depended on outputs of the stomach and the small intestine, respectively, and outputs were formalized with decreasing sigmoid functions. The mathematical formalism is generic but model parameters will be modulated to account for differences related to factors of variation, while factors related to the feed (e.g. composition, technological processes) also need to be considered. This conceptual model is a first step towards the development of a generic predictive tool to determine absorption and excretion of nutrients and to explore how each factor of variation impacts the digestion process and how they interact. This knowledge could contribute to improve the efficiency of animal production and the sustainability of animal production.

Developing win-win outcomes across a range of grassland-based livestock farming systems

B. Dumont[1], M. Benoit[1], L. Delaby[2], M. Duru[3], O. Hughenin-Elie[4], S. Lemauviel-Lavenant[5], J. Ryschawy[3], R. Sabatier[6] and D. Vollet[7]
[1]INRA, UMR Herbivores, 63122 Saint-Genès-Champanelle, France, [2]INRA, UMR Pegase, 35590 Saint-Gilles, France, [3]INRA, UMR Agir, 31324 Castanet-Tolosan, France, [4]Agroscope, Institute for Sustainability Sciences, 8046 Zürich, Switzerland, [5]Caen University, UMR Ecophysiologie Végétale, Agronomie, 14032 Caen, France, [6]INRA, UMR Sadapt, 75231 Paris, France, [7]Irstea, UMR Metafort, 63170 Aubière, France; bertrand.dumont@clermont.inra.fr

In Europe, grassland-based livestock farming systems develop across a wide range of biogeographical conditions. These systems allow producing meat, milk and cheese with limited competition with human food supply. Intensive grazing systems limit greenhouse gas emission per unit of product, but high animal densities lead to nitrate leaching and various other nuisances. Favorable grass-growth conditions combined with the use of grass-legume mixtures allows limiting the inputs needed for production and system environmental footprint. Under more marginal conditions, permanent grasslands and rangelands provide valuable habitats for various forms of biodiversity, and a wide bundle of regulating (e.g. climate regulation via carbon sequestration, pollination, flood regulation, fire prevention, erosion prevention) and cultural (e.g. aesthetic value, product quality, ecotourism) ecosystem services. Provisioning services usually exhibit negative correlations with cultural and regulating services so that trade-offs between stakeholders occur. Though win-wins are the exception, not the rule, case-studies from various European countries illustrate opportunities for win-wins in grassland-based systems. Analyzing the mechanisms involved in grazing management and acting on the spatial arrangement of practices to increase landscape heterogeneity can help reconciling production and biodiversity. In order to create win-win situations, it is also necessary to take into account the point of view of all stakeholders. Importantly, it should not be assumed that provisioning services must always dominate any other ecosystem services. Finally, prospective scenarios offer opportunities to quantify whether shifts in production systems or food choices are likely to lead to win-win solutions.

Multi-indicators approach for the evaluation of efficiency of mountain dairy farms
M. Berton, E. Sturaro, M. Ramanzin and G. Bittante
University of Padova, DAFNAE, viale dell'Università 16, 35020 Legnaro (Padova), Italy; enrico.sturaro@unipd.it

The study aimed to assess the sustainability of the dairy farms in mountainous areas combining the environmental impact computed according to Life Cycle Assessment and the human-edible feed conversion ratio computed as the ratio between the gross energy content in human-edible feedstuffs and the energy content of human-edible animal product (milk) (HeFCR). Data originated from 38 mixed-breed dairy herds located in the Trento Province (North-East Italian Alps).Average herd size was 49±33 cows and mean Fat Protein Corrected Milk (FPCM) production was 22.8±6.6 kg/cow/day. The cradle-to-farm-gate system boundary included the production and use of on and off-farm feedstuffs, of fuels and bedding materials, as well as the herd management phases. The reference unit was the farm, and the functional unit 1 kg of FPCM sold. Milk vs meat multi-functionality was solved using a mass allocation method. Global warming (GW), acidification and eutrophication (EU) potentials, as well as cumulative energy demand and land occupation were assessed as impact categories. Mean environmental impact per 1 kg FPCM and impact category resulted 1.1±0.2 kg CO_2-eq, 20±4 g SO_2-eq, 6±1 g PO_4-eq, 5.1±2.0 MJ and 1.4±0.5 m^2/year respectively. Mean HeFCR was 0.72±0.46 MJ edible feedstuffs/MJ milk, with a favorable output/input energy balance. All the impact categories were negatively correlated with HeFCR (r ranged from -0.37 for EU to -0.48 for GW). The higher was the impact, the lower was the competition for potential human edible resources. The results evidenced that dairy farms in mountainous areas, strongly linked to permanent grasslands, are efficient in terms of competition for potential human edible resources. The sustainability assessment of the dairy sector, in particular that located in disadvantaged mountainous area, has to include different type of indicators in order to take into account the characteristics and environmental conditions as well as the different products and ecosystem services supplied.

Including trees in pasture-based pig systems to improve animal welfare and eco-efficiency
M. Jakobsen and A.G. Kongsted
Aarhus University, Agroecology, Blichers Allé 20, 8830, Denmark; malene.jakobsen@agro.au.dk

In many ways free-range pig production allows the animal to express a wide range of behavioural patterns. However, there are some serious challenges related to animal welfare as well as negative environmental side effects in the current pasture-based systems practiced in Northern Europe. In some countries, sows are snout-ringed, which prevents rooting. This reduces damage of plant cover and consequently the risk of nutrient leaching is decreased. Snout-ringing is problematic as rooting is the preferred explorative behavior of the pig and often referred to as a behavioural need. From practical farming there are reports of high incidences of sunburn in hot seasons. Introduction of trees in pasture-based systems may be one way to reduce these challenges. The overall idea is that trees in a pasture-based system will provide a more natural and stimuli-rich environment for the pig with possibilities to seek shadow in hot periods and shelter during cold periods. Furthermore, the system may lead to reduced snout-ringing due to uptake of nutrients from the trees independent of grass cover. Finally, introduction of trees is expected to improve the overall eco-efficiency of the systems. Behaviour, activity level, location and incidences of sunburn were investigated in 84 lactating sows, housed in individual paddocks, subjected to one of three treatments; (1) pasture + poplar trees and access to trees; (2) pasture + trees but no access to trees; and (3) no trees (control). Preliminary results indicate that on average, the sows rested 57% of the total observation time. Of the total time spent resting, on average, 13% was taking place in the area with trees but with large variation between cold (1-6%) and hot seasons (18-24%). Inclusion of trees in pasture-based systems is suggested to benefit the welfare of sows during hot seasons.

Silvopature is a sustainable intensification option in grassland production systems
J.H. McAdam[1,2]
[1]Queen's University Belfast, School of Biological Sciences, Newforge Lane, BT80 9XL Belfast, United Kingdom,
[2]Agri Food and Biosciences Institute, Grassland and Plant Science, Newforge Lane, BT80 9XL Belfast, United Kingdom; jim.mcadam@afbini.gov.uk

Intensive and semi-intensive temperate grasslands, systems often revolve around landscapes which may suffer from loss of biodiversity, homogeneous community and habitat structure, landscape aesthetic poverty, eutrophication of waterways, soil degradation and rural depopulation. I will review the evidence that the introduction of wide spaced trees in silvopastoral systems can make these grassland landscapes more sustainable, deliver a wide range of ecosystem services and align with a sustainable agriculture land management strategy. Silvopasture can be used to extend the grazing season to help higher grass utilisation and give resilience to grazing during extreme rainfall, while increasing biodiversity, carbon sequestration, reducing water run off and providing renewable fuel. Silvopasture might also be be deemed to be more sustainable than farm woodland because of the intimate spatial integration of trees and agriculture and would include, for example, reduced wind and temperature stress and shelter for animals. There are additional benefits from root differentiation, a reduction in leaching losses of nutrients, faster nutrient cycling in the presence of grazing animals and reduced soil erosion Additionally, silvopasture can make a positive impact on sustainable landscape and rural development, compared to conventional farm woodlands, because of the diversity of employment opportunities created by multi-functional systems. Economic predictions are also encouraging. Favourable farmer surveys, policy inclusion and uptake on commercial farms in Ireland also indicate that farmers have started to consider silvopastoralism as a realistic land-use option which can meet a variety of objectives in future managed landscapes. Appropriate examples of silvopastoral systems in Ireland, UK and Chile will be presented to support the evidence above.

Grassland intensification and inevitable tradeoffs between multiple ecosystem services
D.A. Fornara, E.A. Wasson, C.J. Watson and R. Carolan
Agri-Food & Biosciences Institute (AFBI), Sustainable Agri-Food Sciences Division, Newforge Lane, BT9 5PX, Belfast, United Kingdom; rachael.carolan@afbini.gov.uk

Across Europe grassland management intensification includes regular additions of inorganic and organic nutrients to soils as well as greater livestock density and higher mowing frequency. An important question is how to maintain the ability of these grassland-based livestock farming systems to perform multiple functions and services which benefit our society. For example, if chronic nutrient fertilization greatly increases grass yields, it will also increase nutrient losses from soils while reducing biodiversity. Also soils receiving organic animal manures may have a greater ability to sequester atmospheric CO_2, however, this positive effect can be offset by large increases in GHG emissions from the livestock sector. Here we argue that to make intensive grassland management more sustainable we need a better understanding of the ecological mechanisms that ultimately determine inevitable tradeoffs between different soil and plant functions and services. We first show how long-term (>40 years) intensive grassland management has influenced grass yields, soil carbon and nitrogen sequestration, nutrient availability and nutrient losses from soils, plant root mass and other biogeochemical properties, which are essential to maintain grassland multi-functionality. We then provide evidence to show how different practices (i.e. long-term inorganic vs organic nutrient applications) affect key soil processes. Finally, we discuss how changes in soil processes can inform management decisions in relation to what potential tradeoffs are 'available' among multiple ecosystem services.

Multifunctionality of extensive mountain livestock farming systems

A. Gibon and G. Balent

INRA, UMR 1201 Dynafor, CS 52627, 31326 Castanet Tolosan cedex, France; annick.gibon@toulouse.inra.fr

Extensive grassland-based LFS in European mountainous regions are facing many challenges despite their many functions and services for sustainable regional development. Besides their direct part in economy and social life, they maintain semi-natural grasslands and emblematic cultural landscapes which provide a variety of ecosystem functions and services (environmental protection and mitigation of natural hazards; conservation of a rich biodiversity; cultural services of major value for inhabitants' quality of life and tourism-based economies). European policies failed up to now to secure their future. The lack of understanding of the complex relationships between farmer practices and the ecosystem services they provide is a major impediment to efficient policy support to their sustainability, particularly their economic viability. To help meeting the gap, we ran an integrated assessment of the multi-level relationships between land management and ecosystem services in a valley in the Pyrenees National Park (France), where semi-natural grasslands occupy more than 96% of the agricultural landscape area and livestock farming succeeded reasonably-well up to now to meet its expected functions and services. We outline the social-ecological framework and spatially-explicit method we used to assess the impacts of local land-use system organization and change on landscape services before presenting its main results. These later stress out that local LFS is at a turning point. Most families in the farm population have adaptive strategies rooted in local agropastoral tradition that include practices aimed at the care and long-term stewardship of the various grassland types and the whole landscape. But the biggest ones have or are drifting towards an 'entrepreneurial' strategy prioritizing farm economic performance. Despite detrimental land-use practices to locally-expected services from livestock farming, these farms are the only ones likely to survive at a 10-15 years horizon, due to demographics and the economic context. Indeed, multifunctionality of local LFS up to now relied on widespread adaptive strategies that involved a continuous intensification of family work and a growing share of non-agricultural activities in family income. These strategies have or are reaching their limits.

Research priorities for modelling livestock and grassland systems in Europe under climate change

R.P. Kipling[1], A. Bannink[2], J.C. Van Middelkoop[2] and N.D. Scollan[1]

[1]Aberystwyth University, Institute of Biological, Environmental and Rural Sciences (IBERS), 1st Floor, Stapledon Building, Plas Gogerddan, Aberystwyth, Ceredigion, SY23 3EE, United Kingdom, [2]Wageningen UR, Livestock Research, P.O. Box 338, 6700 AH Wageningen, the Netherlands; rpk@aber.ac.uk

Grassland-based livestock systems provide a range of environmental and cultural benefits in addition to the production of nutritious meat and dairy products. However, climate change poses multi-faceted and complex challenges for the sector. Mathematical modelling plays an important role in understanding and overcoming these challenges. Models can support decision making at farm and policy levels, exploring the impacts of climate change on livestock systems, identifying effective adaptation strategies and improving understanding of complex interactions. Two horizon-scanning exercises were undertaken to identify key challenges and research priorities for grassland modelling and for livestock health and pathogen modelling in the context of climate change. Workshops, questionnaires and literature reviews were used to identify challenges and priorities. These initiatives were initiated within the Modelling European Agriculture with Climate Change for Food Security (MACSUR) knowledge hub and drew together the opinions of 53 modellers and experimental researchers from 30 institutes in 14 countries, including partners of MACSUR and of the Global Research Alliance Animal Health and GHG Emissions Intensity Network. Fifteen challenges for grassland modelling and 20 challenges for livestock health and pathogen modelling were identified. Research priorities common to both fields included capacity building (such as improved data quality and accessibility, stakeholder engagement and fit-for-purpose modelling) and subject-specific priorities (such as modelling overwintering and linking land use change and pathogen spread modelling). It is hoped that the challenges and priorities described will support the development of research structures and agendas for tackling the challenge of climate change.

The effect of stocking rate and prolificacy on profitability of an Irish lowland sheep enterprise

A. Bohan[1,2], L. Shalloo[2], P. Creighton[3], E. Earl[1,3], T.M. Boland[1] and N. McHugh[2]
[1]School of Agriculture & Food Science, UCD, Belfield, Dublin 4, Ireland, [2]Animal & Grassland Research and Innovation Centre, Teagasc, Moorepark, Fermoy, Co. Cork, Ireland, [3]AGRIC, Teagasc, Athenry, Co. Galway, Ireland; alan.bohan@teagasc.ie

Stocking rate and ewe prolificacy have been described as the key drivers of sheep farm profitability in Ireland. The objective of this study was to quantify the effect of stocking rate and prolificacy on farm productivity and subsequent profitability. Three stocking rate and prolificacy scenarios were modelled within the Teagasc lamb production model (TLPM), a computer simulation model of an Irish lowland sheep flock, 7.5 ewes/ha weaning 1.3 lambs per ewe joined (low), 10 ewes/ha weaning 1.5 lambs per ewe joined (medium) and 12 ewes/ha weaning 1.8 lambs per ewe joined (high). Each scenario represented a March lambing lowland flock on a 20 ha grass based system with only stocking rate and prolificacy differing between scenarios. As stocking rate and prolificacy increased grass growth and utilisation were assumed to increase within the model based on flock demand. To assess the effect of maintaining grass growth on profit, sensitivity analysis was undertaken where stocking rate and prolificacy were increased but grass growth was maintained at the low level (7,305 kg/ha) and additional energy was fed through concentrate supplementation. The low, medium and high scenarios had 150, 200 and 240 ewes, respectively, which corresponded to weaning 195 lambs (10/ha), 300 lambs (15/ha) and 432 lambs (22/ha) for the low, medium and high scenarios, respectively. As stocking rate and prolificacy increased individual lamb growth rates and weaning weights decreased, however the overall carcass output per hectare increased from 167 kg (low) to 364 kg (high) carcass/ha. The cost of production per lamb for the low, medium and high scenarios were €102, €82 and €68 respectively and resulted in a net profit per lamb of €5 (€45/ha), €22 (€285/ha) and €32 (€610/ha), respectively. Results from this study show that increasing stocking rate and prolificacy reduced the cost of production per lamb and in turn increased net profit, however increasing stocking rate and prolificacy without increasing grass growth and utilization was counterproductive.

Combined modelling tools to evaluate the impact of flock mobility on ghg emissions in sheep farms

M. Benoit[1], J. Lasseur[2], A. Vigan[3], F. Mouillot[4], M. Eugène[1], L. Mansard[3], M. Vigne[3], P. Lecomte[3] and C. Dutilly[3]
[1]INRA, UMR Herbivores, Theix, 63122 Saint Genès-Champanelle, France, [2]INRA, UMR SELMET, 2 place Viala, 34060 Montpellier, France, [3]CIRAD, UMR SELMET, TA C/112 Baillarguet, 34398 Montpellier, France, [4]IRD/ CNRS, CEFE, Route de Mende, 34000 Montpellier, France; marc.benoit@clermont.inra.fr

Pastoral farming systems have always adapted to the seasonal availability of forage resources by moving flocks. However, the role of flock mobility as a possible mitigating strategy in response to climate change has not been clearly documented. Thus, we investigated: (1) the major methodological challenges linked to the diversity of forage resources used and corresponding methane enteric emissions; (2) the functioning of farming systems based on different combinations of natural resources and their GHG emissions; (3) the impacts of grazing practices on soil and biomass carbon flows. We developed an approach combining two existing models (OSTRAL and CASA) enhanced and adapted. This approach was applied to three French Mediterranean sheep farming systems with different degrees of flock mobility: sedentary (SED), single transhumance (ST) and double transhumance (DT). Results produced by OSTRAL whole farm model showed that, considering GHG emissions, DT is the most efficient system with 27,6 kg CO_2Eq/kc carcass weight vs 35.1 and 30.9 respectively for ST and SED. Indeed, low inputs and high carcass weight compensate the lower ewe productivity. Non-renewable energy consumption (LCA approach) is very low for DT as there is nearly no inputs: 31 MJ/kg CW and 82(ST) and 76 (SED). The CASA model allows us to simulate long term carbon balance scenarios according to land cover dynamics in natural environments relatively to grazing intensities, and so to evaluate contribution of livestock to carbon regulations. Combining both models shows that changing the way of using natural grazed areas, and so their capacity of carbon sequestration, could drastically modify hierarchies between these systems concerning mitigation potential. This underlines the potential interest of coupling tools to enlarge the range of analysis.

Linking deforestation-pasture restoration in Brazil's Intended Nationally Determined Contributions

R. Silva[1], L.G. Barioni[2] and D. Moran[1]
[1]SRUC, West Mains Rd, EH9 3JG, United Kingdom, [2]Embrapa, Campinas-SP, CEP 13083-886, Brazil; rafael.silva@sruc.ac.uk

Brazil has become the first developing country to provide an absolute emissions cut as its Intended Nationally Determined Contributions (INDC), which seeks to reduce total greenhouse gas (GHG) emissions by between 37% and 43% below 2005 levels by 2025 and 2030 respectively. The majority of the INDC mitigation potential is targeted on agriculture and land use change and methods of livestock production. The restoration of 20 Million ha of degraded pastures from 2020-2030 aims to contribute to direct and indirect emissions control, the latter via land sparing and the prevention of deforestation. The INDC also states an ambition to achieve zero net deforestation (ZNDT) in the Amazon by 2030. The INDC ambition on emissions and deforestation are being set as the country seeks to meet an increasing demand for livestock production. This paper shows how the restoration target is derived and how it is sufficient to meet emissions and deforestation targets while simultaneously meeting livestock demand. We evaluated the area restoration requirement using two models. The Demand Constrained Restored Area (DCRA) is a simplified one-equation model that considers two types of grasslands (degraded and not degraded), and the EAGGLE model2, a bottom-up multi-period linear programming model that simulates beef production systems in Brazil subject to demand and pasture area, and estimates the greenhouse gases – including changes in soil carbon stocks. Using both models, we conclude that the required restored pasture to meet baseline demand while accomplishing ZNDT ranges from 15-18.2 Mha. Estimated average costs of direct restoration of degraded pastures, i.e. costs associated with restoration practices (use of inputs and farm operations) are €12.48/ha/yr, €13.87/ha/yr and €12.36 ha/yr, respectively for the Cerrado, Amazon and Atlantic Forest biomes. The total mitigation potential of ZNDT, combined with agricultural intensification through pasture restoration is around 630 million tonnes of CO_2-e relative to the reference scenario, i.e. assuming baseline demand and no accomplishment of the ZNDT. The results are essentially a case study of sustainable agricultural intensification at scale.

Grassland production systems: combing animal species and crossbreeding to strengthen sustainability?

P. Veysset[1], S. Prache[1], K. Vazeille[2] and P. Dhour[2]
[1]INRA, Centre Auvergne-Rhônes-Alpes, UMRH, 63122 Saint-Genès-Champanelle, France, [2]INRA, Centre Auvergne-Rhônes-Alpes, UEMA, 63820 Laqueuille, France; patrick.veysset@clermont.inra.fr

As a response to socio-economic changes over the past decades, sheep and beef systems have become more dependent on external inputs. French beef breeds mature late, forcing farmers to keep animals over 30 months to produce meat from grasslands; farmers prefer to sell males at weaning to specialized Italian fatteners, resulting in a loss of added value for the territory. There are also few pasture-finished lambs in France, as this production is considered as very sensitive to climatic and health hazards; lambs are usually fattened indoor with concentrates. The challenge is to develop low-input farming systems which are both productive and efficient, and which use natural biological regulation services. To address this challenge, we set up an interdisciplinary experiment to compare the functioning and ecosystem services provided by 3 grassland organic farming livestock systems: two specialized (sheep and cattle) systems and one system combining both animal species. The 3 systems are designed to produce meat from grass-fed young animals, they use a cross between a hardy breed dam (Limousine ewes, Salers cows) and an early-maturing breed sire (Suffolk rams, Angus bulls). Each system use the same area (42 ha grassland) with the same number of livestock units (30 LU). These crossbreeding should facilitate grass-based fattening. Combining cattle and sheep (60% bovine LU + 40% ovine LU) should enable to better use forage resources and to better control animal health, therefore to increase system efficiency; however, it may increase workload and complexity of management, and it may affect grassland biodiversity and pollinating insects. For each system, the technical, economic, and environmental performances, together with workload and animal health will be evaluated annually over several years. These 3 systems have been set up on Herbipôle Experimental Unit, INRA Auvergne Rhône Alpes, on the Laqueuille mountainous site.

Tropical grasslands: a pivotal place for a more multi-functional agriculture
M. Boval[1], V. Angeon[2] and T. Rudel[3]
[1]INRA, PHASE, INRA, UMR791 MoSAR, AgroParisTech, 75005 Paris, France, France, [2]INRA, SAD, INRA, UR143, URZ 97170 Petit-Bourg, Guadeloupe, France (F.W.I.);, 97170 Guadeloupe, France, French Southern Terr., [3]the State University of New Jersey, 55 Dudley Road, New Brunswick, NJ 08901., Department of Human Ecology, School of Biological and Environmental Sciences, Rutgers – the State U, 55 Dudley Road, New Brunswick, NJ 08901, USA; maryline.boval@agroparistech.fr

The climate crisis and the current biodiversity make as fallow farmland and meadows are key sites for sustainable intensification efforts. Tropical Grasslands cover a large area of the earth's surface, that can be the support of multifunctional agriculture, i.e. which may allow to produce food and other commodities for the economy in general, to provide services for the ecosystem environment, and to support communities of people who live in these grasslands. Especially as some innovations in animal production (as mixed grazing, practice of composting or other single elementary strategy like the choice of appropriate stage of regrowth), assessed with recent research, and notably through meta-analysis and experiments, can really help improve the multiple functions of grasslands, to meet food security.

Increased foraging in outdoor organic pig production: modeling environmental consequences
M. Jakobsen, T. Preda, A.G. Kongsted and J.E. Hermansen
Aarhus University, Agroecology, Blichers Allé 20, 8830 Tjele, Denmark; malene.jakobsen@agro.au.dk

Consumers' motivations for buying organic products include a wish of acquiring healthy, environmentally friendly products from production systems that also ensure a high level of animal welfare. However, the current Danish organic pig production faces important challenges regarding environmental impact of the system. High ammonia emissions arise from outdoor concrete areas with growing-finishing pigs and sows on pasture possess an increased risk of nitrogen (N) leaching. Direct foraging in the range area is suggested as a way to improve the nutrient efficiency at farm level and to support a more natural behavior of the pig. Thus, by modeling, we investigated the environmental consequences of two alternative scenarios with growing-finishing pigs foraging in the range area and different levels of crops available for foraging; grass-clover (lowest integration of forage) or a combination of lucerne, grass-clover and Jerusalem artichokes (highest integration of forage). It was possible to have growing-finishing pigs on free-range without increasing N leaching compared to the current practice. The alternative system with lucerne, grass-clover and Jerusalem artichokes showed the lowest carbon footprint with 3.12 CO_2 eq/kg live weight pig compared to the current Danish pasture based system with 3.69 kg CO_2 eq/kg live weight pig. Due to positive impact on soil carbon sequestration, the second alternative system based on grass-clover showed a similar carbon foot print compared to current practice with 3.68 kg CO_2 eq/kg live weight pig. It is concluded that in practice there is room for development of organic pig production systems where direct foraging plays a central role. Published in foods: http://www.mdpi.com/2304-8158/4/4/622 The paper has not been presented at any conference.

A mixed-method study on the perception and evaluation of pictures from pig fattening barns

G. Busch[1], S.A. Gauly[2] and A. Spiller[2]
[1]*Free University of Bolzano, Faculty of Science and Technology, Universitätsplatz 5, 39100 Bozen-Bolzano, Italy,*
[2]*Georg-August-University Göttingen, Department für Agrarökonomie und Rurale Entwicklung, Platz der Göttinger Sieben 5, 37073 Göttingen, Germany; gesa.busch@unibz.it*

In order to correct romantic ideas about livestock farming in society and to provoke acceptance about husbandry systems, there is a recent trend in the agricultural industry for using real pictures from barns for communication purposes. Up to now there are only few studies investigating how these pictures are perceived. The current study investigates how such pictures from pig fattening barns are looked at and how they are evaluated to derive recommendations for picture composition. Therefore, a qualitative and quantitative survey is combined with eye-tracking method, where the participants' eye movement is registered via cornea-reflex procedure. A total of 231 students (123 agricultural students and 108 students without connection to agriculture) participated in the study. The study compares three pictures of the same pen with fully slattered floor, taken from different perspective (A=human, B=bird's-eye and C=pig perspective). The results show that all pictures are evaluated negatively by students without connection to agriculture. Different perspectives of the same pen influence picture evaluation and the bird's-eye perspective shows advantages. On picture A and C 20% of those participants wrongly recall straw after seeing the picture. Contrastingly, agricultural students evaluate the pictures more neutral and their evaluation is not influenced by different perspectives. Moreover, they do not recall straw (only 2.5% for picture A). However, the eye movements of both groups when viewing the pictures are very similar. Independent of the perspective, the face regions and bodies of the animals are always looked at first and the longest. The results indicate that agricultural students are quite stable in their evaluation of realistic pictures showing the same circumstance from different perspectives, while for other students, variation in pictures influence evaluations.

The effect of flock size and ewe type on flock management and performance of Irish sheep farms

A. Bohan[1,2], L. Shalloo[2], P. Creighton[3], T.M. Boland[1] and N. McHugh[2]
[1]*School of Agriculture & Food Science, UCD, Belfield, Dublin 4, Ireland, [2]Animal & Grassland Research and Innovation Centre, Teagasc, Moorepark, Fermoy, Co. Cork, Ireland, [3]AGRIC, Teagasc, Athenry, Co. Galway, Ireland; alan.bohan@teagasc.ie*

The objective of this study was to assess the effect of flock size and ewe type on flock management and performance. A detailed survey on Irish sheep farm management and performance was conducted through sheep discussion groups with 718 valid responses received. The survey consisted of 39 questions examining farm size, flock size, ewe type and flock performance. To investigate the effect of flock size and ewe type on flock performance, flock size was divided into four categories: very small (0-61 ewes), small (62-103 ewes), medium (104-190 ewes), and large (>190 ewes); ewe breed was divided into three breed type classes; maternal, terminal and mountain. Higher stocking rates ($P<0.05$), as well as, higher ewe to ram ratio ($P<0.001$) were recorded for larger flocks compared to smaller flocks. Large flocks were 2.40 (95% CI=1.4 to 4.0; $P<0.001$) times more likely to house ewes and were also more likely to body condition score, winter shear and pregnancy scan ewes ($P<0.01$) compared to very small flocks. Larger flocks were more likely to supplement lambs post weaning, whereas the smaller flocks tended not supplement lambs ($P<0.05$). Maternal flocks had higher stocking (+1.11, +3.32; $P<0.001$), scanning (+0.10, +0.30; $P<0.001$) and weaning rates (+0.01, +0.29; $P<0.001$) compared to the terminal and mountain flocks, respectively. Compared to maternal flocks, terminal flocks were more likely to supplement lambs ($P<0.001$). Terminal and mountain flocks were 2.70 (95% CI=1.7 to 4.2; $P<0.001$) and 2.40 (95% CI=1.4 to 4.1; $P<0.01$) times more likely to breed their replacements as hoggets compared to maternal flocks. Terminal flocks and maternal flocks were 4.90 (95% CI=2.5 to 9.6; $P<0.001$) and 3.20 (95% CI=2.1 to 4.9; $P<0.001$) times more likely to house sheep than mountain flocks, respectively. Results from this study show that flock size and ewe type have a significant effect on flock management and performance.

Development of e-learning resources by/for users: a win-win project for teachers and students

J. Faure[1], G. Jalam[1], M. Bonneau[2], C. Aguilera-Charles[1], T. Bodin[1], C. Drieu[1], A. Drouet[1], M. Guillou[1], L. Helary[1], M. Lebouter[1], G. Panaget[1] and L. Montagne[1]
[1]Agrocampus-Ouest, rue de St Brieuc, 35042 Rennes, France, [2]EAAP, IT-00161, Roma, Italy; lucile.montagne@agrocampus-ouest.fr

Post-graduate students are so-called digital native and usual users of numeric tools (web, social networks, software applications) on computers, smartphones, and touch tablets. They have free access anywhere and anytime to a variety of information. Their expectations as students have quickly changed with the numeric innovations in technologies. To match this new generation, learning approaches have to be playful, interactive, and based on visual resources rather than long texts. There is also a need to develop long-life skills. The development of e-learning courses is an adaptive response of teachers to the changes in learning processes. An educational project was proposed to a group of 8 master students in animal science. The objective was to make students actors of the computational development of e-learning resources to make them aware of numeric technologies in addition to knowledge and skills related to animal and livestock production. The project was based on the creation of an e-course linked to the FP7 EU-PLF program. It was expected that the involvement of motivated students in the design of the e-course might result in a creative tool corresponding to their needs. Based on basic knowledge in animal production, students proposed four different learning resources after a period of 60 h working in autonomy with minimal supervision of the teachers. Their resources consisted of texts on livestock farming systems, sketches for dynamic schemes on physiological stages of pig, self-assessment questionnaire, and theoretical growth curve. After validation of the resources by the teacher, the self-assessment questionnaire was tested with other students. The resources were also modified toward a graphic design well adapted to on-line viewing. In a short time this student project management was instructive for them to select and write informative contents, using specific English vocabulary. Background and skills in computational development and design were however a limitation and prevented direct on-line publication. Development of such skills in animal science curriculum is a must.

Shaping the Swedish landscape: a future for the Gotland russ

A. Jansson, L. Edenius, K. Dahlborn, S. Ringmark, A. Skarin, M. Steen, E. Öckinger and C.G. Thulin
Swedish University of Agricultural Sciences, Box 7011, 75007 Uppsala, Sweden; anna.jansson@slu.se

Tree encroachment and abandonment of pastures is an important cause of biodiversity loss in Sweden. In this study, we evaluate the impact of all-year-around grazing by an ancient horse breed, the Gotland russ (GR), on tree encroachment. Twelve GR stallions were kept in 3 enclosures (lat: 59, long: 17, 4 GR/enclosure, 7 ha forest and 3 ha open fields/enclosure) from May 2014 (age 1 yrs) until February 2016. Prior to the study, 3 exclosures (EX, 5×42.5 m) where GR grazing was excluded were centered with the long side at the border between forest and field. The field part of EX were further divided into four 5×5 m sections starting from the forest (1 to 4). Tree type (deciduous, spruce or pine), number and height were registered in the sections and the equivalent, grazed area adjacent to the EX on the 25th of February 2016. Body condition (BC) was assessed weekly using a 9 leveled scoring system. Individuals <4 were withdrawn, fed and reintroduced when BC was >4. Analysis of tree heights and numbers were made by procGLM in SAS. There was no difference in the total number of deciduous trees (21 ± 8 vs 25 ± 8), spruces (2 ± 1 vs 3 ± 1) and pines (0.2 ± 0.2 vs 0.8 ± 0.2) between EX and grazed areas (P>0.05). In the grazed areas, the deciduous trees were shorter compared to the trees in EX (330 ± 57 vs 780 ± 70 mm, P<0.0001). There was no difference in spruce height (EX: 920 ± 110 vs grazed areas: 780 ± 130 mm, P=0.31). In the grazed areas, tree height decreased from quarter 1 to 4 (1: 610 ± 60, 2: 230 ± 110, 3: $140+90$ and 4: 60 ± 100 mm, P<0.0001) whereas in EX, the height increased (1: 740 ± 60, 2: 570 ± 150, 3: 810 ± 90 and 4: 960 ± 100 mm, P<0.02). Four individuals (3 from the same enclosure) were withdrawn temporarily in February 2015 due to BC <4. The study show that grazing by GR may inhibit encroachment by deciduous trees and that BC was maintained in most individuals, despite the semi-feral conditions. The results provide a new use for a threatened, ancient breed in landscape management.

Control of equine diseases

T. Buckley

Irish Equine Centre, Johnstown, Naas, County Kildare, Ireland; tbuckley@equine-centre.ie

The Irish horse industry employs over 25,000 people and is a major contributor to the agricultural sector. As a major exporter, it is important that Ireland remains disease-free from notifiable and exotic diseases, as well as isolating and containing our own infectious diseases. An outline will be given on diseases such as Strangles, foal pneumonia, parasites, enteric infections, reproductive infections and the role of the environment in horse management. The impact of the above diseases on the health and welfare of both the thoroughbred and non-thoroughbred horse population will be given.

Effect of steaming/watering of large hay square bales on particulate matter generation

G. Claußen and E.F. Hessel

Georg-August-Universität Göttingen, Department of Animal Sciences, Division Process Engineering, Gutenbergstraße 33, 37075 Göttingen, Germany; gclauss@gwdg.de

Airway inflammations are a common disease in stabled horses, which are affected by the airborne particulate matter concentration of indoor air. Aim of the study was to investigate the effect of steaming/watering of large square bales on the airborne particulate matter generation of hay. Therefore, three randomly selected hay square bales were watered from above with 150 l and steamed for four hours in a special constructed steaming-box. Before and after the steaming/watering process hay samples reaching the core of the bale were punched out of the top, the back side and the front side of each square bale (n=12/square bale). Moisture contents of the hay samples were determined in a drying cabinet. The generation of airborne particulate matter <10μm (PM10) and <2.5μm (PM2.5) was analysed using an online gravimetric dust measuring instrument (TEOM 1400 AB) under standardized conditions. Statistical analysis was carried out using the GLM procedure of SAS. Within the untreated hay samples the average moisture content was [12.91%; SD 1.67%]. Furthermore, a uniform moisture gradient from the outside to the inside of the bales could be established. After steaming/watering the moisture content increased to [30.91% SD 3.3%]. In two bales an obvious moisture gradient from the outside to the inside was detected [30.66-34.46% and 28.00-36.66%, respectively]. In comparison to the untreated hay, steaming/watering of the square bales led to a significant reduction of PM 10 (-89.93%) and PM 2.5 (-87.59%). Furthermore, an exponential relationship between moisture content of the hay and PM 10 and PM 2.5 generation could be determined. Up to a moisture content of 25% an obvious reduction in the PM 10 and PM 2.5 could be established. In conclusion, the results of this study indicate that vapour distribution within the steamed/watered bales are not satisfying so far. However, a significant reduction of at least 87% of PM 10 and PM 2,5 airborne particular matter has been achieved.

Airborne fungi measurements in bedding hygiene quality assessment
S. Airaksinen[1], A. Salmela[2] and M.-L. Heiskanen[1]
[1]Equine Information Centre, Neulaniementie 5 E, 70210 Kuopio, Finland, [2]University of Eastern Finland, P.O. Box 1627, 70211 Kuopio, Finland; sanna.airaksinen@hevostietokeskus.fi

Bedding is a significant source of fungi in horse stable air with molds exposing horses to respiratory diseases. The purpose of this study was to determine the hygiene quality of six bedding materials with the commercial Mycometer-test. The method detects airborne fungi with the activity of β-N-aminidase acetylhexos (NAHA) and is used for indoor air qualification. Mycometer test does not distinguish fungal species, but it is a rapid method to detect the total fungal biomass. Airborne fungi released from unused and used bedding materials were detected by analyzing 28 air samples from horse boxes (used bedding), 12 samples above unused bedding layer and 15 control samples. Bedding materials studied were peat moss, shavings, sawdust, pelleted straw, pelleted reed canary grass and wood pellets. Air samples were collected from a height 20 cm above the bedding layer or ground (control samples) in 11 different horse stables in eastern Finland during the years 2014-2016 (November – April). The statistical analysis of data was done with the variance analyses (ANOVA). The concentration of airborne fungi released from the unused bedding varied from below the detection limit (LOD 30.31 U/m^3) to 153 U/m^3 and from the used bedding from below the detection limit to 646 U/m^3. Unused peat moss released significantly more molds than other unused bedding materials ($P<0.001$) and used peat moss significantly more molds (Mean 358 U/m^3; SD 226 U/m^3) than other bedding materials in horse boxes. The concentration of fungal biomass in air samples was small above the unused straw pellets, reed canary grass pellets, wood pellets and shavings. Results show that usage in horse box increase the content of airborne fungi released from bedding layer ($P=0.004$) and the mold exposure can be connected the bedding material used ($P=0.001$). The study indicate that peat moss releases more fungal biomass into the air than other studied bedding materials.

Effect of bedding material on air quality of indoor horse stables
A.S. Santos[1,2], C. Bonaldi[2], S. Silva[3], L.M. Ferreira[3] and H. Trindade[2]
[1]EUVG, Escola Universitária Vasco da Gama, Veterinary Department, Lordemão, 3020-210 Coimbra, Portugal, [2]CITAB-UTAD, Agronomy Department, University of Trás-os-Montes and Alto Douro, 5001-801, Vila Real, Portugal, [3]CECAV-UTAD, Animal Science Department, Universidade de Trás-os-Montes e Alto Douro, 5001-801 Vila Real, Portugal; anasofiasantos@euvg.pt

Most of leisure and sports horses are generally kept in riding centres and horse training facilities, where they are stabled in box stalls (approximately 12 m^2/horse). Many horses spend most of their lives (up to 23 h/day) in their stall. The quality of the surrounding air in horse stables is an important factor in maintaining health and welfare of horses. Gases such as ammonia (NH_3), nitrous oxide (N_2O) and methane (CH_4) are produced by transformation processes in the excrement/bedding mixture. Bedding has a key influence on stable ambient conditions in terms of airborne particle generation, water-holding capacity, and ammonia (NH_3) binding. This study aimed to assess the influence of bedding type on air quality of the stable. For this purpose, 5 horses maintained in individual indoor box stalls were used. Two bedding materials were studied: wood shavings (WS) and wheat straw (St). Manure was accumulated for 48 h, during which bedding and faeces were mixed twice/day to ensure homogeneous bedding. Ammonia, CO_2, N_2O, CH_4 and water vapour were measured using a photo acoustic field gas monitor (INNOVA) twice/day, in seven places within the stall, the temperature was measured using a field thermometer. Measurements were taken in duplicate. Results showed that concentrations (ppm) of NH_3, CH_4 and water vapour were higher ($P<0.0001$) for St, and there was a tendency ($P=0.07$) for CH_4 to be higher when St was used as bedding. The temperature within the stall was higher when St was used (26.5 vs 25 °C). This preliminary study points to a higher air quality within the stable when WS was used as bedding material. This work was supported by: European Investment Funds by FEDER/COMPETE/POCI– Operacional Competitiveness and Internacionalization Programme, under Project POCI-01-0145-FEDER-006958 and National Funds by FCT – Portuguese Foundation for Science and Technology, under the project UID/AGR/04033/2013.

Lying behaviour of group-housed horses

E. Mulser[1], A. Zollinger[2], C. Wyss[2] and I. Bachmann[2]
[1]School of Agricultural, Forest and Food Sciences HAFL, Zollikofen, Switzerland, Länggasse 85, 3052 Zollikofen, Switzerland, [2]Swiss National Stud Farm SNSF, Avenches, Switzerland, Les Longs-Prés, 1580 Avenches, Switzerland; esther.mulser@bfh.ch

In recent years, group housing systems have gained importance. However, in such systems, low ranked horses can be kept from lying down. This may result in a REM sleep deficiency because this sleep stage occurs only while the horse is recumbent. The aim of this study was to test whether structuring the lying area lengthens the time horses spend lying down in a group-housed system. A structural element is thought to increase the horses' sense of security because it prolongs their escape time and offers visual privacy. The element used in this study consisted of three walls made of thin, 2 m high rubber mat, fixed in a Y shape. The activities in the lying area of two groups of six mares were recorded continuously on video over 72 h. Both groups were tested once with a structure and once without. The highest ranked mares spent 71 min per day lying down in the non-structured area and 95 min in the structured one. The time the middle-ranking mares spent lying down likewise increased with the structure, from 45 to 72 min per day. The durations recorded in the structured housing were similar to those reported in other studies. However, the structure had no effect on the time most low-ranking horses spent in a recumbent position. Some never lay down during the whole observation time. The Y-shape of the element divided the original entrance into two smaller ones. Two openings are classified as favourable in a group housing system. Nevertheless, we observed more than once a low-ranking horse trapped between higher-ranked members of the group. Sufficient recumbency time is a condition of horse welfare, but the structuring of the lying area tested does not appear to enable the low-ranking horses of a group to adopt a natural lying behaviour. These results showed that more research is required in order to improve the situation of the affected animals. Changing the shapes of the structuring elements, placing the two entrances in opposite corners of the lying area or creating several separate resting places may provide the solution to this welfare problem.

Efficacy of the repellent DEET against tabanid flies on horses evaluated in a field test

C. Herholz[1], C. Kopp[1], M. Wenger[1], A. Mathis[2], S. Wägeli[1], F. Kägi[1] and N. Roth[1]
[1]Bern University of Applied Sciences, School of Agricultural, Forest and Food Science, Länggasse 85, 3052 Zollikofen, Switzerland, [2]Institute of Parasitology, National Centre for Vector Entomology, Vetsuisse Faculty, University of Zurich, Winterthurerstrasse 266a, 8057 Zurich, Switzerland; conny.herholz@bfh.ch

Female tabanid flies (Diptera: Tabanidae) can be a serious nuisance for horses because of their painful bites during blood feeding. They also play a primary role in mechanical transmission of a lentivirus causing Equine Infectious Anemia (EIA), a virus that has spread within Europe in recent years. The aim of the present study was to investigate the efficacy of N,N-diethyl-3-methyl-benzamide (DEET, Brum®, Huebeli-Stud Horse Care AG) in a 15-17% oil-water emulsion against tabanid flies on horses in a field test. Between July and August 2015, four horses were tested on three farms, each on two consecutive days in a cross-over design. Two and a half hours after repellent application the horses were lunged until sweating. Tabanid fly infestations were both photographed and directly counted during five minutes, three and four hours after repellent application, on the right side of the horses in the area from the head to the flank, belly and upper third of the foreleg. Without repellent application, up to 29 tabanid flies were counted on a horse, whereas the maximum for the repellent treated horses was four. In 50% of the treated horses the efficacy of DEET against tabanid flies was 100%, and in all horses the tabanid fly counts were lower than in the control horses with one exemption at 4 h. The efficacy of the DEET repellent was at least 80% and 71% respectively, three or four hours after application (confidence 89%). The design of the present study simulated practical conditions and allowed to quantify the number of tabanids flies and to demonstrate the efficacy of DEET in horses.

Habituation effects to frequent road transport in sport horses
A. Vogt[1] and W. McCormick[2]
[1]University of Göttingen, Department of Animal Science, Albrecht-Thaer-Weg 3, 37075 Göttingen, Germany, [2]The University of Northampton and Moulton College, Department of Animal Welfare and Management, West Street, Moulton, Northampton, NN3 7RR, United Kingdom; anina.vogt@agr.uni-goettingen.de

It has repeatedly been demonstrated that transport is a stressful procedure for horses, but there is only very limited evidence regarding horses which travel regularly and are thus accustomed to transportation. This study aimed to investigate whether the frequent transportation between venues acts as a major stressor for sport horses or if the stress for the animals becomes minimal due to habituation effects. A combination of previously validated behavioural indicators of stress in domestic horses with measurements of salivary cortisol concentrations were employed to assess stress levels of 120 experienced sport horses in response to travelling. Using the Mann-Whitney U-test and Spearman's rank test, results showed that transport experience was positively correlated with the stress levels of horses during transportation (r_s=0.273, P=0.003), whereas the journey length had no significant impact on the stress levels of the horses (r_s=-0.139, P=0.129). For horses that were travelling with very high frequency (≤every 10 days), the measured total stress scores as well as salivary cortisol levels were indicative of no or a low stress response to transportation. Results additionally revealed that the presence of a companion led to significantly lower stress levels in horses that travelled less frequently than every ten days (P=0.033), but that it had no significant effect for very frequent travellers (P=0.066). A high level of transport experience of the horses was also positively correlated with lower stress levels during the loading procedure (r_s=0.301, P=0.004). It was concluded that a high level of transport experience can significantly reduce transport stress in horses. However, individual differences must still be considered in terms of welfare as individual horses showed stress levels that must be of welfare concern despite their travel experience.

Improving equine management by new technologies: drivers and barriers
S. Wägeli, R. Stirnimann, E. Mulser and C. Herholz
Bern University of Applied Sciences, School of Agricultural, Forest and Food Sciences HAFL, Länggasse 85, 3052 Zollikofen, Switzerland; salome.waegeli@bfh.ch

New technologies are usually very helpful to improve stable management and animal welfare. However, many developments do not succeed in the market. In this study, the general affinity for technology of horse stable managers and their willingness to invest in new technologies were analysed. The study aimed to identify drivers for and barriers to investment in new technologies dedicated to equine management. The research was carried out using an example of new technology, in the form of a self-driving robot which automatically gathers horse droppings on pastures. The machine would reduce working hours, thereby saving labour costs, increase pasture quality, improve parasite control and reduce the risk of resistances developing. In an online survey, 1,362 Swiss horse stable managers were questioned and their attitude to new technologies was measured with the Technology Readiness Index (Techqual™). The result showed that this attitude depends both on age and on sex: younger stable managers and men have a significant more positve attitude to new technologies. Willingness to invest is strongly linked to the attitude to new technologies and awareness of animal welfare – and less to economic pressure. On the other hand, willingness to pay is higher when an individual has to act cost-effectively, or owns a stable with more than ten horses. Two groups of stable managers were identified with different drivers for and barriers to investments in new technologies: Stable manager with a farm and in average 11.8 horses, which are cost oriented, and stable managers without a farm, and in average 5.6 horses, which are animal health oriented. These two target groups must be addressed differently in order to ensure the successful establishment of new technologies in the area of equine stable management.

Session 42

Theatre 9

Effect of bedding material on composting temperature of horse manure

A.S. Santos[1,2], C. Bonaldi[2], S. Silva[3], L.M. Ferreira[3] and H. Trindade[2]
[1]EUVG, Veterinary Depratment, Lordemão, 3020-210 Coimbra, Portugal, [2]CITAB-UTAD, Department of Agronomy, University of Trás-os-Montes and Alto Douro, 5001-801, Vila Real, Portugal, [3]CECAV-UTAD, Animal Science Dep., University of Trás-os-Montes and Alto Douro, 5001-801, Vila Real, Portugal; anasofiasantos@euvg.pt

Bedding is used to keep horses clean and for reducing the content of manure gases in the stable. Faeces comprise only 20-30% of the manure while bedding makes up over half of the volume. The choice of bedding has thus a marked effect on the properties of the manure. From an environmental perspective, bedding should have an high capacity to retain nutrients during use and storage but at the same time efficiently release them once applied into soils. Composting provides a way to improve the fertilizer value of manure since it reduces the overall volume and thus increases nutrient concentrations; in addition composting is expected to reduce or completely eliminate pathogenic microorganisms present in fresh manure producing a safe material. This study aimed to assess the temperature evolution on composting horse manure with two different bedding materials: wood shavings (WS) and wheat straw (St). Manure collected from 5 individual horse stalls after a 3 d period was mixed and placed in experimental composters 135 l capacity equipped with a forced ventilation system. A total of 16 composters, 8 with WS manure and 8 with St manure were used. Composting lasted 14 weeks with temperature assessment taken twice/day until constant daily values were obtained (3 days on average), and then daily. Preliminary results show that, although reaching similar values, there is a significant difference ($P<0.001$) in the first 4 days, with St manure presenting a faster and higher temperature rise (69 °C on day 2) than WS manure (68 °C on day 5) also St manure maintained higher temp. for 4 d whereas WS only kept it for 2 d. This faster rise in temperature and duration observed in St manure may eliminate more effectively the pathogenic microorganisms present in fresh manure. This work was supported by: European Investment Funds by FEDER/COMPETE/POCI– Operacional Competitiveness and Internacionalization Programme, under Project POCI-01-0145-FEDER-006958 and National Funds by FCT – Portuguese Foundation for Science and Technology, under the project UID/AGR/04033/2013.

Session 42

Theatre 10

Economical assessments of the saddle horse industry in Europe: what tools for common data?

G. Bigot[1], C. Vial[2], X. Dornier[3], Y. Surry[4] and H. Andersson[4]
[1]Irstea, UMR Métafort, 9 avenue Blaise Pascal, CS 20085, 63 178 Aubière, France, [2]Institut Français du Cheval et de l'Equitation, INRA, UMR 1110 MOISA, 34060 Montpellier, France, [3]Institut Français du Cheval et de l'Equitation, Observatoire Economique et Social, route de Troche, BP 3, 19231 Arnac Pompadour Cedex, France, [4]Swedish University of agricultural Sciences, Box 7013, 750 07 Upssala, Sweden; genevieve.bigot@irstea.fr

Unlike other animal productions mainly oriented towards food production, the horse industry has no common economic databases at the European level and consequently remains invisible to European stakeholders. In order to try solving this problem, we tested a comparative analysis on key variables (numbers of animals, firms, employments…) published by the saddle horse industry between seven European countries: F-France, G-Germany, UK-United Kingdom, B-Belgium, N-Nederland, S-Sweden and I-Ireland. If all countries pointed out the total amount of horses, only three of them (F, G, I) specified the number of horses specialized in sport and leisure, or for breeding. Each country identified the exact number of licensed riders but estimated a total amount of practitioners without specifying methods. The number of firms varied greatly between countries according to administrative statutes and main activities. The assessment of jobs appeared comparable: 10 to 12 horses to create one full-time in leisure activities (I, S) and only 4 to 5 horses in competition activities (G, I). Finally, two methods were mainly used to assess the global economic weight of this industry: the total turnover of all firms from breeding to the various equestrian activities (F, G, B) or the total expenditures spent by horse owners, horse show spectators and public funds (UK, N, I). This analysis pointed out some obstacles for European professionals to elaborate common databases. The use of economic models (such as input-output analysis) should also help to assess the socio-economic impacts of equine activities at different territorial levels.

Tackling the itch: GWAS on insect bite hypersensitivity using IgE levels in two horse populations

L. François[1], H. Hoskens[1], H.F.J. Saverkoul[2], E. Tijhaar[2], B.D. Velie[3], A. Stinckens[1], G. Lindgren[3], S. Blott[4], N. Buys[1], B.J. Ducro[5] and A. Schurink[5]
[1]KU Leuven, Department of Biosystems, Livestock Genetics, 3001, Leuven, Belgium, [2]WU, Cell Biology and Immunology Group, P.O. Box 338, Wageningen, the Netherlands, [3]SLU, Department of Animal Breeding and Genetics, P.O. Box 7023, 75007 Uppsala, Sweden, [4]The University of Nottingham, Reproductive Biology, Faculty of Medicine and Health Sciences, LE12 5RD, Leicestershire, United Kingdom, [5]WU, Animal Breeding and Genomics Centre, P.O. Box 338, Wageningen, the Netherlands; liesbeth.francois@biw.kuleuven.be

The most common skin allergy in horses, insect bite hypersensitivity (IBH), is mainly caused by bites of Culicoides spp. In the Netherlands, Culicoides obsoletus is the most important species involved in the development of IBH. The welfare of allergic horses is reduced as they suffer from severe itch and sometimes secondary infections as a result of self-mutilation in an attempt to relieve the itch. Unt il now, genomic studies have shown regions of interest in several horse breeds but the exact genetic cause of this widespread disease has yet to be identified. While most studies use case/control status or severity scores, we use the IgE levels against C. obsoletus whole body extract and seven recombinant C. obsoletus allergens. The aim is to improve identification of genomic regions and validate previously reported regions. In this study we performed a genome wide association study (GWAS) using IgE levels as phenotypes in Shetland ponies (200 horses) and Icelandic horses (146 horses). The horses were sampled using a matched case/control design and genotyped on the Illumina 70K SNP chip. Serum IgE levels determined by ELISA were available as well as additional information concerning sex, age at sampling, coat colour, and year and month of sampling. GWAS was performed using the GenABEL package in R and the results showed a clear association between a region on ECA20 and IgE levels in Shetland ponies. This region has consistently been reported in former studies with the ELA region as a possible candidate gene. These results are a step closer towards identifying genes involved in IBH and the development of genomic tools to decrease IBH prevalence.

Importance of the design of lying area and optimal barn climate for dairy cattle

F.J.C.M. Van Eerdenburg
Utrecht University, Department of Farm Animal Health, Yalelaan 7, 3584 CL Utrecht, the Netherlands; f.j.c.m.vaneerdenburg@uu.nl

Cows that live in a pasture-based system can get easily over 15 years of age. In barn based systems, however, this is an exception. Apparently, the farm environment is not so healthy for dairy cows. Most farmers try to provide their cows with their needs, because they realize that proper conditions in the barn are crucial for the results of a farm. Two important factors will be discussed here: Freestall design and barn climate. Freestalls should be designed according to the needs and size of the cows. The bedding should be soft, clean and inorganic and cows should have sufficient grip in the stalls. The freestalls must provide cows enough space to stretch their front legs forward, lie on their sides with unobstructed space for their neck and head, rest their heads against their sides without hindrance from a partition and rest with their legs, udders and tails on the platform. Most bedding material is made of organic compounds and thus a risk factor for the growth of bacteria. Sand and ground limestone are the only materials that are inorganic and that can have low bacterial counts. The climatic conditions should be such that there is no dirty smell or toxic gas inside the barn; the temperature inside is not >5 °C above the outside; there is no draught or dead space and the relative humidity in the barn is 50-80%. It is important to realize that the conditions in the various parts of the barn may differ. Climatic conditions have, therefore, to be checked at multiple spots in the barn. Moreover, one cannot be satisfied when the average of the measurements meets the requirements. Every single spot in the barn has to meet the reference values. If not, cows will avoid sites with lower quality, thus reducing their available area. This has not only implications for cow comfort but also for the overall performance of the herd. Cows prefer a relatively cool environment and do not like high air velocities. The optimal temperature range to house high producing dairy cattle is between -5 and +15 °C.

Does herd size affect animal welfare in dairy cattle?

D. Gieseke[1], C. Lambertz[2] and M. Gauly[2]
[1]Georg-August-University Göttingen, Department of Animal Science, Albrecht-Thaer-Weg 3, 37075 Göttingen, Germany, [2]Free University of Bozen, Faculty of Science and Technology, Universitätsplatz 5, 39100 Bozen, Italy; christian.lambertz@unibz.it

Over the last decades the number of cows per farm increased continuously in Europe. Different stakeholders are concerned about the intensification of livestock production, because they perceive large-scale dairy farms with serious animal health and welfare problems. However, it is still scientifically not proven whether there is a direct correla-tion between herd size and these factors. Therefore, the aim of this study was to eva-luate the animal welfare status on 80 conventional dairy cattle farms with different herd sizes (small: <100; medium: 100-299; large: 300-499; very large: ≥500 cows) using animal-based measures of the Welfare Quality® Assessment protocol for dairy cattle (WQP). A mixed model was performed to investigate differences between the four herd size classes. Body condition score (% lean cows), water supply (cm trough per cow) and incidence of milk fever were rated better in farms with more than 300 cows compared with small and medium size farms (P<0.05). In contrast, the frequency of vulvar discharge rose with increasing herd size (P=0.03). The prevalence of mastitis, coughing or diarrhea were found not significantly different in all four classes. Injuries such as severe integument alterations i.e. lesions and swellings (3-71%) as well as severe lameness (0-74%) were detected in each category at a comparable level (P>0.05). The overall welfare score (range 0-100) was low, but did not differ between herd sizes (small: 38.6, medium: 37.2, large: 43.8, very large: 40.5; P>0.05). In conclusion, based on the WQP herd size did not directly affect welfare status of dairy cows.

The impact of husbandry conditions on claw health in Austrian dairy herds

J. Burgstaller[1], D. Feiersinger[1], B. Fuerst-Waltl[2], F. Steininger[3], J. Kofler[1] and C. Egger-Danner[3]
[1]University Clinic for Ruminants, Farm Animals and Veterinary Public Health, Veterinärplatz 1, 1210 Vienna, Austria, [2]University of Natural Resources and Life Science Vienna, Augasse 2-6, 1090 Vienna, Austria, [3]ZuchtData, Dresdnerstraße 89, 1200 Vienna, Austria; johann.burgstaller@vetmeduni.ac.at

The growing human population on earth and the competition for farmland creates a challenging situation for dairy farmers. The Austrian cattle breeders are therefore looking for the efficient cow, which is adapted to the local conditions, has a high lifetime milk production, stays healthy, calves without complications every year and can be fed with farm grown foodstuff. To find those cows, breeds and rearing systems, data were collected from 170 breeding herds consisting of 5,500 dairy cows. The first step included the evaluation of farm specific factors, including animal husbandry, feeding, claw trimming practice and management. At every milk yield recording event, which was performed 8 to 11 times during the year 2014, BCS, lameness scoring, body weight and some body measures were evaluated. During the observation period the ration was analysed and documented, data on animal health and claw trimming was collected. Lameness groups were created, according to the data on lameness scoring, to be able to compare cows that have never been lame, to cows that were slightly, moderately or severely lame. Farm specific data, housing condition, flooring and bedding type, use of pasture or alpine grassland was evaluated concerning its impact on claw health and lameness. The evaluation of housing types gave evidence that cows kept in tied stall barns (which is only allowed by law if combined with access to pasture for at least 90 days) showed the best claw health status with 60% non-lame cows, followed by cows in free stalls with straw bedded cubicles and plane walking alleys (50.6%). Worst results in this investigation were achieved by cows kept in free stall cubicle systems with rubber mattress bedding and slatted floors (39.5%). Herds that used pasture and alpine grassland had a better claw health compared to all year in herds.

The effect of floor type on the performance of growing dairy origin bulls
V.S. Murphy[1,2], D.E. Lowe[1] and F.O. Lively[1]
[1]Agri-Food and Biosciences Institute, Large Park, Hillsborough, BT26 6DR, United Kingdom, [2]Queen's University Belfast, University Road, Belfast, BT7 1NN, United Kingdom; victoria.thompson@afbini.gov.uk

The aim of this study was to evaluate the effect of floor type on the performance of dairy origin bulls during the growing period, following a summer at pasture. Forty dairy origin bulls with a mean initial live weight of 224 kg (SD=29.6 kg) were divided according to live weight and breed into 10 blocks, each of four animals. They were randomly assigned within blocks to one of two floor types, a fully slatted floor or a fully slatted floor covered with rubber strips for a growing period of 100 days. Bulls were at pasture from June 2015, receiving 2.0 kg concentrates/head/day until they were housed in October 2015, at an average age of 8 months. Bulls were offered ad libitum grass silage and supplemented with 2.0 kg concentrates/head/day which was increased by 0.5 kg/week until intake reached 8.5 kg/head/day on day 100. Animal live weights were measured on two consecutive days before allocation to treatment, every 14 days thereafter and finally on two consecutive days at the end of the growing period. Animals were also scanned for back fat depth monthly and cleanliness scored on day 5, 50 and 100. Data were analysed with linear mixed model methodology using REML estimation with Pen as a random effect and Treatment as a fixed effect in all analyses. There was no significant effect of floor type on animal performance. Average daily gain of bulls accommodated on fully slatted flooring was 1.31 kg/day and those accommodated on a fully slatted floor covered with rubber strips was 1.35 kg/day. There was no significant effect of floor type on back fat depth during the growing period or cleanliness of bulls on day 5 and 100 (P>0.05). Bulls accommodated on fully slatted flooring were significantly dirtier than those on fully slatted flooring covered with rubber strips on day 50 (P<0.05). The findings from this study demonstrate that growing bulls reared on fully slatted flooring perform equally to those accommodated on fully slatted flooring covered with rubber strips.

Perch design preferences of commercial broiler chickens reared in windowed houses
C.L. Bailie and N.E. O'Connell
Queen's University, Belfast, Institute for Global Food Security, NITC, Cloreen Park, Malone Road, Belfast, BT9 5HN, United Kingdom; c.bailie@qub.ac.uk

Motivation to perch persists in fast growing commercial broiler chickens, although levels of perching behavior are generally low. The aim of this study was to investigate the perching preferences of these birds. Windowed houses containing Ross 308 broiler chickens were virtually divided into halves (front and back) and 6 perch designs: (1) 'a' frame with incorporated platform; (2) flat topped ramp; (3) curved ramp; (4) fixed wooden bar; (5) suspended wooden bar; and (6) suspended platform, were placed in each half house from Day 1 of the rearing cycle. Perches were provided in 1 house on each of 2 farms, providing 2 replications at 2 initial stocking densities over 2 production cycles (Cycle 1=12 birds/m^2, Cycle 2=17 birds/m^2). Instantaneous scan sampling of video footage was carried out at 4 hour intervals from 00:00 h across a 24 hour period during weeks 1-6 of Cycle 1 and weeks 1-5 of Cycle 2 and the percentage occupancy of component parts (bar, ramp, platform) and overall perches assessed. Continuous counts were made of the numbers of successful and failed perching attempts within 2 min periods at 08:00, 12:00, 16:00 and 20:00 h during weeks 2, 4 and 6 of Cycle 1. Kruskal-Wallis tests were used to compare ranked mean occupancy of perches and components, and successful and failed perching attempts across perches. There was a significant difference in the occupation of different perches (P<0.001); with the suspended platform the most occupied. Platform components were also occupied significantly more than either ramps or bars (P<0.001). These effects were consistent across stocking densities and weeks. The proportion of successful (P<0.001) and failed (P<0.001) perching attempts differed significantly for different perch designs; with most failed attempts at the suspended bar. Results of this study suggest that fast growing broiler chickens display a preference for occupying raised platforms, rather than traditional perches.

Field evaluation of two additional water sources regarding health and welfare aspects in Pekin ducks

L. Klambeck[1], F. Kaufmann[1], J.D. Kämmerling[1], N. Kemper[2] and R. Andersson[1]
[1]University of Applied Sciences Osnabrück, Department of Animal Husbandry and Poultry Sciences, Am Krümpel 31, 49090 Osnabrück, Germany, [2]University of Veterinary Medicine Hannover, Foundation, Institute for Animal Hygiene, Animal Welfare and Farm Animal Behaviour, Bischofsholer Damm 15, 30173 Hannover, Germany; l.klambeck@hs-osnabrueck.de

According to the Council of Europe Recommendations (1999), Pekin ducks without access to open water must be provided with water resources that allow them to take in water with their beak, dip their head and splash water over their bodies. Due to the benificial effects on litter quality and hygine of drinking water standard commercial production systems use nipple drinkers for water supply. This study investigated the effects of two different open water sources (prototypes of a water funnel and troughs) on health and welfare parameters in Pekin ducks under commercial conditions. Water sources affected conditions of foot pads and plumage condition (P≤0.0001). At the end of the fattening periods, the highest foot pad scores were found in birds with access to nipple drinkers only whereas ducks in groups containing nipple drinkers, water funnels and troughs showed the best foot pad conditions (P≤0.0001). Ducks provided with nipple drinkers showed a cleaner overall appearance of the plumage (P≤0.0001), however the body region 'back' was significantly cleaner in groups with open water supply. Water funnels and troughs contained higher bacterial counts and counts of E. coli compared to the nipple drinkers. According to the results of the current study, it is recommended to offer ducks nipple drinkers for water intake and an additional water supply for welfare and behavioural issues. Open water sources may challenge hygiene and health aspects and thus, require an intensive management in commercial production systems. However, further investigations are needed to take hygienic and economical aspects into account.

Commercial comparison of potential dustbathing substrates for intensively farmed broilers

M. Baxter and N.E. O'Connell
Queens University Belfast, Institute for Global Food Security, 18-30 Malone Road, Belfast, BT9 7BL, United Kingdom; mbaxter1004@qub.ac.uk

The aim of this study was to investigate the level of use and types of behaviours performed in different substrates provided to promote foraging and dustbathing in commercial broiler chickens. Five substrates were assessed in a commercial windowed house bedded with sawdust and containing 23,000 broilers (Aviagen Ltd, UK); the trial was repeated for three production cycles between July-Dec 2015. Fifteen steel rings, (0.96 m^2 area, 4″ high), were placed evenly throughout the house. Three rings were filled with either (1) moss-peat, (2) oat husks, (3) straw pellets or (4) clean wood shavings, or (5) were left empty to just contain standard bedding. Rings containing substrates 1-4 were filled on day 10 and were topped up throughout the cycle to maintain their condition. Video footage of ten random rings, two of each substrate, was recorded for an hour (between 12 pm and 3 pm) each on days 15, 22, 29 and 36 of the 40 day cycle. Five scans per hour of footage were performed, scoring the number of birds using each substrate and their behaviour. Data were normally distributed and analysed using one-way ANOVA (IBM SPSS Statistics 23), with post-hoc Tukey tests to investigate significance. Significant substrate effects were found for the number of birds present in substrate ($F_{4,580}$=23.210, P<0.0001), dustbathing ($F_{4,578}$=31.040, P<0.0001), foraging ($F_{4,578}$=89.434, P<0.0001) and sitting inactive ($F_{4,578}$=26.160, P<0.0001). Significantly more birds were present in the moss-peat and dry wood shavings compared to the remaining substrates. The highest levels of dustbathing and foraging were seen in moss-peat, with oat husks also showing high levels compared to the other substrates. Moss-peat and oat husks also recorded the lowest level of inactive sitting, with similarly high levels of sitting recorded in the remaining treatments. This demonstrates that, at a commercial level, birds are capable of performing active behaviours and that the motivation to perform them is affected by bedding material. Parallel studies will look at the benefits of promoting this activity with proven attractive substrates.

Assessing the efficacy of improved animal welfare to control Campylobacter contamination in poultry

S. Messori[1], E. D'Erasmo[2], E. Di Giannatale[1], P. Dalla Villa[1], F. Pomilio[1] and L. Iannetti[1]
[1]*Istituto Zooprofilattico Sperimentale dell'Abruzzo e del Molise, Campo Boario, 64100, Teramo, Italy, [2]Università degli studi di Teramo, Piano D'Accio, 64100, Teramo, Italy; s.messori@izs.it*

Campylobacteriosis is the main zoonoses in the European Union. The strategies currently implemented to control the infection in poultry, mainly based on farm biosecurity and sanitation practices on the production line, often proved insufficient. Compliance to specific microbiological criteria for Campylobacter in poultry meat will be mandatory in EU in the next future. This will pose a huge challenge for the poultry industry, which requires other efficient tools to contain contaminations. There is increasing scientific evidence that animal welfare has an impact on food safety, which may provide a new basis to control this issue. A sample of ten farms from an integrated Italian poultry company, all being positive for Campylobacter infection, was selected. A refined version of the Welfare Quality assessment protocol was used to cluster the farms into two groups based on animal welfare indicators, mainly being animal-based measures. We hypothesised that the two clusters would differ in Campylobacter shedding and, therefore, in the level of carcass contamination. Farm visits occurred 1-2 days prior to slaughtering (mean age=49 days). Mean prevalence for hock burns was 46% (13-52), for footpad dermatitis 67% (43-98) and for lameness (either mild or severe) 12% (5-24). Mortality and culling rates were, on average 2.7 (1.6-5.5) and 2.8% (1.7-3.9) respectively. Significant variations of space allowance were found, with values ranging from 26.8 to 41 kg/m^2. Human-animal relationship was evaluated through touch tests: the mean prevalence of touched birds on the number of those being at arm reach was 58% (52-71). For each measure, the average population values were used as thresholds to create two clusters: farms scoring on average below the threshold were assigned to the 'higher welfare' group, while the others to the 'lower welfare' one. Measurements of the concentration of Campylobacter in caecal contents and carcass skin samples at abattoir, for each of the two clusters, are currently on-going. Preliminary results will be presented and links between animal welfare and food safety will be discussed.

Is tail biting in growing pigs reduced by a prolonged suckling period?

A.F. Naya[1], C. Veit[1], I. Traulsen[1], O. Burfeind[2] and J. Krieter[1]
[1]*Institute of Animal Breeding and Husbandry, Christian-Albrechts-Univisity Kiel, Hermann-Rodewald-Straße 6, 24118 Kiel, Germany, [2]Research Farm Futterkamp, Chamber of agriculture, Gutshof 1, 24327 Blekendorf, Germany; anaya@tierzucht.uni-kiel.de*

Aim of this study was to find out, if a prolonged suckling period combined with a group housing farrowing system has an effect on tail biting. During four batches in total 826 piglets of 72 sows were housed on the research farm Futterkamp Germany. They were split in three experimental groups: Group SH-4 (four weeks suckling period, conventional farrowing pens, n=293), group SH-5 (five weeks suckling period, conventional farrowing pens, n=267) and group GH-5 (five weeks suckling period, group housing system, n=266). After weaning, piglets were moved to conventional rearing pens equipped with jute sacks, wooden bars, chains and piglet bowles filled daily with peat or grass pellets for occupation. Tail posture (curled; lifted but not curled; wagging; hanging; jammed), injuries (unharmed; scratches; small lesions; large lesions) and tail losses (intact; tip loss; partial loss; complete loss) were scored once a week from birth to end of rearing. Tail biting started mostly in the second week after weaning and all pens were affected. Four weeks after weaning more than 60% of the piglets were bitten. Tail posture and injuries seems to correlate because 'curled' and 'lifted' tails decrease when tail biting starts and 'jammed' tails appear. Regarding tail losses, group SH-4 and GH-5 had almost the same percentage of piglets with intact tails at the end of rearing (SH-4:43.49%; GH-5: 42.48%) whilst group SH-5 shows a worse result (SH-5: 27.34% with intact tails). It needs to be mentioned, that there have been infections and health problems in the first batch which might have influenced the investigated effects. However the results indicate that a prolonged suckling period in combination with group housing does not affect tail biting. But the combination of five weeks suckling period with a conventional housing system seems to have a negative effect.

Production of undocked pigs, a survey of farmers' experiences

T. Wallgren, R. Westin and S. Gunnarsson
Swedish University of Agricultural Sciences, Department of Animal Environment and Health, P.O. Box 234, 53223
Skara, Sweden; torun.wallgren@slu.se

Tail biting is a common cause for reduced welfare and production rates within commercial pig production and is more prominent in barren environments. Using enrichment as straw has been shown to reduce tail biting behavior and thus reduce need for tail docking. Implementation of straw in practice has however partly default since it is argued that straw will cause obstruction in the manure handling systems. Sweden has a long tradition of rearing undocked pigs with access to straw due to national legislation banning docking and fully slatted floors while demanding access to manipulable material for pigs. We surveyed 60 randomly selected Swedish nursery and finishing pig farmers' usage of straw and their opinions on straw impact on tail biting and manure handling management. All farmers provided manipulable material, 98% straw. In 50% of nursery and 35% of finishing farms the straw was complemented with material such as wood shavings. Straw rations were 29 g/pig/day (8-85 g) in nursery and 50 g (9-225 g) in finishing farms. Straw was commonly chopped (76%) to a mean length of 6 cm (1-10) in nursery and 8 cm (1-20) in finishing farms. Straw causing problems in the manure handling system occurred in 32% of the farms who experienced this yearly (25%) or monthly (7%). Most common causes were straw making the slurry sluggish, stacked in pivot or blocking slats. The low incidences of problems indicate current systems are able to cope with presented straw rations. Tail biting had been seen at least one time ever in 50% of nursery and 88% of finishing farms. Frequency of observed tail biting was ≤twice/year (78%) 3-6 times/yr (17%) and monthly (4%) by nursey and ≤2 times/yr (21%), 3-6 times/yr (37%), monthly (34%) and weekly (8%) by finishing farmers. An average of 1.6 (0.1-6.5) finishing pigs were reported tail bitten each batch. In partly slatted flooring systems a correlation was found between increased tail biting frequency and percentage of reported tail bitten pigs (r=0.64, P≤0.0001, n=38) (Spearman Rank correlation). The limited tail biting problems indicate that straw usage at this level is enough to prevent major tail biting outbreaks in undocked pigs.

Behaviour and performance of suckling gilts and their piglets in loose single housing

R. Wassmuth[1], C. Biestmann[2] and H. Janssen[2]
[1]University of Applied Sciences, Faculty of Agricultural Sciences and Landscape Architecture, Am Kruempel 31, D49090 Osnabrueck, Germany, [2]Chamber of Agriculture Lower Saxony, Mars-la-Tour-Str. 6, D26121 Oldenburg, Germany; r.wassmuth@hs-osnabrueck.de

The objective was to evaluate suckling performance and behaviour traits of gilts and piglets in two different single housing farrowing systems under practical conditions. Data of 70 crossbred-gilts and their piglets were collected. Gilts of the control group (CG) were fixed during farrowing and suckling (Pro Dromi® 1) and in the experimental group (EG) gilts were fixed for 6 days a.p. only (Pro Dromi® 1.5). Six farrowing crates were included in each group and six replications were carried out. Performance data were collected and gilt's and piglet's behaviour was observed with 10 min scan samples and categorized by standing, walking (only in EG) sitting and lying (side and belly).No significant (P>0.05) differences between CG and EG were found in piglets born alive (13.2 and 13.9, respectively), losses of piglets (1.4 and 1.55, respectively) and weaned piglets including cross-fostering (12.0 and 12.4, respectively). Piglet losses due to crushing were 0.6 (CG) and 0.64 (EG), respectively, with no significant difference (P>0.05). A proportion of 82.5% of all lost piglets due to crushing were lost from farrowing to day 2 a.p. Daily gain of EG piglets was significantly higher than that of CG (205 vs 199 g, respectively) during suckling period (3 weeks). Considering gilt's behaviour significant differences (P≤0.05) were found in sitting duration only (CG 5.8% and EG 4.0%, respectively). CG piglets took more time in the activities lying, sitting and standing (7.4 vs 4.4%, 0.5 vs 0.4%, 9.6 vs 8.4%, respectively, P≤0.05). The reason could be the higher acceptance of the piglet nest in EG. It could be concluded that farrowing crates with loose single housing did not lead to higher piglet losses in the suckling period. The fixation period of the gilt could be finished at day 2 or 3 a.p. (to be tested).

Effect of overstocking during the dry period on the behaviour and welfare of dairy cows
M. Fujiwara[1,2], H.M. Ensor[3], M.J. Haskell[2], A.I. Macrae[1] and K.M.D. Rutherford[2]
[1]University of Edinburgh, Royal (Dick) School of Veterinary Studies, Easter Bush, Midlothian, EH25 9RG, United Kingdom, [2]SRUC, West Mains Road, Edinburgh, EH9 3JG, United Kingdom, [3]Biomathematics and Statistics Scotland, Edinburgh, EH9 3FD, United Kingdom; mayumi.fujiwara@sruc.ac.uk

The level of agonistic social interactions and daily lying times in dairy cows are known to be affected by space allowance in lying and feeding areas as well as regrouping frequency. The aim of this study was to investigate the effects of overstocking and frequent regrouping on the activity and stress levels of cows during the non-lactating (dry) period. Forty-eight Holstein cows were dried off 8-9 weeks before the expected calving date, balanced by parity and allocated to either high (H) or low (L) stocking density groups. During stages1&2 H and L cows had 0.5 vs 1.0 yoke spaces and 1.0 vs 1.5 cubicles/cow respectively. During stage3 they had 0.3 m vs 0.6 m feed face and 6 m^2 vs 12 m^2 lying space/cow respectively. Cubicle sheds was used during experimental stage1 (the first 21 days) and stage2 (from day 22 to 3 weeks before calving), and straw yards were used during stage3 (approximately 3 weeks before calving until calving). Cows' activity (MI: MotionIndex, SC: step count) and lying behaviour (LB: lying bouts, LP: daily proportion of lying time) was recorded throughout the dry period using activity monitors. Faecal samples were collected at 5 time points to measure faecal glucocorticoid metabolites. Data was analysed by Residual Maximum Likelihood (REML) using Genstat. There were no significant differences beween stocking densities in the variables measured. Primiparous cows were more active than multiparous cows (MI: 2,980 vs 2,280, W=10.12, P=0.003, SC: 1,028 vs 847, W=9.00, P=0.004). Main effects of the experimental stage were found on LB (stage1 = 7.78, stage2 = 7.91, stage3 = 9.25, W=74.23, P<0.001), LP (stage1=0.53, stage2=0.52, stage3=0.62, W=265.47, P<0.001) and concentrations of faecal glucocorticoid metabolites (stage1=292.42 ng/g, stage2=299.91 ng/g, stage3=409.26 ng/g, W=19.31, P<0.001). Results suggest that parity and stage of the dry period, rather than stocking density can affect cows' activity level, lying patterns and stress level.

Comparison of cows' and heifers' stress in different automated and manual milking parlours
S. Limberg and U. König V. Borstel
University of Göttingen, A.-Thaer-Weg 3, 37075 Göttingen, Germany; koenigvb@gwdg.de

The aim of the present study was to compare heifers' and cows' stress responses to miliking in different types of automated, semi-automated or manual milking parlours. Behaviour, heart rate and heart rate variability (RMSSD and LF/HF) of a total of 60 Holstein Friesian cows (30 primiparous and 30 multiparous cows) was recorded immediately before, during and after milking in one of the following four milking parlours: herringbone, side-by-side, carousel, or automated milking system. Heifers were observed for their first two milkings after calving, and for each heifer, a multiparous cow was matched and observed during the same day(s) in the same milking system. Expectedly, cows had lower heart rates than heifers throughout milking (e.g. heart rate during milking: 97.4±6.2 vs 122.9±4.1 beats per minute; F-test: P<0.05), and they showed less agitated behaviour (e.g. frequency of kicks: 0.08±0.5 vs 0.96±0.3). However, with the exception of RMSSD assessed after milking before leaving the milking parlour, none of these heart rate variability parameters showed significant differences between primi and multiparous cows either before, during or after milking (all P>0.1). Surprisingly, RMSSD after milking was higher for heifers compared to cows (51.2±16.7 vs 14.9±5.8 ms; P<0.05), indicating that multiparous cows were more stressed during the period waiting for release from the milking parlour compared to primiparous cows. Possibly, the experienced cows expected stressful treatments during this waiting period such as e.g. being driven out of the milking parlour or udder and/or claw treatments as is practiced in some farms after milking when the cows are still in the milking parlour. In addition, there were considerable differences in stress responses to different milking systems, but partial confounding of milking systems with farms, warrants further investigations as to whether differences are due to the milking system or due to overall farm management.

Improving egg production by using split-phase lighting program

H. Al-Khalaifa and A. Al-Nasser
Kuwait Institute for Scientific Research, Shuwaikh, 13109-Safat-Kuwait, Kuwait; hkhalifa@kisr.edu.kw

Egg shell quality and oviposition in laying hens is influenced by a range of factors including strain of birds, age, nutrition, water quality, general stress, heat stress, disease, and lighting program inside houses. A layer experiment was conducted to investigate the effect of split-phase lighting program on egg production efficiency. Four different feeds and average phosphorus (av. P) levels were tested. Diet A was a ration with an av. P level of 0.471%; Diet B was a ration with an av. P level of 0.510%; Diet C contained an av. P level of 0.293%; and Diet D contained an av. P level of 0.327%. The split-phase lighting program tested was one that inserted a 7-hour dark period from 9 am until 4 pm to reduce the heat produced by the feeding increment and physical activity of the hens. Diet B produced significantly more eggs than Diet C or Diet D. Diet A was not significantly different from any of the other diets. Diet B also had the best feed efficiency with the other three diets in the same order and significance as for egg production. Diet D produced eggshells significantly thicker than either Diet A or Diet B. Diet C produced thicker eggshells than Diet B, whose shells were significantly thinner than the other three diets. There were no differences in egg size. From these data, it is apparent that the minimal av. P level for the Lohmann strain of layer in Kuwait is above 0.327%. There was no difference in egg production or eggshell thickness between the split-phase light treatment and the standard light program. There was no difference in oviposition frequency. The split-phase light used 3.66% less feed, however, which was significant. The standard light produced eggs that were significantly heavier (66.30 vs 65.73 g). These results indicate that considerable savings in feed costs could be attained by using split-phase lighting, especially when cooling is not very efficient.

Effect of welfare housing systems and bedding on performance of laying hen in Korea during summer

J.Y. Lee[1], S.E. Woo[1], D.H. Lee[1], D.Y. Choi[1], H.C. Choi[1] and S.J. Ohh[2]
[1]National Institute of Animal Science, Animal Environment Division, 1500, Kongjwipatjwi-ro, Iseo-myeon, Wanju-gun, Jeollabuk-do, 55365, Korea, South, [2]Kangwon National University, Department of Animal Resource Science, 1 Kangwondaehak-gil, Chuncheon-si, Gangwon-do, 24341, Korea, South; andrewlee@korea.kr

This study was conducted to evaluate the effect of different welfare friendly housing systems and litter allowance on the laying performance and eggshell quality in laying hens during the summer season. Two (barn and aviary) rearing systems and two different floor type (w/ or w/o rice husk bedding) were arranged to a total of 800 Lohmann Brown Lite laying hens(30 weeks old) for 8 weeks. The aviary is welfare bestowed housing that allows free locomotion for birds within the barn. In barn and aviary system, welfare facilities and environmental conditions were supplied to meet the guidelines for farm animal welfare certification in Korea. The average temperature and relative humidity in day time in barn and aviary were 30.4°, 31.5° and 72%, 78%, respectively. A commercial layer diet (18% crude protein and 2,750 kcal/kg ME) was supplied manually twice a day to birds in both systems. There were remarkable differences (P<0.01) in hen day egg production (HDEP), egg mass and feed conversion ratio between both rearing systems regardless of bedding allowance during overall period. Daily feed intake of laying hens reared with rise husk litter was lower (P<0.05) than the one reared without litter. During 0~4 weeks, the average percentage of HDEP in aviary and barn were 90.3%, 92.9% and average egg mass were 54.5, 5.0 g, respectively. Those of the remaining 4 weeks (5~8 weeks) were 91.1%, 92.0% and 54.2 g, 55.0 g, respectively. The egg production (10.4%) outside nest in the floor bedded with rice husk was remarkably higher than those (6.7%) in the floor without bedding regardless of welfare friendly housing systems. Overall, exterior egg quality (dirty and cracked eggs) and eggshell quality were not affected by both housing type and litter allowance except for egg yolk color. No difference in the corticosterone contents was observed between treatments. The study implied that the bedding allowance in welfare friendly housing system could be favorable during summer in terms of improved productivity.

Fine dust particles concentration increase during feeding time in a pig barn

A. Costa, F.M. Tangorra, A. Agazzi, G. Savoini and M. Lazzari
Università degli Studi di Milano, Dipartimento di Scienze Veterinarie per la Salute, la Produzione Animale e la Sicurezza Alimentare, Via G. Celoria, 10, 20133 Milan, Italy; giovanni.savoini@unimi.it

In many studies about dust concentrations in pig barns, the strong relationship between animal activity, feeding time and dust concentration was highlighted. The aim of this study was to evaluate the contribution of animal activity to dust concentration peaks in a fattening pig room, with particular regard to fine dust (PM_{10} and smaller particles, up to 0.25 µm), alveolar, thoracic and respirable dust. For this purpose, a fattening room, with 363 animals with a mean age of 131 d and a mean weight of 72 kg, was considered. Animals were reared on concrete slatted floor in 16 boxes. During the trial, dust was measured through a GRIMM Portable Laser Aerosol Spectrometer Model Mini-LAS 11-R, to evaluate dust particles ranging from 0.25 µm to 30 µm, in mass and in counts; and to measure alveolar, thoracic and respirable dust continuously. In the barn, the liquid feeding was released three times a day; a ventilation control system (FANCOM) based on a free running impellers (type Fancom FMS) guaranteed optimal climatic conditions to animals. Measurements, performed at a height of 50 cm, at the respiratory apparatus level of pigs, showed that, soon after feed release corresponding to animal maximum activity level (at visual observation), PM_{10} reached a peak of 1,392 µg/m^3, in agreement with previous studies, creating critical environmental conditions in the animal house. Mean value of inhalable particles was 2,668 µg/m^3, 1,347 µg/m^3 of thoracic particles, 308 µg/m^3 of alveolar particles: a severe environmental situation, occurring for around 40 min before and after feed release. Since the raising in animal activity linked to feed release is similar to the one induced by workers and vets inspection (even more during vaccination), it must be considered as a risk for the health status of animals and workers to be remediated, at barn and management levels.

The effect of birth-weight segregation during the growing period on swine productive features

M. Vázquez-Gómez[1], C. García-Contreras[2], S. Astiz[3], A. Gónzalez-Bulnes[3], C. Óvilo[2] and B. Isabel[1]
[1]UCM, Animal Production, Av. Puerta de Hierro, 28040, Madrid, Spain, [2]INIA, Animal Breeding, Crta. de la Coruña km 7,5, 28040, Madrid, Spain, [3]INIA, Comparative Physiology Group, Crta. de la Coruña km 7,5, 28040, Madrid, Spain; mvgomez@ucm.es

A main factor affecting profitability of modern pork industry is the lack of homogeneity in growth patterns and meat quality. Sex and variability in birth-weight (BW) are principal causes. The aim of the present study was to evaluate if the negative effect of BW variability may be alleviated by segregating pigs during the growing period according to BW. A total of 407 IberianxDuroc piglets were weighed at birth and at weaning (24 days-old) were separated in two groups: segregated (group S, n=240) and non-segregated (group NS, n=80) and the males were castrated. Piglets of the group S were classified, within sex, in Small, Medium or Large by BW, whilst piglets of the group NS were only classified by sex. Their expected slaughter age and weight were 10 months and 150 kg. During the growing period, body weight and average daily gain (ADG) were calculated every 5 weeks. Backfat and loin thickness were measured at weaning and at 15 and 30 weeks-old. Statistical analysis was carried out using the GLM procedure of SAS. The group S reached higher weight at weeks 10 and 15 (P<0.005, for both) and had a higher ADG at week 10 (P<0.0001) and at the end of the growing period (week 35, P<0.005) than the group NS. Effects from sex were found from week 25, when male pigs showed greater weight and ADG than female pigs (P<0.005 for both). At week 35, ADG was higher in females than in male pigs (P<0.005). Regarding backfat and loin thickness, at weaning, females had less fat than males (P<0.005) and at week 30, males reached greater weight and backfat depth (P<0.005) than female pigs. Moreover NS pigs showed higher loin thickness than the S pigs (P<0.01) because of the greatest thickness of NS females. In summary, male pigs showed better growth patterns than female pigs regardless if they were segregated or not. During the growing period, performance was greater in S group than in NS group. However, further studies at slaughterhouse are required to clarify possible effects of segregation by BW on carcass and meat quality and homogeneity.

Improving horse welfare at transport: definition of good practices through a Delphi procedure
S. Messori[1], W. Ouweltjes[2], K. Visser[2], P. Dalla Villa[1], H. Spoolder[2] and W. Baltussen[3]
[1]*Istituto Zooprofilattico Sperimentale dell'Abruzzo e del Molise, Campo Boario, 64100 Teramo, Italy,* [2]*Wageningen UR Livestock Research, P.O. Box 338, 6700 AH Wageningen, the Netherlands,* [3]*LEI Wageningen UR, P.O. Box 29703, 2502 LS The Hague, the Netherlands; s.messori@izs.it*

Slaughter horse transportation is a reality involving thousands of animals every year, and it might have detrimental effects on horse health and welfare. Despite the existence of overarching rules governing animal welfare during transport (i.e. EC Regulation 1/2005), issues with the direct enforcement of the Regulation still exist. In this scenario, the importance of the application of Good Practices (GP) is paramount. Although GP exist, these are often not shared with the community and consensus about them is lacking, making their implementation sparse. In order to obtain GP being agreed by all operators in the sector, a Delphi study was carried out, involving stakeholders (e.g. transporters, NGOs, abattoir, vets) from different countries. The procedure aimed to identify those transport aspects that, if not well managed, would have a major impact on horse welfare, and to define GP that would improve such activities. During Phase 1, participants ranked (on a 1-10 scale) a list of 38 main transport aspects, and drafted a first list of GP that would improve each one of them. Four email rounds followed to reach an agreement on the proposed practices. Scores assigned to transport aspects were weighted to obtain a standardised index score, on the basis of which three main aspects to be targeted were identified: fitness to travel, transport vehicle design and maintenance, and appropriate training. Overall, 23 GP were pre-selected to undergo the Delphi. Experts were asked to score each practice's impact on animal welfare, working conditions of people and profit. The GP selected through this holistic approach will be presented. This EU-funded study represents the first complete guidance to the most challenging phases of horse transportation achieved through a science-based approach. The reached agreement will favour the in-field implementation of these GP, enhancing horse protection during transport.

Performance of dairy cows offered silages produced from grass swards or red clover/grass swards
D.J. Johnston[1,2], A.J. Dale[2] and C.P. Ferris[2]
[1]*Queens University Belfast, University Road Belfast, BT7 1NN, United Kingdom,* [2]*Agri-Food and Biosciences Institute, Large Park, Hillsborough, BT26 6DR, United Kingdom; david.johnston2@afbini.gov.uk*

Due to price volatility and instability in supply, the UK livestock sector is seeking to reduce its reliance on imported protein feedstuffs. The use of locally-grown forage legume crops, including red clover (Trifolium Pratense), may help achieve this objective. This 13 week study (involving 28 dairy cows), examined the performance of cows offered silages produced from three successive harvests of either a pure grass sward (GS) or red clover/grass sward (RCS). Swards were harvested on 12 June, 7 August and 2 October (H1, H2 and H3, respectively) following wilting periods of 48, 72 and 72 hours, respectively. The GS sward received a total of 248 kg N/ha over the season, while no N was applied to the RCS sward. Silages from H1, H2 and H3 were offered for 5, 5 and 3 weeks, respectively, while all cows were offered 8.0 kg concentrate per day throughout the experiment. Data were analysed by ANOVA. Total herbage DM yields (t DM/ha) across H1-3 were 10.4 and 9.9 for grass and red clover/grass swards, respectively. Silage DM intakes for the GS and RCS treatments were 8.8 and 11.7 ($P<0.001$: H1), 9.2 and 10.4 ($P<0.006$: H2) and 12.3 and 11.5 ($P=0.224$: H3) kg/day, respectively, with mean values across H1-H3 being 9.6 and 11.1 kg/day ($P<0.001$), respectively. Milk yield for the GS and RCS treatments were 24.9 and 25.5 ($P=0.221$: H1), 22.1 and 23.1 ($P=0.093$: H2) and 21.8 and 24.7 kg/day ($P<0.05$: H3) respectively, with mean values across H1-3 being 23.2 and 24.4 kg/day ($P<0.05$), respectively. Mean milk fat across H1-3 was 46.6 and 45.7 g/kg ($P=0.439$) for the GS and RCS treatments, respectively, while the respective values for milk protein were 32.3 and 31.5 g/kg ($P=0.201$). Fat plus protein yield for the GS and RCS treatments were 2.60 and 2.55 kg/day ($P=0.648$: H1-3) respectively. In conclusion, intakes were higher with the RCS, but milk solids yield/cow was unaffected. There was a saving in fertiliser costs with the RCS treatment, but milk solids output/ha (assuming 15% in-silo losses with both treatments) was numerically lower with RCS (2,396 and 1,940 kg for GS and RCS, respectively).

Rumen temperature monitoring to assess feed efficiency of dairy cows

A. Fischer[1,2,3] and P. Faverdin[1,2]

[1]Agrocampus-Ouest, UMR 1348 PEGASE, 65 rue de Saint-Brieuc, 35000 Rennes, France, [2]INRA, UMR 1348 PEGASE, domaine de la Prise, 35590 Saint-Gilles, France, [3]Institut de l'élevage, Monvoisin, 35652 Le Rheu, France; amelie.fischer@rennes.inra.fr

Improving feed efficiency can be seen as a solution to increase farms economic and environmental sustainability as it contributes to decrease feed consumption and maintain or increase production. The main issue remains to measure feed efficiency on farm because total feed intake is only available in a few research facilities. However, new monitoring devices offer interesting perspective to find a predictor to assess feed efficiency. Assuming that water intake is highly associated to feed intake, we hypothesized that water efficiency, measured by rumen temperature, could be a predictor of feed efficiency in dairy cows. Feed efficiency and rumen temperature were monitored during the 238 first days of lactation for 60 Holstein cows fed ad libitum a unique TMR. Feed efficiency was defined as the residual energy intake (REI) calculated as the residuals of the linear regression of cumulated net energy intake (UFL, not corrected for digestive interaction due to intake level) on cumulated requirements for fat corrected milk yield, mean metabolic BW, mean BCS, empty BW gain and empty BW loss. Rumen temperature were monitored every 5 min with a rumen thermobolus (San'Phone®, Medria). The daily sum of the difference between the rumen temperature corrected for water intake and raw rumen temperature was used as an indicator of water intake (WI). Water efficiency was defined as the residual water intake (RWI) calculated as the residuals of the linear regression of estimated WI on cumulated milk yield and mean BW. RWI explained 29% of REI variability with a coefficient of correlation of 55%. Feed efficiency assessed with REI can be measured indirectly with RWI which only requires on farm affordable phenotypes: rumen temperature, milk yield and BW. RWI appears as a promising way to include feed efficiency in a genomic selection strategy.

Is PUFA transfer from grass to milk different during grazing compared with indoor feeding?

A. Elgersma

Anjo Elgersma, Independent Scientist, P.O. Box 323, 6700 AH Wageningen, the Netherlands; anjo.elgersma@hotmail.com

A better understanding of factors that influence fatty acid (FA) transfer from forage to milk is essential to enable proper comparisons between effects of grazing and indoor feeding on milk FA composition. There are however methodological constraints in determining FA in fresh herbage in relation to collection and/or handling of plant samples, transport conditions and duration after harvesting, storage conditions and duration, extraction procedures and calculation methods. In a recent survey of FA concentrations, and proportions of α-linolenic acid, linoleic acid, palmitic acid and 'other' FA, the author found large differences in concentrations of 'other' FA among experiments. It may be caused by differences in analytical methods or ways of calculation or data presentation and illustrates methodological constraints when comparing data from various trials and laboratories. If the plant matrix of fresh material affects extraction differently compared with wilted materials, then FA concentrations in fresh herbage samples cannot be directly compared with those in wilted, dried or ensiled herbages; this would pose a methodological constraint when comparing processes during grazing compared with indoor feeding. A fundamental difference between grazing and stall-feeding is that grazing animals eat fresh living herbage while cut forage is more or less wilted, or dead in case of conserved forage. Furthermore, selection possibilities in indoor feeding systems are limited. Besides selection, differences in intake (i.e. rumen fill) and rumination patterns could play an important role in explaining different results found with high-forage dry matter intake in grazing and indoor forage regimens. Intake estimates under grazing are prone to larger errors than values measured indoors. In indoor feeding trials, rations are more constant than under grazing. Thus there is a need for harmonisation of methods to enable comparisons between effects of grazing and indoor feeding on FA in milk.

The effect of supplementing two levels of rumen buffer on rumen and milk parameters in dairy cows
P.S.J. Van Adrichem, P.A. Abrahamse and G.F. Schroeder
Cargill, Innovation Center Velddriel, Veilingweg 23, 5334LD Velddriel, the Netherlands;
peter_van_adrichem@cargill.com

The aim of this study was to evaluate the effects of supplementing dairy cows challenged with mild rumen acidosis with two levels of a rumen buffer. The buffer product tested was Rupromin Balance® (Provimi, the Netherlands), supplied at 0, 150 or 300 g/cow/day (treatments A, B and C respectively). Six rumen cannulated lactating Holstein-Friesian dairy cows (30.6±4.3 l/day; 181±68 days in milk) were used in a replicated 3×3 Latin square experiment with 21-day periods. Cows were housed in tiestalls at the Cargill Innovation Center in Velddriel, The Netherlands and were fed a semi complete ration (52% corn silage, 29% grass silage, 5.4% alfalfa hay, 12.7% protein concentrate and 0.8% premix on DM basis) with additional supplementation of a high-energy concentrate. Milk yield and dry matter intake were measured daily. At the end of each period rumen pH was monitored continuously via indwelling probes for 4 days. Rumen fluid was sampled 3 and 6 h after morning feeding and analysed for VFA and lactate. Milk was analysed for fat content and odd and branched chain fatty acids (OBCFA) profile. Statistical analysis was performed using the MIXED procedure in SAS, a contrast statement was used for testing linear effects. As buffer level increased, rumen acetate:propionate ratio lineally increased ($P<0.05$; A:2.76; B:3.02; C:3.20) and rumen lactate concentration decreased ($P<0.05$; A:3.57; B:3.18; C:2.88 mmol/l). No effect of buffer was observed on rumen pH (mean: 6.06, and 156 min pH<5.6), milk yield (29.5 l/day) and DMI (25.6 kg/day). Milk fat percentage (+0.53 unit %) and yield (+150 g/d) were linearly increased in response to buffer addition ($P<0.05$), resulting in a linear improvement of feed efficiency (+0.06 unit %). Milk OBCFA showed a tendency ($P<0.10$) for lowered t10-C18.1 indicating a shift away from the t10-C18:1 pathway, explaining the observed higher milk fat concentration. In conclusion, buffer supplementation changed rumen fermentation pattern and milk composition, strongly indicating alleviation of rumen acidosis.

Novel selection criteria for rumen microbial genes to improve feed efficiency and methane mitigation
R. Roehe[1], R.J. Dewhurst[1], C.-A. Duthie[1], J.A. Rooke[1], N. McKain[1], D.W. Ross[1], J.J. Hyslop[1], A. Waterhouse[1],
T.C. Freeman[2], M. Watson[2] and R.J. Wallace[3]
[1]Scotland's Rural College, Future Farming Systems, The Roslin Institute Building, EH25 9RG Edinburgh, United Kingdom, [2]University of Edinburgh, The Roslin Institute and Royal (Dick) School of Veterinary Studies, The Roslin Institute Building, EH25 9RG Edinburgh, United Kingdom, [3]University of Aberdeen, Rowett Institute of Nutrition and Health, Bucksburn, AB21 9SB Aberdeen, United Kingdom; rainer.roehe@sruc.ac.uk

Genetic improvement of feed conversion efficiency with simultaneous reduction in methane emissions is of great importance for an economical and environmental sustainable beef industry. However, both traits are difficult and costly to record in large-scale breeding programme. The aim of this study was to investigate the best strategy to identify the most informative rumen microbial genes associated with feed efficiency and methane emissions, as well as to use the abundances of these genes as phenotypes for genetic improvement. Animals selected for low and high methane emissions were used in a metagenomic analysis to determine their microbial gene abundances. A network analysis of the relative abundance of rumen microbial genes resulted in 6 main distinct clusters due to correlated changes in gene abundance. Clusters 2 and 5 or 4 and 6 mainly included microbial genes associated with feed efficiency and methane emissions, respectively. The partial least squares analysis of microbial gene abundances, both within and across these clusters, resulted in small number of genes with known biological mechanisms explaining more than 80% of the variation in feed efficiency and methane emissions. Microbial genes should be chosen using a combination of their relationship with the trait of interest, as well as their biological pathways (networks). Diet effects showed only systematic differences (no interactions) and so could be adjusted for as fixed effects in the genetic model. Because rumen samples can be obtained in the abattoir or from live animals by nasal stomach tube, a functional microbial gene microarray could be a cost-effective way to select cattle without the need to record traits such as feed efficiency and methane emissions.

Isolation and characterization of nutritionally important lactic acid bacteria from cattle gut

S. Ghazanfar[1], M. Qubita[2], F. Hassan[2], A. Muhammad[3], I. Ahmed[1] and I. Muhammad[2]
[1]National Agriculture Research Centre, Park Road, 44000 Islamabad, Pakistan, [2]University Islamabad, Faculty of Biological Sciences, Department of Microbiology, Quaid-i-Azam, 44000 Islamabad, Pakistan, [3]National Agricultural Research Centre, Livestock Research center, Park Road, 44000 Islamabad, Pakistan; shakira_akmal@yahoo.com

Lactic acid bacteria regulate the metabolic pathways and activities of gut micro flora and aids in host resistance mechanism. These also used as natural food preventive to improve food safety. There is still a scarcity of information on the lactic acid bacterial diversity of livestock species such as cattle. So that present study was conducted to highlight the gut LAB diversify of dairy cattle. In the present study, eight LAB strains were isolated from feces of Sahiwal cattle. Results of that study reveal isolates were gram-positive cocci and negative for catalase, oxidase, indole and simmon's citrate. They were non-motile as well as they all lack the ability of gas production. They were positive only for triple sugar iron test and methyl red test. All these properties indicate that the experimental strains belong to lactic acid bacteria family. The phylogenetic analyzes inferred from 16S rRNA gene sequence of these strains demonstrated that these are closely related to species belonging to different genera: Enterococcus (KP256014, KP256015) and Lactococcus (KP256012, KP256013) ssp. This preliminary study showed that the fecal flora of dairy cattle is rich in LAB population, which may be utilized in various industrial applications.

Impact of nutritional grouping on the economics of dairy production efficiency

V.E. Cabrera
Department of Dairy Science, University of Wisconsin-Madison, 1675 Observatory Dr., Madison, WI 53706, USA; vcabrera@wisc.edu

The economic efficiency of nutritional grouping strategies in 5 Wisconsin commercial dairy herds was studied using a daily dynamic stochastic Monte Carlo simulation model. Each month, the clustering method was used to homogeneously regroup cows according to their nutrient concentration requirements. The average NEL and MP+1SD concentration of the group were used to formulate the group diet. The calculated income over feed costs gain (IOFC, $/cow per yr) of having >1 nutritional groups among the herds ranged from $33 to $58, with an average of $39 for 2 groups and from $42 to $58, with an average of $46 for 3 groups. The improved IOFC was explained by increased milk sales and lower feed costs. Higher milk sales were a result of fewer cows having a milk loss associated with low BCS in multi-group scenarios. Lower feed costs were mainly due to less RUP consumption in multi-group scenarios. The percentage of total NEL consumed and captured in milk for >1 nutritional group was slightly lower than that for 1 nutritional group due to better distribution of energy throughout the lactation and higher energy retained in body tissue, which resulted in better herd BCS distribution.

Effect of preweaning plane of nutrition on body size and age at puberty in dairy calves

A. Mendoza[1,2], S. De Trinidad[1], C. Viñoles[2], C. Cajarville[1], T. Morales[2], M. Pla[2], D. Ubilla[1], J. Soutto[1] and E. Garófalo[1]
[1]*Facultad de Veterinaria, Universidad de la República, Ruta 1 km 42, Libertad, San José 80100, Uruguay, [2]Instituto Nacional de Investigación Agropecuaria, Ruta 50 km 11, Colonia 70000, Uruguay; amendoza@inia.org.uy*

To establish the effects of preweaning plane of nutrition on body size and age and puberty, 34 Holstein calves born to multiparous cows were randomly assigned to 2 treatments from days 2 to 56 of life (weaning): daily allowance of 4 (T4) or 8 (T8) L of whole milk, plus starter and water ad libitum. After weaning, all calves were managed as a single group, with access to alfalfa hay and a grain mix. Blood samples were taken every 14 and 30 days during the pre and postweaning (until day 360 of life) periods respectively, to determine IGF-I and insulin concentrations. From day 180 of life follicular dynamics were evaluated by transrectal ultrasonography in alternate days for 21 days, to record the emergence of at least 1 follicular wave. Once heifers reached 195 kg ovaries were examined by ultrasonography twice a week to evaluate the presence of a corpus luteum and determine the onset of puberty. At this moment, live weight and withers height were measured. Statistical analysis was carried out using the MIXED and GLIMMIX procedures of SAS. No significant differences were detected in the total number of follicles, diameter of the dominant and subordinate follicles, regression and growth rates and follicle lifespan. The concentrations of IGF-I were higher in T8 than T4 during the pre (146.3 vs 75.5 ng/ml, SEM=11.6, P<0.001) and postweaning periods (241.3 vs 214.0 ng/ml, SEM=21.6, P=0.03), but treatments did not affect the concentrations of insulin. Calves in T8 reached puberty earlier (259.3 vs 304.0 days, SEM−5.2, P<0.001), but were lighter (253.4 vs 281.0 kg, SEM=3.2, P<0.001) and had a lower hip width (36.1 vs 37.9 cm, SEM=0.3, P<0.001) than T4 calves at this physiological stage. In conclusion, a higher plane of nutrition during the preweaning phase increased the concentrations of IGF-I and reduced the body size and age at puberty in dairy calves.

Relationship between methane emission, residual feed intake and carcass traits of beef cattle

M.E.Z. Mercadante[1], P. Dominguez[2], J.N.S.G. Cyrillo[1], R.C. Canesin[1], L.F. Oliveira[2] and L.G. Albuquerque[2]
[1]*Instituto de Zootecnia, Rod. Carlos Tonani, km 94, Sertãozinho, SP, Brazil, [2]Universidade Estadual Paulista, Via de Acesso Prof. Donato Castellane, Jaboticabal, SP, Brazil; mercadante@iz.sp.gov.br*

To study the inter-relationship between methane emission (mean CH_4 emission = 125±27 g/day), residual feed intake (RFI=from -1.562 to 1.564 kg dry matter/day, SD=0.658 kg/day), average daily gain (mean ADG=0.940±0.258 kg/day) and ultrasound carcass traits (mean LMA=49.6±7.22 cm^2, mean BFT=2.54±1.30 mm and mean RFT=5.45±1.99 mm for longissimus muscle area, backfat and rump fat thickness), data from 88 Nellore bulls and heifers (367±31 days of age and 317±43 kg of BW) were analyzed in a multivariate approach. RFI and ADG were recorded during feedlot performance tests and CH_4, LMA, BFT and RFT were measured at the end of performance tests (after 84 days with a high-roughage diet). Methane emission was estimated using the sulphur hexafluoride traces gas technique over 7-days. Two of the six principal components (PC), considering the Kaiser criterion (eigenvalue greater than 1.0), explained 64% of the total variability. The PC1 explained 45% of the variability and distinguished CH_4 (0.54) and RFI (0.10) from ADG (-0.71), LMA (-0.81), BFT (-0.87) and RFT (-0.72). Nevertheless, CH_4 showed a higher eigen vector coefficient (loading) compared to RFI, hence PC1 had low distinguishing power for RFI. The PC2 explained 19% of the variability and was characterized by RFI with higher loading (0.87) compared to CH_4 (0.15), ADG (0.36), LMA (0.31), BFT (-0.15) and RFT (-0.29). Therefore, CH_4 and RFI were not strongly distinguished in the same PC, but, in both PC, these traits were positively (favorably) correlated. The same tendency was observed for RFI, BFT and RFT, even though they were not strongly distinguished in the same PC. In both PC, they were negatively (unfavorably) correlated. This multivariate approach showed a weak inter-relationship between CH_4, RFI and fat thickness traits. There was no strong evidence that high-efficient cattle (low-RFI) release less enteric methane.

Finishing strategies for dairy bulls slaughtered at 15 months of age
B. Murphy[1,2], A.K. Kelly[1] and R. Prendiville[2]
[1]University College Dublin, Belfield, Dublin 4, Ireland, [2]Teagasc, Animal and Grassland Reseach & Innovation Centre, Grange, Dunsany, Co. Meath, Ireland; brian.murphy@teagasc.ie

The aim was to investigate different finishing regimes for spring born dairy bulls slaughtered at 15 months of age. The production system consisted of 2 periods; at pasture from mid-August to housing (83 days; P1) and indoors on a concentrate ad libitum diet (209 days; P2). Sixty bulls were assigned to one of four treatments; a pasture only diet in P1 followed by concentrate A (800 g/kg barley, 140 g/kg soya bean meal (SBM), 40 g/kg molasses and 20 g/kg minerals) in P2 (T1), pasture plus 3 kg dry matter (DM) of concentrate A per head daily in P1 followed by concentrate A in P2 (T2), pasture plus 3 kg DM of concentrate A per head daily in P1 followed by concentrate B (400 g/kg barley, 400 g/kg maize meal, 140 g/kg SBM, 40 g/kg molasses and 20 g/kg minerals) in P2 (T3) or pasture plus 3 kg DM of concentrate B per head daily in P1 followed by concentrate C (750 g/kg barley, 140 g/kg SBM, 50 g/kg megalac, 40 g/kg molasses and 20 g/kg minerals) in P2 (T4). Data were analysed using PROC MIXED of SAS and pairwise comparisons were determined using a Bonferroni test. Live weight at first housing was similar for all treatments, however average daily gain in P1 was greater (P<0.001) for T4 than for T1 with T2 and T3 intermediate (0.90, 0.96, 1.07 and 1.10 kg/day for T1, T2, T3 and T4, respectively). Live weight at slaughter and carcass weight were similar for T1, T2, T3, and T4; 546, 572, 554 and 552 kg and 283, 296, 288 and 281 kg, respectively. Kill out proportion and conformation and fat scores were similar for all treatments. Total concentrate DM intake was greater for T2 than for T1 with both T3 and T4 intermediate (P<0.001); 1,673; 1,986; 1,790 and 1,866 kg DM per head, respectively. Feed efficiency per kg live weight and per kg carcass weight were similar for all treatments. Results indicated that animal and carcass performance can be maintained by reducing concentrate input and maximising pasture utilisation.

Effects of dietary nitrate and increased lipid concentration on the performance of finishing steers
C.-A. Duthie, J. Rooke, S. Troy, J. Hyslop, D. Ross, A. Waterhouse and R. Roehe
SRUC, Edinburgh, EH9 3JG, United Kingdom; carol-anne.duthie@sruc.ac.uk

Adding nitrate or increasing the lipid concentration in mixed forage/concentrate diets can reduce methane output from cattle without affecting animal performance. The objective was to investigate the combined effects of adding nitrate and increasing the dietary lipid concentration on steer performance and efficiency. The experiment was a 2×4 factorial design consisting of two breeds (AAx, crossbred Aberdeen Angus; LIMx, crossbred Limousin) and 4 treatments (forage:concentrate (g/kg dry matter, DM): 557:443): (1) Control containing rapeseed meal as the main protein source, replaced with either (2) Nitrate (Calcinit, 18 g nitrate/kg diet DM); (3) added lipid (maize dark grains, 37 g lipid/kg diet DM) or (4) Combined nitrate (18 g nitrate/kg diet DM) and lipid (36 g lipid/kg diet DM). Steers (n=80) were group-housed in even numbers of each breed across 4 pens; each treatment allocated to 1 pen. Individual dry matter intake (DMI) was recorded for 56 days. Growth was modelled by linear regression of weight against test date, to describe average liveweight gain (LWG) and mid-test LW. Feed conversion ratio (FCR) and residual feed intake (RFI) were characterised for all animals. Statistical analyses were conducted using the mixed procedure of SAS (SAS Inst. Inc.) with the fixed effects of breed, nitrate and lipid and the random effect of pen (and slaughter batch for carcass traits). Lipid did not adversely affect animal performance. Nitrate addition resulted in poorer LWG (P<0.01) and reduced FCR (P<0.05) compared with diets not containing nitrate. AAx steers achieved greater LWG (P<0.01), but showed greater DMI (P<0.001), greater fat depth (P<0.01) and poorer RFI (P<0.01) than LIMx steers. Adding nitrate to a mixed forage:concentrate diet had a negative impact on growth and efficiency of finishing steers. Increasing the dietary lipid concentration did not adversely affect performance or efficiency and thus can be recommended for use within finishing cattle diets.

In vitro degradability of selected Nigerian forages for their use in ruminant diets
O.J. Bolaji[1], A.S. Chaudhry[1] and J. Dolfing[2]
[1]Newcastle University, School of Agriculture, Food and Rural Development, Newcastle University, Newcastle,
NE1 7RU, United Kingdom, [2]Newcastle University, School of Civil Engineering and Geosciences, Newcastle
University, Newcastle, NE1 7RU, United Kingdom; o.j.bolaji@newcastle.ac.uk

One way ANOVA was used to compare the rumen in vitro dry matter (IVDMD) and organic matter degradability (IVOMD) alongside gas (IVGP), ammonia (NH_3) and methane (CH_4) production of 2 Legumes (L. leucocephala, G. sepium) and 4 grasses (P. purpureum, P. maximum, B. decumbens and A. gayanus) from Nigeria. Among the grasses, A. gayanus recorded the highest IVDMD (522 g/kg), IVOMD (517 g/kg), IVGP (141 l/kg OM), ammonia (147.1 mg/l) and methane (45.7 ml/g OM) whereas B. decumbens recorded the least IVOMD (445 g/kg), IVGP (74 l/kg OM), ammonia (69.1 mg/l) and methane (22 ml/g OM). This was presumably due to the highest crude protein (117 g/kg DM) and least acid detergent fibre (665 g/kg DM) contents in A. gayanus and lowest CP (59 g/kg DM) and higher fibre (714 g/kg DM) contents of B. decumbens. However, the highest nutrient degradability found in A. gayanus was far below the forage of known degradability. Among the legumes, G. sepium had the highest IVDMD (616.0 g/kg), IVOMD (611 g/kg), IVGP (166 l/kg OM), ammonia (161 mg/l) and methane (54.1 ml/g OM) whereas L. leucocephala had the least IVDMD (504 g/kg), IVOMD (529 g/kg), IVGP (87 l/kg OM), ammonia (110 mg/l) and methane (17.6 ml/g OM). This was attributed to higher tannin content of L. leucocephala (91 g/kg DM) than G. sepium (20 g/kg DM) despite the fact that they both had higher CP values (255 and 215 g/kg DM) respectively. The gas production, methane and ammonia tended to increase with increased IVDMD and IVOMD of the forages while increased tannin content in L. leucocephala had a pronounced effect on its nutrient degradability. We conclude that the tested grasses of this study could not meet the energy requirements needed for optimum microbial activity due to low nutrient degradability when compared to a good quality forage. Thus if more rumen by-pass protein is required. L. leucocephala will be the best choice due to its high protein content.

Identification of Genetic Markers Associated with Feeding Efficiency in Holstein fattening Calves
M. Cohen-Zinder[1], A. Asher[1], E. Lipkin[2], R. Agmon[1], A. Brosh[1] and A. Shabtay[1]
[1]Agricultural Research Organization, Beef Cattle Unit, Newe Yaar Research Center, Ramat Yishay, P.O. Box
1021, Ramat Yishay 30095, Israel, [2]Hebrew university of Jerusalem, Department of Genetics, Hebrew university
of Jerusalem, Jerusalem 91904, Israel, Israel; mirico@volcani.agri.gov.il

Ecological and economic concerns drive the need to improve feed utilization by domestic animals. Residual Feed Intake (RFI) is one of the most acceptable measures for feed efficiency (FE). However, phenotyping RFI related traits is complex, expensive, and requires special equipment. Advances in marker technology allow the development of various DNA-based selection tools. In order to assimilate these technologies for the benefit of RFI-based selection, reliable phenotypic measures are prerequisite. In the current study, we used genomic DNA of individuals presenting RFI phenotypic consistency across different ages and diets (stages 1-3), for targeted sequencing of chromosomal regions associated with FE and RFI related traits. Forty-eight top single nucleotide polymorphisms (SNPs), significantly associated with at least one of three stages, were. Identified. Eleven of these SNPs were harbored by the fatty acid binding protein 4 (FABP4). While ten significant SNPs found in FABP4, were common for stage 1 and stage 3, one SNP (FABP4_5; A<G substitution), in the promoter region of the gene, was significantly associated with all three stages. As the three stages reflect changing diets and ages with concomitant RFI phenotypic consistency, the above polymorphisms and in particular FABP4_5, might be considered possible markers for RFI-based selection for FE in the Holstein breed, following a larger scale validation.

Minerals in forage for dairy cows

B. Johansson[1] and M. Åkerlind[2]
[1]Swedish University of Agricultural Sciences, Department of Animal Environment & Health, Box 234, 532 23 Skara, Sweden, [2]Växa Sverige, Box 7024, 750 07 Uppsala, Sweden; birgitta.johansson@slu.se

Cows need minerals for their health and production and a high share of their mineral intake comes from forage. Feeding cows a high proportion of forage is also good for the environment. There are large variations in mineral concentrations (MC) in forages, depending on e.g. plant species composition and concentrations of plant-available minerals in the soil. In Sweden, the MC in soil varies between regions and some dairy farmers report mineral deficiency-related disorders of cows and calves. Dairy cows in Sweden are highly productive and it is possible that the recommended intake does not meet their requirements, at least of certain minerals, in some areas. This study examined whether the MC in forage differed between regions, and between organic and other producers. Large amounts of analytical data on minerals in forages (n=4,872), based on samples taken at harvest and from silages during 2015 and registered by Växa Sverige, were examined. The samples were divided into 10 Swedish regions and 212 were known to be samples from organic farms, while the other samples were mainly from conventional farms. The minerals studied were calcium (Ca), phosphorus (P), potassium (K), sodium (Na), sulphur (S), magnesium (Mg), chlorine (Cl), iron (Fe), manganese (Mn), zinc (Zn) and copper (Cu). The GLM procedure in SAS was used in statistical analyses. All minerals except S and Fe differed between the Swedish regions. In forage samples from northern Sweden, MC were generally lower than in samples from central and southern Sweden. Samples from organic farms had higher concentrations of Ca, P, K and Mg and lower concentrations of S than the other samples. It is known from earlier studies that legumes generally have higher MC than grasses and increased red clover proportion in ley mixtures increases MC in the overall mixed forage. Therefore, the higher MC in forage samples from organic farms may be the result of a higher clover fraction in the forage. It would be beneficial to consider MC differences in forages when composing qow diets.

Effect of new sustained-release non-protein nitrogen on performance of dairy cows

J. Roquet[1], M. Agovino[2] and H. Warren[2]
[1]Alltech Spain, S.L., Can Lletget, 11, 08202 Sabadell, Spain, [2]Alltech Biotechnology Centre, Summerhill Road, Dunboyne, Co. Meath, Ireland; hwarren@alltech.com

This study investigated the effect of sustained-release (SR) ruminal non-protein nitrogen (NPN) source on performance of dairy cows. The milk production and milk composition of 100 Holstein dairy cows has been compared before and after the inclusion of Optisync® (OS, a new generation of SR NPN; Alltech Inc). The diet was based on rye silage, dehydrated alfalfa, corn, barley, soy bean meal and rapeseed meal. OS has been added to the diet replacing part of vegetable protein source, to have the same amount of dry matter and starch and reducing the amount of crude protein (17.35% vs 16.79%). Animals were on treatment for 60 days in average. Milk production and milk quality were recorded. No significant variation on milk quality (fat, protein, urea and SCC) has been established. Average milk production improved by +0,65 liter/head/day after OS inclusion in the diet (37.51 vs 38.16 liter/head/day). Animals between 30 and 100 days in milking recorded an increased production of 1.48 iter/head/day. Dung samples were sieved during the trial period using the Nasco's Digestion Analyser (3 sieves). The digestion analyser has been used as tool to monitor rumen function. A significant reduction of undigested particles in the first sieve was observed and it can explain an improvement of ruminal function and increase of fiber digestion. OS, new generation of sustained-release (SR) ruminal non-protein nitrogen (NPN) source can be used to replace part of vegetable protein source in dairy cows in order to improve milk production.

Continuous measurement of pH in rumen of dairy cows fed with three total mix ratios
R. Loucka, P. Homolka, F. Jancik, P. Kubelkova, Y. Tyrolova, A. Vyborna and M. Michalickova
Institute of Animal Science, Nutrition and feeding of farm animals, Pratelstvi 815, Prague Uhrineves, 104 00,
Czech Republic; michalickova.monika@vuzv.cz

The rumen of dairy cows is a place of complicated digestion processes that increase the total acidity. If the acid contents in the rumen exceeds a specific limit (often set at pH 5.8), acidosis may occur and negatively affect the feed intake, the microbial metabolism, the production and quality of milk and the general health condition of the animals, including limb diseases (laminitis). The risk of acidosis can be reduced by higher proportion of physically efficient neutral detergent fibre (peNDF). The study was assembled from 3 total mix ratios (TMRs) whether the pH of the dairy cow rumen will be different at the same amount of chop particle sizes over 8 mm. The TMRs consisted of alfalfa silage, maize silage, corn cob mix silage, brewer's grains, mashed grains, energy and mineral agent and straw with particles 20 mm long. The main differences between the TMRs consisted in the proportion of brewer's grains (6 kg in TMR1 and TMR2, 10 kg in TMR3) and straw (0 kg in TMR1, 1 kg in TMR2 and TMR3). The experimental period extended 9 weeks. The experiment included 26 Holstein dairy cows, 4 of them with eCOW bolus (Devon, Ltd., Great Britain) in their rumen, implemented per os, to perform pH measurements each 15 minutes. The percentage of chop with particles over 8 mm (pe) of all TMRs amounted to 35%; the calculated peNDF (pe × NDF) of TMR1 amounted to 14.9, that of TMR2 to 15.1, and that of TMR3 to 15.5, which are values below the acidosis limit. The average pH of TMR1 was 6.0 (Standard Deviation 0.24), that of TMR2 6.02 (SD 0.22), and that of TMR 6.11 (SD 0.21). The proportion of time in which the pH values got below 5.8 was 22 for TMR1, 19 for TMR2 and 33% for TMR3. The differences were not statistically significant. Dedication: QJ1510391.

Use of dietary cation-anion difference (DCAD) in formulating dry cow rations
M. Agovino[1], P. Colturato[2] and H. Warren[1]
[1]Alltech Biotechnology Centre, Summerhill Road, Dunboyne, Co. Meath, Ireland, [2]Paragon s.n.c., Via C. degli Uberti, Crema, Italy; hwarren@alltech.com

The transition period (3 weeks before and after calving) is the most critical time in the life of a dairy cow. The cow's metabolism is under severe stress as she transitions to lactation. Fresh cows commonly experience milk fever, retained placenta, displaced abomasum, and mastitis. Hypocalcemia (low blood calcium) resulting from inadequate calcium metabolism is common among fresh cows, and can lead to milk fever and other health disorders. The objective of this field trial was to evaluate the incidence of metabolic disease at calving of dairy cows fed with DCAD diet (Dietary cation-anion difference) during close-up dry period (3 weeks prepartum). A unique formulation of anionic salt (CC, Alltech Calving Care™) containing calcium chloride and magnesium chloride and coated with fat matrix has been fed. The CC contained mannan-derived oligosaccharide components (Actigen®), organic selenium (Sel-Plex®), organic minerals (Bioplex®), and algae (All-G-Rich®). Over a period of 40 days, approximately 21 days before the expected date of calving, 117 multiparous dry cows (from two different dairy farms) were fed with CC. The average feeding rate was 300 grams per head per day. Diet was based on corn and grass silage, wheat straw and rapeseed meal. Urine pH were randomly sampled from 12 animals in average. Relevant clinical, milk production, and fertility data were collected. The urine pH is significantly decreased after 4 days of CC inclusion in the diet from 8.01 to 7.05 ($P<0,05$). Metris, retained placenta and hypocalcemia reduced respectively by 46%, 59% and 41%. In conclusion, the inclusion of CC into the close up diet promote efficient calcium metabolism leading to fewer associated metabolic issues.

Effect of dietary cation–anion differences on rumen fermentation, digestibility and milk production

S. El-Mashed[1] and H. Moussa[2]
[1]*Animal Production Research Institute, Cow's Department, Nadi El-Said street, 12618 Doki, Giza, Egypt,* [2]*Ain Shams University, Faculty of Agiculture, Animal Production Department, 12618 Shubra, Cairo, Egypt; sh.elmashed@hotmail.com*

Twenty Friesian cows were used to study the effect of dietary cation-anion differences (DCAD) on digestibility, rumen fermentation and milk production. Animals were divided into four groups. All groups were fed a basal diet consisting of berseem 30 kg, concentrate feed mixture (CFM) 6 kg, rice straw 6 kg &soya bean meal (SBM) 75 g as a carrier for anionic salts/head/day. Anionic salts were used to control DCAD of the ration at the following levels (groups): (1) Control group was fed the basal diet without anionic salts; (2) DCAD was controlled to be 0 mEq/ kg DM; (3) DCAD was controlled to be negative 150 mEq/kg DM; (4) DCAD was controlled to be negative 150 mEq/kg DM by using Anio-Norel (commercial product). Animals received treatments three weeks before expected calving date and stoped at delivery day. Results obtained showed that: (1) rumen pH was affected by treatment, values were 7.5, 6.9, 6.6 and 7.1 for groups 1, 2, 3 and 4 respectively; (2) organic matter digestibility ranged between 59.43% for group 3 and 57.23% for group 4; (3) CF digestibility was not significantly (P>0.05) affected by treatment; (4) treated groups showed lower 305 days milk yield when compared with control group (P<0.05); (5) treatments didn't affect milk components significantly (P>0.05) except for milk fat where it was lower for treatment 4 compared with other treatments.

Carcass traits of Nelore steers in different Brazil grazing systems

R.R.S. Corte[1], S.L. Silva[1], A.F. Pedroso[2], R.T. Nassu[2], R.R. Tullio[2], A. Berndt[2], L.S. Sakamoto[1], P.H.M. Rodrigues[1] and P.P.A. Oliveira[2]
[1]*University of Sao Paulo, Duque de Caxias Norte, 225, Pirassununga, Brazil,* [2]*Embrapa Southeast Livestock, São Carlos, SP, Brazil; rscorte@usp.br*

This study was carried out to evaluate the effect of different grazing systems on the carcass traits of beef cattle in Brazil. The grazing experiment was conducted for two consecutive years with Nelore steers: (n=24;271±2.2 kg of live weight (LW);15 months old) Animals were allotted to four grazing systems with 2 area replications (block): (1) DP: degraded pasture (1.1 AU/ha; Brachiaria decumbens); (2) IHS: irrigated with high stocking rate (5.9 AU/ha; Panicum maximum); (3) DHS: dryland with high stocking rate (4.9 AU/ha; Panicum maximum); (4) DMS: dryland with moderate stocking rate (3.4 AU/ha; Brachiaria brizantha), at an experimental station of the Brazilian Agricultural Research Corporation (EMBRAPA), Southeast of Brazil. With exception of the degrade pasture, all pastures were manage in a rotational grazing system. Each paddock was grazed by three steers as testers and regulating animals were used to adjust the sward heights. Data were analyzed as completely randomized block design using PROC MIXED and the year was used as a random effect. Animals were slaughtered with approximately 450 kg of LW. Animals from DP have lower final LW (433.9 kg), hot carcass weight (113.6 kg) and dressing percentage (53.3%) (P<0.001) when compared to IHS (530.7 kg; 145.4 kg and 55.8%), DHS (517.3 kg; 141.3 kg and 55.7%) and DMS (527.2 kg; 144.6 kg and 55.9%) grazing systems respectively, with no differences between IHS, DHS and DMS systems. Ribeye areas (REA) were similar among DP (5.9 cm^2) IHS (7.8 cm^2) and DHS (8 cm^2), however, animals from the DMS system presented higher REA (9.6 cm^2, P=0.005) than animals from DP. The backfat thicknesses were similar for all grazing systems (DP=58.4 cm; IHS=64.2 cm; DHS=61.1 cm; DMS=64.7 cm). When compared to IHS, DHS and DMS, the DP grazing system produced smaller carcass, reflecting the inefficiency of the degraded system.

The important and often forgotten role of managing genetic diversity
C. Maltecca, J. Howard and F. Tiezzi
North Carolina State University, Department of Animal Science & Genetics, Campus Box 7621, Raleigh, NC 27695, USA; cmaltec@ncsu.edu

Managing inbreeding and genomic diversity is an often forgotten path to ensure healthy livestock. Management of diversity rests on three pillars. Understanding the basis and consequences of genetic diversity. Controlling the population effective size. Optimize genetic variability deployed through mating plans. Inbreeding management often rely on the implicit assumption that individuals with the same inbreeding share the same genomic load. Marker information allows instead for regions-specific homozygosity causing inbreeding depression to be identified. Yet, these regions are expected to be at a low frequency so that traditional association methods based on estimating dominance effects lack statistical power. Methods that exploit the fact that longs runs of homozygosity (ROH) are enriched with deleterious variants can have greater power in identifying haplotypes linked to inbreeding depression. We utilized an ROH statistic to identify regions that, when in a run of homozygosity, (ROH) impact the number of piglets born alive in a Landrace (LR) and Large White (LW) pig populations. Genotyped dams (n=5,000 LR; n=5,000 LW) with number born alive (NBA) phenotypes were utilized. Regions on SSC1, SSC6 and SSC13 for LR and SSC4, SSC13 and SSC 17 for LW caused a significant reduction (range: -0.64 to -0.34) in NBA. Additionally, metrics that reduce ROH frequency may provide an attractive way to manage the diversity and ensure long-term gains while avoiding inbreeding accumulation. We characterized the frequency of ROH in LR, LW and their cross (LR × LW) through a combination of real and simulated genotypes and to determine the impact of optimizing different inbreeding metrics for nucleus and crossbred populations. Regions of shared high ROH5Mb frequency were found on SS1, SS3 and SS14 and regions with a high ROH5Mb frequency within a breed were found to persist in the crossbreds. ROH and genomic-based relationship matrices decreased the proportion of the overall genome in a ROH by 2.45 and 2.19 fold when compared to pedigree-based relationships.

What do we mean by runs of homozygosity? Assessing effect of parameters involved in their detection
S.T. Rodríguez-Ramilo[1,2] and J. Fernández[1]
[1]INIA, Mejora Genética Animal, Ctra. Coruña Km 7,5, 28040, Spain, [2]INRA, UMR 1388 GenPhySE, BP52627, 31326 Castanet Tolosan, France; jmj@inia.es

Runs of homozygosity (ROH) is a proposed measure of the levels of inbreeding calculated from genomic data. The theoretical definition of ROH is straightforward: long and uninterrupted stretches of homozygosity. However, the empirical identification of ROH depends on several factors. In this study, six parameters involved in the detection of ROH have been investigated. These attributes were: (1) the minimum length of a ROH, (2) the minimum number of SNPs within a ROH, (3) the minimum marker density of a ROH, (4) the maximum distance between two adjacent SNPs, (5) the maximum number of missing genotypes and (6) the maximum number of heterozygous SNPs admitted within a ROH. Individuals from the Spanish Holstein cattle population were genotyped with the Illumina BovineSNP50 BeadChip. After applying filtering criteria, the genomic data included 36,693 autosomal SNPs and 10,569 individuals. Pedigree data for those genotyped animals included 31,203 animals. ROH-based inbreeding was correlated with pedigree-based inbreeding and with the proportion of homozygous SNPs. The results indicate that four factors (the minimum length, the minimum number of SNPs, the minimum marker density, and the maximum distance between two adjacent SNPs) have important effects in the identification of ROH and, thus, in the estimation of ROH-based inbreeding levels. The other two factors seem to have no effect on ROH-based inbreeding. If a high correlation with pedigree-based inbreeding or with the proportion of homozygous SNPs is the desired feature, a sensible combination of parameters to define a ROH could be, for example, a length of a ROH between 0.2-4 Mb, a minimum number of SNPs within a ROH not higher than 80, a minimum density above 1 SNP every 90 Kb, and a maximum distance between two adjacent SNPs higher than 0.4 Mb.

Predicting autozygosity via runs of homozygosity from NGS versus SNP chip data for Nellore bulls

J. Sölkner[1], M. Milanesi[2,3], N. Khayatzadeh[1], A.T. Utsunomiya[3], M. Ferencakovic[4], I. Curik[4], P. Ajmone Marsan[2], J.F. Garcia[3] and Y.T. Utsunomiya[5]
[1]University of Natural Resources and Life Sciences, Vienna, Gregor Mendel Str. 33, 1180, Austria, [2]University Cattolica des Sacro Cuore, Piacenza, Via Emilia Parmense 84, 29122, Piacenza, Italy, [3]UNESP–Univ Estadual Paulista, Rua Clóvis Pestana 793, 16050-680, Araçatuba, Brazil, [4]University of Zagreb, Svetosimunska 25, 10000 Zagreb, Croatia, [5]UNESP–Univ Estadual Paulista, Via de Acesso Prof. Paulo Donato Castelane s/n, 14884-900, Jaboticabal, Brazil; johann.soelkner@boku.ac.at

In the era of high throughput genotyping, inbreeding levels of individuals are often predicted by calculating the proportion of the genome covered by runs of homozygosity (ROH), homozygous chromosomal segments of a particular minimum length. For conservative prediction of autozygosity with bovine 50k SNPchip data, the minimum ROH length should be 4 Mb. With a high density SNP chip with close to 800k, segments longer than 1 Mb may be used and give a more complete picture for common ancestors far back in time. Use of whole genome next generation sequence (NGS) data promises success in further reducing that limit and therefore being able to trace common ancestry even further back. Yet, NGS calls are more prone to genotyping errors than the highly quality controlled SNP chip genotype reads. In this study we used the Illumina HD Bovine SNP chip as well as resequencing data at 10x coverage of 21 important Nellore bulls in Brazil for calling ROH of different minimum lengths (1 Mb, 500 kb and 250 kb). We used two types of frequently used software, PLINKv1.9 and SVS. Of the 27,296,334 NGS raw variant calls, 1,964,474 passed quality control, compared to 498,483 on the SNP chip, therefore providing ~4 fold coverage. The ROH patterns observed depended on the settings used in each software, which we tried to vary from strict to lenient in similar ways. Results were not directly comparable because of very different approaches used for ROH calling. Very long (>10 Mb) ROH segments based on HD chip data were invariably split in several adjacent shorter ROH. We did not succeed to change the settings in a way to systematically avoid this. Therefore, inspection of the results and merging seems to be necessary.

Inbreeding depression for semen quality traits in cattle: estimation and dissection

M. Ferenčaković[1], I. Curik[1], M. Kaps[1] and J. Sölkner[2]
[1]University of Zagreb, Faculty of Agriculture, Department of Animal Science, Svetosimunska 25, 10000, Croatia, [2]University of Natural Resources and Life Sciences Vienna, Department of Sustainable Agricultural Systems, Division of Livestock Sciences, Gregor Mendel Str. 33, 1180, Austria; icurik@agr.hr

The use of artificial insemination (AI) in cattle breeding programs resulted in fewer bulls producing larger numbers of offspring. In such systems, frequent usage of high valued bulls leads to reduction of genetic variability, and to inbreeding and inbreeding depression. In bulls, low concentration of spermatozoa, low volume of sperm, low progressive motility and low number of alive spermatozoa, are considered to be under influence of inbreeding depression. However, inbreeding depression for semen quality traits was, until now, reported only from pedigree inbreeding coefficients, while never using genomic inbreeding coefficient. Here we confirmed the influence of inbreeding on total number of spermatozoa using genomic coefficient on fertility in Simmental bulls. Furthermore, usage of Runs of Homozygosity (ROH) status was explored as potential tool for pinpointing the genome regions that influence quantitative traits of interest. The genomic regions associated with inbreeding depression were shown in 22 significant signals on chromosomes 10, 17, 20 and 27. In total, 31 genes were detected, of which Ribosomal Protein L10-Like (RPL10L), Sperm Flagellar Energy Carrier Protein (SLC25A31), Cadherin 18, Type 2 (CDH18), and Potassium Channel, Subfamily U, Member 1 (KCNU1) were very promising candidates. Thus, it can be confirmed that the inbreeding effects on bull semen quality were detected at SNP level.

Using runs of homozygosity to locate a new recessive mutation in a rare breed of cattle

G.E. Pollott and D.C. Wathes

RVC, PPH, Royal College St., London NW1 0TU, United Kingdom; gpollott@rvc.ac.uk

Segregation analysis of data from a rare breed of dual-purpose cattle indicated the likelihood of an autosomal recessive condition causing early calf mortality. This work describes an attempt to locate the mutation causing this condition using SNP-chip data and a 'runs of homozygosity' (ROH) method of analysis. Tail-hairs were collected from all animals analysed and DNA was extracted from the accompanying follicles. Data from 68 animals (24 dead cases, 44 live controls) was available using the Illumina BovineHD Beadchip. After quality control, 58,071 SNP were available for analysis using the Autozygosity by Difference (ABD) method of ROH analysis and the results expressed as length of ROH in kilobytes. The probability of the result at each SNP was assessed using 10,000 permutations of the data by randomly assigning phenotypes (24 cases, 44 controls) to the 68 sets of SNP-chip data and rerunning the analyses (10,000 times). Large ROH were found on chromosomes 4 (P<0.05; 7,170 kb), 18 (P<0.05; 6.066 kb) and 21 (P<0.05; 10,260 kb) in the cases. However, two of these large ROH were also found in controls (BTA 18 and 21). The ABD methodology finds the difference between cases and controls and so the ROH on BTA4 was suggested as the likely site of the causal mutation of this condition. This region contains the NPY gene which controls a wide range of functions. Further analysis with next-generation sequencing will be applied to this region to fine-map the site of the mutation, and then a test will be developed to eliminate the gene from the population. Applying the ABD methodology has a distinct advantage in this scenario over methods which only score cases, since the controls are also potentially informative. In the case of a small rare breed there is the chance that long ROH are generated 'naturally' and so may become confused with those arising from the mutation. This work demonstrates the advantage of the ABD method and highlights the fact that long ROH may arise for a number of different reasons. Accurate phenotyping is essential since calf mortality clearly has a number of causes which may be confused in the field. The ABD method was extended to try to identify calves dying from the recessive condition as opposed to those dying of other causes.

Investigating Runs of Homozygosity using Whole-Genome Sequence Data in Holstein Bulls

N. Melzer[1,2], M. Sargolzaei[1,3] and C. Baes[1]

[1]University of Guelph, Department of Animal Biosciences, 50 Stone Road East, N1G 2W1 Guelph, Canada, [2]Leibniz Institute for Farm Animal Biology, Institute of Genetics and Biometry, Wilhelm-Stahl-Allee 2, 18196 Dummerstorf, Germany, [3]Semex Alliance, 5653 Highway 6 North, N1H 6J2 Guelph, Canada; melzer@fbn-dummerstorf.de

Due to the rising intensity of artificial selection in dairy populations, inbreeding levels have steadily increased over the past decades. Genomic selection may have also contributed to the increased level of inbreeding by shortening the generation interval. Increased inbreeding levels can detrimentally affect fitness. Hence it is of interest to restrict the increase of inbreeding. To enable more accurate estimation of inbreeding levels, genomic information is commonly used instead of pedigree information. Inbreeding levels can be determined by finding stretches of homozygous loci (runs of homozygosity, ROH) which are identical by descent (IBD) using single nucleotide polymorphism (SNP) arrays or next-generation sequencing (NGS) data. The frequencies of ROH are expected to vary widely within and across chromosomes. The objective of this study was to implement different methods for determination of ROHs. We used the available NGS data from 402 Holstein bulls of the 1000 bull genome project. We compared methods implemented in PLINK, bcftools, and SNP1101 software. PLINK is based on a sliding window approach, bcftools is based on the Hidden-Markov model (HMM), and SNP1101 simply searches for homozygous segments. The HMM approach enables inclusion of SNP positions from the genetic map, which are used to determine the transition probabilities based on likely recombination events between two SNPs. We present results using all approaches. In preliminary tests, we observed that PLINK and bcftools provide similar results for longer ROHs. Applying a HMM revealed additional shorter ROHs than those found with the sliding-window approach. Further analysis of detected ROHs, as well as the relationship of ROHs to pedigree and genomic inbreeding, may also be considered.

Optimum contribution selection for populations with introgression from other breeds

R. Wellmann and J. Bennewitz
University of Hohenheim, Institute of Animal Science, Garbenstraße 17, 70599 Stuttgart,
Germany; r.wellmann@uni-hohenheim.de

The genetic background of many livestock breeds is endangered due to introgression from economically superior breeds. Optimum contribution selection as introduced by Meuwissen aims at maximizing genetic gain while restricting the rate of inbreeding. However, optimum contribution selection often favours the use of animals with high genetic contributions from other breeds as these animals have superior breeding values and have low kinship with the population. Consequently, the application of this method to endangered breeds may further threaten their genetic background. This could be avoided by adding a lower bound for the genetic diversity of alleles originating from native founders and an upper bound for the genetic contribution from other breeds in the offspring. However, until now there is no software available to solve this optimization problem. We introduce the new R package optiSel, which is designed to solve this optimization problem as well as other optimization problems related to optimum contribution selection. These are to maximize genetic gain, to maximize genetic diversity, to minimize genetic contributions from other breeds, or to minimize diversity of alleles originating from native founders while using the remaining criteria as constraints. The methods are applied to a data set consisting of a pedigree of an endangered breed with historic introgression and simulated breeding values that are affected by the contributions from other breeds. It is shown that adding constraints for the genetic contribution from other breeds and for the diversity originating from native founders can lead to completely different selection decisions and considerably reduces the genetic gain achievable with optimum contribution selection but does not further threaten the genetic background of endangered breeds.

The domestication history of sheep and goat

P. Orozco-Terwengel[1], F. Alberto[2], I. Streeter[3], M. Bruford[1] and Nextgen Consortium[2]
[1]Cardiff University, School of Biosciences, Museum Avenue, Cardiff, CF10 3AX, United Kingdom, [2]University of Grenoble, LECA, Domaine Universitaire de Grenoble, 38041 Grenoble Cedex 9, France, [3]European Bioinformatics Insititute, Wellcome Trust Genome Campus, Saffron Walden, CB10 1SD, United Kingdom; orozco-terwengelpa@cardiff.ac.uk

Collecting large genetic datasets (i.e. large sample sizes and number of molecular markers) is becoming an easier and cheaper issue. Moreover, with the advancement in sequencing technologies it has become possible to sequence whole genomes from multiple samples at a not too expensive cost (e.g. ~£2,000 per genome). Making use of these technologies, the NextGen consortium sampled sheep and goats, and their wild ancestors in the Zagros Mountains (Iran) in order to attempt to understand the domestication process of these important livestock species. Sheep and Goat were domesticated in the Middle East around 10,000 years ago, with sheep having been domesticated from wild Mouflon, and goats from Bezoars. The anlayses of these whole genome data enabled reconstructing the demographic history of these species in relation to ecological changes over the last milion years, as well as identify signatures of selection reflecting the domestication process and adaptation to the local environment.

Creation of sheep divergent lines for gastro-intestinal parasitism resistance based on a QTL index

C.R. Moreno[1], S. Aguerre[1], F. Bouvier[2], A. Blanchard[3,4], C. Koch[3,4], J. Cortet[3,4], P. Jacquiet[5] and G. Sallé[3,4]
[1]INRA GenPhySE UMR1388, GA, 24, chemin de Borde-Rouge, Auzeville Tolosane CS 52627, 31326 Castanet Tolosan cedex, France, [2]INRA unité expérimentale de Bourges la Sapinière, 18390 Osmoy, France, [3]Université François Rabelais de Tours, 60 Rue du Plat d'Étain, 37000 Tours, France, [4]INRA UMR1282 ISP Infectiologie et Santé Publique, SA, 37380 Nouzilly, France, [5]ENVT-INRA IHAP, 23 Chemin des Capelles, BP 87614, 31 076 Toulouse Cedex 3, France; carole.moreno@toulouse.inra.fr

Gastro-intestinal nematodes are of major concern to sheep health worldwide. Even if anthelminthic drenching is the main control strategy, several parasite populations have developed resistance to several anthelmintic molecules. Thus, the genetic selection of resistant animals to gastro-intestinal nematodes appears as a complementary control strategy. QTL having an effect on resistance to gastro-intestinal parasites in sheep have been accumulated since 10 years. In the 3SR EU project, the main QTL regions were detected on chromosomes 4, 5, 7, 12, 13, 14 and 21. A dedicated 1000-SNP assay was selected in the 5 to 10 centimorgan around the QTL positions. 277 Romane sheeps were genotyped for these SNP assay and were orally infected twice with 10,000 H. contortus infective larvae. Each experimental challenge lasted a month and fecal egg counts (FEC) were collected at the end of each challenge. SNP genotypes in the QTL regions and FEC at the first and second infestations were used to perform a genomic evaluation. The obtained QTL genomic index was used to select 6 extreme rams among 148 males which were mated respectively to the most and the less resistant females based on the QTL genomic index. In the first generation of divergent lines, 224 lambs were genotyped with the 1000-SNP assay and a genomic selection of 40 resistant and 40 susceptible lambs was performed using parental generation as a reference population. FEC were measured on the 80 lambs following the previous design. After one generation of selection, a strong divergence has been achieved with two genetic standard deviations being observed between resistant and susceptible lines. Consequently, the difference of excretion of fecal egg count between the two lines is as high as 6 times.

Mapping genomic regions associated with resistance to infectious diseases in Ethiopian chicken

A. Psifidi[1], G. Banos[1], O. Matika[1], T.T. Desta[2], J. Bettridge[3], D.A. Hume[1], D. Tadelle[4], R. Christley[3], P. Wigley[3], O. Hanotte[2,4] and P. Kaiser[1]
[1]The Roslin Institute, University of Edinburgh, Easter Bush, Midlothian EH259RG, United Kingdom, [2]School of Life Sciences, University of Nottingham, University Park, Nottingham NG7 2RD, United Kingdom, [3] Institute of Infection & Global Health, University of Liverpool, The Ronald Ross Building, 8 West Derby Street, Liverpool, L69 7BE, United Kingdom, [4]International Livestock Research Institute, Addis Ababa, Ethiopia, P.O. Box 5689, Ethiopia; androniki.psifidi@roslin.ed.ac.uk

Poultry play an important role in the agriculture of many African countries, including Ethiopia. The majority of chickens are indigenous, well adapted to the local environment and raised in scavenging systems. Although these birds are resilient to common diseases, routine vaccination and biosecurity measures are rarely applied and infectious diseases remain a major cause of mortality and reduced productivity. Breeding for increased resistance to infectious diseases offers potentially a sustainable solution. Data were collected for five major infectious diseases: oocyst counts for Eimeria spp and antibody titres for Infectious Bursal disease, Mareks' disease, Fowl Cholera and Fowl Typhoid from indigenous chickens raised in two distinct regions, Horro (n=384) and Jarso (n=376) in Ethiopia. All birds were genotyped with a 580K high density whole-genome DNA array. Principal component analysis (PCA), variance component analyses and genome wide association studies (GWAS) were performed. PCA showed that the chickens from the two regions are genetically distinct. Estimates of heritability were intermediate (0.23-0.45) for all disease phenotypes with the exception of antibody responses to Fowl Typhoid where the estimate was low (0.06). GWAS identified genomic markers significantly associated with response to infectious diseases at genome-wide and chromosome-wide level. Some of the putative QTL regions for antibody responses were common for different diseases. A search for putative candidate genes around the significant markers revealed the presence of many genes involved in immune response. These results underpin the potential of genetic selection for enhanced disease resistance across Ethiopian indigenous chicken ecotypes.

Functional SNP in a polygenic disease induced by high-altitude in fattening Angus steers

A. Canovas[1], R.R. Cockrum[2], D. Brown[3], S. Riddle[3], J. Neary[2], T. Holt[2], G.M. Krafsur[2], J.F. Medrano[4], A. Islas-Trejo[4], M. Enns[2], S. Speidel[2], K. Cammack[2], K.R. Stenmark[3] and M.G. Thomas[2]
[1]University of Guelph, Animal Bioscience, 50 Stone Road E., Guelph, N1G 2W1, Canada, [2]Colorado State University, Animal Science, Fort-Collins, 80523, USA, [3]University of Colorado-Denver, Denver, 80202, USA, [4]University of California-Davis, Animal Science, Sheilds Ave., Davis, 95616, USA; acanovas@uoguelph.ca

High-altitude (>1,800 m) disease is a challenging problem in beef and dairy cattle. The disease is consequential of hypoxia-induced right ventricular (RV) heart failure as per vascular inflammation of the pulmonary artery (PA) and hypertension. The disease has moderate to high heritability ranging from 0.2 to 0.4; however, minimal information exists of the genes involved. The transcriptomes of six tissues (i.e. left and right ventricle, pulmonary artery, aorta, muscle, and lung) were examined in samples harvested from fattening-yearling Angus steers phenotyped to be of low or high pulmonary arterial pressures (LPAP and HPAP; n=10/group). Tissue specific splice variants were identified in RV (n=555), aorta (n=547) and PA (n=152; P<0.01 FC>2) between LPAP and HPAP animals. Most of the splice variants are located in key regulators genes with roles in angiogenesis and cardiomyopathy (NFATC1), movement of leukocytes and neutrophils (OLR1, PLAUR), failure of heart (CTGF), hypertrophy of heart ventricle (TREM1, GATA2) and vascularization (SYVN1). Besides, several SNP variants segregated specifically in either the LPAP or HPAP animals. Among them, 139 SNP were located in key regulator genes involved in the adaptation of high altitude disease. These approaches helped identify splice variants corresponding to key regulator genes in a polygenic disease induced by high-altitude in Angus cattle. The identification of functional SNP associated with high-altitude disease by combining structural and functional genomic data will help to develop more robust approaches for genetic selection in beef cattle.

Genetic parameters for health traits in a multi-breed sheep population

A.C. O'Brien[1,2], N. McHugh[2], E. Wall[3], T. Pabiou[3], K. McDermott[3], S. Randles[3], S. Fair[1] and D.P. Berry[2]
[1]Department of Life Sciences, UL, Limerick, Ireland, [2]Moorepark Research Center, Fermoy, Co. Cork, Ireland, [3]Sheep Ireland, Bandon, Co. Cork, Ireland; aine.obrien@teagasc.ie

The objective of the present study was to quantify the genetic variation in three health-related traits (lameness, dagginess, and mastitis) in an Irish sheep population. Dagginess, a measure of the accumulation of faecal material on the hindquarter of a sheep was measured on a scale of 1 to 5 (1=clean; 5=very daggy). Lameness was assessed on a scale of 0 to 2 (0=not lame; 2=moderately/severely lame). Mastitis was measured (only in adults) by palpation of the udder (0=no mastitis; 1=evidence of (historical) mastitis). Animals were categorised as lambs (<365 days old) or adults (≥365 days old) and were analysed both separately and combined. After edits, 39,315 records from 264 flocks between the years 2009 and 2015 inclusive were analysed. Variance components were estimated using animal linear mixed models. Fixed effects included contemporary group of flock, date of inspection and animal type (i.e. lamb, hogget (i.e. an animal ≥365 days but <730 days old that has not yet had a lamb) or ewe), breed proportion, coefficients of heterosis and recombination, parity/age group and the age difference from the median of each parity/age group. An additive genetic effect and residual effect were both fitted as random terms. Where all data were combined, the heritability of lameness and dagginess was 0.09 (0.01) and 0.15 (0.01), respectively. When age groups were analysed separately, the heritability of lameness and dagginess was 0.06 (0.02) and 0.15 (0.03), respectively in adults and 0.10 (0.01) and 0.14 (0.02), respectively in lambs. Mastitis had a heritability of 0.07 (0.04).The genetic correlation between adult and lamb dagginess was 0.38 (0.12); no correlation existed between adult and lamb lameness. Ample genetic variation exists for all the health traits investigated indicating that genetic improvement in dagginess, lameness and mastitis is possible.

Variation in the prion protein in Dutch goats for selective breeding to eradicate Scrapie

J.J. Windig¹, R.A.H. Hoving¹, J. Priem², L.J.M. Van Keulen², J.P.M. Langeveld² and A. Bossers²
¹Wageningen UR, Animal Breeding and Genomics Centre, P.O. Box 338, 6700 AH Wageningen, the Netherlands,
²Wageningen UR, Central Veterinary Institute, P.O. Box 65, 8400 AB Lelystad, the Netherlands; jack.windig@wur.nl

Scrapie is a neurodegenerative disease occurring in goats and sheep. In sheep genetic selection for resistant genotypes was successful in eliminating the disease. A similar strategy may be followed in goats, but information on genotype frequencies is scarce. Here, we set out to determine frequencies of resistant allelic variants of the PrP gene in the main goat breeds in the Netherlands. Since, scapie in goats is absent in the Netherlands the expectation was that frequencies of resistant alleles will be low in comparison with breeds from SE-Europe where scrapie incidence can be high. Overall frequencies in Dutch goats were determined in 2005, 2008, 2012, derived from random sampling for the national scrapie surveillance without knowledge of the breed. Nine haplotypes were identified in the Dutch breeds. Frequencies for resistant haplotypes were generally low and not in Hardy Weinberg equilibrium. Breed specific frequencies were determined in 2013-2015 for six different breeds. The frequency of the K222 haplotype in the Dutch Toggenburger (29%) and the S146 haplotype in the Nubian and Boer breeds (respectively 20% and 28%).were exceptionally high. The high frequency in the Toggenburger breed is probably the result of genetic drift, while the high frequencies in the Nubian and Boer breed can be explained by their origin from different parts of the world. The frequency in the main Dutch breed, the Saanen derived White Goat was 3.1%, similar to frequencies of other Saanen or Saanen derived breeds around the world. The unexpectedly high frequency of the K222 haplotype in the Dutch Toggenburger, may help to set up selective breeding for scrapie resistance, and underlines the need for conservation of rare breeds in order to conserve genetic diversity rare or absent in other breeds.

Estimation of inbreeding based on runs of homozygosity in Finnish Ayrshire population

K. Martikainen and P. Uimari
University of Helsinki, Department of Agricultural Sciences, Koetilantie 5, 00014 Helsinki, Finland;
katja.martikainen@helsinki.fi

Fertility of dairy cows has decreased over the years due to unfavourable genetic correlation with production traits. Additionally, increased level of inbreeding in dairy cattle populations may have resulted in inbreeding depression in fertility traits. Traditionally inbreeding coefficients are estimated from pedigree. However, pedigree-based estimates may suffer from incomplete pedigree and tend to underestimate the true level of inbreeding. Alternatively, runs of homozygosity (ROH), which are continuous stretches of homozygous genotypes inherited from a common ancestor can be used for genomic based estimates of inbreeding. The objective of this study is to estimate the level of inbreeding based on pedigree (F_{PED}) and ROH (F_{ROH}) analysis and to estimate inbreeding depression in fertility traits in Finnish Ayrshire population. Genotypic data were available for 19 075 Finnish Ayrshire cows that were genotyped with low density panel (9k) and imputed to 50k density. A total of 39 144 SNPs remained for the analysis after removing SNPs with minor allele frequency, MAF<0.05 and Hardy-Weinberg disequilibrium (P<0.0001). F_{ROH} was calculated as the sum of SNP in ROH divided by the total number of SNPs. Average levels of inbreeding of the genotyped cows were 0.063 and 0.024 for F_{ROH} and F_{PED}, respectively. The number of animals with inbreeding coefficient greater than 0.10 was 15 and 1,545 for F_{PED} and F_{ROH}, respectively. Correlation between F_{PED} and F_{ROH} was 0.70. Majority of the found ROH segments were rather small; 48% of the segments were less than 5 Mbp long, 34% of the segments were 5-10 Mbp long, 17% were between 10 Mbp and 50 Mbp and a small proportion (0.2%) were longer than 50 Mbp. Based on these initial results pedigree based and genomic based estimates of inbreeding coefficients are highly correlated but not identical. The next step is the estimate inbreeding depression of fertility traits based on F_{PED} and F_{ROH}.

Analysis of runs of homozygosity and genomic inbreeding in bulls of nine Swiss cattle breeds

H. Signer-Hasler[1], A. Burren[1], M. Neuditschko[2], B. Gredler[3], B. Bapst[3], M. Frischknecht[1,3], D. Garrick[4], C. Stricker[5] and C. Flury[1]
[1]Bern University of Applied Sciences, School of Agricultural, Forest and Food Sciences, Länggasse 85, 3052 Zollikofen, Switzerland, [2]Agroscope, Swiss National Stud Farm, Les Longs-Prés, 1580 Avenches, Switzerland, [3]Qualitas AG, Chamerstrasse 56, 6300 Zug, Switzerland, [4]Iowa State University, Kildee 225, 50011 Ames, USA, [5]agn Genetics, Börtjistrasse 8b, 7260 Davos, Switzerland; heidi.signer@bfh.ch

Pedigree information of 9,214 bulls and 27,612 informative SNP genotypes of the Brown Swiss (BS), Braunvieh (BV), Original Braunvieh (OB), Holstein (HO), Red Holstein (RH), Swiss Fleckvieh (SF), Simmental (SI), Eringer (ER) and Evoléner (EV) breed were used to compare pedigree and markerbased levels of inbreeding. Runs of homozygosity (ROH) were derived with PLINK and consecutively used for the derivation of genomic inbreeding coefficients (F_{ROH}). In addition, pedigree-based inbreeding coefficients (F_{PED}) were calculated for all sires. For 131 bulls no ROH segment could be detected. The largest fraction of animals without ROH was found in SF (9.3%). In total, 82,688 ROH were identified where the average number of ROH segments was largest in BS (14.0) and lowest in ER (4.1). Average F_{ROH} ranged from 2.4% (ER/OB) to 9.6% (BS), whereas F_{PED} varied from 1.2% (EV) to 7.1% (BS). Pearson correlation between genomic and pedigree inbreeding was 0.70 over all 9,214 bulls. Therefore the proportion of the genome present in ROH serves as an indicator of the inbreeding coefficient. Low average genomic inbreeding was found for the local breeds (ER, OB, SF, SI). For ER, OB and SI this result is mainly explained with the ongoing use of natural services and for SF due to its crossbreeding system.

Patterns of runs of homozygosity in four cattle breeds maintained in Poland

T. Szmatoła, A. Gurgul, K. Ropka-Molik, I. Jasielczuk, T. Ząbek and M. Bugno-Poniewierska
National Research Institute of Animal Production, Department of Genomics and Animal Molecular Biology, Krakowska 1, 32-083 Balice, Poland; tomasz.szmatola@izoo.krakow.pl

Runs of homozygosity (ROH) are defined as contiguous homozygous regions of the genome where the two haplotypes inherited from the parents are identical. Their length and frequency may help to describe population history, level of inbreeding, population bottlenecks or signatures of selection. With the use of BovineSNP50 chip, ROH length and frequency in four cattle breeds (Holstein, Polish Red, Limousin and Simmental) maintained in Poland was estimated. Then, the level of autozygosity and genomic regions most commonly associated with ROH were identified. We have observed vast differences between breeds in the length and distribution of homozygous regions: Holstein (mean=290.6 Mb, SD=67.2 Mb) Polish Red (mean=142.8 Mb, SD=67.4 Mb), Limousin (mean=180.5 Mb, SD=79.9 Mb) and Simmental (mean=201.8 Mb, SD=99.4 Mb). Moreover, the level of autozygosity (FROH) varied and ranged from 0.057 for Polish Red to 0.116 for Holstein, while being lower for unselected cattle. In regard to the regions of the genome most commonly associated with ROH, a number of genes potentially being under selection for individual breeds were detected. In general, ROH patterns vary between breeds, with some minor similarities which may reflect common origin of the taurine breeds.

Integration of genetics with Zooarchaeology; Challenges of the MendTheGap
P. Miracle[1], V. Cubric-Curik[2], T. Kivisild[3], S. Radović[4], M. Ferenčaković[2], D. Marchi[5], A. Oroš-Sršen[4], J. Mauh Lenardić[4], D. Radovčić[6], S. Forenbaher[7], A. Džidić[2] and I. Curik[2]
[1]McDonald Institute for Archaeological Research, University of Cambridge, Cambridge, CB2 1TN Cambridge, United Kingdom, [2]University of Zagreb, Department of Animal Science, Svetošimuska 25, 10000 Zagreb, Croatia, [3]University of Cambridge, Division of Biological Anthropology, Cambridge, CB2 1TN Cambridge, United Kingdom, [4]Croatian Academy of Sciences and Arts, Institute for Quaternary Paleontology and Geology, Trg Nikola Šubića Zrinskog 11, 10000 Zagreb, Croatia, [5]University of Pisa, Department of Biology, Lungarno Pacinotti 43/44, 56126 Pisa, Italy, [6]Croatian Natural History Museum, Demetrova 1, 10000 Zagreb, Croatia, [7]Institute for Anthropological Research, Ljudevita Gaja 32, 10000 Zagreb, Croatia; vcubric@agr.hr

'Smart Integration of Genetics with Sciences of the Past in Croatia: Minding and Mending the Gap' (acronym 'MendTheGap' is Multi-Inter-Trans (MIT) disciplinary EU project (the top ranked project in the call H2020-TWINN-2015) performed through the collaboration of the University of Cambridge, University of Pisa and CrEAMA Initiative (MIT disciplinary consortium of Croatian researchers). The big challnge of the MendTheGap project is to answer how large number of already existing archaeological remains from wild and domestic animals in Croatia can contribute to resolving contemporary issues in genetics (domestication, Neolithic migrations, South-East Europe as a glacial refugium). With this poster we present the basic ideas of the MendTheGap project with all activities related to Archaeogenetics. Our presentation aims to stimulate and promote; (1) research collaborations and applications to EU projects; (2) discussions and development of new concepts; (3) the CrEAMA Initiative as a research entity; and (4) transparency of EU-funded projects.

Do seven South African cattle breeds differ genetically?
L. Pienaar-Van Der Westhuizen[1,2], M. Scholtz[1,2], F. Neser[1] and M. Macneil[1,2,3]
[1]University of the Free State, Animal Wildlife and Grassland Sciences, P.O. Box 339, 9300 Bloemfontein, South Africa, [2]ARC, Animal Production Institute, Animal Breeding and Genetics, P Bag X2, 0062 Irene, South Africa, [3]Delta G, Ice Cave Road, MT 59301 Miles City, USA; neserfw@ufs.ac.za

Microsatellite marker data of seven cattle breeds in South Africa were used to investigate the existing genetic relationships amongst the breeds. The landrace cattle breeds used in this study includes the Afrikaner, Bonsmara, Drakensberger and Nguni, while the Brahman, Simmentaler and Angus were used as exotic breeds. The relationships between the different breeds are important not only to determine the genetic distances but also the genetic origin. This is especially important in crossbreeding systems where the amount of heterosis depends on the genetic distance and uniqueness of the participating breeds. Estimates to determine within breed genetic variability included unbiased heterozygosity (Hz), mean number alleles (MNA), allelic richness (Rs) and inbreeding (F_{IS}). Hz, MNA, Rs and F_{IS} ranged between 0.569–0.741, 8.818–11.455, 7.797–10.441 and -0.001–0.050, respectively. From a genetic diversity perspective, all breeds had moderate to high heterozygosity values demonstrating the high degree of genetic variability and low inbreeding levels within each of the breeds. Results from STRUCTURE divided into seven Clusters (K=7), indicating each breed as an unique identity. The second highest probability indicated a total of three genetic Clusters (K=3) and noticeably grouped the separately. When STRUCTURE was forced into three Clusters, breeds were divided into their three different genetic groups (taurine, indicine and Sanga). This was supported by the UPGMA tree and PCoA analysis.

Refining cattle breed proportion using genomic data

T.R. Carthy[1], R.D. Evans[2] and D.P. Berry[1]
[1]Teagasc, Animal and Grassland and Research Innovation Centre, Moorepark, Fermoy, Co Cork, Ireland, [2]Irish Cattle Breeding Federation, Bandon, Cork, Ireland; tara.carthy@teagasc.ie

Precise knowledge of the breed proportion of individual animals is important for herdbook registration but also for properly accounting for breed effects (and non-additive genetic effects) in genetic evaluation models. It is a legal requirement to record the expected breed composition of a calf at birth in Ireland based on the average breed composition of its parents. However, because of mendelian sampling during gametogenesis, it is not possible to truly know the breed proportion of the progeny from a mating which involves at least one crossbred parent. Such inaccuracies can accumulate down the pedigree thus impacting genetic evaluations. The objective of the present study was to correlate the breed proportion of cattle derived from genomic data with expectations based on recorded pedigree data. A total of 14,824 single nucleotide polymorphism data were available on 77,888 Irish cattle. Unsupervised breed predictions was undertaken and herdbook registered animals with a high likelihood of being purebred based on the genomic data were coded as being purebred for subsequent analysis. A total of 5,205 purebred animals were available from the breeds Angus (n=1,097), Charolais (n=822), Hereford (n=685), Holstein-Friesian (n=740), Jersey (n=74), Limousin (n=1,225), Shorthorn (n=126) and Simmental (n=436). The genotypes of these animals were used in a supervised analysis to quantify the proportion of each breed in all remaining animals which were predominantly crossbred. The correlation between the predicted breed proportion of an animal based on the genomic information versus that based on pedigree was, on average, 0.80 varying from 0.73 (Simmental) to 0.87 (Jersey). Although breed proportion of an animal cannot always be predicted in the absence of genomic information, despite this phenomenon and the existence of pedigree errors, a strong correlation still existed between pedigree-predicted and genome-predicted breed proportion of individuals, predominantly crossbred animals.

Genomic characterisation of Pinzgau cattle based on high-throughput molecular information

V. Šidlová[1], N. Moravčíková[1], M. Ferenčaković[2], M. Simčič[3], G. Mészáros[4], A. Trakovická[1], O. Kadlečík[1], I. Curik[2] and R. Kasarda[1]
[1]Slovak University of Agriculture in Nitra, Department of Animal Genetics and Breeding Biology, Tr. A. Hlinku 2, 949 76 Nitra, Slovak Republic, [2]University of Zagreb, Department of Animal Science, Svetošimunska 25, 100 00 Zagreb, Croatia, [3]University of Ljubljana, Department of Animal Science, Jamnikarjeva 101, 1000 Ljubljana, Slovenia, [4]University of Natural Resources and Life Sciences, Division of Livestock Sciences, Gregor-Mendel Str. 33, 1180 Vienna, Austria; veron.sidlova@gmail.com

Pinzgau cattle is a dual purpose Alpine breed, once among the most popular breeds in Austro-Hungarian Empire, whose number is rapidly declining in recent 60 years. Here, we used high-throughput molecular information of 1,040 genotyped individuals; (1) to characterise the population structure and admixture of Pinzgau cattle (Original and Slovak population) in relation to 13 relevant European cattle breeds by Bayesian clustering algorithm implemented in BAPS; and (2) to identify regions in a genome that are specific for the Pinzgau breed based on frequency of autozygosity and Wright's F_{ST} fixation coefficients. With some variations, the analysis from BAPS indicated considerable separation of individuals within their analysed population. Overall, 15 separate populations were observed including Original (Austrian, AP) and Slovakian Pinzgau (SP) populations while Slovenian Cika cattle was very close to Pinzgau populations. The strongest admixture has been observed between AP and SP populations. On chromosomes 2, 4 and 11 we identified genomic regions with highest autozygosity, some previously associated with production and reproduction. Between AP and SP the highest F_{ST} values were observed on the chromosome 6 while some other high F_{ST} values were also observed. Overall, the results obtained will serve in re-evaluation of future breeding strategy of the Pinzgau cattle.

Genetic and molecular characterization of native goat flock of INRA-Tunisia

S. Bedhiaf-Romdhani[1], Y. Ressaissi[2], A. Biao[3], M. Amills[4] and M. Ben Hamouda[5]
[1]INRA-Tunisia, Rue Hédi Karray, 1004, El Menzeh, Tunisia, [2]ISA, Chott Mariem, BP 47, 4042, Tunisia, [3]INAT, Av Charles Nicolles, 1082, Tunisia, [4]CRAG, Campus UAB, Barcelona, Barcelona, 08193, Bellaterra, Spain, [5]INRA-Tunisia, Rue Hédi Karray, 1004, El Menzeh, Tunisia; bedhiaf.sonia@gmail.com

Goats are versatile domesticates that have been integrated into diverse production systems and natural and artificial selection have shaped the variation in this specie. The objective of this research was to characterize a native goat flock selected at INRA-Tunisia since 1970 on growth and adaptive traits. Main results showed that kids, raised under a low input production system, had an average weight at birth (BW) and at weaning (90 days of age (W90)) of 2.6 and 13 kg, respectively. Estimation of genetic parameters of pre-weaning traits showed that direct and maternal heritabilities values' were relatively high for BW and for weight at ten days of age (W10). Molecular analysis of the mitochondrial D-loop region showed that native goats sequenced belong to the A haplo-group with 11 new haplotypes identified, among them four were singletons and seven were composed of 2 to 4 sequences. This D-loop characterization has determined the presence of specific polymorphisms within individuals from the Tunisian D-loop haplotypes raised in the semi-arid. Our study highlights the importance of investigating genome-wide association mapping of indigenous genetic resources as an urgent step.

Genetic diversity of three local chicken breeds using high-density SNP-data

S. Gfeller[1], S. Joller[2], A. Burren[1], M. Neuditschko[3], C. Flury[1], C. Drögemüller[2], E. Kump[4], P. Ammann[4] and A. Spiri[5]
[1]Bern University of Applied Sciences, School of Agricultural, Forest and Food Sciences HAFL, Länggasse, 3052 Zollikofen, Switzerland, [2]University of Bern, Vetsuisse Faculty, Institute of Genetics, Bremgartenstrasse, 3012 Bern, Switzerland, [3]Agroscope, Swiss National Stud Farm, Les Long Prés, 1580 Avenches, Switzerland, [4]ProSpecieRara, Unter Brüglingen, 4052 Basel, Switzerland, [5]ZUN, Bühlstrasse, 9217 Neukirch an der Thur, Switzerland; stgfeller@hotmail.com

Appenzell Spitzhauben (AS), Appenzell Barthuhn (AB) and Schweizerhuhn (SH) are the three remaining local chicken breeds in Switzerland. With less than 200 AS, 300 AB and 600 SH individuals the genetic diversity is expected to be narrow. So far no pedigree information is available and therefore inbreeding coefficients are unknown. The aim of the study was to analyse the genetic diversity within and between the breeds and to estimate genomic inbreeding coefficients. Blood and tissue samples were used to genotype 78 AB, 35 AS and 39 SH with the Affymetrix 600k Axiom chicken SNP array. Thereof 60 genotypes were kindly distributed by the SYNBREED project. After quality control genotypes of 152 chicken and 309'428 SNPs were available. Genomic relationships were estimated using PLINK. ADMIXTURE was applied for structure analysis. Runs of homozygosity were derived (PLINK) to estimate genomic inbreeding coefficients (F_{ROH}). All genome-wide relationships between breeds were zero. This result is supported by population structure analysis, where the three breeds clustered separately. Average genome-wide relationships within breeds ranged between 0.078 (SH), 0.115 (AB) and 0.136 (AS). Within the AS breed remarkable sub-structures were observed between individuals from different feather-colour strains. Average F_{ROH} was highest for AS (0.397) followed by AB (0.238) and SH (0.181). These results are not unexpected considering the current population sizes: The smallest population (AS) showed the highest F_{ROH} whereas SH presented the smallest F_{ROH}. To complete the study selection signatures will be derived and specific regions of the genome with long ROH segments among and within breeds will be investigated. These future analyses are expected to increase the actual knowledge on genomic regions and genes involved in the formation of local Swiss chicken breeds.

Search of selection signatures between fine wool and coarse wool sheep breeds

B. Gutiérrez-Gil[1], I. Gómez-Redondo[1], P. Wiener[2], A. Suárez-Vega[1] and J.J. Arranz[1]
[1]Universidad de León, Department of Animal Production, Faculty of Veterinary Sciences, Campus de Vegazana s/n, 24071, León, Spain, [2]Roslin Institute and R (D) SVS, University of Edinburgh, Division of Genetics and Genomics, Easter Bush, Midlothian EH25 9RG, United Kingdom; asuav@unileon.es

We have performed a search of selection signatures in the sheep genome by comparing three Merino sheep breeds highly specialized in production of fine wool (Australian Industry Merino, Australian Merino and Australian Poll Merino) with three coarse wool breeds (Churra, Altamurana and Chios). For that, considering sliding windows of 9 SNPs, the pooled 50K SNP-chip genotypes of the two breed groups were analyzed to identify genome regions showing extreme genetic differentiation between the two compared groups (top 0.5% of the F_{ST} distribution,) and regions of reduced heterozygosity for each of the analyzed groups (bottom 0.5% of the ObsHtz distribution). The mapping analysis was based on the SNP order and locations according the current version of the sheep genome (Oar_v3.1). Our genetic differentiation analysis highlighted a total of 34 regions distributed across 18 chromosomes, with chromosome 3 being the one showing the larger number of F_{ST}-based candidate regions. A total of 72 genomic regions, distributed across 24 out of the 26 autosomes, were included within the 0.5% bottom of reduced ObsHtz values in the fine-wool group. The same number of regions showing a reduced ObsHtz was identified in the coarse-wool group. Considering the overlapping between the F_{ST} candidate regions and the regions of reduced heterozygosity in the fine-wool group, five regions, located on chromosomes 6, 11, 13, 16 and 25 were identified as convergence selection signals associated to the Merino group. Because the region of OAR13 was coincident with a region of reduced heterozygosity associated to the coarse-breed wool, the four other convergence regions exclusively identified in the Merino group will be considered to be explored in future studies as putative selection signatures associated with traits divergently selected between the two compared groups (wool, growth, meat production/quality traits).

Genetic structure of different Italian and Mexican chicken populations

M.G. Strillacci[1], E. Gorla[1], F. Schiavini[1], M.C. Cozzi[1], S.I. Roman-Ponce[2], F.J. Ruiz-Lopez[2], R.T.M.M. Prinsen[1], V.E. Vega-Murillo[2], S. Cerolini[1] and A. Bagnato[1]
[1]Università degli Studi di Milano, Via Celoria 10, 20133, Italy, [2]Centro Nacional de Investigación en Fisiología y Mejoramiento Animal, Km. 1 Carretera a Colón, 76280 Auchitlán, Mexico; maria.strillacci@unimi.it

This study is part of a Mexico-Italy bilateral project. The goal is the identification of the genetic variability of autochthonous chicken populations to highlight the genetic structure in the Italian populations and Mexican one, where a real breed distinction does not exist. A total of 94 individuals from 6 Italian breeds and 256 Mexican chickens collected from different agro-ecological regions were genotyped with the high-density Affymetrix 600K SNP arrays. Genetic structure was performed using both SNP and copy number variation (CNV) as markers. Using SNP allele frequencies (375,348 SNPs after filtering for MAF and Hardy-Weinberg equilibrium), the genetic relationship among samples was estimated with a principal component analysis. Genetic diversity within breeds was estimated using observed level of heterozygosity (HO), expected heterozygosity (He), Wright's F-statistics (FST and FIS). The Italian and Mexican populations clearly clustered separately. Particularly Italian samples grouped according to breed, while Mexican subjects clustered in a unique group. The individual-based CNV calling, based on LRR and BAF values, was done by the Hidden Markov Model of PennCNV software on autosomes. A hierarchical agglomerative clustering was applied according to the absence or presence of a definite CNV region in each population. The Mexican CNVs map was built on a total of 1,924 CNVs, summarized at the population level into 1,216 CNVRs (226 losses, 959 gains, 31 complex) covering a total of 47 Mb of sequence which corresponds to 5.12% of the chicken assembly autosome. The Italian CNVs map was based on a total of 1,003 CNV that resulted in 564 CNVRs (344 gains, 213 losses and 7 complex) for a total of 9.43 Mb of sequence, 1.03% of the chicken assembly autosome. Acknowledgments: Co-funded by project M01678; Ministry of Foreign affairs of Italy and Mexico.

The effect of Pinzgauer admixture on the coat colour of Cika cattle

M. Simčič, G. Mlakar, D. Bojkovski and D. Kompan
University of Ljubljana, Biotechnical faculty, Jamnikarjeva 101, 1000 Ljubljana, Slovenia;
mojca.simcic@bf.uni-lj.si

The aim of this study was to evaluate the effect of Pinzgauer cattle admixture in the past on the coat colour pattern of Cika cattle, an authentic autochthonous breed in Slovenia. Cika and Pinzgauer cattle share similar red pied sided coat colour pattern. A total of 66 male and 237 female animals, aged from 4 days to 13.1 years were included in the study. The width of the white stripes on the back, on the rump, on the front and on the rear legs were taken with the tape. The red basic coat colour intensity was performed by Minolta CR-300 Chromometer using CIE (L*a*b) colour system at three parts of the body (shoulder, rare ribs, round). All animals were measured in the spring after housing period. The GLM procedure (SAS/STAT) considered fixed effect of sex and the proportion of Pinzgauer admixture from the pedigree of each animal as linear regression. The effect of sex significantly affected the width of white stripes on the back, rump and rear legs as well as the red colour intensity (a*). Females had significantly wider white stripes and more intensively red coat colour than males. The effect of Pinzgauer admixture was found significant related to white stripes on the back and rump as well as in all three coat colour parameters (L*a*b). White stripes on the back and rump significantly increased, while values of L* (lightness), a* (redness) and b* (yellowness) significantly decreased with increased Pinzgauer admixture. Admixed Cika animals with Pinzgauer had wider white stripes on the back and rump as well as darker coat colour with less intensively red and yellow shades. The results comply with an eye evidence of differences in the coat colour pattern according to sex and pedigree data. However, genetic characterisation found admixture with other breeds besides Pinzgauer in some Cika animals which could influence the coat colour as well.

Genetic relationships accross different strata of data according to genomic herd descriptors

S. König and T. Yin
University of Kassel, Animal Breeding, Nordbahnhofstr. 1a, 37213 Witzenhausen, Germany;
sven.koenig@uni-kassel.de

In previous studies, reaction norm or random regression models have been applied to infer genetic parameters in dependency of continuous herd descriptors, e.g. average herd production levels. The aim of the present study was to use high-throughput genotypes (50K SNP chip panel) from 6,616 genotyped cows kept in large-scale contract herds (average 133.5 cows per herd) to create herd descriptors based on genomic architecture measurements: allele frequencies for SNP located close to a functional mutation, linkage disequilibrium measured at r2 for distinct chromosome segments, and average inbreeding coefficients. Furthermore, herd size, the percentage of daughters from specific sire groups, and phenotypic means for production and fertility traits were used as continuous environmental descriptors or as herd classification criteria. Random regression models were used to estimate genetic (co)variance components for test-day production traits and clinical mastitis including 267,393 repeated measurements. Correlations among the herd descriptors ranged between -0.27 (intra-herd inbreeding coefficient with herd non-return rate) and 0.47 (r2 on chromosome 14 with the intra-herd inbreeding coefficient). Most correlations among herd characteristics were close to zero, indicating independency of genetic, genomic and phenotypic herd descriptors. Genetic variances and heritabilities for same traits slightly altered on the continuous herd scale. Genetic correlations in same traits for neighbouring levels of herd descriptors were close to 1, but partly substantially declined for levels in greater distance. Genetic correlations lower than 0.80 in same traits between different levels of a herd descriptor indicate impact of genomic architecture on assessments for genotype by environment interactions. Alterations of genetic values in dependency of herd descriptors suggest utilization of specific sires for different herd structures, allowing new possibilities to improve selection and mating schemes.

Between breed variation in resistance to H. contortus in three indigenous goat breeds in Uganda

R.B. Onzima[1,2], R. Mukiibi[3], B.K. Katali[1], J.A.M. Van Arendonk[2], E. Kanis[2] and H. Komen[2]
[1]*National Agricultural Research Organization (NARO), P.O. Box 421, Kabale, Uganda,* [2]*Wageningen University, Animal Breeding and Genomics Centre, Droevendaalsesteeg 1, 6708 PB, Wageningen, the Netherlands,* [3]*University of Alberta, Department of Agricultural, Food, and Nutritional Science, 410 Agriculture/Forestry Centre Edmonton, AB T6G 2P5, Canada; robert.onzima@wur.nl*

Variation in resistance to the gastrointestinal parasite Heamonchus contortus was studied among three indigenous goat breeds of Uganda: Small East African (SEA), Mubende and Kigezi. Twelve male goats of each breed were purchased at approximately 7 months of age from smallholder farmers in Arua, Mubende and Kabale respectively, and assembled at a research station. At the station, they were dewormed with a combination therapy of the broad spectrum dewormers closantel and albendazole to free the goats from gastrointestinal parasites. During experimentation, the goats were kept indoors and ad libitum fed on clean banana peels. On attaining zero worm egg status, the goats were artificially infected with 18,000 third stage (L3) larvae of H. contortus prepared according Baermann's procedure. Data were collected on faecal egg count (FEC), packed cell volume (PCV) and body weight (BW) on a 2-weekly basis until 12 weeks post infection; and carcass weight and total worm count (WC) in the abomasum at termination of the experiment. The data on FEC, PCV and BW were subjected to repeated measures analysis of variance. Other data were analysed by one-way analysis of variance. FEC between breeds was only significantly different at 12 weeks post infection (P=0.04). There was generally a higher FEC recorded in Kigezi compared to SEA and Mubende. Carcass weight was significantly different among breeds (P<0.05), with Mubende having the highest carcass weight followed by Kigezi and SEA. PCV and daily weight gains were significantly different between breeds (P<0.05). WC was not significantly different between the breeds. FEC, PCV and daily weight gains (DWG) were significantly different between the breeds after artificial infection, suggesting differences in resistance to Heamonchus contortus, with Mubende showing higher resistance. These differences could be useful for designing breeding programs for resistance in goats.

The bovine genome-wide array as a source of novel SNPs in Red deer

R. Kasarda, N. Moravčíková, A. Trakovická and O. Kadlečík
Slovak University of Agriculture in Nitra, Animal Genetics and Breeding Biology, Tr. A. Hlinku 2, 949 76 Nitra, Slovak Republic; radovan.kasarda@uniag.sk

The present study was prepared in order to identify novel SNPs in deer based on cross-species genotyping and to characterize the basic level of genetic variability in population. The genome-wide data were obtained using Illumina BovineSNP50 v2 BeadChip from in total of 27 Red deer (Cervus elaphus). The quality control of genotyping data was conducted to remove any SNPs with more than 10% missing genotype. With respect to the generally accepted criteria for polymorphic markers only loci with minor allele frequency (MAF) higher than 0.01 were retained in following analyses. As expected the decrease of average call rate (61.66%) was found. From in total 54.55% of successfully genotyped SNPs up to 929 loci could be regarded as polymorphic and were informative for genetic variability estimation. In relation to the physical position of each marker in bovine genome most of informative SNPs were located on BTA2 (n=66). Conversely, the lowest proportion of useable loci was detected for BTA23 (n=11). Across all polymorphic markers the average value of MAF at level 0.19±0.10 was detected. Generally, the good level of observed heterozygosity across individuals (0.34) suggested also the value of inbreeding like Wright's F_{IS} coefficient (-0.19) characterizing genetic variability within population. Our study primarily showed that the application of cross-species genotyping can provide the valuable source for novel marker discovery without the need of whole genome sequencing. Despite the decrease of total identified loci number compared to bovine species if only 1% of loci on a 50k genotyping array are polymorphic then those 500 markers would constitute a significant increase of accessible genetic resources for non-model species in which no reference genome is available.

Selecting on fat and protein content instead of milk yield would not improve dairy cows' fertility

N. Bedere[1], C. Disenhaus[1], V. Ducrocq[2], S. Leurent-Colette[3] and L. Delaby[1]
[1]*INRA, AGROCAMPUS OUEST, PEGASE, Saint-Gilles, 35590, France, [2]INRA, AgroParisTech, Université Paris-Saclay, GABI, Jouy-en-Josas, 78352, France, [3]INRA, Domaine Experimental du Pin-au-Haras, Exmes, 61310, France; nicolas.bedere@rennes.inra.fr*

To compare reproductive performances of cows with high genetic merit for fat and protein content vs high genetic merit for milk yield, an experiment was conducted on 500 cows at the INRA farm of Le Pin-au-Haras. The herd was composed of Holstein and Normande cows. In each breed, cows with high genetic merit for milk yield were classified in a 'Milk Group' (MG) and those with high genetic merit for fat and protein content were classified in a 'Content Group' (CG) with equivalent milk solids production over the lactation. Two grazing-based feeding systems were used, a 'High' (maize silage / grazing plus concentrate) and a 'Low' (grass silage / grazing with no concentrate) system. To study the different steps of the reproductive process, milk progesterone information (3/week), oestrous behaviour recording and pregnancy diagnosis (ultrasonography) were combined. Within breeds, cows in the CG had an earlier resumption of luteal activity than cows in the MG (-6 d, P=0.01). Occurrence of prolonged luteal phases was not related to genetic groups. Both genetic groups had similar ovulation detection rate and proportion of standing behaviour. No difference in fertility was found between genetic groups in Normande cows. Holstein cows in the CG showed more non-fertilization/early embryo mortality in the Low system (+31 points, P=0.002) and more late embryo mortality/foetal death in the High system (+18 points, P=0.06). The MG was exporting more energy in early lactation because of a higher milk yield peak and the CG was exporting more energy in mid-lactation because of a better persistency. Trade-off between milk production and reproduction was not found: cows in the MG had a degraded cyclicity and those in the CG had a degraded fertility. Selecting dairy cows on fat and protein content may not be a promising way to maintain good productive and reproductive performances.

Impacts of parentage identification in the genetic evaluation of Nellore animals in a selection herd

L.R. Silva, J.P. Eler, E.C. Mattos and J.B.S. Ferraz
University of Sao Paulo/FZEA, ZMV/NAP-GMABT, Rua Duque de Caxias Norte, 225, Campus da USP, 13635-970 Pirassununga, SP, Brazil; jbferraz@usp.br

The correct assignment of paternity is important to the maintenance of correct pedigree information and consequently for the estimation of breeding values of candidate animals in selection programs. The use of SNP type markers has increased use for genomic selection and in different panels of parentage testing of many species. Parentage identification can be obtained for animals of different ages. This study had evaluated 490,118 Nellore (Bos indicus) beef animals where 7,995 had parentage assigned through DNA test and the greatest improvement of EPD accuracy was shown in younger animals, since they have few, if any, progeny phenotype information and it was up to 21% of increasing in EPD's accuracy. In general, the use of tests such as paternity correction increased the generation of genetic ties between the animals. All animals present in the genetic evaluation benefitted from that. Additionally, dams and sires showed gains in their accuracy of breeding value estimation. Selection conflicts were also analyzed for animals with complete phenotypic data (2011 crop) and showed that there are differences in the animals rank between both assessments. Higher conflicts were observed for weaning weight, but for the other traits, those conflicts fluctuated between 14% and 52% when considering the sum of both difference per trait and animal gender. The results for this study also suggest that a DNA paternity test is a tool that can increase the accuracy of pedigree and increase the performance of genetic evaluation in a beef breeding program. And another benefit can be extracted in a first moment is that it can help in the correction of management system at farm, since mistakes can happen by wrong collecting, annotation or interpretation data, contention of animals in their field and losing of data.

Identification of reproductive trait loci by GWAS of Large White pigs from three breeding population

K. Suzuki[1], R. Otsu[1], K. Katoh[1], Y. Suda[2], E. Kobayashi[3], S. Mikawa[4] and M. Taniguchi[4]
[1]Graduate School of Agricultural Science, Tohoku University, 1-1Tsutsumidori-Amamiyamachi, Aoba-ku, Sendai, 981-8555, Japan, [2]Miyagi University, Sendai, Miyagi, 982-0215, Japan, [3]National Institute of Livestock and Grassland Science, Tsukuba, Ibaragi, 305-0901, Japan, [4]National Institute of Agrobiological Sciences, Tsukuba, Ibaragi, 305-0901, Japan; k1suzuki@bios.tohoku.ac.jp

The purpose of the present study is to identify chromosomal regions and genetic markers that control variation in the reproductive traits with a low heritability such as total number of piglets born (TNB) and total number of piglets born alive (NBA) of the four different pigs population of large white pigs by genome wide association study. Firstly, we estimated heritability and breeding value for reproductive traits such as TNB and NBA using the record of each of the four populations (A, B, D, and D). The farm, parity, year, and month were considered as fixed effects and repeatability animal model was used. Blood was sampled from 475, 453, 323 and 157 sows from each population to extract DNA. Samples of 700-1000 ng with an A260/280 ratio higher than 1.80 and a concentration >20 ng/l were genotyped using the Illumina PorcineSNP60 BeadChip containing 64,232 SNP (Illumina, San Diego, CA). PLINK software was used for the quality control of the identified single nucleotide polymorphisms (SNPs). The exclusion criteria for SNPs were as follows: minor allele frequency, <0.01; call rate, <0.95; and Hardy–Weinberg equilibrium, <0.001. Of the 61,565 SNPs analyzed, 37,234 SNPs were selected for association analyses. The analyses were implemented using PLINK software for each trait. As a result of the genome wide association analysis, a high peak was detected on the chromosome 7 (51.5-53.1 MB, 134.5-134.6 MN) and chromosome 15 (22-23.5 MB) for TNB and NBA in the A population. In this area (134.5-134.6 MB) of the chromosome 7, the GCM1 gene existed. Differences of about 1.6 piglets were admitted between SNP maker genotype. Neither difference between marker genotype nor a candidate gene existed in other population though the area of high peak existed.

Defining a breeding objective for Nile tilapia for Kenyan smallholder production system

S.K. Omasaki[1,2], J.A.M. Van Arendonk[3], A.K. Kahi[1] and H. Komen[2]
[1]Egerton University, Department of Animal Sciences, Egerton University, P.O Box 536, 20115, Kenya, Kenya, [2]Wageningen University, Animal Breeding and Genomic Group, P. O. Box 338, 6700 AH Wageningen, the Netherlands, [3]Hendrix Genetics, Hendrix Genetics Research, Technology and Services, P.O. Box 114, 5830 AC Boxmeer, the Netherlands; simion.omasaki@wur.nl

Selective breeding is an important driving force for increased productivity in livestock and fish, and defining the breeding objective is the first requirement for genetic improvement program. However, developing countries tend to be highly diverse in terms of agro-ecological conditions and market orientation. This makes a formal definition of breeding objective difficult. The aim of the present study was to derive a breeding objective that incorporates and balances attributes of Nile Tilapia (Oreochromis niloticus) that fish farmers consider important given their diverse production systems and economic constraints. A survey was conducted among 100 small-holder Nile tilapia farmers in Kenya to obtain preference values for traits of economic importance, by using multiple pairwise comparisons. Individual, group and consensus preference values were estimated using Analytical Hierachy Process and Weghting Goal Programming. Results: Low and medium income farmers prefered harvest weight (HW) while high income farmers considered growth (GR) and survival (S) as the most important traits. Fingerling producing farmers prefered GR and S than HW. For table size producing farmers, HW was highly prefered, followed by GR and S, while farmers that produced both table size fish and fingerlings prcfcrcd S morc than HW and GR. We concluded that farmers preferences for traits differ depending on income or market objectivc, which needs to be considered in defining breeding goal. We have shown that desired gains can be derived from the estimated consensus values that take into account farmers diverse preferences across the groups to develop a breeding objective for O. niloticus for smallholder production systems.

Genetic parameters for susceptibility to bovine herpesvirus-1 in dairy cattle

S.C. Ring[1,2], R.G. Sayers[1], N. Byrne[1], E. O'Brien[1], M.L. Doherty[2] and D.P. Berry[1]
[1]*Teagasc, Moorepark, Fermoy, Co. Cork, Ireland,* [2]*UCD, School of Veterinary Medicine, Belfield, Dublin 4, Ireland; siobhan.ring@teagasc.ie*

Infectious bovine rhinotracheitis (IBR) is a viral respiratory disease of cattle caused by infection with bovine herpesvirus-1 (BoHV-1). The objective of this study was to quantify the contribution of additive genetics to variability in susceptibility to BoHV-1 in Irish dairy cattle. BoHV-1 antibody test results from two field trials were available for 16,294 female cattle from 81 dairy herds between the years 2010 and 2015, inclusive. Antibody testing was performed on individual blood samples using either an ELISA gB or gE test kit; the majority or tests (70.58%) were analysed using the gE test kit. For comparative analysis of test result values and kits simultaneously, standardised test result values were derived. Only animals born in the herd where they were blood sampled were retained. Within each herd, animals were deemed exposed to BoHV-1 if they were born and tested in the same herd-year as ≥1 animal with a positive BoHV-1 test result. Only the most recent test result per animal with a known sire in a contemporary group (CG) (i.e. herd-year-season of calving or herd-year-season of birth) of ≥5 animals and ≥2 sires were considered further. After all data edits, 6,685 animals in 312 contemporary groups from 60 herds remained. Variance components for BoHV-1 (standardised test value or the binary trait of positive or negative) were estimated using animal linear mixed models. Fixed effects included in the model were CG, parity, age relative to parity median, sample date × test kit interaction, heterosis and recombination coefficients; animal was included as a random effect. The direct heritability (standard error in parenthesis) for susceptibility to BoHV 1 as a binary and as a continuous trait was 0.11 (0.02) and 0.13 (0.03), respectively; the genetic standard deviation for the binary and continuous trait was 0.11 and 0.27 standard deviation units, respectively. This study indicates that genetic variation exists in susceptibility to BoHV-1 and selection of more resistant animals to be parents of the next generation may reduce the susceptibility of their offspring to BoHV-1.

Egg physicochemical and fatty acids composition of Portuguese autochthonous chicken breeds

M.L. Soares[1,2], J.C. Lopes[2], G. Thompson[1,3] and J. Carvalheira[1,3]
[1]*ICETA/CIBIO, University of Porto, Rua Padre Armando Quintas, no. 7, 4485-661 Vairão, Portugal,* [2]*ESAPL-Escola Superior Agrária, Ponte de Lima, 4990-706 Refóios, Portugal,* [3]*ICBAS, University of Porto, Population Studies, Rua de Jorge Viterbo Ferreira no. 228, 4050-313 Porto, Portugal; jgc3@mail.icav.up.pt*

The aim of this study was to determine egg physicochemical (PC) and fatty acids (FA) composition from 3 Portuguese autochthonous chicken breeds located in the Norwest region of continental Portugal (Preta Lusitânica-PL, Amarela-AM and Pedrês Portuguesa-PP), currently classified as rare and particularly endangered. The birds were raised under the traditional production system (pens with open parks), with corn provided ad libitum. Fifty eggs per breed were used to determined pH, protein, moisture and ash in the yolk and albumen, while lipids and FA were determined only in egg yolks. The lipids were determined by Folch method and the FA by direct transmethylation using the Wang method and then, by chromatography analysis. Statistical differences between breeds were inferred using a fixed linear model. This study revealed that the PC parameters of the 3 breeds were within the values referred in the literature. Significant differences (P<0.05) between breeds were found for protein (lipids) content in the yolk and varied between 16.03% (33.48%) for PP and 17.89% (30.35%) for PL, respectively. The breed differences in protein content from the albumen were also significant (P<0.05) and ranged between 10.61% (AM) and 11.03% (PP). Regarding the FA, the oleic acid was the most abundant, followed by the palmitic, stearic and linoleic acids. Polyunsaturated acids of the n-6 series were mostly represented by linoleic and arachidonic acids. The proportion between unsaturated (MUFA+PUFA) and saturated (SFA) FA was 2:1, similar to what is found in general literature. The n-6/n-3 ratio ranged from 14:1, 15:1 and 16:1 for AM, PL, and PP breeds, respectively. As expected, feeding was the key factor in the lipid composition. Diets based on the supplementation of maize alone cannot achieve the best ratios of n-6/n-3, since this cereal is rich in n-6 PUFA and relatively poor in n-3 PUFA.

Providing dairy cow facilities to improve health, welfare, and productive life

T.J. Devries[1], M.A.G. Von Keyserlingk[2] and D.M. Weary[2]
[1]University of Guelph, Animal Biosciences, 50 Stone Road East, Guelph, Ontario, N1G 2W1, Canada,
[2]University of British Columbia, Animal Welfare Program, 2357 Main Mall, Vancouver, BC, V6T 1Z4, Canada;
tdevries@uoguelph.ca

Poorly designed and managed facilities cause injuries and increase the risk of health problems, including lameness and other infectious and metabolic diseases, whose high prevalence is a welfare challenge facing the dairy industry. Despite best efforts, and many scientific improvements, in terms of preventing these problems through genetic selection, it appears that high incidence of these issues are related to cattle housing and management. Housing systems for dairy cattle are provided with the aim of providing a comfortable environment for these animals – one that ensures adequate rest, protection from climatic extremes, and free access to an appropriate, well-balanced diet. Despite these laudable aims, housing systems do not always function well from the perspective of the cow – poorly designed and maintained facilities can reduce feeding and lying time, increase competitive interactions, cause injuries, and increase the risk of health disorders. This review will focus on examples from empirical work on the feeding, standing, and lying areas utilized by dairy cattle, and show how these can be better designed and managed to meet the behavioral needs of dairy cattle, and thus improve welfare, health, longevity, and in some cases, productivity. Much of the research reviewed has evaluated housing systems from the cow's perspective by asking how the housing affects cow health, what elements of the housing environment the cow prefers, and how the these elements affect behavior and welfare.

Timing, costs and incidences of recorded health events on NZ dairy farms

P.R. Amer[1], J.I. Kerslake[1], N.A. Dennis[1], J.R. Roche[2] and C.V.C. Phyn[2]
[1]AbacusBio Ltd, P.O. Box 5585, Dunedin 9058, New Zealand, [2]DairyNZ, Cnr Ruakura & Morrinsville Roads,
Newstead, Hamilton 3240, New Zealand; pamer@abacusbio.co.nz

In dairy herds, suboptimal cow longevity and lifetime productivity due to on-farm mortality (euthanasia and death) and culling is indicative of poor animal health and welfare, and causes financial losses for farmers. This study investigated the timing, incidences and costs of different animal health disorders, culling events and on-farm deaths (as recorded in on-farm software by farmers) in the pasture-based New Zealand dairy industry. Costs of animal removal and death due to various health disorders were quantified using a bioeconomic model, which used derterministic predictions of the age composition within herd to estimate the financial impacts of culling and on-farm mortality of cows at different ages and stages of lactation. The model included assumptions about treatment costs and the cost of discarded milk associated with various health and removal events, The model also accounted for the effect on milk production from high-producing cows being more prone to some health disorders. The costs of health disorders were then combined with incidences of animal removal and disease treatment events to create an aggregate economic cost per disease across the national herd. Cow culling and death were esimtated to cost the industry NZ$1.2B per year, with reproductive failure and unknown reasons contributing to the bulk of this cost (NZ$356M and NZ$402M, respectively) while the cost of culling and death due to health-related disorders was estimated at NZ$295M. The main contributors to the latter were mastitis (17%), udder problems (14%), injured/accident (11%), high somatic cell count (10%), calving trouble (9%) and disease (9%), which totaled NZ$201M per year. Milk production losses and treatment expenses associated with clinical health disorders cost the industry NZ$354M annually although if the estimated effects of health events on delayed calving and survival were included, this cost increased to NZ$519M per year. These estimates of financial loss to the industry support the call for further reseach-led solutions that improve dairy cow health and suvival.

Investigating the use of Mid Infrared spectral data to predict dairy cow cellular immune traits

S.J. Denholm[1], S.L. Smith[1], T.N. McNeilly[2], V. Hicks[3] and E. Wall[1]
[1]SRUC, Animal & Veterinary Sciences, Roslin Institute Building, Easter Bush, Midlothian, EH25 9RG, United Kingdom, [2]Moredun Research Institute, Pentlands Science Park, Bush Loan, Midlothian, EH26 0PZ, United Kingdom, [3]NMR, Fox Talbot House, Chippenham, SN15 1BN, United Kingdom; scott.denholm@sruc.ac.uk

It has been demonstrated that Mid Infra-Red (MIR) spectroscopy of milk samples can be used as an increasingly effective low cost tool for rapid measurement of traditionally expensive and difficult to record phenotypes, such as milk fatty acids, body energy profiles and methane emissions. The objective of the present study was to investigate the capability of using MIR spectroscopy analysis of milk samples as a predictor of cellular immune traits in Holstein-Friesian dairy cows. Cows from the Langhill research herd (n=358) housed at SRUC's Dairy Research Centre (Scotland) were repeatedly sampled over a 3 year period for a range of cellular immune traits. All cows were involved in a long-term selection experiment in a 2×2 approach (genetic line × feeding system). Blood samples were collected on 15 separate occasions and included summer and winter samplings. Samples were analysed by flow cytometry for innate and adaptive immune traits including % T cell subsets, B cells, NK cells and granulocytes. MIR spectral data (from morning milk samples) were aligned to the immune profile of each cow within ±1 week of the milk sampling date, providing a reference dataset. This dataset was analysed by principal component (PCA) and partial least squares (PLS) analysis using SAS 9.3. Prediction accuracy was determined using receiver operating characteristic (ROC) curves carried out using Python 3.4. Positive preliminary results were obtained for CD4$^+$, NKp46$^+$, % eosinophils, % lymphocytes and % PBMC with moderate accuracies of 70%, 68%, 79%, 72%, and 70% respectively. Previously, we found these traits to be associated with fertility, health and welfare events. Results show the potential use of MIR spectroscopy of milk samples in predicting the immune profile of an animal at the individual and population level. If successful this would provide a low-cost, non-invasive tool to aid in the management and improvement of animal health and welfare.

The French observatory on genetic defect: an assessment after 15 years of operation

C. Danchin[1], C. Grohs[2] and the Onab Steering Committee[1,2]
[1]Institut de l'Elevage, 149 rue de Bercy, 75595 Paris Cedex 12, France, [2]INRA GABI, Domaine de Vilvert, 78350 Jouy en Josas, France; coralie.danchin@idele.fr

Livestock breeds have a narrow genetic basis compared with Humans. Therefore, since all animals are carriers of deleterious mutations, the likelihood that offspring are born with a genetic defect is quite high in cattle: breeding organizations had to face multiple crisis due to genetic defects such as CVM or BLAD (Holstein breed). Consequently a French National Observatory of Cattle Abnormalities (ONAB) was set up in 2002. Over a15 year period, more than 5,000 defects were reported to the observatory, leading to the detection of about 20 genetic disorders. Recent discoveries were made on diseases of major economic importance such as Seckel-like syndrome in the Montbéliard breed, ataxia in the Charolais breed or progressive blindness in the Normande breed. About a dozen of other genetic disorders are under current investigation. This number has been increasing recently since the implementation of genomic selection provides very efficient tools to localize haplotypes and mutations. After 15 years of operating, some lessons can be learnt on how to set up an efficient observatory. Our experience showed that it is more efficient to motivate a network to collect data by having them report specific phenotypes instead of random ones: spontaneous declarations on sporadic defects are quite scarce and usually limited to stunning defects such as conjoined twins. Therefore a first step is to have a first level of motivated 'watchmen', such as inseminators, breeders' technician or veterinarian that will report to the observatory the emergence of a new abnormality. Further insights show that this network needs to be organized in the same way that most participative sciences are. For instance, for years, most volunteers that reported cases were quite volatile. We managed to overcome this tendency by disseminating more information through newsletters, a specific website and privileged partnership with our best declarers. In the end, despite a quite low number of cases reported, the ONAB achieved to have successful results thanks to an efficient management of a selected field network, which provides useful DNA samples and accurate phenotypes to researchers, leading to the discovery of new mutations.

A stochastic model to investigate effects of management on calves exposed to Ostertagia ostertagi

Z. Berk[1], Y.C.S.M. Laurenson[2], A.B. Forbes[3] and I. Kyriazakis[1]
[1]School of Agriculture Food and Rural Development, Newcastle University, Newcastle upon Tyne, NE1 7RU, United Kingdom, [2]Animal Science, School of Environmental and Rural Science, University of New England, Armidale, New South Wales, 2351, Australia, [3]Scottish Centre for Production Animal Health and Food Safety, School of Veterinary Medicine, University of Glasgow, Glasgow, G61 1QH, Scotland, United Kingdom; z.berk@newcastle.ac.uk

Predicting the effectiveness of parasite control strategies requires accounting of the responses of individual hosts and the epidemiology of parasite supra and infra-populations. The first objective was to develop a stochastic model that predicted accurately the interaction between a group of first season grazing calves challenged by Ostertagia ostertagi, by considering phenotypic variation amongst the calves and variation in parasite supra-population. The second objective was to assess model behaviour against variation in parasite supra-population and stocking rate. The model showed initial pasture infection level to have little impact on parasitological output traits, such as worm burdens and FEC, or calf performance, whereas increasing stocking rate had a disproportionately large effect on both parasitological output and performance traits. At high stocking rate the intensity of hosts results in lower grass availability and increased total egg excretion, causing a more dramatic rise in pasture contamination. Subsequently, model predictions were validated against published data taken from experiments on common control strategies, such as stocking rates, 'dose and move' or strategic treatment with anthelmintic at specific time points of the grazing season. Model predictions showed a reasonable likeness to observations in most cases, reinforcing model accuracy. The major challenge encountered was in accounting for concurrent infections with Cooperia, although this does not affect parasite infra-populations, there is an effect on bodyweight. A simple consideration was made for this due to a lack of data to validate an additional model component. The stochastic model developed is flexible and can be used to predict the consequences of other nematode control strategies, such as Targeted Selection Treatments, on a group of grazing calves.

Genetic parameters for welfare and disease indicator traits generated in automatic milking systems

L. Santos, K. Brügemann and S. König
University of Kassel, Department of Animal Breeding, Nordbahnhofstraße 1a, 37213 Witzenhausen, Germany; lsantos@uni-kassel.de

The aim of this study was to identify new functional traits reflecting welfare, temperament and health generated in automatic milking systems (AMS), and to infer genetic relationships among those novel and conventional traits. Data recording spanned a period of 30 days, and included 58,821 observations (= entries into the AMS) from 884 Holstein cows kept in three German farms using the same AMS type. AMS traits were combined with test-day records within the 30 day interval, and with non-return rates after 56 (NR56) and 90 days (NR90). AMS behaviour traits included the counts for the daily milking robot visits defined as a binary trait 'VIS3' with VIS3 = 1 for at least three visits, and the binary trait 'VIS2' using a threshold of two daily visits. Additionally, the interval (INT) between two consecutive AMS visits (in hours), average milk flow (AMF in kg milk / minute), duration of the automatic milking process (DUR in minutes), and knock off of the milking device (KO) from one or more udder quarters (binary trait) reflect dairy cow temperament. Milk conductivity (CON) from all udder quarters represents an udder health indicator trait. Linear models and GLMM with a logit link function were applied to Gaussian and binary traits, respectively, in bivariate models for genetic analyses. Heritabilities for CON varied between 0.28 and 0.52, depending on the udder quarter. The heritability for DUR was 0.14, for INT 0.18, for KO 0.17, for VIS3 0.11, and for VIS2 0.16. An increasing no. of cow visits in the milking robot was associated with an increase for KO (r_g=0.18), indicating antagonistic relationships between different types of cow temperament. Genetic correlations between NR56 and NR90 with VIS3 were negative (-0.93 and -0.45, respectively). Cows with higher MY visited the AMS more often (r_g=0.34). An increase of automatically recorded data, combined with a moderate genetic background, justify modifications of dairy cattle breeding goals towards higher emphasis on behaviour and welfare.

Impact of the Friesian POLLED mutation on milk production traits in Holstein Friesian

L. Gehrke[1], D. Seichter[2], I. Ruß[2], I. Medugorac[3], J. Tetens[1] and G. Thaller[1]
[1]*Institute of Animal Breeding and Husbandary, CAU, Kiel University, 24098 Kiel, Germany, [2]Tierzuchtforschung e.V. München, Gendiagnosezentrum, Senator-Gerauer-Str. 23, 85586 Grub, Germany, [3]Chair of Animal Genetics and Husbandry, LMU, Veterinärwissenschaftliches Department, Veterinärstraße 13, 80539 München, Germany; lgehrke@tierzucht.uni-kiel.de*

Serious animal welfare concerns regarding dehorning of calves have led to an increasing interest in breeding genetically hornless (i.e. polled) cattle. Intriguingly, the locus which is supposed to be causative for the polled phenotype seems to influence other traits than horn development as well. Previous studies found differences in performance of polled and horned Holstein Friesian cattle. However, it is unclear if these differences are caused by the POLLED mutations. Aim of this study was to investigate whether a difference in milk production traits of Holstein Friesian can be associated with the Friesian POLLED mutation. For this purpose, genetic material of 478 daughters of eight heterozygous polled sires carrying the Friesian POLLED allele was collected and a direct gene test for polledness was performed. Pearson's Chi Squared test was performed to ascertain that the observed proportion of horned and polled daughters per sire is equal. Yield deviations (YD) of the traits milk, fat and protein yield, fat and protein content and somatic cell score were used as phenotypic data. In addition, the relative estimated breeding value for milk production (RZM) as well as for somatic cell score (RZS) were compared. Using a multi-factor analysis of variances, significant inferiority of polled daughters was found for the traits milk, fat and protein yield. The estimated effects of the polled group were -100, -4.21 and -2.84 kg, respectively. The reason for the correlation between polledness and milk performance remains unclear and the molecular mechanism leading to the polled phenotype is not yet clarified. Therefore, further investigations of the stated negative correlation and of additional possible epistatic interactions between the POLLED locus and production traits will be carried out.

Genetic parameters for F. heptacia in Irish cattle

A.J. Twomey[1,2], R.G. Sayers[2], R.I. Carroll[3], N. Byrne[2], E. O'Brien[2], M.L. Doherty[1], J.C. McClure[4] and D.P. Berry[2]
[1]*School of Veterinary Medicine, UCD, Belfield, Dublin 4, Ireland, [2]Moorepark, Teagasc, Fermoy, Co. Cork, Ireland, [3]AHI, Carrick on Shannon, Co. Leitrim, Ireland, [4]ICBF, Bandon, Co. Cork, Ireland; alan.twomey@teagasc.ie*

The objective of this study was to generate the first international genetic parameter estimates for susceptibility to F. hepatica in dairy cows. Abattoir data on the presence of live F. hepatica or F. hepatica damaged liver without the presence of live F. hepatica were available between the years 2012 and 2015, inclusive. A second dataset was available on cows from 68 selected dairy herds with a blood ELISA test from Autumn 2015. Cows with recorded inter-herd movements >90 days of age were not considered. Cows were defined as exposed by using herd-mate F. hepatica status and only exposed cows were retained for analysis. The abattoir data and study herd data consisted of 16,734 and 6,907 dairy cows, respectively. Variance components for the binary trait of F. hepatica infection in both datasets were estimated using animal linear mixed model. Fixed effects included in the model for both datasets were contemporary group, heterosis, recombination loss, parity, age relative to parity and stage of lactation. An additional fixed effect of abattoir-by-date of slaughter was included in the model for the analysis of the abattoir data. Direct additive genetic effects and a residual effect were included as random effects. The prevalence of F. hepatica in abattoir data and study herd data was 48% and 27%, respectively, prior data edits; the respective prevalence after data edits was 47% and 37%. The heritability (standard error in parenthesis) of F. hepatica in abattoir data and study herd data was 0.03 (0.01) and 0.09 (0.02), respectively. The additive genetic standard deviation for F. hepatica was 0.069 and 0.112 units for abattoir data and study herd data, respectively. Heritable genetic variation exists for susceptibility to F. hepatica in dairy cows.

Estimating heritabilities for fertility disorders using on-farm recorded health data

V. Müller[1], K.F. Stock[2], R. Schafberg[1] and H.H. Swalve[1]
[1]*Martin-Luther-University Halle-Wittenberg, Institute of Agricultural and Nutritional Sciences, Th.-Lieser-Str. 11, 06120 Halle, Germany,* [2]*Vereinigte Informationssysteme Tierhaltung w.V., Heinrich-Schröder-Weg 1, 27283 Verden, Germany; vincent.mueller@landw.uni-halle.de*

The project GKUHplus focusses on the systematic recording and use of producer and/or veterinarian recorded data pertaining to diagnoses of health disorders in dairy herds. The project aims at using the data as an aid in herd management and as basis of genetic-statistical analyses including the estimation of breeding values for health traits for A.I. sires. Data collection is based on on-farm documentation using the German standardized central key for health data recording. For the present study, 741,776 diagnoses from 140 farms were examined. After editing for completeness of recording over time, matching with production data and size of sub-cell within sire and herd-year-season, 22,010 lactations of 17,592 cows from 426 sires and on 72 farms were retained. For the genetic analyses, three fertility traits were defined: A global disorder trait pertaining to any diagnosis (GD), disorders recorded as 'sterility' (SD), and disorders observed post-partum (DPP). Lactation incidences were 45.7% for GD, 36.1% for SD, and 17.6% for DPP. Repeated diagnoses within lactation were treated such that a first diagnosis was treated as one case with a 'sickness-period' of 21 days. Any further diagnosis within this time span was taken as associated with the same health event, and prolonged the 'sickness-period' by ten days when occurring in the second half of the initial 21 d-sickness period. Further diagnoses outside this period were considered as new cases. Heritabilities were estimated using AS-REML 3.0 applying an animal model with parity (1-3) and herd-year-season as fixed effects. Random effects were animal genetic effect and a permanent environmental effect. Estimates for all three traits were low, i.e. from 0.049 to 0.098 under a threshold model and from 0.035 to 0.078 under a linear model. In a next step, random regression models will be applied attempting a better fit for the lactation trend.

Genetic correlations of fighting ability with udder health and longevity in Valdostana cattle

C. Sartori, N. Guzzo and R. Mantovani
University of Padua, Department of Agronomy, Food, Natural Resources, Animals and Environment, Viale dell'Universita' 16, 35020, Italy; cristina.sartori@unipd.it

The ability to fight for resources and establish dominance relationships may be stressful for animals and costly in terms of immune functions and health. Evidences of a depression in health traits and in longevity for more competitive individuals have been found in some species also at genetic level. Valdostana cattle have been selected for fighting ability (FIGH) in traditional competitions for centuries, and antagonistic genetic correlations with milk yield and fertility have been already documented. This study investigated the genetic relationships of FIGH with udder health traits and longevity. Data obtained in 15 years of competitions (referring to 13,000 cows and 24,500 animals in pedigree) were merged with test-day data for somatic cell count (SCC) routinely recorded and morphological scores assigned to primiparous cows. Somatic cells were both measured as normally distributed score (SCS=log2(SCC/100,000)+3) and as a threshold trait for the occurrence of pathological mastitis (SC150=SCC>150,000 cells/ml), whilst longevity was measured as length of productive life from the first calving. Moderate but positive genetic relationships were found (r=0.10) between FIGH and both SCS and SC150, suggesting that greater fighting ability is genetically related to increased somatic cells. The positive genetic relation of longevity and FIGH (r>0.50) reflects the worth of fighting ability for farmers and suggests that fighting ability does not exert a negative effect on this trait. Breeding values for both FIGH and longevity showed a small positive trend in years, whereas somatic cells showed no variations. Thus, the current genetic improvement for fighting ability in addition to traditional milk and meat has not caused a depression in longevity and udder health, but the genetic worth for FIGH has increased the herd life of cows. Nonetheless, the low but negative genetic relationships estimated with udder health traits may warn about possible further detrimental effects on some functional characteristics.

Correlations between health and performance traits in Vorderwaelder and Fleckvieh cattle

H. Hamann[1], S. Gehring[2] and P. Herold[1,2]
[1]*State-Office for Geo-information and Rural Development Baden-Wuerttemberg, Geodata-Center, Stuttgarter Str. 161, 70806 Kornwestheim, Germany,* [2]*University of Hohenheim, Insitute of Animal Science, Farm Animal Genetics and Breeding, Garbenstraße 17, 70599 Stuttgart, Germany; henning.hamann@lgl.bwl.de*

Breeding values for health traits are estimated for Fleckvieh and Brown Swiss cattle in Austria, Bavaria and Baden-Wuerttemberg since 2010 and 2013, repectively. Further research is done on improving quality of available data by integrating observations close to birth (OCB) which are monitored by the farmers on a voluntary basis. Also, the integration of data from regional breeds, like the Vorderwaelder cattle, has to be proven. Additionally, genetic correlations between health and performance traits in the different breeds are of interest. Health data is available from the health monitoring program in Baden-Wuerttemberg. Performance and pedigree information is available from breeding value estimation. Observations close to birth are monitored in the HIT-data base. The analysis consisted of two different steps: (1) comparison of health data and OCB and (2) correlations between health and performance traits based on phenotypic data. The health traits included in the study were mastitis, ovary cysts, milk fever, retained placenta and ketosis. OCB are available for milk fever and retained placenta. Investigated influencing factors are region, farm management (organic/non-organic), herd size, herd performance level. The integration of OCB into the health data analysis is a valuable complementation and improves the data quality. Differences between the two breeds in data recording but also in trait levels could be proven: Fleckvieh cattle showed significantly lower incidence rates in ovary cysts and retained placenta than Vorderwaelder. The estimated heritabilities for the five direct health traits in Fleckvieh were very low as expected, while the corresponding values in Vorderwaelder were slightly higher. In Fleckvieh, genetic correlations between ketosis and traits like fat content, protein content, fat protein ratio and persistency were high and as expected. Therefore, results are promising for application of health monitoring also in smaller populations as contribution for estimation of robustness.

Pleiotropic effects in functional traits in cattle

G. Mészáros, R. Taferner and J. Sölkner
University of Natural Resources and Life Sciences, Vienna, Department of Sustainable Agricultural Systems, Gregor Mendel Str 33, 1180 Vienna, Austria; gabor.meszaros@boku.ac.at

Pleiotropy is a phenomenon of a single gene influencing two or more distinct phenotypes. In this study we have searched for pleiotropic effects between functional traits in cattle, i.e. longevity, fertility, stillbirth and calving ease. Deregressed breeding values with reliability of 0.3 or higher from the German-Austrian Fleckvieh population were used as phenotypes. The data set consisted of 41,889 autosomal SNPs after quality control and 5,673 to 7,384 animals, depending on trait combination. Computations were performed with the R package CAPE, implementing a method of directed interaction networks between genetic variants for predicting the influence of genetic interference on phenotypes. In order to maximize linear independence of phenotypes, singular value decomposition were performed to obtain eigentraits, which were analyzed using multivariate linear regression with intercept, main effects and interaction term for each pair of markers. Variant-to-phenotype effects were derived from significant effects after correction for multiple testing. Several pleiotropic regions were detected for each of the six trait pairs, the most for calving ease and longevity (11 regions), the least for stillbirth and fertility (four regions). Regions around 9.8 Mb, 23-24 Mb, 27 Mb on BTA14, 17.5 Mb on BTA17 and 2-3 Mb on BTA21 were found in multiple trait pairs, with the most prominent representation of region 23-24 Mb on BTA14 in all but two trait pairs (fertility – calving ease, fertility – longevity). Regions with multiple occurrences are hypothesized to have special importance in functional traits, containing genes influencing neural development (KCNQ3), growth and birth weight (XKR4, LYN, SOX17, PLAG1 and others), spermatogenesis and fertility (CLGN), immune response (ELMOD2), thermoregulation (UPC1 and TBC1D9) and fetal growth (UBE3A). The direction of the effects was the same for most pleiotropic regions, although there were also cases of antagonistic relationships between traits.

Producing and marketing quality beef: a meat science perspective
C.R. Calkins
University of Nebraska-Lincoln, Animal Science, Lincoln, NE, USA; ccalkins1@unl.edu

Research has shown that high quality beef can be produced from young animals finished on high-energy diets. In the U.S., high quality beef means meat that is tender, flavorful, and with high marbling. A production system that combines proper management strategies for the live animal with appropriate use of feed resources and selected genetics can lay the foundation for high-value beef. Carcass weight and subcutaneous fat protect against cold shortening (toughening) and marbling instills a rich, savory flavor. Young animals have more soluble connective tissue, leading to better tenderness. Timely supplementation of the calf with a high-energy diet can result in higher marbling content at harvest. Opportunities exist to add value by altering the fabrication process such that muscles can be merchandized for optimal use. Multi-muscle cuts, when the muscles differ widely in their eating quality and composition, drive consumers to cook to the lowest common denominator, reducing overall value. Single-muscle merchandising can improve value and increase consistency in the eating experience. It is well known that ageing beef will improve tenderness. Packaging systems can alter eating quality with time. A consistent production system designed to optimize eating quality can assure consumer satisfaction and improve value.

Supply chain integration; the key to delivering a consistently good product
R. Law
Dunbia, Granville Industrial Estate, Dungannon, BT70 1NJ, United Kingdom; info@dunbiafarmers.com

There are many measures of meat quality but recent research suggests that it is important to deliver a product that is safe and one that consistently delivers a good eating experience to the consumer. Consistency in meat quality is one of the biggest challenges within the industry, especially in the red meat sector. The factors affecting meat quality, and more specifically consistency, are multifactorial and range from processes within primary production and processing to cooking. Failure to link these processes increases the risk of inconsistency and ultimately the risk of a poor eating experience for the consumer. Supply Chain Integration (SCI) is one approach to reducing this risk and involves closer alignment and coordination within the supply chain with the aim of reducing variability at each stage of production. Furthermore, SCI delivers improved data flow within the supply chain and a more coordinated approach to improving economic sustainability. The next stage would be vertical integration whereby the supply chain of a company is actually owned by the company delivering complete control, from sourcing the raw materials to delivering the final product to the consumer. This type of integration within a supply chain introduces high levels of consistency at each stage of supply which inevitably will deliver a much more consistent product to the consumer.

Session 47

Theatre 3

Benefits of an integrated supply chain to the farmer

M. Jones
Trefnant Hall Farm, Berriew, Welshpool, Wales, United Kingdom

At Trefnant Hall farm we are currently involved in two integrated supply chains within in the red meat industry. The first involves the Co-operative supermarket and Dunbia as the processor. Within this scheme calves are sourced directly from the Co-operatives milk producer's with the aim of delivering beef from these calves back to the Co-operative supermarket once finished at approximately 15 months of age. As a calf rearer who rear's calves from 2 weeks of age through to 3 months of age the scheme offers many benefits. Ownership of the calves is retained by the processor; Dunbia, alleviating any capital requirements for calves therefore allowing easy establishment of the business. Furthermore, the contract is performance driven meaning that my business is rewarded for high performance. The business is also provided with an excellent network of support from the processor and retailer which includes training, health and nutrition protocols and a helpline which facilitated high performance, maximising returns for the business. Finally the scheme has allowed dairy farmers to visit the calf rearing units to see their calves, this has allowed feedback to go back to the farmers to encourage best practice. The second supply chain involves the supply of lamb to Sainsbury's (supermarket). Members of the group have to be members of Wales Federation of Young Farmers Clubs. The farmer then receives a premium for the lamb but Wales YFC also receives a premium to aid development of the scheme and to use for education and increasing the efficiency of the suppliers. Dunbia again processes the lamb but it is also involved in running the group between Sainsbury's and Wales YFC. This allows the young farmers to see all stages of the supply chain and educate them to produce the correct product for the retailer. The group also targets efficiency and animal health and welfare with the aim of lifting the skills of the group as a whole.

Session 48

Theatre 1

Developing a 5 tonne sow: Biologically Possible or Commercially Probable?

G.A. Walling[1] and E.F. Knol[2]
[1]JSR Genetics Limited, Genetics and Reproduction, JSR Genetics Limited, Driffield, East Yorkshire, YO25 9ED, United Kingdom, [2]Topigs Norsvin Research Center, Research, Schoenaker 6, 6641 SZ Beuningen, Netherlands Antilles; grant.walling@jsrgenetics.com

With forecasted increasing global demand for pork and a need for producers in developed markets to increase the meat sold to remain profitable, there is significant requirement to increase farm output. Based on current carcase weights (95 kg deadweight) elite sows can produce in excess of 3 tonnes of carcases sold per annum. During the last two decades pig breeding companies have achieved increases in this figure (and can continue doing so) by (1) increasing litter size, however unfavourable correlations between litter size with birth weight and piglet survival mean diminishing returns on output (2) small marginal gains in a number of traits contributing to cycle length such as gestation length where decreasing 1 day increases output by 0.7% (3) rearing more of the piglets that are already born where reducing pre-weaning mortality by 1 percentage point provides an average additional 22 kg of carcase per litter. The impact of these changes to the volume of piglets produced is however small when compared to the effect of increasing carcase weight. Many EU countries still slaughter significantly lighter than the 97 kg average carcase in Austria and the USA has been even higher to offset shortfalls in pig numbers due to disease outbreaks. Increases in output in excess of 10% are relatively simple with modest adjustments to carcase weight. Increasing carcase weights for market already slaughtering heavy (95 kg+) animals is not a complete solution; heavier carcases may fail to meet retail demand for optimal cut size and leanness and will decrease feed efficiency on farm at heavier weights. This coupled with the increasing mortality with higher piglet numbers may mean the 5 tonne sow biologically possible but commercially improbable due to the product, societal and sustainable demands of producers, retailers and consumers.

L-arginine and L-carnitine in gestating sow diets to optimise output and piglet growth

K. Reid[1,2,3], K. O'Driscoll[3], E. Magowan[2], J. O'Doherty[1] and P.G. Lawlor[3]
[1]University College Dublin, Agriculture and Food Science, Belfield, Dublin 4, Ireland, [2]Agri-Food and Biosciences Institute, Large Park, Hillsborough, Co. Down, BT26 6DR, United Kingdom, [3]Teagasc Moorepark, Pig Development Department, Animal and Grassland Research and Innovation Centre, Fermoy, Co. Cork, Ireland; kathryn.reid@teagasc.ie

In an attempt to increase sow output, dam-line genetic improvement has resulted in considerable increases in litter size. However, large litters can result in problems with regard to health and growth performance with one of the major issues for piglets being low birth weight (BW). This study evaluated the effect of supplementing highly prolific multiparous sows with L-Arginine and/or L-Carnitine on sow and piglet performance in a commercial setting. Sows (n=429) were blocked on their breed, AI company who's semen they were served with, assigned feed curve on the farm, parity, back fat, previous numbers born alive and allocated to one of 4 treatments: 1. control (CON), 2. CON plus 25 g/day of L-arginine (ARG), 3. CON plus 0.125 g/day of L-carnitine (CAR), and 4. CON plus 25 g/day of L-arginine plus 0.125 g/day of L-carnitine (ARGCAR); from d28 of gestation to farrowing. The total number of piglets born (TB), born-alive (BA), stillborn (SB), and mummified (MUM) were recorded. Piglets from a subset of 216 litters were weighed at birth (n=3,309) and at weaning (n=1,836). The data was analysed using SAS. There was no interactive effect between CAR and ARG on any variable measured. Arginine tended to reduce TB (14.7±0.4 vs 15.2±0.4; P=0.09) and BA (13.8±0.4 vs 14.4±0.4; P=0.06) compared to non-ARG sows. Piglets from CAR sows were 60 g heavier at birth than those from non-CAR supplemented sows (1.40±0.017 vs 1.34±0.017 kg; P=0.03). There was no treatment effect on average daily gain from birth to weaning, or on weaning weight (P>0.05). In conclusion, Arg tended to reduce numbers born but had no effect on weaning weight, however, CAR was beneficial in increasing piglet birth-weight, although this benefit was not maintained at weaning.

Improving performance of small pigs post weaning using nurse sows or starter nutrition

A. Craig[1,2] and E. Magowan[1,2]
[1]Queens University, Belfast, BT71NN, N. Ireland, United Kingdom, [2]Agri-Food and Biosciences Institute, Hillsborough, BT266DR, N. Ireland, United Kingdom; aleslie02@qub.ac.uk

This study compared the use of nurse sows and post weaning feeding regimes to improve the productivity of small pigs. Pigs (n=245) were used across five treatments: T1: small pigs (av. 5.2 kg at 28 d) were weaned and offered a very high specification dietary regime (S-VH); T2 and T3: small pigs (av. 4.9 kg at 28 d) were reared on a nurse sow until 49 d after which they were offered either a high (S-H) or low (S-L) specification dietary regime; T4 and T5: pigs (av. 8.9 kg at 28 d) were weaned and offered either the High (N-H) or Low (N-L) specification dietary regime. Pigs were weighed at 4, 7, 10 and 12 weeks of age. Feed intake was recorded and average daily feed intake (ADFI) and feed conversion ratio (FCR) calculated. Data was analysed using analysis of variance unbalanced designs with birth weight as a covariate. At 12 weeks of age, the weight of S-VH pigs was similar (33.2 kg) (P>0.05) to that of S-H pigs (32.9 kg) but heavier (P<0.001) than that of S-L pigs (29.6 kg). The N-H pigs were heavier (41.2 kg, P<0.001) than the N-L (36.4 kg) pigs. The intake of S-VH pigs was lower (339 g, P<0.001) than that of N-H (456 g) and N-L pigs (368 g) between days 28 and 49. S-VH had similar FCR (1.20) to that of N-H (1.16) pigs, and better FCR than N-L pigs (1.45, P<0.001) from day 28 to 49. During 7-12 weeks intake differed significantly (P<0.001; SED 0.0164) for each treatment (795, 888, 1,087, 1,162 and 1,224 g for S-L, S-H, S-VH, N-L and N-H respectively). FCR was similar (P>0.05) for S-H and S-L pigs (1.42 and 1.49 respectively) during 7-12 weeks, which was better (P<0.001) than that for S-VH, N-H and N-L pigs (2.15, 1.94, 1.87 respectively). Rearing small pigs for an additional 3 weeks (to 7 weeks old) on a nurse sow and then offering a high specification dietary regime had a similar impact on 12 week weight as weaning small pigs at 4 weeks of age and offering them a very high specification dietary regime. However, the FCR of the nurse fed pigs was better from 7 weeks of age but may be due to them being offered higher specification diets at this time point.

Cross fostering affects the performance of both small and heavy piglets throughout production

A.M.S. Huting[1], I. Wellock[2] and I. Kyriazakis[1]
[1]Newcastle University, School of Agriculture, Food and Rural Development, Agriculture Building, Newcastle upon Tyne, NE1 7RU, United Kingdom, [2]Primary diets, ABAgri, Melmerby Industrial Estate, Melmerby, Ribon, North Yorkshire, HG4 5HP, United Kingdom; a.m.s.huting1@ncl.ac.uk

Major improvements in sow prolificacy have resulted in larger litters, but consequently increased the proportion of piglets born small. Different management strategies have aimed to limit the small piglet contribution to body weight (BW) variability; however, consequences on heavier littermates is often neglected. This study investigated the effect of different litter compositions through cross-fostering, on the performance of piglets born light (L) or normal (N) throughout productive life, and whether the provision of creep feed (CF) could contribute to an improved performance. Within 24 h after birth piglets were cross fostered creating UNIFORM litters with only similar sized piglets (L or N) and MIXED litters that had an equal number of L and N piglets. Half of the litters per treatment were supplemented with CF (≥10 days). The study was replicated over 7 batches to achieve 6-7 litters per treatment. Piglets were weighed at birth, weaning (~d28), grower (~9 weeks), finisher (~3 months), and slaughter (~6 months). Litter type (P=0.0413) and its interaction with birth weight category (P=0.0004) influenced piglets BW at weaning. Piglets born L were ~300 g (6.95 kg) lighter at weaning when reared in MIXED compared to UNIFORM litters (7.28 kg). On the other hand, piglets born N performed considerably better in MIXED litters (8.89 kg in MIXED versus 7.76 kg in UNIFORM). CF provision did not influence performance, however litter type significantly affected daily CF consumption (P=0.0001): N piglets in UNIFORM litters consumed the highest amount of CF. The weight advantages N and L piglets had at weaning when reared in respectively MIXED and UNIFORM litters, was sustained throughout the productive period. Cross fostering; a common farm practise to enhance L pig performance and within litter variation, was beneficial for piglets born small; however piglets born heavier were disadvantaged not only at weaning but up to slaughter.

Quantifying the relation between reproduction and finishing traits

R.H. Vogelzang[1], E.F. Knol[1] and L.L. Janss[2]
[1]Topigs Norsvin Research Center B.V., Schoenaker 6, 6641 SZ Beuningen, the Netherlands, [2]Aarhus University, Department of Molecular Biology and Genetics, P.O. Box 50, 8830 Tjele, Denmark; roos.vogelzang@topigsnorsvin.com

The pork production chain is continuously striving for higher efficiency by increasing number of pigs per litter and their weight gain to finishing and slaughter. Higher litter size is often associated with lower birth weights leading to relatively lower carcass weight but the additional piglets could increase total output per sow. Birth weight is a mainly maternal trait, where phenotypic differences in birth weight have consequences for the weight development of the finishers. We estimated individual weight curves of 2,174 finishing pigs using the Gompertz curve function in a joint Bayesian analysis. The weights during start, middle and end of the finishing period were used. The average weights were 23.2, 123.8, and 70.2 kg, respectively. Average birth weight was 1,391 gram and average litter size was 15.2. An extra piglet born per litter decreased the average birthweight with 30 gram, and increased mortality from farrowing to weaning by 1.76%. Results of this study showed that the effect of birth weight is linear in the range of 1 to 2 kg, and an extra kg of birth weight yields 16.2 kg of live weight at a standardized shipping age of 170 days. This proves that not only the genes of the finisher pig, but also the reproduction potential of the sow is very important for the total weight output per sow at slaughter and that increasing birth weight is highly relevant. Even if it goes at the cost of increased adult weight of the sow. Increasing litter size with 5 piglets resulted in an increased sow output per litter of 2.4 finishers and 247 kg of live weight. The total carcass weight output per sow continues to increase with increase in litter size and there is no evidence that breeding for a further increase in litter size will restrict sow output at slaughter in the near future.

Modelling fattening pig production systems: use of a dynamic, stochastic, mechanistic model

A. Cadéro[1,2,3], A. Aubry[3], L. Brossard[1,2], J.Y. Dourmad[1,2], Y. Salaün[3] and F. Garcia-Launay[1,2]
[1]Agrocampus Ouest, UMR1348 PEGASE, 65 rue de Saint-Brieuc, 35000 Rennes, France, [2]INRA, UMR1348 PEGASE, Domaine de la prise, 35590 Saint-Gilles, France, [3]IFIP-Institut du porc, Domaine de la Motte au vicomte, 35651 Le Rheu, France; alice.cadero@rennes.inra.fr

Pig livestock farming systems face economic and environmental issues. To cope with these issues, models of pig production performance have been developed. However, most of them do not account for the interactions between feeding strategies, management practices and variability of performance and requirements among pigs. Our aim was to develop a model representing the pig fattening unit, taking into account the effects of the interaction between the individual variability of pigs, the farmer's practices and the infrastructure of the farm on the performance of the unit. The farmer's practices included in the model have been chosen from a typology developed using a farm survey that focused on penning, batch management, feeding and delivery practices. The pigs are represented using an individual-based model adapted from the InraPorc model. The model abilities have been illustrated on a typical fattening unit batch management. Feed conversion ratio is 2.73 (±0.06) kg/kg, daily feed gain is 0.856 (±0.02) kg/d. The mean slaughter weight is 119.6 (±0.83) kg with a mean slaughter age of 176.0 (±1.89) days. This predicted performance is in accordance with the mean performance in French farms, considering the same batch management. This model will allows to evaluate the technical performance of a large range of fattening units, in terms of farm size, type of infrastructure and of management. In the near future, this model will be completed with calculation of economic results and environmental impacts of the pig fattening unit.

Fat and protein accretion in grower-finisher pigs fed two protein levels

I. Ruiz-Ascacibar[1,2], P. Stoll[1] and G. Bee[1]
[1]Agroscope, Institute of Livestock Sciences (ILS), Tioleyre 4, 1725 Posieux, Switzerland, [2]ETHZ, Universitätstrasse 2, 8092 Zurich, Switzerland; isabel.ruiz@agroscope.admin.ch

Feeding recommendations are based on growth models, where empty body (EB), protein (P) and fat mass are key variables. The current results are part of a large study aiming to revise the nutrient composition of the EB in order to update the Swiss feeding guidelines for pigs. Sixty-six entire male (EM), 58 castrated (CA) and 66 female (FE) pigs were used for the study. From 20 kg BW, they had either ad libitum access to a control (C) grower-finisher diet or a low P grower-finisher diet (R; 80% of C). The EB composition was determined at birth on 8 EM and 8 FE, at 10 and 20 kg BW, on 2 EM, 2 CA and 2 FE, and at 20 kg intervals from 40 to 140 kg BW on 4 pigs per sex and dietary treatment. At slaughter, organs and empty intestinal tract from each animal were homogenized together. Left carcasses were frozen until grinding. To determine the EB composition, P and fat mass were analyzed in the blood, hair, offal and carcass. Orthogonal contrasts revealed that P mass (g) from birth to 140 kg BW decreased in the following order: C-EM > C-CA; C-FE > R-EM; R-CA; R-FE at each EB weight. Fat mass differed among dietary treatments and sex and decreased as follows: R-CA > R-FE > C-CA > R-EM > C-FE > C-EM. Fat and P mass were fitted to an allometric function ($Y = a \times EB^b + c$), where Y is either fat, or P mass in the EB, a and c are constants and b, the scaling exponent. The R^2 of the function ranged from 0.97 (fat, C-EM) to 0.99 (P, C-EM). In contrast to previous studies, no decay in P accretion (g/kg EB, $Y' = a \times b \times EB^{(b-1)}$) was found in, neither C-EM, nor in R-pigs, with increasing BW. By contrast, P accretion was slower in C-CA and C-FE with increasing BW. The b exponent for fat was over 1.4 in all treatments, implying a marked curvilinear shape. This is especially evident for C-CA and C-FE, who reached at 140 kg BW a similar fat mass as R-CA and R-FE. These novel data obtained from a modern genotype allow improving predictions of the chemical composition of the EB.

Phase feeding groups of pigs representing high or low weight variation

E. Magowan[1], J. Keenan[2] and V.E. Beattie[2]
[1]Agri-Food and Biosciences Institute, Hillsborough, BT26 6DR, United Kingdom, [2]Devenish Nutrition Ltd, Belfast, BT1 3GW, United Kingdom; elizabeth.magowan@afbini.gov.uk

This study investigated the impact of offering the finishing diet as a single diet or through a phased approach on the variation in growth rate and finishing weight of pigs when pigs were in uniform or mixed weight groups. Over eight replicates 640 pigs were assigned to treatment according to weight and gender and penned in groups of 20 from 12 weeks of age. Treatments were arranged in a 2×2 factorial manner. Pens with either high or low weight variation were offered either a single finisher diet (CP 17%, Lysine 1.0%, DE 14.0 MJ/kg) from 40 to 120 kg or a Phase 1 diet (CP18% Lysine 1.1% DE 14.2 MJ/kg) from 40 to 75 kg and Phase 2 diet (CP 16%, Lysine 0.9%, DE 13.8 MJ/kg) from 75 to 120 kg. The SD of the high variation pen weight was aimed to be double that of the low pen weight. Pigs were weighed at 12, 15, 18, 21 & 24 weeks of age. Feed intake was recorded and Feed efficiency (FCR) calculated. Data was analysed by Analysis of variance in Genstat. There was no effect (P>0.05) of single or phase diets, or whether pigs were in pens with high or low weight variation on feed intake, growth rate or FCR, the averages being 2,188 g/d, 905 g/d and 2.42 respectively over the finishing period (41 to 115 kg). At the start of the trial, the co-efficient of variation (CoV) for weight was 0.111 and 0.069 for the high and low weight variation groups respectively. The CoV at 18 weeks of age also differed (P<0.001) but by 21 weeks of age the CoV of weight in 'high' pens had reduced (0.101) whereas it had increased (0.092) in 'low' pens with the result that the CoV for weight at 21 weeks was not different between groups (P>0.05) (averaging 0.097). It is concluded that there was no advantage or disadvantage to pig performance or pen weight variation to adopting a phase feeding approach. Pen weight variance appears to converge to approximately 9.7% as pigs approach slaughter weight regardless of whether there is high (11%) or low (7%) pen weight variation at the start of the finishing period.

The effect of sex and slaughter weight on carcass quality in pigs

A. Van Den Broeke, F. Leen, M. Aluwé, J. Van Meensel and S. Millet
ILVO (Institute for Agricultural and Fisheries Research), Scheldeweg 68, 9090 Melle, Belgium; alice. vandenbroeke@ilvo.vlaanderen.be

The aim of this study was to assess the effect of sex and slaughter weight and their interaction on carcass quality in finishing pigs. In 2 experiments on 2 different farms, approximately 400 pigs per experiment were raised in group pens of the same sex. Gilts, boars, immunocastrates (IC) and barrows were fed ad libitum by phase feeding and were divided in 3 groups of different slaughter weights: 105, 117 and 130 kg. Pigs of both farms were slaughtered at the same slaughterhouse and carcasses were classified using the AutoFom III Classification System. There were no significant interactions between sex, slaughter weight and farm or between sex and slaughter weight on carcass quality parameters. Farm, however, had a significant effect on all carcass quality parameters and was therefore included in the statistical model as fixed factor. Meat percentage did not differ between the three slaughter weight groups. Nevertheless, price per kg carcass was lower for the 105 kg slaughter weight group compared to the 117 kg slaughter weight group (P<0.001). These differences are present because, in addition to meat percentage, also muscle and backfat thickness, ham angle and width and centroid of the carcass determine the price per kg carcass. Muscle thickness of the 105 kg group < 117 kg group < 130 kg group. Ham angle of the 105 kg group was lower compared to the 130 kg group. Ham width and centroid of the carcass of the 105 kg group was lower compared to the 117 and 130 kg groups. Evaluation of sex indicated that meat percentage of boars > IC and gilts > barrows. Backfat thickness, ham angle and width and centroid of the carcass of boars < IC and gilts < barrows and muscle thickness was higher in barrows and gilts compared to IC and boars. Price per kg carcass of gilts > IC and boars > barrows. In conclusion, the farm has a significant effect on carcass quality but overall, gilts of the 117 kg slaughter weight group got the best price per kg carcass. Analysis of the effect of sex and slaughter weight on the obtained margin over feed costs per pig is still in progress.

Performance curves for growing finishing pigs: evaluation of published growth models
F. Leen, A. Van Den Broeke, M. Aluwé, B. Ampe, L. Lauwers, J. Van Meensel and S. Millet
Institute for Agricultural and Fisheries Research (ILVO), Burg. Van Gansberghelaan 115 box, 9820 Merelbeke,
Belgium; frederik.leen@ilvo.vlaanderen.be

A crucial resource for bio-economic optimization of pig production is an accurate growth model. The question is which model suits best to model the commercial growth traject of the pre-maturely slaughtered finishing pig. The objective of this study was twofold: (1) to compare the accuracy of several published growth models; and (2) to provide up-to-date parameters for accurate and applicable growth curves for high-lean pigs of four different sexes (boar, barrow, GnRH-vaccinated boar and gilt).Data from four trials of which two with individually housed and two with commercially group housed pigs were used. Animals were kept ad libitum on a multiphase diet. Body weight and feed intake were recorded from 9 weeks of age until slaughter. The serial body weight data were fitted to two dynamic models (Gompertz and Bridges), describing growth as a function of age and one static model (Monomolecular), describing growth as function of cumulative feed intake, using PROC NLMIXED in SAS. Residual sum of squares (RSS) and root mean squares (RMS) were calculated to evaluate model goodness-of-fit. RSS were used to test significant differences in goodness-of-fit between models. Residual plots were evaluated to investigate under or over prediction. The static monomolecular model showed significant ($P<0.05$) better goodness-of-fit for the four sexes compared to the dynamic models, except for the barrows in one trial. The Gompertz and Bridges models did not significantly differ in goodness-of-fit. Residual plots showed no systematic over or under prediction under 125 kg. However, because of selection bias, due to slaughtering the animals at a fixed weight instead of age, over prediction occurred beyond 125 kg. Application of the current parameters for weight predictions beyond 125 kg is therefore not recommended.

Calculation of economic values for new traits in Norwegian Landrace; lean meat and fat efficiency
K.H. Martinsen[1], D. Olsen[2], J. Ødegård[3] and T.H.E. Meuwissen[1]
[1]Norwegian University of Life Sciences, Department of Animal and Aquacultural Sciences, P.O. Box 5003, 1432
Ås, Norway, [2]Topigs Norsvin, P.O. Box 504, 2304 Hamar, Norway, [3]AquaGen AS, P.O. Box 1240 Sluppen, 7462
Trondheim, Norway; kristine.martinsen@nmbu.no

The breeding goal for Norwegian Landrace consists of traits that are important for the producers, such as growth, feed intake and mothering ability, and for consumers, such as meat quality and animal welfare-related traits. In order to weigh the traits in the breeding goal by their economic importance, the trait's economic value needs to be calculated. Martinsen et al. found genetic variation in two new efficiency traits, lean meat efficiency and fat efficiency. These traits indicate how well the animal utilizes feed for lean meat and fat production. The aim was to calculate the economic value for these two traits in Norwegian Landrace. A profit function for an average slaughter pig was developed. The model included the income and costs related to the slaughter pig from purchased feeder pigs (30 kg) until slaughter (115 kg). The traits included in the model were lean meat efficiency and fat efficiency, which was a part of the feed costs in the profit function. Income was dependent on the animal's slaughter weight and lean meat percentage. Economic values were estimated by improving the mean of the trait by 0.1%. The difference between the original and the improved mean was divided by the change in the trait and represented the marginal economic value of the trait. The marginal economic value was standardized by multiplying with the additive genetic standard deviation for each trait. The relative economic values of lean meat efficiency and fat efficiency were €8.7/σg and €2.8/σg, respectively. These values are rather high compared to both total feed consumption (€2.67/σg) and growth (€1.46/σg), which commonly are the most important traits in pig breeding. The results suggested that lean meat and fat efficiency might be more important to consider for use in a breeding scheme for Norwegian Landrace than growth and feed intake.

Evolution of attitudes towards estimated breeding values in Flemish pig farmers

S. Palmans[1,2,3], M. Van Der Voort[2], S. Millet[2], S. Janssens[3], N. Buys[3] and J. Van Meensel[2]
[1]Agricultural Research and Education Center, Kaulillerweg 3, 3950 Bocholt, Belgium, [2]Institute for Agricultural and Fisheries Research, Burgemeester Van Gansberghelaan 96, 9820 Merelbeke, Belgium, [3]KU Leuven, Livestock Genetics, Kasteelpark Arenberg 30, 3001 Heverlee, Belgium; sander.palmans@pvl-bocholt.be

Estimated Breeding Value (EBV) is a genetic estimation of the performance capacities of an animal. EBVs give pig breeders the opportunity to select sires on specific genetic properties. Pig producers could have economic benefit from an adapted terminal sire for their specific production goals. However EBV is not a common used tool for sire selection by Flemish pig farmers. Therefor we asked some pig farmers to use 5 till 7 specific terminal sires and track the progeny during their fattening period so they could see differences between sires on pigs in their own stable. During the process we questioned the farmers 4 times so we could see the evolution of their attitude towards EBV and sire selection. From the 8 participants 6 already knew something about EBV but only one of them used it for sire selection at the time of the first interview. Others mostly declared that estimation method was not very reliable. Because the farmers fertilize their sows by artificial insemination sires were mostly selected based on advice of the salesman of terminal sire sperm. At the end of the first interview we gave an explanation of the methods to calculate EBVs. One farmer was very impressed and changed his sire selection method to a method based on EBVs. During the fattening period there were two interviews who did not reveal any changes in attitude. The fourth and final interview took place when all pigs were slaughtered. All pig producers declared they were impressed by the agreement of the EBVs and the results in their own farm. One farmer changed his sperm buying strategy by choosing other sperm deliverers. The other farmers changed their sire selection method to an EBV based selection. We conclude that pig farmers mostly are quiet conservative in changing their terminal sire selection strategy but when we showed the possibilities of a tool to them on their own specific farm their belief in EBVs became very strong.

EWPIG: a program to calculate economic weights of traits in pigs

Z. Krupová, E. Krupa, M. Wolfová, J. Wolf and E. Žáková
Institute of Animal Science, Přátelství 815, 10400 Prague Uhříněves, Czech Republic; krupova.zuzana@vuzv.cz

The aim of the present study was to develop a flexible computer program (EWPIG) for calculating the marginal economic values and relative economic weights of traits in pig breeds within various breeding systems. Up to 30 growth, carcass, functional and feed efficiency traits can be evaluated. Calculation of the sow herd structure using Markov chain methodology and calculation of progeny structure are part of the program. To calculate feed costs for young animals in all growth phases and for sows in different reproductive stages (open, lactating, and pregnant sows), the estimation of protein and fat tissues growth, metabolisable energy, feed and water requirements for all animal groups are considered in the program. Based on the feed and non-feed costs and revenues in all production units, the total profit per sow and year is calculated as the main criterion for production efficiency. Using the program, the impact of changes in a great spectrum of input parameters on the trait economic values and on the economic efficiency of the production system for all integrated breeds and crosses can be estimated. Calculating the trait economic weights, the proportions of genes from the individual breeds or lines involved in the given breeding system is considered, but not the time delay when expressing the traits. Estimating economic weights using gene flow methodology is envisaged to be included in the program in near future. Development of the study was supported by project QJ1310109 and MZERO0714 of the Czech Republic.

Length of productive life in Czech Large White and Czech Landrace sows

E. Žáková, E. Krupa and Z. Krupová
Institute of Animal Science, Přáteství 815, 104 00 Praha Uhříněves, Czech Republic; krupova.zuzana@vuzv.cz

Length of sows productive life (LPL) play economic and animal welfare role in pork production systems. This study aimed to characterize the main factors influencing LPL of sows of dam breeds on nucleus and multiplier level in the Czech Republic. The analyzed data contained information of 62 730 Czech Large White (CLW) and of 16 785 Czech Landrace (CL) sows. The Weibull proportional hazard model was fitted on the data using the Survival Kit software. Log likelihood tests were used to determine if the individual effects have a significant impact on LPL. Farm management defined as herd-year-seasons was found as the major factor influencing LPL in both breeds (P<0.001). Other factors significantly (P<0.001) affected LPL were: the stage of reproduction cycle, parity, litter size, age at the first farrowing and the herd level (nucleus or multiplier). The parity and the litter size affected LPL almost equally. The same effect on LPL also had the age at the first farrowing and the level of the herd. Sows at the first parity had higher risk to be culled than sows from second to fifth parities and the same as sows on sixth parity. Sows on second to fourth parity had almost equal risk to be culled. Sows were discarded from herds mainly 29-41 and 41-62 days after farrowing. The risk ratio decreased with increased litter size. Younger CL sows at the first farrowing tended to remain in the herd longer although differences between risk ratios for the first farrowing classes were small. The same pattern was found for CLW sows, but with excessive increase of risk for sows with age on average level. Sows in multiplier herds were discarded later in life (+105 d CLW and + 87 d CL) than sows in nucleus herds. Using animal model, estimates of heritability of LPL was 0.11 and 0.14 for CLW and CL, respectively. This indicated that genetic improvement is feasible, although only a small genetic progress can be expected. This study was supported by the project QJ1310109 of the Ministry of Agriculture of the Czech Republic.

Study of lifetime performance in three pig breeds on a Hungarian commercial farm

Á. Bagi-Hunyadi[1], S.Z. Kusza[1] and P. Balogh[2]
[1]Institute of Animal Science, Biotechnology and Nature Conservation, University of Debrecen, Böszörményi str 138., 4032 Debrecen, Hungary, [2]Institute of Sectoral Economics and Methodology, University of Debrecen, Böszörményi str 138., 4032 Debrecen, Hungary; kusza@agr.unideb.hu

The aim of the present study was to perform lifetime performance analysis in three pig breeds; Hungarian Large White (n=295), Duroc (n=76) and Pietrain (n=91) on a Hungarian commercial farm using analysis of survival sows. We took into consideration the age of sows at the time of their inclusion into breeding, their age at the time of culling, time spent in production, number of mating and parities, parity percentage, intervals between litters, number and mean of piglets born alive and born dead, number of raised piglet litters, number and mean of 21 days old piglets, the weight and mean of raised litter and raise percentage. We carried out the analysis by SPSS 22.0. Single factor analysis of variants, Kaplan-Meier analysis and Cox PH model were used. The determination of the significance of risk rates differences was done by Wald chi square test. Our results showed that the average culling age were 1,056±33.52 days for the Hungarian Large White, 735±73.56 days for Duroc and 818±71.98 days for the Pietrain. The log rank test of the survival analysis indicated a significant difference between the three tested genotypes (x^2 16.981, P<0.001), which means that the survival percentage of the individual breeds varied significantly from one another. In comparison with the Hungarian Large White genotype the Duroc genotype has a 1.6 times higher (P<0.001) culling risk while that of the genotype Pietrain was 1.36 times higher (P<0.001). Our results can be used to compare the breeds kept under the same conditions and to compare the life span of one genotype under different farming conditions. Factors that increase survival and improve the profitability of pig farming can be determined by this method.

Digestible energy intake during gestation and associated sow reproductive performance

A. Lavery[1,2,3], P.G. Lawlor[2], H.M. Miller[1], K. O'Driscoll[2], D.P. Berry[2] and E. Magowan[3]
[1]*University of Leeds, Faculty of Biological Sciences, Leeds, LS2 9JT, United Kingdom,* [2]*Teagasc, Animal & Grassland Research and Innovation Centre, Moorepark, Fermoy, Co. Cork, Ireland,* [3]*Agri-Food and Bioscience Institute, AFBI Hillsborough, Large Park, Hillsborough, Co. Down, BT26 6DR, United Kingdom; anna.lavery@afbini.gov.uk*

Experimental gestation feeding records for 831 multiparous sows on trials at Teagasc Moorepark and AFBI Hillsborough, between the years 2005 and 2014 were used to examine the relationship between gestational digestible energy intake (DEI) and sow reproductive performance. DEI (MJ/Day) for each sow was categorised as: (1) Medium (32.0-33.0, n=257), or Low (29.0-30.0, n=574) during early gestation (day 0-24); (2) High (43.0-46.0, n=85), or Low (29.9-33.0, n=785) during mid gestation (day 25-80); and (3) High (44.0-45.3, n=34), Medium (39.4-40.6, n=32), Low (32.2-34.7, n=395), or Very Low (29.9, n=401) during late gestation (day 81-114). The association between total born (TB) and born alive (BA) with DEI was determined using mixed models, with all stages of gestation included in the model simultaneously. Parity and year of farrowing were included as fixed effects; sow was included as a repeated effect. DEI in early and late gestation was not associated with TB or BA (P>0.05). Mean TB was 12.8, 12.3 (pooled SE=0.546) while mean BA was 11.5 and 10.9 (pooled SE=0.529), respectively for medium and low levels of DEI in early gestation. Mean TB was 12.5, 12.6, 12.8 and 12.3 (pooled SE=0.302) while mean BA was 10.9, 11.4, 11.5 and 11.0 (pooled SE=0.289), respectively for high, medium, low and very low levels of DEI in late gestation. No significant difference existed between categories of DEI in early and late gestation for TB or BA. While a significant statistical association between DEI and both TB and BA was observed during mid gestation, the biological interpretation of this result is difficult due to a possible bias in the data. Increasing DEI intake during early or late gestation was not associated with an improvement in TB or BA.

Relating beef production at farm level to human health, and the state of art of meat substitutes

J.F. Hocquette, D. Gruffat and D. Durand
INRA-VetAgro Sup, UMRH 1213, Herbivore Unit, Theix, 63122 Saint Genes Champanelle, France; jfhocquette@clermont.inra.fr

The healthiness of beef has recently come into question, mainly due to reported levels of saturated fatty acids (FA), its low content in n-3 polyunsaturated FA (n-3 PUFA), and variable trans-FA (TFA) content. These concerns, in addition to concerns regarding sustainability, animal welfare, and projected global food needs have led to increasing interest in meat substitutes, such as soy, wheat, mycoproteins and cell culture meat. However, conventional beef still provides many proteins and micro-nutrients which are beneficial for human health. There is usually a good n-6/n-3 ratio (~3), which is related to a wide range of health benefits due to the high levels of n−3 PUFA (e.g. α-linolenic acid and n-3 long chain PUFA), conjugated linoleic acids (such as the cis-9,trans-11 isomer) and TFA (such as the trans-11 18:1 isomer). Additionally, the n-6/n-3 ratio of beef can be further improved through dietary interventions on cattle with pasture, grass-silage or linseed. However, caution must be taken when increasing PUFA levels as they are susceptible to lipid oxidation result in rancidity, unpleasant tastes and odours, and the formation of toxic compounds which relate to colorectal cancer. These issues are being addressed with further dietary interventions using vitamins, trace elements and vegetable extracts rich in polyphenols which reduce the rate of PUFA oxidation. In this context, the health concerns regarding beef are relatively small when compared with emerging meat replacement technologies. For example, in vitro meat is an untested product and has multiple anticipated health concerns relating to the artificial hormones and growth factors, fungicides, and antibiotics required for cell culture. Therefore, while meat substitutes hold some promise for the future, consumer acceptance is uncertain due to considerable technical, economic and social constraints which mean conventional meat is still a healthy and viable product.

Milk, milk fat and human health: can we alter milk composition on farm, do we want/need to alter it?

A.L. Lock[1] and D.E. Bauman[2]
[1]Michigan State University, East Lansing, Michigan, USA, [2]Cornell University, Ithaca, New York, USA;
allock@msu.edu

Dairy products are nutrient dense foods and public health organizations around the world include milk and other dairy products in recommendations for a healthy, well-balanced diet. Milk fat has been perceived by the public, and much of the scientific community, as a negative component of dairy products and typical dietary advice was to consume only reduced-fat or no-fat dairy products. New research and re-evaluation of previous research is challenging long-held dogma on the relationship between milk fat and human health. We will utilize science-based approaches to address two important questions: (1) Do we need to alter milk fatty acid (FA) composition to improve human health?; and (2) If so, can we modify its composition? Altering the FA content of milk fat requires an understanding of the interrelationship between dietary supply of FA, rumen fermentation, and mammary gland synthesis. Much of this work has involved studies in which whole-scale changes have been the goal, whereby large shifts in the saturated to polyunsaturated FA ratio have been sought-after. Modest changes have been achievable, but these have often led to problems relating to product quality and stability and dairy cow production. Despite concerted efforts there is little evidence that anyone has meaningfully changed the FA composition to an extent that would impact human health. A recent focus has been on the omega-3 FA in milk. Milk fat content of omega-3 FA is generally low (<0.5% total FA), and this is mainly 18:3 n-3. Claims of large percentage increases in omega-3 FA (or another minor FA in milk fat) must be considered with caution due to their initial very low levels. There is also interest in potential effects of different agricultural production practices on product quality and health outcomes. While some small differences in product composition have been reported between production practices, these must be taken within the context of the large impact of diet and the wide range among animals. Opportunities for greater enrichment of specific FA in milk fat will likely rely on post-harvest fortification of milk, if deemed desirable. In conclusion, it is important to emphasize that all dairy products are an excellent source of nutrients for the human population.

Nutrition & health claims for beef, lamb and pork

M. Strong
Agriculture & Horticulture Development Board (AHDB), Stoneleigh Park, Kenilworth, Warwickshire CV8 2TL,
United Kingdom; maureen.strong@ahdb.org.uk

The promotion of lean beef, lamb and pork as part of a health balanced diet continues to be challenged by anti-animal source food (ASF) organisations and some scientists with an interest in demonstrating adverse health concerns about meat consumption. The new legislative framework posed by the implementation of Nutrition and Health (N & H) Claims Regulation 1924/2006 and the Food Information to Consumers (FIC) Regulation 1169/2011 presented an opportunity to clarify how red meat could justifiably be promoted on health grounds within the new legal framework. In 2014 the Agriculture and Horticulture Development Board (AHDB) decided to review all its communications on the promotion of red meat. In line with the EU approved terms and conditions, approved nutrition and health claims that could be made for the three red meats were identified. Documents detailing the justification for the various claims were drafted and used as a point of negotiation with UK regulators. A 'Primary Partnership' arrangement with Buckingham and Surrey Trading Standards (TSI) was formalised. This facilitated an ongoing discussion and agreement on how scientifically worded approved claims could be modified and made more consumer friendly to ensure they were neither misleading to the consumer nor exaggerated the particular health benefit in question. Supported by consumer research the reworded claims were agreed and have subsequently been published on the British Regulatory Delivery Office (BRDO) website as Primary Authority Assured Advice.

Dairy and meat product consumption and cancer risk

E. Kampman
Wageningen University, Human Nutrition, Bomenweg 4, 6703 HD Wageningen, the Netherlands;
ellen.kampman@wur.nl

One out of three people in the Western world will be diagnosed with cancer. About 30% of the most common types of cancer are preventable if we would eat healthy, keep a normal body weight, and be more physically active. The association between foods of animal origin and cancer risk has been mostly evaluated in large observational prospective cohort studies. According to the systematic literature reviews and meta-analyses of the World Cancer Research Fund (WCRF), lastly updated in 2011 (for colorectal cancer) and 2014 (prostate cancer), these studies show that milk is probably associated with a decreased risk of colorectal cancer, while there is limited suggestive evidence that dairy products increase prostate cancer risk. It is hypothesized that milk may protect against colorectal cancer due to it high calcium content, which may bind pro-inflammatory secondary bile acids and ionized fatty acids and may reduce cell proliferation and promote cell differentiation. The potential increased risk of prostate cancer may be mediated through high calcium levels, and by increased blood levels of insulin growth factor-1 (IGF1). Literature reviews and meta-analyses by WCRF and WHO/IARC show that high consumption of red meat (e.g. beef and pork) and processed meat (preserved by smoking, curing, salting, or by the addition of preservatives) is convincingly associated with increased risk of colorectal cancer and (non-cardia) stomach cancer (processed meat). Red meat contains heme iron, which promotes the endogenous formation of carcinogenic N-nitroso compounds and the formation of cytotoxic and genotoxic aldehydes by lipoperoxidation. Red meat cooked at high temperatures or burned during barbecuing, results in the production of carcinogenic heterocyclic amines (HCA) and polycyclic aromatic hydrocarbons (PAH). Processed meat may also contain high levels of nitrite and nitrate. In the stomach, nitrite and nitrate can react with the degradation products of amino acids (from meat) to form N-nitroso compounds. Cured smoked meat may also contain PAHs. The salt included in cooking, processing and preserving meat may cause cancer by damaging the gastric mucosa, leading to inflammation.

Debate - relation between beef and dairy products and human health

Main discussion theme: which actions in the production chain and especially which research is needed to deal with the remarks of the WHO (World Health Organisation) about health risks (including cancer risks) related to beef and milk products

The debate session is based on the next four contributions:
- 'Farm management, milk and human health aspects', Prof. Adam Lock, Animal Science Department, Michigan State University, USA
- 'Dairy and beef product consumption and cancer risk', Prof. Ellen Kampman, Division of Human Nutrition, Wageningen University, the Netherlands
- 'Relating beef production at farm level to human health aspects', Dr. Jean-Francois Hocquette, INRA, Clermont, France
- 'Beef and human health', Dr. Maureen Strong, Agriculture & Horticulture Development Board (AHDB Dairy), Kenilworth, United Kingdom

These contributions are followed by a debate of 1½ hour. This debate is opened by statements of:
- Prof. Herman Barkema, Professor Epidemiology of Infectious Diseases Faculty of Veterinary Medicine and Cumming School of Medicine, University of Alberta, Canada
- Dr. Jean-Louis Peyraud, Directeur de Recherche of INRA, France, and president of the European Animal Task Force (ATF)
- Prof. Partrick Wall, University College Dublin, School of Public Health, Physiotherapy and Population Science, hosting the National Nutrition Surveillance Centre, Dublin, Ireland

Genetic and genomic analyses of enteric methane emissions of dairy cattle

Y. De Haas[1], J.E. Pryce[2], E. Wall[3], S. McParland[4], G. Difford[5] and J. Lassen[5]
[1]Animal Breeding and Genomics Centre of Wageningen UR Livestock Research, P.O. Box 338, 6700 AH Wageningen, the Netherlands, [2]Agribio, Department of Economic Development, Jobs, Transport and Resources and La Trobe University, 5 Ring Road, Bundoora, Victoria 3083, Australia, [3]Animal and Veterinary Sciences, SRUC, Kings Buildings, West Mains Road, Edinburgh EH9 3JG, United Kingdom, [4]Animal and Grassland Research and Innovation Centre, Teagasc, Moorepark, Fermoy, Co. Cork, Ireland, [5]Department of Molecular Biology and Genetics, Faculty of Science and Technology, Aarhus University, P.O. Box 50, 8830 Tjele, Denmark; yvette.dehaas@wur.nl

Climate change is a growing international concern and it is well established that the release of greenhouse gases (GHG) is a contributing factor. Of the various GHG produced by ruminants, enteric methane (CH_4) is the most important contributor. One mitigation strategy is to reduce methane emission through genetic selection. However, successful animal breeding strategies require measurements on a large population of animals. Therefore, we aimed to enlarge the reference population by combining our data on dairy cattle in 5 countries (i.e. Australia, Denmark, Ireland, the Netherlands and UK). The total dataset consists of 3,060 dairy cows. Even though three different measurement equipments (laser, sniffer and SF6) and protocols (measuring for 3 days, 1 week, multiple weeks) were used, these data will be analysed jointly to establish genetic and genomic parameters for enteric methane. The average methane production was 448 g/d in Australia (354 cows); 554 g/d in Denmark (1,769 cows); 381 g/d in IRL (260 cows); 549 g/d in NL (457 cows); and 325 g/d in UK (216 cows). This clearly shows that the populations and diets are different, next to the equipment and protocol. Therefore, a multi-trait approach will be used to analyse the enteric methane emissions. Following the experiences of gDMI, it is expected that each country will benefit for contributing to an international reference set with increased accuracies of the estimates.

Feed efficiency and methane emissions in dairy cattle: overview of the current data recording

A. Butty[1], A. Wilson[1,2], V. Osborne[2], C. Baes[1] and F. Miglior[1,3]
[1]Centre for Genetic Improvement of Livestock, Animal Biosciences, University of Guelph, 50 Stone Road East, Guelph, ON, N1G 2W1, Canada, [2]Centre for Nutritional Modelling, Animal Biosciences, University of Guelph, 50 Stone Road East, Guelph, ON, N1G 2W1, Canada, [3]Canadian Dairy Network, 660 Speedvale Avenue West, Suite 102, Guelph, ON, N1K 1E5, Canada; buttya@uoguelph.ca

Genetic selection of dairy cattle has been successful in improving production and conformation traits over time, but change has not been assessed for feed efficiency and methane emissions. Genetic evaluation of these complex traits is difficult, and requires the use of expensive methods and materials that are not practical on commercial farms. As a result, traditional genetic analysis has not been implemented in these traits. Genomic technologies allow prediction of complex traits requiring fewer phenotypic observations. However, no single country has been able to create a reference population large enough to accurately predict breeding values. To improve the worldwide dairy cattle population more efficiently, gDMI and Genome Canada projects were set up to enhance sharing of dry matter intake and methane emission data between countries. The International Committee for Animal Recording (ICAR) recently approved the creation of the Feed & Gas Working Group. This group aims to provide an overview of the current data available, to facilitate the standardization of recording dry matter intake and methane output in cattle around the world, and to enhance international collaboration by providing technical and methodological tools for data sharing and merging. Together with the Working Group, we developed a survey that was sent to researchers and industry organizations in member countries of ICAR to create an overview about the current or expected recording practices. Questions about the recording methods used or genetic analyses already run within countries will bring together information of realized and planned measurements. Results of the survey will allow assessment of the present situation and definition of standards to record and analyze dry matter intake and methane emission of dairy cattle.

An international initiative to decrease the environmental footprint of dairy cattle using genomics

F. Miglior[1,2], C.B. Baes[1], A. Canovas[1], M. Coffey[3], E.E. Connor[4], M. De Pauw[5], B. Gredler[6], E. Goddard[5], G. Hailu[1], V. Osborne[1], J. Pryce[7], M. Sargolzaei[1,8], F.S. Schenkel[1], E. Wall[3], Z. Wang[5], T. Wright[1,9] and P. Stothard[5]
[1]University of Guelph, Guelph, Ontario, N1G 2W1, Canada, [2]Canadian Dairy Network, Guelph, Ontario, N1K 1E5, Canada, [3]Scottish Rural College, Edinburgh, Scotland, EH9 3FH, United Kingdom, [4]USDA, Animal Genomics and Improvement Laboratory, Beltsville, MD, 20705, USA, [5]University of Alberta, Edmonton, Alberta, T6G 2C8, Canada, [6]Qualitas, Zug, 6300, Switzerland, [7]Dept of Econ Dev, Jobs, Transport and Resources, Melbourne, Victoria, 3001, Australia, [8]Semex Alliance, Guelph, Ontario, N1H 6J2, Canada, [9]Ontario Ministry of Agriculture, Food and Rural Affairs, Guelph, Ontario, N1G 4Y2, Canada; miglior@cdn.ca

We present a large-scale, international project which will bring together research and industry experts to develop effective selection tools for improving feed efficiency (FE) and decreasing methane emissions (ME). To create an appropriately sized reference population, a database for storing individual daily feed intake and ME data for cows and heifers will be developed and implemented integrating international data from the US, UK, Australia, Switzerland and Canada. Also, milk spectral records will be used to further develop predictions of FE and ME. Whole DNA (Genome) and RNA (Transcriptome) sequence information will be used to identify novel markers, which will be incorporated into genomic predictions. Research defining social and economic costs/benefits of selecting for FE and ME, and identifying any barriers to adoption, will also be conducted. Links with industry will allow genomic evaluation services for FE and ME. Since feed is the largest expense in milk production, improving cow, FE will benefit the industry considerably. More efficient animals produce less manure and emit less methane, thereby reducing the industry's environmental footprint.

Genome-wide association study and genomic prediction for methane emission in Danish Holstein cattle

J. Lassen, P. Løvendahl, G. Su and G. Sahana
Centre for Quantitative Genetics and Genomics, Department of Molecular Biology and Genetics, Blichers Alle 20, 8830 Tjele, Denmark; jan.lassen@mbg.au.dk

We carried out a genome wide association study on enteric methane emission in Danish Holstein cows and also investigated if reliable genomic prediction is possible for methane emission. A total of 1,339 cow from 20 heard were individually measured for their methane emission during milking in automatic milking systems fitted with Fourier Transformed Infrared measuring unit. The animals were genotyped with Illumina Bovine SNP50 BeadChip. A SNP-by-SNP analysis association analyses was done using a a linear mixed model approach. We found 44 SNPs showing suggestive significant association with methane emission. The majority SNPs were intragenic, one synonymous, 5 upstream gene variants and 10 intron variants. Genomic predictions were validated using a five-fold cross validation procedure with two validation datasets. In the first dataset, half-sib families were divided randomly into five subsets, and in the second set individuals in the whole data were randomly divided into five subsets. In each fold of cross validation, one of the five subsets was used as the test data where the phenotypic values were masked during genomic prediction, and the remaining four subsets were used as the reference data. Reliability of genomic predicted breeding value (GEBV) and conventional estimated breeding value (EBV) was measured by squared correlation between estimated breeding values and corrected phenotypic value and then divided by heritability of the trait. The reliabilities of GEBVs ranged from 0.17 when using conventional BLUP model to 0.18 when using GBLUPw model combing both marker-based and pedigree-based relationship matrix. Limited power in GWAS and low reliability in genomic prediction necessitate international collaboration and better approaches to collect large scale data in order to implement genomic prediction for scarcely measured traits and to identify genes affecting enteric methane emission.

Genome wide association study (GWAS) of methane emissions in Australian Angus

C.I.V. Manzanilla-Pech[1], Y. De Haas[1], R.F. Veerkamp[1], K.A. Donoghue[2], P.F. Arthur[2] and J.E. Pryce[3]
[1]Wageningen UR Livestock Research, Animal Breeding and Genomics Centre, Droevendaalsesteeg 1, 6700 AH, Wageningen, the Netherlands, [2]Agricultural Research Centre, Department of Primary Industries, NSW 2823, Trangie, Australia, [3]La Trobe University, Agribio, Department of Economic Development, Jobs and Resources, 5 Ring Road, Bundoora, Victoria 3083, Australia; coralia.manzanillapech@wur.nl

One possible solution to reduce methane (CH_4) emissions is to select for animals that produce less methane. For that, accurate genetic parameters and a better understanding of CH_4 traits are needed. The objectives of this study were to perform a GWAS to identify genes associated with several CH_4 traits in Angus beef cattle and evaluate two different residual CH_4 traits and consequently determine if genes associated with CH_4 also control residual CH_4 traits. The Angus population consisted of 1,020 animals with phenotypes on CH_4 production (MeP), dry matter intake (DMI), and weight (WT). Additionally, two residual CH_4 traits: residual genetic methane (RGM) and residual phenotypic methane (RPM) were calculated by adjusting CH_4 for DMI and WT. Animals were genotyped using the 800k Illumina Bovine HD Array, and after editing 632,003 SNPs were available. Estimated heritabilities were 0.30, 0.19 and 0.15 for MeP, RGM and RPM respectively, and genetic correlations of MeP with DMI and WT were 0.83, and 0.80 respectively. Strong associations with MeP were found on chromosomes 4, 12, 14, 20 and 30, at $P<0.0001$ the number of SNPs was 803 and at $P<0.005$, there were 3,304. When SNPs in common between all traits were calculated, DMI had the largest number of significant SNPs ($P<0.005$) in common with MeP (633) followed by WT (627). Based on our results, we conclude that MeP in Angus is dependent on DMI and WT and residual traits could be a good option to include in the breeding goal, to select for lower emitting animals without affect DMI and WT.

Breeding for feed efficiency does not decrease methane production in growing lambs

G. Rose[1], B. Paganoni[2] and A.N. Thompson[1]
[1]Murdoch University, 90 South Street, 6150 Murdoch, Australia, [2]Department of Agriculture and Food Western Australia, 3 Baron-Hay Court, 6151South Perth, Australia; gus@gusrose.com

The Australian sheep industry aims to increase the efficiency of producing more meat by decreasing the amount of feed eaten by lambs. Also, feed intake is related to methane production, and more efficient (low residual feed intake) animals eat less than expected. So we tested the hypothesis that more efficient lambs produce less methane. To test this hypothesis we investigated the genetic correlations between feed intake, residual feed intake and methane with growth and carcass traits. We used 1,870 methane and 670 feed intake measurements from 1,900 Merino lambs during years 2010 to 2015. The lambs were fed a high energy grower pellet for 35 to 45 days ab libitum. Individual feed intake was measured using automated feeders. Methane was measured using portable accumulation chambers up to 3 times during the feed intake period at an average age of 202 days. Live weight was measured at the beginning and end of the feed intake period. We used ASReml to estimate heritabilities and phenotypic and genotypic correlations between traits. Feed intake (0.47) was more heritable than residual feed intake (0.25) and methane (0.09). Feed intake and residual feed intake had high positive phenotypic (0.63) and genetic (0.83) correlations. Feed intake and methane also had a high positive phenotypic correlation (0.51), but almost no genetic correlation (0.05). Feed intake had higher genetic correlation with growth (0.87) and weight (0.74) than methane (0.50 and 0.43). Residual feed intake had a high genetic correlation with growth (0.98) but not live weight (0.17). Therefore, we rejected our hypothesis because feed intake traits were not genetically correlated to methane production. This suggests that selecting for feed efficiency changes different functions that selecting for methane. Additionally, animals that grow the fastest are also the least efficient. This may be because the least efficient lambs eat more. Therefore, it is difficult to breed lambs that grow fast but eat less when fed high energy feed.

Heterogeneity in genetic variation for feed intake and adjusted feed intake between dairy cow breeds

B. Li[1,2], B. Berglund[1], W.F. Fikse[1], J. Lassen[2] and P. Løvendahl[2]
[1]*Swedish University of Agricultural Sciences, Department of Animal Breeding and Genetics, Ulls väg 26, 75007 Uppsala, Sweden, [2]Aarhus University, Department of Molecular Biology and Genetics, Blichers Allé 20, 8830 Tjele, Denmark; bingjie.li@mbg.au.dk*

Dry matter intake (DMI) and adjusted feed intake (AFI) are two traits related to feed efficiency in dairy cattle. This study aimed to estimate genetic variances and genetic parameters for DMI and AFI in cows of two dairy breeds, Holstein and Jersey, by random regression methods (RR). In total, DMI observations in the first 44 lactation weeks from 813 primiparious Holstein cows and 301 primiparious Jersey cows were included. Two RR methods, RR using linear splines and RR using Legendre polynomials, were used to estimate the variance components for DMI and AFI for each breed. The analyses for AFI were carried out by adjusting DMI observations for the cows' milk yields, metabolic body weights (MBW), and the change of body weight during the 44 lactation weeks in the genetic model. The results indicated potential differences between Holstein and Jersey cows in genetic variance for DMI and AFI. The genetic variance for AFI was lower than that for DMI over the 44 lactation weeks. The genetic variance for DMI significantly increased along lactation in each breed, whereas the genetic variance was relatively stable for AFI during the 44 lactation weeks. Both DMI and AFI were moderately heritable, but the heritability for AFI tended to be lower than that for DMI. The heritability for DMI and AFI were less different between breeds compared with genetic variances. In addition, the breed difference was also shown in the regression coefficient of DMI on the change of body weight, whereas the two breeds were similar in their regression coefficients of DMI on milk yields and MBW. The results from the two RR analyses in this study showed high consistency. Based on our current findings, heterogeneity might exist between dairy breeds in genetic variation for feed intake and for AFI, which should be considered in a future multi-breed genetic evaluation for feed efficiency in dairy cattle. Additionally, adjusted feed intake could be a candidate trait for feed efficiency in dairy cattle due to its moderate heritability and its considerations on cows' milk production and body maintenance.

Transcriptomics of feed efficiency in Nordic dairy cattle

S.M. Salleh[1], G. Mazzoni[1], J.K. Höglund[2], D.W. Olijhoek[2,3], P. Lund[2], P. Løvendahl[3] and H.N. Kadarmideen[1]
[1]*University of Copenhagen, Dept. of Large Animal Sciences, SUND, Grønnegårdsvej 7, 1870, Frederiksberg C, Denmark, [2]Aarhus University, Dept. of Animal Science, Blichers Allé 20, 8830 Tjele, Denmark, [3]Aarhus University, Dept. of Molecular Biology and Genetics, Blichers Allé 20, 8830 Tjele, Denmark; surayams@sund.ku.dk*

Feed efficiency (FE) in dairy cattle needs to be improved for sustainable (milk) production. The understanding of the biological mechanisms and the role of gene expression patterns across the whole genome (transcriptomics) may be helpful in order to identify and select animals with high FE. In the present study, the high throughput of RNA sequencing was used to detect genes that are deferentially expressed (DE) between two groups of extreme cows (high and low FE), to detect molecular pathways for FE attributes. Ten Jersey and 9 Holstein cows were used in the experiment. These two groups of cows were divided into two FE groups depending on their FE status. The two FE groups also received two different diets (Control (C) and High Concentrate (HC)) which is contained 70:30 and 40:60 of forage:concentrate, respectively. This enabled us to compare the interaction between FE status and two different diets. mRNA from liver biopsies of the cows analyzed by paired end RNA sequencing resulted in an average of 57 million reads per sample. The RNAseq gene expression data were then analyzed using statistical-bioinformatics pipelines to identify DE genes. On average, 91% of reads per sample were mapped with STAR aligner. The expression of 24,616 genes annotated from Ensembl was quantified with HTSeq-count and analyzed with DESeq2 R package. We compared gene expressions of the FE groups, and identified 61 and 21 DE genes in Holstein and Jersey, respectively. An interaction term (high vs low FE × C vs HC) detected two significantly DE genes in Jersey cows. The biological function of the DE genes was analyzed using the GOSeq R package with the human orthologues of the DE genes. In conclusion, the transcriptomics approach was effective in detecting enrichment of pathways for autoimmune thyroid diseases, metabolism, synthesis of ketone bodies and the immune system being different between High and Low FE cows.

Dissecting purebred-crossbred interaction for feed intake in pigs: feed ingredients

R. Bergsma[1], R.M. Godinho[2] and C.A. Sevillano[1]
[1]TopigsNorssvin Research Center, R&D, P.O. Box 43, 6640 AA Beuningen, the Netherlands, [2]Universidade Federal de Viçosa, Animals science department, 36570-000, Viçosa, MG, Brazil; rob.bergsma@topigsnorsvin.com

In pig breeding traditionally selection takes place in purebreds (PB) on nucleus level while on the commercial level crossbred animals (CB) are used. It has been shown that the realized genetic trend in PB on nucleus level is not fully expressed in CB, especially for feed efficiency traits like daily feed intake (DFI) and average daily gain (ADG). Both a genetic correlation lower than one and a reduced additive variance in CB are responsible. One of the possible causes might be a genotype by environment interaction or more specific a genotype by feed interaction. To test this hypothesis a trial was set up were littermates were split and fed either a corn/soy diet (CS) as is typically fed to grower-finishers in the Americas or a wheat/barley/by-product diet (WB) as is common in Western Europe. Animals fed a WB-diet started off a little slower but consumed around 300 g feed more per day during the finishing phase than animals fed a CS-diet. Because of different energy densities of the diets, the energy intake over the entire grower-finisher phase didn't differ between diets. The results suggest that DFI during the starter phase is limited by volume and DFI during the finishing phase is determined by energy demand and perhaps to a lesser extent by the nutritional history of an animal. Preliminary results show that the genetic correlation (r_g) for ADG is 1.0 between both diets. For DFI the r_g is lower (0.89), although not significantly different from one. Regardless of the diets used, weekly averages of feed intake show genetic correlations differing from 1.0 among each other. Especially DFI during the first month after penning seems a somewhat different trait compared to DFI during the remainder of the growing-finishing period. The different distribution of feed intake over the growing-finishing phase might have caused the somewhat lower r_g for DFI. However, the additive variance for ADG and DFI for animals fed a WB-diet was decreased by around 30% compared to CS-fed animals. Giving rise to different expected genetic trends depending on the selection and production environment.

Partitioning genetic variance of metabolizable energy efficiency in dairy cows

T. Mehtiö, E. Negussie, P. Mäntysaari, E.A. Mäntysaari and M.H. Lidauer
Natural Resources Institute Finland (Luke), Myllytie 1, FI-31600 Jokioinen, Finland; martin.lidauer@luke.fi

Residual energy intake (REI) in dairy cows describes the overall efficiency of an animal in using metabolizable energy. Better understanding of the relationship between the overall efficiency and the different energy pathways is desirable for developing genetic evaluations. In this study our aim was to assess the partitioning of genetic variance of adjusted metabolizable energy (ME) intake, we call it metabolizable energy efficiency (MEE), by four energy pathways: ME use for maintenance, milk production, growth, and ME utilization from tissue. A random regression model was applied, which described the pathways by fixed partial regression coefficients on metabolizable body weight, energy corrected milk, body weight gain and body weight loss, and by two likewise sets of random partial regression coefficients, one set for each, permanent environment and additive genetic effect (pMEE model). Obtained results were compared with estimates from repeatability models for MEE and REI. The conceptual difference between latter two models is that for MEE the cows' energy requirements were fitted directly by the model. All models were tested on data collected from 1998 to 2014 at Luke's experimental farms in Jokioinen. The data included 12 350 weekly MEI and REI records from 495 primiparous Nordic Red dairy cows from lactation week 2 to 40. We found that pMEE fitted the data best and resulted in lowest residual variance, but heritability estimates were higher for MEE (0.26) and REI (0.33). In contrast, overall heritability for MEE, derived from the pMEE model estimates for a cow with average energy requirements, was 0.19. Derived heritability for partial MEE was 0.05, 0.06, and zero for maintenance, milk production, and for the intercept and remaining energy pathways. Genetic correlations between the five regression coefficients varied between -0.02 and 0.03 except for a fairly high negative correlation (-0.61) between the maintenance and milk production pathways.

Live weight prediction and genetic parameter estimation using type traits for Italian Holstein cows

R. Finocchiaro[1], J.B.C.H.M. Van Kaam[1], M. Marusi[1], A. Varotto[1,2] and M. Cassandro[2]
[1]*Italian Holstein Association (ANAFI), Research and Development Office, Via Bergamo 292, 26100 Crepma, Italy,*
[2]*University of Padova, Agripolis, 2Department of Agronomy, Food, Natural resources, Animals and Environment, Viale dell'Università 16, 35020 Legnaro (PD), Italy; jtkaam@anafi.it*

Milk production costs are influenced by feed costs, and therefore there is growing interest in improving feed efficiency. Dry matter intake is an interesting trait for management purposes and could also be included in breeding schemes. However, to directly collect this trait is not easy. In the past several models were developed in order to estimate feed intake starting from other animal traits. Live weight (LWT) is an important trait used in order to maintain cost within herds. It is unfeasible to get routine data collection on LWT, but this can be predicted from type traits. Live weight data for Italian Holstein cows was predicted from different type traits based on actual LWT and type data of 890 first parity cows collected from 30 different herds over a 3-yr period (2013-2015). Actual LWT was collected thanks to availability of milking robots including scales within the herds. Average weight was equal to 598.24±73.00 kg and average age was 30 months at classification. The predictive ability of models was tested in 2-fold cross-validation. Correlation between training and validation data-sets ranged from 0.55 to 0.70. The relationships between actual LWT and selected type traits, including stature, body condition score, chest width, body depth and rump width, were calculated to assess the usefulness of various traits to predict LWT. Model used for genetic parameter estimation included herd-year-season, month of calving, age at classification and interval in days (±30 days) between the day of weighing and the day of type classification. The latter was included in order to increase the number of cows with both LWT and type recorded. Heritability for actual LWT was 0.51±0.06. LWT phenotypic prediction equation is going to be implemented in the national type classification system. Next step will be the estimation of genetic correlations with other traits, the introduction of predicted LWT in a functional index and further the estimation of a predicted feed efficiency index in the Italian Holstein population.

Gompertz model improves breeding value prediction for feed conversion ratio for incomplete weights

V.H. Tran-Huynh, H. Gilbert and I. David
UMR INRA / INPT ENSAT / INPT ENVT, Génétique, Physiologie et Systèmes d'élevage, chemin de borde rouge, 31326, France; van-hung.tran@toulouse.inra.fr

With the development of automatic self-feeders, repeated measurements of feed intake are becoming available in more species. However, corresponding weights are not always recorded which complicate the longitudinal analysis of feed efficiency due to missing values. In our study, male pigs were weekly weighted while females and castrated males were monthly weighted. Our aim is thus to evaluate whether the use of Gompertz model in predicting the missing weights could improve the estimation of breeding values of Feed Conversion Ratio (FCR). For this study, we used 18,387 weekly records of weight and feed intake which were recorded over 18 consecutive weeks for 1,286 male growing pigs. A simulation study was carried out on this dataset to mimic weight missing values according to the pattern of weekly-proportions of incomplete weights observed in females and castrated males. Next, Gompertz model was applied to the data with mimic missing weights and the estimates were used to predict missing records. Then FCR was computed for each week using observed data (FCRo), data with missing weight (FCRm) and data with predicted weight using the Gompertz model (FCRg). Finally, breeding values for FCRo, FCRg and FCRm (EBVo, EBVg and EBVm, respectively) were estimated using Random regression models. We compared the EBV using Pearson correlation coefficient (R^2). Over 100 simulations, the mean of R^2 between EBVo and EBVg was 0.87±0.01 and 0.84±0.02 between EBVo and EBVm. Thus predicting missing weight records using a Gompertz model improved the genetic evaluation of FCR.

Environmental impact of feed optimization: alternative protein sources in pig diets

H.H.E. Van Zanten, P. Bikker, B. Meerburg and I.J.M. De Boer
Wageningen University, De Elst 1, 6708 WD, the Netherlands; hannah.vanzanten@wur.nl

Feed production is responsible for the majority of the environmental impact of livestock production, especially for monogastric animals, such as pigs. Several studies demonstrated that replacing soybean meal (SBM) with alternative protein sources, e.g. waste-fed insects, reduces the environmental impact of pork production. These studies, however, used an attributional life cycle assessment (ALCA), which solely addresses the direct environmental impact of a product. But replacement of SBM, can also have indirect environmental consequences, e.g. impacts related to replacing the original function of the alternative protein source. Food waste to feed insects, for example, can also be used to produce bio-energy. Accounting for indirect environmental consequences might change environmental benefits. This study explored differences in results when including or excluding indirect environmental consequences, related to two case studies: replacing SBM with rapeseed meal (RSM), and replacing SBM with waste-fed larvae meal in diets of finishing pigs. Direct impact results for replacing SBM with RSM showed that global warming potential (GWP) (3%) and energy use (EU) (1%) hardly changed, but land use (LU) was decreased (-16%). Indirect impact results, however, showed that GWP (15%), EU (12%), and LU (10%) increased. Direct impact results for replacing SBM with waste-fed larvae meal showed that EU hardly changed (1%), but GWP (-29%) and LU (-54%) decreased. Indirect impact results, however, showed that GWP (60%) and EU (90%) increased, but LU (-73%) decreased. Environmental benefits from replacing SBM with RSM or larvae meal were more promising when only direct environmental impacts were considered. Indirect environmental impacts results showed that, replacing SBM with RSM or larvae meal resulted in pollution swapping. This would have been overlooked when results were only based on the direct environmental impact. Related to feed optimization, we recommend animal nutritionists to first assess the direct environmental impact to get insight in the environmental impact of their feed. If policy makers want to evaluate the environmental benefits of a mitigation strategy it is recommended to perform also the indirect environmental consequences.

A diet formulation tool designed to reduce the environmental impact of pig production systems

S.G. Mackenzie[1], I. Leinonen[1], N. Ferguson[2] and I. Kyriazakis[1]
[1]School of Agriculture, Food and Rural Development, Agriculture Building, Newcastle University, Newcastle upon Tyne, NE1 7RU, United Kingdom, [2]Trouw Nutrition Canada, 150 Research Ln, Guelph, ON N1G 4T2, Canada; ilias.kyriazakis@newcastle.ac.uk

The objective was to develop a novel methodology that enables pig diets to be formulated explicitly for environmental impact objectives using a Life Cycle Assessment (LCA) approach. To achieve this, the following methodological issues were addressed: (1) account for environmental impacts caused by both ingredient choice and nutrient excretion; (2) formulate diets for multiple environmental impact objectives, and 3) allow flexibility to identify the optimal nutritional composition for each environmental impact objective. An LCA model based on Canadian pig farms was integrated into a diet formulation tool to compare the use of different ingredients in Eastern and Western Canada. By allowing the feed energy content to vary, it was possible to identify the optimum energy density for different environmental impact objectives, whilst accounting for the expected effect of energy density on feed intake. A least cost diet was compared with diets formulated to minimise the following environmental impacts: non-renewable resource use (NRRU), acidification potential (AP), eutrophication potential (EP), global warming potential (GWP) and a combined environmental impact score (using these four categories). The resulting environmental impacts were compared using parallel Monte-Carlo simulations to account for shared uncertainty. When optimising diets to minimise a single environmental impact category, reductions in the said category were observed in all cases. However, this was at the expense of increasing the impact in other categories and higher dietary costs, for e.g. the least GWP diet also reduced NRRU but increased AP and EP in both regions (all P<0.05). The methodology can identify nutritional strategies to minimise environmental impacts, such as increasing the nutritional density of the diets compared to the least cost formulation.

Economic and environmental optimization of feed sequence plans in pig fattening unit
F. Garcia-Launay, L. Brossard, T. Plantevin and J.Y. Dourmad
INRA, Agrocampus Ouest, UMR1348 PEGASE, Domaine de la Prise, 35590 Saint-Gilles, France;
florence.garcia-launay@rennes.inra.fr

Pig livestock farming systems face economic and environmental issues. The pig fattening unit is responsible for most of the feeding cost and nutrient excretion in pig farming. Thus, optimizing the feed sequence plans of growing-finishing pigs is a way to improve the economic and environmental outputs on a commercial farm. We developed a model, called PigOptim that optimizes the feed sequence plan of a population of growing pigs. PigOptim calculates the environmental impacts of pig production by Life Cycle Analysis (LCA). It optimizes the number of feeding phases, the average live weight at the beginning of each phase, the lysine supply (from 70% to 130% of the average requirement of the population) and the formula of each feed with a multi-objective function which integrates gross margin (GM) of the unit and its impact on Climate Change (CC). PigOptim includes response laws of growth performance issued from simulations performed with InraPorc® software for a population. For an optimization of GM only, the best results of GM and CC were obtained for a lysine supply between 90% and 100% of the average requirement of the population, whereas animal performance was maximized at 110% of the average requirement. Indeed, reducing feed costs between 110% and 90% of the average lysine requirement compensates for the reduction in growth performance. This result highlights the interest of models accounting for animal variability when investigating economic and environmental issues. The simultaneous optimization of GM and CC led to a reduction of GM and CC, but some other environmental impacts increased. These first results suggest that PigOptim gives consistent estimations of animal performance, GM and environmental impacts. To our knowledge, PigOptim is the first model optimizing feed sequence plan on criteria calculated by LCA. In the future, PigOptim will have to include various environmental impacts in the objective function and will need an improved optimization process for further integration into a decision support system.

The effect of nutrition on environmental impacts of pig production depends on production context
A.N.T.R. Monteiro[1], F. Garcia-Launay[2], L. Brossard[2], A. Wilfart[3] and J.Y. Dourmad[2]
[1]Animal Science Department, Maringá State University, PR, Brazil, [2]INRA AgroCampus-Ouest, UMR PEGASE, 35590 Saint-Gilles, France, [3]INRA AgroCampus-Ouest, UMR SAS, 35000, Rennes, France;
florence.garcia-launay@rennes.inra.fr

The objective of this study was to evaluate with life cycle assessment (LCA) the effects of different feeding strategies, including phase-feeding, precision-feeding and addition of feed-use amino acids (FU-AA), on the environmental impacts of pig production, in two contrasted geographic contexts, Brazil and France, with soybean from South (SO) or Center West (CW) Brazil. The LCA considered the process of pig fattening, including production and transport of feed ingredients and complete feeds, raising of pigs, and manure management. Impacts were calculated at farm gate and the functional unit was one kg of body weight gain over fattening. Performances of pigs were simulated for each feeding scenario (n=96) using InraPorc population model (2,000 pigs per scenario) and considering between-animal variability. The LCA calculations were performed for each pig according to its own performance and excretion, and the results were subjected to variance analysis. The results indicate that for some impacts there are clear interactions between effects of feeding program, origin of soybean and location of production. For climate change, the interest of phase feeding and incorporation of FU-AA is limited and may even be counterproductive in Brazil with soybean from SO, whereas it appears to be an efficient strategy with soybean from CW, especially in Europe. Similar effects are observed for cumulative energy demand. Terrestrial ecotoxicity is lower and almost not affected by feeding strategy in Brazil, whereas in France this impact is reduced by both increasing the number of feeding phases and including more FU-AA. Conversely, eutrophication and acidification potential impacts are reduced by phase feeding and FU-AA addition in a rather similar way in all situations. In all situations, precision feeding appears the most effective feeding program for reducing life cycle impact of pig fattening.

Greenhouse gas emissions from alternative calf to beef production systems

P. Crosson[1], B. Murphy[1,2] and R. Prendiville[1]
[1]Teagasc, Animal & Grassland Research and Innovation Centre, Grange, Dunsany, Co. Meath., Ireland, [2]Univesity College Dublin, School of Agriculture and Food Science, Belfield, Dublin 4, Ireland; paul.crosson@teagasc.ie

The Irish dairy cow herd has expanded by 21% between 2010 and 2015 with further expansion forecast following the abolition of EU milk quotas. This has led to an increased interest in the production of beef from dairy origin calves. Of particular interest are the opportunities for finishing male calves as bulls rather than steers to avail of the increased inherent efficiencies of bull beef. At the same time, the Irish agricultural sector is faced with the challenge of greenhouse gas (GHG) emissions, with almost one third of total national emissions emanating from that sector. Given the importance of ruminant livestock to the agricultural sector, with almost 70% of the land area in grassland, emissions from cattle enterprises are a significant contributor to agricultural emissions. Thus, the objective of this present study was to quantify GHG emissions generated from alternative dairy calf to beef systems. Biological data from ongoing calf to beef production research was obtained and used to parameterise an existing beef production systems GHG model. Eight alternative production systems were evaluated over two consecutive years. The production system alternatives represented differences in gender, age at slaughter and level of concentrate feeding. Results showed that finishing as steers at 24 and bulls at 16 months of age had the highest (14.36 kg carbon dioxide equivalents (CO_2e)) and lowest (9.41 kg CO_2e), respectively, GHG emissions per kg carcass. In contrast, when GHG emissions were quantified per unit of land used, finishing as bulls at 16 months of age had the highest emissions (31.7 t CO_2e/ kg beef carcass) whereas finishing as steers at 24 months of age were lowest (12.5 t CO_2e/kg beef carcass). Enteric fermentation represented the most important GHG emissions source being responsible for between 51% to 60% of total system emissions. Higher concentrate feeding during finishing for bulls finished at 19 months of age reduced GHG emissions for all metrics measured. There was no effect on GHG emissions of supplementing calves at pasture in their first season with either 1 or 2 kg of concentrates.

Effect of tannins from Birdsfoot trefoil and dietary protein level on nitrogen turnover of lambs

E. Seoni[1,2], G. Battacone[1], M. Corpataux[2], Y. Arrigo[2], F. Dohme-Meier[2] and G. Bee[2]
[1]Università di Sassari, Dipartimento di Agraria, Viale Italia, 39, 07100 Sassari, Italy, [2]Agroscope, Institute for Livestock Sciences, route de la Tioleyre 4, 1725 Posieux, Switzerland; eleonora.seoni@agroscope.admin.ch

The study aimed to investigate the interaction between dietary crude protein (CP) level and condensed tannins (CT) from Birdsfoot trefoil (BT) on nitrogen turnover of growing lambs. The study was carried out with 24 White Alpine ram lambs (21.7±2.7 kg) and arranged according to a 2 × 2 factorial design. The basal diet consisted of BT silage and hay. In addition, the lambs were offered either a concentrate to meet the predicted protein requirements (BP) or a concentrate with a CP level 20% above the requirements (HP). To determine the effect of CT, polyethylene glycol (PEG), known to deactivate the action of CT were used in 2 of the 4 diets: BP+PEG (BP+), HP+PEG (HP+) and BP-, HP-. The diets were offered daily as 2 equal meals at 07:00 and 15:00 for 3 h. After an adaptation period of 14 d, lambs were kept in metabolic cages for 7 d. Feed intake and refusals, feces and urine were collected daily. As expected, N intake and N excretion were greater ($P<0.001$) when the HP- and HP+ compared with BP- and BP+ diets were offered. When expressed per kg $BW^{0.75}$, CT reduced ($P<0.05$) daily N intake and total N excretion by 11%. Lambs fed CT excreted less urinary N ($P<0.05$), whereas neither fecal N excretion nor body N retention differed ($P>0.05$). Total N excretion and body N retention expressed as percentage of N intake were not affected ($P>0.05$) by CT. In addition, fecal N expressed as percentage of total N intake was lower ($P<0.05$) in the HP- and HP+ compared to the BP- and BP+ groups. These results indicate that the inclusion of CT from BT in lambs diet did not improve body N retention.

Effects of breed and concentrate supplementation on nitrogen utilisation in ewes fed fresh grass

Y.G. Zhao[1,2], A. Aubry[1], N.E. O'Connell[2] and T. Yan[1]
[1]*Agri-Food and Biosciences Institute, Hillsborough, Co. Down, BT26 6DR, United Kingdom, [2]Institute for Global Food Security, Queen's University Belfast, BT9 5BN, United Kingdom; yiguang.zhao@afbini.gov.uk*

Nitrogen (N) excretion from sheep production systems is a considerable source of nitrate and nitrous oxide emissions, responsible for groundwater pollution and global warming. This study aimed to investigate the effects of breed and concentrate supplementation on N utilisation efficiency in sheep. Sixteen lowland replacement ewes were used in a 2 breeds (Highlander cross vs Texel cross) × 2 diets (fresh grass vs fresh grass plus 0.5 kg/d (fresh weight) pelleted concentrate) factorial design study. Animals were 14 months old and weighed 61±5.3 kg at the commencement of the study. Fresh perennial ryegrass was harvested daily and offered ad libitum, and provided a high quality forage. The animals were individually housed in pens and fed experimental diets for 19 days before transferred to individual metabolism crates for a further 4 days with N intake and outputs in faeces and urine measured. Data were analysed as a 2 × 2 factorial arrangement using the Analysis of Variance. There were no significant interactions between breed and diet on any variable of N utilisation efficiency. Sheep offered concentrate diets had greater N intake (P<0.001), faecal N (P=0.038), urine N (P=0.002) and manure N (P=0.002) outputs than those given fresh grass only. However, diets had no significant effect on retained N, faecal N/N intake, urine N/N intake, manure N/N intake and urine N/manure N. There was no significant effect of breed on any variable of N intake, outputs and utilisation efficiency. The average manure N/N intake was 74% in the present study, which is in line with those obtained from growing beef cattle. Neither breed nor concentrate supplementation had significant effects on N utilisation efficiency in sheep when fed high quality fresh grass, which has important implications for the development of strategies to reduce pollution from N sources.

Treatment with lactic acid of concentrates alleviates the lack of inorganic P in dairy cow diets

Q. Zebeli, A. Khol-Parisini, E. Humer, H. Harder and E. Mickdam
Innstitute of Animal Nutrition and Functional Plant Compounds, Vetmeduni Vienna, Veterinaerplatz 1, 1210, Austria; qendrim.zebeli@vetmeduni.ac.at

Previous research indicated a potential to improve the availability of organically-bound P by treating cereals with lactic acid (LA) in vitro. The aim of this study was to evaluate the effects of feeding a concentrate with or without inorganic P supplementation and treated in 5% LA on milk production and blood metabolites in early-lactating dairy cows, sampled from d 1 until 37 postpartum. Cows had ad libitum access to a total mixed ration for early-lactating cows that differed with respect to the supplementation of inorganic P and the processing of the concentrates; cows were offered either diets supplemented with 0.8% monocalciumphosphate as inorganic P source, or a diet containing 0% inorganic P sources. The concentrates of the diets were either treated with 5% LA for 24 h prior to feeding or not. Dry matter intake (DMI), milk yield and body weights (BW) were recorded and selected blood metabolites related to glucose and lipid metabolism were determined. Data showed that DMI was lower (P<0.05) in cows receiving LA-treated concentrates, whereby the differences were especially pronounced during the first 3 weeks and disapeared thereafter. Overall, milk yield remained similar among treatment groups. The cows receiving the LA-treated diets had lower (P<0.05) serum non-esterified fatty acids (NEFA), cholesterol and insulin concentrations. Furthermore, serum P was tendentially (P<0.10) enhanced in cows receiving LA-treated concentrates. The BW and BW changes did not differ among the treatment groups. However, milk fat yield was lower (P<0.05) in cows fed the LA diet without P, but not in the LA diet with P. Taken together, except for lowered milk fat yield, the LA-treated diet without inorganic P supplementation did not exert any negative effects on milk production, metabolism, and BW change. Furthermore, the observed modulations in blood metabolites due to the LA-treatments suggest that diets including concentrates treated with 5% LA allow savings of inorganic P supplements and improve the metabolic status of early-lactating dairy cows.

Effect of pure extracts on in vitro CH$_4$ production using rumen fluid of cows fed the same extracts
G. Rossi[1], L. Maccarana[2], V.A. Vieira[3], M. Cattani[2], F. Tagliapietra[1], S. Schiavon[1] and L. Bailoni[2]
[1]*University of Padova, DAFNAE, Viale dell'Università 16, 35020 Legnaro, Italy, [2]University of Padova, BCA, Viale dell'Università 16, 35020 Legnaro, Italy, [3]State University of Sao Paulo, Via de Acesso Prof. Paulo D. Castellane, 14884-900, Jaboticabal, Brazil; giulia.rossi.8@studenti.unipd.it*

This study explored effects of three pure extracts (allyl sulfide, AS; cinnamaldehyde, CI; limonene, LI) on in vitro fermentation of a dairy cow diet, using rumen fluid collected from cows fed the same diet supplemented ('adapted fluid'; AF) or not ('not adapted fluid'; NAF) with the same extracts. In 4 periods of 15 d, 4 cows received a diet supplemented or not with 1 g/d of AS, CI, or LI, according to a 4×4 Latin Square design. On the last day of each period, rumen fluid was collected from each cow to be used for in vitro tests (4 incubations). In each incubation, 10 treatments were tested in 4 replications: NAF (not adapted fluid); AF$_{AS}$, AF$_{CI}$, and AF$_{LI}$ (fluid adapted to AS, CI, or LI); NAF+AS, NAF+CI, and NAF+LI (not adapted fluid with 30 mg in vitro of AS, CI, or LI); AF$_{AS}$+AS, AF$_{CI}$+CI, and AF$_{LI}$+LI (fluid adapted to AS, CI, or LI with 30 mg in vitro of AS, CI, or LI). At the end of incubations, NDF degradability (NDFd), gas, CH$_4$, and volatile fatty acid production were measured. Data (mean of 4 replications) were analyzed considering the treatments and cow as fixed factors, and treatment×cow as random factor. Contrasts were run to compare treatments. Compared to NAF, the NAF+LI reduced strongly (P<0.001) in vitro NDFd (-28%) and CH$_4$ production (-30%). The NAF+AS reduced strongly CH$_4$ production (-32%; P<0.001) without influencing degradability, whereas the NAF+CI showed weak effects. No differences were observed (P>0.05) on in vitro parameters when the four rumen fluids (NAF, AF$_{AS}$, AF$_{CI}$, and AF$_{LI}$) were used as inoculum. Also, the addition of extracts in vitro to the NAF or to the AF provided comparable results (P>0.05). The small effect of pure extracts on rumen fluid activity could be due to a microbial adaptation or to the low dosage used in vivo.

Methane mitigation in dairy cattle with 3-NOP in an on-farm trial
D. Van Wesemael[1], N. Peiren[1], L. Vandaele[1], V. Fievez[2], S. Duval[3] and S. De Campeneere[1]
[1]*Institute for Agricultural and Fisheries Research, Animal Science unit, Scheldeweg 68, 9090 Melle, Belgium, [2]Ghent University, Laboratory for Animal Nutrition and Animal Product Quality, Proefhoevestraat 10, 9090 Melle, Belgium, [3]DSM Nutritional Products, Research Centre for Animal Nutrition and Health, BP 170, 68305 Saint-Louis Cedex, France; dorien.vanwesemael@ilvo.vlaanderen.be*

Nutritional interventions are widely investigated to mitigate methane emissions from ruminants. The methane reducing potential of the feed additive 3-nitrooxypropanol (3-NOP) in dairy cattle was explored. Roughage mixture consisted of 50% maize silage, 40% pre-wilted grass silage and 10% pressed beet pulp on a DM basis. The amount of concentrates (balanced concentrate and (protected) soybean meal), was calculated on an individual basis, to meet the needs of VEM and DPI. Methane emissions of 10 cows (86±31 DIM; on average 21±3 kg DMI/d) were measured with a GreenFeed in the free stall with cubicles at ILVO during a 7 days control period (CP; no treatment) and a 7 days treatment period (TP; from 12 days after first treatment). Treatment involved an additive (1.7 g active compound/day) for 8 cows and a placebo for 2 reference cows. Additive or placebo were mixed in soybean meal and administered via a concentrate feeder. Milk production (MP) and dry matter intake (DMI) were monitored. Daily methane emissions decreased for the treated cows from 482±55 to 390±46 g CH$_4$/day (P<0.01), whereas for the reference cows the daily methane emissions where 462±59 in the CP and 446±53 g CH$_4$/day in the TP (P=0.94). Expressed per kilogram of DMI the additive reduced the methane yield from 23±2 to 19±3 g CH$_4$/kg DMI (P<0.01), where the reference cows produced 23±3 in the CP and 21±1 g CH$_4$/kg DMI in the TP (P=0.48). When MP was considered, no significant change was observed. The reference cows displayed values of 13±2 and 14±2 g CH$_4$/kg MP (P=0.92), whereas the treated cows produced 15±2 and 14±3 g CH$_4$/kg MP (P=0.52), during CP and TP respectively.

Effect of linseed plus nitrate on methane emission in bulls, animal health and residues in offal

M. Doreau, M. Arbre, Y. Rochette, C. Lascoux and C. Martin
INRA, UMRH, Saint-Genès Champanelle, 63122, France; michel.doreau@clermont.inra.fr

The association of linseed and nitrate decreases enteric methane (CH_4) emission in dairy cows, but has not been studied in fattening cattle for animal performance, CH_4 emission, animal health and presence of residues in beef. To address these questions, 16 young Charolais bulls received from weaning at 9 months to 14 months a control diet (C), then were split in 2 groups of 8 balanced for CH_4 emission, dry matter intake (DMI) and liveweight gain (LWG). They received during 2 months either C diet or a diet supplemented with extruded linseed and calcium nitrate (LN); they were then slaughtered. The C diet contained 67% wrapped haylage and 33% concentrate mainly made of maize, wheat and rapeseed meal. In LN diet, rapeseed meal and a part of cereals were replaced by 35% extruded linseed and 6% calcium nitrate; linseed fatty acids and nitrate supply in the diet was 1.9 and 1.0%, respectively. Methane emission was measured continuously using the Greenfeed system. Methaemoglobin was determined in peripheral blood every week for bulls receiving the LN diet. Nitrate and nitrite concentrations were determined in rumen, liver and tongue sampled at slaughter. Total DMI did not differ between diets (11.0 ± 1.2 kg/d on average, P>0.05) but concentrate intake was lower for LN due to refusals, so that LWG was higher for C than for LN (1.60 and 1.26 kg/d, respectively, P=0.013). Daily CH_4 emission was lower by 10% for LN than for C (241 and 269 g/d, respectively; P<0.001), but CH_4 emission per kg DMI or per kg LWG did not vary (P>0.05). Methaemoglobin was under detection limit for most animals (<2% of total haemoglobin), and always lower than 5.6%, suggesting the absence of risk for animal health. Nitrite and nitrate concentrations in offal did not differ between diets. In conclusion, a moderate supply of linseed and nitrate in bulls resulted in a moderate decrease in daily enteric CH_4 emission without adverse effects for animal health and food safety.

Effects of dietary nitrate and increased lipid on methane emissions from beef cattle are independent

S. Troy[1], J.A. Rooke[1], R.J. Wallace[2], C.-A. Duthie[1], J.J. Hyslop[3], D.W. Ross[1], T. Waterhouse[1] and R. Roehe[1]
[1]SRUC, West Mains Road, EH9 3JG Edinburgh, United Kingdom, [2]Rowett Institute of Nutrition and Health, Bucksburn, AB21 9SB Aberdeen, United Kingdom, [3]SAC Commercial Ltd, West Mains Road, EH9 3JG Edinburgh, United Kingdom; shane.troy@sruc.ac.uk

This experiment was designed to investigate the combined effects of adding nitrate or increasing dietary oil on methane (CH_4) emissions from finishing beef cattle fed a mixed 560:440 forage:concentrate diet (DM basis). The experiment had a 2×4 factorial design where the factors were steer genotype (crossbred Aberdeen Angus (AA) or crossbred Limousin (Lim)) with four CH_4 mitigation treatments (control (24 g ether extract (EE) / kg DM, Ctrl); nitrate (19 g nitrate/kg DM, Nit); lipid (37 g EE /kg DM, Lip); combined, (19 g nitrate and 37 g EE / kg DM, Comb). Dietary EE was increased by replacing rapeseed meal with maize distillers dark grains. CH_4 emissions were measured using 6 open-circuit respiration chambers. Steers (18 per treatment; mean live-weight, 659 kg) were assigned to chambers using a randomised block design in 13 weekly periods. Data were analysed using general linear models with fixed effects of breed, nitrate addition and increased lipid and random effects of chamber and block. Lim steers consumed less DM (kg/d 9.3 vs 11.0, P<0.001) and produced more CH_4/kg DM intake (23.2 vs 22.0, P<0.05) than AA steers. Daily CH_4 output (g/d, Ctrl 246; Nit 219; Lip 238; Comb 210) was lower (P<0.001) on nitrate-containing diets than the Ctrl but increasing dietary EE had no significant effect. Similarly, when expressed as g CH_4/kg DM intake (Ctrl 24.6; Nit 22.1; Lip 23.8; Comb 21.0) steers receiving nitrate (Nit and Comb) produced 2.1 g/kg DMI less CH_4 than steers receiving Ctrl, which equated to only 45% of the theoretical maximum CH_4 reduction expected for the amount of nitrate added. In conclusion, there was no interaction between added nitrate and increased lipid concentration on CH_4 emissions and therefore the effects of nitrate or lipid in a mixed forage: concentrate diet were additive. Study supported by EBLEX and Scottish Government.

How does nutrient intake affect methane emission from slurry in pigs?

W. Antezana[1], S. Calvet[1], P. Ferrer[1], P. García-Rebollar[2], C. De Blas[2] and A. Cerisuelo[3]
[1]Universitat Politècnica de València, Camino de Vera, s/n, 46022, Spain, [2]Universidad Politècnica de Madrid, Ciudad Universitaria, s/n, 28040 Madrid, Spain, [3]Instituto Valenciano de Investigaciones Agrarias, Centro de Investigación y Tecnología Animal, Pol. La Esperanza, 100, 12400 Segorbe (Castellón), Spain; cerisuelo_alb@gva.es

The relationship between nutrient intake (g kg $PV^{-0.7}/5$ d^1), nutrient excretion (g kg $PV^{-0.75}/d$) and methane (CH_4, L kg $PV^{-0.75}/d^1$) emission from slurry was investigated using data from three experiments designed to study the effect of feed formulation on gas emissions. A total of 13 diets with different protein, fibre and fat content were assayed in 78 fattening pigs. Individual feed intake and the total amount of urine and faeces produced over seven consecutive days per pig was measured. Feeds and faeces were analysed for dry matter, energy (GE), crude protein (CP), ether extract (EE), soluble fibre (SF), neutral detergent fibre (NDF), acid detergent fibre and acid detergent lignin. Urine was analysed for dry matter and N content. Additionally, potential CH_4 emission from slurry was measured using a vitro batch assay. The whole data set was analysed using the CORR procedure of SAS®. Methane emission was positively related with CP (r: 0.42, P<0.001), GE (0.55, P<0.001), CP-NDF (r: 0.53, P<0.001), EE (r: 0.65, P<0.001) and SF (r: 0.25, P<0.05) intake. Similarly, a positive relationship was found between CH_4 emission and GE, EE and NDF (r: 0.70, r: 0.45 y r: 0.47, respectively; P<0.001) excretion. Therefore, in general, a higher intake of organic matter produces higher CH_4 emission levels. However, once at slurry, GE and, particularly, EE and FND excretion are the components that most determine CH_4 emission. Then, dietary factors affecting energy digestibility in pigs seem to be key factors to control CH_4 emission from slurry. Project funded by the Spanish Ministry of Economy and Competitiveness (AGL2014-56653-C3-3-R).

Isotopic natural abundance as biomarkers of between-animal variation in feed efficiency in ruminants

S.J. Meale[1], I. Ortigues-Marty[1,2], C. Chantelauze[1,2], A.M. Schiphorst[3], R. Robins[3] and G. Cantalapiedra-Hijar[1,2]
[1]INRA, UMR 1213 Herbivores INRA-VetAgro Sup, St Genès Champanelle, 63122, France, [2]Clermont Université, VetAgro Sup, UMR 1213 Herbivores, Clermont-Ferrand, 63000, France, [3]LUNAM Université, University of Nantes, Elucidation of Biosynthesis by Isotopic Spectrometry Group, CEISAM, Nantes, 44322, France; sarah-jade.meale@clermont.inra.fr

Current methods of determining feed efficiency in ruminants are laborious and difficult to measure and consequently, alternative biomarkers are being explored. Based on the idea of isotopic fractionation, we measured the natural abundance of $\delta^{15}N$ and $\delta^{13}C$ (‰) in plasma proteins of 54 Charolais cattle and performed a regression analysis against different feed efficiency indices to determine their potential a as biomarker to predict between-animal variations of feed efficiency. The cattle were examined for feed conversion efficiency (FCE) and residual feed intake (RFI), beginning at 11-13 months of age, over two years (n=20 in 2014 and n=34 in 2015). Despite identical dietary constituents, the crude protein composition varied across the two years (13 vs 15% DM in 2014 and 2015, respectively) and consequently, animals from each year were analyzed separately. The natural abundance of $\delta^{15}N$ in plasma proteins was higher in 2014 cattle ($\delta^{15}N$ av.=6.22), vs 2015 ($\delta^{15}N$ av.=5.54). Whereas, $\delta^{13}C$ was higher in animals from 2015 ($\delta^{13}C$ av.=-24.61) vs 2014 ($\delta^{13}C$ av.=-25.13). A significant negative correlation was observed between $\delta^{15}N$ and FCE in animals from both 2014 and 2015 (R^2=0.62 and R^2=0.52, respectively). Similarly, $\delta^{13}C$ in plasma proteins showed a moderate negative correlation with FCE in 2014 (R^2=0.34), but no correlation with FCE was observed in 2015 (R^2=0.09). Nor was a correlation observed between RFI and either $\delta^{15}N$ (R^2=0.07 and R^2=0.09) or $\delta^{13}C$ (R^2=0.04 and R^2=0.005) in 2014 or 2015. The repeatability of the relationship between $\delta^{15}N$ in plasma protein and feed conversion efficiency in the two groups of cattle indicates its potential as a biomarker between-animal variations of feed efficiency, measured as FCE but not RFI, in ruminants.

Ensiling forage soybean with or without bacterial inoculants on nitrogen and energy retention
B.D. Nkosi and A. Van Niekerk
Animal Production Institute, Agricultural Research Council, Animal Nutrition, Private Bag X2, Irene, 0062, South Africa; dnkosi@arc.agric.za

The study evaluated the effects of ensiling forage soybean (Glycine max (L) Merr.) with or without bacterial inoculants on nitrogen and energy retention in rams. Forage soybean (cv. Link LF 6466) that contains 268 g dry matter (DM); 70 g water-soluble carbohydrates (WSC)/kg DM; 169 g crude protein (CP)/kg DM and pH of 7.04 was harvested at R6 stage, chopped to 15 mm length and ensiled in 210 l drums for 90 d. The forage was treated as follows: (1) control (2 l water/t); (2) Lalsil Dry LB, a heterofermentative LAB inoculant (denoted as LDLB); and (3) Sil-All 4×4® W.S., a homofermentative LAB inoculant (denoted as Sil). After 90 days of ensiling, forage soybean silage (FSS) samples were collected per treatment and analysed for fermentation characteristics and nutritive values. The FSS was top-dressed with 500 g concentrate and fed ad libitum to 18 (6 rams/treatment) matured Damara rams (48.2±2.51 kg live weight) in an N and energy balance study. Rams were placed in individual metabolic crates, and were harnessed with faecal bags prior to urine and faecal samples collection. The trial for 21 days (14 days adaptation plus 7 days sample collection). The intakes of dry matter, N and energy were higher (P<0.05) in rams fed the Sil-treated silage compared to those on the other silages. Further, the retention of both N and energy was improved (P<0.05) in rams fed the Sil-treatment compared to those fed on other silage treatments. It was concluded that the inoculation of forage soybean with Sil was beneficial in reducing the losses of N, hence reduction in environmental pollution.

Nitrogen-use-efficiency of beef cows fed total mixed ration, grass silage or grazed on pasture
H. Scholz[1] and G. Heckenberger[2]
[1]Anhalt University of Applied Sciences, Faculty LOEL, Strenzfelder Allee 28, 06406 Bernburg, Germany, [2]State Institute for Agriculture and Horticulture Saxony-Anhalt, ZTT Iden, Lindenstraße 18, 39606 Iden, Germany; h.scholz@loel.hs-anhalt.de

Nitrogen-use-efficiency (i.e. N-efficiency) is defined as percentage of feed nitrogen (N) that is converted into milk and meat protein. Optimizing N-efficiency has the potential to improve the productivity of livestock systems while reducing their environmental impact. During calving period (2 weeks antepartum and 4 weeks postpartum), 14 beef cows were monitored for daily feed intake. Group 1 received a Total Mix Ration (TMR) containing 52% grass silage, 22% alfalfa silage, 21% corn silage, 4% straw and 1% mineral supplement (all dry matter [DM] basis) Crude protein (CP) and crude fiber content of the TMR was 14% and 27%, respectively. Group 2 received ad libitum grass silage cut to two different chop lengths (15% CP; 27% crude fiber). Mean TMR intake was 17.7 kg DM/d. Intake of grass silage was 14.8 kg DM/d (>10 cm chop length) and 16.9 kg DM/d (≤2 cm chop length), respectively. The percentage share of feces was 25% of dry matter intake and with an amount of N by 26.6 g/kg DM can be calculated N-loss of 27% of N-intake. Mean milk yield of 16.0 kg/d was determined by Weight-Suckle-Weight method. Average milk protein concentration was 3.37%, while milk fat concentration was 3.56%. Milk urea concentration was 237 mg/liter. In this study, N-efficiency from feed to milk was 22%.During the grazing period of suckling cows (June to September) increased N excretion has been observed. Based on estimated N requirements of 85 g/d for maintenance and 93 g/d for milk production there we found N losses in urine and feces by 383 g/d. Feeding TMR or grass silage resulted in an average N loss of 216 g/d (urine and feces combined). Therefore N-efficiency was 18% (calculated based on DM intake of 18 kg/d). It could be shown that there is the N-efficiency of suckler cows at a moderate level compared to dairy cows. However, a more accurate description of N-efficiency of nursing beef cows will require better consideration of N deposition by the suckling calf.

Methanogenesis of a native pasture combined whith trhree protein supplements

M. De J. Marichal, L. Piaggio, R. Comesaña and E. Gomez
Facultad de Agronomía – Universidad de la República, Producción Animal y Pasturas, Garzón 780, 12900
Montevideo, Uruguay; mariadejesus.marichal@gmail.com

The aim of this study was to evaluate the effect of combining forage from a native pasture with protein supplements differing in N degradability on CH_4 production in vitro. Three plots of native pasture (92.7 g CP/kg DM, 41.1% effective degradability of N (NED) and 62.43% IVOMD) were harvested in summer at the Research Center of the Uruguayan Wool Secretariat (S 33°52′, W 55°34′). Protein supplements were sunflower meal (SFM), soybean meal (SBM) and fish meal (FM) (390.4, 515.4 and 676.1 g CP/kg DM and 68.1, 67.2 and 41.7% NED, respectively). Forage was combined with each supplement in such amounts that mixtures contained 120, 140, 160 or 180 g CP/kg DM. Thirteen treatments were evaluated using an in vitro gas production procedure. Rumen contents were collected from two fistulated whethers fed alfalfa hay. Three batches of 3 bottles by treatment, 3 with standard alfalfa and 3 blanks were incubated for 48 h, CH_4 was measured by gas chromatography and expressed per gram of disappeared organic matter (dOM). Analysis of CP levels, testing linear and quadratic effects, was performed by polynomial orthogonal contrasts and CH_4 production at each CP level was analyzed in a complete randomized design with a factorial arrangement of treatments. At each CP level, CH_4 production was lower (P<0.0407) in mixtures with FM than that containing SFM or SBM which registered similar (P>0.1945) values. Forage and all mixtures containing FM produced similar (P=0.497) volume of methane (22.81 ml CH_4/g dOM). When SFM was used a linear (P=0.025) increase in CH_4 production was registered and a trend (P=0.0907) to linearity was observed when SBM was tested. The maximum volumes of CH_4 were registered in mixtures with SFM or SBM and 180 g CP/kg DM, respectively, 25.8 y 27.8 ml/g dOM. Results suggest that protein characteristics supplements would influence CH_4 production. The maximum increments in CH_4 observed in SFM or SBM mixtures may be of low biological significance.

Effect of soybean meal substitution with alternative protein sources on beef cattle carbon footprint

L. Boselli[1], G. Pirlo[2] and L. Migliorati[2]
[1]*Università degli Studi di Milano, Agricultural and Environmental Sciences Department, via Celoria,2 Milano, 20131, Italy,* [2]*Council for Agricultural Research and Economics, Fodder and Dairy Productions Research Centre, via Lombardo,11 Lodi, 26900, Italy; leonardo.boselli@unimi.it*

Alternative protein crops such as Vicia faba var. minor and Pisum sativum, have raised a relevant interest among breeders due to their cheaper market price and less agricultural inputs usage in comparison to soybean. Aim of the present study was to verify if CF of the whole fattening production system can be reduced by the introduction of alternative protein sources in finishing diets. A sample of 13 Italian beef specialized farms, including 10 French – Italian stockers and 3 Piedmontese units, was analyzed. A soybean meal (SB) diet and two virtual diets, in which broad bean (BB) and protein pea (PP) have replaced soybean meal, were tested. Each beef production system was analyzed from cradle to the farm exit-gate, including both calf-to-weanlings and fattening phase, farm inputs and animal transport. Kilogram of live weight (LW) has been adopted as functional unit (FU). Biophysical allocation has been applied to split impacts among the co-products of the analyzed systems. The emissions produced into the F-I stocker phase were 4.15 in SB, 3.76 in BB and 3.69 $kgCO_2$-eq per kg LW in PP respectively. In the Piedmontese fattening system, young bull GHG emissions decreased from 7.50 (SB) to 5.85 (PP) $kgCO_2$-eq per kg LW respectively; while in the case of heifer emissions decreased from 5.47 (SB) to 4.71 (PP) $kgCO_2$-eq. Results are encouraging but in vivo studies are needed to verify effectiveness of alternative protein sources on beef cattle performance and compatibility between economic and environmental aspects should be examined.

The effect of lowering dietary lysine content on broiler performance

M.E.E. Ball

Agri-Food and Biosciences Institute, Agriculture Branch, Hillsborough, BT26 6DR, United Kingdom; elizabeth.ball@afbini.gov.uk

Lysine is the main limiting amino acid in broiler diets and as such, diets are often supplemented with lysine and other essential amino acids in an ideal protein ratio to support optimum performance. However, amino acid supplementation can increase cost of diet production, and if oversupplied, can represent an energy and environmental cost. Therefore, the aim of this trial was to investigate if lysine (and other essential amino acids) could be lowered in diets of broilers without causing any detrimental effect on performance. Eight diets were formulated consisting of two starter (Control (C) 14 g/kg lysine, Low (L) 12 g/kg lysine), two grower (Control (C) 12 g/kg lysine, Low (L) 10 g/kg lysine), and two finisher (Control (C) 9.5 g/kg lysine, Low (L) 7.6 g/kg lysine). The diets were wheat/soya bean meal based and balanced for energy and crude protein. The diets were used in different combinations across the starter, grower and finisher periods to give eight experimental treatments: T1 C-C-C, T2 C-C-L, T3 C-L-C, T4 C-L-L, T5 L-L-L, T6 L-C-C, T7 L-C-C, T8 L-L-C. Each treatment regime was offered to 480 male chicks (Ross 708) in pens of 10 giving six pen replicates per treatment. The starter diet was offered from day 1 to 14, the grower offered from 14 to 21 and finisher from day 21 to 35. Dry matter intake (DMI), liveweight gain (LWG), and feed conversion ratio (FCR) were determined for the starter, grower and finisher periods. Lowering dietary lysine level in the starter phase did not significantly affect performance. However, in the grower phase, LWG and FCR were poorer on the low lysine treatment regimes (LWG=462 vs 440 g (SEM=6.45, $P<0.05$) and FCR=1.51 vs 1.60 (SEM=0.028, P=0.015). For the finisher phase, offering diets low in lysine resulted in significantly ($P<0.001$) poorer performance and body weight at 35 d. DMI was reduced from 2,306 to 1,845 g (SEM=46.4), LWG was reduced from 1,324 to 957 g (SEM=21.8), FCR was less efficient (1.75 vs 2.03, SEM=0.035) and body weight at 35 d was reduced from 2,219 to 1,845 g (SEM=27.5). In conclusion, lowering dietary lysine (and other essential amino acids) to levels below industry standards reduced broiler performance.

Alternative sampling types for assessing the composition of the rumen microbial community in cattle

I. Tapio[1], K.J. Shingfield[1,2], N. McKain[3], A. Bonin[4], D. Fischer[1], A.R. Bayat[1], J. Vilkki[1], P. Taberlet[4], T.J. Snelling[3] and R.J. Wallace[3]

[1]Natural Resources Institute Finland, Green Technology, Myllytie 1, 31600 Jokioinen, Finland, [2]Institute of Biological, Environmental and Rural Sciences, Aberystwyth University, Penglais Campus, SY23 3FL Aberystwyth, United Kingdom, [3]Rowett Institute of Nutrition and Health, University of Aberdeen, Greenburn Road, Bucksburn, AB21 9SB Aberdeen, United Kingdom, [4]Laboratoire d'Ecologie Alpine, CNRS, 2233 rue de la Piscine, 38041 Grenoble, France; ilma.tapio@luke.fi

Ruminant livestock production contributes to global warming primarily due to rumen microbial metabolism. Animal selection for traits, such as lower rumen methanogenesis or improved feed conversion efficiency is theoretically possible but requires rumen microbial phenotypes obtained from large-scale animal studies. Rumen sampling can be carried out by oral intubation, rumenocentesis or via ruminal fistula with the latter being the most reliable but restricted to research facilities. The aim of this study was to test alternative sampling types as proxies for collection of ruminal digesta. In total, 80 samples collected from rumen via fistula, as buccal fluid, regurgitated digesta (bolus) and faeces were analysed by qPCR and by sequencing pro- and eukaryotic rRNA genes and the fungal ITS1 amplicons. Samples were collected from five cows receiving a grass silage based diet without or with four different lipid supplements in a 5×5 Latin square design. Microbial community analysis demonstrated that bacterial, archaeal and fungal communities in faeces were substantially different to rumen, while composition of buccal and bolus samples was much more similar to that of ruminal digesta. Differences between these were mostly related to variation in the relative abundance, rather than the composition of the microbial populations. Furthermore, differences in the microbial communities in rumen content and oral samples were generally consistent between animals and dietary treatments. Results suggest that bolus or buccal swabs may serve as an alternative to rumen sampling, and be of value for screening large animal cohorts.

Enteric methane emissions of cows in two contrasted low input dairy grassland-based systems

M. Eugène[1], A. Danglard[1], F. Fournier[2], Y. Rochette[1], D. Pomiès[1] and A. Farruggia[1]
[1]INRA, UMR1213 Herbivores, Route de Theix, 63122 St Genès-Champanelle, France, [2]INRA, UE 1414, Herbipôle, 15190 Marcenat, France; maguy.eugene@clermont.inra.fr

Livestock contributes to the environmental impact of agriculture with emissions of greenhouse gases. Extensive systems with low milk production per cow are often disadvantaged when emissions are expressed in g per kg of milk, as well as low input systems based on grasslands diet and using low concentrate. In this context, a study was conducted to acquire specific references on methane emissions (CH_4) during the grazing season by measuring in vivo emissions of two dairy cows herds grazing pastures within two contrasted dairy grassland-based systems: an extensive system (Bota) managed with permanent grasslands highly diversified with no concentrate given to cows and an ecologically intensive system (Pepi) conducted with old temporary grasslands and moderate use of mineral fertilization and concentrate. In 2014 at the experimental farm of INRA located in uplands area of central France, two groups of 6 Holstein cows grazing in Pepi and in Bota systems were constituted. Individual measurements of enteric CH_4 emission using the SF6 technique and milk production were performed one week in June (P1) and one in July (P2). Simulated bites were sampled at grazing at the two periods. In P1, CH_4 emission of Bota and Pepi cows, expressed either in daily emission or per milk production, were similar (respectively 402 g/d, 16 g/kg MY). In P2, CH_4 emissions of Bota cows tended to be higher than those of Pepi (400.3 vs 330.5 g/d). The lack of significant difference between the two systems could be related to the low amount of concentrate given to Pepi cows, to the similarity of the botanical family composition of the cows bites and finally to the low difference of milk production between the two systems at this grazing period. These values corresponded to the high range of emission found in the literature, but not when expressed per unit of surface.

DREEM: model on enteric methane emissions in sheep farms

M. Eugène[1], L. Mansard[1], A. Vigan[2], M. Meuret[2], J. Lasseur[3], M. Benoit[1] and P. Lecomte[2]
[1]INRA, UMR1213 Herbivores, Route de Theix, 63122 St Genès Champanelle, France, [2]CIRAD, UMR Selmet, TA C/112 Baillarguet, 34398 Montpellier, France, [3]INRA, UMR Selmet, 2 place Viala, 34060 Montpellier, France; maguy.eugene@clermont.inra.fr

Ruminant livestock systems are significant sources of greenhouse gases. Herd mobility is known as a major adaptation strategy regarding seasonal availability of forage resources. The present study aims at quantifying enteric methane (CH_4) emissions from French Mediterranean sheep farming systems, especially the use of diversified pastoral feed resources, using a simulation model (Diversity of feed REsources and Enteric Methane emissions, DREEM). DREEM model was developed to estimate animal enteric CH_4 emissions from empirical equations and to be subsequently integrated, as a sub-model, to an economic and GHG balance model (OSTRAL) at the whole farm level. Several equations were chosen from literature data to estimate enteric CH_4 emissions. Forage and Feed natures, animal feeding level and performances were referenced according to INRA feeding system and tables. DREEM was applied to the analysis of four case studies covering main contrasted mobility and farming systems situations: from sedentary to highly mobile pastoral systems, in the French Mediterranean area. Overall, equation 1 always gave higher enteric CH_4 estimates than the 3 other ones and small variations were observed between the four equations as the standard deviation within each farming system varied from 2.0 to 3.1 g/kg DMI and from 1.1 and 1.6 kg/head/year. At the individual level, enteric CH_4 emissions (g / d) of ewes in sedentary system are higher than those of ewes in other systems. These differences are mainly due to differences in animal feeding level and feed resources characteristics in these systems. DREEM model could estimate enteric CH_4 emissions at the animal level and be combined to OSTRAL.

Supplementation of Turi or Moringa fodder in crossbred goats diets on methane production
P. Paengkoum[1] and S. Paengkoum[2]
[1]Suranaree University of Technology, School of Animal Production Technology, 111 University Avenue, Muang, Nakhon Ratchasima, 30000, Thailand, [2]Nakhon Ratchasima Rajabhat University, Nakhon Ratchasima Rajabhat University, Muang, Nakhon Ratchasima, 30000, Thailand; pramote@sut.ac.th

Twenty-four crossbred (Thai native × Anglo-Nubian) goats were chosen from a commercial farm on the basis of similar bodyweight (18.5 + 2.5 kg). The goats were randomly allocated to three treatments in randomized complete block design (RCBD). Each goat was given corn silage as roughage plus the respective treatment diets. The diets were iso-nitrogenous and iso-energetic containing cassava pulp, molasses, urea and commercial mineral and vitamin mix. The experimental treatments were: (1) soybean meal (SBM); (2) partial substitution of SBM with Turi (Sesbania grandiflora) fodder; or (3) partial substitution of SBM with Moringa (Moringa oleifera) fodder. Nutrients intakes, ruminal characteristics (pH, ammonia nitrogen and volatile fatty acids), nitrogen balances, plasma urea nitrogen were not significantly different among treatments. The present results indicate that protein foliages locally grown fodders can substitute imported feedstuffs concentrate (e.g. SBM) as protein supplement for goat production.

Effect of plant oil or nitrate supplementation on methane production and rumen fermentation of goats
P. Paengkoum[1], J. Khotsakdee[2] and S. Paengkoum[3]
[1]Suranaree University of Technology, School of Animal Production Technology, 111 University Avenue, Muang, Nakhon Ratchasima, 30000, Thailand, [2]Suranaree University of Technologys, School of Animal Production Technology, 111 University Avenue, Muang, Nakhon Ratchasima, 30000, Thailand, [3]Nakhon Ratchasima Rajabhat University, Faculty of Science and Technology, Agriculture, Muang, Nakhon Ratchasima, 30000, Thailand; pramote@sut.ac.th

The rumen fluid mixed microbes were taken from fistulated crossbred (Thai native and Anglo-Nubian) goats. The dietary treatments were supplemented with 3 levels of sunflower oil 0, 3, 6% DM of total diet and 4 levels of nitrate 0, 1, 2, and 3% DM of total diet. It was found that after incubation the soluble gas fraction, potential of extent of gas production (a+b) and gas production showed were significantly (P<0.05) different among the different in this control group. The gas production was recorded at 24, 48 and 96 hrs of incubation periods and Metabolizable energy (ME), organic matter degradability (OMD) were estimated. There were significant (P<0.05) differences among feedstuffs in terms of OMD and ME. The characteristic of feed samples after incubated was affect to bacteria and NH_3-N increase. Protozoa population was reduced in 6% of sunflower oil, the number of protozoa was ranged from 2.75-3.25 ml^{-1}.

Mitigation of greenhouse gas emissions from two different Emilia-Romagna (Italy) dairy systems

A. Dal Prà[1], L. Valli[1], M.T. Pacchioli[1], R. Chiarini[2] and M. Montanari[2]
[1]Centro Ricerche Produzioni Animali, Viale Timavo 43/2 Reggio Emilia, 42121, Italy, [2]Regione Emilia Romagna, Viale della Fiera 8 Bologna, 40127, Italy; a.dalpra@crpa.it

Farming systems in future will need to reduce greenhouse gas (GHG) emissions per kg of fat and protein corrected milk to continue the development of dairy production while minimizing its influence on climate change. This paper aims to assess GHG emissions in Emilia Romagna (Italy) dairy systems, and to identify farm-specific mitigation options. Twelve dairy farms were surveyed, six 'with silage' (S) as forage base for fresh milk dairy system; six 'without silage' (WS) for the production of the PDO Parmigiano-Reggiano cheese. A specific tool was developed to estimate net GHG emission. The data input required by the model are: herd characteristics, feed composition and supplies, housing and manure management, crop type, fertilizer input and data describing soil's characteristics. Apparent digestibility of DM and nutrients was estimated using uNDF240 fraction as indicator. Data collected over the two years show that the main drivers that reduce the carbon footprint of the milk production are the improvement of farm productivity and the increase of TMR digestibility, achieved both by reducing the number of non-productive animals (through improved performance and increasing the dairy cows' productive life so cutting down on replacement) and by improving forage quality. The most effective and easiest mitigation options to implement were those involving a reduction of inputs, such as low-protein diets, reduction of non-self-produced feed (concentrates) fertilizer use optimization by increasing the manure nitrogen use efficiency (reducing the need for mineral fertilizers) and energy and fuel saving interventions. GHG intensity was estimated in 1.2 kg of CO_2 Eq/kg of milk in the case of fresh milk and 1.3 in the case of milk for Parmigiano-Reggiano cheese. Research supported by LIFE12 ENV/IT/000404 Climate ChangeE-R.

Effect of a lipid-garlic treatment on archaeal community establishment and rumen function in lambs

C. Saro[1,2], M. Bernard[3], D. Graviou[1,2], Y. Rochette[1,2], C. Martin[1,2], M. Doreau[1,2], H. Boudra[1,2], M. Popova[1,2] and D.P. Morgavi[1,2]
[1]Clermont Université, VetAgro Sup, UMR Herbivores, BP 10448, 63000 Clermont-Ferrand, France, [2]INRA, UMRH1213 Herbivores, Site de Theix, 63122 Saint-Genès-Champanelle, France, [3]INRA, UE 1414 Herbipôle, Site de Theix, 63122 Saint-Genès-Champanelle, France; cristina.saro-higuera@clermont.inra.fr

The microbial colonization of the rumen after birth is influenced by several factors that may modulate the microbiota composition and function. We report here the effects of a treatment applied in early life on rumen fermentation and archaeal community structure in lambs from birth up to 4 weeks of age. Sixty lambs and their dams were allocated to two groups. Treatment (T) consisted in linseed oil and garlic essential oil drenched daily during the whole experimental period. T was also applied to dams. Rumen contents were sampled by oral stomach tubing. All lambs were sampled at 1 week, as well as all dams for comparison. A representative set of lambs (n=10) was sampled at 2, 3 and 4 weeks. Volatile fatty acids (VFA) were analysed and 16S rDNA amplicon sequencing was used to explore the archaeal diversity. Total VFA concentration was not affected by the treatment except in 1-week-old lambs (34.4 vs 24.3 mM, for C and T, respectively, P<0.05); VFA profile was mostly affected in T dams with lower proportion of acetic (64.5 vs 71.1%, P<0.001) and higher proportion of propionic acid (20.6 vs 13.8%, P<0.001). In lambs, archaeal diversity increased with age but changes were less marked in T animals; Shannon indices were lower in T 4 week-old lambs and ewes (P<0.001) and numerically lower in T younger lambs. Principal coordinates analysis were constructed independently for each age group using the weighted UniFrac similarity matrix and differentiation between T and C animals increased as age increased. These results indicate that the treatment changed the development of a conventional archaeal rumen community in young lambs without affecting VFA concentration in rumen.

How does nutrient intake affect ammonia emission from slurry in pigs?

W. Antezana[1], S. Calvet[1], P. Ferrer[1], P. García-Rebollar[2], C. De Blas[2] and A. Cerisuelo[3]
[1]Universitat Politècnica de València, Camino de Vera, s/n, 46022, Spain, [2]Universidad Politécnica de Madrid, Ciudad Universitaria, s/n, 28040, Spain, [3]Instituto Valenciano de Investigaciones Agrarias, Pol. La Esperanza, 100, 12400 Segorbe (Castellón), Spain; cerisuelo_alb@gva.es

The relationship between nutrient intake (g kg $PV^{-0.75}$/d), nutrient excretion (g kg $PV^{-0.75}$/d) and ammonia (NH_3, mg kg $PV^{-0.75}$/d) emission from slurry was investigated using data from three experiments designed to study the effect of feed formulation on gas emissions. A total of 13 diets with different protein, fibre and fat content were assayed in 78 fattening pigs. Individual feed intake and the total amount of urine and faeces produced over seven consecutive days per pig was measured. Feeds and faeces were analysed for dry matter, energy, nitrogen (N), ether extract (EE), soluble fibre, neutral detergent fibre, acid detergent fibre (ADF) and acid detergent lignin (ADL). Also, dry matter and N content was determined in urine. Additionally, potential NH_3 emission from slurry was measured using an NH_3 trap system. The whole data set was analysed using the CORR procedure of SAS®. Ammonia emission was low but positively related with crude protein (CP, r: 0.28, $P<0.05$) and ADL (r: 0.29, $P<0.05$) intake and low but negatively related with ADF (r: -0.29, $P<0.01$) intake. On the other hand, NH_3 emission showed a greater positive relation with CP (r: 0.62, $P<0.001$) and ADL (r: 0.31, $P<0.05$) excretion and also a negative relation with EE excretion (r:-0.30, $P<0.01$) in the slurry. Additionally, NH_3 emission was positively related with N content in urine (r: 0.72, $P<0.01$) and negatively related with the ratio N faeces/N urine (r: -0.60, $P<0.05$). Therefore, although N excretion in urine seems a key factor determining NH_3 emission from slurry, other factors that may affect protein digestibility such as fibre fractions and EE might also have a role on NH_3 emission. Project funded by the Spanish Ministry of Economy and Competitiveness (AGL2014-56653-C3-3-R).

In vitro rumen fermentation and methane output of autumn grass fertilised at two rates of nitrogen

A. O'Connor[1], A.P. Moloney[1], P. O'Kiely[1], P. Nolan[1], T. Boland[2] and M. McGee[1]
[1]Teagasc, AGRIC, Grange, Dunsany, Co. Meath, Ireland, [2]University College Dublin, School of Agriculture, Belfield, Dublin 4, Ireland; alan.oconnor@teagasc.ie

Methane is a potent greenhouse gas and the diet consumed by ruminant livestock can affect enteric methane output. The objective of this study was to determine the effects of inorganic nitrogen (N) fertiliser rate on rumen fermentation, including methane production, of autumn grass herbage using an in vitro rumen simulation technique (Rusitec). Experimental herbages were obtained from a rotationally zero-grazed Lolium perenne dominant sward. Herbages were harvested in autumn (August – October) 21 d after receiving either 15 (LN) or 80 (HN) kg N/ha/rotation. An eight-vessel Rusitec was used in a single 18 d incubation (11 d measurement). Un-dried chopped pooled herbage samples were incubated for 48 h. Total gas and methane were measured on d 8-18 using a gas flow meter and a gas analyser, respectively. On d 12 and 15 vessel liquid was sampled at 0 (at grass incubation), 1, 2, 3, 4, 6, 14 and 24 h for rumen fermentation variables. Data were statistically analysed as a completely randomised design with repeated measures on time-related data, using the MIXED models procedure in SAS. In vitro dry matter (DM) digestibility (Tilley and Terry) of LN and HN were 790 (s.d. 10.9) and 804 (s.d. 14.9) g/kg, respectively. Corresponding crude protein concentrations were 143 (s.d. 12.1) and 205 (s.d. 31.7) g/kg DM. There were no ($P>0.05$) treatment × time interactions for rumen fermentation parameters except lactic acid concentrations ($P<0.001$). Rumen pH and ammonia N concentrations were higher ($P<0.05$) for HN compared to LN, whereas volatile fatty acid (VFA) concentrations and molar proportions of VFA did not differ ($P>0.05$) between treatments. Total gas ($P<0.001$) and methane ($P<0.01$) produced per g of feed DM incubated were lower for HN compared to LN. In conclusion, a reduction in in vitro rumen methane output was observed in herbage that received higher fertiliser N application.

The effect of breed on nitrogen utilisation efficiency of first lactation dairy cows

T. Gilfedder and T. Yan

Agri-Food and Biosciences Institute, Ruminant Nutrition, Large Park, Hillsborough, County Down, BT26 6DR, United Kingdom; tim.gilfedder@afbini.gov.uk

Nitrogen (N) pollution from cattle can be reduced by decreasing crude protein levels in total diets. The objective of the present study was to investigate the effects of dairy cow breed and plane of nutrition on N utilisation efficiency of dairy cows. Eight Holstein Fresian (HF) and eight Norwegian (NC) first lactation dairy cattle were used in a 2 breeds (HF vs NC) × 2 nutrition planes (high vs low) factorial design study, with repeated measurements (no changeover) in three periods (28 d/period) at 80, 160 and 240 days of lactation. Animals were offered mixed diets of grass silage and concentrates with concentrate proportions in high vs low plane diets for early, mid and late lactation being 0.60 v. 0.30, 0.50 v. 0.20 and 0.40 v. 0.10 (kg/kg DM), respectively. Nitrogen intake and outputs in faeces, urine and milk were measured in final 6 days of each period. Data were analysed by ANOVA to examine the effect of breed and plane of nutrition on N utilisation in this 2 (breed) × 2 (plane of nutrition) factorial design study. Nitrogen intake for HF was significantly higher than NC in both planes of nutrition ($P<0.001$). Faecal N output was significantly higher from HF than NC in the low plane of nutrition ($P<0.001$). Milk N output was significantly higher for HF than NC across both planes of nutrition ($P<0.001$). No breed effect was observed for retained N although it was higher at the high plane of nutrition for both breeds. The HF had a higher milk N/N intake than NC in the high nutritional plane ($P<0.05$). There was no significant difference between HF and NC for urine N/manure N. In conclusion, HF had higher N intake than NC and the additional N intake mainly partitioned for milk production and body retention. There is no significant difference in the proportion of N intake excreted in urine or faeces between HF and NC.

Effects of breed and concentrate supplementation on methane emissions from ewes fed fresh grass

Y.G. Zhao[1,2], A. Aubry[1], N.E. O'Connell[2] and T. Yan[1]

[1]Agri-Food and Biosciences Institute, Hillsborough, Co. Down, BT26 6DR, United Kingdom, [2]Institute for Global Food Security, Queen's University Belfast, BT9 5BN, United Kingdom; yiguang.zhao@afbini.gov.uk

Breed selection and concentrate supplementation may be appropriate strategies to reduce methane (CH_4) emissions. This study aimed to investigate the effects of breed and concentrate supplementation on CH_4 emissions from sheep. Sixteen lowland replacement ewes were used in a 2 breeds (Highlander cross vs Texel cross) × 2 diets (fresh grass vs fresh grass plus 0.5 kg/d (fresh weight) pelleted concentrate) factorial design study. Animals were 14 months old and weighed 61±5.3 kg at the commencement of the study. Fresh perennial ryegrass was harvested daily and offered ad libitum, and provided a high quality forage. The animals were individually housed in pens and fed experimental diets for 19 days before transferred to individual respiration chambers for a further 4 days with feed intake, total faecal and urine outputs and CH_4 emissions measured. Data were analysed as a 2 × 2 factorial arrangement using the Analysis of Variance. There were no significant interactions between breed and diet on any variable evaluated. Sheep offered concentrate diets tended to have greater dry matter intake (DMI) ($P=0.085$) and greater live weight (LW) ($P=0.056$) than those given fresh grass only. However, diets had no significant effect on CH_4 emissions (g/d), CH_4/DMI or CH_4 energy (CH_4-E)/gross energy intake (GEI). Texel cross ewes produced less CH_4/LW than Highlander cross ewes ($P=0.037$), while breed had no significant effect on CH_4 emission (g/d) or CH_4/DMI. The CH_4-E/GEI obtained in the present study was 5.8% which is less than that of 6.5% currently used in the IPCC guidelines (2006) for sheep. Since concentrate supplementation did not affect CH_4/DMI, it indicates that high quality fresh grass might play a similar role to concentrates in mitigating enteric CH_4 emissions. Furthermore, CH_4-E/GEI for sheep feeding on fresh ryegrass is likely to be less than the default factor currently used by the IPCC.

The effect of by-product based concentrates on feed intake and milk production in dairy cows

J. Karlsson, R. Spörndly, M. Patel and K. Holtenius
Swedish University of Agricultural Science, Dept. of Animal Nutrition and Management, P.O. Box 7024, 750 07
Uppsala, Sweden; johanna.karlsson@slu.se

Large volumes of human-edible products as cereal grains and soya beans are fed to dairy cows in intensive production systems. Given the predicted increase in demand for livestock products and competition of land for food and feed, human inedible feeds are becoming increasingly important. Ruminants have a unique ability to produce high quality food, as milk and meat, from fibrous feed and other products not suitable for human consumption. The aim of this study was to investigate the effects on milk production and feed intake when human-edible feeds as cereal grain and soy beans was completely substituted with different by-products in a high quality forage diet. Twelve multiparous and twelve primiparous dairy cows in mid-lactation, of the breeds Holstein (n=8) and Swedish Red (n=16), were used in a change-over design with four different concentrates. The cows were randomly assigned to one of four groups based on breed, parity and milk yield. The cows were fed grass silage ad libitum. Restricted amounts of concentrate (10 kg dry matter (DM)/day) were offered separately. The four different concentrates were based on: (1) cereal grain and soybean meal (CG-SBM), (2) sugar beet pulp (SBP) and rapeseed meal (RSM), (3) SBP and distiller's grain (DG), (4) SBP, RSM and DG, all with similar content of crude protein (167 g/kg DM) and energy (11.6 MJ ME). The data was analysed by SAS 9.4 PROC MIXED using a change-over model with the effects of treatment, period, order and cow as random variable. Dietary treatments did not affect DM or forage intake. Milk production was higher in diet 2 (32.5 kg/d) compared with diet 3 (30.5 kg/d) and diet 3 had higher fat content (4.7%) than diet 1 (4.4%), while other diets did not differ. In conclusion, replacing concentrate based on CG-SBM with concentrates based on human inedible agricultural by-products in a diet to dairy cows in mid-lactation did not impair feed intake or milk production.

Effects of dietary nitrate on rumen microbial species in beef cattle fed different basal diets

J.A. Rooke[1], R.J. Wallace[2], N. McKain[2], C.-A. Duthie[1], S. Troy[1], J.J. Hyslop[3], D.W. Ross[1], T. Waterhouse[1] and R. Roehe[1]
[1]SRUC, West Mains Road, EH9 3JG, Edinburgh, United Kingdom, [2]Rowett Institute for Nutrition and Health, Bucksburn, AB21 9SB, Aberdeen, United Kingdom, [3]SAC Commercial Ltd, West Mains Road, EH9 3JG, Edinburgh, United Kingdom; john.rooke@sruc.ac.uk

Rumen microbial species including putative nitrate-reducing bacteria were assessed when nitrate was fed to reduce methane (CH_4) emissions from cattle fed two contrasting basal diets. The experiment had a 2×2 factorial design where the factors were basal diet (concentrate, CON (920 g concentrate dry matter (DM)/ kg DM) or forage, FOR (500 g concentrate DM/kg DM)) and presence or absence of nitrate, (20 g/kg DM) giving 4 treatments: CON-C, CON-N, FOR-C, FOR-N. Rumen samples were obtained by stomach tube (10 steers / treatment), DNA extracted and target species measured by quantitative PCR. Data were analysed using general linear models with fixed effects of basal diet and treatment. Neither Veillonella parvula nor Wolinella succinogenes were detected in any sample. The abundance (relative to total bacteria on copy number basis) of archaea (0.028 vs 0.016, sem 0.0115, P<0.05), protozoa (0.135 vs 0.037, sem 0.0.23, P<0.001) were lower on CON than FOR, while the abundance of Selenomonas ruminantium (0.058 vs 0.102, sed, 0.017, P<0.05) was greater on CON than FOR. Nitrate had no effect on microbial populations, nor were there any interactions between nitrate and basal diet. The observed changes in microbial communities are consistent with lower CH_4 emissions (g/kg DM intake) from CON (15.0) than FOR steers (23.0, P<0.001). However, because there was no interaction between basal diet and nitrate addition in microbial species, these data do not explain the ineffectiveness of nitrate in reducing CH_4 on diet CON (g CH_4/kg DM intake, CON-C vs CON-N, 14.7 vs 15.3) compared to diet FOR (FOR-C vs FOR-N, 25.3 vs 20.6). On these diets, S. ruminantium was the only bacterial species capable of reducing nitrate which was identified. Study supported by EBLEX and Scottish Government.

Session 52

Theatre 1

Innovation in equestrian tourism

N. Vaugeois
BC Regional Innovation Chair in Tourism, Sustainable Rural Development, Vancouver Island University, Canada;
nicole.vaugeois@viu.ca

While horses have always been associated to travel as a mode of transport, their use in modern times has evolved, they are now the inspiration for travel experiences. This has resulted in the emergence of a niche form of tourism: equestrian tourism defined as 'travel inspired by the horse, for recreation, leisure and business, encompassing all activity that has the horse as its focus'. The potential to benefit from this niche market of the equestrian tourist is strongest in rural communities many of which have been experiencing shifts in economy due to the realities of globalization. As such, many are integrating alternative forms of economic activity such as tourism to establish greater resilience. These communities have many assets likely to appeal to equestrian tourists including farms, ranches, trails, campgrounds, events and a growing array of accommodations catering to those traveling with a horse. In an effort to capitalize on this market, businesses and destinations have used innovative strategies to design, market and deliver new tourism products. We will highlight how the supply side has been responding to the demand created by equestrian tourists around the world. After describing the niche, we will highlight innovations across the types of equestrian tourism ranging from travel experiences with and without one's horse. These include a range of accommodations (bed, bale and breakfast operations, camping facilities and ranch vacations); unique packages such as girlfriend getaways and trail ride vacations; and special events such as horse artist retreats, wine rides and specialized workshops and clinics combining education, horses and travel. The examples will demonstrate 'within business' innovations where farms and other ventures are adapting practices to serve the niche; 'within industry' innovations where tourism experience providers are collaborating; and 'across sector' innovations where new networks are evolving and technology is being used to enhance the competitive environment for equestrian tourism.

Session 52

Theatre 2

Native Breeds as socio-natural constructs and cultural heritage artefacts

R. Evans
Norwegian University College of Agriculture and Rural Development (HLB), 213 Postvegen, 4353, Klepp Stasjon, Norway; rhys@hlb.no

Horses have been a part of human life for millennia. A study of horses in human history shows a remarkable versatility in terms of the roles they have served. These range from food, to part of the production of food, through industrial and transportation helpmate all the way to therapist, physical activity facilitator and companion. Although horses still fill all those roles somewhere in the world, in the developed world, their role is in the process of adapting to meet the needs of increasingly urbanised life in the 21st Century. In every case, horses have taken up roles which are produced by the state of human life. Changing economies, changing societies and technological change have also taken away the predominant roles held by horses in human life. Thus we have seen the decline of numbers with the advent of motorization due to the loss of horses working in agriculture, forestry and transportation. Throughout history, the potential of the close relationship possible between humans and horses means that they have been companions, and the focus of leisure interests. Now, however, the leisure sector (including fitness, competition, outdoor recreation, tourism, etc.) takes a new role in contemporary society and the economy. Leisure activities are highly significant generators of wealth, and people are willing to pay significantly for them partly because their working and domestic situations generate a need for the things equine activities provide. This paper will explore how these trends in contemporary society is creating new roles for horses and providing new opportunities for the equine economy. At the same time, it will explore some of the challenges this provides and ask a few significant questions about the state of our knowledge and some of the taken-for-granted ideas which need critical assessment.

Automated quality feedback for dressage riders using inertial sensors

R.J. Thompson[1,2], S.M. Matheson[2], T. Ploetz[1], S.A. Edwards[2] and I. Kyriazakis[2]
[1]Newcastle University, Open Lab, Floor 3, 89 Sandyford Road, Newcastle upon Tyne. NE1 8HW, United Kingdom,
[2]Newcastle University, School of Agriculture, Food, and Rural Development, Agriculture Building, Newcastle
University, Newcastle upon Tyne, NE1 7RU, United Kingdom; r.j.thompson3@ncl.ac.uk

The sport of dressage has become very popular not only amongst professional athletes but increasingly also for private horse owners. In well-defined tests, rider and horse execute movements, which demonstrate the strength, endurance, and dexterity of the animal as well as the quality of the interaction between rider and horse. Whilst at a professional level intensive expert coaching to refine the skill set of horse and rider is standard, such an approach to progression is not usually viable for the large amateur population. In this paper we present a framework for automated generation of quality feedback in dressage tests. Using on-body sensing and automated measurement of key performance attributes we are able to monitor the quality of horse movements in an objective way. We validated the developed framework in a large-scale deployment study and report on the practical usefulness of automatically generated quality feedback in amateur dressage.

Eye and rectal temperature, cortisol concentration, heart rate in assessment of stress in racehorses

M. Soroko[1], K. Howell[2], A. Zwyrzykowska[3], K. Dudek[4], P. Zielińska[5] and Z. Dobrzański[3]
[1]Wroclaw University of Environmental and Life Sciences, Department of Horse Breeding and Equestrian Studies,
Kozuchowska 5A, 51-161 Wroclaw, Poland, [2]Royal Free Hospital, Institute of Immunity and Transplantation, Pond
Street, NW3 2QG, London, United Kingdom, [3]Wroclaw University of Environmental and Life Sciences, Department
of Environment Hygiene and Animal Welfare, Chełmońskiego 38C, 50-630 Wroclaw, Poland, [4]Technical University
of Wroclaw, Institute of Machines Design and Operation, Lukasiewicza 7/9, 50-231 Wroclaw, Poland, [5]Wroclaw
University of Environmental and Life Sciences, Department of Surgery, pl. Grunwaldzka 51, 50-366 Wroclaw,
Poland; kontakt@eqma.pl

The aim of the study was to investigate the agreement between eye maximum pixel temperature Tmax (using infrared thermography IRT), and rectal temperature RT in young racehorses, and compare the results with salivary cortisol concentration SCC and heart rate HR, both at rest and in response to exercise. Nineteen healthy horses of two breeds were studied. All horses were undergoing training for flat racing in their first racing season. Tmax, RT, SCC and HR were measured before training (BT), within five minutes of the end of the training session (T+5), and two hours after training (T+120). All measurement techniques were performed on each horse on three days of intensive training scheduled in the training timetable (Tuesday, Friday and the following Tuesday), and the results were expressed as the mean value over the three measurements. Tmax, RT, SCC and HR were all significantly elevated at T+5 compared to BT (all $P<0.001$). At T+120, only Tmax remained significantly elevated compared to BT ($P<0.05$). Bland Altman analysis indicated a poor agreement between Tmax and RT, with the smallest bias between the measurements occurring at T+120 (1.1 °C, with limits of agreement 0.4-1.8 °C). We noted no significant correlations amongst any of the four measurements at any time point, with the exception of Tmax and RT at BT ($r=0.55$, $P=0.01$). In racehorses, Tmax is a poor estimate of core temperature due to its limited agreement with RT. Furthermore, Tmax is not correlated with accepted measures of stress such as SCC and HR.

Could a leisure horse grading system improve horse-rider matching and equine welfare within the UK?
R. Lawson and C. Brigden
Myerscough College, Equine Department, St Michael's Rd, Bilsborrow, Preston, PR3 0RY, United Kingdom;
cbrigden@myerscough.ac.uk

The UK leisure horse market is vast, yet lacks standardised methods of assessing traits of these horses unlike the competition horse. Previous studies validate a demand for an inexpensive and reliable leisure horse assessment method. The purpose of this study was to explore the demand for, and feasibility of, a leisure horse grading system. A self-administered online questionnaire was completed by leisure horse owners (n=157). Mostly closed and ranking / scaling questions were used to gauge general understanding of performance horse grading within the UK and opinions regarding a leisure horse specific grading. Interviews (n=9) were conducted using a simulated leisure horse grading, which allowed participants to rate which aspects work effectively or poorly, identifying feasibility of the concept. Results suggested 60.65% of respondents were entirely positive about the grading concept, with a further 29.03% of respondents unsure. The most influential reason for potentially using a grading was to improve the matching process of horse to rider based on ability. 81.76% of respondents believed the concept would be met with interest within the equine industry. The questionnaire also investigated which proposed components were the most popular for inclusion within the grading, which were personality testing followed by ridden performance and conformation. A Kruskal-Wallace test confirmed that there was a highly significant difference between the components of the grading. Paces were found to be the least important factor for inclusion by a clear margin. This research demonstrates horse owners would have an interest in leisure horse grading and specifically the inclusion of personality testing and riding performance and behaviour. This essentially could aid in the matching of horses to riders, which may help in the reduction of equine wastage and the improvement of equine welfare and rider safety. This could provide considerable benefits across all aspects of the UK leisure industry.

The horse's voice: How trainers and riders make the horse's needs relevant in dressage training
C. Lundgren
Language and Culture, Dept. of Culture and Communication, Linköping University, 58216 Linköping, Sweden;
charlotte.lundgren@liu.se

Welfare issues in horse training, particularly sports dressage training, are the object of heated discussions. Most studies of equine welfare are highly technical, but this paper takes the novel approach of investigating how riders and trainers orient towards equine welfare in everyday training situations in the riding hall. The study is based on multi-modal interaction analysis of 15 hours of video recorded trainings in sports dressage. It departs from the tradition of linguistic ethnography where a dialogical perspective on communication is central. The study rests on the assumption that horses are active participants in the communication during trainings; they contribute to the interaction e.g. by exhibiting behaviours indicating resistance against demands from the rider. However, they are dependent on the humans to interpret and contextualise their behaviour. The interpretations and contextualisations can be examined as they are verbalised in the communication between trainer and rider. The analyses show that trainers and riders discuss the actions of the horses by tying them to the horses' cognitive, emotional and physiological needs. Cognitive needs (e.g. the need for well-timed and correct aids), as well as emotional needs (e.g. motivation and confidence), are made relevant by both riders and trainers, whereas the physiological needs (e.g. rest and muscle strengthening work), are typically made relevant by the trainer. The needs of the horses are used as resources both to explain problems and argue for specific solutions to them. The participating trainers (n=5) are all well-established, professional trainers. The recorded trainings are part of their weekly training schedule and were not arranged for the benefit of the researchers. The riders (n=15) are experienced dressage riders, riding their own horses or horses they have a long-term relation with. The horses (n=15) are Swedish warm-bloods. The data covers everyday dressage trainings with trainers with different educational backgrounds, leisure riders as well competitive riders, and both younger horses and horses educated to the international level.

E-carriage: An opportunity for animal-friendly equine tourism

S. Wägeli, G. Schmidt, A. Scheurer and A. Hochuli
Bern University of Applied Sciences, School of Agricultural, Forest and Food Sciences HAFL, Länggasse 85, 3052
Zollikofen, Switzerland; salome.waegeli@bfh.ch

In Switzerland equine tourism is not very developed so far. However, in neighbouring Austria, which has similar conditions, equine tourism is outstanding and contributes 832 million euros to the economy. Qualitative interviews with experts in equine and nature-based tourism, five each in Austria and Switzerland, showed that there is an interesting market potential for services in equine tourism in Switzerland. Qualitative content analysis following Mayring identified the following trends: 'back to nature', contact with animals, deceleration and higly demanding clients. There is also a growing concern about animal welfare in equine tourism services. This led an innovative Swiss engineer to combine modern technology and horse-power to meet this new demand. He developed a new form of carriage with a battery-powered auxiliary motor, the e-carriage. The system is similar to e-bikes, so the driver is able to control support to the horses by the motor and when the carriage is driving down, the brake energy flows back into the battery. It is a very animal and environmentally-friendly means of transportation. Moreover, e-carriages allow the use of horses even in the Swiss mountain regions. The e-carriage will be introduced in the tourist town of Braunwald, situated 1,250 metres above sea level, in 2016. In the next step of the project, the e-carriage will be embedded in a nature-based tourism concept. The aim is to be a role model for other regions to develop equine tourism – animal and environmentally-friendly.

Wellness and equestrian tourism–innovation and product development

I. Sigurðardóttir
Hólar University College, Rural Tourism, Hólar, IS551, Iceland; inga@holar.is

The importance of enjoying nature and getting away from the strain and stress of everyday life is increasing at the same time as there is a budding interest and increasing acknowledgment of the positive effects of horses on humans. In this research, possibilities of innovation in equestrian tourism towards health/wellness and slow tourism is investigated. Mixed method research was conducted to find out; A: if operators in equestrian tourism in Iceland are currently focusing their product development and marketing towards health, wellness or adventure and B: whether a combination of wellness, slow adventure and outdoor recreation is a realistic option for innovation within the field of equestrian tourism. Content analysis of webpages and open ended interviews with equestrian operators indicate limited systematic use of health tourism concepts in the development of equestrian tourism products. Adventure is important in equestrian tourism but mainly with a focus on 'hard' adventures. However, interviews show congruence in the emphasis of businesses and operator's definitions of future opportunities and the development of slow adventures and health tourism. The research indicates considerable opportunities for future innovation of equestrian tourism towards slow adventures and wellness tourism.

Cluster development and innovation in equestrian tourism in North West of Iceland

I. Sigurðardóttir[1] and R. Steinþórsson[2]
[1]Hólar University College, Rural Tourism, Hólar, IS551, Iceland, [2]University of Iceland, School of Business,
Sæmundargata 1, IS101, Iceland; inga@holar.is

The main purpose of the study is to analyse the extent of regional cluster development and the innovation activities that are manifested in relation to the horse based businesses and the related activities in the North West of Iceland. The main research questions are: A. Can the growth of horse related activity in North West of Iceland be seen as a manifestation of an emerging cluster in the equestrian tourism field in the area? B. If so, how are the cluster externalities, along with the competition and cooperation among the cluster participants, affecting resource utilization, knowledge transfer and innovation? The horse industry in the North West of Iceland was mapped by using secondary data analysis. Twelve open-ended interviews were also conducted with equestrian operators. Findings indicate that horse based businesses and related activities in the research area are both multiple and rich. There are examples of many actors and the relations between them can be interpreted as an emerging cluster with active businesses at all levels of the value chain. Operators claim that variable streams of income are needed to survive. They do emphasize the importance of better utilization of horses, land, buildings, equipment and labor force. They do also stress the importance of collaboration between operators. Specialization is part of the innovation within those businesses. Businesses are developing towards more specialization despite the fact that operators have to be good at multitasking and spotting new opportunities within their businesses. Operators seem in general aware of the importance of flow of knowledge for innovation and competitiveness of businesses and the industry. There are two main streams of knowledge most commonly mentioned. The first one is from the local university. The other stream of knowledge is international communication and going abroad to gain working experience.

Equestrian tourism and innovations: from the adventurer to the connected rider (France)

S. Pickel-Chevalier
ESTHUA, University of Angers, 49, 7 Avenue François Mitterrand 40455, 49004, France;
sylvine.chevalier@univ-angers.fr

The equestrian tourism was born in an informal way in France after the Second World War. It joins in the context of the emergence of the society of tourism and the leisure activities in France. It was the work of adventurers, tired of learning in clubs, practicing an equestrian tourism on the verge of exploitation. The comfort was by definition precarious, within France which entered a process of 'de-equinisation', since the 1930s. The divergent born activity, following the example of the other sports of open air, enters a process of normalization, passing by the creation of professional education and degrees. At the same time, the equestrian tourism becomes a more professional economic activity with the creation of specialized travel agencies: Cheval d'Aventure, Rando-Cheval,…This professionalization accompanies a deep transformation of the needs and practices of the riders. To the former fighters succeeds a clientele of superior middle classes, who look for comfort and safety. This evolution conducts to the necessity to improve the quality and the visibility of paths and facilities. The new technologies are more and more sought (GPS, digital map…). Besides, increases a policy of accommodation labeling. The object of this study is to understand this cultural change in the practice of the equestrian tourism in France. From the history of the equestrian tourism in France, we shall analyze its recent evolution, to know more the practices, the representations and the expectations of the new followers. We shall interrogate the capacity of the digital technologies to transform the equestrian tourism, by adapting it to the contemporary tourist practices. Our methodology is based on a combination of approaches: bibliographical analysis aiming at replacing the history of the equestrian tourism in the emergence of the society of tourism and the leisure activities; exploitation of the statistical data of official authorities allowing to estimate the evolution of the number and the sociological profiles of the followers, and interpretative qualitative survey that we leads in 2015-2016 including: a representative of the FITE; president of CNTE; president of 'Equiliberté'; 1 specialized journalist, and 6 equestrian tourists.

Comparison of snaffle bits made either in steel or titanium in show jumping horses

N. Guzzo[1], C. Sartori[1], C. Stelletta[2], L. Bailoni[3] and R. Mantovani[1]
[1]Dept. of Agronomy Food Natural resources Animals and Environment, Viale dell'Universita', 16, 35020 Legnaro (PD), Italy, [2]Dept. of Animal Medicine, Production and Health, Viale dell'Universita', 16, 35020 Legnaro (PD), Italy, [3]Dept. of Comparative Biomedicine and Food Science, Viale dell'Universita', 16, 35020 Legnaro (PD), Italy; roberto.mantovani@unipd.it

Our aim was to evaluate the effects of snaffle bits made either in steel or in titanium on some well-being indicators in sport horses during show jumping training. The experiment involved 18 rider-horse pairs that alternatively used a traditional steel (S) or a titanium (T) snaffle bit in two subsequent days during a standardized exercise of training, but in a different sequence by applying a crossover design ST|TS, i.e. 9 pairs were randomly assigned to the ST sequence, and the others 9 to the TS. The horses had similar age, training, and normal mouth conditions. Eight skilled riders that trained horses not knowing the bit type assigned were considered. Data recorded during the trial were the length of flat and jumping work, infrared thermographic images at muzzle and neck both before and after the exercise, the temperature of the snaffle bits after training, and the pH of saliva samples after training. Scores on the levels of snaffle bit acceptance, muscle de-contraction, saliva production, and quality of flat work and jumping were also recorded by surveying riders at the end of training. The use of T snaffle bit as compared to S snaffle bit resulted in a lower (P<0.05) temperature change measured via infrared thermal camera (i.e. after and before training) at both muzzle (0.62 vs 1.40 °C for T and S, resp.) and neck (1.95 vs 2.71 °C for T and S, resp.), a lower temperature of the snaffle bit after training (30.2 vs 32.1 °C for T and S, resp.) and in a tendency (P=0.09) for a better snaffle bit acceptance as scored by riders (3.8 vs 3.5 for T and S, resp.). Results indicate that titanium snaffle bits seem to provide a general higher well-being level to the horses during show jumping training.

Analysis of the testing horses for therapeutic purposes in the Czech Republic

B. Hofmanová[1], D. Houdová[1], I. Majzlík[1], H. Vostrá-Vydrová[1] and L. Vostrý[2]
[1]Czech University of Life Sciences Prague, Department of Animal Science and Ethology, Kamýcká 129, 16521 Praha 6, Suchdol, Czech Republic, [2]Czech University of Life Sciences Prague, Department of Genetics and Breeding, Kamýcká 129, 16521 Praha 6, Suchdol, Czech Republic; hofmanova@af.czu.cz

The aim of this paper is to summarize the results of testing horses for therapeutic purposes according to the rules of the Czech Therapeutic Riding Association (CTRA). Therapy horse certification system was started in 2011 with the aim to test the suitability of horses for Equine assisted activities and therapies (EAAT) with the regard to the character and training of the horse. Equine assisted activities and therapies (EAAT) cover several activities of which the above mentioned testing system deals with hippotherapy (HPT) and Equine Facilitated Learning or Psychotherapy (HCT). The horse passing the test could be licensed for the whole EAAT or for its parts only. The therapy horse test is open for any horse, gelding or mare (stallions are excluded from EAAT activities in the Czech Republic) at the age at least 5 years, managed under the 'Rules for the protection of horses at public display during EAAT activities organized by CTRA'. The examination test using scale 0-10 points consists of two parts – first part deals with horse body conformation including movements using linear assessment. Second part deals with horse character and the quality of specific skills required either for EAAT or its parts. The horse has to achieve minimum 6.1 points on average, but on no items of the exam it can get less than 5 points. Since 2011 the examination passed 148 horses of which were 47.29% warmbloods, 25.26% coldbloods and 27% ponies. The group of horses consists of 53.37% mares and 46.62% geldings. It could be concluded, that horses are very good selected and trained as 65% of horses tested for HPT reached full 10 points and all 148 horses passed successfully the test. The paper is further analyzing some partial items of the test (body conformation or handling) and possible relations to the therapeutic performance.

Thoroughbred horse cortisol levels during training session and impact on racing achievements

A. Zwyrzykowska, K. Śpitalniak and R. Kupczyński
Wroclaw University of Evironmental and Life Sciences, Department of Hygiene, Environment and Animal Welfare,
Chelmonskiego 38c, 51-630 Wroclaw, Poland; anna.zwyrzykowska@up.wroc.pl

Cortisol is a natural glucocorticoid hormone produced by the adrenal cortex. Main function of cortisol is energy mobilization in order to protect the homeostasis of essential biochemical processes during effort and stressful situations. Horses trained for races, various competitions are often exposed to different potentially stressful situations for example during transport, change of place of living intensive training, participation in competitions, etc. The aim of this study was to assess the stress level by determination of cortisol concentrations and the influence on racing performance. The experiment was performed on 14 thoroughbred horses on Wrocław Racing Track Partynice. This assessment was carried out in the racing season 2014 (from March to October). Animals were selected by the analogue method. According to the gender, the study group consisted of 7 mares, 7 stallions. Three saliva samples and three blood samples were collected from each horse. Both types of samples were taken when the horse was at rest, immediately after returning from the track and 1 h after the end of exercise. Result shows that the concentration of cortisol in saliva and blood was significantly higher in mares. The difference was evident in all seasons of research, and significant differences were found between I and II part of season ($P<0.01$). In the case of mares it was observed downward trend of cortisol in saliva. This was particularly evident in the retrievals post-exercise ($P<0.05$). Results shows that horses with higher cortisol concentration in all samplings after training, both in saliva and serum had better racing performance.

The effect of Human interactions with horses on physical and emotional wellbeing

A. Šepetkaitė[1], R. Cicėnaitė[1], K. Berškienė[1], A. Vainoras[1], R. Šveistienė[2] and M. Landauskas[3]
[1]Lithuanian University of Health Sciences, Institute of Sport, Tilžės 18, 47181 Kaunas, Lithuania, [2]Lithuanian University of Health Sciences, Institute of Animal Science, R. Žebenkos 12, 82317 Baisogala, Lithuania, [3]Kaunas University of Technology, K. Donelaičio g. 73, 44249 Kaunas, Lithuania; ruta@lgi.lt

For centuries people have talked about the relationship between 'a man and his horse,' and the 'bond that exists between a horse and a rider.' It is found evidence that human interactions with horses provided improved psychological and physiological wellbeing. The aim of this study was to evaluate non riders and rider's heart rate, blood pressure and bio-psychosocial satisfaction during daily activities and interactions with a horse. In this study, 22 people were divided into non riders (n=15), age 27.4±2.3 and riders (n=7), age 26.3±3.2 years, groups. In both groups was measured daily heart rate, blood pressure, measured bio-psychosocial satisfaction, comparing daily activities (such as low, medium and high physical activity and sleep) and interactions with a horse (such as the flattering, brushing, feeding and riding). For heart rate (HR) measurement we have used "first beat'heart rate monitors. Horse flattering during a median heart rate for non riders was 93.5 (72-120; 93.43) beats/min., brushing -97.5 (75-113; 95) beats/min., feeding – 100.5 (83-111; 99) beats/min., riding a horse -93 (82-130; 101) beats/min. Median heart rate for riders was significantly lower in flattering 77 (65-92; 79) beats/min., brushing 85 (72-95; 83) beats/min. and feeding 84 (71-94; 84) beats/min., compared to young age persons ($P<0.05$). In non riders group physical condition after the sessions with the horse increased more than the emotional state ($P<0.05$). HR is higher, when non riders flattering, brushing and feeding the horse, than the low and medium daily physical activity ($P<0.05$). Non riders systolic blood pressure decreased and satisfaction of emotional wellbeing and social interaction increased ($P<0.05$) after sessions with a horse.

Animal behaviour and animal nutrition science working together to support livestock production

S.A. Edwards[1] and H.A.M. Spoolder[2]
[1]Newcastle University Newcastle, School of Agriculture, Food & Rural Development, Agriculture Building, Newcastle upon Tyne NE1 7RU, United Kingdom, [2]Wageningen UR Livestock Research, Animal Welfare dept., De Elst 1, 6708 WD Wageningen, the Netherlands; sandra.edwards@newcastle.ac.uk

Within livestock production and welfare science, many of the interesting and important questions lie at the interface of traditional fields of study and benefit from an interdisciplinary approach. The effects of nutrition on the behaviour of animals have been widely studied. They range from the basic influence of diet quantity and quality on foraging motivation, and its role in modulation of other important behavioural domains and in abnormal behaviour development in restricted environments, through the more nuanced effects of dietary imbalances on food choice and the effect of specific nutrients on mood and cognition. The effects of behaviour on nutritional questions have perhaps been less well appreciated, though the importance of feeding behaviour and intake patterns on the efficiency of nutrient utilisation are receiving increasing study. New precision farming technologies, which allow large-scale automated monitoring of feeding and drinking patterns, not only facilitate such studies, but also the use of knowledge of these behaviours in health monitoring and in optimising feeding systems and grazing management. Another important area of research relates to the behaviourally-mediated social constraints and facilitators of feed intake. This includes not only influences from the immediate social environment, but also learnt responses with long term developmental implications and even transgenerational effects. Given these diverse interactive effects of animal behaviour and animal nutrition on livestock production, collaborations between ethologists and nutritionists will continue to be important for future improvements in both efficiency and animal welfare.

Measurement and analysis of drinking behaviour traits in broilers

J. Rusakovica[1], T. Ploetz[2], V.D. Kremer[3], S. Avendano[3] and I. Kyriazakis[1]
[1]School of Agriculture, Food and Rural Development, Newcastle University, NE1 7RU, Newcastle upon Tyne, United Kingdom, [2]Open Lab, School of Computing Science, Newcastle University, NE1 7RU, Newcastle upon Tyne, United Kingdom, [3]Aviagen, Newbridge, Midlothian, EH28 8SZ, United Kingdom; julija.rusakovica@ncl.ac.uk

According to the European Food Safety Authority (EFSA), there is increased interest to identify risk factors associated with drinking behaviour traits that could improve bird health, such as prevalence of foot-pad dermatitis and overall improvement in bird water use. For this study, a novel system was utilised that automatically measures the individual bird drinking behaviour in large groups. Records of visits to a water station were obtained for two chicken genetic lines (a) line 1 from 1,878 birds from 15-35 days of age (n=1,577,530 events) and (b) line 2 from 2,048 birds from 13-32 days of age (n=2,641,233 events). It was identified that behaviour associated with drinking activity occurs as separate visits to drinkers, which can be clustered into bouts. A bout criterion was estimated by fitting a probability mixture model (PMM) to the natural log transformed interval length between drinking visits. The best PMM was a truncated log normal distribution for within bout intervals and a log normal distribution for between bout intervals. The bout criterion was estimated where the two distributions crossed and resulted in different estimates for the two genetic lines: 846 seconds for line 1 and 566 seconds for line 2. Based on this, eight drinking behaviour traits were identified (medians for line 1 and line 2, Kruskal-Wallis test): (1) number of visits per bout (2.12; 1.99, $P<0.01$), (2) water usage per bout (ml) (22.64; 11.91, $P<0.01$), (3) drinking time per bout (s) (175.30; 132.55, $P<0.01$), (4) bout frequency (12.42; 18.78, $P<0.01$), (5) drinking rate (ml/min) (7.78; 5.51, $P<0.01$), (6) total drinking time per day (min) (36.17; 40.84, $P<0.01$), (7) total bout duration time per day (min) (62.90; 66.91, $P<0.01$), (8) total water usage per day (ml) (282.48; 221.15, $P<0.01$). Overall, birds showed differences in the organisation of drinking behaviour, which can form the basis for further genetic selection analysis.

Feeding and drinking behavior of dairy cows at heat stress
M.A. Karatzia, A. Kalogianni and E.N. Sossidou
Hellenic Agricultural Organization-DEMETER, Veterinary Research Institute, Ktima Thermis, 57001, Thessaloniki, Greece; mkaratz@vet.auth.gr

The issue of environmental impacts on animal production has long been of interest to the dairy industry. However, it is only within the last decade that research efforts have turned to welfare of cows experiencing heat stress. This pilot study attempts to understand the correlation between cows' heat stress and nutritional behavior in 3 distinct time periods (Morning-M:8.00-12.00, Afternoon-A:16.00-18.00 and Evening-E:19.00-20.30). Behavioral aspects of feeding and drinking of 12 healthy Holstein cows were video recorded for 5 months using a surveillance video system of 4 CCTV cameras (720×480, 125 frames/sec video resolution). Two 24 h recordings were analyzed, one within the thermoneutral zone (Controls-C, n=6, T.H.I.=54.6) and one under heat stress conditions (Heat Stressed-HS, n=6, T.H.I.=87.6). Stressed cows spent significantly more (P≤0.05) time feeding in total, while their drinking activity was limited in comparison to the controls. During time period M, HS animals ate significantly more (P≤0.05) and spent 53.4% less time drinking compared to C group. In time period A, when heat stress was acute, HS cows exhibited significantly decreased feeding behavior (-79.15%) and significantly increased drinking behavior (+80.45%), (P≤0.05), in contrast with the controls. Reversed behavior was exhibited during time period E, when feeding duration for the HS group rose to levels similar to the C group and drinking duration decreased, reaching lower levels than in the controls (P≥0.05). Heat stress does seem to affect nutritional behavior of dairy cows however the extent of the situation merits further investigation. This research project was funded under the Action 'Research & Technology Development Innovation Projects'-AgroETAK, MIS453350, in the framework of the Operational Program 'Human Resources Development'. It is co-funded by the European Social Fund through the National Strategic Reference Framework (Research Funding Program 2007-2013) coordinated by the Hellenic Agricultural Organization–DEMETER.

Feeding behaviour, explained by temperament and dominance, impacts feed efficiency in beef cattle
P. Llonch[1], M. Somarriba[2], M.J. Haskell[2] and S.P. Turner[2]
[1]Universitat Autònoma de Barcelona, Department of Animal and Food Science, Campus UAB, Cerdanyola del Vallès, 08193, Spain, [2]Scotland's Rural College, Animal and Veterinary Sciences, West Mains Road, EH9 3JG Edinburgh, United Kingdom; pol.llonch@uab.cat

Feeding behaviour contributes to variations in feed efficiency, but little is known about the underlying traits responsible for this relationship. The aim of this study was to examine the effects of temperament and dominance on feeding behaviour and its contribution to changes in feed efficiency in beef cattle. Eighty-four steers (castrates) (crossbred Charolais or purebred Luing), were used in a 2×2 factorial design, with two breeds and two basal diets (concentrate vs mixed). Over a 56-d period individual animal dry matter intake (DMI; kg/day) and weekly body weight was measured. Ultrasound fat depth was measured on day 56. Average daily gain, food conversion ratio (FCR) and residual feed intake (RFI) were calculated. Dominance was assessed based on agonistic interactions between steers at the feeders using the number of contacts, number of aggressions, and number of displacements per day. Feeding behaviour was described by the number of feeding events, the duration per visit and the total time spent feeding per day. Temperament was assessed using two standardised tests: restlessness when restrained (crush score) and the flight speed on release from restraint. Statistical analysis was performed using multivariate regression models. DMI was lower (P=0.032) in more temperamental animals. The number of meals was higher with a mixed diet (P<0.001) and for more temperamental (P=0.003) and dominant (P=0.017) animals. The time spent eating was greater with the mixed diet (P<0.001), Luings (P=0.009) and dominant steers (P=0.032). Steers that spent more time feeding were more efficient (P=0.039). More time spent eating the mixed diet was associated with improved RFI possibly due to greater secretion of saliva and increased access of microbiota to fibre. In conclusion, temperamental animals have lower DMI despite eating for longer. Dominant animals have more frequent meals and eat for longer. Longer time spent eating improves feed efficiency.

Impact of length of straw by the use of a straw mill on the selective feeding of young cattle and th

H. Scholz[1], B. Fischer[2] and T. Bäthge[2]
[1]Anhalt University of Applied Sciences, Faculty LOEL, Strenzfelder Allee 28, 06406 Bernburg, Germany,
[2]State Institute for Agriculture and Horticulture Saxony-Anhalt, Lindenstraße 18, 39606 Iden, Germany;
h.scholz@loel.hs-anhalt.de

When feeding high quality silage to heifers from the age of two, there is a risk of energy oversupply. Depending on the feeding value or scarce availability of silage or corn silage diets with high proportions of straw is often incorporated. For an energetically standardized young cattle supply of straw proportion can be more than 20% of dry matter. It was investigated whether the grinding of straw with the straw mill selective feeding significantly limits. The investigation has been carried out with young cattle in the second year. 78 animals were kept and fed under similar conditions in two groups. The experimental group (EG) consisted of cattle 12 to 15 months and in the control group (CG), the cattle were 15 to 20 months old. The experimental feeding took place in five days feed distribution and residual feed were weighed. The ration of EG contained ground with the straw mill straw and CG was further fed rotor-cut pressed straw. To determine the selective seizure samples of feed distribution and the remaining food with the particle separator box and the crude protein and energy-content have been determined. The grinding of the straw increased the daily feed intake. In the EG an increase in feed intake was observed by grinding of the straw. Feed intake directly on the day for changing the diet of long on ground straw increased by more than 2.0 kg of DM per animal. In the following days, the feed intake was increased by 0.9 kg DM per animal and day on average (7.4 vs 8.3 kg DM per day). The results of the screen distribution of residual feed point to a differentiated feeding behavior between the groups. In the EG the particle length of the residual feed to a large extent with the template matches. The acid-base-balance (NSBA) values of EG are within normal limits. If straw shares of 25% and more are fed in rations to young cattle (heifers), the particle length of straw has a significant impact on the selective feeding behavior. A particle length of 1.5 cm compared to 7.5 cm long prevented straw certainly a discarding of the straw on the feeding barn. The feed intake increases when short straw is mixed into the TMR.

Effect of antioxidant supplementation on serum α-tocopherol in horses undergoing moderate exercise

E. Velázquez-Cantón, A.H. Ramírez-Pérez, L.A. Zarco, J.C. Ramírez-Orejel and J.C. Ángeles-Hernández
National University of Mexico, Animal Nutrition, Universidad 3000 Col. UNAM-CU, 04510, Mexico;
eliasvelcanton@hotmail.com

Antioxidants as selenium (Se) and vitamin E (E) protect biomembranes against free radicals, which are intensively formed during exercise. Serum tocopherols are the indicators most used to assess the vitamin E status. A trial was carried out to study the effect of both Se (Se-yeast) and E (α-tocopheryl) supplementation on serum α-tocopherol (αE) in horses under moderate exercise conditions. Twenty-four horses of Mexico City's Police (450 kg, 5-15 yr) without physical work during the month prior to our study were used in an 11-week trial. They were individually stabled and randomly assigned to 4 treatments in a factorial arrangement (2×2 Se, E levels) with repeated measures. The experimental treatments were LSeLE, HSeLE, LSeHE and HSeHE [LSe, 0.1; HSe, 0.3 mg Se/kg DM and LE, 1.6; HE, 2 IU vitamin E/kg BW; according to NRC]. Daily ration was poor in Se and E (<2 µg; 14.4 IU /kg DM); hence, Se and E were fully provided using oral supplements. The experimental weeks (W) were: W0 to W4, adaptation period; W5 to W8, exercise period (3 consecutive days/30 min (5-20-5), including: warm up-moderate gallop-cool down). The readaptation to initial conditions period was from W9 to W11. At day 64, supplementation was ended. Once a week, serum samples were obtained. During the exercise period, samples were taken at the end of the 3rd day, after exercise. Serum αE was quantified by HPLC, basal value was considered from W0-1. Serum αE from 2 horses of HSeHE and 1 of LSeLE groups were considered as outliers (±2 SD), therefore, they were not statistically analysed. Data were analysed using PROC MIXED function (SAS, 9.1) where W, E, Se and their interactions were fixed effects while horse nested in treatment was the random effect. Serum αE was affected by Se ($P<0.05$; HSe, 1.27; LSe, 1.34 µg/ml, 0.06 SEM). This result is explained by complementary metabolic role of both Se and E. Week effect was noticeable ($P<0.05$) between basal values (1.19±0.05 µg/ml) and W2 (1.32±0.09 µg/ml). Conclusion: level of supplied Se affects serum αE. This fact is explained by their complementary metabolic functions. Serum αE rose after two weeks of oral supplementation.

High rate monitoring CH$_4$ production dynamics and their link with behavioral phases in cattle

Y. Blaise[1,2,3], F. Lebeau[1,2], A.L.H. Andriamandroso[1,2,3], Y. Beckers[3], B. Heinesch[2] and J. Bindelle[1,3]
[1]University of Liège, Gembloux Agro-Bio Tech, AgricultureIsLife, Passage des deportes 2, 5030 Gembloux, Belgium, [2]University of Liège, Gembloux Agro-Bio Tech, Biose, Passage des deportes 2, 5030 Gembloux, Belgium, [3]University of Liège, Gembloux Agro-Bio Tech, AGROBIOCHEM, Passage des deportes 2, 5030 Gembloux, Belgium; yblaise@ulg.ac.be

Microbial fermentation in the rumen produces methane (CH$_4$) which is a loss of energy for ruminants and also contributes to global warming. While the respiration chamber is the standard reference for CH$_4$ emissions quantification, daily CH$_4$ production dynamics can be measured only by steps of 30 min and measurements on pasture are impossible. The alternative method using SF$_6$ as tracer gas can be applied for grazing animals but provides average CH$_4$ production values over at least several hours, making it impossible to measure short term dynamics of rumen CH$_4$ production with changing animal behavior along the day. Newly developed methods using CO$_2$ as internal tracer gas extrapolate CH$_4$ emissions from few short measurements. However, both CO$_2$ and CH$_4$ emissions fluctuate during the day depending on the behavior and the post-feeding times questioning the validity of this method. Therefore, an innovative device was developed to monitor at a high rate CH$_4$ and CO$_2$ emission dynamics in order to investigate the link between CH$_4$ dynamics and the animal behavior on pasture. Preliminary results showed the ability of the device to record differences in CH$_4$:CO$_2$ ratios and eructation frequencies according to the individual and the behavior. Results from complementary experiments in barn with animals fed contrasting diets regarding CH$_4$ production (with and without linseed) and on pasture with different forage allowance will be presented in order to highlight how post-feeding time and grazing behavior impact CO$_2$ and CH$_4$ emission dynamics along the day.

Non-invasive indicators of rumen function and stress in dairy cows

L.J. Tennant, G.A. White, R. Anand-Ivell and P.C. Garnsworthy
University of Nottingham, Animal Science, School of biosciences, sutton bonington campus, le12 5rd, United Kingdom; sbxlt1@nottingham.ac.uk

Feed efficiency within a dairy herd is likely to be improved by minimising stress. One potential stressor is a change in diet composition, which may also affect rumen function and performance. In this work we investigated effects of diet change on non-invasive indicators of rumen function and stress. Rumen function was assessed through feed intake, milk yield, rumination rate, digestibility and methane emissions. Stress was assessed through faecal glucocorticoid concentration (FGC). Faecal glucocorticoids are metabolites of cortisol, providing a non-invasive alternative to blood cortisol. In the first trial time of day had no effect on FGC, but FGC concentration increased (P<0.001) 10-12 hours after a mild stress event (blood sampling), thereby confirming FGC as an indicator of stress. In the second trial (n=37 cows) concentrate ingredients (protein and energy sources) of 4 total mixed rations (TMR) were changed every 3 weeks in a 4×4 Latin square design. Diet composition and diet change did not affect rumen function. Diet composition did not affect FGC, but mean FGC increased (P<0.001) with successive diet changes across periods, suggesting a cumulative mild stress response. In the third study (n=39 cows), forage components of TMR were changed in 4 periods. In the first 3 periods only the batch of grass silage was changed and, although each diet change increased (P<0.001) FGC transiently, there was no period effect on mean FGC. In the fourth period, however, wheat straw replaced molasses in the TMR and mean FGC increased (P<0.001) after diet change. Results of this work suggest that minor changes in diet ingredients do not affect rumen function or stress indicators. Major changes in ingredients, however, can elicit mild stress responses even in the absence of changes in rumen function, and these responses may be cumulative. Increased stress induced by diet change were not considered to be welfare concerns, but future studies will indicate if this level of stress affects feed efficiency.

Accuracy of a commercially available activity monitor to record feeding and rumination time

E. Kennedy[1], M. Ryan[2], E. Harty[2], L. Mullane[2] and D.P. Berry[1]
[1]Teagasc, AGRIC, Moorepark, Fermoy, Co. Cork, Ireland, [2]Dairymaster, Causeway, Co Kerry, Ireland; emer. kennedy@teagasc.ie

Commercially available behaviour monitoring technology gives farmers the ultimate management tool, provided information is accurate. In this study feeding and rumination duration data of a MooMonitor+ activity meter (MM+; Dairymaster) was compared to IGER recorder data. IGER recorders have been validated to measure feeding and ruminating behaviour and are considered the gold standard; however they are expensive and not user-friendly. In total 20 recordings were taken from six Holstein-Friesian cows in 2014 and 2015 at the Teagasc Moorepark research farm. IGER recorders and MM+ meters were fitted for a 24-hour period. Individual files were analysed using 'Graze' analysis software. Data from the MM+ was summarised into 15 minute intervals; only data corresponding to the 24-hour period coinciding with when the IGER recorders were fitted was analysed. Cows were offered fresh feed in 24-hour allocations during the trial. The ability to predict IGER-recorded feeding and ruminating using the MM+ was quantified using logistic regression. The dependent variability was the binary trait of whether or not the cow was feeding during the measured 15 minute period as dictated by the IGER and the independent variable was the respective binary trait as dictated by the MM+. The same approach was used to evaluate the precision of detecting ruminating. A strong (P<0.001) predictive ability was evident. The sensitivity of the MM+ compared to IGER was 0.95 for both feeding and ruminating time. The specificity was 0.97 for feeding and 0.96 for ruminating when the MM+ was compared to IGER. The results from this study indicate an excellent concordance between 24 hour feeding and ruminating duration estimated from the IGER recorders and MM+ meters. The availability of this technology enables a more precise level of continuous real-time behaviour and performance monitoring of individual animals not previously possible.

Ruminating and feeding behaviour in grazing dairy herds

A. Spengler Neff[1], J.K. Probst[1], F. Leiber[1] and S. Ivemeyer[2]
[1]Research Institute of Organic Agriculture (FiBL), Ackerstrasse 113, 5070 frick, Switzerland, [2]Uni-Kassel, Nordbahnhofstr. 1a, 37213 Witzenhausen, Germany; anet.spengler@fibl.org

The aim of the present study was to investigate environmental and individual characteristics influencing feeding and rumination behaviour and whether individual differences of these behaviours are linked to health. 225 dairy cows in 8 herds in Switzerland and southern Germany were observed 3 times for 3 days (72 hours) during pasture season (spring, summer and fall, respectively). During each observation event all animals were equipped with Rumiwatch®-halters. No intervention on feeding management took place but only empirical observation. Using linear mixed models, feeding and ruminating time per day, ruminating speed, number of rumination boles per day and bites per bolus during rumination were analysed and related to forage analyses, cows' age and lactation stage as well as milk production. Relations of calving interval (CI) and somatic cell scores (SCS) to feeding and ruminating behaviour as well as to body condition scores (BCS) were investigated. Younger cows used more time per day for feeding and less for ruminating, they chewed faster and with more bites per bolus and they showed more activity changes (especially during the night) than older cows. With ongoing lactation, the number of activity changes decreased. High producing cows ruminated longer than low producing ones. Forage with high fibre (ADF) contents caused a longer ruminating time per day and tended to cause more bites per bolus. Forage with a high energy density caused less ruminating time per day, less bites per bolus, and a tendency to lower chewing speed. Forage with a high crude protein content tended to decrease ruminating time per day. Cows with high CI performed shorter feeding times per day and had higher BCS-ranges. Cows with high SCS showed less chewing bites per bolus. Older cows and cows with lower milk production showed higher SCS. This study showed that there are influences of cows' age, lactation stage and production and of feed contents and forage structure on feeding and ruminating behaviour. There are relations between feeding and ruminating behaviour and health and fertility parameters. Reasons for those relations have to be investigated in further studies including adaptation of behaviour to feed changes.

Adaptation of dairy cows to grazing after TMR feeding in early lactation on hepatic gene expression

A.L. Astessiano[1], P. Chilibroste[1], D.A. Mattiauda[1], A. Meikle[2] and M. Carriquiry[1]
[1]Facultad de Agronomia, Departamento de Produccion Animal y Pasturas, Av. Garzos 780, 12900, Uruguay,
[2]Facultad de Veterinaria, Alberto Lasplaces 1550, 11600, Uruguay; lauaste@gmail.com

Multiparous cows (n=18) were used in a randomized complete block design to study metabolic profile and hepatic gene expression related to glucose and fatty acid metabolism on the adaptation to grazing after TMR feeding during early lactation. During the first 60 days postpartum (DPP), cows were assigned to: (1) TMR (30 kgDM/d offered; 45% forage, 55% concentrate); and (2) 50% pasture in one (am) grazing session (6 h; pasture allowance 4 cm above ground level =15 kgDM/d) + 50% TMR (15 kgDM/d offered) (G1). At 61 DPP and during 21 days, all cows were managed as G1 group determining 2 treatments (TREAT): PostTMR and G1. Plasma and liver biopsies were collected pre (+55 DPP) and post (+75 DPP) dietary change to measure glucose, NEFA and BHB concentrations by colorimetric assay and mRNA abundance of 10 genes related with glucose and fatty acid metabolism by SYBR-Green real time RT-PCR. Data were analyzed as repeated measures with a mixed model that included TREAT, DPP and its interaction and means were considered to differ when P<0.05. Milk energy output and BCS from 61 to 82 DPP did not differ between TREAT (27.5 and 25.7±1.4 Mcal/d and 2.8 and 2.7±0.1 units for PostTMR and G1, respectively). Glucose, non-esterified fatty acids and beta-hydroxybutyrate concentrations were maintained during the experimental period in both TREAT. Regardless of the early lactation feeding strategy, hepatic PC, PDK, PDH1A, ACADVL and ACoAoX mRNA had a 1.5 to 2.5 fold change from +55 to +75 DPP. Expression of G6PC, PCK1A, CPT1A and CS mRNA were not affected by TREAT, DPP or its interaction. However, HMGCS2 mRNA was greater in G1 than PostTMR throughout the experimental period (pre and post dietary change; 0.81 and 2.03±0.3 for PostTMR and G1, respectively). Results indicate that hepatic expression of fatty acid oxidation increased regardless of the early lactation feeding strategy probably due to increased energy requirements from +55 to +75 DPP associated with an increase in milk yield in G1 cows or increased activity in PostTMR cows during this period.

Overstocking dairy cattle during the dry period affects DHEA and cortisol secretion

M. Fustini[1], A. Palmonari[1], G. Galeati[1], G. Gabai[2], D. Bucci[1] and A. Formigoni[1]
[1]University of Bologna, DIMEVET, Via Tolara di Sopra, 73, 40064 Ozzano Emilia, Italy, [2]University of Padova, BCA, Via dell'Università, 16, 35020 Padova, Italy; mattia.fustini3@unibo.it

Stressful situations trigger a number of changes such as the secretion of cortisol (C) and dehydroepiandrosterone (DHEA) from the adrenal cortex, in response to adrenocoticotropic hormone (ACTH). The aim of this study was to verify whether overstocking during the dry period affects DHEA and C secretion and behaviour in Holstein Friesian cows Twenty-eight cows were randomly divided into two groups (14 animals each), balanced for number of lactations, BCS (body condition score) and expected date of calving, organized in seven replicates. Animals from 21 days to the expected calving till calving were housed in pens with the same size but in different crowding conditions due to the introduction into the pen of heifers (interference animals). Control condition (CTR) had 2 animals per pen with 11.9 m^2 each, while the overstocked condition (OS) had three interference animals in the same pen with 4.8 m^2 for each animal. On days -30, -21, -15, -10, -5 before and 10, 20, 30 after calving blood samples were collected from each cow for the determination of plasma DHEA and C concentrations by RIA. Rumination time, activity (steps/h), lying time (min) and lying bouts were measured daily. Data were analysed by ANOVA for repeated measures using the MIXED model of SAS. In both groups, there was an increase in DHEA before calving and after parturition the concentration declined rapidly. Overstocking significantly (P<0.05) increased DHEA concentration compared to CTR group at day -10 (1.79±0.09 vs 1.24±0.14 pmol/ml) while an increase of C was observed (P<0.05) at day -15 (3.64±0.52 vs 1.64±0.46 ng/ml). No relationship was found between DHEA and C. OS group showed significantly higher activity (step/hour), compared with CTR group. Daily lying bouts tended to be higher for OS group compared with CTR group in the first week of treatment. The overall results of this study show that overstocking during the dry period is associated with changes in DHEA and cortisol. Work supported by a MIUR-COFIN grant (prot. 2010YBP4LZ_003).

Effect of lactation number on the respiratory rate of dairy cows on hot days

S. Pinto[1], W. Pauw[1], C. Ammon[1], J. Heinicke[1], W. Heuwieser[2] and T. Amon[1,3]
[1]Leibniz Institute for Agricultural Engineering Potsdam-Bornim (ATB), Engineering for Livestock Management, Max-Eyth-Allee 100, 14469, Germany, [2]Freie Universität Berlin, Department of Veterinary Medicine, Animal Reproduction Clinic, Königsweg 65, 14163 Berlin, Germany, [3]Freie Universität Berlin, Institute of Animal Hygiene and Environmental Health, Department of Veterinary Medicine, Robert-von-Ostertag-Str. 7-13, 14163 Berlin, Germany; tamon@atb-potsdam.de

It was the aim of this study to assess the effect of lactation number on the respiratory rate of dairy cows on high temperature days. The respiratory rate was measured in thirty lactating cows (1st to 5th lactation), housed in a naturally ventilated barn in Brandenburg, Germany within on 36 hot days between June and August 2015. Measurements took place in the morning from 07:00 h to 10:00 h and in the afternoon from 11:00 h to 14:00 h (GMT + 01:00 h). The cows were divided in three different groups (1, 2 and 3 or more lactations). The respiratory rate was measured visually by counting flank movements for thirty seconds (breaths/30 sec). At the same time, the body posture of the cows was noted (laying and standing). Simultaneously, air temperature (AT) and relative humidity (RH) in the barn were recorded. Data were analyzed for differences between factor levels with a repeated measurements linear mixed model at an overall significance level of 0.05 ($P<0.05$). The mean values of AT and RH were 19.2 °C and 77.1% in the morning and 25.1 °C and 57.4% in the afternoon, respectively. The respiratory rate of both morning and afternoon was higher ($P<0.05$) in cows with three or more lactations (20.8 and 23.0 breaths/30 sec) than in second lactation cows (17.9 and 20.9 breaths/30 sec). Furthermore, cows in lying position had a significantly ($P<0.05$) higher respiratory rate (20.1 and 23.5 breaths/30 sec) than standing cows (17.2 and 20.1 breaths/30 sec) in the morning and afternoon, respectively. In conclusion, the respiratory rate is influenced by lactation number and body posture of cows on hot days. These data will serve as a basis for developing an automatic sensor of the respiratory rate of dairy cows to evaluate heat stress.

Technical note: Validation of a device for automatically measuring of lying behaviour in dairy cows

J.C.S. Henriksen and L. Munksgaard
Animal Science, Aarhus University, Animal nutrition and physiology/Behaviour and stress biology, Blichers allé 20, 8830 Tjele, Foulum, Denmark; juliec.henriksen@anis.au.dk

Using electronic devices for measuring lying behaviour, is time saving compared to direct observations. However, the device needs to be validated by comparing with direct observations or similar devices that have already been validated. Since the AfiTagII system is relatively new on the market, the objective was to validate the lying time and frequency of lying bouts by direct observations. The validation included 40 cows with 20 lactating cows on slatted floor, and 20 dry cows on deep bedding. The activity pedometers were attached to the hind leg. Both Danish Jersey (n=21) and Danish Holstein (n=19) were represented including primiparous and multiparous cows. Each cow was observed for 2 hours in their home pen. A proc mixed procedure was conducted to test if there was any effect of breed, parity and housing system. To find the correlation between the activity pedometers and direct observations a spearman correlation was calculated. No difference was found for breed, parity or housing system. Lying time showed high positive correlation between direct observations and the AfiTagII system for the total observation period (r=0.99). The lying bouts in each observation period ranged between 0-3 bouts. On the slatted floor there was only one observation where the lying bout number deviated from the activity pedometers (PPV=0.96), and on the deep bedding lying bout number deviated with 1 bout between direct observations and activity pedometers for three observations (PPV=0.85). In conclusion the AfiTagII measures lying time with high accuracy.

Air flow patterns and gas concentration distribution in naturally ventilated barns

M. Koenig[1], D. Janke[1], S. Hempel[1], C.K. Saha[2], M. Fiedler[1], W. Berg[1] and T. Amon[1,3]
[1]Leibniz Institute for Agricultural Engineering Potsdam-Bornim (ATB), Engineering for Livestock Management, Max-Eyth-Allee 100, 14469, Germany, [2]Bangladesh Agricultural University, Department of Farm Power and Machinery, Main Road, 2202 Mymensingh, Bangladesh, [3]Freie Universität Berlin, Institute of Animal Hygiene and Environmental Health, Robert-von-Ostertag-Str. 7-13, 14163 Berlin, Germany; tamon@atb-potsdam.de

Airflow through naturally ventilated barns (NVB) links the outdoor environment and the buildings microclimate. Understanding the principles of air motion and predicting the air exchange rate (AER) and indoor climate parameters is crucial to support animal welfare and to mitigate gaseous emissions. We present high-resolution wind and gas concentration measurements under on-farm conditions in a NVB in Northeast Germany and identify typical patterns for different meteorological boundary conditions. The loss of accuracy by neglecting measurement points for calculating a characteristic CO_2 value is investigated. Computational fluid dynamics simulations are conducted to assess the influence of wind direction and surroundings on the AER. Further simulations compare Reynolds-average Navier-Stokes (RANS) approaches with an unsteady Large Eddy Simulation (LES) approach considering the cross-section of the NVB. The simulations are compared to measurements. Large spatial deviations in wind speed lead to high fluctuating gas concentrations at multiple measurement locations affecting the AER estimation from measurements. In addition, simulations imply real differences in the AER up to 77% depending on the wind direction. The unsteady behaviour reproduced with LES also affects the simulated AER. For further research, we combine field measurements, physical modelling in a boundary layer wind tunnel and numerical simulations focussing on LES in order to significantly improve our knowledge of the emission source „dairy barn'. Field measurements are crucial to validate models and to assess the effect of air flow on emission rates. Modelling permits to study the influence of boundary conditions based on high-resolution data.

The welfare of dairy cows in continuously housed versus pasture-based production systems

G. Arnott[1], C.P. Ferris[2] and N.E. O'Connell[1]
[1]Institute for Global Food Security, School of Biological Sciences, Queen's University Belfast, 97 Lisburn Road, Belfast, BT9 7BL, United Kingdom, [2]Agri-Food and Biosciences Institute, Large Park, Hillsborough, BT26 6DR, United Kingdom; g.arnott@qub.ac.uk

There is increasing interest in the use of continuous housing systems for dairy cows, with various reasons put forward to advocate such systems. A review was conducted with the aim of interrogating the existing scientific literature to compare the welfare, including health, of dairy cows in continuously housed and pasture-based systems. Using a validated systematic approach, this review summarises existing work, while also identifying knowledge gaps and directions for future research. The scope of the review is broad, examining relevant topics under two main headings; 'health' and 'other welfare indicators'. For health, the topics examined included; lameness, hock lesions, mastitis, uterine disease, other infectious disease, endoparasites and mortality. Overall in relation to the health topics examined, cows were generally healthier within pasture-based systems, with the unsurprising exception of being at greater risk of endoparasite infection. The 'other welfare indicators' category considered; behaviour, cow preference for pasture, physiology / biomarkers, thermal stressors, and the use of a holistic approach to welfare assessment. This revealed benefits of pasture access for dairy cow behaviour. Moreover, when given the choice between pasture and indoor housing, cows showed an overall preference for pasture, particularly at night. However, potential areas for concern in pasture-based systems included physiological indicators of more severe negative energy balance, and in some situations the potential for comprised welfare with exposure to unpredictable weather conditions. Finally, evidence using a holistic welfare index score indicated improved welfare when cows are at pasture. In summary, the results from this review highlight that there remain considerable welfare benefits from incorporating pasture access into dairy production systems.

Simulation of the indoor climate of livestock buildings: a tool to assess climate change scenarios

G. Schauberger

University of Veterinary Medicine, Department for Biomedical Sciences, WG Environmental Health, Veterinärplatz 1, 1210, Austria; gunther.schauberger@vetmeduni.ac.at

In the future, the production of pig and poultry may be severely affected by heat stress. In temperate climate regions like Central Europe these animals are predominantly kept in confined housing systems. Therefore the impact of climate change scenarios on these systems cannot be assessed only by ambient meteorological parameters, as they are modified by the confined livestock building and the livestock itself. This means that the indoor climate, which is the environment relevant for a large share of farm animals, has to be considered by simulation models, describing the interaction between animals which release sensible and latent (water vapour) heat, the insulation of the building to capture the sensible heat, and the ventilation system. The ventilation system is the most effective link to the outside, but inevitably differences will occur between climatic conditions outside and inside livestock buildings. The indoor climate will in turn impact on animal health and welfare, productive, reproductive and economic performance. The features as well as the limitations of such simulation models will be presented and discussed. On the basis of such model calculations the husbandry conditions can be simulated to reveal the effectiveness of adaptation measures (i.e. adaptive capacity), which sums up to livestock and farm vulnerability. The thermal environment is a major parameter as it directly impacts animal welfare and health. The productivity of farm animals can be investigated by parameters which are sensible to the thermal environment: For fattening pigs and broilers this can be described by daily weight gain and feed conversion, egg production for laying hens and the reproductive performance of sows (litter size, number and body weight of weaned piglets, etc.). Furthermore, an increase of the airborne emissions (NH_3, odour) can be anticipated, leading to increased ambient concentrations, a reduced indoor air quality, and increased separation distances to protect the neighbours from odour nuisance.

Bayesian Networks to assess potential barn adaptations to climate change in dairy farms

E. Galán[1], A. Del Prado[1], E. Sanchís[2], F. Estellés[2] and S. Calvet[2]

[1]Basque Centre for Climate Change (BC3), Alameda Urquijo 4, 4ª, 48008 Bilbao, Spain, [2] Universitat Politècnica de València (UPV), Institute of Animal Science and Technology, Camino de Vera, s/n, 46022 Valencia, Spain; elena.galan@bc3research.org

The information about the effects of climate change in dairy cow farms is made using different scales, methods and regions. This mismatch challenges the reliability of the assessments to implement cost-effective measures of adaptation in farms. Therefore, integration of the available information and uncertainties associated is needed to design effective farm-scale adaptation strategies. In the framework of the ERANET+ project OptiBarn, focused on the adaptation of naturally ventilated barns of permanently housed dairy cows to the effects of climate change, we present a risk assessment for three potential impacts at farm-scale: welfare, economic costs and emissions. We use a Bayesian Belief Network (BBN) to relate quantitatively the information of the effects of climate change described in literature and apply it to conduct scenarios of sub-optimal and optimal housing design of barns in three countries: Germany, Israel and Spain. Using BBN allow us to take into account the biases of the information either due to methodological design, mismatch of scales, unavailable data, unclear causal relationships, etc., and therefore to determinate which is the minimum information needed to take decisions based on our risk assessment. To control for other socio-economic impacts such as milk prices, concentrate prices and non-market values concerning welfare, we perform a sensitivity analysis based on regional data.

Nighttime cooling improves milk yield of dairy cattle exposed to moderate heat stress

U. Serbester, S. Goncu and M. Gorgulu
Cukurova University Agriculture Faculty, Deparment of Animal Science, Cukurova University Agriculture Faculty
Deparment of Animal Science Balcalı, 01330 Adana, Turkey; ugurserbester@gmail.com

The present study was conducted to investigate the effects of daytime and nighttime cooling on performance, milk composition and physiological parameters of dairy cattle. Two experiments were conducted. In the first experiment, 28 lactating Holstein dairy cows, averaging parity 2.6±0.84 (mean±standart deviation), daily in milk 123.9±37.64, body weight 552.00±46.9, milk yield 29.0±4.08, were divided into 4 groups and were exposed to 1 of 4 groups: Control (without cooling), Sprinklers, Sprinklers+Fans, and Fans for daytime (between 10 am – 5 pm) cooling. Treatments were identical in the second experiments but cooling was applied between 10 pm – 5 a.m. Dairy cattle used in the second experiments averaged parity 2.9±1.24, daisy in milk 141.3±52.51, body weight 528.4±54.08, milk yield 22.7±3.68. Both of the experiments were lasted 6 weeks and dairy cows were fed total mixed ration containing corn silage, alfalfa, wheat straw and concentrate (18% crude protein and 2,650 kcal/kg metobolizable energy). Milk yield and dry matter intake were measured daily in both of the experiments. Milk samples were collected 2 times per week and analyzed for components. Body weight and body condition score were recorded weekly. Respiration rate, heart rate, rectal temperature and skin temperature were measured 2 times (on 10.30 a.m., 1.30 p.m. and 3.30 p.m.) per week. Data were analyzed using PROC MIXED of SAS with repeated measures. In the experiment I, cows received mild heat stress, on the other hand, in the second experiment, cows received moderate heat stress. As a conclusion, daytime cooling did not effect performance and physiological parameters of dairy cows, whereas night cooling and especially sprinklers treatment increase daily milk yield while decreasing of dry matter intake.

Comparison of three techniques for measuring enteric methane emissions by ruminants

M. Arbre, C. Martin, Y. Rochette, C. Lascoux, M. Eugène and M. Doreau
INRA, UMRH, Saint-Genès Champanelle, 63122, France; michel.doreau@clermont.inra.fr

The open-circuit chamber (OC), the SF_6 tracer technique (SF_6) and the GreenFeed system (GF) are three techniques for measuring enteric methane (CH_4) emissions by ruminants. These methods estimate individual daily CH_4 emissions, from eructed and expired gas samples collected either continuously over 24 h, in a canister for SF_6 and in chamber for OC, or several times a day during short-term periods (3-8 min) when cattle visit an automated feeder fitted with a head chamber for GF. The objective of this work is to compare these three techniques during a 46-day period. Methane emissions were measured on 8 dry cows fed the same diet (70% hay + 30% concentrate) in restricted amounts, constant among cows throughout the experiment: 13.3±0.5 kg/day dry matter intake (DMI). Measurements in OC were performed twice for 4 days at 3-week interval, and SF_6 technique was performed in free-stalls barn (SF_6-FS) twice for 4 days at 3-week interval and once for 4 days in chambers (SF_6-C). GreenFeed was used continuously for 30 days and data were split in two 15-day periods. Data were analyzed using PROC MIXED of SAS. Methane emissions in g/kg DMI averaged for the two periods: 24.2±4.9 for SF_6-FS, 27.7±3.3 for OC and 23.7±3.3 for GF. Analyses of variance showed an effect of techniques, OC resulting in higher CH_4 emissions in g/kg DMI than the two other techniques (P<0.001). Neither an effect of period nor a technique × period interaction was shown (P>0.05). Emissions using SF_6 in chambers (SF_6-C) were 23.7±2.4 g/kg DMI. The difference between CH_4 measurement techniques can be explained in part because GF and SF_6 record only eructed and expired gases whereas OC records in addition gases from flatulence and CH_4 from feces fermentation. Pearson correlation coefficients for individual methane emissions (n=8) were 0.80, 0.85, 0.30 and 0.37 between OC and SF_6-FS, OC and SF_6-C, OC and GF, SF_6-FS and GF, respectively.

Genetic control of environmental variation: Different structures of residual variance in DHGLM model
M. Felleki[1], A. Gilmour[2] and L. Bodin[3]
[1]Swedish University of Agricultural Sciences, Department of Animal Nutrition and Management, The Beijer Laboratories in Uppsala, Box 7024, 750 07 Uppsala, Sweden, [2]CargoVale, Cargo, NSW, Australia, [3]INRA, UMR1388 / INPT ENSAT / INPT ENVT GenPhySe, 24 Chemin de Borde Rouge, 31326 Castanet-Tolosan, France; majbritt.felleki@slu.se

Models of genetic control of environmental variation have been broadly studied with different analysis methods which were applied to different structures of the residual variance. Modeling the environmental variability with fixed, permanent environmental and additive genetic effects has been done firstly for the logarithm of the environmental variance, later for the environmental variance itself and finally for the environmental standard deviation. Up to now the method of Double Hierarchical Generalized Linear Models (DHGLM), had been considered only for the logarithmic scale. We have extended the use of the method of DHGLM to the other structures by the use of new link functions in the DHGLM settings: identity (additive structure of the residual variance), and square root (additive structure of the residual standard deviation). Examples of simulated and empiric data sets have been analyzed with these three link functions. The problem of possible negative residual variances and standard deviations when using identity and square root links has been addressed, as well as the interpretation of the results among structures. It has been found that the use of DHGLM with identity and square root links is feasible, thus the repertoire of easy to use tools for analyzing genetic control of environmental variation has been extended. Comparing differently structured residual variances of quantitative traits might lead to an increased understanding of the connection between genetic control of the mean level and the variation of a trait, as one structure could show up to generally better describe this feature than do the others.

Genetics of carcass condemnations and relationships with growth, backfat and uniformity in pigs
H.A. Mulder[1], P.K. Mathur[2], R.H. Vogelzang[2] and E.F. Knol[2]
[1]Animal Breeding and Genomics Centre, Wageningen University, P.O. Box 338, 6700 AH Wageningen, the Netherlands, [2]Topigs Norsvin BV, P.O. Box 43, 6640 AA Beuningen, the Netherlands; han.mulder@wur.nl

Carcass condemnations recorded during meat inspection at slaughter plants are indications of reduced health and welfare of pigs and farms differ in incidence. These records offer opportunities for breeding to increase health and welfare of pigs. However, in most cases pedigree data of finisher pigs are lacking. In this study, we used data of 56,319 finisher pigs from three farms with known pedigree and that were scored for 69 carcass condemnations at meat inspection at the slaughter plant. These three-way crossbred pigs were offspring of Pietrain sires. The main aim of this study was to test the hypothesis that uniform animals are more robust and have lower disease incidence. We used a trivariate double hierarchical generalized linear model in ASReml 4. Among the carcass condemnations, incidences were highest for bursitis (14%), pneumonia (14%) and lower for pericarditis (2.5%) and pleuritis (3.4%). The heritabilities for these traits ranged from 0.014 to 0.073 on the observed scale. The genetic variance for residual variance on the log scale was 0.05 for growth rate an 0.20 for backfat thickness showing good opportunities for genetic improvement of uniformity in these traits. The genetic correlations between carcass condemnations and growth rate, backfat thickness and their residual variances were all small and most of them were not significantly deviating from zero. Therefore, the hypothesis that selection for increased uniformity would improve disease resistance and welfare was rejected. However, these results show good opportunities to utilize carcass condemnations at meat inspection in genetic evaluations to improve health and welfare of pigs while improving uniformity in growth rate and backfat at the same time.

Selection for intramuscular fat in rabbits: direct and correlated responses

M. Martínez Álvaro, P. Hernández Pérez and A. Blasco Mateu
Universitat Politècnica de València, Institute for Animal Science and Technology, Camino de vera s/n, 46022 Valencia, Spain; mamaral9@upv.es

After seven generations of divergent selection for intramuscular fat (IMF) in rabbits, the response to selection and the correlated responses in carcass and meat quality traits were studied. Selection criterion was the phenotypic value of IMF measured at 9 wk of age in 2 full sibs of the candidate. Direct response in IMF and correlated responses in reference carcass weight, dissectible fat content, scapular and perirrenal fat weights, carcass and meat color, pH, and protein and fatty acid composition of Longissimus dorsi muscle were estimated as differences between the divergent lines in the seventh generation. Response to selection was 0.39 g/100 g, about 2.6 standard deviations of the trait. The high line showed higher carcass dissectible fat content than the low line. Selection on IMF changed the fatty acid profile of the meat, the high line showing lower PUFA and higher MUFA percentages than the low line. Protein content was higher in the high than in the low line, whereas the rest of the traits did not show any substantial change due to selection.

A deficient protein supply could be affecting selection for growth rate in rabbits

P.J. Marín-García, E. Blas, C. Cervera and J.J. Pascual
Instituto de Ciencia y Tecnología Animal, Animal Science, Camino de Vera S/N, Universidad Politécnica de Valencia, Valencia 46022, Spain; pabmarg2@doctor.upv.es

The use of non-specific diets during the fattening period could be affecting the expression of the genetic potential in parental lines of rabbits with high growth rate. To test this hypothesis, a total of 175 growing rabbits from three genetic lines (two maternal and one paternal) were used to cover a wide range of daily growth rates. Fifteen animals per line were slaughtered at weaning (28 day of age). The rest were fed with a common growing diet, containing 111 g digestible protein (DP) per kg dry matter (DM), until they were slaughtered at 63 days of age. Protein and amino acids retention during the growing period was determined from empty bodies. Body weight and feed intake were also controlled to evaluate how this diet could be affecting protein supply. Animals from the paternal line showed better feed conversion ratio (2.36±0.04) than those coming from maternal lines (2.67±0.05; P<0.05). Animals showing a high growth rate had slightly lower digestible protein intake to that expected from their empty body weight (EBW) gain. This fact, coupled to the greater maintenance requirements, could lead to lower DP and amino acids available to be retained in the body. In fact, when the amount of protein retained in the EBW of growing rabbits was represented in function of the daily EBW gain, animals with an EBW gain up to 45 g/day showed a lower protein retention to that expected. Therefore, a rabbit with an average daily gain of 70 g/d would need to increase its protein retention in 1.75 g/d to maximize its growth rate, being needed a diet with at least 125 g DP/kg DM, or an adequate adjustment of the limiting amino acids. These results reveal a possible protein deficit on growing rabbits with a high growth rate, when common diets are used, affecting their genetic expression and hindering the selection process. Determining limiting amino acids requirements in function of growth rate will be need to develop specific growing diets for paternal lines.

Investigating genes and enzymes related to intramuscular fat content in pig meat

M. Zappaterra[1], M. Deserti[2], P. Zambonelli[1] and R. Davoli[1]
[1]University of Bologna, Department of Agricultural and Food Sciences (DISTAL), Viale G. Fanin 44, 40127 Bologna, Italy, [2]University of Bologna, Department of Experimental, Diagnostic and Specialty Medicine, Via Massarenti 9, 40138 Bologna, Italy; martina.zappaterra2@unibo.it

Technological and nutritional qualities of pig meat are affected by several aspects and among them intramuscular fat (IMF) content plays a crucial role. Nevertheless, despite the increasing interest towards IMF composition and deposition, the knowledge regarding molecular and biological mechanisms behind IMF metabolism is still incomplete. Therefore, we decided to study in two groups of Large White pigs divergent for IMF deposition the transcription levels of PLIN1, PLIN2, PLIN3, PLIN4, PLIN5, ATGL, LIPE, LPL, MGLL, FASN and SCD, genes involved in lipids metabolism. Furthermore, FASN, LIPE, SCD and LPL protein levels were quantified through commercially available ELISA kits. The genes and proteins expression levels were then compared, and were assessed the correlations to identify co-expression patterns. The aim of the study was to test the associations between IMF deposition and the transcription and/or translation levels of the considered genes. PLIN1, PLIN4, PLIN5, ATGL, FASN, LPL, MGLL genes showed to be more expressed in high IMF pigs, while PLIN2 was the only gene presenting higher mRNA levels in low IMF samples. The results of protein quantitation suggest the involvement of LIPE enzyme in IMF metabolism, as their amounts significantly differed between divergent IMF pigs. Additionally, a poorly known correlation was found between LIPE and FASN enzymes only in female individuals, suggesting the existence of a sex-related co-expression pathway linking the translation of these two proteins. The obtained results provide new information for the understanding of IMF deposition and indicate that LIPE protein quantification may be an useful parameter indicating the pig attitude towards IMF deposition.

Genetic analysis of stillbirth, birth weight and fertility traits in Landrace and Large White pigs

E. Heuß[1], C. Große-Brinkhaus[1], M.J. Pröll[1], H. Henne[2], A.K. Appel[2], K. Schellander[1] and E. Tholen[1]
[1]Institute of Animal Science, University of Bonn, Endenicher Allee 15, 53115 Bonn, Germany, [2]BHZP GmbH, An der Wassermühle 8, 21368 Dahlenburg-Ellringen, Germany; esther.heuss@uni-bonn.de

Aim of this study was to investigate the potential of improving piglet survival in German Landrace (LR) and Large White (LW) populations via genomic selection (GS). Piglet mortality does not only have a negative impact on animal welfare but also decreases the economic productivity of piglet producers. Therefore, variance components were estimated for stillbirth (SB) and individual birth weight (BW) for 32.627 LW and 38.042 LR piglets using ASReml 4.0. In addition heritabilities (h^2) for reproductive traits for their 3.476 LW and 2.594 LR dams were evaluated. SB and BW were analyzed with a threshold using a generalized linear mixed model (GLMM) and a logit link function. Reproductive traits were investigated with a univariate animal model. Classical breeding values were obtained and used in a genome-wide association study (GWAS) (n=553 (LW), n=454 (LR)) of SB, BW and number of piglets born alive (NBA). GWAS was conducted with GenABEL and the estimation of genomic breeding values (GEBV) with rrBLUP. The h^2 estimated for SB were 0.057 (LR) and 0.051 (LW). For BW estimates were 0.146 and 0.137 in LR and LW, respectively. The traits exhibited negative genetic (-0.570 (LR), -0.400 (LW)) and phenotypic (-0.189 (LR), -0.208 (LW)) associations indicating that piglets with lower birth weights have a higher probability to be stillborn. For the reproductive traits in both populations low to medium h^2 were estimated. GWAS revealed 4 promising QTL regions inhibiting potential candidate genes for SB. GS presented encouraging accuracies and will be the method of choice. To gain a more in-depth picture of the relationships between reproduction traits of sows and piglet survival until weaning a multivariate analysis of piglet survival was conducted and the application of GWAS and GS methods was expanded. The results will be presented.

Comparison of male selection strategies in Finnish blue fox population using stochastic simulation

J. Peura[1], A.C. Sørensen[2] and L. Rydhmer[1]
[1]SLU, Department of animal breeding and genetics, Ulls väg 26, 75007 Uppsala, Sweden, [2]Aarhus University,
Department of molecular biology and genetics, Center for quantitative genetics and genomics, Blichers Allé 20,
8830 Tjele, Denmark; jussi.peura@slu.se

The selection in current Finnish blue fox breeding scheme is based on traditional BLUP evaluation. Among fur farmers, there is increasing discussion if genomic selection should be taken into use also in blue fox breeding. Breeding value evaluation is done across farms but selection is done mainly within farms. Currently, selection of blue fox breeding animals is in some extend inefficient and uncontrolled. Because selection of breeding males have most pronounced effect to the population, this study compared three optional future blue fox breeding male selection strategies using genomic information: (1) male selection within farm using truncation selection; (2) Male selection across farms using truncation selection; and (3) Male selection across farms using optimal contribution selection. In all three scenarios two optional female/male ratio (10/1 and 25/1) and two optional proportions of genotyped male pups (10 and 50%) was compared. The breeding goal in all scenarios included litter size, pelt size, pelt quality, feed efficiency and leg conformation. Study based on stochastic simulation. Genomic selection is simulated using pseudo-genomic simulation and three optional levels of accuracies of genomic breeding values were tested (low, medium and high). Across farms selection resulted higher total economic gain than within farm selection. Moreover, scenarios with higher females/male ratio and higher proportion of genotyped male pups resulted higher genetic gain than low female/male ratio and low proportion of genotyped male pups.

Genetic relationship between polledness and other relevant traits in Holstein dairy cattle

H. Dressel, M. Wensch-Dorendorf and H.H. Swalve
Martin-Luther-University Halle-Wittenberg, Institute of Agricultural and Nutritional Sciences, Th.-Lieser-Str. 11,
06120 Halle, Germany; holger.dressel@landw.uni-halle.de

Previous studies found in the literature have indicated substantial differences between the genetic level of polled and horned Holstein bulls. Recently, these differences appear to become smaller. The objective of the present study was to evaluate the genetic relationships between polled status and other important traits. In previous research, predominately estimated breeding values of A.I. sires were used for such an analysis. However, bulls available for artificial insemination represent a preselected sample. Thus, in the present study genomic estimated breeding values of unselected cows from commercial dairy herds were used to estimate the genetic correlation between polled status and functional traits as well as production traits. Data consisted of a random sample of 4,937 genotyped and phenotyped Holstein cows born between 2011 and 2013. In this sample, 32 cows were heterozygous polled. The polled status was converted into a binary trait (0 = polled, 1 = horned). For the bivariate analysis of the polled status on the logistic scale and any estimated breeding value of a trait (total merit index, milk yield, conformation, reproduction) on the normal scale an animal model with genomic relationship matrix was used. The 'heritability' for polled status was estimated as 0.56. Genetic correlations between polled status and other traits were between -0.004 and 0.016. Only the two correlations between polled status and total merit index and between polled status and milk yield were significantly different from zero.

Genome-wide association and prediction studies using a Bayesian dominance model
J. Bennewitz[1], T.H.E. Meuwissen[2] and R. Wellmann[1]
[1]Institute of Animal Science, University Hohenheim, Garbenstrasse 17, 70599 Stuttgart, Germany,
[2]Institute of Animal and Aquacultural Science, Norwegian University of Life Science, 1450 Aas, Norway;
j.bennewitz@uni-hohenheim.de

In recent SNP-based studies it was repeatedly shown that dominance variance explains a non-negligible proportion of the phenotypic variance in important cattle and pig breeding traits. Moreover, it is well known that dominance and additive gene effects depend on each other in a complicated manner. This complex relationship was used in a recent study to develop priors for Bayesian genomic dominance models. These so-called BayesD models are extensions of BayesC. The present study investigated the power and precision of BayesC and BayesD in genome-wide association and prediction studies by means of stochastic simulations. During the simulation process special emphasis was on the joint distribution of additive and dominance effects of causative mutations. For genome-wide association analysis a sliding window approach and the R. Fernando window posterior probability of association were used for inference purpose. The power to map segregating causal effects and the mapping precision were assessed for various marker densities with up to full sequence information. Prediction accuracy was assessed by cross validation. The results showed an improved power between -2 and 8% for causative genes explaining more than 2.5% of the genetic variance. Mapping precision was only marginally affected by the choice of the model. The accuracy of genomic predictions increased in previous simulations by about 15%. The improvement with BayesD is due to the use of dominance effects, which are modelled by diplotype marker information using priors that were supported by the data. Currently we are in progress of analysing a dairy cattle data set in order to validate the superiority of BayesD if dominance is present in a real data set.

Increased prediction accuracy using a genomic feature model including information on QTLs
P. Sarup[1], J. Jensen[1], T. Ostersen[2], M. Henryon[2] and P. Sørensen[1]
[1]Center for Quantitative Genetics and Genomics, Aarhus, Department of Molecular Biology and Genetics, Blichers Allé 20, 8830 Tjele, Denmark, [2]SEGES Danish Pig Research Centre, Axeltorv 3, 1609 København V, Denmark;
pernille.sarup@mbg.au.dk

In animal breeding, genetic variance for complex traits is often estimated using linear mixed models that incorporate information from single nucleotide polymorphism (SNP) markers using a realized genomic relationship matrix. In such models, individual genetic markers are weighted equally and genomic variation is treated as a 'black box.' This approach is useful for selecting animals with high genetic potential, but it does not generate or utilise knowledge of the biological mechanisms underlying trait variation. Here we propose a linear mixed-model approach that can evaluate the collective effects of sets of SNPs and thereby open the 'black box.' The described GFBLUP model has two components that are defined by genomic features. We analysed data on average daily gain, feed efficiency, and lean meat percentage from 3,085 Duroc boars, along with genotypes from a 60 K SNP chip. In addition information on known quantitative trait loci (QTL) from the animal QTL database was integrated into the GFBLUP as a genomic feature. Our results showed that the most significant QTL categories were indeed biologically meaningful. Additionally, for high heritability traits, prediction accuracy was improved by the incorporation of biological knowledge in prediction models. A simulation study using the real genotypes and simulated phenotypes demonstrated challenges regarding detection of causal variants in low to medium heritability traits. The GFBLUP model showed increased predictive ability when enough causal variants were included in the genomic feature to explain over 10% of the genomic variance, and when dilution by non-causal markers was minimal.

Effect of selection for residual variance of litter size on biochemical parameters in rabbits

M.L. García[1], K. Zbynovska[2], P. Petruska[2], I. Bovdišová[2], A. Kalafová[2], M. Capcarova[2], A. Blasco[3] and M.J. Argente[1]
[1]*Universidad Miguel Hernández de Elche, Departamento Tecnología Agroalimentaria, Ctra Beniel km3.2, 03312 Orihuela, Spain,* [2]*Slovak University of Agriculture in Nitra, Tr. A. Hlinku 2, 949 76 Nitra, Slovak Republic,* [3]*Institute for Animal Science and Technology. Universitat Politècnica de València, P.O. Box 22012, 46071 Valencia, Spain; mariluz.garcia@umh.es*

An experiment of divergent selection for residual variance of litter size (Ve) was carried out on rabbits. The Ve was estimated as the phenotypic variance of litter size within female after correcting litter size for year-season and lactation status. The aim of this study was to examine the correlated response on health status in female. We analysed the Ve, the litter size at birth, and the concentration of total bilirubin, conjugated bilirubin, cholesterol, triglycerides, glucose, Ca, K, Cl, Na, P, urea and creatinine. Data were analyzed using Bayesian methods. After eight generations of selection, the High (H) line showed a higher Ve, being the difference between lines D=1.36 kits^2, with a probability of being positive P=100%. Selection for increasing Ve decreased total number of kits born (D=-0.45 kits, P(D<0) = 92%) in this line. Besides, the H line showed higher total bilirubin (D=7.66 µmol/l, P(D>0)=90%), conjugated bilirubin (D=5.06 µmol/l, P(D>0)=93%), cholesterol (D=0.44 mmol/l, P(D>0)=99%), triglycerides (D=0.28 mmol/l, P(D>0)=98%), glucose (D=0.56 mmol/l, P(D>0)=86%), Ca (D=0.52 mmol/l; P(D>0)=93%)) and K (D=2.33 mmol/l, P(D>0)=76%)) than L line. The concentration of Cl, Na, P, urea and creatinine was similar in both lines. In conclusion, selection for reducing litter size variability appears to lead to a better health status in females, which would allow them for better adaptation to harsh environmental conditions.

Effect of selection for residual variance of litter size on body condition in rabbits

E.W. Calle[1], M.L. García[2], A. Blasco[1], M.E. García[2] and M.J. Argente[2]
[1]*Institute for Animal Science and Technology. Universitat Politècnica de València, P.O. Box 22012, 46071 Valencia, Spain,* [2]*Universidad Miguel Hernández de Elche, Departamento de Tecnología Agroalimentaria, Ctra de Beniel Km 3.2, 03312 Orihuela, Spain; mariluz.garcia@umh.es*

A divergent selection experiment for residual variance of litter size has been performed in rabbits at University Miguel Hernández of Elche in Spain. After seven generations of selection, litter size in the Low (H) line was 30% more homogenous than in the High (H) line comparing residual variance of litter size with the original mean. The objective of this study was to analyze the effect of selection for residual variance of litter size on female body condition. Body condition was measured as body weight (BW), perirenal fat thickness (PFT), basal non-esterified fatty acids (NEFA$_b$), response in NEFA to isoproterenol (NEFA$_r$) and lipolytic potential of fat reserves (LPF = NEFA$_r$ – NEFA$_b$) at second mating, delivery and 10th day of lactation. Data were analysed using Bayesian methodology. Selection process did not affect BW. For PFT, the L line showed higher PFT than the H line at delivery and 10th d of lactation, being the difference between lines (D$_{L-H}$) of 0.16 mm and 0.17 mm respectively (P(D$_{L-H}$>0) = 86%). When body condition was estimated using NEFA$_b$, NEFA$_r$ and LPF, the L line displayed lower concentration in NEFA$_b$ (DL-H =-0.04 mmol/l, P(D$_{L-H}$<0)=76%) and higher concentration in NEFA$_r$ (D$_{L-H}$ =0.05 mmol/l, P(D$_{L-H}$<0)=78%) and LPF (D$_{L-H}$ =0.09 mmol/l, P(D$_{L-H}$<0)=99%) than the H line at delivery. These findings show that the L line had higher body fat depot at delivery. This line seems to mobilize more efficiently body fat reserves in physiological status under stronger energetic demands, such as delivery. In conclusion, a decrease in litter size variability showed a favourable effect on body condition.

Survival and pregnancy differences between lines divergently selected for environmental variability

N. Formoso-Rafferty[1], I. Cervantes[1], M. De La Flor[1], N. Ibáñez-Escriche[2,3] and J.P. Gutiérrez[1]
[1]Universidad Complutense, Producción Animal, Avda. Puerta de Hierro s/n, 28040 Madrid, Spain, [2]University of Edinburgh, The Roslin Institute, Edinburgh, Edinburgh, United Kingdom, [3]Centre IRTA Lleida, Genètica i Millora Animal, Lleida, Lleida, Spain; n.formosorafferty@ucm.es

A divergent selection experiment for environmental variability has been successfully conducted during 12 generations in mice. Animals of the low variability selected line have been shown to be more robust in the sense of having higher litter size and survival of animals at weaning in a common breeding environment. However, they had not been faced to an environmental challenge. The objective of this work was to compare the selected lines according to the influence that feeding restriction would have on growth and reproductive function. A total of 40 females (four full-sib females from 10 random different litters) were chosen within high or low selected line to be split in four groups of 10 according to the combination of restriction or not in two periods, from weaning at 3 weeks to 13 weeks and one week before mating to the 2nd partition. A food restriction of 75% of the normal intake of the animals was applied, leading to unexpected surprising mortality, particularly in the high line (hypothesized as less robust animals). Survival and number of parties were compared based on single statistical Chi-square tests. Under feeding restriction, survival was significantly (P<0.01) higher in the low line (95% vs 65%). During reproductive period, feeding restriction had also an effect on having parity (P<0.05). Only 33% of the restricted females in the high line had at least one litter in contrast with the 72% of the females under the same conditions in the low line. According to these results, the line selected for low environmental variance would be less affected by the food restriction. It would indicate a higher robustness of the low line for this environmental challenge.

Widespread third order linkage disequilibrium in the swine genome

L. Gomez-Raya, M.C. Rodriguez, L. Silió, C. Barragán and W.M. Rauw
Instituto Nacional de Investigación y Tecnología Agraria y Alimentaria, Departamento de Mejora Genética Animal, Crta. de la Coruña Km. 7,5, 28040, Spain; rauw.wendy@inia.es

The objective of this work is to investigate the scope and distribution of third order linkage disequilibrium in the swine genome. A novel algorithm to compute third order linkage disequilibrium is developed and tested on 305 Iberian sows typed for the 60K Illumina BeadChip array. After filtering, just 26,310 SNPs remained. It is established that the third order linkage disequilibrium ranges between -0.125 and +0.125. Estimation of third order linkage disequilibrium was carried out for each of three consecutive SNPs using a sliding window, after choosing markers in each triplet separated from each other by 0, 5, 25, 50, 100, and 200 SNPs. The average absolute value of third order linkage disequilibrium for the three consecutive SNPs across the genome was 0.023 (SD=0.023), 0.016 (SD=0.017), 0.012 (SD=0.013), 0.007 (SD=0.008), and 0.006 (SD=0.007) with un average distance between the two markers at the extremes of 0.185, 0.831, 4.515, 9.098, 18.716, and 28.732 Mb, respectively. Therefore, third order disequilibrium decays following a similar pattern that second order linkage disequilibrium. Implications of third order linkage disequilibrium on the haplotype structure of swine populations are discussed.

Exact calculation of the Mendelian sampling covariability and their application in mating decisions

S. Bonk[1], M. Reichelt[1], F. Teuscher[1], D. Segelke[2] and N. Reinsch[1]

[1]Leibniz-Institut for Farm Animal Biology, Institute of Genetics and Biometry, Wilhelm-Stahl-Allee 2, 18196 Dummerstorf, Germany, [2]Vereinigte Informationssysteme Tierhaltung w.V., Heideweg 1, 27283 Verden, Germany; bonk@fbn-dummerstorf.de

In mating decision not only the expected mean of the breeding value of full sibs but also the genetic variability is of interest. This variance is known as Mendelian sampling variance (MSV) and can be estimated by simulating gametes from parents with known diplotypes, recombination rates and additive marker effects. In this contribution we present an exact method for the calculation of the additive and dominant MSV. In contrast to established simulation methods, our novel approach is not affected by Monte Carlo error. The method is based on parent specific covariance matrices which are derived from parental diplotypes and recombination rates. These matrices exactly quantify the within-family (co-)variability of the coded genotypes. Combining covariance matrices with additive marker effects provides the Mendelian sampling (co-)variances of (estimated) breeding values for several traits as well as for the aggregate genotype. We have benchmarked our new method by comparing it to an established simulation method, showing a good agreement between both approaches. The exact method is therefore well suited to calculate the MSV and can, in combination with optimization techniques, ultimately be used in practical mating decisions.

New developments in grassland to make grazing management easier

M. O'Donovan, A. Geoghegan, M. O'Leary and L. Shalloo
Teagasc, Animal and Grassland Innovation Centre, Fermoy, Co Cork, Ireland; michael.odonovan@teagasc.ie

Northwestern Europe, and in particular Ireland, brings the natural advantage of temperate climate to food production which favours ruminant grass based production systems. There are major improvements needed in the areas of grass production and utilisation if we are to achieve these targets from a low cost base. While every farm situation is unique with varying soil types, local climatic conditions, stocking rates and farmer management capabilities, grass production is limiting on most farms. If the levels of expansion predicted in recent dairy industry surveys are realised and they will be going by current trends, then farm grass production will have to increase substantially from current levels to meet additional cow requirements. Ultimately, the optimum stocking rate for an individual farm is that which gives sustainable profitability and is dependent on the individual farm's grass growth capability. Many Irish farms, for example, are only producing 50% of their grass growth capability. Large increases in grass production can be achieved. Dairy expansion must come from utilising more grass, and not from importing supplementary feed. At present, Irish dairy farmers are utilising 7.5 t DM/ha (from NFS data) during a 210 day grazing season, from a milking platform stocked at 1.8 livestock units (LU)/ha. In 2014, an analysis of matched farms completing grassland measurements in PastureBase Ireland and Profit Monitor analysis were used to compare the effect of increasing milk output by increasing grass utilisation or by increasing purchased feed. These farms were, on average, stocked at 2.35 cows/ha, producing 404 kg MS/cow and 950 kg MS/ha and were utilising 9.6 t DM/ha. The analysis included an allowance for the farmers own land and labour. The effect on milk output of providing 1 tonne of purchased feed per cow was lower than the effect of 1 extra tonne of grass DM utilised per cow. Increasing grass utilised by 1 t DM/ha increased net profit by €267/ha. This example shows the benefits of increasing farm grass production. Options to improve grass production in northwestern Europe, e.g. via optimising spring grazing management, will be provided.

Combining automatic milking and grazing using detailed cow information

A.H. Ipema[1], G. Holshof[1], F.W. Oudshoorn[2], C. Allain[3] and F. Flessire[4]
[1]Wageningen UR Livestock Research, De Elst 1, 6708 WD Wageningen, the Netherlands, [2]SEGES P/S, Agro Food Park 15, 8200 Aarhus N, Denmark, [3]Institut de l'Elevage, 149 rue de Bercy, 75595 Paris Cedex 12, France, [4]Université de Liège, Avenue de Cureghem 7b, 4000 Liège, Belgium; bert.ipema@wur.nl

Combining automatic milking with grazing is an additional challenge for management of both herd and individual cow. How do feeding and grazing strategies affect cow traffic and what are the consequences for milk yield and for the algorithms for detecting cows in oestrus or with health problems? During the grazing period robot and sensor data were collected on four research and two private farms in the Netherlands, Belgium, Denmark and France. The accessibility to the pastures ranged from 7 to 24 hours per day. On farms with part-time grazing cows received additional feed during their stay in the barn. The analysed data consisted of the data of the visits of the cows to the milking robot and the associated milk yields of all cows. In addition, data from activity and eating sensors were used. Increasing grazing resulted in a lower milking frequency and lower milk production per cow. On three farms, the eating time was recorded with two types of neck sensors. If the proportion of grass in the forage diet decreased, total eating time was lower. With 40%, 65% and 100% grass in the diet, cows needed almost 6 hours, 7 hours and 9 hours, respectively, to consume their diet. Oestrus detection is usually based on the principle that a cow in oestrus exhibits an increased activity over a given period of time. Applied activity sensor brands use a variety of operating technologies and principles in which activity is measured and recorded. In addition, the sensors can be attached to various body parts. Measured activity values therefore differ in level and variation throughout the day. Oestrus detection models are usually based on the principle that a cow in oestrus exhibits an increased activity over a given period of time. In order to cope with the different characteristics of activity measurements, different approaches for oestrus detection models were used. Oestrus detection models could be developed with a sensitivity of more than 80% and a specificity of more than 96%.

Analysis of behaviour of grazing cattle based on GPS and accelerometer data

J. Maxa, S. Thurner, M. Kaess and G. Wendl
Bavarian State Research Centre for Agriculture, Institute of Agricultural Engineering and Animal Husbandry, Voettinger Str. 36, 85354 Freising, Germany; jan.maxa@lfl.bayern.de

Global positioning systems (GPS) have been applied in numerous studies focusing on wildlife and recently on domesticated animals, especially livestock. Among GPS other motion sensors such as accelerometers are nowadays used to monitor behavior and pasture use of grazing animals. The main objective of this study was to analyze behavior of grazing cattle based on information obtained from GPS tracking collars and an accelerometer sensor for potential use of such a technique on alpine pastures. A four days trial was conducted in 2015 on a pasture in Bavaria, Germany where six heifers of Limousine and Simmental breed were equipped with collars collecting GPS position (1 Hz) and 3-axis accelerometer data (3 Hz). Several active and inactive behaviors were recorded based on continuous sampling observations of random animals for a max. of 8 hours per day. Collected GPS and accelerometer data were merged with behavioral observations and aggregated into intervals of 2 and 10 seconds. For each interval the mean values with standard deviation (SD) were calculated for each axis to access information about position of the neck and activity of the observed animal. Furthermore, turn angle and walking speed of heifers were calculated from GPS data. Differences among the mean values of behavioral classes were detected using mixed effects regression. Frequency histograms and density functions of means and SD were used to observe the distribution of different behaviors and to investigate thresholds among animals' activity levels. The first results showed significant differences ($P<0.05$) among most of behaviors based on SD obtained from accelerometer measurements. Furthermore, different activity levels of observed animals based on information from visualized accelerometer data could be distinguished. The results will be implemented for further investigations of behavior of grazing cattle and for a possible establishment of classification algorithms.

The use of a virtual electric fence for dairy cow grazing

P.H. Hogewerf, P. Koene and A.H. Ipema
Wageningen UR Livestock Research, Animal Welfare, De Elst 1, 6708 WD Wageningen, the Netherlands;
pieter.hogewerf@wur.nl

There is a growing demand for products produced in 'natural' production systems. On dairy farms, cows should have the possibility to graze during a significant part of the day. Current grazing systems are labour intensive and inefficient, resulting in a growing number of farmers that keep cows inside throughout the year. Tools for improving grazing efficiency and reducing the labour-intensity could be helpful to reverse this trend. In the Amazing Grazing project (www.amazinggrazing.eu) the possibility of optimizing grazing behaviour by controlling cow movement with Virtual Electric Fencing was evaluated. A non-GPS system with underground wires that remotely can activate a cow collar sensor in which the boundary signal is received and from which a warning signal could be given to the cow approaching the boundary and a correction signal crossing the boundary. Questions are whether the virtual fence works as expected and what the impact of the signals is on the behaviour and welfare of cows. A group of four dairy cows in a virtual fenced area was compared with a group in an electric fenced area. This was replicated once with two different groups (total n=16). During 12 days the groups were compared in several aspects. One of the aspects was grazing on a single strip (80×10 m) compared to double strip (80×20 meter). The behaviour, position, activity were monitored during 5 hours per day. Cows learned the virtual fence system in the same pace as the electric fence. During the 12 day period all cows stayed in the strips with for the virtual fence only few temporary boundary crossings were observed. The cows in the electric fence system came closer to the visible fence while the virtual fence cows kept more distance to the fence (P<0.01). The behaviour of the cows was significantly influenced by the strip size. In a 1-strip area the virtual fence cows showed more standing (34% vs 8.5%; P<0.001) and less lying (4% vs 14%; P<0.01). No difference was found for standing (8.5% vs 13%) and lying (25% vs 25%) in the 2-strip area. Our preliminary conclusion is that a virtual fence is effective and has some effects on the behaviour of cows different from an electric fence. More detailed analysis will show whether the welfare of the cows is affected.

Satellite imagery to support sheep management in Tierra del Fuego, Chile

S. Radic[1], M. Villa[2], C. Moraga[3], R. Muñoz[1] and J. McAdam[4]
[1]Universidad de Magallanes, Facultad de Ciencias, Avda Bulnes 01855, Punta Arenas, 6200000, Chile,
[2]Organización de las Naciones Unidas para la Alimentación y la Agricultura, Proyecto Apoyo al Mecanismo Conjunto en Acciones de Mitigación al Cambio Climático en los Bosques de Bolivia, La Paz., Bolivia, [3]School of Natural Resources and the Environment, and Wildlife Ecology and Conservation Department, U, 354 Newins-Ziegler Hall, Gainesville, FL 32611, USA, [4]Agri Food and Biosciences Institute, Newforge Lane Belfast, BT9 5PX, United Kingdom; sergio.radic@umag.cl

Tierra del Fuego is characterized by sheep extensive production with animal movement between wintering and summering grounds, also the weather seasonality and abundant snowfall during winter that determines different problems in sheep management. The aim of this study was to evaluate the critical dates for sheep grazing management, from wintering and summering grounds, in ranches of central Tierra del Fuego, Chile. The analysis considered a 12 year (2000-2011) series of the Normalized Difference Vegetation Index (NDVI), snow cover, and Land Surface Temperature (LST); to analyze sheep movements in four sheep ranches of central Tierra del Fuego. Results indicate that wintering grounds presented higher NDVI from May 8 to July 27. Snow cover is notoriously higher in summering grounds during autumn and winter, specifically from March 29 to September 5. The LST critical period occurs between March 21 to September 29. The actual movement of sheep to wintering grounds generally occurs in end of April, as soon as snow precipitation begins. In relation with these results the critical period considering all parameters analyzed, should be between March 21 to September 29, and a prohibitive period is from May 8 until September 13. These findings may be incorporated into sheep management planning, also it can help in decision making for a public organisms when the climatic conditions are extreme.

Grazing behaviour in dairy cows as a predictor of grass intake

R.L.G. Zom, J.T.N. Van Der Werf, B. Timmer, D. Hoeksma and C.G. Van Reenen
Wageningen UR Livestock Research, P.O. Box 338, 6700 AH Wageningen, the Netherlands; ronald.zom@wur.nl

At present, there are no practically feasible methods to reliably estimate the grass intake of individual cows kept on pasture. The aim of the present study was to examine whether cow behaviours continuously recorded with the use of sensors, in combination with other potential predictors related to the pasture and to production characteristics of the cow, could be used to estimate grass intake by grazing dairy cows. Behavioural and grass intake data were collected during a one-week experimental period in 60 cows involved a grazing trial. All cows were equipped with two commercially available 3D accelerometers: an 'IceQube' (IceRobotics, UK) attached to the leg, and a Smarttag Neck (NEDAP, Groenlo) attached to the neck, recording standing/lying/number of steps and grazing behaviour, respectively. Grass intake (kg DM/day) was determined in individual cows with the use of the n-alkane method. Grass intake was considered as response variable (y), and stepwise multiple regression analysis was used to identify the best fitted models based on percentage of adjusted variance explained (adj R Square). The best model to predict individual grass intake (adj R^2=84%) had the following predictors: grazing area/cow, days in milk (DIM), milk protein, milk urea, milk production (kg) and time spent grazing (min/hour spent on pasture). When milk protein and urea were left out of the analysis, the best model to predict grass intake (adj R^2=74%) had the following predictors: milk production (kg), DIM, grazing area/cow, time spent grazing, and the total number of steps while on pasture. These findings demonstrate that grass intake of individual cows on pasture can be reliably estimated from a number of behavioural and other predictors. These models may represent potential management tools in grazing systems.

Cut and Carry: Investigating higher proportions of fresh grass in the diets of high yielding cows

P.R. Hargreaves[1], M.A. Lee[2] and D.J. Roberts[1]
[1]SRUC, Future Farming Systems, Hestan House, Crichton Campus, DG1 4TA, Dumfries, United Kingdom,
[2]Royal Botanic Gardens, Royal Botanic Gardens, Kew, Richmond, TW9 3AB, Richmond, United Kingdom;
paul.hargreaves@sruc.ac.uk

Cut and carry is an approach to dairy farming in which grass is cut in the field and delivered to housed cows. Benefits of this system are; grass is cheaper to produce than a total mixed ration (TMR) and grass is not subject to the same fluctuations in production costs. Two studies, each study consisting of three groups of 16 cows considered aspects of increasing the proportion of fresh grass; the first study considered the replacement of two amounts of fresh cut grass (25% and 50%) for TMR on a dry matter basis and found that the inclusion of 25% cut grass reduced the milk yield by 5.5 kg/cow/d (P<0.05) but only 4.3 kg/cow/d (P<0.05) for 50% cut grass compared to a 100% TMR-fed group, with no significant difference in milk yield between the 25% or 50% diets. Fresh weight intakes averaged 89, 72 and 54 kg/d for the 50%, 25% and TMR-fed groups. With DM intakes the TMR group consumed the most (20 kg/d) compared to 18 kg/d for the 50% and 19 kg/d and for the 25% groups, respectively. In a second study a cut and carry group that was grass fed during the day was compared with a housed TMR-fed group and group grazed for two windows; again the TMR group produced a mean milk yield greater (P<0.001) than both the cut and carry or the grazing groups by 1.39 and 1.61 kg/cow/d, respectively. However, the cut and carry group did show a mean increase of milk protein (33. 7 g/kg cow/d (P<0.001)) compared to the grazed (31.5 g/kg cow/d) and TMR (32.8 g/kg cow/d) groups. Inclusion of fresh cut grass in the diets of high yielding cows did show a reduction of milk yield and changes in milk composition compared to TMR-fed housed cows.

Effect of ewe prolificacy & stocking rate on lamb growth, production efficiency and carcass output

E. Earle[1,2], N. McHugh[2], T.M. Boland[1] and P. Creighton[2]
[1]*School of Agriculture and Food Science, University College Dublin, Belfield, Dublin 4, Ireland,* [2]*Teagasc, Animal & Grassland Research and Innovation Centre, Athenry, Galway, Ireland; elizabeth.earle@teagasc.ie*

Ewe prolificacy potential (PP; predicted number of lambs born/ewe/year) and stocking rate (SR; ewes/ha) are two vital factors influencing output in grass-based lamb production systems in temperate grazing regions. Grazed grass is the most economic source of nutrition available to ruminants and has the potential to supply up to 95% of the energy requirements of sheep. The objective of this study therefore was to investigate the effect of PP, SR and their interaction on lamb performance, production efficiency (ratio of live weight weaned ewe^{-1} & hectare^{-1} to live weight mated) and carcass output (kg hectare) from a grass-based lamb production system. The study was conducted over three production years and had a 2×3 factorial design, consisting of two PP levels and three SR's incorporating 360 ewes (180 medium prolificacy potential (MP – Suffolk crossbred ewes; target weaning rate 1.5 lambs/ewe) and 180 high prolificacy potential (HP – Belclare crossbred ewes; target weaning rate 1.8 lambs/ewe)). Stocking rates were low (LSR; 10 ewes/ha), medium (MSR; 12 ewes/ha), and high (HSR: 14 ewes/ha). The effect of PP, SR and their interaction on lamb performance, production efficiency and carcass output were modelled using linear mixed models in PROC HPMIXED (SAS Inst. Inc., Cary, NC, USA) with ewe included as a random effect. As SR increased lamb average daily gain (ADG; grams/day) decreased (P<0.001). Despite lower individual lamb performance levels, carcass output/ha increased as PP (P<0.01) and SR (P<0.001) increased. Production efficiency was greatest for HP ewes +0.05 ewe^{-1} (0.70 vs 0.65; P<0.01) and +0.10 hectare^{-1} (0.86 vs 0.76; P<0.05). It can therefore be concluded that greater carcass output is achievable through increasing ewe PP and SR and that increasing ewe PP will increase production efficiency in a grass-based lamb production system.

Use of grazing in a dairy goat farm to design sustainable production systems in France

H. Caillat[1], B. Ranger[1], E. Bruneteau[1], C. Paraud[2] and R. Delagarde[3]
[1]*INRA, UE1373 FERLus, 86600 Lusignan, France,* [2]*ANSES, 60, rue de Pied de Fond, 79000 Niort, France,* [3]*INRA-Agrocampus Ouest, UMR PEGASE, 35590 Saint-Gilles, France; hugues.caillat@lusignan.inra.fr*

France is the first producer of goat milk in European Union with about 550 millions liters collected. Since years 2000, goat farms moved gradually towards intensive farming, significantly increasing their need for purchased inputs. To increase productivity, the grazing has been stopped and the indoor breeding has been developed. Today, feed self-sufficiency of French goat systems are only 55%, increasing economic risks and feeding costs. PATUCHEV is an experimental device of Inra (UE FERLus) set up to assess and propose innovative, low input and sustainable goat farming systems. Before 2012, goats of this flock were bred in sexual season and fed indoors with concentrates and straw. In 2013, the feeding system was changed to grazing on multi-species grasslands and to mix crops for concentrate self-production. During three years before and after the change, feed intake, milk yield and health data were recorded monthly. Gastrointestinal parasitism was controlled every month by measuring strongyle faecal egg excretion on faecal group samples. Since 2013, on a fixed lactation period of 200 days, the annual mean milk production decreased by 16% per goat (682 kg/goat vs 812). Milk fat (35 g/kg) and protein (32 g/kg) concentrations were unaffected. With grazing, concentrate supplementation was 242 kg per goat per year, which is 62% lower than in the previous feeding system (343 g of concentrate/l milk vs 644). The global feed self-sufficiency is now greater than 75%. Gastro-intestinal infection by strongyles appeared in September 2013 but an integrated management enabled to maintain a low infection level. Annual mean flock excretions were 198 and 310 eggs per gram of faeces in 2014 and 2015, respectively. An economic study is in preparation to complete these results and to confirm that good economic results are possible despite the decrease of milk production per goat.

Influence of a virtual fence on synchronicity and social grouping of dairy cows
P. Koene
Wageningen University & Research, Animal Welfare, De Elst 1, 6708 WD Wageningen, the Netherlands;
paul.koene@wur.nl

Dairy cows may graze more efficiently when allowed grazing in predefined areas with predictable sward height and grass quality. Moving the animals daily or preferably more frequent is needed. Using traditional fencing so-called strip grazing causes high labor requirements. Undesirable effects of electric fencing might be prevented by the use of virtual fencing. We used a non-GPS system with underground wires signaling a boundary that could be partly and differentially activated by remote control together with a collar in with the boundary signal is received. A warning signal could be given to the cow when approaching the boundary and a correction signal when crossing the boundary. One of the questions is whether the virtual fence influences the social behavior and welfare of cows. Dairy cows – grazing in a group of four – in the virtual fenced area were compared with cows in an electric fenced area. This was replicated once with two different groups (n=16). During 12 days the groups were compared in several aspects. One of the aspects was grazing on a single strip (80×10 m) compared to a double strip (80×20 meter). The behaviour, position and activity were monitored during 5 hours per day. During the 12 day period all cows stayed in the strips with only few temporary boundary crossings. The cows in the virtual fence system showed lower nearest neighbor distances than the cows in the electric fence system (P<0.01). The social activity of the cows was significantly influenced by strip size. In a 1-strip area the virtual fence cows were standing closer together. Also behavioral synchronization of the group in the virtual fence condition measured during standing and lying was weaker (P<0.001). Our preliminary conclusion is that a virtual fence has some effects on the social behavior of cows different from social behavior of electric fenced cows. More detailed analysis will show whether the social organization and welfare of the cows is also affected.

The effects of early lactation pasture allowance and duration on milk yield
E. Kennedy[1], L. Delaby[2], J.R. Roche[3], B. Horan[1] and E. Lewis[1]
[1]Teagasc, AGRIC, Moorepark, Fermoy, Co. Cork, Ireland, [2]INRA, AgroCampus Ouest, UMR Pegase, 35590 Saint-Gilles, France, [3]Down to Earth Advice Ltd, P.O. Box 12520, Chartwell, Hamilton, New Zealand; emer.kennedy@teagasc.ie

Higher stocking rates in the post-quota era, due to increased herd sizes may result in feed deficits, particularly in early spring. The objective of this experiment was to investigate if different pasture allowances (PA) offered to early lactation grazing dairy cows for varying time durations influenced milk yield (MY). The study was carried out over a two year period; from March 25 to November 27, 2014 and March 9 to November 23, 2015. Each year 96 cows were assigned to a randomised complete block design experiment with a 4×2 factorial arrangement of treatments. Cows were randomly assigned from within block to one of four PA (60%, 80%, 100% or 120% of estimated intake capacity; IC) for either 2 or 6 weeks. Once the 2 and 6-week time durations had elapsed, the treatments were offered 100% of estimated IC. Milk yield was recorded daily. Data were analysed using covariate analysis and mixed models in SAS v9.3. Terms for year, parity, breed, PA, duration and the interaction of PA and duration were included. Pre-experimental values were used as covariates in the model. During the first 2 weeks of the experiment there was a linear increase in MY with increasing PA (19.9, 20.9, 22.3 and 24.0 kg/cow/day; 60%, 80%, 100% and 120% PA, respectively). At the end of six weeks there was an interaction between PA and duration (P<0.001) on cumulative six-week MY. There was no difference between any of the two week treatments (935 kg/cow), they were similar to the 100×6 treatments (985 kg/cow) but different all other six-week treatments. The 120×6 cows had a higher MY than all other treatments; the 60×6 and 80×6 treatments had similar MY but they were lower than all other treatments. When cumulative MY (33 weeks) was examined there was a PA effect (P=0.05), the 120% treatment (4,277 kg/cow) was higher than all other treatments (3,977 kg/cow). There was no effect of duration. Although early lactation MY is reduced when PA is restricted there is little effect when cumulative lactation yield is considered, however offering 120% of IC for six weeks in early lactation tended to increase MY.

Effect of grazing rotation length on dairy cow performance grazing at two herbage allowances

L.A. Pérez-Prieto[1], H. Gonzalez-Verdugo[1] and R. Delagarde[2]
[1]Universidad de Chile, Facultad de Ciencias Agronómicas, Estación experimental Oromo, P.O. Box 161, 5380000, Purranque, Chile, [2]INRA, UMR 1348, Pegase, Saint-Gilles, 35590, Saint-Gilles, France; lucio.perez.prieto@gmail.com

At the grazing season level, pasture quality and quantity is mainly determined by grazing rotation length and grazing severity (e.g. herbage allowance). The results presented here correspond to the first 3 months (spring, from September 28, 2015, to January 1, 2016) of a grazing season experiment which will end on May 30, 2016. This study was funded by the Chilean government (Fondecyt project 11140341). The objective was to enlarge the knowledge of the effect of GRL (short: 17 d vs long: 24 d), and its possible interaction with HA (medium: 25 vs high: 33 kg DM/d above 3 cm), on dairy cow performance. Forty eight spring-calving Holstein cows, producing 29 kg/d at the beginning of the experiment, were used in a 2×2 factorial arrangement of the treatments. The effect of GRL and HA on animal and pasture data were analysed as a randomised block design, including the fixed effects of GRL, HA, week, and the interaction between GRL and HA. At the beginning of the experiment, pre-grazing herbage mass, platemeter height, and OMD of the pasture were similar between treatments and averaged 2,233 kg DM/ha, 9.3 cm, and 0.89, respectively. At the end of the period (week 14), pre-grazing herbage mass and platemeter height were greater with long than short GRL: 2,841 vs 1,068 kg DM/ha (P<0.001), 9.0 vs 5.5 cm (P<0.001), respectively. The interaction between GRL and HA was significant (P<0.10) for total milk production. Total milk production was similar between low and high HA for long GRL (average=2,299 kg), and was lower at low than high HA with short GRL (2,223 vs 2,451 kg). The effect of GRL on total milk production, milk fat concentration, milk protein concentration, liveweight, and animal condition score, was not significant. This result was consistent with previous short term studies reporting the effect of pre-grazing HM above 3 cm on animal performance and may be in accordance with previous long term experiments reporting the effect of GRL during the entire grazing season.

Different energy sources with or without additives for pasture finished beef heifers

C.W. Cruywagen, M.N. Engelbrecht, L.C. Hoffman and W.H. Hoffman
Stellenbosch University, Department of Animal Sciences, Stellenbosch, 7600, South Africa; cwc@sun.ac.za

Sixty Bonsmara heifers on planted pasture were used to evaluate the effect of two energy sources with, or without additives, on body weight gain. Two supplementary feeds with dried apple pulp (A) or maize (M) as energy source were formulated on an iso-nutrient basis. The energy supplements further contained either an oregano oil extract (OE) or monensin (MO) as additive, or no additive, designated as placebo (PL). Animals were stratified according to initial weight in ten blocks and treatments were allocated randomly to each block. The study was conducted during spring in the Western Cape Province of South Africa. The pasture consisted of a perennial grass-legume mixture. A rotational grazing system was applied and animals were moved to new paddocks once a week. Based on falling plate meter readings, the heifers consumed a calculated mean amount of 4.48 kg DM/day over the entire experimental period. A fixed amount of 4 kg/head (air dry basis) of the respctive supplements were offered daily during the first 42 days, followed by 5 kg/day until 66 days. Animals were weighed bi-weekly and average daily gain (ADG) was calculated. Production data were subjected to a two way factorial ANOVA. Least Square Mean tests were applied and Shapiro Wilks tests were used to test for normality. Significance was declared at P<0.05. The mean ADG of the six treatment groups was 1.44 kg/day. No interactions occurred between the energy sources and additives. The apple pulp based energy source supplements resulted in a higher (P<0.02) ADG (1.54 kg/day) than the maize based supplements (1.33 kg/day). There were no differences between any of the additives (EO, MO or PL). Mean ADG values (kg/day) of the additive effects were 1.44 (PL), 1.49 (MO) and 1.38 (EO). All the heifers were slaughtered at the end of the trial. Carcas weight and dressing percentage did not differ between energy sources or additives. The mean dressing percentage was 52.5%. It was concluded that dried apple pulp as energy source is not only less expensive than maize, but also resulted in higher ADG values in beef heifers finished on cultivated pastures.

Pasture allocation time and tannins in partial mixed ration: intake and performance of dairy cows

C.A. Pozo[1], C. Cajarville[2], J.L. Repetto[2], M. Cuffia[3], A. Ramirez[2] and G.V. Kozloski[1]
[1]Universidade Federal de Santa Maria, Departamento de Zootecnia, Avenida Roraima 1000, 97105-900, Santa Maria, RS, Brazil, [2]Universidad de la República, Facultad de Veterinaria, Ruta 1, km 42.5, 80100, San José, Uruguay, [3]Universidad Nacional del Litoral, Facultad de Ciencias Agrarias, Kreder 2805, 3080 Esperanza, Santa Fe, Argentina; claudioapozo@gmail.com

The aim of this study was to evaluate the effect of Acacia mearnsii tannin extract (TE) inclusion in a partial mixed ration (PMR) and the time of pasture allocation on intake (DMI) and performance of grazing dairy cows. Nine Holstein cows averaging 197±12 days in milk were arranged on a triplicate 3×3 Latin square design, conducted through three experimental periods of 21 days. The three treatments consisted of: morning pasture allocation and afternoon PMR supply (AM), morning pasture allocation and afternoon PMR supply with 15 g/kg of TE (AMt) and morning PMR supply and afternoon pasture allocation (PM). Individual DMI was measured during 5 days of each period. Pasture DMI was determined in individual paddocks as the difference in herbage mass before and after grazing. Individual milk production was recorded daily during 5 days, and milk samples were collected during 2 days of each period for composition analysis. Statistical analysis was carried out using the MIXED procedure of SAS. Although DMI was similar for all treatments (19.9±1.2 kg/d), the proportion of pasture in the diet was higher in PM than AM and AMt (0.43 vs 0.41, P<0.05). In AM and AMt diets PMR DMI was higher than PM (11.7 vs 11.4 kg/d, P<0.05), while pasture DMI in PM tended to be higher than AM and AMt (8.6 vs 8.1 kg/d, P=0.06). Neither milk production (21.7±3.3 kg/d) nor milk composition were affected by treatments. These results suggest that afternoon grazing could increase pasture DMI in PMR without affecting total DMI, milk production and composition. At the level used in this study, TE supply was not effective to increase milk production nor induce changes in milk composition.

Effect of prolificacy potential & stocking rate on ewe body weight & body condition score

E. Earle[1,2], N. McHugh[2], T.M. Boland[1] and P. Creighton[2]
[1]School of Agriculture and Food Science, University College Dublin, Belfield, Dublin 4, Ireland, [2]Teagasc, Animal & Grassland Research and Innovation Centre, Athenry, Galway, Ireland; elizabeth.earle@teagasc.ie

During the production year, ewe body weight (BW; kg) and body condition score (BCS; 1-5 point scale) fluctuates depending on the stage of production and energy demand and supply of the ewe. Ewe prolificacy potential (PP; predicted number of lambs born/ewe/year) and stocking rate (SR; ewes/ha) are two key drivers of production in grass-based lamb production systems in temperate grazing regions. The objective of this study therefore was to investigate the effect of PP, SR and their interaction on ewe BW and BCS and change of both during the production year. The study was conducted over three production years and had a 2×3 factorial design, consisting of two PP levels and three SR's incorporating 360 ewes (180 medium prolificacy potential (MP – Suffolk crossbred ewes; target weaning rate 1.5 lambs/ewe) and 180 high prolificacy potential (HP – Belclare crossbred ewes; target weaning rate 1.8 lambs/ewe)). Stocking rates were low (LSR; 10 ewes/ha), medium (MSR; 12 ewes/ha), and high (HSR: 14 ewes/ha). The effect of PP, SR and their interaction on ewe BW, BCS and the change in BW and BCS were modelled using linear mixed models in PROC HPMIXED (SAS Inst. Inc., Cary, NC, USA) with ewe included as a random effect. Ewe BW decreased as PP (P<0.05) and SR (P<0.001) increased, although MSR and HSR ewes did not significantly differ from each other. From weaning to mating HP ewes gained more BW (4.3 vs 1.9 kg; P<0.001). There were significant PP by SR interactions for ewe BCS and BCS change, with HP ewes at the HSR having a lower BCS at mating (-0.2 BCS; P<0.01) and greater BCS loss (-0.2 BCS; P<0.01) from pregnancy scanning to lambing. It can be concluded that increasing PP and SR decreases ewe BW and BCS, and impacts on the level of change in ewe BW and BCS throughout the production year in a grass-based lamb production system.

Factors affecting the grazability of grass varieties

N. Byrne[1,2], M. O'Donovan[2], L. Delaby[3] and T.J. Gilliland[1,4]
[1]Institute for Global Food Security, School of Biological Sciences, Queen's University, Belfast, N Ireland, United Kingdom, [2]Teagasc, Animal and Grassland Research and innovation centre, Moorepark, Fermoy, Co, Cork, Ireland, [3]INRA, UMR Production du Lait, 35590, St-Gilles, France, [4]Agri-food Biosciences Institute, Hillsborough, Co. Down, BT26 6DR, N Ireland, United Kingdom; nicky.byrne@teagasc.ie

Selection of grass varieties as a means of improving the 'grazability' of swards in order to achieve higher levels of herbage utilisation is of common occurrence on many Irish farms. The level of utilisation is a trait which has drastic effects on profitability, but due to its complexity it goes unexamined within simulated grazing evaluations. The objective of this study was to use information available for recommended list varieties to make judgements on their 'grazability' thus providing indications of their levels of utilisation. This study is comprised of 59 grass varieties across three species, namely festulolium, hybrid ryegrass and perennial ryegrass. These varieties were sown in plots (21 m^2) managed under commercial farm conditions being grazed on eight occasions across the year with lactating dairy cows, receiving nitrogen applications totalling 250 kg/N/ha. Plots were grazed on a 21-30 day rotation at an estimated herbage mass of 1,300-1,600 kg DM/ha. At each grazing event pre and post grazing compressed sward heights were recorded using a plate meter. From these records the proportion of herbage removed was calculated. This study is a randomised block design with three replicates per variety. A mixed procedure of SAS was used to analysis these data. From the initial year of this study heading date (P<0.0001), ploidy (P=0.0021) and variety (P=0.0060) all had significant effects on the proportion of herbage removed (indicator of utilisation). Varieties do differ in terms of their grazability. Variety selection based on ploidy and heading date can be used as a means of achieving higher levels of pasture utilisation on-farm.

Effects of two types of pastures on the performance of lambs during autumn in Western Patagonia

H.F. Elizalde[1], M. Tapia[1], R. Morales[2] and M. Silva[1]
[1]Instituto de Investigaciones Agropecuarias, INIA Tamel Aike, P.O. Box 296, Coyhaique, Chile, [2]Instituto de Investigaciones Agropecuarias, INIA Remehue, P.O. Box 24-0, Osorno, Chile; helizald@inia.cl

Lamb production in Western Patagonia (Chile) is based on grass, matching the growth curve of pastures. The main crop of lambs is harvested from extensive range farms at weaning time, i.e. early summer, however the tail of the lamb crop always are below the minimum weight to be slaughtered for either the local or export market and hence there is a need to finished them on grass during the rest of the grass growing season. A special window for fattening them is during autumn, taking advantage of the end of the normal summer draught and before winter. The aim of this study was to evaluate the performance of Corriedale lambs raised with two distinct types of pastures. The treatments were: (1) lambs grazing on a legume based pasture of alfalfa (Medicago sativa), n=12; (2) lambs grazing on permanent pasture, n=12; Permanent improved pasture was based mainly on ryegrass, cocksfoot and white clover. With an initial LW of 29.1 kg, lambs were maintained in the experiment for 77 d when they were slaughtered. Lamb live weight gain was higher (P<0.001) with alfalfa compared to permanent pasture, therefore, alfalfa grazed lambs averaged 8.4 kg heavier body weight (P<0.001) than permanent pasture lambs. The two treatments presented different (P<0.001) hot carcass weight, with values of 18.6 and 14.0 kg for the alfalfa and the permanent pasture raised lambs, respectively. Effects of type of pasture on cold carcass weight followed a similar pattern (P<0.001). The carcass length and GR measurements were also affected by type of pasture (P<0.001). Most commercial cuts of lambs were significantly affected by the type of pasture utilized, since permanent pasture raised smaller lambs with lighter carcass than when animals are fattening on alfalfa. Results indicate that lamb raised on alfalfa regrowth during autumn, showed higher productive performance than lambs fed on permanent improved pastures. However, the use of improved permanent pastures, although less advantageous, it is also an option because of the high costs involved when sowing alfalfa in Western Patagonia.

Assessment of artificial neural networks to predict digestibility of ryegrass

G.A. Burns[1], A. Byrne[2], D. Johnston[1] and T.J. Gilliland[3]
[1]Agri-Food and Biosciences Institute, Horticulture, Plant Breeding & Land Use Unit, 4 Manor House, Loughgall, BT61 8JB, Co. Armagh, United Kingdom, [2]Agri-Food and Biosciences Institute, Bacteriology, 12 Stoney Road, Belfast, BT4 3SD, Co. Down, United Kingdom, [3]Agri-Food and Biosciences Institute, Agriculture Branch, Large Park, Hillsborough, BT26 6DR, Co. Down, United Kingdom; trevor.gilliland@afbini.gov.uk

Grass digestibility is the most frequently utilised forage quality trait on evaluation and breeding programmes to provide an indication of animal performance and is routinely determined using near infrared reflectance spectroscopy (NIRS). This technique utilises predictive calibration models to relate spectral data to reference laboratory data. Most calibration models published in the literature utilise multiple linear regression (MLR) techniques, however machine learning algorithms, such as artificial neural networks (ANN), offer several advantages over MLR techniques; including the ability to model non-linear relationships, the use of non-spectral predictor variables (e.g. species) and ability to detect interactions between all predictor variables. Despite these advantages, ANNs are an under-explored area of the forage analysis. The current study evaluated ANNs to predict the digestibility of perennial ryegrass (Lolium perenne L.) and hybrid ryegrass (Lolium boucheanum Kunth) samples. The inclusion of non-spectral traits (e.g. species) as dummy variables in the ANNs resulted in increased accuracy of the calibration models. Previous research assessing NIRS as a rapid and cost-effective means of predicting animal intake potential from forage samples displayed promising but limited results. Results from the current study suggest that the use of ANNs with the inclusion of non-spectral variables (e.g. sward morphology) could be combined with spectral data and improve upon previous MLR models to predict animal intake potential. This could potentially provide a technique for the rapid assessment of forages to predict animal intake potential for grass evaluation and breeding programmes.

Comparative grazing behaviour of lactating dairy crossbred suckler and beef suckler cows

M. McManus[1,2], R. Prendiville[1], T. Boland[2], S. McCabe[1] and P. Creighton[3]
[1]Teagasc, Animal and Grassland Research and Innovation Centre, Dunsany, Grange, Co. Meath, Ireland, [2]University College Dublin, School of Agriculture and Food Science, College Dublin, Belfield, Dublin 4, Ireland, [3]Teagasc, Animal and Grassland Research and Innovation Centre, Mellows Campus, Athenry, Co. Galway, Ireland; michelle.mcmcmanus@teagasc.ie

The aim was to investigate differences in grazing behaviour (GB) among lactating dairy crossbred suckler (F_1) and beef suckler (S) cows. A total of 186 records were available across two stages of lactation from 103 cows; 109 F_1 and 77 S. Seventeen F_1 and 10 S were in parity one and the remainder in parity two. Cows were 159 and 178 days in milk and GB was determined over a 24 hour period using IGER GB headset recorders. Average pre and post-grazing sward heights were 112 mm and 48 mm, respectively. Data were analysed using the MIXED procedure of SAS. Live weight (LW) was 62 kg greater (P<0.001) and body condition score (BCS) was 0.45 units greater (P<0.001) for S than F_1; 653 kg and 591 kg LW and 3.05 and 2.60 BCS, respectively. No significant differences were observed among the breed groups for grazing time, number of grazing bouts or grazing bout duration. Grazing time and number of grazing bouts were similar for F_1 and S cows; 533 minutes, 540 minutes and 7.0 and 7.6 bouts, respectively. Grazing bout duration was 83 and 78 minutes for F_1 and S cows, respectively. Total number of bites and grazing mastications were similar for F_1 and S cows; 28,460 and 28,792 bites and 9,197 and 9,010 grazing mastications, respectively. No differences in ruminating time, number of ruminating bouts or ruminating bout duration were observed between the breed groups. Ruminating time was 407 and 409 minutes, numbers of ruminating bouts were 10.6 and 11.0 bouts and ruminating bout duration was 40 and 39 minutes for F_1 and S cows, respectively. Ruminating mastications were similar for F_1 and S cows; 26,517 and 27,068, respectively. No difference was observed between breeds for any of the GB variables investigated. Results from the this study suggests that despite differences in LW and BCS no difference in grazing or ruminating time, number of bouts, bout duration, mastications or total number of bites were observed between the breed groups.

(Some) slurry is good for (some) earthworms

A.K. Murchie[1], R.P. Blackshaw[2], A.W. Gordon[1] and P. Christie[1]
[1]*Agri-Food & Biosciences Institute, Sustainable Agri-Food Sciences, Newforge Lane, BT9 5PX Belfast, United Kingdom,* [2]*Blackshaw Research and Consultancy, Chudleigh, TQ13 0JF Devon, United Kingdom; archie.murchie@afbini.gov.uk*

The Agri-Food and Biosciences Institute maintains experimental plots that have been continually treated with animal slurry since 1970, representing the longest–running experiment of this kind in the world. The experiment consists of a ryegrass sward divided into 48 plots, each measuring 8.5×3.5 m. These plots are treated three times a year (March, May and October) with cattle and pig slurries and inorganic fertiliser. Animal slurries are applied at three rates (50, 100 and 200 m^3 per ha per yr) and inorganic fertiliser as 200 kg N, 32 kg P and 160 kg K per ha per yr. We sampled earthworms prior to treatment inputs on three annual occasions (March, May and October) for four consecutive years, by chemical extraction. Earthworm biomass data were analysed using a factorial arrangement of treatments and year using a split plot in time ANOVA according to the experimental design. Twelve earthworm species were recovered by chemical extraction. There were highly significant responses of earthworms to fertiliser inputs but this varied with treatment and earthworm species. Epigeic earthworms and the endogeic Allolobophora chlorotica responded strongly and positively to cattle slurry inputs. For Lumbricus rubellus, their biomass was five times greater in the highest input cattle slurry plots than in the untreated controls. Other endogeic earthworms responded positively to inorganic fertiliser treatment. Juveniles Aporrectodea spp. had significantly greater biomass in the inorganic fertiliser treatment. Pig slurry had least effect on earthworm biomass. Pig feed is often enriched with copper, which is known to have a variety of sub-lethal effects on earthworm growth and reproduction. Despite this, earthworm biomass in pig slurry plots was no worse than in the controls. Whilst applications of slurry can kill earthworms close to the soil surface, in general slurry inputs are beneficial to earthworms as they add organic matter and nutrients to the soil. The invasive New Zealand flatworm, Arthurdendyus triangulatus, was present at this site, and does represent a significant threat to earthworm biodiversity in grassland.

Quality and Management of Horse Pastures in Switzerland

F. Kägi[1], B. Reidy[1], H. Hertzberg[2] and C. Herholz[1]
[1]*Bern University of Applied Sciences, School of Agricultural, Forest and Food Science, Laenggasse 85, 3052 Zollikofen, Switzerland,* [2]*Institute of Parasitology, Vetsuisse Faculty, Winterthurerstrasse 266a, 8057 Zurich, Switzerland; franziska.kaegi@bfh.ch*

Considering the physiological and ethological needs of horses such as species-appropriate forage, exercise and social contact pastures provide the most natural habitat for horses. The objective of the study was to identify parameters that affect the quality of a horse pasture and to define quality criteria for adapted plant populations. Vegetation surveys (Daget-Poissonet) were conducted on 20 randomly selected pastures on horse farms in Switzerland. Sward height (Rising Plate Meter), sward density and the prevalence of undesired plants were assessed. In a survey, farmers were questioned on various parameters affecting the pasture management. Management factors with a potential impact on the quality of the pastures were statistically analyzed (NCSS). In total 54 different plant species could be identified, perennial ryegrass (Lolium perenne L.) and white clover (Trifolium repens L.) being the most abundant plant species. Dandelion (Taraxacum officinale Weber), smooth meadowgrass (Poa pratensis L.), orchard grass (Dactylis glomerata L.) and creeping buttercup (Ranunculus repens L.) could be found frequently. Grasses were predominant, followed by forbs and legumes. Based on the palatability and trampling-tolerance, the different plant species were classified as 'suitable' or 'unsuitable'. Multi-species grazing had a significant (P=0.0176) beneficial effect on the prevalence of suitable plant species. Furthermore, the grazing intensity significantly influenced the sward density (P=0.0319). In conclusion, the study revealed that Swiss horse pastures are largely dominated with typical pasture species such as perennial ryegrass and white clover. Trampling-tolerant species become frequently stand-forming. Multi-species grazing and an adapted grazing density are key factors to improve the sward density and the proportion of suitable plants for horse pastures.

Circulating microRNA expression changes during grazing in cattle

S. Muroya[1], H. Ogasawara[2] and M. Hojito[2]
[1]Animal Products Research Division, NARO Institute of Livestock and Grassland Science, Ikenodai 2, Tsukuba, Ibaraki 305-0901, Japan, [2]Field Science Center, School of Veterinary Medicine, Kitasato University, Yakumo-cho, Hokkaido 049-3121, Japan; muros@affrc.go.jp

Circulating microRNAs (c-miRNAs) are associated with physiological adaptation in mammals. Under the hypothesis that grazing have some influence on circulating miRNA in cattle, we conducted profiling of miRNA expression in exosomes from the plasma of both grazing and housed Japanese Shorthorn cattle. Microarray analysis of the c-miRNAs resulted in detection of a total of 231 bovine exosomal miRNAs in the plasma, with a constant expression level of let-7g across the duration and cattle groups. Expression of muscle-specific miRNAs such as miR-1, miR-133a, miR-206, miR-208a/b, and miR-499 were undetectable, suggesting the mildness of grazing movement as exercise. According to validation by quantitative RT-PCR, the circulating miR-150 level in the grazing cattle normalized by the endogenous let-7g level was down-regulated after 2 and 4 months of grazing (P<0.05), and then its levels in housed and grazing cattle equalized when the grazing cattle were returned to a housed situation. Likewise, the levels of miR-19b, miR-142-5p, miR-148a, miR-221, miR-223, miR-320a, miR-361, and miR-486 were temporarily lowered in the cattle at 1 and/or 2 month of grazing compared to those of the housed cattle (P<0.05). In contrast, the miR-451 level was up-regulated in the grazing cattle at 2 months of grazing (P=0.044). These results revealed influence of grazing on exosomal c-miRNAs in cattle, suggesting that the roles of miRNA in molecular events of physiological response to grazing cattle with association with endocytosis, focal adhesion, axon guidance, and a variety of intracellular signaling, as predicted by bioinformatic analysis.

Automating measurements of grassland production

A. Van Den Pol-Van Dasselaar[1,2] and M.W.J. Stienezen[2]
[1]CAH Vilentum University of Applied Sciences, De Drieslag 4, 8251 JZ Dronten, the Netherlands, [2]Wageningen UR Livestock Research, De Elst 1, 6708 WD Wageningen, the Netherlands; agnes.vandenpol@wur.nl

Grazing has always been an efficient and cheap method to harvest and utilise grass. However, due to increasing herd sizes and increasing average milk production, grazing has become more complicated. In this era of automation, management tools could be automated to support the farmer. An important issue in grazing management is to match the grassland production with the animal needs. Insight into grass allowance is therefore needed. The objective of the research was to support grazing by automating management tools for determining grass allowance. Grass allowance can be estimated, but more precise data are gathered when the grass supply is measured. The standard for measuring grass heights in the Netherlands is a simple calibrated rising plate metre. Measurements with this metre were compared with a more user-friendly automated rising plate metre, that transfers data on the spot to a mobile phone. Measurements were taken weekly on a commercial dairy farm in the Netherlands in spring and early summer 2015. Regression analysis showed that the relation between the data of both metres was good. The automated rising plate metre saved time and was easier to use. Results of grass height are usually presented in a FeedWedge, which allow farmers to get insight in the expected grass allowance in the coming weeks. The next steps in the research are (1) to automatically upload the data of the automated rising plate metre in the FeedWedge progam available for all farmers on the internet and (2) to predict grass growth based on indicators like soil temperature and soil moisture content and incorporate this information in the FeedWedge. These steps will be taken in the near future (2016-2017) in the Dutch research project 'Amazing Grazing 2.0'. Further automation of management tools for determining grass allowance will facilitate farmers that practise grazing and thereby support grazing.

On-farm technologies to understand variation in growth performance

W. Thomson

Harbro Ltd., Markethill Road, Turriff AB53 4PA, Scotland, United Kingdom; willie.thomson@harbro.co.uk

Conventional manual weighing of animals has a limited use when attempting to monitor variation in growth over short periods of time. New precision technologies, including video imaging growth monitors and automatic step-on weighers, provide the opportunity to automatically capture multiple weight recordings from individuals, or groups, on a daily basis. Compared with manual systems, these systems offer reduced labour and minimal animal disturbance. The use of these new technologies generates a flow of detailed production data including daily weights – but also potentially including records of eating and drinking behaviour. This rich data stream offers the prospect of a new understanding of the growth and behaviour of livestock. Whilst average weight gain is still an important monitor of performance, the new systems create a greater understanding of both the pattern of growth and the variation of performance in groups of animals. It is now possible to closely monitor growth through key stages of production, and to understand in more depth the impact of health or nutritional changes on performance. The new understanding of variation will allow greater targeting of animals which either fall below target, or out-perform their peers. There is great scope to utilise the new data to identify animals in early stages of disease, potentially before symptoms are readily identifiable, to more precisely target health interventions. Statistical monitoring of group variation may also provide sensitive guides to health risk or diet suitability. New growth monitoring technologies offer the potential to redefine our understanding of livestock performance. They offer the potential to precisely target nutrition and health interventions and will provide a much greater understanding of animal response to management interventions.

Use of new sensor derived data in Australian dairy systems

C.E.F. Clark and S.C. Garcia

Dairy Science Group, School of Life and Environmental Sciences, Faculty of Veterinary Science, The University of Sydney, Camden, NSW 2570, Australia; cameron.clark@sydney.edu.au

Consumers are increasingly demanding high quality dairy products, produced at very high standards of animal welfare. In this same environment, cow numbers per farm continue to increase in systems where farmers are already working long (>60 hours per week) hours. Systems that focus farmer time towards priority tasks to solve problems and exploit opportunities, or even reduce the time that a farmer is required to spend on farm, will become increasingly important and will increase the attractiveness of the dairy industry for the next generation of farmers – who never knew a world without the internet. Over the last two decades, a large number of new technologies and sensors have been developed that enable monitoring of soil, plant and animal functions with increased accuracy. Whilst data on animal-based (primarily activity) sensor deployment across the Australian dairy industry is scarce, it is estimated that such sensors (primarily used for oestrus detection) are now on over 100,000 dairy cows with the rate of deployment increasing. Our group has shown new, additional uses of these and other milking system data for decision making on farm. Sensor derived rumination data now helps identify calving for farmers as rumination levels decline rapidly (~30%) in the days pre-partum. We are now advancing this research to substantially increase the health, welfare and productivity of the cow and calf by optimising the timing of farmer assistance for calving using detailed, multiple sensor derived behaviours. Whilst the potential application of this new technology is exciting, simple existing data such as milking order, associated with the timing of pasture access for each cow, is now being used to optimise individual cow/group feed allocations as pasture is a diverse nutrient resource, changing in chemical composition from leaf to stem. Together, these new sensor data and methods to convert data in actions will enable Australia's current, and next generation, dairy farmers to meet and exceed the food quality and animal welfare challenges set by themselves and the market.

Mixed ration and free cow traffic in automatic milking: effects on production and milking frequency
M. Patel, H. Driscoll and E. Spörndly
Swedish University of Agricultural Sciences, Department of Animal Nutrition and Management, P.O. Box 7024,
75007 Uppsala, Sweden; mikaela.patel@slu.se

One of the most important factors to maintain high production level in automatic milking systems is well-functioning cow traffic and frequent visits to the milking unit. Low milking frequencies may jeopardize udder health and milk yield. The aim of this study was to evaluate the effect of different feeding strategies on high yielding dairy cows. Thirty-eight cows were randomly allotted to either a mix of grass/clover silage and concentrate (MIX; 12.1 MJ ME) or separate ration (SR) of silage (11.4 MJ ME) and concentrate during a ten week experiment in a free cow traffic system. The feeds were from the same silos/batches and the mix and silage was fed ad libitum. The concentrate ration in the SR treatment was continuously adjusted in relation to silage intake to ensure that the silage/concentrate ratio was the same in both treatments. Daily feed intake, milk yield and milking frequencies were recorded automatically and the milk constituents analysed fortnightly. The statistical model included treatment, parity and days in milk using a mixed model with repeated measurements and cow as random variable. The results revealed no significant differences in somatic cell count or milk production with 35.0 and 35.4 kg ECM/day in the MIX and SR groups, respectively but feed intake in the MIX group was significantly higher ($P<0.05$) (26.8 vs 24 kg DM/day). Milking frequency was also higher in the MIX compared with the SR group, 2.6 and 2.3 milkings/day, respectively, which was unexpected since nutrient dense mixed feed are often claimed to lower milking frequency ('lazy cow syndrome'). The same pattern remained when fetched visits were excluded from the analyses. In conclusion, this study showed no evidence of lower milk frequencies or milk yield when dairy cows are fed a nutrient dense mixed ration in a free cow traffic system.

Increasing milking frequency and milking yield in robotic milking using an automatic herding system
U. Drach[1,2], A. Degani[2], T. Pnini[3], I. Izhaki[4] and I. Halachmi[1]
[1]Institute of Agricultural Engineering, Agricultural Research Organization (A.R.O.), The Volcani Center, Bet Dagan, Bet Dagan, Israel, [2]Faculty of Civil and Environmental Engineering, Technion, Israel Institute of Technology, The Division of Environmental, Water and Agricultural Engineering, Haifa, Haifa, Israel, [3]Pnini Shiran Milk Corporation Ltd., Tel Adashim, Tel Adashim, Israel, [4]University of Haifa, Depart. of Evolutionary and Environmental Biology, Haifa, Haifa, Israel; halachmi@volcani.agri.gov.il

Cow motivation to be milked is a key factor in milking robot utilization, if a cow does not voluntarily attend a robot stall fetching work is required. Feeding, cooling and behavioral studies tried to solve this problem by willingly attracting the cows to the milking robot. None of them achieved full success while the fetching work was still required. Therefore, this research suggests a new concept, herding all the cows to the milking robot using an Automatic Herding System (AHS). The AHS was built as a system of slow moving mobile fences controlled by an industrial controller. The AHS herds all the cows to the milking robot. The AHS was built in a commercial farm with two milking robots, and the experiment was conducted for three month. The farm was divided into a control group (43 cows) and an experiment group (39 cows). The AHS was installed only in the experiment group. For statistical analysis, two control groups were used: (1) the group without the AHS, (2) the group with AHS before installation of the system. Milking frequency increased in the experiment group by 47% (1.88 vs 2.75 milkings/d), while there was no major change in the milking frequency in the control group 2% (2.38 vs 2.39 milkings/d). Milk yield increased in the experiment group 17% (35 vs 41 kg/d). There was no major change in the yield in control group 4% (31 vs 29.8 kg/d). The results were statistically tested by mixed model ANOVA and were not rejected ($\alpha<0.005$). The system appeared to be a complementary system to the automatic milking system.

Using automation to identify and prevent health problems in dairy cattle

T.J. Devries
University of Guelph, Animal Biosciences, 50 Stone Road East, Guelph, Ontario, N1G 2W1, Canada;
tdevries@uoguelph.ca

The dairy industry has an increasing availability of commercially available equipment that is readily available for the automation of management tasks, including milking and feeding, as well as the monitoring of dairy cow behaviour. Such automation not only has the ability to improve production and time efficiency on farm, but also increases our ability to detect health disorders, both in occurrence and in advance of clinical symptoms. This, in turn, allows producers to identify and implement prevention and treatment protocols at earlier time points. Thus, this review will focus on how we can utilize automation in precision farming to keep cows healthy, productive, and efficient. Particular focus will be on measures of behaviour that may be collected automatically, including feeding, rumination, lying/standing, and milking behaviour, and the association these behaviours have with various health issues, including metabolic and infectious transition disease, lameness, and mastitis. It is anticipated that through precision monitoring of such behaviours, particularly with the aid of automation, those cows at risk or experiencing illness can be identified. This, in turn, will allow producers to identify and implement prevention and treatment protocols at earlier time points.

Modeling individual voluntary feed intake based on feeding behavior

I. Richter[1], I. Halachmi[1,2] and Y. Parmet[1]
[1]Ben-Gurion University of the Negev, Department of Industrial Engineering and Management, P.O.B. 653, Beer-Sheva 84105, Israel, [2]Institute of Agricultural Engineering, ARO, Volcani Center, P. O. Box 6, Bet Dagan, Israel;
ilanarichter@gmail.com

Dairy cow's dry matter intake (DMI) is an important variable in dairy management. As such, it can contribute greatly in various economic decisions relating this field. For example, combining the DMI with other measured data such as milk yield and milk components, allows us to assess the economic value of the individual cow feed. However, Existing models are still unable to measure the voluntary food intake of a single cow with high accuracy. In this research we developed a feed intake model for the individual dairy cow. The model includes available data in commercial farms and incorporates not yet available data of feeding behavior. As technology develops, low-cost feeding-behavior sensors will soon be available for commercial use in dairy farms. Sensors of feeding behavior, i.e. time spent in the feeding lane, duration and frequency of meals through both day and night, are currently being developed by numerous commercial companies. To develop the model, data of DMI and indicators of 73 cows during 98 days of lactation were obtained. The subjects were from the ARO research farm in Bet-Dagan, Israel. Two models were developed, the first assesses dry matter intake on a daily basis and the second assesses single meals. In both models we used information of eating time to predict the DMI. Linear mixed model was fitted to consider the random effect of both cows and days. Using the cows as a random effect allowed us to transform the model results to precision farming. Results show that the meals DMI model, for a cow that is measured for the first time, the coefficient of determination (R2) is equal to 0.78 and for cows with previous data in the model, the R2 is 0.88. For the daily DMI model, a cow that is measured for the first time, has R2 equal to 0.67 and cows with previous data in the model the coefficient of determination is 0.94. The result suggests that first calibration is of importance.

Rumen telemetry as an aid to managing water intake

T.T.F. Mottram[1,2] and D. Bradley[1]
[1]eCow Devon Ltd, King St Business Centre, EX1 1BH, United Kingdom, [2]Royal Agricultural University, School of Agriculture Food and Environment, Stroud Road, Cirencester, GL7 6SJ, United Kingdom; toby.mottram@rau.ac.uk

Wireless rumen telemetry has revealed the nature of daily changes in rumen pH and how this differs between systems. The features of the daily rumen profile on individual animals have demonstrated that we are able to monitor the daily routine of animals. The rumen temperature profile features sudden drops indicating drinking events and these can be related to changes in pH. However, the monitoring of drinking behaviour with a rumen bolus has not been reported and may be a useful addition in farm management. This paper will describe a case study where observations of both daily rumen pH and trough visits were used to inform changes of management in an individual animal, leading to a better understanding of the demands of differing systems and the animals within them. These data and a literature review identified that the interaction between drinking behaviours, rumen pH and sub-clinical dehydration is not well understood in commercial cattle.

A multi-Kinect system for monitoring and measuring functional traits in dairy cows

J. Salau, J.H. Haas, W. Junge and G. Thaller
Kiel University, Institute of Animal Breeding & Husbandry, Hermann-Rodewald-Str. 6, 24118 Kiel, Germany; jsalau@tierzucht.uni-kiel.de

Animal health, productivity, and fertility are important cornerstones of modern dairy cow breeding and husbandry. Health monitoring and conformation recording based on manual or visual evaluation is time consuming and prone to subjectivity. Camera based systems are a progressive way to overcome the need to examine animals by assessors and deliver precise and objective measurements. This study presents a system, in which 6 Microsoft Kinect cameras are mounted in a frame with ≈2 m in height and width. While cows walk freely through the frame, 3D data covering almost the complete cow surface is collected. Software for recording, synchronizing multiple cameras, automated system calibration, and registering 3D data from different viewpoints has been developed. After a fixed installation on a research farm, 8 runs of 4 Holstein Frisian cows were gathered as test data set. Shoulders and front teats were marked manually for each run of each cow. From these coordinates, shoulder height and teat lengths were calculated and compared between the runs of each cow to test the system's measuring precision. Mean shoulder heights ranged between cows from 1.36 m to 1.46 m with standard deviations 0.03 m to 0.08 m and standard errors between 0.01 m and 0.03 m. Mean teat lengths varied between cows from 40.4 mm to 44.6 mm (right) and 32.1 mm to 47.3 mm (left) with standard deviations 4.3 mm to 9.0 mm and 5.2 mm to 8.3 mm and standard errors between 1.8 mm and 3.2 mm, respectively, 1.8 mm and 2.9 mm. Next steps are to evaluate the system's accuracy, i.e. comparing system and physically taken measurements on the cow, and to automate the determination of body parts of interest. To achieve the latter, a database for models of cows' body parts was set up, and software for the recognition of 3D objects via a comparison with those models was implemented. Information concerning conformation, age, and lactation of the cow the model was taken from was linked to the model in the database. Exemplarily, this makes it possible to find similarities between those cows with the body part of interest matching the database model. In this way, a solid data basis for analyzing relationships between functional traits and genetics could be provided.

Spatial, inter and intra-cow repeatability of thermal imaging

D.T. Byrne[1,2], H. Esmonde[1], D.P. Berry[2] and N. McHugh[2]
[1]*Animal & Grassland Research and Innovation Centre, Teagasc, Moorepark, Fermoy, Co. Cork, Ireland,* [2]*Dublin City University, Glasnevin, Dublin 9, Co. Dublin, Ireland; thomas.byrne@teagasc.ie*

Infrared thermography (IRT) is a quick, non-invasive method for quantifying the energy radiating from an object and thereby facilitates estimation of the temperature of an object. High repeatability between replicate IRT images has been documented in medical studies; however in an agricultural environment many factors can adversely affect the repeatability of IRT. The objective of this study was to quantify the inter and intra-cow repeatability of IRT measurements on three anatomical regions (i.e. udder, eyes and hooves) of Holstein-Friesian cattle and to quantify the number of replicate images required to gain a certain level of precision. All thermal images were captured using a FLIR T430sc thermal camera with a spectral range of 7.5 to 13 µm. The camera resolution was 320×240, the thermal sensitivity was <0.03 °C, and the range within which the true measurement lies (accuracy) was ±2 °C. A set of 30 replicates were manually taken of the ventral face of the udder, lateral face of the left eye, lateral face of the right eye, palmar face of the front hooves and plantar face of the back hooves on 15 cows. Images were manually cropped to extract maximum, minimum and average temperatures. Cow variance (intra) and error variance (inter) were calculated using mixed models where temperature was the dependent variable and cow was included as a random effect. The repeatability (t) and the co-efficient of variation (CV) were calculated for each extracted temperature from all anatomical areas. The precision associated with differing number of images was calculated as the range within which the extracted temperature lies relative to the gold standard (i.e. average of 30 measurements). Maximum temperature had the highest cow variance and CV, while average temperature had the lowest error variance and greatest precision for hoof images. However the opposite was true for eye and udder images. Across all temperature variables, hoof images were the most repeatable, with the greatest repeatability (0.95 to 0.99). If an average of 5 images were captured a precision of ±0.25 °C could be achieved for hoof images.

Classifiers based on 3D-head acceleration data used for lameness detection

Y. Link[1], J. Salau[1], S. Karsten[2] and J. Krieter[1]
[1]*Institute of Animal Breeding & Husbandry,Christian-Albrechts-University, Olshausenstraße 40, 24098 Kiel, Germany,* [2]*TiDaTier und Daten GmbH, Bosser Str. 4c, 24259 Westensee, Germany; jsalau@tierzucht.uni-kiel.de*

Lameness in dairy cow husbandry is a great health and welfare problem. Moreover it is time consuming to detect. Therefore, an automated monitoring system is of great interest for scientists and farmers. A new approach is using 3D-head acceleration data to identify lame dairy cows. Between January 2014 and February 2015, 70 cows at the research farm Futterkamp were equipped with 10 Hz or 1 Hz ear tags. For a reference data set locomotion scores were recorded three times a week. Sensor data covering three weeks prior to each lameness was used to create classifiers to detect lameness on a day to day basis. For pre-processing, nine different methods (i.e. sum or mean) were applied to condense the data within 100 equally long time sections, respectively. Further on, 22 daily features (i.e. mean, variance or skewness) were calculated from the condensed vectors. To define ranges for cow individual thresholds the data of the first 10 days was used. These ranges were used to classify cows as lame. The Area Under ROC (receiver operating characteristic) Curve (AUC) analysis was used to evaluate the effectiveness of each classifier. ROC curves for all classifiers were plotted and AUC values calculated as measure for classifier performance. For the 10 Hz dataset the best classifier was based on the pre-processing method 25% quantile and the feature mean-median with an AUC value of 0.71. The same combination of pre-processing method and feature for the 1 Hz dataset generated an AUC value of 0.74. Though, the highest AUC values (0.80) for this dataset was observed for the pre-processing method kurtosis and the feature standard deviation respectively variance. These results indicate that 3D-head acceleration data can be used to detect lameness in dairy cows.

Electricity and water consumption on Irish commercial dairy farms
P. Shine[1], M.D. Murphy[1] and T. Scully[2]
[1]Cork Institute of Technology, Department of Process, Energy and Transport Engineering, Bishopstown, T12 T66T Cork, Ireland, [2]Cork Institute of Technology, Department of Computing, Bishopstown, T12 T66T Cork, Ireland; philip.shine@mycit.ie

Understanding variances in resource consumption on dairy farms is essential for farmers and regulatory bodies from both an environmental and economic perspective to ensure the future sustainability of the Irish dairy industry. With this, an exploratory analysis of electricity and direct water consumption has been carried out through a random sample of 58 pasture-based, Irish commercial dairy farms, remotely monitored throughout 2014-2015 along with associated milk production, cow numbers and infrastructural data. The sub-metering of electricity and water equipment allowed for 15 minute resolution, cumulative consumption measurements to be recorded for each process involved in the on-farm production of milk and autonomously sent via a 3G network to a receiver on site at Teagasc, Moorepark, Animal & Grassland Research and Innovation Centre, Fermoy. Annual electricity and water key performance indicators of 203 kWh/cow (42 Wh/Lm) and 29.6 m3/cow (6.3 Lw/Lm) were established. Dairy farms incorporating an ice bank within its milk cooling system consumed on average 21% greater total electricity consumption per cow in comparison to farms employing a direct expansion with pre-cooling system. While consuming 22% less electricity during daytime hours due to load shifting capabilities of ice bank cooling systems, this increased night time consumption failed to offset the greater electricity consumption whereby ice bank dairy farms had a 7% increase in the cost per litre of milk. Employing ground water for milk pre-cooling through a plate heat exchanger, on average, decreased electricity consumption for milk cooling per litre by 20% compared to standard direct expansion systems. This resulted in a population increase of 5% in litres of water consumed per litre of milk; however, no significant difference between mean water consumption efficiencies suggests dairy farmers in Ireland recycle pre-cooler water effectively throughout the farm. Results pertaining from this analysis may facilitate the development of adaptive predictive and optimisation methodologies for agriculture resource consumption.

Strategies of the Walloon dairy breeders faced to the uncertain dairy future
A.-C. Dalcq[1], T. Dogot[1], H. Soyeurt[1], Y. Brostaux[1], F. Vanwindekens[2], E. Froidmont[2], P. Rondia[2], B. Wyzen[3], A. Masure[4], C. Bauraind[5] and Y. Beckers[1]
[1]ULg-GXABT, Passage des déportés, 2, 5030 Gembloux, Belgium, [2]CRA-W, Rue de Liroux, 8, 5030 Gembloux, Belgium, [3]AWE, Rue des Champs Elysées, 4, 5590 Ciney, Belgium, [4]FWA, Chaussée de Namur, 47, 5030 Gembloux, Belgium, [5]Collège des Producteurs, Avenue Comte de Smet de Nayer 14, 5000 Namur, Belgium; anne-catherine.dalcq@ulg.ac.be

This study observes the strategies, and their determinants, of the Walloon dairy breeders faced to the post quota perspective through the realization of 245 surveys, conducted from November 2014 to February 2015. Three kinds of strategical variables were defined and related to the evolution of milk production (MP) [the breeders who increase MP (HighMP) vs keep constant MP (ConstantMP) vs stop MP]; the valorisation of MP [alternative (ValMP) vs classical] and the diversification of activities [with (DivMP) vs without such activities]. The relationships between the chosen strategies and the quantitative technical variables were studied using generalized linear models. The independence between qualitative technical variables and the strategical variables was tested using Chi Square test. HighMP and ConstantMP bredeers represent 38.4% and 53.9% of respondents, respectively. HighMP breeders were significantly more declared as legal entity (P=0.03), had more family members on the farm (p-value<0.01), larger agricultural area in property (P=0.03) and higher MP quota (P=0.01) compared to ConstantMP breeders. Only 9.8% of respondents decide to valorise differently MP. ValMP breeders tend to have more employees (P=0.08) and an agricultural area less fragmented (P=0.07) than classical breeders. A total of 7.8% of respondents decide to develop other activities. DivMP breeders tend to have more employees (P=0.10), more agricultural area in property (P=0.03) and a more recent year of installation (P<0.01). Finally, 44.9% of ConstantMP breeders do not want to start an alternative valorisation of MP and diversify their activities. In conclusion, a relationship exists between, amongst others, the legal status, workforce available, characteristics of the agricultural area, the dairy production and the strategy chosen by the Walloon dairy breeders.

Potential effects of herd expansion on cow welfare in a pasture-based dairy industry

J. Marchewka[1,2], J.F. Mee[2] and L. Boyle[2]
[1]Institute of Genetics and Animal Breeding, PAS, Department of Animal Behaviour, ul. Postepu 36A, Jastrzębiec,
05-552 Magdalenka, Poland, [2]Teagasc, Animal and Grassland Research and Innovation Centre, Moorepark,
Fermoy, Co. Cork, Ireland; laura.boyle@teagasc.ie

Due to the abolition of milk quota (2015) Irish pasture-based dairy farms have increased herd sizes and yields of milk solids per hectare, coupled with increasing stocking density at pasture and in housing, and/or the size and fragmentation of the grazing platform. Unless accompanied by investments in infrastructure such intensification could have negative consequences for cow welfare. This study investigated changes in herd size, associated investments occurring over the last 3 years in Irish dairy herds, as well as the perceptions of key stakeholders: dairy farmers (F; n=115), cattle veterinarians (V; n=60) and Teagasc dairy advisors regarding related risks to cow welfare. A Chi-Square Fisher test was used to investigate differences in response frequencies between groups (PROC FREQ, SAS). 77% of F increased cow numbers in the past 3 years. More farmers who increased cow numbers invested in their farm (85.4%) than those that did not expand (14.6%) ($X^2 = 4.46$; P=0.035) and invested in the modernization or enlargement of the milking parlour (93.5%) than those that did not expand (6.5%) ($X^2 = 8.3$; P=0.004). There was no difference in handling facilities, housing or roadways investments (P>0.05) between expanding and non-expanding farms. Frequencies of A, F and V indicating intensification-related advantages or disadvantages for cow welfare were the same, with the majority in all groups expressing concern (87.0 vs 79.1 vs 86.4%, respectively). Encouragingly, investment in milking parlours indicates that cow welfare is not likely to be adversely affected by inefficiencies in the milking process, causing prolonged standing on concrete. However, the lack of investment in housing and roadways in expanded herds poses risks to cow welfare, particularly lameness. Proactive steps should be taken to protect cow welfare and the sustainability of Irish pasture-based dairying systems.

Understanding the mechanism of impacts of agricultural science. A case study: Dairy Fertility Index

S. Fowler and P. Midmore
Aberystwyth University, School of Management and Business, Rheidol Bldg, Llanbadarn Campus, Aberystwyth,
SY23 3AL, United Kingdom; suf@aber.ac.uk

Increasing pressure on research budgets to demonstrate impact has kindled new interest in the relationship between original science and its contribution to solving socio-economic and environmental policy challenges. Whilst it can be demonstrated that agricultural research, after sometimes long lags, produces substantial benefits, much less is known about the mechanisms involved at the level of the individual project. Study of this tier of impact can give important insights into the framing of projects and the overall development of programmes to enhance the efficiency of public support for agricultural research. This paper describes the multiple impact pathways involved in a single project which used performance recording to develop the Dairy Fertility Index (FI). Using predominantly qualitative case study analysis, we studied the targeting of research activity and the quality and relevance of results and their dissemination, within the overall political and cultural context in which the research system is embedded. The approach included documentary analysis, in-depth interviews with key stakeholders in science, extension and the dairy industry, and three workshops with farmers and advisers in different parts of the UK. The analysis reveals a unique set of circumstances which enabled the project to be brought to fruition, despite initial and some continuing opposition. Price pressure on dairy farmers created an environment where short term survival was in doubt and longer term issues were deemed unimportant. Enabling factors arose from a strong basis of established personal relationships, and a public service ethos that endured even after the MMB and NMR became privatised, enabling provision of data that benefited the dairy industry as a whole. In particular, the MAFF Sustainable Livestock research programme and LINK funding mechanism that required engagement with commercial companies benefited dissemination and adoption routes, although progress in building awareness was slow and required sustained effort. We conclude that it is much less likely that such a project would have succeeded in current circumstances and make recommendations that might help to improve the overall impact of dairy science research.

Knowledge management on Austrian dairy farms

T. Gehmair[1], B. Fuerst-Waltl[1], F. Sturmlechner[2], M. Drillich[3], M. Wöckinger[4], C. Egger-Danner[5] and M. Wurzinger[1]
[1]University of Natural Resources and Life Sciences Vienna (BOKU), Gregor-Mendel-Str 33, 1180 Vienna, Austria, [2]ZAR, Association of Austrian Cattle Breeders, Dresdner Str. 89, 1200 Vienna, Austria, [3]University of Veterinary Medicine, Veterinärplatz 1, 1210 Vienna, Austria, [4]Chamber of Agriculture Upper Austria, Auf der Gugl 3, 4021 Linz, Austria, [5]ZuchtData, Dresdner Str. 89, 1200 Vienna, Austria; theresa.gehmair@gmx.at

Nowadays Austrian dairy farmers are confronted with increasing challenges such as the end of the milk quota and the increasing tendency of entering global markets. In this struggle, having access to the latest information about new technologies that concern the various fields of a farm, from feeding to economic data, is important. The aim of this study was to investigate the current needs and demands of the Austrian dairy farmers concerning knowledge management and acquisition. Based on these results a new training concept will be developed. In 2015, an on-line survey was conducted and 9,992 dairy farmers were invited to participate. After six weeks 983 questionnaires were completed. Statistical analysis was performed using SAS software. The results showed that print media such as agricultural magazines and magazines of breeding organisations are more popular than meetings and electronic media. The majority (80%) preferred simple internet search or visit of homepages of various relevant organisations. Online courses were less popular (30% of respondents) at the moment, but were given high priority for future use. A clear result was the importance of veterinarians as contact persons for the farmer considering animal health (95%). Farmers ranked fertility disorders, feeding management and udder health as key topics, where they would require additional information. The best time for training courses is during the winter months due to the absence of work peaks. During the day, the farmers prefer the evening for further training. Limiting factors for attending trainings are inconvenient times or dates and unappealing programmes. The factor of costs was ranked on last position. The openness of farmers concerning on-line trainings (on-line courses, webinars) seems to be a new avenue to be further developed in new training programmes.

Sources of information for small-holder cattle farmers in North West Province, South Africa

D.M. Motiang[1] and E.C. Webb[2]
[1]Agricultural Research Council, Animal Production Institute, Private Bag X2 Irene, 0062, South Africa, [2]University of Pretoria, Department of Animal and Wildlife Sciences, University of Pretoria, 0002, South Africa; dan@arc.agric.za

In South Africa, role players use various modalities to complement public extension and improve productivity. The purpose of this study was to identify sources of information used by small-holder cattle enterprises and evaluate their influence on productivity of those enterprises. A random sample of 308 cattle producers from Dr RSM District was interviewed between May and July 2012 using a semi structured questionnaire. A ranking scale was used to measure the extent to which producers used various sources of information where 1= never and 4= always. Data were analysed using IBM SPSS statistics 22 (2013). The GLM multivariate analysis was performed to test effect of farming area and farm level variables on herd mortality and off-take rates. Means were separated using least significant differences (LSD) tests. Correlation analysis was performed to measure associations between demographics, information sources and herd off-take. Public extension is the main (85-87%) information source for small-holder cattle producers, followed by radio (52%) and other farmers (47-48%). Educated farmers prefer magazine and TV to extension officers while women prefer office visits to magazine, TV and commercial farmers (P<0.05). Older farmers tend not to rely on magazine, radio, TV, commercial farmers and veterinarian while owners of larger herds prefer these sources to office visits (P<0.05). Source commonly used by more educated farmers, men, younger farmers and owners of large herds such as radio, TV and commercial farmers, suppressed the sale of cattle and small stock (P<0.05). Reliance on other small-holder farmers suppressed calving rates while other farmers lowered the sale of small stock (P<0.05). Sources preferred by women such as office visits and telephone stimulated the sale of sheep and small stock in general. Results show that age, gender, education and herd size determine extension channels.

Participatory famers trials for increasing productivity of smallholders dairy production in Nepal

D. Sapkota
IDE, Kathmandu, 00977, Nepal; dr.deepak2011@yahoo.com

Dairy farming is an integral part of the farming system in Nepal and is a major source of household income. The substantial growth of dairy sub-sector in recent years is due to increasing demand of milk and milk products in local and regional markets. However, smallholder farmers are not keen towards commercial promotion of dairy animals because of high cost of production and low return. Animal feed alone constitutes around 65% total production cost (Better to be abstract based on study rather than range?). Majority of these farmers own breeds that produce less milk. Traditional animal feeding practices and lack of knowledge in ruminant nutrition further lessen productivity. Forum for Rural welfare and Agricultural Reform for Development (FORWARD Nepal) in partnership with International Livestock Research Institute (ILRI) implemented the project entitled 'Improvement of Livestock Production System through Feed Based Interventions in Chitwan District, Nepal' and introduced different options of residue based feeding for reducing milk production costs. Activities conducted under this project included Feed Assessment Study (FEAST), nutritional gap analysis and cost-benefit study of milk production based on current feeding practices. A 350 kg Jersey cow was assessed as optimum feeding of Dry Matter (DM) (10.87 kg) but the crude protein (CP) intake was 1.27 kg which was very high. Analysis showed that farmers can reduce costs by reducing the quantity of costly compound feed and effectively utilizing crop residues. Based on the existing feeding pattern, the average cost of production for one liter milk was NRs. 14, whereas average benefit was approx NRs 13 per liter of milk irrespective to other expenses incurred in dairy animal farming.

Using farming system approach for creating new occupations for small holder farmers in Thailand

A. Molee[1], W. Molee[1], B. Likitdejcharoj[1], T. Chormai[2], C. Hormta[1] and P. Mernklatoak[1]
[1]Suranaree University of Technology, School of Animal Production Technology, 111 University Avenue, Muang District, Nakhon Ratchasima Province, 40000, Thailand, [2]Department of Livestock Development, Kabin Buri District, Prachin Buri, 25110, Thailand; amonrat@sut.ac.th

A new meat chicken breed for small holder farmer's occupation is needed since it can be an option for farmers who prefer to have freedom of occupation, that means without contract farming with a private company. The aim of this study was to apply the farming system approach to design an appropriate so called Korat chicken for the group of small holder farmers. The five processes of the approach are (1) target area selection (2) analysis of problem or need of farmer (3) appropriated technology designing (4) on farm trial, and (5) results reporting and transfer of the technology to another area than was used for the study. A group of farmers at the Srikhiew District, Nakhon Ratchasima Province was selected, since this area was convenient for travelling of the researchers to the project. A collateral meeting was held between researchers and farmers' representatives. The aim of the meeting was to acquire information regarding market demands of chicken meat quantity and quality. High end marketing was surveyed. The breeding goal for the Korat chicken was formulated and designed. It contained a better growth rate than the indigenous chicken, while taste and springy texture of the meat is maintained at the same level. The knowledge of animal breeding was applied. When the 1st generation of the chicken was generated, its performances were tested in SUT's farm and followed by an on farm trial by farmer. The performances of the chickens, which were tested in 2 places, were compared. Similar performances were shown. The new breed of chicken could be accepted by the group of farmers. The results were reported to the other communities, and the chicken was transferred to more than 15 groups of farmers in Thailand. The acceptance proves that Korat chicken is an appropriate technology for the farmer his occupation. However, for the sustainability of the occupation it is needed to evaluate the new breed as now used by the various farmers groups.

Breeding ewe replacements to lamb at 1 year of age: effects of ewe genotype and live weight at join
T.W.J. Keady and J.P. Hanrahan
Teagasc, Grassland Science, Athenry, Co Galway, Ireland; tim.keady@teagasc.ie

Breeding ewe replacements to lamb at 1 year of age can increase lifetime lamb carcass output and may reduce replacement rate, and thus cost. The objective of this study was to evaluate the effects of ewe genotype and live weight at joining on the performance of ewe replacements joined to lamb at 1 year, and the performance of their progeny. A total of 230 ewe replacements [84 Belclare (B), 65 B × Suffolk (S) and 81 '>75%' S; mean (range) live weight (LW) at joining 45.3 (36.5-56.5), 46.1 (37.9-55.6) and 48.3 (41.9-59.0) kg, respectively] were joined with Charollais rams. The ewe replacements were exposed to vasectomized rams for 48 h (to advance oestrus via the 'ram effect') at day 14 prior to joining, which lasted for 42 days. The ewes were housed in mid-December, shorn and offered a diet based on grass silage until lambing in March/April. Ewes and their lambs were put to pasture within 48 h of parturition. Ewes rearing singles received no concentrate supplementation; those rearing twins received 0.5 kg daily for 5 weeks post lambing. Lambs had access to a maximum of 300 g concentrate daily until weaning at 14 weeks. Litter size, number of lambs reared per ewe joined, lamb birth weight (kg), lamb weaning weight (kg) and lamb LW gain (g/d) from birth to weaning were: 1.65, 1.41 and 1.26 (s.e. 0.067; P<0.001); 1.19, 1.05 and 0.82 (s.e. 0.081; P<0.01); 4.2, 4.4 and 4.4 (s.e. 0.11; P>0.05); 29.0, 31.9 and 31.0 (s.e. 0.54; P<0.001); and 255, 279 and 269 (s.e. 5.0; P<0.01) for the progeny of the B, B × S and '>75%' S ewes, respectively. To have a 90% chance of rearing at least one live lamb B, B × S and '>75%' S ewes needed to be 48, 51 and 60 kg at joining; equivalent to 62, 64 and 72% of their expected mature LW, respectively. It is concluded that target LW at joining ewe replacements to lamb at 1 year of age depends on ewe genotype. The productivity of Belclare ewes lambing at 1 year of age, based on the number of lambs reared per ewe joined, was equivalent to that achieved from the national lowland ewe flock in Ireland.

Genetic management without pedigree: effectiveness of a breeding circle in a rare sheep breed
J.J. Windig[1], M. Verweij[1], M. Maurice-Van Eijndhoven[2] and K. Oldenbroek[1,2]
[1] Wageningen University and Research, Animal Breeding and Genomics Centre, P.O. Box 338, 6700 AH Wageningen, the Netherlands, [2] SZH, Dutch Rare Breed Trust, Dreijenlaan 2, 6703 HA Wageningen, the Netherlands; jack.windig@wur.nl

Rotational mating can be an effective tool to reduce inbreeding rates especially in populations without pedigree recording. Theory is well developed but the effectiveness in practice has never tested due to a lack of pedigree recording. The Dutch rare sheep breed Veluws Heideschaap is an example where a breeding circle has been maintained for over 20 years. To do this we sampled age classes of 1 year old (rams) and of 6 year old (ewes), about 2.4 generations apart and typed on a 10K SNP chip in the 8 flocks participating in the breeding circle. Average heterozygosity was higher for 6 year animals (0.400) than for 1 year old animals (0.384). This corresponded to an inbreeding rate of well over 1% per generation in each of the flocks. Mathematical derived inbreeding rates based on the number of animals per flock were about 5 times lower, andsowre inbreeding rates from computer simulations. Possibly dominant rams leading to an unequal distribution of number of offspring across sires, increased inbreeding rates. Simulations, however, indicate that dominant rams can only partly explain the increased inbreeding rates. Selection of related animals for breeding and differential survival of more and less heterozygous individuals may further have increased inbreeding rates estimated from DNA.

Selection of Norwegian White Sheep rams for AI on EBV and ram circle connectedness

J.H. Jakobsen[1], T. Blichfeldt[1] and R.M. Lewis[2]

[1]Norwegian Association of Sheep and Goat Breeders, Box 104, 1431 Ås, Norway, [2]University of Nebraska, Animal Science Department, Lincoln, NE 68583, USA; jj@nsg.no

In Norway, flocks are genetically connected within ram circles by circulating natural service test and elite rams. Across ram circles, connections are established by use of artificial insemination (AI). The aim of the current study was to investigate the strength of connectedness within and across ram circles in Norwegian White Sheep, as an indication of risk of selection bias at different stages of ram selection. Performance data on weaning weight on 2,376,830 lambs born in 1996 to 2015 were extracted from the Sheep Recording database along with 2,541,608 pedigree records. An animal model was fitted with fixed effects of flock-year and interaction of litter size at tagging × litter size at weaning × age of dam × sex. The data contained 966 flocks in 132 ram circles. Pairwise connectedness between flocks was determined as the average prediction error correlation of mean flock EBV. Values of this statistic above 0.10, and between 0.05 and 0.10, have been shown to correspond with strong and moderate levels of connectedness, respectively. Average connectedness among flocks within a given ram circle with at least two members ranged from 0.050 to 0.676, with an average of 0.318, indicating strong connectedness among flocks within ram circles. Selection of test and elite rams based on the EBVs thus should be subject to limited bias. Since AI rams are selected across ram circles, unbiased comparison of their EBV across ram circles is key. Average connectedness of flocks within a ram circle to all other ram circle flocks ranged from 0.011 to 0.063 with an average of 0.034. This suggests considerable uncertainly when comparing EBV among candidate AI rams. In the future, a strategy that considers both EBV and ram circle connectedness to all other breeding flocks may be adopted when selecting AI rams. In addition, ram circles with low connectedness values can be encouraged to increase their use of AI in order to improve the strength of genetic links.

An across breed comparison of Irish versus New Zealand ovine genetics

F.M. McGovern[1], P. Creighton[1], H. Walsh[1], K. McDermott[2], E. Wall[2] and N. McHugh[1]

[1]Teagasc, Animal and Bioscience, Animal & Grassland Research and Innovation Centre, Teagasc, Athenry, Co. Galway, Ireland, 0000, Ireland, [2]Sheep Ireland, Highfield House, Shinagh, Bandon, Co. Cork, Ireland, 0000, Ireland; fiona.mcgovern@teagasc.ie

Sheep genetic indexes have been available for sheep farmers in New Zealand for over 20 years with large increases in sheep farm productivity and profitability being achieved. The rate of genetic progress is three times higher for the New Zealand sheep industry compared to the Irish sheep industry. The aim of this study is to compare elite Irish sheep genetics to elite New Zealand genetics. Three groups of animals (n=60), comprising of two main breeds (Texel and Suffolk), were assembled; representing high genetic merit New Zealand ewes (NZ), high genetic merit Irish ewes (Irish High), and low genetic merit Irish ewes (Irish Low). The experiment commenced in October 2015 when ewes were mated within group, using artificial insemination (A.I.). Ewes were housed and pregnancy scanned in late December and began lambing on the 29th February 2016. The effect of ewe strain on ewe body condition score (BCS), live weight was analysed using PROC HPMIXED SAS (SAS Inst. Inc., Cary, NC, USA) where breed (Texel or Suffolk), age of ewe and strain (NZ, Irish High or Irish Low) were included as fixed effects. The log of the odds of a ewe conceiving was modeled using logistic regression in PROC GENMOD (SAS Inst. Inc., Cary, NC, USA) assuming a binomial distribution of the errors with breed, age, strain and A.I type (fresh or frozen) included as fixed effects. New Zealand ewes had significantly lighter live-weights (P<0.05) and higher BCS (P<0.01) when compared to Irish ewes. Ewe breed significantly effected both ewe BCS (P<0.01) and scanned litter size (P<0.01), with Texel ewes having a greater scanned litter size and higher BCS than Suffolk ewes. Texel ewes were 2.69 times more likely to hold to first service compared to Suffolk ewes (P<0.01). In conclusion, ewe type affects both mature live-weight and BCS however ewe breed has a greater influence on scanned litter size and conception to first service than ewe type.

RFI divergent lines in sheep and relationship with TMR RFI: first results

D. Francois[1], J.L. Weisbecker[1], E. Ricard[1], J.F. Bompa[1], D. Marcon[2], F. Bouvier[2], C. Moreno[1], C. Marie-Etancelin[1] and F. Tortereau[1]
[1]INRA, UMR 1388 GenPhySE, 24 chemin de Borde Rouge, Auzeville, 31326 Castanet Tolosan, France, [2]INRA, UE 0332, Domaine de la Sapinière, 18390 Osmoy, France; dominique.francois@toulouse.inra.fr

Meat sheep selection programmes include traits such as growth and body composition. One way to improve sheep industry may be to increase the efficiency of feed transformation. Concentrated diet feed intake has been measured with automatic feeders on young rams as they are undergone to individual test for selection on liveweight (LW), growth and body composition. The test is following a national process and lasts 8 weeks from about 100 to 156 days of age. LW and ultrasound scan (US) are recorded at start (LW only), mid (both LW and US) and end (both LW and US) of the test. Feeders provided detailed data for each ram meal (identity number, intake, duration) allowing computation of daily feed intake (DFI). Since DFI is strongly correlated with LW, growth and body composition, linear regression of DFI on these traits was performed and residual feed intake (RFI) was expressed as the residue of the regression. Breeding Values have been estimated for RFI, LW, Average Daily Gain, Body composition (Conformation and Backfat). Based on RFI EBV, two divergent lines have been established in 2014 with 5 rams and 60 ewes per line. A similar feeder but for total mixed ration (TMR) has been developped by the institute and feed intake measurements have started with the lambs of the two lines born in 2015. DFI was measured first on concentrate diet of both males and females (between 100 and 150 days of age for males and between 175 and 225 days of age for females) and then measured by TMR feeders (between 175 and 225 days of age for 32 males and between 200 and 250 days of age for 32 females). Correlation between RFI concentrate and RFI TMR was 0.41 for males and 0.17 for females suggesting that DFI of both diets should be phenotyped to improve genetics of feed efficiency.

Relations between farmers and cheese-makers in dairy ewe sector: what room for sector coordination?

M. Millet and F. Casabianca
INRA, Science for Action and Development, Quartier Grossetti, 20250 CORTE, France; fca@corte.inra.fr

In a context of chronic crisis, a solution for local actors would be to create sector coordination integrating their common interests and ensuring sector sustainability. To what extent such joint trade organization can build a collective strategy? For answering our question, we compared dairy ewe sectors in Western Pyrenees (WP) and Corsica Island (CI) in Southern France. After a period of milk provision for Roquefort, various modes of relations between farmers and cheese makers are now coexisting in each basin (expectations on milk quality, milk prices, technical support). Our aim is to retrace the trajectory of relations for both cases in order to better understand: (1) how the present diversity was built; and (2) what room was left to sector coordination. To do so, we realized a set of interviews with firms (12) and farmers (50). We completed our data with narratives (retired actors) and archives. We split the evolution of relations into distinct periods. First, a unique regime was under external decision center (Roquefort Era). Then, as Roquefort firms drew back (1970-1980) in both basins, new firms implemented (national group or local initiatives) and farmers were considered as equal to firms being the condition to their success. After stabilization, crisis occurred inducing a new set of relations (1990s in WP; 2000s in CI): national groups demanding milk quality, local initiatives growing, some others staying small. Finally, in WP only, new initiatives emerged in 2000s based on raw milk. During the 90s, joint trade organizations were created in each basin for managing dairy ewe sector. Such tools regulate various interests and undertake missions as milk quality and management of production (volumes), technical support for dairy ewe development (CI) or sanitary management (WP). However, as their missions grew, joint trade organizations failed in building a collective strategy. Therefore, sector coordination is necessary but not enough as the involvement of farmers and firms remains crucial.

World sheepmeat production: a meta-analysis

L.E. Robles Jimenez[1], A. Partida De La Peña[2], O.A. Castelan Ortega[1], S. Radic[3] and M. Gonzalez Ronquillo[1]
[1]Universidad Autónoma del Estado de México, Departamento de Nutricion Animal., Instituto Literario Ote. No. 100, Col. Centro, 5000, Toluca, Mexico, [2]Centro Nacional de Investigacion Disciplinaria en Fisiologia y Mejoramiento Animal, INIFAP, Km. 1 Carretera a Colón, 76280, Ajuchitlan-Queretaro, Mexico, [3]Universidad de Magallanes, Escuela de Ciencias y Tecnología en Recursos Agrícolas y Acuícolas, Manuel Bulnes 01855, 6210427, Punta Arenas, Chile; lizroblez@hotmail.com

The free trade agreements among the world's countries are generating greater openness and balance, which is being reflected in the production of heavier sheep carcasses for export cuts worldwide. The aim of this study was to conduct a meta-analysis to identify differences with regard to the weight of sheep carcasses and their main characteristics obtained in production systems around the world, according to their genotype and geographic location. A meta-analysis of the types of sheep carcasses produced worldwide (64 studies, n=36,583 animals) was performed, using the databases Scopus and Web of Science. The data analysis was performed using the methodology of mixed method. The Kruskal-Wallis and Mann-Whitney tests were applied. The average daily gain (g/d) among genotypes varied ($P<0.05$), being higher for wool vs hair genotypes; the hot and cold carcass weight (kg), the hot carcass yield (%) and backfat thickness (mm) were higher ($P<0.05$) in animals supplemented with concentrate; the initial live weight (kg) was higher ($P<0.05$) in animals in the feedlot system, which also had a higher cold carcass weight. The free-range system ($P<0.05$) achieved a higher hot carcass weight compared with the feedlot system. The average weight gain, hot carcass weight and longissimus dorsi area (cm^2) was higher for animals situated in the southern hemisphere ($P<0.05$) compared with those in the northern hemisphere and intertropical area. We conclude that the world market shows a homogeneous production of carcasses, which have increased in weight regardless of the production and feeding system.

Species, breeds and potential for improvement in animal fibre production in Europe

H. Galbraith, M. Antonini, M. Gerken, J.P. Gutierrez Garcia, R. Niznikowski, C. Renieri, M. Wurzinger and D. Allain
EAAP Animal Fibre Working, Group (AFWG), c/o INRA, UMR 1388 GenPhySE, CS52627, 31326 Castanet-Tolosan, France; h.galbraith@abdn.ac.uk

Keratin-based fibre is a natural product of a range of animal species routinely farmed in Europe. The major species are wool-producing sheep, which include 62 m breeding ewes (Eurostat) with production (EU28), of both fleece and skin wool, of 186,000 tonnes (FAO), and small numbers of goats (Angora; Cashmere), South American camelids (SAC: Alpaca) and Angora rabbits. There is considerable variation in sheep populations, raw wool produced and profitability in different countries. Trends in wool production indicate reductions, including those noted in Central and Eastern Europe following political change, and frequently poor economic returns. While knowledge on the genetics and underpinning biology of keratin-fibre production has increased, there is limited production of fine fibre as a primary product. Sheep are kept for meat (mainly) or milk. In excess of 150 breeds have been identified with, for example, 73 breed societies listed by the British Wool Marketing Board. Sheep wools vary in colour, lustre and diameter: the finest Merino superfine at 16-21 micron to the coarser-wooled Blackface at 35+ micron. Concomitant uses range, with reducing fineness, from fine knitwear to knitting yarns, wool blends for apparel, blankets, Japanese futons, floor carpets and insulation products. Utilisation may be small-scale craft (including goat, rabbit and SAC fibres), or at larger scale industrial level. Some wool of poor quality may be discarded. Up to 90% of better quality may go for export. Finer qualities are typically imported from other countries and processed into high value products such as suiting and knitwear. Strategies to improve production, quality and utilisation in Europe will be explored.

Heritabilities for preweaning and its genetic correlations with other traits in Peruvian alpacas

A. Cruz[1], I. Cervantes[2], A. Burgos[1], R. Morante[1] and J.P. Gutiérrez[2]
[1]INCA TOPS S.A, Miguel Forga 348, Arequipa, Peru, [2]Universidad Complutense de Madrid, Departamento de Producción Animal, Avda/ Puerta de Hierro s/n, 28040 Madrid, Spain; gutgar@vet.ucm.es

The aim of the study was to estimate genetic parameters for preweaning traits and their genetic relationship with reproductive, productive and morphological traits in alpacas. The data was collected from 2001 to 2015 by Pacomarca experimental farm for Huacaya (HU) and Suri (SU) ecotypes. Birth Weight (BW), Birth Withers Height (BH), Weaning Weight (WW), Weaning Withers Height (WH), average Daily Gain (DG) and Survival to Weaning (SW) were analyzed. The genetic models included additive, maternal and residual random effects for all traits, and sex, color, number of calving and month-year as fixed effects. Age at weaning were used as linear covariate for WW and WH. The estimated direct heritabilities were, respectively in HU and SU, 0.097 and 0.131 for BW, 0.070 and 0.162 for BH, 0.497 and 0.336 for WW, 0.364 and 0.659 for WH, 0.447 and 0.201 for DG and 0.038 and 0.150 for SW. These figures for HU and SU maternal heritabilities were 0.248 and 0.376 for BW, 0.179 and 0.324 for BH, 0.289 and 0.392 for WW, 0.192 and 0.263 for WH, 0.265 and 0.355 for DG and 0.020 and 0.061 for SW. Since alpacas could be use for fiber or meat production, the genetic correlation values suggested that WW would be recommended to be used directly as a selection criterion under meat production objetive. However, under a scenario of fiber production it would be necessary to use a genetic index as a selection criterion to take into account the genetic correlations with fiber traits. Direct genetic correlations of preweaning traits with reproductive traits were favorable but unfavorable for maternal effects, especially for SU. Genetic correlations with morphological traits were high and positive for SU, but not important for HU, which suggests a spurious association for SU ecotype in the tested population.

Lifetime production of Istrian pramenka sheep in Slovenia

D. Kompan
University of Ljubljana, Biotechnical faculty, Department of Animal Science, Jamnikarjeva 101, 1000 LJUBLJANA, Slovenia; drago.kompan@bf.uni-lj.si

Istrian pramenka is the Slovenian autochthonous breed of sheep. It is used for milk production and lamb (dairy breed). The breed was developed in the area of the Karst and Istria, where shepherds kept the local, indigenous sheep, which they named istrijanka, Karst ewe, coastal sheep and even the coastal sheep of raw wool. In 2009, Slovenia had 1,150 animals of Istrian pramenka, among which 813 were ewes and 36 breeding rams. Production data, collected according to the ICAR rules by Istrian pramenka sheep born between 1994 and 2012 were used for the analysis of its lifetime production. Dataset included records of 1 290 ewes (5 565 lactations) from 10 flocks. The studied traits were lifetime milk yield production, lifetime fat yield production, lifetime protein yield production, number of lactations completed, number of born lambs in lifetime, age at first lambing and age at culling. Preparation of the data and statistical analysis were done by the statistical programme R and Excel. All ewes born to 1999 were culled. On average, these ewes had 3.4 lactations completed in their lifetime, 4.4 born lambs, 559.5 kg of milk yield, 36.4 kg of fat yield and 30.1 kg of protein yield. Ewes were on average 23.3 months old at first lambing and 76.6 months old at culling. Ewes born after 1999, had on average 3.0 completed lactations, 3.5 born lambs, 386.9 kg of milk yield, 26.8 kg of fat yield and 21.9 kg of protein yield during their lifetime. On average, ewes were 27.7 months old at first lambing and 70.1 months old at culling. Almost a half of the ewes were culled in the first and in the second parity. Even though the autochthonous Istrian Pramenka have relatively low lifetime production, the breed is important for our country due to its modesty, resistance and adaptability to relatively rough surroundings where it is bred.

Birthcoat type and lamb survival in the French Romane sheep breed

D. Allain[1], Y. Boudillon[2], D. Marcon[2] and D. François[1]
[1]INRA, UMR1388, GenPhySE, CS52627, 31326 Castanet Tolosan, France, [2]INRA, UE0332, Domaine expérimental de Bourges, La Sapinière, 18390 Osmoy, France; daniel.allain@toulouse.inra.fr

Birth coat type is an important component of lamb survival for sheep raised under harsh environment. At birth two types of coat were observed in the Romane breed: a long hairy coat or a short woolly one. It was shown that hairy coat lambs are more adapted to survive around lambing time due to a better coat protection with less heat losses at coat surface and show better growth performances up to the age of 10 days than woolly coat lambs. The aim of the present work was to quantify lamb survival and growth performances in relation to birth coat type in the French Romane breed raised under standard or conventional indoor housing from birth. A total of 8,657 lambs were used in a 6 years experiment from 2011. Birth coat type, lamb survival and growth performances were registered from birth to weaning. At birth two types of coat were observed: a hairy coat (39.2% of lambs) or a woolly one (60.8% of lambs). It was shown that hairy-bearing coat lambs had a slight better survival from birth to 50 days of age (3.0 vs 4.1%, 5.5 vs 6.4%, 6.8 vs 7.6% and 7.4 vs 8.2% for total mortality rate at 1, 10, 30 and 50 days, respectively) but only the difference in lamb survival at 1 day was significant (P<0.01). If at birth, any differences in weight were not observed between both birth coat types, significant differences (P<0.001) were thereafter observed on the live body weight. Woolly-bearing coat lambs were significantly heavier than hairy ones: +250 g and +710 g at 30 days of age and weaning (about 65 days) respectively. If selection to obtain a hairy birth coat type had advantages when the Romane sheep breed is raised under harsh and permanent outdoors conditions, it seems to have some significant reverse effects under conventional indoors housing. A hairy coat lamb survived a little better (with a significant survival rate around birth) than a wool coat one but had a significant lower weight at weaning. It could be suggested a genotype×environment effect involving birth coat type on both lamb survival and lamb growth when comparing standard housing and harsh outdoors conditions.

Genetic loci involved in antibody response to Mycobacterium avium ssp. paratuberculosis in sheep

F. Palazzo[1], G. Marras[1], E. Nicolazzi[1], E. Pieragostini[2], F. Petazzi[2], G. Gandini[3], G. Pagnacco[3], P. Crepaldi[3], J.L. Williams[4] and G. Minozzi[1,3]
[1]Parco Tecnologico Padano, Via Einstein, 26900 Lodi, Italy, [2]University of Bari, Piazza Umberto I, Bari, 70121 Bari, Italy, [3]University of Milan, DIVET, Via Celoria 10, 20133 Milan, Italy, [4]University of Adelaide, School of Animal and Veterinary Sciences, Faculty of Sciences, Roseworthy, 5371 Adelaide, Australia; giulietta.minozzi@unimi.it

Mycobacterium avium ssp. paratuberculosis (MAP) causes chronic enteritis in different ruminant species. In sheep, MAP causes a chronic disease called Johne's disease, or paratuberculosis. There have been several genome wide association studies over the last years that have identified genetic loci putatively associated with MAP susceptibility in cattle. The aim of this work was to perform a case-control study in the ovine species, using the 50K SNP panel (Illumina OvineSNP50 BeadChip) to unravel the genes underlying the mechanisms of susceptibility to Johne's disease in the ovine species. During 2011-2015 in the regions of Sicily and Apulia in Italy, more than 69.000 sheep have been tested for antibodies against MAP and blood sampled. In herds with high occurrence of Johne's diseases, three cohorts of animals designed for genome wide association studies, corresponding to three dairy sheep breeds (Sarda, Valle del Belice and Comisana) have been selected. All samples were classified based on the serum antibodies produced in response to MAP. This analysis refers to the first cohort, composed by the Sarda breed sampled in Apulia, consisting of a total 300 samples, 150 MAP ELISA positive and 150 negative. Whole genome association analysis was performed using the R package GenABEL with the Grammar-CG approach. The analysis identified several chromosomal regions associated with the disease status on chromosomes 1, 3, 5, and 13 in position 203971381, 70227376, 35140513 and 24755379 with high significance (P<5×10[-8]). These results provide evidence for genetic loci involved in the antibody response to MAP in sheep. Adding further knowledge of genetic variations related to susceptibility in all the other breeds, will facilitate the improvement of the health status in the ovine species.

Predicting intake from faecal chemical composition in pre-weaned lambs consuming milk and pellets

A.S. Danso[1,2], P.C.H. Morel[1,3], P.R. Kenyon[1,2] and H.T. Blair[1,2]
[1]Gravida, National Centre for Growth and Development, The University of Auckland, 1142 Grafton, Auckland, New Zealand, [2]International Sheep Research Centre, IVABS, Massey University, 4442 Tennet Drive, Palmerston North, New Zealand, [3]Monogastric Research Centre, IVABS, Massey University, 4442 Tennet Drive, Palmerston North, New Zealand; a.s.danso@massey.ac.nz

Predicting feed intake is difficult in suckling lambs consuming both milk and pasture. Feed intake values are typically predicted from values derived from those fed either milk only or solid feed only. Faecal chemical composition has been used to predict nutrient intake and digestibility in grazing lambs. The aim of the present study was to investigate if the feed intake of lambs fed a combination of milk and pellets can be predicted with sufficient accuracy from faecal chemical composition under controlled milk and pellet intake. A total of 34 lambs with live weight (LW) ranges from 9 kg to 18 kg were used in the study. Fifty-four faecal samples with detailed information about their chemical composition and the chemical composition of the milk and pellets consumed by pre-weaned lambs from two digestibility trials were used in the study. The lambs were bottle-fed milk replacer with or without access to pellets and kept in metabolic cages for four days. Records of feed intake, LW of the lambs and their total faecal collection were used to develop the prediction equations. Pellet dry matter intake (DMI_p, g/d) was predicted (R^2=0.72; RPE=39.28%; CCC=0.83) from faecal Neutral Detergent Fiber (NDF, %), pellet NDF (%), DM (%) and LW (kg). Milk dry matter intake (DMI_m, g/d) was predicted (R^2=0.96; RPE=3.48%; CCC=0.98) from faecal Nitrogen (%) and LW. Milk and pellet DMI and their metabolisable energy (ME) content were combined to predict DMI / d (R^2=0.88; RPE=9.93%; CCC=0.93), and ME intake / d (R^2=0.93; RPE=6.53%; CCC=0.96). The results indicated that the equations developed can be used with enough precision to estimate milk intake and pellet intake in pre-weaned lambs consuming both milk and pellets, thus providing a means for farmers to develop feeding strategies for young lambs.

Effect of three rearing systems on ewe milk production and lamb growth in dairy sheep

M. Margetin[1,2], M. Oravcová[2], O. Bučko[1], J. Margetínová[2] and J. Huba[2]
[1]Slovak University of Agriculture Nitra, Tr. Andreja Hlinku 2, 949 76 Nitra, Slovak Republic, [2]National Agricultural and Food Centre, Research Institute for Animal Production Nitra, Hlohovecká 2, 951 41 Lužianky, Slovak Republic; margetin@vuzv.sk

The objective of this study was to analyze the growth intensity of lambs as well as milk traits (weaning, milking period and lactation) of ewes of synthetic Slovak Dairy sheep population in dependence on three rearing systems. A total of 211 ewes and 314 lambs were included. First, lambs were weaned at an age of two to three days postpartum. Lambs were raised artificially; ewes were machine milked twice daily (ARM2). Second, lambs were reared traditionally in nurseries. Since two weeks postpartum lambs were not allowed to suckle twelve hours before milking of ewes, however, were allowed to access the mammary gland in the remaining day time i.e. in regular intervals for a limited period of time. Ewes were machine milked once daily at 6.30 a.m. (TRM1). Third, lambs were reared traditionally in nurseries and allowed to access the mammary gland in regular intervals for a limited period of time daily. Ewes were not milked (TRM0). Weaning milk yield (55.1±1.80 l), fat yield (2.69±0.080 kg), protein yield (2.72±0.080 kg) and fat content (4.83±0.086%) in ARM2 ewes were significantly higher (P<0.0001) than milk yield (20.4±1.63 l), fat yield (0.70±0.080 kg), protein yield (1.15±0.076 kg) and fat content (3.20±0.084%) in TRM1 ewes. Differences in protein content and somatic cell score were insignificant. Milking period yield and daily milk yield were insignificant when comparing ARM2, TRM1 and TRM0 ewes. The total of 215 d lactation milk yield (TMY) and useful dry matter (UDM) (200.0±5.17 l and 23.80±0.657 kg) in ARM2 ewes, the 219 d TMY and UDM (178.6±5.55 l and 22.21±0.644 kg) in TRM1 ewes and 220 d TMY and UDM (153.9±5.3 l and 18.58±0.673 kg) in TRM0 ewes differed significantly (P<0.05). Weight at weaning (17.35±0.292 kg) and average daily gain (0.264±0.0053 kg) in TRM1 lambs were insignificantly lower (P>0.05) than weight at weaning (18.03±0.290 kg) and average daily gain (0.275±0.0052 kg) in TRM0 lambs. The study was supported by the projects APVV 0458-10 and VEGA 1/0364/15.

Sexually-active rams and bucks as sustainable tools for out-of-season reproductive control

J.A. Delgadillo[1], J.A. Abecia[2], M. Keller[3], L. Zarazaga[4], H. Hernández[1], M. Chasles[3] and P. Chemineau[3]
[1]CIRCA, Univ Aut Agr Antonio Narro, 27054 Torreón, Mexico, [2]IUCA, Prod Animal, Univ Zaragoza, 50013 Zaragoza, Spain, [3]Phys Reprod Comport, INRA Univ Tours, 37380 Nouzilly, France, [4]C Agroforestales, Univ Huelva, 21819 Huelva, Spain; philippe.chemineau@tours.inra.fr

Sheep and goats from temperate and subtropical latitudes display reproductive seasonality, which is mainly controlled by photoperiod. Hormonal treatments are used to induce out-of season reproduction, although these hormonal treatments raise questions about their effects on environmental pollution and animal health and welfare. In addition, there are no free-hormone techniques to advance puberty or reduce the postpartum anestrus (PPA) duration, limiting the productivity of females. Here we describe cheap, simple and sustainable methods for controlling out-of-season reproduction, based on the presence of males rendered sexually active by appropriate photoperiodic treatments. Sexually active bucks were much more efficient than those in seasonal sexual rest in inducing ovulation and estrous behavior in seasonally anovulatory goats (Mexico: 1; France: 2; Spain: 3). Goats in permanent presence of vasectomized sexually active males cycled permanently during two consecutive anestrous seasons (4), whereas Mediterranean ewes prolonged ovarian activity and increased estrous activity in spring (5). The presence of sexually active males advanced puberty in goat-kids (6) and autumn-born ewe-lambs (7), and reduced PPA in adult females during anestrus (ewes: 8) or the breeding season (goats: 9). In conclusion, the presence of sexually active rams and bucks is a very efficient way to override the photoperiodic control of female sexual activity. These findings provide a new and sustainable approach to the reproductive control of small-ruminant flocks, without the use of hormonal treatments.

Mechanistic model to analyse the effect of body weight and milk production in dairy sheep

J.C. Angeles Hernandez[1,2], S.S. Radic Schilling[3], M.A. Vera Arias[3], R.A. Echeverria Perez[3], O.A. Castelan Ortega[4], A.H. Ramirez Perez[1] and M. Gonzalez Ronquillo[4]
[1]UNAM, Depto. de Nutrición Animal y Bioquimica, Ciudad Universitaria, 04510, Mexico, [2]UNAM, Programa de Maestría y Doctorado en Ciencias de la Salud y la Producción Animal, Ciudad Universitaria, 04510, Mexico, [3]Universidad de Magallanes, Depto. de Ciencias Agropecuarias y Acuícolas, Avenida Bulnes, 01855, Chile, [4]FMVZ, UAEMex, Depto. de Nutricion Animal, Instituto Literario Ote. No. 100, Toluca, México, 50000, Mexico; sergio.radic@umag.cl

During late pregnancy and early lactation utilization of nutrients is less efficient and feed intake decreases, hence metabolic reserves are paramount to successful lactation. The aim of the current study was to analyse the effect of sheep body weight at mating (MAT), mid pregnancy (MP), lambing (LAM), early lactation (ELAC) and late lactation (LLAC) on milk yield using a mechanistic model of lactation curve. Fifty-two multiparous East Friesian ewes from an experimental flock at University of Magallanes in Chile, were used. Ewes were assigned to three different groups according to body weight at each productive stage: Low (LBW, n=13), Moderate (MBW, n=26) and High body weight (HBW, n=13). Lactations were fitted using the model described by Pollott (2000). Total milk yield (TMY), peak yield (PY) and time at peak yield (TPY) were also calculated. Ewes with HBW showed significantly higher values (P<0.001) of TMY than LBW and MBW when measured at MAT (145.0 vs 113.2 and 77.6 l), MP (144.4 vs 11.8 and 78.6 l), LAM (143.5 vs 121.2 and 81 4 l), ELAC (132.5 vs 115.3 and 82.2 l) and LLAC (129.4 vs 115.6 and 107.1 l, respectively). Ewes with HBW had higher PY at MAT (1.39 vs 0.97 and 1.17 l), MP (1.52 vs 0.96 and 1.15 l) and LAM (1.35 vs 0.95 and 1.26 l, respectively) compared to LBW and MBW ewes (P<0.05). The TPY occurred significantly earlier (P=0.04) in HBW (29.6) than LBW (33.4) and MBW ewes (37.9 days). HBW ewes showed the highest values of maximum secretion parameter of Pollott model (2.37 vs 1.57 LBW and 1.69 l MBW), which could partially explain the higher milk yields in HBW ewes. The ewes's body weight in pregnancy and early lactation is useful indicator at farm level to improve the milk yield performance in dairy sheep.

Intake and total phenolic content of milk of goat as affected by the forms of two-stage olive cake

G. Keles[1], F. Akgul Yıldız[2] and S. Unubol Aypak[3]
[1]*Adnan Menderes University, Faculty of Agriculture, Deparment of Animal Science, 09100, Aydın, Turkey,* [2]*Adnan Menderes University, Faculty of Agriculture, Deparment of Dairy Technology, 09100, Aydin, Turkey,* [3]*Adnan Menderes University, Faculty of Veterinary Medicine, Deparment of Biochemistry, 09100, Turkey; gurhan.keles@adu.edu.tr*

Quality of dairy goat milk may be improved through feeding two-stage olive cake (SOC) that is rich in total phenolic compounds and ether extract. Feeding value and potential benefit of SOC on milk quality may vary depending on form of SOC used in diets. The present study investigated if silage or dried forms of SOC can affect the performance and milk quality of lactating dairy goats. A total of 15 Saanen goats (mid-lactation) divided into three groups and were individually fed total mixed rations (TMR) that contain either dried or silage forms of SOC at DM proportions of 0.0 (C, control), 0.17 (DSOC, dried SOC) and 0.17 (SOCS, silage of SOC). The SOC was replaced with concentrate in TMRs contained at DM proportions of 0.45 roughages. The TMRs were isonitrogenous, but the energy levels were decreased by inclusion of SOC in TMRs. The experiment continued four weeks and experimental data was taken last week. The data were analyzed in a completely randomized design using a model that accounted for the main effects of form of SOC. Dry matter intake (DMI) of goats fed SOCS was greater (P<0.05) than goats fed DSOC, but dairy goat on either SOCS or DSOC had higher (P<0.001) NDF intake. Dairy goats that were on either SOCS or DSOC had lower (P<0.05) digestible DM and NDF values than the control group, but intake of digestible NDF per 100 kg body weight was similar (P>0.05) among the three groups. Milk yield decreased (P<0.05) by feeding either forms of SOC, but energy corrected milk yield did not differ (P>0.05) among groups. Total phenol content of milk was correlated to their dietary level, namely increased (P<0.001) by feeding either forms of SOC. The results indicated that total phenolic content of milk of dairy goat can be improved by feeding two-stage SOC, regardless of its form, but using either SOCS or DSOC that increases the NDF in TMRs fed to dairy goat could adversely affect the intake as goats consumed similar digestible NDF.

Use of olive polyphenols in the diet of dairy sheep: effects on milk and rumen liquor FA composition

A. Cappucci[1], S. Alves[2], R. Bessa[2], G. Conte[1], A. Serra[1] and M. Mele[1]
[1]*University of Pisa, Dipartimento di Scienze Agrarie, Alimentari e Agro-ambientali, Via del Borghetto 80, 56124 Pisa, Italy,* [2]*University of Lisbon, Department of Animal Production and Food safety, Polo universitario do Alto da Ajuda, 1300-477 Lisboa, Portugal; alice.cappucci@for.unipi.it*

During the next 30 years, meat and milk production demand will increase. At the same time, livestock systems are requested to reduce the use of natural resources and the consumption of human food. A valuable approach to this problem may be an increase of the use of agro-industrial by-products, which are sources of important nutrients and plant metabolites. In the Mediterranean area, the olive pomace represents an interesting opportunity for ruminant feeding. However, it contains significant amounts of polyphenols, which are able to interfere with rumen microbes. Aim of the study was to evaluate the effect of increasing amounts of polyphenols from olive pomace (OP) in the diet of dairy sheep on rumen liquor and milk fatty acid (FA) composition. Twenty dairy ewes were allotted into four groups after a baseline period of three weeks, when the animals received the same dietary regimen based on alfalfa hay and concentrate feed. The four experimental diets were based on alfalfa hay and 800 g/head/day of a concentrate feed containing: linseed (L diet) or linseed plus different quantity of OP to provide 0.4 (L0.4), 0.8 (L0.8) or 1.2 (L1.2) g/kg DM of polyphenols. The use of OP at the maximum dose, resulted in a significant increase of linoleic and alpha-linolenic acid in milk fat by 18% and 24%, respectively. At the same time, milk fat from ewe fed L1.2 diet was lower in t11 18:1, CLA content (-38% for both FA) and other intermediates of the biohydrogenation process. The same patter of FA composition was also observed in the rumen fluid. Moreover, in the rumen fluid a significant increase of dimethylacetal (DMA, alk-1-enyl (vinyl) ether chains), which are plasmalogenic lipids contained in the cell membranes of rumen bacteria, was observed. Further studies are needed in order to evaluate the role of DMA as a marker of rumen microbiome composition.

Production performance of ram and castrated hill lambs offered a range of diets

A. Aubry[1], D. Irwin[1], M. Connolly[1], J. McKinnie-Hill[2], T. Hagan[2], L. Farmer[2], S. Watson[2] and F. Monahan[3]
[1]Agri-Food and Biosciences Institute, Agriculture branch, Large Park, Hillsborough, BT26 6DR, United Kingdom,
[2]Agri-Food and Biosciences Institute, 18a Newforge Lane, Belfast, BT9 5PX, United Kingdom, [3]University College
Dublin, School of Agriculture and Food Science, Belfield, Dublin 4, Ireland; aurelie.aubry@afbini.gov.uk

There is much debate within the sheep meat industry about the relative merits of keeping male lambs entire or having them castrated. The objective of this study was to determine the role of gender and finishing diet outside of the main grazing season (i.e. October-January) on animal growth and carcass characteristics. A total of 138 hill lambs of two breed types (Blackface × Swaledale cross or Swaledale × Blackface cross (BS x) vs SuffTex cross lambs (SuffTex x)) and 2 genders (castrated at birth or entire) were finished on three outdoor lowland diets (grazed grass, stubble turnip, forage rape) and three indoor diets for BS × lambs only, fed ad lib (grass silage or two different concentrates plus silage at 250 g/lamb/day) from October to slaughter (slaughtered at 50, 64 and 92 days). Similar growth rates were observed between breed types (P>0.05), but kill out percentages were lower (P<0.001) for BS × lambs (40.5%) than SuffTex × lambs (43.9%). Clear gender effects were found, with greater live weight gains (P<0.05) for ram lambs (142 g/day) than castrated lambs (120 g/day), but higher kill out percentages for castrated lambs (P<0.001), resulting in similar carcass weights for both genders. Fat grades were lower for ram lambs (P<0.001) but conformation scores were similar for both genders. Highest growth rates were achieved in lambs fed concentrates (P<0.001) (196 and 216 g/day for each of the two concentrate diets, BS × lambs) as well as best carcass attributes, followed by those fed grass silage, stubble turnip or forage rape (111 to 117 g/day). The clear differences found between breeds, genders and diets can help to develop cost-effective management strategies best suited to specific production systems. In particular, the study indicates that performance of lambs grazing brassicae can be similar to lambs fed indoors on good grass silage-based diets.

Effect of eCG and GnRH administration on reproductive performance of Khalkhali goat

V. Vahedi[1] and H. Abdi Benemar[2]
[1]Moghan College of Agriculture and Natural Resources, University of Mohaghegh Ardabili, Department of Animal Science, Ardabil, 5697194781, Iran, [2]College of Agriculture Science, University of Mohaghegh Ardabili, Department of Animal Science, ardabil, 1136756199, Iran; vahediv@uma.ac.ir

Increment of ovulation rate in farm animals to improve reproductive performance by increasing in litter size and fecundity rate. The objective of this study was the improvement of reproductive performance of Khalkhali goats used GnRH and equine chorionic gonadotrophin (eCG) administration during estrous synchronization in breeding season. A total of 150 goats within 2-5 years old and weighting 37-39 kg in completely randomized design were allocated into three groups. In order to synchronize estrous cycle, all does were treated for 16 days with a controlled internal sponge devise. The control group (T1) exposed to Khalkhali bucks without any treatments for mating after sponge removal. Goats in second (T2) and Third (T3) groups were injected intramuscularly 400 IU eCG and 400 IU eCG with 2.5 ml GnRH respectively in sponge removal time and then the does were mated. The data of reproductive parameters were analyzed by ProcGenmod and data collected for born weight were analyzed using the GLM procedure of SAS. There was no effect of hormone treatment (eCG and eCG+GnRH) on the kidding rate and pregnancy rate (P>0.05) in breeding season. Control group had lower Fecundity rate (106%) than the treatment groups (158 and 188% in second and third groups respectively). Lambing rate in 1 to 3 treatments were calculated 113, 175 and 177% respectively that the treatment groups had higher (P<0.05) Lambing rate compared to the control. Litter size in groups 2 and 3 (49% and 56% respectively) were greater (P<0.05) than the control as well. There was no significant difference in born weight of kids between three groups (P>0.05). In conclusion, giving eCG and GnRH concurrent with removal of the sponge containing progesterone increased fecundity rate, kid production efficiency and twinning rate. Therefore, using of synthetic hormone caused improvement of reproductive efficiency of Khalkhali goat in breeding season.

Effect of polyethylene glycol 4000 supplementation on the performance of indigenous Pedi goats
D. Brown, J.W. Ng'ambi and D. Norris
University of Limpopo, Animal Production, Private Bag X1106, Sovenga, 0727, South Africa; db4010396@gmail.com

Acacia karroo is widely spread and abundantly available during the dry season and has potential to increase productivity of indigenous Pedi goats feeding on low quality roughages. The only limitation to its use is the presence of tannins in their foliage. The negative effects posed by high tannin contents must be overcome to improve the feeding value of the feeds. An experiment was carried out to determine the effect of polyethylene glycol (PEG) 4000 supplementation on the performance of indigenous Pedi goats. Eighteen yearling male indigenous Pedi goats with mean live weight of 21.76 ± 3.1 kg were randomly allocated to six treatments in a 2 (Acacia karroo leaf meal levels) × 3 (levels of PEG 4000) factorial arrangement in a completely randomised design for a period of 21 days. Each treatment had three replicates with one goat per replicate. Each goat was supplemented with 0, 23 or 30 g of PEG 4000/litre of water per day in addition to dietary mixture of S. verticillata hay and A. karroo leaf meal inclusion levels as follows: $A_{20}S_{80}P_0$, $A_{20}S_{80}P_{23}$, $A_{20}S_{80}P_{30}$, $A_{50}S_{50}P_0$, $A_{50}S_{50}P_{23}$ and $A_{50}S_{50}P_{30}$. Goats supplemented with 23 g of PEG 4000 and fed a dietary mixture of 20% A. karroo and 80% S. verticillata grass hay had higher (P<0.05) dry matter (DM), organic matter (OM), neutral detergent fibre (NDF) and acid detergent fibre (ADF) intakes than those on a mixture of 50% A. karroo and 50% S. verticillata grass hay and supplemented with either 23 or 30 g of PEG 4000. Dry matter, OM, NDF and ADF intakes per goat were optimized at PEG 4000 supplementation levels of 19.62, 19.62, 19.61 and 19.53 g/goat/day, respectively, for diets with a mixture of 20% A. karroo and 80% S. verticillata grass hay. Goats supplemented with either 23 or 30 g of PEG 4000 and fed 20% A. karroo and 80% S. verticillata grass hay had higher (P<0.05) DM, OM, CP and NDF digestibility values than those on a mixture of 50% A. karroo and 50% S. verticillata grass hay and supplemented with 23 g of PEG 4000. Crude protein digestibility was optimized at a PEG 4000 supplementation level of 15.78 g/goat/day. Results indicate that PEG 4000 has potential to improve the feeding value of A. karroo and can be used in the feeding systems for ruminants.

Intake and digestibility of lambs fed cassava silage with dry tamarind residue
C.M. Souza[1], R.L. Oliveira[1], M.S. Borja[1], C.B. Pellegrini[1], N.J.A. Dos Santos[1], D.R. Menezes[2] and T.V. Voltolini[3]
[1]Federal University of Bahia, Av. Adhemar de Barros 500, 40170110 Salvador Bahia, Brazil, [2]Federal University of São Francisco Valley, Rodovia 407, 12, 6300000, Petrolina, Pernambuco, Brazil, [3]Embrapa Semi-Arid, BR-428, Km 152, 56302970 Petrolina, Pernambuco, Brazil; ronaldooliveira@ufba.br

This study was conducted to evaluate the effects of cassava silage added to different levels of dry tamarind residue in the intake and digestibility in lambs. The Experiment lasted 72 days, were used 40 castrated male lambs, vaccinated and wormed. Lambs were six months of age, of an undefined breed, and had an average body weight of 22.09 ± 2.87 kg. were fed silage made from the upper third of cassava as an additive with dried tamarind residue and a concentrate composed of soybean meal and corn meal. Mineral salt and water were available ad libitum. The treatments were characterized by the inclusion of 0, 10, 20 or 30% dry tamarind residue levels included with the upper third of cassava as roughage. The chopped forage was in the form of meal ground through a 3-mm sieve. The silage from the upper third of cassava contained the additive. The experiment uses a completely randomized design that had 4 treatments and 10 repetitions. The inclusion of dry tamarind residue in the silage from the upper third of cassava resulted in a linear increase in the intake of DM (P=0.0034), CP (P=0.0035), NDF (P<0.0001) and NFC (P=0.0229). In contrast, there was a reduction in the EE intake (P=0.0056) with the inclusion of dry tamarind residue. A linear trend was observed (P=0.0775) in the intake of NDT with the inclusion of dry tamarind residue in cassava silage. There was an increasing linear effect on DM (P=0.0005) and NDF (P<0.0001) intake relative to BW. There was a positive quadratic effect on the digestibility of dry matter (P=0.0017), crude protein (P=0.0057) and NDF (P=0.0203). The digestibility coefficients of NFC (P=0.1643) and EE (P=0.3208) were not influenced by levels of dry tamarind residue inclusion. The 30% inclusion level is recommended because it contributes to the increase in dry matter intake and, consequently, an increase in the intake of other nutrients. It also increases the digestibility coefficients.

The Icelandic heritage goat breed: conservation supported by utilization

O.R. Dyrmundsson
Agricultural Consultant, Jorusel 12, 109 Reykjavik, Iceland; oldyrm@gmail.com

The Iceland breed of goats is an isolated, native breed of Nordic origin, dating back to the settlement of Iceland over 1,100 years ago. The goats have been kept, mainly in small numbers, often as pets, in all parts of the country. The national population reached the maximum of 3,000 head in 1930. They were on the verge of extinction in the late 19th century and again by 1960 when just over 100 breeding goats remained. Since 1965 annual state conservation grants have been available for individually recorded goats, now amounting to 40 euros per head. This subsidy, one of the first of its kind in Europe, has beyond doubt contributed substantially to both increased goat and herd numbers, 900 and 90, respectively. This development has been supported by the Farmers Association of Iceland, for example, through advisory and recording services and by the publication of a poster illustrating the diversity of colours and other breed characteristics. Furthermore, since 1991, the Goat Breeders Society of Iceland has promoted goat products such as meat, milk, skins and fine fibres of Cashmere quality. In recent years the Agricultural Genetic Resources Committee has also contributed to the revival of this endangered heritage breed, for example, by arranging AI services. Now, in 2016, special efforts will be made through a new State Agricultural Support Scheme to promote goat production as a market orientated farm enterprise. Although the population is highly inbred there is some scope for genetic selection and together with better management productivity is likely to be enhanced. Recently the Iceland goat breed received Slow Food recognition (Ark of Taste, Presidium). It is hoped that the generation of income from the marketing of quality goat products will strengthen utilization of the breed and thus continued conservation efforts in the future.

Comparative analysis of growth curves of Awassi lambs fed different protein diets

P.Y. Aad[1], R. Makdessi[2], R. Ghorayeb[2], K. Houchaymi[3] and S. Abi Saab[2]
[1]Notre Dame University-Louaize, Sciences, Zouk Mosbeh, Lebanon, [2]Holy Spirit University of Kaslik, Agricultural Sciences, Lebanon, [3]Lebanese Agricultural Research Institute, Terbol, Lebanon; paad@ndu.edu.lb

Awassi sheep, the local fat-tail breed of the Mediterranean region, is used for its ability to withstand long dry summers. The purpose of this study was to observe the response of lambs from weaning to puberty under harsh or enriched conditions. Therefore 96 newly-weaned male and female Awassi lambs were divided into 3 groups with low (LP), normal (MP) or high (HP) crude protein (CP) levels in isocaloric rations for 4 months. Body Weight (BW), body circumference (BC) and height (BH) at the point of the withers and fat-tail circumference (FC) and volume (FV) were measured bi-weekly. Data were analyzed using SPSS20 and presented as Means ± SD. Results showed a similar growth trend in both males and females of all groups, with males reaching 1.2 fold higher ($P<0.05$) weights by week 13 than females, especially those in HP and MP. Further, feed restriction showed a decrease in growth and fat reserves in both males and females, with HP and MP groups doubling ($P<0.05$) their BC by week 16. While LP BC remained the same as week 0. Interestingly, both male and females groups stored reserves in their fat tails from week 0 to 5, afterwhich HP and MP females showed a decrease ($P<0.05$) in the fat circumference and volume until week 13 when these reserves were restored until week 16. HP and MP males showed a decrease ($P<0.05$) in fat reserves until week 15 and slightly restored reserves until week 16. These events align with the puberty onset in the HP and MP groups, showing the high energy requirements and the importance of the fat tail in supplementing the feeding of growing lambs. Unfortunately, LP lambs showed a constant decrease in body weight and fat tail volume, to a stronger extent in males than in females. Most of these lambs showed a delayed puberty onset, many of them not showing any of these signs by week 16. In conclusion, low pasture during growth period of lambs under transhumance will not promote puberty onset in that same season and will further delay productivity and reproductive capacity.

Factors influence the behavioural parameters displayed during milking in dairy ewes

P. Simitzis, I. Seferlis, M. Goliomytis, I. Bizelis and I. Politis
Agricultural University of Athens, Animal Breeding and Husbandry, 75 Iera Odos, 11855, Greece;
pansimitzis@aua.gr

The interaction of farmer with dairy animals during milking could influence their behaviour, physiology and productivity. An experiment was therefore conducted to determine the effects of breed, age, milking frequency and human handling on behavioural parameters displayed during milking in dairy ewes. Thirty-two ewes of Chios and Karagouniko breed were used; half of them was primiparous and the rest had already conducted two gestations (multiparous). Animals were assigned to two groups; milked twice or once daily for 2.5 months after lamb weaning. The feeding regime and the handlings were similar between the groups throughout the experiment. Milk yield, milking rate and number of kick responses during milking were weekly recorded. Each milking was randomly conducted by an aversive or a friendly handler. The position in milking parlour was also recorded. At the end of the first and second month, an isolation – flight distance test was also implemented to estimate the emotional state of animals. Results indicated that Chios had higher milk yield but lower milking rate compared to Karagouniko ewes. Ewes milked twice per day also produced more milk than those milked once. No effect of ewe age was demonstrated. The presence of an aversive handler decreased milk yield and increased the number of kick responses during milking. A preference of ewes for a specific position in the milking parlour was also recorded. During the isolation – flight distance test, Chios ewes bleated more, but had shorter flight distances compared to Karagouniko ones. Flight distance was also shorter in multiparous than primiparous ewes. Finally, ewes milked twice per day displayed more jumps and less bleats compared to those milked once. The significant effects found in the present study indicate the possibility of applying specific treatments with the intention to reduce fear responses and improve productivity in dairy ewes.

Effect of grouping hour on agonistic behaviour of goats after introduction of unfamiliar individuals

M. Almeida[1], J. Magalhães[2], C. Araujo[2], J. Almeida[1], G. Stilwell[3] and S. Silva[1]
[1]*University of Trás-os-Montes and Alto Douro, CECAV, Animal and Veterinary Research Centre, P.O. Box 1013, 5000-801 Vila Real, Portugal,* [2]*University of Trás-os-Montes and Alto Douro, Biology Student, P.O. Box 1013, 5000-801 Vila Real, Portugal,* [3]*Faculty of Veterinary Medicine, Technical University of Lisbon, Animal Behaviour and Welfare Lab., CIISA, Avenida da Universidade Técnica, 1300-477 Lisboa, Portugal; m_dbalmeida@hotmail.com*

This study hypothesised that unfamiliar goats introduced in a stable group in the morning or at night show different levels of agonistic behaviours. Fifty-six Serrana breed goats were randomly organised into four groups of 9 animals (G9) and four groups of 5 animals (G5). Each group was housed in similar deep-bedded straw pens and had ad libitum access to hay offered twice daily. To verify the hypothesis, the goats were introduced at 10:00 h (GH10) and at 18:00 h (GH18). For both grouping hours, a G5 was grouped with an established group G9. The experiment lasted for 5 days: two days previous to grouping (D-2 and D-1); grouping day (D0); and two days after grouping (D+1 and D+2). During this period, the goats' behaviour was continuously video recorded and five agonistic behaviours were noted: frontal clashing-FC, butting-B, chasing-C, threat-T and avoiding-A. The agonistic behaviours were analysed using a model of analysis of variance with grouping hour and day as factors. FC, B, C behaviours were more intense at GH10 than at GH18 ($P<0.05$). Regarding day, all agonistic behaviours were affected by this factor ($P<0.001$). Agonistic behaviours increased 3.5 to 10.4 times and 1.5 to 2.4 times the days before and after grouping, respectively, comparing with D0. It was also observed a significant interaction between grouping hour and day for all behaviours ($P<0.001$), except A ($P=0.125$). From the results obtained in the present study, we can conclude that the introduction of unfamiliar goats into a stable group at night leads to less frequent agonistic behaviours.

Is the Qualitative Behaviour Assessment a suitable method to assess positive emotions in pigs?

I. Czycholl and J. Krieter
Institute of Animal Breeding and Husbandry, Christian-Albrechts-University, Olshausenstraße 40, 24118 Kiel, Germany; iczycholl@tierzucht.uni-kiel.de

Positive emotions constitute an important part of animal welfare. However, it is probably also the most difficult part to assess. In the Welfare Quality® protocols for cattle and pigs which aim at evaluating animal welfare, a Qualitative Behaviour Assessment (QBA) with a fixed list of adjectives was included as the animal-based parameter for the assessment of positive emotions. Studies concerning the reliability of the finally included versions of the QBA are rare. Therefore, the aim of the present study was to assess the reliability of the QBA as included in the Welfare Quality® protocol for growing pigs in its practical implication. First, interobserver (IR) and retest reliability (RR) were analysed based on 109 video sequences: For the IR, three trained observers completed QBAs on these sequences and for the RR, one of the observers assessed the sequences twice. Afterwards, reliability was assessed on-farm: For the evaluation of IR, two observers completed 19 QBAs on German growing pig farms at the same time and on the same animals, but completely independent of each other. RR was evaluated by the repeated assessment of 22 farms in two consecutive growing periods. The results were compared between the observers and repeated observations, respectively. The statistical analysis was carried out in accordance with Welfare Quality® and literature by means of a principal component analysis. Thereby, two principal components (PC) were extracted explaining for the greatest part of variance. The factor loadings of PC1 and 2 were compared between the observers or repeated observations by the calculation of the Spearman's rank correlation coefficient (RS). The results proved a sufficient reliability in the study based on video sequences (IR: RS: PC1: 0.67, PC2: 0.60, RR: RS: PC1: 0.54, PC2: 0.47), but not in the on-farm use of the QBA (IR: RS: PC1: -0.16, PC2: 0.13, RR: RS: PC1: -0.28, PC2: 0.03).

Oxytocin candidate genes and maternal behaviour of sows

L. Rydhmer and E. Jonas
Swedish University of Agricultural Sciences, Dept Animal Breeding and Genetics, Box 7023, 75598 Uppsala, Sweden; lotta.rydhmer@slu.se

Oxytocin affects calmness, milk let-down and maternal behaviour. But measuring the hormone level in blood is difficult as frequent sampling is required. In this study we therefore decided to investigate candidate genes for oxytocin and behaviour, instead of measuring blood levels of oxytocin. One hundred Swedish Yorkshire sows in pens without crates were studied during farrowing and lactation at an experimental farm. Behaviour traits were recorded with a questionnaire answered by the staff at d 1, d 5 and d 21. It included nest building, sows' carefulness and reaction to piglets, anxiety, fear of humans and aggression. Farrowing and nursing were recorded with video cameras. Sow body weight and body condition during lactation, piglet growth and survival were also recorded. Nine genes related to maternal behaviour and the regulation of oxytocin were identified in the literature: Fosb and Peg3 on chromosome (SSC) 6, CD38 on SSC8, AVPR1B and Grb10 on SSC9, Oxyr on SSC13, Oxy and AVP on SSC17 and Mest on SSC18. Fragments from these genes were sequenced to identify SNP markers. These SNP and some known SNP located close to the candidate genes were used to genotype the sows. There was genetic variation in all candidate genes; 31 out of 45 markers segregated. Associations between markers and behavioural traits have been analysed using generalised linear models. There is phenotypic variation between sows with regard to nest building, being careful with newborn piglets, anxiety, and reaction to piglets when returning from weighing. Aggression towards humans and anxiety are repeated from day 1 to day 21. It was found that 29% of the sows had no colostrum in the udder on the day before farrowing. We have identified significant associations between some genes and some behavioural traits. Further investigations are ongoing.

The added value of molecular phenotypes: towards the identification of animal welfare proxies

L. Fontanesi[1], S. Bovo[1], G. Schiavo[1], G. Mazzoni[1], A. Ribani[1], V.J. Utzeri[1], S. Dall'Olio[1], F. Bertolini[1], F. Fanelli[2], M. Mazzullo[2], G. Galimberti[3], D.G. Calò[3], P.L. Martelli[4], R. Casadio[4], P. Trevisi[1], P. Bosi[1] and U. Pagotto[2]
[1]University of Bologna, Department of Agricultural and Food Sciences, Division of Animal Sciences, Viale Fanin 46, 40127 Bologna, Italy, [2]University of Bologna, Department of Surgical and Medical Sciences, Endocrinology Unit, via Massarenti 9, 40138 Bologna, Italy, [3]University of Bologna, Department of Statistical Sciences 'Paolo Fortunati', Via delle Belle Arti 41, 40126 Bologna, Italy, [4]University of Bologna, Department of Biological, Geological, and Environmental Sciences, Biocomputing Group, Via San Giacomo 9/2, 40126 Bologna, Italy; luca.fontanesi@unibo.it

Metabolomics can detect and quantify hundreds of metabolites that constitute internal (or molecular) phenotypes that could be useful to dissect complex and difficult traits, like welfare state and disease resistance of the animals. Metabotypes are referred as phenotypes defined by the level of metabolites in a biological fluid. Metabotypes can be influenced by environmental and genetic factors. In this study we showed that genetic variants affect the level of many metabolites in pigs, leading from minor (even if detectable) to relevant modifications of metabolomic profiles. Genetically influenced metabotypes or GIM were used to dissect at the genome level biochemical pathways that may contribute to define behaviour of the animals or to cope with stressful conditions. We determined about 200 plasma metabotypes in two performance tested pig populations: Italian Large White (~1000 animals) and Italian Duroc pigs (~400 animals). These animals were genotyped with the Illumina PorcineSNP60 BeadChip. Genome wide association studies identified a large number of significant regions affecting metabotypes in both breeds. Several metabotypes from the same metabolite families were influenced by the same genomic regions providing useful information to characterize unknown metabolomic pathways. Metabolites relevant in behaviour were affected by mutations in genes coding for proteins or enzymes involved in neurological functions. Several metabolites could be important proxies to define animal welfare states and to design new breeding concepts indirectly including welfare biomarkers in selection plans.

Genetic Analysis of Feather Pecking and Mortality in Laying Hens

A.R. Sharifi[1], C. Reimer[1], N.T. Ha[1], M. Erbe[2], D. Cavero[3], R. Preisinger[3] and H. Simianer[1]
[1]Animal Breeding and Genetics Group, Department of Animal Sciences, Albrecht-Thaer-Weg 3, 37075 Göttingen, Germany, [2]Bavarian State Research Centre for Agriculture, Vöttinger Straße 38, 85354 Freising, Germany, [3]Lohmann Tierzucht, Am Seedeich 9, 27472 Cuxhaven, Germany; rsharif@gwdg.de

Feather pecking and cannibalism are major welfare problems in laying hens resulting in high mortality and an adverse effect on the egg-laying performance particularly in non-cage housing systems. Genetic improvement of socially-affected traits via individual selection may result in unfavorable selection response. Therefore, a breeding strategy suitable for such traits could be a group selection based on group average plumage condition as auxiliary trait, rather than direct selection for feather pecking which is not easily measurable. The aim of this study was to estimate the genetic parameters for plumage condition, mortality and egg performance in laying hens, housed in a group-cage housing systems. A total number of 32'565 brown layer hens from 689 sires (group-cage per sire) were used for estimation of heritabilities and genetic correlations. A sire model was fitted on sire-group averages of plumage condition score at 45 and 70 weeks, mortality at 70 weeks and laying performance. Estimated heritability and genetic correlations were generally low, with the estimated genetic correlation between mortality and plumage condition being negative. Taking into consideration the very high reproductive ability and very short generation interval in chicken, selection for reducing feather pecking and cannibalism appears promising despite the low heritabilities found. A GWAS based on a genome-wide chicken array was performed using estimated breeding values as phenotypes. Significant signals were annotated using Ensembl genes.

Foot pad health and mortality as part of 'controlling' in commercial Turkey Production

K. Toppel[1], H. Schoen[1], F. Kaufmann[1], M. Gauly[2] and R. Andersson[1]
[1]University of Applied Sciences Osnabrueck, Department of Animal Husbandry and Poultry Sciences, Am Kruempel 31, 49090 Osnabrueck, Germany, [2]Free University of Bozen, Faculty of Science and Technology, Piazza Università 5, 39100 Bozen-Bolzano, Italy; k.toppel@hs-osnabrueck.de

For a continuous improvement of animal welfare and health in turkey production the applicability of indicators as part of the on-farm controlling is required. In the present study mortality and foot pad (FP) health related data were collected on 13 conventional farms during rearing and fattening period in two cycles. Starting in the first week of life until slaughtering, FPs of 60 randomly selected birds per flock and barn were scored in four week intervals. FPs of birds were scored individually according to the five-point scale from Hocking et al. Results of mortality indicate a positive correlation between seven-day mortality and 35-day mortality (rearing period) (r=0.677; P<0.01). Weekly mortality (median) of toms was 0.05% higher in summer when compared to winter cycle. Time frames for mortality risk were evaluated. The average FP score (median of pooled data) from week 12 until slaughter was two, depending on season and sex (P<0.05). Thus, risk of FP lesions has to be considered higher during the first eight weeks of life. Furthermore, FP health is influenced by the amount of litter and litter dispersing interval (P<0.05). The correlation coefficient (Spearman) between right and left FP of individual birds ranged between r=0.252 to r=1.000. The results of the histological examinations of the FP lesions showed differences between macroscopic score and microscopic manifestation. In order to evaluate FP health as an indicator for animal welfare it is recommended to consider the worst foot (highest score) of a bird. Mortality and FP health need to be monitored and evaluated in frequent intervals within a production cycle. Furthermore, farm specific risk factors, seasonal effects and different husbandry systems can cause different target values and corridors which then need to be taken into account when applying a controlling system.

Results of a desk study on best practices for animal transport

M.A. Mitchell[1] and H.A.M. Spoolder[2]
[1]SRUC, The Roslin Institute Building, Easter Bush, Midlothian, EH25 9RG Edinburgh, United Kingdom, [2]Wageningen UR Livestock Research, Department of Animal Welfare, P.O. Box 338, 6700 AH, Wageningen, the Netherlands; malcolm.mitchell@sruc.ac.uk

The EU has provides a harmonised legal framework for animal transport. Part of that framework is the EU adopted Regulation (EC) 1/2005. The content and impact of the Regulation has been the subject of a Scientific Opinion by EFSA in 2011 followed in 2012 by an impact report from the Commission to the European Parliament and the Council. Three key recommendations were formulated the latter of which stated 'As regards the gap between the requirements of the legislation and available scientific evidence the Commission sees that this is best addressed by the adoption of guides to good practice'. It is expected that the development of Guides to Good and Best Practice can improve the welfare of animals during transportation, particularly by reducing national and regional differences in interpretation of the requirements of the Regulation. The Transport Guides project was commissioned by DG SANTE to produce Guides to Good Practice for cattle, sheep, pigs, horses and poultry in the European Union (http://www.animaltransportguides.eu). The project has been undertaken by a research consortium comprised of 16 representative organisations from 9 Member States and a stakeholder platform with 10 member organisations. The first phase of this project involved the examination of a wide range of information sources and literature to identify and evaluate available recommendations for good and best practices during all stages of transport. Good practices are those that reflect the requirements in the legislation. Best Practices are defined as providing additional guidance on how to exceed legally defined minimum welfare requirements. The first results of this project include analyses of the collected information for each of the species: each species section includes analyses of the practices identified, followed by an overview of all available practices presented in tabulated form. These findings are the basis for the development of Guides to Good and Best Practices in the next steps of the project, and will be presented at the meeting.

Sow removal patterns in Norwegian and US farms

S.L. Thingnes, T. Aasmundstad, L.T. Bogevik, J. Kongsro and E. Grindflek
Topigs Norsvin, Storhamargata 44, 2317 Hamar, Norway; signe-lovise.thingnes@norsvin.no

High removal rates in lower parities have a negative influence on production economy and can also negatively impact the consumer's perception of the pig industry. In general, an improvement in sow longevity is beneficial for overall herd productivity, health, economics and animal welfare. As part of a project where the overall aim was to identify available phenotypes that could help in describing the innate robustness of pigs more accurately, patterns of sow removal in Norwegian and US farms were investigated. Recent records of removal for more than 60 000 sows in Norway and USA were collected from the litter recording scheme Ingris. Statistical analyses were carried out using the LOGISTIC procedure in SAS 9.3. Preliminary data analysis shows that involuntary removal rates (dead/euthanized sows) are higher in US farms (24.6%) than in Norwegian farms (7.7%) (P<0.001), but for both countries most unplanned removals occur after the first or second parity. Sow removal due to poor body condition, conformation traits, farrowing difficulty, health, prolapses and reproductive disorders were more frequent in US herds compared to Norwegian herds (P<0.001), while removal due to injury (probably due to different housing system) or management practices were more frequent in Norway compared to US herds (P<0.001). No difference was found with regards to litter rearing ability. Regardless of country, the top three reasons for removal were reproductive failure, management practices and conformation traits. In both countries more than 10% of sow removal was categorized as unknown. Further analysis will aim at a more in-depth study of removal pattern and look at possible influential factors (e.g. age at first mating, production performance, and housing). These results show that there is a high proportion of unplanned removal among young sows, and that removal patterns differ between countries. Although sow longevity is a multifactorial trait, identifying removal patterns (at least at herd level) should be an important step towards improving longevity.

Welfare assessment of Low Atmospheric Pressure Stunning in chickens

J.E. Martin[1,2], K. Christensen[3], Y. Vizzier-Thaxton[3], M.A. Mitchell[1] and D.E.F. McKeegan[4]
[1]*SRUC, Animal Behaviour and Welfare, Animal and Veterinary Science Group, Easter Bush, Edinburgh, EH25 9RG, United Kingdom,* [2]*University of Edinburgh, R(D)SVS and The Roslin Institute, Easter Bush, Midlothian, EH25 9RG, United Kingdom,* [3]*University of Arkansas, Centre of Excellence for Poultry Science, Fayetteville, AR 72701, USA,* [4]*University of Glasgow, College of Medical, Veterinary & Life Sciences, Bearsden Road, Glasgow, G61 1QH, United Kingdom; jessica.martin@ed.ac.uk*

A novel pre-slaughter stunning method for chickens has been developed, where birds are rendered unconscious by progressive hypobaric hypoxia. Low Atmospheric Pressure Stunning (LAPS), involves gradual decompression (280s) according to a prescribed curve. We examined responses to LAPS in two trials by recording behaviour, electroencephalogram (EEG) and electrocardiogram (ECG) in broilers, and interpreted in regards to welfare impact. Trial 1: characterised responses of broilers to LAPS in 30 triplets. One bird was instrumented for recording EEG+ECG and all three underwent behavioural observation. Trial 2: examined the influence of illumination of the chamber, plus responses to the chamber without LAPS applied (sham) in a 2×2 factorial design. Twenty pairs per treatment; one bird instrumented for recording EEG+ECG, with behaviour of both recorded. Birds showed a consistent order of behaviours in LAPS (ataxia, loss of posture, clonic convulsions). Durations of conscious behaviours (e.g. stand) were predictably higher in sham. Illumination increased activity and dark induced sleep. During LAPS, progressive decreases in median frequency of EEG were observed along with corresponding increases in total power, followed later by decreases in total power as all birds died (isoelectric). There was an increase in total power at ~55s into LAPS corresponding to dominance of high amplitude slow waves, indicating loss of consciousness. ECG showed pronounced bradycardia ~45s into LAPS, but was absent in sham. Collectively, these results add to evidence that behavioural and EEG responses to LAPS are consistent and indicative of a process that is equivalent to controlled atmosphere stunning. This evidence has been presented to EFSA to facilitate approval of the method in the EU regulatory framework.

Inter-observer reliability of scoring of skin and other welfare-related lesions in finisher pigs
N. Van Staaveren[1,2], A. Diana[1,2], L. Boyle[2] and J.A. Calderón Díaz[2,3]
[1]*University College Dublin, Belfield, Dublin 4, Ireland,* [2]*Teagasc Animal and Grassland Research and Innovation Centre, Moorepark, Fermoy, Co. Cork, Ireland,* [3]*Institute of Genetics and Animal Breeding, Polish Academy of Sciences, ul. Postepu 36A, Jastrzębiec, 05-552 Magdalenka, Poland; nienke.vanstaaveren@teagasc.ie*

Animal-based measures like skin lesions or tail/ear/flank lesions are used to assess pig welfare. Visual inspection and scoring of these lesions rely on training and experience of the observers. This study examined inter-observer reliability in the identification and scoring of skin and other welfare-related lesions in pigs between two inexperienced observers. Ninety-nine finisher pigs were observed in their pens on a commercial farm and visually scored once for lesions by two observers. Skin lesions were scored according to severity (0: no evidence of lesions – 6: >1 extensive lesion) for neck, shoulder, flank and hindquarter on both sides the body. Ear lesions (0-3) and tail lesions (0-4) were scored on a scale where 0 indicated no evidence of lesions and score 3 or 4 indicated partial loss of ears and tail, respectively. Presence of flank biting lesions was recorded. Due to the low number of pigs receiving extreme values, observations were reclassified as presence/absence of lesions. A kappa statistics test was used to measure the level of agreement between observers. Moderate to substantial levels of agreement were found for presence of skin lesions on the neck (Kappa=0.41, $P<0.001$), shoulder (Kappa=0.45, $P<0.001$), flank (Kappa=0.62, $P<0.001$) and hindquarter (Kappa=0.52, $P<0.001$). In contrast, levels of agreement were lower for the welfare-related lesions. Ear lesions showed moderate levels of agreement (Kappa=0.52, $P<0.001$), but observers did not agree on the presence of tail (Kappa=0.26, $P<0.01$) and flank lesions (Kappa=0.39, $P<0.001$). Although inexperienced with the scoring system, moderate to substantial levels of agreement between observers were found for the presence of skin lesions. Lower levels of agreement for tail and flank lesions could be explained by difficulty of scoring these lesions due to dirtiness and movement by the animals. It is possible that in better conditions and with more training the level of agreement for these lesions could be improved.

Determination of some biochemical parameters in the blood of camels in Algeria
N. Sahraoui, O. Douadji, A. Doudou, M. Brahim Errahmani, D. Guetarni and J.L. Hornick
Blida university, Blida university, 09400, Algeria; nasahraoui@gmail.com

A survey of 45 female camels has been conducted over a 1-year period in Algeria to determine their metabolic profiles and to study the correlations between this profile and the feeding and health status. The following parameters were measured: total protein, urea, glucose, free fatty acid, minerals (Ca, Mg). The values obtained were similar to those reported in desert areas throughout the world, but the standard deviation was generally higher. This important variation might be due to the large variability of feeding conditions: total protein (64.1±5.1 g/l), urea (32.0±16.6 mg/100 ml), glucose (95.0±7.2 mg/100 ml), calcium (9.4±5.5 mg/100 ml), magnesium (3.3±0.5 mg/100 ml). The health status had a significant effect on the metabolic profile of the camels.

Use of haemoglobin analysis to evaluate anaemia status of piglets after weaning

E. Royer[1], F. Barbé[2], D. Guillou[2], Y. Rousselière[1,2] and E. Chevaux[2]
[1]Ifip-institut du porc, 34 bd de la gare, 31500, France, [2]Lallemand SAS, 19 rue des Briquetiers, 31702, France; eric.royer@ifip.asso.fr

The weaning of piglets results in a negative nutrient balance which could lead to anaemia. A monitoring was undertaken to evaluate the anaemia status after weaning and to assess the effects of a stress factor and of a nutritional supplementation. A total of 120 castrate male and female (LWxLd)xPietrain weaned piglets (8.2±0.5 kg) were blocked, with 2 single-sex pens per treatment and 15 pigs per pen. All piglets had received 200 mg iron by IM injection at 3 days of age. At weaning, piglets were vaccinated or not against PCV2, and received from weaning to day 14, NRC (2012) levels for vitamin E and selenium or extra supplementation in vitamin E, organic selenium and SOD-rich melon pulp concentrate in a 2×2 factorial design. Within each pen, 5 pigs close to the initial mean weight of the pen were chosen and their blood level for haemoglobin was measured at days 1, 5, 9 and 34 after weaning (Hg1 to 34). Analysis were carried out in the barn using a point-of-care quantitative haemoglobin reader (HemoCue Hb201+). The average haemoglobin level of piglets at day 1 was 10.09±1.45 g/dl and was significantly influenced by the lactation litter (P=0.03), but not by the birth litter (P=0.28) or the weaning weight (P=0.58). Hg5, Hg9 and Hg34 were 10.18±1.56, 10.38±1.05 and 10.35±0.82 g/dl, respectively. At days 5 and 9, step regressions show decreasing influences of Hg1 (or Hg5), weaning weight, daily gain (measured from day 0 to 14), diet and sow parity number. At day 34, haemoglobin level was influenced by day 0 to 14 daily gain, sex, parity number, Hg1 and vaccination stressor. Fewer piglets had anaemia (<9 g/dl) with the anti-oxidant supplementation than with the standard diet at days 5 (15 vs 35%) and 9 (0 vs 20%). As a conclusion, haemoglobin monitoring may be useful to evaluate the individual nutritional status of weaned piglets.

vitality measured at birth is a better indicator of survival than measured 3 hours post-partum

H. Davis, A.E. Taylor and H.M. Miller
University of Leeds, Faculty of Biological Science, LS2 9JT, Leeds, United Kingdom; h.m.miller@leeds.ac.uk

Piglet vitality is an indicator of it's future growth and survival. The aim of this study was to determine whether piglet survival was better predicted by vitality assessed at birth or at 3 hours post-partum. Seventy-two (Large White × Landrace) sows were moved into the farrowing house on day 112 of gestation and kept in conventional farrowing crates measuring 4.2 m[2]. At birth and 3 hours post-partum, piglets (n=585) were given vitality scores and their temperatures taken using a tympanic ear thermometer. Deaths were recorded. Piglets were sorted according to birth temperature (BT) and split into three groups; the same was done for 3 hour temperature (3T). Vitality scores at birth and 3 hours were grouped into separate categories (LOW, MEDIUM, HIGH). Pearson's Chi squared tests were used to analyse mortality in each category and assess if there was an association between vitality at birth and at 3 hours, and mortality. Post hoc tests were used to determine where the association was. Piglets with a BT below 34.8 C had a higher chance of mortality than piglets with a BT above this (P<0.05) and piglets with a 3T below 35.8 C had a higher chance of mortality than piglets with a 3T above this (P<0.01). Piglets with birth vitality scores of LOW had a higher chance of mortality than piglets with higher vitality scores (P<0.001) and piglets with 3 hour vitality scores of LOW had a higher chance of mortality than piglets with vitality scores above this (P<0.001). However, piglets that had a LOW vitality at birth and a LOW vitality at 3 hours had a higher chance of mortality than piglets that had a HIGH vitality at birth and a LOW vitality at 3 hours (P<0.001). In summary piglets with a BT below 34.8 C and a 3T below 35.8 C and piglets with LOW vitality scores at birth and 3 hours had reduced chances of survival compared to piglets with temperatures and vitality scores above this, demonstrating they are both indicators of vitality. Overall, piglets with LOW birth vitality scores had a higher chance of mortality, indicating that birth vitality gave a better indicator of survival.

Genome-wide association study of the Eimeria maxima response in Cobb500 broilers

E. Hamzić[1,2], M.-H. Pinard-Van Der Laan[2], A.J. Buitenhuis[1], R. Hawken[3] and B. Bed'hom[2]
[1]Aarhus University, Department of Molecular Biology and Genetics, Blichers Allé 20, 8830 Tjele, Denmark, [2]Université Paris-Saclay, GABI AgroParisTech / INRA, Domaine de Vilvert, 78352 Jouy-en-Josas, France, [3]Cobb-Vantress Inc., US-412, AR 72761 Siloam Springs, USA; marie-helene.pinard@jouy.inra.fr

Coccidiosis is one of the most common and costly diseases in poultry industry. The current control of coccidiosis, based on the use of anticoccidial drugs and vaccination, faces serious difficulties such as drug resistance and the high costs for the development of efficient vaccines, respectively. Therefore, the present control programs must be expanded with additional approaches such as the use of genetics to improve the host response to coccidiosis. Recently, we performed a large-scale challenge study on Cobb500 broilers using Eimeria maxima. We investigated variability of measured traits in response to the challenge and performed a genome-wide association study (GWAS) to identify genomic regions underlying variability of the measured traits in the response to the challenge. Furthermore, we performed a post-GWAS functional analysis to increase our biological understanding of the underlying response to E. maxima challenge. We identified 22 single nucleotide polymorphisms (SNPs) with q value<0.1 distributed across five chromosomes. The highly significant SNPs were associated with body weight gain (three SNPs on GGA5, one SNP on GGA1 and one SNP on GGA3), plasma coloration measured as optical density at wavelengths in the range 465–510 nm (10 SNPs all on GGA10) and the percentage of β2-globulin in blood plasma (15 SNPs on GGA1 and one SNP on GGA2). The post-GWAS functional analysis, which combined the biological pathway analysis (KEGG) and network-based analysis (Ingenuity Pathway Analysis), indicated that the genes and biological pathways involved in tissue repair, general robustness as well as the primary immune response might play a key role during the early stage of E. maxima infection.

Genome wide association analysis for cell mediated immune response in chicken

V. Raeesi[1], A. Ehsani[1], R. Vaez Torshizi[1], M. Sargolzaei[2], A.A. Masoudi[1] and R. Didehban[1]
[1]Tarbiat Modares University, Animal Science, Jalal Ale Ahmad Highway, Tehran, Iran, 14115-111, Iran, [2]University of Guelph, Animal Biosciences, 50 Stone Road East Guelph, Ontario, N1G 2W1 Canada, N1G 2W1, Canada; alireza.ehsani@modares.ac.ir

Availability of genetic map for farm animals has enabled animal breeding community to seek for associations between genomic variables and phenotypes of complex traits and diseases, commonly called genome wide association studies (GWAS). Application of genomic variables for low heritable or hard to measure traits such as immunity related traits is an extra advantageous of genomic selection. Cell mediated immunity (CMI) refers to various immune cells, including neutrophils and cytotoxic T cells. A GWAS performed to identify genomic regions associated with cell-mediated immunity in chicken using chicken 60k high density single nucleotide polymorphism (SNP) array. Phenotypes of 198 animal for Cell mediated immunity was available in a F2 population. In order to account for multiple testing a chromosomal false discovery rate was applied as significant thresholds. Based on genomic inflation factor, the power of fixed and mixed linear models (MLM) compared. Since MLM had better inflation rate, its results used for subsequent analysis. In order to account for multiple comparison, P-values were adjusted based on chromosomal false discovery rate of 5 and 10% as significant and suggestive threshold respectively. 3 significantly associated SNPs on chromosome 24 and linkage group E22C19W28_E50C23 and 3 suggestively associated SNPs on chromosome 1, 5 and 16 were identified. Pathway analysis showed that 2 biological pathways which are related to immune responses were strongly associated with the candidate genes surrounding identified SNPs and their influences were mostly on antigen processing and presentation and cellular structure. This study showed that it might be possible to predict genomic estimated breeding values and use it to improve immune response in poultry populations through genomic selection and marker assisted selection.

Identification of bovine respiratory disease causative bacteria using an amplicon sequencing assay

D. Johnston[1,2], M.S. McCabe[2], P. Cormican[2], G. Murray[3], M. McGee[2], S.M. Waters[2], D.A. Kenny[2], A.K. Kelly[1] and B. Earley[2]
[1]*University College Dublin, College of Agriculture Food Science and Veterinary Medicine, Belfield, Dublin 4, D04 W6F6, Ireland,* [2]*Teagasc Grange, Animal & Grassland Research and Innovation Centre, Dunsany, Co. Meath, C15 PW93, Ireland,* [3]*Regional Veterinary Laboratory, Department of Agriculture, Food and the Marine, Sligo, F91 XRW7, Ireland; sinead.waters@teagasc.ie*

Currently used methods for identification of bacteria associated with bovine respiratory disease (BRD) will not identify unknown or unculturable pathogens. Therefore, the objective of this study was to develop a single universal assay, based on the 16S ribosomal RNA (rRNA) gene PCR amplicon sequencing, with potential to identify all bacteria in post-mortem lung and lymph node tissue samples from fatal BRD. Thirty-eight lung (apical lobe) and thirty-two corresponding mediastinal lymph node post-mortem tissues were collected from beef and dairy calves confirmed with BRD by veterinary laboratory pathologists, and from twenty healthy Holstein-Friesian calves. Gene amplicon libraries were prepared by extracted DNA from these tissues and subjecting it to two rounds of PCR amplification. The libraries were subsequently sequenced on an Illumina MiSeq. Quantitative Insights Into Microbial Ecology was used to determine the operational taxon units (OTU)s which corresponded to the 16S rRNA gene sequences. All calves had unique lung and lymph node microbiomes. The read counts were lower for the OTUs in healthy compared to BRD-affected calves. Leptotrichiaceae, Pasteurellaceae, and Fusobacterium were the most abundant OTUs identified from BRD-affected calves. Leptotrichiaceae, Fusobacterium, and Mycoplasma had greater relative abundances in lung samples and Leptotrichiaceae and Pasteurellaceae had greater relative abundances in lymph node samples, collected from BRD-affected dairy calves, compared with healthy calves. However, the Leptotrichiacea OTU was not identical to known gerera and Leptotrichiaceae species are not associated with BRD. In conclusion, we have developed a 16S rRNA gene amplicon sequencing assay which can detect the bacteria present in a tissue sample. Using it, we have identified a family of bacteria which appears to contain a novel species member implicated in BRD.

Phenotypic Profile of Enterobacteriaceae in Poultry Faeces, Litters and Water to farms in Nigeria

O.O. Adebowale[1] and O.K. Adeyemo[2]
[1]*Federal University of Agriculture Abeokuta, Veterinary Public Health and Reproduction, Alabata, Abeokuta, Ogun State, Nigeria,* [2]*University of Ibadan, Veterinary Public Health and Preventive Medicine, Ibadan, Oyo State, Nigeria; oluwawemimo1@yahoo.com*

The family Enterobacteriaceae comprises commensal intestinal bacteria of humans and animals, and diverse environmental sources. Many species belonging to this group are opportunistic and food borne pathogens responsible for wide range of infections in humans. This study investigated the diversity of bacteria in poultry faeces, litter and water sources to farms in Ogun State, Nigeria. Samples were collected from 113 commercial laying hen farms, and bacteria identified using culture and biochemical tests (Microbact GNB 24E, Oxoid UK). A total of 129, 88 and 18 isolates were identified from poultry faeces, litter and water samples respectively. Bacteria identified from faeces included Aeromonas hydrophilia (30.2%), Actinobacillus spps (10.1%), Providencia stuartii (8.5%), Burkholderia cepacia (8.5%), S. marcescens (7.8%), Citrobacter diversus (5.4%), Klebsiella oxytoca (4.7%), and Enterobacter georgiviae (4.7%). Other bacteria were Enterobacter agglomerans (3.1%), Escherichia coli (2.3%), Plesiomonas shigelloides (2.3%), Vibro alginolyticus (2.3%), Morganella morgani (2.3%), Vibrio mimicus (1.5%), Pseudomonas aeruginosa (1.5%), Burkholderia pseudomallei (1.5%), Salmonella arizonae (0.8%), Klebsiella pneumonia (0.8%), Acinetobacter iwoffii (0.8%), Vibrio vulnificus (0.8%), Shewanella putrefaciens (0.8%), Proteus mirabilis (0.8%) and Pseudomonas vulgaris (0.8%). Similar bacteria phenotypes were isolated from poultry litters except additional strains as Flavobacterium memingosepticum, Hafnia alvei and Moraxella spp. Aeromonas hydrophilia (21. 6%), Pseudomonas aeruginosa (11.4%), and Klebsiella oxytoca (11.4%) were commonly isolated. Water samples tested positve for Aeromonas hydrophilia, B. cepacia, Citrobacter freundii, Salmonella, Acinetobacter iwoffii and Enterobacter cloacae. Overall, study showed a high recovery rate for Aeromonas hydrophilia from the three types of samples investigated, and close similarity in poultry faeces and litter bacteria phenotypic profiles. There is the potential risk of Aeromonas and other opportunistic pathogens to humans in close contact with the birds, and the general public.

Effect of prolonged storage of duckeggs with warming system on reproduction, growth and carcass
M. Waehner[1], H. Pingel[2] and H. Sun[1]
[1]*Anhalt University of Applied Sciences, Agriculture, Strenzfelder Allee 28, 06406 Bernburg, Germany,* [2]*University of Halle-Wittenberg, Animal Breeding, Th.-Lieser Str. 11, 06120 Halle/Saale, Germany; m.waehner@loel.hs-anhalt.de*

In pedigree breeding it would be necessary to prolong the storage time for hatching eggs to have more uniform offspring. Sometimes it is necessary to store hatching eggs up to 3 or 4 weeks. In 2 experiments it was compared the effect of storage duck eggs with periodical warming up to 37.4 °C for 3 hours at the 3rd day and for 2 hours at days 8, 11, 14, 19, 21, 23 25 and 27 with storage duck eggs without warming procedure on hatching rate. The negative effect of storage duration on hatchability was more pronounced in the group without periodic warming. In exp. 1 hatchability was 89.1, 87.0, 80.8 and 71.1% with warming and 85.7, 86.4, 70.0 and 64.2% without warming for storage duration of 1, 2, 3 and 4 weeks. In exp. 2 the results were: 83.4, 81.0, 77.7 and 70.9% with warming procedure and 79.9, 80.2, 77.8 and 61.4% without warming procedure, respectively. The reason for decreased hatchability was increasing of early embryonic mortality, of dead in shell and of delayed hatching, but pronounced in the group without warming. The hatched ducklings (1,644 in exp. 1; 360 in exp. 2) were tested in single boxes (3.-6. week) their growthl, carcass quality and feed intake. All 6 weeks old ducks were slaughtered. The storage of hatching eggs with warming resulted higher 2-week bodyweight than of hatching eggs without periodic warming (585 to 545 g exp. 1; 603 to 572 g exp. 2). Also the 6-week bodyweight was higher in exp. 2 in the group with warming procedure (3,173 to 2,928 g), but not in exp. 1 (3,102 to 3,094 g). Feed efficiency was improved in the groups with periodic warming in both experiments significantly (exp. 1:442 to 437 g; exp. 2: 403 to 377 g). The breast filet percentage was higher by periodic warming in experiment 1, but not in exp. 2. (exp.1: 21.8 to 21.4%, exp. 2: 22.6 to 22.7%.). The was no significant effect of duration of storage. Also interactions between duration of storage and storage conditions were unimportant.

Towards mixed models focusing on the animal nature of body condition: Reproductive rabbit does
A. Arnau-Bonachera[1], D. Savietto[2] and J.J. Pascual[1]
[1]*Institute for Animal Science and Technology/ Universitat Politècnica de Valencia, Camino de Vera, S/N. Valencia, Spain, 46022, Spain,* [2]*INRA UMR 1388 Génétique, Physiologie et Système d'Élevage, Castanet-Tolosan, 32326, France; alarbo@upv.es*

Life traits are dynamic. Consequently, the analysis of body condition throughout the reproductive life requires a proper design of the (co)variance matrix for the random effects, which also could improve the knowledge of its sources of variability (i.e. animal or environment). To achieve this goal, we assessed perirenal fat thickness (PFT) on 203 rabbit females at three states of their lactation throughout 5 consecutive reproductive cycles (RC). All PFT data was analysed in a common base model, further modified to contemplate four different ways of modelling the variance-covariance structure matrix for the random effects: M1 considered PFT as a single trait with a permanent effect of the animal and no correlation among the residuals; M2 considered decreasing correlations with increasing the lag-time between measures for the residuals; M3 considered PFT at different states of lactation as being different correlated traits that were assessed at each RC; M4 considered each single measure as different correlated traits (3 states of 5 lactation = 15 traits). The fact that Akaike's information criteria (AIC) value for the unstructured model (M4) did not decreased (in fact it increased an average of +65 points respect to the other models), together with the observed (co)variance behaviour of models (constant for M1 and M2, cyclical for M3 and unstructured for M4), suggests that each record of PFT should not be analysed as a different trait. Given the assumption of no flexibility of PFT through time for M1 and M2, and no relevant distinction of their AIC with respect to M3, we suggest a better biological description of PFT dynamics with M3. Using M3 we observed an average repeatability value of 0.308, residual correlations close to 0 and permanent animal correlations close to 1. These results, suggests a relevant effect of the animal. However, results also suggest small flexibility among animals for pattern of PFT use through the RC.

Preliminary Association analysis of SLC6A4 SNPs with behavioural traits in PRE dressage competitions

S. Negro[1], M. Solé[1], M.J. Sánchez[1], E. Bartolomé[2], A. Molina[3] and M. Valera[1]
[1]*University of Seville, Agro-Forestal Sciences, Seville, 41013, Seville, Spain,* [2]*Universidade Lusófona de Humanidades e Tecnologias, Facultade de Medicina Veterinária, Lisboa, 1749-024, Lisboa, Portugal,* [3]*University of Córdoba, Genetics, Córdoba, 14071, Córdoba, Spain; z12neras@uco.es*

Animal welfare is nowadays a core value in equestrian disciplines. New methods, such as eye temperature (ET) using infrared thermography (IR) or heart rate (HR), have been proved to be good indicators of stress in horses during competitions. The equine serotonin transporter gene (SLC6A4; 5-HTT) plays a central role in the circuitry of cognition and emotions. Therefore, this study elucidates whether equine SLC6A4 variants (g.43864387C>T, g.43864288A>C, AC1615-1616GT, g.43864091C>A; and haplotypes) are involved in welfare-behavioural traits recorded in 115 Pura Raza Español (PRE) young horses (aged from 4 to 6 years old) during dressage competitions. For that, ET and HR measures taken 3 hours before the competition (BC), just after the competition (JAC) and 3 hours after (AC), jointly with the submission scores (indirect measure of temperament and performance) on the dressage test (awarded points on a scale of 1-10 by a panel of judges) were analysed. Two interval measurements (ETI and HRI) corresponding to basal (Initial) and resting (Final) phases were also assessed for each parameter. An association analysis based on ML-Chi2 and General Linear Model procedure (GLM) revealed that the allele A at g.43864091C>A was associated to HR_AC (P=0.029). This polymorphism (P=0.016) and AC1615-1616GT (two consecutive SNPs in linkage disequilibrium which provoke the change from Threonine to Valine) (P=0.021) were associated to initial ETI. The C allele at g.43864288A>C was associated to submission score (P−0.047), as well as the haplotype of SLC6A4 gene (P=0.014). These preliminary results showed that SLC6A4 gene appears to be involved in welfare-behavioural traits in horses. It denotes its potential use as a molecular tool for developing marker-assisted selection strategies for improving behaviour in dressage PRE horses.

Effect of yeast based interventions on reducing Campylobacter spp., colonisation broilers

A. Corrigan and R. Murphy
Alltech, European bioscience centre, Meath, A86X006, Ireland; acorrigan@alltech.com

Campylobacteriosis is considered to be the most important cause of zoonotic illness globally, with poultry being the main source of infection. Reducing the load of Campylobacter spp., colonisation in broilers entering the processing unit should make an effective contribution at reducing the incidence of zoonotic transmission of this pathogen. With the increase in antibiotic resistance spread and a reduction in antibiotic use it is essential to search for new, natural and sustainable strategies to reduce the incidence of this bacterium in the food chain. The aim of this study was to examine if dietary supplementation of broilers with 3 different yeast based products (T1, T2 and T3) reduced the level of natural Campylobacter spp., colonisation in the caecum. The study was carried out using 500 male broilers divided amongst four different experimental groups using a randomised complete block design. Birds were allowed to naturally acquire Campylobacter spp., colonisation from the environment and were tested at days 21 and 35 post hatch for Campylobacter spp., colonisation levels using plate counts of caecal swabs and at day 35 using quantitative PCR (qPCR). Results from plate count analysis at day 21 showed that Campylobacter spp., colonisation had successfully occurred in the broilers and at day 35 showed that the level of colonisation was highest in the control group with decreasing levels in T1, T2 and T3 respectively. No significant differences were noted using plate counts at day 35. Further analysis of the caecal content by qPCR showed that both T2 and T3 significantly reduced the levels of Campylobacter spp., colonisation in the broiler caecum. While T1 did not significantly reduce the level of Campylobacter spp., colonisation, the levels were still lower than those in the control group. In conclusion, this study showed yeast based dietary interventions were effective at significantly reducing the level of natural Campylobacter spp., colonisation in the broiler caecum using qPCR.

Detection of faecal and gut contamination on chicken carcasses using a chlorophyll-based marker

S.A. Morgan, V.J. Theobald, D. Leemans, A.P. Gay and N.D. Scollan
Institute of Biological, Environmental and Rural Sciences, Plas Gogerddan, Aberystwyth University, Aberystwyth,
Wales, United Kingdom; sam61@aber.ac.uk

A sizeable proportion of foodborne illness is associated with the consumption of undercooked or raw chicken contaminated with pathogens such as Campylobacter or Salmonella. One of the sources of this bacterial contamination is transfer of faecal or gut material onto carcasses at the processing plant. Image-based tools are currently being investigated as a means to detect surface faecal and gut contamination on chicken carcasses by detecting fluorescence from natural fluorophores such as chlorophyll found in faecal material. However, poultry diets typically contain low levels of these natural fluorophores reducing the reliability of faecal fluorescence as a means of contamination detection. To alleviate this problem a number of studies have been carried out to assess the efficacy and practical delivery of a chlorophyll-based marker to chickens in order to improve our ability to detect faecal and gut contamination using fluorescence techniques. Factors including delivery of the marker using feed versus water, the time scale of appearance of faecal fluorescence, the minimum dose required to produce fluorescence and the effects of feed and water withdrawal prior to slaughter on faecal fluorescence were explored. Encouraging results have been found which establish the optimum timing and amounts of chlorophyll-based marker supplementation required before slaughter to ensure adequate faecal fluorescence post-slaughter. This will enhance our ability to detect contamination of carcasses by faecal and gut material using fluorescence techniques.

Comparative study of inactivation of B. sporthermodurans by hydrostatic pressure and heat treatment

C. Aouadhi[1,2], H. Simonin[3], A. Maaroufi[1], M. Djemali[2] and S. Mejri[2]
[1] *Institute Pasteur of Tunisia (IPT), Tunisia., Laboratory of Microbiology, Bioprocess group, BP 74, 13 place*
Pasteur, 1002 Belvédère Tunis, Tunisia, [2]INAT, Labo Ressources Génétiques animales et alimentaires, 43 Av.
Charles Nicolle, 1082 Tunis Mahrajène, Tunisia, [3]ONIRIS-UMR-CNRS, ONIRIS, UMR CNRS, BP 82225, France,
44322 Nantes Cedex 3, France; mejri.slah1980@gmail.com

The presence of Bacillus sporothermodurans spores, are known to be extremely heat-resistant, in milk has emerged as an important problem in dairy industry because this bacterium may contribute to reduce their quality. It thus became necessary to develop more efficient processes to inactivate these spores completely and ensure milk commercial sterility. In this work, a central composite experimental design was used to evaluate the effect of pressure (300-500 MPa), temperature (30-50 °C), and pressure-holding time (10-30 min) on the inactivation of spores in distilled water and skim milk. The inactivation observed was shown to fit well with the values predicted by the quadratic equation, since $Radj2$ were 0.970 and 0.977 in distilled water and milk, respectively. By analyzing the obtained results, the inactivation was shown to be higher in distilled water than in milk under all the conditions tested. This was probably due to a protective effect of milk against inactivation by pressure. The optimum process parameter values for a 5-log cycle reduction of spores were calculated as 477 MPa/48 °C for 26 min and 495 MPa/49 °C for 30 min in water and in milk, respectively. This study shows the efficiency of hydrostatic pressure in combination with moderate temperature to inactivate B. sporothermodurans spores. Such treatments could be applied by the dairy industry to ensure the commercial sterility of UHTmilk.

The components of litter size in rabbits: effect of Ano-Genital Distance

R. Belabbas[1], M.L. García[2], H. Ainbaziz[3], A. Berbar[1], G.H. Zitouni[1], N. Benali[1,3], M. Lafri[1] and M.J. Argente[2]
[1]Biotechnology Laboratory of Animal Reproduction (LBRA), Institute of Veterinary Sciences, University of Saad Dahleb Blida I, 09000 Blida, Algeria, [2]Departamento de Tecnología Agroalimentaria, Universidad Miguel Hernández de Elche, 03312 Alicante, Spain, [3]Health and Animal Production, National Veterinary School, 16200, Algeria; r_belabbas@yahoo.fr

The aim of this work was to test in rabbit of local Algerian population, the effect of Ano-Genital Distance (AGD) before the mating on, sexual behavior, litter size and its components (ovulation rate and prenatal mortality) and sex ratio (ratio of male pups to females pups at birth). In total, 64 multiparous rabbit does were used in this experiment. At the moment of mating, the AGD was measured by three operators and the behavior of the females was noted. At 12 d post coïtum, an endoscopy was realized on the pregnant females in order to measure the ovulation rate (number of non-hemorrhagic corpora lutea) and the number of implanted embryos (live and resorbed). At kindling, the number of pups (live and dead) and their sex were noted. The females with larger AGD were more aggressive (25,4% vs 8%; P<0,01) but presented similar receptivity rate when compared to the females with shorter AGD (82 vs 86%; P>0,05). At 12 d of pregnancy, the effect of the AGD was not significant on the ovulation rate (9,22 vs 9,35; P>0,05). However, the females with larger AGD presented higher early embryonic and fetal mortalities (+45% and +57% respectively; P<0,01). The females with larger AGD gave birth to almost 62% male pups. Conversely, the females with sorter AGD gave birth to about 41% male pups. In conclusion, the AGD in rabbits has influenced the majority of the traits related to reproduction and more investigations are necessary in order to understand more the origin of results obtained.

COST Action FA1401-European network on the factors affecting the gastro-intestinal microbial balance

P. Trevisi
University of Bologna, DISTAL, V. Fanin, 40127, Bologna, Italy; paolo.trevisi@unibo.it

COST Actions are an efficient networking instrument for scientists, engineers and scholars to cooperate and coordinate nationally funded research activities. In 2015 the COST Action – FA1401 'European network on the factors affecting the gastro-intestinal microbial balance and the impact on the health status of pigs (PiGutNet) started. Today, the network involves 49 European intuitions and several companies from 22 European countries. Moreover, partners from Canada, China and Australia collaborate in the network activities. The main objective of PiGutNet is to increase the knowledge about the effect/interaction of environmental and genetic factors on the composition of the microbiota in the gastrointestinal tract of pigs and to improve the risk management associated with antibiotic resistance in pig production. In order to disentangle the factors involved in the gut microbial balance, four main topics were identified: i. Molecular microbiology; ii. Environment, host genetics and epigenetic approach; iii. Feeding strategy; iv. Antibiotic resistance. According to the 4 main research areas above described 5 working groups (WGs) were defined: WG1. Functional and genetic characterization of microbial communities in the gastrointestinal tract of pigs; WG2. Genetic and environmental factors to understand dysbiosis including their interaction (epigenetics); WG3. Feeding strategy to maintain/restore the gut homeostasis; WG4. Antibiotics as a factor of dysbiosis and spread of antibiotic resistance genes; WG5. Knowledge and management exchange. The PiGutNet network is open to accept new partners that aim to share information to progress in the field of the gut microbial balance of pigs. In order to have more information on the network, please, visit the webpage http://www.cost.eu/COST_Actions/fa/Actions/FA1401 and www.pigutnet.eu.

Microbes, diet and host: how do they interact in newborn piglets ?
T. Thymann
University of Copenhagen, Veterinary Clinical and Animal Sciences, 68 Dyrlægevej, 1870 Frederiksberg C,
Denmark; thomas.thymann@sund.ku.dk

Following birth the gastrointestinal tract in pigs is quickly populated with bacteria, viruses and parasites. At the one hand this may induce severe pathological changes in the gut during the first few days after birth, but at the other hand proper gut colonization is also a prerequisite for development of important functions like digestion and innate and adaptive immunity. Gut microbial composition in early life is a reflection of which microbes are present in the local environment as well as factors related to the diet and the host. We have in a long series of experiments studied early life gut colonization in newborn pigs housed individually under very standardized laboratory conditions. Although this represents a much different colonization pattern than seen in commercial pig production, it offers a possibility to study intervention strategies and host responses without the influence of confounding factors. On the host side we have shown that the gut microbiota is influenced by the gestational age at birth, as preterm pigs display a different microbial profile up to one month after birth relative to pigs born at term, given the same diet and housing conditions. Although it is not clear to which extent newborn pigs under farming conditions display characteristics of prematurity, the high litter size and morbidity and mortality in pigs relative to other farm animals, makes it a plausible working hypothesis. Ingestion of milk diets is known to influence gut microbial composition during the first week of life, but we have shown that these diet-induced gut microbial effects may not persist if the diet is changed after the first week. Despite the unstable and transient nature of the gut microbiota in early life, ongoing activities in our lab are now trying to identify if induced differences during the first week of life associates with a persistent change in the epigenetic fingerprint in gut mucosal cells. Further basic and applied research is needed to determine how a proper gut colonization can be secured in early life, and to what extent this influences gut health in later life.

The porcine gut microbiota: composition and links with host's genetics and phenotypes
J. Estellé and C. Rogel-Gaillard
INRA, UMR1313 Génétique Animale et Biologie Intégrative (GABI), Domaine de Vilvert, Bat. 320, 78350 Jouy-
en-Josas, France; jordi.estelle@jouy.inra.fr

Microbiomes and their effects on hosts are emerging as outstanding factors to study in the animal production field. In fact, the microbiome of the intestinal tract may be considered a new host organ that plays a major role in health and well-being. In our laboratory we are developing a research line that targets the pig's gut microbiota with the objective to study the interplay with its host for shaping host's phenotypes. To this end, a cohort of French Large White piglets ranging from 14 to 70 days old was assessed for fecal microbiota composition by pyrosequencing the 16S rRNA gene. All animals were weaned at 28 days and measured for immunity and production traits. Bacteroidetes, Firmicutes, and Proteaobacteria phyla were predominant at all ages, while specific microbial groups (e.g. Lactobacillus) were more represented in the youngest animals. In this sense, a temporal trajectory of bacterial communities in 31 piglets revealed a stratification of piglets in two main groups after weaning that were primarily distinguished by the levels of unclassified Ruminococcaceae and Prevotella, respectively. This results were confirmed in phylogenetic network and clustering analyses in 518 60-days old pigs. A Dominance of Prevotella was positively correlated to increased concentrations of luminal secretory IgA, average daily gain and body weight. In parallel, the genetic parameters of the gut microbiota composition were estimated and, among a set of 63 genera, 7 had low (0.1<h2<0.2), 15=" 8=" and=" h2=" high–" medium–">0.4) heritabilities for abundance variation. Finally, regularized canonical correlations and sparse Partial Least Squares analyses highlighted both positive and negative correlations between health traits (e.g. monocytes, eosinophils, platelets) and genera such as Prevotella, Roseburia and Dialister. Thus, the gut microbiota composition is both influenced by the host's genetics and linked to health and growth traits, which confirms the relevance of this ecosystem for the porcine production.</h2<0.2),>

Delineating spatio-temporal processes in the gut mucosa of pigs

B. Hulsegge[1], J.M.J. Rebel[1], D. Schokker[1] and M.A. Smits[1,2]
[1]Wageningen Livestock Research, Droevendaalsesteeg 1, 6708 PB, the Netherlands, [2]Central Veterinary Institute, 8219PH, Lelystad, the Netherlands; ina.hulsegge@wur.nl

Biological tissues, like intestine, are highly dynamic and develop in time. To investigate this intestinal development we have used whole genome analyses. Temporal gene expression patterns provide an important characterization of gene function. Identifying such gene expression patterns provide much greater insight into their biological functions and timing of certain biological processes compared to individual genes because they often share similar expression patterns. To get insight into the dynamics of biological processes in the gut mucosa of pigs we used a set of jejunal gene expression data. To identify such reference gene expression pattern of control pigs, data of 9 microarray experiments were combined. In total 98 arrays from 2 platforms were used, measuring expression levels at 17 different time points (range 0-63 days post-natal). A set of 8,069 genes were common across both platforms was used as input to extract nine reference patterns of expression over the time course. In order to identify whether these clusters of genes showed similar functions, a Reactome pathway enrichment analysis was performed for each cluster. Decreasing expression patterns over time were enriched for 'extracellular matrix', 'metabolism', 'platelets' and 'kainate receptors'. Whereas an increasing pattern over time was associated to 'immune system' and 'cell proliferation'. Genes in the relatively flat pattern over the whole time course were related to 'gene expression'. In conclusion, by performing a meta-analysis on jejunal gene expression patterns of (control) piglets from multiple experiments, we obtained insight into the time-dependent fluctuations of biological processes in this tissue. This insight may be exploited to modulate particular processes by changes in management, nutrition, or genetic background.

The effect of host genetics factors on shaping pig gut microbiota

M. Maushammer[1], A. Camarinha-Silva[1], M. Vital[2], R. Wellmann[1], S. Preuss[1] and J. Bennewitz[1]
[1]University of Hohenheim, Institute of Animal Science, Farm Animal Genetics and Breeding, Garbenstr. 17, 70599 Stuttgart, Germany, [2]Helmholtz Centre for Infection Research, Inhoffenstraße 7, 38124 Braunschweig, Germany; maria.maushammer@uni-hohenheim.de

Studies on humans suggests that host genotype plays an important role in the diversity of the gut microbial community. Currently, there is a lack of research in pigs regarding the genetic influences on gut microbial community. The aim of this project is to address how host-specific genotype background influences the composition of the pig gut microbiota and how this composition influences growth performance. Colon microbiota of 210 Piétrain sows was characterized by 16S Illumina amplicon sequencing. Sows were genotyped with PorcineSNP60 BeadChip. Phylogenetic analysis assessed using RDP pipeline. Univariate and bivariate genomic mixed linear models were used to estimate genetic parameters (heritabilities and genetic correlations). Next to the random pig effect, a random slaughter day effect and several fixed effects were included. The covariance structure of the pig effect was modelled by a SNP-based genomic relationship matrix. All pigs showed a microbial community similarity of 35%. Animals more colonized with Firmicutes presented a 68% dissimilarity from those more colonized with Bacteroidetes. Phylotypes contributing for this variation were Unc. Clostridium, Lactobacillus amylovorus, Streptococcus alactolyticus. Prevotella showed negative correlations with Clostridium sensu stricto, Clostridium XI and Ruminococcaceae and a positive one with Lachnospiraceae. Spirochaetales was the most highly heritable group of microbes in the pig gut (0.52), followed by the genera Alloprevotella, Blautia, Catenibacterium, Lactobacillus and Spirochaetes (~0.36). Lactobacillus and feed consumption showed the highest negative genetic correlation (-0.5) followed by Alloprevotella (-0.46). The genetic correlations between daily gain and bacteria ranged between -0.26 and 0.19. To conclude, the microbial composition in the gut seems to be a heritable trait of the pig, which might become part of the breeding goal to improve feed efficiency.

FUT1 gene polymorphism: impact on gut microbiota, immune response and metabolomic profile of piglets

C. Lauridsen[1], A.S.R. Poulsen[1], D. Luise[2], S. Sugiharto[3] and N. Canibe[1]
[1]Aarhus University, Blichers Alle 20, P.O. Box 50, 8830 Tjele, Denmark, [2]University of Bologna, Viale G. Fanin 50, 40127 Bologna, Italy, [3]Diponegoro University, Semarang, 50275 Central Java, Indonesia; charlotte.lauridsen@anis.au.dk

A single guanine-to-adenine mutation at nucleotide 307 in the α-(1,2)-fucosyltransferase (FUT1) gene is determinant for the susceptibility of piglets to Escherichia coli F18-diarrhea. We studied the influence of FUT1 gene variants on the intestinal luminal microbiota, mucosal immune responses, binding of E. coli F18, and plasma metabolomic profile of weaners. Two E. coli F18-sensitive sows (heterozygous, FUT1-M307GA) were mated with a resistant boar (homozygous, FUT1-M307AA). Offspring were raised with their dams until weaning on d 28. Piglets were sacrificed (d 34) and gastrointestinal digesta collected. Distal jejunal tissue was obtained to determine the binding of E. coli F18 to intestinal tissue ex vivo, using a porcine intestinal organ culture model. Mucosa was sampled from the cultured intestinal tissue for IgA and IgM determination. Feces and plasma were collected during the study. Sensitive piglets weighed more than resistant piglets on day 28 (P=0.003) and 34 (P=0.01) of age. The ex vivo adherence of E. coli to intestinal tissue and the numbers in intestinal content were higher (P=0.05) in sensitive piglets than in resistant piglets. No effects on concentration of immunoglobulins in mucosa were measured. In susceptible pigs, numbers of lactic acid bacteria tended (P=0.06) to be higher in faecal samples and at 34 days of age, the fecal number of hemolytic bacteria was higher (P<0.003). A higher number of Enterobacteriaceae (P<0.02) in the distal small intestine, caecum and mid-colon; and of hemolytic bacteria (P=0.02) along the gastrointestinal tract were found in susceptible piglets. The concentration of acetic acid was higher in the colon of sensitive piglets (P=0.01). Minor differences between the genotypes were obtained with regard to the plasma metabolic profile. In conclusion, our results indicate that FUT1 genotype might influence the gastrointestinal colonization of other bacterial groups than E. coli F18, and may also influence growth and exert some minor influence on the metabolism of the host.

The A0 blood groups effect on the porcine gut microbiota colonization

V. Motta, D. Luise, M. Colombo, P. Trevisi and P. Bosi
University of Bologna, DISTAL, V. Fanin, 40127, Bologna, Italy; vincenzo.motta2@unibo.it

Intestinal epithelium glycocalix plays a role in bacterial-host interaction. The porcine histo-blood group A0 system affected the jejunal mucosa glycomic pattern profile, and thus may influence the gut microbiota colonization events. The present study verified the influence of A and 0 blood groups on the porcine gut microbiota and tested the resilience of the bacterial community after weaning. Two sows with A blood group and two with 0 blood group were selected. Three piglets per sow with blood group identical to the mother were chosen. Fecal samples were collected from the piglets at 7 (tI) and 14 (tII) days after birth and 2 weeks after weaning (tIII). Faeces from the sows were also collected at tI and tII in order to check the microbiota similarity between mother and piglets. Bacterial DNA was extracted with QIAGEN DNA stool mini kit, V3-V4 region of 16S rRNA gene was sequenced on Illumina MiSeq and the resulting data were analyzed in QIIME (v1.9.1) with open-reference OTU strategy using default settings. No significant difference in microbiota composition and diversity were reported between A and 0 genotypes. The dominant phylum was Firmicutes, followed by Bacteroidetes and Proteobacteria, both in sows and in piglets. It is visible a cline in which the Bacteroidetes rise from tI piglets samples (9%) to the mature microbiota (21% in sows samples), whereas Proteobacteria decrease (8% to 3%).The sows showed stable and higher values of alpha diversity, instead microbiota diversity in piglets increased along time, reaching after weaning (tIII) values close to those of sows (mature microbiota), furthermore, a linear correlation between tI and tIII in piglets, for Shannon (r=0.64 P=0.02) and Chao1 (r=0.62 P=0.02) indices, was notable. Results show the absence of A and 0 blood group effect on the porcine gut microbiota and suggest that the manipulation of gut microbiota in the first days of life can influence the microbial community composition of the gastrointestinal tract even after weaning.

Session 61

Theatre 8

Impact of high-wheat bran diet on sows' microbiota, performances and progeny's growth and health

J. Leblois[1], J. Bindelle[1], F. Dehareng[2], S. Massart[1], B. Li[1], H. Soyeurt[1], J. Wavreille[2], Y. Beckers[1] and N. Everaert[1]

[1]ULg-GxABT, Passage des Déportés, 2, 5030 Gembloux, Belgium, [2]CRA-W, Rue de Liroux, 9, 5030 Gembloux, Belgium; julie.leblois@ulg.ac.be

Finding alternatives to antimicrobial growth promoters is part of the goal of improving sustainability in pig production. Dietary fibres are considered as health-promoting substances acting on pigs' microbiota. This study aimed to investigate whether the enrichment of sows' diet with high levels of wheat bran (WB) could impact the performances of sows and piglets' health. Seven sows were fed a control diet (CON) and 8 sows a WB diet from day 43 of gestation (WB 240 g/kg DM) until the end of the lactation period (WB 140 g/kg DM). Diets were formulated to be iso-energetic and iso-nitrogenous by changing the proportions of some ingredients. Faeces were sampled at different time points (before treatment, during treatment: in gestation and lactation) to determine microbiota composition (sequencing with Illumina MiSeq). Milk was sampled weekly to determine lactose, fat and protein concentration by mid-infrared technology and IgA and IgG contents by ELISA. Before weaning (d26-27), piglets were euthanized, intestinal contents and tissues sampled for further analyses. Zootechnical performances of sows and piglets were recorded. Statistical analyses were performed using the SAS MIXED procedure and repeated measurements. Treatment never impacted piglets' weight (P=0.51). Sows' ingestion during the lactation period was comparable between both treatments until the last 4 days of lactation where the percentage of target ingestion was significantly (P<0.001) lower for the WB (66%) compared to the CON group (89%). No effect on sows' backfat and weight changes was observed. An increased abundance of Lactobacillus spp. in feces of the WB group was observed in gestation before and after diet change (8.8% vs 15.1% of total bacteria). However, for the overall genera changes between treatments, it only seems to occur for minor groups of bacteria. Milk protein, fat, IgG and IgA were not affected by treatment, but a time-effect (P<0.001) was observed while treatment impacted (P<0.05) lactose content. In conclusion, sows' performances were not affected by the high WB diet and more research on the piglets' samples is foreseen.

Session 61

Theatre 9

Effects of dietary protein sources on intestinal and systemic responses of pigs

S.K. Kar[1], A.J.M. Jansman[2], D. Schokker[2], L. Kruijt[2] and M.A. Smits[1,2]

[1]Wageningen University, Host-Microbe Interactomics, De Elst 1, 6708 WD Wageningen, the Netherlands, [2]Wageningen UR Livestock Research, De Elst 1, 6708 WD Wageningen, the Netherlands; soumya.kar@wur.nl

Previously we provided evidence that experimental diets prepared with different protein sources and administered for 3 weeks to mice, greatly affected host (immune and metabolic parameters) and gut microbiota. In the present study we used pigs to investigate the effects of five experimental diets each containing one of five different protein sources on a range of physiological and immunological parameters. Pigs aged six weeks were fed for four weeks with experimental diets containing one of the following protein sources providing a dietary protein level of 160 g/kg: soybean meal (SBM), wheat gluten meal (WGM), rapeseed meal (RSM), spray dried plasma protein (SDPP), or black solider fly (BSF). To discern changes in intestinal mucosal gene expression, intestinal microbiota composition and systemic immunity, genome-wide gene expression profiling in jejunal and ileal tissue, 16S rRNA gene sequencing of intestinal microbiota (on-going), and multiplex detection of cytokines and chemokines in serum were used. Gene set enrichment analysis (GSEA) was performed on the transcriptomic data to identify biological pathways and processes affected in small intestine by the various diets. Jejunum showed the highest response in transcriptome analysis and therefore we focused on jejunum. We observed higher (FDR, q<0.05) expression of genes related to barrier function and immune signalling in duodenum of pigs fed with SBM compared to SDPP and BSF, and these response were higher in the former comparison. Moreover, we have observed lower (q<0.05) expression of genes in duodenum of pigs related to metabolism of bio-molecules (xenobiotics, retinol and tryptophan) in SBM compared to RSM and WGM. Further, overall responsiveness to these diets showed no significant (P<0.05) effects on nine measured blood immune parameters. From the knowledge gained from our previous mice study we expect different effects on the composition and/or diversity of small intestinal microbiota. The results of these studies will be presented and discussed with regard to the potential use of alternative protein sources to replace the traditional protein sources in pig diets.

Effect of protected benzoic acid supplementation in nursery diets on piglet growth performance
S. Keller[1], M. Blanch[1], P. Buttin[1] and J. Morales[2]
[1]*Novus Europe SA/NV, rue Neerveldstraat 101-103, 1200 Brussels, Belgium,* [2]*PigChamp Pro Europa S.L., Calle Santa Catalina, 10, 40003 Segovia, Spain; sven.keller@novusint.com*

Delivering benzoic acid to the intestinal tract by using an embedding technology, improves growth performance of pigs as a result of higher nutrients digestibility, inhibition of pathogenic microorganisms and maintenance of intestinal microecological balance. This study evaluated the effect of protected benzoic acid (Provenia®, Novus International Inc., USA) in nursery diets on growth performance and also assessed the compatibility of its combination with zinc oxide (ZnO) used at therapeutic dose (3,000 ppm). A total of 288 weaned piglets (7.5 kg body weight; 28 d of age) were used and distributed in a 2×2 factorial design, based on the administration or not of two additives, embedded benzoic acid (2.5 kg/t) and ZnO (3,000 ppm). Pigs were fed a pre-starter diet (d 1 to 14) and a starter diet (d 15 to 35), and ZnO was administered only in pre-starter feed in those treatments containing it, while Provenia was added during the whole experimental period. Average daily gain (ADG), feed intake and feed conversion ratio (FCR) were controlled and daily health and faecal score were recorded. There was no interaction between benzoic acid and ZnO dietary supplementation during the experimental period. Benzoic acid supplementation tended to increase ADG in the starter phase (535 vs 510 g/d; P=0.09) and improved FCR in the global nursery period (1.36 vs 1.43 g/g; P=0.01). Zinc oxide did not affect growth performance, both in pre-starter and total nursery period. However, ZnO reduced the incidence of diarrhea compared with benzoic acid. In conclusion, compared to ZnO supplemented at therapeutic dose in the pre-starter phase, supplementation of nursery feed with protected benzoic acid improved growth performance of piglets.

Transcriptome analysis of porcine mesenchymal stem cells subjected to epigenetic modulation
A. Gurgul, J. Opiela, K. Pawlina, J. Romanek, K. Żukowski, T. Szmatoła and M. Bugno-Poniewierska
National Research Institute of Animal Production, Krakowska 1, 32-083 Balice, Poland; artur.gurgul@izoo.krakow.pl

Up to now, practically little is known about gene expression profile and its regulation in porcine mesenchymal stem cells (MSCs). Moreover, nothing is known about the impact of epigenetical modifications on porcine MSCs transcriptome. Our previous results showed that using epigenetically modified MSC as donor cells for somatic cell cloning in pigs significantly increased the efficacy of the procedure. Therefore, in this initial study, a comparative gene expression analysis was carried out between porcine MSC treated with Histone Deacetylase (HDAC) inhibitor – Trichostatin A (TSA) and control cells, to address the regulation of the epigenomic transformation and its impact on gene expression profile and to define which sets of genes may be responsible for increased usefulness of TSA-modified MSCs. By using high throughput mRNA sequencing (Illumina) we characterized the transcriptome of in vitro cultured pig bone-marrow-derived MSCs, both treated (24 h) and untreated with TSA. The expression of positive and negative surface markers specific for MSC was confirmed by flow cytometry and immunofluorescence analyses. A TopHat/Cufflinks pipeline allowed for identification of 209 differentially expressed genes at a genome wide level (q<0.05). Of the genes, 129 were upregulated in TSA-modified cells of which the vast majority was associated with cellular processes, such as: cell communication and cell cycle or primary metabolic processes. Molecular functions of the encoded proteins were mainly associated with antigen binding and transferase, hydrolase or enzyme activity. A large number of the downregulated genes had molecular functions connected with protein and nucleic acid binding and participated in variety of biological processes involving e.g. cell communication, developmental processes and protein metabolism. A deeper analysis of the obtained results will provide more insights into the mechanisms of impact of epigenetic modulation on properties of cultured MSCs, especially their usefulness for different downstream applications including somatic cell nuclear transfer.

Effects of FUT1 and MUC4 genotypes on microbiota and gene expression in the jejunum of healthy pigs
D. Luise, V. Motta, P. Bosi and P. Trevisi
University of Bologna, DISTAL, V. Fanin, 40127, Bologna, Italy; paolo.trevisi@unibo.it

The gastrointestinal microbiota is relevant to animal health and performance and is partially affected by host genetic factors. Pigs genotype is related to some E. coli infections. A G/A mutation at 307pb of α(1,2)-fucosyltransferase (FUT1) gene affects the properties or the quantity of the mature enzyme, inducing the expression of the E. coli F18 receptor. The g.8227 G/C mutation of mucin 4 (MUC4) gene is linked with the causative gene for the intestinal susceptibility to E. coli F4ac. The impact of FUT1 and MUC4 genotypes on the gastrointestinal microbiota and on the intestinal expression of some genes related to inflammation were tested. Seventy weaned pigs were reared for six weeks and then gastrointestinal digesta and mucosa tissue from jejunum were collected. FUT1 and MUC4 genotypes were tested by PCR-RFLP. Total bacteria DNA was extracted using the QIAamp DNA Stool Mini Kit and the v3-v4 regions of 16S rRNA gene were sequenced using the Illumina MiSeq. Microbiota sequencing data were analysed using QIIME's with open-reference based on OTU strategy. Total RNA was extracted using the Trizol reagent and reverse transcribed (ImProm-II Reverse Transcription System) and the qPCR of IL8, GPX2 and REG3G was performed in a Roche Light Cycler instrument. The effect of the combinations between FUT1 and MUC4 genotypes on the data was tested using the CONTRAST option of the GLM procedure of SAS. The allele frequencies distribution of FUT1 locus was 0.29 (A) and 0.71 (G), while for MUC4 locus was 0.78 (G) and 0.22 (C). No pig with the genotype AA (FUT1) – CC (MUC4) was found. The pigs with the combination of FUT1 AG genotype with MUC4 GG or CG had more variability in microbiota composition (PD_whole_tree variation index, chao index), more abundant presence of Streptococci and Clostridium, and a higher expression of GPX2, vs GG_GG and GG_CG (P<0.05). Contrasts between MUC4 genotypes were not significant. FUT1 genotype may affect the adaptive response to the bacterial colonization in heathy pigs.

Dietary organic acids, prebiotic and probiotic on gut microflora and lymphocyte populations in pigs
I. Skoufos[1], A. Tzora[1], A. Karamoutsios[1], G.K. Papadopoulos[1], I. Giannenas[2], A. Tsinas[1], E. Christaki[2] and P. Florou-Paneri[2]
[1]TEI of Epirus, Agriculture Technology, Division Animal Production, Kostakioi Artas, 47100, Greece, [2]Aristotle University of Thessaloniki, Veterinary Medicine, Thessaloniki, 54124, Thessaloniki, Greece; jskoufos@teiep.gr

The objective of this study was to evaluate whether a combination of natural feed additives consisted of Enteroccocus faecium 2×1,010 cfu/g, at 35 mg/kg feed, benzoic acid at 5 g/kg feed and mannan-oligosacharides at 1 g/kg feed could affect growth performance, microflora composition and immunocyte populations on fattening pigs. One hundred ninety two crossbred pigs were allocated into two experimental groups from 114 until 165 day of life. Control group was fed a basal diet, whereas the other group received the diet with the combination of feed additives. Intestinal samples were collected from the jejunum, caecum and mid colon. Total aerobes, total anaerobes, Clostridium perfringens, Enterococci, Enterobacteriaceae spp, Lactobacilli spp and Bifidobacteria spp were estimated by conventional microbiological techniques, using selective agar media. Immunocyte population analysis was also performed via four-color flow cytometry using the following monoclonal antibodies: cytoplasmic CD3 (pan-T-cells), CD79a (B-cells), SLA-DR (swine MHC-II), CD4a (T-helper cells) and CD8a (T-cytotoxic cells). The group fed the diet supplemented with the aforementioned additives showed improved (P<0.05) body weight on slaughter age, increased (P<0.05) counts of Enterococci and Lactobacilli in the jejunum, Enterococci and Bifidobacteria in the caecum, and Enterococci, Lactobacilli and Bifidobacteria in the mid colon, along with decreased counts of Enterobacteriaceae spp. Flow cytometric analysis showed no differences in the total B and T-helper and cytotoxic cell populations. The tested combination of feed additives provides promising results on fattening pigs in the absence of immune challenge, in order to improve growth performance and establish a beneficial microflora.

ECO FCE: Effect of early artificial rearing and milk replacer supplementation on growth of pigs
J.G. Madsen[1,2], M. Kreuzer[1] and G. Bee[2]
[1]*ETH Zurich, Universitätstrasse 2, 8092 Zurich, Switzerland, [2]Agroscope, Route de la Tioleyre 4, 1725 Posieux, Switzerland; johannes.madsen@agroscope.admin.ch*

Low birth weight piglets (L-BtW; 0.8-1.2 kg BtW) from large litter sows show impaired pre and post-weaning growth compared to their heavier littermates. The objective of this study was to compare survival and pre and post-weaning growth performance of conventionally (CON; n=18) and artificially (ART; n=18) reared L-BtW piglets born from large litter sows (>15 pigs born/litter). Artificially reared piglets were allocated to two dietary treatments where the milk replacer either remained unsupplemented (C) or was supplemented with L-carnitine (0.05% of DM) and L-arginine (1.67% of DM) (CarArg). The ART piglets were housed in pairs and offered either C or CarArg milk replacer ad libitum in rescue decks from d 7-28 of age. The CON piglets were nursed by the sow from birth to d 28. On d 28, one C and CarArg piglet per pair was sacrificed and their organs were weighed. The other piglet was weaned together with a CON piglet and offered ad libitum the same standard weaning and grower diet for 60 d post weaning. Pre-weaning survival rate and growth performance in the pre and post-weaning period was assessed. Except for the greater (P<0.05) liver (27.6%) and stomach weight (26.7%) of C compared to CarArg piglets, no differences in other phenotypic traits were found between dietary treatments. All artificially reared piglets survived the nursing period, which is in strong contrast to a 56% mortality rate of L-BtW CON pigs reared in loose house sow system. Weaned ART and surviving CON piglets displayed weaning weights (5.0 kg) which were below herd average of 6.6 kg. At 88 d of age, BW was again comparable between ART and CON piglets (25.8 vs 26.0 kg). These preliminary results clearly show that survival rate of L-BtW piglets is markedly improved using rescue decks, without compromising the subsequent growth performance in the early post-weaning period. Thus, overall production efficiency of pig production with hyperprolific sows is enhanced by artificial rearing.

ECO-FCE: Phytase, xylanase, and protease use in a wheat DDGS and rapeseed diet for finisher pigs
A. Torres-Pitarch[1], U. McCormack[1], V. Beattie[2], E. Magowan[3], G.E. Gardiner[4] and P.G. Lawlor[1]
[1]*Teagasc, Pig Development Department, Moorepark, Fermoy, Co Cork, Ireland, [2]Devenish Nutrition Ltd., Belfast, Northern Ireland, United Kingdom, [3]Agri-Food and Bioscience Institute, Belfast, Northern Ireland, United Kingdom, [4]Waterford Institute of Technology, Dept. of Science, Waterford, Ireland; alberto.torrespitarch@teagasc.ie*

Exogenous enzyme use has been suggested as a means of increasing feed efficiency in pigs. This study aimed to determine the effect of enzyme supplementation to wheat distiller's dried grains with solubles (DDGS) and rapeseed-based diets on growth and carcass quality of finisher pigs. A total of 162 pigs (38±2.9 kg) housed in same sex pairs were allocated to 1 of 9 dietary treatments: (1) Positive control (PC) diet formulated to exceed nutritional requirements (9.9 NE MJ/kg, 8.3 g/kg of available Lys); (2) Negative control (NC) diet formulated to 95% of the NE and amino acid content of PC; (3) NC+phytase (Phyzyme 5000 XP TPT, Danisco Animal Nutrition); (4) NC+xylanse+β-glucanase (XB; Rovabio Spiky, Adisseo); (5) NC+protease (Ronozyme ProAct, DSM); (6) NC+phytase+protease; (7) NC+phytase+XB (8) NC+XB+protease; (9) NC+phytase+XB+protease. Diets including phytase (3, 6 and 9) were formulated to allow a sparing effect from phytase of 0.15 and 0.10 g/kg for digestible P and Ca, respectively. The experiment lasted 63 days and growth and feed intake were recorded. Pigs on all treatments had similar average daily gain (ADG) (1,051±23.0 g/day) and final live weight (106.8±1.42 kg). Pigs fed the NC diet had higher average daily feed intake (ADFI) and poorer feed conversion ratio (FCR; P<0.01) than pigs fed the PC diet. The sparing effect of phytase was effective, as ADG, ADFI and FCR were similar for diets formulated with low P and Ca and supplemented with phytase to diets formulated with adequate Ca and P (P>0.05). XB or protease supplementation did not improve FCR (P>0.05). Dietary treatment had no effect on cold carcass weight, muscle depth, fat depth or lean meat yield. Supplementation of phytase, XB and protease did not improve FCR when used alone or in combination.

ECO-FCE: Composition of the intestinal microbiota varies between high and low RFI pigs

U.M. McCormack[1,2], T. Curiao[2], P. Cotter[2], O. O'Sullivan[2], F. Crispie[2], S. Buzoianu[2], D. Berry[2], B. Metzler-Zebeli[3], G.E. Gardiner[1] and P.G. Lawlor[2]
[1]Waterford Institute of Technology, Waterford, Waterford, Ireland, [2]Teagasc, Moorepark, Fermoy, Co Cork, Ireland, Fermoy, Co Cork, Ireland, Ireland, [3]University of Veterinary Medicine, Vienna, Vienna, Ireland; tania.curiao@teagasc.ie

The intestinal microbiota plays an important role in energy harvest for the host and therefore could impact feed efficiency. The aim here was to examine the variation of microbial profiles in pigs divergent for residual feed intake (RFI). High throughput 16S rRNA gene sequencing (Illumina) was performed on faeces at weaning, days 42 and 139 post weaning (pw) and on caecal and ileal digesta from pigs of high (HRFI; n=10), medium (MFRI; n=10) and low RFI (LRFI; n=12). Sequence data were analysed using a QIIME pipeline and significant differences were determined. Bacterial richness was highest in the caecum; however, no significant differences between RFI ranks were found. Bacterial diversity clustered by sample/time point but not by RFI rank. Firmicutes and Bacteroidetes were the most abundant phyla in the faeces and caecum, while Firmicutes and Actinobacteria predominated in the ileum. The relative abundance of 18 taxa varied significantly with RFI rank. These included those detected at high relative abundance i.e. decreased Actinobacteria (Rhodococcus spp) and increased Fusobacteria in the ileum of LRFI pigs, and increases/decreases of Firmicutes members (Clostridiales) in the faeces at day 139pw. The low abundance taxa of Clostridiales were higher in LRFI pigs at day139 and Erysipelotrichales and Lactobacillales were lower in LRFI pigs at weaning and day 42pw respectively. Bacteroides spp. (from Bacteroidetes) increased by 0.14% at day 139pw in LRFI pigs. Overall, the intestinal microbiota matured over time, but bacterial richness and diversity did not vary with RFI rank. However, compositional shifts occurred between HRFI and LRFI pigs. The increase of Bacteroides spp. and some Clostridiales suggests a higher metabolic efficiency in LRFI pigs and these could be potential biomarkers for feed efficiency.

Impact of amino acid formulation strategy on finishing pig performance

E. Magowan[1], K.J. McCracken[2], S. Smyth[3], F.J. Gordon[3], J. Kennan[2] and V.E. Beattie[2]
[1]Agri-Food and Biosciences Institute, Hillsborough, BT26 6DR, United Kingdom, [2]Devenish Nutrition Ltd, Belfast, BT1 3BG, United Kingdom, [3]John Thompson and Sons Ltd, Belfast, BT15 3GW, United Kingdom; elizabeth.magowan@afbini.gov.uk

This study investigated the performance of pigs offered diets formulated to a range of amino acid target levels. 480 pigs (PIC 337 × (LRxLW)) were allocated to one of four treatments based on weight and gender (boar and gilt) and were penned in groups of 10 across 8 time periods from 12 weeks of age to slaughter (112 kg). Diets were formulated to contain 13.85 MJ/kg DE. The control diet contained 17.5%CP and the remaining three diets contained 15% CP. Diet 2 was formulated to M+C, Thr, Trp, Val, Iso and Arg, diet 3 to M+C, Thr and Trp and diet 4 to M+C and Thr. The total (analysed) levels of lysine, Met, Cys, Thr, Trp, Val, Isoleu and Arg were (% of fresh diet) 1.05, 0.32, 0.32, 0.66, 0.14, 0.77, 0.66 and 1.06 for diet 1 (control) respectively; 1.06, 0.32, 0.27, 0.67, 0.11, 0.71, 0.57 and 0.86 respectively for diet 2; 1.06, 0.31, 0.29, 0.67, 0.11, 0.65, 0.53, 0.86 respectively for diet 3 and 1.04, 0.36, 0.30, 0.66, 0.11, 0.65, 0.53, 0.87 respectively for diet 4. Pigs were weighed at 12, 15, 18, 21 weeks of age and slaughter. Feed intake and feed efficiency (FCR) were recorded and data was analysed by ANOVA in Genstat. Pig start weight averaged 40 kg. Daily gain was improved (P<0.05, SEM 21.4) between 12 and 15 weeks of age when diets 2 and 3 were offered (786 and 755 g/day respectively) compared with diet 1 (682 g/day) with diet 4 being intermediate (742 g/day). FCR was also improved between 12 and 15 weeks of age (P<0.05; SEM 0.071) when diet 2 was offered (2.08) compared with diet 1 (2.41) with diets 3 and 4 being intermediate (2.20 and 2.22 respectively). There was no effect (P>0.05) of diet on the overall finishing pig performance between 40 and 112 kg. It is concluded that it is not necessary to formulate diets to six amino acids but formulating to Lys, Met + Cys, Thr and Trp does have benefits for finishing pigs, especially in the early finish period when adopting the levels of amino acids used in this trial.

Performance of male layer hybrids fed different dietary protein sources as fattening cockerels

S. Ammer[1], N. Quander[1], I.D.M. Gangnat[2], V. Maurer[1] and F. Leiber[1]
[1]*Research Institute of Organic Agriculture (FiBL), Department of Livestock Sciences, Ackerstrasse 113, 5070 Frick, Switzerland, [2]ETH Zurich, Institute of Agricultural Sciences, Universitätsstrasse 2, 8092 Zurich, Switzerland; stefanie.ammer@fibl.org*

We investigated the performance and meat quality of male layer hybrids as fattening cockerels under feeding schemes with different protein sources. A total of 90 male Lohmann Brown (LB), 90 male Lohmann Selected Leghorn (LSL) and 90 mixed-sex chicken of an organic fattening strain (Hubbard JA-757 (HUB)) were randomly divided into three feeding groups per genotype: (1) commercial feed for organic chicken fattening containing 25.5% soybean cake (Control); (2) control diet, supplemented with extra alfalfa meal ad libitum (Alf-ext); and (3) control diet with a 12% replace of soybean cake by alfalfa meal (Alf-int). The feed intake was recorded group-based while daily weight gains, live weights and meat quality were measured individually. Fattening periods took 63, 91 and 104 days for HUB, LB and LSL, respectively. Final weights were influenced by the interaction of genotype and feed (P=0.02) and were greater for HUB (1,843 g) and LB (1,871 g) than for LSL (1,631 g). Highest daily weight gains were measured for HUB (29.3 g), followed by LB (20.6 g) and LSL (20.7 g), influenced by the genotype (P<0.001) but not by the diet. For LB and LSL, the proportions of leg muscle were higher than for HUB. LB had lower proportions of breast muscles compared to HUB and LSL (P<0.001). Diet did not affect these parameters (P>0.05). Regarding meat quality, the maximum shear force of the right breast muscle was lower for HUB and LSL compared to LB (P<0.001) and was not influenced by the diet (P>0.05). Breast raw protein and raw fat contents were higher for chicken fed Alf-int than Alf-ext (P<0.001). In conclusion, male LB layer hybrids may be used as fattening cockerels, but longer fattening periods are required. Additionally, soybean cake can be replaced partially by alfalfa meal achieving similar performance and meat quality.

Ability to digest explains part of the between cow feed efficiency variability

A. Fischer[1,2,3], R. Delagarde[1,2] and P. Faverdin[1,2]
[1]*Agrocampus-Ouest, UMR 1348 PEGASE, 65 rue de Saint-Brieuc, 35000 Rennes, France, [2]INRA, UMR 1348 PEGASE, domaine de la Prise, 35590 Saint-Gilles, France, [3]Institut de l'élevage, Monvoisin, 35652 Le Rheu, France; amelie.fischer@rennes.inra.fr*

Improving feed efficiency may be a solution to increase farms economic and environmental sustainability, through decreasing feed consumption and maintaining or increasing milk production. The main issue is to understand why inefficient cows need to consume more feed than the most efficient ones independently of their energy requirements level. Feed inefficiency could result from a lower dry matter (DM) diet digestibility: inefficient cows need to consume more feed than the efficient ones to cover their energy requirements. The aim of this study was to assess the relationship between feed efficiency and DM diet digestibility in dairy cows. Sixty Holstein cows were fed ad libitum a unique TMR during the 238 first days of lactation. Feed efficiency was defined as the residual energy intake (REI). The 60 individual REI were calculated as the residuals of the linear regression of cumulated net energy intake (UFL) not corrected for digestibility differences on cumulated requirements for fat corrected milk yield, metabolic BW, BCS, empty BW gain and empty BW loss. The cows were dosed with ytterbium oxide as an external marker during 2 periods of 14 days (2 × 5 days of rectal faecal sampling) to assess individual mean DM digestibility (known individual DM intake). Intake level (kg DMi/kg BW) explained 14% of REI variation and DM digestibility explained 9% of REI variation. Intake level explained 15% of DM digestibility variation. The more efficient the cow is, the lower the intake level is and the higher its digestibility is. Further investigation is planned to test the hypothesis that inefficient cows would be able to increase their feed efficiency through restricting their feeding level.

Transcriptional profile of bovine rumen papillae in response to diet restriction and re-alimentation

K. Keogh[1], S.M. Waters[1], P. Cormican[1], A.K. Kelly[2], E. O'Shea[1] and D.A. Kenny[1]
[1]Teagasc, Animal and Bioscience Research Department, Grange, Dunsany, Co. Meath, Ireland, [2]University College Dublin, School of Agriculture and Food Science, Belfield, Dublin 4, Ireland; kate.a.keogh@teagasc.ie

The objective of this study was to examine the effect of dietary restriction and subsequent re-alimentation induced compensatory growth (CG) on ruminal epithelial papillae. For this, Holstein Friesian bulls (n=60) were assigned to one of two groups: (1) restricted feed allowance (RES; n=30) for 125 days (Period 1) followed by ad libitum access to feed for 55 days (Period 2); or (2) ad libitum access to feed throughout (ADLIB; n=30). At the end of each period 15 animals from each treatment were slaughtered and rumen papillae harvested. Messenger RNA (mRNA) was isolated from all papillae samples collected. cDNA libraries were then prepared from rumen papillae mRNA and sequenced. Quality analysis of resultant sequence reads was performed using FASTQC. Reads were then aligned to the bovine genome using TopHat. Mapped reads were counted and differentially expressed genes (DEG) identified using EdgeR. DEGs are defined as having a Benjamini-Hochberg P value of <0.05. Animals undergoing re-alimentation displayed CG, growing at 1.8 times the rate of ADLIB animals in Period 2. Following a period of dietary restriction at the end of Period 1, 64 genes were identified as differentially expressed, with only one gene differentially expressed at the end of Period 2. When analysed within treatment (RES, Period 2 vs Period 1), 392 genes were differentially expressed in animals undergoing re-alimentation induced CG compared to dietary restriction. Genes identified as differentially expressed included those involved in processes including transport, cellular interactions and organisation, protein folding and gene expression. This study provides an insight into the molecular mechanisms underlying the expression of CG in rumen papillae in cattle; however the results suggest that most of the CG has occurred by day 55 of re-alimentation.

Concentrate supplementation during the close-up dry period: dairy cow production and immune function

M. Little[1,2], N. O'Connell[1] and C. Ferris[2]
[1]IGFS, Queens University Belfast, BT9 5BN, United Kingdom, [2]Agri-Food and Biosciences Institute, Hillsborough, BT26 6DR, United Kingdom; mark.little@afbini.gov.uk

The transition period (last three weeks of gestation and the first three weeks of lactation) is the period of greatest disease risk during the dairy cow's production cycle, with negative energy balance (NEB) contributing to immune dysfunction. Thus there is interest in nutritional strategies which minimise NEB at this time. This two treatment study involved multiparous (n=28) and primiparous (n=22) Holstein Friesian dairy cows. All cows were offered a medium quality grass silage from dry-off until four weeks before the predicted calving date, when half the cows (treatment S+C) were moved to a mixed ration comprising grass silage plus concentrates (60:40 DM ratio, mean daily concentrate intake of 4.5 kg DM/cow/day), while the remaining cows (treatment SO) continued to be offered the grass silage diet only until calving. Postpartum, all cows were offered a common diet consisting of grass silage and concentrates (40:60 on a DM basis) for ten weeks. During the four weeks prepartum cows on S+C had a higher total DMI (P<0.001), gained more liveweight (P=0.027), tended to gain more body condition score (P=0.060), had a lower serum NEFA concentration (P=0.023) and tended to have a higher serum BHBA concentration (P=0.074) than cows on SO. Offering concentrates prepartum had no effect (P>0.1) on total DMI, mean daily milk yield, milk composition, liveweight, body condition score, and serum NEFA or BHBA concentrations postpartum. Similarly, treatment had no effect (P>0.1) on the phagocytic activity of neutrophils at day 14 prepartum or at days 3, 7, 14 or 21 postpartum, or on the oxidative burst activity of neutrophils at day 14 prepartum or at days 3, 7 or 21 postpartum. However, cows on S+C tended (P=0.078) to have a higher oxidative burst index at day 14 postpartum, than cows on SO. In summary, concentrate supplementation of a grass silage diet during the close-up dry period had no beneficial effects on production, metabolic or immune function postpartum.

Improving the prediction of Amino Acid Digestible in the Intestine through meta-analysis

S. Lemosquet[1], C. Panzuti[1], L. Bahloul[2], G. Cantalapiedra-Hijar[3], P. Chapoutot[4], P. Nozière[3], D. Sauvant[4] and J. Guinard-Flament[1]
[1]PEGASE, Agrocampus Ouest, INRA, 35590 Saint-Gilles, France, [2]ADISSEO SAS, CERN, 03600 Commentry, France, [3]UMR H, INRA, Vetagro Sup, 63122 Saint Genès-Champanelle, France, [4]MoSAR, AgroParisTech, INRA, 75015 Paris, France; sophie.lemosquet@rennes.inra.fr

The metabolisable protein (PDI) supplies were revised in the new INRA feeding system (2013). Consequently, the digestible amino acid (AA) supply in the intestine (AADI, in % of PDI) for cattle was to revise since it was developed in 1998. A meta-analysis was performed to estimate the AADI using a close approach to the past method. The new database is larger than the past one (30 publications and 133 diets from 1967 to 1992) since it includes 64 publications (from 1967 to 2013) on cattle and 243 dietary treatments mainly focussed on nitrogen supply (amount or quality). Each dietary AA supply in the duodenum (g/kg DM) was calculated using a linear combination of the rumen undegraded (RUP), the microbial and the endogenous protein flows in the intestine (PI) with the individual feed AA profiles and the two fixed microbial and endogenous profiles, respectively. It was then expressed in % of total AA (i.e. 16 AA). Each relationship between the measured and the calculated AA supply was analysed using a linear model of covariance with a fixed publication effect. The new vs the past models were for the 4 AA recommended in the INRA system: $0.80 + 0.90 \times X$ (RMSE=0.29) vs $1.90 + 0.76 \times X$ (RMSE=0.38) for Lys; $1.03 + 0.45 \times X$ (RMSE=0.17) vs $0.32 + 0.73 \times X$ (RMSE=0.26) for Met; $0.57 + 0.79 \times X$ (RMSE=0.24) vs $0.93 + 0.58 \times X$ (RMSE=0.22) for His; $1.78 + 0.86 \times X$ (RMSE=0.32) vs $2.55 + 0.76 \times X$ (RMSE=0.45) for Leu, respectively. The new INRA PI estimations led to lower biases between the measured and the calculated AA supplies compared to the past models since lower RMSE were observed for 15 AA models in addition to a slope closer to one and an intercept closer to zero for 12 AA models. Further analyses will describe the interfering factors influencing the relationships observed for Met, Phe, Arg, Gly, and the RMSE for His. It will be possible to improve the AADI calculations by using these new models to estimate the AA in RUP and microbial PI and the digestibilities of RUP and microbial PI.

Insect's based extruded feed: process optimization and impact on in vitro digestibility

M. Ottoboni[1], T. Spranghers[2], J. Michiels[2], P. De Clercq[2], S. De Smet[2], W. De Jaeghere[2], V. Dell'Orto[1], F. Cheli[1], L. Pinotti[1] and M. Eeckhout[2]
[1]University of Milan, Department of Health, Animal Science and Food Safety (VESPA), Via Celoria 10, 20134 Milano, Italy, [2]Gent University, Faculty of Bioscience Engineering, Valentin Vaerwyckweg 1, 9000 Gent, Belgium; matteo.ottoboni@unimi.it

This study investigated the inclusion of Hermetia illucens (HI) material in an experimental extruded feed, and the impact of extrusion on in vitro digestibility. For this purpose, premixes of HI larvae or prepupae (wet material) and wheat flour in a ratio of 25:75 with or without soy oil addition, have been formulated as follows: prepupae+wheat (no oil); prepupae+wheat (low oil); prepupae+wheat (medium oil); prepupae+wheat (high oil); larvae+wheat (no oil). Ether extract (EE) content on wet basis prior to extrusion of the respective formulations was: 3.2, 3.9, 4.6, 5.4 and 4.6%. Mixtures were homogeneous in term of moisture and protein content (23.89 and 11.26% on wet basis respectively). Feed mixtures were then extruded using a co-rotating, conical twin-screw mini extruder (HAAKE™ MiniLab II). In order to evaluate extrusion features net torque value (NTV) was measured for each HI-wheat blend. On the best performing mixture, selected according to NTV, further extrusion test have been done at four barrel temperatures (60, 70, 80 and 90 °C) at the same screw speed (100 rpm). In vitro organic matter digestibility (OMD) and protein digestibility (CPD) of these extrudates were measured in order to evaluate the effect of technological treatments on their nutritional value. Increasing the blend EE content (up to 5.4%), the NTV was reduced by four times (<100 Ncm) compared to the 3.2 and 3.9% EE mixtures. In this EE range, the best performing mixture was larvae+wheat (no oil, 4.6% EE). Processing the mixture by extrusion, increased OMD by 13%, but not CPD, compared to the untreated (no extrusion) mixture. By contrast, extrusion temperature did not affect OMD nor CPD. Concluding, fresh insect larvae material con be included in extruded feed material; EE content in the mixture is a key variable that must be considered; extrusion can contribute to increasing OMD in insect containing feed blends. The present results obtained on lab scale need to be confirmed in a large-scale pilot plant.

Effect of olive cake feeding and Saccharomyces cerevisiae on performance of Awassi lambs
B.S. Obeidat
Jordan University of Science and Technology, Animal Production, P.O. Box 3030, Irbid 22110, Jordan;
bobeidat@just.edu.jo

A 63-d study was conducted to determine the influence of feeding olive cake (OC) and Saccharomyces cerevisiae (SC) supplementation on growth performance, nutrient intake and digestibility of Awassi lambs. Thirty lambs (initial body weight (BW) = 25.0±1.91 kg) were randomly assigned into 3 dietary treatments; fed the conventional diet (CON; n=10) or OC containing diets (n=20). Within the OC diet, 10 of the lambs received a diet containing 15% dietary dry matter (DM) OC (OC; n=10) and the other 10 received OC diet plus 0.5 g/head/d SC (OCSC; n=10). Diets were formulated to be isonitrogenous and to meet nutrient requirements for growing lambs. Lambs were offered ad libitum access to the assigned diet until the end of the study. Before the morning feeding, lamb BW was measured on d 0, 21, 42, and 63 to evaluate ADG, feed:gain ratio, and other performance efficiency. Dry matter, crude protein (CP), and ether extract (EE) intake was greater for lambs fed OC containing diets than for the CON diet. However, no significant differences were observed in intake of neutral detergent fiber (NDF) and acid detergent fiber (ADF) among dietary diets. Digestibility of DM, CP, NDF and ADF decreased (P≤0.04) in OC and OCSC diets vs the CON diet. However, digestibility of EE increased (P<0.0001) in OC containing diets than the CON diet. No significant differences were noticed in total gain, ADG and feed efficiency among dietary diets. In all measured variables, SC supplementation did not show any differences when included in the diet of OC compared with OC diet alone. In summary, results of the current study indicated that feeding olive cake for growing Awassi lambs had beneficial effects without adverse effect and could replace the conventional feeds; whereas Saccharomyces cerevisiae supplementation did not show any benefits.

Evaluation of hay for dairy cow produced in Mediterranean Area
R. Grazioli, N. Musco, M.I. Cutrignelli, R. Tudisco, F. Infascelli and S. Calabrò
University of Napoli Federico II, Veterinary Medicine and Animal Production, via F. Delpino, 1, 80137, Italy;
serena.calabro@unina.it

The influence of hay quality on milk yield and nutritional characteristics is widely demonstrated. Moreover, producing high quality hay is one of the most important economic considerations for livestock managers; any forage deficiency of nutrients have to be supplemented by concentrates, which increases the total feeding cost. In the hilly area of Campania Region, large amounts of forage are used in dairy cow ratio producing high quality milk. The aim of this study was to evaluate the hay quality in seven mixed hays produced in this area. In particular, the samples were subjected to sensory evaluation (including color, odor, consistency, leafiness, presence of powder, Leguminous/Gramineae ratio, botanical species) and proximate analysis (crude protein, NDF, ash). Moreover, the fermentation characteristics were studied using the in vitro gas production technique incubating the hay samples under anaerobic condition, for 120 h, using bovine rumen fluid as inoculum; organic matter degradability (dOM) and cumulative volume of gas related to incubated OM (OMCV) were determined. Moreover, methane production was evaluated at 24 h using gas-chromatograph. To test the differences between the hay the GLM of SAS was used. Few significant differences appear between the tested hays. As a whole, the forage resulted poor in their chemical composition (mean values 116±19.2 and 514±46.8 g/kg DM, for CP and NDF, respectively) and low in OM degradability (mean value 72.7±3.27%) and gas production (OMCV: 264±8.80 ml). A significant (P<0.01) correlation was found between parameters of chemical analysis and sensory evaluation. Regarding methane production, the values obtained (16.0±1.13% of total gas) resulted similar compared the mean value reported in literature for forage of similar composition. The results show that in this Mediterranean Area it is necessary improve the haymaking technique in order to valorize their quality and use in livestock nutrition.

Mycorrhized maize grain: effect on rumen environment and degradability
A. Chiariotti, D. Meo Zilio, G. Contò, S. Di Giovanni and C. Tripaldi
Consiglio per la ricerca in agricoltura e l'economia agraria, CREA-PCM, via salaria 31, 00015, Italy;
antonella.chiariotti@entecra.it

The effect of MICOSAT F® microbial treatment on maize grain degradability and rumen environment was evaluated. The trial was conducted on 14 Holstein cows divided into two groups of 7 heads each and lasted 100 days. Besides six cannulated cows were used for rumen microbial analyses in a cross-over design and three of them were used for in situ degradability trial. Rations included 5 kg maize grain, mycorrhized (E) and not mycorrhized (C), administered as unifeed (0.91 Milk FU/kg Dry matter (DM), 155 g/kg DM crude protein) and ad libitum basis. Chemical characteristics of feedstuffs and diets were determined (AOAC, 1995; Goering and Van Soest 1970). Dry matter intake (DMI) and average daily gain (ADG) were measured. qPCR quantifications of total bacteria, and the main cellulolytic (R. Albus, R. Flavefaciens, F. succinogenes) were performed on rumen content. Dry matter degradability was predicted according to Ørskov and McDonald. Milk yield recording and sampling were carried on every two weeks and fat, protein, urea, somatic cells were determined. Data were analysed by GLM procedure using the mono-factorial model. Maize grain proximate composition showed no difference. Immediately degradable fraction of dry matter was higher in E compared to C, while the slowly degradable fraction was significantly higher in C than in E. In the rumen fluid, protozoa counts were significantly higher in E diet (+15.6%, P<0.05) and total bacterial behaved accordingly (6.91 vs 6.19 log10 pg/g DM, P<0.01).Milk yield was the same in the two groups, but E group showed an increased milk protein content. This is likely due to a higher DMI (22.35 vs 21.11 kg/d, P=0.015) for E group, which also showed a tendency to a higher ADG (272.21 vs 124.72 g/d). In conclusion, the use of mycorrhized maize seems to affect rumen microflora and some animal performances.

Effects of Silosolve bacterial inoculant on chemical composition and fermentation of ensiled Oats
R.S. Thomas[1], B.D. Nkosi[1,2], T. Langa[1] and S. Modiba[1]
[1]Agricultural Research Council: Animal Production Institute, Animal Nutrition, P/Bag x2, Irene, 0062, South Africa, [2]University of the Free State, Centre for Sustainable Agriculture, P.O. Box 339, Bloemfontein, 9300, South Africa; ronaldt@arc.agric.za

The present study evaluated the effects of adding Silosolve bacterial inoculant on chemical composition and fermentation characteristics of Oats (Avena sativa L.) silage. Oats forage was harvested at 24% dry matter (DM) and wilted to achieve 48% DM, and was chopped to 5 cm length. The forages were treated with or without inoculant, as follows: CON (no additive) and Silosolve, a heterofermentative LAB inoculant that contains strain of Lactobacillus buchneri (DSM 16568), was applied to obtain at least 6×10^5 cfu/g fresh material and ensiled in 210 litter drums. After 90 days of ensiling, drums were opened and sampled for chemical composition and fermentation characteristics analyses. Silosolve inoculation reduced (P<0.05) silage pH and neutral detergent fibre compared to the control. In addition, the Silosolve treatments had higher (P<0.05) contents of lactic acid, water-soluble carbohydrates and dry matter compared to control. It was concluded that the addition of Silosolve to forage oats at ensiling improve the fermentation characteristics. Further work to determine the aerobic stability of the silage and nutrient digestion using rams is warranted.

ECO FCE: Phosphorus nutritional conditioning affects SLC34A2 expression in broiler chickens

M. Ballester[1], M. Francesch[2], O. González[1], D. Torrallardona[2] and R. Quintanilla[1]
[1]IRTA, Animal Breeding and Genetics, Torre Marimon, E08140 Caldes de Montbui, Spain, [2]IRTA, Monogastric Nutrition, Mas de Bover, E43120 Constantí, Spain; raquel.quintanilla@irta.cat

Nutritional conditioning (NC) can improve lifetime performance and nutrient utilization. A trial was conducted with 96 male broiler chickens (Cobb 500 FF) to test if dietary phosphorus (P) conditioning results in a better utilisation of P later in life. Solute carrier family 34 (type II sodium/phosphate cotransporter) member 2 (SLC34A2) is considered to be the major Na-P cotransporter, primarily expressed in the brush-border of the small intestinal epithelium. For the NC experiment, a deficient P diet during the first week of age was offered to 48 animals (conditioned animals) vs a conventional diet to the other 48 (non-conditioned animals). At 21 d of age, after a period (8-21 d) with a common balanced diet, 24 animals from each group were maintained on a balanced diet whereas the other 24 fed a P deficient diet. This resulted in four treatments according to a 2×2 factorial design with NC and late deficiency (LD) as factors. At d 30, 12 birds per treatment were killed, ash content in tibia was measured, and the expression of SLC34A2 in duodenum and jejunum was analysed by RT-qPCR. LD impaired ($P<0.05$) the birds' growth whereas no NC or NCxLD effects were observed. A relevant NCxLD interaction ($P<0.01$) was observed for tibia ash, being the NC associated to an increased tibia ash only in conditioned birds subsequently fed a balanced diet. SLC34A2 was more expressed in duodenum than in jejunum epithelium. In duodenum, both NC and LD increased significantly SLC34A2 expression, and interaction was non-significant. Conversely, in jejunum there was a suggestive ($P<0.1$) LD×NC interaction: although SLC34A2 expression was increased in both LD treatments, this increase was significantly larger for non-conditioned animals. Further studies of the gut transcriptome are being performed to identify genes and pathways involved in metabolic efficiency regarding animal feed utilization within each treatment.

ECO-FCE: Serum metabolite profiles as predictors for feed efficiency in broiler chickens

B.U. Metzler-Zebeli[1], E. Magowan[2], P.G. Lawlor[3] and Q. Zebeli[1]
[1]University of Veterinary Medicine Vienna, Veterinaerplatz 1, 1210 Vienna, Austria, [2]Agri-Food and and Biosciences Institute, Large Park, Hillsborough, BT26 6DR, United Kingdom, [3]Teagasc, Moorepark, Fermoy, Co. Cork, P61 C996, Ireland; barbara.metzler@vetmeduni.ac.at

To reduce feed costs, selecting chickens for enhanced feed efficiency (FE) is a way of improving profitability. As knowledge about physiological differences in birds of diverging FE advances, the present objective was to assess if peripheral serum metabolite profiles and acute-phase-proteins (APP) can be used to predict FE in chickens. At 5 weeks of age, female and male chickens were selected according to their residual feed intake (RFI) at 2 locations (L1: Austria; L2: UK; n=9/ low, medium and high RFI and location). Blood was collected from the jugularis vein at L1 and heart at L2. Serum glucose, lipids, and urea were biochemically determined and ovotransferrin (OVT) and alpha-1-acid glycoprotein (AGP) using chicken-specific ELISA. Effects of RFI, location and their two-way interaction were analyzed in SAS. Chickens at L1 had a 15%-greater body weight gain than at L2 ($P<0.05$) but similar feed intake. We observed location effects for serum glucose, urea, cholesterol and NEFA in females, and for serum glucose and triglycerides in male chickens. Serum urea, NEFA ($P<0.05$), cholesterol and triglycerides ($P<0.1$) linearly increased from low, medium to high RFI in females, whereas in males cholesterol ($P<0.05$) and triglycerides ($P<0.1$) showed the same linear trend from low, medium to high RFI. Serum AGP linearly increased by 35% from low to high RFI in male chickens at L1 ($P<0.05$), whereas males at L2 showed similar values. Also, APP levels in females were similar, only OVT was higher in females at L1 than at L2 ($P<0.05$). Overall, serum metabolites may be used as FE predictors, particularly in female chickens, whereas APP appeared less useful. Location effects may render it difficult to set uniform thresholds for low, medium and high RFI which may be related to chicken's growth performance and systemic nutrient assimilation.

Effect of methionine supplementation in processed Mucuna sloanei seed meal in broiler chickens diet

A.H. Akinmutimi[1] and O.O. Ezeigbo[2]
[1]*Michael Okpara University of Agriculture, Umudike, Animal Nutrition and Forage Science, MOUA, Umudike. PMB 7267, Umuahia, Abia State, Nigeria, 440001, Nigeria, [2]Holy Cross Anglican Church, Pastorial, Ezeleke, Umuopara, Umuahia, Nigeria, 440001, Nigeria; henryakinmutimi2@yahoo.com.au*

The objective of this study was to develop a high quality animal protein at reduced cost through the use of alternative feedstuff (Mucuna sloanei seed meal). One hundred and fifty day-old Marshal Broiler chicks were used to assess the quantitative replacement of soybean meal with soaked and boiled Mucuna sloanei seed meal (SBMSSM) with or without methionine supplementation. Ten birds per treatment were replicated thrice in a completely randomized design. Diet 1 was control (0% SBMSSM), diet 2 had 6% SBMSSM quantitatively replacing soybean meal but without methionine supplementation, Diets 3 to 5 had 6% SBMSSM quantitatively replacing soybean meal but with 0.1, 0.2 and 0.3% methionine supplementation, respectively. Feed and water were given ad-libitum for 56 days. The crude protein (23.89%) and gross energy value (4.56 kcal/g) of SBMSSM makes it a potential feedstuff. It contains 2.58 mg/kg HCN, 0.24% saponin and 3.11% L-Dopa. Growth performance showed significant ($P<0.05$) differences for all the parameters measured, except for feed intake/bird/day. The final weight, weight gain/bird, weight gain/bird/day supported diet 5. The feed conversion ratio also favoured diet 5 (2.61) among others (diet 1, 2.68; diet 2, 3.89; diet 3, 2.73; and diet 4, 2.67, respectively). For cut-parts, there were significant differences ($P<0.05$) for all parameters measured with diet 5 comparing favourably with the control diet. All the organ weight parameters showed no significant difference ($P>0.05$). The economics of the diet revealed that diet 5 had the least cost/kg weight gain, highest revenue and gross margin, making diet 5 an economically viable diet. Conclusively, diet 5 (0.3% methionine supplementation) enhanced a high quality animal protein production at reduced cost.

Effect of sow diet trace mineral source on colostrum and milk composition and growth of piglets

C. Rapp[1] and J. Morales[2]
[1]*Zinpro Corporation, Akkerdistel 2E, 5831 PJ Boxmeer, the Netherlands, [2]PigChamp Pro Europa S.L., Santa Catalina, 10, 40003 Segovia., Spain; christof@zinpro.com*

Trace minerals are essential nutrients that might limit performance of the lactating sow and thus decrease growth rate of suckling pigs. This study evaluated the effect of sow diet trace mineral source on composition of colostrum and milk as well as piglet weaning weight. At weaning, a total of 40 sows were allocated to one of two dietary treatments: (1) 110 ppm Zn as ZnO, 40 ppm Mn as MnO and 15 ppm Cu as CuSO4 (ITM); or (2) 50 ppm Zn, 20 ppm Mn and 10 ppm Cu from complexed sources (Availa®Zn, Availa®Mn, Availa®Cu, Zinpro Corporation, Eden Prairie, USA; CTM) replacing equal amounts of trace minerals in inorganic form. Animals remained on dietary treatments for an entire gestation and lactation period. Lactation feed intake, back fat thickness, sow body weight, and numbers of total pigs born, pigs born alive and pigs weaned were not affected by dietary treatment ($P>0.44$). The number of mummies was lower ($P=0.04$) and stillbirths tended to be decreased ($P=0.10$) in the CTM treatment (0.0 vs 0.3 and 0.4 vs 0.9 pigs/litter for CTM and ITM, respectively). Litter weaning weight was increased (69.5 vs 61.7 kg, $P=0.05$) and homogeneity of pig weaning weights tended to be higher (85.5 vs 83.3%; $P=0.11$) in the CTM treatment. Protein, fat, and lactose concentrations in colostrum and milk were not affected by dietary treatment ($P>0.25$). Overall somatic cell count in colostrum and milk was lower ($P=0.05$) for the CTM treatment (6.90 vs 7.81 ln/ml). IgG concentration in colostrum and piglet serum were numerically higher in the CTM treatment (52.0 vs 41.7, $P=0.35$; 43.6 vs 40.6 mg/ml, $P=0.22$, for CTM and ITM, respectively). In conclusion, supplementing sow diets with complexed trace minerals is a means to improve sow reproductive performance, decrease somatic cell count and increase litter weaning weight.

Effect of trace mineral source in sow and nursery diets on nursery pig growth performance
C. Rapp[1] and J. Morales[2]
[1]Zinpro Corporation, Akkerdistel 2E, 5831 PJ Boxmeer, the Netherlands, [2]PigChamp Pro Europa S.L., Santa Catalina, 10, 40003 Segovia, Spain; christof@zinpro.com

Zinc, copper and manganese are essential nutrients for swine. This study evaluated the effect of trace mineral source in sow and nursery diets on growth performance in the nursery period using a two by two factorial arrangement of treatments with a total of 216 weanling pigs (7.1 kg body weight; 28 d of age). Each combination of treatment factors comprised nine replicate pens with six pigs each. Sow diets were either supplemented with 110 ppm Zn, 40 ppm Mn and 15 ppm Cu in inorganic form (ITM) or 50 ppm Zn, 20 ppm Mn and 10 ppm Cu from complexed sources (Availa®Zn, Availa®Mn, Availa®Cu, Zinpro Corporation, Eden Prairie, USA) replacing equivalent amounts of inorganic forms (CTM). Diets fed in the nursery phase were either supplemented with 160 ppm Cu and 110 ppm Zn from sulfates (ICZ) or equal amounts of Cu and Zn from complexed sources (Availa-Cu, Availa-Zn; CCZ). Pigs were fed a pre-starter diet (d 1 to 17) and a starter diet (d 18 to 37). There was no interaction (P>0.3) between dietary treatments in the sow and nursery phases. Average daily gain (ADG), body weight (BW) and average daily feed intake (ADFI) were not affected (P>0.27) by dietary treatment of sows. Feed to gain ratio (F:G) from d 18 to 37 tended to be increased (P=0.06) in pigs from sows fed CTM (1.56 vs 1.51 for CTM and ITM, respectively) but F:G from d 1 to 37 was not affected (P>0.31) by sow diet trace mineral source. Dietary treatment in the nursery phase did not affect ADFI (P>0.11). Average daily gain was increased and F:G was decreased (P<0.02) from d 18 to 37 and d 1 to 37 in pigs consuming CCZ compared to ICZ (0.494 vs 0.440 and 0.341 vs 0.314 kg/d; 1.47 vs 1.60; 1.42 vs 1.52 kg/kg). Day 35 BW of pigs receiving CCZ was increased compared to ICZ (P=0.02; 19.8 vs 18.8 kg). In conclusion, compared to inorganic sources of the same trace minerals, supplementing complexed Cu and Zn in the nursery phase improved growth performance of nursery pigs.

Effect of xylanase and β-glucanase on growth of weaned pigs fed wheat-barley diets: a meta-analysis
L. Payling, M.C. Walsh and S. Allan
Danisco Animal Nutrition, DuPont Industrial Biosciences, Marlborough, Wiltshire, SN8 1XN, United Kingdom; laura.payling@dupont.com

Exogenous feed enzymes have been demonstrated to negate some of the negative impacts of weaning on the growth performance of weaned pigs. The objective of this meta-analysis was to investigate the efficacy of a xylanase and β-glucanase combination on the growth performance of weaned pigs in seven different trials, using a total of 692 piglets (initial BW: 7.4±0.1 kg). Diet formulations were based on wheat, barley, and soybean meal, and were fed in two phases, each lasting 21 days. There were 4 treatments consisting of a negative control diet (NC) formulated to be deficient in digestible energy (~200 kcal/kg) and the NC diet supplemented with Axtra® XB at 0.05 kg/tonne (to supply 610 U/kg xylanase, 76 U/kg β-glucanase), 0.1 kg/t (to supply 1,220 and 152 U/kg) or 0.2 kg/t (to supply 2,440 and 304 U/kg). Experiments were conducted using a randomized complete block design with 8-10 replicates per treatment per study. Average daily feed intake (ADFI), average daily gain (ADG) and feed conversion ratio (FCR) were calculated for each phase and overall. Data were analysed using the fit model platform in JMP 11.0, using trial as a random effect. Means separation was conducted using Student's t-test, differences were considered significant at P≤0.05. Xylanase + β-glucanase at 0.05 kg/t reduced FCR relative to the NC in Phase 2 (P=0.02), and in the overall period by 7 points (1.65 vs 1.58; P=0.05). xylanase + β-glucanase at 0.1 and 0.2 kg/t also reduced FCR compared to the NC, by 10 and 9 points, respectively (1.55 and 1.56 vs 1.58; P=0.05). There were no significant effects on ADG and ADFI. In conclusion, xylanase + β-glucanase at 0.05, 0.1 and 0.2 kg/t were found to improve growth performance in weaned pigs fed wheat-barley based diets.

Effects of dietary Glibenclamide and Sucrose on performance and carcass characteristics of broilers

M. Bouyeh, A. Poorhassan Talemi and A. Bouyeh
Department of Animal Science Rasht Branch, Islamic Azad University, Rasht, 41939-63115, Iran;
mbouyeh@gmail.com

The present study was conducted to investigate the effects of different levels of Glibenclamide (GLI) as a stimulator of insulin release in pancreas, and Sucrose (SUC) on performance and some carcass parameters of broilers. A total of 240 one-day-old male chicks Ross 308 were used for 6 weeks based on a completely randomized design in a factorial (4×2) arrangement (4 levels of GLI include 0, 5, 10 and 15 mg/kg × 2 levels of SUC include 0 and 5 g/kg diet) with 3 replicates and 10 chicks per replicate in floor boxes. Food and water was provided ad-libitum and lighting system was provided according to Ross 308 recommendations. The data were collected weekly or at the end of experiment. SAS software was used to analyze variances and the Duncan test was used to compare the means on a value of $P<0.05$. The results indicated that SUC levels could not significantly affect the studied parameters ($P>0.05$) except for liver weight, and feed consumption in the first 3 weeks, for which SUC 5 g/kg was the highest ($P<0.05$). There was no significant effect of GLI on some parameters such as heart, small intestine, gizzard and crop weight ($P>0.05$). GLI had significant effects on European Production Efficiency Factor (EPEF), feed consumption, carcass efficiency, feed conversion ratio (FCR), liver, kidney, pancreas, breast and femur weights ($P<0.01$), so that kidney, liver, femur and breast weights increased parallel with GLI level (significant difference with control group), but in addition the highest performance belonged to the treatment group 5 mg/kg GLU+ 0 g/kg SUC with the best FCR (1.69 vs 1.81 in control group) and EPEF (364 vs 327 in control group) ($P<0.05$).

Evaluation of Ceriporiopsis subvermispora on improving the nutritive value of wheat straw

N. Nayan[1,2], J.W. Cone[2], A.S.M. Sonnenberg[3] and W.H. Hendriks[2]
[1]Universiti Putra Malaysia, Department of Animal Science, Kompleks Agrobio, 43400 UPM Serdang, Malaysia,
[2]Wageningen University, Animal Nutrition Group, Department of Animal Science, De Elst 1, 6708WD Wageningen,
the Netherlands, [3]Wageningen University, Plant Breeding Group, Droevendaalsesteeg 1, 6708PB Wageningen,
the Netherlands; nazri.nayan@wur.nl

The white-rot fungus, Ceriporiopsis subvermispora produces enzymes which effectively degrade lignin, increases the cellulose accessibility and enhances the degradability of wheat straw in rumen fluid. In this study, two strains of C. subvermispora, CS1 and CS2, were evaluated for their potential in improving the nutritive value of the wheat straw for ruminants. Wheat straw was inoculated with the fungi under solid state fermentation for 7 weeks. Weekly samples were analyzed for in vitro gas production, ergosterol and ligninolytic enzyme activities. Data were analyzed as a function of fungal treatment, incubation period and their interactions. Results showed a higher total in vitro gas production of CS2-treated straw compared to CS1. Both strains showed high activities of laccase and manganese peroxidase (MnP) especially during the first week of inoculation. No activity of lignin peroxidase was detected in both strains. At week 1, CS2 showed a higher laccase activity (1.16 vs 0.47 U/l), but lower MnP activity (0.25 vs 0.52 U/l) compared to CS1. High activities of these ligninolytic enzymes indicate that the fungi start colonizing the wheat straw by degrading lignin which subsequently increase the accessibility of the cellulose for rumen microbes. This also explains a higher gas production upon incubation in rumen fluid with CS2-treated wheat straw. CS2 produced more fungal biomass compared to CS1, measure as ergosterol. The different performance of the two strains indicate that enzyme activities and colonization (formation of fungal biomass) are possible key-elements in selective removal of lignin and thus increasing digestibility of wheat straw and targets for fungal strain improvements.

The effect of barley quality and exogenous enzyme supplementation on performance in finisher pigs

L.C. Clarke[1], E. Curley[2], P.O. McAlpine[1] and J.V. O'Doherty[1]
[1]University College Dublin, School of Agriculture and Food Science, Belfield, Dublin 4, Ireland, Ireland, [2]National University of Ireland, Galway, Botany and Plant Science, Plant and AgriBioscience Centre, NUI Galway, Ireland; louise.clarke.4@ucdconnect.ie

An experiment was conducted to investigate the interaction between barley quality (hectolitre weight, HL) and non-starch polysaccharide (NSP) enzyme inclusion on growth performance in finisher pigs. The experiment was set up as a 2 (61.3 and 56.6 kg/HL of barley) × 2 (with or without NSP enzyme) factorial arrangement. The four dietary treatments were as follows: T1) high HL barley-based diet (HB); T2) high HL barley-based diet plus enzyme supplementation (HBE) T3) low HL barley-based diet (LB) and T4) low HL barley-based diet plus enzyme supplementation (LBE). The enzyme used contained beta-glucanase and beta-xylanase. The inclusion rate of barley (var. Sebastian) was 500 g/kg and the diets were formulated to contain similar levels of digestible energy and lysine (13.1 MJ/kg and 10.1 g/kg respectively). The high HL barley had a significantly (P<0.01) higher thousand grain weight (50.2 and 35.6 g/kg, sem 0.94) and lower screening percentage (1.2 and 6.7 sem 0.220) compared to low HL barley. There was a significant (P<0.01) difference in the crude protein (CP) and dry matter (DM) content with low HL barley having a higher CP (134.0 and 115.8 g/kg, sem 1.17) and DM content (888.8 and 864.8 g/kg, sem 0.26) compared to the high HL barley. Ninety six pigs (44.6 kg; SD 4.8 kg) were offered the experimental diets for 28 days. Pigs offered the high HL barley had a higher (P<0.01) average daily gain (ADG) (1.13 vs 1.06 kg, sem 0.019) and feed intake (FI) (2.63 vs 2.47 kg, sem 0.025) compared to the low HL barley, however there was no significant difference between barley quality on feed efficiency. Enzyme inclusion increased FI (P<0.01) (2.63 vs 2.47, sem 0.025) but had no effect on ADG or feed efficiency. In conclusion, the HL weight of barley positively increased ADG and FI of pigs during the finisher stage.

Probiotic fermentation using orange vinasse and molasses as feed substrates for animal production

J.E. Miranda Yuquilema[1], A. Marin Cardenas[1] and D. Sánchez Macías[2]
[1]Universidad Central Marta Abreu de Las Villas, Cuba, Facultad de Ciencias Agropecuarias, Km 5 1/2 Carretera a Camajuani, 54830, Santa Clara, Cuba, [2]Universidad Nacional de Chimborazo, Producción Animal e Industrialización, Facultad de Ingeniería, Avda. Antonio José de Sucre s/n, 060150, Riobamba, Ecuador

The use of bacteria as beneficial biological agents being used as food ingredients or as active components of food supplements dates back to the early 1900s. The objective of this study was to prepare a probiotic for animal feed using microorganism with previously determined probiotic capacity and residues from sugar cane and orange processing industry. A selection of LAB for feed fermentation (Lactobacillus acidophilus ATCC® 4356™, Lactobacillus bulgaricus ATCC® 11842™, and Streptococcus salivarius subsp. thermophilus ATCC® 19258™) and the yeasts Saccharomyces cerevisiae ATCC® 2601™ and Kluyveromyces marxianus subsp. fragilis ATCC® 12424™ were used to produce the probiotic food. Molasses (as carbohydrates source) and orange by-products vinasse (as protein source) were chosen as feed substrates to obtain high numbers of microorganisms and levels organic acids. The microorganisms were inoculated (2%) in previously sterile low-fat milk for their activation and incubated during 24 h at 37 °C. After that, 500 ml of vinasse and 300 g of molasses were mixed with 125 ml of the milk with the microorganisms. At that moment, pH was 4.4, and after 72 h of fermentation at ambient temperature, pH was 3.8. The final probiotic food color was, in the CIELab* system: L 31.85, a 11.48, b 24.52, C 27.08, and H 64.91, similar to the HTML code #61382B. The odor and taste was sweet. The gross chemical composition was: 18% DM, 3.3% ash, 19% crude protein, 12% net protein, 3.2% EE, and 0.75% organic acids. The microorganism counts were 9×10^9 cfu/ml, and the viability was 95%. The results showed that by-products as vinasse and molasses are good and economic feed substrates for probiotic growth and obtain an acceptable probiotic for animal feed.

Prediction of ruminal starch degradability of maize forage

J. Peyrat[1], E. Meslier[2], A. Le Morvan[3], A. Ferard[2], R. Baumont[3], B. Deroche[2], P.V. Protin[2] and P. Noziere[3]
[1]ARVALIS Institut du végétal, Ferme expérimentale des Bordes, 36120 Jeu Les Bois, France, [2]ARVALIS Institut du végétal, Station expérimentale de La Jaillière, 44370 La Chapelle-Saint-Sauveur, France, [3]INRA, UMR 1213 Herbivores, Site de Theix, 63122 Saint-Genes-Champanelle, France; a.ferard@arvalisinstitutduvegetal.fr

Maize silage is the main source of energy in the diet of high-yielding ruminants. The current French system to evaluate its energetic value provides knowledge on energetic value of the whole plant. However, there is no indication on the nature of its available energy for animals resulting in rate and extent of ruminal degradation of starch and cell-wall fraction. In this study, we investigated different parameters of maize whole plant chemical composition in order to predict ruminal starch degradability. Nineteen degradability trials with fistulated cows (from 1996 to 2013) were selected to build up the database. The data of the 115 samples of ensiled forage maize were collected from samples initially dried during 72 h at 60 °C and ground at 4 mm. The methodology involved measuring ruminal starch degradability at different incubation times using nylon bags. Effective starch degradability (ED6) was calculated with a step by step model and assuming a particulate passage rate of 0.06 h^{-1}. Models were fitted by taking into account within-hybrid and within-harvest year effects. Starch ED6 is closely related to the association between dry matter (DM) at harvest and starch content (g/kg DM) with the equation: ED6starch = 118.9 − 0.125×DM + 0.022×starch (R^2adjusted=0.89; RSD=4.7; Ntotal=175; Nexp=19). Other chemical composition parameters did not significantly improve the prediction model of effective starch degradability (P>0.05). A validation step is planned before use with forage maize harvested in 2015 (n=28). These results indicate that the use of criteria characterizing whole plant maturity stage and chemical composition could be relevant to estimate starch ED6. The prediction model of starch ED6 will be included in the future French feed evaluation system of nutritive and integrated in the future infrared laboratory calibration.

The effect of wheat quality and enzyme or mycotoxin binder inclusion on growth performance in pigs

L.C. Clarke[1], E. Curley[2], P.O. McAlpine[1] and J.V. O'Doherty[1]
[1]University College Dublin, School of Agriculture and Food Science, Belfield, Dublin 4, Ireland, [2]National University of Ireland, Galway, Botany and Plant Science, NUI, Galway, Ireland; louise.clarke.4@ucdconnect.ie

Two experiments were carried out to investigate the interaction between wheat quality (hectolitre weight, HL) and non-starch polysaccharide (NSP) enzyme or mycotoxin binder inclusion on growth performance of pigs. In experiment 1, 96 pigs (11.59 kg; SD 0.973) were used in a 3 (73.79, 70.17 and 66.42 kg/HL of wheat) × 2 (with or without NSP enzyme) factorial arrangement for 28 days. In experiment 2, 96 pigs (38.65 kg; SD 3.477) were used in a 3 (73.79, 70.17 and 66.42 kg/HL of wheat) × 2 (with or without a mycotoxin binder) factorial arrangement for 42 days. The enzyme used contained beta-glucanase and beta-xylanase while the mycotoxin binder contained silicoglycidol. The inclusion rate of wheat (var. JB Diego) was 500 g/kg and the diets were formulated to contain similar levels of digestible energy and lysine. In experiment 1, pigs offered the low HL wheat had a lower (P<0.01) average daily gain (ADG) (0.541 vs 0.655 vs 0.651, sem 0.0117) and a poorer feed conversion ratio (FCR) (2.46, vs 1.74, vs 1.83, sem 0.0798) compared to the high and medium HL wheat. Enzyme inclusion decreased (P<0.05) feed intake (FI) (1.12 vs 1.17, sem 0.015) and improved FCR (1.916 vs 2.111, sem 0.065). Pigs offered the high HL wheat had higher (P<0.01) total tract digestibility of gross energy and nitrogen compared to pigs on the low HL wheat while the inclusion of enzymes had no significant effect on nutrient digestibility. In experiment 2, an interaction between the mycotoxin binder and wheat quality was observed for ADG (P<0.05) and FI (P<0.05) with the mycotoxin binder increasing ADG and FI with the low HL wheat, with no effects observed with medium and high HL wheat. In conclusion, the HL of wheat positively predicted pig performance during the weaner stage. NSP enzyme inclusion improved feed efficiency in grower pigs. The addition of a mycotoxin binder to low quality wheat improved ADG and FI.

The INRA feed tables and prediction equations for forage quality evaluation

R. Baumont, G. Maxin and P. Noziere
INRA, UMR1213 Herbivores, Centre Auvergne Rhône-Alpes, Site de Theix, 63122 Saint-Genès-Champanelle,
France; pierre.noziere@clermont.inra.fr

Accurate evaluation of nutritive value and voluntary intake potential of forages produced from grasslands and annual forage crops is a key feature of their efficient utilization in diets. INRA feed tables for forages are based on a database of more than 2000 in vivo measurements of digestibility and voluntary intake in sheep accumulated over several decades with a standard protocol. The tables cover a large variety of grassland and annual forage crop species (3 types of permanent pastures, 8 grass species, 6 forage cereals species, 5 forage legume species and 8 other species). A second database for in sacco nitrogen degradability (n=452) is used for estimation of protein value. Feed values in net energy, in digestible protein in the intestine and in fill unit are provided for fresh forages, as well as for conserved ones from statistical models accounting for the modification of feed value induced by ensiling and making hay. Prediction equations of feed value components established on this databases and from enzymatic digestibility measurements (n=375 for grasslands, n=290 for maize) allow calculations from laboratory analyses. A new version of forage feed tables will come out with updated net energy and digestible protein calculations taking into account the effect of feeding level on digestion and an improved estimation of organic matter fermented in the rumen ('Systali' project). It will include several novelties concerning fatty acid content of forages, estimation of energy loss in urine and as methane emission and estimation of starch degradability of maize silage. This will allow to evaluate forage quality also in relation to other dimensions than nutritive value, as impact on product quality, animal health and emissions in the environment. Finally the reference values provided by French tables show a good consistency with those provided in Netherland (CVB) and in Nordic countries (Norfor). This is promising for interoperability of feed tables and possible harmonization at European level.

The use of laser diffraction to assess diet effects on faecal particle size of steers

J.M. Bowen[1,2], S. Sherriff[3], S.J. Lister[1] and R.J. Dewhurst[4]
[1]Aberystwyth University, IBERS, Gogerddan, Ceredigion, SY23 3EB, United Kingdom, [2]Teagasc, Grange, Dunsany,
Co Meath, Ireland, [3]Teagasc, Johnstown Castle, Co Wexford, Ireland, [4]SRUC, West Mains Rd, Edinburgh, EH9
3JG, United Kingdom; richard.dewhurst@sruc.ac.uk

Differences in faecal particle sizes can show effects on digestive tract processes related to changes in passage rates through the tract. Traditional wet sieving techniques for assessing faecal particle size distribution are laborious and time consuming. The aim of this study was to assess the use of laser diffraction (LD) as a rapid technique to assess faecal particle size in cattle offered diets with a series of additives designed to alter methane emissions. Faecal samples were collected from 75 steers at the end of a 56 day period of measuring feed efficiency, expressed as residual feed intake (RFI). Diets were: Control (TMR based diet, n=18), and 3 diets in which rapeseed meal was replaced with Nitrate (18 g nitrate/kg DM in the form of calcium nitrate, n=19), Oil (increased by 12 g/kg DM by use of maize dark grains, n=20), Nitrate/Oil (a combination of Nitrate and Oil, n=18). Faecal samples were sieved to remove particles larger than 2 mm, diluted, and assessed using LD with 27 pre-set particle size categories ranging from 0.01 to 2,000 μm. Particle size distributions were analysed using Principal Component Analysis (PCA) using all samples and subsequently individual diets. One way ANOVA was used to evaluate diet effects on Principal Component (PC) scores. Separation of diets was observed in the PCA plot containing all 4 diets. Diet had significant effects on PC1 (P<.001) and PC2 (P<.001) for all samples. PCA plots for individual diets showed differences in particle size were not related to RFI level. This study shows the potential for LD as a rapid technique to assess faecal particle size. Inclusion of nitrate and oil additives both affected faecal particle size distribution and it is interesting to note that the nitrate exerted effects in the absence of changes in feed particle distribution, whilst the Oil treatment effects may have resulted from changes in diet ingredients.

Efficiency of novel probiotic complexes in a cattle diet

R.V. Nekrasov[1], M.G. Chabaev[1], A.A. Zelenchenkova[1], V.V. Dzhavakhiya[2], E.V. Glagoleva[2] and M.I. Kartashov[2]
[1]L.K. Ernst Institute of Animal Husbandry, Department of Farm Animal Feeding and Feed Technology, Moscow region, Podolsk district, Dubrovicy, 142132, Russian Federation, [2]OOO «Fermlab», Moscow, 123592, Russian Federation; nek_roman@mail.ru

There is an ongoing search for promising supplements as alternative to antibiotics. Potential alternatives are probiotics, prebiotics, enzymes and their complexes in rations. The supplementation of the ration seems most relevant for calves during period of maternal suckling, when development of gastrointestinal tract and enzyme systems of the body is not yet completed, as well as for fresh cows in the beginning of lactation, when increase in availability of nutrients and energy is necessary. Therefore we studied the effects of Bacillus subtilis VKM B-2998D, Bacillus licheniformis VKM B-2999D, Bacillus subtilis (natto) VKPM B-12079 (at least 5×10^9 cfu/g) blend and the blend combined with enzyme in the rations of suckling calves and fresh cows. The results of the study showed an increase in average daily gain of 5.9-8.5% (700.5 & 717.6 vs 661.2 g/d, P>0.05) (blend & blend+enzyme vs control) in calves. Similarly, milk yield was 11.0 and 6.8% higher (30.46 & 29.29 vs 27.42 kg/d, P>0.05) (blend & blend+enzyme vs control) in fresh cows during the experiment. The analysis of experimental data showed improvements in nutrient digestibility (CF), increase in total blood serum protein, phagocytic index and phagocytic number in the blood of calves with probiotic supplementation (P<0.05, for all parameters). Also, overall CFA content in feces of calves increased with 53.6 and 25.1% (blend and blend+enzyme, P>0.05) whereas the number of hemolytic α-streptococci decreased (P<0.05). Thus, the use of complex probiotic supplements improves the growth of calves and milk production of cows.

The accuracy of Near Infra-red Spectroscopy analysis when used on clover-grass silages in the UK

A. Thomson, C. Reynolds, C. Rymer and D. Humphries
University of Reading, School of Agriculture, Policy and Development, P.O. Box 237, Earley Gate, RG6 6AR, United Kingdom; a.l.thomson@reading.ac.uk

Near Infra-red Spectroscopy (NIRS) is a quick and inexpensive technique, routinely used to provide nutritional analysis of silage for use in livestock diet formulation. However, obtaining accurate results requires robust calibrations. The purpose of this study was to ascertain whether NIRS calibrations based on predominantly pure grass silage samples, could accurately predict the composition of mixed grass-clover silage samples. The hypothesis was that there would be differences between the predicted composition of grass-clover silages (using NIRS) and the observed composition (using standard chemical analysis) and that any differences would be greater at higher clover concentrations. To test the hypothesis, a diverse set of 90 grass-clover silages (ranging in clover content from 40 to 1000 g/kg as fed) were analysed for chemical composition using reference techniques, in vivo organic matter digestibility (OMD, in sheep) and by NIRS (at AFBI, Northern Ireland). Predicted and observed results were compared using Student's t-test and Lin's concordance coefficient (ρ_c). NIRS under-predicted crude protein (CP) by 17 g/kg (P=0.001, ρ_c=0.71), acid detergent fibre by 35 g/kg (P=0.001, ρ_c=0.51) and ash by 6 g/kg (P=0.024, ρ_c=0.55) on a dry matter (DM) basis. In each case, the difference between predicted and observed measures increased with greater clover concentration. Differences might be partly attributed to the silages being analysed wet for NIRS, which is standard UK practice, rather than dried which can reduce heterogeneity. Prediction of DM, volatile content, pH and OMD was acceptably accurate (ρ_c>0.71, no significant difference between analytical techniques). It was concluded that, for clover-grass silages, a revised prediction equation or correction factor is required to improve the accuracy of NIRS prediction for some components that are important for accurate diet formulation. As CP is under-predicted, the current NIRS analysis may result in excess CP being fed, particularly where the silage has a high clover content, and correcting this would reduce the need for expensive bought-in feed products.

Effect of diet supplemented with Euphorbia heterophylla on lipid parameters in the thighs of rabbits

Y.D. Ahongo[1], D.V. Kouakou[1], E.M. Angbo[1], N.E. Assidjo[1] and M. Kouba[2]
[1]Institut National Polytechnique Félix Houphouet Boigny, BP 1093, Yamoussoukro,
Cote d'Ivoire, [2]INRA agrocampus ouest, phase, 65 rue de Saint Brieuc, 35042 Rennes cedex, France;
maryline.kouba@agrocampus-ouest.fr

Euphorbia heterophylla is a major weed of crops whose α-linolenic acid content (ALA; 18:3n-3) can induce the enrichment in omega 3 fatty acids of guinea pig carcasses and egg yolks of laying hens. A total of 27 male weaned rabbits (Oryctolagus cuniculus), from a local Ivorian breed were allocated to three dietary treatments (9 animals per diet) to test the effect of the duration of the diet supplementation of leaves and stems of Euphorbia heterophylla on the lipid parameters of the thighs of the animals. A basal commercial diet was used as a control diet (C) and was distributed 90 days. This basal diet was supplemented by 50% of Euphorbia heterophylla and was either distributed 90 days (diet C 90) or during 30 days (diet C 30) (from the day 60 to the day 90 of the study). The animals were killed on the last day of the experiment and the lipid parameters were analysed on a sample from their thighs. Lipid contents were not influenced by the diet in contrast to cholesterol whose contents decreased by 42 and 59% respectively after 30 days and 90 days of supplementation (P≤0.001), compared to the control diet. The supplementation of Euphorbia heterophylla led to a significant increase in the α-linolenic fatty acid contents which were multiplied by 2 and 4, respectively after 30 and 90 days of supplementation, compared to the control animals. The ratios of linoleic/α-linolenic acids were 12.67 (SD=1.97), 5.13 (SD=0.83) and 2.21 (SD=0.15) for the thighs of the rabbits submitted to diets C, C 30 and C 90, respectively. The incorporation of leaves and stems of Euphorbia heterophylla in the diet during 30 days in the rabbit is sufficient to obtain a meat in accordance with nutritional recommendations for polyunsaturated fatty acids.

Mixed enzymes supplementation in laying hens fed dried cassava pulp

S. Khempaka, P. Maliwan and W. Molee
School of Animal Production Technology, Institute of Agricultural Technology, Suranaree University of Technology,
Nakhon Ratchasima, 30000, Thailand; khampaka@sut.ac.th

The cassava starch industry generates a large amount of waste in the form of cassava pulp annually. The pulp contains a lot of starch (50-70%). However, the high fiber content of cassava pulp, which is mostly present in the form of insoluble fiber, is an obvious concern. Therefore, this study evaluates the effect of mixed enzymes supplementation in diets containing dried cassava pulp (DCP) on the nutrient digestibility and retention in laying hens. A total of 45 laying hens (Isa Brown) aged 45 weeks were placed in individual cages and randomly allocated to 9 groups with 5 replicates using a Completely Randomized Design (CRD). Nine dietary treatments were given as follows: control and DCP at 20, 25, 30, and 35% supplemented with mixed enzymes (cellulose, glucanase and xylanase) at 0.10 and 0.15%. Feed and water were provided ad libitum for 10 days. Excreta were collected on the last 4 days of the experimental period. Dried excreta and diets were analyzed for dry matter, crude fiber, crude protein and ash to calculate their digestibilities and retention. The results showed that the use of DCP at 20-35% added with enzymes (0.10 and 0.15%) had no negative effects on dry matter and crude fiber digestibilities (P>0.05). While organic matter digestibility and nitrogen retention decreased as DCP was increased in diets, but there were no significant differences in the laying hen group which received DCP up to 20%. The supplementation of mixed enzymes at 0.10 and 0.15% showed similar results on nutrient digestibility and retention (P>0.05). In conclusion, the supplementation of mixed enzymes can improve nutrient digestibility in laying hen fed DCP diets.

The use of faecal NIRS to assess diet effects on digestive processes in steers

J.M. Bowen[1,2], S.J. Lister[1], R. Sanderson[1] and R.J. Dewhurst[3]
[1]Aberystwyth University, IBERS, Gogerddan, Ceredigion, SY23 3EB, United Kingdom, [2]Teagasc, Grange, Dunsany, Co Meath, Ireland, [3]SRUC, West Mains Rd, Edinburgh, EH9 3JG, United Kingdom; richard.dewhurst@sruc.ac.uk

Faecal near infrared spectroscopy (faecal NIRS) is a rapid technique used to assess chemical composition of faeces, with chemical differences in faeces showing differences in digestion processes between animals. The aim of this study was to assess the use of faecal NIRS as a rapid technique to assess chemical differences of faeces of cattle offered diets with a series of additives designed to alter methane emissions. Faecal samples were collected from 72 steers at the end of a 56 day period of measuring feed efficiency, expressed as residual feed intake (RFI). Diets were: Control (TMR based diet, n=18), and 3 diets in which rapeseed meal was replaced with Nitrate (18 g nitrate/kg DM in the form of calcium nitrate, n=19), Oil (increased by 12 g/kg DM by use of maize dark grains, n=19), Nitrate/Oil (a combination of Nitrate and Oil, n=16). Faecal samples were dried and milled before being scanned in duplicate at 2 nm intervals between 1,100 and 2,500 nm. Averaged spectra were transformed using SNV and Detrend, before being assessed using PCA, using all samples and subsequently individual samples based on diet. Separation of diet in to 3 clusters was observed in the PCA plot containing all diets (1; Control and Oil, 2; Nitrate, 3: Nitrate/Oil). Diet had a significant effect on PC-1 (P=0.007) but not for PC-3 (P=0.092), with wavelengths associated with protein, lipids and cellulose/sugars showing highest loadings. PCA plots for individual diets showed no separation by RFI level. This study shows the potential for NIRS as a rapid technique to assess chemical differences of faeces associated with differences in digestion processes, with the presence or absence of Nitrate having a large effect.

Using of probiotic complexes in feeding of fresh cows

V.V. Dzhavakhiya[1], E.V. Glagoleva[1], M.I. Kartashov[1], R.V. Nekrasov[2], M.G. Chabaev[2] and A.A. Zelenchenkova[2]
[1]OOO «Fermlab», Moscow, 123592, Russian Federation, [2]L.K. Ernst Institute of Animal Husbandry, Department of Farm Animal Feeding and Feed Technology, Moscow region, Podolsk district, Dubrovicy, 142132, Russian Federation; nek_roman@mail.ru

Young animals and lactating dairy cows, especially early in their lactation benefit from use of probiotics in their rations. Probiotics modulate anti-infective, immunnological responses in the body and increase productivity through improving the stomach barrier function, stimulating digestion and increased absorption of nutrients. A goal of the current study was to conduct extensive research on the effeciency of new probiotic complex (a combination of strains of Bacillus subtilis, Bacillus licheniformis and their combination after fermentation) in a ration of fresh cows at 14 factory farms (n=410). We conducted our studies in Altai and Vologda regions, which are among the leading regions for the milk production in the Russian Federation. The results demonstrated that addition of the probiotics to the ration of lactating cows increased daily milk production of standardized milk on 15.0%, in some cases through higher fat content of the milk in experimental groups. We also analyzed qualitative characteristics of the milk, its biochemical and hematological blood parameters, immunoresistance and quantitative composition of intestinal microflora in experimental animals. In most cases, the use of probiotic complexes optimized metabolism in the cows. Probiotic supplemetation increased phagocytic index and phagocytic number in the blood of fresh cows (P<0.05). Similarly, the number of bifidobacteria and lactobacilli increased in the the colon of the animals (P<0.05). Thus, the use of studied strains of microorganisms with/without enzyme treatment enhanced productive potential of the animals while maintaining their health.

Influence of extrusion on nitrogen degradability of faba bean or lupine blends including urea
P. Chapoutot[1], O. Dhumez[1], A. Germain[2] and G. Chesneau[2]
[1]UMR MoSAR INRA-AgroParisTech, 16 rue Claude Bernard, 75005 Paris, France, [2]Valorex, La Messayais, 35210 Combourtillé, France; patrick.chapoutot@agroparistech.fr

The reduction in nitrogen (N) ruminal degradability of cereals, proteaginous or oilseeds by extrusion has already been widely described, but studies on proteaginous or oilseed mixtures including urea are rare. Moreover, precise technological parameters were seldom reported in these trials. In this study, different processes of extrusion were applied just after grinding to four blends, composed of 90% either faba bean (F) or lupine (L) and 10% linseed, combined or not with 7%urea. Compared to controls, the treatments varied in the exit temperature (T) of the product and how this T inside the extruder was obtained (mechanical efforts vs external heating): high T (140 °C) obtained either with high pressure (P) due to mechanical constraints (HTHP) or with low P but exogenous heat input (HTLP), and low T (110 °C) due to low mechanical constraints without exogenous heat input (LTLP). Many technological parameters were recorded during the treatments. In sacco N ruminal degradation was measured on feeds and used to calculate N effective degradability (NED, %). All the main experimental effects and their double interactions appeared statistical significant (P<0.001) by analysis of variance carried on degradation parameters. N effective degradability was higher for L than F blends (91.5 vs 87.9%) and the presence of urea logically increased NED, but to a greater extent for F compared to L (+9.5 vs +7.5 pts). In comparison with controls, only the most severe treatment (HTHP) led to a significant and lower degradability, but this effect was stronger for L than for F (-9.5 vs -4.3 pts). Moreover, processes had no effect on blends including urea. All the variations of NED can be precisely predicted by in vitro N degradability. When related to the technological criterion of extrusion, NED reduction is more linked to strong mechanical constraints than to external heat input.

Effects of hybrid and dietary mussel meal on broiler performance during 10 weeks of rearing
L. Jönsson[1], S. Gunnarsson[2], J. Yngvesson[2], M. Rezaei[1] and A. Wallenbeck[1,2]
[1]SLU, Department of Animal Breeding and Genetics, Box 7023, 750 07 Uppsala, Sweden, [2]SLU, Department of Animal Environment and Health, Box 234, 532 23 Skara, Sweden; anna.wallenbeck@slu.se

Organic broiler production is characterised by a long rearing period (>10 weeks) and diets based on locally available feedstuff. Until recently, conventional fast growing broilers hybrids have been used exclusively in organic broiler production in Sweden. Rearing fast growing hybrids during long periods has resulted in e.g. ascites and leg disorders and the need for alternative slow growing hybrids for these production systems has been outspoken. Moreover, the nutrient composition (especially protein quality) in the organic diets does not meet the high demands of the conventional fast growing hybrid. This study investigated broiler production performance in the fast growing hybrid Ross 308 (R) and the slower growing hybrid Rowan Ranger (RR) fed either a standard organic diet (O, 11.3 MJ ME and 164 g protein/kg feed) or an organic diet with mussel meal as protein source (M, 11.2 MJ ME and 156 g protein/kg feed) during a 10 weeks rearing period. The study included in total 389 birds housed in groups of 20-21 birds per group divided over the treatment combinations; 5 RO, 4 RM, 5 RRO and 4 RRM groups. During the rearing period, 22 birds were culled due to leg weakness and 16 died or were culled due to other reasons. The statistical analyses were performed with linear mixed models in SAS. Preliminary results indicated that R birds grew faster when fed the mussel meal diet while this difference in the response to protein quality was less pronounces in RR birds (4,586a±55.3, 3,961b±49.5, 2,919c±55.3 and 2,891c±49.5 g live weight at 71 d in the RM, RO, RRM and RRO treatment respectively, LSM±SE; different subscript letters indicate pairwise differences of P<0.05). A significantly higher proportion of RM birds were culled due to leg weakness compared to the other treatments (P<0.05). Dressing percentage was higher in R than RR birds (P<0.05), but there were no effect of diet treatment.

Improved in vitro digestibility of protein feedstuffs with a bacterial endo-xylanase

C. Boudry[1], F. Philippe[2], B. Kambashi[3] and J. Bindelle[3]
[1]Puratos, Rue Bourrie 12, 5300 Andenne, Belgium, [2]JEFO Europe, 2 rue Claude Chappe, Espace Performance La Fleuriaye, 44481 Carquefou Cedex, France, [3]Gembloux Agro-Bio Tech, ULG, Precision Livestock and Nutrition Unit, Passage des Deportes 2, 5030 Gembloux, Belgium; cboudry@puratos.com

Xylanases are mainly used in broilers and pigs to reduce the viscosity and increase the digestibility of cereal grains rich in arabinoxylans. Fibrous protein feedstuffs may also contain significant concentrations of arabinoxylans, mainly in an insoluble form, which can enclose nutrients. The aim of this study was to investigate the effect of a bacterial endo-xylanase on the in vitro digestibility of protein feedstuffs and diversified pig diets. Five feedstuffs (Wheat and Corn DDGS, palm cake, hulled and unhulled sunflower cake) and three different pig diets (Soya diet – Rapeseed diet – Rapeseed and sunflower diet) were hydrolyzed in vitro with pepsin (pH 3, 2 h) and pancreatin (pH 6.8, 4 h) at 39 °C. The incubation was performed with and without a bacterial endo-xylanase at a level of 10 mIU/g which is the minimum dose recommended for broiler and pig diets (E1606). Residues were collected through filtration on a 42-μm Nylon cloth, dried at 60 °C and analyzed for DM, Ash, protein, NDF, Gross Energy and Starch content, as well as the initial samples. To perform the analyses, residues of 12 hydrolyses were pooled to get enough samples. Statistics were performed for the in vitro DM digestibility (n=12) with SAS (Proc GLM, material and enzyme: fixed factors). The in vitro DM digestibility was increased by addition of the enzyme for 7 of the 8 materials. The increase was in average of 1% but was not statistical (P>0.05). Looking to the different chemical components, as expected, the highest effect was observed on the in vitro NDF digestibility which was increased by 40% in average (13.1% without vs 18.2% with enzyme). The highest increase was observed for palm cake (10.2% without vs 22% with enzyme). This led to an increase in gross energy and protein digestibility (respectively by 2 and 2.7%). Results indicate that the bacterial endo-xylanase used in this screening can also improve the digestibility of other feedstuffs than cereal grains and byproducts.

Fungi from Antarctica as possible producers of polyunsaturated fatty acids for feed additives

L. Skofic[1], P. De Rossi[2], A. Del Fiore[2], C. Nobili[2], R. Gatti[2], A. Ambrico[3], R. Albergo[4], M. Trupo[3], R. Balducchi[1,3], O. Maggi[1], A.M. Persiani[1], S. Zjalic[5], S. Urbanelli[1], D. Poretta[1], A. Ceci[1], C. Fanelli[1] and M. Reverberi[1]
[1]University Sapienza, Department of environmental biology, P. le Aldo Moro 5, 00100 Roma, Italy, [2]ENEA, Research Centre Casaccia, via Anguilarese 301, 00123, Italy, [3]ENEA, Research Centre Trisaia, SSPT-BIOAG, SS 106 Jonica, km 419,500, 75206 Rotondella (Mt), Italy, [4]ENEA, Research Centre Trisaia, DTE-BBC, SS 106 Jonica, km 419,500, 75206 Rotondella (Mt), Italy, [5]University of Zadar, Department of Ecology, Agronomy and Aquaculture, trg kneza Viseslava 9, 23000 Zadar, Croatia; patrizia.derossi@enea.it

The benefits of polyunsaturated fatty acids (PUFAs), particularly ω3-ω6-ω9, as regulators of cholesterol levels in the blood and beneficial for the cardiovascular system are well known. The introduction in the feed of polyunsaturated fatty acids could have positive effects on both animal health and quality of their products. The purpose of this study was to investigate the fungal sources for producing PUFAs that are poorly represented in conventional foods. The interest of this work is focused in particular on the phenotypic plasticity of Fungi isolated from Antarctica material compared with PUFA such as adaptive response to low temperatures. These fungi, to maintain a state of semi-fluid membranes during the lowering of the temperature, increase the proportion of unsaturated fatty acids. The fungi were identified molecularly (ITS1-4) at species level while the quali-quantitative characterization of the lipidic component was carried out by means of gas chromatography techniques and by liquid chromatography techniques coupled to mass spectrometry. For a selected and limited number of fungal strains that showed a particular lipid profile of potential nutritional interest, a fermentation process from batch to scale up was developed. The results showed that all samples are rich in saturated (stearic acid i.e.) and mono and polyunsaturated (oleic acid, linoleic, linolenic and arachidonic acid) fatty acids. The results confirmed that the ratio of saturated/unsaturated fatty acids was strongly influenced by environmental parameters; this has allowed the identification of the conditions to promote the production of the lipids of interest for use as feed additives.

Selenium bioaccessibility in commercial selenium yeast products

S. Fagan[1], R. Owens[2], P. Ward[1], S. Doyle[2] and R. Murphy[1]
[1]Alltech, Sarney, Summerhill Rd, Dunboyne, Co. Meath, Ireland, [2]Maynooth University, Department of Biology, Maynooth, Co. Kildare, Ireland; sfagan@alltech.com

The compounds available for use as selenium (Se) supplements in animal nutrition include the inorganic forms, sodium selenite and sodium selenate, and the organic forms, L-Selenomethionine (SeMet) and Se-enriched yeast. Se nutritional status is greatly influenced by the bioavailability of Se, therefore the quality of the supplement will depend directly on the amount of Se present that is ultimately available for the organism. In addition, bioavailability of Se varies according to the chemical form of the element, being significantly higher for organic forms. Animal feed that contain only organic Se mostly supply it in form of SeMet, which is actively absorbed in the intestine as an amino acid employing similar processes as methionine (Met). However, it is important to make the distinction between total SeMet and free SeMet content, as only the free form of this selenoamino acid is available for non-specific incorporation into proteins in place of Met, leading to Se-containing proteins, representing a biological pool of Se. Characterisation of Se enriched yeast products has traditionally been made by quantifying total SeMet content. A disadvantage of this approach however, is that it does not consider the effects of Se deposition in yeast on subsequent digestive availability. In this study, Se-yeast products were processed through a poultry in vitro digestion model, using pepsin and pancreatin as digestive enzymes. The objective of this work was to examine in vitro, the bioaccessibility of Se in commercially available Se-enriched yeast feed supplements. Using HPLC-ICP-MS, products were assessed for total Se and SeMet content. Results indicate that Se yeast products vary considerably in terms of total SeMet and SeMet as a % of total Se. A difference also exists between gastric and intestinal release of SeMet which may have implications for the bioavailability of individual products.

Effect of Buttiauxella phytase on grower pig performance fed wheat/barley diets without inorganic P

Y. Dersjant-Li[1], A.L. Wealleans[1], K. Schuh[2] and G. Dusel[2]
[1]Danisco Animal Nutrition, DuPont Industrial Biosciences, P.O. Box 777, Marlborough, SN8 1XN, United Kingdom, [2]University of Applied Sciences Bingen, FB1, Life Sciences, Bingen am Rhein, Germany; alexandra.wealleans@dupont.com

This study determined the effect of increasing doses of Buttiauxella phytase up to 1000 FTU/kg on production performance in grower pigs fed a European type wheat/corn/barley based diet without inorganic phosphorus. Five dietary treatments were tested: a nutritionally adequate positive control (PC); a negative control (NC) without inorganic P and reduced Ca (-0.12%) and ME (-34 kcal/kg) content; NC supplemented with Buttiauxella sp. phytase at 250, 500 or 1000 FTU/kg. A total of 100 pigs (initial mean BW of 30 kg, Topig × Pi genetic) was used, with 20 replicates per treatment (50% gilts and 50% barrows). Pigs were individually housed in pens and randomly allocated on the basis of BW and gender to the 5 treatments. Feed (in mash form) and water were offered ad libitum for 54 days. Basal diets had an endogenous phytase level of ~300 FTU/kg. Mean daily weight gain (ADG), mean daily feed intake (ADFI) and feed conversion ratio (FCR) were measured and calculated. Treatment means were compared using Tukey's HSD, linear response was determined using JMP 11 (SAS Institute Inc., USA). Overall mortality was very low and did not differ between treatments. Increasing phytase dose resulted in a linear increase in ADG ($P<0.05$). Phytase supplementation at 500 and 1000 FTU/kg to NC diet improved ADG by 6.9 and 7.6% respectively compared to NC ($P<0.05$) and by 4.6 and 5.4% respectively compared to PC ($P>0.05$). The estimated feed cost saving per kg BWG was 4.46 and 4.92% for phytase at 500 and 1000 FTU/kg compared to PC. The data suggest that it is beneficial to increase phytase dose up to 1000 FTU/kg in grower pigs up to 85 kg BW.

Response of finishing steers to graded levels of sorghum dry distiller's grains in the diet

A. Simeone, V. Beretta, J. Franco, E. Martínez and F. Zanetti
University of the Republic, Ruta 3, km 363, 60000 Paysandu, Uruguay; asimeone@adinet.com.uy

The present study evaluated the effect of level of inclusion of dry distiller's grains plus soluble (DDGS) from sorghum in the diet on the performance and carcass traits of lot-fed finishing steers. Twenty four Hereford steers (312,8±27.1 kg, 18 months old) were randomly allocated to four total mixed rations differing in the level of DDGS from sorghum (28% CP, 35% NDF, 12.5% fat): 0, 10, 20 or 30% of ration DM. Diets were balanced for minimum CP=14%, ME=2.5 Mcal/kg and 30% NDF across treatments, by varying sorghum grain (SG) sunflower meal (SFM) and wheat bran (WB) with respect to control diet (0% DDGS, 66.4% SG, 8.4% SFM, 12.6% WB, 1% urea, 1.5% molasses, 8.3% rice hulls, and 1.8% mineral-vitamin premix). Animals were gradually introduced to diets and then fed ad libitum in individual pens outdoor during 10 weeks. Liveweight (LW) was recorded every 14 days, and dry matter intake (DMI) was determined daily. Feed to gain ratio (FGR) was calculated based on mean values for the experimental period. Steers were all slaughtered on same date and carcass and meat traits were measured. The experiment was analyzed according to a randomized plot design with repeated measures for LW and DMI. Linear and quadratic effects associated to DDGS level were tested. No differences were observed in LW gain due to DDGS level (0% 1.44, 10% 1.74, 20% 1.87, 30% 1.84 kg/d, SE 0.18; P>0.10). DMI expressed as kg/100 kg LW, increased linearly with DDGS level (DMI = 2.93+0.020 DDGS; P<0.01 R2=0.95) while no differences were observed in FGR (0% 8.8, 10% 7.8, 20% 7.9, 30% 8.4 kg/kg, SE 0.34; P>0.10). Carcass weight (266 kg SE 3.4), yield (59.3% SE 0.31) and subcutaneous back fat were not affected by DDGS level (P<0.10), while rib eye area increased linearly (RA, cm^2=45.1+0.234 DDGS; P<0.01). Meat quality parameters (L. dorsi 48 h pH and color, lightness L* and brightness *a) did not differ between treatments (P>0.10). Results suggest that including DDGS from sorghum in up to 30% of DM in high grain finishing rations would not affect steer's feedlot performance, carcass or meat traits.

The effect of phytase, time and freezing method on gastric inositol phosphate concentration in pigs

S. Laird[1], I. Kühn[2], C. Brearley[3], H. Whitfield[3] and H.M. Miller[1]
[1]University of Leeds, The Faculty of Biological Sciences, LS2 9JT, United Kingdom, [2]AB Vista, Woodstock Court, SN8 4AN, United Kingdom, [3]University of East Anglia, University of East Anglia, NR4 7TJ, United Kingdom; bs09sl@leeds.ac.uk

Phytate (myo-inositol hexakisphosphate, InsP$_6$) and its degradation products are frequently measured in the digesta of monogastrics as a means of determining phytase efficacy. However, there is no standardised method for the collection and processing of the digesta in such studies. This experiment set out to determine the effect of time and freezing temperature on gastric inositol phosphate (InsP$_{2-6}$) concentration in pigs fed diets containing different levels of phytase. A total of 40 pigs were fed 1 of 4 diets ad libitum for 28 days. The diets comprised a nutritionally adequate diet (PC) with no added phytase, a similar diet but with Ca and P reduced by 1.6 and 1.24 g/kg respectively (NC), and this NC + 500 or 2,000 FTU/kg. At the end of the experiment the pigs were euthanised and the stomach chymus was collected and mixed thoroughly. Two subsamples of the chymus were frozen immediately; 1 on dry ice (-78 °C) and 1 in the freezer (-26 °C). Thereafter the remaining chymus was left to sit for 5 minutes before a further 2 subsamples were collected and frozen as above. This process was repeated every 5 minutes for a total of 15 minutes from the time of mixing. The stomach chymus samples were freeze dried and milled prior to InsP$_{2-6}$ determination by HPIC. Data were analysed using the GLM procedure of SPSS with time and freezing method as repeated measurements. In pigs fed diets with phytase, gastric InsP$_6$ was 82.4% and InsP$_5$ was 75.7% lower than those fed diets with no added phytase (P<0.001), the level of Ca and P did not influence phytate degradation. Gastric InsP$_6$ (P<0.001) and InsP$_5$ (P<0.001) were also influenced by freezing method with lower levels being detected in samples frozen at -26 °C than at -78 °C. Further, gastric InsP$_6$ was also influenced by time (P<0.01), with levels of InsP$_6$ linearly decreasing (P<0.001) over time. These results suggest that the rapid collection and freezing of chymus is necessary in order to minimise phytate degradation after sampling.

Productive and cecal microbial counts of broilers feeding with dry whey or whey protein concentrate

C. Pineda-Quiroga, R. Atxaerandio, R. Ruiz and A. García-Rodríguez
Neiker-Tecnalia, Animal Production, Campus Agroalimentario de Arkaute, 01080, Vitoria-Gasteiz, Spain;
cpineda@neiker.eus

Whey is a liquid co-product of the dairy industry remaining during the production of cheese. However, whey has about 50% of nutrients present in milk as lactose, serum proteins, minerals, vitamins and small amount of fat. Dry whey (W) is rich in lactose content (70% of dry matter). In poultry, lactose is not digested in the foregut but could be fermented in the hindgut, promoting the growth of intestinal acid lactic bacteria and acting as prebiotic. Otherwise, whey protein concentrate (P, 30 to 85% crude protein) is rich in proteins that have higher nutritional and biological value. The aim of the present trial was to evaluate the effect of supplementing corn-soybean based diets with W and P formulated to provide the same lactose content (42 g/kg) but different protein source over productive performance and cecal microbial counts. Nine hundred day-old broiler chicks Ross 308 were allocated in pens of 2.75 m^2. Three treatments with 9 replicates of 30 chickens were evaluated: control diet (0W0P), 60 g W/kg (60W0P) and 80 g P/kg (0W80P). Animals were fed with starter diet from day 0 to 21 and then were switched to a grower-finisher diet until day 42. Soybean meal was replaced at 20% and 11% on 0W80P starter and grower-finished diets respectively, related to 0W0P. Thus, protein supplied by P represented 14% of the total protein provided. Body weight (BW), daily weight gain (DWG), feed intake (FI) and feed conversion ratio (FCR) were calculated on a weekly basis. Last day of experiment, 5 chickens per treatment were randomly selected and slaughtered to remove cecum for Bifidobacterium spp., Lactobacillus spp., Clostridium perfringens and E. coli counts. At day 42, animals fed with 0W80P and 60W0P had higher BW (2,231 and 2,090 vs 1,882 g) DWG (52 and 74 vs 41 g) and FI (91 and 85, vs 78 g) without differences on FCR respect to 0W0P (P<0.001). Non differences were determined between 0W80P and 60W0P for productive parameters measured. Non-effect was observed of experimental diets on cecal bacterial counts. The presented results showed favourable results of dry whey or whey protein concentrate in diets for broilers on productive performance without affecting the cecal counts of monitored bacteria.

Supplementing ruminally protected phenylalanine to dairy diets containing high levels of canola meal

N. Swanepoel[1], P. Robinson[1] and L. Erasmus[2]
[1]Univ of California, Dept Animal Science, Davis, CA 95616, USA, [2]Univ of Pretoria, Animal and Wildlife Sciences, Lynnwood Rd, Pretoria 0001, South Africa; lourens.erasmus@up.ac.za

Limited research identified Phe as a limiting amino acid (AA) in corn silage based diets, after Lys and Met. A previous study fed 7.5 g/cow/d of intestinally absorbable (IA) Phe, and suggested that it was insufficient to increase milk yield since it was primarily used to support increased body condition score (BCS, AFST 205, 10). Our objective was to determine if supplementing 15 g/cow/d of IA ruminally protected (RP) Phe (HCMP) to a diet containing 17% CM (HCM) would increase milk yield after fulfilling its apparent 1st priority of restoring previously mobilized peptides to muscle protein synthesis, thereby regaining animal performance possibly lost with higher dietary CM inclusion levels (i.e. 13.0% (LCM) to 16.5% (HCM)) based upon Swanepoel et al. To avoid a possible Met limitation 2.0 g/cow/d of IA RP Met was added. The experimental design was a 3×3 Latin square using 3 pens of ~315 early lactation cows/pen with three 21 d periods. Dry matter intake was not affected (avg: 27.5±0.5 kg/d) and there was no impact on milk and component yields, except a reduced lactose concentration (P=0.02) with Phe addition. Even though plasma Phe levels only differed numerically among treatments, supplementation resulted in energy being diverted towards BCS, as in Swanepoel et al., but this time the change was much higher and not at the expense of milk protein and fat components, possibly suggesting that the higher Phe supplementation level was successful in supplying enough Phe to replace mobilized muscle protein while maintaining milk yield. In Swanepoel et al., total NE output was numerically reduced with Phe supplementation to the high CM diet but, in this study, it was increased to the level of the LCM diet, when Phe was added to the HCM diet. This suggests that supplementation of Phe to HCM regained animal performance that was lost when CM inclusion was increased with the HCM diet. It also seems possible that a further increase of supplemental Phe may have additional benefits.

Biohydrogenation of linoleic acid by the rumen bacterium Propionibacterium acnes as affected by pH

M.R.G. Maia[1,2], A.R.J. Cabrita[2], R.J.B. Bessa[3], A.J.M. Fonseca[2] and R.J. Wallace[4]
[1]REQUIMTE, LAQV, FC, Universidade do Porto, Rua do Campo Alegre s/n, 4169-007 Porto, Portugal, [2]REQUIMTE, LAQV, ICBAS, Universidade do Porto, Rua de Jorge Viterbo Ferreira 228, 4050-313 Porto, Portugal, [3]CIISA, FMV, Universidade de Lisboa, Av. da Universidade Técnica, 1300-477 Lisboa, Portugal, [4]Rowett Institute of Nutrition and Health, University of Aberdeen, Greenburn road, Bucksburn, Aberdeen AB21 9SB, United Kingdom; ajfonseca@icbas.up.pt

The ruminal biohydrogenation (BH) of unsaturated fatty acids is a much more complex process than was once thought. In fact, the diversity of BH intermediates found in the rumen is not fully explained and might be a result of changes in the microbial population or of metabolic adaptations to environmental stress stimuli. Our aim was to evaluate if Propionibacterium acnes, a ruminal bacterium with biohydrogenating activity, might modify its linoleic acid (LA) BH pattern when challenged by low pH. A type strain of P. acnes (DMS 1897) was inoculated into Hungate tubes with the liquid form of Hobson's M2 medium adjusted at pH 5.5 or 6.5, and supplemented with 50 mg/l LA. Tubes were incubated at 39 °C for 0, 6, 12, 18, 24, 30, 36, 48, 72, 96, and 120 h in triplicate. At each time point, growth was determined by turbidity at 650 nm, pH measured and fatty acids methyl esters analysed by GC-FID equipped with a 100-m capillary column. Growth of P. acnes was affected by pH. The maximum optical density measured was over two fold higher at pH 6.5 (1.1) compared to pH 5.5 (0.5). However, no differences were observed in the lag phase, which was of 6 h for both pH levels. pH had no effect on the LA BH pathway. Regardless of medium pH, LA was isomerised to four conjugated linoleic acid (CLA) isomers, namely 18:2 trans-10,cis-12, 18:2 trans-10,trans-12, 18:2 cis-10,cis-12, and 18:2 cis-10,trans-12, corresponding to 53 and 52, 37 and 38, 6 and 5, and 4 and 5% of total CLA for pH 5.5 and 6.5, respectively. In agreement with previous studies, P. acnes was unable to reduce the CLA isomers to monoenoic acids, neither at pH 5.5 nor 6.5. Although the BH pathway was the same, the BH rate differed, being slower at pH 5.5 than 6.5. In conclusion, although growth is affected, the pathway of LA BH by P. acnes is unchanged by pH stress.

Fermentation characteristics of milling by-products of durum wheat contaminated by mycotoxins

P.P. Danieli[1], N. Musco[2], R. Primi[1], M.I. Cutrignelli[2], R. Tudisco[2], B. Ronchi[1], F. Infascelli[2] and S. Calabrò[2]
[1]University of Tuscia, Agricultural and Forestry Sciences, Via S. Camillo de Lellis, 01100 Viterbo, Italy, [2]University of Napoli Federico II, Veterinary Medicine and Animal Production, via F. Delpino 1, 80137 Napoli, Italy; serena.calabro@unina.it

The use of cereal by-products in livestock nutrition requires careful evaluation in terms of mycotoxins contamination that can cause health problems in animals and contamination of animal products. Ruminants are considered to be less susceptible to the negative effects of mycotoxins than single-stomached, due to the rumen microflora ability to degrade/deactivate mycotoxins. Nevertheless, there are evidences that pure mycotoxins can have some in vitro effects on fermentation parameters, i.e. gas and volatile fatty acids (VFA) production, dry matter degradability and ammonia (NH_3-N) release. At EU level, ochratoxin A (OTA) and fusarium-toxins in feed have not specific law limits yet: there are only recommended guidance values (Comm. Rec. 2006/576/EC), which are conceived to avoid direct health problem in animal, without considering the impact on productivity. Aim of the study was to test the effect on in vitro fermentation characteristics of three milling by-products (bran, shorts, middlings) of durum wheat (Triticum durum L.) naturally contaminated by OTA and the tricothecenes deoxynivalenol (DON), its metabolites and nivalenol. The milling by-products contaminated at high level (HL) and the correspondents samples contaminated at low level (LL) (4.07 vs 0.55 mg/kg, 88% DM, as sum of mycotoxins) were incubated under anaerobic conditions for 96 h with bovine inoculum. Organic matter degradability (dOM), gas, VFA and NH_3-N production and fermentation kinetics were determined. GLM of SAS (2000) was used to test the effect of contamination level within the three milling by-products. The contamination level significantly (P<0.01) affected many fermentation parameters as by-products type. In particular, the HCL reduced gas (329 vs 312 ml/g) and VFA (6.44 vs 5.88 mM/g) in all by-products; less clear was the influence on dOM and NH_3-N. Mycotoxins also affected the fermentation kinetics in terms of rate, which appears higher in the LL by-products (17.9 vs 12.5 ml/h). The practical perspectives of these finding deserves further careful evaluations.

The nutritional value for dairy cow of wheat grain and wheat co-products from bioethanol production

P. Dagah and K. Theodoridou
Queen's University Belfast, Institute for Global Food Security, Clooreen Park, Malone Road, BT95HN, BELFAST,
United Kingdom; k.theodoridou@qub.ac.uk

The rapid expansion of bioethanol production in recent years results in producing large amount of co-products, such as dried distillers grains with solubles (DDGS). Due to starch removal, the remaining chemical components in DDGS products concentrate approximately threefold compared to the parent grain. Although, DDGS have been widely used in animal feed and livestock production industry, limiting research has been conducted on wheat and the co-products (w-DDGS). Apart from the chemical composition, the nutrient structure may also influence protein quality and availability. However, the inherent protein molecular structures of wheat and w-DDGS have not extensively been studied. Protein molecular structure profiles were identified by Fourier-transform infrared attenuated total reflection spectroscopy (FT/IR-ATR). Mineral composition was carried out on a X-ray fluorescence spectrometry. Protein supply to dairy cattle predicted with the PDI French system. Statistical analyses were performed using the MIXED procedure of SAS. This study aimed to evaluate the: (1) chemical profiles; (2) metabolic characteristics; and (3) protein molecular structures of wheat and w-DDGS from the same wheat-based bioethanol processing plant. Wheat had lower ash (1.4 vs 7.8%, $P<0.001$), crude protein (9.6 vs 33.6% $P<0.001$), ether extract (1.9 vs 4.1%, $P<0.001$), and nitrogen digestibility ($P<0.001$) content than w-DDGS. For carbohydrate profiles wheat had significant lower fiber (3.2 vs 5.9%), neutral detergent fiber (12.7 vs 83.1%), acid detergent fiber (3.1 vs 8.3%) but higher starch (56.1 vs 11.5%; $P<0.001$) than the w DDGS. The Mg, P, K, Ca, Mn, Fe, Zn was significant lower for wheat than for w-DDGS. DDGS exhibited significant differences in their protein molecular profiles which may result from the bioethanol process. The spectral intensity of β-sheet was significant higher for w-DDGS, which may cause low access to gastrointestinal digestive enzymes, resulting in a low protein availability. The findings indicate that w-DDGS has a great potential as livestock feed due to its high protein and fiber content and the knowledge of protein molecular structure may help to improve the quality of DDGS by optimizing bioethanol processing.

Energy-protein supplement affects fatty acid profile of liver and adipose tissue of cows

A.M. Brzozowska, K. Rutkowska, M. Lukaszewicz and J.M. Oprzadek
Institute of Genetics and Animal Breeding of the Polish Academy of Sciences, Department of Animal Improvement,
Postępu 36A, 05-552 Jastrzebiec, Poland; j.oprzadek@ighz.pl

The aim of this study was to remove the hypothesised effects of lactation stage (100-120, 120-200, 200<) and parity (1st and later) on fatty acids in blood, liver and adipose tissue of dairy cows while testing the effect of energy-protein feed additive (EPS – whole golden flaxseed, protected rapeseed cake, calcium soaps of flaxseed and fish oils, wheat bran, Blattin Lacto-Fett, mineral-vitamin additive). The study covered 24 Polish Holstein cows fed diet based on corn or grass silage. Liver, adipose tissue and blood samples were collected in adequate intervals and their fatty acid composition was determined. ANOVA accounted for the effects of the stage of lactation, parity and energy-protein supplement. Stage of lactation affected significantly ($P<0.05$) the contents of C16:0, C16:1 c9, C18:0, C18:1 c9, C18:2n-6, C18:3 n-3, C20:3 n-3, and C20:4 n-6 fatty acids, in the blood. Palmitic, palmitoleic, oleic and arachidonic acids reached highest values in early lactation (up to 120 days) – 15.59, 0.39, 12.58 and 1.18%, respectively. At the later stages of lactation their contents set up at constant lower levels (14.37, 0.28, 10.46 and 1.03%, respectively). Lactation stage affected also palmitic, stearic, oleic, linoleic and docosapentaenoic acids in liver and c9, t11 CLA in the adipose tissue. Parity of cows had no effect on any of the fatty acids in blood while palmitoleic acid in liver and C14:1 with C16:1 in the adipose tissue, increased with parity. EPS significantly affected the content of most fatty acids in blood (e.g. oleic and alpha-linolenic acid) and liver (alpha-linolenic acid, both isomers of CLA and n-3 fatty acids derived from fish oil), but it did not affect the profile of the adipose tissue of cows. EPS significantly ($P<0.05$) changed content of the most fatty acids in blood and liver, but it did not affect the profile of the adipose tissue. Most of the observed changes are beneficial from the human health point of view. In addition to nutrition also stage of lactation and parity influenced fatty acid content of the tissues.

Dietary effects of sodium bentonite on growth performance in broiler chickens

O. Besseboua[1], A. Ayad[2] and H. Benbarek[1]
[1]Department of Agricultural Sciences, Faculty of Life and Nature Sciences, University M. Istambouli, Mascara, 29000, Mascara, Algeria, [2]Department of Environment and Biological Sciences, Faculty of Life and Nature Sciences, University A. Mira, 06000, Bejaia, Algeria; besseboua.omar@gmail.com

The present experiment was conducted to investigate the effect of supplementing poultry feed with graded levels of Algerian sodium bentonite (Na-B) on growth performance and the development of villus height in jejunum and some biochemical parameters during 50 days in broiler chickens. Four hundred twenty 1-day-old broiler chicks (Arbor acres) were obtained from a commercial hatchery. The birds were randomly allocated into 6 groups (A, B, C, D, E and F). The treatments were 0 (control), 1, 2, 3, 4 and 5% of Algerian Na-B levels. Statistical analysis was performed using t-test to compare between different groups ($P<0.05$). The results obtained indicate clearly that weight gain in the chickens fed treatments containing 3% Na-B had greater weight gain than the chickens fed different treatments (0, 1, 2, 4 and 5% Na-B). Feed conversion rate was lower birds supplemented with NaB 1% (1.66) than control group (2.12) ($P<0.05$). Maximum feed consumption was observed in the birds' control (2,224 g), while the lowest was noted in the chickens with diet added 4% Na-B (1,856 g). The weight of duodenum, jejunum and ileum was decreased for the Algerian Na-B supplemented group compared with the control group ($P<0.05$). The villus height was affected by dietary treatments (1, 2, 3 and 5%) at day 18 and 50. feeding the supplemented graded levels Na-B resulted in increase in plasma cholesterol, triglyceride and HDL concentrations at 50 days of age compared to the control group. These results showed clearly that the Na-B from Algeria can improve the growth performance in broiler chickens. Thus, dietary inclusion of Na-B had positive effect on plasma triglyceride, cholesterol and HDL values in broiler chickens at the end experiment.

Can Seaweeds be used as protein source in organic dairy farms?

U. Tayyab and L. Sinclair
Harper Adams Univeristy, Animal Production, Welfare and Veterinary Sciences, Edgmond, Newport Shropshire, TF10 8NB, United Kingdom; utayyab@harper-adams.ac.uk

Use of seaweeds in animal feed is not new, Norway was first EU member who started producing seaweed meal in 70s for animal feeding. North Ronaldsay sheep in UK have been grazing exclusively on seaweed for some time in a year. To decrease the dependence of imported soya, the focus has shifted on locally producing legumes (e.g. Lucerne) and other alternative protein sources (seaweeds). Red and green seaweeds are high in protein in comparison to brown seaweeds. Uses of seaweeds also help to bring back nutrients to land, that leached to marine waters. Recently, few studies have focused in feeding value of seaweeds for dairy cows and concluded the potential use as protein source. Increased awareness in product quality has compelled farming to switch from conventional dairy systems to organic farming. Lack of enough high organic protein feedstuffs are one big hurdle in organic production. However, seaweeds production in marine waters believed to be organic and potentially can be used to feed ruminants in organic system.

Effects of chemical additives on fermentation characteristics of pressed sugar beet pulp

H. Scholz[1], C. Potthast[2], E. Kramer[3] and H. Auerbach[4]
[1]Anhalt University of Applied Sciences, Faculty LOEL, Strenzfelder Allee 28, 06406 Bernburg, Germany, [2]Südzucker AG Mannheim/Ochsenfurt, Marktbreiter Straße 74, 97199 Ochsenfurt, Germany, [3]ISF GmbH, An der Mühlenau 4, 25421 Pinneberg, Germany, [4]ISC, Thomas-Müntzer-Straße 12, 06193 Wettin-Löbejün, Germany; h.scholz@loel.hs-anhalt.de

As the knowledge of the effects of silage additives on pressed beet pulp (PBP) quality is still limited, 50 t of fresh PBP per treatment was pressed into plastic tubes of 3 m diameter and 6-8 m length on December. During filling, the material in each bag was treated with one of three chemical additives (KOFASIL STABIL: 1.5 l/t; Silostar liquid HD: 2.0 l/t; NOVIBAC: 2.7 l/t), which differed in composition, by an applicator mounted on the bagger. Untreated PBP served as control. All bags were placed outside, covered with a bird-protection silo net and opened in June. Subsequently, the bags were exposed to air for 7 days and the same surface ('old') sampled again, followed by removal of 1 m of PBP silage and sampling from the new surface ('fresh'). This procedure was repeated twice at weekly intervals. The trial was terminated after 4 weeks when only 'old' surfaces were left for taking samples. All silages were individually analyzed for dry matter, fermentation pattern, fungal counts and aerobic stability. Fresh PBP contained typical concentrations of nutrients and energy. Fungal counts averaged log 1.5/g for moulds and log 5.3/g for yeasts, respectively. All PBP silages taken on the day of bag opening were well fermented, with only minor differences between treatments. Additives increased aerobic stability of PBP taken from 'fresh' surfaces (n=72) from 81 hours (control) to 248 hours (P<0.001), which can be explained by the reduction in yeast (log 2.8 vs log 1.8, P<0.001) and mould (log 3.3 vs log 2.1, P<0.001) counts. Aerobic stability of PBP silages was negatively correlated with the number of yeasts (rs=-0.65, P<0.001) and moulds (rs=-0.61, P<0.001). It is concluded that chemical silage additives effectively improve hygienic quality and aerobic stability of PBP silages stored in plastic tubes.

Protein requirement of Korat chickens from 0 to 3 weeks of age

P. Maliwan, S. Khempaka and W. Molee
Institute of Agricultural Technology, Suranaree University of Technology, School of Animal Production Technology, 111 University Avenue, Muang, Nakhon Ratchasima, 30000, Thailand; praphot382@gmail.com

The Korat (KR) chicken is a Thai indigenous crossbred (50%) meat chicken, which has a better growth performance while meat qualities and consumer perceptions are similar to indigenous chickens. The demand for KR chickens by domestic consumers is increasing because KR chicken's meat has a unique taste, texture, less fat and high collagen. However, there is limited information on nutrient requirements, especially protein, for KR chickens. Therefore, this study was conducted to evaluate the protein requirements of KR chickens from 0-12 wk of ag, divided into 4 phases (0-3, 3-6, 6-9 and 9-12 wk of age). Since this research is not yet completed, only results of age 0-3 wk are presented. Five dietary protein levels were composed: 19, 20, 21, 22 and 23%, respectively. Nine hundred 1-d-old mixed-sex KR chicks were randomly divided over the 5 dietary treatments, each containing 6 replicate pens (30 birds per pen), and using completely randomized design (CRD). These birds were raised until 3 wk of age. Body weight (BW), BW gain, feed intake (FI), feed conversion ratio (FCR), feed cost per kilogram of BW gain, protein efficiency ratio (PER), energy efficiency ratio (EER) and blood urea nitrogen (BUN) of chickens from each pen were measured. The results found that there were significant differences (P<0.01) in BW, BW gain, protein intake and PER among treatments. As dietary protein increased from 19 to 23%, the BW, BW gain and protein intake were increased (P<0.01), while PER were decreased (P<0.01). However, FI, FCR, feed cost per weight gain 1 kg, ME intake, EER and BUN did not differ significantly (P>0.05) among treatments. According to broken-line regression analysis, protein requirements of KR chickens for optimal BW gain was 21.26% and the regression equation predicting the protein requirement for optimal BW gain was y = 229.00 – 5.5508 X (21.26 – x) [P<0.01, R^2=0.48].

A robust approach to modelling in vitro rumen fermentation data

N.G. Bootes, B.J. Leury and F.R. Dunshea
The University of Melbourne, Faculty of Veterinary and Agricultural Sciences, Royal Parade, Parkville, Victoria 3010, Australia; fdunshea@unimelb.edu.au

The gas production profiles from in vitro fermentation of substrate in rumen microflora typically exhibit a curve-linear sigmoidal relationship. The end-point data in and of itself does not represent the dynamic nature of substrate utilisation unless each data measurement is regressed against time. Due to the nature of the time-dependent data equation modelling is needed so that variability within data sets be limited and a cross-comparison between individual units possible. The choice of model is critical and poorly understood or applied models can lead to false interpretations and conclusions. Early models, due to the simplicity of the data, were minimalistic and oversimplified. The Gompertz equation is a commonly used equation to fit growth patterns that have lower and upper limiting stages and has been used extensively to model in vitro gas production rates. However, the Gompertz equation can sometimes lead to nonsensical estimates of gas production, particularly early in the fermentation. In this study the alternative Hill curve which is a modified version of the Generalised Michaelis-Menten curve to the Gompertz curve was analysed. This equation has three parameters, A b and C depict the change in gas volume (G) at time (t) as is seen in the following equation where $G = At^b/(C^b + t^b)$. A cross comparison of goodness-of-fit between Gompertz and Hill models over 33 experiments and 1,291 individual in vitro gas fermentation profiles indicated that in 31 of 33 experiments the Hill model had lower residual error. If model parameters led to impossible estimates (e.g. negative fermentation rates) they were deemed to be a 'fail'. 100% of the Hill model parameters were successful whereas up to 55% of the experiments had at least one fail for the Gompertz model parameters. Of those experiments that contained failed model fitting, up to 39% of the individual profiles are deemed as failed. With the exception of two experiments, the cumulative residual sum of squares was higher for data that was regressed against the Hill equation than Gompertz. These findings indicate that the modified Hill equation should be applied to in vitro gas production curves where gas production is measured in close to real time.

From vegetable food waste to earthworms as novel food/feed: microbial hazard identification

D. Tedesco, J. Parisi, M. Castrica and C. Balzaretti
Department of Health, Animal Science and Food Safety, University of Milan, Via Celoria 10, 20133 Milano, Italy; doriana.tedesco@unimi.it

Earthworms grown on fruit and vegetable waste can be a future alternative, and sustainable, food/feed source, contributing to waste disposal efficiency, responding to global food demand and reducing GHG (Greenhouse Gases) emission. In this study we evaluated the earthworms (Eisenia foetida) when reared on food vegetable waste, in order to establish critical limits and critical control points (HACCP) of this rearing system, for the prevention, elimination or reduction of identified microbial hazards. Our aim is to verify the safety of these new food / feed, regarding the presence of pathogenic micro-organisms, other than those provided by the reg. (EU) 2073/2005. In our first HACCP evaluation, the earthworms from this rearing process were characterized by levels of microflora ranged above common benchmark and threshold values to be considered a safety product. In this step, continuing on the HACCP system and Critical Control Point (CCP) identification, we evaluated the microbial contamination of earthworms (Eisenia foetida) reared on food vegetable waste that belong to the category 'fruit and vegetables to consume raw, uncut and unprocessed'. The earthworms samples from this rearing process were analyzed according to AFNOR and ISO methods for Total Bacterial Count, Enterobacteriaceae, coagulase-positive Staphylococci, Bacillus cereus, Total Coliforms, Listeria Monocytogenes, sulphite reducing Clostridia, Escherichia coli and Salmonella spp. Bacterial counts were expressed as Log cfu/g sample. Conforming to the Reg. (EU) 2073/2005, Salmonella spp. and Listeria monocytogenes were absent in 25 g samples; Escherichia coli and coagulase-positive Staphylococci were below the detection limit (1 log cfu/g). Nevertheless, samples were characterized by unsatisfactory levels of Enterobacteriaceae and total coliforms, whereas Bacillus cereus and sulphite reducing Clostridia were slightly higher than the acceptable safety range. Considering that fruit and vegetable waste substrate, represents the primary source of microbiological contamination for earthworms and consequently a relevant CCP, other detailed studies are needed to reduce the hazard and guarantee the safety of this new nutritional source.

From vegetable food waste to earthworms as novel food/feed: nutritional characteristics

D. Tedesco[1], A. Tava[2] and J. Parisi[1]
[1]Department of Health, Animal Science and Food Safety, University of Milan, Via Celoria 10, 20133, Milano, Italy, [2]Consiglio per la Ricerca in Agricoltura e l'Analisi dell'Economia Agraria, viale Piacenza 29, 26900 Lodi, Italy; aldo.tava@entecra.it

Earthworms grown on fruit and vegetable waste can be a future alternative and sustainable food/feed source, contributing to waste disposal efficiency, responding to global food demand and reducing greenhouse gases (GHG) emission. As a part of studies on the feasibility to transform vegetable food wastes in this novel protein source, we evaluated the nutritional characteristic of earthworms (Eisenia foetida) reared on food vegetable waste from different rearing cycle. The focus was to evaluate the protein and fat content and the amino acids and fatty acids profile. Briefly, approximately 800 earthworms were separated from each substrate, washed in running controlled water to remove dirt from the body surface, kept in saline solution at room temperature changing the solution until digestive system were clean, and then freeze – dried for further analyses. Samples of earthworms were analyzed for crude protein (CP) and total amino acids content, determined by the High Performance Liquid Chromatography (HPLC) analysis. Amino acids were quantified according to calibration curves made with pure reference compounds, using the internal standard method. The ether extract was evaluated by solvent extraction, fatty acids profile was determined using reference fatty acid methyl esters (FAME) by gaschromatography flame ionization detector (GC/FID) analysis. The CP content ranged from 64.4% to 67.15% of the dry matter. The earthworm samples contains all the essential amino acids, with high lysine, threonine and arginine level. Total fats were quantified from 9.3% to 10.84% dry weight, while the fatty acids composition revealed a great amount of polyunsaturated fatty acids (PUFA), quantified from 40.32% to 43.91% of the total. Among PUFA, the ω3 and ω6 series were steadily well represented, ranged from 7.5% to 8.22% of the total amount. From these results, earthworms reared on vegetable food waste, showed an attractive nutritional profile for food/feed applications.

Genetic analysis of temperament traits assessed by riders and judges at Icelandic horse field tests

H. Sigurðardóttir[1], S. Eriksson[1] and E. Albertsdóttir[2]
[1]Swedish University of Agricultural Sciences, Dept. of Animal Breeding and Genetics, P.O. Box 7023, 75007 Uppsala, Sweden, [2]The Farmers Association of Iceland, Bændahöllinni við Hagatorg, 107 Reykjavik, Iceland; elsa@bondi.is

Temperament is an important trait in the breeding goal for Icelandic horses. The aim is a very willing, even fiery, horse that is easy to handle and tries to please the rider. In breeding field tests judges assess conformation and riding ability including the trait 'spirit'. The aim of this study was to give more insight in how well the trait spirit describes different aspects of the temperament that riders seek in Icelandic horses. A questionnaire about temperament of the horse was distributed to judges and riders at breeding field tests in Iceland in 2014 and 2015. A seven-point scale was used and the trait spirit was included together with total suppleness of the horse derived from the sub-traits rein contact, nerve strength, top line, head carriage and cooperation. In 2015 the total suppleness was based on rein contact and nerve strength only. In total judges answered the questionnaire for 2,088 horses, whereas the response from riders was lower with answers for only 451 of the horses. Genetic parameters were estimated using REML animal model in the DMU package. Estimated heritabilities for temperament traits were low to moderate and differed depending on whether assessments by judges or riders were used. Higher heritabilities were estimated for nerve strength and suppleness when assessed by riders (0.4) compared with by judges (0.1-0.2). Estimated genetic correlations with spirit were 0.8-1.0 for all temperament traits assessed by riders. When assessed by judges, the genetic correlations with spirit were in the range 0.5-1.0, with the lowest correlations estimated between spirit and nerve strength, and between spirit and rein contact. Genetic correlations between judge and rider assessments of the same traits were high except for spirit (0.6) and rein contact (0.5), but estimated with high standard errors. We conclude that the current assessment of temperament work rather well but could be further improved. Future studies of behavior of Icelandic horses in their daily environment and of temperament faults as reason for culling are planned.

Analysis of different definitions of performance traits in riding horses

V. Welker[1], K. Schöpke[2], M. Wensch-Dorendorf[1] and H.H. Swalve[1]
[1]Martin-Luther-University Halle-Wittenberg, Institute of Agricultural and Nutritional Sciences, Th.-Lieser-Str. 11, 06120 Halle, Germany, [2]Sächsische Gestütsverwaltung, Schloßallee 1, 01468 Moritzburg, Germany; viktoria.welker@landw.uni-halle.de

Genetic parameters for sport performance of horses recorded between 1995 and 2014 in Germany were estimated from 4,582,191 and 9,997,439 observations of 221,132 and 253,945 horses, for show jumping and dressage, respectively. The aim of the study was to define and check alternative trait definitions. Three traits were tested: Firstly, the transformation of the rank by using the formula: TR=mean of the ranks-rank, secondly, the highest level of competition achieved, and thirdly, the accumulated life points. Data were analyzed using SAS procedures and genetic parameters were estimated via ASReml. Heritabilities of TR were estimated separately for competition performance classes of increasing difficulty (A, L, M, S). The statistical model for TR comprised the combined fixed effects for age and sex, performance class of the rider and year, event, and the random effects for permanent environment and an additive genetic effect. For the highest level achieved the model included the combined fixed effects for age and sex, year of the competition, the random effect for rider and an additive genetic effect. Finally, the model for accumulated life points involved the fixed effects of years in competition, sex, birth year, and an additive genetic effect. Heritability estimates of TR in dressage and show jumping performance classes ranged from 0.02 to 0.14; for dressage and show jumping, estimates for class A were 0.07±0.007 and 0.02±0.003, class L 0.05±0.005 and 0.03±0.002, class M 0.07±0.006 and 0.04±0.004, class S 0.14±0.01 and 0.06±0.005. Results for highest level achieved and life points showed heritabilities of 0.32±0.001 and 0.16±0.005 for dressage, and 0.32±0.001 and 0.25±0.005 for show jumping. Further research will focus on analyses of estimates of breeding values and multivariate evaluations.

Genetic composition of Nordic riding horse populations and joint breeding value of stallions

S. Furre[1], Å. Viklund[2], J. Philipsson[2], B. Heringstad[1] and O. Vangen[1]
[1]Norwegian University of Life Sciences, Arboretveien 4, 1430 Aas, Norway, [2]Swedish University of Agricultural Sciences, P.O. Box 7070, 750 07 Uppsala, Sweden; siri.furre@gmail.com

Riding horse studbooks has increased the use of foreign stallions steadily during the last decades thus challenging the traditional prediction of breeding values (EBV) based on national data as stallions might have too few offspring for the EBV to be published based on national data only. Genetic composition and genetic similarity of the 4 Nordic sport horse studbooks, Danish (DWB), Finnish (FWB), Norwegian (NWB) and Swedish Warmblood (SWB), were estimated using data from young horse test (YHT), 41 216 records, competition (COMP), 73 962 records, and 229 163 horses in pedigree file. The largest genetic contributors originated from Germany with an average genetic contribution of 50% of the genetic composition of the Nordic studbooks, while horses of Thoroughbred origin contributed between 7.3% (FWB) and 31.7% (DWB). Genetic similarity between populations was from 13% (FWB and NWB) to 57% (DWB and SWB). On average, the Holstein contributed most to the similarity with a contribution ranging from 15.5% (FWB and NWB) to 28.1% (DWB and NWB). SWB had the single largest contribution between SWB and FWB (51.7%), and SWB and NWB (43%), while DWB-stallions contributed between 12.4 to 18.5% to the similarity between DWB and the other studbooks. Correlation between YHT traits ranged from 0.68 (walk) to 0.93 (canter). For a stallion to have official EBVs in DWB or SWB, at least 15 offspring must have test records within the population. There were 631 stallions having offspring with YHT records in at least two ≥15 offspring in total). Including test records of offspring from the other populations increased the number of stallions with publishable EBVs from 167 to 255 (DWB), 1 to 94 (FWB), 3 to 113 (NWB) and 175 to 305 (SWB). Accuracy of EBVs for stallions increased from an average of 0.36 to 0.54 over all studbooks. Joint EBV is beneficial and recommended for the Nordic sport horse populations.

Implications of across-studbook genetic correlations between linear traits for sport horse breeding

F. Sperrle[1,2], Å. Viklund[3], E. Thorén-Hellsten[4], W. Schulze-Schleppinghoff[5] and K.F. Stock[1,2]
[1]*University of Veterinary Medicine Hannover (Foundation), Inst. for Anim. Breed. & Genet., Bünteweg 17p, 30559 Hannover, Germany,* [2]*IT Solutions for Animal Production (vit), Heinrich-Schröder-Weg 1, 27283 Verden, Germany,* [3]*Swedish University of Agricultural Sciences (SLU), Dept. of Anim. Breed. & Genet., Box 7023, 75007 Uppsala, Sweden,* [4]*Swedish Warmblood Association, Box 2, 24729 Fylinge, Sweden,* [5]*Oldenburger Pferdezuchtverband e.V., Grafenhorststr. 5, 49377 Vechta, Germany; friederike.katharina.stock@vit.de*

Linear systems have been introduced for routine assessment of conformation and performance in several studbooks for riding horses, with the aim to strengthen the breeding programs by refined phenotype recording. Genetic parameters for linear traits have indicated the potential of targeted improvement of certain characteristics of conformation, gaits and jumping, so systematic linear profiling of young horses and routine breeding applications based on linear data are expected to gain in importance. Knowledge about genetic correlations between different linear traits assessed within and across studbooks helps to correctly interpret and make best use of the new breeding tool. For this study, results of genetic evaluations (GE) for linear traits from the Oldenburg Horse Breeding Association (linear data from 2012-2015; 6,018 foals, 3,622 mares and stallions) and the Swedish Warmblood Association (linear data from 2013-2014; 1,889 3-year-olds) were available. Estimated breeding values (EBV) of stallions with linearly described progeny provided the basis of genetic correlation analyses. Of the 130 stallions with progeny records in both GE, only 17 had 5 or more offspring in each of the GE. Pearson correlation coefficients and Spearman rank correlations between analogous linear traits were significant ($P<0.01$), ranging between 0.4 and 0.7. With the highest values found for trot and jumping traits, the pattern of correlations plausibly reflected both the different heritability levels and the similarity of definitions of the linear traits. Transparency of linear profiling practices with reference to agreed trait definitions facilitates the interpretation and use of EBV while similarity of breeding goals for sporthorses may warrant sufficient overlap of linear systems.

A genome-wide association study of young horse test traits in Swedish Warmblood

S. Eriksson, Å. Viklund and S. Mikko
Swedish University of Agricultural Science, Dept. of Animal Breeding and Genetics, P.O. Box 7023, 75007 Uppsala, Sweden; susanne.eriksson@slu.se

Tests of three-year-old Swedish Warmblood (SWB) horses include traditional subjective assessments on a ten point scale of the quality of the horses in relation to the breeding goal. In 2013 a complementary linear scoring of the horses on a nine point scale from one extreme to the other was introduced. The horses are assessed for traits related to conformation, gaits and jumping ability. Also, deviations from normality are indicated for some traits (0/1). In this study we aimed to find genomic regions associated with these test traits. In total 380 SWB horses tested as three-year-olds in 2013 or 2014 were genotyped using the Affymetrix 670k SNP bead chip. After quality control the data set included 379 horses and 467,606 SNPs. Counting also traits recorded in only one of the years, the data included 97 traits. A fast score test for association in the R software GenAbel was used for a first set of analyses. The uncorrected lambda value was >1 for most traits so genomic control was used, and the model included year and place of the test and sex of the horse. Preliminary results validate findings in other horse breeds of a region on ECA3 highly significant for height at withers. SNP-associations significant after Bonferroni correction were also found for traditionally scored type of the horse, and linear measures of type (light – heavy built). Because the data was particularly stratified for jumping traits, analyses were also done within clusters of genomic relationship. Within one of the two clusters, significant associations were also found for jumping technique. Suggestive significant associations were found for several traits, and continued analyses including polygenic effects are in the pipeline.

No signs of decline in genetic variation in racing performance traits in Swedish standardbred

T. Árnason
IHBC AB, Knubbovagen 34, 74494 Morgongava, Sweden; info@ihbc.se

The BLUP methodology has been applied to the genetic evaluation of Swedish standardbred trotters since the early eighties. Substantial genetic progress in racing performance traits has been achieved. In 30 years the population mean for racing performance has increased by more than 2 phenotypic SD, and in the last decade the estimated rate of annual genetic progress is 7% of σ_P. The population, which has the actual size of about 4,000 broodmares, has a small founder base, the current level of inbreeding is exceeding 8% and the rate of increase in F correspond to N_e of 40 animals. In previous studies an inbreeding depression in racing performance has been estimated to be 0.4% of σ_P/yr in this population. The main objective of this study was to evaluate presumed changes in genetic variation in the performance traits due to inbreeding and selection. The material was divided into 3 periods according to birth-years (78-90; 91-00; 01-09). Each data set contained ca 30 000 horses with racing records and ca 50 000 horses in complete 5 generations pedigree files. The genetic analyses were performed with AI-REML using DMU software applying the same AM-model to the 3 data sets. Heritability estimates for ranks, earnings and time were in the range 0.30-0.37 and for earnings per race 0.46-0.48. No decline in heritability or in genetic variance was observed over time. Inclusion of maternal linage in the analyses indicated that mitochondrial DNA contributed very little (0 to 0.5% of σ^2_p) to racing performance in Swedish standardbred trotters relative to the contribution of nuclear DNA (reflected by h^2). The results indicate that the actual size of the population allows for sufficient creation of new genetic variation balancing the loss due to the inbreeding. Constant selection intensity might have created an equilibrium (steady state) between reduction in genetic variance due to directional selection and the variation rebuilt by recombination (the Bulmer effect).

Estimation of heritabilities of conformation traits in Slovenian population of Lipizzan horse

K. Potočnik[1], M. Mesarič[2], V. Štepec[1], J. Krsnik[1], A. Kaić[3] and M. Simčič[1]
[1]University of Ljubljana, Biotechnical Faculty, Department of Animal Science, Jamnikarjeva 101, 1000 Ljubaljana, Slovenia, [2]University of Ljubljana, VeterinaryFaculty, Gerbičeva 60, 1000 Ljubaljana, Slovenia, [3]University of Zagreb, Faculty of Agriculture, Department of Animal Science and Technology, Svetošimunska cesta 25, 10000 Zagreb, Croatia; klemen.potocnik@bf.uni-lj.si

Scoring of conformation traits as the main breeding criteria is described in the breeding program for Slovenian population of Lipizzan horses. At the moment, the quality of the horse is assessed in relation to the phenotypically scored traits. The aim of this study was to estimate genetic parameters in the population of Lippizan horse in Slovenia for the first time and to check their suitability for the genetic evaluation. Data from 291 horses born from 1996 to 2011 were analysed. The pedigree file contained 945 individuals. There were 12 conformation traits in total (3 measured and 9 scored) and 2 gait traits. Traits are subjectively evaluated on a 10 point scale by a team of experts after horses achieved sexual maturity. Covariance components were estimated in the VCE 6.0 package with an animal model including age at scoring, birth year and sex as fixed effects. Estimated heritabilities ranged from 0.07 for breed type to 0.64 for cannon bone circumference. The highest heritabilities were estimated for the neck (0.59), chest girth (0.58), front part of the body (0.47), and overall score (0.45). Estimated heritabilities for front legs (0.39), middle part of the body, gait correctness (0.31) and height in wither (0.30) were moderate, while rear legs (0.26), rear part of the body (0.25), head (0.21) and gait efficiency (0.16) had lower heritabilities. The results show that it is possible to use the conformation traits for genetic evaluation, but the low number of scored animals considered for the covariance components estimation needs to be taken in the account. To increase the number of scored animals for the prediction of breeding values for conformation traits, one of the options may be to merge data from populations from different countries.

Population structure and recent migration in closely related Old Kladruber and Lipizzan populations

N. Moravčíková[1], L. Vostrý[2], H. Vostrá-Vydrová[2], O. Kadlečík[1] and R. Kasarda[1]
[1]Slovak University of Agriculture in Nitra, Department of Animal Genetics and Breeding Biology, Tr. A. Hlinku 2, 949 76 Nitra, Slovak Republic, [2]Czech University of Life Sciences Prague, Department of Genetics and Breeding, Kamycka 129, 165 21 Praha 6-Suchdol, Czech Republic; nina.moravcikova@uniag.sk

The study was prepared in order to examine the population subdivision, genetic relationship and recent migration events between historically related Old Kladruber and Lipizzan populations. A total of 270 Old Kladruber horses representing two colour varieties (175 white and 95 black) and 418 Lipizzan horses originating from Slovak (61) and Slovenian (357) studs were evaluated. A set of 13 microsatellite markers has been used for analysis of genetic variability. All loci used in this study reached good level of polymorphism (PIC=0.67±0.09) and therefore considered as informative. As expected due to the origin of each analysed subpopulation the genetic diversity parameters at intra-population levels showed only small differences. Observed heterozygosity (0.65-0.69) and inbreeding coefficient (-0.02-0.08) indicated the sufficient proportion of heterozygotes within each subpopulation. Despite the fact that the degree of differentiation estimated based on F_{ST} and genetic distances were low, the differences obtained at intra-population level of Old Kladruber horses could relate to segregation of colour variants. However, the genetic distances computed at individual level indicated formation of only three main clusters according to breed origin. Bayesian assignment analysis showed that the average log-likelihood was maximized at K=3. The membership probability outputs suggested that the frequencies of alleles varied across the three regions that confirmed the evidence of genetic clusters in relation to the horse breeding history. Following, analysis of relative migration rate showed mostly consistent results with patterns of detected population differentiation which means low migrations rate across the whole population. The results of this study will be useful for the development of breeding strategies, which consider classical horse breeding as well as recent achievements of population and conservation genetics.

High level and amateur competition are compatible objectives for jumping horses

A. Ricard[1,2]
[1]IFCE, Développement et Recherche, Jumenterie du Pin, 61310 Exmes, France, [2]INRA, UMR 1313, Domaine de vilvert, 78352 Jouy-en-Josas, France; anne.ricard@toulouse.inra.fr

The objective of this study was to estimate genetic correlation between young and adult performances and amateur and high level in jumping competition. French Equestrian Federation provided results of competition from 2009 to 2013. There were 3,254,266 results in competition performed by 96,478 horses and 160,949 relatives. Five traits were distinguished: JEUNE AMATEUR (4-6 years, normal competition) JEUNE LIBRE (4-6 years, special young horse competition for amateur riders), JEUNE PRO (4-6 years, special young horse competition for professional riders), ADULTE AMATEUR (>6 years, <1.30 m) and ADULTE PRO (>6 years, ≥1.30 m). Performances were measured by the annual sum of points (allowed according to rank and level of the event) by number of starts with log transformation. A multiple traits animal model was used. Heritabilities were moderate for JEUNE LIBRE (0.19) to high for ADULTE PRO (0.35). The genetic correlations were all highly positive (0.78 to 0.96). They were especially high for ADULTE AMATEUR and ADULTE PRO (0.95) and JEUNE AMATEUR and ADULTE AMATEUR (0.96) and lower with JEUNE LIBRE. The phenotypic correlations were lower, but still moderately high for ADULTE (0.66 to 0.74), and considerably lower between JEUNE and ADULTE (0.39 to 0.56). In conclusion, production of a leisure sport horse appears to be compatible with the choice of high level stallions, and performances in amateur competition provide useful information to produce elite horses.

Cross-validation analysis of genetic evaluation models for ranking in endurance horses

S. García-Ballesteros[1], L. Varona[2], J.P. Gutiérrez[1] and I. Cervantes[1]
[1]Departamento de Producción Animal, Universidad Complutense de Madrid, Avda. Puerta de Hierro s/n, 28040, Madrid, Spain, [2]Departamento de Genética, Universidad de Zaragoza, C/ Miguel Servet, 177, 50013, Zaragoza, Spain; icervantes@vet.ucm.es

The ranking in the race is one of the most common traits used as measure of performance in sport horse competitions. Different models can be found in the literature to predict breeding values such as linear and threshold models. However recent studies showed that breeding value for sport performance could be better predicted using a Thurstonian model. The objective of this study was to compare prediction abilities of linear, threshold and Thurstonian methodologies for genetic evaluation of ranking in endurance horses. For this purpose, eight genetic models were used for each methodology according to different combinations of random effects: rider, rider-horse and environmental permanent effect. All models included gender, age and race as systematic effects. The dataset contained 4,065 ranking records from 966 horses. The pedigree contained 8,733 animals (47% Arab horses). A cross-validation approach was used to evaluate the models' prediction abilities. The average of the correlation between real and predicted performances across models was 0.24 for threshold, 0.58 for linear and 0.60 for Thurstonian method. Despite no significant differences were found between models within methodology, the best model included the permanent environmental, rider and rider-horse random effects for threshold and Thurstonian methods, and only rider and permanent effects for linear methodology. The correlations of predicted breeding values between models were higher between threshold and Thurstonian: 0.92, 0.87 and 0.80 for all animals, top 20% and top 5% animals, respectively. For the rank correlations respective figures were 0.87, 0.75 and 0.74. Lower values were seen between linear and threshold methods (0.69, 0.37 and 0.30). We can conclude that Thurstonian method could be recommended for being used in the routine genetic evaluations for ranking in endurance horses.

Multi-trait animal model estimate of genetic parameters for morphometric traits in the Murgese horse

G. Bramante[1], A.M. Cito[1], E. Ciani[2] and E. Pieragostini[1]
[1]University of Bari, Department of Emergency and Organ Transplant, Division of Veterinary Clinics and Animal Production, Via Amendola 165/A, 70126 Bari, Italy, [2]University of Bari, Department of Biosciences, Biotechnologies and Biopharmaceutics, Via Amendola 165/A, 70126 Bari, Italy; knowout@gmail.com

Murgese, a baroque-type Southern Italian horse breed, is experiencing a renovated interest due to its valuable aptitudes. Selection based on estimated breeding values (EBVs) may expedite genetic progress compared to selection based on phenotypic values currently performed in the breed, for which a studbook has been recently established. For genetic evaluation accurate estimates of genetic parameters are needed. This work aimed at: (1) characterizing the morphological evolution and realized genetic trend in the Murgese population over the last four decades; and (2) estimating the genetic parameters for three routinely measured biometric traits. An observational study was carried out examining a data set, including measurements for withers height (WH), chest girth (CG) and cannon bone circumference (CBC) on 2,366 individuals approved for breeding use in the period 1974-2010. A multiple-trait animal model with fixed effects (herd, sex, birth year) was adopted. Genetic parameters and EBVs were estimated using restricted maximum likelihood. Trends in genetic gain were evaluated by linear regression of EBVs on birth years. The results showed significant increase of WH (males), and decrease of CG and CBC over the last four decades, while no significant trend in genetic gain was observed for these traits. Heritability estimates were 0.27 (WH), 0.31 (CG) and 0.24 (CBC). Genetic correlations ranged from 0.641 (CG × WH) to 0.709 (CG × CBC). According to our results, phenotypic selection has not had so far a large impact on the Murgese morphology, whose evolution over the last four decades seems more likely attributable to improvement in management practices. However, the obtained heritability estimates and the favorable genetic correlations indicate that there is ground for quantitative genetic selection to be applied and give successful results for the considered morphometric traits.

Linear conformation traits of Domestic hilly horse breed in Montenegro
B. Marković, R. Mumović and M. Marković
University of Montetnegro, Biotechnical Faculty, Livestock Science, Mihaila Lalica 1, 81000, Montenegro;
bmarkovic@t-com.me

The specific natural condidtions and the extensive system of rearing for centuries have shaped and adapted the population of Domestic hilly horse in Montenegro that belongs to the wider population of the Balkan hilly horse. It is the most common horse breed in Montenegro. In the past, horses in Montenegro had always been very important working animals, while today they are more used for recreation and tourism purposes. A horse`s conformation is of significant importance for its use and genetic value. The aim of this work was to assess body conformation of autochthonous Domestic hilly horse by using linear scoring system. The research was carried out on 118 adult animals (76 males and 42 females). Linear scoring (scale of 1 to 9) was applied to 12 conformation tratis (head profile, neck topside line, neck muscularity, correctness of back line, croup shape, hind croup profile, abdomen line, body muscularity, forelegs set front view, forelegs set side view, hind legs set side view, hind legs set rear view). The averages of linear conformational traits of the studied horse population ranged from 4.09 to 5.69. The lowest average scores was for hind legs set and forelegs set side view (4.09 and 4.13), while the highest average score were assigned for body muscularity, head profile and neck muscularity (5.69, 5.68 and 5.61, respectively). Males had a higher total average score for the conformational properties (5.20) than females (4.86), but the differences between the groups were not statistically significant (P>0,05). As the first of this kind, these results are the important contribution to description of Domestic hilly horse breed in Montenegro. Having in mind the importance of this breed, research should be extended to the other relvant breed characteristics and work abilities.

Yearling conformation and radiographic findings related to Swedish Standardbred trotter performance
I. Clemensson Lindell[1], Å. Viklund[2], C. Olsson[1] and S. Eriksson[2]
[1]The Swedish Trotting Association, Hästsportens hus, 16189 Stockholm, Sweden, [2]Swedish University of Agricultural Sciences, Dept. of Animal Breeding and Genetics, P.O. Box 7023, 75007 Uppsala, Sweden., Sweden;
susanne.eriksson@slu.se

When breeding Swedish Standardbred trotters, the goal is well-tempered, sustainable and competitive racing horses. Conformation traits are believed to be important for a high performance and a long career, but conformation has not routinely been assessed in young Swedish Standardbreds. To enable studies of the relationship between conformation, leg health and later trotting performance, in total 1,125 Swedish Standardbred trotters born in 2011 and 2012 were assessed at Swedish yearling auctions. The conformation traits were subjectively described on a linear scale between two extremes. Results from radiographic examinations of fetlocks and hocks of most horses were also available. After data editing, 1,028 horses remained in the dataset. The conformation assessment scores and radiographic findings were statistically analyzed together with the performance of the horses as three-year-olds using SAS GLM. For horses born in 2011 also records of performance as four-year-olds were analyzed. The performance measures used were best racing time, earnings per start, percentage of placings (1-3) and number of starts. Radiographic findings of osteochondrosis and palmar/plantar osseous fragments were categorized according to a scale of severity and if the horse had had surgical treatment or not. Conformation traits assessed on yearlings with significant effects on their performance as three-year-olds were: body size, development, body direction, proportions (height vs length of body), muscularity of the neck, length and height of the withers, thoracic depth, front leg cannon bone length, and front and hind leg conformation from the side. Radiographic findings in yearlings significantly affected best racing time, earnings, number of starts and placings as three-year-olds (P<0.05). Our results suggest that conformation scores and radiographic findings in yearling Standardbred trotters could be useful indicators of trotting performance later in life.

A genome-wide association study (GWAS) for Equine Recurrent Airway Obstruction (RAO)

D. Schnider[1,2], S. Rieder[1], T. Leeb[3], V. Gerber[2] and M. Neuditschko[1]
[1]*Agroscope, Swiss National Stud Farm, Research group for animal breeding and genetics, Les Longs-Prés, 1580 Avenches, Switzerland, [2]Swiss Institute for Equine Medicine, University of Bern, Department of Clinical Veterinary Science, Vetsuisse Faculty, Länggasstrasse 124, 3012 Bern, Switzerland, [3]Institute of Genetics, University of Bern, Vetsuisse Faculty, Bremgartenstrasse 109a, 3012 Bern, Switzerland; debora.schnider@vetsuisse.unibe.ch*

Equine recurrent airway obstruction (RAO), also known as heaves, is an asthma-like, strongly debilitating disease. It is the most common chronic lower airway disorder of mature horses in the Northern hemisphere affecting 10-20% of the adult horse population. Typical clinical signs of an affected horse are chronic cough, nasal discharge, increased respiratory effort and exercise intolerance. RAO is related to sensitization and exposure to moldy hay and has a familial basis with a complex mode of inheritance. Apart from the environmental factors contributing to the development of RAO it has been shown in several previous studies that there is a link to genetics. However, so far no causative QTL could be identified based on a case-control study including 666 warmblood horses genotyped for 54'602 SNPs. In order to improve the quality of this existing dataset, pedigrees were reviewed thoroughly and complemented wherever possible. Only horses with at least 5 complete generations were considered for further investigation. Quality control based upon marker and pedigree-information revealed a coefficient of determination of 0.68 which can be considered as a good value. Following this quality control 356 horses (184 cases, 172 controls) were selected and genotyped with the high-density 670K SNP chip. Based on the pruned dataset a genome-wide association study for RAO will be performed. It is our goal to detect additional QTLs significantly associated with RAO. We will use a mixed-model approach considering population stratification of the sampled horses as implemented in the function mmscore (R package GenABEL). In order to preclude false positive association signals QQ-plots will be examined for inflation of small p-values. We will use a Bonferroni corrected significance threshold to determine genome-wide significant associations.

Genetic parameters of cryptorchidism in Friesian stallions

A. De Jong[1], A. Schurink[1], J.R. De Nooij[2], I. Hellinga[3] and B.J. Ducro[1]
[1]*Animal Breeding and Genomics Centre, Wageningen University, P.O. Box 338, 6700 AH Wageningen, the Netherlands, [2]Dierenartsencombinatie Gorredijk, De Klok 22, 8401 CM Gorredijk, the Netherlands, [3]Koninklijke Vereniging 'Het Friesch Paarden-Stamboek', P.O. Box 624, 9200 AP Drachten, the Netherlands; bart.ducro@wur.nl*

Cryptorchidism occurs in several species such as pigs, dogs, horses and humans. In a cryptorchid stallion one (unilateral) or both (bilateral) of its testes do not descend into their normal position in the scrotum which can result in infertility and aggressive behaviour. Our aim was to estimate genetic parameters of cryptorchidism in Friesian stallions. In 2009 to 2012 1,327 colts with known pedigrees were scored for cryptorchidism as 0 (no cryptorchidism) or 1 (cryptorchidism). In 2010 to 2012 testis size (in centimetres) of all unaffected foals was measured. Genetic parameters were estimated in ASReml4. Fixed effects included in the model were age, location, year and month of inspection. 14.2% of the stallions were cryptorchid, where 12.5% (n=166) were unilateral and 1.7% (n=22) were bilateral. Heritability of cryptorchidism was 0.14 (SE=0.06). Data indicated that foals should not be inspected at a very young age, as the heritability increased when only older cryptorchid colts were considered. Heritabilities of testis size were 0.24 (SE=0.09) (left testis) and 0.27 (SE=0.09) (right testis). Genetic correlation between left and right testis size was 0.77 (SE=0.15) and therefore not significantly different from 1. The genetic correlation between cryptorchidism and left testis size was -0.96 (SE=0.14), while with right testis size -0.63 (SE=0.22). More data seem to be needed to be able to conclude whether this correlation indeed differs from -1. From the results we can conclude that cryptorchidism is a hereditary train in Friesian horses and selection against this disorder is possible and recommended.

Are the breeders of native horse breeds interested in preserving the animal genetic resources?

G.M. Polak
National Research Institute of AnimalProduction, National Focal Point, Wspolna Str. 30, 00-930, Poland;
grazyna.polak@izoo.krakow.pl

The aim of the study conducted in 2008-2014 was to analyze the native types of cold-blooded horses, which participate in the genetic resources conservation programs of Sztumski and Sokolski horse in Poland, for the number and value of offspring and for stallion use. Herds of two local, native types, Sztumski and Sokólski, were considered. The herds were covered by payments under the Common Agricultural Policy. The results of the study showed that over the 7 years the number of eligible mares increased from 567 to 1,934 (total for both types), number of herds increased from 217 to 459, which means that breeders were interested in increasing the number of herds. In the genetic resources conservation programs of Sztumski and Sokolski horse, a total of 958 stallions were used for mating, but 598 produced no more than 1 offspring, and only 7 stallions (2 Sokólski and 5 Sztumski) produced >40 offspring. In period from 2008 to 2014 6,611 foals were born, but only 13% of them have been involved in conservation programs. Most stallions produced offspring whose conformation was scored to be 15 (adequate) and 16 points (good) on a scale of 20, and the least stallions were rated at 12 and 13 points (poor) and 19 and 20 points (excellent), which means that breeders chose stallions of medium value and were not after producing the best offspring. An additional survey showed that the breeders are interested in selling their foals for slaughter regardless of their quality (conformation score). This indicates that most breeders are not interested in the choice of most typical individuals. The implementation of conservation objectives should be therefore put greater emphasis on their education.

Maternal genetic diversity among Estonian local horse breeds

E. Sild, S. Värv and H. Viinalass
Estonian University of Life Sciences, Fr. R. Kreutzwaldi 1, 51014 Tartu, Estonia; sirje.varv@emu.ee

The maternal genetic diversity in Estonian horse breeds was assessed based on mtDNA variation in D-loop sequence between nucleotides 15343-15852. Hair and blood samples (n=110) of Estonian local horse breeds including the Estonian Native (n=40), the Estonian Heavy Draught (n=30) and the Tori (n=40) were analysed. To detect the nucleotide substitution model fitting to the population and to construct phylogeny tree of the mtDNA haplotypes, MEGA v5.2 was used. The number of different haplotypes, Tajima's D, K, and π were calculated using DnaSP v5.1. Of the total 38 haplotypes identified in Estonian horse population, five were common over all three breeds with the whole population frequency of 13%. The highest number of haplotypes occurred in a single breed in a single or few individuals, comprising 60% of the sample sequences. The highest haplotype diversity was observed in Tori (96.9), also having the highest number of unique haplotypes (25). The nucleotide diversity ranged from π=11.4 in Estonian Heavy Draught to π=16.5 in Estonian Native Horse. All breeds considered under neutral evolution, as the results were statistically insignificant for the Tajima's D-test. On the haplogroup level, the most characteristic for Estonian local horse breeds were D3, X2b and X3c1 of the determined 17 haplogoups in terms of Cieslak et al. definition. The within-breed diversity estimates varying among the studied breeds showed less matrilines in Estonian Heavy Draught and in Estonian Native Horse, indicating loss of variation due to genetic drift and bottleneck effect(s) after breed formation. The results showed that the maternal variation of the ancestral population of Estonian horse breeds has been maintained by rare haplotype distribution within Tori.

Runs of homozygosity in four different horse breeds

A. Burren[1], H. Signer-Hasler[1], M. Neuditschko[2], M. Hunziker[1], V. Gerber[3], D. Schnider[3], T. Leeb[4], S. Rieder[2] and C. Flury[1]
[1]*Bern University of Applied Sciences BFH, School of Agricultural, Forest and Food Sciences HAFL, Länggasse 85, 3052 Zollikofen, Switzerland,* [2]*Agroscope, Haras national Suisse, Les Long Prés, 1580 Avenches, Switzerland,* [3]*University of Bern, Institut suisse de médecine équine ISME, Vetsuisse Faculty, Länggassstrasse 124, 3012 Bern, Switzerland,* [4]*University of Bern, Institute of Genetics, Vetsuisse Faculty, Bremgartenstrasse 109a, 3012 Bern, Switzerland; markus.neuditschko@agroscope.admin.ch*

Franches-Montagnes (FM) and Warmblood horses build the majority of living equids in Switzerland, whilst Shetland ponies (SP) represent the most prominent pony breed, and Tinker (TI) horses are considered as a niche with lacking pedigree information. For these four breeds we investigated the properties of runs of homozygosity (ROH) using filtered data from a total of 1,345 horses (196 WB, 1,077 FM, 48 SP and 24 TI) and 37,075 autosomal SNPs. ROH are long, consecutive homozygous stretches of the genome, which can be applied to derive genomic inbreeding. Genomic inbreeding coefficients (FROH) were derived for all horses. In total 14,578 ROHs were detected. The fraction of horses with missing ROH ranged from 0.00% (SP and TI) to 0.51% (WB). Average FROH within breed was 3.30% (WB), 5.00% (TI), 6.10% (FM) and 9.30% (SP), respectively. Eleven out of 48 SP horses (23%) showed a FROH>12%, which could be a result of half-sib matings. This is not surprising as few imported breeding stallions and mares are widely used to fulfill the demand of hobby-breeders. Using the pedigree information for all 196 WB and 1,077 FM horses the correlation between pedigree-based inbreeding coefficients (FPED) and FROH was 0.50 (WB) and 0.62 (FM), respectively. Including the parents in the fitted regression equation the correlation further increased (to r=0.72 in WB and r=0.85 in FM). Based on our preliminary results it can be concluded that FROH can be applied to accurately derive inbreeding coefficients of horses. This might be especially useful for breeds with incomplete or lacking pedigree information. The study will be completed by further investigating specific regions of the genome with long ROH segments among and within breeds.

Estimation of segregation variance in Spanish Trotter metapopulation by hierarchical Bayes model

I. Cervantes[1], M. Valera[2], M.D. Gómez[3], F.F. Cardoso[4] and A. Molina[3]
[1]*Departamento de Producción Animal, Universidad Complutense de Madrid, Avda. Puerta de Hierro s/n, 28040, Madrid, Spain,* [2]*Departamento de Ciencias Agroforestales, Universidad de Sevilla, Ctra. de Utrera km 1, 41013, Sevilla, Spain,* [3]*Departamento de Genética, Universidad de Córdoba, Ctra. Madrid-Córdoba km 396a, 14071, Córdoba, Spain,* [4]*EMBRAPA, Empresa Brasileira de Pesquisa Agropecuária, Parque Estação Biológica s/n, CEP 70770-901, Brasilia, Brazil; icervantes@vet.ucm.es*

The Spanish Trotter Horse (STH) can be considered as a composite of the main World trotter populations, because this breed often uses imported breeding stock from the main Trotter Horse Breeds, mainly American and French Trotters. The objective of this study was to estimate additive genetic variances, direct and due to segregation, in the STH. The dataset contained 14,614 annual earnings (calculated from a total of 247,378 records) from 4,406 participants. The pedigree contained 10,946 animals. Although different countries contributed to the STH population, only the genetic contribution of the countries USA, France (FR), Spain (SPA) and Italy (ITA) was computed as the sum over all known ancestors of the different countries of the terms computed as the sum of $(1/2)^n$. The genetic model included sex, year of birth and year of racing as fixed effects. The number of starts was included as covariate. The permanent environmental effect was included as random effect, besides the animal and error. The software INTERGEN v. 1.2 was used to run the analyses, using chain length of 500,000 MCMC rounds with 50,000 burning cycles. The additive genetic variance was 0.18, 0.16, 0.72 and 0.82 for USA, FR, SPA and ITA genetic lines, respectively. The corresponding additive genetic variance due to segregation ranged between 0.02 and 0.08. The highest values were for SPA-ITA and USA-SPA combinations, whereas the segregation variance FR-SPA had the lowest value. Despite STH is not a multi-breed population, the differentiation of trotter horses by genetic origin lines could produce the existence of segregation variance. More analyses are necessary to confirm this hypothesis.

Pedigree analysis of the indigenous Tori Horse breed

K. Rooni and H. Viinalass
Estonian University of Life Sciences, Kreutzwaldi 1, 51014, Tartu, Estonia; krista.rooni@emu.ee

The aim of this study was to describe the dynamics of the population and the present state of inbreeding of the multi-purpose Tori horse breed. The Tori horse was originally bred as a universal horse suitable for different agricultural work and transportation. Nowadays, there are three types in the Tori horse breed: Types TA and Old-T representing the older universal type with only pure breeding allowed, and type TB representing the more modern sport type where cross-breeding with related breeds is allowed. The Tori types TA and Old-T have the status of an endangered breed. The current size for the Tori breed is 1,252 horses, among them 395 horses used for breeding, including 340 mares and 55 stallions. For the pedigree analysis, the full pedigree of the Tori breed with the oldest known ancestors dating back to the 19th century was used. The data set consisted of the pedigree data from 27,176 Tori horses. The data were analysed with the software package Eva Interface 2.0. The analysis showed that during the breed's history there have been considerable changes in the size of the population. The highest number of breeding horses per year of birth (649) was registered in 1939. Starting from the 1950s, the number of breeding horses per year of birth has gradually declined, reaching an absolute low point at present with less than 50 horses registered per year of birth in the past two decades. At the same time, the number of inbred horses has increased with most of the current breeding population being more or less inbred. For the past two decades, the inbreeding coefficient (F) per year of birth has ranged from 0.01 to 0.04 (F=0.01-0.07 when excluding type TB). Since the population numbers are small and only pure breeding is allowed for the Tori types TA and Old-T, it is necessary to monitor the breeding schemes in order to minimise the inbreeding and its accompanying negative effects on the population.

Genetic diversity loss in populations Czech draft horse breeds

L. Vostry[1,2], H. Vostra-Vydrova[1], B. Hofmanova[2], Z. Vesela[1], J. Schmidova[1,2] and I. Majzlik[2]
[1]Institute of Animal Science, Pratelstvi 815, 10401 Prague, Uhrineves, Czech Republic, [2]Czech University of Life Sciences Prague, Faculty of Agrobiology, Food and Natural Resources, Kamycka 129, 165 21 Prague, Czech Republic; vostry@af.czu.cz

In the Czech Republic, three draught horse breeds are kept: Silesian Noriker, Noriker and Czech-Moravian Belgian horse. The objective of this study was to analyse the genetic diversity and identify the major causes of loss of genetic diversity in these three breeds. Genetic diversity loss was analysed based on pedigree information. The loss of genetic diversity in the reference population due to genetic drift or unequal founder contribution was derived from effective number of founders and founder genome equivalent. The total loss of genetic diversity caused by genetic drift or bottleneck, the loss of genetic diversity due to the unequal founder contribution and the loss of genetic diversity caused by genetic drift across the generations of non-founders were analysed. The average relative genetic diversity loss in the reference population was 6% (Silesian Noriker), 3% (Noriker) and 5% (Czech-Moravian Belgian), respectively. The loss of genetic diversity during the last 53 years due to genetic drift accumulated over non-founder generations was more important than the loss due to unequal contribution of founders in all breeds. The highest loss of genetic diversity has been recorded between 1960 and 1980. The loss of genetic diversity can be seen in all breeds also in last year, which points to an increase in genetic diversity loss in future generations to be expected. Supported by the project QJ1510141.

Carcass characteristics and growth parameters of foals slaughtered at 13 months of age

M.V. Sarriés[1], J.M. Lorenzo[2], S. Crecente[3], N. Díaz[3], M. Ruiz[1], D. Franco[2], M.J. Beriain[1], R. Dominguez[2], K. Insausti[1], R. Agregán[2] and A. Purroy[1]
[1]*Universidad Pública de Navarra, Producción Agraria, Etsia, Edif. Los Olivos-Campus de Arrosadía, Pamplona, Navarra, Spain, 31006, Spain, [2]Centro Tecnolóxico da Carne (CTC)), Rua Galicia, 4, Parque Tecnológico de Galicia, San Cibrao das Viñas, Ourense, Spain, 32900, Spain, [3]Centro de Investigacións Agrarias de Mabegondo (CIAM), Estrada Betanzos, Mesón do Vento, km 7, Abegondo, A Coruña, Spain, 15318, Spain; vsarries@unavarra.es*

The aim of this work was to study the effect of sex and type of fodder supplementation on carcass characteristics and growth parameters of the Galician Mountain and Burguete crossbreeding foals slaughtered at 13 months of age. Twenty-two foals were used in this experiment (6 males and 16 females). Foals were reared at pasture with their mothers and after weaning, foals at 6-7 months-old were fed on pasture until slaughter. Previous to it, one group of eleven foals (3 males and 8 females) was supplemented with 2 kg of conventional fodder (CF) per foal and day, increasing live weight from 271.4±7.6 kg to 333.7±10.6 kg during 87 days. The other group of eleven foals (3 males and 8 females) was supplemented with 2 kg of linseed-rich fodder (LF) per foal and day, increasing live weight from 270.9±10.0 kg to 343.7±14.5 kg during 97 days. Both conventional and linseed-rich fodder, were isoenergetic and isoproteic. There were no significant sex or fodder supplementation effect in the percentage distribution of the conformation and fatness degree of the carcasses (P>0.05), since 27.3% of the foals were scored in category B of conformation and 36.4% of the foals scored as fatness degree 2. Neither sex nor the type of fodder supplementation traits affected significantly (P>0.05) any of the growth parameters reaching similar values in the four groups in dressing percentage (53.1±0.5%), carcass length (108.0±1.0 cm), loin length (64.3±0.6 cm), leg width (22.41+0.35 cm), leg perimeter (99.9±1.2 cm), carcass thickness at 9th rib (2.4±0.1 cm) and compactness index (1.7±0.1). In conclusion, the type of fodder and sex had no effect on the carcass characteristics and growth parameters, which contributed to obtain similar results.

Equine viral arteritis in breeding stock: a quantitative estimation of the surveillance sensitivity

J.-P. Amat[1,2,3], T. Vergne[4], A. Hans[3], B. Ferry[5], P. Hendrikx[6], J. Tapprest[3], B. Dufour[1] and A. Leblond[2]
[1]*Ecole vétérinaire d'Alfort, EpiMAI USC Anses, U. Paris Est, Maisons-Alfort, France, [2]INRA, UR346, Saint-Genès-Champanelle, France, [3]Anses, Laboratory for Equine Diseases, Dozulé, France, [4]Royal Veterinary College, N Mymms, London, United Kingdom, [5]IFCE, 75013, Paris, France, [6]Anses, 69007, Lyon, France; jean-philippe.amat@anses.fr*

Equine viral arteritis (EVA) is a cause of abortion and neonatal deaths. In France, surveillance of EVA is based mostly on serological testing of all or part of breeding horses, depending on the studbooks' regulations. Hence, its incidence is not precisely known in the overall breeding population. Moreover, the exact number of detected cases remains unknown due to difficulties in interpretation of certain series of titers. Our goals were to establish suitable rules for identifying seroconversion in order to estimate the number of cases and outbreaks detected by the breeding stock surveillance (BSS) in France between 2006 and 2013, and to assess the BSS's sensitivity by estimating the total number of outbreaks that occurred during this period, including unreported outbreaks. Data from mares which had at least one positive result in serology were used for analysis (n=1,645), i.e. annual antibody titers and location (towns) of mares. A consensus among multidisciplinary experts was reached to consider seroconversion as a change in antibody titer from negative to at least 32 or a three-fold or greater increase. The number of seroconversions was counted for each town and modeled using a zero-truncated binomial (ZTB) capture-recapture model with R software. Using the defined rules, 239 cases located in 177 towns (outbreaks) were identified from 2006 to 2013. The sensitivity of BSS was estimated as the ratio of the number of detected outbreaks to the total number of outbreaks that occurred estimated using the ZTB model. The total number of outbreaks was estimated at 215 (95% credible interval 195-249) and sensitivity of BSS at 82% (CrI[95%] 71-91), which is relatively high and supports the relevance of BSS to prevent EVA spreading. The rules proposed for identifying seroconversion could be applied to analyze EVA surveillance data collected from other surveillance systems or, after adjustment to local environment, in other countries.

Horizon 2020 and european partnerships
L. Mahy
European Commission, DG Agriculture and Rural Development, Brussels, 1049, Belgium;
louis.mahy@ec.europa.eu

Horizon 2020, the EU's flagship programme for research and innovation, is by definition an international collaborative programme. It aims for a first class knowledge base built on excellent science, and an environment conducive to innovation in Europe. It also includes the agricultural sector and aims at jobs, growth and investment in rural areas. To filter the most important needs from the sector down to a limited number of research topics, the process via which the latter are formed involves an intensive dialogue with the stakeholders. This presentation will give an outline of some key aspects of this dialogue. An overview will be given of the different channels via which the dialogue takes place and on how livestock research community and industry can voice their needs. Also the broader context in which these research needs are to be expressed, will be discussed. What is to be expected from the agricultural research community? How does the political context, and more specifically the priorities related to the H2020 programme, relate to them? How does this translate into research topics? What is the relation to the multi-actor approach? In short, this presentation aims at informing the livestock research community about how to further integrate into the H2020 programme.

Effective knowledge transfer through discussion groups
T. O'Dwyer
Teagasc, Animal & Grassland Research and Innovation Centre, Moorepark, Fermoy, Co. Cork, Ireland;
tom.odwyer@teagasc.ie

Dairy farmers face many challenges to ensure the on-going success of their farm businesses. While research has identified the key technologies for success, practice adoption will only happen once knowledge is acquired, skills are learned and attitudes are changed. An effective discussion group will help improve both the farm performance and lifestyle of everyone in the group. High functioning discussion groups typically are well organised with regular, scheduled meetings; engage in a range of well-planned activities with a clear purpose; feature a strong ownership of the group's direction by the farmer members; and are effectively facilitated by a committed adviser/ facilitator. Currently Teagasc Dairy Advisers facilitate approximately 330 dairy discussion groups with 5,000 farmer members. Teagasc research has shown that, on average, the adoption of a range of technologies is higher by discussion group members than by non-discussion group members. Members achieved a higher net margin of up to 3 cent per litre or €470 per hectare in nominal terms (2 cent per litre or €240 per hectare higher gross margin when farm and farmer characteristics are controlled for; 2011 data). The 'Dairy Efficiency Programme' (DEP) was a three year programme which provided a financial incentive to dairy farmers to join a discussion group. Funded by the Department of Agriculture, Food and the Marine (DAFM), its main motivation was to stimulate technology adoption, improve efficiency and promote profitable expansion. While DEP attracted a new cohort of farmers to discussion groups, it is not clear whether those new members benefitted to the same degree as earlier members. DAFM are currently rolling out the Knowledge Transfer Groups (KT Groups) Programme under the 2014-2020 RDP Programme to build on the DEP.

KT Using on-farm research and the multi-actor approach to boost effectiveness of knowledge exchange

J. Birnie
Dunbia, Dungannon, Co Tyrone, BT70 1NJ Northern Ireland, United Kingdom; info@dunbiafarmers.com

Livestock farming efficiency and output in the United Kingdom and Ireland is well below its potential level. There are many practical, research based solutions to this lack of efficiency, but uptake is painfully slow. The author spent 14 months on a Nuffield scholarship trying to understand the reasons for this and to identify solutions. The main reasons for failing to attempt change were found to be a lack of knowledge, a lack of confidence and a lack of vision (being able to visualise the result of change), while the reasons for failing to consistently implement change were identified as a poor feedback (primarily a lack of data collection) and the absence of accountability. Without addressing these issues holistically, the author argues that effective change is at best unlikely, and at worst impossible. People learn in different ways and many lack confidence in their own ability to learn. A solid understanding of different learning systems can allow tailored learning systems to be designed and appropriate support to be given. Technology has a vital role in enabling and driving change. The notable lack of data collection on beef and sheep farms must be addressed through simple to use technology. Data can and must be used to assess a farm's current performance and assess the effectiveness of any changes made.

DAIRYMAN: An exemplar of effective knowledge transfer and exchange

J. Bailey
Agri-Food and Biosciences Institute (AFBI), 18A Newforge Lane, Belfast, BT9 5PX, United Kingdom

The DAIRYMAN (INTERREG IVB) project was established to investigate differences in ecological, economic and social performance between dairy sectors in different regions of North West Europe and to see how performance might be improved via stakeholder cooperation. The project involved 14 partner organizations from 10 of the main dairy farming regions of North West Europe: Brittany, Pays de la Loire, Nord-Pas de Calais, Ireland, Northern Ireland, Flanders, Wallonia, Baden-Württemberg, Luxembourg and the Netherlands. A key objective was to test out a multi-actor strategy for knowledge exchange between farmers, advisors and researchers within and between regions, and assess its effectiveness in addressing economic, environmental and social policy issues. In each region, a stakeholder network was established, which included between 9 and 30 pilot dairy farms, where innovative technologies were trialed, a local agricultural research institute or school where the innovations were developed in response to regional policy requirements, and a farm advisory or knowledge transfer centre, which acted as the hub to facilitate knowledge exchange between researchers, advisors and farmers. These stakeholder networks, together, formed a larger inter-regional network, which facilitated exchange visits by groups of pilot farmers, advisors and researchers to other regions, and resulted in the sharing of ideas, skills and experiences, and the exchange of management tools of proven worth in optimising resource use efficiency. Knowledge transfer to a wider stakeholder group, which included policy makers and legislators as well as farmers, advisors and researchers, was delivered via publications, media articles, and regional and interregional workshops. Overall, the strategy proved to be very effective in delivering positive changes in farming practices to meet a range of environmental, economic and social objectives, including reduced potential for methane emission, lowered energy and feed costs, and increased vacation time for farmers and their families.

Low emission slurry application technology: a research, adviser's, and farmer's perspective
M. Mulholland
Cafre, Greenmount Campus, Antrim, Co. Antrim, United Kingdom; martin.mulholland@dardni.gov.uk

An experimental programme was conducted at the Agri-Food and Biosciences Institute (AFBI) Hillsborough from 2002 to 2003 to determine the yield response and nitrogen efficiency of cattle slurry applied to grassland by splash plate (SP), trailing-shoe (TS) and band spreader (BS). A total of 23 harvests were taken. Overall, the BS and TS treatments resulted in increased dry matter yields of 18 and 26% respectively, compared to SP. These responses were similar across all seasons. Comparisons with inorganic fertilizer treatments were made to determine the fertilizer equivalent of the slurry applied. The inorganic fertilizer equivalents of slurry applied (mean 55 t/ha) by SP, BS and TS were 25, 58 and 69 kg/ha N respectively for herbage yield. A major knowledge transfer event jointly organised by AFBI and the College of Agriculture, Food and Rural Enterprise (CAFRE) was held at AFBI, Hillsborough in June 2006 attended by 490 farmers who observed working demonstrations of a range of slurry application technologies. Over 20 small on-farm practical demonstrations of TS technology were then held on commercial farms across Northern Ireland from 2007 to 2009 attended by over 800 farmers. A capital grant scheme (the Manure Efficiency Technology Scheme) was developed giving 40% grant aid to farmers purchasing band spreading, trailing shoe or shallow injection slurry application equipment. This equipment has the potential to spread 20% of slurry produced in N. Ireland.

Effective KT: a farmers perspective
D. McConnell
Omagh, County Tyrone, Northern Ireland, United Kingdom; jason@agrisearch.org

Drew is a leading dairy farmer from Omagh, County Tyrone and has had a long involvement in research having been involved in a number of on-farm research projects. He will outline experiences from this and being a 'Focus Farmer' under 2007-2013 Northern Ireland Rural Development Programme. Drew is currently Chair of the AgriSearch Dairy Advisory Committee and a former member of DairyCo's Research and Development Advisory Forum so well placed to understand the the farmers view of KT transfer.

Knowledge Integration and Knowledge Exchange: broadening the concepts and tools that we use
S. Hoste
Quantech Solutions, Howe Farm, YO17 6RG, United Kingdom; sam@qtsuk.co.uk

Knowledge Integration here is taken to be the 'combination of specialised, differentiated, but complementary knowledge' in contrast to KE or KT as the 'sharing or transferring of knowledge'. KE/KT has possibly become reified within agricultural to the exclusion of other approaches, for example approaches made by environmental scientists who also have dialogue and interactions with farmers. Grant (1996) suggested that 'transferring knowledge is not an efficient approach to integrating knowledge' (italics in original). If production requires the integration of many people's specialist knowledge, the key to efficiency is to achieve effective integration while minimising knowledge transfer through cross-learning by organizational members. There follows a description of two management scientists and the inefficiency of them trying to understand their respective disparate disciplines, rather than establishing a mode of interaction between them to integrate knowledge between them. An example in animal science could thus be characterised by the genomics knowledge domain and the animal nutrition knowledge domain: tt is not necessary that either party transfers their knowledge to the other, but that there is some means to integrate these two knowledge domains. The general problem highlighted in the literature on knowledge integration is as a consequence of knowledge specialisation (e.g. genomics and nutrition) with most authors agreeing with Steven Postrel's general characterisation: 'the traditional problem of the division of labour is to trade off the superior task of efficiency of specialisation against its inferior coordination properties, the fundamental tension in the division of knowledge is between the superior learning efficiency of specialisation and its inferior integration properties'. Agriculture is increasingly necessitating the integration of multiple disciplines (genomics, precision agriculture, animal and human health and nutrition to name but a few) where knowledge integration is arguably required and not knowledge exchange. The paper will describe what the author sees as the challenges, the need for research, and some potential mechanisms to assist in knowledge integration.

Improving beef production efficiency through growth monitoring tools
F.O. Lively[1], S.J. Morrison[1], F. Titterington[1], J. Freeburn[2], A. Johnston[2], N. Weatherup[2] and J. Rankin[3]
[1]Agri Food and Bioscience Institute, Agriculture Branch, Large Park, Hillsborough, BT26 6DR, United Kingdom, [2]CAFRE, Beef and Sheep Development, Greenmount Campus, Antrim, BT41 4PS, United Kingdom, [3]AgriSearch, Innovations Centre, Large Park, Hillsborough, BT26 6DR, United Kingdom; francis.lively@afbini.gov.uk

Industry data in Northern Ireland highlighted two major inefficiencies for beef production, high first calving age for suckler replacements (31 months) and high slaughter age of prime beef cattle (26-28 months). An on-farm research study was undertaken to develop an online growth monitoring tool for beef producers. The online growth monitoring tool enables producers to select a production system (24 month calving or 16 month bull beef production or 24 month steer beef production) for particular animals and by inputting the animal's live weight the tool reports the animal's growth rate and predicts further growth requirements if the target end-point was is to be achieved. This predicted growth rate requirement enables producers to alter the plane of nutrition offered to the animals to ensure the target is achieved. The tool was demonstrated on 12 beef farms (6 suckler farms rearing suckler replacements and 6 farms rearing dairy-origin beef cattle). Over the duration of the study the suckler farms reduced calving age from 28 months to 24 months, whilst beef finishing farms reduced the average slaughter age of their cattle. Each of the farms involved hosted a farm walk to demonstrate to their neighbours and colleague farmers how they had improved the efficiency of their businesses through monitoring performance and altering nutrition to ensure set performance targets were achieved.

Perceptions of French private veterinary practitioners on their role in organic dairy farms

J.E. Duval[1], N. Bareille[1], C. Fourichon[1], A. Madouasse[1] and M. Vaarst[2]
[1]LUNAM Université, Oniris, INRA, UMR1300 BioEpAR, Atlanpôle, La Chantrerie, CS 40706, 44307 Nantes, France, [2]Aarhus University, Department of Animal Science, Blichers Allé 20, 8830 Tjele, Denmark; julie.duval@oniris-nantes.fr

Veterinarians could be farmers' sparring partners in reaching organic principles such as promoting animal health and welfare, by exchanging and converting their knowledge into relevant management practices with farmers. However, in the past, organic dairy farmers have not always considered veterinarians to be their pertinent advisors. The objectives of this study are (from private veterinary practitioners' point of views): (1) to describe the roles of veterinarians today in organic dairy farmers' animal health promotion strategies; (2) to identify factors related to organic farming which determine that role; and (3) to identify opportunities for improvement of veterinarians' advisory services for these farms. Fourteen veterinarians were interviewed, using qualitative semi-structured research interviews. A modified approach to Grounded Theory was used for data collection and analysis. Many veterinarians had only contact with organic dairy farmers in cases of individual ill animals or acute herd health problems. Even though certain veterinarians experienced animal health and welfare situations and practices not meeting their standards, they were not always able to establish themselves in an advisory role supporting farmers in improving this. Indeed, organic production principles, regulations and farmers' health approaches challenged veterinarians' values on animal health and welfare and their perceptions of 'good veterinary practices'. Some veterinarians found that there was no direct economic interest for them in the organic dairy sector and that could diminish their willingness to invest in this sector. Opportunities for improvement were identified; propose proactively advice, adapt advisory services and/ or dissociate veterinarians' curative role from their advisory role in disease prevention.

Knowledge transfer within a retail supply chain: The Tesco Supplier Network

E. Fardy[1], C. Maddocks[1] and F.G. Roberts[2]
[1]Tesco Stores (UK) Ltd, Progress House, The Boulevard Shire Park, Welwyn Garden City, Hertfordshire, AL7 1RZ, United Kingdom, [2]Integra Food Secure Ltd, 2Integra Food Secure Ltd., Hanborough Business Park, Long Hanborough, Oxfordshire, OX29 8SJ, United Kingdom; fiona@foodsecure.co.uk

The vision of the Tesco Supplier Network is to build strong partnerships between Tesco, suppliers and producers to secure the best affordable, sustainable products for our customers by: 1. Sharing knowledge and expertise 2. Building a more sustainable supply chain 3. Innovating together The network was established in January in 2015. It now has over 5,000 engaged suppliers, across 50 countries and 16 product categories. The Supplier Network incorporates an on-line platform which allows suppliers and producer visibility of communication (using formats familiar to social media users). A resource library allows suppliers and producers to access relevant information and then use the network to discuss issues and concerns with a wide range of contributing experts. These insights range from price cutting case studies, technical audit presentations, and best practice guidelines on managing bio-security, to customer and market trends updates. After 12 months a survey was launched to review the effectiveness and value of the network. The survey responses showed that 87% of those who responded are satisfied or extremely satisfied with the network. 89% would recommend the network to colleagues and 94% feel the network is helping to develop their relationship with Tesco. Additionally and to underpin the network approach we have developed Tesco Sustainable Farming Groups which are an end-to-end supply chain initiative established to drive productivity, sustainability and efficiency on farm. Developed with aligned farmers, growers and processors, the groups are governed and directed via a committee of representatives from each stage in the supply chain from conception to consumption. Using this two phase approach the vision is to create a consistent model and message which extends throughout our supply base to encourage best practise, knowledge transfer and continuous improvement.

Welfare training in emerging livestock Sectors: Asian tropical prawn supply base of Tesco (UK) Ltd

F.G. Roberts[1], C. Maddocks[2] and A.J. Lucas[1]
[1]Integra Food Secure Ltd, Hanborough Business Park, Long Hanborough, Oxfordshire OX29 8SJ, OX29 8SJ, United Kingdom, [2]Tesco Stores (UK) Ltd, Progress House, The Boulevard Shire Park, Welwyn Garden City, Hertfordshire, AL7 1RZ, United Kingdom; fiona@foodsecure.co.uk

Validation of conditions of production and slaughter have been carried by Integra Food Secure Ltd (an independent, ISO 17020 accredited body) across the Tesco global supply base for terrestrial protein sources since 1998. The instigation of comparable inspection schedules across commercial aquatic sectors has been a phased process with the most recent, Tropical Prawns commencing in 2013. Inspection outcomes have indicated that the invertebrate sector, has struggled with embedding the concept of animal welfare into production practices i.e. the number and nature of non-conformances identified appeared consistent between subsequent inspections between and within supply chains. The extent to which this is a consequence of species-specific and/or cultural understanding is unclear; but the inability to access relevant training was identified as a repeated concern during the inspection process. Thus, a 2-day, targeted training programme was commissioned by Tesco stores for delivery by industry experts in both Thailand and Vietnam, to the Tropical Prawn supply base. Overall there was an improvement in number and nature of non-conformances and associated rating (Red, Amber, Green) from June 2015 (post-training) with fewer major welfare issues e.g. associated with slaughter and handling, resulting in a reduction in Amber (from 62% to 28%) and Red (from 8% to 0%) inspection outcomes. There was a net increase in suppliers achieving Green status (from 31% pre-training to 72% post training) importantly, participant feedback indicated improved understanding and positive engagement with the content of the retailer standard. This demonstrates the value of structured training, particularly in emerging sectors where the importance of animal welfare and the influence on management practice has historically been less well understood.

Session 65   Theatre 1

Tools and efforts for improved health in Swedish dogs

S. Malm
Svenska Kennelklubben, Rinkebysvängen 70, 163 85 Spånga, Sweden; sofia.malm@skk.se

The Swedish Kennel Club (SKK) dedicates substantial resources to helping breeders breed dogs in a way that benefits both the individual dog and the breeds. There are about 780,000 dogs in Sweden, out of which around 70% are pedigree dogs registered with SKK. About 50,000 puppies are registered yearly. The SKK database contains information on pedigree, coefficients of inbreeding, health programs, behaviour assessments and other official tests and trials. All records are publically available on the web, including breeding statistics for both individual dogs and each breed as a whole. Hence, the database constitutes a valuable source of information, for breed clubs, breeders and dog owners. It also offers unique possibilities for epidemiological, behavioural and genetic studies in dogs. New results are added daily and SKK puts emphasis on including all test results, positive as well as negative. About 15,000 and 11,000 records for hip and elbow dysplasia, and 18,000 records from eye examinations are registered yearly. BLUP EBVs for hip and/or elbow dysplasia are currently predicted weekly in 20 breeds. The database also contains a large number of records from behaviour assessments and EBVs have been introduced for some behaviour traits, based on data from a standardized and validated behaviour assessment. Every breed in Sweden should have a breed-specific breeding strategy including all aspects relevant for the breeding goal of each breed. The strategy should constitute an overall plan for the breed and act as a guideline to breeders. Moreover, SKK applies breed-specific instructions (BSI) for show judges, aiming at identifying breed-specific areas at risk for exaggerations. Dog breeding is an international activity. Thus, international collaboration with respect to health and breeding is of utmost importance for a long-term sustainable breeding of dogs. International Partnership for Dogs (IPFD, a non-profit organisation) is a ground-breaking collaboration between cynological organisations, health registries, research and veterinary organizations and other stakeholders in dog health, with the mission to enhance the health, well-being and welfare of dogs. IPFD and the internet platform dogwellnet.com represents an exciting step in the collaboration and sharing of information and resources within the global dog community.

Developments in dog genetics: a UK perspective

T.W. Lewis
The Kennel Club, Clarges Street, London, W1J 8AB, United Kingdom; thomas.lewis@thekennelclub.org.uk

Tremendous progress has been made in the last 10-15 years in the area of canine genetics. From the emergence of DNA tests for simple Mendelian diseases which frequently occur in pedigree dog populations, through sequencing of the canine genome in 2005, and to the application of quantitative genetic techniques to pedigree populations, canine genetics research has advanced quickly. The result has been the generation of a plethora of useful information to assist breeders seeking to both select for health (and other desirable traits) while also attempting to maintain sustainable levels of genetic diversity. However, while in the field of livestock breeding tools exist to simultaneously select for a range of traits (selection index) while minimising the rate of inbreeding (optimisation), big differences in 'industry' structure mean these methods are ill-suited to pedigree dog breeding. A pertinent challenge, therefore, is how to collate and present the information relevant to individual breeds in an accessible format to allow a disparate group of breeders to make informed breeding decisions with the common aim of improving health and maintaining sustainability. The UK Kennel Club is attempting to achieve this goal by producing tailored 'Breeding for Health Strategy' reports for each individual breed, comprising: (1) epidemiological data on the most prevalent diseases in each breed, and those with the greatest impact on welfare; (2) results of pedigree/population analysis detailing the rate of inbreeding and effective population size, alongside data on actual population size and information on the extent of popular sire usage; (3) trends in EBVs (where available), describing the response to selection against a particular condition; (4) summary of health testing/screening results, including rates of participation and results among breeding animals; (5) summary of DNA testing results, including rates of participation and trends in estimated disease allele frequency; (6) results of data from dog shows reporting minor faults (such as poor dentition) and from vet checks for over-exaggeration in conformation. It is hoped that the reports will enable individual breeders to gain a more holistic picture of the challenges facing their breed, assist in clarifying and ranking of overall breeding objectives, and encourage the best use of the wide range of tools available to breed sustainably for health.

The genetics of gun-dog trial results in Retrievers

G.E. Pollott and C. Wilkinson
RVC, Royal College Street, NW1 0TU London, United Kingdom; gpollott@rvc.ac.uk

Retriever dogs are used extensively as gun dogs in the UK and also participate in gun-dog trials run by local clubs. This study was designed to investigate the genetics of gun-dog traits and the factors which influence them in gun-dog trials. The trait analysed was the ranking of 1,002 dogs in all UK gundog trials in 2013-15. Raw place data (1st, 2nd, etc.; PL) was converted to a trait on an underlying continuous scale and adjusted for size of competition (PLund). The various trials were not all of equal standing so PL was also computed to allow for these differences (PLeq). These two adjustments were combined into one measure (PLcomb) and log values taken to achieve normality. Various features of all dogs were available from Kennel Club records including 5-generation pedigree, inbreeding coefficient, age, colour, sex and EBV for hip and elbow scores, to be used as fixed effects or covariates in a mixed-model analysis. All traits (PL, PLund, PLeq, PLcomb and their log values) were analysed with ASReml with the A-matrix used to fit the animal effect and an identity matrix used to model repeated animal measurements. Both PL and PLund (untransformed and log transformed) were analysed and no significant effects or heritability was found for these traits. However, for PLeq and PLcomb there were both significant effects and genetic parameters. Taking log PLcomb as an example, age affected this trait such that older dogs had higher scores and the more highly placed dogs had a higher level of inbreeding. Sex was also found to be a significant effect. Males had higher mean score than females. The heritability and repeatability of log(PLcomb) were 0.25 ± 0.06 and 0.51 ± 0.04 respectively but the trait was not correlated to hip score or elbow score EBV. This is the first report of the genetics of gun-dog competition results although one report from the USA found the heritability of gundog traits ranged between 0.08 and 0.35 for different of breeds. This study supports the view that gun dog traits are heritable and could be included in more formal breeding programmes if so desired by the breeders.

Genetic relationships between hunting traits of leashed and non-leashed traits in Norwegian Elkhound

M. Wetten[1], K. Lysaker[2], T. Aasmundstad[3,4] and O. Vangen[3]
[1]Aninova AS, at. Norsvin, 2317 Hamar, Norway, [2]GENINOVA AS, at Norsvin, 2317 Hamar, Norway, [3]Norwegian University of Life Sciences, Animal and Aquacultural Sciences, P.O. Box 5003, 1432 Ås, Norway, [4]NORSVIN, Norwegian Pig Breeders Org., Storhamargata 44, 2317 Hamar, Norway; odd.vangen@nmbu.no

Norwegian Elkhound Grey is a dog breed used for hunting elk in Nordic countries, and is one of the national breeds of Norway. The breed is used both for leashed and non-leashed tracking. For the future breeding program it is important to know the genetic relationships between leashed and non-leased groups of traits. The study provides estimates of heritabilities of traits registered on hunting trials with both leashed and non-leashed dogs, as well as genetic correlations between them. Performance test of leashed dogs includes recording of performance of five different traits, while in non-leached dogs there are ten traits included. In total, 1,991 animals (5,393 records from the years 1995-2014) on lashed performance, and 2,070 animals (5,405 records from 2005-2014) on non-leashed performance were included in the dataset. No dogs had registrations on both groups of traits. The pedigree file included 8,623 animals. Data was analyzed using a BLUP animal model, with permanent environment and month-year as random effects and sex and test number within dog as fixed effects. Age was included as regression variable. Heritabilities were generally low in non-leashed traits (0.01-0.14) and in leashed traits (0.06-0.16). Genetic correlations between performance traits in the two groups were generally negative and unfavourable (from -0.44 to -0.71). Due to the relative small population size (900 pups born annually) it is important not to split the population in two separate breeding populations. This must be taken into consideration when the breeding goal is defined. Furthermore, the avoidance of inbreeding should be an important aspect of the breeding program.

Non genetic factors affecting hunting ability in Italian Scent Hound dog

S. Riganelli, S. Antonini, M. Gubbiotti, A. De Cosmo, A. Valbonesi and C. Renieri
University of Camerino, Via Gentile III Da Varano s/c, 62032, Camerino (MC), Italy; stefano.antonini@unicam.it

The aptitude for hare hunting in Italian Scent Hound (ISH) is estimated by scoring 7 traits: morphology, breed style, search, approach, voice, ability to flush out the prey and chasing of prey. Analysis was based on ISH field trials results from competitions held in North-Central Italy and isle of Elba, from December 2014 to October 2015. The total trial data had 1,406 records from 525 males and 881 females. Dog was tested as individuals, pairs and pack (from 6 to 12 components). The effect of five non-genetic factors (sex, coat colour, coat types, competition judges, type of trial) was carried out by ANOVA analysis using the SPSS 12.0 predictive analysis software. The results indicate that there is no effect on traits for sex. Coat colour (tawny or seal black) has a significant effect on voice ($P<0.001$). Coat type (rough or short haired) has a significant effect on breed style ($P<0.05$) and search ($P<0.05$). Type of trial (individual, pair and pack) has a statistically significant effect on all the seven hunting traits ($P<0.001$). Judges factor has a significant effect on all the seven hunting traits ($P<0.001$). Coat colour, coat type, competition judges and type of trial affect the seven tested traits in Italian Scent Hound. These data are the basis to improve our knowledge about the genetic quantification of aptitudes in scent hounds. Also, they provide insight in assessing the link between standard breed traits and functional aptitudes.

Heritability of mentality traits in Swedish dogs

E. Strandberg[1], K. Nilsson[1] and K. Svartberg[2]
[1]Swedish University of Agricultural Sciences, Department of Animal Breeding and Genetics, P.O. Box 7023, 75007 Uppsala, Sweden, [2]Svartbergs Hundkunskap, Voxome 102, 74791 Alunda, Sweden; erling.strandberg@slu.se

The dog mentality assessement (DMA) has been used in Sweden since 1989 to describe mentality in dogs. DMA was developed mainly for working dogs and there was a demand for a mentality assessment suitable also for other breeds. With this aim, the new behaviour and personality assessment in dogs (BPH) was created and assessments were started in 2012. The aim of this study was to estimate heritabilities (h^2) for the created behaviour traits for 5 breeds with the highest number of observations: Rhodesian Ridgeback (RR, n=493), Labrador Retriever (LR, n=491), Nova Scotia Duck Tolling Retriever (NSDTR, n=425), Staffordshire Bullterrier (SBT, n=343), and American Staffordshire Terrier (AST, n=298). From the original 242 markings on the score sheet, 6 summarizing behavioural traits were defined across the 5 breeds using exploratory factor analysis. Average h2 for all traits within breed varied from 0.29-0.38. Hostility had the highest h^2 across breeds but also the largest variation (0.38, range 0.18-0.62). Confidence had more consistently high h^2 (0.36, range 0.22-0.45). The other 4 traits (Playfulness, Sociability, Curious and Confident, and Positive and Energetic) had average h^2 of around 0.3, with range from 0.10 to 0.59. Standard errors ranged from around 0.10 for the largest breeds to about 0.15 for the smallest. These results show that the traits defined in the new BPH have at least as high h2 as the traits in DMA.

Syringomyelia in Cavalier King Charles spaniels: new detailed phenotypes for the genetic evaluation

K. Wijnrocx[1], W. Eggelmeijer[2], L. Van Bruggen[2], N. Buys[1], S. Janssens[1] and P. Mandigers[2]
[1]KU Leuven, Department of Biosystems, Livestock Genetics, Kasteelpark Arenberg 30, 3001 Leuven, Belgium, [2]University of Utrecht, Department of Clinical Sciences of Companion Animals, Yalelaan 108, 3584 Utrecht, the Netherlands; katrien.wijnrocx@biw.kuleuven.be

Syringomyelia (SM) and chiari-like-malformation (CM) are two neurologic conditions occurring frequently in small toy breeds such as the Cavalier King Charles Spaniel (CKCS) and Griffon Bruxellois. In these breeds, the formation of fluid-containing cavities (syringes) in the spinal chord results in the syringomyelia phenotype. A previous study found a moderate heritability thereby showing that selection against syringomyelia is possible. The frequency of CM and SM in the CKCS is estimated to be as high as 97% and 50% respectively. As a result breeders in Belgium and the Netherlands require an examination of SM, by means of a MRI scan, before using their dog as a breeding animal. Until recently SM was scored from 0 to 2, however, after re-evaluation of all MRI scans, the measurement of the exact width of the syrinx was proposed as a more precise measure amongst other detailed phenotypes. In order to replace current selection based on phenotypes, we studied this new information in a set of 900 individuals. Several mixed linear animal models using Misztal's BLUPF90 family of programs were used in order to evaluate relevant non-genetic factors on the traits such as age at scanning, examining veterinarian, sex and color.

Population analysis and conservation options of the Norwegian Lundehund

A. Kettunen[1], M. Daverdin[2], T. Helfjord[3] and P. Berg[1]
[1]The Nordic Genetic Resource Center, NordGen, Farm Animals, P.O. Box 115, 1431 Ås, Norway, [2]NTNU Museum, Norwegian University of Science and Technology, Erling Skakkes Gate 47a, 7491 Trondheim, Norway, [3]Norsk Hestesenter, Starumsveien 90, 2850 Lena, Norway; anne.kettunen@nordgen.org

The Norwegian Lundehund is a highly endangered dog breed, native to Norway, and traditionally used to retrieve puffins (Fratercula arctica) from their nests on the steep mountainsides. Although, Lundehund are currently used as companion dogs, there is great interest in conserving the traditional characteristics of the breed. Since the 1940s the Lundehund has experienced several severe bottlenecks. Molecular genetic studies have confirmed extremely low genetic variability in the Lundehund and close relatedness between the individuals. Both reduced litter size and high frequency of intestinal lymphangiectasia in Lundehund are most likely consequences of high levels of inbreeding. Our objective was to assess the genetic diversity in the current Lundehund population based on in-depth pedigree analysis from 5,433 individuals born between 1930 and 2015. Using optimal contribution selection (OCS) we evaluated the possibilities of conservation and sustainable management of this rare dog breed. Average relatedness and inbreeding in the last cohort was 0.76 and 0.38, respectively. The effective population size based on individual increases in inbreeding/coancestry was estimated to be 13, whereas regression based methods gave higher and highly variable estimates. Parameters based on the probability of the gene origin indicated that 38% of the genetic variation in the base population was lost, mainly due to random genetic drift (78.5%). Results indicate that the current Lundehund population is based on a few related individuals. Due to the extremely high relatedness, OCS alone facilitates no improvement in the current situation in the Lundehund. Due to more or less depleted genetic diversity, cross-breeding with a foreign breed(s) is the only option to introduce new genetic variation to the Lundehund. We quantify and discuss the effects of cross-breeding with one or several breeds on the average parental relatedness and genetic diversity.

Genetic variability of the short-haired and long-haired Segugio Italiano

S. Pallotti[1], A.M. De Cosmo[1], A. La Terza[1], D. Pediconi[1], I. Pazzaglia[1], C. Nocelli[1], M. Gubbiotti[2] and C. Renieri[3]
[1]University of Camerino, School of Bioscience and Veterinary Medicine, Via Gentile III da Varano, 62032 Camerino, Italy, [2]Universitá Telematica Marconi, Via Plinio 44, 00193 Roma, Italy, [3]University of Camerino, School of Pharmacy, Piazza dei Costanti 4, 62032 Camerino, Italy; carlo.renieri@unicam.it

The short-haired and long-haired Segugio Italiano are two breeds of Italian scenthound dogs classified under the Italian Kennel Club (ENCI) group 6 and largely diffused throughout Italy. Four thousands four hundreds ninety dogs were registered in 2015. The breeds differ by the type of the coat which can be short-haired or long-haired. The aim of this study is to evaluate the genetic variability of both breeds and the genetic distance with the Italian related breeds, Segugio dell'Appennino and Segugio Maremmano. 126 one year old unrelated dogs (56 long-haired and 70 short-haired) were sampled during the meetings organized by the Prosegugio Italian Society throughout the country. For each dog, molecular characterization of 21 microsatellite markers was carried out. Results were analyzed by GenALEx software v.6.5 in order to estimate the main parameters of genetic variability among breeds. Number of alleles ranged from 5 to 11 and from 5 to 12 for the long-haired and for the short-haired breed respectively (total mean 7.76). The mean Fis, Fit and Fst were 0.048, 0.061 and 0.013 respectively. The observed and expected heterozigosity were 0.68 and 0.72 for the long-haired breed whereas were 0.68 and 0.71 for the short-haired breed. All the loci were in HWE except for 2 and 3 loci for the long-haired and the short-haired breed respectively. Results from AMOVA showed that the larger portion of the variation was whitin individuals. The low value of genetic distance (0.072) between the long-haired and the short-haired breeds indicated that there was no substantial variability among the two breeds. Similar result was observed respect the genetic distance whit the related breeds (Segugio dell'Appennino and Segugio Maremmano).These results provide a basis for the development of breeding programs which tend to avoid the loss of genetic variability of the two breeds.

Campylobacter posing new challenges and opportunities to the food supply chain

N. Corcionivoschi, C. Kelly and M. Linton
Agri-Food and Biosciences Institute, 18a Newforge Lane, Belfast BT9 5PX, United Kingdom;
nicolae.corcionivoschi@afbini.gov.uk

Food-borne pathogens, including Campylobacter and E. coli, are a concern for the poultry industry. Campylobacter jejuni, a microaerophilic bacterium, is well known for its ability to cause severe gastroenteritis and life-threatening diseases in humans and is considered a commensal in poultry. The main source of infections in humans is considered to be chicken meat. A new virulence factor, the type six secretion system (T6SS), has previously been reported in Campylobacter spp. with the hcp gene being the key component. We now know that 56.1% of our C. coli isolates and 28.8% of the C. jejuni are positive for this novel virulence factor. Overall, it was discovered that hcp$^+$ C. coli and C. jejuni isolated from retail chicken isolates posses genetic and phenotypic properties associated with enhanced virulence. In view of the significant burden caused by campylobacteriosis resulting from the preparation and consumption of infected poultry meat, new approaches are required to control this pathogen in food production systems. We will report how the ability to differentiate between pathogens, in relation to their pathogenic abilities, can enable us to quantify the effect of in farm biosecurity or management measures in applying genetic pressure that affect the survival and growth of the most virulence bacteria.

Food safety, integrity and animal welfare challenges for M&S protein supply

P. Kennedy
Marks & Spencer plc., Waterside House, 35 North Wharf Road, London W2 1NW, United Kingdom;
peter.kennedy@marks-and-spencer.com

Marks & Spencer (M&S) has built a unique brand position over the last 130 years and much of this is down to customer trust. An own-label retailer, with more than 1,330 stores across the globe, M&S sells food, clothing and home products through its stores and online, both in the UK and internationally. M&S is renowned for providing consumers with unique, innovative products sourced and marketed with an industry-leading approach to provenance, ethics and environmental standards. This position has evolved over decades, and the business is focused on protecting that position by always acting with the integrity its customers have come to know and expect. It goes without saying that food safety is critically important to the customer and M&S has long-standing codes of practice for suppliers, and an extensive audit programme, to ensure food safety is maintained throughout its supply chain. This is continuously developed with innovative projects to drive real changes in food safety, for example, M&S has been instrumental in developing a new approach to minimise Campylobacter levels in fresh chicken. When it comes to integrity, M&S introduced its Select Farm standards and audit process some 20 years ago to ensure that proteins are sourced from farms it knows and trusts and which operate to specific M&S production standards. M&S was one of the few retailers that wasn't affected by the horsegate scandal, but since then has implemented changes to supply chain structures and DNA testing of all beef to further protect integrity. Recent projects have also focused on liquid egg free-range supply chains and prawn sourcing, among many others. The retail environment remains challenging and customers are only getting more demanding in terms of provenance, integrity and farm animal welfare. M&S continues to evolve its approach to continue to exceed customer expectations and maintain brand position.

Impact of information load on the centrality parameters of a pig trade network in Northern Germany
K. Büttner and J. Krieter
Institute of Animal Breeding and Husbandry, Christian-Albrechts-University, Olshausenstr. 40, 24098, Germany;
kbuettner@tierzucht.uni-kiel.de

Dealing with network analysis of trade networks faces always the challenge of incomplete data sets due to unavailability of information, e.g. caused by country borders or different producer communities. These various information loads can influence the outcome of the network analysis and, thus, also the disease control strategies based on centrality parameters. The aim of the study was to evaluate how the centrality parameters varied between different network scenarios created based on available information. Therefore, the centrality parameters (in and out-degree, betweenness, ingoing and outgoing closeness) of three different scenarios of a pig trade network in Northern Germany (2013-2014) were compared. The first scenario contains all trade contacts with information about the purchaser, the supplier as well as the truck (A: 978 nodes, 2,280 edges). In the second scenario only those contacts stayed in the data set with full geographical information (B: 866 nodes, 1,884 edges) and in the last scenario only contacts with additional information about the farms, e.g. farm type, were kept (C: 188 nodes, 625 edges). The information load increased continuously from network A to C, whereas the number of nodes and edges decreased. In order to evaluate the differences of the centrality parameters between these three network scenarios a Spearman Rank Correlation was carried out. For network A and B the correlation coefficients of all centrality parameters ranged between 0.91 and 0.99. Both scenarios were least different, in particular regarding the size of the network. But in the comparison of network B and C (A and C) the correlation coefficient of the parameters in-degree, betweenness and ingoing closeness decreased clearly to 0.68 to 0.78 (0.63 to 0.74). Especially for control strategies based on centrality parameters it is of great relevance to know about the influence of the information load and, thus, of the resulting network scenarios.

Nutritional benefits of animal products
R.K. Price, A. Yeates, L.K. Pourshahidi and M.B.E. Livingstone
University of Ulster, Northern Ireland Centre for Food and Health, Cromore Road, Coleraine, BT52 1SA, N.
Ireland, United Kingdom; rkprice@ulster.ac.uk

Animal products such as red meat, poultry, eggs, milk and dairy products, and to a lesser extent fish, form an integral component of the habitual UK diet. However they are often the dietary component which evoke the widest array of complex scientific, economic, environmental and political issues. Despite these controversies it is indisputable that animal products make a significant and valuable contribution to human health, at all life stages, through the provision of high quality protein and highly absorbable essential nutrients. Moreover, scientific evidence increasingly demonstrates the beneficial role that animal products can have in preventing and combating obesity and certain non-communicable diseases related to over-nutrition. This presentation will (1) explore the contribution of animal products to nutrient intakes within the UK diet, (2) compare reported intakes with current dietary recommendations, and (3) discuss consumption trends and comparisons with other EU countries. The talk will also provide an overview of a number of human intervention trials recently undertaken at the Northern Ireland Centre for Food and Health (NICHE), Ulster University, investigating the link between the intake of animal products and nutritional status. These studies include the use of novel fortification and bio-fortification techniques to positively manipulate the nutrient profile of animal products for the benefit of consumer health.

Sustainability of cattle supply in the United Kingdom
J. Birnie
Dunbia, Granville Industrial Estate, Dungannon, Co. Tyrone, BT70 1NJ, Northern Ireland, United Kingdom;
info@dunbiafarmers.com

Cattle demographics have changed substantially across the world for a number of different reasons. Demand has risen and fallen, different consumption patterns have changed purchasing, and market intervention has influenced breed usage and production types. Europe in particular has been strongly influenced by the actions of first the EEC then the EU. Arguably support for suckler cattle production has enabled larger suckler herds than would otherwise have been possible while intervention in the 1980's and 1990's strongly influenced the cattle breeds which were used in the UK, with a wholesale move from traditional British breeds to Limousin, Charolais and other continental sires. This was a move which the rest of the world did not follow. There has been a clear pattern of legislative and 'production led production' as opposed to demand led production. The end consumer of beef has been broadly ignored by the primary producer, who has been led by the examples demonstrated at agricultural shows throughout the country. The sustainability of cattle supply in the UK is dependent on being consumer led and efficiency driven. There is a clear need for the cattle breeder to understand what the customer is buying and to change the animals they breed.

Further development of temporal parameters for directed networks
K. Büttner, J. Salau and J. Krieter
Institute of Animal Breeding and Husbandry, Christian-Albrechts-University, Olshausenstr. 40, 24098, Germany;
kbuettner@tierzucht.uni-kiel.de

Previous studies dealing with network theory focused mainly on the static aggregation of edges over specific time window lengths which were analysed separately. Thus, most of the temporal information (e.g. chronological order of animal movements in trade networks) got lost. In order to assess the quality of this static aggregation the temporal correlation coefficient can be calculated which measures the overall possibility for an edge to persist between two consecutive time steps. Up to now, this measure is only defined for undirected networks, meaning the edge direction is neglected. However, for some issues the edge direction has a great impact on the investigated system, e.g. trade networks or agonistic interactions. Here, the nodes of the network can clearly be distinguished between initiator and receiver. Thus, the aim of the study was to develop a formula for the temporal correlation coefficient of directed networks which enables the differentiation between ingoing and outgoing edges. The newly developed parameters were then calculated for a real pig trade network to emphasize the importance of considering the edge direction. Here, the farm types at the beginning of the pork supply chain, i.e. multipliers and farrowing farms, showed clearly higher values for the outgoing temporal correlation coefficient compared to the farms at the end of the pork supply chain. Contrarily, these farm types, i.e. finishing and farrow-to-finishing farms showed higher values for the ingoing temporal correlation coefficient. Considering the yet known dependencies and issues in dealing with the analysis of temporal networks, the temporal correlation coefficient is a valuable tool to understand the structural dynamics of these systems, as it assesses the consistency of the edge configuration. The adaption of this measure to directed networks may help to preserve meaningful additional information about the investigated network that might get lost if the edge directions are neglected.

RNA-Seq transcriptomics analyses to identify biomarkers for in vitro production of embryos in cows

G. Mazzoni[1], S.M. Salleh[1], K. Freude[2], H.S. Pedersen[3], L. Stroebech[4], H. Callesen[3], P. Hyttel[2] and H.N. Kadarmideen[1]
[1]University of Copenhagen, Animal breeding, Quantitative Genetics and Systems Biology Group, IPH-SUND, Grønnegårdsvej 7, 1870 Frederiksberg C, Denmark, [2]University of Copenhagen, Anatomy and Biochemistry, IKVH-SUND, Grønnegårdsvej 7, 1870 Frederiksberg C, Denmark, [3]Aarhus University, Department of Animal Science, Blichers Alle 20, 8830 Tjele, Denmark, [4]EmbryoTrans BioTech, Bregnevang 6, 4690 Haslev, Denmark; gianluca.mazzoni@sund.ku.dk

Genomic Selection (GS) reduces the cost of traditional progeny and performance tests and increases the accuracy of estimated breeding values. For these reasons GS already has a substantial impact on genetic improvement of livestock production. The combination of GS with Ovum Pick Up–In vitro Production (OPU-IVP) of embryos can further speed up the genetic improvements in cattle. This is foreseen to have a huge impact in cattle production since a single donor cow can produce many calves by using IVP-based embryo transfer. Our objectives are to identify biomarkers in cows for IVP related traits as well as to provide a better understanding of the biological mechanisms that play a major role in the IVP process allowing to make a widespread large scale application of GS-OPU-IVP possible. In this study we aspirated oocytes together with follicular cells from the ovaries of 24 sacrified cows. Using Illumina HISeq2500, we sequenced the 24 RNA samples extracted from follicular cells. We preprocessed the raw reads, we aligned them using STAR and finally we quantified the expression of genes annotated in Ensembl. The oocytes from each animal were kept individually and were matured, (IVM), fertilised (IVF) and cultured (IVC) until blastocyst stage (day 8). We computed a score for each donor cow based on the blastocyst rate and on the morphology and kinetic parameters. The score (donor score) represents the quality of animal as donor cow for IVP. Using DESeq2, we identified 17 genes whose expression in follicular cells resulted to be significantly related to the donor score representing potential candidate biomarkers for donor cows. We identified respectively 2 and 3 genes involved specifically with morphology and kinetic parameters which gave us information about biological process controlling blastocyst development.

Transcriptome Profiling to Predict Piglet Early Development and Vitality from Umbilical Cord Blood

L. Canario[1], V. Voilet[1], N. Iannucelli[1], Y. Lippi[2], Y. Billon[3], J. Bailly[3], M. San Cristobal[1] and L. Liaubet[1]
[1]INRA, Animal Genetics, INRA UMR1388 / INPT ENSAT / INPT ENVT GenPhyse, 31326 Castanet-Tolosan, France, [2]INRA, Animal Genetics, UMR1331 Toxalim, 31076 Toulouse, France, [3]INRA, Animal Genetics, UE1372 GENESI, 17700 Surgères, France; laurianne.canario@toulouse.inra.fr

The objective was to identify molecular predictors of piglet level of vitality and development at birth. The parental genetic influences on early-life traits were analyzed with comparison of purebred and crossbred piglets born in the same litter, from Meishan (MS) or Large White (LW) sows. We used mixed semen from the 2 breeds to inseminate first parity sows. Phenotypes included growth, body mass indexes and indicators of vitality, i.e. reactivity scores, with repeated measurements. The study included litters from 8 LW sows and 9 MS sows producing a total of 245 piglets. Blood samples were collected at birth from the umbilical cord in PAXgene Blood RNA Tube. Transcriptome analyses were carried out with the Agilent 60K microarray on 3 to 8 newborn piglets from each litter, resulting in a total of 21 LWLW, 19 MSLW, 22 LWMS and 19 MSMS samples. Statistical analyses were carried out with restricted maximum likelihood methodology, differential and discriminant analyses. Crossbred piglets from LW sows had a higher early growth than their purebred counterparts (P<0.01). No significant differences were observed between purebreds and crossbreds born from MS sows. Among the 2,247 differentially expressed genes that were identified with a significance threshold of 1%, 1,962 genes were expressed differentially according to genotype. Co-variation between phenotypes and whole blood transcriptome profiling were investigated by multivariate approaches. Clusters of genes that co-vary with early growth were identified, in favour of crossbreds from LW sows. We are refining the analyses. Samples from LW piglets dead in early-life were added to the data base. The identification of genes biomarkers for early-life development and vitality is in progress and confirmed by RT-qPCR.

A large X-chromosomal haplotype is associated with small body size of minipigs

C. Reimer[1], A.R. Sharifi[1], C.J. Rubin[2], S. Weigend[3], K.H. Waldmann[4], O. Distl[4], S.D. Pant[5], M. Fredholm[5] and H. Simianer[1]
[1]*University of Göttingen, Albrecht-Thaer-Weg 3, 37075 Göttingen, Germany, [2]Uppsala University, Husargatan 3, 75123 Uppsala, Sweden, [3]Friedrich-Loeffler-Institut, Höltystr. 10, 31535 Neustadt, Germany, [4]University of Veterinary Medicine, Bünteweg 2, 30559 Hannover, Germany, [5]University of Copenhagen, Groennegaardsvej 3, 1870 Frederiksberg C, Denmark; creimer@gwdg.de*

Domestic pigs display a huge variety in body sizes, with miniature pigs as smallest representatives. They are genetically up to ten times smaller than commercial fattening pigs. Expected Heterozygosity and F_{ST} was calculated on whole genome sequencing data of Göttingen Minipigs (GMP), Berlin Minipigs and various large breeds to screen for selection signatures underlying small body size. We found an outstanding signal of decreased variation on chromosome X (52-100 Mb) carrying only three major haplotypes. Focusing on the region from 52 to 61 Mb, there were only two haplotypes, one carried by minipigs and South Chinese wild boars only. SNP array data from 175 GMP and 55 other pure-bred pigs were used to find SNPs associated with either the minipig or the large-pig-haplotype. Associated SNPs were used to determine the haplotypic state of 83 F_1 and 447 F_2-animals from a GMP boar × large pig (Duroc and Yorkshire) sow intercross. Using a linear model, we estimated the effect of the haplotype on body length and height at two different ages (Ø: 63 and 240 days). Sex, age, and the breed of the maternal granddam were used as covariates. Three SNPs around 57 Mb were associated with either GMP or large pigs. The distribution of the haplotypes in the F_2 animals followed expectations. 243 boars and sows carried the large pig haplotype only, 90 sows were heterozygous and 114 boars carried the minipig haplotype. Boars carrying the minipig haplotype had significant lower body height (3%) at both ages and were shorter (3%) at the older age. In conclusion, the analyzed region was shown to contain genetic variation significantly contributing to the miniaturization of minipigs.

Genome-wide association studies for production traits in pooled pig F2 designs

I. Blaj[1], J. Tetens[1], S. Preuß[2], R. Wellmann[2], J. Bennewitz[2] and G. Thaller[1]
[1]*Institute of Animal Breeding and Husbandry, Christian-Albrechts-University, Hermann-Rodewald-Straße 6, 24118 Kiel, Germany, [2]Institute of Animal Husbandry and Breeding, University of Hohenheim, Garbenstraße 17, 70599 Stuttgart, Germany; iblaj@tierzucht.uni-kiel.de*

Gene mapping experiments in livestock primarily aim to unravel the genetic architecture of quantitative traits and to identify genetic markers to facilitate breeding progress. In swine, several F2 resource populations have been established and analysed in the past. Despite their high power, the mapping resolution achieved is limited by the number of individuals, i.e. of meioses included in the analysis. Thus, the objective of this study is to combine data from two experimental F2 crosses between various types of founder breeds with the purpose to structurally identify short chromosome regions harboring causal mutations for the following quantitative traits: daily gain, back fat depth and meat fat ratio. The founder breeds of the first cross are Piétrain × Large White/(Large White × Landrace) and Meishan/Wild boar × Piétrain and Wild boar × Meishan in the second design. A total of 2,984 animals consisting of the F2 individuals and the respective F1 and F0 ancestors were genotyped with Illumina PorcineSNP60 Bead-Chip. The study focuses on gene variants segregating in the Piétrain breed due to its relevance from a breeding perspective and also because the highest mapping resolution is expected for the Piétrain related causal mutations. Results of genome-wide association studies (single and multi-marker approach) carried out in each individual design as well as in the joint design will be presented.

Estimation of recombination rates in a dairy cattle population

A. Hampel, F. Teuscher and D. Wittenburg
Leibniz Institute for Farm Animal Biology, Genetics and Biometry, Wilhelm-Stahl-Allee 2, 18196 Dummerstorf, Germany; hampel@fbn-dummerstorf.de

Half-sib family is a typical population structure in dairy cattle. The typical population structure has influence on population parameters such as linkage disequilibrium (LD) and recombination rate between pairs of genomic markers. Estimates of LD can be incorporated in extended models for genome-based phenotype prediction. The paternal contribution to LD can be determined from the recombination rate. The maternal part of LD requires estimates of the maternal haplotype frequencies. It is possible to estimate LD and recombination rates from progeny genotypes, for instance, using a complex expectation-maximization algorithm which requires a high computation time ('old' model). In this study, we aimed at developing a new model based on a minimization approach with less computation time. The new model takes parameters such as allele frequencies of progeny genotypes and empirical covariances between genotype codes for additive and dominant effects at two loci into account. The old and new models were applied to an empirical dataset comprising 1,295 half sibs from multiple families of Holstein-Friesian cattle with 40,317 SNP genotypes. The estimates were obtained for each autosome. The comparison of results showed that both models had a similar distribution of the maternal LD but the distribution of the recombination rate differed. A verification of the model's accuracy was performed in simulated half-sib families with various population sizes, maternal haplotype frequencies and paternal recombination rates. The results revealed that the new model had more accurate estimates for higher recombination rates (larger than 0.2) and is three times computationally faster than the old model. For smaller sample sizes (e.g. n=100), the old model, however, outperformed the new approach in terms of accuracy of estimates.

Implementation of genomic selection in three French regional dairy cattle breeds

M.P. Sanchez[1], D. Jonas[1,2], A. Baur[1,2], V. Ducrocq[1], C. Hozé[1,2], R. Saintilan[1,2], F. Phocas[1], S. Fritz[1,2], D. Boichard[1] and P. Croiseau[1]
[1]GABI INRA, AgroParisTech, Université Paris-Saclay, Domaine de Vilvert, 78350 Jouy en Josas, France, [2]Allice, 149 rue de Bercy, 75012 Paris, France; david.jonas@jouy.inra.fr

Genomic selection was implemented in three large dairy breeds (Holstein, Montbéliarde, Normande) in 2009 and in Brown Swiss in 2014 based on the Intergenomics reference population. In smaller breeds, assembling large reference populations is challenging due to the limited number of progeny tested bulls. An across breed genomic evaluation using HD markers was tested for the Abondance, Tarentaise and Vosgienne breeds but led to accuracies too low for practical selection. Hence, within breed reference populations were created: the AI bulls reference population was extended by genotyping cows representative of the breed genetic diversity and with complete phenotypes. The first genotypes were on the HD chip (60 to 250 bulls per breed) while the others were from the 50K chip. In total, the size of the reference population was 3,158, 1,892 and 1,237 in Abondance, Tarentaise and Vosgienne, respectively. The model used was similar to the one used for the other breeds, with 250 to 1000 QTL and about 9,000 SNP to fit the residual polygenic part. According to the breeds, 40 to 46 traits with a polygenic evaluation on production, conformation, fertility, mastitis resistance and longevity, as well as some composite traits, were included in this genomic evaluation. Phenotypes were weighted according to their accuracy by defining an equivalent number of independent own performances. The reliability of candidates without phenotype varied from 20 to 59% depending on the traits and the breeds. The first official genomic evaluation was released in March 2016 while unofficial indicators have been weekly computed since January 2016. Furthermore, the imputation accuracy from the EuroG10k LD to the 50k chip was found to vary from 99.56 to 99.64% of concordance rate. Hence, the LD chip is now also used in these breeds. The authors acknowledge the financial support of ANR and APIS-GENE through the GEMBAL and G2R projects.

Systematic genotyping of randomly sampled cows to improve reliability of GBVs of young candidates

L. Plieschke[1], C. Edel[1], E.C.G. Pimentel[1], R. Emmerling[1], J. Bennewitz[2] and K.-U. Götz[1]
[1]Bavarian State Research Center for Agriculture, Institute of Animal Breeding, Prof.-Dürrwaechter-Platz 1, 85586 Poing-Grub, Germany, [2]University Hohenheim, Institute of Animal Science, Garbenstraße 17, 70599 Stuttgart, Germany; laura.plieschke@lfl.bayern.de

In this investigation we explored the potential to increase reliabilities of breeding values of young selection candidates by genotyping first-crop daughters of each AI bull in a balanced and regular system of genotyping. We simulated a base scenario resembling the dual-purpose Fleckvieh cattle with respect to important key parameters. In the base scenario the reference population comprised only genotyped bulls. We extended this reference population in a stepwise manner by assuming that an increasing number of daughters was genotyped and included in the reference population. We studied the effects on model derived reliabilities, validation reliabilities and regression slopes as measures of bias for selection candidates. For the largest design with two generations of sires each having 200 genotyped daughters, the validation reliabilities for candidates reached 80% and higher. We demonstrate that even a moderate number of 25 genotyped daughters per sire will lead to considerable improvements of 16% to 19% for the reliability of selection decisions in candidates compared to the base scenario. Although the sires in the last generation were heavily pre-selected based on their Mendelian sampling term, genomic breeding values for candidates could be estimated with high reliability when genotyped cows were integrated in the reference population. The concept proposed is based on genotyping and phenotyping a random sample of first-crop daughters of each AI sire. We found that within-family selected daughter samples lead to strongly diminished benefits with respect to validation reliabilities and to biased genomic breeding values of selection candidates.

Cow genotyping strategies for genomic selection in small dairy cattle population

J. Jenko[1], G.R. Wiggans[2], T.A. Cooper[2], S.A.E. Eaglen[1], W.G. De L. Luff[3], M. Bichard[4], R. Pong-Wong[1] and J.A. Woolliams[1]
[1]The Roslin Institute and Royal (Dick) School of Veterinary Studies, The University of Edinburgh, Easter Bush, Midlothian, EH25 9RG, United Kingdom, [2]Animal Improvement Programs Laboratory, BARC, USDA-ARS, Beltsville, MD 20705, USA, [3]World Guernsey Cattle Federation, The Hollyhocks, 10 Clos des Goddards, Rue des Goddards, Castel, GY5 7JD, United Kingdom, [4]English Guernsey Cattle Society, 12 Southgate Street, Launceston, Cornwall, PL15 9DP, United Kingdom; janez.jenko@roslin.ed.ac.uk

This study compares how cow genotypes and different cow genotyping strategies increase the accuracy of genomic estimated breeding values (GEBVs) in dairy cattle breeds with low numbers. For this purpose Guernsey cattle from England and Guernsey Island were used. The prediction accuracies of GEBVs were estimated for milk (MY), fat (FY), and protein yield (PY). Altogether 1,637 genotypes were available: 197 bulls and 1,440 cows and GEBVs were predicted using three different methods: (1) genomic BLUP using only bulls; (2) univariate genomic BLUP using bulls and cows; and (3) bivariate genomic BLUP using bull and cow training set as different traits. Genotyping cows with phenotypes and using their data for the prediction of SNP effects increased the correlation between phenotypes and GEBVs. The increase with univariate GBLUP when using bulls and cows genotypes as compared to scenario where only bulls were used was: from 0.215 to 0.376 for MY, from 0.237 to 0.336 for FY, and from 0.217 to 0.331 for PY. Bivariate genomic BLUP gave further increases in the correlation. The increase when using bivariate compared to univariate GBLUP model was: 0.015 for MY, 0.019 for FY, and 0.017 for PY. It was always the best to genotype all the cows, but if only half of the cows were genotyped a top-bottom selection strategy was better than either random or top strategy. The research leading to these results has received funding from the EU 7th FP for research, technological development and demonstration [G.A. 289592] – Gene2Farm.

The value of genotyping in commercial farms for effective use of genotype-environment interactions

P.K. Mathur[1], Ø. Nordbø[2,3], T. Aasmundstad[3] and E. Grindflek[3]
[1]Topigs Norsvin Research Center, BV, Schoenaker 6, 6641SZ Beuningen, the Netherlands, [2]Geno SA, Storhamargata 44, 2317 Hamar, Norway, [3]TopigsNorsvin, Storhamargata 44, 2317 Hamar, Norway; pramod.mathur@topigsnorsvin.com

In major pig breeding programs around the world, selection for creating genetic gain is carried out in high health nucleus environments while the actual commercial production take place in less optimal environments. Estimation of the magnitude of genotype-environment interaction (G×E), estimation of its effect on predictive ability of breeding values for actual phenotypes in commercial farms, and development of suitable approaches for enhancing genetic gain are therefore valuable to make more effective use of the G×E. Genotyping of pigs from commercial farms provides even further opportunities. In this study, the additional gain of using genomic information in estimation of G×E interactions and in breeding value predictions were evaluated. This was attempted by using 151,918 maternal ability records from nucleus farms in Europe, 20,170 records of the same traits from a large commercial farm in USA, genotypes of 2,997 pigs from the commercial farm and 12,210 pigs from nucleus farms. Addition of genotypes from the commercial customer farm highly increased the accuracy of the breeding values for the customer from 0.58 to 0.77, averaged over seven traits, showing that genotyping in commercial farms increases the potential for more directional gain in customer herds. The accuracy in nucleus increased, as well, from 0.64 to 0.66 in average. The genetic correlations between nucleus and commercial farm ranged from 0.49 to 0.88 suggesting moderate G×E for some of the traits. The heritability was higher for those traits in the commercial farm compared to the European nucleus farms due to G×E. In view of these results, opportunities for optimum use of genotyping information and characteristics underlying the G×E phenomenon are discussed.

Joint genetic analysis of Holstein-Friesian dairy cows performing in Sub Saharan Africa

O. Abejide[1], M. Chagunda[1], R. Mrode[1], G. Banos[1], J. Ojango[2], C. Banga[3] and G. Simm[1]
[1]Scotland's rural college (SRUC), Animal breeding and genetics, Roslin institute building (RIB), Easter bush, EH25 9RG, Edinburgh, United Kingdom, [2]International livestock research institute (ILRI), Animal breeding strategies department, P.O. Box 30709, Nairobi, Kenya, [3]Agricultural research council (ARC), Animal production institute, 1134 Park Street, Hatfield, P.O. Box 8783, Pretoria, South Africa; oluyinka.abejide@sruc.ac.uk

Currently, there are a few countries that carry out genetic evaluation for dairy cattle in Sub Saharan Africa. Where this is done, it is at individual country level. Hypothesis was that joint genetic evaluation would result in robust and accurate genetic parameters and hence, improve genetic progress. This study used data of Holstein-Friesian cows performing in Kenya and South Africa. Test interval method (ICAR 2003) was used to determine 305-day milk yield from test day records for Holstein-Friesian cattle (n=358,327) obtained from the Kenya Livestock Breeders Organisation. In South Africa, 305-day milk yield records (n=10,767,516) were obtained from Agricultural Research Council. Data were from cows between first and fifth lactation from 1996 to 2012. There were 326 sires with an average of 20 daughters per sire from Kenya and 924 sires with 19 daughters per sire from South Africa. Of these, 191 sires had daughters in both countries. Variance component estimation was performed fitting a bivariate mixed linear model using ASReml. Cows in Kenya calved (31.0 SD=7.21 months) for the first time at a slightly older age than in South Africa (28.0 SD=4.10). Cows in Kenya had relatively lower milk yield (MY) than cows in South Africa (4,730 SD=1,916 vs 8,887 SD=2,593). Heritability for MY was 0.47 (s.e.=0.02) in Kenya and 0.15 (s.e.=0.01) in South Africa. MY heritability from the joint evaluation was 0.15 (s.e.=0.013). Repeatability for MY was 0.59, s.e.=0.007, for Kenya; 0.32, s.e.=0.01 for South Africa; and 0.30, s.e.=0.007 from the joint analysis. For calving interval, heritability values were 0.08 (s.e.=0.01) for Kenya, 0.03 (s.e.=0.01) for South Africa and 0.05 (s.e.=0.007) from joint analysis. Results indicate that joint genetic evaluation is feasible and offers potential for robust and accurate genetic parameters especially where there are insufficient individual country data.

AlphaSim: fast and flexible simulation and evaluation of breeding programs

S. Gonen, A.-M. Faux, R.C. Gaynor, M. Battagin, S.M. Edwards, D.L. Wilson, G. Gorjanc and J.M. Hickey
The Roslin Institute and Royal (Dick) School of Veterinary Studies, Genetics and Genomics, Easter Bush,
Midlothian, Scotland, EH25 9RG, United Kingdom; serap.gonen@roslin.ed.ac.uk

Simulation offers a quick and inexpensive alternative to empirical testing for measuring the impact, cost and benefit of alternative breeding program designs. AlphaSim is a software package for simulating, testing and optimising animal and plant breeding programs with a high degree of flexibility. It simulates breeding programs in a series of steps: (1) simulate haplotype sequences and initial pedigree; (2) drop haplotypes into the base generation and select marker and causal loci; (3) simulate allele substitution effects; (4) drop haplotypes into the burn-in generations and generate breeding values and phenotypes; and (5) perform selection and simulate new generations. The program is flexible in terms of historical population structure and diversity, recent pedigree structure, trait architecture and selection strategy. It allows the user to simulate multiple traits and multiple environments, specify recombination hot-spots and cold-spots, specify gene jungles and deserts, perform genomic predictions, and apply optimal contribution selection. AlphaSim can incorporate the use of new and existing biotechnologies, including genome editing, gene drives and the manipulation of recombination. Individuals can be selected as candidates for breeding based on phenotypes, progeny testing schemes, genomic selection using SNP chips of different densities, and optimal contributions with the aim of minimising inbreeding and maximising genetic gain. AlphaSim includes restart functionalities, which increase its flexibility by allowing the simulation process to be paused so that the parameters can be changed or to import an externally created pedigree, trial design or results of an analysis of previously simulated data. By combining the options a user can simulate simple or complex breeding programs with several generations, variable population structures and variable breeding decisions over time. In summary, AlphaSim is an efficient and flexible software with a diverse array of functionalities enabling the design, testing, optimisation and comparison of different breeding designs, decisions and goals prior to the application within real settings.

Social genetic effects for growth in Danish Landrace pigs increase with group size

H.M. Nielsen[1], B. Ask[1], J. Krogsdal[1], O.F. Christensen[2], L. Janss[2] and P. Madsen[2]
[1]SEGES Pig Research Centre, Axeltorv 3, 1609 Copenhagen V, Denmark, [2]AU Foulum, Blichers Alle 20, 8830
Tjele, Denmark; hmni@seges.dk

The aim of this study was to test if social genetic effects for growth in pigs depend on group size, and if they are reduced with increasing group size. Records included 70.080 pigs from 16 nucleus Landrace herds in Denmark. At 30 kg the pigs entered the performance test, and were assigned to pens varying from 8 to 15 pigs with approximately the same density. All pigs were fed ad libitum from weaning until they ended the test at an average age and weight of 146 days and 93.7 kg. A total of 6.364 pens were included with an average genetic relationship of 0.18 within the pens. Average daily gain from birth until end of the test was 644 g per day. Daily gain was analysed for the total data set with eight different group sizes, and for four subsets of this data with different group sizes (8-9, 10-11, 12-13, and 14-15). Variance components were first estimated using a conventional animal model including fixed effect of sex, contemporary compartment, and age and age squared at end of the test as covariates in addition to random effects of animal, pen and litter. Next, variance components were estimated using a social genetic model including a social genetic effect in addition to effects in the conventional model. For the total dataset with all group sizes, the social genetic variance was 3.55±1.29, and the total genetic variance was 0.163. When analysed separately for different group sizes, social genetic variances of 0.41±3.34, 4.86±3.65, 7.76±4.63, and 10.52±4.28 were found for group sizes 8-9, 10-11, 12-13, and 14-15, yielding total genetic variances of 0.25, 0.41, 0.81, and 1.15. The direct-social co-variances were negative for all group sizes indicating competition. In conclusion, social genetic effects for growth in pigs depend on group size, but contrary to expectations, our results indicate that the social genetic effect increases with increasing group size.

Estimating genetic parameters for direct and indirect effects on survival time in crossbred layers

E.D. Ellen, T. Brinker and P. Bijma
Wageningen University, Animal Breeding and Genomics Centre, Droevendaalsesteeg 1, 6708 PB Wageningen, the Netherlands; esther.ellen@wur.nl

Genomic information is widely used to improve accuracy of estimated breeding values and increase response to selection. For socially affected traits, however, benefits of genomic selection are still largely unknown. Estimation of genetic parameters for socially affected traits requires data where interacting individuals come from different families. However, in poultry breeding, hens are often kept in sire family groups, which is known as recurrent testing and particularly common for crossbred offspring. With groups of family members, direct and indirect genetic effects are fully confounded. Thus little is known of indirect genetic effects in crossbred laying hens. Here we evaluate the benefit of genomic information for estimating genetic parameters and breeding values for direct (DGE) and indirect genetic effects (IGE) on survival time of crossbred laying hens kept in sire-family groups. In total, 8,220 laying hens from four different crosses were used. For the crosses, one sire line and four dam lines were used. Birds had intact beaks. Two models were used; a sire model with pedigree information (sire-BLUP) and a direct-indirect genetic effects model with genomic information (ssGBLUP). With family groups, sire-BLUP captures the joint effect of the DGE and IGE, known as the total breeding value (TBV). ssGBLUP yields estimates for both DGE and IGE, from which TBVs can be calculated. Cross validation was used to compare both models. Total heritable variance, expressed as proportion of phenotypic variance (T^2), ranged from 3 to 28% with sire-BLUP and from 33 to 62% with ssGBLUP. ssGBLUP allowed estimation of genetic parameters and breeding values for DGE and IGE separately. IGE were significant for only one of the four crosses, probably reflecting the limited data and power with family groups. ssGLUP captured more genetic variation than sire BLUP, suggesting greater genetic variance in the dam lines than in the sire line. First results of cross validation suggest that ssGBLUP improves the accuracy of estimated TBV relative to sire-BLUP. Results show that prediction of breeding values for survival time in layers can be improved with ssGBLUP.

Ignoring genetic service sire effect for litter size in sire and dam lines is unjustified

B. Ask, B. Nielsen and T. Ostersen
SEGES Pig Research Centre, Breeding, Axeltorv 3, 1609 Copenhagen V, Denmark; bas@seges.dk

We investigated whether the genetic service sire effect for litter size was significant in both sire and dam lines in pigs. For sire lines, the genetic service sire effect on litter size is often ignored, because male fertility is assumed to be important for piglet survival rather than litter size. For dam lines, the genetic service sire effect is often ignored, because male fertility has a negligible economic value from the farmer's point of view. We show that this is not justified. Genetic parameters were estimated for litter size in Danish Duroc (D), Landrace (L), and Yorkshire (Y). Litter size traits were total number born (TNB) in D, and live piglets at day 5 (LP5) in L and Y. Two animal models, either excluding or including a genetic service sire effect, were applied to data from 2012-15 including 10,093 purebred litters in D, and 111,017 and 100,745 purebred and crossbred litters in L and Y. According to the likelihood ratio test, the model including genetic service sire effect fitted better for all breeds. Sow heritabilities were 0.09, 0.06, and 0.05, and heritabilities of the service sire effect were 0.06, 0.04, and 0.03 in D, L, and Y, respectively. The genetic correlations between the sow and service sire effects were 0.37, 0.76, and 0.17 in D, L, and Y, respectively. Including the genetic service sire effect improved both predictive ability and bias, though in Y only for purebred litters. For D, the large genetic service sire effect on TNB and its correlation to dam fertility suggests that male fertility may account for more than merely piglet survival. Combined with its economic importance for pig producers, the genetic service sire effect on TNB therefore becomes important in the sire line. For the dam lines, the prediction of the sow effect on LP5 is improved due to early and accurate information on top boars resulting from the combination of a large genetic service sire effect and, in L, a surprisingly high correlation to the sow effect.

Selective DNA pool-seq approach on bitter taste receptor genes in different pig populations
A. Ribani, F. Bertolini, G. Schiavo, E. Scotti, V.J. Utzeri, S. Dall'Olio, P. Trevisi, P. Bosi and L. Fontanesi
University of Bologna, Department of Agricultural and Food Sciences (DISTAL), viale Fanin 46, 40127, Italy;
anisa.ribani2@unibo.it

Taste perception in mammals influences feed intake and may affect performance and production traits in livestock. One of the most important taste is bitter which is sensed by G-coupled protein receptors encoded by a family of TAS2R genes. In this research, using a DNA pool-seq approach, coupled with next generation semiconductor based target re-sequencing, we analysed 9 porcine TAS2R genes (TAS2R1, TAS2R3, TAS2R4, TAS2R7, TAS2R9, TAS2R10, TAS2R16, TAS2R38 and TAS2R39) in order to identify variability and, at the same time, estimate single nucleotide polymorphism (SNP) allele frequencies in several pig populations. Allele frequency estimation obtained with this approach were also used to test differences in extreme and divergent groups of pigs for back fat thickness estimated breeding value (BFT-EBV) using a selective DNA-pool-seq approach. Equimolar DNA pools were prepared for five pig breeds (Italian Duroc, Italian Landrace, Pietrain, Meishan and Casertana) and wild boars (5-10 individuals each) and for two groups of Italian Large White pigs having extreme and divergent BFT (50 positive + 50 negative pigs). We obtained about 1.8 million of reads by sequencing amplicons generated from these pools. A total of 125 SNPs were identified, of which 37 were missense mutations. Three of them (p.I53F and p.W85L in TAS2R4; p.L37S in TAS2R39) could have important effects on the receptors function according to in silico prediction. A lower level of variability was found in wild boars than in domestic breeds as a potential result of selective pressure in the wild towards defensive bitter taste perception (bitter taste is usually associated with dangerous food protection). Thirteen SNPs in 5 TAS2R genes were significantly associated with BFT-EBV. These results might be important to understand the complexity of bitter taste perception and their effects on production and performance traits that could be useful to develop nutrigenetics approaches in pig breeding and nutrition.

Relationship between selection criteria and kits survival in the INRA 1777 rabbit line
H. Garreau[1], E. Bamisse[2], J. Ruesche[1], J. Hurtaud[3], M. Maupin[3], L. Canario[1] and I. David[1]
[1]INRA, Animal Genetics, Génétique, Physiologie et Systèmes d'élevage, 31326 Castanet tolosan, France, [2]INRA, Animal Genetics, PECTOUL, 31326 Castanet tolosan, France, [3]Hypharm, La Corbière Roussay, 49450, France; laurianne.canario@toulouse.inra.fr

The INRA 1777 line has been selected for a breeding goal (BG) that has included number of kits alive per litter (NA) and two components of weaning weight: Direct effects (WWD), influence of the young rabbit genotype on its growth, and maternal effects (WWM), incidence of mother's genotype on young rabbit growth. The objective of this study was to estimate responses to selection for the three criterion and to evaluate the relationship between the selection criterion and the survival of young rabbits from birth to weaning. The study included 1,948 litters dedicated to renew each generation (4th and 5th AI) from 1,158 does during 12 generations of selection. For each selection criteria and for the breeding goal, females were divided into 4 groups according to their estimated genetic value (-, -, +, ++). For stillbirths, the effect of genetic group of each of the four criteria was significant: The stillbirth rate is even lower for high genetic merit does. For mortality between birth and weaning, the effect of genetic group was significant for NA, WWM and BG criteria with lower mortality in groups + and ++ than in groups - and -. In conclusion, the breeding goal of the INRA 1777 line is not opposed to improvement of survival but puts to select individuals from litters with low mortality.

Estimation of maternal and cytoplasmic genetic variance for visual scores in Nellore Cattle
L. Grigoletto[1], E.C. Mattos[1], M.H.A. Santana[1], L.G.G. Figueiredo[2], F. Baldi[3], J.P. Eler[1] and J.B.S. Ferraz[1]
[1]College of Animal Science and Food Engineering, USP, 225 Duque de Caxias Norte Av., 13635-900 Pirassununga, Sao Paulo, Brazil, [2]National Association of Breeders and Researchers, ANCP, 463 João Godoy St., 14020-230 Ribeirão Preto, Sao Paulo, Brazil, [3]School of Agricultural Science and Veterinarias, UNESP, Acess way Prof. Paulo Donato Castellane s/n, 14884900 Jaboticabal, Sao Paulo, Brazil; jbferraz@usp.br

Visual scores are frequently included as selection criteria in the majority breeding programs in Nellore cattle in Brazil. However, the role maternal effects are not included in the models for genetic analysis because of the low heritability. The aim of this study was to estimate (co)variance components and genetic parameters of maternal and cytoplasmic effects for conformation (C), precocity (P) and muscle (M) visual scores of Nellore animal, measured at 18 months to analyze the impact of those effects on genetic evaluation. The data came from Nellore herds and were recorded from 1984 to 2013 with total of 142,290, 144,399 and 166,522 records by C, P and M, respectively. The cytoplasmic effect is a maternally inherited through the transmission of mitochondrial DNA (mtDNA) through maternal generations on a clonal way, therefore, to trace the maternal effect the founder LinMat software was used. For C, P and M were found 7,477, 11,337 and 8,800 cytoplasmic lines, respectively. Single-trait models were used for the estimation of (co)variance components, direct and maternal heritability coefficients and the genetic correlation between direct and maternal additive effects, the permanent environmental variance and cytoplasmic lineage variance as a percentage of the phenotypic variance were reported, using REML method by BLUPF90 program. Direct and maternal heritability coefficients averaged 0.15 and 0.01 respectively for conformation, 0.21 and 0.01 for precocity, 0.20 and 0.01 for muscle. The components of cytoplasmic lineage variance were 0.2×10^{-3}, 0.3×10^{-3} and 0.3×10^{-3}, respectively to C, P and M. The present study confirms the low heritability of maternal and cytoplasmic effects and, also, that those effects did not affect the estimation of (co)variance components and prediction of breeding values.

Genetic variability of hand reared Northern Bald Ibis Geronticus eremita released in Spain
I. Cervantes[1], J.P. Gutiérrez[1], M. De La Flor[1] and M. Cuadrado[2]
[1]Universidad Complutense de Madrid, Producción Animal, Avda. Puerta de Hierro s/n, 28040, Madrid, Spain, [2]ZooBotánico de Jerez, Departamento Técnico, Madreselva s/n, 11408 Jerez de la Frontera (Cádiz), Spain; gutgar@vet.ucm.es

The Northern Bald Ibis (Geronticus eremite) is listed in the IUCN Red List as 'Critically Endangered' specie on a global scale. Since 2003, the ZooBotánico Jerez in cooperation with Andalusia government has carried out a reintroduction program where hand-reared birds were released into wildness. The objective of this study was to analyze the genetic variability of the free-ranging population being alive in February 2014, by using its genealogical information. The pedigree included 480 individuals. A reference population of 88 birds was defined including the individuals released at any time. ENDOG program was used for the analyses. The mean equivalent complete generations was 1.4 for the reference population. Regarding the probability origin of genes, the effective number of founders was 35, while the effective number of ancestors was 33. The founder genome equivalent, which accounts for all possible causes of losses of diversity, became 23. Mean inbreeding resulted 0.4% and mean average relatedness 2.7%. The effective population sizes based on individual increase in inbreeding and on increase in coancestry were, respectively, 244 and 28. Given the scarceness of pedigree, Ne based in increase in coancestry becomes more informative. Furthermore, given that mating of captive birds were not directed, the effective population size based on coancestries would provide the expected Ne under random mating. Preliminary results suggest that the genetic variability of released population was good enough to favor the establishment of a healthy self-sustained population of this species in wildness. However, danger has not disappeared, so that management should continue or even improve if possible to ensure the stable presence of this species in its natural habitat.

Effect of the polymorphism in GPX5 gene on reproductive performance traits in Large White × Landrace

D. Polasik[1], A. Terman[1], M. Kumalska[1], O. Chomczyńska[1], P. Urbański[2] and G. Żak[3]
[1]West Pomeranian University of Technology in Szczecin, Al. Piastów 45, 70-311 Szczecin, Poland, [2]Institute of Genetics and Animal Breeding, ul. Postępu 36 A, 05-552 Magdalenka, Poland, [3]National Research Institute of Animal Production, Sarego 2, 31-047 Krakow, Poland; grzegorz.zak@izoo.krakow.pl

The aim of the experiment was to detect polymorphism in GPX5 gene encoding glutathione peroxidase 5 and to determine associations between individual genotypes and following reproductive traits in Polish Large White × Landrace sows crossbreed (n=442): total number of piglets born (TNB), number of piglets born alive (NBA) and number of piglets weaned (NW). The polymorphism in GPX5 gene was detected using PCR-RFLP (Restriction Fragment Length Polymorphism Analysis of PCR-Amplified Fragments) method with specific primers and restriction enzyme HinfI. Two different alleles of GPX5 gene were identified in the analyzed population – 1B (0.42) and 2B (0.58). Genotypes distribution was in a state of Hardy-Weinberg equilibrium. The association analysis showed statistically significant ($P \leq 0.01$) differences between sows carrying different genotypes and TNB, NBA NW. 1B1B genotype was favourable for all analyzed reproductive traits in first parity. In later parities similar tendency was observed, but differences were smaller and statistically not significant.

Association of LIF polymorphisms with fattening and slaughter traits of Polish Landrace pigs

A. Mucha, K. Ropka-Molik, K. Piórkowska and M. Tyra
National Research Institute of Animal Production, ul. Sarego2, 31-047 Kraków, Poland; aurelia.mucha@izoo.krakow.pl

Leukemia inhibitory factor (LIF) is a cytokine that serves many functions in the body. It is essential for embryo development and implantation, and plays a crucial role during early pregnancy. LIF regulates the differentiation and proliferation of many cell types, affects calcium and bone metabolism, contributes to the pathogenesis of inflammatory conditions, and stimulates myoblast and osteoblast proliferation. The aim of the study was to determine if LIF1 and LIF3 polymorphisms influence fattening and slaughter traits in pigs. The study involved 132 Polish Landrace gilts which were kept at the Pig Performance Testing Station under the same housing and feeding conditions. The animals were slaughtered after reaching 100 kg of body weight. The fattening and slaughter traits included: daily gain 30-100 kg (g), mean backfat thickness from 5 measurements (cm), loin eye area (cm^2), weight of meat of primal cuts (kg), and carcass meatiness (%). The polymorphisms were genotyped by PCR-RFLP and PCR-SSCP methods. Highest frequency in the LIF1 and LIF3 polymorphisms was found for the AB genotype (0.57 and 0.58) and lowest for the BB genotype (0.16) of the LIF1 and for the AA genotype (0.17) of the LIF3. For the LIF1 polymorphism, the highest values of all analysed traits were observed for pigs with BB genotype, and lowest for those with AA genotype (except for daily gain). For the LIF3 polymorphism, the highest values were characteristic of the animals with AA genotype (except for backfat thickness), and lowest of those with BB genotype. However, no statistically significant differences were found. It is concluded from the study that LIF1 and LIF3 polymorphisms have no effect on fattening and slaughter traits.

Predictive ability of a single-step terminal-cross model

L. Tusell Palomero[1], H. Gilbert[1], J. Riquet[1], M.J. Mercat[2], A. Legarra[1] and C. Larzul[1]
[1]INRA, UMR1388 / INPT ENSAT / INPT ENVT GenPhySE, 24 Chemin de Borde Rouge Auzeville Tolosane, CS 52627, 31326 Castanet Tolosan Cedex, France, [2]IFIP/BIOPORC, La Motte au Vicomte, 35651 Le Rheu, France; llibertat.tusell-palomero@toulouse.inra.fr

Selection goal in purebred pig lines is to improve crossbred performance although information from crossbreds is not accounted for in routine genetic evaluations. This study evaluates predictive ability of a bivariate single-step terminal-cross model (GEN) that jointly accounts for purebred (PB) and crossbred (CB) performance in a 6-fold cross-validation. The GEN model decomposes the additive genetic effect of a crossbred individual into the allelic contribution effects of its purebred sire (correlated with the additive genetic effect of the PB individual) and dam plus a Mendelian sampling confounded with the residual. A matrix combining both genomic and pedigree relationship matrices is used. PED model is of the same form as GEN but accounting only for pedigree information. Two univariate models containing same effects as the GEN model for either the PB or CB performance are also assessed. Data consisted in measurements of growth rate, individual feed intake, lean meat, pH longissimus dorsi, drip loss and intramuscular fat of approximately 650 purebred and 700 crossbred descendants of 90 sires. The predictive ability was evaluated using the average Pearson's correlation between yet-to-be observed and predicted phenotypes. Accounting for genomic information increases the predictive ability with respect to using only pedigree-based relationships among all traits (correlations for PB and CB performance are respectively 0.02 to 0.11 and 0.03 to 0.05 higher with the GEN than with the PED model). Nevertheless, the joint analysis of PB and CB performance in a single-step terminal-cross model did not increase the predictive ability with respect to single–step univariate analyses, possibly due to the high genetic correlation estimated between PB and CB performance for the analyzed traits (>0.68). Current results will be compared with predictive ability of alternative single-step terminal-cross models.

Association analysis of c.105G>A SNP in CAPN1 gene with carcass and meat quality traits in goose

S. Negro[1], M. Solé[1], A. Membrillo[2], F. Peña[3], V. Domenech[3], I. Verona[4], M. Rubí[4], M. Valera[1] and A. Molina[2]
[1]Universidad de Sevilla, Departamento de Ciencias-Agroforestales, Sevilla, 41013, Sevilla, Spain, [2]Universidad de Córdoba, Departamento de Genética, Córdoba, 14071, Córdoba, Spain, [3] Universidad de Córdoba, Departamento de Producción Animal, Córdoba, 14071, Córdoba, Spain, [4]Inddeco S.L., Ganso Ibérico de Dehesa, Madrid, 28003, Madrid, Spain; z12neras@uco.es

Meat quality is an important feature for the poultry industry. Domestic geese usually present a though meat and it is necessary to select them to improve meat tenderness. The relation of the calpain 1 (CAPN1) gene with the tenderness process of meat post-mortem has been demonstrated in several species. Thus, the objective of the present study was to identify polymorphisms in this gene and to determine for first time the possible association between these polymorphisms and related economic traits in goose raised in a dehesa ecosystem. For the analysis, forty geese of two different genotypes (20 Embden-Anser anser, EE; 20 Toulouse-Anser anser, TT) were studied. A novel SNP was found in the CAPN1 gene, c.105GàA. This Polymorphism was statistically associated with different carcass and meat quality traits as tight quality (circumference, P=0.020; and length, P=0.026) and b* (meat color, P=0.024) parameters for the global goose population. The association of this gene with meat tenderness (WBSF) was only confirmed in the case of female individuals of the Toulouse breed (P=0.043), where the G allele of Calpain 1 gene contributes to obtain a tender meat. The results suggest the possibility of using molecular markers in CAPN1 gene as a promising tool for the improvement of carcass and meat quality traits in poultry breeding programs.

Accuracy of genomic selection to improve litter traits in the French Landrace pig population

A. Bouquet[1], T. Bellec[2], L. Flatres Grall[3], B. Ligonesche[4] and C. Larzul[5]
[1]IFIP, BP 35104, 35651 LE RHEU Cedex, France, [2]ADN, 15 Rue M. de Tresiguidy, 29190 PLEYBEN, France,
[3]Gène+, 12 Rue du Moulin, 62134 ERIN, France, [4]Nucleus, 7 Rue des Orchidées, 35650 LE RHEU, France, [5]INRA
UMR1388 GenPhySE, 24 chemin de Borde Rouge, 31326 Castanet-Tolosan, France; alban.bouquet@ifip.asso.fr

The objective of this study was to assess the gain in selection accuracy obtained due to integration of genomic information in genetic evaluation models. In total, 579 boars (born>2002) and 504 sows (born>2010) were genotyped with the Porcine60K SNP panel. After edition, 1,067 animals were available for analysis. Four traits were analyzed: the number of piglets born alive (NBA), the number of weaned piglets (NW), the mean birth weight of piglets (MBW) and the within-litter standard deviation of piglet birth weights (SDBW). The predictive ability of the conventional BLUP and single-step GBLUP model was assessed with a validation study in which the genomic dataset was divided in a reference and a validation population. For the validation study, performances of individuals born after 12/31/2013 were removed. The reference population comprised 491 boars with daughters and 355 sows with own performances. The validation population was made up of 46 boars born in 2013 having at least 15 daughters with reproduction data on 09/30/2015. The predictive ability was estimated for validation boars as the correlation between the EBV estimated based on offspring performance and the EBV / GEBV estimated as candidates with performances recorded up to 12/31/2013. We used the same fixed and random effects and genetic parameters as in the routine genetic evaluations. With conventional BLUP, the predictive ability of the model was 0.46 for NBA, 0.45 for NW, 0.37 for MBW and 0.26 for SDBW. With GBLUP model, the predictive ability was 0.55 for NBA, 0.63 for NW, 0.45 for MBW and 0.26 for SDBW. Thus, genomic evaluation improves the accuracy of selection for all maternal traits.

Implementation of BLUP for breeding value estimation in Russian Federation

A.A. Kudinov[1], J. Juga[2], P. Uimari[2], E.A. Mäntysaari[3], I. Strandén[3], K.V. Plemyashov[1], E.I. Saksa[1] and M.G.
Smaragdov[1]
[1]Russian Reserch Institute of Farm Animal Genetics and Breeding, Moscovscoe shosse 55A, Tyarlevo, Saint-
Petersburg, 196601, Russian Federation, [2]University of Helsinki, Helsinki, FI-00014, Finland, [3]Natural Resources
Institute Finland (Luke), Jokioinen, FI-31600, Finland; kudinov_aa@list.ru

Current prediction of breeding values for dairy bulls in Russian Federation is based on contemporary comparison. This method is old and ineffective. Thus, there is a tremendous need to modernize the system including up-to-date modeling of the data and utilizing genomic information. The first step is to implement BLUP animal model to predict breeding values for dairy cows and bulls. Leningrad district is one of the most productive and well recorded dairy region in Russia. Data from 47 large herds (>700 cows) with pure bred Russian Black and White and Holstein cattle were used to develop the animal model BLUP. Pedigree records of 415,145 animals born in 1990-2015 were available together with milk records from 146,098 cows, including 305 d milk yield, fat yield and protein yield in kg. Milk yields and contents were measured once a month during lactation (~10 measurements). Average 305 d milk yield was 8,267 kg (SD=1,766), average fat yield was 306 kg (SD=69), and average protein yield was 263 kg (SD=54). A repeatability animal model with three lactations was used for variance component estimation and for breeding value (BV) estimation up to five lactations were used. The statistical model included herd-year-season and age at calving × days open as fixed effects. Outliers with records ±3 SD from the mean were excluded from the analysis. DMU software package was used to estimate variance components. Heritability and repeatability estimates were for milk $h^2=0.19$ and r=0.3, for fat $h^2=0.16$ and r=0.34, and for protein $h^2=0.14$ and r=0.34. With data from first lactations only, the heritability estimates were $h^2=0.23$, $h^2=0.20$, and $h^2=0.18$ for milk, fat and protein yield, respectively. MiX99 was used to estimate breeding values. Average genetic trend for milk yield for cows born from 2002 to 2015 was +43 kg per year.

Comparison of single and multi-trait animal models for genetic evaluation of milk production traits

T.K. Belay and T. Ådnøy
Norwegian University of Life Sciences, Department of Animal and Aquacultural Sciences, P.O. Box 5003, 1432 Ås, Norway; tormod.adnoy@nmbu.no

This study aimed to compare use of single and multi-trait methods. Genetic parameters and trends of test-day (TD) milk yield and percentage of fat, protein and lactose were estimated using 758,996 TD records from 90,065 cows kept in 1,015 herds. (Co)-variance components (additive genetic, permanent environmental, herd-test day and residual) and the corresponding heritabilities, correlations and breeding values estimated by WOMBAT. Breeding values of cows for a trait averaged by and regressed on the birth years were used to estimate genetic trends for the periods 2001 to 2005, 2006 to 2011 and 2001 to 2011. Estimates of heritability obtained under the single and multi-trait model were similar and had moderate values. Variance due to herd-test day effects was an important source of variation for fat percentage under both types of analysis. Genetic correlations among the traits were moderately positive or negative, except for correlations between lactose and the remaining traits, where the genetic and phenotypic correlations were close to zero. Results of this study showed that the difference between the estimates derived by both methods were small and that the implemented genetic program has had positive impact on percentage traits and negative impact on milk yield for the studied period (2001-2011) in Norwegian dairy cattle.

Genetic parameter estimations for 'out of season' breeding in UK dairy goats

S. Desire[1], S. Mucha[1], R. Mrode[1,2], M. Coffey[1] and J. Conington[1]
[1]SRUC, Animal & Veterinary Sciences, Roslin Institute Building, Easter Bush, Midlothian, EH25 9RG, United Kingdom, [2]International Livestock Institute, Animal biosciences, 30709 Naivasha Rd, Nairobi, Kenya; suzanne.desire@sruc.ac.uk

Consistent, year-round dairy production is important when supplying fresh milk, but is difficult in species with seasonal breeding cycles, such as goats. Historical efforts to overcome this problem have focused on extending lactation length, however milk yields typically decrease over the course of the lactation period, affecting herd milk output. This study sought to estimate genetic parameters for 'out of season' breeding, as a first step towards developing genetic evaluations for this trait. Date of first kidding was recorded for 17,366 dairy goats from 2 sites managed by the same UK producer. Goats were a composite of 3 breeds: Alpine, Saanen, and Toggenburg and 1 herd was based on emigration of animals from the other. Kidding dates spanned from 1989 to 2015. Peak kidding season was defined as the four weeks of the year where the highest average number of births occurred across all years. Each doe was assigned a value from 0 to 24, depending on how many weeks either side of peak kidding season she kidded. Contemporary groups (n=92), defined as farm, year and season of birth were created. Groups containing fewer than 50 animals were excluded from the dataset. The 'out of season' phenotype was defined as week of kidding relative to peak breeding season. A univariate analysis was used to estimate breeding values using the following animal model: $y = Xb + Za + e$, where the incidence matrices X and Z relate phenotypic records contained in vector y to fixed (b) and additive genetic (a) effects, and vector e represents residual error. Contemporary group and age at first kidding were fitted as fixed effects ($P<0.001$). Heritability was 0.19 (s.e 0.02). The results of this study indicate that selection for out of season breeding is feasible. The next step will be to evaluate the relationship between this trait and other traits of economic importance.

Longitudinal analysis of pig growth using Legendre polynomial equations

J.M. Coyne[1,2,3], K. Matilainen[2], D.P. Berry[3], M.-L. Sevon-Aimonen[2], E.A. Mantysaari[2], J. Juga[1], T. Serenius[4] and N. McHugh[3]
[1]*University of Helsinki, Dept. of Agricultural Science, PL 28 (Koetilantie 5), Helsingin Ylioposto, Finland,* [2]*Natural Resources Institute Finland (Luke) Green technology, Biometrical Genetics, Finland, Jokioinen, Finland,* [3]*Teagasc Moorepark, Animal & Bioscience Research Department, AGRIC, Teagasc Moorpark, Fermoy, Co. Cork, Cork, Ireland,* [4]*Figen Ltd, FI-01301, Seinäjoki, Finland; noirin.mchugh@teagasc.ie*

The objective of this study was to estimate (co)variance components for pig weight across an animal's life. A total of 51,893 live weight records from 10,201 animals between the years 2006 and 2012, inclusive, were available from the Finnish pig breeding company Figen Oy. Genetic variance components for pig weight across ages were estimated using random regression Legendre polynomial models in ASREML, with gender and contemporary group (date of entry to test station) included as fixed effects. The animal genetic effect was modelled using random regression Legendre polynomials across age measured in days; a litter environmental effect was also included as a random effect. The most parsimonious order of the random regression to adequately model the data was determined by comparing the Akaike information criterion of the higher order random regression model and the immediately lower order fit. Residual variances were estimated across nine defined age groups. Within block, residual variances were assumed homogenous but were allowed to vary between blocks. The quadratic random regression describing the additive genetic component of pig weight fit the data significantly better than the linear model. The genetic variance for pig weight varied from of 13 kg^2 at 76 days of age to 226.76 kg^7 at 197 days of age. The heritability of pig weight varied from 0.43 (76 days) to 0.72 (120 days). In general the genetic correlations between weight measured across the age trajectory weakened as the interval between measures lengthened and ranged from -0.189 (between days 64 and 143) to 0.999 (between days 65 and 66). The results indicate ample genetic variation exists for pig weight and modelling the data using Legendre polynomials better accounts for the longitudinal nature of the data.

Genome-wide association analyses confirm a major QTL for fatty acid composition in porcine backfat

M. Van Son, E. Gjerlaug-Enger, H. Hamland and E. Grindflek
TopigsNorsvin, Storhamargata 44, 2317, Norway; eli.grindflek@norsvin.no

The composition of fatty acids is a nutritional quality trait that influences human health as well as the technical and sensory quality of pork. Fatty acids vary in carbon length and degree of saturation. Healthier meat could be produced by higher levels of monounsaturated fatty acids (MUFA) and polyunsaturated fatty acids (PUFA), lower levels of saturated fatty acids (SFA), and decreased ratio of n-6/n-3 PUFA. The aim of this study was to detect loci affecting fatty acid composition in the Norwegian Duroc and Landrace breeds. For this purpose, nine different fatty acid traits (C16:0, C16:1n-7, C18:0, C18:1n-9, C18:2n-6, C18:3n-3 and total MUFA, PUFA and SFA) were measured on boars at dissection line by NIRS technology. Altogether, 454 Duroc and 659 Landrace boars were phenotyped, and thereafter genotyped with the porcine 60K Illumina SNP chip. The NIRS predicted fatty acid compositions are estimated to be, in most cases, highly heritable. Given the same data, heritabilities for C16:0, C18:1 and C18:2 in Landrace were 0.21, 0.42 and 0.62 (SE 0.12; n=646), respectively, and for Duroc the heritabilities were 0.86, 0.52 and 0.68 (SE 0.17; n=444), respectively. Association analyses were performed using the R package GenABEL where a mixed model was run using the polygenic function. A highly significant genomic region on SSC14 (p<1.0-6) was found in Duroc for the six de novo synthesized fatty acids and their families (C16:0, C16:1n-7, C18:0, C18:1n-9, MUFA and SFA). The QTL is additive for all traits and is explaining 55-83% of the genetic variance. In Landrace, this QTL is fixed but a Landrace specific QTL was obtained on SSC8 for C16:1n-7. The QTL region on SSC14 in Duroc is a confirmation of previous results on Duroc and Asian breeds, and the causative mutation has been identified in the stearoyl-CoA desaturase (SCD). Also the SSC8 QTL region has previously been identified in several breeds and a promoter SNP in ELOVL6 has been suggested to be the causative variant.

Organic pig breeding programs with limited trait registration should adopt genomic selection

T.O. Okeno and A.C. Sørensen
Aarhus University, Center for Quantitative Genetics and Genomics, Department of Molecular Biology and Genetics, P.O. Box 50, 8830, Tjele, Denmark; tobias.okenootieno@mbg.au.dk

Organic pig producers rely on genetic materials from conventional breeding programs. However, a separate organic breeding program with intensive trait registration has been recommended. In this study we consider the same traits in conventional and organic breeding goals: growth rate from birth to 30 kg, growth rate from 30-100 kg, lean meat percentage, back and leg strength, feed efficiency, live piglets at 5 days, slaughter loss and sow longevity. These traits differ in relative economic importance in the two breeding goals and the genetic correlation between the two breeding goals is 0.83. In addition, the organic farmers are also interested piglet mortality and number of functional teats. The current organic pig breeding herds are characterized by limited pedigree and performance trait registration. This makes it difficult to realize high response to selection. We envision that genomic selection (GS) could improve response to selection when phenotypic information is limited. GS will provide additional information source that can increase prediction accuracies and accurately estimate relationship between selection candidates. We used stochastic simulation to test this hypothesis. We compared economic and genetic gain per year and rate of inbreeding per generation realised in the traditional breeding scheme (TS) with limited phenotypes with GS scheme with intensive (GSI) and limited (GSL) genotyping intensities. The change in economic response per year realised were €1.59, €1.88 and €1.78 for TS, GSI and GSL, respectively. The GSI and GSL outperformed TS in genetic gain for individual traits. The GSI resulted to lower rate of inbreeding per generation 0.037 compared to 0.051 for TS and GSL. Our findings show that adoption of GS in organic pig breeding program with limited trait registration is beneficial.

Comparison between animal model and polynomial Legendre test-day model in dairy cattle

M.C. Rotar, H. Grosu, M.A. Gras and R.S. Pelmus
National Research Development Institute for Animal Biology and Nutrition (IBNA), Laboratory of Animal Biology, Calea Bucuresti No.1, Balotesti, 077015, Ilfov, Romania; rotar.mircea.catalin@gmail.com

The objective of this study is comparison between traditional genetic evaluation system and test-day models, in two dairy cattle breeds from Romania. A total of 175 animals (consisting of cows, their parents and their descendants), with 523 records, were evaluated based on performances of dairy cows from National Research Development Institute for Animal Biology and Nutrition. Dairy cows belong to Montbeliarde (31 cows) and Holstein (38 cows) breeds. Data were filtered and animals with less than 3 test-days were eliminated. Also, cows under 200 or over 400 days in milk were not taken into account. Breeding values were estimated for all the animals from data set, all estimates being adjusted for all other effects from the models, all effects are simultaneous estimated and predicted each other. Genetic correlations between test days were very high (0.96) between two consecutive tests, but decrease when tests were more distant from each other (0.36). Heritability increased at the lactation extremes due to artefacts generated by Legendre polynomial procedure. Between 35 and 275 days in milk, heritability is trending normally. Spearman rank correlation (ρ=0.80) shown an 80% analogy between ranks. A comparison of both sets of breeding values shows some differences between that two models. The test-day model offer possibility to estimate more precise breeding values, with a better correction of environmental effects along lactation. High correlation ranks show a pronounced similarity between models, but we recommend using test-day model in dairy cattle genetic evaluation.

Effect of different management systems on performance and carcass traits of crossbred chickens

W. Molee, P. Mernkrathoke, S. Khempaka and A. Molee
Suranaree University of Technology, School of Animal Production Technology, Institute of Agricultural Technology,
111 University Avenue, Muang District, Nakhon Ratchasima 30000, Thailand; wittawat@sut.ac.th

Feed costs have a major impact on the profitability of chicken enterprises. Feeding and rearing systems will help in reducing feed costs. The objective of this study was to determine the effect of different feeding and rearing management systems on growth performance and carcass composition of Thai indigenous crossbred chickens. Two hundred forty 42-d-old of the 'Korat Meat Chicken' genotype were assigned to 3 different management systems: housing in an indoor pen (8 birds/m^2) with ad libitum feeding (control group); housing in an indoor pen (8 birds/m^2), subjected to a feeding restriction equal to 80% of ad libitum feeding, and access to grass provided in separate feeder (FR-Grass); and housing in an indoor pen (8 birds/m^2), subjected to a feeding restriction equal to 80% of ad libitum feeding, with access to a grass paddock (bird/m^2) (FR-Outdoor). Each treatment was represented by 4 pens containing 20 birds each. All birds were raised until slaughter at 70 days of age. The results showed that body weight (BW) was not different among groups (P>0.05). However, FR-outdoor group had better feed conversion ratio (FCR) than control group (P<0.05). There were no significant differences between the 3 groups for eviscerated carcass, breast meat and thigh meat (P>0.05). However, FR-Outdoor and FR-Grass groups had lower abdominal fat than control group (P<0.05). The feed cost of FR-Outdoor group had lower than FR-Grass and control group, respectively. It can be concluded that the results of this study may suggest that a feeding restriction with access to a grass paddock or access to grass provided in separate feeder could significantly reduce feed costs and abdominal fat, but with no effect on growth performance and meat composition of Thai indigenous crossbred chicken.

Social genetic effects for daily gain and correlation to conformation traits in two pig breeds

J.J. Krogsdahl, B. Ask and H.M. Nielsen
Pig Research Centre, SEGES, Axeltorv 3, 1609 Copenhagen V, Denmark; juk@seges.dk

Including social genetic effects in pig breeding programs may increase welfare of pigs for example through a reduction in aggression between pigs. Yorkshire pigs are known to be more aggressive than Landrace pigs. Therefore, the first aim of the study was to test if social genetic effects for daily gain differed between the two breeds. Conformation traits explain the robustness of a pig and can be associated with how pigs interact with each other. Therefore, the second aim was to estimate the genetic correlation between the direct effect for conformation traits and the social genetic effect for daily gain. The data were from the performance test at Danish breeding herds in the period of 30 to 100 kg and data were from 13 Landrace herds and 13 Yorkshire herds. Data consisted of 60,939 Landrace pigs across 5,512 groups and 62,753 Yorkshire pigs across 5,362 groups. Group size varied from 8 to 15 pigs. Average daily gain was 962 g for Yorkshire and 988 g for Landrace. Variance components for daily gain were estimated using: (1) a classical animal model; and (2) a social genetic model. Genetic correlations between the social genetic effect of daily gain and the direct genetic effect of conformation will be estimated using a bivariate analysis using model (1) for conformation traits and model (2) for daily gain. Social genetic effects were present for daily gain for both breeds with higher total heritabilities compared to classical heritabilities for both Landrace and Yorkshire pigs (0.26±0.01 vs 0.24±0.02, and 0.24±0.01 vs 0.21±0.02 respectively). For both breeds the genetic correlation between the direct and social genetic effects for daily gain was negative, which indicates competitive interactions. Contrary to expectations, there was no significant difference between the breeds. We expect a negative genetic correlation between direct genetic effects for conformation and social genetic effects for daily gain because we assume that aggressive pigs have to have a good conformation to dominate other pigs.

Genome wide association studies for dry-cured ham quality traits in two Italian heavy pig breeds

G. Schiavo[1], G. Galimberti[2], S. Bovo[1], F. Bertolini[1], M. Gallo[3], V. Russo[1], L. Buttazzoni[4] and L. Fontanesi[1]
[1]University of Bologna, Department of Agricultural and Food Sciences, Division of Animal Sciences, Viale Fanin 46, 40127, Bologna, Italy, [2]University of Bologna, Department of Statistical Sciences 'Paolo Fortunati', Via delle Belle Arti 4, 40126, Bologna, Italy, [3]Associazione Nazionale Allevatori Suini, Via L. Spallanzani 4, 00161, Roma, Italy, [4]Centro di Ricerca per la Produzione delle Carni e il Miglioramento Genetico, Consiglio per la Ricerca in Agricoltura e l'Analisi dell'Economia Agraria, Monterotondo, 00015, Roma, Italy; luca.fontanesi@unibo.it

In Italy, about 70% of pigs are raised for protected designation of origin (PDO) productions. The most important Italian PDO products are Parma and San Daniele dry-cured hams that are obtained from legs of heavy pigs slaughtered at about 160 kg live weight. Tights reach the optimal dryness after at least 12 months of seasoning with the combined action of the added salt and the right temperature and humidity of the processing plants. Two dry-cured ham meat quality traits has been considered for breeding purposes: ham weight loss at first salting (HWLFS) and visible intermuscular fat (VIF). HWLFS, measured within the first 7 days of seasoning, is highly correlated to the total weight loss of the hams at the end of the seasoning period. An excess level of VIF in the hams decreases consumers' acceptance. In this work, we report the results of genome wide association studies (GWAS) in Italian Large White (ILW) and Italian Duroc (IDU) pigs for these traits. A total of 1,388 ILW and 419 IDU were performance tested by the National Pig Breeder Association (ANAS) and genotyped with the Illumina PorcineSNP60 BeadChip. Random residuals (RR) for HWLFS and VIF were used as trait values in GWAS. A total of 31 and 20 QTLs were identified for HWLFS and VIF, respectively. Different QTLs for both traits segregate in the two investigated breeds. These results highlight the complexity of these meat quality traits for dry-cured ham production and pave the way to include information of these QTLs in genomic selection programs for heavy pigs.

Session 68   Theatre 1

Amazing grazing in periods of drought

M.R. Mosquera-Losada, J.J. Santiago-Freijanes, M.P. González-Hernández, R. Romero-Franco, N. Ferreiro-Domínguez, J.L. Fernández-Lorenzo and J.L. Rigueiro-Rodríguez
University of Santiago de Compostela, Crop Production, EPS, 27002 Lugo, Spain; mrosa.mosquera.losada@usc.es

Grasslands are one of the main types of area in southern European countries. Due to the different pedoclimatic conditions, grasslands are grazed with sheep, pigs, goats, horses, dairy cows and beef cows. The grazing season is limited by different periods of drought during the summer going from 2 to 6 months of housing period. Strategies have been developed in the past to increase the feed use produced within the farm systems and avoid inputs from outside. Most of the strategies were based on the use of woody vegetation, as this type of vegetation has the ecological trait that enables them to survive to the long summer period. So, the introduction of woody vegetation (trees or shrubs) is usually employed to feed animals and overcome in a cheap way the restrictive period. On the other hand, ecological traits linked to the use of annual species self-seed are also successful to supply feed to animals, while high levels of biodiversity are maintained. All these strategies are really very useful to enhance adaptation to climate change in northern latitudes. This is of high relevance taking into account the inclusion of LULUCF as a commitment for reducing greenhouse gas emissions by 40% in 2030 to that produced in 1990 for the Kyoto Protocol rules. Links to the Common Agricultural Policy will be further evaluated.

Ten years-data of milk production in beef cattle under South American grazing conditions
M. Lewdiukow[1,2], O.M. Lema[1], J.I. Velazco[1] and G. Quintans[1]
[1]*National Institute for Agricultural Research, Ruta 8 km 281, 33000, Uruguay, [2]Facultad de Agronomía, UdelaR, Garzon 780, 12900, Uruguay; gquintans@inia.org.uy*

The aim of this study was to characterize milk production in multiparous beef cows (4 to 10 years old) under grazing conditions in Uruguay. For a preliminary analysis British crossbred cows (n=114) were selected from a database of 660 over 10 years. They belonged to different experiments, where milk production was measured under identical protocol. Cows calved in spring and were managed on native pastures with a forage allowance between 8 and 12 kg DM/kg LW (8-10% CP and 45-55% of digestibility). Milk yield was assessed between 20 and 40 d postpartum (pp) and monthly until weaning (average 155 d pp), using a milking machine after oxytocin injection. Milk samples were analysed for fat, protein and lactose. LW and Body Condition Score (BCS; scale 1 to 8 u) at calving were 420.0±4.22 kg and 3.9±0.04 u. Calves LW was registered at birth and monthly until weaning (35.9±0.52 and 155.7±1.45 kg for birth and weaning, respectively). Milk yield and calves performance were analysed with lineal splines with 3 knots at 30, 60 and 150 d using PROC GLIMMIX (SAS). The model included individual dam as random effect, experiment and calves sex as fixed effects, and BCS at calving as covariable. Milk production for the entire lactation period was, on average, 826±15.2 l. Milk fat, protein and lactose were 2.15±0.06, 3.07±0.02 and 4.94±0.01% respectively. The curve showed a peak at 45 d (7.33±0.23 l) decreasing until weaning with a milk average production of 3.98±0.11 for this final period. Calves daily LW gain was 946±19.3 g from birth until 45 d, decreasing to 758±12.4 g until weaning. This milking protocol applied to cows under grazing conditions allows us to characterize the beef milk curve. It was consistent with calves LW gain evolution under the conditions of this analyses. The consolidation of a larger data set of beef milk production will enable us to reinforce the parameters to better characterize the curve shape in the production systems described above.

Herbage and milk production from perennial ryegrass and white clover swards in a grazing system
M. Egan and D. Hennessy
Teagasc, Grassland Department, Animal & Grassland Research and Innovation Centre, Moorepark, Fermoy, Co. Cork, Ireland; michael.egan@teagasc.ie

White clover (Trifolium repens L.; clover) can increase the sustainability of grass based dairy systems and has the potential to increase milk production. This experiment compared milk production from a perennial ryegrass (Lolium perenne L.; PRG) sward receiving 250 kg N/ha/yr (Gr250), a PRG/clover sward receiving 250 kg N/ha/yr (Cl250) and a PRG/clover sward receiving 150 kg N/ha/yr (Cl150) in an intensive grazing system over two grazing seasons (2013 and 2014). Three groups of cows were allocated to graze each sward in 2013 (n=14) and 2014 (n=19) from February to November at a stocking rate of 2.74 livestock units/ha. Fresh herbage was offered daily and 350 kg concentrate were fed per cow. Pre-grazing herbage mass, sward clover content and milk production were measured. Herbage dry matter intake (DMI) was estimated on three occasions each year. Sward clover content was greater (P<0.001) on Cl150 (266 g/kg DM) compared to Cl250 (225 g/kg DM). Sward clover content increased (P<0.05) from year 1 to year 2. Treatment did not effect (P>0.05) cumulative herbage production (13,504 kg DM/ha). There was a significant (P<0.05) treatment × week interaction on daily milk yield and daily milk solids (MS) yield; in the second half of the year, the cows on the Cl250 and Cl150 treatments had a greater daily milk yield and daily MS yield compared to cows on the Gr250 treatment. The clover treatments had greater milk solids compared to the Gr250 treatment (477 and 447 kg/cow, respectively). The cows grazing had the greatest (P<0.01) DMI compared to Gr250, and the Cl150 was intermediate, 17.2, 16.2 and 16.8 kg DMI/cow, respectively. In conclusion, the lower N fertiliser application to the Cl150 treatment increased sward clover content; however there was no effect on herbage production. Inclusion of clover increased daily milk yield and daily MS yield as a result of greater DMI on both clover treatment compared to the cows grazing the Gr250 treatment.

The effect of grass yield on economic and environmental performance of dairy farms

C.W. Klootwijk[1], C.E. Van Middelaar[1], A. Van Den Pol-Van Dasselaar[2,3], P.B.M. Berentsen[4] and I.J.M. De Boer[1]

[1]Wageningen University, Animal Production Systems group, De Elst 1, 6700 AH Wageningen, the Netherlands, [2]CAH Vilentum University of Applied Sciences, De Drieslag 4, 8251 JZ Dronten, the Netherlands, [3]Wageningen UR Livestock Research, De Elst 1, 6700 AH Wageningen, the Netherlands, [4]Wageningen University, Business Economics group, Hollandseweg 1, 6700 EW Wageningen, the Netherlands; cindy.klootwijk@wur.nl

In this study we explored the importance of grass yield for the economic and environmental performance of dairy farms that apply grazing. A whole-farm optimization model was used to analyse the effect of grass yield on labour income, nitrogen (N) and phosphate (P_2O_5) losses and greenhouse gas emissions (summed in kg CO_2-equivalents per kg fat-and-protein-corrected milk (FPCM)). The model contains all relevant activities and constraints present on Dutch dairy farms and represents a typical farm in 2016. The objective function of the model maximizes labour income. The model was used to optimize farm management for a dairy farm of 50 ha with different grass yields. Data on grass yields were derived from 23-26 Dutch dairy farms that apply grazing for the years 2012-2014. Grass yield varied between farms from 34.5 to 69 MJ NE_L/ha (NE_L = net energy for lactation) at the same fertilization level of about 225 kg N/ha. The model results show that, based on the lowest and highest grass yield, labour income varied from -€9,300 to €23,700 yr[-1]. A higher grass yield led to a higher labour income mainly due to lower costs for feed purchases. A higher grass yield also resulted in lower on-farm surpluses, i.e. 181 kg N/ha and 7 kg P_2O_5/ha (highest grass yield), compared to 245 kg N/ha and 22 kg P_2O_5/ha (lowest grass yield), and lower greenhouse gas emissions, i.e. 970 kg CO_2-eq/kg FPCM (highest grass yield) compared to 1,047 kg CO_2-eq/kg FPCM (lowest grass yield). Results show the importance of grass yield to improve the economic and environmental performance of dairy farms. To identify options to increase grass yields further research will focus on quantifying relations between grassland management and grass growth, utilization and intake on modern dairy farms.

Substitution rate between forage supplement and grazed pasture in dairy cows: a meta-analysis

M.F. Miguel[1], H.M.N. Ribeiro Filho[2] and R. Delagarde[1]

[1]INRA-Agrocampus Ouest, UMR Pegase, 35590 Saint-Gilles, France, [2]UDESC, Centro de ciencias Agroveterinarias, Lages, SC, Brazil; remy.delagarde@rennes.inra.fr

Grazing dairy cows are often supplemented with a conserved forage when pasture amount is not sufficient. Milk production response to forage supplementation (MPR, in kg of milk/kg DM supplement) and forage/pasture global substitution rate (GSR) are highly variable. A meta-analysis was carried out to quantify the within-experiment effects of forage supplementation in rotationally grazing dairy cows. We hypothesized that GSR and MPR are mainly affected by grazing conditions (pasture allowance, PA) and forage type (maize silage, pasture silage or hay). Comparisons of supplementation levels at same PA or when supplemented cows receive lower PA than unsupplemented cows were considered separately. At same PA (61 comparisons, 4.6 kg DM/d of supplementary forage), the GSR averaged 0.39, and was lower for maize or pasture silage than for hay (0.36 and 0.31 vs 0.51). The MPR averaged 0.41, and was higher for maize silage than for pasture silage and hay (0.53, 0.30, 0.22, respectively). The GSR increased by 0.018 per kg DM of PA and was unaffected by forage supplementation level. At lower PA (52 comparisons, 7.2 kg DM/d of supplementary forage, -16 kg DM/d of PA), the GSR averaged 0.79. Higher GSR was observed with maize and pasture silage than for hay (1.28, 1.15 and 0.58, respectively). Feeding supplementary forage while reducing PA leaded to a decrease of milk production. The average MPR was negative (-0.13), and was affected by forage type (-0.22, -0.47 and -0.03 for maize silage, pasture silage and hay, respectively). On both databases, any increase in GSR of 0.1 decreased MPR by 0.06. The most effective forage for increasing milk production was maize silage. It is concluded that GSR is largely affected by PA, and that it is lower when supplemented cows receive the same PA than unsupplemented cows than when they received a lower PA (increased stocking rate). As for concentrate supplementation, MPR is highly and negatively correlated with GSR.

Effects of substitution of kikuyu forage by oat silage on milk production and quality in dairy cows

S. Yuste[1], W. Sánchez[2], A. De Vega García[1] and J.A. Guada[1]
[1]*Universidad de Zaragoza, Producción Animal y Ciencia de los Alimentos, Miguel Servet, 177, 50013 Zaragoza, Spain, [2]Instituto Nacional de Innovación y Transferencia en Tecnología Agropecuaria, Sabana Sur, de Canal 7 300 m Sur, Antiguo Colegio La Salle-MAG, 382 CENTRO COLÓN, Costa Rica; susanay2@gmail.com*

This work aimed to determine the effects of substitution of kikuyu forage by oat silage on milk production and quality, and on urinary excretion of purine derivatives as an index of microbial protein synthesis in the rumen, in grazing dairy cows. Eight dairy cows, four Jersey and four crossbreed, with an average body weight (BW) of 350 kg in the 12th week of lactation were assigned to two treatments (supplementation with oat silage or kikuyu forage) in a cross-over design with two periods (7 days of adaptation and 5 days of sampling per period) and four cows (two Jersey and two crossbreed) per treatment and period. Animals were daily fed, at milking, 1.5 kg dry matter (DM)/day of either oat silage or mature kikuyu forage plus a mixture of 0.9 kg DM/day of Citrocón (a commercial concentrate) and variable amounts (between 1.75 and 3.5 kg DM/day, depending on milk production) of Vapp Feed (also a commercial concentrate). Moreover, cows strip grazed kikuyu pasture (one day of grazing and 30 resting days for regrowth) and had free access to fresh water throughout the experiment. Estimated pasture intake was numerically higher (5%; $P>0.1$) in cows supplemented with oat silage than in those supplemented with kikuyu forage, and this led to a tendency ($P=0.0925$) to increase milk production from 14.8 to 15.6 kg/day, without variation in its composition. Daily production (kg) of protein ($P=0.0429$) and lactose ($P=0.0205$) was higher in cows supplemented with oat silage. Urinary excretion of purine derivatives was not affected ($P>0.1$) by the type of forage supplemented. As a conclusion, substitution of kikuyu forage by oat silage (on a DM basis) seems to be an advisable practice for dairy milk producers in the highlands of Costa Rica.

The effect of tetraploid and diploid swards sown with and without white clover on milk production

B. McCarthy[1], M. Dineen[1,2], C. Guy[1,2], F. Coughlan[1] and T. Gilliland[2,3]
[1]*Teagasc, Animal and Grassland Research and Innovation Centre, Moorepark, Fermoy, Co. Cork, Ireland, [2]Institute of Global Food Security, Queens University Belfast, Belfast, United Kingdom, [3]Agri-food Biosciences Institute, Plant testing Station, Crossnacreevy, Belfast, BT5 7QJ, United Kingdom; brian.mccarthy@teagasc.ie*

There is renewed interest in the use of forage legumes, and white clover (WC) in particular, in pasture-based dairy production systems. An experiment to investigate the impact of tetraploid and diploid perennial ryegrass swards sown with and without WC on the productivity of spring milk production systems was established in Clonakilty Agricultural College, west Cork, Ireland, in 2012 and 2013. Four separate grazing treatments were sown for the experiment: a tetraploid only sward, a diploid only sward, a tetraploid + WC sward and a diploid + WC sward. Four diploid cultivars (Tyrella, Aberchoice, Glenveagh and Drumbo) and four tetraploid cultivars (Aston Energy, Kintyre, Twymax and Dunluce) were chosen and sown as monocultures with and without WC. Thirty cows were randomly assigned to each treatment after calving in February 2014 and 2015. All treatments were stocked at 2.75 cows/hectare (ha) and received 250 kg of nitrogen fertiliser per ha annually. There was no difference in milk or milk solids yield (kg fat + protein) between the tetraploid only (4,972 and 420 kg/cow) and diploid only (4,994 and 423 kg/cow) treatments. However, incorporating WC into perennial ryegrass swards resulted in 15.7% and 13.8% greater ($P<0.001$) milk and milk solids yield, respectively. Tetraploid + WC and diploid + WC treatments produced 5,783 and 5,750 kg milk per cow and 481 and 478 kg milk solids per cow, respectively. Pasture dry matter production was 12.2% greater ($P<0.001$) on the grass WC swards (tetraploid + WC and diploid + WC; 17.4 t DM/ha) compared with the grass only swards (tetraploid only and diploid only; 15.5 t DM/ha). Incorporating WC into perennial ryegrass swards increased milk and pasture production over the first two years of this experiment. Further research as to the long term persistency of WC in intensive pasture-based milk production systems is required.

Are physical and feeding activities at pasture impacted by cattle breed and feeding restriction?
A. De La Torre[1], F. Anglard[1], M. Barbet[2], A. Le Morvan[1], J. Agabriel[1] and R. Baumont[1]
[1]INRA, UMR1213 Herbivores, Theix, 63122 Saint-Genès-Champanelle, France, [2]INRA, UE1414 Herbipôle, Theix, 63122 Saint-Genès-Champanelle, France; anne.delatorre@clermont.inra.fr

Recording locomotion and feeding activities coupled with animal performances could contribute to characterize the adaptive responses of animals subjected to disturbances. In this study physical and feeding activities at pasture of 16 suckling cows subjected to a previous indoor feed restriction have been characterized. The experimental design combined 2 breeds (Charolais, CH and Salers, SA) and 2 indoor feeding levels (Control, C: 120% of requirements and Restricted, R: 70% of requirements) from calving to turnout (110 days). Physical activities as standing, walking and lying and feeding activities as eating and ruminating activities were recorded with the Rumiwatch® system over one week immediately after the turning out (Period 1, P1) and 8 weeks later (Period 2, P2). Body weight recovery at pasture was significantly higher in R than in C cows and these changes were greater in CH than in SA cows (40 vs 12 kg in CH-R and CH-C; 25 vs 2 kg in SA-R and SA-C respectively, $P<0.001$). These differences were related to changes in feeding and physical activities especially during P1. The time spent in an upright position (included either standing or walking) was greater in R than in C cows in both breeds (12.7 vs 11.8 hours, $P=0.07$). These results are in accordance with the time spent eating which was higher in CH than in SA cows in both feeding levels (in average 7.08 vs 6.22 hours/day, $P<0.08$). In contrast, the ruminating time was higher in SA than in CH cows especially as they were underfed (11.3 vs 9.8 hours/day). With time (P2), differences in feeding and physical activities between breeds and previous feeding levels disappeared. Physical activity and time spent eating of suckling cows at turnout are significantly increased after an indoor feed restriction in accordance with a higher weight gain. In addition, SA cows were characterized by longer rumination time at pasture than CH cows.

Effects of grazing previously abandoned grassland on performance in sheep
L. Grøva[1], H. Steinshamn[1], E. Brunberg[2] and U. Støbet Lande[1]
[1]NIBIO, Gunnarsvei 7, 6630 Tingvoll, Norway, [2]NORSØK, Gunnarsvei 7, 6630 Tingvoll, Norway; lise.grova@nibio.no

Large areas of farmland are annually abandoned in Norway; In Mid-Norway 8 600 ha farmland has been abandoned since 2005. Knowing that access to spring and autumn pastures is a limiting factor for sheep farmers, this study aims at testing the effect of introducing abandoned farmland into sheep production. One sheep flock with 83(88) ewes(lambs) in 2014 and 77(106) ewes(lambs) in 2015 was each year assigned with respect to age of ewe and number of lambs born into three treatments: (1) control; common farm procedure with short spring grazing period before summer grazing on range pasture; (2) spring extended; 4 weeks extended spring grazing period on abandoned cultivated grassland before summer grazing on range pasture; (3) whole season grazing on abandoned grassland. Weight gain from birth to autumn as well as slaughter weight were significantly ($P<0.05$) higher in lambs assigned to treatment 2 with four weeks extended spring grazing period (259 g/day and 15.7 kg) compared to treatment 1 (238 g/day and 14.3 kg) and treatment 3 (216 g/day and 13.2 kg). The use of abandoned cultivated grassland for extended spring grazing improved weight gain and slaughter weight, while whole season grazing on abandoned grassland were the least productive.

Does access to pasture affect welfare in dairy cows?

L. Armbrecht[1], C. Lambertz[2], D. Albers[3] and M. Gauly[2]
[1]University of Göttingen, Department of Animal Science, Albrecht-Thaer-Weg 3, 37075 Göttingen, Germany, [2]Free
University of Bozen, Faculty of Science and Technology, Piazza Universitá 5, 39100 Bozen, Italy, [3]Chamber of
Lower Saxony, Field Testing Station for Grassland Management and Cattle Farming, Albrecht-Thaer-Straße 1,
26939 Ovelgönne, Germany; linda.armbrecht@agr.uni-goettingen.de

In the animal welfare discussion, dairy cows are perceived positively by consumers when compared with other species. This is mainly due to the fact, that the majority of consumers associate dairy farming with pasture, which in turn they relate with benefits for animal health and welfare. However, overall scientific assessments of effects of pasturing on animal welfare are rare. Hence, it was the aim of the study to estimate the animal welfare level in 61 dairy farms by using the Welfare Quality® Protocol (WQP). The WQP includes 32 indicators, which are aggregated to 12 criteria and four principles. Data were collected at the end of the pasture season (April to November) in 2014 and at the end of the barn (winter) season in 2015. Farms were classified based on animals pasture access during pasture season: group 1 (G1) >10 hours, group 2 (G2) 6-10 hours, group 3 (G3) ≤6 hours pasture access per day, group 4 (G4) without pasture. The average herd size was 129 Holstein-Friesian or Red-Holstein cows (58 to 527). Statistical analyses were computed using the GLIMMIX procedure. While groups did not differ for the principle 'good feeding' ($P>0.05$), pasture groups got better results than G4 for the principle 'good husbandry' at the end of the pasture ($P<0.05$), though not at the end of the barn season ($P>0.05$). Compared with the two other groups, G1 and G2 reached superior scores at both visits for 'good health' ($P<0.05$), mainly due to better ratings for 'absence of injuries'. The principle 'appropriate behavior' did not differ ($P>0.05$). This study shows positive effects for some welfare indicators when cows have access to pasture. However, differences occurred predominantly during pasture season.

Wolf predation risk in summer farms of Italian Alps

G. Faccioni[1], F. Marucco[2], A. Menzano[2], E. Sturaro[1], S. Calderola[3] and M. Ramanzin[1]
[1]University of Padova, DAFNAE, viale dell'Università 16, 35020 Legnaro, Italy, [2]Centro Gestione e Conservazione
Grandi Carnivori, Parco naturale Alpi Marittime, P. zza Regina Elena 30, 12010 Valdieri (CN), Italy, [3]Regione
del Veneto, Sezione Caccia e Pesca, via Torino 110, 30172 Mestre (VE), Italy; georgia.faccioni@studenti.unipd.it

This study is part of the project LIFE WolfAlps and analyzed the wolf predation risk due to grazing management in 6 areas exposed to different wolf presence in the Italian Alps. Data on herd composition, grazing management, use of damage preventive methods, and farmers' willingness to adopt preventive measures were collected on 806 summer farms (temporary units used for summer grazing in highlands; 44% with cattle, 26% with sheep and goats, and 31% with mixed ruminants). Use of preventive measures (mostly electrified fences and/or guardian dogs) varied across areas from 0 to 64%, from 3 to 100%, and from 13 to 93% of farms with cattle, with sheep and goat, and with mixed ruminants respectively, and was much more frequent in areas with long term stable wolf presence than in areas recently recolonized or without predators. Preventive measures were used in 58% of farms with continuous presence of shepherd and in 57% of farms using rotational grazing, but only in 16% and 32% of farms with discontinuous shepherd presence and free grazing. Willingness of farmers to adopt preventive measures varied greatly across core areas, being higher in long term stable wolf presence areas and for small ruminant shepherds than for cattle farmers. The use of unguarded free-grazing, most frequent in summer farms with cattle and in areas where wolf presence is discontinuous or absent, is the main obstacle to the use of preventive measures. Refunding farmers for the costs of acquiring protection measures contributes little to modify their negative attitudes, because adopting grazing practices compatible with prevention would increase labour and costs. The necessary implementation of suitable grazing practices and prevention methods should be integrated into a wider approach to innovate and sustain the multifunctionality of livestock farming in mountainous areas.

Nutrient composition of faeces in feral Giara horses in relation to seasonal availability of feed

M.G. Cappai[1], G. Arru[1], M. Di Marco[2], S. Poddesu[1], D. Bergero[2] and W. Pinna[1]
[1] University of Sassari, Agraria, via De Nicola 39, 07100, Italy, [2]University of Turin, Veterinary Sciences, l. go Paolo Braccini 2, 10095, Italy; gisella.arru@gmail.com

Feeding habits and nutrient composition of individual faeces were determined in a group of feral horses, composed of 11 specimens living on the Giara tableland in Sardinia Island. The peculiarity of these horses is represented by the strict dependence of their survival on natural feeding stuffs, exclusively composed of wild flora, in a well defined geographical area, identified by natural limits. Variations in the nutritional value of plants were hypothesised to occur in relation to season and the composition of nutrients in faeces were determined with the aim to obtain basic data about the grazing behaviour and selection of feedstuffs by Giara horses, in comparison to nutrient composition of faeces from 6 horses, experimentally fed a hay based diet. At the analysis of data, the differences in crude ash, ether extract and crude fiber content of faeces turned out to be statistically significant ($P<0.05$) according to seasons. This finding is suggestive of the grazing activity in relation to seasonal availability of feeding stuffs as well as of the selected organs or parts of plants ingested. The extremely high crude ash content of faeces from feral Giara horses suggests a high content of mineral sources in the diet, which did not appear to be responsible for sand colic syndromes in these animals. Worth of note, the seasonal ether extract content of faeces from Giara horses could be related to the crude fat content of selected organs of different plant species, observed to be moderately grazed. On the basis of our results, the nutrient profile of faeces from Giara horses allows to establish the highly specialised selection of plant species by grazing feral horses on Giara tableland, which can successfully compose their diet on feeding stuffs naturally available in the different seasons of the year.

Recent advances in pasture-based automatic milking systems

C.E.F. Clark
Dairy Science Group, School of Life and Environmental Sciences, Faculty of Veterinary Science, The University of Sydney, Camden, New South Wales 2570, Australia; cameron.clark@sydney.edu.au

Automatic milking systems (AMS) present an opportunity for dairy farmers to not only improve their lifestyle and conditions of work, but also save on labour costs and/or increase the time available to focus on overall farm management. However, the viability of Australian AMS into the future will largely rely on producing milk at a low cost which is particularly relevant in the current environment of high international milk price volatility. Our group has focused on two factors to produce milk from AMS at low cost. The first is the ability to achieve high levels of pasture utilisation in AMS by following well established pasture management principles and the second is the implementation of grazing systems to fully utilise milking robots across the 24 hours of a day. Similar levels of pasture utilisation (mean 13,500 kg DM/ha) were recorded between AMS and CMS at the same site highlighting the ability to follow established grazing management principles and achieve high levels of pasture utilisation on pasture-based AMS farms. Data collected from elite Australian AMS farmers also shows the ability to achieve consistently high levels of milking robot utilisation throughout 24 hours in grazing AMS, with the typical drop in early morning robot utilisation eradicated. Ongoing work is now isolating the factors driving consistently high levels of milking robot utilisation in a research setting as a first step towards providing simple changes in management to drive more efficiency out of grazing AMS.

Precision Livestock Farming in Europe: EU-PLF project

D. Berckmanns

KU Leuven, Department of Biosystems, Division M3-BIORES: Measure, Model & Manage Bioresponses, Kasteelpark Arenberg 30, 3001 Heverlee, Belgium; daniel.berckmans@biw.kuleuven.be

The worldwide demand for animal products is increasing strongly while the number of farmers is decreasing which leads to bigger groups of animals per farm. The question is how to achieve high-quality, sustainable and safe meat production in such big groups of animals. At the same time there are several concerns about animal health in relation to food safety and human health, how to guarantee an improved animal welfare, reducing the environmental impact of the livestock sector. The European Commission has subsidized the EU-PLF project to analyse how the technology of Precision Livestock Farming (PLF) can create value for the farmer and his animals beside other stakeholders in the food chain. Twenty farms (fattening pigs, broilers, milking cows) spread over Europe were equipped with PLF technology (real time image analysis, sound analysis, sensor signals) to work over 60 production cycles for each species. The objective is to give an overview of the most interesting results from the project that used continuous, fully automatic monitoring of animals in 20 European farms and how this technology should be applied to create value. We will present results collected on the farms, results from the monitoring systems and testimonies from farmers. Finally we will give a proposal of business models to be used for PLF technology in connection with the Internet of Things.

Validation of portable pig weighing system on Iberian pigs based on image analysis and cloud service

I. Amat-Roldan and D. Verde

YMAGING, Plaza Urquinaona 6 12A, 08010, Spain; david@ymaging.com

Pig weighing determines pig value at slaughter-house which is generally performed by visual inspection or direct scaling. Visual inspection is a cost effective approach but highly inaccurate, which is penalised by the slaughter. Direct scaling is an intensive task which can be only done before slaughtering, but becomes impractical for management of livestock growth. A prototype of PigWei was presented last year as a potential solution for pig weighing ubiquosly by means of a portable device that captures a sequence of images of a pig and sends them to a central server which replies the weight in few seconds. PigWei has been now further developed into a compact device weighing less than 900 g, specific algorithms for Iberian breed and a realtime cloud service operating in less than 8 seconds on a 3G network. PigWei has been tested in different farms of Iberian pigs ranging 140 to 160 kg and compared to scales. This range is critical for this breed as optimal slaughtering is at 148 kg according to Spanish Law. We acquired a total of 178 image sequences out of 45 pigs in two different farms. Farm #1 scale had an average error of 19 kg on 15 pigs, whereas farm #2 had an average error of 0.8 kg in 30 pigs. We had normally found average errors around 5 kg in scales of different farms. PigWei processed 65% of image sequences with an average error of 1.9%, which corresponds to 3.0 kg. PigWei also allows for a high resolution mode, which was computed off-line to test its performance. High resolution mode has lower acceptance and only processed 11% of images but obtained 0.9% of average error, corresponding to 1.4 kg. This results will be further improved in the following months and we expect that errors will be halved in our next release with no changes in hardware. Farm tests on white pigs will be also performed in the following months. In summary, PigWei is an attractive alternative to other weighing methods due to its cost and accuracy.

Automatic image analysis to predict the onset of farrowing in sows

I. Traulsen[1], J. Brünger[2], M. Kardel[2], O. Burfeind[3] and J. Krieter[1]
[1]Institute of Animal Breeding and Husbandry, Christian-Albrechts-University, Olshausenstr. 40, 24098 Kiel, Germany, [2]Institute of Computer Science, Christian-Albrechts-University, Olshausenstr. 40, 24098 Kiel, Germany, [3]Chamber of Agriculture Schleswig-Holstein, Gutshof, 24327 Blekendorf, Germany; itraulsen@tierzucht.uni-kiel.de

The aim of the present study was to detect the onset of farrowing with help of automatic video image analysis. From December 2013 until March 2014, in total 30 sows (three batches) in the farrowing unit of the Futterkamp research farm, Germany, were videotaped by color cameras starting at the day of housing-in into the farrowing unit (conventional farrowing pens) until two days after farrowing. The exact point in time of the onset of farrowing (birth of the first piglet) was determined by manual video analysis. For the automatic analysis video material from 48 hours before the onset of farrowing until the onset of farrowing was used. Activity of the sows was evaluated by measuring the amount of change in pixel color (chPix) using two automatic image analysis algorithms: namely background subtraction (BG) and optical flow (FL). For chPix of both algorithms the parameters mean, standard deviation, variance, maximum and 1st, 2nd and 3rd p-variation per half-hour showed an increase before the onset of farrowing. A sow individual CUSUM control chart was used to monitor these parameters. For all parameters on average almost 70% of the sows could be detected (BG: 64.3%, FL: 69.6%). Optimal parametrization of the chart increased the detection rate up to more than 80% (BG: 81.7%, FL: 83.0%). The signal was produced on average 23.5 (BG)/ 21.0 (FL) hours before the onset of farrowing. The parameter 2nd p-variation produced most reliable results. The onset of farrowing can be detected using automatic image analysis and provide beneficial information to the famer to improve farrowing management.

Can an accelerometer based labelling method replace a human: postural profile of farrowing sows?

M. Oczak[1,2], K. Maschat[2], D. Berckmans[1], E. Vranken[1] and J. Baumgartner[2]
[1]M3-BIORES: Measure, Model, Manage Bioresponses, Kasteelpark Arenberg 30, 3001 Leuven, Belgium, [2]Institute of Animal Husbandry and Animal Welfare, Veterinärplatz 1, 1210 Vienna, Austria; i106pc03@staff.vetmeduni.ac.at

The objective of this research was to develop a method for automated classification of postures of sows on the basis of accelerometer data. The second objective was to evaluate if the technique can be used in a reliable way instead of human labelling to automatically create postural profiles of sows on new unlabelled data sets. The experiment took place in the research farm of University of Veterinary Medicine Vienna, using a herd of 120 Large White sows. Data were collected from sows housed in three types of farrowing pens with temporary crating. The behaviour of 14 sows was video recorded and labelled in six periods: day of entering the farrowing rooms, 24 h before the start of farrowing, farrowing, first day after farrowing, 4th day after farrowing and 6th day after farrowing. The focus of labelling was on three types of postural behaviours: active, resting in lateral position (RLP) and resting in sternal position (RSP). Each sow had an ear tag with 3 axis accelerometer sensor. Acceleration was measured at a frequency of 10 Hz. From raw accelerometer data 156 parameter vector was extracted for further analysis. Linear discriminant analysis (LDA) was used to classify the three labelled postural behaviours on the basis of parameters extracted. The overall accuracy of classification of postural behaviours with the developed method was 84% in the cross-validation and 86% in the training set. Comparison of labelling and classification results revealed that the accuracy was good enough to detect the effect of crating and pen type on active behaviour of the animals. The developed method could be further tested for automated monitoring of health and welfare status of sows.

Activity and occupation analyses as indicators of broiler welfare risk

A. Peña Fernández[1], T. Norton[1], V. Exadaktylos[1], E. Vranken[2] and D. Berckmans[1]
[1]KU Leuven, Biosystems Department, Animal and Human Helath Engineering division, M3-BIORES Laboratory, Kasteelpark Arenberg 30, Room 01.35, 3001 Leuven, Belgium, [2]Fancom B.V., Research Department, P.O. Box 7131, 5980 AC Panningen, the Netherlands; daniel.berckmans@biw.kuleuven.be

The activity and occupation patterns of broiler chickens are measured to check the performance and status of the flock using Precision Livestock Farming (PLF) technologies. The aim of this study is to illustrate how monitoring flock activity and occupation, provides retrievable information on their habits and activity patterns, which can be used to check the flock health and welfare status. Four top view cameras are used in a commercial broiler house in The Netherlands to visualize the floor area, continuously recording images during the light cycle periods which are translated into activity and occupation indexes per minute. The monitoring of these indexes through successive light periods reveals how behavioral patterns of the flock evolve during the cycle. By using the previous three light periods, it is possible to estimate the activity and occupation levels for the following period and evaluate deviations (15% or higher). These deviations are expressed as the percentage of time birds spent in an alert situation. This percentage shows a statistical relevant correlation ($P<0.05$) with the welfare quality assessment scores. Thus, performing these analyses along every growing cycle is possible to indicate the level of risk for broilers being under a potential welfare problem. Farmers can use this information to support their management decisions.

Precision tools for Brazilian beef production systems

P.P. Pires
EMBRAPA, Avenida Rádio Maia, 830, Zona Rural, CEP 791006-550, Campo Grande, MS, Brazil; pedropaulo.pires@embrapa.br

Precision Livestock Farming (PLF) aims to provide solutions to the needs of businesses, regardless of the conditions. It can be used to improve the management of land and animals from various perspectives: agronomy (optimising pasture management and fertilizer use); animal management (e.g. in planning vaccinations); environmental (reduction of livestock impacts); and economic (increasing production and/or reducing costs). Another benefit for the producer is the accurate and reliable record of their production practices and performance, which matches increasing market requirements for traceability. For PLF to have its maximum benefit for the meat industry, there us a need for analysis of data that will add value to the final product. Animal production is an increasingly competitive business and these combinations of sensors and softwares allow farmers to make more intelligent decisions. There is a close relationship between the concept of precision and the question of productive efficiency. PLF requires effective monitoring and control of all production stages. Interest in PLF in Brazil started following failure of traditional veterinary approaches to deal with an outbreak of cattle deaths in Central Brazil (1986 to 2000). The ensuing research programme involved a lot of animal sampling and recording that was not easy for either animals or researchers. At around the same time, Dr Alain Moreau, was looking for rapid and reliable solutions to control the herd of cattle owned by the Kadiwéu Indians. A partnership was initiated and we searched for identification methods using radio frequency (RFID). The first antenna for reading these transponders was developed from aviation equipment. Over the following years, Embrapa developed the first 'Digital Corral', with all necessary hardware and software for the automatic management of cattle. A series of developments included identification transponders, temperature recording (and disease prediction), supporting softwares, and a 'walk-over' weigher system. Developments continue with work on new sensors linked to information management strategies and decision software development.

Ultrasonography as a method to determine body composition in cattle

C. Staub[1], J.-L. Touze[2], C. Mouaze[1], D. Egal[3], J.-M. Giraud[4], E. Briant[1], A. Touchard[1], N. Muller[1], F. Vincent[1], S. Leurent-Colette[5], D. Dozias[5], G. Renand[6], A. De La Torre[4] and J. Agabriel[4]
[1]INRA, UEPAO, Domaine de l'Orfrasière, 37380 Nouzilly, France, [2]INRA, UMRPRC, Site INRA de Tours, 37380 Nouzilly, France, [3]INRA, UE Herbipôle, Domaine de Laqueuille, 63122 Saint-Genes-Champanelle, France, [4]INRA, UMRH, Site INRA de Theix, 63122 Saint-Genes-Champanelle, France, [5]INRA, UEDEP, Domaine de Borculo, 61310 Le Pin-au-Haras, France, [6]INRA, UMRGABI, Domaine de Vilvert, 78530 Jouy-en-Josas, France; christophe.staub@tours.inra.fr

Estimation of body composition in live cattle is difficult and usually performed using subjective approaches (visual and handling evaluation). Today, there is no reference methodological framework for the standardization of the determination of body fatness in cattle by ultrasonography. The objective of this study was to compare ultrasonographic measurements of fat and muscle subcutaneous tissue thickness at the buttock, lumbar, back and rib level with the conventional body condition scoring method. We performed ultrasonography on 160 cows from 5 different breeds (Normande, Holstein, Montbeliarde, Charolaise, Salers) by selecting populations of animals with the widest range of body condition scores possible (from 0.5 to 4.5 on a 0-5 scale). The study was carried out with four experimenters working with two different ultrasound devices and repeating the measurements two times. Low variation between assessors was observed at the lumbar and the rib sites. Measurements performed on dairy cows allow good correlation between body scores and ultrasound indicators at the buttock (fat, r=0.709), lumbar (fat and muscle, r=0.829) and rib (fat and muscle, r=0.707)) sites. The best result was obtained with the measurement of fat thickness + muscle at the lumbar site on the Normandy cows (R=0.856, P<0.001). A combination of these ultrasound indicators provides an accurate and objective assessment of body composition on dairy cows. However, no relationship was found between body scores and ultrasound indicators on beef cows. Ultrasonography is an easy, objective and non-invasive way to determine body composition in dairy cows. This work is offering an opportunity to standardize this technique and use it routinely on a large number of animals in the field.

Management application of activity data monitoring in grazing beef cows

R. Gabrieli
Ministy of Agriculture and Rural development, Extension Service Beef cattle, Hamacabim st. Rishon le Zion, p.o box 28 Beit Dagan, 50250, Israel; ragav@shaham.moag.gov.il

Activity patterns reflect the measure of freedom each cow has, to choose her own behaviour within the herd. Assuming that establishing low energy cost routine is beneficial for each cow, identifying cows that are unable to do so can help minimize social stress and improve production. Previous study showed that high and fluctuated activity graphs characterized low ranking cows. The current study tested effects of group composition on social behaviour. Activity was monitored hourly during 170 days by a wireless pedometric system. 148 cows, (78 heifers and 70 adults) were tagged. 46 of them were kins. 18 of the adult cows were monitored continuously for 150 days prior to the study period and identified as low ranking. Observations were conducted daily and interactions recorded. Interactions were grouped as 'positive' (grooming, proximity) or 'negative' (pushing, isolation). Pedometric data showed high variability in daily activity (145+-80 steps per hour). Lowest variability was found in activity data between 6:00-7:00 am (217+-41 steps per hour) implying the importance of these hours for data analysis. Number of positive interactions had a weak but significant negative correlation to 06:00-0:700 activity (R^2=0.13; P=0.034). Heifers showed higher activity than adult cows (262+-32.6 and 194+-18; P<0.001) respectively, suggesting an advantage to adult cows in a mixed age group. Similarly, the group of previously low ranking cows, once regrouped with heifers showed a marked reduction in activity and graph amplitude, implying this grouping strategy can serve as a management tool to improve welfare and production of low ranking cows. No interactions whatsoever were observed between mothers and daughters, though there were more positive interactions and lower 6:00-7:00 activity within the group of kins (108+-4.4 and 62+-3.7; P=0.001; 253+-12 and 290+-18 steps per hour; P=0.078) kins and none respectively, implying a positive effect of grouping kins. This intriguing nature of kinship indirect effects need to be further studied. Data was analyzed using 'jmp' software.

EU-PLF e-course: a free on-line e-course to understand concepts of precision livestock farming

J. Faure[1,2], M. Bonneau[3], L. Montagne[1,2], Y. Le Cozler[1,2], G. Jalam[2], A. Verbrugge[4] and D. Berckmans[4]
[1]INRA, La Prise, 35590 St Gilles, France, [2]Agrocampus-Ouest, rue de St Brieuc, 35042 Rennes, France, [3]EAAP, IT-00161, Roma, Italy, [4]KU Leuven, 3001, Leuven, Belgium; justine.faure@agrocampus-ouest.fr

Precision Livestock Farming (PLF) technologies in farms offer a lot of new opportunities to improve animal welfare, health and efficiency and reduce the environmental impact of animal production. Continuous monitoring of animals throughout the use of cameras, microphones and sensors, generates data that need to be analysed to transform data into valuable information, helping farmers to take a better care of their animals. The EU-PLF e-course has been created to disseminate the main outputs of the FP7 Smart Farming for Europe project (EU-PLF) which aims to bring PLF tools from the lab to end-users in dairy, pig and poultry farms in order to create value for the farmers and other stakeholders. The animal science team in AGROCAMPUS OUEST used results created by the EU-PLF project and contents of the advanced course provided by CIHEAM to design educational on-line resources. Three learning pathways were defined (animal science students, suppliers of PLF technologies, professionals in animal production) to answer the specific need of the various categories of stakeholders. The e-course contains basic knowledge, including examples and methods required to analyse a study case. Seven units were generated for an equivalent of a two-day long training (16 h) but learners may spend more time depending on their needs. Dynamic schemes, self-assessment and videos of farmers, scientists and professionals make the course user-friendly. The course will be available at the following address: http://plf.agrocampus-ouest.fr. Its free accessibility will allow a wide appropriation by individual learners but it can also be used as a teaching tool for educational and professional trainers. Quick changes and innovations in PLF technologies will require a regular update of the e-course content via continuous interaction with end-users, scientists, teachers, informatics engineers and PLF providers.

Discrepancy between expected and actual benefits of automatic heat detectors in commercial herds

C. Disenhaus[1], C. Allain[2], R. Courties[2,3], Y. Quiniou[1] and N. Bareille[3]
[1]UMR 1348 INRA-Agrocampus-Ouest, 65 rue de Saint-Brieuc, 35000 Rennes, France, [2]IDELE, Monvoisin, 35650 Le Rheu, France, [3]UMR 1300 Oniris-INRA, 102 route de Gachet, 44300 Nantes, France; catherine.disenhaus@agrocampus-ouest.fr

This work aims to get a feedback on the use and the benefits of automatic heat detectors (AHD) in commercial dairy farms in France. The survey covered mainly the motivations to an AHD, the past and current methods of heat detection and the perceived efficiency and impact on labour. A semi-directive survey was led beside 32 commercial dairy farms, chosen to enhance the diversity of the dairy systems. The main characteristics were: (1) location from an intensive plain area (Brittany) to a mountain area (Jura); (2) size from 38 to 194 cows; (3) workers from 1 to 8 people; (4) productivity from 6,000 to 11,800 l/cow/year v) standard (19) vs automatic milking system (AMS) (13). 22 farms own a standalone AHD vs 10 get AHD included in the AMS. The main motivation was to increase detection and/or reproductive performances (22) especially when sexed semen was used (16). Only 6 farms had the main purpose to relieve the work, mainly in Brittany (5). 6 had no expectation as AHD was provided with the AMS. After equipment, heat detection was more or less delegated to AHD. 6 farms used systematic visual heat detection, confirmed alert by visual detection and examination on computer. On the opposite, heat detection was entirely delegated to AHD without any confirmation (4). Intermediary strategies were used by 22 farms. The main perceived benefit was neither the saving of time (5) nor the improvement of the detection rate (6) but the comfort felt at services (21). This comfort was pointed for multiple reasons: confidence in detection at services, better transmission between different workers, serenity in case of absence of the main reproduction manager, fewer constraints of schedule. Finally, 29 farms were fully satisfied with their AHD even though they mainly didn't get what they expected. This suggests that labour comfort has to be included in decision-making tools, which lead to further research to characterise this comfort.

Productivity evaluation using precision livestock farming vs conventional methodologies

S. Ocak[1] and O. Yilmaz[2]
[1]Nigde University, Faculty of Agricultural Sciences and Technologies, Department of Animal Production & Technologies, Nigde University, 51240, Turkey, [2]Adnan Menderes University, Faculty of Agriculture, Deperment of Animal Science, 09010, Aydın, Turkey; sezenocak1@gmail.com

For thousands of years farmers have cultivated crops, grazed and mated animals using trial and error, relying on self accomplished wisdom. Precision Livestock Farming on the other hand is about collecting reliable data and managing variations in the field accurately to make decision that are more resource efficient in the face of rising production costs and growing environmental constraints. By using Radio-Frequency Identification (RFID) to monitor and track animal activity and specifics accurately, a farmer can more effectively tackle the complex issue of managing both production and resource use in an integrated manner by making sustainable management decisions. A typical sheep farm with120 sheep with lambs at foot were divided into two groups of 60 ewes each; the Conventional Farming Group (CF) and the Precision Farming Group (PF). Decisions regarding Group CF were based purely on subjective visual appraisal and the decision for the PF Group was made based on information provided to the manager from a software that collected the relevant data on a daily basis. The objective of the study was to compare and evaluate the total operational costs and profitability between the two systems. Simple procedures such as automatic drafting, walk-over weighing was conducted without labour intensive human intervention, as well as providing a physical association between ewes and their lambs to record lamb survival and assign maternal pedigree without costly mothering-up procedures. Results showed that Precision Farming can reduce on farm labour costs by 43%, reduce feed loss by 24%. With the ability to link abattoir carcass feedback with on-farm data for individual animals the study also showed significant improved benefits along the lamb supply chain by avoiding unnecessary fat production equating to a cost saving of 31%.

Genetics of alternative energy efficiency definitions in grazing lactating dairy cows

A.M. Hurley[1,2], N. Lopez-Villalobos[1], S. McParland[2], E. Lewis[2], E. Kennedy[2], M. O'Donovan[2], J.L. Burke[1] and D.P. Berry[2]
[1]Massey University, Institute of Veterinary, Animal and Biomedical Science, Palmerston North, 4410, New Zealand, [2]Teagasc, Animal & Grassland Research and Innovation Centre, Moorepark, Fermoy, Co. Cork, Ireland; alan.hurley@teagasc.ie

This study aimed to estimate genetic parameters for alternative measures of energy conversion efficiency (ECE), energy balance (EB) and production traits within lactation stages. Individual measurements of net energy intake (NEI) (n=7,675) from 2,445 lactations on 1,245 grazing cows were available. Residual energy intake (REI) was defined as NEI minus predicted energy requirements; residual energy production (REP) was defined as net energy of lactation (NEL) minus predicted energy requirements. Energy conversion efficiency was defined as NEL divided by NEI; EB was defined as the difference between NEI and energy required for both maintenance and lactation. Lactation was divided into three stages (8-90, 91-180, and >180 days in milk [DIM]). Genetic and phenotypic (co)variances were estimated using univariate and bivariate animal repeatability models. The models included the fixed effects of contemporary group (treatment and test-date), parity, DIM, and a random additive genetic effect of animal, a within-lactation stage and across lactation random permanent environmental effect. Heritability for NEI, REI, REP, ECE and EB across lactation stages varied from 0.13 (8-90 DIM) to 0.28 (91-180 DIM), from 0.03 (8-90 DIM) to 0.11 (>180 DIM), from 0.16 (8-90 DIM) to 0.24 (>180 DIM), from 0.04 (8-90 DIM) to 0.18 (>180 DIM), and from 0.04 (8-90 DIM) to 0.10 (91-180 and >180 DIM). A strong genetic association between REI and EB in the present study was particularly noticeable in mid and late lactation (0.91 and 0.87, respectively) when live-weight change was close to zero. This correlation is expected because REI and EB are mathematically equivalent when live-weight change is zero.

Genetic parameters for a multiple-trait linear model conception rate evaluation

K. Muuttoranta[1], A.-M. Tyrisevä[1], E.A. Mäntysaari[1], J. Pösö[2], G.P. Aamand[3], J.-Å. Eriksson[4], U.S. Nielsen[5] and M.H. Lidauer[1]
[1]Natural Resources Institute Finland (Luke), Myllytie 1, FI-31300 Jokioinen, Finland, [2]Faba Co-op, P.O. Box 40, FI-01301 Vantaa, Finland, [3]Nordic Cattle Genetic Evaluation, Agro Food Park 15, 8200 Aarhus N, Denmark, [4]Växa Sweden, Box 288, 75105 Uppsala, Sweden, [5]SEGES Cattle, Agro Food Park 15, 8200 Aarhus N, Denmark; kirsi.muuttoranta@luke.fi

Genetic evaluation for female fertility in Finnish, Danish, and Swedish dairy cows has been upgraded to multiple trait animal model evaluation, where heifer and cow fertility up to third parity are considered as four different traits. In next step, non-return rate will be replaced by conception rate (CR). The aim of this study was to estimate required variance components for the implementation of a CR model using outcomes of all available inseminations. A multiple trait multiple parity sire model with 11 traits was used to estimate simultaneously the variance components for CR, interval from first to last service (IFL) and interval from calving to the first service (ICF) in the different parities. Within each lactation the CR observations were treated as repeated observations. The data consisted of Nordic Red dairy cattle heifers and cows (n=101 315) from Swedish milk recording data. The heritabilities of CR were 0.02 in all parities and ranged from 0.02 to 0.05 for the interval traits in different parities. The within-trait genetic correlations between heifer and cow traits varied from 0.4 to 0.7 in CR and in IFL, and were 0.9 or higher among all the cow traits in the later parities. CR are binomial observations with a variance structure depending on the number of repeated trials service period. Proper modeling of CR observations was found to be crucial to avoid biased estimates for heritability and genetic correlations between traits. Female fertility traits are low-heritable and benefit from multiple trait analysis where all available information is used.

Genetic parameters of body weight and body measurements of Austrian dairy cattle

C. Pfeiffer[1], B. Fuerst-Waltl[1], F. Steininger[2] and C. Egger-Danner[2]
[1]University of Natural Resources and Life Sciences (BOKU), Department of Agricultural Systems, Gregor-Mendel-Strasse 33, 1180 Vienna, Austria, [2]ZuchtData EDV-Dienstleistungen GmbH, Dresdner Strasse 33, 1200 Vienna, Austria; christina.pfeiffer@boku.ac.at

During the last decades the breeding goal mainly focused on increased milk production. As a consequence dairy cows got larger. Concerning resource efficiency, body weight (BW) is of increasing interest, as maintenance costs do have an impact on efficiency. Changes in BW may also be used as predictors for health problems. Currently, the Austrian breeding organizations aim at improving efficiency of dairy cows. BW could be a trait to select for; however, weighing animals reliably is cost and time intensive. Therefore, some indirect traits have to be identified for routine genetic evaluation. The aim of this study was to estimated heritabilities, repeatability and genetic correlations between BW, waist circumference (WC), chest circumference (CC), muscularity (MU) and body condition score (BCS) of Austrian Fleckvieh (FV, dual purpose Simmental), Brown Swiss (BS) and Holstein Friesian (HF). In total, 33,347, 18,552 and 10,766 records of FV, BS and HF, respectively, were used to estimate genetic parameters. A bivariate animal model was fitted using VCE 6.0. Heritabilities were similar for all breeds and in the range of 0.22 for BCS (FV) to 0.49 for CC (BS). Traits showed a moderate repeatability (0.51 for MU (BS) to 0.80 for BW (FV)). Genetic correlations between BW and WC, CC, MU, and BCS were 0.76, 0.85, 0.32 and 0.37 for FV, 0.90, 0.81, 0.51 and 0.56 for BS and 0.84, 0.74, 0.59 and 0.52 for HF, respectively. Among body measurements significant and positive correlations were observed (ranging from 0.23 to 0.89). Especially WC and CC are genetically closely linked to BW and can therefore be used as indirect traits to select for an optimal BW.

Holstein-Friesian vs 3-breed crossbred dairy cows within a low and moderate concentrate input system

C.P. Ferris[1], P.J. Purcell[1], A.W. Gordon[1], T. Larsen[2] and M. Vestergaard[2]
[1]Agri-Food and Biosciences Institute (AFBI), Largepark, Hillsborough, BT26 6DR, United Kingdom, [2]Aarhus University, Department of Animal Science, Blichers Alle 20, 8830 Tjele, Denmark; conrad.ferris@afbini.gov.uk

While the use of '3-breed' rotational crossbreeding programmes are often advocated, there is relatively little information available on the performance of 3-breed crossbred dairy cows. This experiment examined the performance of spring calving Holstein-Friesian (HF, n=34) and Swedish Red × (Jersey/Holstein-Friesian) crossbred (SR×, n=34) cows within low (Low) and Medium (Medium) concentrate input grassland-based systems. The experiment encompassed one full lactation. Cows were offered diets containing grass silage and concentrates (70 : 30 dry matter (DM) ratio, and 40 : 60 DM ratio, for Low and Medium, respectively) until turnout, followed by grazed grass plus either 1.0 or 4.0 kg concentrate/d during the grazing period (Low and Medium, respectively). Full lactation concentrate DM intakes were 739 and 1,844 kg/cow for systems Low and Medium, respectively. There were no significant Genotype × System interactions for any of the feed intake or full lactation milk production data examined. While HF cows had a higher milk yield than SR× cows, the latter produced milk with a higher fat and protein content, thus fat + protein yield was unaffected by genotype. Somatic cell score was highest with the SR× cows. Throughout lactation HF cows were on average 30 kg heavier than SR× cows, while the SR× cows had a higher body condition score. Overall pregnancy rate was unaffected by either genotype or concentrate input system, although there was an increased incidence of ovarian disorders with the HF cows. HF cows had an increased incidence of mastitis, as supported by N-acetyl-β-D-glucosaminidase and β-glucoronidase activity in milk (measured during weeks 1-6 of lactation), although SCC measurements suggested a low level of mastitis in both genotypes at this time. Three-breed crossbred cows had similar fat + protein yields as pure HF cows, while having fewer health problems.

Biological characterization of the estrous cycle in lactating Holstein cows

Y. Montanholi[1], S. Bourgon[1], A. Macdonald[2], P. Park[2], S. Lam[2], M. Kozak[2], B. Potvin[2], K. Colliver[2], L. Haas[2], A. Rocha[2] and S. Miller[2,3]
[1]Dalhousie University, 58 River Road, B2N5E3, Canada, [2]University of Guelph, 50 Stone Road E, N1G2W1, Canada, [3]AgResearch, Private Bag 50034, Mosgiel, 9053, New Zealand; yuri.r.montanholi@gmail.com

Estrus detection in dairy cows is an emerging challenge due to the increasing herd size and to the difficulties in assessing estrus in certain biotypes. The aim was to measure biological parameters throughout the estrus cycle and their potential as a proxy for estrus onset. In total 24 lactating Holstein cows were monitored during the estrus cycle in a tie-stall facility. Cows were estrus synchronized by inserting a controlled internal drug release (CIDR) progesterone on day -7, followed by an injection of prostaglandin on day -1. On day 0 the CIDR was removed and estrus signs were monitored for up to 3 days. The study was carried out with six groups of cows, where 7 cows were submitted to estrous synchronization at each time with the aim to identify 4 cows in estrus. Evaluations included daily feed intake and twice daily analysis of milk including yield, chemical (i.e. lactose and fatty-acids profile) and physical properties, as well as, levels of cortisol and progesterone in the milk and fecal extract. Continuous assessments included contact body surface temperature (ear and neck), standing and lying patterns using an accelerometer, heart rate using a heart girth electrode device and thermographs (i.e. feet, udder and vulva). Progesterone served to indicate estrus onset. Variables were modelled in relation to the estrus cycle as repeated measures and tested for their predictability of estrous onset. Preliminary analysis are indicating variation in milk freezing point in response to estrus onset. Body surface temperatures appears to fluctuate with the estrous cycle, with the highest temperature on the day of estrus. In light of the technological support for enhancing farming practices, this data may serve to stimulate the development of complementary devices to detect estrus in cows to cope with the demand of assessments of estrus due to herd size and diversity in estrus behavior.

The cow for the system: limiting milk yield and body condition loss to ensure reproduction on time

N. Bedere[1], L. Delaby[1], S. Leurent-Colette[2] and C. Disenhaus[1]
[1]INRA, AGROCAMPUS OUEST, PEGASE, Saint-Gilles, 35590, France, [2]INRA, Domaine Experimental du Pin-au-Haras, Exmes, 61310, France; nicolas.bedere@rennes.inra.fr

To study reproductive performances of Holstein and Normande cows managed under a compact calving system (3 months), 2 contrasted grazing-based feeding systems (FS) are used since 2006 in a trial at the INRA farm of Le Pin-au-Haras. The High FS enables cows to produce more milk (MY) while limiting their body condition loss whereas the Low FS limits MY while inducing a large body condition loss throughout lactation (High vs Low: +2,355 kg MY for Holstein, +1,402 kg MY for Normande and +0.40 body condition score in both breeds, P<0.001). It was possible to study the different steps of the reproductive process by combining milk progesterone information (sampled 3 times a week) with intensive oestrous behaviour recording and pregnancy diagnosis (using ultrasonography). Cyclicity of dairy cows was not affected by FS. Normande cows had an earlier resumption of ovarian activity (28.0 vs 31.5 d) and a higher proportion of normal cyclicity patterns than Holstein cows (+22 points, P<0.001). Estruses were more intense in the Low FS than in the High one (+13 points standing to be mounted, P<0.001). For Holstein cows, fertility problems were different in each FS: in the Low FS a higher proportion of inseminations were not-fertilizing or resulted in early embryo mortality (+14 points, P<0.05) whereas in the High FS a higher proportion of inseminations resulted in late embryo mortality (+9 points, P<0.05). Finally, Normande had a higher re-calving rate than Holstein (+19 points, P<0.001), and the High FS tended to be positively associated with re-calving rate (+6 points, P<0.10). The Low FS was very restrictive in nutrients. By limiting their MY to this nutritive constraint Normande do not experience too severe negative energy balance while Holstein tried to maintain the highest MY possible. This resulted in preserved reproduction for Normande cows, and thus animals more robust to compact calving and grazing-based dairy systems.

Application of a milk MIR methane prediction equation to Swiss dairy cattle population data

F. Grandl[1], A. Vanlierde[2], M.-L. Vanrobays[3], C. Grelet[2], F. Dehareng[2], N. Gengler[3], H. Soyeurt[3], M. Kreuzer[4], A. Schwarm[4], A. Münger[5], F. Dohme-Meier[5] and B. Gredler[1]
[1]Qualitas AG, Chamerstr. 56, 6300 Zug, Switzerland, [2]Walloon Agricultural Research Centre, Valorization of Agricultural Products Department, Chée de Namur, 24, 5030 Gembloux, Belgium, [3]University of Liège, Gembloux Agro-Bio Tech, Passage des Déportés, 2, 5030 Gembloux, Belgium, [4]ETH Zürich, Institute of Agricultural Sciences, Universitätstr. 2, 8092 Zürich, Switzerland, [5]Agroscope, Institute for Livestock Sciences ILS, Tioleyre 4, 1725 Posieux, Switzerland; florian.grandl@qualitasag.ch

To make use of genetic variation of methane (CH_4) emissions in dairy cattle, a large number of individual CH_4 emissions across entire populations are needed. As direct measurement of animal CH_4 production is very laborious, alternatives like the prediction of CH_4 emissions are needed. A promising approach for this purpose is the use of milk mid-infrared (MIR) spectral data, which is readily available from routine milk recording analyses. A published MIR prediction equation was applied to spectral data from routine milk recording in Switzerland from August to October 2015. Swiss spectral data has been standardised to reduce spectral variability between instruments. In total, 933,307 spectral data were available from the time period. In a first step only data from Swiss Holstein cows in their 1st to 4th lactation and with less than 306 days in milk were considered in the study, which resulted in 88,974 predicted daily CH_4 production values from 40,082 cows. Semiparametric regression analysis using regression splines was carried out using the R package mgcv. Without any correction for diet, animal, and management influences, increasing CH_4 production from lactation 1 to 4 was found and changing emissions during lactation were identified. Combining the predicted CH_4 emissions with additional herd book data allows a more detailed population-wide screening for potentially high or low emitting animals. The information can be used for management purposes and genetic analyses. In a joint project, the analyses will be extended to other breeds and adding new reference measurements will increase accuracy and robustness of the prediction equation.

Effect of stocking rate and animal genotype on milk production in spring-calving dairy cows

E.L. Coffey[1], S. Fitzgerald[1], K.M. Pierce[2] and B. Horan[1]

[1]Teagasc Moorepark, Fermoy, Co. Cork, Ireland, [2]School of Agriculture and Food Science, UCD, Dublin 4, Ireland; emmalouise.coffey@teagasc.ie

Successful grazing systems are dependent on dairy cows achieving large intakes of high quality grass and efficiently converting the feed to high-value milk solids (MS). The objective of this experiment was to investigate the effect of stocking rate (SR) and genotype (B) on milk production in spring calving grass-based systems. Four hundred and seventeen dairy cows (68 Holstein-Friesian (HF) and 71 Holstein-Friesian × Jersey crossbred (Jx) cows in 2013-2015, respectively) were randomly assigned, within B, to one of three SR treatments based on achieving a similar live weight per hectare (BW/ha) for each B. The three SRs were; low (LSR; 1,200 kg BW/ha), medium (MSR; 1,400 kg BW/ha) and high (HSR; 1,600 kg BW/ha). Milk yield was recorded daily and milk constituents weekly. Least squares means for each SR and B were estimated using linear mixed models. The dependent variables of interest included milk and MS yield per cow and per hectare. Fixed effects in the model included year, calving date, SR, B and parity. Cow was included as a random effect. Low SR achieved the greatest ($P<0.001$) milk and MS yield per cow, while HSR was least. In contrast, HSR achieved the greatest ($P<0.001$) milk and MS yield per hectare while LSR was least. At similar BW/ha, HF cows produced greater ($P<0.001$) milk yield per cow. In contrast and based on a lower daily herbage allowance, JX produced greater ($P<0.001$) MS yields per cow and per hectare at each SR. The absence of an SR×B interaction indicates that at similar BW per hectare, the milk production response to a change in SR was similar for HF and JX. In conclusion, MS production per hectare is maximized in Jx cows at higher SR in Irish grass based systems.

Effect of different regrowth grass silages on dairy cow performance

D. Pang, P. Huhtanen and S. Krizsan

Swedish University of Agricultural Sciences, Department of Agricultural Research for Northern Sweden, Skogsmarksgränd, 90734, Sweden; degong.pang@slu.se

Grass is the most important feed resource for ruminants in the Nordic countries. The objective of this study were to evaluate regrowth grass silages from three different 2-cut systems and a 3-cut system when fed to lactating dairy cows. The regrowth grass was harvested at an early and late stage after an early respectively late first cut, and additionally a third cut silage was harvested after the early first and second cut. Thirty lactating Swedish Red cows averaging 74 days in milk and 32 kg milk/d pretrial were assigned to six blocks in a Latin square design with four 21-d periods. Treatments were the five different regrowth grass silages supplemented with equal amounts of crimped barley and rapeseed meal on dry matter (DM) basis. Time of harvest significantly ($P<0.01$) affected DM intake with the greatest difference of 2.7 kg/d observed between silages harvested after early vs late first cut compared with 0.9 kg/d difference between early and late harvested second cut silages. Yield of energy corrected milk (ECM) was significantly ($P<0.01$) greater for regrowth silages harvested after the early compared to the late first cut (3.6 kg/d), for early harvested second cut grass silage compared to the later harvest (3.8 kg/d), and for third cut compared to second cut silages (2.2 kg/d). Total tract digestibility of nutrients were greater ($P<0.01$) for cows fed regrowth silages harvested after early vs late first cut, for early vs late second cut silages, and for third vs second cut silages. Methane emissions related to ECM yield of 13.7 g/kg was lowest ($P<0.01$) for cows fed third cut silages, not affected ($P=0.29$) by timing of first cut, and greater ($P=0.02$) for late compared to early second cut silages of 15.6 and 14.5 g/kg, respectively. The results from this study support a 3-cut harvesting strategy for improved dairy cow performance and more efficient utilization of nutrient resources.

Variation in social organization in cattle breeds determined by social network analyses
P. Koene
Wageningen University & Research, Animal Welfare, De Elst 1, 6708 WD Wageningen, the Netherlands;
paul.koene@wur.nl

Many domestic animals tend to be large, non-selective feeders occupying open habitats. They are socially organized non-territorial species, typically occurring in relatively large groups in their natural environments. A strong social network is functional in maintaining the group and coping with many challenges. Social relations of individuals determine social stress, social support and even disease transmission and are important for health and welfare. In cattle management the focus is mainly on individual performance and measurements despite the fact that social information is shown to be important in managing introductions and removals, estrus, lameness, cow traffic with respect to feeding (fence) and milking (robot). Social influence can be both positive and negative. Relevant behavioral information includes interactions, rank order, nearest neighbors, farthest neighbors, friends, approaches, avoidances, time spend together and many social behavioural elements, such as vocalizations. In this review, the social organization of cattle in different contexts is measured and interactions and/or associations are analyzed using Social Network Analysis (SNA). Natural social organizations – using SNA – are shown in Heck cattle (fenced and free-range), Galloways and Scottish Highland cattle (free-range), while social organizations of dairy cows, beef bulls and calves are shown in a farm setting. Also, a practical application of the use of SNA is shown in the effect of fencing on heifer social behaviour. The presented information shows a functional and evolutionary interpretation of the social organization of cattle. Using the social information in the field will improve actual management by more precise interpretation of performance measurements and the opportunity to guide the behavior in cows, bulls and calves. Furthermore, assessments of individual and group welfare are possible, facilitating benchmarking. The conclusion is that social networks are important for the health and welfare management of cattle. Furthermore, the automatic measurement of location, movements and/or nearest neighbors for management purposes using GPS is feasible and automatic recording and follow-up management actions will probably become an essential part of cattle management in the near future.

Technical and economic analysis of using ensiled or wrapped grass for young bulls finishing
A. Ferard[1], Y. Carel[2], P. Kardacz[3] and J. Peyrat[4]
[1]ARVALIS Institut du Végétal, Station expérimentale de la Jaillière, 44370 La Chapelle Saint Sauveur, France,
[2]ARVALIS Institut du Végétal, Station expérimentale de Boigneville, 91720 Boigneville, France, [3]ARVALIS Institut du Végétal, Ferme expérimentale professionnelle Lorraine, 55160 St HILAIRE EN WOEVRE, France, [4]ARVALIS Institut du Végétal, Ferme expérimentale des Bordes, 36120 Jeu Les Bois, France; a.ferard@arvalisinstitutduvegetal.fr

To improve the feed autonomy of cattle fattening activity, an increase in the grass use could be an efficient and cheap solution according to her characteristics. 18 diets were tested for fattening young bulls (YB). These trials checked the technical and economic performances of diets including ensiled or wrapped grass in comparison to control diets based on maize forage or cereals without grass. The feeding costs per YB produced, calculated for each diet and for 5 prices contexts: from 2010-2011 to 2014-2015. In cereals diets, introducing grass ad libitum to YB doesn't increase the finishing duration. The introduction of ensiled or wrapped grass at 35% of the total dry matter intake of a maize forage diet increases finishing period by 21 days. Average daily gain is however maintained over 1,400 g/d. In any type of diets, to produce carcasses of 427 kg, grass intake is on average 850 kgDM per YB produced. The grass consumption allows saving 40% of oil-cake and a reduction of 3 to 6% of N and P_2O_5 effluents of a farm with a maize-based fattening diet. On average, feed costs per YB produced is 33€ higher (+6%) in grass+maize based diets and is reduced by 29 € (-7%) in the diets based on grass+cereals compared to controls. Using SIMULBOX software, the simulation at farm scale showed that the change of land use from 1 ha of wheat (sold as grain) + 1 ha of maize forage to 2 ha of a temporary grass pasture leads to maintain EBITDA and farm income. The decrease in costs of oil-cake feed, fertilisers, seeds and phytosanitary products offsets the decrease in cereals sales and the increase in mechanisation and straw bedding due to the longer fattening duration of maize-based diets. In grass farming systems, YB finishing with grass-based diets can answer to environmental rules and social attempts while benefiting to farmers in increasing the economic sustainability of their systems.

Heterogeneity of variance for milk, fat and protein a small cattle population

N. Guzzo, C. Sartori and R. Mantovani
Dept. of Agronomy Food Natural resources Animals and Environment, Viale dell'Universita', 16, 35020 Legnaro (PD), Italy; nadia.guzzo@unipd.it

The aims of this study were to identify possible heterogeneity of variance for productive traits in a small cattle population. The Rendena breed, accounting for about 4,000 registered cows, was used as case study. On the basis of the average milk yield recorded in Rendena farms under functional control in years 2004-2013 (101 farms), 2 productive groups were identified via cluster analysis. The first group (Cl1) consisted in 80 herds with an average milk yield of 4,464 kg/year/cow, the second one (Cl2), in 21 herds with an average milk yield of 6,142 kg/year/cow. A sample of 49 herds with the same numerical proportion as in Cl1 and Cl2 was obtained, and 73,916 test-day records for milk (M), fat (F) and protein (P) yields belonging to 4,582 cows were used to estimate heritability and genetic correlations between groups, i.e. treating M, F and P yield as different traits. The effects of herd test day within lactation (up to the third lactation), gestation, age at parity within lactation, month of parity within lactation, permanent environment, and additive genetic component were taken into account in a repeatability test-day model. The shape of the lactation curve was described by 4th order Legendre polynomials for the fixed effects of the age at parity and month of parity. A Gibbs sampling method was used to estimate (co)variances components. The estimated genetic (σ_a) and permanent environmental variance (σ_{pe}) resulted higher in Cl2 than in Cl1, i.e. σ_a was doubled and σ_{pe} was increased of about 60% for both M, F and P yields. Heritability values obtained were 0.16 vs 0.30 for M, 0.15 vs 0.22 for F and 0.15 vs 0.25 for P in Cl1 and Cl2, respectively. Genetic correlations between productive groups were 0.90 for M and 0.88 for F and P yield. The rank correlation between EBVs of bulls that had daughters in both groups resulted 0.98, suggesting that selection for M, F, and P is not biased although in the presence of a large heterogeneity of variance.

Effects of chemical additive applied into the rumen of dairy cows in form of bolus

R. Loucka, P. Homolka, F. Jancik, P. Kubelkova, Y. Tyrolova, A. Vyborna and J. Schmidova
Institute of Animal Science, Nutrition and feeding of farm animals, Pratelstvi 815, Prague Uhrineves, 104 00, Czech Republic; schmidova.jitka@vuzv.cz

The application of different boluses per os into the dairy cow rumen is becoming almost commonplace in practice. Chemical additives are usually applied short after birth in order to boost the rumen functions and to reduce acidosis. The study was aimed at assessing the efficiency of the bolus applied to Holstein dairy cows at about 190 days of lactation. Both in the first and in the second period (3 weeks) of the experiment, the dairy cows received the same total mix ratio. At the beginning of the 2nd period, boluses containing sodium bicarbonate, niacin, inactivated yeast and choline were applied to the dairy cows of the experimental group (P), next group (K) was without bolus. Each group included 2 dairy cows with eCOW bolus (Devon, Ltd., Great Britain) in the rumen and their rumen pH was recorded in 15-minute intervals. The rumen fluid, taken with the probe, showed significant difference in pH (pH 6.87 vs 6.67) and in lactic acid content (0.52 vs 0.65 mM) between the P and K group. At the beginning of the experimental period, the dairy cows of group P with eCOW boluses showed pH values of 6.16 pH and the dairy cows of group K without boluses showed pH values of 6.09. At the end of the experiment, the pH of group P was 0.11 higher (6.27), while that of group K was only 0.02 higher (6.11). The proportion of time in which the pH values got below 5.8 was 5.7% for group P and 16.8% for group K. The development of pH values in group P with the bolus had 2 tops, which consists with the manufacturer's statement on gradual release of active substances of the agent. The effect of the bolus on milk production and quality parameter improvement was not proved. Dedication: MZE00027014.

Technical note: Use of internal transcribed spacer for ruminal yeast identification in dairy cows

E. Vargas-Bello-Pérez[1], N. Cancino-Padilla[1], C. Ramírez[2] and J. Romero[2]
[1]Pontificia Universidad Católica de Chile, Departamento de Ciencias Animales, Facultad de Agronomía e Ingeniería Forestal, Casilla 306. C.P. 6904411. Santiago, Chile, [2]Instituto de Nutrición y Tecnología de los Alimentos (INTA)., Laboratorio de Biotecnología, Universidad de Chile, Santiago, Chile; evargasb@uc.cl

Molecular techniques are important tools for microbiological studies in different habitats, and the internal transcribed spacer (ITS) has been proved to be useful for analyzing fungal diversity. The aim of this study was to use the ITS region to generate ruminal yeast profile and to identify ruminal yeast. DNA from ruminal digesta of three fistulated non-lactating dairy cows was extracted to amplify the ribosomal ITS region. The profile from the PCR products was visualized and the excised bands from the profile were identified as the genera Millerozyma, Pichia, Rhizomucor and Hyphopichia. There are some aspects that need to be considered when using the ITS approach, for example, the ITS results can be a preliminary step for the quantification of yeast species in rumen. In future studies the use of qPCR with the same primers used in this study could provide more accurate results if the objective is to determine the total concentration of ruminal yeast. Also, findings from this study showed that the ITS approach is not accurate enough to discern among yeast species and therefore this molecular technique is more suitable to study genera. Additionally, it is recommended to use reference sequences from GenBank RefSeq or UNITE repositories, which are databases with validated sequences to avoid misleading or incorrect sequences. Overall, the ITS approach used in this study allowed the identification of ruminal yeast in an easier, faster and economic procedure, and could provide new insights into the diversity and ecology of many different groups of fungi. In this study, the use of ITS allowed the identification of Millerozyma and Hyphopichia (phylum Ascomycota) which have not been previously described in ruminants. This study was funded by a research grant from FONDECYT 11121142 and 1140734.

Effect of high and medium weight gain on puberty onset in early-weaned heifers

M.V.C. Ferraz Junior[1], A.V. Pires[1,2], M.H. Santos[1], R.G. Silva[1], M.V. Biehl[2], G.B. Oliveira[1], D.M. Polizel[1], A.A. Miszura[1], E.M. Moreira[1], A.B. Bertoloni[1], M.F. Westphalen[2] and M.L. Day[3]
[1]University of São Paulo, FMVZ / VNP, Pirassununga, SP, 13635-900, Brazil, [2]University of São Paulo, Esalq, Piracicaba, SP, 13418-900, Brazil, [3]University of Wyoming, Laramie WY, 82071, USA; ferrazmvc@gmail.com

The aim of this study was to evaluate the effects of timing of high weight gain on puberty. Forty beef heifers (½Angus × ½Nellore) were weaning (86±3 d of age and 91±1 kg of BW), and allocated in 4 treatments arranged in 2×2 factorial design. Heifers were submitted to high (H) and median (M) ADG in two period (first period was of 3 to 7 mo, and second was 7 mo to puberty), forming four treatment HH, HM, MH and MM. Diet in first period (PB 19%, NDF 24% and TDN 79%), was composed by hay (20%DM bases), corn (57.5%), SB meal (21%) and mineral (1.5%). Diet in second period (PB 14%, NDF 24% and TDN 78%), was composed by sugar cane bagasse (20%), corn (62.5%), SB meal (16%) and mineral (1.5%). Diets were the same for all treatments, but the amount was controlled to get the ADG wanted. Total dry matter intake (tDMI) was calculated by sum of daily intake from 3 mo to puberty. Weekly, heifers were weighed and puberty status assessed by US (CL). Statistical analysis was done by Proc Mixed (SAS 9.3). The ADG was 0.989±0.06 kg in H group and 0.509±0.03 kg in M group in first period. In second period, the ADG was 0.810±0.03 kg for HH, 0.509±0.03 kg for HM, 1.022±0.04 kg for MH and 0,697±0.03 kg for MM treatment. Heifers submitted to high AGD (330±8 kg) on second period reached puberty heavier (P=0.037) than heifers in median ADG (306±7 kg). There was interaction between factors in age at puberty (P=0.009) and tDMI (P=0.002). Heifers from MM (14.4±0.4 mo) treatment reached puberty older than other treatments (HH=12.1±0.3 mo; HM=12.5±0.2 mo; MH=12.3±0.1 mo). However, heifers in MM (1.1±53 t) treatment had the lowest tDMI. Heifers from HH (1.8±62 t) treatment had the highest total tDMI, but this did not anticipated the age at puberty when to compare with HM and MH that feeding 1.3±29 and 1.3±31 t, respectively. In conclusion, heifers from HH, HM and MH treatments reached puberty in similar age, and MM reached puberty about 2 mo later. However, this treatment had the lowest tDMI.

Trial of construction of the in vitro model to analysis placenta separation at delivery

H. Kamada[1], S. Matoba[1] and H. Takahashi[2]
[1]*NARO Institute of Livestock and Grassland Science, Reproduction, Ikenodai-2, Tsukuba, Ibaraki, 305-0901, Japan,*
[2]*NARO Headquarters, 3-1-1 kannondai, Tsukuba, Ibaraki, 305-8517, Japan; kama8@affrc.go.jp*

The mechanism of fetus release at delivery is well understood; however, there is little information about the process of fetal membrane (FM) release. We discovered that 12-oxoETE is a candidate of signal for FM release after delivery and succeeded in FM release by its injection to PG-induced delivery cows. The present study tried to construct the in vitro multilayer model of uterine epithelial cells and trophoblast cells. Bovine uterine epithelial cells, placenta-derived fibroblast cells, blastocyst and blastocyst-derived trophoblast cells were used. Blastocyst did not attach on the monolayer of uterine epithelial cells in normal culture dish; however, it attached on the monolayer of the mixture of uterine epithelial cells and fibroblasts derived from placenta. Subsequently, the colony of trophoblast cells was formed by pushing aside the epithelial cells, and then generated small floating trophoblastic vesicles. This floating vesicle generated secondary small vesicles several days ago in another culture dish seeded uterine epithelial cells, and then it attached on the monolayer of uterine epithelial cells and formed trophoblastic cell colony; however, the epithelial cells under the colony was lost. On the other hands, blastocyst attached on the monolayer of the mixture of uterine epithelial cells and fibroblasts in the Transwell culture dish and formed trophoblastic cell colony. Addition of oxoETE into the medium lead to the exfoliation of trophoblast cell colony. In this case, cell sheets existed under the exfoliated colony (confirmation of duplex structure). This in vitro model may use to analysis the mechanism of placenta separation after delivery.

Effect of CIDR reused by 4 or 6 time in Nellore heifers submitted to short Fixed time AI protocol

M.H. Santos[1], A.V. Pires[1,2], M.V.C. Ferraz Junior[1], M.V. Biehl[2], R.G. Silva[1], A.A. Miszura[1], G.B. Oliveira[1], D.M. Polizel[1], A.B. Bertoloni[1], E.M. Moreira[1], M.F. Westphalen[2] and M.L. Day[3]
[1]*University of São Paulo, FMVZ/VNP, Pirassununga, SP, 13635-900, Brazil,* [2]*University of São Paulo, Esalq, Piracicaba, SP, 13418-900, Brazil,* [3]*University of Wyoming, Laramie WY, 82071, USA; ferrazmvc@gmail.com*

In Brazil, the most common fixed time AI protocols use progesterone (P_4) device by 8 or 9 d in beef cattle. The device is used 3 time in cows and the 4th time (27 to 34 d) in heifers because Bos indicus heifers are more sensitive to P_4 concentration than Bos taurus. The aim of this study was to evaluate the effectiveness of two CIDR reused on Nellore heifers submitted to short protocol. Ninety two puberty heifers (BW 315±2,1 kg; BCS 2.86±0.01 – scale 1-5; Age 25±1 mo; means ± SE) were used. Fixed time AI protocol was composed by insertion of a device with 1.9 g de P4 (CIDR®) previously used by 21 (CIDR4) or 35 (CIDR6) d and administration of 2.0 mg of estradiol benzoate on day 0. Seven days after, CIDR was removed and were administrate 300 IU of eCG, 0.6 mg of estradiol cipionate and 150 µg of D – cloroprostenol in both treatments. Estrus detection was performed by visual observation and tag (Estrotec™). AI was realized 55 h after CIDR removal, and at this time was performed US in the heifers to access the diameter of ovulatory follicle. Heifers were distributed randomly in treatments. Continuous variables was analyzed by Mixed procedure and binomial variables was analyzed by Glimmix procedure (Both procedure SAS 9.3). At time of CIDR removal, the diameter of ovulatory follicle was not influenced (P=0,83) by CIDR4 (7.43±0.37 mm) or CIDR6 (7.55±0.41 mm). In the same way, diameter of ovulatory follicle was not influenced (P=0.37) by CIDR4 (10.22±0.25 mm) or CIDR6 (9.83±0.35 mm) at the time of AI. As seen in follicle diameter, estrus detection (88.0 and 94.1%; P=0.37) and pregnancy (42.1 and 47.0%; P=0.56) rates were not influenced by CIDR4 or CIDR6, respectively. In conclusion, there was no difference between use of CIDR4 or CIDR6 in Nellore heifers, submitted to 7 d protocol. These data imply that it is possible the use CIDR by up to 42 d, reducing cost with Fixed time AI.

Improvement of udder conformation by culling cows not suitable for milking once-a-day

N. Lopez-Villalobos, J.F. Rocha, R.A. Harris, F. Lembeye, N.W. Sneddon and D.J. Donaghy
Massey University, Private Bag 11222, 4442 Palmerston North, New Zealand; n.lopez-villalobos@massey.ac.nz

In 2013 Massey University established a mixed breed herd milked once-a-day (OAD) consisting of 70 Holstein-Friesian (F), 65 Jersey (J) and 127 F×J crossbred cows. Milk traits and traits other than production (TOP) were recorded for the last three years. This study aimed to evaluate the change in udder traits over time and across breeds after culling cows not suitable for OAD milking. There was a significant increase in udder support (US: 5.94 vs 6.40 P=0.0004) and udder overall (UO: 5.85 vs 6.15, P=0.0218) over the three years. Jersey cows had higher US scores than F×J cows (6.37 vs 5.99, P<0.05) with F cows (6.17) similar to J or F×J cows. Jersey cows had the greatest UO score (6.27), F and F×J were not significantly different (5.91 and 5.79, respectively). Younger cows (<3 years old) had higher US and UO scores than older cows (>3 years old) (6.46 vs 5.90 and 6.23 vs 5.76, respectively). There has been significant improvement in US scores for J (6.13 vs 6.65) and F×J (5.73 vs 6.14) cows between 2013 and 2016 (P<0.02) while F cows trended for an improvement (P=0.0683). Heifers starting the year 2016 had higher US scores than those starting in 2013 (6.10 vs 6.67, P=0.0072), older cows have also improved over the same period (5.77 vs 6.13, P=0.0116). All breeds experienced a decline in US score as they aged (F 6.42 vs 5.92, P=0.0175; J 6.61 vs 6.14, P=0.0080; F×J 6.36 vs 5.63, P<0.0001) but the decline in F×J was the greatest. Udder support was negatively correlated with cow stature (-0.158, P<0.0001) and cow live weight (-0.152, P<0.0001), and positively correlated with front and rear teat position (0.64 and 0.61, P<0.0001). Repeatabilities were between 0.18 and 0.53 for TOP traits, and 0.33 and 0.42 for milk traits. This study has shown that active culling cows not suitable for OAD milking over three years had increased the udder trait score across breeds and ages, with heifers having greater scores than those in the first year.

Pilot study to evaluate the use of sorted semen within a synchronisation programme in beef herd

D. Anderson[1], F.O. Lively[2] and D.E. Lowe[2]
[1]AgriSearch, Large Park, Hillsborough, BT26 6DR, United Kingdom, [2]Agri Food and Bioscience Institute, Agriculture Branch, Large Park, Hillsborough, BT26 6DR, United Kingdom; francis.lively@afbini.gov.uk

Beef farms operate under tight financial margins. Increasing herd output, in terms of additional kilograms of live weight from the same resources, with minimal additional cost is a mechanism for improving financial margins. It is well defined that male progeny perform better than female progeny in terms of live weight gain so maximising the number of male calves born could be a mechanism for improving herd output. Using a maternal sire on the suckler herd for breeding replacement heifer's results in the production of both heifer and bull calves. The bull calves have lower genetic potential for growth and carcass characteristics relative to bull calves sired by terminal bulls. The use of female sorted semen to breed maternal female replacements and male terminal semen to breed male calves could have huge potential at increasing the output from suckler enterprises. A pilot study was undertaken during 2015 on the AFBI suckler herd to evaluate the use of either conventional or sorted (both male and female) semen on herd fertility. The study involved 56 heifers synchronised and AI'd to either conventional (n=22) or female sorted (n=34) semen and 98 cows synchronised and AI'd to either conventional (n=26), female sorted (n=38) or male sorted (n=34) semen. The heifer synchronisation programme involved 4 handlings (Day 0 insertion of progesterone device, Day 7 prostaglandin-F2α injection, Day 8 progesterone device removal and 56 hours later (Day 10) fixed time AI. The cow synchronisation programme involved 5 handlings (Day 0 insertion of progesterone device plus GnRH injection, Day 7 prostaglandin-F2α injection, Day 8 progesterone device removal, Day 10 (36 hours after progesterone device removal) GnRH injection and Day 11 (56 hours after progesterone device removal) fixed timed AI. The conception to first service was 0.59 for the heifers regardless of semen type and was 0.58, 0.61 and 0.68 for the conventional, female sorted and male sorted semen, respectively, when used on the cows, none of which differed significantly (P>0.05). These results suggest that sorted semen may be a viable option for increasing the suckler herd output in the future.

Interactions of alfalfa hay provision and physical form of starter in calves

M.R. Nilieh

Department of Animal Science, Faculty of Agriculture, Arak University, Arak, Iran, Department of Animal Science, Faculty of Agriculture, Arak University, Arak, Iran, Isfahan no.112 mehregan st. 24 st. behdari bus stop modares av., 8149615941, Iran; mohammadrezanilieh@yahoo.com

Forty-four 3-d-old Holstein dairy calves with a meanstarting BW of 39.9 kg (SD 2.3) were used in a 2 × 2 factorial arrangement with the factors dietary forage level (0 or 15% on DM basis) and physical form of starter(mashed vs pellet). Individually housed calveswere randomly assigned (n=11 calves per treatment:6 males and 5 females) to 4 treatments: (1) a mashedstarter (MS) feed with no ALF (MS-NALF); (2) a MS feedwith ALF (MS-ALF); (3) a pellet starter (PS) feed with no ALF (PS-NALF); and (4) a PS feed with ALF (PS-ALF).The milk consumption was similar for all calves throughout the study. The calves were weaned on 60 d but study lasted until 74 days of age. The mixed model in SAS was used for statistical analysis. The results of the present study showed that alfalfa provision did not affect starter intake, however pelleted starter decreased intake compared with mash one (902 and 678 g/d for MS and PS, respectively) (P<0.05). The average daily gain (ADG) was tended to be significant (P=0.09)for interaction of alfalfa provision and starter form (ADG was 541, 585, 545 and 466 g/d for MS-NALF, MS-ALF, PS-NALF and PS-ALF, respectively). The results show that in post weaning time feed efficiency (gain/feed) was tended to be significant (P=0.08). The feed efficiency was the lowest in pellet form with no inclusion of alfalfa forage in starter diet. Regardless the forage inclusion in starter diets, mash form of starter may be recommendable for dairy calves compared to pellet form based on the results of the present study. In conclusion it has been found that provision of alfalfa forage in to pelleted starter rather than mash starter diet may have more positive effect on Holstein dairy calves performance.

Biogas production from livestock waste: potential of bovine e buffalo rumen inocula

A. Chiariotti[1], M. Calì[1], G. Lembo[2] and A. Signorini[2]

[1]Consiglio per la ricerca in agricoltura e l'economia agraria. CREA-PCM, via salaria 31, 00015, Italy, [2]ENEA DTE-BBC-BBE, via Anguillarese 301, 00123, Italy; antonella.chiariotti@entecra.it

Anaerobic digestion of agricultural wastes allows for simultaneous recovery of hydrogen, methane and digestate as fertilizers. Rumen microorganisms are able to digest a variety of biomass, including agricultural residues. To investigate hydrogenic and methanogenic activity, two different batch experiments were carried out at 39 °C using three different inocula: buffalo rumen fluid (BU), bovine rumen fluid (BO) and buffalo sludge (BS). Total Solid (TS), Volatile Solid (VS) were determined at the beginning and at the end of the trials. Methane and Hydrogen yield (ml/gVS), biogas composition (% v/v), volatile fatty acids (g/l) and Lactic acid (LA) were scored. The Hydrogen Experiment (HE) was conducted at pH 5.5 using sterile buffalo sludge and low protein cheese whey (LPCW at 40 and 60%) as substrate and 15% of inoculum. Methane Experiment (ME) was performed at pH 7 using sterile buffalo sludge as substrate and three different inoculum/substrate ratio (1:5; 1:2.5; 1:1 v:v). Until the 7th day of HE no hydrogen was produced, because the archea population consumed H_2 to produce methane. Afterwards hydrogen was produced only in samples containing 40% of LPCW when BU e BO inocula were used. Acetic acid concentration increased (3.15 e 3.95 g/l BU e BO, respectively) and the highest H_2 yield was obtained with BU inoculum (15.4 ml/gVS) with a 30% concentration of H_2 in biogas. When 60% of LPCW was used, pH dropped due to LA rapid increase (9.83 BS, 7.40 BO and 8.00 g/l BU). In ME experiment the highest CH_4 yield (356 ml/gVS) was obtained by BU, followed by BO and BS inocula (324 and 229 ml/gVS, respectively) at inoculum/substrate ratio of 1:1. Moreover, BO and BU gave about 75% of methane in biogas compared to BS (69%). Rumen inocula achieved better hydrogen and methane results, especially buffalo, being suitable for anaerobic digestion processes.

Incorporating meat quality in sheep breeding programmes: potential of non-invasive technologies

N.R. Lambe[1], N. Clelland[1], R. Roehe[1], K.A. McLean[1], J. Gordon[1], D. Evans[2] and L. Bunger[1]
[1]Scotland's Rural College, Animal and Veterinary Science Group, Roslin Institute Building, Easter Bush, Midlothian, EH25 9RG, Scotland, United Kingdom, [2]Wm Morrisons Supermarkets PLC, Hilmore House, Gain Lane, Bradford, BD3 7DL, England, United Kingdom; nicola.lambe@sruc.ac.uk

Genetic selection for sheep meat quality is rare, due to difficulties in measuring these traits within practical breeding programmes. Non-invasive methods to predict lamb meat quality in vivo and post-mortem have been investigated, and results indicate scope for their commercial implementation. UK research has examined relationships between meat quality traits and parameters resulting from x-ray computed tomography (CT) scanning, which is routinely implemented in UK terminal sire breeding. CT is poor at predicting mechanical tenderness in vivo (R^2<15%), across various sheep populations. However, CT can predict intramuscular fat (IMF) with accuracies ranging from 33-70% in live lambs of different breeds, providing a unique in-vivo predictor of meat quality. In vivo CT-predicted IMF is moderately heritable (h^2 0.31) and its genetic control differs from total carcass fat (r_g 0.68), suggesting potential for its improvement within a multi-trait selection index. Novel methodologies have also been tested using CT and visible and near-infrared spectroscopy (NIR) to analyse chilled and vacuum-packed meat samples post-mortem, allowing their return to the food chain. Lamb loin cuts from commercial carcasses (n=303), varying in fat and conformation grades, were scanned by CT and NIR in vacuum packs, then tested for IMF, tenderness and by a trained UK taste panel. CT parameters predicted IMF with moderate accuracy (R^2 36%), but did not accurately predict shear force or sensory traits. NIR predicted IMF in unpackaged meat with moderate accuracy, but predictions were poor in vacuum-packed meat. Samples predicted by CT as having >3% IMF scored significantly higher for sensory eating quality than those predicted as <3% IMF. Work is underway to incorporate CT-predicted IMF, as a proxy for meat quality, into UK breeding programmes for terminal sire sheep, in which IMF is known to be low as a consequence of selection for lean carcasses.

Novel analytical technologies of Quality in the Sheep & Goat Dairy Sector

G. Psathas and O. Tzamaloukas
Cyprus University of Technology, Department of Agricultural Sciences, Biotechnology and Food Science, P.O. Box 50329, 3603 Limassol, Cyprus; gpsathas@cytanet.com.cy

France, Greece, Italy, and Spain the major sheep and goat milk producers in Europe, got the initiative through an interim of 30 years (1985-2015), to organize – through IDF – 7 symposiums aiming at the upgrade of the Dairy Sector. One major interest was the milk quality. Stepwise, the efforts followed cow milk structural development in organizing the testing laboratory, the definition of quality in composition (fat, protein, lactose, total solids), in hygienic (total bacteria count, somatic cell count, antibiotics residues and aflatoxin M1 contamination), adulteration with water and detection of milk species. EU, by Decision 2002/657/EC, defined the rules for the Directive 96/23/EC, enumerated the testing methods, the demands in sampling procedures, the min. limits of performance, the quality control (ISO 17025) and the interpretation of results. The equivalency of results from different laboratories, through ISO 17025, demanded the application of ISO analytical standards. In a period of 5 years these standards have been reformed in order to cover the needs of sheep and goat milk analysis. Novel development in analytical instrumental methods, like FTIR interferometer working with mid-Infrared region of the spectrum from 3-10 μm corresponding to 1000-5,000 cm^{-1}, gives opportunities of collecting data by making measurements for new parameters in complex dairy products, like Urea in milk, FFA profile, casein, pH, etc. Flowing Cytometry for Individual Bacteria Count and Somatic Cells Count is opening the road for applications on testing for raw milk in hygienic control of the sheep and goat dairy sector. New developed screening methods are applied to give rapid results for antibiotics or aflatoxin M1 residues. Milk quality in the Central Testing Laboratory of today, is upgraded and promoted through rapid, trustful and traceable systems of LIMS in payments and herd management. Milk species mixtures can today be analysed in milk and dairy products with screening and reference methods. PDO products need today testing methods for authenticity of animals' breed and geographical region of production.

Managing and improving meat quality through the chain

M. McWhinney
Dunbia, Granville Industrial Estate, Dungannon, Co. Tyrone, BT70 1NJ, Northern Ireland, United Kingdom;
info@dunbiafarmers.com

The quality of red meat is high, but variable. This is an ongoing concern for our key customers, who monitor and feed information back on customer complaints. In order of importance customers are firstly concerned with tenderness and then flavour. This value on tenderness means that which is tender on one occasion but less tender (or even tough) on another occasion discourages repeat purchasing. This variability in quality means that sales are lost against other meats, particularly chicken and pork. The red meat chain must engage and address this problem more effectively. However, managing quality throughout the chain is very difficult as there are so many variables. Influencing factors include genetics, management, diet type, pre-loading transport, transport style, pre-slaughter handling, post slaughter management (electrical stimulation, suspension method, chill regime, maturation) and packaging style. In general if these elements are well managed, the quality of the meat will be at least acceptable. The most pressing need within the industry is for a reliable quality measurement which will allow feedback to farm level and measure the effectiveness of other quality interventions throughout the chain. In reality, this is a long way off, and until such a measurement exists, quality must be managed through a series of controlled inputs rather than outputs. Different techniques are used to manage and improve quality and the author will discuss the use of different techniques within the different supply lines of a large red meat processor.

Effects of electrical stimulation on pH/ temperature decline

R. Law[1], J. Birnie[1], S. Leitch[1], G. Williamson[1], D. Devlin[2], D. Farrell[2] and L. Farmer[2]
[1]Dunbia, Agriculture, Dunbia, Granville Industrial Eatate, Dungannon, BT70 1NJ, United Kingdom, [2]AFBI, Food Research Branch, 18A Newforge Lane, Belfast, Co. Antrim BT95PX, United Kingdom; gwilliamson@dunbia.com

pH fall and temperature decline are two of the main quality determinants in processing muscle to meat. In order to control pH fall, carcasses are electrically stimulated (low voltage electrical stimulation) according to customer specification and in line with the MLC blueprint. However, the current stimulation programme generates variable pH decline in carcasses that can create inconsistency in meat quality. Four electrical stimulation (ES) treatments were evaluated on 120 carcasses to assess their effects on; rate of pH decline, pH/ temperature decline, ultimate pH, drip and cooking loss and meat tenderness. The four treatments assessed were; (1) No ES (negative control), (2) Low voltage ES at start of the line (current practice), (3) Low voltage ES (50 V) at the end of line and (4) medium voltage ES (300 V) at the end of line. Results identified significant differences ($P<0.001$) in muscle pH parameters between carcases stimulated at the start and end of line. Electrical stimulation at the end of the line reduced the variation in muscle pH and resulted in significantly lower ($P<0.001$) ultimate pH values. However, ES at the end of line (both low and medium voltage) greatly affected pH/ temperature declines and increased the incidence of heat shortened carcasses. This detrimental effect was confirmed by Warner Bratzler Shear Force (WBSF) analysis and tenderness evaluation using a MIRINZ tenderometer. Thaw loss and cooking loss values further support the heat shortening effect caused by ES at the end of line. Results show that end of line ES encourages lower and more consistent ultimate pH values. However, when used in conjunction with the present chilling blueprint, the incidence of heat shortening increases. Further consideration should therefore be given to revising post slaughter chilling programmes to complement end of line ES.

Relationship between flavour volatiles and eating quality of lamb

J. McKinnie-Hill[1], A. Aubry[1], T. Hagan[1], L. Farmer[1] and F. Monahan[2]
[1]Agri-Food and Biosciences Institute (AFBI), Food Research Branch, 18a Newforge Lane, Belfast, BT9 5PX,
United Kingdom, [2]University College Dublin, School of Agriculture and Food Science, University College Dublin,
Belfield, Dublin 4, Ireland; linda.farmer@afbini.gov.uk

In lamb production on the island of Ireland there is a trend towards raising ram lambs as opposed to castrated lambs, for welfare reasons and for improved efficiency because rams usually reach slaughter weight faster. However meat from ram lambs is often perceived by the industry to be of lower quality due to anecdotal evidence of a ram taint. It is important to determine whether this perceived taint is real and to identify the cause with a view to developing mechanisms to minimise it. Previous studies have shown that a range of odour compounds may be responsible for the off-odours associated with lamb meat. A joint project between the Agri-Food & Biosciences Institute (AFBI), University College Dublin and Teagasc aims to establish the factors affecting the occurrence of any off-odour in entire male lambs. This paper presents the results of an experiment, conducted at AFBI to determine the effect of forage diet, breed and gender of the lambs on the flavour quality of the meat. Sensory profiling analyses, using a trained panel, on the loins from 272 lambs, reared on a range of forage and concentrate diets are currently being conducted at AFBI. Results from Year 1 identified a number of significant attributes from animals across these treatments. These included a 'greasy' and 'meaty' aroma associated with breed, tenderness related to gender, and also gender:breed:diet interactions. Gas chromatography-olfactometry analysis on selected treatments has also been conducted. Initial analysis has indicated possible compounds that may contribute to a less desirable odour of lamb meat from some treatments.

The type of condensed tannins affected differently growth and meat lipid oxidation of light lambs

S. Lobón, A. Sanz, G. Ripoll, M. Joy and M. Blanco
Centro de Investigación y Tecnología Agroalimentaria de Aragón. Instituto Agroalimentario de Aragón-, Avda.
Montañana 930, 50059 Zaragoza, Spain; slobon@cita-aragon.es

The aim of this study was to assess the effect of the feeding during lactation and the inclusion of condensed tannins (CT) in the concentrate during the fattening period on productive parameters and on meat lipid oxidation of light lambs. At parturition, 63 Rasa Aragonesa ewe-lamb pairs were randomly distributed in 3 treatments. During the lactation period, one group was housed indoors and received a total mixed ration (TMR), a second group was stocked on alfalfa (Medicago sativa) paddocks (Alfalfa) and the third group was stocked on sainfoin (Onobrychis viciifolia) paddocks (Sainfoin). At day 42, the lambs were weaned and half of the lambs of each feeding treatment was fed a commercial concentrate (Control; 11.9 MJ/kg FM, 17.5% crude protein) or a commercial concentrate with 5% of Quebracho (QUE, SYLVAFEED ByPro Q, Spain, with 75% of CT; 11.7 MJ/kg FM, 17.5% crude protein). When lambs reached the target slaughter weight (22-24 kg BW), they were slaughtered. Carcass characteristics were registered and samples of the Longissimus thoracis et lumborum muscle were obtained to study the intramuscular fat content and the lipid oxidation of the meat. The feeding treatment during lactation slightly affected weight gains and BW at slaughter of lambs. Sainfoin treatment tended to improve the BW at slaughter (P=0.09). TMR lambs had the heaviest carcasses and greater dressing percentage and kidney fat depots, Sainfoin lambs intermediate, and Alfalfa lambs the lowest. Regarding lipid oxidation of meat, Sainfoin lambs presented the lower level from 5 d until 14 d of storage (P<0.05). The inclusion of Quebracho in the concentrate tended to improve the weight gains during fattening period and the BW at slaughter (P<0.1) at the same age (P>0.05) but did not affect lipid oxidation.

Effect of ageing and hanging method on flavour precursors in beef

D. Farrell[1], T.D.J. Hagan[1], C. McIlroy[1,2], J. Birnie[3] and L.J. Farmer[1]
[1]Agri-Food and Biosciences Institute, Food Research Branch, Newforge Lane, Belfast BT9 5PX, United Kingdom,
[2]Queen's University Belfast, University Road, Belfast BT7 1NN, United Kingdom, [3]Dunbia, Granville Industrial
Estate, Dungannon, Co Tyrone BT70 1NJ, United Kingdom; david.farrell@afbini.gov.uk

Tenderstretch hanging of carcases and extended ageing are two methods frequently used to improve the tenderness of beef muscles. Tenderstretch beef has increased sarcomere length in many beef muscles, while ageing causes proteolytic breakdown of the muscle proteins. Ageing of beef can also improve flavour and this is probably due to increased formation of flavour precursors by proteolysis to give amino acids and by glycolysis and other pathways to release free sugars. However, it is generally assumed that these precursors are not affected by hanging method. Six steers were slaughtered, and the two sides were Achilles or tenderstretch hung. After two days, the longissimus dorsi were removed, sliced into 25 mm steaks, vacuum packed and aged for 3, 7, 14 and 21 days. Samples were analysed for flavour precursors, (sugars, sugar phosphates, ribonucleotides and free amino acids) using a balanced experimental design. Tenderness was assessed by Warner Bratzler Shear Force. Differences in Warner Bratzler shear force were small, probably due to the small number of animals sampled. However, there were significant differences in flavour precursors. Ageing very highly significantly increased the concentrations of the sugars while reducing the concentration of most sugar phosphates. Significant interactions between hanging method and ageing were observed for fructose ($P<0.01$), fructose-6-phosphate ($P<0.05$) and glucose-6-phosphate ($P<0.05$). Ageing had no effect on the AMP concentration, but there was a significant ($P<0.001$) decrease in IMP and increases in hypoxanthine and inosine. There were significant increases ($P<0.001$) with ageing in the levels of all of the free amino acids studied. There were no interactions between hanging method and ageing but, there was a tendency for all amino acids to form more slowly in the tenderstretch beef. This study confirmed that there are consistent increases in both sugars and amino acids during ageing. Surprisingly, there was evidence that hanging method can influence the concentrations of some flavour precursors.

Sensory benchmarking of commercial sirloin steaks

T.D.J. Hagan, L.H. Majury and L.J. Farmer
Agri-Food and Bioscience Institute, Food Research Branch, 18a Newforge Lane, Belfast, BT9 5PX, United Kingdom; terence.hagan@afbini.gov.uk

The Agri-Food Biosciences Institute was contracted by a major food retailer to both externally benchmark standard sirloin steaks with three of their competitors and to internally benchmark three of their own processors. Standard sirloin steaks were purchased from outlets in Great Britain and Northern Ireland weekly over a three week period and were stored at 4 °C until sensory evaluation. Samples were grilled to an internal temperature of 70 °C using an agreed protocol in a commercial oven. Sensory profiling was conducted on each delivery of sirloin steaks by a trained panel (8 assessors) using 31 sensory attributes generated over two training sessions. These encompassed aroma, appearance, texture, flavour and aftertaste. Each attribute was scored on a 0-100 line scale. Statistical analysis by Analysis of Variance (ANOVA) identified significant differences in a range of both 'positive' attributes such as 'steak' and 'beefy' for both aroma and flavour and also 'negative' attributes such as 'sour', 'farmyard' flavour and 'greasy' aftertaste. The steak from two suppliers generally received lower scores for the positive attributes and higher scores for the negative attributes. Principal Component Analysis (PCA) demonstrated that steak from these suppliers tended to have a 'sour',' bitter' 'farmyard' flavour while steak from a third supplier tended to demonstrate more of the positive attributes such as 'sweet' flavour 'beefy' and 'steak' aroma. The variation for each competitor between each week was minimal. The three internal suppliers showed a difference in product consistency. One processor continually supplied a consistent product with high positive scores for 'steak' and 'beefy' attributes. A second was less consistent but for two of the three weeks the steaks generally received positive attributes. However there was a wide variation over the three weeks for the third processor. These results demonstrate that the consistency of eating quality of beef is variable and that the causes are not always understood.

Dried Distillers Grain byproduct improves the fatty acid profile of ewe milk in Chios sheep

O. Tzamaloukas[1], D. Sparaggis[2], C. Michael[1], C. Constantinou[1], M. Orford[1] and C. Papacrhistoforou[1]
[1]Cyprus University of Technology, Department of Agricultural Sciences, Biotechnology and Food Science, P.O. Box 50329, 3603 Lemesos, Cyprus, [2]Agricultural Research Institute of Cyprus, Animal Production Section, P.O. Box 22016, 1516 Nicosia, Cyprus; ouranios.tzamaloukas@cut.ac.cy

The objective of this research was to examine the effect of Dried Distillers Grains with Solubles (DDGS), as an alternative protein feed, on milk production and milk quality of Chios sheep. A total of 45 ewes of the Chios breed were allocated to three groups of 15 animals each and fed three iso-nitrogenous and iso-energetic mixtures of concentrates with different inclusion rates of DDGS, replacing part of the soybean, as follows: G0 no inclusion, G10 with 10% and G20 with 20% inclusion of DDGS on DM basis. The experiment lasted for 9 weeks and milk samples, collected once weekly, were analysed by gas-chromatography mass-spectrometry for fatty acid (FA) profile. The effect of the DDGS was tested by ANOVA for repeated measurements. Results showed no significant differences between the three groups in milk yield, protein, lactose or solids non-fat content of milk. Fat content, though, was affected by the inclusion of DDGS with lower values observed in milk from the G20 group (total means of 6.0, 5.9 and 5.0 for the D0, D10 and D20 group, respectively, $P<0.001$). FA composition of fat was also affected. Expressed in g/100 g of fat (for the D0, D10 and D20 groups, respectively), milk from ewes fed DDG had less saturated FA (means of 67.5, 65.7 and 60.2, $P<0.001$), more mono-unsaturated (23.5, 25.2 and 28.2, $P<0.001$) and more poly-unsaturated FA (6.0, 6.6, and 8.2, $P<0.001$) compared with the control milk. Health related FA, such as the conjugated linoleic (18:2 c9, t11) and linolenic (C18: 2 c9, c12, c15) acids were also increased ($P<0.01$) throughout the experiment in G10 and G20 groups. Overall, the inclusion of DDGS, as an alternative protein feed, reduced the total fat content of milk, improved the FA profile of milk fat and had no effect on other milk constituents or milk yield.

Investigation into the impact of age and mixing with ewes on eating quality of rams

T.D.J. Hagan[1], J. Birnie[2], D.J. Devlin[1], I.J. Tollerton[1], N.F.S. Gault[1], B.W. Moss[1] and L.J. Farmer[1]
[1]Agri-Food and Bioscience Institute, Food Research Branch1, 18a Newforge Lane Belfast, BT44 0AQ, United Kingdom, [2]Dunbia, Agriculture and Research Group, Granville Industrial Estate, BT701NJ, United Kingdom; terence.hagan@afbini.gov.uk

The adverse impact of entire males on the eating quality of lamb has been reported widely in the literature. Lambs were reared on the same farm, of the same breed and same diet and were born in late March/early April and weaned in August. 'Mixed' ram lambs were held with ewe lambs while the 'unmixed' ram lambs were segregated from females after weaning. They were slaughtered at the same meat plant and whole loins were boned out and frozen 7 days after slaughter. Two hundred consumers participated in the consumer panels Assessors scored the samples for their liking of appearance, liking of aroma and overall liking of the entire sample (lean and fat). However, they were asked to only judge liking of flavour and liking of texture for the lean lamb meat. Experienced taste panellists (n=7) developed a lamb profile encompassing appearance, aroma, flavour, texture and aftertaste of cooked lamb. The data was statistically analysed by REML variance components analysis, internal and external preference mapping and hierarchical cluster analysis. The loins from animals slaughtered at 7.75 months were less acceptable than older and younger lambs with one third deemed 'unsatisfactory' and there were significant differences for the sensory profiling attributes. Consumer liking for meat from rams at 10.75 months was less than that for rams at 5.5 or 6.5 months, sometimes significantly so, again with differences in sensory attributes. Whether rams are reared with ewes or separately had no significant effect on consumer scores and no clear effect on profiling scores. The relationship of consumer liking with specific sensory attributes will be discussed.

Milk protein profile: measure from mid infrared spectra and identification of influence factors

M. Gelé[1], M. Ferrand-Calmels[1], G. Miranda[2], N. Ballot[3], L. Bianchi[3], M. Brochard[1,2] and P. Martin[2]
[1]Institut de l'Elevage, 149 rue de Bercy, 75 595 Paris cedex 12, France, [2]INRA, UMR1313 GABI, Domaine de Vilvert, 78 350 Jouy en Josas, France, [3]CNIEL, 42, rue de Châteaudun, 75 314 Paris cedex 09, France; marine.gele@idele.fr

Protein composition is an important factor of cheese-making properties (milk clotting and cheese yield) that dairy processors and milk producers should take into account to improve their competitiveness. Monitoring protein content and composition requires: (1) the possibility to measure these parameters in routine; and (2) the knowledge of their influence factors and genetic determinism. All conditions have been met concerning protein content, but it is not the case for the protein composition. This study aims to develop a method to measure concentrations of major milk proteins in routine and to identify nutrition and physiology-related factors that influence this composition. Prediction equations were developed to estimate protein profile from mid infrared (MIR) spectra. 280 milk samples were analyzed both by MIR and liquid chromatography coupled with mass spectrometry. Total casein, αs1-casein, αs2-casein, β-casein, κ-casein and β-lactoglobulin contents in milk can be estimated by using MIR spectrometry. Those equations were applied on 293,525 spectra from Holstein, Montbeliarde and Normande cows milk samples (PhénoFinlait program). Statistical analysis performed with SAS software highlighted some findings concerning the main factors of variance of milk protein composition. Proteins mostly affected by non-genetic factors are β-casein and β-lactoglobulin. The lactation stage of the cows impacts the major protein proportions, particularly during the first month. This study pointed out the very limited influence of the nutrition-related factors on milk protein composition, which was more sensitive to the physiological status of the cows. This program received financial support from ANR, Apis-Gène, CASDAR, CNIEL, FranceAgriMer, France Génétique Elevage and Ministry of Agriculture.

Effect of genetic group on sheep milk composition and lamb development

C. Salgado[1], J.C.A. Hernandez[1], A.A. Lugo Galindo[2], A. Lizarazo[3], R. Angulo[3], N. Suarez[4] and A. Ramirez[4]
[1]UNAM, FMVZ, 04510, Mexico, [2]Fundacion Hidalgo Produce A.C., Hidalgo, 42186, Mexico, [3]UNAM, FMVZ, Centro de Enseñanza Práctica e Investigación en Producción y Salud Animal, 04510, Mexico, [4]UAM-X, Depto de Producción Agrícola y Animal, DF, 14387, Mexico; mvzjuancarlos@comunidad.unam.mx

Quality of sheep milk is determinate by their capacity to be transformed in high quality dairy products; hence, manipulation of genetic and management factors have been carried out to improve sheep milk composition. A total of 116 milk samples of ½ East Friesian × ½ Pelibuey (EF×PL) and Katahdin (KT) ewes were used to analyze the effect of genetic group (Gen) on milk composition; also, the growing of lambs was analyzed. The trial was carried out during the first 60 days of post-lambing; milk samples were weekly obtained and analyzed to fat (F), protein (P), lactose (L) and solids not-fat (SNF)). EF×PL ewes remained with their lambs the first 30 days post-lambing; after 31 days post-lambing the ewes were separated from their lambs during the evening, and the ewes were milked once daily, lambs were weaned at day 60. KT ewes were not milked and their lambs were kept with them for the first 60 days post-lambing. Lambs were weighted at lambing (LWL), 30 days post lambing (LW30) and at weaning (LWW). Repeat measures analysis was used to analyze the effect of Gen, week of measure (WM) and their interaction on LWL, LW30, LWW, F, P, L and SNF. F content varied in relation to WM (P<0.01), decreased of 5.1 to 2.7% from the third to fourth week. WM×Gen interaction was significant (P=0.01) to P content, largest difference between Gen was at the first WM (EF×PL 4.5 vs KT 3.8%). EF×PL ewes showed higher levels of L content (4.7 vs 4.5%). There was a trend in the interaction (P–0.09) WM×Gen to SNF with the higher differences in the first WM (EF×PL, 10.4 vs KT, 8.7%). There were no differences (P>0.05) between EF×PL and KT sheep to LWL (4.2 vs 4.4), LW30 (10.03 vs 10.06) and LWW (14.2 vs 15.8 kg, respectively). The higher levels of L content of EF×PL could be associated with a better ability to milk synthesis; also, the milking not affected the lamb development. A deeper analysis must be carry out to explain the milk fat depression that shown both Gen, but can be associate with the feed management.

In vivo estimation of lamb carcass quality using ultrasound measurements

E. Ghita, R. Pelmus, C. Lazar, M.C. Rotar, M. Habeanu, M. Gras and M. Ropota
National Research Development Institute for Animal Biology and Nutrition, Animal Biology, 1 Calea Bucuresti,
077015, Balotesti, Ilfov, Romania; elena.ghita@ibna.ro

With a stock of 8.8 million sheep and with large pasture areas, Romania has an important potential for sheep meat production. Since much of the stock of sheep consists of local breeds, with low meat production, the purpose of the research was to evaluate in vivo the carcass quality of the lambs from two local breeds: Palas Merino (PM – dual purpose breed, for fine wool and meat) and Palas Meat Breed (PMB – breed specialised in meat production, recently homologated). We used ultrasound measurements on live animals to determine the properties of Longissimus Dorsi muscle (depth, area and perimeter) and the thickness of the subcutaneous fat layer, on a number of 45 PM lambs aged 165 days and 32.29 kg average weight, and on 45 PMB lambs, aged 210 days and 42.38 kg average body weight. The ultrasound measurements were performed with an Echo blaster 64 using LV 7.5 65/64 probe, supplied by Telemed ultrasound medical systems. The ultrasound images were recorded using Echo Wave II software version 1.32/2009. The first measurement point was 5 cm from the spine, at the 12th rib; the second measuring point was between 3rd and 4th lumbar vertebrae. The average values of the subcutaneous fat layer thickness, and of LD muscle depth, area and perimeter were 2.21 mm; 22.73 mm; 10.71 cm^2 and 135.53 mm for PM lambs, and of 2.45 mm; 26.59 mm; 11.89 cm^2 and 139.26 mm, for PMB lambs. There were significant phenotypic correlations between the body weight and the properties of the LD muscle (depth, area and perimeter), the correlation coefficients ranging between 0.44 and 0.66. The results revealed very significant differences (P<0.001) between the lambs from the two breeds. The LD muscle characteristics from these two breeds have been further compared with another local breed, with the same body weight, and the comparison has shown that PMB lambs had the best performance, followed by PM lambs. This conclusion supports the possibility of using Palas Meat Breed rams to improve the meat production of other local breeds.

Comparison of meat quality of lambs raised on two types of pastures on Western Patagonia

L. Bravo-Lamas[1,2], H.F. Elizalde[3], I. Subiabre[1], B. Martinez[1] and R. Morales[1]
[1]Instituto de Investigaciones Agropecuaris, INIA Remehue, P.O. Box 24-0, Osorno, Chile, [2]University of the
Basque Country (UPV /EHU), Department of Pharmacy and Food Sciences, Lascaray Research Centre, 01006
Vitoria-Gasteiz, Spain, [3]Instituto de Investigaciones Agropecuarias, INIA Tamel Aike, P.O. Box 296, Coyhaique,
Chile; helizald@inia.cl

Lamb production in Western Patagonia (Chile) is mainly carried out on direct grazing of permanent pastures. Legumes are used for increasing the carcass weight of lambs. However, information of the meat quality of lamb produced under these conditions is unknown. The aim of this study was to compare meat quality of Corriedale lambs raised with two fattening strategies. Treatments were: (1) lambs grazing on permanent pasture, n=12; (2) lambs grazing on a legume based pasture of alfalfa (Medicago sativa), n=12. Permanent improved pasture is based on ryegrass, cocksfoot and white clover. The initial live weight was 29.1 kg and lambs were maintained in the experiment for 77 d when they were slaughtered. Instrumental color, shear force, pH and compositional chemical analysis were measured in longissimus dorsi muscle. The instrumental color in the subcutaneous fat of chops was also measured. No differences (P>0.05) were observed between permanent pasture and alfalfa for shear force, pH and lean meat color, although subcutaneous color from permanent pasture showed higher L* values than those produced on alfalfa (P≤0.05). Regarding meat composition, meats from permanent pasture had more protein (alfalfa 82.1% vs permanent pasture 86.2%; P≤0.01) but less intramuscular fat (alfalfa 13.9% vs permanent pasture 8.48%; P≤0.001) than those produced on alfalfa. Thus, dry matter value was higher for alfalfa meat samples (alfalfa 27.5% vs permanent pasture 25.7%; P≤0.01). These data were related with the higher significate carcass weight (alfalfa 18.6 kg vs naturalized pasture 14.0 kg; P≤0.01). Results shown that meat quality from lambs produced on permanent pastures were leaner whereas, meat produced in alfalfa pastures allowed an increase of intramuscular fat, conferring desirable sensorial attributes. Both types of pastures get an interesting meat quality attributes, characteristics that made them suitable for the export market.

A preliminary investigation on the variation of milk composition during lactation of ten pure bred A

A. Mohammed, A. Rolingson, A. Edwards, A. Khan and A. Griffith
The University of Trinidad and Tobago, Biosciences Agriculture and Food Technologies, Caroni North Bank Road,
01 Centeno Arima, Trinidad and Tobago; aphzal.mohammed@utt.edu.tt

The quality of processed goat milk products depends on its main components protein, fat and lactose on milk composition which is influenced by factors such as season, farming systems and farming methods. This study represents a first report on milk composition of lactating Anglo Nubian females of early weaned kids of the 'up the hills' farm. Milk samples were analysed over a seven week period from 10 pure bred Anglo Nubian goats without evidence of subclinical mastitis, intensively reared on an 18% Crude protein diet and at varying stages of lactation. Composition of the milk from individual goats were analysed using a lactoscan SP (Nova Zagora, 8900) and the pooled data subdivided into early (0-90 d; n=42), mid (90-180 d; n=23) and late lactation (>180 d; n=5). A one way ANOVA revealed lactose, Solids Non Fat, protein and salt composition was significantly higher (P<0.05) in early lactation compared with values in mid or late lactation. However, neither fat, temperature, salt concentrations nor the freezing point of the milk (P>0.05) varied with stage of lactation. Overall means for fat, lactose, Solids Non Fat, protein were 4.9, 4.8, 8.7 and 3.2%, respectively. Average daily milk yields from a single milking of the does were between 1.3 to 1.5 litres. Milk composition from these pure bred Anglo Nubian goats would be suitable for the production of hard and soft cheeses and other dairy products. The low average milk yields reported need to be improved through a better system of feeding and management. Studies are under way to determine nutritional input pertaining to individual milk yields and milk composition of pure bred Anglo Nubian, Saanen and Alpine breeds of lactating goats from Trinidad.

Effects of the rearing system, sex and age on the fatty acid profile of three muscles in Chios breed

E. Tsiplakou[1], G. Papadomichelakis[1], D. Sparaggis[2], K. Sotirakoglou[1], M. Georgiadou[1] and G. Zervas[1]
[1]Agricultural University of Athens, Nutritional Physiology and Feeding, Iera Odos 75, GR11855, Greece,
[2]Agricultural Research Institute, Nicosia, 22016, 1516, Cyprus; gzervas@aua.gr

The effects of the rearing system, sex and age on the intramuscular fatty acid (FA) profile was studied using samples of the Semimembranosus proprius (SP), Longissimus dorsii (LD) and Triceps brachii (TB) muscles of 40 lambs of Chios breed. Two groups of 10 male and 10 female lambs, balanced for body weight, were used in this experiment: lambs nursed by their ewes (n=20) and lambs fed a commercial milk replacer ad libitum (n=20). Five male and five female suckling lambs from each group were weighted prior to slaughter at about 35 days of age. The remaining lambs were weaned at 35 days and then slaughtered at the age of about 98 days. During that period, solid feed in pelleted form was offered ad libitum and barley hay was restricted to 0.1 kg per animal per day. Body weight and dressing percentage at the age of 98 days were similar for both groups. Analysis of variance was performed by a General Linear Model (GLM) procedure for the analysis of the fatty acids composition. The model included fixed effects ascribed to sex, rearing system, age and their interactions. The results confirm that the intramuscular fatty acid composition of young lambs depends to a great degree on the fatty acid composition of the milk source (maternal or milk replacer). However, there are variations in the FA composition among the three muscles examined (SP, LD and TB). Lamb sex was found to have little effect on the FA profile of muscles while, age of slaughter has a significant effect on the proportions of the majority of the fatty acids in all three muscle tissues examined. Lambs slaughtered at weaning age have a more desirable FA profile in their muscles from the perspective of human nutrition, than lambs slaughtered at an older age. It can be concluded that type of rearing and age of slaughter have a considerable effect on the fatty acid composition of lamb meat, but not evenly in all the three muscles.

Kappa casein gene polymorphism in Iranian local Goats breed

N. Hedayat Evrigh and V. Karimi Ouri
University of Mohaghegh Ardabili, animal science, Daneshghah, 5619911367, Iran; hedayatuma@gmail.com

The kappa casein fraction of goat milk plays a crucial role in the formation, stabilization and aggregation of casein micelles and thus affects on milk compositions. In the recent years, due to relationship between milk protein and economic traits in livestock animals, investigation of genetic of milk protein was interested. In this study blood samples were taken from 100 Khalkhali Iranian goat for screen SNP of kappa casein gen in Khalkhali goat. DNA extracted using mammals blood kit and amplify using specific primer. The PCR products of primers (405 bp) were genotype using PCR-SSCP methods. The genotype determined using different pattern and were sequence with sanger sequencing methods. For comparison with other livestock's animals, the sequences obtained from the NCBI database and analyzed using bioinformatics software. The results of Analysis showed two different pattern and the sequencing was showed single nucleotide polymorphism in 248 nt position that changed C to T in exon 4 of kappa casein gene. This substitution caused changing in amino acids and make changes Serine to Proline with the codons ACC to ATG respectively. The result of biodiversity analysis in Khalkhali goats showed low diversity. Sequence analysis of livestock species showed 41 polymorphic loci with 6 and 35 singleton and parsimony polymorphisms respectively, in overall 405 sequence. Using this mutations we identified 21 haplotype. Haplotype diversity in overall was 0.83 with 1.5899 Tajima D that was not significant. The funding of network analysis of haplotypes showed that sheep and goat located in one haplogroup and bus Indicus and Bos Taurus were in another groups but buffalo was in different haplogroup between two groups. This study indicated that the κ-Casein genetic variants may be used as a genetic aid through increasing the frequency of desired genotypes to improve the quality of production of this herd.

Effect of STAT5A/Eco81I gene polymorphism on milk yield in Garganica goat breed

M. Selvaggi[1], A.G. D'Alessandro[2], F. Pinto[1], G. Martemucci[2] and C. Dario[1]
[1]University of Bari, Department of DEto, strada prov. le per Casamassima km 3, 70010 Valenzano, Italy; [2]University of Bari, Department of Agro-Environmental and Territorial Sciences, Via Amendola, 70125 Bari, Italy; maria.selvaggi@uniba.it

Signal transducers and activators of transcription (STATs) are latent cytoplasmic transcription factors. They belong to a 7-member family (STAT1-4, STAT5A, STAT5B, and STAT6) playing a prominent role as mediators of many hormone and cytokines within target cells. STAT5 is known as the main mediator of growth hormone action on target genes; it plays a key-role as mediator of prolactin signalling and can activate transcription of milk protein genes in response to prolactin. A single STAT5 gene was identified in sheep, subsequently two forms, STAT5 and STAT5B have been identified. The STAT5A gene has been mapped to chromosome 19 in goat. Many polymorphisms at STAT5A gene has been investigated in bovine species as candidate markers affecting quantitative traits. To date, only few SNPs at STAT5A gene were detected in goat species and few studies have been conducted. The aims of the present study were to investigate the STAT5A/Eco81I polymorphism in a sample of Garganica goats, an Italian autochthonous breed, and to investigate the possible association between this polymorphism and milk production. Most of Italian goats breeds are represented by small populations with peculiar characteristics and variable productive traits. Garganica breed showed an exceptional ability to adapt to difficult environments and to use poor pasture. This breed originates in the Gargano promontory as a result of a crossbreeding with west European goats. Nowadays, the Garganica goat breed is included in the list of Italian endangered breeds drawn up by the Italian Department for Environment, Food, and Rural Affairs. In the investigated Garganica population, only two out of three possible genotypes were identified, moreover the association between milk yield and different genotypes was investigated.

Study of the external morphometry and carcass traits in rabbit of local population

A. Berbar[1], K. Lounaouci[1], N. Benali[2], Z. Boumahdi[1] and R. Belabbas[1]
[1]Biotechnology Laboratory of Animal Reproduction (LBRA), Institute of Veterinary Sciences, University of Saad Dahleb Blida I, 09000, Algeria, [2]Health and Animal Production, National Veterinary School, Algeria, 16200, ALGIERS, Algeria; berbarali@yahoo.fr

This work was undertaken in order to identify the rabbits of local Algerian population by the study of its external morphometry and carcass characteristics. In total 220 rabbits from both sex and at different ages (from 1 to 6 months) were used for the morphometry study. 12 morphological traits were measured and used to compare between young and adult in one hand and male and female in other hand. 60 male and female (30 per group) were sacrificed at the age of 6 months in order to study the traits of the carcass. The rabbit of local Algerian population is characterized by several colors of phenotypes. The weight of the rabbit at the adult age was 2,622±142 g allowing the classification of this population in category of small breed. The majority of the morphometric parameters measured in this study were influenced significantly by the age of the animal. There were no differences between males and females for the majority of the traits except for, the distance between the eyes and the length of the head (+15% for the male; P<0,05). At the slaughtering, the females were slightly heavier than male (+8%; P<0,05) without any effect on the carcass yield. However, they presented more adiposity than males (+10% for perirenal and inter scapular fat). This study is considered as preliminary research in order to characterize the rabbit of local Algerian population and more investigations by molecular methods seem necessary.

Lipid mobilization, immune system and vitamin E in transition cows: an old story revisited

I. Politis and G. Theodorou
Agricultural University of Athens, Animal Science and Aquaculture, 75 Iera Odos, 11855, Athens, Greece; i.politis@aua.gr

The multitude of disorders that dairy cows face during the transition to lactation can be classified in three main categories; disorders related to energy metabolism (ketosis, fatty liver, acidosis), disorders related to mineral metabolism (milk fever) and disorders related directly or indirectly to impaired immune function (mastitis, retained placenta). Among the many physiological changes during the transition period, perhaps the most crucial, is an increase in the concentration of plasma non-esterified fatty acids (NEFA) which can be partially attributed to an energy deficit, known as negative energy balance. An observational study performed in 4 commercial farms located in southeastern Europe found a strong negative correlation between blood concentrations of NEFA and α-tocopherol but not between β-hydroxybutyrate and α-tocopherol. An in vitro experiment investigated whether a cross-talk between adipocytes and monocytes/neutrophils is mediated by fatty acids. Results indicated that all fatty acids tested (myristic, palmitic, palmitoleic, stearic and oleic) up-regulated expression of various pro-inflammatory genes (urokinase plasminogen activator, u-PA; u-PA receptor, u-PAR; intercellular adhesion molecule 1 and inducible NO synthase) by both monocytes and neutrophils. The longer the carbon chain, the most potent was the effect. Pretreatment of cells with α-tocopherol did not modify the effect of fatty acids on expression of pro-inflammatory genes. Thus, fatty acids compromise the function of two immunocompetent cells. Vitamin E is unable to block this effect. This finding should not be interpreted as vitamin E having no role at all. Data will be presented from many studies showing that vitamin E supplementation of dairy cows during the periparturient period improves the antioxidant status, restores partially proper immune function and affects in a positive manner milk quality.

Lactic acid treatment and P levels in the diet impact microbial dynamics in lactating dairy cows
R.M. Petri, B.U. Metzler-Zebeli and Q. Zebeli
Vetmeduni, Animal Nutrition and Functional Plant Compounds, Veterinärplatz 1, 1210, Vienna, Austria;
renee.petri@vetmeduni.ac.at

Recent in vitro data indicates that there are beneficial effects of treating grain with lactic acid (LA) on specific ruminal microbes, fermentation profiles and fiber degradation in low phosphorus (P) diets. This study used lactating dairy cows (n=18) to access the impact of LA treated diet with either non-supplemented (low) or supplemented (high) inorganic P content on fecal microbial populations compared to an LA untreated control diet. Extracted fecal microbial DNA was sequenced using the Illumina MiSeq platform. Sequenced DNA was then stitched, cleaned, analyzed using QIIME software. Significant OTUs were blasted using the NCBI database. A total of 2,268 OTUs were identified as representing a minimum of 1% of the total population. The treatment of diet with LA significantly impacted 10 OTUs whereas the variation in P significantly impacted only three OTUs from the fecal microbiota. Of the OTUs impacted by the LA treatment, 6 OTUs had an increased abundance in the control diet. From those 6 OTUs, two showed an additional significant effect of P. These two OTUs were classified as being members of the RC4-4 genus and the Ruminococcaceae family respectfully. Comparison of these OTUs to the NCBI database gave identification as Peptococcus niger and Defluviitalea saccharophila with 89 and 88% respective similarity. A total of 15 of the 18 OTUs impacted by LA treatment or P level belong to one of the following Clostridium clusters: i, iii, iv, xi, or xv. These results show a large representation of fecal bacteria from the Ruminococcaceae family. Additional qPCR analysis, performed using Ruminococcus genus primers, showed no significant differences based on LA treatment or P levels in the diet. The lack of qPCR response and the low percent identification provided by curated databases indicates that there are many members from the ruminococci group which remain unknown. By increasing organic P availability from feedstuffs and decreasing supplemented inorganic P without adversely impacting the rumen microbial community, it may be possible to reduce the amount of P excretion in commercial dairy production.

Recording of feed efficiency under on-farm conditions
F. Steininger[1], M. Ledinek[2], L. Gruber[3], B. Fuerst-Waltl[2], K. Zottl[4] and C. Egger-Danner[1]
[1]ZuchtData EDV-Dienstleistungen GmbH, Dresdner Str. 89/19, 1200 Vienna, Austria, [2]University of Natural Resources and Life Sciences, Department of Sustainable Agricultural Systems, Division of Livestock Sciences, Gregor-Mendel-Str. 33, 1180 Vienna, Austria, [3]Agricultural Research and Education Centre Raumberg-Gumpenstein, Raumberg 38, 8952 Irdning, Austria, [4]LKV Austria Qualitätsmanagement GmbH, Dresdner Str. 89/19, 1200 Vienna, Austria; steininger@zuchtdata.at

During recent years discussions about greenhouse gas emissions, higher costs for concentrates and decreasing producer milk prices resulted in increased interest in feed efficiency in dairy production. Calculation of measures for feed efficiency is challenging as it is influenced by several factors. Several possibilities to use feed efficiency data for a genetic evaluation exist: (1) to record the feed intake precisely for a small group of animals on station; (2) to estimate feed intake of many animals on farm based on animal and diet information with impact on the feed intake or (3) to work with further auxiliary traits like mid-infrared-spectra. Within the Austrian project 'Efficient Cow' composition of diets, feed analyses and amounts of concentrates based on automatic transponder feeders were recorded for approximately 5,400 cows, i.e. 3,100 Fleckvieh (FL – dual purpose Simmental), 1,300 Brown Swiss (BS), 1,000 Holstein (HF), kept on 167 farms. Based on parameters breed, live weight, milk yield, number of lactation, days in milk, proportion of concentrates and net energy content of forage the dry mater intake (DMI) for every single cow was estimated using the model of Gruber et al. To make the results comparable between animals a standardisation for the 100th day in milk and without pregnancy was done. The means for standardized DMI were 20.0 (FL), 19.4 (BS), and 21.3 (HF) kg dry matter with a standard deviation of 2.9 (FL), 2.9 (BS), 3.1 (HF). On average, the cows produced 0.23 (FL), 0.22 (BS), and 0.24 (HF) kg milk per MJ NEL intake on their 100th day in milk. The proportion of milk produced out of forage (grass, hay, silage or maize silage) was in the range of 20 to nearly 100% with a mean of 60.1% (FL), 63.6% (BS), and 59.9% (HF).

Growth stage and ensiling effects on ruminal degradability of whole-crop oats

S. Stirling[1], C. Cajarville[1], J.L. Repetto[1], J.E. Díaz[2], M. Sosa[1], J. Benkendorf[1] and M. Pla[3]
[1]Facultad de Veterinaria, Universidad de la República, Nutrición Animal y Bovinos, Ruta 1, km 42.5, San José,
Uruguay, [2]Instituto Nacional de Investigación Agropecuaria, Mejoramiento Genético de Cultivos, La Estanzuela,
Colonia, Uruguay, [3]Instituto Nacional de Investigación Agropecuaria, Unidad de Lechería, La Estanzuela,
Colonia, Uruguay; sofiastirlings@gmail.com

This study evaluated the effects of growth stage and ensiling of whole-crop oats on in situ ruminal dry matter (DM) degradability. Grain-type oats were ensiled at boot, heading, water ripe, early milk, early dough and grain ripe stages of growth (i.e.: growth stage (GS) 45, 59, 69, 73, 83, and 91 respectively). Field plots were arranged in a randomized complete design with 3 replicates per growth stage. DM disappearance of fresh and ensiled forage was measured in situ for 3, 6, 12, 24, 48, 72 and 96 h (3 cows and 2 incubation series). Data was fitted to the model $d = a + b (1 - e^{-kd×t})$ ('d': material lost from the bag at time 't'; 'a': soluble fraction; 'b': degradable fraction; 'kd': fractional degradation rate of 'b' (/h); 'u': undegradable fraction). Effective degradability (ED) was estimated as $ED = a + (b×kd)/ (kd+kp)$, assuming a rumen particle outflow rate (kp) of 0.03 (ED03) and 0.06 /h (ED06). Effects of growth stage, ensiling and their interaction were analysed using the SAS MIXED procedure, performing linear (L) and quadratic (Q) regressions for growth stage. With increasing maturity, fraction 'a' decreased from GS 45 to 69, increasing thereafter (Q, $P<0.05$); fraction 'b' declined (L, $P<0.01$), while fraction 'c' increased (L, $P<0.01$; Q, $P=0.03$). Ruminal 'kd' decreased with maturity for fresh (L, $P<0.01$; Q, $P<0.01$) and ensiled (L, $P<0.01$) forage. ED03/06 decreased sharply (L, $P<0.01$; Q, $P<0.01$) from GS 45 to 69, remaining stable thereafter. Fraction 'b' increased due to the ensiling process (0.38 vs 0.41; $P<0.01$), while fraction 'c' (0.208 vs 0.188; $P=0.02$) and ruminal 'kd' (0.029 vs 0.023; $P<0.01$) decreased. ED03/06 was lower after ensiling ($P<0.01$) at GS 45, 59 and 91. Growth stage had a negative effect on DM degradability kinetics of whole-crop oats from boot to grain ripe stages, while ensiling increased the degradable fraction but decreased degradation rate, decreasing ED at some stages of growth.

Oxidative stability of meat from lambs fed silage mixtures of timothy grass, red clover and sainfoin

G. Luciano[1], A. Priolo[2], B. Valenti[1], S. Mattioli[1], M. Pauselli[1], G. Copani[3], C. Ginane[3] and V. Niderkorn[3]
[1]University of Perugia, Dipartimento DSA3, Borgo XX Giugno 74, 06123, Perugia, Italy, [2]University of Catania,
Dipartimento Di3A, Via Valdisavoia 5, 95123, Catania, Italy, [3]INRA, UMR1213 Herbivores, Site de Theix, 63122,
Saint-Genès-Champanelle, France; giuseppe.luciano@unipg.it

Red clover (RC) and sainfoin (SF) may be conveniently used in silage mixtures with grass. These legumes also contain compounds able to affect the quality of meat from ruminants, as it has been reported for intramuscular fatty acids. The effects of feeding silage mixtures of grass, RC and SF on other meat quality traits, such as oxidative stability, are still poorly understood. In this study, 5 groups of lambs (n=8) were fed for 10 weeks a restricted amount of barley and straw and one of the following silages ad libitum: timothy grass (T, control), T+SF (1:1), T+RC (1:1), T+SF+RC (2:1:1), or SF+RC (1:1). The concentration of vitamin E was measured in the longissimus dorsi muscle. Lipid oxidation (TBARS values) was measured over time of refrigerated storage in raw and cooked muscle slices (7 days and 4 days, respectively) and in muscle homogenates incubated for 90 min with Fe^{3+} and ascorbate. The data were analyzed with a repeated measures statistical model. The concentration of vitamin E in muscle decreased with the inclusion of RC in the diets ($P<0.001$). TBARS values increased regardless of the storage/incubation treatment ($P<0.001$). The dietary treatment did not affect TBARS values in raw meat. Nevertheless, in cooked meat, the inclusion of legume forages in the mixtures reduced the extent of lipid oxidation ($P<0.001$). Interestingly, the T-RC group displayed the lowest TBARS values, despite the lower vitamin E content in muscle. Under more pro-oxidant conditions (incubation with Fe^{3+}/ascorbate), TBARS values followed the order: T-RC < T-SF-RC < SF-RC < T-SF < T. The results highlight the advantage of mixing grass and legume forages and the need of further investigation to understand their impact on the oxidative status of meat.

Differences in digestibility between cows contribute to explain differences in feed efficiency

A. De La Torre[1], D. Andueza[1], R. Baumont[1], G. Renand[2], S. Rudel[3], L. Genestoux[1], G. Cantalapiedra-Hijar[1] and P. Nozière[1]
[1]*INRA, UMR 1213 Herbivores, Theix, 63122 Saint-Genès-Champanelle, France, [2]INRA, UMR1313 Gabi, Domaine de Vilvert, 78352 Jouy-en-Josas, France, [3]INRA, UE1414 Herbipôle, Theix, 63122 Saint-Genès-Champanelle, France; anne.delatorre@clermont.inra.fr*

Residual Feed Intake (RFI) is a feed efficiency (FE) trait phenotypically independent of body weight and level of production. It reflects inherent between-animal variations in biological determinants of FE. Among these determinants, differences in digestion have been suggested but it may rely on the type of diets used. We tested the hypothesis of a higher digestibility in more efficient cows (low RFI) regardless of the type of diets. The objective of this study was to determine the apparent digestibility of dry matter (DMd) of high (NDF=62%) and low fibre (NDF=39%) diets in simultaneously 2 RFI divergent groups of 8 non-pregnant non-lactating Charolais cows (RFI-: -0.73±0.59, 705±96 kg and RFI+: 1.02±0.34, 711±78 kg). RFI was measured when cows were heifers (21 months old) during 12 weeks on herb silage (cultivated fescue) diet distributed ad libitum. Four periods (P) of digestion trials were carried out successively using either high (100% hay, P1 and P2) or low fibre diet (maize silage/concentrate, 67/33, P3 and P4). Each period lasted 2-3 weeks ad libitum followed by 7 days of total faeces collection at 95% ad libitum. DMd was calculated using dry matter intake (DMI) and feacal dry matter excretion. The intra period individual range of variation of DMI and DMd were closed to 4 kg/d and 6 to 8 g/100 g for both diets respectively. Differences between RFI groups were significant for DMd (P=0.024) and not for DMI (P=0.27). RFI-cows presented higher DMd than RFI+ cows for diets (61.0±1.81% vs 60.0±2.1% for high-fibre diet and 69.6±1.96% vs 68.3±1.82% for low-fibre diet). This study showed that on average feed efficient cows (RFI-) had higher digestive efficiency than non-efficient cows (RFI+) regardless of the type of diets used.

Maternal dietary interventions affect piglet intestinal development in different ways

A. De Greeff[1], J. Allaart[2], D. Schokker[3], C. De Bruijn[2], S. Vastenhouw[1], P. Roubos[2], M. Smits[1,3] and J. Rebel[3]
[1]*Central Veterinary Institute, part of Wageningen UR, Edelhertweg 15, 8219PH, the Netherlands, [2]Trouw Nutrition R&D, Veerstraat 38, 5831JN Boxmeer, the Netherlands, [3]Wageningen Livestock Research, Droevendaalsesteeg 1, 6708PB Wageningen, the Netherlands; astrid.degreeff@wur.nl*

Evidence is accumulating that the early life environment of piglets is critical for microbial colonization of the gut and the concomitant maturation of their immune system. It is known that sows contribute to the microbial colonization of the piglet's gut through microbial transmission. Here we used three maternal dietary interventions in order to investigate the effect on microbial transmission from the sow to the piglets, on microbial colonization and on intestinal immune development of piglets. Sows (24) were divided into four groups of 6 that received the following dietary interventions: medium chain fatty acids (MCFA), beta-glucans (BG), galactooligosaccharides (GOS) or a control diet. MCFA possess bactericidal activity and may induce positive effects. BG can modulate the immune status of the host, and GOS is a substrate for Bifidobacteria and Lactobacilli. At day 1 and 31 (3 days post-weaning) intestinal tissue samples and digesta from 6 piglets per group were collected. The digesta samples were used to determine the microbiota composition and diversity, tissue samples were used to profile intestinal gene expression patterns. Only small differences were observed in microbial diversity and composition between maternal dietary interventions. Contrary, gene expression data showed many differentially expressed genes for all three interventions on day 31. On day 1 only GOS induced a gene expression effect. The GOS intervention resulted in the modulation of many pathways related to immunity and barrier function. Similar findings were observed for the BG group at day 31. The pathways affected by the MCFA intervention were mainly associated to fat metabolism. In conclusion, maternal dietary interventions slightly affect the intestinal microbiota in the offspring. Maternal GOS and BG affect intestinal processes in the offspring related to immune competence, whereas MCFA modulates intestinal processes in the offspring related to fat metabolism.

Ruminal protected arginine supplementation in cattle and sheep
J.S. Caton[1], J.L. Peine[1], L.P. Reynolds[1] and A.M. Meyer[2]
[1]North Dakota State University, Animal Sciences, Dept 7630, P.O. Box 6050, Fargo, ND 58102, USA, [2]University of Missouri, Division of Animal Science, 15B Animal Science Research Center, Columbia, MO 65211, USA; joel.caton@ndsu.edu

To investigate the effects of rumen protected arginine (RPA) supplementation in cattle and sheep, 3 experiments were conducted. In Exp.1, four ruminal and intestinal cannulated steers were used in a 4×4 Latin square. Steers were fed grass hay (7.2% CP) for ad libitum intake with either no added arginine (CON), 54 mg L-Arg/kg BW injected intravenously (Arg-INJ), 180 mg RPA/kg BW daily (RPA180), or 360 mg RPA/kg BW daily (RPA360). In Exp. 2, Rambouillet ewes (n=60) were used in a dose titration study. Treatments were: control (CON) and groups receiving 90 mg RPA/kg BW (RPA90), RPA180, or RPA360. In Exp. 3, multiparous, Rambouillet ewes (n=32) were randomly allocated to 3 treatments at d 54 of gestation. Dietary treatments were 100% of requirements (REQ), 60% of REQ (RES), or RES plus RPA180. At parturition, lambs were removed from dams and reared independently. At 54 ± 3 d of age, lamb internal organs were harvested and weighed. In Exp. 1, carotid artery hemodynamics indicated greater (P<0.05) tissue perfusion in steers fed RPA180 compared with CON and Arg-INJ. Duodenal arginine (ARG) flow and small intestinal ARG disappearance (g/d) were least (P<0.05) in CON, intermediate in RP180, and greatest in RPA360. In Exp. 2, all ewes fed RPA had greater (P=0.02) distal tissue blood perfusion than CON. Ewes receiving RPA180 had greater (P<0.05) serum ARG than ewes receiving RPA90, and RPA180 were similar to RPA360 fed ewes (P≥0.55). In Exp. 3, RPA180 did not recover ewe BW or BCS when fed to RES ewes. Providing RPA180 to RES ewes did not enhance circulating ARG in ewes at parturition, but did increase (P<0.04) lamb circulating ARG compared with RES at birth. Lamb birth weight was greater (P=0.04) in REQ than RES ewes. On d 19, lambs from REQ and RP180 ewes both had greater (P≤0.04) BW than lambs from RES ewes. On d 54, lambs from RPA180 ewes were longer (P=0.003) than lambs from RES ewes. Livers from lambs born to RP180 ewes weighed more (P=0.05) than those from lambs born to RES ewes. Additional, research is needed to further delineate potential benefits of RPA supplementation in ruminants.

The effect of linseed expeller supplementation on carcass traits and meat colour of finishing gilts
E. Garcia-Hernandez, M. Tor, D. Villalba and J. Alvarez-Rodriguez
University of Lleida, Animal Science, Av. Alcalde Rocira Roure 191, CP 25198. Lleida, Spain; esther.garcia@udl.cat

A total of 82 lean gilts (Landrace × Large-White × Pietrain), with initial body-weight (BW) of 92.5 ± 2 kg (168 ± 3 days of age), were used to study the effect of top-dressing linseed expeller supplementation for 21 days (0.5 kg/day) on their carcass and meat traits. After slaughter, technological meat quality characteristics of caudal loins were assessed (control n=24, supplement n=17). The colour attributes (CIELab scale and Japanese colour score standard) were measured in loin slices stored in modified atmosphere package (MAP, 70% O_2, 30% CO_2) at 4 °C for 1, 4, 8 or 12 days. Although linseed supplement slightly increased daily gains (689 vs 739 ± 47 g, P>0.05), the final BW was greater in linseed than in control (105.2 vs 109.9 kg, P<0.05). Thus, the slaughter BW was accounted as a covariate in post-mortem data analyses. Carcass weight and dressing did not differ between groups (average 83.1 ± 0.4 kg and 75.5 ± 0.3%, P>0.05), but back-fat depth increased due to linseed supplement (22.8 vs 24.6 ± 0.6 mm at 3rd-4th last rib and 10.9 vs 12.5 ± 0.5 mm at last rib, P<0.05). The muscle pH at 45 minutes and at 24 hours post-mortem were similar across groups (average 6.42 ± 0.07 and 5.61 ± 0.05, respectively; P>0.05). The colour attributes of loin were not affected by linseed supplement (P>0.05). Compared to day 1, lightness increased after 4 days in MAP (46.4 vs 49.9 ± 0.6; P<0.05), while chroma increased from day 1 to days 4 and 8 (10.0 vs 13.8 and 13.3 ± 0.2; P<0.05) but decreased at 12 days (11.8 ± 0.2; P<0.05). The estimated metmyoglobin content increased linearly from day 1 to 12 of MAP storage (1.12 to 1.35 ± 0.007 $K/S_{572/525}$; P<0.05). The visual meat colour evaluation was not affected by the linseed supplement, except at 12 days of MAP storage, when the loins from linseed group were darker than their control counterparts (2.7 vs 3.2 ± 0.14; P<0.05). Linseed supplement increased carcass adiposity without negatively affecting the muscle pH or meat colour during at least 8 days of display storage.

High phytase supplementation levels improve performance in piglets fed P-adequate diets
H. Graham and P. Wilcock
AB Vista, 3 Woodstock Court, Marlborough, Wilts SN8 4AN, United Kingdom; hadden.graham@abagri.com

Phytate has been shown to be an anti-nutrient and that feeding high levels of phytase can breakdown phytate, improving nutrient utilization and nursery pig performance. Weaned pigs (n=88; mean initial weight 5.76 kg) were allotted to 2 diets (11 pen replicates per diet, with 4 pigs per pen) to evaluate the effect of phytase supplemention. Diets were as follows: (1) Control and (2) Control + phytase (Quantum Blue) at 2,500 FTU/kg. Pigs were fed a three-phase feeding program; P1 (d1-7), P2 (d7-21) and P3 (d21-42) with P1 and P2 diets based on corn, soyabean meal, fishmeal and whey, and P3 based on corn and soyabean meal. All diets were formulated to meet the pig's nutrient requirements, including P and Ca, and phytase was added to the control diet with no nutritional changes made. Body weight and feed consumption were recorded to calculate growth performance. During phase 1 of the experiment, faecal scores of pigs were assessed visually on a pen basis, and recorded each day by 2 independent evaluators, with the score ranging from 1 to 5 (1 = normal feces and 5 = watery diarrhoea). Data was analysed by ANOVA using JMP Pro in a randomized complete design, with pen as the experimental unit. Overall (d1-42), pigs fed phytase supplemented diets tended to have a higher ADG (378 vs 413; P<0.15), resulting in an extra 1.42 kg extra gain (P<0.05) at 42 d. Phytase supplementation also improved G/F (0.65 vs 0.70; P<0.05) but had no effect of ADFI. There was no effect of phytase on post-wean diarrhoea score from d1-7. In conclusion, the addition of 2,500 FTU/kg phytase to nutrient adequate piglet diets significantly improved G/F and tended to improve gain, even in a P-adequate diet. This indicates that this performance response may be associated with the breakdown of anti-nutritive phytate rather than P release.

The effect of xylanase and phytase supplementation on performance and egg quality in laying hens
A.E. Taylor[1], M.R. Bedford[2] and H.M. Miller[1]
[1] University of Leeds, Faculty of Biological Sciences, Leeds, LS2 9JT, United Kingdom, [2] AB Vista Feed Ingredients, Marlborough, Wiltshire, SN8 4AN, United Kingdom; a.e.taylor@leeds.ac.uk

The aim of this study was to investigate the interactive effects of xylanase and phytase on layer hen performance, egg quality, phytate breakdown and blood concentration of PYY. A total of 240 Bovan Brown hens (22 weeks of age) were allocated to 60 enriched colony cages with 4 birds per cage. Hens were offered one of six diets in a 2×3 arrangement with phytase (0, 500 or 1,500 FTU/kg Quantum Blue) and xylanase (0 vs 12,000 BTU/kg Econase XT) as factors. Diets were wheat barley based and formulated to meet or exceed the nutrient requirements of laying hens. Hen body weight, feed intake and egg production were recorded. At the end of the 24 week trial digesta and blood samples were collected. Data were analysed by ANOVA according to a general linear model procedure (SPSS 22). Addition of phytase or xylanase had no effect on weight gain or plasma PYY concentration. Addition of phytase increased the number of eggs laid (P<0.05) and daily egg mass (P<0.05) irrespective of phytase level. A phytase × xylanase interaction was observed for adjusted FCR (P<0.05) with xylanase improving feed efficiency when phytase was fed at 500 FTU/kg but tending to reduce feed efficiency when fed with 0 FTU/kg phytase. Inositol phosphate (InsP)6 and InsP5 concentrations in the gizzard were reduced at 1,500 FTU/kg phytase (P<0.05) whereas InsP4 and myo-inositol (MYO) concentrations increased (P<0.05). Ileal InsP6 and InsP5 concentrations were reduced in response to phytase treatment (P<0.05) whereas InsP3 and MYO concentrations increased (P<0.05). Xylanase reduced InsP6 concentration in the gizzard, however increased InsP6 concentration in the ileum (P<0.05). Xylanase had limited effect on laying hen productivity and egg quality. Phytase improved egg production and reduced phytate concentration in the gizzard and ileum. The only interactive effect observed was improved feed efficiency when 12,000 BTU/kg xylanase and 500 FTU/kg phytase were both present.

In vivo and in vitro assessment of two copper sources for broilers

M. Hamdi[1], D. Solà-Oriol[1], R. Franco-Rosselló[1], A. Romeo[2] and J.F. Perez[1]
[1]Animal Nutrition and Welfare Service (SNiBA), Departament de Ciència Animal i dels Aliments, Universitat Autònoma de Barcelona, 08193 Bellaterra, Spain, [2]Animine, 335 Chemin du Noyer, 74330 Sillingy, France; aromeo@animine.eu

Copper supplementation (125 to 250 mg Cu/kg) in poultry diets is a common practice in many non-EU countries to enhance bird health, but high doses of Cu may affect animal performance and interact with phytate. However, these effects could directly depend on the Cu source. The objective of the trial was to compare effects of copper sulphate ($CuSO_4$) and dicopper oxide (Cu_2O, CoRouge®) at 3 levels in the diets (15, 150, 300 ppm) following a 2×3 factorial arrangement. A total of 576 one-day male broilers (Ross 308) were randomly distributed into 6 groups (8 pens per treatment, 12 birds per pen). Bodyweight (BW) and feed intake (FI) were recorded weekly up to d35. On d35, 1 bird per pen was euthanized, skin fat and breast muscle were sampled, and liver and kidney were collected and weighted for Cu determination. Cu sources were also evaluated in vitro to measure Cu solubility, PP (phytic phosphorus) hydrolysis by phytase at pH 2.5, 4.5 and 6.5. The use of 300 ppm Cu from $CuSO_4$ decreased (P<0.01) BW on d14, 28 and 35 and increased (P=0.04) liver Cu content in comparison with the use of 300 ppm Cu from Cu_2O (7.91 vs 4.63 µg/g). Feed conversion ratio increased (P<0.01) for chicken fed with 300 ppm Cu from $CuSO_4$ (2.19) in comparison to those fed with 300 ppm Cu from Cu_2O (1.84). The use of high level of Cu (300 ppm), either from Cu_2O or $CuSO_4$, also increased (P<0.01) Cu level in kidney and breast muscle in comparison to other groups. Solubility of Cu from Cu_2O and $CuSO_4$ was higher at pH=2.5 but it decreased at higher pHs. Including a level of 300 ppm Cu from $CuSO_4$ reduced PP solubility (68.66%) in comparison to Cu_2O (97.41%). Increasing the levels of Cu with both sources reduced PP hydrolysis by phytase at pH=4.5 and 6.5. It can be concluded that dietary levels of 150 and 300 ppm Cu from Cu_2O are adequate to ensure broiler growth performance and limit organ accumulation in comparison to $CuSO_4$.

Effect of the diet on the consumer acceptability of fresh loin from Iberian pigs

M. Gispert[1], J. García-Gudiño[1], I. Blanco-Penedo[1], J. Perea-Muñoz[2], J. Garcia Casco[3], E. Gonzalez[4], A. Brun[1] and M. Font-I-Furnols[1]
[1]IRTA, Product Quality, Finca Camps i Armet s/n, 17121 Monells, Spain, [2]UCO, Campus Universitario de Rabanales, Ctra.N-IV, km 396, 14014 Córdoba, Spain, [3]INIA, Centro de I+D en Cerdo Iberico, INIA, Ctra. Ex101, km 4,7, 06300 Zafra, Spain, [4]UEX, Ctra de Cáceres s/n, 06071 Badajoz, Spain; marina.gispert@irta.es

It is of interest to achieve a reduction of the growth rate and a higher meat quality of crossbred Iberian pigs fattened under intensive regimen in order to benefit the sustainability of the pork meat production. The aim of the present work was to evaluate the influence of three feeding regimes: control diet (CO), diet with low protein (LP) and sustainable-type diet using a local sub-product from the olive named 'alperujo' (AF), on the acceptability of fresh loin of Iberian pigs by consumers. Twenty-eight castrated male pigs (CO and AF n=8; LP n=10) were slaughtered in commercial conditions at 154.6+10.28 kg live weight. A sample of loin muscle with 5 mm of subcutaneous fat was frozen until the day before the sensory evaluation. A total of 80 consumers representative of the Spanish population evaluated 3 samples of meat, one of each feeding regimen, in a Hall Test. Meat samples were cooked in an oven pre-heated at 200 °C until reach the internal temperature of 76 °C and distributed to the consumers in a monadic way and following a design to avoid the first sample and carry-over effect. Statistical analysis was performed with SAS software. The mixed procedure was applied considering feeding regimen as fixed effect, session as blocking variable and consumer as a random effect. Tukey test was used to compare means. Results show that consumers' overall and tenderness acceptability of meat was significantly (P<0.05) higher in meat from CO compared with AF, being meat from LP in between. Flavour acceptability of AF meat was significantly lower (P<0.05) than LP. No differences were found in odour acceptability (P=0.51) between treatments. In the conditions of this experiment, it can be concluded that AF diet seems to decrease consumers' loin acceptability while LP does not affect it.

The effect of curcumin on the oxidative stability of pork

O. Bučko, A. Lehotayová, J. Petrák, P. Juhás, M. Margetín and O. Debrecéni
Slovak University of Agriculture in Nitra, Department of Animal Husbandry, Tr. A. Hlinku 2, 949 76, Nitra, Slovak Republic; milan.margetin@uniag.sk

The aim of the experiment was to consider the effect of the addition of curcumin (Curcuma longa powder) on the parameters of the oxidative stability of pork. The experiment included 29 pigs of Large White breed. The pigs were divided into a control group of 9 pigs and two experimental groups of 10 pigs. Experimental groups were fed compound feed mixture enriched with curcumin. The first experimental group has fed supplemented by the addition of 200 mg curcumin per kg of live weight and second by 400 mg per kg. Oxidative stability of pork was determined from the sample longissimus thoracis muscle by the method of determination the level of malondialdehyde (MDA) on the day of the slaughter and 3, 5 and 7 days post mortem. The two-way analysis of variance was used for determining the differences of the monitored parameters. Results showed increasing amount of oxidative stability of pork the indicator MDA in meat with the number of days post mortem with the rising addition of curcumin in the feed. The lowest increase during the days post mortem was recorded in the control group. Difference among the average amount MDA at day of slaughter and at monitored days post mortem was not statistically significant. At the 3. day post mortem were detected significant differences between the control and the second experimental group (P<0,001) and between experimental groups (P<0,05). Statistically significant differences were found at the 5. day post mortem between the control and the second experimental group (P<0,001) and between the control and the first experimental group and experimental groups at P<0,05. Statistically significant differences were found at 7. day post mortem between control and the first experimental group and experimental groups at P<0,05. Between the control and experimental group was detected a statistical difference at P<0,001. It can be concluded that the rising addition of curcumin in the feed increased the amount of MDA in meat during the process of post mortem. This work was supported by projects VEGA 1/0364/15 and KEGA 035SPU-4/2015.

Colostral transfer of immunoglobulins in dairy calves

E.M. Andrée O'Hara, I. Olsson and K. Holtenius
Swedish University of agricultural sciences, Department of animal nutrition and management, Box 7024, 75007 Uppsala, Sweden; lisa.ohara@slu.se

The concentration of immunoglobulin G (IgG) and total protein in plasma (TP) are both related to absorption of immunoglobulins from colostrum after birth in calves. Numerous studies have shown that failure to absorb immunoglobulins are related to increased mortality. This study included 153 calves of Swedish Holstein (SH; n=70) and Swedish Red breeds (SR; n=83). A blood sample was taken at the age of 2-9 days and plasma was harvested. Birth weight, volume of colostrum fed at the first feeding, age when fed first colostrum and density of the colostrum were recorded. TP was determined by refractometry and the IgG concentration in plasma was determined using a two-sited enzyme-linked immunoassay. A general linear regression model (SAS software 9.3) was used to investigate the associations between density of the first colostrum, volume of colostrum at first feeding, time after birth at first feeding, birth weight, breed and gender and the levels of TP and IgG. TP was higher in SH calves (55.3 g/l) compared to calves of the SR breed (52.6 g/l, P=0.003). The same trend was observed for the plasma IgG concentration (SH 10.0 g/l, SR 8.5 g/l, P=0.08). There was no gender effect on IgG or TP. Neither did birth weight or age at first feeding have an effect on IgG or TP, which could have been due to calves being fed within 12 hours and all but three calves were fed within 6 hours. Both the density and the amount of colostrum were correlated to TP (P=0.003, P=0.021) and IgG (P<0.0001, P<0.0001). The results of the present study highlight the importance of feeding sufficient amounts of colostrum with a high density to neonatal calves.

Metabolomic profile of early lactating dairy cow receiving rumen-protected choline

C. Giromini[1], A. Baldi[1], A.A.K. Salama[2], G. Caja[2], M. Tretola[1], V. Caprarulo[1], G. Invernizzi[1] and L. Pinotti[1]
[1]University of Milan, Health, Animal Science and Food Safety, via celoria 10, 20133, Italy, [2]Universitat Autonoma de Barcelona, Department of Animal and Food Sciences, Campus de la UAB, 08193 Bellaterra, Barcelona, Spain; antonella.baldi@unimi.it

In dairy cows, choline has been proposed as limiting nutrient, especially at the onset of lactation when silage-based diets are fed. Choline plays a major role in lipid metabolism and traffic. The aim of the current study was to evaluate the effect of rumen-protected choline supplementation on plasma metabolomic profile of early lactating dairy cow receiving a hay-based diet. Fourteen Italian Holstein multiparous cows (28 DIM) fed hay-based diet (50% hay and 50% concentrates) were divided in 2 groups for an 8-weeks experimental period: control group (CTR) receiving no choline, and treated group (RPC) receiving 20 g/day of choline in rumen protected form (Balchem, US). Dry matter intake and milk yield were measured daily, whereas blood plasma samples were collected at weeks 4, 5, 6, 8, 12. Plasma samples were analyzed with ^{1}H NMR spectroscopy operating at a frequency of 600 MHz. Principal component analysis (PCA) and partial least square–discriminant analysis (PLS-DA) were used to identify possible metabolite markers in blood plasma. Although RPC supplementation did not significantly affect plasma metabolomics, RPC cows showed a greater plasma concentration of choline, indicating that choline, in the rumen-protected from, was bio-available. When progress of lactation has been considered, both PCA and PLS-DA revealed distinct metabolomics profiles. Lipids (mainly LDL/VLDL) were greater at week 12 compared to week 4; whereas proline and tyrosine were significantly reduced over the weeks of lactation. In conclusion, metabolomics changes over weeks of lactation were observed, while rumen-protected choline impact in hay-based diet deserves further investigations.

Evaluation of untreated field pea (Pisum sativum L.) as a protein source for laying hens

G. Ciurescu[1], M. Hăbeanu[1] and C.O. Pana[2]
[1]National Research Development Institute for Animal Biology and Nutrition, Animal Nutrition, Balotesti, Calea Bucuresti no. 1, Ilfov, 077015, Balotesti, Romania, [2]University of Agronomic Science & Veterinary Medicine, 59 Marasti Street, Bucharest, 011464, Romania; mihaela.habeanu@ibna.ro

Price and availability of soybean meal on global markets may change rapidly, thereby stimulating interest in maximizing the use of locally produced protein sources, like seed legumes. This study was designed to establish the effect on laying performance and egg quality resulting from substitution of soybean meal with untreated field pea (Pisum sativum L.) and enzyme supplementation in the diet of laying hens. A total of 270 Lohmann Brown Classic pullets, 18 wks of age were randomly allocated to five experimental groups of 54, hens each. Each treatment had six replicates of 9 hens with similar levels of production at the beginning of the trial and fed 35 wks (between 20 and 54 wks of age). The dietary treatments were as follows: (1) corn/soybean meal (soybean diet); (2) corn/pea (pea diet); (3) pea diet supplemented with Kemzyme HF® (100 mg/kg diet); (4) pea diet supplemented with Allzyme®Vegpro (100 mg/kg diet); (5) pea diet supplemented with enzyme combination [Kemzyme HF® (50 mg/kg diet) plus Allzyme® Vegpro (50 mg/kg diet)]. Statistical analysis was carried out using the General Linear Models procedure of SPSS. The substitution of soybean meal with untreated pea did not significantly ($P>0.05$) change the bird's live weight as well as feed conversion ratio; egg production (rate of lay and egg mass) and none of egg quality traits examined (Haugh unit, specific weight, shell strenght) were influenced, except for yolk color that was higher ($P<0.05$) in hens fed the pea diet. Our results show that the addition of Allzyme® Vegpro or the enzyme combination (Kemzyme HF® plus Allzyme®Vegpro), respectively to pea-based diets significantly improved laying performance; however, the effect of Kemzyme HF® was not statistically significant. No effect of enzyme supplementation, were observed on egg quality traits.

Effects of dietary n-3 PUFA-rich camelina cake on fatty acids distribution in some tissue of pigs

M. Hăbeanu, M. Ropota, N.A. Lefter, G. Ciurescu, E. Ghita and T. Mihalcea
National Research Development Institute for Animal Biology and Nutrition, Animal Nutrition, Balotesti, Calea
Bucuresti no. 1, Ilfov, 077015, Balotesti, Romania; mihaela.habeanu@ibna.ro

Camelina cake (CC) is a by-product of oil extraction industry that can add value to the diet by its essential
fatty acids (FA) and amino acids composition. The objective of this study was to assess FA composition of
brain, heart and liver tissue by the dietary addition of CC in comparison with sunflower meal or rapeseed
meal. The FA concentration of these tissues may affect the health. The trial was conducted on 36 finishing
pigs, 72.22±6.33 kg, randomly assigned to three experimental treatments containing either 12% sunflower
meal, or 12% rapeseed meal or 12% CC. The FAs were determined by gas chromatography. The data were
submitted to variance analysis with SPSS, version 20. The CC used in our study had 39.61% crude protein,
11.03% ether extractives on dry matter basis and 61.38% polyunsaturated FA of total FA ester methyl, of which
31.50% were α-linolenic. The α-linolenic concentration in rapeseed meal and sunflower meal was 6.52% and
0.16% respectively. The α-linolenic FA was affected significantly by both CC inclusion and tissue type. The
increase of α-linolenic FA concentration by dietary addition of CC was associated with an increase of n-3 FA,
irrespective of the tissue. The concentration of total n-3 PUFAs was more pronounced in the brain (9.46 vs
1.68% in heart and 3.08% in liver). CLA distribution was 3.34 times higher in the brain than in the heart and
8.0 times higher in brain than in liver. Long-chain n-3 FA (EPA, DPA and DHA, whose health properties are
known) showed a higher (P<0.001) deposition in the liver of the group fed CC but the highest concentration
was noticed in the brain, irrespective of the diet. N-6:n-3 ratio was significantly lower in the brain (1.49%)
than liver (13.87%) and heart (22.77%). We conclude that the supplemental CC favoured mostly the retention
of α-linolenic FA. This effect is associated with the presence of EPA, DPA and DHA, particularly in the brain,
and with lower n-6:n-3 ratio irrespective of the tissue, beneficial to health.

Evaluating homogeneity of TMR in dependence on feed mixing times and different loads of a TMR wagon

F.M. Tangorra, A. Agazzi, A. Costa, A. Pilotto and G. Savoini
Università degli Studi di Milano, Dipartimento di Scienze Veterinarie per la Salute, la Produzione Animale e la
Sicurezza Alimentare, Via G. Celoria, 10, 20133 Milan, Italy; giovanni.savoini@unimi.it

The aim of the trial was to determine the influence of cutting and mixing time on homogeneity and particle size
distribution of a dairy cow total mixed ration (TMR), considering different loading levels of the mixing wagon.
The TMR particle size distribution, intended as homogeneity of the diet, and the chemical correspondence
to the designed diet where investigated applying a Central Composite Design (CCD) in a randomized block
design generated by JMP 11 Pro software considering: (1) three mixing wagon loads (30%, 40 and 70% of
the maximum nominal load, 21 m^3); (2) three cutting times (4, 5 and 6 min); and (3) three mixing times (4, 5
and 6 min). The mixing wagon was a two-screws vertical type, and the diet was based on corn silage (57%)
and hay (14.7%). Samples of TMR were collected at the beginning, in the middle and at the end of the feeding
alley for dry matter (DM) content, ether extract (EE), crude protein (CP), neutral detergent fibre (NDF),
and ash content determination. At the same time TMR particle size distribution was determined in triplicate
for each sampling point by Penn Particle Separator. The homogeneity of the diet was evaluated in terms of
variation coefficient (CV, %) of DM (% wet basis); CP (% DM); NDF (%ss) to evaluate the multiple effect
of loading level, cutting and mixing time. The CCD allowed the 3D response surfaces for CV of DM, CP,
and NDF on loading levels and mixing times, since cutting time was not a significant parameter. Data were
also analysed by a Principal Component Analysis PCA,, keeping into account TMR chemical composition
of samples collected along the alley and the chemical composition of the designed diet. The higher level of
similarity of the real and the designed diet was reached with a 70% nominal loaded mixing wagon, together
with 5 minutes of cutting and 5 minutes of mixing.

Expander processed maize increases digestibility of nutrients and hence alters the lysine to energy

R. Puntigam, K. Schedle, C. Schwarz, E.M. Lechner and M. Gierus
University of Natural Resources and Life Sciences, Institute of Animal Nutrition, Livestock Products and Nutrition Physiology, Muthgasse 11, 1190 Vienna, Austria; martin.gierus@boku.ac.at

Expander processing, known as high temperature short time (HTST) technology is widely used in the feed industry. Especially the high specific energy inputs (kWh/t) used for processing single components like maize before mixing leads to a pronounced starch gelatinization and disruption of cell walls. The aim of the present study was to determine the effect of an intensive expander treatment of maize on apparent total tract digestibility (ATTD), and carcass quality. After grinding, maize remains HTST unprocessed (control diet (C)), and HTST processed maize for treatments SC, LC and LC+AA. In these treatments, maize was short (60 s, SC) and long-term (1,080 s, LC) conditioned (80 °C) and subsequently intensively expanded (~45 kWh/t). In treatment LC+AA, AA content was increased by 10% to maintain the ideal protein. The experiment used 60 crossbreed barrows (30.7±0.3 kg). Mash feed and water were provided ad libitum. Fattening and slaughter parameters were recorded. Data were analysed using ANOVA as randomized block design, with Tukey-Kramer test for LS-mean separation (P<0.05). Expanded maize improved ATTD of DM and EE and DE content. Within expander treatments, SC maize had higher ATTD for DM and DE content, followed by LC and unprocessed maize (C). Considering the maize proportion in diets, the amount of DE/kg DM of maize reached 8.4%, 5.1% and 4.1% in SC, LC and LC+AA respectively, compared to the control. Supplemented AA did not affect ATTD-parameters, but reduced feed intake and feed conversion rate (FCR). Expanded maize declined loin depth and lean percentage, which was reversed in LC+AA treatment, while no influence of the average daily gain (ADG) was observed. There is a potential to improve the nutrient digestibility of individually expander treated feeds for pig diets. However, the Lys:energy ratio should be considered in diet formulation, to avoid higher fat proportion in the carcass.

Effect of feeding level on body composition of pure and crossbred lambs finished in feedlot

W. Henrique[1], G. Aferri[1], I.H.S. Fuzikawa[2], E.A. Oliveira[2], S.N. Esteves[2] and M.M. Alencar[2]
[1]APTA, Rodovia Washington Luiz, km 445, 15082-000 São José do Rio Preto, SP, Brazil, [2]EMBRAPA/CPPSE, Rodovia Washington Luiz, km 234, 13560-970 São Carlos, SP, Brazil; wignez@terra.com.br

The objective was to evaluate the effect of feeding level on body composition of lambs, confined after weaning at 90 days of age, of seven genetic groups: 15 White Dorper (WDO), 15 Ile de France (IF), 11 Texel (Tx), 18 Santa Inês (SI), 18 ½WDo+½SI, 18 ½IF+½SI and 18 ½Tx+½SI. There were three weaning dates, spaced 15 days. The animals were kept in individual pens for 42 days, after 15 days of adaptation. All animals were shearing during adaptation period. They received a complete pelleted diet consisting of 10% alfalfa hay and 90% concentrate, at the following levels: ad libitum (leftovers around 10%, adjusted daily), 0.75 and 0.63 g/kg of metabolic weight (MW), both readjusted weekly after weighing the animals without fasting. At the end, the animals were weighed after solid fasting for 14 hours and slaughtered in commercial abattoir. There were separated and weighed the following body components: blood, head + feet + leather, empty viscera and carcass. It was considered a randomized block design (dates of weaning as block) in factorial scheme 7×3 (genetic groups × feeding levels), the initial weight of the animals was used as a covariate and means compared by Tukey test at 5% probability. The interaction among factors was not significant for any variable. The feeding level affected the weight of blood, viscera, head + legs + leather, carcass and empty body; the highest values were observed for the treatment ad libitum: 1.096, 6.71, 6.95, 18.95 and 33.71 kg, respectively. The genetic group also affected head + legs + leather, viscera and carcass weights; highest values (7.41, 6.59 and 20.35 kg, respectively) were achieved by Tx lambs. The lighter carcass weights were obtained by the SI and ½WDo+½SI animals (15.27 and 15.63 kg, respectively). The hot carcass yield was influenced by genetic group, ranging from 45.57 to 49.77%, and IF and SI groups had the lowest yields. Body composition of lambs is independently altered by genetic group and feeding level in the finishing phase.

Body composition of purebred and crossbred lambs finished in feedlot

G. Aferri[1], W. Henrique[1], I.H.S. Fuzikawa[2], E.A. Oliveira[2], S.N. Esteves[2] and M.M. Alencar[2]
[1]*Apta, Rodovia Deputado Leônidas Pacheco Ferreira km 304, Pouso Alegre Baixo, 17206-697 Jaú, SP, Brazil,*
[2]*Embrapa Pecuária Sudeste, Rodovia Washington Luiz km 234, Fazenda Canchim, 13560-970 São Carlos, SP,*
Brazil; gabriela@apta.sp.gov.br

The objective was to evaluate the body composition of lambs, confined after weaning at 90 days of age, of seven genetic groups: five White Dorper (WDo), five Ile de France (IF), four Texel (Tx), six Santa Inês (SI), six ½WDo + ½SI, six ½IF + ½SI and six ½Tx + ½SI. There were three weaning dates, spaced 15 days among them. The animals were kept in individual pens with wooden slatted floor for 42 days, after 15 days of adaptation. All animals were shearing during adaptation period. They received a complete pelleted diet balanced with 10% alfalfa hay and 90% concentrate, allowing leftovers around 10%. At the end of the experimental period, the animals were weighed after solid fasting by 14 hours and slaughtered in commercial abattoir. At slaughter, there were separated and weighed the following body components: blood, head + feet + leather, empty viscera and carcass. The statistical analysis considered a randomized block design (considering the dates of weaning as block) with seven treatments (genetic groups), the weight of the animals at the start of feedlot considered as a covariate and the averages compared by Tukey test at 5% of probability. There was no difference between genetic groups for the weight of blood and viscera, average of 1.10 and 6.72 kg, respectively. The Tx, IF and ½IF + ½SI had the highest head weight + feet + leather, with the following values, respectively: 7.73; 7.90 and 7.68 kg. The carcass weight was higher to Tx group with 21.8 kg and the lowest was 15.95 kg for SI. The hot carcass yield ranged from 46.52 to 52.11%, and the highest value was obtained by ½Tx + ½SI. Tx lambs showed the highest empty body weight (37.91 kg), calculated as the addition of the parts, and the lowest weight obtained by SI and ½WDo + ½SI groups (30.83 and 29.38 kg, respectively). Animals obtain from crossing different breeds of sheep in feedlot not always have the best carcass weight and empty body. Texel lambs proved to be more promising in southeastern Brazil confinement system.

Performance of purebred and crossbred lambs finished in feedlot

G. Aferri[1], W. Henrique[1], I.H.S. Fuzikawa[2], E.A. Oliveira[2], S.N. Esteves[2] and M.M. Alencar[2]
[1]*Apta, Rodovia Deputado Leônidas Pachêco Ferreira km 304, Pouso Alegre Baixo, 17206-697 Jaú, SP, Brazil,*
[2]*Embrapa Pecuária Sudeste, Rodovia Washington Luiz km 234, Fazenda Canchim, 13560-970 São Carlos, SP,*
Brazil; gabriela@apta.sp.gov.br

The objective was to evaluate the feedlot performance of lambs, newly weaned with 90 days old, of seven genetic groups: five White Dorper (WDo), five Ile de France (IF), four Texel (Tx), six Santa Inês (SI), six ½WDo+½SI, six ½IF+½SI and six ½Tx+½SI. There were three weaning dates, spaced 15 days. The animals were kept in individual pens for 42 days, after 15 days of adaptation. They received a complete pelleted diet consisting of 10% alfalfa hay and 90% concentrate, allowing leftovers around 10%. At the beginning and end of the experimental period, the animals were weighed after solid fasting for 14 hours. It was used a randomized block design (dates of weaning) with seven treatments (genetic groups), the initial weight of the animals considered as a covariate and means were compared by Tukey test at 5% probability. No differences among genetic groups were observed, except for feed efficiency. Texel lambs had the lowest weight at the beginning of feedlot. The dry matter intake was different in daily amount, as a percentage of body weight and in relation to metabolic weight; Tx lambs did not differ from IF (2.90 and 3.26% of body weight), but consumed less than the others (3.68% on average). The intake of crude protein and ether extract was different among genetic groups. This nutrients intake but did not follow the pattern of dry matter ingested, despite of it had been used a ration which make difficult the selection of food. Despite of lower food intake in relation to body weight, Tx animals showed the greatest weight gain (0.860 kg/day), different from all the other genetic groups, and ½WDo+½SI had the lowest weight gain (0.299 kg/day), not different from WDo (0.440 kg/day). Thus, the highest final weight was achieved by Tx, ½IF+½SI, IF and ½Tx+½SI lambs, with respectively 43.9, 43.4, 42.0 and 40.5 kg. Animals obtained by crossing different breeds of sheep have not always the best performance in fattening feedlot in southeastern Brazil, and Texel lambs shown promise in these conditions.

Effect of feed restriction on performance of pure and crossbred lambs finished in feedlot

W. Henrique[1], G. Aferri[1], I.H.S. Fuzikawa[2], E.A. Oliveira[2], S.N. Esteves[2] and M.M. Alencar[2]
[1]APTA, Rodovia Washington Luiz, km 445, 15082-000 São José do Rio Preto, SP, Brazil, [2]EMBRAPA/CPPSE, Rodovia Washigton Luiz, km 234, 13560-970 São Carlos, SP, Brazil; wignez@terra.com.br

The objective was to evaluate the effect of feed restriction on feedlot performance of lambs weaned at 90 days of age, of seven genetic groups: 15 White Dorper (WDo), 15 Ile de France (IF), 11 Texel (Tx), 18 Santa Inês (SI), 18 ½WDo+½SI, 18½IF+½SI and 18 ½Tx+½SI. There were three weaning dates, spaced 15 days. The animals were kept in individual pens for 42 days, after 15 days of adaptation. They received complete pelleted diet consisting of 10% alfalfa hay and 90% concentrate, at the following levels: ad libitum (leftovers around 10%, adjusted daily), 0.75 and 0.63 g/kg metabolic weight (MW), both readjusted weekly after weighing. Animals were weighed after solid fasting by 14 hours at the beginning and end. It was considered a randomized block design (considering the dates of weaning as block) in a factorial scheme 7×3 (genetic groups × feed supply levels), the initial weight of the animals considered as a covariate and means compared by Tukey test at 5% probability. The feed supply levels and genetic groups affected the daily weight gain and feed efficiency. The body weight gain decreased with the lower supply of food, but feed efficiency was not different between animals that received feeding level ad libitum or 0.75 g/kg MW, but was lower in animals level 0.63 g/kg MW. High weight gains were achieved by Tx, IF and ½Tx+½SI lambs, respectively 0.654, 0.645 and 0.604 kg/day; and the crossbred lambs ½WDo+½SI showed the lower daily gain (0.242 kg). On the other hand, this genetic group, along ½IF+½SI, improved feed efficiency. The final body weight of the lambs was affected by the feed supply level, and the animals that received ad libitum feed showed the highest slaughter weight (38.12 kg) than animals fed restrictive levels (35.45 and 32.32 kg, respectively to 0.75 and 0.63 g/kg MW). Offer restricted amount of food for lambs in termination is not a good strategy because it reduces feed efficiency and slaughter weight.

Effect of L-Selenomethionine supplementation in starter broilers

J. Michiels[1], J. Degroote[1], M. Majdeddin[1,2,3], A. Golian[3], S. De Smet[2], M. Rovers[4], A.B.J. Van Dalen[4] and L. Segers[4]
[1]Ghent University, Department of Applied Biosciences, Valentyn Vaerwyckweg 1, Ghent, Belgium, [2]Laboratory for animal Nutrition and Animal Product Quality, Ghent University, Department of Animal Production, Proefhoevestraat 10, 9090 Melle, Belgium, [3]ferdowsi University of Mashhad, Centre of Excelence in the Animal Science, P. O. Box 91775-1163, Mashhad, Iran, [4]Orffa, Vierlinghstraat 51, 4251 LC, the Netherlands; dalen@orffa.com

Amongst different stress factors, the onset of broilers has a significant impact on broiler production. The starter period of broiler production may cause stress and the relation between selenium (Se) and oxidative stress is described in literature. It is hypothesized that above normal feed Se levels, extra dietary supplementation with L-Selenomethionine (L-SeMet) could improve the performance of starter broilers. In this trial the addition of 0,2 ppm Se in form of L-SeMet (Excential Selenium4000) to the basal starter (0-10 d) and grower diets (11-25 d) containing respectively 0.5 and 0.4 mg Se/kg was evaluated. Diets contained 11.5 and 10.5 g dig lysine/kg respectively. Dietary treatments were replicated in 18 and 13 pens respectively with 20 Ross308 birds each. In the starter period (0-10 d) significant differences between treatments were observed. Average daily gain (ADG) improved 3.6% ($P<0.05$) by supplementation with L-SeMet (22.32 vs 23.12 g/d) resulting in a higher body weight ($P<0.05$) for the supplemented group (269 vs 277 g). Feed conversion ratio (FCR) was 1.215 in control and 1.195 in L-SeMet supplemented group. Growth, feed intake and feed conversion were not significantly different in the grower period. Overall mortality (d1-25) was not statistical different. In conclusion, supplementation of broiler diets with L-Selenomethionine could be a nutritional tool to optimize broiler performance during stressful periods like the onset of broilers.

Changes in body condition and serum metabolites in endurance horses fed stabilized rice bran

A. Agazzi[1], G. Savoini[1], G. Farina[1], E. Mariani[1], M. Doronzo[1], C. Valentinelli[2] and C. Belladelli[2]
[1]Università degli Studi di Milano, Dipartimento di Scienze Veterinarie per la Salute, la Produzione Animale e la Sicurezza Alimentare, Via Celoria, 10, 20133 Milan, Italy, [2]Riso Scotti Ingredients, Riso Scotti spa, Via A. Scotti, 2, 27100 Pavia, Italy; greta.farina@unimi.it

The objective of the trial was to study the effects of dietary stabilized rice bran on body condition and serum lipid and carbohydrates metabolites in endurance horses. Four subjects (419±5.98 kg) in training received 500 g/d of stabilized rice bran (Risosprint®, Riso Scotti Ingredients, Italy) in a 85F:15C diet, fed at 2.5% body weight (23 Mcal/d DE and 760 g/d crude protein). Dietary treatment lasted 7 months. Feed intake was recorded daily, while body weight, body condition score, heart rate, capillary refill, and mucous membrane colour were evaluated monthly, at rest. Blood samples were collected for ALT, AST, glucose, cholesterol and triglycerides serum content before starting the dietary treatment, monthly, and one month after the administration of stabilized rice bran. Data were analysed by MIXED procedure of SAS considering the effect of time response to the dietary treatment. Feed intake, body weight and body condition score were not affected during time as heart rate, capillary refill, and mucous membrane colour. Significantly reduced ALT, cholesterol and triglycerides serum content were found mainly at the end of the trial, when compared with samples collected on months three and four ($P<0.05$). Glucose serum content showed more variable trends in the first four months of the trial, with significant increased levels at 30 days of treatment than in the subsequent two months (85.75 vs 77.00 and 76.35 mg/dl respectively; $P<0.01$) and in the rest of the trial ($P<0.05$). The inclusion of stabilized rice bran in the diet of endurance horses in training exerted lowering effects on ALT, cholesterol and triglycerides basal serum content with a long-time administration effect.

Effect of hydroxy copper and zinc and L-selenomethionine on claw health of sows

A.B.J. Van Dalen and I. Eising
Orffa, Vierlinghstraat 51, 4251 LC Werkendam, the Netherlands; dalen@orffa.com

Trace minerals (TM) copper (Cu) and zinc (Zn) influence claw health positively. Selenium (Se) plays an important role in inflammation and immunity. On a practical farm with 750 sows, currently used TM sources were replaced with higher bioavailable TM sources to investigate the effect of these higher bioavailable TM sources on claw health over time. In the gestation diet 14 ppm Cu from CuSO4, 70 ppm Zn from ZnSO4+37 ppm Zn from Zn-chelate and 0,2 ppm Se from Se-yeast were replaced by 14 ppm Cu from hydroxy-Cu, 75 ppm Zn from hydroxy-Zn and 0,2 ppm Se from L-selenomethionine (L-SeMet). In the lactation diet 16 ppm Cu from CuSO4, 80 ppm Zn from ZnSO4+43 ppm Zn from Zn-chelate and 0,2 ppm Se from Se-yeast were replaced by 16 ppm Cu from hydroxy-Cu, 85 ppm Zn from hydroxy-Zn and 0,2 Se from ppm L-SeMet. Claws of sows in the farrowing room were monthly scored. The claws were scored on five different parameters: cracks in heel area; dew claw length; inner & outer toe length; cracks in horn wall; skin damages above claw. Scores 1 – 4 were given, 1 being good/healthy, 4 being extremely deviating. The claw score method developed by Hoofs (2006) was used. Scores from 1 month before start and start of the trial were averaged (control) and compared to average scores of 7, 8 and 9 months after TM replacement (treatment). Differences were observed between control and treatment for the amount of severe claw problems (scores 3 and 4). Score 3 and 4 for cracks in heel area was 4.32% in control vs 0.19% in treatment. Dew claw length score 3 and 4 in control was 0.61 vs 1.37% in treatment. Inner and outer toe length score 3 and 4 were 1.79% in control vs 1.77% in treatment. Score 3 and 4 for cracks in horn wall was 4.88% in control vs 0.78% in treatment. Score 3 and 4 for skin damages above claw was 10.65% in control vs 1.35% in treatment. From this trial it was concluded that trace elements have an effect on claw health. Replacing Cu, Zn and Se by these better available sources might be a useful tool to improve claw health problems in practical situations.

Nutritional value of Proglobulin® in piglets

X. Guan[1], F. Molist[1], C. Van Vuure[2] and P.J. Van Der Aar[1]
[1]Schothorst Feed Research, Swine Nutrition, Meerkoetenweg 26, 8200 AM Lelystad, the Netherlands, [2]Darling Ingredients International, Sonac, Kanaaldijk Noord 20-21, 5690 AA Son, the Netherlands; xguan@schothorst.nl

Protein is the most expensive nutrient in diets for weaned piglets. Type and quality of protein sources can influence the performance and health of weaned piglets. Proglobulin® is plasma powder which is produced from porcine blood from veterinarian approved pigs. The objective of the experiment was to determine the ileal protein and amino acid digestibility of Proglobulin® in weaned piglets. Ten piglets (barrows) of around 18 days of age and averaged body weighed of 8.1 kg were surgically fitted with a T-cannula. After surgery, piglets remained with the sow for 7 days. Afterwards, piglets were weaned and housed individually in metabolic cages and received a standard weaner diet for 4 days. Thereafter, piglets were divided into 2 experimental treatments a negative control (NC) diet and the same NC diet supplemented with 15% Proglobulin®. Piglets received the diets during two consecutive rounds of 14 days. Piglets were fed 3.2 times maintenance and had ad-libitum access to drinking water. On days 13 and 14 of the experiment ileal samples were collected. Diets and ileal samples were analysed for dry matter, ash, crude protein, amino acids and marker (TiO2). Digestibility's of Proglobulin® were calculated by subtracting the contribution of the basal diet from the experimental diet, using the difference method. Apparent ileal digestibility (± SEM) of Proglobulin® nutrients were as follows: dry matter (110.7±8.4), ash (114.1±22.3), organic matter (104.3±10.6), crude protein (92.9±4.4), Lysine (94±1.8), Methionine (93±3.7), Cysteine (77±5.0), Threonine (92±3.2), Tryptophan (94±2.5), Valine (93±2.9), Arginine (97±2.4), Histidine (92±2.7), Isoleucine (93±3.6), Leucine (94±2.6), Phenylalanine (94±2.6), Tyrosine (98±2.6), Alanine (93±3.7), Aspartic Acid (87±4.3), Glutamic Acid (83±5.5), Glycine (93±8.5), Proline (91±19.2), Serine (91±3.1). In conclusion the apparent ileal nutrient digestibility of Proglobulin® in cannulated piglets was higher than 90% for most of the nutrients.

Effect of dietary selenium source on selenium deposition in broiler muscle tissue

M. Rovers[1], L. Segers[1], I. Eising[1] and G. Du Laing[2]
[1]Orffa, Vierlinghstraat 51, 4251 LC Werkendam, the Netherlands, [2]Ghent University, Faculty of Bioscience Engineering, Coupure Links 653 Bl. B, 9000 Ghent, Belgium; eising@orffa.com

Aim of this preliminary study was to investigate the effect of different dietary selenium (Se) sources on the selenium deposition in broiler muscle tissue. For this purpose, 4 pens with 30 Ross 308 male birds each were fed from day 0 to 14 one of the 4 treatment diets, which differed only in the added selenium source. Treatment 1 contained no added Se. Diets of treatments 2, 3 and 4 were supplemented with 0.2 mg/kg Se from respectively sodium selenite (NaSe), L-selenomethionine (SeMet) and seleno-hydroxy-methionine (SeOH). Samples of breast meat (whole left filet) were taken from 6 birds from each treatment on day 7 and also from 6 birds on day 14. Initial Se level in breast meat was determined on 6 birds euthanized on day 0. Breast meat samples were analyzed for selenium content by HLPC ICP-MS. Se content of broiler muscle tissue on day 0 was 197±10 µg/kg. On day 7 the Se content of the non-supplemented birds decreased and reached a level of 84±6 µg/kg. Birds fed supplementary Se as sodium selenite also had a decreased Se (130±7 µg/kg) on day 7. However, treatment with SeOH and SeMet resulted at day 7 in increased Se levels of 229±14 and 269±24 µg/kg, respectively. On day 14, the Se content in muscle tissue was 68±2 µg/kg in non-supplemented diets and 121±6, 240±12 and 271±28 µg/kg for NaSe, SeOH and SeMet supplemented diets, respectively. Results from this trial suggest that Se deposition in broiler muscle tissue is dependent of the supplemented Se source. Supplementing inorganic Se (NaSe) results within 7 days in a decrease of Se in broiler muscle, while supplementing organic sources results in an increased level of Se in broiler muscle and these levels are stable over time. These findings provide a good indication for future research of the effect of Se sources on the selenium status in poultry.

Modifying the feeding strategy of lightest pigs helps to partly catch up their bigger counterparts

S. López-Vergé[1], D. Solà-Oriol[1], J. Bonet[2], J. Coma[2] and J. Gasa[1]

[1]Animal Nutrition and Welfare Service, Department of Animal and Food Sciences, Universitat Autònoma de Barcelona, Bellaterra, 08193, Barcelona, Spain, [2]Vall Companys Group, Polígono Industrial El Segre, 25191 Lleida, Spain; sergio.lopez.verge@uab.cat

Maximize the lightest pig's body weight (BW), constitutes a huge issue for pig producers. The standard feeding programs implemented during the growing-fattening period use to treat all the animals of a batch as a unit and change from one feed to the following at a fixed day, although other approaches may be implemented. A total of 1,067 males and females crossbreed pigs [Pietrain × (Landrace × Large White)] were used and classified, when leaving the nursery at 63 days of age, as Heavy (Hp, n=524) and Low (Lp, n=543) pigs. Pigs were individually weighed every three weeks until 125 d of age. Along this growing period, Hp and half Lp pigs were fed with three consecutive feeds following a standard feeding program (Std). Alternatively half Lp pigs were fed 'by budget', changing the two first feeds on the basis of an equivalent feed consumption instead of age (Sp). It is expected that Lp eat the two first feeds for more days, therefore favoring its startup and growth. Data were analysed with ANOVA using the GLM procedure of SAS. The two treatments started with the same BW at d 63 (LSp: 18.4 vs LStd: 18.4 kg, P>0.05). Higher BW was observed for animals corresponding to the LSp treatment at 83 (LSp: 29.5 kg vs LStd: 29.0 kg, P=0.052), 104 (LSp: 45.4 kg vs LStd: 43.5 kg, P<0.0001) and 125 (LSp: 63.8 kg vs LStd: 60.8 kg, P<0.0001) d of age respectively. BW of pigs HStd were always higher (P<0.0001) than the two Lp pigs (22.8, 35.2, 51.3 and 69.8 kg at 63, 83, 104 and 125 d). At 125 d of age LStd and LSp were, respectively, a 16.1% and 9.1% lighter than the HStd pigs. Pigs assigned to LSp treatment decreased around 44% the differences in BW between HStd and LStd pigs. Results suggest that Lp subjected to the Sp feeding strategy at the start of the growing period increase their growth rate and partially catch up their bigger counterparts.

Relation between selenomethionine in dietary selenium sources and selenium deposition in broiler

S. Van Beirendonck[1], B. Driessen[1], M. Rovers[2], L. Segers[2], I. Eising[2], A. Ruttens[3] and G. Du Laing[4]

[1]KU Leuven, Faculty of Engineering Technology, Kleinhoefstraat 4, 2440 Geel, Belgium, [2]Orffa, Vierlinghstraat 51, 4251 LC Werkendam, the Netherlands, [3]CODA-CERVA-VAR (Veterinary and Agricultural Research Centre), Leuvensesteenweg 17, 3080 Tervuren, Belgium, [4]Ghent University, Faculty of Bioscience Engineering, Coupure Links 653 Bl. B, 9000 Ghent, Belgium; eising@orffa.com

The aim of this study was to investigate the relation between the selenomethionine (SeMet) content in two different selenized yeast (SeYeast) products and L-selenomethionine (L-SeMet) on selenium (Se) deposition in broiler muscle tissue. Two different commercially available selenized yeast products were analyzed for SeMet content with HPLC-ICP-MS, after twofold enzymatic extraction (protease/lipase at pH 7.5) in a water bath (37 °C). Se in the form of SeMet was 26% in SeYeast A and 69% in SeYeast B. Male broilers were fed one of 4 treatment starter diets. All treatments had 4 pens with 5 animals per pen. Treatment 1 was supplemented with 0.2 mg/kg total Se from sodium selenite (NaSe). Treatment 2 and 3 were supplemented with 0.2 mg/kg total Se from SeYeast products A or B, which resulted in a supplemented Se in the form of SeMet of 0.052 mg/kg and 0.138 mg/kg for SeYeast A and B, respectively. Treatment 4 was supplemented with L-SeMet (Excential Selenium4000) at a dosing of 0.2 mg/kg total Se, which equals 0.2 mg/kg Se in the form of SeMet. Representative samples of the left breast of 3 broilers per pen were taken on d14 and analyzed for Se content by ICP-MS. Results show Se content in broiler muscle for treatment 1 (NaSe) of 133 µg/kg Se. Treatment 2 (SeYeast A) and 3 (SeYeast B) showed 161 and 267 µg/kg Se, respectively. Treatment 4 (L-SeMet) showed the highest Se content in muscle with 337 µg/kg Se. The data show that the Se deposition in muscle is linearly correlated with the added Se as SeMet. Linear regression of the data shows a fit with a R2 of 0.7186. This study shows that Se deposition in broiler muscle tissue is related to the added SeMet in the diet (P<0.0001). SeYeast with high SeMet content results in higher Se deposition compared to SeYeast with low SeMet content (P<0.0001). L-SeMet supplementation results in the highest Se deposition.

The effect of wheat microdochium content on broiler performance

M.E.E. Ball[1], J. Sloss[2] and A. Catlett[2]
[1]Agri-Food and Biosciences Institute, Agriculture Branch, Hillsborough, BT26 6DR, United Kingdom, [2]Moy Park Ltd., Coolhill, Killyman Road, BT71 6LN, Dungannon, United Kingdom; elizabeth.ball@afbini.gov.uk

It has been well established that different batches of wheat result in varying levels of broiler performance which is of huge economic importance. There is some concern that the microdochium content of wheat (from the non-mycotoxin fungi infection of Microdochium nivale) may reduce the nutritive value through lower grain weight but the relationship between microdochium content and broiler performance has never been investigated before. The aim of this study was to examine the effect of microdochium content on broiler performance. Microdochium content was determined on 20 wheat samples from a range of sources. Starter, grower and finisher broiler diets were formulated using each wheat sample (approx. 650 g/kg inclusion). The diets were balanced for metabolisable energy, crude protein, lysine, methionine, threonine and tryptophan. For each wheat sample, the inclusion of ingredients were adjusted slightly on the basis of the determined amino acid analysis of the wheat. Six hundred male chicks (Ross 308) were obtained on day of hatch, weighed, divided into 20 and arranged throughout the room according to a predetermined randomisation to give six pen replicates/treatment. The starter diet was offered from day 1 to 14, the grower offered from day 14 to 21 and the finisher from day 21 to 35. Dry matter intake (DMI), liveweight gain (LWG), and feed conversion ratio (FCR) were determined. Wheat microdochium contents were; 0.45, 1.04, 1.56, 3.10, 6.59, 7.88, 8.50, 15.52, 18.92, 23.91, 25.70, 25.82, 27.62, 30.23, 34.92, 38.09, 38.10, 40.48, 41.06 and 60.66 pg/ng. Wheat microdochium content had a significant linear effect on DMI in the grower stage ($R^2=0.16$, $P=0.047$), on finisher FCR ($R^2=0.17$, $P=0.006$) and on overall FCR ($R^2=0.19$, $P=0.048$). While the relationships were not strong, they were significant, and point to microdochium content as a possible indicator of how the bird will perform. However, more work is required on additional samples before firm conclusions can be drawn.

Effects of different levels of wDDGS and enzyme inclusion on fatty acid concentrations of broiler br

N.B. Rano, A.S. Chaudhry and S.A. Edwards
Bayero University Kano, Department of Animal Science, Faculty of Agriculture, P.M.B.3011, Kano, 700241, Kano, Nigeria; nbrano.asc@buk.edu.ng

This study determined the effects of different levels of wheat distillers dried grains with solubles (wDDGS) with or without enzyme inclusion on fatty acid composition of broiler breast meat. A completely randomized design with a 2×3 factorial arrangement comprising 2 wDDGS levels (0 and 15%) and 3 enzyme levels (no enzyme=NE and with enzyme A =EA or enzyme B =EB) was used. One hundred and eighty day-old Ross broiler chicks were individually weighed; wing tagged and assigned to 6 groups of 30 chicks per treatment. Each group had 5 sub-groups of 6 birds which were housed in floor pens covered with wood shavings. The birds received diets and drinking water ad libitum. Each enzyme was separately included at 0.5 kg /tonne of a relevant diet. At 45 days of age, 2 birds per pen as a replicate were killed, skinned and eviscerated. The breast muscles were excised, minced and extracted to obtain lipids which were analysed for fatty acid concentrations. The data were statistically compared for the effects of wDDGS, Enzyme and wDDGS × Enzyme interaction at $P<0.05$. The results showed that inclusion of 15% wDDGS supported higher palmitic acid concentration in breast muscles. While the diet containing EA caused increased palmitic acid concentration, the EB based diet increased the arachidonic acid concentration. It appeared that the n-6:n3 ratio was more affected by 15% wDDGS with EA ($P<0.05$) than 15% wDDGS with EB. It is concluded that wDDGS based diets supplemented with specific enzymes can positively affect the concentration of fatty acids of broiler breast meat and hence its quality.

Health and Performance of preweaned calves fed Alltech Blueprint milk replacer

M. Agovino[1], A.C. Berge[2] and F. Aubry[1]
[1]Alltech Biotechnology Centre, Summerhill Road, Dunboyne, Co. Meath, Ireland, [2]Berge Veterinary Consulting BVBA, Oulstbergstr 24, 1570 Vollezele, Belgium; magovino@alltech.com

The objective of this field trial was to evaluate the performance and health of calves fed with a new milk replacer (MR) (BP, Alltech Blueprint Milk Replacer) compared to MR currently used on the farm (CO, Pingvin Milk Replacer). The CO MR was a 20% protein, 15% fat MR made from whey, plant oils, sweet whey and wheat gluten. The BP MR was a 21.5% protein, 18% fat MT made from skim milk, whey, buttermilk, plant oils, and wheat gluten. The BP MR, contained a 5% inclusion of mannan-derived oligosaccharide components (Actigen®), organic selenium (Sel-Plex®), nucleotides (Nupro®), organic minerals (Bioplex®), and algae (All-G-Rich®). Day-old heifer calves were randomly enrolled into BP MR (18 animals) or CO MR (17 animals) group in June and July 2015. Calves were fed 3 liters milk replacer at 12,5% concentration, twice daily and starter grain and water ad lib. Serum total proteins in blood, milk intake, weight at birth, 4 weeks and 6 weeks, and incidence of diseases were measured. General linear models were developed controlling for serum total proteins and birth weight to evaluate difference in average daily gains between the milk replacers. The BP MR calves gained 0,607 kg/d compared to the control calves that gained 0,472 kg/d. The heifer calves fed the BP MR had significantly higher daily weight gain (0,134 kg/d higher, P-value=0,02) the first four weeks compared to the calves fed CO MR. At 6 weeks of age, the calves fed BP MR had a marginally significant higher daily weight gain (0,093 kg/d higher, P-value=0,06)) compared to the CO MR. The BP MR calves gained 0,662 kg/d compared to the CO MR calves that gained 0,569 kg/d. All the calves had high serum total proteins level and no calf had failure of passive transfer of immunity. The average serum total protein (STP) was 6.8 g/dl in the BP calves and 6.9 g/dl in the CO calves and the minimum STP recorded in BP and CO group was 5.5 and 5.4 g/dl respectively indicating that no calf in this study suffered from failure of passive transfer of immunity from colostrum. In conclusion, the BP MR, containing the Alltech supplements and the optimized ingredient composition most likely optimize performance in the calves compared to CO MR calves.

Effect of vitamin E in the diet of pigs on the fatty acids of muscles during frozen storage

T. Popova, M. Ignatova and P. Marinova
Institute of Animal Science, Pochivka str., 2232 Kostinbrod, Bulgaria; tlpopova@yahoo.com

The effect of vitamin E supplementation in the diet on the fatty acid composition of m. Longissimus dorsi (m. LD) and m. Semimembranosus (m. SM) during frozen storage was studied in 20 hybrid pigs. The animals were divided in two groups – one receiving a control diet and the other a diet containing vitamin E in amount of 400 mg/kg feed during the finishing period. Each group contained 10 animals. After slaughter, samples of the two muscles were taken, frozen and stored at -20 °C. The fatty acid composition was analysed in the fresh tissue and after frozen storage of 3 and 6 months. Data were statistically assessed by two-way ANOVA, the model including fixed effects ascribed to the treatment, storage duration and their interaction on the fatty acid profile of the muscles. Vitamin E influenced considerably the fatty acid composition of m. LD. The content of the saturated C14:0 and C16:0 was significantly increased (P<0.001) while that of C18:0 was decreased (P<0.01) in the pigs of the supplemented group. The proportion of C16:1 was higher in both m. LD (P<0.001) and m. SM (P<0.01) in the pigs receiving vitamin E. Significant effect of vitamin E on the content of the individual polyunsaturated fatty acids was observed only in m. LD. The pigs from the supplemented group displayed significantly lower content of C18:2, C18:3n-6, C20:4 (P<0.01). The duration of the storage had less influence on the fatty acid composition in the pig muscles. In m. LD the content of C14:0 decreased over time (P<0.001), while significant increase was observed for C18:0 (P<0.001) and total amount of the saturated fatty acids (P<0.05) in the pigs from the control and supplemented group. In m. SM the frozen storage affected only the content of C18:3n-3 (P<0.05) which increased at the end of the storage period in both groups.

Seed yield, oil content and fatty acid composition of new genotypes of spineless safflower

C. Stamigna[1], D. Chiaretti[1], E. Chiaretti[1], P.P. Danieli[2], R. Primi[2] and B. Ronchi[2]
[1]ENEA, Centro Ricerche Casaccia, Via Anguillarese, 301, 00123 Roma, Italy, [2]Università degli Studi della Tuscia, Dipartimento di Scienze Agrarie e Forestali, Via S. Camillo de Lellis, snc, 01100 Viterbo, Italy; ronchi@unitus.it

Safflower (Carthamus tinctorius L.) is increasingly used as a feed for ruminants for its nutritional characteristics and because it is reach of linoleic acid. In fact, several researches demonstrated that including safflower seed or oil in the ruminants feed, increases the presence of health-promoting fatty acids (cis-9, trans-11 conjugated linoleic acid; CLA) content in body and milk. Our study was aimed to establish new genotypes of spineless safflower with increased seed productivity and oil content (linoleic acid in particular) to be grown in the Mediterranean area. Starting from 36 spineless genotypes with highly variable flower colour, 8 lines with proven stability of morpho-physiological characters were selected, and in 2012 they were grown in central Italy for testing seed productivity, oil content and fatty acid composition. Average seed production was 2.46 t/ha (1.40-3.48) and oil content was 29.79% (26.64-33.53). The oil contained 78.95% linoleic (77.70-79.70), 11.84% oleic (11.37-12.37), 6.24% palmitic (5.63-6.84), 2.31% stearic (1.92-2.77), 0.26% begenic (0.14-0.26), 0.17% linolenic (0.14-0.21), 0.10% miristic (0.09-0.11), 0.09% arachinic (0.06-0.16) and 0.05% palmitoleic (0.02-0.18) acids. From this preliminary study, 3 promising lines with higher production of seeds and content of linoleic and oleic acids were selected and sowed in January 2016. In order to assess the possibility of utilization for animal feeding, appropriate tests should be performed.

Effect of probiotic mixes on growth performance, mortality and incidence of diarrhea of piglets

J.E. Miranda Yuquilema[1], D. Sánchez Macías[2], A. Marin Cardenas[1] and L. Marrero[1]
[1]Universidad Central Marta Abreu de Las Vilas, Facultad de Ciecias Agropecuarias, km 5 1/2 carretera Camajuani, Santa Clara., 54830, Santa Clara, Cuba, [2]Universidad Nacional de Chimborazo, Producción Animal e Industrialización, Avda. Antonio Jose de Sucre s/n, 060150, Riobamba, Ecuador

The aim of this study was evaluate probiotic mixes on growth performance, mortality and incidence of diarrhea of newborn piglets. 159 piglets (Yorkshire-Hampshire breed), 80 males and 79 females, were randomly selected for 5 treatments: I. Control; II. Single dose every 7 d during 33 days of 3 ml of L. acidophilus ATCC® 4356™ and Kluyveromyces ssp fragilis ATCC® 12424™ grew on molasses and and torula yeast; III. Single dose every 7 d during 33 days L. acidophillus ATCC® 4356™ grew on los-fat milk and molasses; IV. L. acidophillus ATCC® 4356™ grew on low-fat milk and molasses, but administrate into the growing pellet for piglets; V. single dose of L. bulgaricus ATCC® 11842™ + L. acidophillus ATCC® 4356™ + Kluyveromyces ssp fragilis ATCC® 12424™ grew in molasses and torula yeast. All the animals were growing and feeding under the same system production and management. From 7 to 15 days postpartum, the groups III, IV and V obtained the highest weight gain (2.1, 2.2 and 2.2 kg, respectively), and the II group had the lowest weight gain (1.3 kg). On the other hand, during the 16 to 30 days postpartum period, groups IV and V (4.9 and 5 kg), followed by II piglets group (4.6 kg), had the highest weight gain than control and III group (4.3 kg). From 31 to 33 days, there not were statistical differences among the groups. During the first week of the experiment, the highest diarrhea incidence was in the V group (9.6%), while the lowest percentage was found in control group (6.4%). However, during the rest of the experiment with probiotics, the lowest diarrhea incidence was presented in the piglets groups treated with the different probiotics mixes (aprox. 4% vs 9%). Also, during the experiment, the control group presented higher mortality percentage when compared with the probiotic groups. In conclusion, administration of L. acidophilus in the growing piglets is better than single dose administration. On the other hand, the probiotic mix including L. bulgaricus increases significantly the weight gain of piglets.

Replacing dietary corn with sugar beet pulp on ruminal function and performance of Egyptian buffalo

H.M. El-Zaiat[1], H.M. Abu-Zeid[1], A.S. Morsy[2], M.F. Attia[1], M.A. Abaza[1] and S.M.A. Sallam[1]
[1]Faculty of Agriculture, Alexandria University, Animal and Fish Production Department, Aflaton St., El-Shatby, Alexandria, 21545, Egypt, [2]Arid Land Cultivation Research Institute, City of Scientific Research and Technological Applications, Livestock Research Department, Universities and Research Center District, New Borg El-Arab, 21934, New Borg El-Arab, Egypt; hani.elzaiat@alexu.edu.eg

In Egypt, scarce information is available regarding the impacts of substituting increasing levels of dried pelleted sugar beet pulp (SBP) for corn in buffalo diets. So, the objectives of this study were to evaluate the effects of replacing corn with SBP on rumen fermentation and performance of buffalo growing calves. Forty calves male Egyptian buffalo (237.2 ± 24.46 kg) were allotted randomly to 1 of 4 experimental diets: Control (0% SBP) fed basal diet containing 60% corn, SBP33% fed basal diet containing 40% corn; SBP67% fed basal diet containing 20% corn, SBP100% fed basal diet containing 0% corn for 130 days. Animals were weighed biweekly and DMI was measured daily. Rumen liquor samples were collected every month before the morning feeding using a stomach tube for measuring the rumen pH, NH_3-N and short chain fatty acids. The apparent nutrients digestibility coefficients were determined using lignin contents as an internal marker. The Acid detergent lignin was determined according to Van Soest et al. The DMI, SCFA and NH_3 concentrations were not affected ($P>0.05$) by treatments. Substitution of corn by 33% and 67% of SBP increased ($P<0.05$) ADG compared with SBP0% and SBP100% diets (1.21, 1.81, 1.08 and 1.11, respectively). Ruminal pH values ($P=0.001$) was increased with increasing levels of SBP (6.89 with SBP0%, SBP33% to 7.03, 7.19 with SBP66% and SBP100%, respectively). The complete substitution of corn by SBP improved ($P<0.05$) the digestibility of NDF, ADF, cellulose and hemicelluloses values. In conclusion, SBP is a good candidate for partially substituting corn in ruminants diets without adverse effects on animal performance.

Effect of a novel silage inoculant on fermentation profile and aerobic stability of ryegrass silage

V.J. Taha[1], P. Townley[2] and J.A. Huntington[3]
[1]Harper Adams Univ., Edgmond, TF10 8NB, United Kingdom, [2]Univ. Duhok, Kurdistan region, Iraq, [3]EM Effective Micro-Organisms, Devon, EX16 7HN, United Kingdom; jhuntington@harper-adams.ac.uk

The effect of a mixed inoculant containing L. plantarum, L. casei plus S. cerevisiae (EM silage) on silage fermentation and aerobic stability was studied. First cut perennial ryegrass (PRG) was treated with either EM Silage (80 ml/t) or water equivalent as control (C). Treated forage was ensiled in minisilos (1.5 kg) with 6 reps for 100 days. Samples of fresh and ensiled material were analysed for proximate analysis and fermentation characteristics. Aerobic stability of the ensiled material was determined by temperature difference to ambient. Samples were contained in insulated boxes open to atmosphere at 18-20 °C for 20 days. EM treated PRG had similar DM, CP and WSC to those observed for the C (355, 357; 70.2, 71.1; 43.7, 41.3 g/kgDM for EM and C treatments respectively). Yeast and mould counts were found to be similar for both treatments (2.7, 2.9; 2.35, 2.65 log10cfu/g for EM and C respectively), however lactobacilli (log10cfu/g) were higher in the EM treated silage (2.45, 1.65 log10cfu/g for EM and C respectively). Treating forage with EM inoculant had no effect ($P>0.05$) on silage pH or NH_3-N %TN (4.13, 4.15; 1.77, 1.66 for EM and C respectively). EM treatment reduced ($P<0.001$) lactic acid content (13%) but increased $P<0.001$ volatile fatty acid content of the EM silage compared to C by 34% largely due to elevate acetic acid ($P=0.002$) concentration (12.1, 8.8 g/kg EM and C respectively). Total alcohol concentration was elevated by 228% in EM treated silage resulting from elevated concentration of ethanol (143%), propane-1-ol (294%), and propane-1,2-diol (389%). EM treated silage was found to be more aerobically stable than C. Control treatment reached 3 °C above ambient room temperature (18-20 °C) within 6.8 days whereas EM Silage took a further 7 days to reach the same temperature difference. Treating PRG with EM Silage inoculant was found to enhance the fermentation characteristics and aerobic stability of the PRG silage.

Lasalocid can manipulate macro-mineral utilisation in dairy cows

D.M. McNeill[1], J. Olm[1], M. Whybird[1] and T. Cowper[2]
[1]*University of Queensland, School of Veterinary Science, Gatton Campus, 4343 Gatton, Australia, [2]Zoetis, 6/5 Rider Boulevard, 2138 Rhodes, Australia; d.mcneill@uq.edu.au*

Ionophores are known to manipulate the organic outputs of the ruminal microbiome. Yet their capacity to transport mineral ions across membranes, including mammalian, has received little attention despite it being an action that defines ionophores. Our aim was to show that a calcium-ionophore, lasalocid, influences the utilisation of dietary minerals associated with milk, bone and electrolyte balance, in lactating cows. Ten Holstein-Friesian multiparous, mid-lactation (about 30 l milk/day) cows, n=5 per treatment, in individual pens except at milking, were fed morning and afternoon 21.5 kg DM/day of a total-mixed-diet including a 1.0 kg DM/day separate 'carrier' mix (± lasalocid sodium, 300 mg/cow/day) for 6 weeks. For the last 10 days, indigestible markers (Chromium chloride (Cr) and Cobalt-EDTA (Co)) were added daily to the morning carrier-mix and for the last 6 days faecal grab-samples collected to determine ratios of Calcium (Ca), Magnesium (Mg), and Phosphorus (P) to Cr or Co excreted relative to the ratios consumed, to estimate the digestibility of Ca, Mg and P. Blood was collected on the final day to measure plasma concentrations of total Ca, Mg, P, Sodium (Na), Potassium (K), and Chloride (Cl). Responses to lasalocid were tested by ANOVA. In response to lasalocid the concentration of total Ca in plasma tended to increase (2.27 v. 2.13 mmol/l, P=0.06) and Cl declined (101 v. 104 mmol/l, P<0.05), whilst plasma Mg, P, Na and K were unaffected. Digestibility declined for Mg (36.2 v. 41.9%, P<0.05), but not for Ca or P. Limited data are available on the capacity of ionophores to manipulate mineral utilisation in lactating cows. The decline observed in Mg digestion was unexpected. However, since plasma Mg remained within a healthy range and others have found lasalocid to improve Mg utilisation, it would be premature to conclude that Mg digestibility is a problem. The potential improvement in plasma Ca and decline in plasma Cl could benefit feed intake and reduce the risk of metabolic disease such as milk fever.

Assessment of various forms of metals on the antioxidant activity of a selection of feed components

M.J. Concarr, L.A. Byrne and R.A. Murphy
Alltech, Research, Alltech Bioscience Centre, Summerhill Road, Sarney, Dunboyne, Co. Meath, Co. Meath, Ireland; rmurphy@alltech.com

Antioxidants play a vital role in animal feed preservation as they inhibit the oxidation, damage and subsequent destruction of various important premix constituents. One antioxidant commonly added to animal feed to limit oxidative damage is the synthetic antioxidant butylated hydroxytoluene (BHT). The aim of this study was to determine whether there was a significant difference in the oxidation rate of BHT in the presence of various organic and inorganic forms of iron, copper, zinc and manganese. This was performed by utilising the β-carotene bleaching assay. BHT and various forms of a selection of trace minerals were added to a β-carotene solution and the absorbance was measured immediately. The samples were then subject to pro-oxidative conditions for 20, 40 and 60 minutes. The decrease in absorbance corresponds to the destruction of the β-carotene. The degradation rate of the β-carotene in the presence of the minerals was calculated and compared to a positive sample which contained no trace minerals. The difference between these degradation rates corresponds to the inhibitory effect that the trace minerals have on the antioxidant activity of BHT. Results indicated that the inorganic minerals generally inhibited the antioxidant activity of BHT at a higher rate than organic minerals. Observable differences between the commercial organic mineral products were also detected indicating not all organic mineral forms are equally effective. Commercially available premix samples containing vitamins and organic or inorganic mineral sources were also assessed for antioxidant activity over a 20 day storage period using a DPPH assay. The results showed that the antioxidant activity of the premix sample containing inorganic minerals decreased by 20.04% while the premix sample containing organic trace minerals only decreased by 5.99%. These results indicate that by incorporating specific organic trace mineral products in livestock premixes in place of their inorganic counterparts will prolong their storage life by decreasing the rate of antioxidant destruction which enhances the quality of the feedstuff.

Influence of trace minerals on feed grade phytase activity
M. Gaffney, R. O'Rourke, L. Byrne and R. Murphy
Alltech, Sarney, Summerhill Road, Dunboyne, Co Meath, A86 X006, Ireland; mgaffney@alltech.com

Monogastric diets are largely comprised of plant-based materials, of which phytic acid is a common constituent. While representing an abundant source of organic phosphorous, phytic acid also acts as an anti-nutritional factor, adversely affecting protein digestibility. Monogastrics are unable to synthesise the enzyme required to hydrolyse phytic acid, so the inclusion of exogenous phytase enzymes to respective diets is common practice. Consequently, the potential for antagonistic interactions between exogenous enzymes and other feed components becomes an important consideration for premix formulations, particularly in terms of efficacy and productivity. The present study investigated the influence of commonly supplemented trace minerals (iron, copper, zinc) on phytase activity. Dietary minerals come in a variety of forms, from inorganic salts to complexes and chelates, generically referred to as organic trace minerals. Mineral form and structure can have a pronounced effect on phytase function, potentially interacting with the enzyme's active and binding sites. Microbial phytase sources differ in function and specificity based on botanical origin. In this instance, a number of commercial phytase preparations were exposed to both organic and inorganic minerals, at inclusion levels reflective of poultry premixes. Following interaction, residual enzyme activity was determined by colorimetric assay. Results indicated that mineral source, type and concentration had a pronounced effect on enzyme function. Significant reductions in phytase activity were detected for each of the minerals tested (Fe, Cu, Zn), whereby interactions with inorganic minerals equated to a 2-3 fold reduction in phytase activity relative to their organic mineral counterparts ($P \leq 0.05$). Of the minerals assessed, zinc had the least antagonistic effect on enzyme function, compared to iron and copper which had strong inhibitory potentials on phytase activity. Negative interactions between high cost feed ingredients such as enzymes and trace minerals carry real and often underappreciated financial costs. Given the effects of mineral type and concentration on enzyme activity, such interactions may contribute to the varied animal responses observed in vivo following phytase supplementation.

Effect of replacing soybean meal with field pea on meat quality in the diets for turkeys
R. Juodka, V. Juskiene, R. Nainiene and R. Juska
Lithuanian University of Health Science, Institute of Animal Science, R. Zebenkos 12, 82317, Baisogala, Lithuania; violeta@lgi.lt

A study was carried out to determine the efficiency of soybean oil-meal replacement with increasing amount of peas (from 10% to 40%) in the diets for turkeys on anatomic carcass dissection and chemical indicators of breast, shoulder, thigh and shin muscles, physical indicators of breast muscles and the content of tryptophan and oxyprolin. The anatomic dissection of carcasses indicated that male turkeys fed peas had from 0.28 to 0.61% ($P>0.05–P<0.05$) less shoulder muscles, whereas female turkeys of the same groups had from 0.09 to 0.31% ($P>0.05–P<0.05$) more abdominal fat compared with the turkeys the diets of which did not contain any peas. The composition of the feed did not affect breast, thigh, shin muscles, internal edible body parts and the dressing percentage of both genders and also shoulder muscle content in females and abdominal fat content in males. The chemical analysis of meat showed that feeding the diets with the increasing amount of peas resulted in from 0.69% to 0.82% lower amount of fat in female shoulder muscle ($P>0.05–P<0.05$). Soybean oil-meal replacement with field peas in broiler turkey diets had no significant influence on the other physicochemical indicators of meat as well as the content of trypthophan and oxyprolin in the muscles. Our study indicated that soybean oil-meal replacement with peas up to 20% from 0 to 4 weeks, 25% from 5 to 8 weeks, 30% from 9 to 12 weeks and 40% from 13 to 16 weeks, and with 100% soybean oil-meal replacement at the final stage had no negative influence on the carcass qualities of broiler turkeys.

Transcriptomic signature of selenium versus zinc and iodine supplementation in sheep

F. Palazzo[1], R. Elgendy[2], F. Castellani[1], L. Grotta[1], S. Marchetti[1], M. Giantin[2], M. Dacasto[2] and G. Martino[1]
[1]University of Teramo, Faculty of Bioscience and Technology for Food, Agriculture and Environment, Via C. Lerici 1, 64023 Mosciano Sant'Angelo, Italy, [2]University of Padua, Department of Comparative Biomedicine and Food Science, Viale dell'Università 16, 35020 Legnaro, Italy; fpalazzo@unite.it

Selenium (Se), zinc (Zn) and iodine (I) are significant elements in ruminant diet for their role on the immune system, fertility and hormone activity. In cattle and sheep, deficiencies of these elements have been confirmed under natural grazing conditions in many countries where soils produces pasture grasses and legumes low in content. Three different groups of 10 sheep each were administered for 40 days diets containing high supplementations (almost reaching the highest tolerated amounts according to EC restrictions) of Se (1.4 mg/animal/day), Zn (212 mg/animal/day) and I (28 mg/animal/day). The effects of supplementation on whole blood transcriptome were investigated using microarray technology and a sheep custom microarray slides. In the group receiving Se, 942 and 244 mRNAs were shown to be up and down-regulated, respectively (1.5 $\geq$ FC $\leq$1.5; FDR<0.05). The functional annotation of these differentially expressed genes (DEGs) showed an involvement of immune system-related biological processes (lymphocyte and leukocyte activation, cytokine binding, T-cell differentiation) and pathways (cytokine and interleukin signalling). Then, we verified if the transcriptional effects induced by Se supplementation (in term of DEGs) were mirrored by the other two micronutrients. When we compared the transcriptional portrait of sheep given Zn and Se, a total of 231 and 20 mRNAs appeared to be commonly up-and down-regulated, respectively. Likewise, 166 and 13 mRNAs were commonly up and down-regulated in sheep provided with I and Se, respectively. In conclusion, these results show that micronutrient supplementation seems to affect the sheep blood transcriptome; moreover, some DEGs were proved to be similarly affected, particularly when we compared animals given Se and Zn. Clearly, further studies are needed to investigate more in depth the consequences of these effects on animal health and the possible presence of potential crosstalks between each micronutrient-related molecular pathways.

Inclusion of palm kernel expeller meal in barley-based concentrates for beef cattle

D. Magee[1,2], A.P. Moloney[2], A. Kelly[1], E.G. O'Riordan[2] and M. McGee[2]
[1]University College Dublin, School of Agriculture, Belfield, Dublin 4, Ireland, [2]Teagasc, Animal & Grassland Research and Innovation Centre, Grange, Dunsany, Co. Meath, Ireland; mark.mcgee@teagasc.ie

This study evaluated the inclusion of palm kernel expeller meal (PK) in barley-based concentrates on the performance of beef cattle. Late-maturing crossbred steers (n=57: 460 kg, s.d. 35.0) were blocked on breed and live weight and randomly allocated to one of five concentrate treatments (C). The control C comprised 862 g/kg rolled barley (RB), 60 g/kg soya bean meal (SBM), 50 g/kg molasses and 28 g/kg minerals/vitamins. For the other C, 80+20, 160+40, 240+60 and 340+60 g of RB+SBM were replaced with 100, 200, 300, and 400 g PK (crude protein (CP) 158, and oil B 72, g/kg dry matter (DM)) /kg fresh weight, respectively. Steers were individually offered 3.0 kg DM of the respective C as a supplement to grass silage (DM digestibility, 640 g/kg; CP, 126 g/kg DM) offered ad libitum during a 71-d growing study (GS). At the end of GS they were offered silage-only ad libitum for 21 d, and then blocked on weight and GS treatment and randomly re-assigned to the one of the same five C. Following a 29 d dietary adaption period, C were offered ad libitum plus 3 kg silage (fresh weight) daily for a 70-d finishing study (FS). Data were statistically analysed using mixed models with C as fixed and block as random, effects. Linear and quadratic response terms were included. During the GS, DM intake was lower (P<0.01) for PK200 and PK400 compared to other C, whereas average daily gain (ADG) did not differ (P>0.05) between C. During the FS there was no effect (P>0.05) of C on DM intake, whereas ADG and slaughter weight declined linearly (P<0.05) with increasing PK inclusion. Carcass weight, kill-out proportion, conformation score and, colour of subcutaneous fat and M. longissimus dorsi muscle, did not differ (P>0.05) between C. Carcass fat score was lower for PK100 than PK400 (P<0.05), with other C intermediate (P>0.05). In conclusion, PK can be included in a barley-based C at up to 400 g/kg when offered as a supplement to grass silage and up to 100 g/kg when offered ad libitum.

Impact of incorporating Moringa oleifera seed meal as protein source in growing lambs rations

S.I. El-Naggar[1], G.A. Abou-Ward[1], M.A. Tawila[1], S.M. Gad[1] and A.M. Ali[2]
[1]National Research Centre, Animal Production, Dokki, Giza, 12311, Giza, Egypt, [2]Cairo University, Animal Production, Giza, 12613 Giza, Egypt; soadelnaggar75@gmail.com

Thirty growing Rahmani lambs (27.25±0.25 kg average body weight 6-7 months old) were used to evaluate the impact of incorporating Moringa oleifera seed meal as protein source in a total mixed ration (TMR) for 90 days feeding trial followed by 21 day digestion. Moringa oleifera seed meal (MSM) was used for partial substitution of ration protein by (0, 25 and 50%) in three experimental rations, control (R1), R2 and R3, respectively. Results were statistically analyzed via general liner model for one way analysis and indicated that the best nutrients digestibility and nutritive value were recorded with feeding R1 compared with R2 or R3. Also the highest average daily gain recorded with R1 (222.2 g) followed by R3(208.3 g) and R2 (203 g). Thus it could be concluded that MSM can use as protein source in a TMR for growing lambs rations up to 50%. And more studies were needed to evaluate different levels replacement for MSM as protein source in the rations of ruminant.

Effect of diet supplementation on tenderness development and on proteomic changes in lamb meat

A. Della Malva[1], M. Albenzio[1], G. Annicchiarico[2], M. Caroprese[1], A. Santillo[1], A. Sevi[1] and R. Marino[1]
[1]University of Foggia, Department of Agricultural Food and Environmental Sciences, Via Napoli,25, 71121, Italy, [2]Council for Research and Experimentation in Agriculture (CRA), Borgo Segezia, 71020, Italy; antonella.dellamalva@unifg.it

Although it is established that diet affect meat quality, a better understanding of proteomics changes due to different dietary strategies may be useful to improve tenderness. We investigate the effect of supplementing linseed and quinoa seed and their combination on proteomic changes associated with tenderness in lamb meat. 32 Altamurana weaned lambs were divided in 4 groups with different diet supplementation: control (CO), quinoa (QS), linseed (LS) and combination of quinoa and linseed (QS+LS). After 50 days of supplementation, animals were slaughtered and longissimus thoracis muscle were removed. Warner Bratzler shear force (WBSF) and changes of myofibrillar proteins with SDS-PAGE and Two-Dimensional Gel Electrophoresis (2DE) were estimated. All data were subjected to an analysis of variance using the GLM procedure including fixed effect due to dietary supplementation. Based on WBSF values, LS meat showed lower values (P<0.01) than CO, QS and QS+LS. Diet supplementation significantly affected the amount of several myofibrillar proteins bands: QS, LS and QS+LS groups showed lower values of myosin heavy chain (P<0.001) and higher values of bands from 190-130 kDa (P<0.05) than control group. LS meat showed the lowest values of desmin (P<0.05), Troponin T (P<0.001), Troponin I (P<0.001) and Troponin C (P<0.001) intact proteins, together with the highest Myosin Light Chain 2 (MLC2; P<0.01) and 32-25 kDa (P<0.001) bands. 2DE image analyses showed significant differences in the total number of spots among experimental groups with the highest value in LS meat. Proteins separation also revealed differences in the spot expression patterns of MLC2 isoforms. Results indicate that LS dietary supplementation may have the potential to induce changes in the tenderness mechanisms and on proteolysis of structural myofibrillar proteins.

In vitro evaluation of plants vegetable waste: effects on rumen fermentation

D. Tedesco, J. Parisi and L. Garavaglia
Department of Health, Animal Science and Food Safety, University of Milan, Via Celoria 10, 20133 Milano, Italy;
doriana.tedesco@unimi.it

The plants vegetable waste originating from the processing of food and phytopharmaceutical products, still contain valuable phytochemicals. These waste could be recovered as feed or feed additives in ruminant nutrition. In an anaerobic batch culture rumen system, the activities of 18 vegetable waste were evaluated on microbial growth, pH, ammonia-nitrogen (NH_3-N), volatile fatty acid concentration (VFA) and methane production. Of the waste were tested the raw material and their extracts after water, ethanol and heptane extraction. The waste, raw and extracts, in two replicates, were added into the bottles at three different concentrations. Data on pH, concentration and proportion of VFA, NH_3-N and total bacterial count (after logarithmic transformation), were statistically analysed using the PROC GLM of SAS. Considering the effect on energy metabolism, among the screened waste, mango peel, larch sawdust and thyme waste, for all the dosage tested, exhibited an increase in total VFA concentration, associated with a slightly increase in methane production. Considering the effect on rumen nitrogen metabolism mango peel and larch sawdust decreased the NH_3-N concentration (P<0.05). Of the plant waste with a detrimental effect on rumen fermentation, olive fruit waste decreased the VFA concentration (P<0.05). In conclusion, mango peel exhibited positive effects on rumen activity followed by larch sawdust and thyme, and they can be utilised to improve the rumen environment and nutrient use efficiency. Furthermore, the following beneficial effects were highlighted: mitigation of the environmental impact of livestock by improving the efficiency of diets and by the complete and efficient use of plant waste.

Effect of early introduction of milk replacer feeding on pre and post-weaning weight gain

J. Barry[1,2], J.P. Murphy[2], E. Bokkers[1] and E. Kennedy[2]
[1]Wageningen UR Livestock Research, Wageningen University P.O. Box 338, 6700 AH, Wageningen, the Netherlands,
[2]Teagasc, AGRIC, Moorepark, Fermoy, Co. Cork, Ireland; john.barry@teagasc.ie

Early introduction of milk replacer feeding could be a future strategy to reduce transmission of diseases to neonatal dairy calves. This study aimed to investigate the effect on pre and post weaning body weight (BW) of dairy heifer calves introduced to MR feeding at different timings. A study population of 120 dairy heifer calves (Holstein Friesian (HF) and HF × Jersey) was used. Calves were weighed immediately after birth (mean BW at birth 33.2 (S.D 3.34) kg) and received 8.5% of their birth BW in colostrum, obtained from the first milking of the dam within two hours of birth, and fed via oesophageal tube. Calves were then randomly assigned to one of three treatment groups: (1) fed milk replacer at the feed following colostrum feeding (MR); (2) fed milk replacer following colostrum feeding and four feeds of transition milk (TM); and (3) fed milk replacer following colostrum feeding, four feeds of transition milk and ten feeds of whole milk (WM). Milk replacer (26% crude protein) was fed at a rate of 15% of birth BW and prepared at 15% concentration. Weight recording was carried out weekly from birth to weaning, and at two week intervals post weaning, to monitor the effect of the treatments on BW. There was no difference in BW at birth (33.2 kg) or at weaning (93.2 kg) between the treatments. There was no difference in time to reach target weaning weight (90.2 days), and no difference existed in average daily gain (ADG) from birth to weaning (0.66 kg/day) between treatments. Body weight at 140 and 240 days old did not differ (129 and 215 kg, respectively). Similarly there was no difference in ADG from birth to 240 days old (0.75 kg/day) and weaning to 240 days old (0.81 kg/day) between treatments. Results indicate that early introduction of MR feeding does not affect BW or ADG of dairy heifer calves pre and post-weaning. This suggests that MR can be offered immediately after colostrum feeding to reduce the risk of disease transmission to neonatal calves; however colostrum feeding needs to adhere to best practice.

Effects of enhanced distillers' grains on caecal microbiome composition in growing broiler chicks

A.A. Tonks, C. Rymer and D.T. Juniper
Animal, Dairy and Food Chain Science, School of Agriculture, Policy and Development, University of Reading, Reading, RG6 6AR, United Kingdom; a.tonks@pgr.reading.ac.uk

Wheat distillers' grains (WDG), a cereal by-product of alcohol distillation, is potentially a feedstuff that will become increasingly abundant with the expansion of bioethanol production. There may be scope for removing some components of these grains for industrial use, leaving a residue that may still have value as an animal feed. WDG, without the addition of solubles, were subjected to a series of procedures in which first lipid was extracted, using supercritical CO_2, followed by protein by sodium hydroxide and ethanol. The remaining carbohydrate fraction was finally extracted using steam explosion and fermentation with yeast. This experiment aimed to investigate the effects of replacing rapeseed meal with WDG products on the gut microbiome of 10-day old chicks. Male Ross 301 broiler chicks were fed one of five diets, 77 g/kg of which consisted of: rapeseed meal (CON), fresh WDG (FDG), dried WDG (RDG), WDG following treatment as above (EDG) or RDG following treatment as above (ERDG). Diets were isoenergetic and had equal concentrations of limiting amino acids. At 10 days of age, birds were sacrificed and pooled caecal contents (n=2 per pool, four samples per treatment) were taken for DNA sequencing on the Roche 454 platform. The microbiome of each treatment was similar though Erysipelotrichales (P=0.003) and Lactobacillales (P=0.067) were significantly higher in RDG than CON. RDG was associated with significantly increased Lactobacillales, Clostridiales and Erysipelotrichales, whilst decreasing Anaeroplasmatales (P<0.05). ERDG significantly reduced Erysipelotrichales (P=0.039) and increased Anaeroplasmatales (P=0.059). RDG also showed the smallest range between samples, perhaps suggesting that feeding dried WDG stabilises the gut microbiome but that alteration of distillers' grains can perturb the gut microbiota.

Effect of 4 feed additives on daily salivation and ruminal pH dynamic in lactating dairy cows

S. Stephan
Phode Sciences, ZI Albipole, 81150 terssac, France; sstephan@phode.fr

Latent acidosis is common in high productive dairy herds because milking cows are fed with high energy diets to meet production requirements in early lactation. In ruminants, the most important amount of saliva buffers is produced during rumination. This study investigates the effect of different sensory feed additives supplementation on the daily saliva production and the ruminal pH dynamic. The three feed additives evaluated are oleoresins from pungent spices. They are classified as botanically defined aromatic substances regarding the regulation (EC) N°1831/2003. Four experimental concentrates are compared (A control, and 3 additives B, C, D) over a balanced Latin square design (4 dairy cows and 4 experimental 14-d). Feed additves are incorpated and fed via 3 kg of concentrates. Diet is formulated to be acidogenic. Concentrate is fed twice a day at 9.00 am and 5.00 pm. Total amount of diet is limited to 95% ad lib. Saliva amount produced during the day (estimated from duodenal liquid flow marked with Co-EDTA) and continuous ruminal pH are measured from 8.30 am during 24 hours. Statistical analysis was carried out using ANOVA with procedure Mixed of SAS. Comparison of adjusted means B, C and D feed additives to the control is done using a Dunett's test. Diet formulation and its distribution mode lead to sub-clinical acidosis. Daily rumen pH averages 6.17 with a pH<6.0 duration of 7.4 h/d, matching the criteria of latent acidosis defined by Sauvant et al. No significant effect of feed additves on feed intake and milk production is observed. Feed additive B tends to increase the liquid volume in the rumen (103.0 vs 82.0; P=0.06) and the water flow in the duodenum (11.01 vs 9.16 l/h; P=0.10) compared to control group. No general effect of feed additives is observed on the pH dynamic (P>0.05) even if there is a numerical trend with feed additive B. Feed additive B tends to increase the numerical daily saliva production (l/day), average pH and tends to decrease the daily pH duration under 6 value (7.31 vs 6.45 h/day; NS).

Developing and implementing technological wellbeing evaluation in dairy animals: COST DairyCare

C.H. Knight[1], A. Castro-Costa[2] and G. Caja[2]
[1]*University of Copenhagen, IKVH, Dyrlægevej 100, 1870 Frb C, Denmark,* [2]*Universitat Autonoma de Barcelona, G2R, Faculty of Veterinary Science, 08193 Bellaterra, Spain; chkn@sund.ku.dk*

The problem faced by modern animal production systems is to maintain the best possible standards of animal health and welfare ('wellbeing') together with high productivity and a minimal environmental impact in an era of greatly increased demand for high-quality livestock products, larger livestock units and decreased contact between animals and husbandry staff. This requires better connection between two complementary but distinct 'states-of-art' concerning the monitoring and managing of animal wellbeing. On the one hand, the technological approach, employing automated sensing and the power afforded by novel proteomics and metabolomics. On the other hand, the observational approach of combining a number of relevant behavioural measures into a Welfare Index. Animal wellbeing is multifactorial, and achieving the best solutions will require inputs from animal scientists, ethologists, veterinarians and agricultural engineers working in partnership with SMEs and dairy equipment companies to create technologies that are then integrated into best practice blueprints that are then channeled through dairy consultancies to achieve better dairy animal welfare on the farm. By enabling networking, exchange of skills, capacity building through enhanced collaboration and targeted dissemination, this COST Action aims to achieve exactly what is required. In this presentation we will review the achievements of the Action's first two years with focus on lameness detection, metabolic diseases, stress biomarkers and assessment of social behaviour. Then we will preview the activities that are planned for the remaining two years of the Action.

Indicators of early lactation disease in the dairy cow from the previous lactation and dry period

G.L. Smith[1], M.G.G. Chagunda[1], C.J. Ashworth[2] and N.C. Friggens[3]
[1]*Scotland's Rural College, Future Farming Systems, West Mains Road, EH9 3JG, Edinburgh, United Kingdom,* [2]*University of Edinburgh, The Roslin Institute, Easter Bush Campus, EH25 9RG, Edinburgh, United Kingdom,* [3]*INRA/AgroParisTech, MoSAR, 16 rue Claude Bernard, Paris 05, France; gls90@hotmail.co.uk*

Identification of different indicators of cow health is necessary for the development of precision farming tools which aid in the detection of disease. This study aimed to identify indicators of early lactation disease from data recorded at the end of lactation and throughout the dry period preceding a disease event. Retrospective analyses were conducted on 460 cow-lactations from the long term genetics × environment study in the SRUC's Langhill herd. Each cow-lactation was assigned to one of four health groups based on predominant disease incidence in the first 30 days after calving; no clinical disease (NCD) (n=339), reproductive (REP) (n=69), subclinical mastitis (SCM) (n=37) or metabolic (MET) (n=15). Weekly body weight and body condition score data from the day of dry-off to the day of calving were used to calculate weekly body energy content (MJ) throughout the dry period. Dry-off milk yield, milk yield and fertility measures in the lactation of disease were also recorded. SCM cows had greater milk yield at dry-off than REP or MET cows (P<0.01). REP cows had a greater rate of change in body energy content during the first 15 days of the dry period compared to NCD cows; -18.26±7.44 MJ per day vs 0.6±5.11 MJ per day, respectively (P<0.01). In the first 30 days of lactation in which disease occurred, REP and MET cows had average daily yields 3.0 and 5.2 litres lower than NCD cows respectively. The results indicate that cows which developed different diseases in the first thirty days of lactation had different characteristics in their physiology and production traits during the transition from being lactating cows to dry cows and during the dry and early lactation periods. This highlights the importance of identifying risk factors for production diseases that occur after calving, before the transition period starts.

Effect of abrupt weaning at housing on welfare biomarkers in beef suckler calves

B. Earley, E. Lynch and M. McGee
Teagasc, AGRIC, Grange Research centre, Dunsany, Co. Meath, Ireland; bernadette.earley@teagasc.ie

Altered leukocyte populations and elevated acute phase protein concentrations are evident following abrupt weaning in beef suckler calves. There is no research examining the effect of weaning stress on the proportions of leukocytes expressing selected differentiation antigens and the functional activity of neutrophils. The objective was to examine the effect of abrupt weaning at housing on circulating leukocyte subsets and phagocytic activity of neutrophils in beef suckler calves. At weaning (day (d) 0), 16 spring-born, single-suckled, previously grazed calves (mean age and weight (s.d.): 227 (18.2) d and 310 (31.1) kg, respectively) of Limousin × Holstein-Friesian and Simmental × Holstein-Friesian dams, were either abruptly weaned (AW, n=8) and housed in a slatted floor shed and offered grass silage ad libitum, or not weaned (control) (NW, n=8) and housed with their dam and offered the same diet. Jugular blood was collected on d -7, 0 (weaning), 2, 7, and 14. Percentages of positive-staining CD4+, CD8+, WC1+ lymphocytes, MHC Class II+ cells and G1+ neutrophils were recorded. Surface expression of CD62L was recorded as mean fluorescence intensity (MFI) of neutrophils gated on the basis of their light-scattering properties. Phagocytic activity of neutrophils was examined in heparinised blood using a commercial assay (PhagoTest®). On d 2, neutrophil number increased (P<0.001) in AW ($6.7\pm0.36\times10^3$ cells/μl) compared with d 0 ($2.6\pm0.36 \times 10^3$ cells/μl), whereas neutrophil number in NW did not differ (P>0.05) from d 0 ($2.9\pm0.36 \times 10^3$ cells/μl) or throughout the experimental period. On d 2, % CD4+ and CD8+ lymphocytes decreased (P<0.001) in AW and did not differ (P>0.05) in NW, compared with d 0. The % WC1+ cells, CD62L MFI and phagocytic activity decreased (P<0.05) on d 2 in AW and NW, however the decrease was greater (P<0.001) in AW. Abrupt weaning at housing resulted in increased neutrophil number but the functionality of these cells to potentially traffic and phagocytose efficiently was impaired. Together with the greater changes in lymphocyte subsets and % MHC Class II+ cells this suggests that there was a greater transitory reduction in immune function in abruptly weaned than non-weaned beef suckler calves.

Validating activity sensors for measuring lying behaviour in dairy calves

G. Finney, S.J. Morrison, A. Gordon and G. Scoley
Agri-Food and Biosciences Institute, Hillsborough, Northern Ireland, BT26 6DR, United Kingdom; graham.finney@afbini.gov.uk

The objective of this study was to assess whether activity sensors (tri-axial, 4 Hz accelerometer: IceRobotics, Edinburgh, Scotland) would be a suitable alternative to making manual observations of dairy calf lying behaviour and whether a correction model is necessary to improve their accuracy. Thirteen straw housed calves (55±5 days old) were studied over 4 days. Recordings of lying bout duration were made according to two methodologies. In the automated method each calf wore an activity sensor on the right hind ankle. In the manual method calves were video recorded and watched back in full by two observers with 94.4% agreement and a kappa of 0.91. All lying behaviour was calculated into lying bouts in a format which matched the automated data, this facilitated direct comparison of the two methodologies. Both observers' datasets were compared and any discrepancies rechecked against the videos, this generated a 'gold standard' of greater accuracy for comparison with the automated dataset. A total of 1,084 lying bouts were compared, in the manual method 4.6% of these bouts showed discrepancy between observers, all such discrepancies owed to data inputting errors. Bouts which did not match the 'gold standard' were deemed 'sensor error' and totalled 7.6% of bouts recorded. Sensor error owed entirely to incorrectly registered bouts during nested behaviours (play & feeding) (5.6%), and missed standing/lying events (1.6%). All possible correction models were trialled with the optimal model found to be the exclusion of bouts ≤8 s in duration; this reduced sensor error to 3.7% (i.e. to within 95% confidence) consisting of 0.2% miss-classified bouts, 1.6% missed standing/lying events and 1.8% true bouts removed by the model. In summary, these sensors are a reliable alternative to making manual lying behaviour observations of calves, and when used with calves a correction model is necessary to calibrate their data outputs to within 95% confidence interval.

Processing automatic tracking data to identify interactions between Holstein-Friesian cows
S. Medisetti, B. Foris, S. Trißl, J. Langbein and N. Melzer
Leibniz Institute for Farm Animal Biology, Wilhelm-Stahl-Allee 2, 18196 Dummerstorf, Germany;
medisetti@fbn-dummerstorf.de

Indoor tracking of farm animals using Ultra Wide Band (UWB) technology is getting more into focus to study their spatial behaviour and to gain deeper insight into social processes of group-housed animals. In contrast to video observation, which has to be analyzed manually, data from UWB technology may be analyzed automatically. This would allow for continuous monitoring of a group of animals and thus may contribute to monitoring animal welfare. These measurements, however, contain uncertainty and missing values as transmitter (tag) fastened to an animal may not correctly or not at all detected by the sensors due to metallic barn equipment, missing coverage, or tracking system malfunction. In order to remove random fluctuations several smoothing methods are described in literature, such as the Kalman filter, weighted least squares regression, or outlier removal using a sliding window approach. Up to date it has not been investigated how well proximity between animals can be determined based on such kind of data to reveal the real interactions between animals kept in a barn. Hence, to detect interactions between two animals it is necessary to be able to reliably determine the distance between two tags. In a preliminary test, we prepared an array of nine tags in a row attached to a wooden pole. We could already observe that the method of smoothing has an effect on the measured distance between tags and thus also on the reliability to determine the distance between two tags. Therefore, we compared different methods of smoothing on UWB tracking data recorded over a period of 3 days on a group of 15 Holstein-Friesian cows in a barn area of 21.5×8 m. To enable verification of interactions between cows, we also acquired video data covering the same time period. We describe our proposed workflow to smooth data and show results of comparative analyses.

Assessment of novel technologies as indicators of stress in dairy origin calves
G.E. Scoley[1,2] and S.J. Morrison[2]
[1]Queen's University Belfast, School of Biological Sciences, University Road, Belfast, Co. Antrim, BT7 1NN, United Kingdom, [2]Agri-Food and Biosciences Institute, Hillsborough, Large Park, Hillsborough, Co. Down, BT26 6DR, United Kingdom; gillian.scoley@afbini.gov.uk

Traditional methods of measuring animal stress are typically invasive, thus an increasing importance has been placed on the development and application of non-invasive techniques for measuring stress. A main objective of this study was to examine the use of novel technologies to assess stress caused by commercial farm practices in dairy origin calves. Forty eight Holstein Friesian calves were allocated to one of three milk feeding treatments: i. Once daily feeding commencing at day 14 (OAD14), ii. Once daily feeding commencing at day 28, iii. Twice daily feeding (TAD). Calves were weaned abruptly at 56 days of age and remained in individual pens until 63 days of age at which point they were transferred to a group pen. A representative number of calves from each treatment were fitted with heart rate monitors and accelerometers during periods of milk step-down, weaning and transfer to the group pen. There was no significant effect of treatment on total daily lying time or number of lying bouts during periods of milk step-down from twice to once daily feeding, however, TAD calves tended to have a decreased total daily lying time (P=0.095) at weaning. Calves across all treatments displayed an increased number of lying bouts (P≤0.001) which were of shorter duration (P=0.088) during the period of transfer to the group pen. There was no significant effect of treatment on heart rate variability as determined by the root mean square of successive differences (RMSSD) of inter-beat intervals (IBIs) during periods of milk step-down or at weaning, however, TAD calves showed significantly lower RMSSD values during the period of transfer to the group pen. The data from the present study suggests that the sensors were able to pick up physiological and behavioural changes during the grouping period that are indicative of increased stress load. We cannot conclusively say that calves did not perceive milk step down or weaning as a stressor, but increased stress levels were not observed within the measures used.

Can infra red thermography be used to identify ill-health and stress in pre-wean calves?

G.E. Scoley[1,2] and S.J. Morrison[2]
[1]*Queen's University Belfast, School of Biological Sciences, University Road, Belfast, Co. Antrim, BT7 1NN, United Kingdom, [2]Agri-Food and Biosciences Institute, Hillsborough, Large Park, Hillsborough, Co. Down, BT26 6DR, United Kingdom; gillian.scoley@afbini.gov.uk*

Infrared thermography (IRT) has previously been shown to be a viable, non-invasive method of detecting alterations in radiated temperature which are linked to changes in emotional state as a result of stress or ill-health. However, little research has investigated the use of IRT in detecting illness and stress in pre-wean calves. A main objective of this study was to examine how infra red (IR) temperature develops over time and to consider the extent to which it can be used as an indicator for ill health or stress. Forty eight individually penned Holstein Friesian calves underwent thermal imaging of the eye and rectal area over 3 time periods: 7-21 days of age (Period 1), 21-35 days of age (Period 2) and 49-62 days of age (Period 3). Internal rectal temperature was also recorded at the time of thermal imaging. There was a highly significant effect ($P \leq 0.001$) of age, with IR eye temperature decreasing as age increased. No significant relationship was found between IR eye temperature and internal rectal temperature during Periods 1 ($P=0.587$) and 2 ($P=0.227$), however there was a significant relationship in Period 3 ($P \leq 0.001$). No significant relationship was found between IR rectal temperature and internal rectal temperature during Period 1 ($P=0.152$), however, a significant relationship was found during Periods 2 ($P=0.033$) and 3 ($P \leq 0.001$). The current data set demonstrates the possibility of using IR temperature as a non-invasive indicator of ill-health in pre-wean calves. It also demonstrates the possibility that IR temperature could be indexed to internal body temperature; however, there is a need to expand the data set of calves in early life to examine if correlation between the two methods can be improved, and also to extend the data set past the point of weaning to examine how IR temperature is affected during the post wean period.

Expression of 11b-HSD1 mRNA in adipose tissue of dairy cows during the periparturient period

A. Alizadeh[1,2,3], H. Sadri[1], J. Rehage[4], S. Dänicke[5] and H. Sauerwein[1]
[1]*Institute of Animal Science, Physiology and Hygiene Unit, University of Bonn, 53115 Bonn, Germany, [2]Islamic Azad University, Department of Animal Science, Saveh Branch, Islamic Azad University, Saveh, Iran, [3]Royan Institute for Reproductive Biomedicine, Department of Embryology, Reproductive Biomedicine Research Center, Biomedicine, ACECR, Tehran, Iran, [4]University of Veterinary Medicine, Clinic for Cattle, University of Veterinary Medicine, Hannover, Germany, Hannover, Germany, [5]Institute of Animal Nutrition, Institute of Animal Nutrition, Friedrich-Loeffler-Institute, Braunschweig, Germany; sauerwein@uni-bonn.de*

The key enzyme for cortisol metabolism is 11b-hydroxysteroid dehydrogenase type 1 (11b-HSD1). Although the pivotal roles of 11b-HSD1 in human adipose tissue (AT) were shown, limited information exists on 11b-HSD1 gene expression in AT of dairy cows. Thus, our objective was to determine how calving influences 11b-HSD1 gene expression in subcutaneous tail head AT (scAT) and retroperitoneal AT (rpAT) of dairy cows during late gestation and early lactation. scAT and rpAT biopsies of twenty Holstein cows were taken and quantified by qPCR for mRNA abundance of 11b-HSD1 on day 42 antepartum (a.p.) and day 1, 21 and 100 postpartum (p.p.). The 11b-HSD1 mRNA abundance was measured relative to the expression of EMD, LRP10, and POLR2A in scAT and EIF3K, LRP10, MARVELD, EMD, and Pol2 in rpAT as reference genes. Data were analyzed using the MIXED procedure of SAS. The mRNA abundance of 11b-HSD1 in scAT was 2-fold greater on day 1 p.p. than other sampling days ($P<0.001$). Similarly, calving led to a greater expression of 11b-HSD1 in rpAT on day 1 p.p. compared to other time-points ($P<0.01$). Expression of 11b-HSD1 mRNA in rpAT on day 42 a.p. and 100 p.p. was comparable. Expression of 11b-HSD1 mRNA was detected in rpAT and scAT of dairy cows during late gestation and early lactation. Moreover, the evidence of high 11b-HSD1 mRNA abundance in scAT and rpAT on day 1 p.p. confirms that it may be triggered by calving in AT which warrants further studies related to metabolic disorders.

Effects of weaning conditions on metabolic parameters, growth and health of piglets

A. Buchet[1,2,3], C. Belloc[1], N. Le Floc'h[3] and E. Merlot[3]
[1]*ONIRIS INRA UMR 1300 BioEpAR, BP 40706, 44307 Nantes, France,* [2]*COOPERL ARC Atlantique, BP 60238, 22403 Lamballe, France,* [3]*INRA, UMR 1348 PEGASE, 35590 Saint Gilles, France; arnaud.buchet@rennes.inra.fr*

Social, nutritional and spatial changes at weaning generate very high adaptive requirements for piglets. This study investigated the effects of management practices and age at weaning on metabolic status, growth and health of piglets. Piglets born from 12 litters were weaned at either 21 (W21) or 28 (W28) days of age in optimal (OC) or deteriorated (DC) conditions (n=16 per group). In OC post-weaning pens, animals originating from 2 litters were housed by 4 (0.39 m²/pig) in cleaned and disinfected pens. In DC post-weaning pens, animals originating from 4 litters were housed by 8 (0.20 m²/pig) in dirty pens after a waiting time of 4 h at 20 °C. One week after weaning, DC pigs were mixed again. From 12 to 61 days of age, blood was collected and pigs were weighed weekly. Diarrheas were recorded daily. Plasma concentrations of metabolites were measured. Body weight (BW) was significantly lower from 47 days of age in DC compared to OC (P<0.001) and, from 54 days of age, for W28 compared to W21 (P<0.001). Average daily gain of the week before blood sampling influenced positively non-esterified fatty acids (NEFA) and urea concentrations (P<0.05) and negatively creatinine concentration (P<0.001). After weaning, diarrhea frequency was greater in DC than OC (P<0.05) and negatively influenced by BW at weaning (P<0.05). Whatever the age and conditions at weaning, NEFA and urea concentrations increased at weaning (P<0.05) whereas glucose concentration consistently decreased with time (P<0.001). Five days after weaning, creatinine concentration was higher in DC than OC pigs (P<0.01), and urea concentration in W28 than W21 pigs (P<0.001). The increase in NEFA, urea and creatinine concentrations after weaning probably reflects lipolysis and muscle catabolism. Whatever the age at weaning, DC increase diarrhea occurrence and protein and lipid mobilization, which consistently slows down piglet growth.

Characterization of central, systemic and peripheral thermoregulation in the dromedary camel

H. Al-Tamimi[1] and M. Daradka[2]
[1]*Jordan University of Science and Technology, Animal Science, College of Agriculture, 22110 Irbid, Jordan,* [2]*Jordan University of Science and Technology, Clinical Veterinary Medical Sciences, College of Veterinary Medicine, 22110, Jordan; hosamt@gmail.com*

The Arabian camel is well known for its adaptation to desert and arid climates, mainly due to its unique thermophysiological processes. However, immense contradictory reports regarding the thermal lability of the camel's core exist in literature. A couple of principal shortcomings in these reports stem either from the very limited time window in thermal monitoring of the animals, or the highly restricted space provision (i.e. in radiotelemetric studies). Therefore, we chronically (110 days) instrumented five freely-roaming adult male camels using miniaturized thermo-loggers, to gain parallel understanding of central, systemic and peripheral body temperatures, coupled with ambient summer climatic conditions in Jordan, measured at 30-minute intervals. Mean temperatures of the implanted sites (meningeal; M, intraperitoneal; IP, peri-rectal; Re, hoof; H, intra-scrotal; Sc; and subdermally within the hump; Hsq, skull; Ssq; and sternal midline; Stsq regions) were 37.9, 37.5, 37.2, 36.7, 34.3, 37.1, 37.4 and 36.6±0.6 °C, respectively. Thermal lability was most pronounced (P<0.01) peripherally (20.1 °C), than systemically (7.1 °C) and then centrally (5.8 °C). Scrotal temperature was consistently kept about 3.2 °C below that of the core, which is in agreement with previous work indicating a 2-6 °C deficit below IP, seemingly a critical thermoregulatory process essential for optimum spermatogenesis in mammals. The greatest (P<0.01) correlation for IP was found with Re, reiterating the legitimacy for considering the rectal temperature in mammals as a good representation of the core's. Ambient temperature and solar radiation had the highest correlation (r=0.74 and 0.76; respectively, P<0.001) associated with Ssq, far than all other body sites. Patterns of central temperatures indirectly suggest whole body arousal is initiated by the brain, as evidenced by onset of thermal rise in the meningeal region prior to all other monitored body sites. The current trial provides a highly practical, intensive and chronic means of body temperature-monitoring in camels, comfortably while free-roaming.

Acute stress response in sheep of four different identity profiles

M. Pascual-Alonso[1], G.C. Miranda-De La Lama[2], W. Sepúlveda[1], L. Aguayo-Ulloa[1], M. Villarroel[3], J.A. Abecia[1] and G.A. María[1]
[1]*Universidad de Zaragoza, Producción Animal y Ciencia de los Alimentos, Facultad de Veterinaria. Miguel Servet, 177, 50013 Zaragoza, Spain,* [2]*Universidad Autónoma Metropolitana Unidad Lerma, Av. Hidalgo Pte. 46, 52006 Lerma de Villada, Mexico,* [3]*Universidad Politécnica de Madrid, Ciencia Animal, Av. Puerta de Hierro, 2, 28040 Madrid, Spain; alf@unizar.es*

The strategy used to deal with life in a social environment (the adoption of an identity profile) will lead to differences in stress response. The aim of this study was to identify the existence of identity profiles based on the use of social strategies in a stable flock of Roya Bilbilitana sheep and its relation to acute stress response provoked by shearing. Four different profiles were identified: aggressive, affiliative, evasive and passive in a group of 36 adult sheep housed indoors under a conventional management system. Blood samples were collected before shearing, immediately after shearing, 4 and 24 hours after shearing (May 2015). Stress response parameters were assessed (cortisol, glucose, lactate, CK, NEFA and haptoglobin). Internal body temperature was also measured using internal e-botton data loggers placed in a sponge device introduced into the vagina of the ewe and compared with infrared thermograph pictures of the eye of the ewe. Data were analyzed using a mixed model with repeated measures including the animal as a random effect and the identity profile (four levels) as the fixed effect. A hierarchical cluster analysis was performed to identify differences in social interactions to detect the identity profiles clusters. Affiliative ewes showed the lowest (P≤0.05) cortisol peak after shearing (39.8±15 nmol/l), while aggressive profile ewes showed the highest (69.2±14 nmol/l). The profiles of the stress response variables were corroborated by the internal body temperature recorded from the e-botton data logger every minute for 24 hours. Overall results indicate that the ewes with an affiliative strategy were better able to adapt to the shearing stress.

Effects of genetic selection and production system on sheep reproduction rates and welfare

D. Gavojdian, I. Patras, G. Flutur and S. Kusza
Research and Development Station for Sheep and Goats, Caransebes, 325400, Romania;
gavojdian_dinu@animalsci-tm.ro

Objective of the current study was to evaluate the health, reproductive rates and fitness indicators in three Tsigai sheep populations, divergently selected for milk (Sombor), meat (Cokan) and unimproved (Covasna, mountain type). A total of 588 ewe-year units were observed, with ewes (ranging from 1.5 to 8 years of age) managed under intensive (meat population), semi-intensive (dairy selected) and extensive (unimproved flock) rearing conditions for a period 12 months. The Main Effect ANOVA analysis of variance was applied. Clinical mastitis incidence was significantly lower (P≤0.01 and P≤0.001) in Covasna population (3.8±0.13%) compared to that of Sombor (10.8±0.21%) and Cokan (15.4±0.27%) genotypes, respectively. Significant differences (P≤0.05) for lameness were found between Cokan and Covasna populations, with occurrence rates of 9.7±1.98% and 4.5±1.58%, respectively. Conception, abortion and attrition rates were not affected (P>0.05) by selection pressure among the three populations. Litter size was significantly lower (P≤0.001) in Covasna Tsigai than for Cockan and Sombor populations, of 121.0±2.93, 142.3±4.07 and 145.1±4.06, respectively. For the two selected specialized Tsigai genotypes (Sombor and Cokan) it would be recommended to include fitness traits into the breeding selection schemes, with special focus on traits such as genetic resistance to mastitis and ewe stayability, in order to improve animal welfare and overall productivity.

Heat stress and PUFA in the diet of sheep: an ex vivo study

M.G. Ciliberti, M. Albenzio, C. Inghese, R. Mancino, R. Marino, A. Santillo and M. Caroprese
University of Foggia, Department of Agricultural Food and Environmental Sciences, Via Napoli 25, 71121, Italy;
maria.ciliberti@unifg.it

The effects of the diet enriched with polyunsaturated fatty acid (PUFA) on ex vivo immunological responses of sheep under high ambient temperatures were studied. Four balanced groups each with four Cominsana ewes were assigned to one of four dietary treatments in a 2 by 2 factorial arrangement. The experimental groups were: the FS group supplemented with whole flaxseed (250 g/ewe/d); the AG group supplemented with seaweed Ascophyllum nodosum (pelleted concentrate incorporating 5% Ascophyllum nodosum), the FS+AG group supplemented with a combination of flaxseed and Ascophyllum nodosum. The fourth group (CON group) was a control. Blood samples from each ewe were collected for isolation of peripheral blood mononuclear cells (PBMC) by density gradient centrifugation. The cells were stimulated with phytohemagglutinin (PHA, 5 µg/ml) (Stimulated Cells, SC), or not (NSC) for 96 h at 37 °C and 5% CO_2. After incubation, the supernatant free-cells was collected and stored at -20 °C until ELISA to measure cytokines production; whereas the cells were incorporated with Bromodeoxyuridine (BrDU) for 18 h in order to test lymphocyte proliferation. All the data were analysed using ANOVA for mixed models using the PROC MIXED of SAS (2013). PBMC from AG sheep registered an impairment of cells proliferation and IL-6 production after PHA stimulation compared to PBMC from FS+AG sheep. In addition, PBMC from AG sheep displayed a reduction of cell proliferation compared to cells from CON sheep. FS+AG sheep produced a lower level of IL-10 than CON cells, and higher IL-6 than AG and CON cells. Based on previous results, the immunological responses of sheep under high ambient temperatures can be influenced by supplementation of PUFA from different sources in the diet. In particular, the simultaneous administration in the sheep diet of PUFA from flaxseed and Ascophyllum nodosum exerted an ex vivo synergistic mechanism on activation of inflammation responses.

Influence of metabolic status to milk yield

M. Vlček and R. Kasarda
Slovak University of Agriculture in Nitra, Department of Animal Genetics and Breeding Biology, Tr. A. Hlinku 2,
949 76 Nitra, Slovak Republic; radovan.kasarda@uniag.sk

The aim of the study was to evaluate the influence of fat to protein ratio (F/P ratio) as the indicator of metabolic status to milk yield. 201 Holstein cows from two dairy farms between 2012 and 2015 were observed. In total 529 records of month milking were studied. Statistical analysis was performed by the SAS. 16.26% of observations were in risk of acidosis and 12.85% of observations were in risk of ketosis. The selected effects of breed, parity, year and season of calving, days in milk and fat to protein ratio described the milk yield on 41% and the averge milk yield was 38.15±9.45 kg/day. Main influences to milk yield were observed in number of lactation (P<0.001), days in milk (P<0.001) and in F/P ratio (P<0.001). The effect of the F/P ratio was in depressed milk yield in 6.22 kg/day. It is possible to use the F/P ratio like the noninvasive method of prediction of metabolic disorders to eliminate the negative effect of metabolic disorders to milk yield.

Diseases of the older horse from a genetic point of view

A. Pacholewska and V. Gerber
University of Bern and Agroscope, Swiss Institute of Equine Medicine, Länggassstrasse 128, 3012 Bern,
Switzerland; alicja.pacholewska@vetsuisse.unibe.ch

Among farm animals, horses are the most likely to experience senescence, since they are often treated as companions and a large proportion is now living beyond 20 years of age. Thus, horses develop age-related diseases. Some of these have a genetic background, which is not easy to identify as they typically are multi-factorial and multigenetic, and show a complex mode of inheritance. The genetically best studied disease of mature horses is equine asthma (recurrent airway obstruction, RAO) that affects horses across the Northern hemisphere, with a relatively late onset in life. It has been shown that RAO has a heterogeneous genetic background and the offspring of an RAO-affected parent is more likely to develop the disease. As in other species, the horse's immune response to vaccines, as well as to parasitic infections decreases with age, e.i. immunosenescence. Also, it has been shown that there is an age-related increase in the gene expression of pro-inflammatory cytokines in horses. In humans, this so called inflamm-aging has been shown to have a genetic component. Moreover, most grey horses will develop melanomas, usually in their late teens. It is noteworthy that older age and a genetic predisposition do not necessarily lead to the pathologic condition in all cases. Environmental factors can either increase or decrease the risk of developing a disease. RAO is an excellent example for this relationship: excluding hay, which is considered as major RAO-inducer, from the horse's diet and environment may suppress the clinical signs completely. Therefore, proper education of horse owners and breeders about genetic and environmental factors of late-onset diseases may help to improve horse welfare and reduce the risk of disease development.

Nutrition of old horses

S.L. Ralston
Rutgers, The State University of New Jersey, Department of Animal Sciences, School of Environmental and
Biological Sciences, 84 Lipman Drive, New Brunswick, NJ 08901, USA; sarah.ralston@rutgers.edu

With increased longevity has come a change in the perception of older horses. In 1989 only 2% of the equine animals admitted to a university veterinary hospital in the USA were over 20 years old. By 2003 that number had risen to over 20%. Now almost all feed companies produce a 'senior' product. However the digestive alterations on which most of the senior feed formulations are based were reported in aged horses studied in the mid-1980's It has since been theorized that the changes seen (reduced digestion of protein, fiber and phosphorus) may been due to chronic damage done to the intestinal mucosa by intestinal parasites. More recent studies of digestion in old horses did not show the same problems, suggesting that improved anthelminthic and dental care received by horses born in the 1980's and 1990's have reduced the need for special feeds in otherwise healthy old horses. The definition of 'aged' should now apply to horses over 20 years of age and 'geriatric' only to aged horses with health problems. Just because a horse is aged does not mean it needs a 'senior' formulation. If you see an aged horse in poor body condition or with chronic infections, colic, etc. there is usually something other than age alone causing the problem. In such cases the social environment, quality and amount of feeds offered, dentition, pituitary, renal and hepatic function should be carefully evaluated. This talk will give a practical approach to the nutritional management of the common problems seen in truly geriatric equine animals.

Body condition and location of fat stores in the aging horse
S.J. Wood, E. Allen, A. Corr, M. Lee and A. Magee
College of Agriculture, Food and Rural Enterprise, Greenmount Campus, 45 Tirgracy Road, Antrim, Co. Antrim,
BT41 4PS, United Kingdom; stephanie.wood@dardni.gov.uk

The effect of age on body fat stores is of interest as horses are living longer and it is unclear at what age losses in fat stores and body weight (BW) become a potential health issue. This study aimed to compare the body condition scores (BCS) and location of fat stores in horses of different ages. A random sample of 103 equines of varying breed, age, sex and BW from three livery yards in Northern Ireland was used in the study. Body condition score of eight points on the horse's body (crest of neck, base of neck, behind shoulder [Forehand BCS], ribs, base of withers [Body BCS], point of hip, loins, tail head [Hindquarter BCS]) on a scale of 0-5 (very poor-very fat) was assessed by a trained researcher. Mean BCS was calculated for overall BCS (all eight points), forehand, body and hindquarters and all BCS data was normally distributed. To determine the effect of age on BCS and location of fat stores horses were categorised by age (young, ≤ 6 years, n=25; adult, 7-20 years, n=63; old, 21-26 years, n=9; geriatric, ≥ 27 years, n=6) and analysed using one-way ANOVA (Minitab17). Horses aged from 2 to 35 years and from overall BCS 2 to score 5 (mean 3.6 ± 0.06). Young, adult and old horses were of statistically similar BCS for overall, forehand, body and hindquarter assessments (P>0.05). Geriatric horses had the lowest overall BCS and for forehand, body and hindquarters, but were only statistically different to adult horses for overall, forehand and hindquarter BCS (P<0.05). Location of fat stores did not differ between age categories although there was a trend for all horses to have higher BCS in the body region (withers and ribs) compared to forehand and hindquarters. Old horses had their lowest BCS on their hindquarters whilst geriatric horses had low BCS on both their neck and hindquarters (means 2.9). This change in fat storage location indicates that as horses enter the geriatric years there is increasing difficulty in maintaining adequate fat stores. Results also indicate that maintaining BW does not become difficult until horses reach their late 20's, most likely due to improved dental and veterinary care and the availability of feeds specifically designed for the aging horse.

Long-term changes of bone ultrasound measurements and metabolic indicators in the Lusitano horse
M.J. Fradinho, L. Mateus, R.J.B. Bessa, R.M. Caldeira and G. Ferreira-Dias
CIISA-Faculdade de Medicina Veterinária, ULisboa, Av. Universidade Técnica, 1300-477 Lisboa, Portugal;
amjoaofradinho@fmv.ulisboa.pt

In the equine athlete, the success and longevity of sport career depend greatly on the soundness of the locomotor system. Thus, a better knowledge of the factors involved in the skeletal development is fundamental for breeders and users. A longitudinal field study was performed in order to evaluate long-term changes of ultrasound measurements, biochemical markers, growth factors and other metabolic variables related with bone metabolism in the Lusitano horse. Twenty-seven Lusitano foals (11 colts, 16 fillies), born and raised at four stud-farms, were periodically assessed from birth to 36 months of age. Blood samples were collected for determination of osteocalcin, bone alkaline phosphatase, insulin-like growth factor I (IGF-I), leptin, insulin, glucose, parathyroid hormone, calcium, phosphorus and magnesium plasma concentrations. Ultrasound measurements were performed on the dorsal and lateral regions of the third metacarpal bone with a quantitative ultrasound device developed to mesure the speed of sound (SOS) in axial transmission mode. The effects of age, gender and foaling season were analyzed. SOS values increased with age (P<0.0001) for both regions, but this increase was influenced by an interaction with gender (P<0.01) with the highest values observed for colts. Bone biochemical markers, IGF-I, glucose and plasma mineral concentrations changed markedly with age (P<0.01), and some interactions with gender and foaling season were also observed (P<0.01). The present study provided information on a sequence of events during the first three years of life of the same foals, under field conditions. Therefore, this detailed characterization of variables involved in skeletal development during such a long period, may become useful for further investigation, also in other breeds.

Performing good death for animal companions
N. Schuurman
Karelian Institute, University of Eastern Finland, P.O. Box 111, 80101 Joensuu, Finland;
nora.schuurman@uef.fi

With pet burials, cremation services and cemeteries becoming common phenomena in the Western world, the culture and practices around companion animal death bear significant resemblance to those of humans. They also illustrate the contemporary status of animals as companions with a position as a friend and a family member. This status, combined with recent advances in veterinary knowledge and care, result in the possibility for animals to live well into old age. However, the deaths of animals living with humans are nearly always undertaken by humans, in the form of animal euthanasia. Ideally, this is carried out at the moment when the animal is no longer enjoying a good life due to serious illness, injury or extreme old age. The aim is to avoid excessive suffering, but what consists of excessive suffering, is to be culturally and contextually negotiated. In this presentation I explore the question of good death in the context of animal aging and euthanasia. The presentation draws on discussions on the themes of death, care, liminality and the human–animal boundary, as well as on empirical studies on the practices and conceptions concerning the aging and death of horses and pets. Ensuring a good death for an animal requires skill, experience and expertise on the part of the human, especially with respect to the major task of determining the ideal time for the animal to be euthanized. Understood in the Aristotelian sense, however, as a good ending to one's life, a good death starts before the actual death. Performing good death, therefore, involves not only controlling the process of aging and the approaching death, but also managing the relationship between the human and the animal. From this perspective, places such as horse retirement yards and veterinary clinics can be understood as spaces where the culturally defined ending for a human–animal relationship is actively pursued. The performance of good death in these spaces is constructed on the actions, choreographies and mutual communication of both humans and animals, and on human expertise.

Equine passports and end of life decisions: have we produced a welfare crisis?
S. Schofield and C. Brigden
Myerscough College, Equine Department, St Michael's Rd, Bilsborrow, Preston, PR3 0RY, United Kingdom;
cbrigden@myerscough.ac.uk

The horsemeat scandal and increased equine abandonment raise doubt over the effectiveness of the passport. Permanent removal of horses from the food chain is inconsistent with agriculture and reduces options for owners at the end of equine life (EoL). The study aim was to explore opinions regarding the passport and its influence on EoL. A self-administered online survey was distributed via relevant social media sites and equine forums. The survey contained 25 closed and open questions, covering experiences of passport enforcement and opinions regarding its influence on EoL. Responses (n=159) were analysed with χ^2 test of association and content analysis. Most owners (88.1%) have considered their options for EoL, but this was positively associated with increased length of ownership ($\chi^2$3=9.606; P<0.05). Most horses (71.3%) were signed out of the food chain; in 24.3% this was the previous owner's choice. A proportion (20.9%) of owners may or would like the option to reverse this decision. 32.1% of owners would like the chance to dispose of their horses as meat should they be unable to afford other options. A similar proportion of owners had a positive and negative attitude to the use of horses as meat. Attitudes were driven by some mutual concerns, such as breeding and welfare, but some were exclusive to a particular viewpoint. For example, the ethics of horsemeat featured only within negative or neutral responses. The majority of this horse population has been excluded from the meat trade by their passport, limiting options for current or future owners. Some owners were open to using the meat trade for disposal, but many would be prohibited by passport regulations. Almost half of owners saw potential benefits in the use of horse meat, relating to the potential outlet for unwanted horses. Other negativity towards horsemeat linked to risks of breeding for meat purposes and general ethical standpoints. Amendment of passport regulations to allow the reversal of the meat chain decision could help to reduce equine suffering at EoL.

Usefulness of fallen equine data in describing quantitative mortality in France

J. Tapprest[1], M. Borey[1], X. Dornier[2], E. Morignat[3], D. Calavas[3], P. Hendrikx[4], B. Ferry[2] and C. Sala[3]
[1]French Agency for Food, Environmental and Occupational Heath & Safety (Anses), Laboratory for Equine Diseases, F14430 Goustranville, France, [2]French horse and riding institute (IFCE), 83-85 boulevard Vincent Auriol, F75013 Paris, France, [3]Anses, Epidemioogy Unit, 31 avenue Tony Garnier, F69364 Lyon cedex 07, France, [4]Anses, Scientific Directorate for Laboratories, 31 avenue Tony Garnier, F69364 Lyon cedex 07, France; jackie.tapprest@anses.fr

Information about quantitative mortality is scarce while it could be of great value for improving knowledge of the equine population health and welfare. In France, a potential source of quantitative mortality data is represented by the Fallen Stock Data Interchange (FSDI) database managed by the French Ministry of Agriculture. In fact, all equine cadavers have to be collected by one of the fallen stock companies and since 2011, data from rendering plants have been comprehensively transmitted via computerized data transfer to the FSDI database. The objective of our research was to evaluate the quality of the FSDI database in the context of its use to provide quantitative mortality information on the French equine population. The main data collected in the FSDI database are the date of removal request, the date of removal, the zip code of the removal location, the number of animals collected and their age/breed category, the individual identification number and an estimation of global cadaver weight. The quality of FSDI equine data from 2011 to 2014 was assessed by evaluating the completeness and reliability of the important fields for 139,726 removals of dead equines and by quantifying the accuracy of their age/breed categorization in the FSDI. For that, we used base data registered in the French equine census database, SIRE, managed by the IFCE horse and riding institute. Despite a perfectible quality, the FSDI database proved to be a very valuable source of quantitative equine mortality data and useful for thoroughly describing the spatial distribution of deaths for different age and breed categories. However, improvements in the FSDI database are needed, in particular regarding the registration of animal identification numbers, in order to detail equine mortality for epidemiological purposes.

Survey on equine hoof costs in Switzerland

A. Estermann, S. Wägeli, C. Kopp, B. Durgiai, F. Kägi and C. Herholz
Bern University of Applied Sciences, School of Agricultural, Forest and Food Science, Laenggasse 85, 3052 Zollikofen, Switzerland; conny.herholz@bfh.ch

'No hoof, no horse' is an old saying. Hoof health is a vital precondition to the performance of the horse. Despite the high importance of hoof health little data are available with regard to hoof management costs. The goal of the present study was to evaluate the total hoof costs of horse owners in Switzerland. In an online questionnaire 523 horse owners were asked for hoof costs on services like shoeing, trimming, veterinary care, but also on feed supplements for hoof quality and growth, hoof care products and other. Statistical analysis was performed using R, version 3.0.2, and a regression analysis was performed to evaluate factors which influenced hoof costs. It was shown that 59% of the horse owners shoe, 40% have the horses' feet trimmed. More than 30% of the horse owners used hoof grease, 14% a product against thrush and 12% a feed supplement to improve hoof horn quality. Hoof ointment (7.6%) and hoof tar (2.8%) were used to a lesser extend. Remarkably 51% of the horse owners were not able to report on the brand of hoof grease they are using. The costs for shoeing were between 46 and 642 Euro, horse owners spent in average 198 Euro per shoeing of their horses' feet. Hoof trimming costs were in average 80 Euro. Routine hoof care, including hoof trimming and shoeing was most frequently performed in an interval between 7-8 weeks. It was calculated that total costs for shoeing, trimming and care products are on average 1,160 Euro per year and horse, respectively 96 Euro per month and horse. Additionally horse owners spent on average 116 Euro for protective hoof products like overreach boots or hoof boots. A statistically significant connection (P<0.05) between the amount of hoof costs and age, breed, intensity of usage per week and primary use for spot versus leisure was found. Extrapolated to the total population of 107'619 equids in Switzerland in 2015, the total hoof costs exceeded 124 Mio. Euro demonstrating the importance for the equine business in Switzerland.

T waves and QT intervals in the ECG of Standardbred horses
E. Preiss and P.W. Physick-Sheard
Ontario Veterinary College, Population Medicine, 50 Stone Road, Guelph, Ontario, N1G 2W1, Canada;
epreiss@uoguelph.ca

In electrocardiography, the human T wave (T) and QT interval (QT) are well characterized, and long and short QT and dispersion have been associated with malignant ventricular arrhythmias and sudden death. Factors influencing QT in humans include heart rate (HR), autonomic tone, age, sex, and some drugs. Manual and automatic QT measurement methods have been proposed, along with HR correction equations. In horses, significance has been attached to QRS duration (QRS), T morphology, and QT, but clinical associations are not well supported. Sudden death, a problem in racehorses, with negative effects on welfare, safety, perception and economics, may be associated with QT changes, but no standardized approach has been proposed for automated measurement and systematic description. While acknowledging inherent limitations of the body surface ECG in measuring ventricular repolarization, this research attempts to describe the equine QT and automate measurement. Sample populations for the study were apparently healthy Standardbred racehorses that were holter monitored during racing, and clinically normal Standardbred horses at rest at the Ontario Veterinary College. Convenience samples of 10 horses with artifact-free recordings were selected from each population. A total of 545 ECG segments of 12 RR intervals each were studied. Segments were selected to cover a wide range of HR and QT conformations. The algorithm successfully identified 97.0% of key waveform points, and all points in 403/545 files. QT onset and offset were reliably identified 99.8 and 94.5% of the time, respectively. At this time, the algorithm appears to be robust but predictably maintains some sensitivity to noise. This study phase was not designed to define normal QT or T, but several relationships were revealed. Parameters, including QT, appeared to vary consistently with HR, but a change in relationship occurred at about 60 bpm. This specific HR may be individual and was found to be most strongly associated with QT and T morphology. There was more intra than inter-horse variation in this sample. Longitudinal studies are required to investigate further, but with the aid of a reliable automated algorithm, it will be possible to establish a normal equine QT and accurate HR correction.

Author index

Bayraktaroglu, A.G.	248	Bessa, R.J.B.	524, 569, 681
Beattie, V.E.	121, 442, 547, 548	Besseboua, O.	571
Beckers, Y.	481, 512, 544	Bettridge, J.	418
Bedere, N.	428, 630	Beudou, J.	237
Bedford, M.R.	652	Biagini, D.	293, 294
Bedhiaf-Romdhani, S.	170, 424	Bianchi, L.	643
Bed'hom, B.	535	Biao, A.	424
Bednarczyk, M.	117, 118	Bichard, M.	602
Bee, G.	119, 181, 322, 441, 457, 547	Bieber, A.	104, 109, 111, 306, 307, 344
Behrendt, A.	380	Biehl, M.V.	634, 635
Bekurtz, J.	123	Bielas, W.	232
Belabbas, R.	182, 540, 647	Biemans, F.	190
Belay, T.K.	611	Biermann, A.D.M.	215
Bellabes, R.	186	Biestmann, C.	400
Belladelli, C.	660	Bigot, G.	394
Bell, B.	335	Bijma, P.	190, 605
Bellec, T.	610	Bikker, P.	455
Bell, M.	299, 300	Billon, D.	196
Belloc, C.	317, 677	Billon, Y.	180, 316, 599
Ben Abdallah, I.	164, 170	Bilton, T.P.	107
Benali, N.	182, 186, 540, 647	Bindelle, J.	481, 544, 565
Benaouda, M.	329	Binder, M.	325
Benbarek, H.	571	Birnholz, C.A.	325
Ben Hamouda, M.	424	Birnie, J.	587, 598, 639, 641, 642
Benkendorf, J.	649	Bishop, S.C.	165, 187, 188
Bennewitz, J.	282, 417, 492, 542, 600, 602	Bittante, G.	382
Benoit, M.	381, 385, 465	Bizelis, I.	528
Benzie, J.	260	Blackshaw, R.P.	505
Berard, J.	290	Blair, H.T.	522
Berbar, A.	182, 540, 647	Blaise, Y.	481
Berckmanns, D.	622	Blaj, I.	600
Berckmans, D.	268, 623, 624, 626	Blake, D.P.	318
Berentsen, P.B.M.	115, 617	Blanchard, A.	418
Beretta, V.	142, 146, 215, 567	Blanch, M.	545
Berge, A.C.	664	Blanco, A.E.	175
Bergero, D.	621	Blanco, M.	640
Berglund, B.	287, 311, 452	Blanco-Penedo, I.	653
Berg, P.	595	Blanquefort, P.	189
Bergsma, R.	282, 340, 453	Blasco, A.	493
Berg, W.	485	Blasco Mateu, A.	489
Beriain, M.J.	585	Blas, E.	489
Berk, Z.	433	Blavy, P.	375
Bermingham, M.	188	Blichfeldt, T.	517
Bernard, M.	467	Blonk, R.J.W.	256
Bernau, M.	223, 270	Blott, S.	286, 395
Berndt, A.	298, 413	Blunk, I.	216
Bernués, A.	237	Bluyssen, H.	246
Berri, C.	239	Bodin, L.	488
Berry, D.P.	129, 130, 163, 164, 167, 171, 194,	Bodin, T.	389
	199, 217, 303, 304, 305, 335, 342, 343, 348, 351,	Boekhorst, J.J.M.	154
	419, 423, 430, 434, 446, 482, 511, 548, 612, 627	Boekschoten, M.V.	242
Berškienė, K.	477	Bogevik, L.T.	532
Bertolini, F.	317, 530, 606, 615	Bogucka, J.	117
Bertoloni, A.B.	634, 635	Bohan, A.	385, 388
Berton, M.	209, 382	Boichard, D.	187, 189, 601
Bertozzi, C.	291, 298	Boison, S.A.	284

Boissy, A.	363	Brauning, R.	107, 161
Bojkovski, D.	426	Bravo-Lamas, L.	644
Bokkers, E.	671	Brearley, C.	567
Bolaji, O.J.	410	Breitsma, R.	320
Boland, T.M.	366, 385, 388, 468, 499, 502, 504	Briant, E.	625
Bompa, J.F.	518	Brien, F.D.	362
Bonaldi, C.	391, 394	Briens, M.	245
Bonaudo, T.	154	Brigden, C.	473, 682
Bonet, J.	662	Brinker, T.	605
Bonilha, S.F.M.	140	Brinkmann, J.	306
Bonin, A.	464	Brlek, V.	297
Bonin, M.N.	146	Brocard, V.	115
Bonk, S.	495	Brochard, M.	643
Bonneau, M.	389, 626	Brockmann, G.A.	123
Bonny, S.P.F.	133, 212	Bromley, C.	178
Boonanuntanasarn, S.	257, 260	Brosh, A.	410
Bootes, N.G.	573	Brossard, L.	441, 456
Borey, M.	683	Brostaux, Y.	512
Borg, R.C.	202	Brotherstone, S.	187
Borja, M.S.	526	Brown, A.	371
Borowska, A.	203, 342	Brown, D.	419, 526
Borrisser-Pairó, F.	224	Bruckmaier, R.M.	285, 373, 378
Boselli, L.	463	Bruegemann, K.	307
Bosi, P.	530, 543, 546, 606	Bruford, M.W.	111, 417
Bosman, L.	139, 214	Brügemann, K.	309, 433
Bossers, A.	420	Brun, A.	653
Boudillon, Y.	521	Brunberg, E.	619
Boudon, A.	292	Bruneteau, E.	499
Boudra, H.	467	Brünger, J.	623
Boudry, C.	565	Brüssow, K.-P.	228
Boumahdi, Z.	182, 647	Brzozowska, A.M.	570
Bouquet, A.	610	Bucci, D.	483
Bourgon, S.	181, 629	Buchet, A.	317, 677
Boussaha, M.	190, 191	Buckley, F.	271
Boutinaud, M.	269, 274, 296	Buckley, T.	390
Bouvarel, I.	239	Bučko, O.	522, 654
Bouvier, F.	418, 518	Bueno, R.S.	146
Bouwman, A.C.	107	Bugno-Poniewierska, M.	345, 421, 545
Bouyeh, A.	557	Buitenhuis, A.J.	288, 290, 303, 535
Bouyeh, M.	557	Bujko, J.	355
Boval, M.	387	Bunger, L.	638
Bovdišová, I.	493	Bunnom, R.	331
Bovenhuis, H.	272, 290, 301	Buono, N.	324
Bovo, S.	530, 615	Burfeind, O.	221, 399, 623
Bowen, J.M.	560, 563	Burggraaf, V.	182
Boyd, A.	178	Burgos, A.	520
Boylan, P.	177	Burgstaller, J.	347, 396
Boyle, L.	172, 221, 227, 513, 533	Burke, J.L.	627
Bozkurt, Y.	130	Burns, G.A.	504
Bradley, D.	510	Burren, A.	166, 313, 421, 424, 583
Brahim Errahmani, M.	533	Busch, G.	273, 388
Bramante, G.	579	Busch, J.	249
Branco, R.H.	140	Bussemas, R.	223
Brändle, J.	292	Butler, G.	360
Brandt, H.	223	Buttazzoni, L.	615
Brassington, A.H.	244	Buttin, P.	545

Chen, X.	328	Cone, J.W.	557
Cherrane, M.	186	Conington, J.	162, 363, 611
Chesneau, G.	564	Connolly, M.	525
Chevaux, E.	240, 534	Connolly, S.M.	130
Cheype, A.	364	Connor, E.E.	450
Chiaretti, D.	665	Constantinou, C.	642
Chiaretti, E.	665	Conte, G.	524
Chiarini, R.	467	Contò, G.	553
Chiariotti, A.	553, 637	Contreras-Jodar, A.	126
Chilibroste, P.	483	Cooper, R.	301
Choi, D.Y.	402	Cooper, T.A.	602
Choi, H.C.	402	Copani, G.	649
Chomczyńska, O.	608	Coppens, K.	243
Chormai, T.	331, 515	Corcionivoschi, N.	596
Christaki, E.	546	Cordell, H.J.	122
Christen, A.M.	347, 350	Cormican, P.	185, 536, 550
Christensen, K.	532	Corner, R.	178
Christensen, O.F.	604	Corpataux, M.	457
Christie, P.	505	Corr, A.	681
Christley, R.	418	Corraze, G.	255
Christodoulou, V.	254	Corrigan, A.	323, 538
Ciani, E.	579	Corte, R.R.S.	298, 413
Cicėnaitė, R.	477	Cortes, C.	196
Ciliberti, M.G.	679	Cortet, J.	418
Ciobanu, D.	317	Cosenza, G.	294
Cirot, C.	296	Coso, L.	297
Cito, A.M.	579	Costa, A.	403, 656
Ciurescu, G.	655, 656	Cotter, P.	118, 199, 548
Clark, C.E.F.	507, 621	Cottrell, J.J.	251
Clarke, L.C.	558, 559	Coughlan, F.	618
Clarke, S.	161	Courties, R.	626
Clarke, S.M.	107	Cowper, T.	667
Claußen, G.	390	Coyne, J.M.	612
Clelland, N.	638	Cozzi, M.C.	425
Clemensson Lindell, I.	580	Craig, A.	439
Clements, D.N.	124	Crecente, S.	585
Clerc, J.	146	Creighton, P.	385, 388, 499, 502, 504, 517
Cleveland, M.A.	206	Crepaldi, P.	168, 521
Cloete, J.J.E.	165	Crispie, F.	118, 199, 548
Cloete, S.W.P.	165	Cristobal-Carballo, O.	367, 368
Cockrum, R.R.	419	Croiseau, P.	191, 601
Coffey, E.L.	631	Cromie, A.	147, 193
Coffey, M.	162, 188, 193, 450, 611	Cromie, A.R.	130, 335
Coffey, M.P.	129, 187	Crooijmans, R.P.M.A.	286
Cohen-Zinder, M.	241, 410	Crosson, P.	206, 209, 211, 293, 457
Coizet, B.	168	Croué, I.	283
Coleman, J.	356	Crowe, M.A.	299, 300
Colinet, F.G.	291, 298	Cruywagen, C.W.	501
Colliver, K.	629	Cruz, A.	520
Colombo, M.	543	Cruz Hidalgo, D.	190
Colonna, M.A.	259	Cuadrado, M.	607
Colturato, P.	412	Cubric-Curik, V.	422
Coma, J.	662	Cuffia, M.	502
Combs, G.F.	150	Cummings, B.	371
Comesaña, R.	463	Curiao, T.	118, 199, 548
Concarr, M.J.	667	Curik, I.	125, 415, 422, 423

Curley, E.	558, 559	De Greeff, A.	650
Curtasu, M.	227	Degroote, J.	250, 659
Cutrignelli, M.I.	552, 569	De Haas, Y.	204, 449, 451
Cyrillo, J.N.S.G.	140, 408	Dehareng, F.	291, 298, 299, 300, 544, 630
Cziszter, L.T.	353	De Jaeghere, W.	551
Czycholl, I.	221, 529	De Jong, A.	581
		De Jong, G.	195, 349
D		De Jong, M.C.M.	190
Dacasto, M.	669	Dekkers, J.C.M.	225, 318
Dagah, P.	570	De Koning, D.J.	287
Daghigh Kia, H.	252	Delaby, L.	381, 428, 500, 503, 630
Dahlborn, K.	389	Delacroix-Buchet, A.	272
Dahmani, Y.	186	De La Flor, M.	494, 607
Dakheel, M.M.	322	Delafosse, A.	189
Dalcq, A.-C.	512	De La Foye, A.	252
Dale, A.J.	404	Delagarde, R.	499, 501, 549, 617
D'Alessandro, A.G.	646	De La Torre, A.	619, 625, 650
Dalla Villa, P.	399, 404	Del Fiore, A.	565
Dall'Olio, S.	530, 606	Delgadillo, J.A.	523
Dal Prà, A.	467	Della Malva, A.	670
Danchin, C.	432	Dell'Orto, V.	359, 551
Danglard, A.	465	De L. Luff, W.G.	602
Dänicke, S.	241, 245, 268, 676	Del-Pozo, J.	165
Danieli, P.P.	569, 665	Del Prado, A.	486
Daniel, J.B.	380	Delval, E.	363
Daniels, V.	347	De Marchi, M.	304, 305
Dankowiakowska, A.	117	Demirezer, H.	364
Danso, A.S.	522	Denholm, S.J.	432
Daradka, M.	677	Dennis, N.A.	431
Dardenne, P.	299, 300	De Nooij, J.R.	581
Dario, C.	646	De Pauw, M.	450
Das, G.	228, 321	De Paz, R.	258
Daverdin, M.	595	D'Erasmo, E.	399
Davergne, A.	189	Derno, M.	267
Davey, M.	165	Deroche, B.	559
David, I.	454, 606	De Rossi, P.	565
David, X.	264	Dersjant-Li, Y.	566
Davière, J.-B.	187	Deserti, M.	490
Davies, D.R.	358	Deshpande, P.D.	190
Davies, H.	365	Desire, S.	162, 611
Davies, S.	255	De Smet, S.	551, 659
Davis, H.	534	Desta, T.T.	418
Davoli, R.	490	De Tonnac, A.	231
Dawson, S.	143	De Trinidad, S.	408
Day, M.L.	634, 635	De Vega García, A.	618
De Blas, C.	461, 468	Devlin, D.J.	131, 132, 133, 148, 639, 642
De Boer, I.	114	Devries, T.J.	431, 509
De Boer, I.J.M.	115, 153, 234, 455, 617	Dewhurst, R.J.	406, 560, 563
Debrecéni, O.	361, 654	De Wolf, P.L.	153
De Bretagne, T.	187	Dezetter, C.	196
De Bruijn, C.	650	Dhour, P.	386
De Campeneere, S.	459	Dhumez, O.	564
De Clercq, P.	551	Diana, A.	172, 533
De Cosmo, A.M.	593, 595	Diaz, C.	108
Deeb, N.	318	Díaz, J.E.	649
Degani, A.	508	Díaz, N.	585

Dibbits, B.W.	286	Drillich, M.	514
Didehban, R.	535	Driscoll, H.	508
Difford, G.F.	126, 285, 449	Drögemüller, C.	285, 424
Digiacomo, K.	251	Drouet, A.	389
Di Giannatale, E.	399	Druet, T.	106
Di Giovanni, S.	553	Duclos, A.	326
Di Lello, S.	324	Duclos, M.J.	381
Dillon, P.	261	Ducro, B.J.	286, 395, 581
Di Marco, M.	621	Ducrocq, V.	190, 283, 428, 601
Dineen, M.	618	Dudek, K.	472
Disenhaus, C.	428, 626, 630	Duffy, S.K.	358, 359
Diskin, M.G.	354	Dufour, B.	585
Distl, O.	600	Duggan, B.M.	124
Dixon, R.M.	243, 288	Du Laing, G.	661, 662
Djemali, M.	164, 169, 170, 312, 330, 539	Dumont, B.	381
Djikeng, A.	161	Dunislawska, A.	118, 246
Dobbin, D.	114	Dunn, A.	267
Dobrzański, Z.	472	Dunshea, F.R.	251, 573
Dodds, K.G.	107, 161	Durand, D.	252, 446
Do, D.N.	229	Durgiai, B.	683
Doeschl-Wilson, A.	189, 260, 318	Duru, M.	381
Dogan, C.	130	Dusurt, L.	239
Dogot, T.	512	Dusel, G.	566
Doherty, M.L.	348, 430, 434	Duthie, C.-A.	406, 409, 460, 470
Dohme-Meier, F.	457, 630	Dutilly, C.	385
Dolezal, M.A.	344	Duval, J.E.	590
Dolfing, J.	410	Duval, S.	459
Domenech, V.	609	Dwyer, C.M.	362, 365
Domig, K.J.	292	Dwyer, J.	333
Domingues, J.P.	154	Dyrmundsson, O.R.	186, 527
Dominguez, P.	408	Dzama, K.	141, 351
Dominguez, R.	184, 585	Dzhavakhiya, V.V.	561, 563
Dominguez-Romero, E.	248	Džidić, A.	422
Donaghy, D.J.	636		
Donnellan, T.	198	**E**	
Donoghue, K.A.	451	Eaglen, S.A.E.	602
Döpfer, D.	192, 347	Earl, E.	385
Doreau, M.	460, 467, 487	Earle, E.	499, 502
Dorenlor, V.	228	Earley, B.	143, 217, 293, 536, 674
Dornier, X.	394, 683	Eberspächer, E.	230
Doronzo, M.	660	Echeverria Perez, R.A.	523
Dos Santos, N.J.A.	526	Edel, C.	338, 602
Douadji, O.	533	Edenius, L.	389
Doudou, A.	533	Eding, H.	349
Dourmad, J.Y.	441, 456	Edwards, A.	645
Downey, G.	147	Edwards, D.	308, 312
Doyle, S.	566	Edwards, S.A.	122, 314, 315, 316, 472, 478, 663
Dozias, D.	625	Edwards, S.M.	604
Drach, U.	508	Eeckhout, M.	551
Drag, M.	229	Egal, D.	625
Drake, C.	322	Egan, M.	616
Draper, J.	138, 211	Egea, M.	224
Dressel, H.	192, 491	Eggelmeijer, W.	594
Dreux, D.	207	Egger-Danner, C.	108, 171, 173, 194, 330, 347, 348, 396, 514, 628, 648
Driessen, B.	662		
Drieu, C.	389	Egri, E.	147

Ehsani, A.	535	Fagan, S.	566
Eising, I.	660, 661, 662	Fahey, A.G.	300
El Balaa, R.	251	Fair, S.	171, 185, 419
Eler, J.P.	144, 146, 169, 428, 607	Fanelli, C.	565
Elgendy, R.	669	Fanelli, F.	530
Elgersma, A.	405	Fangmann, A.	281
Elizalde, H.F.	503, 644	Fang, Z.H.	272
Eljarah, A.H.	250	Faouen, A.	222
Ellen, E.D.	204, 605	Fardy, E.	590
El-Mashed, S.	413	Farina, G.	359, 660
Elminowska-Wenda, G.	117	Farmer, L.J.	131, 132, 133, 148, 212, 222, 291,
El-Naggar, S.I.	670		525, 639, 640, 641, 642
El-Zaiat, H.M.	666	Farouk, M.H.	275
Emmerling, R.	103, 338, 602	Farquharson, C.	165
Emsen, E.	364	Farrell, D.	639, 641
Emsenhuber, C.	246	Farruggia, A.	465
Engblom, L.	374	Fasching, C.	269
Engelbrecht, M.N.	501	Faure, J.	389, 626
Engelhard, T.	276, 278	Faux, A.-M.	604
English, A.M.	185	Faverdin, P.	405, 549
Englishby, T.M.	129	Fearon, A.	291
Enns, M.	419	Feiersinger, D.	396
Ensor, H.M.	401	Fekete, A.A.	360
Eono, F.	228	Felleki, M.	488
Erasmus, L.	568	Fenelon, M.A.	305
Erazo Solines, J.	274	Ferard, A.	559, 632
Erbe, M.	103, 281, 530	Ferenčaković, M.	415, 422, 423
Eriksson, J.-Å.	628	Ferguson, N.	455
Eriksson, S.	574, 576, 580	Fernández, J.	258, 414
Ertl, P.	238	Fernández-Lorenzo, J.L.	615
Escribano, A.J.	289	Fernández Pierna, J.A.	291
Esmonde, H.	511	Ferrand-Calmels, M.	643
Espagnol, S.	157	Ferrandi, C.	108
Espinoza Castro, J.	274, 275	Ferraz, J.B.S.	144, 146, 169, 218, 286, 428, 607
Estellé, J.	541	Ferraz Jr., M.V.C.	634, 635
Estellés, F.	486	Ferreira-Dias, G.	681
Estermann, A.	683	Ferreira, G.	183
Esteves, S.N.	657, 658, 659	Ferreira, J.G.	177, 178
Estévez, J.	120	Ferreira, L.M.	391, 394
Ettema, J.F.	377	Ferreiro-Domínguez, N.	615
Eudier, S.	228	Ferrer, J.	145
Eugène, M.	385, 465, 487	Ferrer, P.	461, 468
Evans, D.	638	Ferris, C.P.	291, 299, 300, 327, 371, 404, 485, 550,
Evans, G.	151		629
Evans, R.D.	335, 423, 471	Ferry, B.	585, 683
Eveno, E.	228	Fidelis, H.A.	140
Everaert, N.	544	Fiedler, A.	346, 347
Exadaktylos, V.	624	Fiedler, M.	485
Ezelgbu, O.O.	555	Fievez, V.	459
		Figueiredo, L.G.G.	144, 607
F		Fikse, W.F.	452
Fablet, C.	228	Filioussis, G.	175
Fàbrega, E.	119, 122	Finan, S.	334
Facciolongo, A.M.	259	Finlayson, H.A.	165
Faccioni, G.	620	Finney, G.	674
Fagan, C.C.	361	Finocchiaro, R.	454

Fiorelli, J.L.	157	Fuerst, C.	310, 348, 366
Firat, M.Z.	346	Fuerst-Waltl, B.	306, 307, 310, 330, 347, 348, 366,
Firth, C.L.	171, 173		396, 514, 628, 648
Fischer, A.	405, 549	Fujiwara, M.	401
Fischer, B.	480	Füller, I.	144
Fischer, D.	464	Fuller, R.	207
Fischer, R.	184	Furre, S.	575
Fitzgerald, S.	631	Fustini, M.	483
Flanders, D.	220	Fuzikawa, I.H.S.	657, 658, 659
Flatres Grall, L.	610		
Fleming, H.R.	358	**G**	
Flessire, F.	496	Gabai, G.	483
Fletcher, M.T.	243, 288	Gabler, N.K.	225
Florou-Paneri, P.	546	Gabrieli, R.	625
Flury, C.	104, 109, 111, 341, 421, 424, 583	Gabrielle, B.	154
Flutur, G.	678	Gac, A.	326
Flynn, P.	167	Gad, S.M.	670
Fogh, A.	311	Gaffney, M.	668
Foldager, L.	299	Gaillard, C.	369, 375, 376
Fonseca, A.J.M.	569	Galama, P.A.	153
Fonseca, M.	146	Galán, E.	486
Fontanesi, L.	530, 606, 615	Gulbraith, H.	519
Font-I-Furnols, M.	653	Galeati, G.	483
Forbes, A.B.	433	Galimberti, G.	530, 615
Forenbaher, S.	422	Gallo, L.	209
Foris, B.	675	Gallo, M.	615
Formayer, H.	330	Gameiro, A.H.	154
Formigoni, A.	483	Gandini, G.	521
Formoso-Rafferty, N.	125, 494	Gangnat, I.D.M.	549
Fornara, D.A.	383	Garavaglia, L.	671
Forrest, J.	301	García-Ballesteros, S.	579
Foulquié, D.	363	Garcia Casco, J.	653
Fourichon, C.	189, 590	García-Contreras, C.	403
Fournier, F.	465	Garcia-Cortes, L.A.	257
Fowler, S.	513	García De Salazar, A.	248
Fradinho, M.J.	681	García-Gudiño, J.	653
Francesch, M.	119, 120, 122, 554	Garcia-Hernandez, E.	651
Franco, D.	184, 585	Garcia, J.F.	125, 415
François, D.	518, 521	Garcia-Launay, F.	152, 441, 456
François, L.	286, 395	García, M.E.	493
Franco, J.	567	García, M.L.	182, 493, 540
Franco-Rosselló, R.	653	García-Rebollar, P.	461, 468
Frąszczak, M.	108, 110, 194	García-Rodríguez, A.	247, 568
Frattini, S.	168	Garcia, S.C.	507
Fredholm, M.	600	Gardiner, G.E.	118, 121, 199, 547, 548
Freeburn, J.	589	Gardner, G.E.	133, 212
Freeman, T.C.	406	Garnsworthy, P.C.	481
Freude, K.	599	Garófalo, E.	408
Frevert, H.	215	Garreau, H.	606
Fries, R.	104, 109, 111	Garrick, D.	104, 109, 111, 218, 421
Friggens, N.C.	673	Garrido, M.D.	224
Frischknecht, M.	104, 109, 111, 341, 421	Garrido-Varo, A.	295
Fritz, S.	187, 191, 601	Gasa, J.	662
Froidmont, E.	512	Gatti, R.	565
Frössling, J.	374	Gault, N.F.S.	131, 132, 133, 148, 291, 642
Fuchs, K.	171, 173	Gauly, M.	321, 352, 396, 531, 620

Gauly, S.A.	388	Goddard, E.	450
Gautier, J.M.	364, 365	Godinho, R.M.	453
Gautier, M.	106	Golian, A.	659
Gavojdian, D.	168, 353, 678	Goliomytis, M.	528
Gay, A.P.	539	Golzar Adabi, S.H.	248
Gaynor, R.C.	604	Gomarasca, M.	324
Gazbar, F.	170	Gomes Ferreira, J.	176
Gazzarin, C.	166	Gómez-Cabrera, A.	295
Gebreyesus, G.	290	Gomez, E.	463
Gehmair, T.	514	Gómez, M.D.	583
Gehring, S.	436	Gomez-Raya, L.	257, 494
Gehrke, L.	434	Gómez-Redondo, I.	425
Geibel, J.	124	Goncu, S.	487
Gelasakis, A.I.	302	Gondret, F.	315
Gelé, M.	292, 300, 643	Gonen, S.	105, 604
Genestoux, L.	650	Gónzalez-Bulnes, A.	403
Gengler, N.	291, 298, 299, 300, 304, 377, 630	Gonzalez, E.	653
Geoghegan, A.	348, 495	Gonzalez-Garcia, E.	152
Georgiadou, M.	645	González-Hernández, M.P.	615
Geraert, P.A.	243, 245	González, O.	554
Gerber, V.	581, 583, 680	González-Prendes, R.	112, 233
Gerken, M.	519	Gonzalez-Rivas, P.A.	251
Germain, A.	564	González-Ronquillo, M.	173, 329, 519, 523
Gfeller, S.	424	Gonzalez-Verdugo, H.	501
Ghaderi, H.	110	Goodwin, K.	243, 288
Ghazanfar, S.	287, 407	Gordon, A.	267, 371, 674
Ghita, E.	644, 656	Gordon, A.W.	136, 137, 210, 505, 629
Ghorayeb, R.	527	Gordon, F.J.	548
Giannakou, R.	302	Gordon, J.	638
Giannenas, I.	321, 546	Gorgulu, M.	487
Giannico, F.	259	Gorjanc, G.	105, 206, 604
Giannico, K.	110	Gorla, E.	425
Giannico, R.	108	Görs, S.	228, 331
Giantin, M.	669	Goselink, R.M.A.	195
Gibon, A.	384	Goti, L.	237
Gierus, M.	657	Götz, K.U.	103, 338, 602
Gieseke, D.	396	Gous, R.M.	320
Gilbert, H.	316, 319, 454, 609	Govignon-Gion, A.	191
Gilfedder, T.	469	Graczyk, M.	203, 343
Giller, K.	290	Graham, H.	379, 652
Gilliland, T.J.	503, 504, 618	Grandison, A.S.	361
Gillon, A.	377	Grandl, F.	372, 630
Gil, M.	224, 313	Gras, M.	644
Gilmore, B.F.	279	Gras, M.A.	613
Gilmour, A.	488	Graviou, D.	467
Ginane, C.	649	Grayaa, M.	377
Giraldo, P.A.	251	Grazioli, R.	552
Girard, M.	322	Gredler, B.	104, 109, 111, 125, 344, 421, 450, 630
Giraud, J.-M.	625	Green, S.	243
Giromini, C.	360, 655	Grelet, C.	299, 300, 630
Gispert, M.	122, 653	Griffith, A.	645
Givens, D.I.	263, 360, 361	Grigoletto, L.	144, 286, 607
Gjerde, B.	258	Grimberg-Henrici, C.G.E.	221
Gjerlaug-Enger, E.	612	Grindflek, E.	532, 603, 612
Glagoleva, E.V.	561, 563	Groeneveld, E.	103
Glass, E.J.	188	Grohs, C.	189, 432

Große-Brinkhaus, C.	490	Hanrahan, J.P.	516
Grossen-Rösti, L.	373	Hans, A.	585
Gross, J.J.	285, 373, 378	Ha, N.T.	285, 378, 530
Grosu, H.	613	Hanzen, C.	377
Grotta, L.	297, 669	Harder, H.	458
Grøva, L.	619	Harding, J.	317
Grubelic, M.	297	Hargreaves, P.R.	498
Gruber, L.	269, 648	Harris, R.A.	636
Gruffat, D.	252, 446	Harstad, O.M.	208, 209
Guada, J.A.	618	Harty, E.	482
Guan, L.	208	Haskell, M.J.	401, 479
Guan, X.	249, 661	Hassan, F.	287, 407
Guatteo, R.	189	Haugaard, K.	216
Gubbiotti, M.	593, 595	Havard, T.	292
Guérin, S.	248	Hawken, R.	203, 343, 535
Guerrero-Ginel, J.E.	295	Hayes, B.J.	107
Guetarni, D.	533	Hayes, H.	121, 356
Guggenbichler, S.	347	Hazard, D.	363
Guillaume, F.	187	Headon, D.J.	165
Guillevic, M.	231	Hébrard, W.	180
Guillou, D.	240, 534	Heckenberger, G.	462
Guillou, M.	389	Hedayat Evrigh, N.	646
Guinard-Flament, J.	271, 551	Heinesch, B.	481
Gunnarsson, G.G.	186	Heinicke, J.	484
Gunnarsson, S.	374, 400, 564	Heise, J.	375
Guo, J.	361	Heiskanen, M.-L.	391
Gurgul, A.	345, 421, 545	Helary, L.	389
Gutierrez Garcia, J.P.	519	Helfjord, T.	595
Gutiérrez-Gil, B.	284, 425	Hellinga, I.	581
Gutiérrez, J.P.	125, 494, 520, 579, 607	Hely, F.	193
Gutzwiller, A.	322	Hempel, S.	485
Guy, C.	618	Hendriks, W.H.	557
Guzzo, N.	435, 476, 633	Hendrikx, P.	585, 683
		Henne, H.	490
		Hennes, N.	367
H		Hennessy, D.	616
Haas, J.H.	510	Hennessy, T.H.	198
Haasken, C.	325	Hennies, M.	321
Haas, L.	629	Henri, J.	248
Hăbeanu, M.	644, 655, 656	Henriksen, J.C.S.	484
Hagan, T.D.J.	222, 525, 640, 641, 642	Henrique, W.	657, 658, 659
Hager-Theodorides, A.	166	Henryon, M.	492
Haikukutu, L.	139	Henry, W.	233
Hailu, G.	450	Herholz, C.	392, 393, 505, 683
Hajibabaei, A.	248	Heringstad, B.	192, 347, 350, 575
Halachmi, I.	508, 509	Herlin, M.	258
Hamann, H.	216, 436	Hermansen, J.E.	387
Hamdi, M.	653	Hermans, K.	299, 300
Hamill, R.M.	147, 200	Hernández, H.	523
Hamland, H.	612	Hernandez, L.L.	372
Hammami, H.	304, 377	Hernández Maya, C.	274, 275
Hampel, A.	601	Hernández Pérez, P.	489
Hamrouni, A.	311	Herold, P.	436
Hamzaoui, S.	126	Herrero, M.	262
Hamzić, E.	535	Hertzberg, H.	505
Hanlon, A.	221	Hervé, L.	269, 274, 296
Hanotte, O.	418		

Hessel, E.F.	390	Hummel, J.	380
Heuß, E.	490	Humphries, D.J.	361, 561
Heuwieser, W.	484	Huntington, J.A.	270, 272, 666
Hickey, J.M.	105, 206, 604	Hunziker, M.	583
Hicks, V.	301, 432	Hurley, A.M.	627
Hill, D.L.	327	Hurtaud, C.	289, 292, 296
Hille, K.	352	Hurtaud, J.	606
Hill, T.R.	320	Hussein, A.	253
Hjortø, L.	308, 312	Hutchings, M.R.	328
Hochuli, A.	474	Húth, B.	144, 147, 149
Hocking, P.M.	124	Huting, A.M.S.	440
Hocquette, J.F.	133, 212, 446	Huuskonen, A.	134, 135
Hodgson, C.	358	Hyslop, J.J.	207, 406, 409, 460, 470
Hoeksma, D.	498	Hyttel, P.	599
Hoffman, L.C.	501		
Hoffmann, I.	160	**I**	
Hoffman, W.H.	501	Iannetti, L.	399
Hofmanová, B.	476, 584	Iannucelli, N.	599
Hogewerf, P.H.	497	Ibáñez-Escriche, N.	125, 260, 494
Höglund, J.K.	299, 300, 452	Icken, W.	175
Höinghaus, K.	223	Ignatova, M.	664
Hojito, M.	506	Ilie, D.E.	168, 353
Hollmann, M.	199, 201	Imran, M.	287
Holló, G.	144, 147, 149	Inca Guerrero, V.	274, 275
Holló, I.	147, 149	Infascelli, F.	552, 569
Holshof, G.	496	Inghese, C.	679
Holtenius, K.	470, 654	Ingvartsen, K.L.	299
Holt, T.	419	Insausti, K.	585
Homolka, P.	412, 633	Invernizzi, G.	359, 655
Hooiveld, G.J.	242	Ioannidou, M.	296
Horan, B.	117, 500, 631	Ipema, A.H.	496, 497
Hormta, C.	331, 515	Irwin, D.	525
Hornick, J.L.	533	Isabel, B.	403
Horodyska, J.	200	Islas-Trejo, A.	419
Hoskens, H.	395	Ismael, A.	311
Hostens, M.	299, 300	Itenge, T.O.	139
Hoste, S.	220, 589	Ivarsson, E.	226
Houchaimi, K.	251	Ivemeyer, S.	306, 307, 482
Houchaymi, K.	527	Izhaki, I.	508
Houdijk, J.G.M.	226, 328		
Houdová, D.	476	**J**	
Houston, R.D.	189	Jack, C.	333
Hoving, R.A.H.	420	Jacquiet, P.	418
Howard, J.	349, 414	Jaeger, M.	307
Howell, K.	472	Jagoe, A.	334
Hozé, C.	601	Jagusiak, W.	374
Huba, J.	522	Jakobsen, J.H.	517
Huber, K.	179, 268, 369	Jakobsen, M.	382, 387
Huber, M.	313	Jalam, G.	389, 626
Hughenin-Elie, O.	381	Jamrozik, J.	350
Hugh, N. Mc	171, 210, 213, 419, 511	Jancik, F.	412, 633
Huhtanen, P.	631	Janke, D.	485
Hulsegge, B.	542	Jansman, A.J.M.	544
Hulzebosch, A.	301	Janssen, H.	400
Hume, D.A.	418	Janssens, S.	444, 594
Humer, E.	458	Janss, L.L.	290, 440, 604

Jansson, A.	389	Kämmerling, J.D.	398
Jardim, J.	191	Kampman, E.	448
Jarosz, B.	232	Kanis, E.	427
Jasielczuk, I.	345, 421	Kanitz, E.	228
Jawasreh, K.I.Z.	132	Kaps, M.	415
Jemmali, B.	170	Karaman, E.	346
Jenet, A.	324	Karamoutsios, A.	546
Jenko, J.	206, 602	Karatzia, M.A.	296, 479
Jensen, A.	136	Kardacz, P.	632
Jensen, J.	127, 203, 204, 492	Kardel, M.	623
Joerg, H.	166, 313	Kargo, M.	127, 308, 311, 312
Johansen, M.	136	Karhapää, M.	319
Johansson, B.	411	Karimi, H.	252
Johnston, A.	210, 589	Kariuki, C.M.	310
Johnston, D.	504, 536	Karlsson, J.	470
Johnston, D.J.	404	Kar, S.K.	544
Joki-Tokola, E.	135	Karsten, S.	511
Joller, S.	424	Kartashov, M.I.	561, 563
Joly, A.	189	Kasarda, R.	423, 427, 578, 679
Jonas, D.	601	Kaseja, K.	363
Jonas, E.	341, 529	Kasprzak, M.M.	226
Jondreville, C.	248	Katali, B.K.	427
Jones, G.	112	Katoh, K.	429
Jones, M.	438	Kato, K.	109
Jönsson, L.	564	Kaufmann, F.	398, 531
Jordana, J.	233	Kaufmann, T.	325
Jordan, C.	177	Kause, A.	257
Joy, M.	640	Kaushik, S.J.	254
Juanes, X.	156	Keady, T.W.J.	516
Judge, M.M.	194, 217	Kearney, J.F.	194, 217, 335
Juga, J.	610, 612	Keatley, P.	327
Juhás, P.	654	Keenan, J.	442
Julián, R. San	183	Keles, G.	524
Junge, W.	510	Kelleher, M.M.	217
Jungnickel, R.	192	Keller, M.	523
Juniper, D.T.	672	Keller, S.	545
Juodka, R.	668	Kelly, A.K.	139, 141, 142, 185, 206, 211, 212, 217,
Juska, R.	668		224, 244, 359, 409, 536, 550, 669
Juskiene, V.	668	Kelly, C.	596
		Kelly, D.	167
K		Kelton, D.	350
Kaart, T.	276, 277	Kemp, B.	370
Kadarmideen, H.N.	229, 452, 599	Kempeneers, A.	363
Kaddour, R.	186	Kemper, N.	398
Kadlečík, O.	423, 427, 578	Kenéz, Á.	268
Kaess, M.	496	Kennan, J.	548
Kägi, F.	392, 505, 683	Kennedy, A.E.	174
Kahi, A.K.	310, 429	Kennedy, E.	482, 500, 627, 671
Kaić, A.	577	Kennedy, J.T.	222
Kaiser, P.	418	Kennedy, P.	596
Kalafová, A.	493	Kenny, D.A.	185, 217, 240, 244, 354, 536, 550
Kalogianni, A.	479	Kenyon, P.R.	522
Kals, J.	256	Keogh, K.	185, 244, 550
Kamada, H.	635	Kerslake, J.I.	431
Kamal, M.M.	180	Kettunen, A.	595
Kambashi, B.	565	Khalfaoui, H.	169

Khalifa, M.	312
Khan, A.	367, 645
Khan, M.A.	368
Khatkar, M.S.	341
Khayatzadeh, N.	125, 415
Khempaka, S.	257, 331, 562, 572, 614
Khol-Parisini, A.	458
Khotsakdee, J.	466
Kidd, L.J.	242, 243, 288
Kightley, S.P.J.	226
Kijas, J.W.	165
Kipling, R.P.	384
Kirkpatrick, B.W.	194
Kivisild, T.	422
Klaffenböck, M.	269
Klambeck, L.	398
Kliem, K.E.	361
Klootwijk, C.W.	115, 617
Klopčič, M.	113, 234
Klopp, C.	284
Knaus, W.	238
Kneifel, W.	292
Knight, C.H.	673
Knol, E.F.	340, 438, 440, 488
Knol, F.W.	367, 368
Knudsen, M.T.	137
Kobayashi, E.	109, 429
Koch, C.	418
Koeck, A.	350
Koene, P.	497, 500, 632
Koenig, M.	485
Kofler, J.	347, 396
Kogelman, L.J.A.	229
Koivula, M.	337
Kok, A.	370
Kok, M.G.	154
Kolodziejski, P.	118
Komen, H.	310, 427, 429
Kominakis, A.	302
Komolka, K.	145
Kompan, D.	426, 520
Kongsro, J.	532
Kongsted, A.G.	382, 387
König, S.	215, 305, 307, 309, 426, 433
König Von Borstel, U.	306, 352, 401
Koolaard, J.	182
Koops, W.	219
Kopke, G.	192
Kopp, C.	392, 683
Kortelainen, T.	319
Kouakou, D.V.	562
Kouba, M.	562
Kovač, M.	234
Kozak, M.	629
Kozloski, G.V.	502
Kozłowska, A.	158

Krafsur, G.M.	419
Kramer, E.	572
Krattenmacher, N.	331
Kraus, A.-S.	270
Kremer-Rücker, P.V.	223
Kremer, V.D.	478
Kreuzer, L.S.	223
Kreuzer, M.	119, 290, 372, 547, 630
Kreuzer-Redmer, S.	123
Krieter, J.	221, 399, 511, 529, 597, 598, 623
Kristensen, I.S.	137
Kristensen, T.	137, 369, 370, 377
Kristjánsson, T.	107
Krizsan, S.	631
Kroeger, C.	346
Krogsdahl, J.J.	614
Krogsdal, J.	604
Kronberg, S.L.	150
Krpálková, L.	352
Krsnik, J.	577
Kruger, A.C.M.	165
Kruijt, L.	544
Krupa, E.	355, 444, 445
Krupová, Z.	352, 355, 444, 445
Kubelkova, P.	412, 633
Kudinov, A.A.	610
Kuhla, B.	267, 331
Kühne, P.	276, 278
Kühn, I.	567
Kuipers, A.	113, 114, 174
Kumalska, M.	344, 608
Kump, E.	424
Kupczyński, R.	232, 278, 477
Kusza, S.Z.	168, 445, 678
Kutluca Korkmaz, M.	364
Kyriazakis, I.	314, 316, 318, 320, 433, 440, 455, 472, 478

L

Ladinig, A.	225
Lafri, M.	182, 540
Lainé, A.	304
Laird, S.	567
Laloe, D.	190
Lambe, N.R.	368, 638
Lamberton, P.	269, 274
Lambertz, C.	396, 620
Lam, S.	181, 208, 629
Landauskas, M.	477
Langa, T.	553
Langbein, J.	675
Langeveld, J.P.M.	420
Larroque, H.	271
Larsen, T.	299, 629
Larzul, C.	341, 609, 610
Lascoux, C.	460, 487

Lassen, J.	126, 127, 285, 288, 449, 450, 452	Lema, O.M.	616
Lasseur, J.	155, 385, 465	Lemauviel-Lavenant, S.	381
La Terza, A.	595	Lembeye, F.	636
Launay, A.	191	Lembo, G.	637
Laurenson, Y.C.S.M.	433	Lemme, A.	249
Lauridsen, C.	227, 543	Le Morvan, A.	559, 619
Lauwers, L.	443	Lemosquet, S.	551
Lavery, A.	446	Lencart-Silva, J.D.	177
Lavery, U.	356	Lenehan, C.	141, 142
Lawlor, P.	200, 202	Leonard, N.	172
Lawlor, P.G.	118, 121, 199, 200, 201, 439, 446,	Lescoat, P.	239
	547, 548, 554	Le, T.H.	378
Lawlor, P.J.	121	Leurent-Colette, S.	428, 625, 630
Lawlor, T.J.	339	Leury, B.J.	251, 573
Law, R.	437, 639	Lewdiukow, M.	616
Law, R.A.	371	Lewis, C.	266
Lawrence, P.	143, 217	Lewis, E.	500, 627
Lawson, R.	473	Lewis, H.	131, 132
Lazar, C.	644	Lewis, R.M.	517
Lazzari, M.	403	Lewis, T.W.	592
Lazzaroni, C.	293, 294	Lherm, M.	209, 326
Lazzarotto, V.	255	Liaubet, I.	180, 599
Leal, L.N.	242	Li, B.	452, 544
Leal Yepes, F.	371	Lidauer, M.H.	453, 628
Lebeau, F.	481	Liebig, M.A.	150
Leblois, J.	544	Liegard-Vanhecke, D.	228
Leblond, A.	585	Ligda, C.	157
Lebouter, M.	389	Ligonesche, B.	610
Lechartier, C.	196	Liiva, M.	276, 277
Lechner, E.M.	657	Likitdejcharoj, B.	331, 515
Leclerc, H.	187	Lillehammer, M.	163, 284
Lecomte, P.	385, 465	Limberg, S.	401
Le Cozler, Y.	626	Linares, M.B.	224
Ledinek, M.	648	Lind, C.E.	260
Leeb, T.	341, 581, 583	Lindegaard, J.Ø.	227
Lee, D.H.	402	Lindena, T.	116
Lee, J.Y.	402	Lindgren, G.	286, 395
Lee, M.	681	Link, Y.	511
Lee, M.A.	498	Linscott, A.	135
Leemans, D.	539	Linton, M.	596
Lee, M.R.F.	358	Lipkin, E.	410
Leen, F.	442, 443	Lippi, Y.	599
Lefaucheur, L.	315	Lisle, A.T.	242
Lefebvre, R.	191	Lister, S.J.	560, 563
Le Floc'h, N.	222, 677	Little, M.	550
Lefter, N.A.	656	Liu, Z.	283
Le Gall, A.	115	Lively, F.	291
Legarra, A.	338, 609	Lively, F.O.	136, 137, 143, 207, 210, 214, 397,
Legrand, I.	133, 212		589, 636
Lehmann, J.O.	369, 370, 377	Livingstone, M.B.E.	597
Lehotayová, A.	654	Lizarazo, A.	643
Leiber, F.	360, 482, 549	Lizardo, R.	121
Leidinger, D.	330	Llonch, P.	479
Leinonen, I.	455	Lobley, M.	332
Leitch, S.	639	Lobón, S.	640
Lelyon, B.	326	Lock, A.L.	447

Maroto-Molina, F.	295	Mayer, M.	128, 216
Marras, G.	521	May, K.	309
Marren, D.	129, 134	Mazzoni, G.	452, 530, 599
Marrero, L.	665	Mazzullo, M.	530
Martelli, P.L.	530	McAdam, J.H.	383, 497
Martemucci, G.	646	McAlpine, P.O.	358, 558, 559
Martikainen, K.	420	McBride, S.H.	188
Martin, C.	460, 467, 487	McCabe, M.S.	536
Martínez Álvaro, M.	489	McCabe, S.	210, 213, 504
Martinez, B.	644	McCarthy, B.	618
Martínez, E.	567	McCarthy, J.	335
Martínez, P.	258	McClure, J.C.	434
Martinez, R.	111	McClure, M.	335, 342, 343
Martin, G.	155, 237	McCoard, S.A.	182, 367, 368
Martin, J.E.	532	McConnell, D.	588
Martin, O.	152	McConnell, D.A.	197
Martino, G.	297, 669	McCormack, U.M.	118, 199, 547, 548
Martin, P.	272, 643	McCormick, W.	393
Martinsen, K.H.	443	McCracken, K.J.	548
Martín-Tereso, J.	242	McDermott, A.	304, 305
Märtlbauer, E.	270	McDermott, K.	163, 164, 167, 171, 419, 517
Martyniuk, E.	158	McDonald, R.	117
Marucco, F.	620	McEvoy, T.G.	368
Marusi, M.	454	McEwan, J.	161
Mascarell, J.	120	McEwan, J.C.	107
Maschat, K.	623	McGee, M.	129, 134, 139, 141, 142, 143, 148, 206,
Mason, S.	324		211, 217, 293, 468, 536, 669, 674
Masoudi, A.A.	110, 535	McGonigle, C.	177
Massart, S.	544	McGovern, F.M.	366, 517
Massart, X.	300	McHugh, N.	163, 164, 167, 385, 388, 499, 502,
Masuda, Y.	336, 338, 339		517, 612
Masure, A.	512	McIlroy, C.	641
Matek Saric, M.	297	McKain, N.	406, 464, 470
Mateus, L.	681	McKeegan, D.E.F.	532
Matheson, S.M.	314, 315, 316, 472	McKenna, C.E.	240, 244
Mathis, A.	392	McKinnie-Hill, J.	525, 640
Mathur, P.K.	488, 603	McLaren, A.	363
Matika, O.	165, 188, 351, 418	McLean, K.A.	638
Matilainen, K.	337, 340, 612	McLean, S.	177
Matoba, S.	635	McManus, M.	504
Matthews, E.	300	McMenamin, K.	139, 148
Mattiauda, D.A.	483	McMullen, P.	233
Mattioli, S.	649	McNeill, D.M.	242, 243, 288, 667
Mattos, E.C.	144, 169, 428, 607	McNeilly, T.N.	432
Mauch, E.D.	225	McParland, S.	128, 271, 303, 304, 305, 449, 627
Mauh Lenardić, J.	422	McRae, A.	301
Maupin, M.	606	McRoberts, C.	291
Maurer, V.	360, 549	McTeir, L.	165
Maurice-Van Eijndhoven, M.	516	McWhinney, M.	639
Maushammer, M.	542	Meale, S.J.	461
Mausse, F.	139	Méda, B.	152, 239, 248
Maxa, J.	496	Médale, F.	255
Maxin, G.	560	Medisetti, S.	675
Mayasari, N.	370	Medrano, J.F.	419
Mayer, E.	246	Medugorac, I.	434
Mayerhofer, M.	173	Mee, J.F.	513

Meerburg, B.	455	Millet, S.	442, 443, 444
Meese, S.	267	Miltiadou, D.	166
Mehtiö, T.	453	Minozzi, G.	108, 110, 168, 521
Meikle, A.	483	Miracle, P.	422
Mejri, S.	539	Miranda-De La Lama, G.C.	678
Mekkawy, W.	260	Miranda, G.	272, 643
Mekuriaw, G.	161	Miranda Yuquilema, J.E.	558, 665
Mele, M.	524	Mishra, N.	211
Melzer, N.	416, 675	Misztal, I.	336, 338, 339
Membrillo, A.	609	Miszura, A.A.	634, 635
Mena, M.A.	325	Mitchell, M.A.	531, 532
Méndez, J.	184	Mlakar, G.	426
Mendoza, A.	408	M'naouer, D.	311
Menezes, D.R.	526	Modiba, S.	553
Menghi, A.M.	116	Mogensen, L.	137, 369, 370, 377
Menzano, A.	620	Mohamed, M.	155
Meo Zilio, D.	553	Mohammed, A.	645
Mercadante, M.E.Z.	140, 408	Molee, A.	331, 515, 614
Mercat, M.J.	341, 609	Molee, W.	331, 515, 562, 572, 614
Mercier, Y.	245	Molenaar, A.	182
Merda, K.	354	Molina, A.	538, 583, 609
Merlot, E.	315, 317, 677	Molist, F.	249, 661
Mernklatoak, P.	331, 515	Moloney, A.P.	129, 134, 139, 141, 142, 148, 468,
Mernkrathoke, P.	614		669
Merrifield, D.	255	Monahan, F.J.	129, 134, 525, 640
Mesarič, M.	577	Montagne, L.	316, 319, 389, 626
Meslier, E.	559	Montanari, M.	467
Messori, S.	399, 404	Montanholi, Y.	181, 208, 629
Mészáros, G.	125, 269, 423, 436	Monteiro, A.N.T.R.	456
Metges, C.C.	228, 331	Moore, H.	177, 178
Metzler-Zebeli, B.U.	118, 199, 200, 201, 202, 225,	Moore, K.L.	129, 193, 363
	230, 548, 554, 648	Moraga, C.	497
Meuret, M.	465	Morales, J.	545, 555, 556
Meuwissen, T.H.E.	103, 104, 163, 303, 443, 492	Morales, R.	503, 644
Meyer, A.M.	651	Morales, T.	408
Mezgebo, G.B.	129, 134	Mora, M.A.	325
Michael, C.	642	Moran, A.	193
Michaličková, M.	352, 355, 412	Moran, D.	386
Michiels, J.	250, 551, 659	Moran, L.	148
Mickdam, E.	458	Morante, R.	520
Midmore, P.	513	Moravčíková, N.	423, 427, 578
Mielczarek, M.	108, 110	Moreira, E.M.	634, 635
Migliorati, L.	463	Morek-Kopec, M.	374
Miglior, F.	106, 350, 449, 450	Morel, P.C.H.	522
Miguel, M.F.	617	Moreno, C.R.	418, 518
Mihalcea, T.	656	Morgan-Davies, C.R.	368
Mikawa, S.	429	Morgan, R.	361
Mikko, S.	576	Morgan, S.A.	138, 539
Miksza-Cybulska, A.	345	Morgavi, D.P.	467
Milanesi, M.	415	Morignat, E.	683
Milerski, M.	367	Mormède, P.	180
Millán, A.	258	Morrison, S.J.	143, 210, 267, 327, 589, 674, 675,
Miller, C.	333		676
Miller, H.M.	446, 534, 567, 652	Morsy, A.S.	666
Miller, S.	181, 208, 629	Mosnier, C.	326
Millet, M.	518	Mosquera-Losada, M.R.	615

Moss, B.W.	133, 148, 642	Neamt, R.	353
Mote, B.	317	Neary, J.	419
Motiang, D.M.	514	Neciu, F.	353
Motta, V.	543, 546	Neeteson-Van Nieuwenhoven, A.M.	265
Mottram, T.T.F.	510	Negro, S.	538, 609
Mouaze, C.	625	Negussie, E.	453
Mouillot, F.	385	Neil, M.	226
Moulin, C.H.	152, 155, 157	Nekrasov, R.V.	561, 563
Mourot, J.	231	Nemecek, T.	236
Mousaoumai, E.	251	Neser, F.	422
Moussa, H.	413	Neuditschko, M.	341, 421, 424, 581, 583
Mrode, R.	161, 162, 188, 603, 611	Newman, M.A.	225, 230
Mucha, A.	608	Nfor, B.M.	138
Mucha, S.	162, 363, 611	Ng'ambi, J.W.	526
Mueller-Harvey, I.	322	Nicholas-Davies, P.K.	151
Muetzel, S.	368	Nicholas, P.	157
Muhammad, A.	287, 407	Nicolazzi, E.L.	108, 110, 521
Muhammad, I.	407	Nicol, C.J.	262
Mühlbach, S.	267	Niderkorn, V.	649
Muir, P.	182	Nielsen, B.	378, 605
Mukiibi, R.	427	Nielsen, H.M.	604, 614
Mulder, H.A.	189, 318, 488	Nielsen, M.B.F.	227
Muley, N.S.	249	Nielsen, N.I.	137
Mulholland, M.	588	Nielsen, P.	347
Mullane, L.	482	Nielsen, U.S.	628
Müller, K.E.	347	Niemann, H.	265
Muller, N.	316, 319, 625	Ni, G.	103
Müller, V.	435	Nilieh, M.R.	637
Mulser, E.	392, 393	Nilsson, K.	378, 594
Mumović, R.	580	Nishiura, A.	376
Münger, A.	630	Niznikowski, R.	519
Munksgaard, L.	484	Nkosi, B.D.	462, 553
Muñoz, R.	497	Nobili, C.	565
Munro, J.	181, 208	Nocelli, C.	595
Murani, E.	200, 201, 203	Noel, S.J.	288
Murchie, A.K.	505	Nolan, P.	468
Muroya, S.	506	Norberg, E.	378
Murphy, B.	212, 409, 457	Nordbø, Ø.	603
Murphy, J.P.	671	Norris, A.	266
Murphy, M.D.	273, 512	Norris, D.	526
Murphy, R.A.	280, 323, 538, 566, 667, 668	Norton, T.	268, 624
Murphy, V.S.	137, 397	Notenbaert, A.M.	325
Murray, G.	536	Novotný, L.	355
Musco, N.	552, 569	Noziere, P.	559, 560
Mustière, C.	269	Nozière, P.	379, 380, 551, 650
Muuttoranta, K.	628	Nunes, J.P.	176, 177
Mwacharo, J.	161	Nydam, D.	371

N

Nainiene, R.	668	**O**	
Napoleone, M.	155	Obeidat, B.S.	552
Narcy, A.	381	O'Brien, A.	163, 164, 167
Nascimento Jr., N.G.	150	O'Brien, A.C.	171, 419
Nassu, R.T.	298, 413	O'Brien, E.	348, 351, 430, 434
Naya, A.F.	399	Obritzhauser, W.	171, 173
Nayan, N.	557	Ocak, S.	627
		Öckinger, E.	389

Pascual-Alonso, M.	678	Phuong, H.N.	375
Pascual, J.J.	489, 537	Phyn, C.V.C.	431
Pastorelli, H.	315	Physick-Sheard, P.W.	684
Patel, M.	470, 508	Piaggio, L.	183, 463
Patience, J.F.	225	Piasentier, E.	138
Patras, I.	168, 678	Pickel-Chevalier, S.	475
Pauciullo, A.	294	Piegas, J.	142
Paul, B.K.	325	Pienaar-Van Der Westhuizen, L.	422
Pausch, H.	103, 104	Pieragostini, E.	521, 579
Pauselli, M.	649	Pierce, K.	117
Pauw, W.	484	Pierce, K.M.	271, 631
Pawlina, K.	545	Piette, D.	268
Payling, L.	556	Piles, M.	123
Pazzaglia, I.	595	Pilotto, A.	656
Pedersen, H.S.	599	Pimentel, E.C.G.	338, 602
Pedersen, M.L.M.	227	Pinard-Van Der Laan, M.-H.	535
Pediconi, D.	595	Pineda-Quiroga, C.	247, 568
Pedroso, A.F.	298, 413	Pingel, H.	184, 537
Peine, J.L.	651	Pinna, W.	621
Peiren, N.	459	Pinotti, L.	551, 655
Peixoto, M.	191	Pinto, F.	646
Pellegrini, C.B.	526	Pinto, S.	484
Pelmus, R.	644	Piórkowska, K.	608
Pelmus, R.S.	613	Pires, A.V.	634, 635
Peña, F.	609	Pires, P.P.	624
Peña Fernández, A.	624	Pirlo, G.	314, 463
Penasa, M.	304	Pla, M.	408, 649
Perea-Muñoz, J.	653	Plantevin, T.	456
Père, M.C.	180, 315	Plastow, G.	317
Pérez, F.	215	Pleijte, M.	153
Perez, J.F.	653	Plemyashov, K.V.	610
Pérez-Marín, D.C.	295	Plichta, D.F.	288
Pérez-Prieto, L.A.	501	Plieschke, L.	602
Pérez-Vendrell, A.M.	119, 120, 122	Ploetz, T.	314, 472, 478
Perini, M.	138	Plötz, T.	316
Perrot, C.	115	Plowiec, A.	246
Persiani, A.M.	565	Pnini, T.	508
Perucho, L.	157	Poccard-Chappuis, R.	152
Pesonen, M.	134, 135	Pocrnic, I.	336, 338
Petazzi, F.	521	Poddesu, S.	621
Peter, S.	372	Pogoda-Sewerniak, K.	278
Peters, A.R.	218	Polak, G.M.	582
Pethick, D.W.	133, 212	Polák, P.	149
Petrák, J.	654	Polasik, D.	344, 608
Petridou, E.	175	Pol, F.	228
Petri, R.M.	200, 648	Politis, I.	254, 528, 647
Petruska, P.	493	Polizel, D.M.	634, 635
Peura, J.	491	Polkinghorne, R.J.	133, 212
Peyrat, J.	559, 632	Pollott, G.E.	302, 416, 592
Peyraud, J.L.	289, 379	Pomiès, D.	465
Pfeiffer, C.	310, 628	Pomilio, F.	399
Philau, S.	274	Ponchunchoovong, S.	259
Philippe, F.	565	Pong-Wong, R.	602
Philippe, M.	187	Ponsuksili, S.	200, 201, 203
Philipsson, J.	575	Poorhassan Talemi, A.	557
Phocas, F.	601	Popova, M.	467

Ribani, A.	530, 606	Roquet, J.	411
Ribeiro Filho, H.M.N.	617	Rose, G.	451
Ribeiro, R.D.X.	150	Rose, N.	228
Ribikauskas, V.	306	Rosner, F.	192, 267
Ricard, A.	578	Ross, D.	220, 409
Ricard, E.	518	Ross, D.W.	406, 460, 470
Richardson, I.	138, 211	Rossi, G.	459
Richardson, R.I.	129, 134, 358	Rossoni, A.	108, 344
Richter, I.	509	Rota, G.	168
Riddle, S.	419	Rotar, M.C.	613, 644
Rieder, S.	341, 581, 583	Roth, N.	392
Riganelli, S.	593	Rothschild, M.F.	317
Riggio, V.	165, 188, 351	Rotondi, P.	259
Rigueiro-Rodríguez, J.L.	615	Roubos, P.	650
Rinell, E.	350	Rouillé, B.	292
Ringmark, S.	389	Rouleau, B.	320
Ring, S.C.	348, 351, 430	Rousing, T.	306
Ripoll, G.	640	Rousselière, Y.	240, 534
Riquet, J.	180, 341, 609	Rovers, M.	659, 661, 662
Risso, S.	215	Rowe, S.	161
Robert, F.	315, 316, 319	Rowland, R.	318
Roberts, D.J.	498	Royer, E.	240, 248, 534
Roberts, F.G.	590, 591	Ruane, E.	157
Robinson, P.	568	Rubí, M.	609
Robins, R.	461	Rubin, C.J.	600
Robles Jimenez, L.E.	519	Rudel, S.	650
Rocco, J.	146	Rudel, T.	387
Rocha, A.	629	Ruesche, J.	316, 606
Rocha, J.F.	636	Ruiz-Ascacibar, I.	441
Rocha, T.C.	150	Ruiz-Lopez, F.J.	425
Roche, J.R.	431, 500	Ruiz, M.	585
Rochette, Y.	460, 465, 467, 487	Ruiz, R.	237, 247, 568
Rodiles, A.	255	Rusakovica, J.	478
Rodrigues, P.H.M.	298, 413	Ruß, I.	104, 109, 111, 434
Rodriguez, M.C.	494	Russo, V.	615
Rodríguez-Ramilo, S.T.	414	Rutherford, K.M.D.	401
Rodríguez-Sánchez, J.A.	145	Rutkowska, K.	570
Roehe, R.	406, 409, 460, 470, 638	Ruttens, A.	662
Rogel-Gaillard, C.	541	Ryan, M.	482
Roger, L.	316, 319	Rydhmer, L.	491, 529
Roger, S.	381	Rymer, C.	176, 322, 561, 672
Rolingson, A.	645	Ryschawy, J.	150, 151, 237, 381
Romanek, J.	545	Rzewuska, K.	329
Roman-Ponce, S.I.	425		
Romeo, A.	250, 653	**S**	
Romero-Franco, R.	615	Saavedra, R.	324
Romero, J.	634	Sabatier, R.	381
Romero Oiartzun, M.	371	Sabljak, I.	297
Rommelfanger, E.	215	Sadeghi Sadegh Abad, F.	252
Ronchi, B.	569, 665	Sadri, H.	241, 245, 676
Rondia, P.	512	Saha, C.K.	485
Rooke, J.	409	Sahana, G.	191, 450
Rooke, J.A.	328, 406, 460, 470	Sahar, A.	147
Rooni, K.	584	Sahraoui, N.	533
Ropka-Molik, K.	421, 608	Saibi, L.	245
Ropota, M.	644, 656	Saidi, R.	172

Seichter, D.	434	Skinkyte-Juskiene, R.	229
Seioudy, A.F.	275	Skofic, L.	565
Selvaggi, M.	646	Skoufos, I.	321, 546
Seoni, E.	457	Skoufos, S.	321
Šepetkaitė, A.	477	Skuce, R.A.	188
Sepúlveda, W.	678	Slagboom, M.	308, 312
Serbester, U.	487	Slawinska, A.	118, 246
Serenius, T.	612	Sleator, R.D.	217
Serra, A.	524	Sloss, J.	663
Service, M.	177, 178, 179	Smaragdov, M.G.	610
Sevane, N.	111	Smith, G.L.	673
Sevi, A.	670	Smith, S.L.	300, 301, 432
Sevillano, C.A.	282, 453	Smits, M.A.	542, 544, 650
Sevón-Aimonen, M.-L.	319, 612	Smyth, S.	548
Shabtay, A.	241, 410	Sneddon, N.W.	636
Shalloo, L.	385, 388, 495	Snelling, T.J.	464
Sharifi, A.R.	124, 175, 281, 380, 530, 600	Soares, M.L.	430
Sherriff, S.	560	Soberon, F.	242
Shetty, N.	303	Sobolewska, A.	117
Shine, P.	512	Solà-Oriol, D.	653, 662
Shingfield, K.J.	464	Solberg, T.	108
Shirali, M.	127, 203, 204	Solé, M.	538, 609
Sibony, V.	241	Sölkner, J.	104, 109, 111, 125, 269, 415, 436
Šidlová, V.	423	Somarriba, M.	479
Siegerstetter, S.-C.	200	Sonesson, A.K.	163, 284
Signer-Hasler, H.	104, 109, 111, 341, 421, 583	Sonnenberg, A.S.M.	557
Signorini, A.	637	Sørensen, A.C.	308, 312, 491, 613
Sigurðardóttir, H.	574	Sorensen, J. T.	306
Sigurðardóttir, I.	474, 475	Sorensen, M.T.	299, 300
Sild, E.	582	Sørensen, P.	204, 492
Silió, L.	494	Sorg, D.	267
Siljander-Rasi, H.	319	Soroko, M.	472
Silva, L.R.	428	Sosa, M.	649
Silva, M.	503	Sossidou, E.N.	175, 296, 479
Silva, R.	386	Sotirakoglou, K.	645
Silva, R.G.	634, 635	Sousanna, J.F.	261
Silva, S.	391, 394, 528	Soutto, J.	408
Silva, S.L.	298, 413	Souza, C.M.	526
Silva, V.H.	286	Soyeurt, H.	300, 512, 544, 630
Silveira, X.	142	Sparaggis, D.	254, 642, 645
Silvestre, T.	182	Speidel, S.	419
Simantke, C.	306, 307	Spengler, A.	166, 307
Simčič, M.	423, 426, 577	Spengler Neff, A.	306, 482
Simeone, A.	142, 146, 215, 567	Sperrle, F.	576
Simianer, H.	103, 124, 281, 285, 375, 378, 530, 600	Spiers, J.G.	243, 288
		Spiller, A.	273, 388
Simitzis, P.	528	Spiri, A.	424
Simm, G.	603	Śpitalniak, K.	232, 278, 477
Simonin, H.	539	Spoolder, H.A.M.	404, 478, 531
Sinclair, K.D.	244	Spörndly, E.	508
Sinclair, L.A.	244, 270, 272, 571	Spörndly, R.	470
Sindic, M.	291, 298	Spranghers, T.	551
Singh, A.	211	Stadnicka, K.	117
Siwek, M.	118, 246	Staines, K.A.	165
Skarin, A.	389	Stalgiene, A.	113
Skiba-Cassy, S.	255	Stamigna, C.	665

Stanton, A.V.	357	Sveinsson, I.	186
Stark, F.	152	Šveistienė, R.	149, 477
Staub, C.	625	Svendsen, M.	192
Steele, M.A.	208, 242	Sviben, M.	231, 232
Steen, M.	389	Svitakova, A.	367
Steiner, A.	347	Swalve, H.H.	192, 267, 435, 491, 575
Steiner, K.	166	Swaminathan, M.	190
Stein, H.	230	Swanepoel, N.	568
Steinig, E.J.	341	Sweeney, T.	147, 224
Steininger, F.	330, 347, 396, 628, 648	Symeou, S.	166
Steinshamn, H.	619	Szałański, M.	126
Steinwidder, A.	238, 269	Szczerba, A.	117
Steinþórsson, R.	475	Szmatoła, T.	345, 421, 545
Stelletta, C.	476	Szumny, A.	232
Stenmark, K.R.	419	Szwaczkowski, T.	203, 342, 343
Štepec, V.	577	Szyda, J.	108, 110, 194
Stephan, S.	672		
Stergiadis, S.	328, 360	**T**	
Stevens, D.	367, 368	Taberlet, P.	464
Stewart, S.	291	Tadelle, D.	418
Stienezen, M.W.J.	506	Taferner, R.	436
Stilwell, G.	528	Tagliapietra, F.	459
Stinckens, A.	286, 395	Taha, V.J.	666
Štípková, M.	352	Takahashi, H.	635
Stirling, S.	649	Takeda, H.	376
Stirnimann, R.	393	Talenti, A.	168
Støbet Lande, U.	619	Tänavots, A.	276
Stock, K.F.	347, 375, 435, 576	Tangorra, F.M.	403, 656
Stoll, P.	441	Taniguchi, M.	429
Stoop, W.M.	195	Tanpol, N.	331
Stothard, P.	450	Tapia, M.	503
Strabel, T.	329	Tapio, I.	464
Strandberg, E.	311, 594	Tapprest, J.	585, 683
Strandén, I.	336, 337, 339, 340, 610	Taskinen, M.	339
Streeter, I.	417	Tava, A.	574
Stricker, C.	104, 109, 111, 307, 421	Taveewannaboon, K.	259
Strillacci, M.G.	344, 425	Tawila, M.A.	670
Stroebech, L.	599	Taylor, A.E.	534, 652
Strong, M.	447	Taylor, M.	177
Strube, C.	309	Taylor-Pickard, J.	272
Sturaro, E.	209, 382, 620	Taylor, R.F.	211
Sturmlechner, F.	514	Taylor, U.	207
Suarez, N.	643	Tayyab, U.	571
Suárez-Vega, A.	284, 425	Tedeschi, L.O.	140
Subagiyo, I.	154	Tedesco, D.	573, 574, 671
Subiabre, I.	644	Temim, S.	186
Suchocki, T.	194	Tenghe, A.M.M.	287
Suda, Y.	109, 429	Tennant, L.J.	481
Su, G.	450	Ten Napel, J.	281
Sugiharto, S.	543	Terler, G.	269
Sukkel, W.	153	Terman, A.	344, 608
Sun, H.	537	Tetens, J.	331, 434, 600
Surry, Y.	394	Teuscher, F.	128, 495, 601
Sutcliffe, E.	122	Thaller, G.	331, 434, 510, 600
Suzuki, K.	109, 429	Thanner, S.	322
Svartberg, K.	594	Thenard, V.	155

Theobald, V.J.	539	Tretola, M.	655
Theodoridou, K.	570	Trevisi, P.	530, 540, 543, 546, 606
Theodorou, G.	254, 647	Triboulet, P.	155
Thingnes, S.L.	532	Tribout, T.	191
Thirlwell, D.	207	Trindade, H.	391, 394
Tholen, E.	490	Tripaldi, C.	553
Thomasen, J.R.	308, 312	Trißl, S.	675
Thomas, G.	347	Troch, T.	298
Thomas, M.G.	419	Troy, S.	409, 460, 470
Thomas, R.S.	553	Trupo, M.	565
Thompson, A.N.	451	Tsairidou, S.	187, 189
Thompson, G.	430	Tsinas, A.	321, 546
Thompson, R.J.	314, 316, 472	Tsiplakou, E.	254, 645
Thomson, A.	561	Tudisco, R.	552, 569
Thomson, W.	507	Tullio, R.R.	298, 413
Thorén-Hellsten, E.	576	Turini, T.	235
Thornton, P.	262, 324	Turner, S.P.	479
Thulin, C.G.	389	Tusell Palomero, L.	609
Thurner, S.	496	Tussel, L.	341
Thymann, T.	541	Tuzun, C.G.	130
Tichit, M.	154, 235	Twomey, A.J.	348, 351, 434
Tielemenans, M.	291	Tyra, M.	344, 608
Tiengtam, N.	257	Tyrisevä, A.-M.	628
Tiernan, K.	293	Tyrolova, Y.	412, 633
Tiezzi, F.	349, 414	Tzamaloukas, O.	166, 638, 642
Tighiouart, H.	186	Tzora, A.	321, 546
Tijhaar, E.	395		
Till, B.E.	272	**U**	
Timmer, B.	498	Ubilla, D.	408
Titterington, F.	589	Uimari, P.	420, 610
Titterington, F.M.	137, 143, 210	Ulbrich, S.E.	290
Tixier-Boichard, M.	159	Ule, A.	234
Todd, C.G.	293	Unubol Aypak, S.	524
Tolkamp, B.J.	328	Urbanelli, S.	565
Tolland, E.	131, 132, 133, 148	Urbański, P.	344, 608
Tollerton, I.J.	642	Ureña Ureña, E.	274, 275
Tollerton, J.	131, 132, 133, 148	Utsunomiya, A.T.	415
Tomka, J.	149, 229, 361	Utsunomiya, Y.T.	125, 415
Tonks, A.A.	176, 672	Utzeri, V.J.	530, 606
Toppel, K.	531		
Tor, M.	651	**V**	
Toro, M.A.	258	Vaarst, M.	590
Torrallardona, D.	119, 120, 121, 122, 323, 554	Vaez Torshizi, R.	535
Torres-Pitarch, A.	121, 547	Vahedi, V.	525
Tortereau, F.	364, 518	Vailati, M.	126
Tosser-Klopp, G.	284	Vainoras, A.	477
Toteda, F.	259	Valbonesi, A.	593
Touchard, A.	625	Valenti, B.	649
Touze, J.-L.	625	Valentinelli, C.	660
Townley, P.	666	Valera, M.	538, 583, 609
Trajano, J.S.	150	Vallas, M.	277
Trakooljul, N.	201	Valli, L.	467
Trakovická, A.	423, 427	Van Adrichem, P.S.J.	406
Tran-Huynh, V.H.	454	Van Amburgh, M.E.	242
Traulsen, I.	399, 623	Van Arendonk, J.A.M.	310, 427, 429
Travel, A.	248	Van Beirendonck, S.	662

Vanbergue, E.	289, 296	Vazeille, K.	386
Van Bruggen, L.	594	Vázquez-Gómez, M.	403
Vandaele, L.	459	Veer, A. De	114
Van Dalen, A.B.J.	659, 660	Veerkamp, R.F.	287, 451
Van Den Broeke, A.	442, 443	Vega-Murillo, V.E.	425
Vandenplas, J.	105, 281, 282	Veit, C.	399
Van Den Pol-Van Dasselaar, A.	506, 617	Velazco, J.I.	183, 616
Vandepitte, J.	300	Velázquez-Cantón, E.	480
Van Der Aar, P.J.	249, 661	Veldkamp, T.	204
Van Der Hoek, R.	325	Velie, B.D.	286, 395
Van Der Mheen, H.W.	256	Venneman, J.G.B.	159, 263
Van Der Voort, M.	444	Ventura, R.V.	146
Van Der Werf, J.T.N.	498	Vera Arias, M.A.	523
Van Der Westhuizen, R.R.	214	Verbrugge, A.	626
Van Duinkerken, G.	195	Verde, D.	622
Van Eerdenburg, F.J.C.M.	395	Vergne, T.	585
Van Eetvelde, M.	180	Verloop, J.	153, 157
Vangen, O.	575, 593	Vermeer, I.J.P.	195
Van Hal, O.	153	Vernet, J.	140
Van Hoeij, R.	370	Vernunft, A.	228
Van Kaam, J.B.C.H.M.	454	Verona, I.	609
Van Keulen, L.J.M.	420	Verreth, J.A.J.	256
Van Knegsel, A.T.M.	370	Verweij, M.	516
Vanlierde, A.	299, 300, 630	Vesela, Z.	584
Van Marle-Köster, E.	139, 214	Veselko, D.	291, 298
Van Meensel, J.	442, 443, 444	Vestergaard, M.	136, 369, 373, 376, 629
Van Middelaar, C.E.	115, 617	Veysset, P.	386
Van Middelkoop, J.C.	384	Vial, C.	394
Van Milgen, J.	381	Vicario, D.	108
Van Niekerk, A.	462	Vieira, V.A.	459
Van Noten, N.	250	Vigan, A.	385, 465
Van Reenen, C.G.	498	Vigne, M.	385
Vanrobays, M.-L.	630	Vigors, S.	224
Van Son, M.	612	Viinalass, H.	276, 277, 582, 584
Van Staaveren, N.	221, 533	Viklund, Å.	575, 576, 580
Van Stijn, T.C.	107	Vilkki, J.	464
Van Straten, M.	263	Villalba, D.	651
Van Troos, K.	324	Villa, M.	497
Van Vuure, C.	249, 661	Villanueva, B.	189, 258
Van Wesemael, D.	459	Villarroel, M.	678
Vanwindekens, F.	512	Vincent, F.	625
Van Wyk, J.B.	165	Viñoles, C.	408
Van Zanten, H.H.E.	153, 455	Virtanen, J.K.	357
Vargas-Bello-Pérez, E.	634	Visentin, G.	304, 305
Varley, P.	204	Visker, M.H.P.W.	272
Varley, P.F.	127, 203, 342	Visser, B.	160
Varona, L.	108, 579	Visser, C.	139, 214
Varotto, A.	454	Visser, K.	404
Varshney, R.K.	206	Vitale, M.	313
Värv, S.	582	Vital, M.	542
Vaseghi, H.	252	Vizzier-Thaxton, Y.	532
Vašíček, D.	229	Vlček, M.	679
Vašíčková, K.	229	Vogelzang, R.H.	440, 488
Vasilopolou, D.	361	Vogt, A.	352, 393
Vastenhouw, S.	650	Voilet, V.	599
Vaugeois, N.	471	Voller, S.	255